Statistical Mechanics: Theory and Molecular Simulation

Statistical Mechanics: Theory and Molecular Simulation

SECOND EDITION

MARK E. TUCKERMAN

Department of Chemistry and Courant Institute of Mathematical Sciences
New York University

OXFORD
UNIVERSITY PRESS

Great Clarendon Street, Oxford, OX2 6DP,
United Kingdom

Oxford University Press is a department of the University of Oxford.
It furthers the University's objective of excellence in research, scholarship,
and education by publishing worldwide. Oxford is a registered trade mark of
Oxford University Press in the UK and in certain other countries

First Edition published in 2010
Second Edition published in 2023

Published in the United States of America by Oxford University Press
198 Madison Avenue, New York, NY 10016, United States of America

British Library Cataloguing in Publication Data
Data available

Library of Congress Control Number: 2023930434

ISBN 978-0-19-882556-2

DOI: 10.1093/oso/9780198825562.001.0001

Printed and bound by
CPI Group (UK) Ltd, Croydon, CR0 4YY

Links to third party websites are provided by Oxford in good faith and
for information only. Oxford disclaims any responsibility for the materials
contained in any third party website referenced in this work.

To Jocelyn and all my group members, past, present, and yet to join

Preface

The field of statistical mechanics is evolving with stunning rapidity. Practitioners are actively developing theoretical and computational tools for solving complex problems and those tools are being deployed in increasingly novel applications to real systems of physical, chemical, biological, and engineering import. The first edition of this book provided a solid foundation in the theoretical underpinnings and computational technologies that allow the classical and quantum statistical mechanics of particles to be formally understood and practically implemented. These core concepts will forever remain central to the field and indispensable learning for anyone wishing to enter it. However, the intervening ten-plus years have witnessed advances of such significance that I felt a new edition of the book was needed to incorporate these developments into the book's framework. These incorporations and revisions caused the book's length to swell by over 150 additional pages.

In the last decade, the application of models and methods from the subarea of artificial intelligence known as *machine learning* has transformed the landscape of computational statistical mechanics. Large-scale computer simulations in statistical mechanics often generate output data sets of such enormity and heterogeneity that they could be regarded as "big data" within a "chemical" scope (broadly speaking). These data sets often represent multiple systems containing heterogeneous environments evolving under a variety of external conditions. The techniques of machine learning allow these data sets to be mined for hidden patterns, patterns that can be further leveraged to construct simplified models of the data. Owing to their low computational overhead, these models can be employed in subsequent simulations that reach longer length and time scales, reveal meaningful reaction coordinates for specific processes, and even drive rare-event simulations leading to highly featured free-energy hypersurfaces. Because machine learning has become such an impactful tool in statistical mechanics, a substantial new chapter (Chapter 17) has been added that introduces basic machine learning concepts and model types, including kernel-ridge methods, support-vector machines, neural networks, weighted-neighbor models, and data-clustering algorithms, and demonstrates the application of these models, in both regression and classification modes, in statistical mechanical simulation problems.

Other important developments have shaped the second edition significantly. The discussions of collective variables and free-energy based rare-event sampling techniques in Chapter 8 have been overhauled to capture the impressive innovations in these methods and to highlight how novel techniques such as well-tempered metadynamics and driven adiabatic free-energy dynamics are connected and can be synergistically combined, both with each other and with simpler methods like umbrella sampling. The treatment of Feynman path integrals in Chapter 12 has been enhanced to clarify the meaning of imaginary time, to include methods for simulating systems of N identical bosons and fermions via path integrals and to incorporate an elegant technique for reducing the computational overhead of path integral simulations. The discussion of approaches for approximating quantum time correlation functions in Chapter 14 has been augmented to include a formally exact open-chain formulation. Functionals also make a more prominent appearance in the second edition, being used to construct a classical entropy maximization principle in Chapters 4 through 6 to derive ensemble distributions, and an additional appendix on the calculus of functions has also been provided. Proofs of the Henderson Theorem on radial distribution functions and the Potential Distribution Theorem on excess chemical potentials can be found in these chapters as well. Resonances in multiple time-scale integration are dealt with in much greater depth in Chapters 4 and 15 and algorithms for circumventing these numerical artifacts and allowing very large time steps in molecular dynamics simulations are presented. Other added material includes an expanded discussion of numerical integrators for deterministic and stochastic thermostats in molecular dynamics, more exactly solvable models and numerical applications, resonance-free multiple time-scale algorithms, and more end-of-chapter exercises. The overall structure of the book has not been changed from the first edition: equilibrium classical statistical mechanics precedes equilibrium quantum statistical mechanics, and both precede discussions of classical and quantum time-dependent statistical mechanics. Stochastic dynamics, discrete models, and machine learning make up the final three chapters of the book. As in the original edition, computational methods are presented side-by-side with the theoretical developments, and, to the extent possible, mathematical complexity gradually increases within each chapter.

As I did in the first edition, I wish to close this preface with a list of acknowledgments. I am, as ever, truly grateful to all of the teachers, mentors, colleagues, and coworkers acknowledged in the first edition's preface, which will immediately follow. In addition, content for the second edition would not have been possible without highly fruitful collaborations with Benedict Leimkuhler, Charlles Abreu, Jutta Rogal, Ondrej Marsalek, Serdal Kirmizialtin, and Joseph Cendagorta. I must also express my continued thanks to the National Science Foundation, the U.S. Department of Energy, and the Army Research Office for the ongoing support I have received from these agencies. Finally, I owe, once again, a tremendous debt of gratitude to my wife Jocelyn Leka who lent me her considerable talents as a substantive editor. As with the first edition, her skills were employed only for the textual parts of the book; any mathematical errors remain mine and mine alone.

M.E.T.
New York
December, 2022

Preface to the first edition

Statistical mechanics is a theoretical framework that aims to predict the observable static and dynamic properties of a many-body system starting from its microscopic constituents and their interactions. Its scope is as broad as the set of "many-body" systems is large: as long as there exists a rule governing the behavior of the fundamental objects that comprise the system, the machinery of statistical mechanics can be applied. Consequently, statistical mechanics has found applications outside of physics, chemistry, and engineering, including biology, social sciences, economics, and applied mathematics. Because it seeks to establish a bridge between the microscopic and macroscopic realms, statistical mechanics often provides a means of rationalizing observed properties of a system in terms of the detailed "modes of motion" of its basic constituents. An example from physical chemistry is the surprisingly high diffusion constant of an excess proton in bulk water, which is a single measurable number. However, this single number belies a strikingly complex dance of hydrogen bond rearrangements and chemical reactions that must occur at the level of individual or small clusters of water molecules in order for this property to emerge. In the physical sciences, the technology of molecular simulation, wherein a system's microscopic interaction rules are implemented numerically on a computer, allow such "mechanisms" to be extracted and, through the machinery of statistical mechanics, predictions of macroscopic observables to be generated. In short, molecular simulation is the computational realization of statistical mechanics. The goal of this book, therefore, is to synthesize these two aspects of statistical mechanics: the underlying theory of the subject, in both its classical and quantum developments, and the practical numerical techniques by which the theory is applied to solve realistic problems.

This book is aimed primarily at graduate students in chemistry or computational biology and graduate or advanced undergraduate students in physics or engineering. These students are increasingly finding themselves engaged in research activities that cross traditional disciplinary lines. Successful outcomes for such projects often hinge on their ability to translate complex phenomena into simple models and develop approaches for solving these models. Because of its broad scope, statistical mechanics plays a fundamental role in this type of work and is an important part of a student's toolbox.

The theoretical part of the book is an extensive elaboration of lecture notes I developed for a graduate-level course in statistical mechanics I give at New York University. These courses are principally attended by graduate and advanced undergraduate students who are planning to engage in research in theoretical and experimental physical chemistry and computational biology. The most difficult question faced by anyone wishing to design a lecture course or a book on statistical mechanics is what to include and what to omit. Because statistical mechanics is an active field of research, it comprises a tremendous body of knowledge, and it is simply impossible to treat the entirety of the subject in a single opus. For this reason, many books with the words "statistical mechanics" in their titles can differ considerably. Here, I have attempted to bring together topics that reflect what I see as the modern landscape of statistical mechanics. The reader will notice from a quick scan of the table of contents that the topics selected are rarely found together in individual textbooks on the subject; these topics include isobaric ensembles, path integrals, classical and quantum time-dependent statistical mechanics, the generalized Langevin equation, the Ising model, and critical phenomena. (The closest such book I have found is also one of my favorites, David Chandler's *Introduction to Modern Statistical Mechanics*.)

The computational part of the book joins synergistically with the theoretical part and is designed to give the reader a solid grounding in the methodology employed to solve problems in statistical mechanics. It is intended neither as a simulation recipe book nor a scientific programmer's guide. Rather, it aims to show how the development of computational algorithms derives from the underlying theory with the hope of enabling readers to understand the methodology-oriented literature and develop new techniques of their own. The focus is on the molecular dynamics and Monte Carlo techniques and the many novel extensions of these methods that have enhanced their applicability to, for example, large biomolecular systems, complex materials, and quantum phenomena. Most of the techniques described are widely available in molecular simulation software packages and are routinely employed in computational investigations. As with the theoretical component, it was necessary to select among the numerous important methodological developments that have appeared since molecular simulation was first introduced. Unfortunately, several important topics had to be omitted due to space constraints, including configuration-bias Monte Carlo, the reference potential spatial warping algorithm, and semi-classical methods for quantum time correlation functions. This omission was not made because I view these methods as less important than those I included. Rather, I consider these to be very powerful but highly advanced methods that, individually, might have a narrower target audience. In fact, these topics were slated to appear in a chapter of their own. However, as the book evolved, I found that nearly 700 pages were needed to lay the foundation I sought.

In organizing the book, I have made several strategic decisions. First, the book is structured such that concepts are first introduced within the framework of classical mechanics followed by their quantum mechanical counterparts. This lies closer perhaps to a physicist's perspective than, for example, that of a chemist, but I find it to be a particularly natural one. Moreover, given how widespread computational studies based on classical mechanics have become compared to analogous quantum investigations (which have considerably higher computational overhead), this progression seems to be both logical and practical. Second, the technical development within each chapter is graduated, with the level of mathematical detail generally increasing from chapter start to chapter end. Thus, the mathematically most complex topics are reserved for the final sections of each chapter. I assume that readers have an understanding of calculus (through calculus of several variables), linear algebra, and ordinary differential equations. This structure hopefully allows readers to maximize what they take away from each chapter while rendering it easier to find a stopping point within each chapter. In short, the book is structured such that even a partial reading of a chapter allows the reader to gain a basic understanding of the subject. It should be noted that I attempted to adhere to this graduated structure only as a general protocol. Where I felt that breaking this progression made logical sense, I have forewarned the reader about the mathematical arguments to follow, and the final result is generally given at the outset. Readers wishing to skip the mathematical details can do so without loss of continuity.

The third decision I have made is to integrate theory and computational methods within each chapter. Thus, for example, the theory of the classical microcanonical ensemble is presented together with a detailed introduction to the molecular dynamics method and how the latter is used to generate a classical microcanonical distribution. The other classical ensembles are presented in a similar fashion as is the Feynman path integral formulation of quantum statistical mechanics. The integration of theory and methodology serves to emphasize the viewpoint that understanding one helps in understanding the other.

Throughout the book, many of the computational methods presented are accompanied by simple numerical examples that demonstrate their performance. These examples range from low-dimensional "toy" problems that can be easily coded up by the reader (some of the exercises in each chapter ask precisely this) to atomic and molecular liquids, aqueous solutions, model polymers, biomolecules, and materials. Not every method presented is accompanied by a numerical example, and in general I have tried not to overwhelm the reader with a plethora of applications requiring detailed explanations of the underlying physics, as this is not the primary aim of the book. Once the basics of the methodology are understood, readers wishing to explore applications particular to their interests in more depth can subsequently refer to the literature.

A word or two should be said about the problem sets at the end of each chapter. Math and science are not spectator sports, and the only way to learn the material is to solve problems. Some of the problems in the book require the reader to think conceptually while others are more mathematical, challenging the reader to work through various derivations. There are also problems that ask the reader to analyze proposed computational algorithms by investigating their capabilities. For readers with some programming background, there are exercises that involve coding up a method for a simple example in order to explore the method's performance on that example, and in some cases, reproduce a figure from the text. These coding exercises are included because one can only truly understand a method by programming it up and trying it out on a simple problem for which long runs can be performed and many different parameter choices can be studied. However, I must emphasize that even if a method works well on a simple problem, it is not guaranteed to work well for realistic systems. Readers should not, therefore, naïvely extrapolate the performance of any method they try on a toy system to high-dimensional complex problems. Finally, in each problem set, some problems are preceded by an asterisk (*). These are problems of a more challenging nature that require deeper thinking or a more in-depth mathematical analysis. All of the problems are designed to strengthen understanding of the basic ideas.

Let me close this preface by acknowledging my teachers, mentors, colleagues, and coworkers without whom this book would not have been possible. I took my first statistical mechanics courses with Y. R. Shen at the University of California Berkeley and A. M. M. Pruisken at Columbia University. Later, I audited the course team-taught by James L. Skinner and Bruce J. Berne, also at Columbia. I was also privileged to have been mentored by Bruce Berne as a graduate student, by Michele Parrinello during a postdoctoral appointment at the IBM Forschungslaboratorium in Rüschlikon, Switzerland, and by Michael L. Klein while I was a National Science Foundation postdoctoral fellow at the University of Pennsylvania. Under the mentorship of these extraordinary individuals, I learned and developed many of the computational methods that are discussed in the book. I must also express my thanks to the National Science Foundation for their continued support of my research over the past decade. Many of the developments presented here were made possible through the grants I received from them. I am deeply grateful to the Alexander von Humboldt Foundation for a Friedrich Wilhelm Bessel Research Award that funded an extended stay in Germany where I was able to work on ideas that influenced many parts of the book. I am equally grateful to my German host and friend Dominik Marx for his support during this stay, for many useful discussions, and for many fruitful collaborations that have helped shape the book's content. I also wish to acknowledge my long-time collaborator and friend Glenn Martyna for his help in crafting the book in its initial stages and for his critical reading of the first few chapters. I have also received many helpful suggestions from Bruce Berne, Giovanni Ciccotti, Hae-Soo Oh, Michael Shirts, and Dubravko Sabo. I am indebted to the excellent students and postdocs with whom I have worked over the years for their invaluable contributions to several of the techniques presented herein and for all they have taught me. I would also like to acknowledge my former student Kiryn Haslinger Hoffman for her work on the illustrations used in the early chapters. Finally, I owe a tremendous debt of gratitude to my wife Jocelyn Leka, whose finely honed skills as an editor were brought to bear on crafting the wording used throughout the book. Editing me took up many hours of her time. Her skills were restricted to the textual parts of the book; she was not charged with the onerous task of editing the equations. Consequently, any errors in the latter are mine and mine alone.

M.E.T.
New York
July, 2010

Contents

1
Classical mechanics

1.1 Introduction

The first part of this book is devoted to the subject of classical statistical mechanics, which is founded upon the fundamental laws of classical mechanics as originally stated by Newton. Although the laws of classical mechanics were first postulated to study the motion of planets, stars and other large-scale objects, they turn out to be a surprisingly good approximation at the molecular level (where the true behavior is correctly described by the laws of quantum mechanics). Indeed, an entire computational methodology, known as *molecular dynamics*, is based on the applicability of the laws of classical mechanics to microscopic systems. Molecular dynamics has been remarkably successful in its ability to predict macroscopic thermodynamic and dynamic observables for a wide variety of systems using the rules of classical statistical mechanics to be discussed in the next chapter. Many of these applications address important problems in biology, such as protein and nucleic acid folding, in materials science, such as surface catalysis and functionalization, in the structure and dynamics of glasses and their melts, and in nanotechnology, such as the behavior of self-assembled monolayers and the formation of molecular devices. Throughout the book, we will be discussing both model and realistic examples of such applications.

In this chapter, we will begin with a discussion of Newton's laws of motion and build up to the more elegant Lagrangian and Hamiltonian formulations of classical mechanics, both of which play fundamental roles in statistical mechanics. The origin of these formulations from the action principle will be discussed. The chapter will conclude with a first look at systems that do not fit into the Hamiltonian/Lagrangian framework and the application of such systems in the description of certain physical situations.

1.2 Newton's laws of motion

In 1687, the English physicist and mathematician Sir Isaac Newton published the *Philosophiae Naturalis Principia Mathematica*, wherein three simple and elegant laws governing the motion of interacting objects are given. These may be stated briefly as follows:

1. In the absence of external forces, a body will either be at rest or execute motion along a straight line with a constant velocity $\mathbf{v}$.

2. The action of an external force $\mathbf{F}$ on a body produces an acceleration $\mathbf{a}$ equal to the force divided by the mass m of the body:

$$\mathbf{a} = \frac{\mathbf{F}}{m}, \qquad\qquad \mathbf{F} = m\mathbf{a}. \qquad\qquad (1.2.1)$$

3. If body A exerts a force on body B, then body B exerts an equal and opposite force on body A. That is, if $\mathbf{F}_{AB}$ is the force body A exerts on body B, then the force $\mathbf{F}_{BA}$ exerted by body B on body A satisfies

$$\mathbf{F}_{BA} = -\mathbf{F}_{AB}. \qquad\qquad (1.2.2)$$

In general, two objects can exert attractive or repulsive forces on each other, depending on their relative spatial location, and the precise dependence of the force on the relative location of the objects is specified by a particular *force law*.[1]

Although Newton's interests largely focused on the motion of celestial bodies interacting via gravitational forces, most atoms are massive enough that their motion can be treated reasonably accurately within a classical framework. Hence, the laws of classical mechanics can be approximately applied at the molecular level. Naturally, there are numerous instances in which the classical approximation breaks down, and a proper quantum mechanical treatment is needed. For the present, however, we will assume the approximate validity of classical mechanics at the molecular level and proceed to apply Newton's laws as stated above.

The motion of an object can be described quantitatively by specifying the Cartesian position vector $\mathbf{r}(t)$ of the object in space at any time t. This is tantamount to specifying three functions of time, the components of $\mathbf{r}(t)$,

$$\mathbf{r}(t) = (x(t), y(t), z(t)). \qquad\qquad (1.2.3)$$

Recognizing that the velocity $\mathbf{v}(t)$ of the object is the first time derivative of the position, $\mathbf{v}(t) = d\mathbf{r}/dt$, and that the acceleration $\mathbf{a}(t)$ is the first time derivative of the velocity, $\mathbf{a}(t) = d\mathbf{v}/dt$, the acceleration is easily seen to be the second derivative of position, $\mathbf{a}(t) = d^2\mathbf{r}/dt^2$. Therefore, Newton's second law, $\mathbf{F} = m\mathbf{a}$, can be expressed as a second-order differential equation

$$m\frac{d^2\mathbf{r}}{dt^2} = \mathbf{F}. \qquad\qquad (1.2.4)$$

(We shall henceforth employ the overdot notation for differentiation with respect to time. Thus, $\dot{\mathbf{r}} = d\mathbf{r}/dt$ and $\ddot{\mathbf{r}} = d^2\mathbf{r}/dt^2$.) Since eqn. (1.2.4) is a second-order equation, it is necessary to specify two initial conditions, these being the initial position $\mathbf{r}(0)$ and initial velocity $\mathbf{v}(0)$. The solution of eqn. (1.2.4) subject to these initial conditions uniquely specifies the motion of the object for all time.

The force $\mathbf{F}$ that acts on an object is capable of doing *work* on the object. In order to see how work is computed, consider Fig. 1.1, which shows a force $\mathbf{F}$ acting on a

---

[1]Throughout the book, vector quantities will be designated using boldface type. Thus, in three spatial dimensions, a vector $\mathbf{u}$ has three components u_x, u_y, and u_z, and we will represent the vector as the ordered triple $\mathbf{u} = (u_x, u_y, u_z)$. The vector magnitude $u = |\mathbf{u}| = \sqrt{u_x^2 + u_y^2 + u_z^2}$ will be denoted using normal type.

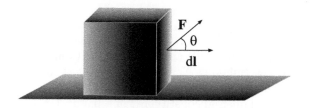

Fig. 1.1 Example of mechanical work. Here $dW = \mathbf{F} \cdot \mathbf{dl} = F\cos\theta dl$.

system along a particular path. The work dW performed along a short segment $\mathbf{dl}$ of the path is defined to be

$$dW = \mathbf{F} \cdot \mathbf{dl} = F\cos\theta dl. \tag{1.2.5}$$

The total work done on the object by the force between points A and B along the path is obtained by integrating over the path from A to B:

$$W_{\mathrm{AB}}(\text{path}) = \int_A^B \mathbf{F} \cdot \mathbf{dl}. \tag{1.2.6}$$

In general, the work done on an object by a force depends on the path taken between A and B. For certain types of forces, called *conservative forces*, the work is independent of the path and only depends on the endpoints of the path. We shall describe shortly how conservative forces are defined.

Note that the definition of work depends on context. Equation (1.2.6) specifies the work done *by* a force $\mathbf{F}$. If this force is an intrinsic part of the system, then we refer to this type of work as work done *by the system*. If we wish to calculate the work done *against* such a force by some external agent, then this work would be the *negative* of that obtained using eqn. (1.2.6), and we refer to this as work done *on the system*. An example is the force exerted by the Earth's gravitational field on an object of mass m. If the mass falls under the Earth's gravitational pull through a distance h, we can think of the object and the gravitational force as defining the mechanical system. In this case, the system does work, and eqn. (1.2.6) would yield a positive value. Conversely, if we applied eqn. (1.2.6) to the opposite problem of raising the object to a height h, it would yield a negative result. This is simply telling us some external agent must do work *on* the system against the force of gravity in order to raise it to a height h. Generally, it is obvious what sign to impart to work, yet the distinction between work done on and by a system will become important in our discussions of thermodynamics and classical statistical mechanics in Chapters 2 through 6.

Given the form of Newton's second law in eqn. (1.2.4), it can be easily shown that, in a flat or Euclidean space, Newton's first law is redundant. According to Newton's first law, an object initially at a position $\mathbf{r}(0)$ moving with constant velocity $\mathbf{v}$ will move along a straight line described by

$$\mathbf{r}(t) = \mathbf{r}(0) + \mathbf{v}t. \tag{1.2.7}$$

This is an example of a *trajectory*, that is, a specification of the object's position as a function of time and initial conditions. If no force acts on the object, then, according to Newton's second law, its position will be the solution of

$$\ddot{\mathbf{r}} = 0. \tag{1.2.8}$$

The straight line motion of eqn. (1.2.7) is, in fact, the unique solution of eqn. (1.2.8) for an object whose initial position is $\mathbf{r}(0)$ and whose initial (and constant) velocity is $\mathbf{v}$. Thus, Newton's second law embodies Newton's first law.

Statistical mechanics is concerned with the behavior of large numbers of objects that can be viewed as the fundamental constituents of a particular microscopic model of the system, whether they are individual atoms or molecules, or even groups of atoms in a macromolecule (for example, the amino acids in a protein). We shall, henceforth, refer to these constituents as "particles" (or, in some cases, "pseudoparticles"). The classical behavior of a system of N particles in three dimensions is given by the generalization of Newton's second law to the system. In order to develop the general form of Newton's second law, note that particle $i, i \in [1, N]$, will experience a force $\mathbf{F}_i$ due to all of the other particles in the system and possibly the external environment or external agents as well. Denoting the position vectors of the N particles as $\mathbf{r}_1, ..., \mathbf{r}_N$, the forces $\mathbf{F}_i$ are generally functions of these positions, and if frictional forces are present, $\mathbf{F}_i$ could also be a function of the particle's velocity $\dot{\mathbf{r}}_i$. We denote this functional dependence as $\mathbf{F}_i = \mathbf{F}_i(\mathbf{r}_1, ..., \mathbf{r}_N, \dot{\mathbf{r}}_i)$. For example, if the force $\mathbf{F}_i$ depends only on individual contributions from every other particle in the system, we say that the forces are *pairwise additive*. In this case, the force $\mathbf{F}_i$ can be expressed as

$$\mathbf{F}_i(\mathbf{r}_1, ..., \mathbf{r}_N, \dot{\mathbf{r}}_i) = \sum_{j \neq i} \mathbf{f}_{ij}(\mathbf{r}_i - \mathbf{r}_j) + \mathbf{f}^{(\text{ext})}(\mathbf{r}_i, \dot{\mathbf{r}}_i). \tag{1.2.9}$$

The first term in eqn. (1.2.9) describes forces that are intrinsic to the system and are part of the definition of the mechanical system, while the second term describes forces that are entirely external to the system. For a general N-particle system, Newton's second law for particle i takes the form

$$m_i \ddot{\mathbf{r}}_i = \mathbf{F}_i(\mathbf{r}_1, ..., \mathbf{r}_N, \dot{\mathbf{r}}_i). \tag{1.2.10}$$

These equations, referred to as the *equations of motion* of the system, must be solved subject to a set of initial positions, $\{\mathbf{r}_1(0), ..., \mathbf{r}_N(0)\}$, and velocities, $\{\dot{\mathbf{r}}_1(0), ..., \dot{\mathbf{r}}_N(0)\}$. In any realistic system, the interparticle forces are highly nonlinear functions of the N particle positions so that eqns. (1.2.10) possess enormous dynamical complexity, and obtaining an analytical solution is hopeless. Moreover, even if an accurate numerical solution could be obtained, for macroscopic matter, where $N \sim 10^{23}$, the computational resources required to calculate and store the solutions for each and every particle at a large number of discrete time points would exceed by many orders of magnitude all those presently available, making such a task equally untenable. Given these considerations, how can we ever expect to calculate physically observable properties of realistic systems starting from a microscopic description if the fundamental equations governing the behavior of the system cannot be solved?

The rules of statistical mechanics provide the necessary connection between the microscopic laws and macroscopic observables. These rules, however, cannot circumvent the complexity of the system. Therefore, several approaches can be considered for dealing with this complexity: A highly simplified model for a system that lends

itself to an analytical solution could be introduced. Although often of limited utility, important physical insights can sometimes be extracted from a clever model, and
it is usually possible to study the behavior of the model as external conditions are
varied, such as the number of particles, containing volume, applied pressure, and so
forth. Alternatively, one can consider a system, not of 10^{23} particles, but of a much
smaller number, perhaps 10^2–10^9 particles, depending on the nature of the system,
and solve the equations of motion numerically subject to initial conditions and the
boundary conditions of a containing volume. Fortunately, many macroscopic properties are well-converged with respect to system size for such small numbers of particles!
The rules of statistical mechanics are then used to analyze the numerical trajectories
thus generated. This is the essence of the technique known as *molecular dynamics*. Although the molecular dynamics approach is very powerful, a significant disadvantage
exists: in order to study the dependence on external conditions, a separate calculation
must be performed for every choice of these conditions, hence a very large number of
calculations is needed, for example, in order to map out a phase diagram. In addition, the "exact" forces between particles cannot be determined and, hence, models
for these forces must be introduced. Usually, the more accurate the model, the more
computationally intensive the numerical calculation, and the more limited the scope
of the calculation with respect to time and length scales and the properties that can
be studied. Often, time and length scales can be bridged by combining models of different accuracy, including even continuum models commonly used in engineering, to
describe different aspects of a large, complex system, and devising clever numerical
solvers for the resulting equations of motion. Numerical calculations (typically referred
to as *simulations*) have become an integral part of modern theoretical research, and
since many of these calculations rely on the laws of classical mechanics, it is important that this subject be covered in some detail before advancing to a discussion of
the rules of statistical mechanics. The remainder of this chapter will, therefore, be
devoted to introducing the concepts from classical mechanics that will be needed for
our subsequent discussion of statistical mechanics.

1.3 Phase space: visualizing classical motion

Newton's equations specify the complete set of particle positions $\{\mathbf{r}_1(t), ..., \mathbf{r}_N(t)\}$ and,
by differentiation, the particle velocities $\{\mathbf{v}_1(t), ..., \mathbf{v}_N(t)\}$ at any time t, given that
the positions and velocities are known at one particular instant in time. For reasons
that will be clear shortly, it is often preferable to work with the particle momenta,
$\{\mathbf{p}_1(t), ..., \mathbf{p}_N(t)\}$, which, in Cartesian coordinates, are related to the velocities by

$$\mathbf{p}_i = m_i \mathbf{v}_i = m_i \dot{\mathbf{r}}_i. \tag{1.3.1}$$

Note that, in terms of momenta, Newton's second law can be written as

$$\mathbf{F}_i = m\mathbf{a}_i = m_i \frac{\mathrm{d}\mathbf{v}_i}{\mathrm{d}t} = \frac{\mathrm{d}\mathbf{p}_i}{\mathrm{d}t}. \tag{1.3.2}$$

Therefore, the classical dynamics of an N-particle system can be expressed by specify-
ing the full set of $6N$ functions, $\{\mathbf{r}_1(t), ..., \mathbf{r}_N(t), \mathbf{p}_1(t), ..., \mathbf{p}_N(t)\}$. Equivalently, at any

instant t in time, all of the information about the system is specified by $6N$ numbers (or $2dN$ in d dimensions). These $6N$ numbers constitute the *microscopic state* of the system at time t. That these $6N$ numbers are sufficient to characterize the system follows entirely from the fact that they are all that is needed to seed eqns. (1.2.10), from which the complete time evolution of the system can be determined.

Suppose, at some instant in time, the positions and momenta of the system are $\{\mathbf{r}_1, ..., \mathbf{r}_N, \mathbf{p}_1, ..., \mathbf{p}_N\}$. These $6N$ numbers can be regarded as an ordered $6N$-tuple or a single point in a $6N$-dimensional space called *phase space*. Although the geometry of this space can, under certain circumstances, be nontrivial, in its simplest form, a phase space is a Cartesian space that can be constructed from $6N$ mutually orthogonal axes. We shall denote a general point in the phase space as

$$\mathbf{x} = (\mathbf{r}_1, ..., \mathbf{r}_N, \mathbf{p}_1, ..., \mathbf{p}_N) \tag{1.3.3}$$

also known as the *phase-space vector*. (As we will see in Chapter 2, phase spaces play a central role in classical statistical mechanics.) Solving eqns. (1.2.10) generates a set of functions

$$\mathbf{x}(t) = (\mathbf{r}_1(t), ..., \mathbf{r}_N(t), \mathbf{p}_1(t), ..., \mathbf{p}_N(t)) \equiv \mathbf{x}_t, \tag{1.3.4}$$

which describe a parametric path or *trajectory* in the phase space. Therefore, classical motion can be described by the motion of a point along a trajectory in phase space. Although phase-space trajectories can only be visualized for a one-particle system in one spatial dimension, it is, nevertheless, instructive to study several such examples.

Consider, first, a free particle with coordinate x and momentum p, described by the one-dimensional analog of eqn. (1.2.7), *i.e.*, $x(t) = x(0) + (p/m)t$, where p is the particle's (constant) momentum. A plot of p vs. x is simply a straight horizontal line starting at $x(0)$ and extending in the direction of increasing x if $p > 0$ or decreasing x if $p < 0$. This is illustrated in Fig. 1.2. The line is horizontal because p is constant for all x values visited on the trajectory.

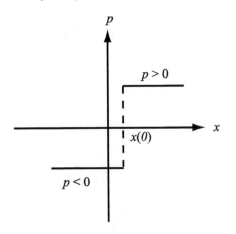

Fig. 1.2 Phase space of a one-dimensional free particle.

Another important example of a phase-space trajectory is that of a simple Harmonic oscillator, for which the force law is given by Hooke's law, $F(x) = -kx$, where k is a constant known as the *force constant*. In this case, Newton's second law takes the form

$$m\ddot{x} = -kx. \tag{1.3.5}$$

For a given initial condition, $x(0)$ and $p(0)$, the solution of eqn. (1.3.5) is

$$x(t) = x(0)\cos\omega t + \frac{p(0)}{m\omega}\sin\omega t, \tag{1.3.6}$$

where $\omega = \sqrt{k/m}$ is the natural frequency of the oscillator. Equation (1.3.6) can be verified by substitution into eqn. (1.3.5). Differentiating once with respect to time and multiplying by the mass gives an expression for the momentum

$$p(t) = p(0)\cos\omega t - m\omega x(0)\sin\omega t. \tag{1.3.7}$$

Note that $p(t)$ and $x(t)$ are related by

$$\frac{(p(t))^2}{2m} + \frac{1}{2}m\omega^2(x(t))^2 = C, \tag{1.3.8}$$

where C is a constant determined by the initial condition according to

$$C = \frac{(p(0))^2}{2m} + \frac{1}{2}m\omega^2(x(0))^2. \tag{1.3.9}$$

(This relation is known as the *conservation of energy*, which we will discuss in greater detail in the next few sections.) From eqn. (1.3.8), it can be seen that the phase-space plot, p vs. x, specified by $p^2/2m + m\omega^2 x^2/2 = C$ is an ellipse with axes $(2mC)^{1/2}$ and $(2C/m\omega^2)^{1/2}$ as shown in Fig. 1.3. The analysis also indicates that different initial

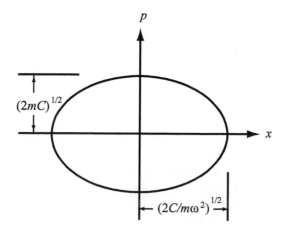

Fig. 1.3 Phase space of the one-dimensional harmonic oscillator.

conditions give rise to different values of C, which changes the size of the ellipse. Changing the mass and frequency changes the shape of the ellipse. Phase-space plots determine the values of position and momentum the system will visit along a trajectory for a given set of initial conditions. These values constitute the *accessible phase space*. For a free particle, the accessible phase space is unbounded since x lies in the interval $x \in [x(0), \infty)$ for $p > 0$ or $x \in (-\infty, x(0)]$ for $p < 0$. The harmonic oscillator, by contrast, provides an example of a phase space that is bounded.

Consider, finally, the example of a particle of mass m rolling over a hill under the influence of gravity, as illustrated in Fig. 1.4(a). (This example is a one-dimensional idealization of a situation that should be familiar to anyone who has ever played miniature golf and also serves as a paradigm for chemical reactions.) We will assume that the top of the hill corresponds to a position $x = 0$. The force law for this problem is non-linear, so that a simple, closed-form, analytical solution to Newton's second law is not readily available. However, an analytical solution is not needed in order to visualize the motion using a phase-space picture. Several kinds of motion are possible depending on the initial conditions. First, if the particle is not rolled quickly enough, it cannot roll completely over the hill. Rather, it will climb part way up the hill and then roll back down the same side. This type of motion is depicted in the phase-space plot of Fig. 1.4(b). Note that the plot only shows the motion in a region close to the hill. A full phase-space plot would extend to $x = \pm\infty$. On the other hand, if the initial speed is high enough, the particle can reach the top of the hill and roll down the other side as depicted in Fig. 1.4(d). The crossover between these two scenarios occurs for

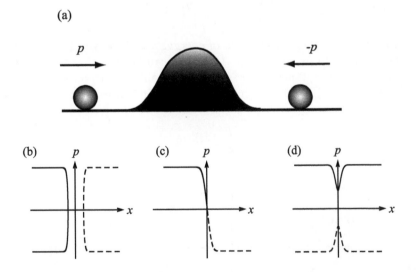

Fig. 1.4 Phase space of a one-dimensional particle subject to the "hill" potential: (a) Two particles approach the hill, one from the left, one from the right. (b) Phase-space plot if the particles have insufficient energy to roll over the hill. (c) Same if the energy is just sufficient for a particle to reach the top of the hill and come to rest there. (d) Same if the energy is greater than that needed to roll over the hill.

one particular initial rolling speed in which the ball can just climb to the top of the hill and come to rest there, as is shown in Fig. 1.4(c). Such a trajectory clearly divides the phase space between the two types of motion shown in Figs. 1.4(b) and 1.4(d) and is known as a *separatrix*. If this example were extended to include a large number of hills with possibly different heights, then the phase space would contain a very large number of separatrices. Such an example is paradigmatic of the force laws that one encounters in complex problems such as protein folding, one of the most challenging computational problems in biophysics.

Visualizing the trajectory of a complex many-particle system in phase space is not possible due to the high dimensionality of the space. Moreover, the phase space may be bounded in some directions and unbounded in others. For formal purposes, it is often useful to think of an illustrative phase-space plot, in which some particular set of coordinates of special interest are shown collectively on one axis and their corresponding momenta are shown on the other with a schematic representation of a phase-space trajectory. This technique has been used to visualize the phase space of chemical reactions in an excellent treatise by De Leon *et al.* (1991). In other instances, it is instructive to consider a particular cut or surface in a large phase space that represents a set of variables of interest. Such a cut is known as a *Poincaré section* after the French mathematician Henri Poincaré (1854–1912), who, among other things, contributed substantially to our modern theory of dynamical systems. In this case, the values of the remaining variables will be fixed at the values they take at the location of this section. The concept of a Poincaré section is illustrated in Fig. 1.5.

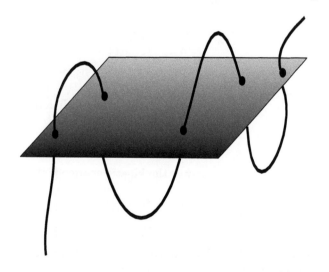

Fig. 1.5 A Poincaré section. The dark line represents a trajectory, and the collection of points at which it crosses the plane is the Poincaré section.

1.4 Lagrangian formulation of classical mechanics: A general framework for Newton's laws

Statistical mechanics is concerned with characterizing the number of microscopic states available to a system and, therefore, requires a formulation of classical mechanics that is more closely connected to the phase-space description than the Newtonian formulation. Since phase space provides a geometric description of a system in terms of positions and momenta, or equivalently in terms of positions and velocities, it is natural to look for an algebraic description of a system in terms of these variables. In particular, we seek a "generator" of the classical equations of motion that takes the positions and velocities or positions and momenta as its inputs and produces, through some formal procedure, the classical equations of motion. The formal structure we seek is embodied in the *Lagrangian* and *Hamiltonian* formulations of classical mechanics (Goldstein, 1980), named for Joseph-Louis Lagrange (1736–1813) and William Rowan Hamilton (1805–1865), respectively. The introduction of such a formal structure places some restrictions on the form of the force laws. Specifically, the forces are required to be *conservative*. Conservative forces are defined to be vector quantities that are derivable from a scalar function $U(\mathbf{r}_1, ..., \mathbf{r}_N)$, known as a *potential energy function*, via

$$\mathbf{F}_i(\mathbf{r}_1, ..., \mathbf{r}_N) = -\nabla_i U(\mathbf{r}_1, ..., \mathbf{r}_N), \tag{1.4.1}$$

where $\nabla_i = \partial/\partial \mathbf{r}_i$. Consider the work done by the force $\mathbf{F}_i$ in moving particle i from points A to B along a particular path. This work is

$$W_{AB} = \int_A^B \mathbf{F}_i \cdot d\mathbf{l}. \tag{1.4.2}$$

Since $\mathbf{F}_i = -\nabla_i U$ is conserved, the line integral simply becomes the difference in potential energy between path endpoints A and B, $W_{AB} = U_A - U_B$, independent of the path taken. Because the work is independent of the path, it follows that along a closed path

$$\oint \mathbf{F}_i \cdot d\mathbf{l} = 0. \tag{1.4.3}$$

Given the N particle velocities, $\dot{\mathbf{r}}_1, ..., \dot{\mathbf{r}}_N$, the kinetic energy of the system is given by

$$K(\dot{\mathbf{r}}_1, ..., \dot{\mathbf{r}}_N) = \frac{1}{2} \sum_{i=1}^N m_i \dot{\mathbf{r}}_i^2. \tag{1.4.4}$$

The *Lagrangian* $\mathcal{L}$ of a system is defined as the difference between the kinetic and potential energies expressed as a function of positions and velocities:

$$\mathcal{L}(\mathbf{r}_1, ..., \mathbf{r}_N, \dot{\mathbf{r}}_1, ..., \dot{\mathbf{r}}_N) = K(\dot{\mathbf{r}}_1, ..., \dot{\mathbf{r}}_N) - U(\mathbf{r}_1, ..., \mathbf{r}_N). \tag{1.4.5}$$

The Lagrangian serves as the generator of the equations of motion via the *Euler-Lagrange* equation:

$$\frac{d}{dt}\left(\frac{\partial \mathcal{L}}{\partial \dot{\mathbf{r}}_i}\right) - \frac{\partial \mathcal{L}}{\partial \mathbf{r}_i} = 0. \tag{1.4.6}$$

It can be easily verified that substitution of eqn. (1.4.5) into eqn. (1.4.6) gives eqn. (1.2.10):

$$\frac{\partial \mathcal{L}}{\partial \dot{\mathbf{r}}_i} = m_i \dot{\mathbf{r}}_i$$

$$\frac{\mathrm{d}}{\mathrm{d}t} \left(\frac{\partial \mathcal{L}}{\partial \dot{\mathbf{r}}_i} \right) = m_i \ddot{\mathbf{r}}_i$$

$$\frac{\partial \mathcal{L}}{\partial \mathbf{r}_i} = -\frac{\partial U}{\partial \mathbf{r}_i} = \mathbf{F}_i$$

$$\frac{\mathrm{d}}{\mathrm{d}t} \left(\frac{\partial \mathcal{L}}{\partial \dot{\mathbf{r}}_i} \right) - \frac{\partial \mathcal{L}}{\partial \mathbf{r}_i} = m_i \ddot{\mathbf{r}}_i - \mathbf{F}_i = 0, \tag{1.4.7}$$

which is just Newton's second law of motion.

As an example of the application of the Euler-Lagrange equation, consider the one-dimensional harmonic oscillator discussed in the previous section. The Hooke's law force $F(x) = -kx$ can be derived from a potential

$$U(x) = \frac{1}{2}kx^2, \tag{1.4.8}$$

so that the Lagrangian takes the form

$$\mathcal{L}(x, \dot{x}) = \frac{1}{2}m\dot{x}^2 - \frac{1}{2}kx^2. \tag{1.4.9}$$

Thus, the equation of motion is derived as follows:

$$\frac{\partial \mathcal{L}}{\partial \dot{x}} = m\dot{x}$$

$$\frac{\mathrm{d}}{\mathrm{d}t} \left(\frac{\partial \mathcal{L}}{\partial \dot{x}} \right) = m\ddot{x}$$

$$\frac{\partial \mathcal{L}}{\partial x} = -kx$$

$$\frac{\mathrm{d}}{\mathrm{d}t} \left(\frac{\partial \mathcal{L}}{\partial \dot{x}} \right) - \frac{\partial \mathcal{L}}{\partial x} = m\ddot{x} + kx = 0, \tag{1.4.10}$$

which is the same as eqn. (1.3.5).

It is important to note that when the forces in a particular system are conservative, then the equations of motion satisfy an important conservation law, namely the conservation of energy. The total energy is given by the sum of kinetic and potential energies:

$$E = \sum_{i=1}^{N} \frac{1}{2}m_i \dot{\mathbf{r}}_i^2 + U(\mathbf{r}_1, ..., \mathbf{r}_N). \tag{1.4.11}$$

In order to verify that E is a constant, we need only show that $dE/dt = 0$. Differentiating eqn. (1.4.11) with respect to time yields

$$
\begin{aligned}
\frac{dE}{dt} &= \sum_{i=1}^{N} m_i \dot{\mathbf{r}}_i \cdot \ddot{\mathbf{r}}_i + \sum_{i=1}^{N} \frac{\partial U}{\partial \mathbf{r}_i} \cdot \dot{\mathbf{r}}_i \\
&= \sum_{i=1}^{N} \dot{\mathbf{r}}_i \cdot \left[m_i \ddot{\mathbf{r}}_i + \frac{\partial U}{\partial \mathbf{r}_i} \right] \\
&= \sum_{i=1}^{N} \dot{\mathbf{r}}_i \cdot \left[m_i \ddot{\mathbf{r}}_i - \mathbf{F}_i \right] \\
&= 0,
\end{aligned}
\qquad (1.4.12)
$$

where the last line follows from the fact that $\mathbf{F}_i = m_i \ddot{\mathbf{r}}_i$.

The power of the Lagrangian formulation of classical mechanics lies in the fact that the equations of motion in an arbitrary coordinate system, which might not be easy to write down directly from Newton's second law, can be derived straightforwardly via the Euler-Lagrange equation. Often, the standard Cartesian coordinates are not the most suitable coordinate choice for a given problem. Suppose, for a given system, there exists another set of $3N$ coordinates, $\{q_1, ..., q_{3N}\}$, that provides a more natural description of the particle locations. These coordinates are known as *generalized coordinates,* and they can be related to the original Cartesian coordinates, $\mathbf{r}_1, ..., \mathbf{r}_N$, via a *coordinate transformation*

$$
q_\alpha = f_\alpha(\mathbf{r}_1, ..., \mathbf{r}_N), \qquad \alpha = 1, ..., 3N. \qquad (1.4.13)
$$

Thus, each coordinate q_α is generally a function of the N Cartesian coordinates, $\mathbf{r}_1, ..., \mathbf{r}_N$. It is assumed that the coordinate transformation eqn. (1.4.13) has a unique inverse

$$
\mathbf{r}_i = \mathbf{g}_i(q_1, ..., q_{3N}), \qquad i = 1, ..., N. \qquad (1.4.14)
$$

In order to determine the Lagrangian in terms of generalized coordinates, eqn. (1.4.14) is used to compute the velocities via the chain rule:

$$
\dot{\mathbf{r}}_i = \sum_{\alpha=1}^{3N} \frac{\partial \mathbf{r}_i}{\partial q_\alpha} \dot{q}_\alpha = \sum_{\alpha=1}^{N} \frac{\partial \mathbf{g}_i}{\partial q_\alpha} \dot{q}_\alpha, \qquad (1.4.15)
$$

where $\partial \mathbf{r}_i / \partial q_\alpha$ is computed from $\partial \mathbf{g}_i / \partial q_\alpha$. Substituting eqn. (1.4.15) into eqn. (1.4.4) gives the kinetic energy in terms of the new velocities $\dot{q}_1, ..., \dot{q}_{3N}$:

$$
\tilde{K}(q, \dot{q}) = \frac{1}{2} \sum_{\alpha=1}^{3N} \sum_{\beta=1}^{3N} \left[\sum_{i=1}^{N} m_i \frac{\partial \mathbf{r}_i}{\partial q_\alpha} \cdot \frac{\partial \mathbf{r}_i}{\partial q_\beta} \right] \dot{q}_\alpha \dot{q}_\beta
$$

$$\equiv \frac{1}{2} \sum_{\alpha=1}^{3N} \sum_{\beta=1}^{3N} G_{\alpha\beta}(q_1, ..., q_{3N}) \dot{q}_\alpha \dot{q}_\beta, \tag{1.4.16}$$

where

$$G_{\alpha\beta}(q_1, ..., q_{3N}) = \sum_{i=1}^{N} m_i \frac{\partial \mathbf{r}_i}{\partial q_\alpha} \cdot \frac{\partial \mathbf{r}_i}{\partial q_\beta} \tag{1.4.17}$$

is called the *mass metric matrix* or *mass metric tensor*. The matrix $G_{\alpha\beta}$ is a function of some or all of the generalized coordinates, depending on the form of the transformations in eqn. (1.4.14). The Lagrangian in generalized coordinates then takes the form

$$\mathcal{L} = \frac{1}{2} \sum_{\alpha=1}^{3N} \sum_{\beta=1}^{3N} G_{\alpha\beta}(q_1, ..., q_{3N}) \dot{q}_\alpha \dot{q}_\beta - U(\mathbf{g}_1(q_1, ..., q_{3N}), ..., \mathbf{g}_N(q_1, ..., q_{3N})), \tag{1.4.18}$$

where the potential U is expressed as a function of the generalized coordinates through the transformation in eqn. (1.4.14). Substitution of eqn. (1.4.18) into eqn. (1.4.6), the Euler-Lagrange equation, gives the equations of motion for each generalized coordinate, q_γ, $\gamma = 1, ..., 3N$:

$$\sum_{\beta=1}^{3N} G_{\gamma\beta}(q_1, ..., q_{3N}) \ddot{q}_\beta + \sum_{\alpha=1}^{3N} \sum_{\beta=1}^{3N} \left[\frac{\partial G_{\gamma\beta}}{\partial q_\alpha} - \frac{1}{2} \frac{\partial G_{\alpha\beta}}{\partial q_\gamma} \right] \dot{q}_\alpha \dot{q}_\beta = -\frac{\partial U}{\partial q_\gamma}. \tag{1.4.19}$$

Here, $\partial U / \partial q_\gamma$ can be obtained from the chain rule

$$\frac{\partial U}{\partial q_\gamma} = \sum_{i=1}^{N} \frac{\partial U}{\partial \mathbf{r}_i} \cdot \frac{\partial \mathbf{r}_i}{\partial q_\gamma}. \tag{1.4.20}$$

Equation (1.4.19) can be recast in a form that manifestly reveals the geometric structure of generalized coordinate space. Noting that the dyad $\dot{q}_\alpha \dot{q}_\beta$ is symmetric with respect to an exchange of α and β, we can rewrite the second term on the left in eqn. (1.4.19) as

$$\frac{\partial G_{\gamma\beta}}{\partial q_\alpha} - \frac{1}{2} \frac{\partial G_{\alpha\beta}}{\partial q_\gamma} = \frac{1}{2} \left[\frac{\partial G_{\gamma\beta}}{\partial q_\alpha} + \frac{\partial G_{\gamma\alpha}}{\partial q_\beta} - \frac{\partial G_{\alpha\beta}}{\partial q_\gamma} \right]. \tag{1.4.21}$$

If we substitute eqn. (1.4.21) into eqn. (1.4.19), multiply through by the inverse of the mass-metric tensor $G_{\lambda\gamma}^{-1}$, and sum over γ, we obtain an equation of motion of the form

$$\ddot{q}_\lambda + \sum_{\alpha=1}^{3N} \sum_{\beta=1}^{3N} \Gamma_{\lambda\alpha\beta} \dot{q}_\alpha \dot{q}_\beta = -\sum_{\gamma=1}^{3N} G_{\lambda\gamma}^{-1} \frac{\partial U}{\partial q_\gamma}, \tag{1.4.22}$$

where

$$\Gamma_{\lambda\alpha\beta} = \sum_{\gamma=1}^{3N} \frac{1}{2} G_{\lambda\gamma}^{-1} \left[\frac{\partial G_{\gamma\beta}}{\partial q_\alpha} + \frac{\partial G_{\gamma\alpha}}{\partial q_\beta} - \frac{\partial G_{\alpha\beta}}{\partial q_\gamma} \right] \tag{1.4.23}$$

is known as the *affine connection* in the generalized coordinate space. On a general manifold, an affine connection is a geometric structure that provides a way to connect nearby tangent spaces. In the absence of forces, eqn. (1.4.22) gives the equations of motion of geodesics (free-particle motion) in terms of the generalized coordinates. The remainder of this section will present several examples of the use of the Lagrangian formalism.

1.4.1 Example: Motion in a central potential

Consider a single particle in three dimensions subject to a potential $U(\mathbf{r})$ that depends only on the particle's distance from the origin. This means $U(\mathbf{r}) = U(|\mathbf{r}|) = U(r)$, where $r = \sqrt{x^2 + y^2 + z^2}$ and is known as a *central potential*. The most natural coordinates are not the Cartesian coordinates (x, y, z) but rather spherical polar coordinates (r, θ, ϕ) given by

$$r = \sqrt{x^2 + y^2 + z^2}, \qquad \theta = \tan^{-1} \frac{\sqrt{x^2 + y^2}}{z}, \qquad \phi = \tan^{-1} \frac{y}{x}, \qquad (1.4.24)$$

which can be inverted to give

$$x = r \sin \theta \cos \phi, \qquad y = r \sin \theta \sin \phi, \qquad z = r \cos \theta. \qquad (1.4.25)$$

The mass metric tensor is a 3×3 diagonal matrix given by

$$\begin{aligned} G_{11}(r, \theta, \phi) &= m \\ G_{22}(r, \theta, \phi) &= mr^2 \\ G_{33}(r, \theta, \phi) &= mr^2 \sin^2 \theta \\ G_{\alpha\beta}(r, \theta, \phi) &= 0 \qquad \alpha \neq \beta. \end{aligned} \qquad (1.4.26)$$

Returning to our example of a single particle moving in a central potential, $U(r)$, we find that the Lagrangian obtained by substituting eqn. (1.4.26) into eqn. (1.4.18) is

$$\mathcal{L} = \frac{1}{2} m \left(\dot{r}^2 + r^2 \dot{\theta}^2 + r^2 \sin^2 \theta \dot{\phi}^2 \right) - U(r). \qquad (1.4.27)$$

In order to obtain the equations of motion from the Euler-Lagrange equations, eqn. (1.4.6), derivatives of $\mathcal{L}$ with respect to each of the variables and their time derivatives are required. These are given by:

$$\frac{\partial \mathcal{L}}{\partial \dot{r}} = m\dot{r}, \qquad\qquad \frac{\partial \mathcal{L}}{\partial r} = mr\dot{\theta}^2 + mr \sin^2 \theta \dot{\phi}^2 - \frac{dU}{dr}$$

$$\frac{\partial \mathcal{L}}{\partial \dot{\theta}} = mr^2 \dot{\theta}, \qquad\qquad \frac{\partial \mathcal{L}}{\partial \theta} = mr^2 \sin \theta \cos \theta \dot{\phi}^2$$

$$\frac{\partial \mathcal{L}}{\partial \dot{\phi}} = mr^2 \sin^2 \theta \dot{\phi}, \qquad\qquad \frac{\partial \mathcal{L}}{\partial \phi} = 0. \qquad (1.4.28)$$

Note that in eqn. (1.4.28), the derivative $\partial \mathcal{L}/\partial \phi = 0$. The coordinate ϕ is an example of a *cyclic* coordinate. In general, if a coordinate q satisfies $\partial \mathcal{L}/\partial q = 0$, it is called cyclic. It is also possible to make θ a cyclic coordinate by recognizing that the quantity

$l = \mathbf{r} \times \mathbf{p}$, called the *orbital angular momentum*, is a constant ($l(0) = l(t)$) when the potential only depends on r. (Angular momentum will be discussed in more detail in Section 1.11.) Thus, the quantity l is conserved by the motion and, therefore, satisfies $dl/dt = 0$. Because l is constant, it is always possible to simplify the problem by choosing a coordinate frame in which the z axis lies along the direction of l. In such a frame, the motion occurs solely in the xy plane so that $\theta = \pi/2$ and $\dot{\theta} = 0$. With this simplification, the equations of motion become

$$m\ddot{r} - mr\dot{\phi}^2 = -\frac{dU}{dr}$$

$$mr^2\ddot{\phi} + 2mr\dot{r}\dot{\phi} = 0. \tag{1.4.29}$$

The second equation can be expressed in the form

$$\frac{d}{dt}\left(\frac{1}{2}r^2\dot{\phi}\right) = 0, \tag{1.4.30}$$

which expresses another conservation law known as the conservation of *areal velocity*, defined as the area swept out by the radius vector per unit time. Setting the quantity, $mr^2\dot{\phi} = \lambda$, where λ is constant, the first equation of motion can be written as

$$m\ddot{r} - \frac{\lambda^2}{mr^3} = -\frac{dU}{dr}. \tag{1.4.31}$$

Since the total energy

$$E = \frac{1}{2}m\left(\dot{r}^2 + r^2\dot{\phi}^2\right) + U(r) = \frac{1}{2}m\dot{r}^2 + \frac{\lambda^2}{2mr^2} + U(r) \tag{1.4.32}$$

is conserved, eqn. (1.4.32) can be inverted to give an integral expression

$$dt = \frac{dr}{\sqrt{\frac{2}{m}\left(E - U(r) - \frac{\lambda^2}{2mr^2}\right)}}$$

$$t = \int_{r(0)}^{r} \frac{dr'}{\sqrt{\frac{2}{m}\left(E - U(r') - \frac{\lambda^2}{2mr'^2}\right)}}, \tag{1.4.33}$$

which, for certain choices of the potential, can be integrated analytically and inverted to yield the trajectory $r(t)$.

1.4.2   Example: Two-particle system

Consider a two-particle system with masses m_1 and m_2, positions $\mathbf{r}_1$ and $\mathbf{r}_2$, and velocities $\dot{\mathbf{r}}_1$ and $\dot{\mathbf{r}}_2$ subject to a potential U that is a function of only the distance

$|\mathbf{r}_1 - \mathbf{r}_2|$ between them. Such would be the case, for example, in a diatomic molecule. The Lagrangian for the system can be written as

$$\mathcal{L} = \frac{1}{2}m_1\dot{\mathbf{r}}_1^2 + \frac{1}{2}m_2\dot{\mathbf{r}}_2^2 - U(|\mathbf{r}_1 - \mathbf{r}_2|). \tag{1.4.34}$$

Although such a system can easily be treated directly in terms of the Cartesian positions $\mathbf{r}_1$ and $\mathbf{r}_2$, for which the equations of motion are

$$m_1\ddot{\mathbf{r}}_1 = -U'(|\mathbf{r}_1 - \mathbf{r}_2|)\frac{\mathbf{r}_1 - \mathbf{r}_2}{|\mathbf{r}_1 - \mathbf{r}_2|}$$

$$m_2\ddot{\mathbf{r}}_2 = U'(|\mathbf{r}_1 - \mathbf{r}_2|)\frac{\mathbf{r}_1 - \mathbf{r}_2}{|\mathbf{r}_1 - \mathbf{r}_2|}, \tag{1.4.35}$$

a more natural set of coordinates can be chosen. To this end, we introduce the *center-of-mass* and *relative* coordinates defined by

$$\mathbf{R} = \frac{m_1\mathbf{r}_1 + m_2\mathbf{r}_2}{M}, \qquad \mathbf{r} = \mathbf{r}_1 - \mathbf{r}_2, \tag{1.4.36}$$

respectively, where $M = m_1 + m_2$. The inverse of this transformation is

$$\mathbf{r}_1 = \mathbf{R} + \frac{m_2}{M}\mathbf{r}, \qquad \mathbf{r}_2 = \mathbf{R} - \frac{m_1}{M}\mathbf{r}. \tag{1.4.37}$$

When eqn. (1.4.37) is substituted into eqn. (1.4.34), the Lagrangian becomes

$$\mathcal{L} = \frac{1}{2}M\dot{\mathbf{R}}^2 + \frac{1}{2}\mu\dot{\mathbf{r}}^2 - U(|\mathbf{r}|), \tag{1.4.38}$$

where $\mu = m_1m_2/M$ is known as the *reduced mass*. Since $\partial\mathcal{L}/\partial\mathbf{R} = 0$, we see that the center-of-mass coordinate is cyclic, and only the relative coordinate needs to be considered. After elimination of the center of mass, the reduced Lagrangian is $\mathcal{L} = \mu\dot{\mathbf{r}}^2/2 - U(|\mathbf{r}|)$ which gives a simple equation of motion

$$\mu\ddot{\mathbf{r}} = -U'(|\mathbf{r}|)\frac{\mathbf{r}}{|\mathbf{r}|}. \tag{1.4.39}$$

Alternatively, one could transform $\mathbf{r}$ into spherical-polar coordinates as described in the previous subsection, and solve the resulting one-dimensional equation for a single particle of mass μ moving in a central potential $U(r)$.

We hope that the reader is now convinced of the elegance and simplicity of the Lagrangian formulation of classical mechanics. Primarily, it offers a framework in which the equations of motion can be obtained in any set of coordinates. Beyond this, in Section 1.8, we will see how it connects with a more general concept, the action extremization principle, which allows the Euler-Lagrange equations to be obtained by extremization of a particular mathematical form known as the *classical action*, an object of fundamental importance in quantum statistical mechanics to be explored in Chapter 12.

1.5   Legendre transforms

We shall next derive the Hamiltonian formulation of classical mechanics. Before we can do so, we need to introduce the concept of a *Legendre transform*.

Consider a simple function $f(x)$ of a single variable x. Suppose we wish to express $f(x)$ in terms of a new variable s, where s and x are related by

$$s = f'(x) \equiv g(x) \tag{1.5.1}$$

with $f'(x) = \mathrm{d}f/\mathrm{d}x$. Can we determine $f(x)$ at a point x_0 given only $s_0 = f'(x_0) = g(x_0)$? The answer to this question, of course, is no. The reason, as Fig. 1.6 makes clear, is that s_0, being the slope of the line tangent to $f(x)$ at x_0, is also the slope of $f(x) + c$ at $x = x_0$ for any constant c. Thus, $f(x_0)$ cannot be uniquely determined

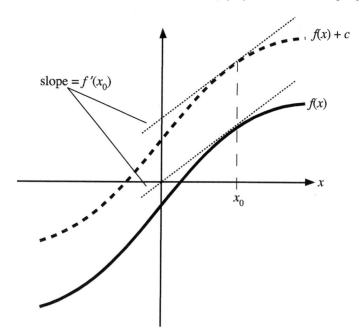

Fig. 1.6 Depiction of the Legendre transform.

from s_0. However, if we specify both the slope, $s_0 = f'(x_0)$, and the y-intercept, $b(x_0)$, of the line tangent to the function at x_0, then $f(x_0)$ can be uniquely determined. In fact, $f(x_0)$ will be given by the equation of the line tangent to the function at x_0:

$$f(x_0) = f'(x_0)x_0 + b(x_0). \tag{1.5.2}$$

Equation (1.5.2) shows how we may transform from a description of $f(x)$ in terms of x to a new description in terms of s. First, since eqn. (1.5.2) is valid for all x_0, it can be written generally in terms of x as

$$f(x) = f'(x)x + b(x). \tag{1.5.3}$$

Then, recognizing that $f'(x) = g(x) = s$ and $x = g^{-1}(s)$, and assuming that $s = g(x)$ exists and is a one-to-one mapping, it is clear that the function $b(g^{-1}(s))$, given by

$$b(g^{-1}(s)) = f(g^{-1}(s)) - sg^{-1}(s),\tag{1.5.4}$$

contains the same information as the original $f(x)$ but expressed as a function of s instead of x. We call the function $\tilde{f}(s) = b(g^{-1}(s))$ the *Legendre transform* of $f(x)$. The function $\tilde{f}(s)$ can be written compactly as

$$\tilde{f}(s) = f(x(s)) - sx(s),\tag{1.5.5}$$

where $x(s)$ serves to remind us that x is a function of s through the variable transformation $x = g^{-1}(s)$.

The generalization of the Legendre transform to a function f of n variables $x_1, ..., x_n$ is straightforward. In this case, there will be a variable transformation of the form

$$s_1 = \frac{\partial f}{\partial x_1} = g_1(x_1, ..., x_n)$$

$$\vdots$$

$$s_n = \frac{\partial f}{\partial x_n} = g_n(x_1, ..., x_n).\tag{1.5.6}$$

Again, it is assumed that this transformation is invertible so that it is possible to express each x_i as a function $x_i(s_1, ..., s_n)$ of the new variables. The Legendre transform of f will then be

$$\tilde{f}(s_1, ..., s_n) = f(x_1(s_1, ..., s_n), ..., x_n(s_1, ..., s_n)) - \sum_{i=1}^{n} s_i x_i(s_1, ..., s_n).\tag{1.5.7}$$

Note that it is also possible to perform the Legendre transform of a function with respect to any subset of the variables on which the function depends.

1.6 Generalized momenta and the Hamiltonian formulation of classical mechanics

For a first application of the Legendre transform technique, we will derive a new formulation of classical mechanics in terms of positions and momenta rather than positions and velocities. The Legendre transform will appear again numerous times in subsequent chapters. Recall that the Cartesian momentum of a particle $\mathbf{p}_i$ is just $\mathbf{p}_i = m_i\dot{\mathbf{r}}_i$. Interestingly, the momentum can also be obtained as a derivative of the Lagrangian with respect to $\dot{\mathbf{r}}_i$:

$$\mathbf{p}_i = \frac{\partial \mathcal{L}}{\partial \dot{\mathbf{r}}_i} = \frac{\partial}{\partial \dot{\mathbf{r}}_i}\left[\sum_{j=1}^{N} \frac{1}{2}m_j\dot{\mathbf{r}}_j^2 - U(\mathbf{r}_1, ..., \mathbf{r}_N)\right] = m_i\dot{\mathbf{r}}_i.\tag{1.6.1}$$

For this reason, it is clear how the Legendre transform method can be applied. We seek to derive a new function of positions and momenta as a Legendre transform of the Lagrangian with respect to the velocities. Note that, by way of eqn. (1.6.1), the velocities can be easily expressed as functions of momenta, $\dot{\mathbf{r}}_i = \dot{\mathbf{r}}_i(\mathbf{p}_i) = \mathbf{p}_i/m_i$. Therefore, substituting the transformation into eqn. (1.5.7), the new Lagrangian, denoted $\tilde{\mathcal{L}}(\mathbf{r}_1, ..., \mathbf{r}_N, \mathbf{p}_1, ..., \mathbf{p}_N)$, is given by

$$\tilde{\mathcal{L}}(\mathbf{r}_1, ..., \mathbf{r}_N, \mathbf{p}_1, ..., \mathbf{p}_N) = \mathcal{L}(\mathbf{r}_1, ..., \mathbf{r}_N, \dot{\mathbf{r}}_1(\mathbf{p}_1), ..., \dot{\mathbf{r}}_N(\mathbf{p}_N)) - \sum_{i=1}^{N} \mathbf{p}_i \cdot \dot{\mathbf{r}}_i(\mathbf{p}_i)$$

$$= \frac{1}{2} \sum_{i=1}^{N} m_i \left[\frac{\mathbf{p}_i}{m_i} \right]^2 - U(\mathbf{r}_1, ..., \mathbf{r}_N) - \sum_{i=1}^{N} \mathbf{p}_i \cdot \frac{\mathbf{p}_i}{m_i}$$

$$= -\sum_{i=1}^{N} \frac{\mathbf{p}_i^2}{2m_i} - U(\mathbf{r}_1, ..., \mathbf{r}_N). \tag{1.6.2}$$

The function $-\tilde{\mathcal{L}}(\mathbf{r}_1, ..., \mathbf{r}_N, \mathbf{p}_1, ..., \mathbf{p}_N)$ is known as the *Hamiltonian* $\mathcal{H}$:

$$\mathcal{H}(\mathbf{r}_1, ..., \mathbf{r}_N, \mathbf{p}_1, ..., \mathbf{p}_N) = \sum_{i=1}^{N} \frac{\mathbf{p}_i^2}{2m_i} + U(\mathbf{r}_1, ..., \mathbf{r}_N). \tag{1.6.3}$$

The Hamiltonian is simply the total energy of the system expressed as a function of positions and momenta and is related to the Lagrangian by

$$\mathcal{H}(\mathbf{r}_1, ..., \mathbf{r}_N, \mathbf{p}_1, ..., \mathbf{p}_N) = \sum_{i=1}^{N} \mathbf{p}_i \cdot \dot{\mathbf{r}}_i(\mathbf{p}_i) - \mathcal{L}(\mathbf{r}_1, ..., \mathbf{r}_N, \dot{\mathbf{r}}_1(\mathbf{p}_1), ..., \dot{\mathbf{r}}_N(\mathbf{p}_N)). \tag{1.6.4}$$

The momenta given in eqn. (1.6.1) are referred to as *conjugate* to the positions $\mathbf{r}_1, ..., \mathbf{r}_N$.

The relations derived above also hold for a set of generalized coordinates. The momenta $p_1, ..., p_{3N}$ conjugate to a set of generalized coordinates $q_1, ..., q_{3N}$ are given by

$$p_\alpha = \frac{\partial \mathcal{L}}{\partial \dot{q}_\alpha}, \tag{1.6.5}$$

and the Hamiltonian becomes

$$\mathcal{H}(q_1, ..., q_{3N}, p_1, ..., p_{3N}) = \sum_{\alpha=1}^{3N} p_\alpha \dot{q}_\alpha(p_1, ..., p_{3N})$$

$$- \mathcal{L}(q_1, ..., q_{3N}, \dot{q}_1(p_1, ..., p_{3N}), ..., \dot{q}_{3N}(p_1, ..., p_{3N})). \tag{1.6.6}$$

Now, according to eqn. (1.4.18), since $G_{\alpha\beta}$ is a symmetric matrix, the generalized conjugate momenta are

$$p_\alpha = \sum_{\beta=1}^{3N} G_{\alpha\beta}(q_1, ..., q_{3N}) \dot{q}_\beta. \tag{1.6.7}$$

Inverting this, we obtain the generalized velocities as

$$\dot{q}_\alpha = \sum_{\beta=1}^{3N} G_{\alpha\beta}^{-1}(q_1, ..., q_{3N})p_\beta, \tag{1.6.8}$$

where the inverse of the mass-metric tensor is

$$G_{\alpha\beta}^{-1}(q_1, ..., q_{3N}) = \sum_{i=1}^{N} \frac{1}{m_i} \left(\frac{\partial q_\alpha}{\partial \mathbf{r}_i}\right) \cdot \left(\frac{\partial q_\beta}{\partial \mathbf{r}_i}\right). \tag{1.6.9}$$

It follows that the Hamiltonian in terms of a set of generalized coordinates is

$$\mathcal{H}(q_1, ..., q_{3N}, p_1, ..., p_{3N}) = \frac{1}{2} \sum_{\alpha=1}^{3N} \sum_{\beta=1}^{3N} p_\alpha G_{\alpha\beta}^{-1}(q_1, ..., q_{3N})p_\beta$$

$$+ U(\mathbf{r}_1(q_1, ..., q_{3N}), ..., \mathbf{r}_N(q_1, ..., q_{3N})). \tag{1.6.10}$$

Given the Hamiltonian (as a Legendre transform of the Lagrangian), one can obtain the equations of motion for the system from the Hamiltonian using

$$\dot{q}_\alpha = \frac{\partial \mathcal{H}}{\partial p_\alpha}, \qquad\qquad \dot{p}_\alpha = -\frac{\partial \mathcal{H}}{\partial q_\alpha}. \tag{1.6.11}$$

Equations (1.6.11) are known as *Hamilton's equations of motion*. Whereas the Euler-Lagrange equations constitute a set of $3N$ second-order differential equations, Hamilton's equations constitute an equivalent set of $6N$ first-order differential equations. When subject to the same initial conditions, the Euler-Lagrange and Hamiltonian equations of motion must yield the same trajectory.

Hamilton's equations must be solved subject to a set of initial conditions on the coordinates and momenta, $\{q_1(0), ..., q_{3N}(0), p_1(0), ..., p_{3N}(0)\}$. Equations (1.6.11) are completely equivalent to Newton's second law of motion. In order to see this explicitly, let us apply Hamilton's equations to the simple Cartesian Hamiltonian of eqn. (1.6.3):

$$\dot{\mathbf{r}}_i = \frac{\partial \mathcal{H}}{\partial \mathbf{p}_i} = \frac{\mathbf{p}_i}{m_i}$$

$$\dot{\mathbf{p}}_i = -\frac{\partial \mathcal{H}}{\partial \mathbf{r}_i} = -\frac{\partial U}{\partial \mathbf{r}_i} = \mathbf{F}_i(\mathbf{r}). \tag{1.6.12}$$

Taking the time derivative of both sides of the first equation and substituting the result into the second yields

$$\ddot{\mathbf{r}}_i = \frac{\dot{\mathbf{p}}_i}{m_i}$$

$$\dot{\mathbf{p}}_i = m_i \ddot{\mathbf{r}}_i = \mathbf{F}_i(\mathbf{r}_1, ..., \mathbf{r}_N), \tag{1.6.13}$$

which shows that Hamilton's equations reproduce Newton's second law of motion. The reader should check that the application of Hamilton's equations to a simple harmonic oscillator, for which $\mathcal{H} = p^2/2m + kx^2/2$, yields the equation of motion $m\ddot{x} + kx = 0$.

Hamilton's equations conserve the total Hamiltonian:

$$\frac{d\mathcal{H}}{dt} = 0. \tag{1.6.14}$$

Since $\mathcal{H}$ is the total energy, eqn. (1.6.14) is just the law of energy conservation. In order to see that $\mathcal{H}$ is conserved, we simply compute the time derivative $d\mathcal{H}/dt$ via the chain rule in generalized coordinates:

$$\frac{d\mathcal{H}}{dt} = \sum_{\alpha=1}^{3N} \left[\frac{\partial \mathcal{H}}{\partial q_\alpha} \dot{q}_\alpha + \frac{\partial \mathcal{H}}{\partial p_\alpha} \dot{p}_\alpha \right]$$

$$= \sum_{\alpha=1}^{3N} \left[\frac{\partial \mathcal{H}}{\partial q_\alpha} \frac{\partial \mathcal{H}}{\partial p_\alpha} - \frac{\partial \mathcal{H}}{\partial p_\alpha} \frac{\partial \mathcal{H}}{\partial q_\alpha} \right]$$

$$= 0, \tag{1.6.15}$$

where the second line follows from Hamilton's equation, eqns. (1.6.11). We will see shortly that conservation laws, in general, are connected with physical symmetries of a system and, therefore, play an important role in the analysis of the dynamics.

Hamilton's equations of motion describe the unique evolution of the coordinates and momenta subject to a set of initial conditions. In the language of phase space, they specify a trajectory $x_t = (q_1(t), ..., q_{3N}(t), p_1(t), ..., p_{3N}(t))$ in the phase space starting from an initial point x_0. The energy conservation condition

$$\mathcal{H}(q_1(t), ..., q_{3N}(t), p_1(t), ..., p_{3N}(t)) = \text{const}$$

is expressed as a condition on a phase-space trajectory. It can also be expressed as a condition directly on the coordinates and momenta $\mathcal{H}(q_1, ..., q_{3N}, p_1, ..., p_{3N}) = \text{const}$, which defines a $6N - 1$ dimensional surface in the phase space on which a trajectory must remain. This surface is known as the *constant-energy hypersurface* or simply the *constant-energy surface*. An important theorem, known as the *work-energy theorem*, follows from the law of conservation of energy. Consider the evolution of the system from a point x_A in phase space to a point x_B. Since energy is conserved, the energy $\mathcal{H}_A = \mathcal{H}_B$. But since $\mathcal{H} = K + U$, it follows that

$$K_A + U_A = K_B + U_B, \tag{1.6.16}$$

or

$$K_A - K_B = U_B - U_A. \tag{1.6.17}$$

The right side expresses the difference in potential energy between points A and B and is, therefore, equal to the work, W_{AB}, done on the system in moving between these two points. The left side is the difference between the initial and final kinetic energy. Thus, we have a relation between the work done on the system and the kinetic energy difference

$$W_{AB} = K_A - K_B. \tag{1.6.18}$$

Note that if $W_{AB} > 0$, net work is done on the system, which means that its potential energy increases, and its kinetic energy must decrease between points A and B. If

$W_{AB} < 0$, work is done by the system, its potential energy decreases, and its kinetic energy must, therefore, increase between points A and B.

In order to understand the formal structure of a general conservation law, consider the time evolution of any arbitrary phase-space function, $a(\mathbf{x})$. Viewing $\mathbf{x}$ as a function of time $\mathbf{x}_t$, the time evolution can be analyzed by differentiating $a(\mathbf{x}_t)$ with respect to time:

$$\frac{da}{dt} = \frac{\partial a}{\partial \mathbf{x}_t} \cdot \dot{\mathbf{x}}_t$$

$$= \sum_{\alpha=1}^{3N} \left[\frac{\partial a}{\partial q_\alpha} \dot{q}_\alpha + \frac{\partial a}{\partial p_\alpha} \dot{p}_\alpha \right]$$

$$= \sum_{\alpha=1}^{3N} \left[\frac{\partial a}{\partial q_\alpha} \frac{\partial \mathcal{H}}{\partial p_\alpha} - \frac{\partial a}{\partial p_\alpha} \frac{\partial \mathcal{H}}{\partial q_\alpha} \right]$$

$$\equiv \{a, \mathcal{H}\}. \tag{1.6.19}$$

The last line is known as the *Poisson bracket* between $a(\mathbf{x})$ and $\mathcal{H}(\mathbf{x})$. The general definition of a Poisson bracket between two functions $a(\mathbf{x})$ and $b(\mathbf{x})$ is

$$\{a, b\} = \sum_{\alpha=1}^{3N} \left[\frac{\partial a}{\partial q_\alpha} \frac{\partial b}{\partial p_\alpha} - \frac{\partial a}{\partial p_\alpha} \frac{\partial b}{\partial q_\alpha} \right]. \tag{1.6.20}$$

Note that the Poisson bracket is a statement about the dependence of functions on the phase-space vector and no longer refers to time. This is an important distinction, as it will often be necessary for us to distinguish between quantities evaluated along trajectories generated from the solution of Hamilton's equations and quantities that are evaluated at arbitrary (static) points in the phase space. From eqn. (1.6.19), it is clear that if $a(\mathbf{x})$ is a conserved quantity, then $da(\mathbf{x}_t)/dt = 0$ along a trajectory, and, therefore, $\{a(\mathbf{x}), \mathcal{H}(\mathbf{x})\} = 0$ in the phase space. Conversely, if the Poisson bracket between any quantity $a(\mathbf{x})$ and the Hamiltonian of a system vanishes, then the function $a(\mathbf{x}_t)$ is conserved along a trajectory generated by Hamilton's equations.

As an example of the Poisson bracket formalism, suppose a system has no external forces acting on it. In this case, the total force $\sum_{i=1}^{N} \mathbf{F}_i = 0$, since all internal forces are balanced by Newton's third law. The condition $\sum_{i=1}^{N} \mathbf{F}_i = 0$ implies that

$$\sum_{i=1}^{N} \mathbf{F}_i = -\sum_{i=1}^{N} \frac{\partial \mathcal{H}}{\partial \mathbf{r}_i} = 0. \tag{1.6.21}$$

Now, consider the total momentum $\mathbf{P} = \sum_{i=1}^{N} \mathbf{p}_i$. Its Poisson bracket with the Hamiltonian is

$$\{\mathbf{P}, \mathcal{H}\} = \sum_{i=1}^{N} \{\mathbf{p}_i, \mathcal{H}\} = -\sum_{i=1}^{N} \frac{\partial \mathcal{H}}{\partial \mathbf{r}_i} = \sum_{i=1}^{N} \mathbf{F}_i = 0. \tag{1.6.22}$$

Hence, the total momentum $\mathbf{P}$ is conserved. When a system has no external forces acting on it, its dynamics will be the same no matter where in space the system lies.

That is, if all of the coordinates were translated by a constant vector **a** according to $\mathbf{r}'_i = \mathbf{r}_i + \mathbf{a}$, then the Hamiltonian would remain invariant. This transformation defines the so-called *translation group*. In general, if the Hamiltonian is invariant with respect to the transformations of a particular group $\mathcal{G}$, there will be an associated conservation law. This fact, known as *Noether's theorem*, is one of the cornerstones of classical mechanics and also has important implications in quantum mechanics.

Another fundamental property of Hamilton's equations is known as the condition of *phase-space incompressibility*. To understand this condition, consider writing Hamilton's equations directly in terms of the phase-space vector as

$$\dot{\mathbf{x}} = \eta(\mathbf{x}), \tag{1.6.23}$$

where $\eta(\mathbf{x})$ is a vector function of the phase-space vector x. Since

$$\mathbf{x} = (q_1, ..., q_{3N}, p_1, ..., p_{3N}),$$

it follows that

$$\eta(\mathbf{x}) = \left(\frac{\partial \mathcal{H}}{\partial p_1}, ..., \frac{\partial \mathcal{H}}{\partial p_{3N}}, -\frac{\partial \mathcal{H}}{\partial q_1}, ..., -\frac{\partial \mathcal{H}}{\partial q_{3N}} \right). \tag{1.6.24}$$

Equation (1.6.23) illustrates the fact that the general phase-space "velocity" $\dot{\mathbf{x}}$ is a function of x, suggesting that motion in phase space described by eqn. (1.6.23) can be regarded as a kind of "flow field" as in hydrodynamics, where the flow pattern of a fluid is described by a velocity field $\mathbf{v}(\mathbf{r})$. Thus, at each point in phase space, there will be a velocity vector $\dot{\mathbf{x}}(\mathbf{x})$ equal to $\eta(\mathbf{x})$. In hydrodynamics, the condition for *incompressible* flow is that there be no sources or sinks in the flow, which means that the flow field is divergence free: $\nabla \cdot \mathbf{v}(\mathbf{r}) = 0$. In phase-space flow, the analogous condition is $\nabla_{\mathbf{x}} \cdot \dot{\mathbf{x}}(\mathbf{x}) = 0$, where $\nabla_{\mathbf{x}} = \partial/\partial \mathbf{x}$ is the phase-space gradient operator. Hamilton's equations of motion guarantee that the incompressibility condition in phase space is satisfied. To see this, consider the compressibility in generalized coordinates

$$\nabla_{\mathbf{x}} \cdot \dot{\mathbf{x}} = \sum_{\alpha=1}^{3N} \left[\frac{\partial \dot{p}_\alpha}{\partial p_\alpha} + \frac{\partial \dot{q}_\alpha}{\partial q_\alpha} \right]$$

$$= \sum_{\alpha=1}^{3N} \left[-\frac{\partial}{\partial p_\alpha} \frac{\partial \mathcal{H}}{\partial q_\alpha} + \frac{\partial}{\partial q_\alpha} \frac{\partial \mathcal{H}}{\partial p_\alpha} \right]$$

$$= \sum_{\alpha=1}^{3N} \left[-\frac{\partial^2 \mathcal{H}}{\partial p_\alpha \partial q_\alpha} + \frac{\partial^2 \mathcal{H}}{\partial q_\alpha \partial p_\alpha} \right]$$

$$= 0, \tag{1.6.25}$$

where the second line follows from Hamilton's equations of motion.

One final important property of Hamilton's equations that merits comment is the so-called *symplectic structure* of the equations of motion. Given the form of the vector function, $\eta(\mathbf{x})$, introduced above, it follows that Hamilton's equations can be recast as

$$\dot{x} = M \frac{\partial \mathcal{H}}{\partial x}, \tag{1.6.26}$$

where M is a matrix expressible in block form as

$$M = \begin{pmatrix} 0 & I \\ -I & 0 \end{pmatrix}, \tag{1.6.27}$$

where **0** and **I** are the $3N \times 3N$ zero and identity matrices, respectively. Dynamical systems expressible in the form of eqn. (1.6.26) are said to possess a symplectic structure. Consider a solution x_t to eqn. (1.6.26) starting from an initial condition x_0. Because the solution of Hamilton's equations is unique for each initial condition, x_t will be a unique function of x_0, that is, $x_t = x_t(x_0)$. This dependence can be viewed as defining a variable transformation on the phase space from an initial set of phase space coordinates x_0 to a new set x_t. The Jacobian matrix J of this transformation, whose elements are given by

$$J_{kl} = \frac{\partial x_t^k}{\partial x_0^l}, \tag{1.6.28}$$

satisfies the following condition:

$$M = J^T M J, \tag{1.6.29}$$

where J^T is the transpose of J. Equation (1.6.29) is known as the *symplectic property*. We will have more to say about the symplectic property in Chapter 3. At this stage, however, let us illustrate the symplectic property in a simple example. Consider, once again, the harmonic oscillator $\mathcal{H} = p^2/2m + kx^2/2$ with equations of motion

$$\dot{x} = \frac{\partial \mathcal{H}}{\partial p} = \frac{p}{m} \qquad \dot{p} = -\frac{\partial \mathcal{H}}{\partial x} = -kx. \tag{1.6.30}$$

The general solution to these for an initial condition $(x(0), p(0))$ is

$$x(t) = x(0) \cos \omega t + \frac{p(0)}{m\omega} \sin \omega t$$
$$p(t) = p(0) \cos \omega t - m\omega x(0) \sin \omega t, \tag{1.6.31}$$

where $\omega = \sqrt{k/m}$ is the frequency of the oscillator. The Jacobian matrix is, therefore,

$$J = \begin{pmatrix} \frac{\partial x(t)}{\partial x(0)} & \frac{\partial x(t)}{\partial p(0)} \\ \frac{\partial p(t)}{\partial x(0)} & \frac{\partial p(t)}{\partial p(0)} \end{pmatrix} = \begin{pmatrix} \cos \omega t & \frac{1}{m\omega} \sin \omega t \\ -m\omega \sin \omega t & \cos \omega t \end{pmatrix}. \tag{1.6.32}$$

For this two-dimensional phase space, the matrix M is given simply by

$$M = \begin{pmatrix} 0 & 1 \\ -1 & 0 \end{pmatrix}. \tag{1.6.33}$$

Thus, performing the matrix multiplication $J^T M J$, we find

$$J^T M J = \begin{pmatrix} \cos \omega t & -m\omega \sin \omega t \\ \frac{1}{m\omega} \sin \omega t & \cos \omega t \end{pmatrix} \begin{pmatrix} 0 & 1 \\ -1 & 0 \end{pmatrix} \begin{pmatrix} \cos \omega t & \frac{1}{m\omega} \sin \omega t \\ -m\omega \sin \omega t & \cos \omega t \end{pmatrix}$$

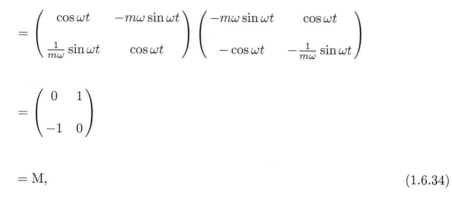

$$= \left(\begin{array}{cc} \cos\omega t & -m\omega\sin\omega t \\ \frac{1}{m\omega}\sin\omega t & \cos\omega t \end{array}\right) \left(\begin{array}{cc} -m\omega\sin\omega t & \cos\omega t \\ -\cos\omega t & -\frac{1}{m\omega}\sin\omega t \end{array}\right)$$

$$= \left(\begin{array}{cc} 0 & 1 \\ -1 & 0 \end{array}\right)$$

$$= \mathrm{M}, \tag{1.6.34}$$

showing that the symplectic condition is satisfied.

1.7 A simple classical polymer model

Before moving on to more formal developments, we present a simple classical model for a free polymer chain that can be solved analytically. This example, known as the *Rouse model* (Rouse, 1953), not only serves as a basis for more complex models of biological systems presented later but will also reappear in our discussion of quantum statistical mechanics. The model is illustrated in Fig. 1.7. It consists of a set of N point particles connected by nearest neighbor harmonic springs for which the Hamiltonian is

$$\mathcal{H} = \sum_{i=1}^{N} \frac{\mathbf{p}_i^2}{2m} + \frac{1}{2}\sum_{i=1}^{N-1} m\omega^2(|\mathbf{r}_i - \mathbf{r}_{i+1}| - b_i)^2, \tag{1.7.1}$$

where b_i is the equilibrium bond length. For simplicity, all of the particles are assigned

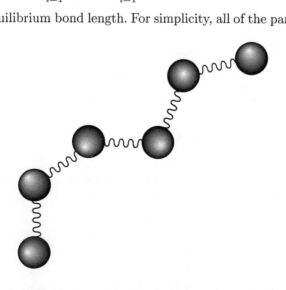

Fig. 1.7 The harmonic polymer model.

the same mass, m. Consider a one-dimensional analog of eqn. (1.7.1) described by

$$\mathcal{H} = \sum_{i=1}^{N} \frac{p_i^2}{2m} + \frac{1}{2} \sum_{i=1}^{N-1} m\omega^2 (x_i - x_{i+1} - b_i)^2. \tag{1.7.2}$$

In order to simplify the problem, we begin by making a change of variables of the form

$$\eta_i = x_i - x_{i0}, \tag{1.7.3}$$

where $x_{i0} - x_{(i+1)0} = b_i$. The Hamiltonian in terms of the new variables and their conjugate momenta is given by

$$\mathcal{H} = \sum_{i=1}^{N} \frac{p_{\eta_i}^2}{2m} + \frac{1}{2} \sum_{i=1}^{N-1} m\omega^2 (\eta_i - \eta_{i+1})^2. \tag{1.7.4}$$

The equations of motion obeyed by this simple system can be obtained directly from Hamilton's equations and take the form

$$\begin{aligned}
\dot{\eta}_i &= \frac{p_{\eta_i}}{m} \\
\dot{p}_{\eta_1} &= -m\omega^2 (\eta_1 - \eta_2) \\
\dot{p}_{\eta_i} &= -m\omega^2 (2\eta_i - \eta_{i+1} - \eta_{i-1}), \qquad i = 2, ..., N-1 \\
\dot{p}_{\eta_N} &= -m\omega^2 (\eta_N - \eta_{N-1}),
\end{aligned} \tag{1.7.5}$$

which can be expressed as second-order equations

$$\begin{aligned}
\ddot{\eta}_1 &= -\omega^2 (\eta_1 - \eta_2) \\
\ddot{\eta}_i &= -\omega^2 (2\eta_i - \eta_{i+1} - \eta_{i-1}), \qquad i = 2, ..., N-1 \\
\ddot{\eta}_N &= -\omega^2 (\eta_N - \eta_{N-1}).
\end{aligned} \tag{1.7.6}$$

In eqns. (1.7.5) and (1.7.6), it is understood that the $\eta_0 = \eta_{N+1} = 0$, since these have no meaning in our system. Equations (1.7.6) must be solved subject to a set of initial conditions $\{\eta_1(0), ..., \eta_N(0), \dot{\eta}_1(0), ..., \dot{\eta}_N(0)\}$.

The general solution to eqns. (1.7.6) can be written in the form of a Fourier series

$$\eta_i(t) = \sum_{k=1}^{N} C_k a_{ik} e^{i\omega_k t}, \tag{1.7.7}$$

where ω_k is a set of frequencies, a_{ik} is a set of expansion coefficients, and C_k is a complex scale factor. Substitution of this ansatz into eqns. (1.7.6) gives

$$\sum_{k=1}^{N} C_k \omega_k^2 a_{1k} e^{i\omega_k t} = \omega^2 \sum_{k=1}^{N} C_k e^{i\omega_k t} (a_{1k} - a_{2k})$$

$$\sum_{k=1}^{N} C_k \omega_k^2 a_{ik} e^{i\omega_k t} = \omega^2 \sum_{k=1}^{N} C_k e^{i\omega_k t} (2a_{ik} - a_{i+1,k} - a_{i-1,k})$$

$$\sum_{k=1}^{N} C_k \omega_k^2 a_{Nk} e^{i\omega_k t} = \omega^2 \sum_{k=1}^{N} C_k e^{i\omega_k t} (a_{Nk} - a_{N-1,k}). \tag{1.7.8}$$

Since eqns. (1.7.8) must be satisfied independently for each function $\exp(i\omega_k t)$, we arrive at an eigenvalue equation of the form:

$$\omega_k^2 \mathbf{a}_k = \mathbf{A}\mathbf{a}_k. \tag{1.7.9}$$

Here, $\mathbf{A}$ is a matrix given by

$$\mathbf{A} = \omega^2 \begin{pmatrix} 1 & -1 & 0 & 0 & 0 & \cdots & 0 & 0 \\ -1 & 2 & -1 & 0 & 0 & \cdots & 0 & 0 \\ 0 & -1 & 2 & -1 & 0 & \cdots & 0 & 0 \\ & & & \cdots & & & & \\ 0 & 0 & 0 & 0 & 0 & \cdots & -1 & 1 \end{pmatrix}, \tag{1.7.10}$$

and the ω_k^2 and $\mathbf{a}_k$ are the eigenvalues and eigenvectors, respectively. The square roots of the eigenvalues are frequencies that correspond to a set of special modes of the chain known as the *normal modes*. By diagonalizing the matrix $\mathbf{A}$, the frequencies can be shown to be

$$\omega_k^2 = 2\omega^2 \left[1 - \cos\left(\frac{(k-1)\pi}{N}\right) \right]. \tag{1.7.11}$$

Moreover, the orthogonal matrix U whose columns are the eigenvectors $\mathbf{a}_k$ of $\mathbf{A}$ defines a transformation from the original displacement variables η_i to a new set of variables ζ_i via

$$\zeta_i = \sum_k \eta_k U_{ki}, \tag{1.7.12}$$

known as *normal mode variables*. By applying this transformation to the Hamiltonian in eqn. (1.7.4), it can be shown that the transformed Hamiltonian is given by

$$\mathcal{H} = \sum_{k=1}^{N} \frac{p_{\zeta_k}^2}{2m} + \frac{1}{2} \sum_{k=1}^{N} m\omega_k^2 \zeta_k^2. \tag{1.7.13}$$

(The easiest way to derive this result is to start with the Lagrangian in terms of $\eta_1, ..., \eta_N, \dot{\eta}_1, ..., \dot{\eta}_N$, apply eqn. (1.7.12) to it, and then perform the Legendre transform to obtain the Hamiltonian. Alternatively, one can directly compute the inverse of the mass-metric tensor and substitute it directly into eqn. (1.6.10).) In eqn. (1.7.13), the normal modes are decoupled from each other and represent a set of independent modes with frequencies ω_k.

Note that independent of N, there is always one normal mode, the $k = 1$ mode, whose frequency is $\omega_1 = 0$. This zero-frequency mode corresponds to overall translations of the entire chain in space. In the absence of an external potential, this translational motion is free, with no associated frequency. Considering this fact, the solution of the equations of motion for each normal mode

$$\ddot{\zeta}_k + \omega_k^2 \zeta_k = 0 \tag{1.7.14}$$

can now be solved analytically:

$$\zeta_1(t) = \zeta_1(0) + \frac{p_{\zeta_1}(0)}{m} t$$

$$\zeta_k(t) = \zeta_k(0) \cos \omega_k t + \frac{p_{\zeta_k}(0)}{m \omega_k} \sin \omega_k t \qquad k = 2, ..., N, \qquad (1.7.15)$$

where $\zeta_1(0), ..., \zeta_N(0), p_{\zeta_1}(0), ..., p_{\zeta_N}(0)$ are the initial conditions on the normal mode variables, obtainable by transformation of the initial conditions of the original coordinates. Note that $p_{\zeta_1}(t) = p_{\zeta_1}(0)$ is the constant momentum of the free zero-frequency mode.

In order to better understand the physical meaning of the normal modes, consider the simple case of $N = 3$. In this case, there are three normal mode frequencies given by

$$\omega_1 = 0, \qquad \omega_2 = \omega, \qquad \omega_3 = \sqrt{3}\omega. \qquad (1.7.16)$$

Moreover, the orthogonal transformation matrix is given by

$$U = \begin{pmatrix} \frac{1}{\sqrt{3}} & \frac{1}{\sqrt{2}} & \frac{1}{\sqrt{6}} \\ \frac{1}{\sqrt{3}} & 0 & -\frac{2}{\sqrt{6}} \\ \frac{1}{\sqrt{3}} & -\frac{1}{\sqrt{2}} & \frac{1}{\sqrt{6}} \end{pmatrix}. \qquad (1.7.17)$$

Therefore, the three normal mode variables corresponding to each of these frequencies are given by

$$\omega_1 = 0: \qquad \zeta_1 = \frac{1}{\sqrt{3}} (\eta_1 + \eta_2 + \eta_3)$$

$$\omega_2 = \omega: \qquad \zeta_2 = \frac{1}{\sqrt{2}} (\eta_1 - \eta_3)$$

$$\omega_3 = \sqrt{3}\omega: \qquad \zeta_3 = \frac{1}{\sqrt{6}} (\eta_1 - 2\eta_2 + \eta_3). \qquad (1.7.18)$$

These three modes are illustrated in Fig. 1.8. Again, the zero-frequency mode corresponds to overall translations of the chain. The ω_2 mode corresponds to the motion of the two outer particles in opposite directions, with the central particle remaining fixed. This is known as the *asymmetric stretch* mode. The highest frequency ω_3 mode corresponds to symmetric motion of the two outer particles with the central particle oscillating out of phase with them. This is known as the *symmetric stretch* mode. On a final note, a more realistic model for real molecules should involve additional terms beyond just the harmonic bond interactions of eqn. (1.7.1). Specifically, there should be potential energy terms associated with bend angle and dihedral angle motion. For now, we hope that this simple harmonic polymer model illustrates the types of techniques used to solve classical problems. Indeed, the use of normal modes as a method for efficiently simulating the dynamics of biomolecules has been proposed (Sweet *et al.*,

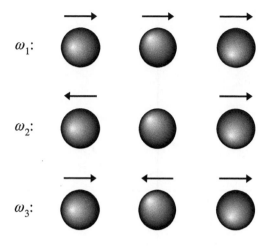

$\omega_1:$

$\omega_2:$

$\omega_3:$

Fig. 1.8 Normal modes of the harmonic polymer model for $N = 3$ particles.

2008). In general, for anharmonic potentials, one can only obtain local normal modes obtained by diagonalizing the Hessian of the potential $H_{\alpha\beta} = \partial^2 U/\partial q_\alpha \partial q_\beta$ in the vicinity of a point in configuration space.

1.8   The action integral

Having introduced the Lagrangian formulation of classical mechanics and derived the Hamiltonian formalism from it using the Legendre transform, it is natural to ask if there is a more fundamental principle that leads to the Euler-Lagrange equations. In fact, we will show that the latter can be obtained from a variational principle applied to a certain integral quantity, known as the *action integral*. At this stage, however, we shall introduce the action integral concept without motivation because in Chapter 12, we will show that the action integral emerges naturally and elegantly from quantum mechanics. The variational principle to be laid out here has more than formal significance. It has been adapted for actual trajectory calculations for large biological macromolecules by Olender and Elber (1996) and by Passerone and Parrinello (2001).

In order to define the action integral, we consider a classical system with generalized coordinates $q_1, ..., q_{3N}$ and velocities $\dot{q}_1, ..., \dot{q}_{3N}$. For notational simplicity, let us denote by Q the full set of coordinates $Q \equiv \{q_1, ..., q_{3N}\}$ and $\dot{Q}$ the full set of velocities $\dot{Q} \equiv \{\dot{q}_1, ..., \dot{q}_{3N}\}$. Suppose we follow the evolution of the system from time t_1 to t_2 with initial and final conditions $(Q_1, \dot{Q}_1)$ and $(Q_2, \dot{Q}_2)$, respectively, and we ask what path the system will take between these two points (see Fig. 1.9). We will show that the path followed renders stationary the following integral:

$$A = \int_{t_1}^{t_2} \mathcal{L}(Q(t), \dot{Q}(t)) \, dt. \qquad (1.8.1)$$

The integral in eqn. (1.8.1) is known as the *action integral*. We see immediately that the action integral depends on the entire trajectory of the system. Moreover, as specified, the action integral does not refer to one particular trajectory but to *any* trajectory

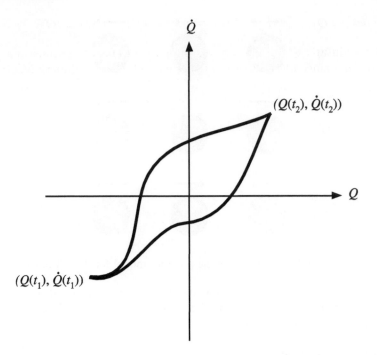

Fig. 1.9 Two proposed paths joining the fixed endpoints. The actual path followed is a stationary path of the action integral in eqn. (1.8.1.)

that takes the system from $(Q_1, \dot{Q}_1)$ to $(Q_2, \dot{Q}_2)$ in a time $t_2 - t_1$. Each trajectory satisfying these conditions yields a different value of the action. Thus, the action can be viewed as a "function" of trajectories that satisfy the initial and final conditions. However, this is not a function in the usual sense since the action is really a "function of a function." In mathematical terminology, we say that the action is a *functional* of the trajectory, which is specified by writing A as $A[Q]$. A functional is a quantity that depends on all values of a function between two points of its domain. (For readers unfamiliar with functionals, an introduction is provided in Appendix B.) Here, the action is a functional of trajectories $Q(t)$ leading from Q_1 to Q_2 in time $t_2 - t_1$. In order to express the functional dependence, the notation $A[Q]$ is commonly used. Also, since at each t, $\mathcal{L}(Q(t), \dot{Q}(t))$ only depends on t (and not on other times), $A[Q]$ is known as a *local functional* in time. Functionals will appear from time to time throughout the book, so it is important to become familiar with these objects.

Stationarity of the action means that the action does not change to first order if a small variation of a path is made keeping the endpoints fixed. In order to see that the true classical path of the system is a stationary point of A, we need to consider a path $Q(t)$ between points 1 and 2 and a second path, $Q(t) + \delta Q(t)$, between points 1 and 2 that is only slightly different from $Q(t)$. If a path $Q(t)$ renders $A[Q]$ stationary, then to first order in $\delta Q(t)$, the variation δA of the action must vanish. This can be shown by first noting that the path $Q(t)$ satisfies the initial and final conditions:

$$Q(t_1) = Q_1, \quad \dot{Q}(t_1) = \dot{Q}_1, \quad Q(t_2) = Q_2, \quad \dot{Q}(t_2) = \dot{Q}_2. \tag{1.8.2}$$

Since all paths begin at Q_1 and end at Q_2, the path $Q(t) + \delta Q(t)$ must also satisfy these conditions, and since $Q(t)$ already satisfies eqn. (1.8.2), the function $\delta Q(t)$ must satisfy

$$\delta Q(t_1) = \delta Q(t_2) = 0, \qquad \delta \dot{Q}(t_1) = \delta \dot{Q}(t_2) = 0. \tag{1.8.3}$$

The variation in the action is defined to be the difference

$$\delta A = \int_{t_1}^{t_2} \mathcal{L}(Q(t) + \delta Q(t), \dot{Q}(t) + \delta \dot{Q}(t)) \, \mathrm{d}t - \int_{t_1}^{t_2} \mathcal{L}(Q(t), \dot{Q}(t)) \, \mathrm{d}t. \tag{1.8.4}$$

This variation must vanish to first order in the path difference $\delta Q(t)$. Expanding to first order, we find:

$$\delta A = \int_{t_1}^{t_2} \mathcal{L}(Q(t), \dot{Q}(t)) \, \mathrm{d}t + \int_{t_1}^{t_2} \sum_{\alpha=1}^{3N} \left[\frac{\partial \mathcal{L}}{\partial q_\alpha} \delta q_\alpha(t) + \frac{\partial \mathcal{L}}{\partial \dot{q}_\alpha} \delta \dot{q}_\alpha(t) \right] \, \mathrm{d}t$$

$$- \int_{t_1}^{t_2} \mathcal{L}(Q(t), \dot{Q}(t)) \, \mathrm{d}t$$

$$= \int_{t_1}^{t_2} \sum_{\alpha=1}^{3N} \left[\frac{\partial \mathcal{L}}{\partial q_\alpha} \delta q_\alpha(t) + \frac{\partial \mathcal{L}}{\partial \dot{q}_\alpha} \delta \dot{q}_\alpha(t) \right] \, \mathrm{d}t. \tag{1.8.5}$$

We would like the term in brackets to involve only $\delta q_\alpha(t)$ rather than both $\delta q_\alpha(t)$ and $\delta \dot{q}_\alpha(t)$ as it currently does. We thus integrate the second term in brackets by parts to yield

$$\delta A = \sum_{\alpha=1}^{3N} \frac{\partial \mathcal{L}}{\partial \dot{q}_\alpha} \delta q_\alpha(t) \bigg|_{t_1}^{t_2} + \int_{t_1}^{t_2} \sum_{\alpha=1}^{3N} \left[\frac{\partial \mathcal{L}}{\partial q_\alpha} - \frac{\mathrm{d}}{\mathrm{d}t} \left(\frac{\partial \mathcal{L}}{\partial \dot{q}_\alpha} \right) \right] \delta q_\alpha(t) \, \mathrm{d}t. \tag{1.8.6}$$

The boundary term vanishes by virtue of eqn. (1.8.3). Then, since $\delta A = 0$ to first order in $\delta q_\alpha(t)$ at a stationary point, and each of the generalized coordinates q_α and their variations δq_α are independent, the term in brackets must vanish independently for each α. This leads to the condition

$$\frac{\mathrm{d}}{\mathrm{d}t} \left(\frac{\partial \mathcal{L}}{\partial \dot{q}_\alpha} \right) - \frac{\partial \mathcal{L}}{\partial q_\alpha} = 0, \tag{1.8.7}$$

which is just the Euler-Lagrange equation. The implication is that the path for which the action is stationary is that which satisfies the Euler-Lagrange equation. Since the latter specifies the classical motion, the path is a classical path.

There is a subtle difference, however, between classical paths that satisfy the endpoint conditions specified in the formulation of the action and those generated from a set of initial conditions as discussed in Section 1.2. In particular, if an initial-value problem has a solution, then it is unique, assuming smooth, well-behaved forces. By contrast, if a solution exists to the endpoint problem, it is not guaranteed to be a unique solution. However, it is trivial to see that if a trajectory with initial conditions $Q(t_1)$ and $\dot{Q}(t_1)$ passes through the point Q_2 at $t = t_2$, then it must also be a solution of the endpoint problem. Fortunately, in statistical mechanics, this distinction is not very important, as we are never interested in the unique trajectory arising from one particular initial condition, and in fact, initial conditions for Hamilton's equations are generally chosen at random (*e.g.*, random velocities). Typically, we are interested in the behavior of large numbers of trajectories all seeded differently. Similarly, we are rarely interested in paths leading from one specific point in phase space to another as much as paths that evolve from one *region* of phase space to another. Therefore, the initial-value and endpoint formulations of classical trajectories can often be two routes to the solution of a particular problem.

The action principle suggests the intriguing possibility that classical trajectories could be computed from an optimization procedure performed on the action given knowledge of the endpoints of the trajectory. This idea has been exploited by various researchers to study complex processes such as protein folding. As formulated, however, stationarity of the action does not imply that the action is minimum along a classical trajectory, and, indeed, the action is bounded neither from above nor below. In order to overcome this difficulty, alternative formulations of the action principle have been proposed which employ an action or a variational principle that leads to a minimization problem. The most well known of these is Hamilton's *principle of least action*. The least action principle involves a somewhat different type of variational principle in which the variations are not required to vanish at the endpoints. A detailed discussion of this type of variation, which is beyond the scope of this book, can be found in Goldstein's *Classical Mechanics* (1980).

1.9 Lagrangian mechanics and systems with constraints

In mechanics, it is often necessary to treat a system that is subject to a set of externally imposed constraints. These constraints can be imposed as a matter of convenience, e.g. constraining high-frequency chemical bonds in a molecule at fixed bond lengths, or as true constraints that might be due, for example, to the physical boundaries of a system or the presence of thermal or barostatic control mechanisms.

Constraints are expressible as mathematical relations among the phase-space variables. Thus, a system with N_c constraints will have $3N - N_c$ degrees of freedom and a set of N_c functions of the coordinates and velocities that must be satisfied by the motion of the system. Constraints are divided into two types. If the relationships that must be satisfied along a trajectory are functions of only the particle coordinates $q_1, ..., q_{3N}$ and possibly time, then the constraints are called *holonomic* and can be expressed as N_c conditions of the form

$$\sigma_k(q_1, ..., q_{3N}, t) = 0, \qquad k = 1, ..., N_c. \qquad (1.9.1)$$

If they cannot be expressed in this manner, the constraints are said to be *nonholonomic*. A class of a nonholonomic constraints consists of conditions involving both the particle positions and velocities:

$$\zeta(q_1, ..., q_{3N}, \dot{q}_1, ..., \dot{q}_{3N}) = 0. \qquad (1.9.2)$$

An example of a nonholonomic constraint is a system whose total kinetic energy is kept fixed (thermodynamically, this would be a way of fixing the temperature of the system). The nonholonomic constraint in Cartesian coordinates would then be expressed as

$$\frac{1}{2}\sum_{i=1}^{N} m_i \dot{\mathbf{r}}_i^2 - C = 0, \tag{1.9.3}$$

where C is a constant.

Since constraints reduce the number of degrees of freedom in a system, it is often possible to choose a new system of $3N - N_c$ generalized coordinates, known as a *minimal set* of coordinates, that eliminates the constraints. For example, consider the motion of a particle on the surface of a sphere. If the motion is described in terms of Cartesian coordinates (x, y, z), then a constraint condition of the form $x^2 + y^2 + z^2 - R^2 = 0$, where R is the radius of the sphere, must be imposed at all times. This constraint could be eliminated by choosing the spherical polar angle θ and ϕ as generalized coordinates. However, it is not always convenient to work in such minimal coordinate frames, particularly when there is a large number of coupled constraints. An example of this is a long hydrocarbon chain in which all carbon-carbon bonds are held rigid (an approximation, as noted earlier, that is often made to eliminate the high frequency vibrational motion). Thus, it is important to consider how the framework of classical mechanics is affected by the imposition of constraints. We will now show that the Lagrangian formulation of mechanics allows the influence of constraints to be incorporated into its framework in a transparent way.

In general, it would seem that the imposition of constraints no longer allows the equations of motion to be obtained from the stationarity of the action, since the coordinates (and/or velocities) are no longer independent. More specifically, the path displacements δq_α (cf. eqn. (1.8.6)) are no longer independent. In fact, the constraints can be built into the action formalism using the method of *Lagrange undetermined multipliers*. However, in order to apply this method, the constraint conditions must be expressible in a differential form as:

$$\sum_{\alpha=1}^{3N} a_{k\alpha} dq_\alpha + a_{kt} dt = 0, \qquad k = 1, ..., N_c, \tag{1.9.4}$$

where $a_{k\alpha}$ is a set of coefficients for the displacements dq_α. For a holonomic constraint as in eqn. (1.9.1), it is clear that the coefficients can be obtained by differentiating the constraint condition

$$\sum_{\alpha=1}^{N} \frac{\partial \sigma_k}{\partial q_\alpha} dq_\alpha + \frac{\partial \sigma_k}{\partial t} dt = 0 \tag{1.9.5}$$

so that

$$a_{k\alpha} = \frac{\partial \sigma_k}{\partial q_\alpha}, \qquad a_{kt} = \frac{\partial \sigma_k}{\partial t}. \tag{1.9.6}$$

Nonholonomic constraints cannot always be expressed in the form of eqn. (1.9.4). A notable exception is the kinetic energy constraint of eqn. (1.9.3):

$$\sum_{i=1}^{N} \frac{1}{2} m_i \dot{\mathbf{r}}_i^2 - C = 0$$

$$\sum_{i=1}^{N} \frac{1}{2} m_i \dot{\mathbf{r}}_i \cdot \left(\frac{\mathrm{d}\mathbf{r}_i}{\mathrm{d}t} \right) - C = 0$$

$$\sum_{i=1}^{N} \frac{1}{2} m_i \dot{\mathbf{r}}_i \cdot \mathrm{d}\mathbf{r}_i - C \mathrm{d}t = 0 \tag{1.9.7}$$

so that

$$\mathbf{a}_{1i} = \frac{1}{2} m_i \dot{\mathbf{r}}_i, \qquad a_{1t} = C \tag{1.9.8}$$

($k = 1$ since there is only a single constraint).

Assuming that the constraints can be expressed in the differential form of eqn. (1.9.4), we must also be able to express them in terms of path displacements δq_α in order to incorporate them into the action principle. Unfortunately, doing so requires a further restriction, since it is not possible to guarantee that a perturbed path $Q(t) + \delta Q(t)$ satisfies the constraints. The latter will hold if the constraints are integrable, in which case they are expressible in terms of path displacements as

$$\sum_{\alpha=1}^{3N} a_{k\alpha} \delta q_\alpha = 0. \tag{1.9.9}$$

The coefficient a_{kt} does not appear in eqn. (1.9.9) because there is no time displacement. The equations of motion can then be obtained by adding eqn. (1.9.9) to eqn. (1.8.6) with a set of Lagrange undetermined multipliers, λ_k, where there is one multiplier for each constraint, according to

$$\delta A = \int_{t_1}^{t_2} \sum_{\alpha=1}^{3N} \left[\frac{\partial \mathcal{L}}{\partial q_\alpha} - \frac{\mathrm{d}}{\mathrm{d}t} \left(\frac{\partial \mathcal{L}}{\partial \dot{q}_\alpha} \right) + \sum_{k=1}^{N_c} \lambda_k a_{k\alpha} \right] \delta q_\alpha(t) \, \mathrm{d}t. \tag{1.9.10}$$

The equations of motion obtained by requiring that $\delta A = 0$ are then

$$\frac{\mathrm{d}}{\mathrm{d}t} \left(\frac{\partial \mathcal{L}}{\partial \dot{q}_\alpha} \right) - \frac{\partial \mathcal{L}}{\partial q_\alpha} = \sum_{k=1}^{N_c} \lambda_k a_{k\alpha}. \tag{1.9.11}$$

It may seem that we are still relying on the independence of the displacements δq_α, but this is actually not the case. Suppose we choose the first $3N - N_c$ coordinates to be independent. Then, these coordinates can be evolved using eqns. (1.9.11). However, we can choose λ_k such that eqns. (1.9.11) apply to the remaining N_c coordinates as well. In this case, eqns. (1.9.11) hold for all $3N$ coordinates provided they are solved subject to the constraint conditions. The latter can be expressed as a set of N_c differential equations of the form

$$\sum_{\alpha=1}^{3N} a_{k\alpha} \dot{q}_\alpha + a_{kt} = 0. \tag{1.9.12}$$

Equations (1.9.11) together with eqn. (1.9.12) constitute a set of $3N + N_c$ equations for the $3N + N_c$ unknowns, $q_1, ..., q_{3N}, \lambda_1, ..., \lambda_{N_c}$. This is the most common approach used in numerical solutions of classical-mechanical problems.

Note that, even if a system is subject to a set of *time-independent* holonomic constraints ($a_{kt} = 0$), the Hamiltonian is still conserved. In order to see this, note that eqns. (1.9.11) and (1.9.12) can be cast in Hamiltonian form as

$$\dot{q}_\alpha = \frac{\partial \mathcal{H}}{\partial p_\alpha}$$

$$\dot{p}_\alpha = -\frac{\partial \mathcal{H}}{\partial q_\alpha} - \sum_{k=1}^{N_c} \lambda_k a_{k\alpha}$$

$$\sum_{\alpha=1}^{3N} a_{k\alpha} \frac{\partial \mathcal{H}}{\partial p_\alpha} = 0. \tag{1.9.13}$$

Computing the time-derivative of the Hamiltonian, we obtain

$$\frac{d\mathcal{H}}{dt} = \sum_{\alpha=1}^{3N} \left[\frac{\partial \mathcal{H}}{\partial q_\alpha} \dot{q}_\alpha + \frac{\partial \mathcal{H}}{\partial p_\alpha} \dot{p}_\alpha \right]$$

$$= \sum_{\alpha=1}^{3N} \left[\frac{\partial \mathcal{H}}{\partial q_\alpha} \frac{\partial \mathcal{H}}{\partial p_\alpha} - \frac{\partial \mathcal{H}}{\partial p_\alpha} \left(\frac{\partial \mathcal{H}}{\partial q_\alpha} + \sum_{k=1}^{N_c} \lambda_k a_{k\alpha} \right) \right]$$

$$= \sum_{k=1}^{N_c} \lambda_k \sum_{\alpha=1}^{3N} \frac{\partial \mathcal{H}}{\partial p_\alpha} a_{k\alpha}$$

$$= 0. \tag{1.9.14}$$

From this, it is clear that no work is done on a system by the imposition of holonomic constraints.

1.10   Gauss's principle of least constraint

The constrained equations of motion (1.9.11) and (1.9.12) constitute a complete set of equations for the motion subject to the N_c constraint conditions. Let us study these equations in more detail. To keep the notation simple, let us consider just a single particle in three dimensions described by a Cartesian position vector $\mathbf{r}(t)$ subject to a single constraint $\sigma(\mathbf{r}) = 0$. According to eqns. (1.9.11) and (1.9.12), the constrained equations of motion take the form

$$m\ddot{\mathbf{r}} = \mathbf{F}(\mathbf{r}) + \lambda \nabla \sigma$$

$$\nabla \sigma \cdot \dot{\mathbf{r}} = 0. \tag{1.10.1}$$

These equations will generate classical trajectories of the system for different initial conditions $\{\mathbf{r}(0), \dot{\mathbf{r}}(0)\}$ provided the condition $\sigma(\mathbf{r}(0)) = 0$ is satisfied. If this condition is true, then the trajectory will obey $\sigma(\mathbf{r}(t)) = 0$. Conversely, for each $\mathbf{r}$ visited along the trajectory, the condition $\sigma(\mathbf{r}) = 0$ will be satisfied. The latter condition defines a surface on which the motion described by eqns. (1.10.1) must remain. This surface is called the *surface of constraint*. The quantity $\nabla\sigma(\mathbf{r})$ is a vector that is orthogonal to the surface at each point $\mathbf{r}$. Thus, the second equation (1.10.1) expresses the fact that the velocity must also lie in the surface of constraint, hence it must be perpendicular to $\nabla\sigma(\mathbf{r})$. Of the two force terms appearing in eqns. (1.10.1), the first is an "unconstrained" force which, alone, would allow the particle to drift off of the surface of constraint. The second term must, then, correct for this tendency. If the particle starts from rest, this second term exactly removes the component of the force perpendicular to the surface of constraint as illustrated in Fig. 1.10. This minimal projection

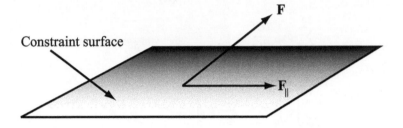

Fig. 1.10 Representation of the minimal force projection embodied in Gauss's principle of least constraint.

of the force, first conceived by Karl Friedrich Gauss (1777-1855), is known as *Gauss's principle of least constraint* (Gauss, 1829). The component of the force perpendicular to the surface is

$$\mathbf{F}_{\perp}(\mathbf{r}) = [\mathbf{n}(\mathbf{r}) \cdot \mathbf{F}(\mathbf{r})]\,\mathbf{n}(\mathbf{r}), \qquad (1.10.2)$$

where $\mathbf{n}(\mathbf{r})$ is a unit vector perpendicular to the surface at $\mathbf{r}$; $\mathbf{n}$ is given by

$$\mathbf{n}(\mathbf{r}) = \frac{\nabla\sigma(\mathbf{r})}{|\nabla\sigma(\mathbf{r})|}. \qquad (1.10.3)$$

Thus, the component of the force parallel to the surface is

$$\mathbf{F}_{\parallel}(\mathbf{r}) = \mathbf{F}(\mathbf{r}) - \mathbf{F}_{\perp}(\mathbf{r}) = \mathbf{F}(\mathbf{r}) - [\mathbf{n}(\mathbf{r}) \cdot \mathbf{F}(\mathbf{r})]\,\mathbf{n}(\mathbf{r}). \qquad (1.10.4)$$

If the particle is not at rest, the projection of the force cannot lie entirely in the surface of constraint. Rather, there must be an additional component of the projection which can project any free motion of the particle directed off the surface of constraint. This additional term must sense the curvature of the surface in order to affect the required projection; it must also be a minimal projection perpendicular to the surface.

In order to show that Gauss's principle is consistent with the Lagrangian formulation of the constraint problem and find the additional projection when the particle's velocity is not zero, we make use of the second of eqns. (1.10.1) and differentiate it once with respect to time. This yields:

$$\nabla\sigma \cdot \ddot{\mathbf{r}} + \nabla\nabla\sigma \cdot \cdot\,\dot{\mathbf{r}}\dot{\mathbf{r}} = 0, \qquad (1.10.5)$$

where the double dot-product notation in the expression $\nabla\nabla\sigma \cdot\cdot\dot{\mathbf{r}}\dot{\mathbf{r}}$ indicates a full contraction of the two tensors $\nabla\nabla\sigma$ and $\dot{\mathbf{r}}\dot{\mathbf{r}}$. The first of eqns. (1.10.1) is then used to substitute in for the second time derivative appearing in eqn. (1.10.5) to yield:

$$\nabla\sigma \cdot \left[\frac{\mathbf{F}}{m} + \frac{\lambda\nabla\sigma}{m} \right] + \nabla\nabla\sigma \cdot\cdot\dot{\mathbf{r}}\dot{\mathbf{r}} = 0. \tag{1.10.6}$$

We can now solve for the Lagrange multiplier λ to yield the analytical expression

$$\lambda = -\frac{\nabla\nabla\sigma \cdot\cdot\dot{\mathbf{r}}\dot{\mathbf{r}} + \nabla\sigma \cdot \mathbf{F}/m}{|\nabla\sigma|^2/m}. \tag{1.10.7}$$

Finally, substituting eqn. (1.10.7) back into the first of eqns. (1.10.1) yields the equation of motion:

$$m\ddot{\mathbf{r}} = \mathbf{F} - \frac{\nabla\nabla\sigma \cdot\cdot\dot{\mathbf{r}}\dot{\mathbf{r}} + \nabla\sigma \cdot \mathbf{F}/m}{|\nabla\sigma|^2/m}\nabla\sigma, \tag{1.10.8}$$

which is known as *Gauss's equation of motion*. Note that when $\dot{\mathbf{r}} = 0$, the total force appearing on the right is

$$\mathbf{F} - \frac{(\nabla\sigma \cdot \mathbf{F})\nabla\sigma}{|\nabla\sigma|^2} = \mathbf{F} - \left(\frac{\nabla\sigma}{|\nabla\sigma|} \cdot \mathbf{F} \right) \frac{\nabla\sigma}{|\nabla\sigma|}, \tag{1.10.9}$$

which is just the projected force in eqn. (1.10.4). For $\dot{\mathbf{r}} \neq 0$, the additional force term involves a curvature term $\nabla\nabla\sigma$ contracted with the velocity-vector dyad $\dot{\mathbf{r}}\dot{\mathbf{r}}$. Since this term would be present even if $\mathbf{F} = 0$, this term clearly corrects for free motion off the surface of constraint.

Having eliminated λ from the equation of motion, eqn. (1.10.8) becomes an equation involving a velocity-dependent force. This equation, alone, generates motion on the correct constraint surface, has a conserved energy, $E = m\dot{\mathbf{r}}^2/2 + U(\mathbf{r})$, and, by construction, conserves $\sigma(\mathbf{r})$ in the sense that $d\sigma/dt = 0$ along a trajectory. However, this equation cannot be derived from a Lagrangian or a Hamiltonian and, therefore, constitutes an example of *non-Hamiltonian* dynamical system (see also Section 1.12). Gauss's procedure for obtaining constrained equations of motion can be generalized to an arbitrary number of particles or constraints satisfying the proper differential constraints relations.

1.11 Rigid body motion: Euler angles and quaternions

The discussion of constraints leads naturally to the topic of rigid body motion. Rigid body techniques can be particularly useful in treating small molecules such as water or ammonia or large, approximately rigid subdomains of large molecules, in that these techniques circumvent the need to treat large numbers of explicit degrees of freedom. Imagine a collection of n particles with all interparticle distances constrained. Such a system, known as a *rigid body*, has numerous applications in mechanics and statistical mechanics. An example in chemistry is the approximate treatment of small molecules with very high frequency internal vibrations. A water molecule (H_2O) could be treated as a rigid isosceles triangle by constraining the two OH bond lengths and the distance between the two hydrogens for a total of three holonomic constraint conditions. An ammonia molecule (NH_3) could also be treated as a rigid pyramid by fixing the three NH bond lengths and the three HH distances for a total of six holonomic constraint conditions. In a more complex molecule, such as the alanine dipeptide shown in Fig. 1.11, specific groups can be treated as rigid. Groups of this type are shaded in Fig. 1.11.

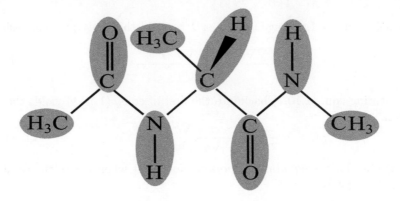

Fig. 1.11 Rigid subgroups in a large molecule, the alanine dipeptide.

Of course, it is always possible to treat these constraint conditions explicitly using the Lagrange multiplier formalism. However, since all the particles in a rigid body move as a whole, a simple and universal formalism can be used to treat all rigid bodies that circumvents the need to impose explicitly the set of holonomic constraints that keep the particles at fixed relative positions. Before discussing rigid body motion, let us consider the problem of rotating a rigid body about an arbitrary axis in a fixed frame. Since a rotation performed on a rigid body moves all of the atoms uniformly, it is sufficient for us to consider how to rotate a single vector **r** about an arbitrary axis. The problem is illustrated in Fig. 1.12. Let **n** designate a unit vector along the axis of rotation, and let **r**' be the result of rotating **r** by an angle θ clockwise about the axis. In the notation of Fig. 1.12, straightforward geometry shows that **r**' is the result of a simple vector addition:

$$\mathbf{r}' = \overrightarrow{OC} + \overrightarrow{CS} + \overrightarrow{SQ}. \tag{1.11.1}$$

Since the angle CSQ is a right angle, the three vectors in eqn. (1.11.1) can be expressed in terms of the original vector **r**, the angle θ, and the unit vector **n** according to

$$\mathbf{r}' = \mathbf{n}(\mathbf{n} \cdot \mathbf{r}) + [\mathbf{r} - \mathbf{n}(\mathbf{n} \cdot \mathbf{r})] \cos\theta + (\mathbf{r} \times \mathbf{n}) \sin\theta, \tag{1.11.2}$$

which can be rearranged to read

$$\mathbf{r}' = \mathbf{r} \cos\theta + \mathbf{n}(\mathbf{n} \cdot \mathbf{r})(1 - \cos\theta) + (\mathbf{r} \times \mathbf{n}) \sin\theta. \tag{1.11.3}$$

Equation (1.11.3) is known as the *rotation formula*, which can be used straightforwardly when an arbitrary rotation needs to be performed.

In order to illustrate the concept of rigid body motion, consider the simple problem of a rigid homonuclear diatomic molecule in two dimensions, in which each atom has

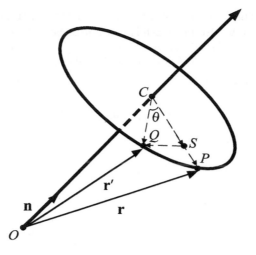

Fig. 1.12 Rotation of the vector **r** to **r**′ about an axis **n**.

a mass m. We shall assume that the motion occurs in the xy plane. Let the Cartesian positions of the two atoms be $\mathbf{r}_1$ and $\mathbf{r}_2$ and let the molecule be subject to a potential of the form $U(\mathbf{r}_1 - \mathbf{r}_2)$. The Lagrangian for the molecule can be written as

$$\mathcal{L} = \frac{1}{2}m\dot{\mathbf{r}}_1^2 + \frac{1}{2}m\dot{\mathbf{r}}_2^2 - U(\mathbf{r}_1 - \mathbf{r}_2). \tag{1.11.4}$$

For such a problem, it is useful to transform to center-of-mass $\mathbf{R} = (\mathbf{r}_1 + \mathbf{r}_2)/2$ and relative $\mathbf{r} = \mathbf{r}_1 - \mathbf{r}_2$ coordinates, in terms of which the Lagrangian becomes

$$\mathcal{L} = \frac{1}{2}M\dot{\mathbf{R}}^2 + \frac{1}{2}\mu\dot{\mathbf{r}}^2 - U(\mathbf{r}), \tag{1.11.5}$$

where $M = 2m$ and $\mu = m/2$ are the total and reduced masses, respectively. Note that in these coordinates, the center of mass has an equation of motion of the form $M\ddot{\mathbf{R}} = 0$, which is the equation of motion of a free particle. As we have already seen, this means that the center-of-mass velocity $\dot{\mathbf{R}}$ is a constant. According to the principle of *Galilean relativity*, the physics of the system must be the same in a fixed coordinate frame as in a coordinate frame moving with constant velocity. Thus, we may transform to a coordinate system that moves with the molecule. Such a coordinate frame is called a *body-fixed frame*. The origin of the body-fixed frame lies at the center of mass of the molecule, and in this frame the coordinates of the two atoms $\mathbf{r}_1 = \mathbf{r}/2$ and $\mathbf{r}_2 = -\mathbf{r}/2$. It is, therefore, clear that only the motion of the relative coordinate $\mathbf{r}$ needs to be considered. Note that we may use the body-fixed frame even if the center-of-mass velocity is not constant in order to separate the rotational and translational kinetic energies of the rigid body. In the body-fixed frame, the Lagrangian of eqn. (1.11.5) becomes

$$\mathcal{L} = \frac{1}{2}\mu\dot{\mathbf{r}}^2 - U(\mathbf{r}). \tag{1.11.6}$$

In a two-dimensional space, the relative coordinate $\mathbf{r}$ is the two-component vector $\mathbf{r} = (x, y)$. However, if the distance between the two atoms is fixed at a value d, then there is a constraint in the form of $x^2 + y^2 = d^2$. Rather than treating the constraint via a Lagrange multiplier, we could transform to polar coordinates according to

$$x = d\cos\theta \qquad y = d\sin\theta. \tag{1.11.7}$$

The velocities are given by

$$\dot{x} = -d(\sin\theta)\dot{\theta} \qquad \dot{y} = d(\cos\theta)\dot{\theta} \tag{1.11.8}$$

so that the Lagrangian becomes

$$\mathcal{L} = \frac{1}{2}\mu\left(\dot{x}^2 + \dot{y}^2\right) - U(x,y) = \frac{1}{2}\mu d^2\dot{\theta}^2 - \tilde{U}(\theta), \tag{1.11.9}$$

where the notation, $U(\mathbf{r}) = U(x,y) = U(d\cos\theta, d\sin\theta) \equiv \tilde{U}(\theta)$, indicates that the potential varies only according to the variation in θ. Equation (1.11.9) demonstrates that the rigid body has only one degree of freedom, namely, the single angle θ. According to the Euler-Lagrange equation (1.4.6), the equation of motion for the angle is

$$\mu d^2\ddot{\theta} = -\frac{\partial\tilde{U}}{\partial\theta}. \tag{1.11.10}$$

In order to understand the physical content of eqn. (1.11.10), we first note that the origin of the body-fixed frame lies at the center-of-mass position $\mathbf{R}$. We further note the motion occurs in the xy plane and therefore consists of rotation about an axis perpendicular to this plane, in this case, about the z-axis. The quantity $\omega = \dot{\theta}$ is called the *angular velocity* about the z-axis. The quantity $I = \mu d^2$ is a constant known as the *moment of inertia* about the z-axis. Since the motion is purely angular, we can define an angular momentum, analogous to the Cartesian momentum, by $l = \mu d^2\dot{\theta} = I\omega$. In general, angular momentum, like the Cartesian momentum, is a vector quantity given by

$$\mathbf{l} = \mathbf{r}\times\mathbf{p}. \tag{1.11.11}$$

For the present problem, in which all of the motion occurs in the xy-plane (no motion along the z-axis), $\mathbf{l}$ has only one nonzero component, namely l_z, given by

$$l_z = xp_y - yp_x$$

$$= x(\mu\dot{y}) - y(\mu\dot{x})$$

$$= \mu(x\dot{y} - y\dot{x})$$

$$= \mu d^2(\dot{\theta}\cos^2\theta + \dot{\theta}\sin^2\theta)$$

$$= \mu d^2\dot{\theta}. \tag{1.11.12}$$

Equation (1.11.12) demonstrates that although the motion occurs *about* the z-axis, the direction of the angular momentum vector is *along* the z-axis. Since the angular momentum l_z is given as the product of the moment of inertia I and the angular velocity ω ($l_z = I\omega$), the angular velocity must also be a vector whose direction is along the z-axis. Thus, we write the angular velocity vector for this problem as $\boldsymbol{\omega} = (0, 0, \dot{\theta})$ and $\mathbf{l} = I\boldsymbol{\omega}$. Physically, we see that the moment of inertia plays the role of "mass" in angular motion; however, its units are mass$\times$length^2. The form of the moment of inertia indicates that the farther away from the axis of rotation an object is, the greater will be its angular momentum, although its angular velocity is the same at all distances from the axis of rotation.

It is interesting to calculate the velocity in the body-fixed frame. The components of the velocity $\mathbf{v} = (v_x, v_y) = (\dot{x}, \dot{y})$ are given by eqn. (1.11.8). Note, however, that these can also be expressed in terms of the angular velocity vector, $\boldsymbol{\omega}$. In particular, the velocity vector is expressible as a cross product

$$\mathbf{v} = \dot{\mathbf{r}} = \boldsymbol{\omega} \times \mathbf{r}. \tag{1.11.13}$$

Since $\boldsymbol{\omega} = (0, 0, \dot{\theta})$, the cross product has two nonvanishing components

$$v_x = -\omega_z y = -d(\sin\theta)\dot{\theta}$$
$$v_y = \omega_z x = d(\cos\theta)\dot{\theta}, \tag{1.11.14}$$

which are precisely the velocity components given by eqn. (1.11.8). Equation (1.11.14) determines the velocity of the relative position vector $\mathbf{r}$. In the body-fixed frame, the velocities of atoms 1 and 2 would be $-\dot{\mathbf{r}}/2$ and $\dot{\mathbf{r}}/2$, respectively. If we wish to determine the velocity of, for example, atom 1 at position $\mathbf{r}_1$ in a space-fixed frame rather than in the body-fixed frame, we need to add back the velocity of the body-fixed frame. To do this, write the position $\mathbf{r}_1$ as

$$\mathbf{r}_1 = \mathbf{R} + \frac{1}{2}\mathbf{r}. \tag{1.11.15}$$

Thus, the total velocity $\mathbf{v}_1 = \dot{\mathbf{r}}_1$ is

$$\dot{\mathbf{r}}_1 = \dot{\mathbf{R}} + \frac{1}{2}\dot{\mathbf{r}}. \tag{1.11.16}$$

The first term is clearly the velocity of the body-fixed frame, while the second term is the velocity of $\mathbf{r}_1$ relative to the body-fixed frame. Note, however, that if the motion of $\mathbf{r}_1$ relative to the body-fixed frame is removed, the remaining component of the velocity is just that due to the motion of the body-fixed frame, and we may write

$$\left(\frac{d\mathbf{r}_1}{dt}\right)_{\text{body}} = \frac{d\mathbf{R}}{dt}. \tag{1.11.17}$$

Since $\dot{\mathbf{r}} = \boldsymbol{\omega} \times \mathbf{r}$, the total time derivative of the vector $\mathbf{r}_1$ becomes

$$\left(\frac{d\mathbf{r}_1}{dt}\right)_{\text{space}} = \left(\frac{d\mathbf{r}_1}{dt}\right)_{\text{body}} + \boldsymbol{\omega} \times \frac{1}{2}\mathbf{r}$$

$$\left(\frac{d\mathbf{r}_1}{dt}\right)_{\text{space}} = \left(\frac{d\mathbf{r}_1}{dt}\right)_{\text{body}} + \boldsymbol{\omega} \times \mathbf{r}_1, \tag{1.11.18}$$

where the first term in the second line is interpreted as the velocity due solely to the motion of the body-fixed frame and the second term is the rate of change of $\mathbf{r}_1$

in the body-fixed frame. A similar relation can be derived for the time derivative of the position $\mathbf{r}_2$ of atom 2. Indeed, eqn. (1.11.18) applies to the time derivative of any arbitrary vector $\mathbf{G}$:

$$\left(\frac{d\mathbf{G}}{dt}\right)_{space} = \left(\frac{d\mathbf{G}}{dt}\right)_{body} + \boldsymbol{\omega} \times \mathbf{G}. \qquad (1.11.19)$$

Although it is possible to obtain eqn. (1.11.19) from a general consideration of rotational motion, we content ourselves here with this physically motivated approach.

Consider, next, the force term $-\partial U/\partial \theta$. This is also a component of a vector quantity known as the *torque* about the z-axis. In general, $\boldsymbol{\tau}$ is defined by

$$\boldsymbol{\tau} = \mathbf{r} \times \mathbf{F}. \qquad (1.11.20)$$

Again, because the motion is entirely in the xy-plane, there is no z-component of the force, and the only nonvanishing component of the torque is the z-component given by

$$\tau_z = x F_y - y F_x$$

$$= -d\cos\theta \frac{\partial U}{\partial y} + d\sin\theta \frac{\partial U}{\partial x}$$

$$= -d\cos\theta \frac{\partial U}{\partial \theta}\frac{\partial \theta}{\partial y} + d\sin\theta \frac{\partial U}{\partial \theta}\frac{\partial \theta}{\partial x}, \qquad (1.11.21)$$

where the chain rule has been used in the last line. Since $\theta = \tan^{-1}(y/x)$, the two derivatives of θ can be worked out as

$$\frac{\partial \theta}{\partial y} = \frac{1}{1 + (y/x)^2}\frac{1}{x}$$

$$= \frac{x}{x^2 + y^2} = \frac{\cos\theta}{d}$$

$$\frac{\partial \theta}{\partial x} = \frac{1}{1 + (y/x)^2}\left(-\frac{y}{x^2}\right)$$

$$= -\frac{y}{x^2 + y^2} = -\frac{\sin\theta}{d}. \qquad (1.11.22)$$

Substitution of eqn. (1.11.22) into eqn. (1.11.21) gives

$$\tau_z = -\frac{\partial U}{\partial \theta}\left(d\cos\theta\frac{\cos\theta}{d} + d\sin\theta\frac{\sin\theta}{d}\right)$$

$$= -\frac{\partial U}{\partial \theta}\left(\cos^2\theta + \sin^2\theta\right)$$

$$= -\frac{\partial U}{\partial \theta}. \tag{1.11.23}$$

Therefore, we see that the torque is simply the force on an angular coordinate. The equation of motion can thus be written in vector form as

$$\frac{d\mathbf{l}}{dt} = \boldsymbol{\tau}, \tag{1.11.24}$$

which is analogous to Newton's second law in Cartesian form

$$\frac{d\mathbf{p}}{dt} = \mathbf{F}. \tag{1.11.25}$$

A rigid diatomic, being a linear object, can be described by a single angle coordinate in two dimensions or by two angles in three dimensions. For a general rigid body consisting of n atoms in three dimensions, the number of constraints needed to make it rigid is $3n-6$ so that the number of remaining degrees of freedom is $3n-(3n-6) = 6$. After removing the three degrees of freedom associated with the motion of the body-fixed frame, we are left with three degrees of freedom, implying that three angles are needed to describe the motion of a general rigid body. These three angles are known as the *Euler angles*. They describe the motion of the rigid body about three independent axes. Although several conventions exist for defining these axes, any choice is acceptable.

A particularly convenient choice of the axes can be obtained as follows: Consider the total angular momentum of the rigid body, obtained as a sum of the individual angular momentum vectors of the constituent particles:

$$\mathbf{l} = \sum_{i=1}^{n} \mathbf{r}_i \times \mathbf{p}_i = \sum_{i=1}^{n} m_i \mathbf{r}_i \times \mathbf{v}_i. \tag{1.11.26}$$

Now, $\mathbf{v}_i = d\mathbf{r}_i/dt$ is measured in the body-fixed frame. From the analysis above, it follows that the velocity is just $\boldsymbol{\omega} \times \mathbf{r}_i$ so that

$$\mathbf{l} = \sum_{i=1}^{n} m_i \mathbf{r}_i \times (\boldsymbol{\omega} \times \mathbf{r}_i). \tag{1.11.27}$$

Expanding the double cross product, we find that

$$\mathbf{l} = \sum_{i=1}^{n} m_i \left(\boldsymbol{\omega} r_i^2 - \mathbf{r}_i (\mathbf{r}_i \cdot \boldsymbol{\omega}) \right), \tag{1.11.28}$$

which, in component form, becomes

$$l_x = \omega_x \sum_{i=1}^{n} m_i (r_i^2 - x_i^2) - \omega_y \sum_{i=1}^{n} m_i x_i y_i - \omega_z \sum_{i=1}^{n} m_i x_i z_i$$

$$l_y = -\omega_x \sum_{i=1}^{n} m_i y_i x_i + \omega_y \sum_{i=1}^{n} m_i (r_i^2 - y_i^2) - \omega_z \sum_{i=1}^{n} m_i y_i z_i$$

$$l_z = -\omega_x \sum_{i=1}^{n} m_i z_i x_i - \omega_y \sum_{i=1}^{n} m_i z_i y_i + \omega_z \sum_{i=1}^{n} m_i (r_i^2 - z_i^2). \qquad (1.11.29)$$

Equation (1.11.29) can be written in matrix form as

$$\begin{pmatrix} l_x \\ l_y \\ l_z \end{pmatrix} = \begin{pmatrix} I_{xx} & I_{xy} & I_{xz} \\ I_{yx} & I_{yy} & I_{yz} \\ I_{zx} & I_{zy} & I_{zz} \end{pmatrix} \begin{pmatrix} \omega_x \\ \omega_y \\ \omega_z \end{pmatrix}. \qquad (1.11.30)$$

The matrix elements are given by

$$I_{\alpha\beta} = \sum_{i=1}^{n} m_i \left(r_i^2 \delta_{\alpha\beta} - r_{i,\alpha} r_{i,\beta} \right), \qquad (1.11.31)$$

where $\alpha, \beta = (x, y, z)$ and $r_{i,\alpha}$ is the αth component of the ith position vector in the body-fixed frame. The matrix $I_{\alpha\beta}$ is known as the *inertia tensor* and is the generalization of the moment of inertia defined previously. The inertia tensor is symmetric $(I_{\alpha\beta} = I_{\beta\alpha})$, and can therefore be diagonalized via an orthogonal transformation. Consequently, it has real eigenvalues denoted I_1, I_2 and I_3. The eigenvectors of the inertia tensor define a new set of mutually orthogonal axes about which we may describe the motion. When these axes are used, the inertia tensor is diagonal. Since the angular momentum is obtained by acting with the inertia tensor on the angular velocity vector, it follows that, in general, l is not parallel to $\boldsymbol{\omega}$ as it was in the two-dimensional problem considered above. Thus, $\boldsymbol{\omega} \times \mathbf{l} \neq 0$ so that the time derivative of l in a space-fixed frame obeys eqn. (1.11.19):

$$\left(\frac{\mathrm{d}\mathbf{l}}{\mathrm{d}t} \right)_{\text{space}} = \left(\frac{\mathrm{d}\mathbf{l}}{\mathrm{d}t} \right)_{\text{body}} + \boldsymbol{\omega} \times \mathbf{l}. \qquad (1.11.32)$$

Accordingly, the rate of change of l in the space-fixed frame will be determined simply by the torque according to

$$\left(\frac{\mathrm{d}\mathbf{l}}{\mathrm{d}t} \right)_{\text{space}} = \boldsymbol{\tau}. \qquad (1.11.33)$$

Expressing this in terms of the body-fixed frame (and dropping the "body" subscript) eqn. (1.11.32) yields

$$\frac{\mathrm{d}\mathbf{l}}{\mathrm{d}t} + \boldsymbol{\omega} \times \mathbf{l} = \boldsymbol{\tau}. \qquad (1.11.34)$$

Finally, using the fact that $\mathbf{l} = \mathbf{I}\boldsymbol{\omega}$ and working with a set of axes in terms of which I is diagonal, the equations of motion for the three components ω_1, ω_2, and ω_3 of the angular velocity vector become

$$I_1 \dot{\omega}_1 - \omega_2 \omega_3 (I_2 - I_3) = \tau_1$$

$$I_2 \dot{\omega}_2 - \omega_3 \omega_1 (I_3 - I_1) = \tau_2$$
$$I_3 \dot{\omega}_3 - \omega_1 \omega_2 (I_1 - I_2) = \tau_3. \tag{1.11.35}$$

These are known as the *rigid body equations of motion*. Given the solutions of these equations of motion for $\omega_i(t)$, the three Euler angles, denoted (ϕ, ψ, θ), are then given as solutions of the differential equations

$$\omega_1 = \dot{\phi} \sin\theta \sin\psi + \dot{\theta} \cos\psi$$
$$\omega_2 = \dot{\phi} \sin\theta \cos\psi - \dot{\theta} \sin\psi$$
$$\omega_3 = \dot{\phi} \cos\theta + \dot{\psi}. \tag{1.11.36}$$

The complexity of the rigid body equations of motion and the relationship between the angular velocity and the Euler angles renders the solution of the equations of motion a nontrivial problem. (In a numerical scheme, for example, there are singularities when the trigonometric functions approach 0.) For this reason, it is preferable to work in terms of a new set of variables known as *quaternions*. As the name suggests, a quaternion is a set of *four* variables that replaces the three Euler angles. Since there are only three rotational degrees of freedom, the four quaternions cannot be independent.

In order to illustrate the idea of the quaternion, let us consider the analogous problem in a smaller number of dimensions (where we might call the variables "binarions" or "ternarions" depending on the number of angles being replaced). Consider again a rigid diatomic moving in the xy plane. The Lagrangian for the system is given by eqn. (1.11.9). Introduce a unit vector

$$\mathbf{q} = (q_1, q_2) \equiv (\cos\theta, \sin\theta). \tag{1.11.37}$$

Clearly, $\mathbf{q} \cdot \mathbf{q} = q_1^2 + q_2^2 = \cos^2\theta + \sin^2\theta = 1$. Note also that

$$\dot{\mathbf{q}} = (\dot{q}_1, \dot{q}_2) = (-(\sin\theta)\dot{\theta}, (\cos\theta)\dot{\theta}) \tag{1.11.38}$$

so that

$$\mathcal{L} = \mu d^2 \dot{\mathbf{q}}^2 - U(\mathbf{q}), \tag{1.11.39}$$

where $U(\mathbf{q})$ indicates that the potential depends on $\mathbf{q}$ since $\mathbf{r} = d\mathbf{q}$. The present formulation is completely equivalent to the original formulation in terms of the angle θ. However, suppose we now treat q_1 and q_2 directly as the dynamical variables. If we wish to do this, we need to ensure that the condition $q_1^2 + q_2^2 = 1$ is obeyed, which could be achieved by treating this condition as a constraint (in Section 3.12, we shall see how to formulate the problem so as to avoid the need for an explicit constraint on the components of $\mathbf{q}$). In this case, $\mathbf{q}$ would be an example of a "binarion." The "binarion" structure is rather trivial and seems to bring us right back to the original problem we sought to avoid by formulating the motion of a rigid diatomic in terms of the angle θ at the outset! We must wait until Section 3.12 to see how to sidestep the constraint. For a diatomic in three dimensions, the rigid-body equations of motion would be reformulated using three variables (q_1, q_2, q_3) satisfying $q_1^2 + q_2^2 + q_3^2 = 1$ so that they are equivalent to $(\sin\theta\cos\phi, \sin\theta\sin\phi, \cos\theta)$.

For a rigid body in three dimensions, we require four variables, (q_1, q_2, q_3, q_4), the quaternions, that must satisfy $\sum_{i=1}^{4} q_i^2 = 1$ and are, by convention, formally related to the three Euler angles by

$$q_1 = \cos\left(\frac{\theta}{2}\right) \cos\left(\frac{\phi + \psi}{2}\right)$$

$$q_2 = \sin\left(\frac{\theta}{2}\right) \cos\left(\frac{\phi - \psi}{2}\right)$$

$$q_3 = \sin\left(\frac{\theta}{2}\right) \sin\left(\frac{\phi - \psi}{2}\right)$$

$$q_4 = \cos\left(\frac{\theta}{2}\right) \sin\left(\frac{\phi + \psi}{2}\right). \tag{1.11.40}$$

From eqn. (1.11.40), it is straightforward to verify that $\sum_i q_i^2 = 1$. The advantage of the quaternion structure is that it leads to a simplification of the rigid-body motion problem. First, note that at any time, a Cartesian coordinate vector in the space fixed frame can be transformed into the body-fixed frame via a rotation matrix involving the quaternions. The relations are

$$\mathbf{r}^{(\text{body})} = \mathbf{A}(\theta, \phi, \psi)\mathbf{r}^{(\text{space})} \qquad \mathbf{r}^{(\text{space})} = \mathbf{A}^{\mathsf{T}}(\theta, \phi, \psi)\mathbf{r}^{(\text{body})}. \tag{1.11.41}$$

The rotation matrix is the product of individual rotations about the three axes, which yields

$$\mathbf{A}(\theta, \phi, \psi) = \begin{pmatrix} \cos\psi\cos\phi - \cos\theta\sin\phi\sin\psi & \cos\psi\sin\phi + \cos\theta\cos\phi\sin\psi & \sin\theta\sin\psi \\ -\sin\psi\cos\phi - \cos\theta\sin\phi\cos\psi & -\sin\psi\sin\phi + \cos\theta\cos\phi\cos\psi & -\sin\theta\cos\psi \\ \sin\theta\sin\phi & -\sin\theta\cos\phi & \cos\theta \end{pmatrix}. \tag{1.11.42}$$

In terms of quaternions, the matrix can be expressed in a simpler-looking form as

$$\mathbf{A}(\mathbf{q}) = \begin{pmatrix} q_1^2 + q_2^2 - q_3^2 - q_4^2 & 2(q_2 q_3 + q_1 q_4) & 2(q_2 q_4 - q_1 q_3) \\ 2(q_2 q_3 - q_1 q_4) & q_1^2 - q_2^2 + q_3^2 - q_4^2 & 2(q_3 q_4 + q_1 q_2) \\ 2(q_2 q_4 + q_1 q_3) & 2(q_3 q_4 - q_1 q_2) & q_1^2 - q_2^2 - q_3^2 + q_4^2 \end{pmatrix}. \tag{1.11.43}$$

It should be noted that in the body-fixed coordinate, the moment of inertia tensor is diagonal. The rigid-body equations of motion, eqns. (1.11.35), can now be transformed into a set of equations of motion involving the quaternions. Direct transformation of these equations leads to a new set of equations of motion given by

$$\dot{\mathbf{q}} = \frac{1}{2}\mathbf{S}(\mathbf{q})\boldsymbol{\omega}$$

$$\dot{\omega}_x = \frac{\tau_x}{I_{xx}} + \frac{(I_{yy} - I_{zz})}{I_{xx}}\omega_y\omega_z$$

$$\dot{\omega}_y = \frac{\tau_y}{I_{yy}} + \frac{(I_{zz} - I_{xx})}{I_{yy}}\omega_z\omega_x$$

$$\dot{\omega}_z = \frac{\tau_z}{I_{zz}} + \frac{(I_{xx} - I_{yy})}{I_{zz}}\omega_x\omega_y. \tag{1.11.44}$$

Here, $\boldsymbol{\omega} = (0, \omega_x, \omega_y, \omega_z)$ and

$$\mathbf{S(q)} = \begin{pmatrix} q_1 & -q_2 & -q_3 & -q_4 \\ q_2 & q_1 & -q_4 & q_3 \\ q_3 & q_4 & q_1 & -q_2 \\ q_4 & -q_3 & q_2 & q_1 \end{pmatrix}. \tag{1.11.45}$$

These equations of motion must be supplemented by the constraint condition $\sum_i q_i^2 = 1$. The equations of motion have the conserved energy

$$E = \frac{1}{2}\left[I_{xx}\omega_x^2 + I_{yy}\omega_y^2 + I_{zz}\omega_z^2\right] + U(\mathbf{q}). \tag{1.11.46}$$

Conservation of the energy in eqn. (1.11.46) can be shown by recognizing that the torques can be written as

$$\boldsymbol{\tau} = -\frac{1}{2}\mathbf{S(q)}^\mathsf{T}\frac{\partial U}{\partial \mathbf{q}}. \tag{1.11.47}$$

1.12 Non-Hamiltonian systems

There is a certain elegance in the symmetry between coordinates and momenta of Hamilton's equations of motion. Up to now, we have mostly discussed systems obeying Hamilton's principle, yet it is important for us to take a short detour away from this path and discuss more general types of dynamical equations of motion that cannot be derived from a Lagrangian or Hamiltonian function. These are referred to as *non-Hamiltonian* systems.

Why might we be interested in non-Hamiltonian systems in the first place? To begin with, we note that Hamilton's equations of motion can only describe a conservative system isolated from its surroundings and/or acted upon by an applied external field. However, Newton's second law is more general than this and could involve forces that are non-conservative and, hence, cannot be derived from a potential function. There are numerous physical systems that are characterized by non-conservative forces, including systems subject to frictional forces and damping effects or the celebrated Lorenz model that catalyzed major interest in the study of chaotic dynamics. We noted previously that Gauss's equations of motion (1.10.8) constitute another example of a non-Hamiltonian system.

In order to understand how non-Hamiltonian systems may be useful in statistical mechanics, consider a physical system in contact with a much larger system, referred to as a *bath*, which regulates some macroscopic property of the physical system such as its pressure or temperature. Were we to consider the microscopic details of the system plus the bath together, we could, in principle, write down a Hamiltonian for the entire system and determine the evolution of the physical subsystem. However, we are rarely interested in all of the microscopic details of the bath. We might, therefore, consider treating the effect of the bath in a more coarse-grained manner by replacing its microscopic coordinates and momenta with a few simpler variables that couple to the physical subsystem in a specified manner. In this case, a set of equations of motion describing the physical system plus the few additional variables used to represent the action of the bath could be proposed which generally would not be Hamiltonian in form because the true microscopic nature of the bath had been eliminated. For this reason, non-Hamiltonian dynamical systems can be highly useful and it is instructive to examine some of their characteristics.

We will restrict ourselves to dynamical systems of the generic form

$$\dot{x} = \xi(x), \tag{1.12.1}$$

where x is a phase space-vector of n components and $\xi(x)$ is a continuous, differentiable n-dimensional vector function. A key signature of a non-Hamiltonian system is that it can have a nonvanishing phase-space compressibility:

$$\kappa(x) = \sum_{i=1}^{n} \frac{\partial \dot{x}_i}{\partial x_i} = \sum_{i=1}^{n} \frac{\partial \xi_i}{\partial x_i} \neq 0. \tag{1.12.2}$$

When eqn. (1.12.2) holds, many of the theorems about Hamiltonian systems no longer apply. However, as will be shown in Chapter 2, some properties of Hamiltonian systems can be generalized to non-Hamiltonian systems provided certain conditions are met. It is important to note that when a Hamiltonian system is formulated in non-canonical variables, the resulting system can also have a nonvanishing compressibility. Strictly speaking, such systems are not truly non-Hamiltonian since a simple transformation back to a canonical set of variables can eliminate the nonzero compressibility. However, throughout this book, we will group such cases in with our general discussion of non-Hamiltonian systems and loosely refer to them as non-Hamiltonian because the techniques we will develop for analyzing dynamical systems with nonzero compressibility factors can be applied equally well to both types of systems.

A simple and familiar example of a non-Hamiltonian system is the case of the damped forced harmonic oscillator described by an equation of motion of the form

$$m\ddot{x} = -m\omega^2 x - \zeta \dot{x}. \tag{1.12.3}$$

This equation describes a harmonic oscillator subject to the action of a friction force $-\zeta\dot{x}$, which could arise, for example, by the motion of the oscillator on a rough surface. Obviously, such an equation cannot be derived from a Hamiltonian. Moreover, the microscopic details of the rough surface are not treated explicitly but rather are modeled grossly by the simple dissipative term in the equation of motion for the physical subsystem described by the coordinate x. Writing the equation of motion as two first order equations involving a phase-space vector (x, p), we have

$$\dot{x} = \frac{p}{m}$$

$$\dot{p} = -m\omega^2 x - \zeta \frac{p}{m}. \tag{1.12.4}$$

It can be seen that this dynamical system has a non-vanishing compressibility

$$\kappa(x,p) = \frac{\partial \dot{x}}{\partial x} + \frac{\partial \dot{p}}{\partial p} = -\frac{\zeta}{m}. \tag{1.12.5}$$

The fact that the compressibility is negative indicates that the effective "phase-space volume" occupied by the system will, as time increases, shrink and eventually collapse onto a single point in the phase space ($x = 0, p = 0$) as $t \to \infty$ under the action of the damping force. All trajectories regardless of their initial condition will eventually approach this point as $t \to \infty$. Consider an arbitrary volume in phase space and let all of the points in this volume represent different initial conditions for eqns. (1.12.4). As these initial conditions evolve in time, the volume they occupy will grow ever smaller until, as $t \to \infty$, the volume tends toward 0. In complex systems, the evolution of such a volume of trajectories will typically be less trivial, growing and shrinking in time as the trajectories evolve. If, in addition, the damped oscillator is driven by a periodic forcing function, so that the equation of motion reads

$$m\ddot{x} = -m\omega^2 x - \zeta \dot{x} + F_0 \cos \Omega t, \tag{1.12.6}$$

then the oscillator will never be able to achieve the equilibrium situation described above but rather will achieve what is known as a *steady state*. The existence of a steady state can be seen by considering the general solution

$$x(t) = e^{-\gamma t} \left[A \cos \lambda t + B \sin \lambda t \right] + \frac{F_0}{\sqrt{(\omega^2 - \Omega^2)^2 + 4\gamma^2 \Omega^2}} \sin(\Omega t + \beta) \tag{1.12.7}$$

of eqn. (1.12.6), where

$$\gamma = \frac{\zeta}{2m} \quad \lambda = \sqrt{\omega^2 - \gamma^2} \quad \beta = \tan^{-1} \frac{\omega^2 - \Omega^2}{2\gamma\Omega}, \tag{1.12.8}$$

and A and B are arbitrary constants set by the choice of initial conditions $x(0)$ and $\dot{x}(0)$. In the long-time limit, the first term decays to zero due to the $\exp(-\gamma t)$ prefactor, and only the second term remains. This term constitutes the *steady-state* solution. Moreover, the amplitude of the steady-state solution can become large when the denominator is a minimum. Considering the function $f(\Omega) = (\omega^2 - \Omega^2)^2 + 4\gamma^2 \Omega^2$, this function reaches a minimum when the frequency of the forcing function is chosen to be $\Omega = \omega \sqrt{1 - \gamma^2/(2\omega^2)}$. Such a frequency is called a *resonant* frequency. Resonances play an important role in classical dynamics when harmonic forces are present, a phenomenon that will be explored in greater detail in Chapter 3. Driven systems and steady states will be discussed in greater detail in Chapter 13.

1.13   Problems

1.1. Solve the equations of motion given arbitrary initial conditions for a one-dimensional particle moving in a linear potential $U(x) = Cx$, where C is a constant, and sketch a representative phase-space plot.

*1.2. A particle of unit mass moves in a potential of the form

$$U(x) = -\frac{\omega^2}{8a^2}\left(x^2 - a^2\right)^2.$$

a. Show that the function

$$x(t) = a\tanh[(t - t_0)\omega/2]$$

is a solution to Hamilton's equations for this system, where t_0 is an arbitrary constant.

b. Let the origin of time be $t = -\infty$ rather than $t = 0$. To what initial conditions does this solution correspond?

c. Determine the behavior of this solution as $t \to \infty$.

d. Sketch the phase-space plot for this particular solution.

1.3. Determine the trajectory $r(t)$ for a particle of mass m moving in three dimensions subject to a central potential of the form $U(r) = kr^2/2$. Verify your solution for different values of l and given values of m and k by numerically integrating eqn. (1.4.33). Discuss the behavior of the solution for different values of l.

1.4. Repeat problem 3 for a potential of the form $U(r) = \kappa/r$.

*1.5. Consider Newton's equation of motion for a one-dimensional particle subject to an arbitrary force, $m\ddot{x} = F(x)$. A numerical integration algorithm for the equations of motion, known as the velocity Verlet algorithm (see Chapter 3), for a discrete time step value Δt is

$$x(\Delta t) = x(0) + \Delta t \frac{p(0)}{m} + \frac{\Delta t^2}{2m} F(x(0))$$

$$p(\Delta t) = p(0) + \frac{\Delta t}{2}\left[F(x(0)) + F(x(\Delta t))\right].$$

By considering the Jacobian matrix

$$\mathbf{J} = \begin{pmatrix} \frac{\partial x(\Delta t)}{\partial x(0)} & \frac{\partial x(\Delta t)}{\partial p(0)} \\ \frac{\partial p(\Delta t)}{\partial x(0)} & \frac{\partial p(\Delta t)}{\partial p(0)} \end{pmatrix},$$

show that the algorithm is symplectic, and show that $\det[J] = 1$.

1.6. A water molecule H_2O is subject to an external potential. Let the positions of the three atoms be denoted $\mathbf{r}_O$, $\mathbf{r}_{H_1}$, $\mathbf{r}_{H_2}$, so that the forces on the three atoms can be denoted $\mathbf{F}_O$, $\mathbf{F}_{H_1}$, and $\mathbf{F}_{H_2}$. Consider treating the molecule as completely rigid, with internal bond lengths d_{OH} and d_{HH}, so that the constraints are

$$|\mathbf{r}_O - \mathbf{r}_{H_1}|^2 - d_{OH}^2 = 0$$
$$|\mathbf{r}_O - \mathbf{r}_{H_2}|^2 - d_{OH}^2 = 0$$
$$|\mathbf{r}_{H_1} - \mathbf{r}_{H_2}|^2 - d_{HH}^2 = 0.$$

a. Derive the constrained equations of motion for the three atoms in the molecule in terms of undetermined Lagrange multipliers.

b. Show that the forces of constraint do not contribute to the work done on the molecule in moving it from one spatial location to another.

c. Determine Euler's equations of motion about an axis perpendicular to the plane of the molecule in a body-fixed frame whose origin is located on the oxygen atom.

d. Determine the equations of motion for the quaternions that describe this system.

1.7. Calculate the classical action for a one-dimensional free particle of mass m. Repeat for a harmonic oscillator of spring constant k.

1.8. A simple mechanical model of a diatomic molecule bound to a flat surface is illustrated in Fig. 1.13. Suppose the atom with masses m_1 and m_2 carry electrical charges q_1 and q_2, respectively, and suppose that the molecule is subject to a constant external electric field E in the vertical direction, directed upwards. In this case, the potential energy of each atom will be $q_i E h_i$, $i = 1, 2$ where h_i is the height of the atom i above the surface.

a. Using θ_1 and θ_2 as generalized coordinates, write down the Lagrangian of the system.

b. Derive the equations of motion for these coordinates.

c. Introduce the small-angle approximation, which assumes that the angles only execute small amplitude motion. What form do the equations of motion take in this approximation?

1.9. Use Gauss's principle of least constraint to determine a set of non-Hamiltonian equations of motion for the two atoms in a rigid diatomic molecule of bond length d subject to an external potential. Take the constraint to be

$$\sigma(\mathbf{r}_1, \mathbf{r}_2) = |\mathbf{r}_1 - \mathbf{r}_2|^2 - d^2.$$

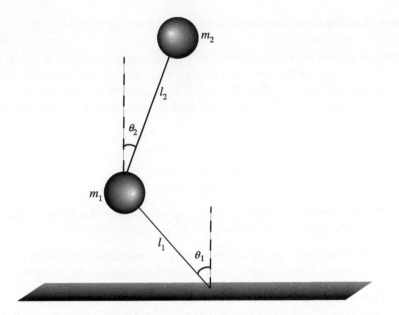

Fig. 1.13 Schematic of a diatomic molecule bound to a flat surface.

Determine the compressibility of your equations of motion.

1.10. Consider the harmonic polymer model of Section 1.7 in which the harmonic neighbor couplings all have the same frequency ω but the masses have alternating values m and M, respectively. For the case of $N = 5$ particles, determine the normal modes and their associated frequencies.

1.11. The equilibrium configuration of a molecule is represented by three atoms of equal mass at the vertices of a right isosceles triangle. The atoms can be viewed as connected by harmonic springs of equal force constant. Find the normal mode frequencies of this molecule, and, in particular, show that the zero-frequency mode is triply degenerate.

1.12. A particle of mass m moves in a double-well potential of the form

$$U(x) = \frac{U_0}{a^4} \left(x^2 - a^2\right)^2.$$

Sketch the contours of the constant-energy surface $\mathcal{H}(x,p) = E$ in phase space for the following cases:
a. $E < U_0$.
b. $E = U_0 + \epsilon$, where $\epsilon \ll U_0$.
c. $E > U_0$.

*1.13. The Hamiltonian for a system of N charged particles with charges q_i, $i = 1, ..., N$ and masses m_i, $i = 1, ..., N$, positions, $\mathbf{r}_1, ..., \mathbf{r}_N$ and momenta $\mathbf{p}_1, ..., \mathbf{p}_N$ interacting with a static electromagnetic field is given by

$$\mathcal{H} = \sum_{i=1}^{N} \frac{(\mathbf{p}_i - q_i \mathbf{A}(\mathbf{r}_i)/c)^2}{2m_i} + \sum_{i=1}^{N} q_i \phi(\mathbf{r}_i) + \sum_{i=1}^{N-1} \sum_{j=i+1}^{N} \frac{q_i q_j}{|\mathbf{r}_i - \mathbf{r}_j|},$$

where $\mathbf{A}(\mathbf{r})$ and $\phi(\mathbf{r})$ are the vector and scalar potentials of the field, respectively, and c is the speed of light. In terms of these quantities, the electric and magnetic components of the electromagnetic field, $\mathbf{E}(\mathbf{r})$ and $\mathbf{B}(\mathbf{r})$ are given by

$$\mathbf{E}(\mathbf{r}) = -\nabla \phi(\mathbf{r}), \qquad \mathbf{B}(\mathbf{r}) = \nabla \times \mathbf{A}(\mathbf{r}).$$

  a. Derive Hamilton's equations for this system, and determine the force on each particle in terms of the electric and magnetic fields $\mathbf{E}(\mathbf{r})$ and $\mathbf{B}(\mathbf{r})$, respectively. This contribution to the force due to the field is known as the *Lorentz force*. Express the equations of motion in Newtonian form.

  b. Suppose $N = 1$, that the electric field is zero everywhere ($\mathbf{E} = 0$), and that the magnetic field is a constant in the z direction, $\mathbf{B} = (0, 0, B)$. For this case, solve the equations of motion for an arbitrary initial condition and describe the motion that results.

1.14. Prove that the energy in eqn. (1.11.46) is conserved making use of eqn. (1.11.47) for the torques.

1.15. Let $q_1, ..., q_{3N} \equiv q$ denote the $3N$ Cartesian coordinate components in a system, such that $(q_1, q_2, q_3) = \mathbf{r}_1$, $(q_4, q_5, q_6) = \mathbf{r}_2$, *etc.* Similarly, let $p_1, ..., p_{3N} \equiv p$ denote the $3N$ Cartesian momentum components. Consider a Hamiltonian of the form

$$\mathcal{H}(q, p) = \gamma \sum_{\alpha=1}^{3N} \ln \cosh \left(\frac{p_\alpha}{\sqrt{\gamma m_\alpha}} \right) + U(q_1, ..., q_{3N}),$$

where γ is an arbitrary parameter and $\{m_\alpha\}$ are the masses associated with each degree of freedom.

  a. Derive Hamilton's equations of motion from this Hamiltonian.

  b. Show that the equations of motion imply that the speed of any degree of freedom $|\dot{q}_\alpha|$ is restricted such that $|\dot{q}_\alpha| = \sqrt{\gamma/m_\alpha}$, even though $|p_\alpha| \in [0, \infty)$.

*1.16 (For amusement only): Consider a system with coordinate q, momentum p, and Hamiltonian

$$H = \frac{p^n}{n} + \frac{q^n}{n},$$

where n is an integer larger than 2. Show that if the energy E of the system is chosen such that $nE = m^n$, where m is a positive integer, then no phase-space trajectory can ever pass through a point for which p and q are both positive integers.

2
Theoretical foundations of classical statistical mechanics

2.1 Overview

The field of thermodynamics began in precursory form with the work of Otto von Guericke (1602–1686) who designed the first vacuum pump in 1650 and with Robert Boyle (1627–1691) who, working with von Guericke's design, discovered an inverse proportionality between the pressure and volume of a gas at a fixed temperature for a fixed amount of gas. This inverse proportionality became known as *Boyle's Law*. Thermodynamics matured in the nineteenth century through the seminal work of R. J. Mayer (1814–1878) and J. P. Joule (1818–1889), who established that heat is a form of energy, of R. Clausius (1822–1888) and N. L. S. Carnot (1796–1832), who originated the concept of entropy, and of numerous others. This work is neatly encapsulated in what we now refer to as the laws of thermodynamics (see Section 2.2). As these laws are based on experimental observations, thermodynamics is a phenomenological theory of macroscopic matter, which has, nevertheless, withstood the test of time. The framework of thermodynamics is an elegantly self-consistent one that makes no reference to the microscopic constituents of matter. If, however, we believe in a microscopic theory of matter, then it must be possible to rationalize thermodynamics based on microscopic mechanical laws.

In Chapter 1, we presented the laws of classical mechanics and applied them to several simple examples. The laws of classical mechanics imply that if the positions and velocities of all the particles in a system are known at a single instant in time, then the past evolution of the system leading to that point in time and the future evolution of the system from that point forward are known. The example systems considered in Chapter 1 consisted of one or a small number of degrees of freedom with simple forces, and we saw that the past and future of each system could be worked out from Newton's second law of motion (see, for example, eqn. (1.2.10)). Thus, classical mechanics encodes all the information needed to predict the properties of a classical system at any instant in time.

In order to provide a rational basis for thermodynamics, we should apply the microscopic laws of motion to macroscopic systems. However, this idea immediately meets with two serious issues. First, macroscopic systems possess an enormous number of degrees of freedom (1 mole consists of 6.022×10^{23} particles); second, real-world systems are characterized by highly nontrivial interactions. Hence, even though we should be able, in principle, to predict the microscopic detailed dynamics of any classical system knowing only the initial conditions, we quickly realize the hopelessness of this effort.

The highly nonlinear character of the forces in realistic systems means that an analytical solution of the equations of motion is not available. If we propose, alternatively, to solve the equations of motion numerically on a computer, the memory requirement to store just one phase-space point for a system of 10^{23} particles exceeds what is available both today and in the foreseeable future. Thus, while classical mechanics encodes all the information needed to predict the properties of a system, the problem of extracting that information is seemingly intractable.

In addition to the problem of the sheer size of macroscopic systems, another, more subtle, issue exists. The second law of thermodynamics prescribes a direction of time, namely, the direction in which the entropy increases. This "arrow" of time is seemingly at odds with the microscopic mechanical laws, which are inherently reversible in time.[1] This paradoxical situation, known as *Loschmidt's paradox*, seems to pit thermodynamics against microscopic mechanical laws.

The reconciliation of macroscopic thermodynamics with the microscopic laws of motion required the development of a new field, *statistical mechanics*, the main topic of this book. Statistical mechanics began with ideas from Clausius and James C. Maxwell (1831–1879) but grew principally out of the work of Ludwig Boltzmann (1844–1906) and Josiah W. Gibbs (1839–1903). (Other significant contributors include Henri Poincaré, Albert Einstein, and later, Lars Onsager, Richard Feynman, Ilya Prigogine, Kenneth Wilson, and Benjamin Widom, to name just a few.) Early innovations in statistical mechanics derived from the realization that the macroscopic observable properties of a system do not depend strongly on the detailed dynamical motion of every particle in a macroscopic system but rather on gross averages that largely "wash out" these microscopic details. Thus, by applying the microscopic mechanical laws in a statistical fashion, a link can be provided between the microscopic and macroscopic theories of matter. Not only does this concept provide a rational basis for thermodynamics, it also leads to procedures for computing many other macroscopic observables. The principal conceptual breakthrough on which statistical mechanics is based is that of an *ensemble*, which refers to a collection of systems that share common macroscopic properties. Averages performed over an ensemble yield the thermodynamic quantities of a system as well as other equilibrium and dynamic properties.

In this chapter, we will lay out the fundamental theoretical foundations of *ensemble theory* and show how the theory establishes the link between the microscopic and macroscopic realms. We begin with a discussion of the laws of thermodynamics and a number of important thermodynamic functions. Following this, we introduce the notion of an ensemble and the properties that an ensemble must obey. Finally, we will describe, in general terms, how to use an ensemble to calculate macroscopic properties. Specific types of ensembles and their use will be detailed in subsequent chapters.

---

[1]It can be easily shown, for example, that Newton's second law retains its form under a time-reversal transformation $t \rightarrow -t$. Under this transformation, $d/dt \rightarrow -d/dt$, but $d^2/dt^2 \rightarrow d^2/dt^2$. Time-reversal symmetry implies that if a mechanical system evolves from an initial condition x_0 at time $t = 0$ to x_t at a time $t > 0$, and all the velocities are subsequently reversed ($v_i \rightarrow -v_i$), the system will return to its initial microscopic state x_0. The same is true of the microscopic laws of quantum mechanics. Consequently, it should not be possible to tell if a "movie" made of a mechanical system is running in the "forward" or "reverse" direction.

2.2   The laws of thermodynamics

Our discussion of the laws of thermodynamics will make no reference to the microscopic constituents of a particular system. Concepts and definitions we will need for the discussion are described below:

i. A *thermodynamic* system is a macroscopic system. Thermodynamics always divides the universe into the system and its surroundings. A thermodynamic system is said to be *isolated* if no heat or material is exchanged between the system and its surroundings and if the surroundings produce no other change in the thermodynamic state of the system.

ii. A system is in *thermodynamic equilibrium* if its thermodynamic state does not change in time.

iii. The fundamental thermodynamic parameters that define a thermodynamic state, such as the pressure P, volume V, the temperature T, and the total mass M or number of moles n are measurable quantities assumed to be provided experimentally. A thermodynamic *state* is specified by providing values of all thermodynamic parameters necessary for a complete description of a system.

iv. The *equation of state* of a system is a relationship among the thermodynamic parameters prescribing how these parameters vary from one equilibrium state to another. Thus, if P, V, T, and n are the fundamental thermodynamic parameters of a system, the equation of state takes the general form

$$g(n, P, V, T) = 0. \tag{2.2.1}$$

As a consequence of eqn. (2.2.1), there are in fact only three independent thermodynamic parameters in an equilibrium state. When the number of moles remains fixed, the number of independent parameters is reduced to two. An example of an equation of state is that of an *ideal gas*, which is defined (thermodynamically) as a system whose equation of state is

$$PV - nRT = 0, \tag{2.2.2}$$

where $R = 8.315$ J·mol^{-1}·K^{-1} is the *gas constant*. The ideal gas represents the limiting behavior of all real gases at sufficiently low density $\rho \equiv n/V$.

v. A *thermodynamic transformation* is a change in the thermodynamic state of a system. In equilibrium, a thermodynamic transformation is effected by a change in the external conditions of the system. Thermodynamic transformations can be carried out either *reversibly* or *irreversibly*. In a reversible transformation, the change is carried out slowly enough that the system has time to adjust to each new external condition imposed along a prescribed thermodynamic path, so that the system can retrace its history along the same path between the endpoints of the transformation. If this is not possible, then the transformation is irreversible.

vi. A *state function* is any function $f(n, P, V, T)$ whose change under any thermodynamic transformation depends only on the initial and final states of the transformation and *not* on the particular thermodynamic path taken between these states (see Fig. 2.1).

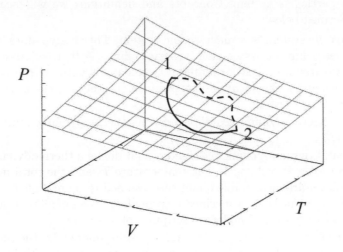

Fig. 2.1 The thermodynamic state space defined by the variables P, V, and T with two paths (solid and dashed lines) between the state points 1 and 2. The change in a state function $f(n, P, V, T)$ is independent of the path taken between any two such state points.

vii. In order to change the volume or the number of moles, work must be performed on a system. If a transformation is performed reversibly such that the volume changes by an amount dV and the number of moles changes by an amount dn, then the work performed on the system is

$$dW_{\mathrm{rev}} = -PdV + \mu dn. \tag{2.2.3}$$

The quantity μ is called the *chemical potential*, defined to be the amount of work needed to add 1.0 mole of a substance to a system already containing that substance.

viii. In order to change the temperature of a system, heat must be added or removed. The amount of heat dQ needed to change the temperature by an amount dT in a reversible process is given by

$$dQ_{\mathrm{rev}} = CdT. \tag{2.2.4}$$

The quantity C is called the *heat capacity*, defined to be the amount of heat needed to change the temperature of 1.0 mole of a substance by 1.0 degree on a chosen scale. If heat is added at fixed pressure, then the heat capacity is denoted C_P. If heat is added at fixed volume, it is denoted C_V.

2.2.1 The first law of thermodynamics

The first law of thermodynamics is a statement of conservation of energy. We saw in Section 1.6 that performing work on a system changes its potential (or internal) energy (see Section 1.4). Thermodynamics recognizes that heat is also a form of energy. The first law states that *in any thermodynamic transformation, if a system absorbs an amount of heat ΔQ and has an amount of work ΔW performed on it, then its internal energy will change by an amount ΔE given by*

$$\Delta E = \Delta Q + \Delta W. \tag{2.2.5}$$

(Older books define the first law in terms of the heat absorbed and work done *by* the system. With this convention, the first law is written $\Delta E = \Delta Q - \Delta W$.) Although neither the heat absorbed ΔQ nor the work ΔW done on the system are state functions, the internal energy E is a state function. Thus, the transformation can be carried out along either a reversible or irreversible path, and the same value of ΔE will result. If E_1 and E_2 represent the energies before and after the transformation respectively, then $\Delta E = E_2 - E_1$, and it follows that an exact differential dE exists for the energy such that

$$\Delta E = E_2 - E_1 = \int_{E_1}^{E_2} dE. \tag{2.2.6}$$

However, since ΔE is independent of the path of the transformation, ΔE can be expressed in terms of changes along either a reversible or irreversible path:

$$\Delta E = \Delta Q_{\text{rev}} + \Delta W_{\text{rev}} = \Delta Q_{\text{irrev}} + \Delta W_{\text{irrev}}. \tag{2.2.7}$$

Suppose that reversible and irreversible transformations are carried out on a system with a fixed number of moles, and let the irreversible process be one in which the external pressure drops to a value P_{ext} by a sudden volume change ΔV, thus allowing the system to expand rapidly. It follows that the work done on the system is

$$\Delta W_{\text{irrev}} = -P_{\text{ext}}\Delta V. \tag{2.2.8}$$

In such a process, the internal pressure $P > P_{\text{ext}}$. If the same expansion is carried out reversibly (slowly), then the internal pressure has time to adjust as the system expands. Since

$$\Delta W_{\text{rev}} = -\int_{V_1}^{V_2} PdV, \tag{2.2.9}$$

where the dependence of the internal pressure P on the volume is specified by the equation of state, and since P_{ext} in the irreversible process is less than P at all states visited in the reversible process, it follows that $-\Delta W_{\text{irrev}} < -\Delta W_{\text{rev}}$, or $\Delta W_{\text{irrev}} > \Delta W_{\text{rev}}$. However, because of eqn. (2.2.7), the first law implies that the amounts of heat absorbed in the two processes satisfy

$$\Delta Q_{\text{irrev}} < \Delta Q_{\text{rev}}. \tag{2.2.10}$$

Equation (2.2.10) will be needed in our discussion of the second law of thermodynamics.

Of course, since the thermodynamic universe is, by definition, an isolated system (it has no surroundings), its energy is conserved. Therefore, any change ΔE_{sys} in a system must be accompanied by an equal and opposite change ΔE_{surr} in the surroundings so that the net energy change of the universe $\Delta E_{\text{univ}} = \Delta E_{\text{sys}} + \Delta E_{\text{surr}} = 0$.

2.2.2 The second law of thermodynamics

Before discussing the second law of thermodynamics, it is useful to review the *Carnot cycle*. The Carnot cycle is the thermodynamic cycle associated with an ideal device or "engine" that takes in heat at a high temperature T_h, expels heat at a low temperature T_l, and over the cycle, delivers a net amount of useful work. The ideal engine provides an upper bound on the efficiency that can be achieved by a real engine.

The thermodynamic cycle of a Carnot engine is shown in Fig. 2.2, which is a plot of the process in the P-V plane. In the cycle, each of the four transformations (curves

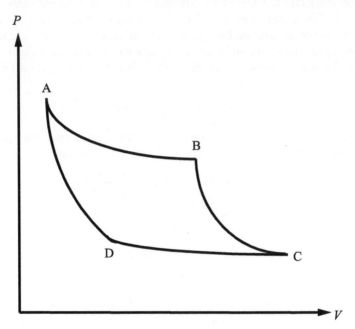

Fig. 2.2 The Carnot cycle.

AB, BC, CD and DA in Fig. 2.2) is assumed to be performed reversibly on an ideal gas. The four stages of the cycle are defined as follows:

- **Path AB**: An amount of heat Q_h is absorbed at a high temperature T_h, and the system undergoes an isothermal expansion at this temperature.
- **Path BC**: The system further expands adiabatically so that no further heat is gained or lost.
- **Path CD**: The system is compressed at a low temperature T_l, and an amount of heat Q_l is released by the system.
- **Path DA**: The system undergoes a further adiabatic compression in which no further heat is gained or lost.

Since the cycle is closed, the change in the internal energy in this process is $\Delta E = 0$. Thus, according to the first law of thermodynamics, the net work output by the Carnot engine is given by

$$-W_{\text{net}} = \Delta Q = Q_h + Q_l. \tag{2.2.11}$$

The efficiency of any engine ϵ is defined as the ratio of the net work output to the heat input

$$\epsilon = -\frac{W_{\text{net}}}{Q_h}, \tag{2.2.12}$$

from which it follows that the efficiency of the Carnot engine is

$$\epsilon = 1 + \frac{Q_l}{Q_h}. \tag{2.2.13}$$

On the other hand, the work done on (or by) the system during the adiabatic expansion and compression phases cancels, so that the net work comes from the isothermal expansion and compression segments. From the ideal gas law, eqn. (2.2.2), the work done on the system during the initial isothermal expansion phase is simply

$$W_h = -\int_{V_A}^{V_B} P dV = -\int_{V_A}^{V_B} \frac{nRT_h}{V} dV = -nRT_h \ln\left(\frac{V_B}{V_A}\right), \tag{2.2.14}$$

while the work done on the system during the isothermal compression phase is

$$W_l = -\int_{V_C}^{V_D} \frac{nRT_l}{V} dT = -nRT_l \ln\left(\frac{V_D}{V_C}\right). \tag{2.2.15}$$

However, because the temperature ratio for both adiabatic phases is the same, namely, T_h/T_l, it follows that the volume ratios V_C/V_B and V_D/V_A are also the same. Since $V_C/V_B = V_D/V_A$, it follows that $V_B/V_A = V_C/V_D$, and the net work output is

$$-W_{\text{net}} = nR(T_h - T_l) \ln\left(\frac{V_B}{V_A}\right). \tag{2.2.16}$$

The internal energy of an ideal gas is $E = 3nRT/2$, and therefore the energy change during an isothermal process is $\Delta E = 0$. Hence, for the initial isothermal expansion phase, $\Delta E = 0$ and $W_h = -Q_h = nRT_h \ln(V_B/V_A)$. The efficiency can also be expressed in terms of the temperatures as

$$\epsilon = -\frac{W_{\text{net}}}{Q_h} = \frac{nR(T_h - T_l)\ln(V_B/V_A)}{nRT_h \ln(V_B/V_A)} = 1 - \frac{T_l}{T_h}. \tag{2.2.17}$$

Equating the two efficiency expressions, we have

$$1 + \frac{Q_l}{Q_h} = 1 - \frac{T_l}{T_h}$$
$$\frac{Q_l}{Q_h} = -\frac{T_l}{T_h}$$
$$\frac{Q_h}{T_h} + \frac{Q_l}{T_l} = 0. \tag{2.2.18}$$

Equation (2.2.18) indicates that there is a quantity $\Delta Q_{\mathrm{rev}}/T$ whose change over the closed cycle is 0. The "rev" subscript serves as a reminder that the Carnot cycle is carried out using reversible transformations. Thus, the quantity $\Delta Q_{\mathrm{rev}}/T$ is a state function, and although we derived this fact using an idealized Carnot cycle, it turns out that this quantity is always a state function. This means that there is an exact differential $\mathrm{d}S = \mathrm{d}Q_{\mathrm{rev}}/T$ such that S is a state function. The quantity ΔS, defined by

$$\Delta S = \int_1^2 \frac{\mathrm{d}Q_{\mathrm{rev}}}{T}, \tag{2.2.19}$$

is therefore independent of the path over which the transformation from state "1" to state "2" is carried out. The quantity S is the entropy of the system.

The second law of thermodynamics is a statement about the behavior of the entropy in any thermodynamic transformation. From eqn. (2.2.10), which implies that $\mathrm{d}Q_{\mathrm{irrev}} < \mathrm{d}Q_{\mathrm{rev}}$, we obtain

$$\mathrm{d}S = \frac{\mathrm{d}Q_{\mathrm{rev}}}{T} > \frac{\mathrm{d}Q_{\mathrm{irrev}}}{T}, \tag{2.2.20}$$

which is known as the *Clausius inequality*. If this inequality is now applied to the thermodynamic universe, an isolated system that absorbs and releases no heat ($\mathrm{d}Q = 0$), then the *total* entropy $\mathrm{d}S_{\mathrm{tot}} = \mathrm{d}S_{\mathrm{sys}} + \mathrm{d}S_{\mathrm{surr}}$ satisfies

$$\mathrm{d}S_{\mathrm{tot}} \geq 0. \tag{2.2.21}$$

That is, *in any thermodynamic transformation, the total entropy of the universe must either increase or remain the same.* $\mathrm{d}S_{\mathrm{tot}} > 0$ pertains to an irreversible process while $\mathrm{d}S_{\mathrm{tot}} = 0$ pertains to a reversible process. Equation (2.2.21) is the second law of thermodynamics.

Our analysis of the Carnot cycle allows us to understand two equivalent statements of the second law. The first, attributed to William Thomson (1824–1907), known later as the First Baron Kelvin or Lord Kelvin, reads: *There exists no thermodynamic transformation whose sole effect is to extract a quantity of heat from a high-temperature source and convert it entirely into work.* In fact, some of the heat absorbed at T_h is always lost in the form of *waste heat*, which in the Carnot cycle is the heat Q_l released at T_l. The loss of waste heat means that $-W_{\mathrm{net}} < -W_h$ or that the net work done by the system must be less than the work done during the first isothermal expansion phase.

Now suppose we run the Carnot cycle in reverse so that an amount of heat Q_l is absorbed at T_l and released as Q_h at T_h. In the process, an amount of work W_{net} is *consumed* by the system. Thus, the Carnot cycle operated in reverse performs as a refrigerator, moving heat from a cold source to a hot source. This brings us to the second statement of the second law, attributed to Clausius: *There exists no thermodynamic transformation whose sole effect is to extract a quantity of heat from a cold source and deliver it to a hot source.* That is, heat does not flow spontaneously from cold to hot; moving heat in this direction requires that work be done.

2.2.3 The third law of thermodynamics

As with any state function, it is only possible to measure changes in the entropy, which does not inherently require an absolute entropy scale. The third law of thermodynamics defines an absolute scale of entropy: *The entropy of a system at the absolute zero of temperature is a universal constant, which can be taken to be zero.* Absolute zero of temperature is defined as $T = 0$ on the Kelvin scale; it is a temperature that can never be physically reached. The unattainability of absolute zero is sometimes taken as an alternative statement of the third law. A consequence of the unattainability of absolute zero temperature is that the ideal (Carnot) engine can never be 100% efficient, since this would require sending $T_l \to 0$ in eqn. (2.2.17), which is not possible. As we will see in Chapter 10, the third law of thermodynamics is actually a macroscopic manifestation of quantum mechanical effects.

2.3 The ensemble concept

We introduced the laws of thermodynamics without reference to the microscopic origin of macroscopic thermodynamic observables. Without a microscopic basis, thermodynamics must be regarded as a phenomenological theory. We now wish to provide this microscopic basis and establish a connection between the macroscopic and microscopic realms. As we remarked at the beginning of the chapter, we cannot solve the classical equations of motion for a system of 10^{23} particles with the complex, nonlinear interactions that govern the behavior of real systems. Nevertheless, it is instructive to pose the following question: If we could solve the equations of motion for such a large number of particles, would the vast amount of detailed microscopic information generated be necessary to describe macroscopic observables?

Intuitively, we would answer this question with "no." Although the enormous quantity of microscopic information is certainly *sufficient* to predict any macroscopic observable, there are many microscopic configurations of a system that lead to the same macroscopic properties. For example, if we connect the temperature of a system to an average of kinetic energy of the individual particles composing the system, then there are many ways to assign the velocities of the particles consistent with a given total energy such that the same total kinetic energy and, hence, the same measure of temperature is obtained. Nevertheless, each assignment corresponds to a different point in phase space and, therefore, a different and unique microscopic state. Similarly, if we connect the pressure to the average force per unit area exerted by the particles on the walls of the container, there are many ways of arranging the particles such that the forces between them and the walls yields the same pressure measure, even though each assignment corresponds to a unique point in phase space and hence, a unique microscopic state. Suppose we aimed, instead, to predict macroscopic time-dependent properties. By the same logic, if we started with a large set of initial conditions drawn from a state of thermodynamic equilibrium, and if we launched a trajectory from each initial condition in the set, then the resulting trajectories would all be unique in phase space. Despite their uniqueness, these trajectories should all lead, in the long time limit, to the same macroscopic dynamical observables such as vibrational spectra, diffusion constants, and so forth.

The idea that the macroscopic observables of a system are not sensitive to precise microscopic details is the basis of the *ensemble* concept originally introduced by Gibbs. More formally, an *ensemble* is a collection of systems described by the same set of microscopic interactions and sharing a common set of macroscopic properties (e.g. the same total energy, volume, and number of moles). Each system evolves under the microscopic laws of motion from a different initial condition so that at any point in time, every system has a unique microscopic state. Once an ensemble is defined, macroscopic observables are calculated by performing averages over the systems in the ensemble. Ensembles can be defined for a wide variety of thermodynamic situations. The simplest example is a system isolated from its surroundings. However, ensembles also describe systems in contact with heat baths, systems in contact with particle reservoirs, systems coupled to pressure control mechanisms such as mechanical pistons, and various combinations of these influences. Such ensembles are useful for determining static properties such as temperature, pressure, free energy, average structure, *etc.* Thus, the fact that the systems in the ensemble evolve in time does not affect properties of this type, and we may freeze the ensemble at any instant and perform the average over the ensemble at that instant. These ensembles are known as *equilibrium ensembles*, and we will focus on them up to and including Chapter 12. Finally, ensembles can also be defined for systems driven by external forces or fields for the calculation of transport coefficients and other dynamical properties. These are examples of non-equilibrium ensembles, which will be discussed in Chapters 13 and 14.

In classical ensemble theory, every macroscopic observable of a system is directly connected to a microscopic function of the coordinates and momenta of the system. A familiar example of this comes from the kinetic theory, where the temperature of a system is connected to the average kinetic energy. In general, we will let A denote a macroscopic equilibrium observable and $a(\mathbf{x})$ denote a microscopic phase-space function that can be used to calculate A. According to the ensemble concept, if the ensemble has $\mathcal{Z}$ members, then the "connection" between A and $a(\mathbf{x})$ is provided via an averaging procedure, which we write heuristically as

$$A = \frac{1}{\mathcal{Z}} \sum_{\lambda=1}^{\mathcal{Z}} a(\mathbf{x}_\lambda) \equiv \langle a \rangle. \tag{2.3.1}$$

This definition is not to be taken literally, since the sum may well be a continuous "sum" or integral. However, eqn. (2.3.1) conveys the notion that the phase-space function $a(\mathbf{x})$ must be evaluated for each member of the ensemble at that point in time when the ensemble is frozen. Finally, A is obtained by performing an average over the ensemble. (The notation $\langle a \rangle$ in eqn. (2.3.1) will be used throughout the book to denote an ensemble average.)

Let us recall the question we posed earlier: If we could solve the equations of motion for a very large number of particles, would the vast amount of detailed microscopic information generated be necessary to describe macroscopic observables? Previously, we answered this in the negative. However, the other side can also be argued if we take a purist's view. That is, all of the information needed to describe a physical system is encoded in the microscopic equations of motion. Indeed, there are many physical and chemical processes for which the underlying atomic and molecular mechanics are of significant interest and importance. In order to elucidate these, it is necessary to know how individual atoms and molecules move as the process occurs. Experimental techniques such as ultrafast laser spectroscopy can resolve processes at increasingly short time scales and thus obtain important insights into such motions. (The importance of such techniques was recognized by the award of the 1999 Nobel Prize in chemistry to the physical chemist Ahmed Zewail (1946–2016) for his pioneering work in their development.) While we cannot expect to solve the equations of motion for 10^{23} particles, we actually can solve them numerically for systems whose particle numbers range from 10^2 to 10^9, depending on the complexity of the interactions in a particular physical model. The technique of solving the equations of motion numerically for small representative systems is known as *molecular dynamics*, a method that has become one of the most important theoretical tools for solving statistical mechanical problems. Although the system sizes currently accessible to molecular dynamics calculations are not truly macroscopic, they are large enough to capture the macroscopic limit for certain properties. Thus, a molecular dynamics calculation, which can be viewed as a kind of detailed "thought experiment" performed *in silico*, can yield important microscopic insights into complex phenomena including the catalytic mechanisms of enzymes, details of protein folding and misfolding processes, formation supramolecular structures, and many other fascinating phenomena.

We will have more to say about molecular dynamics and other methods for solving statistical mechanical problems throughout the book. For the remainder of this chapter, we will focus on the fundamental underpinnings of ensemble theory.

2.4 Phase-space volumes and Liouville's theorem

As noted previously, an ensemble is a collection of systems with a set of common macroscopic properties such that each system is in a unique microscopic state at any point in time as determined by its evolution under some dynamical rule, *e.g.*, Hamilton's equations of motion. Given this definition, and assuming that the evolution of the collection of systems is prescribed by Hamilton's equations, it is important first to understand how a collection of microscopic states (which we refer to hereafter simply as "microstates") moves through phase space.

Consider a collection of microstates in a phase-space volume element dx_0 centered on the point x_0. The "0" subscript indicates that each microstate in the volume element serves as an initial condition for Hamilton's equations, which we had written in eqn. (1.6.23) as $\dot{x} = \eta(x)$. The equations of motion can be generalized to the case of a set of driven Hamiltonian systems by writing them as $\dot{x} = \eta(x,t)$. We now ask how the entire volume element dx_0 moves under the action of Hamiltonian evolution. Recall that x_0 is a complete set of generalized coordinates and conjugate momenta:

$$x_0 = (q_1(0), ..., q_{3N}(0), p_1(0), ..., p_{3N}(0)). \tag{2.4.1}$$

(We will refer to the complete set of generalized coordinates and their conjugate momenta collectively as the *phase-space coordinates*.) If we follow the evolution of this volume element from $t = 0$ to time t, dx_0 will be transformed into a new volume element dx_t centered on a point x_t in phase space. The point x_t is the phase-space point that results from the evolution of x_0. As we noted in Section 1.6, x_t is a unique function of x_0 that can be expressed as $x_t(x_0)$. Since the mapping of the point x_0 to x_t is one-to-one, this mapping is equivalent to a coordinate transformation on the phase space from initial phase-space coordinates x_0 to phase-space coordinates x_t. Under this transformation, the volume element dx_0 transforms according to

$$dx_t = J(x_t; x_0)dx_0, \tag{2.4.2}$$

where $J(x_t; x_0)$ is the Jacobian of the transformation, the determinant of the matrix J defined in eqn. (1.6.28), from x_0 to x_t. According to eqn. (1.6.28), the elements of the matrix are

$$J_{kl} = \frac{\partial x_t^k}{\partial x_0^l}. \tag{2.4.3}$$

We propose to determine the Jacobian in eqn. (2.4.2) by deriving an equation of motion it obeys and then solving this equation of motion. To accomplish this, we start with the definition,

$$J(x_t; x_0) = \det(J), \tag{2.4.4}$$

analyze the derivative

$$\frac{d}{dt}J(x_t; x_0) = \frac{d}{dt}\det(J), \tag{2.4.5}$$

and derive a first-order differential equation obeyed by $J(x_t; x_0)$.

The time derivative of the determinant is most easily computed by applying an identity satisfied by determinants

$$\det(J) = e^{\mathrm{Tr}[\ln(J)]}, \tag{2.4.6}$$

where Tr is the trace operation: $\mathrm{Tr}(J) = \sum_k J_{kk}$. Equation (2.4.6) is most easily proved by first transforming J into a representation in which it is diagonal. If J has eigenvalues λ_k, then $\ln(J)$ is a diagonal matrix with eigenvalues $\ln(\lambda_k)$, and the trace operation yields $\mathrm{Tr}[\ln(J)] = \sum_k \ln \lambda_k$. Exponentiating the trace yields $\prod_k \lambda_k$, which is just the determinant of J. Substituting eqn. (2.4.6) into eqn. (2.4.5) gives

$$\frac{d}{dt}J(x_t; x_0) = \frac{d}{dt}e^{\mathrm{Tr}[\ln(J)]}$$

$$= e^{\mathrm{Tr}[\ln(J)]}\mathrm{Tr}\left[\frac{dJ}{dt}J^{-1}\right]$$

$$= J(x_t; x_0)\sum_{k,l}\left[\frac{dJ_{kl}}{dt}J_{lk}^{-1}\right]. \tag{2.4.7}$$

The elements of the matrices J^{-1} and dJ/dt are easily seen to be

$$\frac{dJ_{kl}}{dt} = \frac{\partial \dot{x}_t^k}{\partial x_0^l}. \qquad J_{lk}^{-1} = \frac{\partial x_0^l}{\partial x_t^k}. \tag{2.4.8}$$

Substituting eqn. (2.4.8) into eqn. (2.4.7) gives

$$\frac{\mathrm{d}}{\mathrm{d}t} J(\mathrm{x}_t; \mathrm{x}_0) = J(\mathrm{x}_t; \mathrm{x}_0) \sum_{k,l} \left[\frac{\partial \dot{\mathrm{x}}_t^k}{\partial \mathrm{x}_0^l} \frac{\partial \mathrm{x}_0^l}{\partial \mathrm{x}_t^k} \right]. \tag{2.4.9}$$

The summation over l of the term in square brackets, is just the chain-rule expression for $\partial \dot{\mathrm{x}}_t^k / \partial \mathrm{x}_t^k$. Thus, performing this sum yields the equation of motion for the Jacobian:

$$\frac{\mathrm{d}}{\mathrm{d}t} J(\mathrm{x}_t; \mathrm{x}_0) = J(\mathrm{x}_t; \mathrm{x}_0) \sum_k \frac{\partial \dot{\mathrm{x}}_t^k}{\partial \mathrm{x}_t^k}. \tag{2.4.10}$$

The sum in the last line of eqn. (2.4.10) is easily recognized as the phase-space compressibility $\nabla_{\mathrm{x}_t} \cdot \dot{\mathrm{x}}_t$ defined in eqn. (1.6.25), where $\nabla_{\mathrm{x}_t} = \partial / \partial \mathrm{x}_t$. Equation (1.6.25) also revealed that the phase compressibility is 0 for a system evolving under Hamilton's equations. Thus, the sum on the right side of eqn. (2.4.10) vanishes, and the equation of motion for the Jacobian reduces to

$$\frac{\mathrm{d}}{\mathrm{d}t} J(\mathrm{x}_t; \mathrm{x}_0) = 0. \tag{2.4.11}$$

This equation of motion implies that the Jacobian is a constant for all time. The initial condition $J(\mathrm{x}_0; \mathrm{x}_0)$ on the Jacobian is simply 1, since the transformation from x_0 to x_0 is an identity transformation. Thus, since the Jacobian is initially 1 and remains constant in time, it follows that

$$J(\mathrm{x}_t; \mathrm{x}_0) = 1. \tag{2.4.12}$$

Substituting eqn. (2.4.12) into eqn. (2.4.2) yields the volume element transformation condition

$$\mathrm{d}\mathrm{x}_t = \mathrm{d}\mathrm{x}_0. \tag{2.4.13}$$

Equation (2.4.13) is an important result known as *Liouville's theorem* (named for the nineteenth-century French mathematician Joseph Liouville (1809–1882)). Liouville's theorem is essential to the claim made earlier that ensemble averages can be performed at any point in time.

  If the motion of the system is driven by highly nonlinear forces, then an initial hypercubic volume element $\mathrm{d}\mathrm{x}_0$, for example, will distort due to the chaotic nature of the dynamics. Because of Liouville's theorem, the volume element can spread out in some of the phase-space dimensions but must contract in other dimensions by an equal amount so that, overall, the volume is conserved. That is, there can be no net attractors or repellors in the phase space. This is illustrated in Fig. 2.3 for a two-dimensional phase space.

2.5 The ensemble distribution function and the Liouville equation

Phase space consists of all possible microstates available to a system of N particles. However, an ensemble contains only those microstates that are consistent with a given set of macroscopic observables. Consequently, the microstates of an ensemble are either a strict subset of all possible phase-space points or are clustered more densely in certain

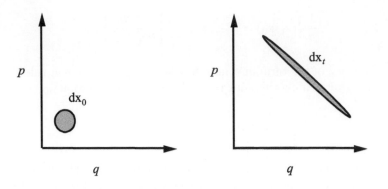

Fig. 2.3 Illustration of phase-space volume conservation prescribed by Liouville's theorem.

regions of phase space and less densely in others. We, therefore, need to describe quantitatively how the systems in an ensemble are distributed in the phase space at any point in time. To do this, we introduce the *ensemble distribution function* or *phase-space distribution function* $f(\mathrm{x}, t)$. The phase-space distribution function of an ensemble has the property that $f(\mathrm{x}, t)\mathrm{dx}$ is the fraction of the total ensemble members contained in the phase-space volume element dx at time t. From this definition, it is clear that $f(\mathrm{x}, t)$ satisfies the following properties:

$$f(\mathrm{x}, t) \geq 0$$
$$\int \mathrm{dx}\, f(\mathrm{x}, t) = 1. \tag{2.5.1}$$

Therefore, $f(\mathrm{x}, t)$ is a probability density.

When the phase-space distribution is expressed as $f(\mathrm{x}, t)$, we imagine ourselves sitting at a *fixed* location x in the phase space and observing the ensemble distribution evolve around us as a function of time. This perspective is known as the *Eulerian view* of the ensemble evolution and is analogous to an Eulerian frame in fluid mechanics, in which an observer remains at a fixed location and observes the motion of the fluid at that point. In order to determine the number of ensemble members in a small element dx at our location, we could simply "count" the number of microstates belonging to the ensemble in the volume element dx at any time t, determine the fraction $f(\mathrm{x}, t)\mathrm{dx}$, and build up a picture of the distribution by moving from one observation point x to another until we have exhausted every point in the phase space. On the other hand, the ensemble consists of a collection of systems evolving in time according to Hamilton's equations of motion. Thus, we can also let the ensemble distribution function describe how a bundle of trajectories in a volume element dx_t centered on a trajectory x_t is distributed at time t. This will be given by $f(\mathrm{x}_t, t)\mathrm{dx}_t$. The latter, "co-moving" perspective, known as the *Lagrangian view* of the ensemble evolution (analogous to a Lagrangian frame in fluid mechanics), more closely fits the originally stated definition of an ensemble and will, therefore, be employed in the subsequent derivation to determine an equation satisfied by f in the phase space.

The fact that f has a constant normalization means that there can be neither sources of new ensemble members nor sinks that reduce the number of ensemble

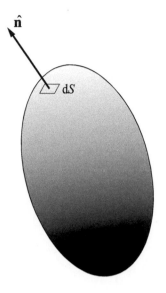

Fig. 2.4 An arbitrary volume in phase space. $\mathrm{d}S$ is a hypersurface element and $\hat{\mathbf{n}}$ is the unit vector normal to the surface at the location of $\mathrm{d}S$.

members—the number of members remains constant. This also means that any volume Ω in the phase space with a surface S (see Fig. 2.4) contains no sources or sinks. Thus, the rate of decrease (or increase) of ensemble members in Ω must equal the rate at which ensemble members leave (or enter) Ω through the surface S. The fraction of ensemble members in Ω at time t can be written as

$$\text{Fraction of ensemble members in } \Omega = \int_{\Omega} \mathrm{d}\mathbf{x}_t \, f(\mathbf{x}_t, t). \tag{2.5.2}$$

Thus, the rate of decrease of ensemble members in Ω is related to the rate of decrease of this fraction by

$$-\frac{\mathrm{d}}{\mathrm{d}t} \int_{\Omega} \mathrm{d}\mathbf{x}_t \, f(\mathbf{x}_t, t) = -\int_{\Omega} \mathrm{d}\mathbf{x}_t \, \frac{\partial}{\partial t} f(\mathbf{x}_t, t). \tag{2.5.3}$$

On the other hand, the rate at which ensemble members leave Ω through the surface can be calculated from the flux, which is the number of ensemble members per unit area per unit time passing through the surface. Let $\hat{\mathbf{n}}$ be the unit vector normal to the surface at the point $\mathbf{x}_t$ (see Fig. 2.4). Then, as a fraction of ensemble members, this flux is given by $\dot{\mathbf{x}}_t \cdot \hat{\mathbf{n}} f(\mathbf{x}_t, t)$. The dot product with $\hat{\mathbf{n}}$ ensures that we count only those ensemble members actually leaving Ω through the surface, that is, members whose trajectories have a component of their phase-space velocity $\dot{\mathbf{x}}_t$ normal to the surface. Thus, the rate at which ensemble members leave Ω through the surface is obtained by integrating over S:

$$\int_{S} \mathrm{d}S \, \dot{\mathbf{x}}_t \cdot \hat{\mathbf{n}} f(\mathbf{x}_t, t) = \int_{\Omega} \mathrm{d}\mathbf{x}_t \nabla_{\mathbf{x}_t} \cdot (\dot{\mathbf{x}}_t f(\mathbf{x}_t, t)). \tag{2.5.4}$$

The right side of eqn. (2.5.6) follows from the divergence theorem applied to the hypersurface integral. Equating the right sides of eqns. (2.5.4) and (2.5.3) gives

$$\int_{\Omega} d\mathbf{x}_t \nabla_{\mathbf{x}_t} \cdot (\dot{\mathbf{x}}_t f(\mathbf{x}_t, t)) = -\int_{\Omega} d\mathbf{x}_t \frac{\partial}{\partial t} f(\mathbf{x}_t, t), \tag{2.5.5}$$

or

$$\int_{\Omega} d\mathbf{x}_t \left[\frac{\partial}{\partial t} f(\mathbf{x}_t, t) + \nabla_{\mathbf{x}_t} \cdot (\dot{\mathbf{x}}_t f(\mathbf{x}_t, t)) \right] = 0. \tag{2.5.6}$$

Since the choice of Ω is arbitrary, eqn. (2.5.6) must hold locally, so that the term in brackets vanishes identically, giving

$$\frac{\partial}{\partial t} f(\mathbf{x}_t, t) + \nabla_{\mathbf{x}_t} \cdot (\dot{\mathbf{x}}_t f(\mathbf{x}_t, t)) = 0. \tag{2.5.7}$$

Finally, since $\nabla_{\mathbf{x}_t} \cdot (\dot{\mathbf{x}}_t f(\mathbf{x}_t, t)) = \dot{\mathbf{x}}_t \cdot \nabla_{\mathbf{x}_t} f(\mathbf{x}_t, t) + f(\mathbf{x}_t, t) \nabla_{\mathbf{x}_t} \cdot \dot{\mathbf{x}}_t$, and the phase-space divergence $\nabla_{\mathbf{x}_t} \cdot \dot{\mathbf{x}}_t = 0$, eqn. (2.5.7) reduces to

$$\frac{\partial}{\partial t} f(\mathbf{x}_t, t) + \dot{\mathbf{x}}_t \cdot \nabla_{\mathbf{x}_t} f(\mathbf{x}_t, t) = 0. \tag{2.5.8}$$

The quantity on the left side of eqn. (2.5.8) is just the total time derivative of $f(\mathbf{x}_t, t)$, which includes both the time dependence of the phase-space vector $\mathbf{x}_t$ and the explicit time dependence of $f(\mathbf{x}_t, t)$. Thus, we obtain finally

$$\frac{df}{dt} = \frac{\partial}{\partial t} f(\mathbf{x}_t, t) + \dot{\mathbf{x}}_t \cdot \nabla_{\mathbf{x}_t} f(\mathbf{x}_t, t) = 0, \tag{2.5.9}$$

which states that $f(\mathbf{x}_t, t)$ is conserved along a trajectory. This result is known as the *Liouville equation*. The conservation of $f(\mathbf{x}_t, t)$ implies that

$$f(\mathbf{x}_t, t) = f(\mathbf{x}_0, 0), \qquad \qquad \cdot \tag{2.5.10}$$

and since $d\mathbf{x}_t = d\mathbf{x}_0$, we have

$$f(\mathbf{x}_t, t)d\mathbf{x}_t = f(\mathbf{x}_0, 0)d\mathbf{x}_0. \tag{2.5.11}$$

Equation (2.5.11) states that the fraction of ensemble members in the initial volume element $d\mathbf{x}_0$ is equal to the fraction of ensemble members in the volume element $d\mathbf{x}_t$. Equation (2.5.11) ensures that we can perform averages over the ensemble at any point in time because the fraction of ensemble members is conserved. Since $\dot{\mathbf{x}}_t = \eta(\mathbf{x}_t, t)$, eqn. (2.5.9) can also be written as

$$\frac{df}{dt} = \frac{\partial}{\partial t} f(\mathbf{x}_t, t) + \eta(\mathbf{x}_t, t) \cdot \nabla_{\mathbf{x}_t} f(\mathbf{x}_t, t) = 0. \tag{2.5.12}$$

Writing the Liouville equation this way allows us to recover the "passive" view of the ensemble distribution function in which we remain at a fixed location in phase space.

In this case, we remove the t label attached to the phase-space points and obtain the following partial differential equation for $f(\mathbf{x}, t)$:

$$\frac{\partial}{\partial t} f(\mathbf{x}, t) + \eta(\mathbf{x}, t) \cdot \nabla_{\mathbf{x}} f(\mathbf{x}, t) = 0, \tag{2.5.13}$$

which is another form of the Liouville equation. Since eqn. (2.5.13) is a partial differential equation, it can only specify a class of functions as solutions. Specific solutions for $f(\mathbf{x}, t)$ require input of further information; we will return to this point again as specific ensembles are considered in subsequent chapters.

Finally, note that if we use the definition of $\eta(\mathbf{x}, t)$ in eqn. (1.6.24) and apply the analysis leading up to eqn. (1.6.19), it is clear that $\eta(\mathbf{x}, t) \cdot \nabla_{\mathbf{x}} f(x, t) = \{f(\mathbf{x}, t), \mathcal{H}(\mathbf{x}, t)\}$, where $\{..., ...\}$ is the Poisson bracket. Thus, the Liouville equation can also be written as

$$\frac{\partial}{\partial t} f(\mathbf{x}, t) + \{f(\mathbf{x}, t), \mathcal{H}(\mathbf{x}, t)\} = 0, \tag{2.5.14}$$

a form we will employ in the next section for deriving general equilibrium solutions.

2.6   Equilibrium solutions of the Liouville equation

In Section 2.3, we argued that thermodynamic variables can be computed from averages over an ensemble. Such averages must, therefore, be expressed in terms of the ensemble distribution function. If $a(\mathbf{x})$ is a microscopic phase-space function corresponding to a macroscopic observable A, then a proper generalization of eqn. (2.3.1) is

$$A = \langle a(\mathbf{x}) \rangle = \int d\mathbf{x}\, f(\mathbf{x}, t) a(\mathbf{x}). \tag{2.6.1}$$

If $f(\mathbf{x}, t)$ has an explicit time dependence, then so will the observable A, in general. However, we also remarked earlier that a system in thermodynamic equilibrium has a fixed thermodynamic state. This means that the thermodynamic variables characterizing the equilibrium state do not change in time. Thus, if A is an equilibrium observable, the ensemble average in eqn. (2.6.1) must yield a time-independent result, which is only possible if the ensemble distribution of a system in thermodynamic equilibrium has *no* explicit time dependence, *i.e.*, $\partial f/\partial t = 0$. This will be the case, for example, when no external driving forces act on the system, in which case $\mathcal{H}(\mathbf{x}, t) \to \mathcal{H}(\mathbf{x})$ and $\eta(\mathbf{x}, t) \to \eta(\mathbf{x})$.

When $\partial f/\partial t = 0$, the Liouville equation eqn. (2.5.14) reduces to

$$\{f(\mathbf{x}), \mathcal{H}(\mathbf{x})\} = 0. \tag{2.6.2}$$

The general solution to eqn. (2.6.2) is any function of the Hamiltonian $\mathcal{H}(\mathbf{x})$:

$$f(\mathbf{x}) \propto \mathcal{F}(\mathcal{H}(\mathbf{x})). \tag{2.6.3}$$

This is as much as we can say from eqn. (2.6.2) without further information about the ensemble. In order to ensure that $f(\mathbf{x})$ is properly normalized according to eqn. (2.5.1), we write the solution as

$$f(x) = \frac{1}{\mathcal{Z}} \mathcal{F}(\mathcal{H}(x)), \tag{2.6.4}$$

where $\mathcal{Z}$ is defined to be

$$\mathcal{Z} = \int dx \, \mathcal{F}(\mathcal{H}(x)). \tag{2.6.5}$$

The quantity $\mathcal{Z}$, referred to as the *partition function*, is one of the central quantities in equilibrium statistical mechanics. The partition function is a measure of the number of microscopic states in the phase space accessible within a given ensemble. Each ensemble has a particular partition function that depends on the macroscopic observables used to define the ensemble. We will show in Chapters 3 to 6 that the thermodynamic properties of a system are calculated from the various partial derivatives of the partition function. Other equilibrium observables are computed according to

$$A = \langle a(x) \rangle = \frac{1}{\mathcal{Z}} \int dx \, a(x) \mathcal{F}(\mathcal{H}(x)). \tag{2.6.6}$$

Note that the condition $f(x_0)dx_0 = f(x_t)dx_t$ implied by eqn. (2.5.11) guarantees that the equilibrium average over the systems in the ensemble can be performed at any point in time.

Equations (2.6.5) and (2.6.6) constitute the essence of equilibrium statistical mechanics. As Richard Feynman remarks in his book *Statistical Mechanics: A Set of Lectures*, eqns. (2.6.5) and (2.6.6) embody "the summit of statistical mechanics, and the entire subject is either the slide-down from this summit, as the [principles are] applied to various cases, or the climb-up where the fundamental [laws are] derived and the concepts of thermal equilibrium . . . [are] clarified" (Feynman, 1998).[2] We shall, of course, embark on both, and we will explore the methods by which equilibrium ensemble distributions are generated and observables are computed for realistic applications.

2.7   Problems

2.1. Consider n moles of an ideal gas in a volume V at pressure P and temperature T. The equation of state is $PV = nRT$ as given in eqn. (2.2.2). If the gas contains N molecules, so that $n = N/N_0$, where N_0 is Avogadro's number, then the total number of microscopic states available to the gas can be shown (see Section 3.5) to be $\Omega \propto V^N(kT)^{3N/2}$, where $k = R/N_0$ is known as Boltzmann's constant. The entropy of the gas is defined via Boltzmann's relation (see Chapter 3) as $S = k \ln \Omega$. Note the total energy of an ideal gas is $E = 3nRT/2$.

   a. Suppose the gas expands or contracts from a volume V_1 to a volume V_2 at constant temperature. Calculate the work done on the system.

---

[2]This quote is actually made in the context of quantum statistical mechanics (see Chapter 10); however, the sentiment applies equally well to classical statistical mechanics.

b. For the process in part a, calculate the change of entropy using Boltz-mann's relation and using eqn. (2.2.19). Show that these two approaches yield the same entropy change.

c. Next, suppose the temperature of the gas is changed from T_1 to T_2 under conditions of constant volume. Calculate the entropy change using the two approaches in part a and show that they yield the same entropy change.

d. Finally, suppose that the volume changes from V_1 to V_2 in an adiabatic process ($\Delta Q = 0$). The pressure also changes from P_1 to P_2 in the process. Show that

$$P_1 V_1^\gamma = P_2 V_2^\gamma,$$

and find the numerical value of the exponent γ.

2.2. A substance has the following properties:
   i. When its volume is increased from V_1 to V_2 at constant temperature T, the work done in the expansion is

$$W = RT \ln\left(\frac{V_2}{V_1}\right).$$

   ii. When the volume changes from V_1 to V_2 and the temperature changes from T_1 to T_2, its entropy changes according to

$$\Delta S = k \left(\frac{V_1}{V_2}\right)\left(\frac{T_2}{T_1}\right)^\alpha,$$

   where α is a constant. Find the equation of state and Helmholtz free energy of this substance.

2.3. Reformulate the Carnot cycle for an ideal gas as a thermodynamic cycle in the T-S plane rather than the P-V plane, and show that the area enclosed by the cycle is equal to the net work done by the gas during the cycle.

2.4. The equation of state for one mole of a substance takes the general form $g(V, P, T) = 0$, which can be expressed equivalently as $V = F(P, T)$. Suppose we are able to perform experiments that yield the following information about the P and T dependence of the volume:

$$\frac{1}{V}\left(\frac{\partial V}{\partial T}\right)_P = \frac{1}{F}\left(\frac{\partial F}{\partial T}\right)_P = A\left(1 - aP\right)$$

$$\frac{1}{V}\left(\frac{\partial V}{\partial P}\right)_T = \frac{1}{F}\left(\frac{\partial F}{\partial P}\right)_T = -K\left[1 + b(T - T_0)\right],$$

where a, b, K, and T_0 are constants. (These two relations define two thermo-dynamic variables known as the *volume expansion coefficient* and the *isother-mal compressibility*, respectively.)

a. Let $h(P,T) = \ln V = \ln F(P,T)$. Show that

$$\left(\frac{\partial h}{\partial P}\right)_T = \frac{1}{V}\left(\frac{\partial V}{\partial P}\right)_T = \frac{1}{F}\left(\frac{\partial F}{\partial P}\right)_T,$$

$$\left(\frac{\partial h}{\partial T}\right)_P = \frac{1}{V}\left(\frac{\partial V}{\partial T}\right)_P = \frac{1}{F}\left(\frac{\partial F}{\partial T}\right)_P,$$

and, therefore, that

$$dh = -K\left[1 + b(T - T_0)\right]dP + A\left(1 - aP\right)dT.$$

b. Using the identity for $h(P,T)$

$$\frac{\partial^2 h}{\partial T \partial P} = \frac{\partial^2 h}{\partial P \partial T},$$

show that $Aa = Kb$.

c. By integrating $\left(\frac{\partial h}{\partial P}\right)_T$ with respect to P, show that

$$h(P,T) = -K\left[1 + b(T - T_0)\right]P + f(T),$$

where $f(T)$ is an arbitrary function of T.

d. Use the derivative $\left(\frac{\partial h}{\partial T}\right)_P$ to show that $f(T) = AT + C$, where C is an arbitrary constant.

e. Suppose we know that at $T = T_0$ and at a pressure $P = P_0$, the volume has a value $V = V_0$. Given this data, determine the constant C, and find the functional form of $F(P,T)$, thereby obtaining the equation of state for this system.

2.5. Consider the thermodynamic cycle shown in Fig. 2.5. Compare the efficiency of this engine to that of a Carnot engine operating between the highest and lowest temperatures of the cycle in the figure. Which one is greater?

2.6. Consider an ensemble of one-particle systems, each evolving in one spatial dimension according to an equation of motion of the form

$$\dot{x} = -\alpha x,$$

where $x(t)$ is the position of the particle at time t and α is a constant. Since the compressibility of this system is nonzero, the ensemble distribution function $f(x,t)$ satisfies a Liouville equation of the form

$$\frac{\partial f}{\partial t} - \alpha x \frac{\partial f}{\partial x} = \alpha f$$

(see eqn. (2.5.7)). Suppose that at $t = 0$, the ensemble distribution has a Gaussian form

$$f(x,0) = \frac{1}{\sqrt{2\pi\sigma^2}} e^{-x^2/2\sigma^2}.$$

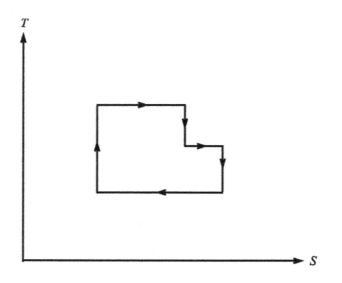

T

S

Fig. 2.5 Thermodynamic cycle.

a. Find a solution of the Liouville equation that also satisfies this initial distribution.

   Hint: Show that the substitution $f(x,t) = e^{\alpha t}\tilde{f}(x,t)$ yields an equation for a conserved distribution $\tilde{f}(x,t)$. Next, try multiplying the x^2 in the initial distribution by an arbitrary function $g(t)$ that must satisfy $g(0) = 1$. Use the Liouville equation to derive an equation that $g(t)$ must satisfy and then solve this equation.

b. Describe the evolution of the ensemble distribution qualitatively and explain why it should evolve this way.

c. Show that your solution is properly normalized in the sense that

$$\int_{-\infty}^{\infty} dx f(x,t) = 1.$$

d. Repeat the problem for the case that the initial distribution is a Lorentzian

$$f(x,0) = \frac{1}{\pi}\frac{\sigma}{\sigma^2 + x^2}$$

*2.7. An alternative definition of entropy was proposed by Gibbs, who expressed the entropy in terms of the phase-space distribution function $f(x,t)$ as

$$S(t) = -k\int dx\, f(x,t) \ln f(x,t).$$

Here, $f(x, t)$ satisfies the Liouville equation eqn. (2.5.13). The notation $S(t)$ expresses the fact that an entropy defined this way is an explicit function of time.

a. Show that for an arbitrary distribution function, the entropy is actually constant, *i.e.*, that $dS/dt = 0$, $S(t) = S(0)$, so that $S(t)$ cannot increase in time for any ensemble. Is this in violation of the second law of thermodynamics?

Hint: Be careful how the derivative d/dt is applied to the integral!

b. The distribution $f(x, t)$ is known as a "fine-grained" distribution function. Because $f(x, t)$ is fully defined at every phase-space point, it contains all of the detailed microstructure of the phase space, which cannot be resolved in reality. Consider, therefore, introducing a "coarse-grained" phase space distribution $\bar{f}(x, t)$ defined via the following operation: Divide phase space into the smallest cells over which $\bar{f}(x, t)$ can be defined. Each cell C is then subdivided into small subcells such that each subcell of volume Δx centered on the point x has an associated probability $f(x, t)\Delta x$ at time t (Waldram, 1985). Assume that at $t = 0$, $f(x, 0) = \bar{f}(x, 0)$. In order to define $\bar{f}(x, t)$ for $t > 0$, at each point in time, we transfer probability from subcells of C where $f > \bar{f}$ to cells where $f < \bar{f}$. Then, we use $\bar{f}(x, t)$ to define a coarse-grained entropy

$$\bar{S}(t) = -k \int dx \, \bar{f}(x, t) \ln \bar{f}(x, t),$$

where the integral should be interpreted as a sum over all cells C into which the phase space has been divided. For this particular coarse-graining operation, show that $\bar{S}(t) \geq \bar{S}(0)$ where equality is only true in equilibrium.

Hint: Show that the change in $\bar{S}$ on transferring probability from one small subcell to another is either positive or zero. This is sufficient to show that the total coarse-grained entropy can either increase in time or remain constant.

2.8. Consider a single particle moving in three spatial dimensions with phase-space vector (p_x, p_y, p_z, x, y, z). Derive the complete canonical transformation to spherical polar coordinates (r, θ, ϕ) and their conjugate momenta (p_r, p_θ, p_ϕ) and show that the phase-space volume element $d\mathbf{p}d\mathbf{r}$ satisfies

$$dp_x dp_y dp_z dx dy dz = dp_r dp_\theta d\phi dr d\theta d\phi.$$

3

The microcanonical ensemble and introduction to molecular dynamics

3.1 Brief overview

In the previous chapter, it was shown that statistical mechanics provides the link between the classical microscopic world described by Newton's laws of motion and the macroscopic observables that are actually measured in experiments, including thermodynamic, structural, and dynamical properties. One of the great successes of statistical mechanics is its provision of a rational microscopic basis for thermodynamics, which otherwise is only a phenomenological theory. We showed that the microscopic connection is provided via the notion of an ensemble—an imaginary collection of systems described by the same Hamiltonian with each system in a unique microscopic state at any given instant in time.

In this chapter, we will lay out the basic classical statistical mechanics of the simplest and most fundamental of the equilibrium ensembles, that of an isolated system of N particles in a container of volume V and a total energy E corresponding to a Hamiltonian $\mathcal{H}(x)$. This ensemble is known as the *microcanonical ensemble*. The microcanonical ensemble provides a starting point from which all other equilibrium ensembles are derived. Our discussion will begin with the classical partition function, its connection to the entropy via Boltzmann's relation, and the thermodynamic and equilibrium properties that it generates. Several simple applications will serve to illustrate these concepts. However, it will rapidly become apparent that in order to treat any realistic system, numerical solutions are needed, which will lead naturally to a discussion of the numerical simulation technique known as molecular dynamics (MD). MD is a widely used, immensely successful computational approach in which the classical equations of motion are solved numerically and the trajectories thus generated are used to extract macroscopic observables. MD also permits direct "visualization" of the detailed motions of individual atoms in a system, thereby providing a "window" into the microscopic world. Although such animations of MD trajectories should never be taken too seriously, they can be useful as a guide toward understanding the mechanisms underlying a given chemical process. At the end of the chapter, we will consider a number of examples that illustrate the power and general applicability of molecular dynamics to realistic systems.

3.2   Basic thermodynamics, Boltzmann's relation, and the partition function of the microcanonical ensemble

We begin by considering a system of N identical particles in a container of volume V with a fixed internal energy E. The variables N, V, and E are all macroscopic thermodynamic quantities referred to as *control variables*. Control variables are simply quantities that characterize the ensemble and that determine other thermodynamic properties of the system. Different choices of these variables lead to different system properties. In order to describe the thermodynamics of an ensemble of systems with given values of N, V, and E, we seek a unique state function of these variables. We will now show that such a state function can be obtained from the first law of thermodynamics, which relates the energy E of a system to a quantity Q of heat absorbed and an amount of work W done on the system:

$$E = Q + W. \tag{3.2.1}$$

The derivation of the desired state function begins by examining how the energy changes if a small amount of heat dQ is added to the system and a small amount of work dW is done on the system. Since E is a state function, this thermodynamic transformation may be carried out along any path, and it is particularly useful to consider a reversible path for which

$$dE = dQ_{\text{rev}} + dW_{\text{rev}}. \tag{3.2.2}$$

Note that since Q and W are not state functions, it is necessary to characterize their changes by the "rev" subscript. The amount of heat absorbed by the system can be related to the change in the entropy ΔS of the system by

$$\Delta S = \int \frac{dQ_{\text{rev}}}{T}, \qquad\qquad dS = \frac{dQ_{\text{rev}}}{T}, \tag{3.2.3}$$

where T is the temperature of the system. Therefore, $dQ_{\text{rev}} = TdS$. Work done on the system is measured in terms of the two control variables V and N. Let $P(V)$ be the pressure of the system at the volume V. Mechanical work can be done on the system by compressing it from a volume V_1 to a new volume $V_2 < V_1$:

$$W_{12}^{(\text{mech})} = -\int_{V_1}^{V_2} P(V)dV, \tag{3.2.4}$$

where the minus sign indicates that work is positive in a compression. A small volume change dV corresponds to an amount of work $dW_{\text{rev}}^{(\text{mech})} = -P(V)dV$. Although we will typically suppress the explicit volume dependence of P on V and write simply, $dW_{\text{rev}}^{(\text{mech})} = -PdV$, it must be remembered that P depends not only on V but also on N and E. In addition to the mechanical work done by compressing a system, chemical work can also be done on the system by increasing the number of particles. Let $\mu(N)$ be the chemical potential of the system at particle number, N (μ also depends on V

and E). If the number of particles is increased from N_1 to $N_2 > N_1$, then chemical work

$$W_{12}^{(\text{chem})} = \sum_{N=N_1}^{N_2} \mu(N) \qquad (3.2.5)$$

is done on the system. Clearly, the number of particles in a system can only change by integral amounts, ΔN. However, the changes we wish to consider are so small compared to the total particle number ($N \sim 10^{23}$) that they can be regarded approximately as changes dN in a continuous variable. Therefore, the chemical work corresponding to such a small change dN can be expressed as $dW_{\text{rev}}^{(\text{chem})} = \mu(N)dN$. Again, we suppress the explicit dependence of μ on N (as well as on V and E) and write simply $dW_{\text{rev}}^{(\text{chem})} = \mu dN$. Therefore, the total reversible work done on the system is given by

$$dW_{\text{rev}} = dW_{\text{rev}}^{(\text{mech})} + dW_{\text{rev}}^{(\text{chem})} = -PdV + \mu dN, \qquad (3.2.6)$$

so that the total change in energy is

$$dE = TdS - PdV + \mu dN. \qquad (3.2.7)$$

By writing eqn. (3.2.7) in the form

$$dS = \frac{1}{T}dE + \frac{P}{T}dV - \frac{\mu}{T}dN, \qquad (3.2.8)$$

it is clear that the state function we are seeking is just the entropy of the system, $S = S(N, V, E)$, since the change in S is related directly to the change in the three control variables of the ensemble. However, since S is a function of N, V, and E, the change in S resulting from small changes in N, V, and E can also be written using the chain rule as

$$dS = \left(\frac{\partial S}{\partial E}\right)_{N,V} dE + \left(\frac{\partial S}{\partial V}\right)_{N,E} dV + \left(\frac{\partial S}{\partial N}\right)_{V,E} dN. \qquad (3.2.9)$$

Comparing eqn. (3.2.9) with eqn. (3.2.8) shows that the thermodynamic quantities T, P, and μ can be obtained by taking partial derivatives of the entropy with respect to each of the three control variables:

$$\frac{1}{T} = \left(\frac{\partial S}{\partial E}\right)_{N,V}, \qquad \frac{P}{T} = \left(\frac{\partial S}{\partial V}\right)_{N,E}, \qquad \frac{\mu}{T} = -\left(\frac{\partial S}{\partial N}\right)_{V,E}. \qquad (3.2.10)$$

We now recall that the entropy is a quantity that can be related to the number of microscopic states of the system. This relation was first proposed by Ludwig Boltzmann in 1877, although it was Max Planck who actually formalized the connection. Let Ω be the number of microscopic states available to a system. The relation connecting S and Ω states that

$$S(N, V, E) = k \ln \Omega(N, V, E). \qquad (3.2.11)$$

Since S is a function of N, V, and E, Ω must be as well. The constant, k, appearing in eqn. (3.2.11) is known as Boltzmann's constant; its value is $1.3806505(24) \times 10^{-23}$

J·K^{-1}. That logarithmic dependence of the entropy on $\Omega(N, V, E)$ will be explained shortly. Assuming we can determine $\Omega(N, V, E)$ from a microscopic description of the system, eqn. (3.2.11) then provides a connection between this microscopic description and macroscopic thermodynamic observables.

In the last chapter, we saw that the most general solution to the equilibrium Liouville equation, $\{f(\mathbf{x}), \mathcal{H}(\mathbf{x})\} = 0$, for the ensemble distribution function $f(\mathbf{x})$ is any function of the Hamiltonian: $f(\mathbf{x}) = F(\mathcal{H}(\mathbf{x}))$, where $\mathbf{x}$ is the phase-space vector. The specific choice of $F(\mathcal{H}(\mathbf{x}))$ is determined by the conditions of the ensemble. The microcanonical ensemble pertains to a collection of systems in isolation obeying Hamilton's equations of motion. Recall, however, from Section 1.6, that a system obeying Hamilton's equations conserves the total Hamiltonian

$$\mathcal{H}(\mathbf{x}) = E \tag{3.2.12}$$

with E being the total energy of the system. Conservation of $\mathcal{H}(\mathbf{x})$ was demonstrated explicitly in eqn. (1.6.15). Moreover, the ensemble distribution function $f(\mathbf{x})$ is static in the sense that $\partial f/\partial t = 0$. Therefore, each member of an equilibrium ensemble is in a single unique microscopic state. For the microcanonical ensemble, each unique state is described by a unique phase-space vector $\mathbf{x}$ that satisfies eqn. (3.2.12). It follows that the choice of $F(\mathcal{H}(\mathbf{x}))$ must be consistent with eqn. (3.2.12). That is, $F(\mathcal{H}(\mathbf{x}))$ must restrict $\mathbf{x}$ to those microscopic states for which $\mathcal{H}(\mathbf{x}) = E$. A function that achieves this is the Dirac δ-function

$$F(\mathcal{H}(\mathbf{x})) = \mathcal{N}\delta(\mathcal{H}(\mathbf{x}) - E) \tag{3.2.13}$$

expressing the conservation of energy condition. Here, $\mathcal{N}$ is an overall normalization constant. For readers not familiar with the properties of the Dirac δ-function, a detailed discussion is provided in Appendix A. Since eqn. (3.2.12) defines the constant-energy hypersurface in phase space, eqn. (3.2.13) expresses the fact that, in the microcanonical ensemble, all phase-space points must lie on this hypersurface and that all such points are equally probable; all points not on this surface have zero probability. The notion that $\Omega(N, V, E)$ can be computed from an ensemble in which all accessible microscopic states are equally probable is an assumption that is consistent with classical mechanics, as the preceding discussion makes clear. More generally, we assume that for an isolated system in equilibrium, all accessible microscopic states are equally probable, which is known as the *assumption of equal a priori probability*. The quantity $1/\Omega(N, V, E)$ is a measure of the probability of randomly selecting a microstate in any small neighborhood of phase space anywhere on the constant-energy hypersurface.

The number $\Omega(N, V, E)$ is a measure of the amount of phase space available to the system. It must, therefore, be proportional to the fraction of phase space consistent with eqn. (3.2.12), which is proportional to the $(6N - 1)$-dimensional "area" of the constant-energy hypersurface. This number can be obtained by integrating eqn. (3.2.13) over the phase space, as indicated by eqn. (2.6.5).[1] An integration over the

---

[1]If we imagined discretizing the constant-energy hypersurface such that each discrete patch contained a single microstate, then the integral would revert to a sum that would represent a literal counting of the number of microstates contained on the surface.

entire phase space is an integration over the momentum $\mathbf{p}_i$ and position $\mathbf{r}_i$ of each particle in the system and is, therefore, a $6N$-dimensional integration. Moreover, while the range of integration of each momentum variable is infinite, integration over each position variable is restricted to that region of space defined by the containing volume. We denote this region as $D(V)$, i.e., the spatial domain defined by the containing volume. For example, if the container is a cube of side length L, lying in the positive octant of Cartesian space with a corner at the origin, then $D(V)$ would be defined by $x \in [0, L]$, $y \in [0, L]$, $z \in [0, L]$ for each Cartesian vector $\mathbf{r} = (x, y, z)$. Therefore, $\Omega(N, V, E)$ is given by the integral

$$\Omega(N, V, E) = M \int \mathrm{d}\mathbf{p}_1 \cdots \int \mathrm{d}\mathbf{p}_N \int_{D(V)} \mathrm{d}\mathbf{r}_1 \cdots \int_{D(V)} \mathrm{d}\mathbf{r}_N \, \delta(\mathcal{H}(\mathbf{r}, \mathbf{p}) - E), \quad (3.2.14)$$

where M is an overall constant whose value we will discuss shortly. Equation (3.2.14) defines the partition function of the microcanonical ensemble. For notational simplicity, we often write eqn. (3.2.14) in a briefer notation as

$$\Omega(N, V, E) = M \int \mathrm{d}^N\mathbf{p} \int_{D(V)} \mathrm{d}^N\mathbf{r} \, \delta(\mathcal{H}(\mathbf{r}, \mathbf{p}) - E), \quad (3.2.15)$$

or more simply as

$$\Omega(N, V, E) = M \int \mathrm{d}\mathbf{x} \, \delta(\mathcal{H}(\mathbf{x}) - E), \quad (3.2.16)$$

using the phase-space vector. However, it should be remembered that these shorter versions refer to the explicit form of eqn. (3.2.14).

In order to understand eqn. (3.2.14) somewhat better and define the normalization constant M, let us consider determining $\Omega(N, V, E)$ in a somewhat different way. We perform a thought experiment in which we "count" the number of microstates via a "device" capable of determining a position component, say x, to a precision Δx and a momentum component p to a precision Δp. Since quantum mechanics places an actual limit on the product $\Delta x \Delta p$, namely Planck's constant h (this is Heisenberg's uncertainty relation to be discussed in Chapter 9), h is a natural choice for our thought experiment. Thus, we can imagine dividing phase space up into small hypercubes of volume $\Delta x = (\Delta x)^{3N}(\Delta p)^{3N} = h^{3N}$, such that each hypercube contains a single measurable microstate. Let us denote this phase-space volume simply as Δx. We will also assume that we can only determine the energy of each microstate to be within E and $E + E_0$, where E_0 defines a very thin energy shell above the constant-energy hypersurface. For each phase-space hypercube, if the energy of the corresponding microstate lies within this shell, we increment our counting by 1, which we represent identically as $\Delta x / h^{3N}$. We can therefore write Ω as

$$\Omega(N, V, E) = \sum_{\substack{\text{hypercubes} \\ E < \mathcal{H}(\mathbf{x}) < E + E_0}} \frac{\Delta x}{h^{3N}}, \quad (3.2.17)$$

where the summand $\Delta x / h^{3N}$ is added only if the phase-space vector $\mathbf{x}$ in a given hypercube lies in the energy shell. Since the hypercube volume Δx is certainly extremely small compared to all of phase space, eqn. (3.2.17) can be very well approximated by an integral

$$\Omega(N, V, E) = \frac{1}{h^{3N}} \int_{E < \mathcal{H}(\mathbf{x}) < E + E_0} d\mathbf{x}. \qquad (3.2.18)$$

Finally, since eqn. (3.2.18) is measuring the volume of a very thin $6N$-dimensional shell in phase space, we can approximate this volume by the $(6N - 1)$-dimensional area of the surface defined by $\mathcal{H}(\mathbf{x}) = E$ times the thickness E_0 of the shell, leading to

$$\Omega(N, V, E) = \frac{E_0}{h^{3N}} \int d\mathbf{x}\, \delta(\mathcal{H}(\mathbf{x}) - E), \qquad (3.2.19)$$

which is eqn. (3.2.16) with $M = E_0/h^{3N}$. In principle, eqn. (3.2.19) should be sufficient to define the microcanonical partition function. However, remember we assumed at the outset that all particles are identical, so that exchanging any two particles does not yield a uniquely different microstate. Unfortunately, classical mechanics is not equipped to handle this situation, as all classical particles carry an imaginary "tag" that allows them to be distinguished from other particles. Thus, in order to avoid overcounting, we need to include a factor of $1/N!$ in M for the number of possible particle exchanges that can be performed. This factor can only be properly derived using the laws of quantum mechanics, which we will discuss in Chapter 9. Adding the $1/N!$ in by hand yields the normalization factor

$$M \equiv M_N = \frac{E_0}{N! h^{3N}}. \qquad (3.2.20)$$

Note that the constant E_0 is irrelevant and will not affect any thermodynamic or equilibrium properties. However, this normalization constant renders $\Omega(N, V, E)$ dimensionless according to eqn. (3.2.16).[2]

Since $\Omega(N, V, E)$ counts the number of microscopic states available to a system with given values of N, V, and E, the thermodynamics of the microcanonical ensemble can now be expressed directly in terms of the partition function via Boltzmann's relation:

$$\frac{1}{kT} = \left(\frac{\partial \ln \Omega}{\partial E}\right)_{N,V}, \qquad \frac{P}{kT} = \left(\frac{\partial \ln \Omega}{\partial V}\right)_{N,E}, \qquad \frac{\mu}{kT} = -\left(\frac{\partial \ln \Omega}{\partial N}\right)_{V,E}. \qquad (3.2.21)$$

Moreover, an observable A is obtained from the ensemble average of a phase-space function $a(\mathbf{x})$ according to

$$A = \langle a \rangle = \frac{M_N}{\Omega(N, V, E)} \int d\mathbf{x}\, a(\mathbf{x}) \delta(\mathcal{H}(\mathbf{x}) - E) = \frac{\int d\mathbf{x}\, a(\mathbf{x}) \delta(\mathcal{H}(\mathbf{x}) - E)}{\int d\mathbf{x}\, \delta(\mathcal{H}(\mathbf{x}) - E)}. \qquad (3.2.22)$$

---

[2]Of course, in a multicomponent system, if the system contains N_A particles of species A, N_B particles of species B,...., and N total particles, then the normalization factor becomes $M_{\{N\}}$, where

$$M_{\{N\}} = \frac{E_0}{h^{3N}\, [N_A! N_B! \cdots]}.$$

Throughout the book, we will not complicate the expressions with these general normalization factors and simply use the one-component system factors. However, the reader should always keep in mind when the factors M_N, C_N (canonical ensemble), and I_N (isothermal-isobaric ensemble) need to be generalized.

The form of the microcanonical partition function shows that the phase-space variables are not all independent in this ensemble. In particular, the condition $\mathcal{H}(\mathrm{x}) = E$ specifies a condition or constraint placed on the total number of degrees of freedom. An N-particle system in the microcanonical ensemble therefore has $6N - 1$ independent phase-space degrees of freedom (or $2dN - 1$ in d dimensions). In the limit where $N \to \infty$ and $V \to \infty$ such that $N/V = \mathrm{const}$, a limit referred to as the *thermodynamic limit*, we may approximate, $6N - 1 \approx 6N$ so that the system behaves approximately as if it had $6N$ phase-space degrees of freedom.

Equation (3.2.22) now allows us to understand Boltzmann's relation $S(N, V, E) = k \ln \Omega(N, V, E)$ between the entropy and the partition function. From eqn. (3.2.10), it is clear that quantities such as $1/T$, P/T, and μ/T, which are themselves macroscopic observables, must be expressible as ensemble averages of phase-space functions, which we can denote as $a_T(\mathrm{x})$, $a_P(\mathrm{x})$ and $a_\mu(\mathrm{x})$. Consider, for example, $1/T$, which can be expressed as

$$\frac{1}{T} = \frac{M_N}{\Omega(N, V, E)} \int d\mathrm{x}\, a_T(\mathrm{x})\delta(\mathcal{H}(\mathrm{x}) - E) = \left(\frac{\partial S}{\partial E}\right)_{N,V}. \tag{3.2.23}$$

We seek to relate these two expressions for $1/T$ by postulating that $S(N, V, E) = CG(\Omega(N, V, E))$, where G is an arbitrary function and C is an arbitrary constant, so that

$$\left(\frac{\partial S}{\partial E}\right)_{N,V} = CG'(\Omega(N, V, E)) \left(\frac{\partial \Omega}{\partial E}\right)_{N,V}. \tag{3.2.24}$$

Now,

$$\left(\frac{\partial \Omega}{\partial E}\right)_{N,V} = M_N \int d\mathrm{x}\, \frac{\partial \delta(\mathcal{H}(\mathrm{x}) - E)}{\partial E}$$

$$= M_N \int d\mathrm{x}\, \delta(\mathcal{H}(\mathrm{x}) - E)\frac{\partial \ln \delta(\mathcal{H}(\mathrm{x}) - E)}{\partial E}. \tag{3.2.25}$$

Thus,

$$\left(\frac{\partial S}{\partial E}\right)_{N,V} = CG'(\Omega(N, V, E))M_N \int d\mathrm{x}\, \delta(\mathcal{H}(\mathrm{x}) - E)\frac{\partial \ln \delta(\mathcal{H}(\mathrm{x}) - E)}{\partial E}. \tag{3.2.26}$$

Equation (3.2.26) is in the form of a phase-space average as in eqn. (3.2.23). If we identify $a_T(\mathrm{x}) = (1/k)\partial[\ln \delta(\mathcal{H}(\mathrm{x}) - E)]/\partial E$, where k is an arbitrary constant, then it is clear that $G'(\Omega(N, V, E)) = k/\Omega(N, V, E)$, which is only satisfied if $G(\Omega(N, V, E)) = k \ln \Omega(N, V, E)$, with the arbitrary constant identified as Boltzmann's constant. In the next few sections, we shall see how to use the microcanonical and related ensembles to derive the thermodynamics for several example problems and to prove an important theorem known as the *virial theorem*.

3.3 The classical virial theorem

In this section, we present a microcanonical ensemble proof of the classical virial theorem. Consider a system with Hamiltonian $\mathcal{H}(\mathrm{x})$. Let x_i and x_j be specific components of the phase-space vector x. The classical virial theorem states that

$$\left\langle x_i \frac{\partial \mathcal{H}}{\partial x_j} \right\rangle = kT\delta_{ij}, \tag{3.3.1}$$

where the average is taken with respect to a microcanonical ensemble.

To prove this theorem, we begin with the following ensemble average:

$$\left\langle x_i \frac{\partial \mathcal{H}}{\partial x_j} \right\rangle = \frac{M_N}{\Omega(N,V,E)} \int dx \, x_i \frac{\partial \mathcal{H}}{\partial x_j} \delta(E - \mathcal{H}(\mathbf{x})). \tag{3.3.2}$$

In eqn. (3.3.2), we have used the fact that $\delta(x) = \delta(-x)$ to write the energy conserving δ-function as $\delta(E - \mathcal{H}(\mathbf{x}))$. It is convenient to express eqn. (3.3.2) in an alternate way using the fact that the Dirac δ-function $\delta(x) = d\theta(x)/dx$, where $\theta(x)$ is the Heaviside step function, $\theta(x) = 1$ for $x \geq 0$ and $\theta(x) = 0$ for $x < 0$. Using these relations, eqn. (3.3.2) becomes

$$\left\langle x_i \frac{\partial \mathcal{H}}{\partial x_j} \right\rangle = \frac{M_N}{\Omega(N,V,E)} \frac{\partial}{\partial E} \int dx \, x_i \frac{\partial \mathcal{H}}{\partial x_j} \theta(E - \mathcal{H}(\mathbf{x})). \tag{3.3.3}$$

The step function restricts the phase-space integral to those microstates for which $\mathcal{H}(\mathbf{x}) < E$. Thus, eqn. (3.3.3) can be expressed equivalently as

$$\left\langle x_i \frac{\partial \mathcal{H}}{\partial x_j} \right\rangle = = = \frac{M_N}{\Omega(N,V,E)} \frac{\partial}{\partial E} \int_{\mathcal{H}(\mathbf{x})<E} dx \, x_i \frac{\partial \mathcal{H}}{\partial x_j}$$

$$= \frac{M_N}{\Omega(N,V,E)} \frac{\partial}{\partial E} \int_{\mathcal{H}(\mathbf{x})<E} dx \, x_i \frac{\partial(\mathcal{H}(\mathbf{x}) - E)}{\partial x_j}, \tag{3.3.4}$$

where the last line follows from the fact that E is a constant. Recognizing that $\partial x_i/\partial x_j = \delta_{ij}$ since all phase-space components are independent, we can express the phase-space derivative as

$$x_i \frac{\partial(\mathcal{H}(\mathbf{x}) - E)}{\partial x_j} = \frac{\partial}{\partial x_j} [x_i(\mathcal{H}(\mathbf{x}) - E)] - \delta_{ij}(\mathcal{H}(\mathbf{x}) - E). \tag{3.3.5}$$

Substituting eqn. (3.3.5) into eqn. (3.3.4) gives

$$\left\langle x_i \frac{\partial \mathcal{H}}{\partial x_j} \right\rangle = \frac{M_N}{\Omega(N,V,E)}$$

$$\times \frac{\partial}{\partial E} \int_{\mathcal{H}(\mathbf{x})<E} dx \left\{ \frac{\partial}{\partial x_j} [x_i(\mathcal{H}(\mathbf{x}) - E)] + \delta_{ij}(E - \mathcal{H}(\mathbf{x})) \right\}. \tag{3.3.6}$$

The first integral is over a pure derivative. Hence, when the integral over x_j is performed, the integrand $x_i(\mathcal{H}(\mathbf{x}) - E)$ must be evaluated at the limits of x_j which lie on the constant-energy hypersurface $\mathcal{H}(\mathbf{x}) = E$. Since $\mathcal{H}(\mathbf{x}) - E = 0$ at the limits, the first term vanishes leaving

$$\left\langle x_i \frac{\partial \mathcal{H}}{\partial x_j} \right\rangle = \frac{M_N}{\Omega(N,V,E)} \delta_{ij} \frac{\partial}{\partial E} \int_{\mathcal{H}(\mathbf{x})<E} dx \, (E - \mathcal{H}(\mathbf{x})). \tag{3.3.7}$$

If we now carry out the energy derivative in eqn. (3.3.7), we obtain

$$\left\langle x_i \frac{\partial \mathcal{H}}{\partial x_j} \right\rangle = \frac{M_N \delta_{ij}}{\Omega(N,V,E)} \int_{\mathcal{H}(\mathbf{x})<E} d\mathbf{x}$$

$$= \frac{\delta_{ij}}{\Omega(N,V,E)} M_N \int d\mathbf{x}\,\theta\left(E - \mathcal{H}(\mathbf{x})\right).$$

$$= \frac{E_0 \delta_{ij}}{\Omega(N,V,E)} C_N \int d\mathbf{x}\,\theta\left(E - \mathcal{H}(\mathbf{x})\right), \qquad (3.3.8)$$

where $C_N = 1/(N!h^{3N})$. The phase-space integral appearing in eqn. (3.3.8) is the partition function of an ensemble that closely resembles the microcanonical ensemble, known as the *uniform ensemble*:

$$\Sigma(N,V,E) = C_N \int d\mathbf{x}\,\theta\left(E - \mathcal{H}(\mathbf{x})\right). \qquad (3.3.9)$$

The partition function of this ensemble is related to the microcanonical partition function by

$$\Omega(N,V,E) = E_0 \frac{\partial \Sigma(N,V,E)}{\partial E}. \qquad (3.3.10)$$

As noted previously, the uniform ensemble requires that the phase-space integral be performed over a phase-space volume for which $\mathcal{H}(\mathbf{x}) < E$, which is the volume enclosed by the constant-energy hypersurface. While the dimensionality of the constant-energy hypersurface is $6N-1$, the volume enclosed has dimension $6N$. However, in the thermodynamic limit, where $N \to \infty$, the difference between the number of microstates associated with the uniform and microcanonical ensembles becomes vanishingly small since $6N \approx 6N - 1$ for N very large. Thus, the entropy $\tilde{S}(N,V,E) = k \ln \Sigma(N,V,E)$ derived from the uniform ensemble and that derived from the microcanonical ensemble $S(N,V,E) = k \ln \Omega(N,V,E)$ become very nearly equal as the thermodynamic limit is approached. Substituting eqn. (3.3.10) into eqn. (3.3.8) gives

$$\left\langle x_i \frac{\partial \mathcal{H}}{\partial x_j} \right\rangle = \delta_{ij} \frac{\Sigma(E)}{\partial \Sigma(N,V,E)/\partial E}$$

$$= \delta_{ij} \left(\frac{\partial \ln \Sigma(E)}{\partial E} \right)^{-1}$$

$$= k \delta_{ij} \left(\frac{\partial \tilde{S}}{\partial E} \right)^{-1}$$

$$\approx k \delta_{ij} \left(\frac{\partial S}{\partial E} \right)^{-1}$$

$$= kT \delta_{ij}, \qquad (3.3.11)$$

which proves the theorem. The virial theorem allows for the construction of microscopic phase-space functions whose ensemble averages yield macroscopic thermodynamic observables.

As an example of the use of the virial theorem, consider the choice $x_i = p_i$, a momentum component, and $i = j$. If $\mathcal{H} = \sum_i \mathbf{p}_i^2/2m_i + U(\mathbf{r}_1, ..., \mathbf{r}_N)$, then according to the virial theorem,

$$\left\langle p_i \frac{\partial \mathcal{H}}{\partial p_i} \right\rangle = kT$$

$$\left\langle \frac{p_i^2}{m_i} \right\rangle = kT$$

$$\left\langle \frac{p_i^2}{2m_i} \right\rangle = \frac{1}{2}kT.$$

Thus, at equilibrium, the kinetic energy of each particle must be $kT/2$. By summing both sides over all $3N$ momentum components, we obtain the familiar result:

$$\sum_{i=1}^{3N} \left\langle \frac{p_i^2}{2m_i} \right\rangle = \sum_{i=1}^{3N} \left\langle \frac{1}{2}m_i v_i^2 \right\rangle = \frac{3}{2}NkT. \tag{3.3.12}$$

3.4   Conditions for thermal equilibrium

Another important result that can be derived from the microcanonical ensemble and that will be needed in the next chapter is the equilibrium state reached when two systems are brought into thermal contact. By thermal contact, we mean that the systems can exchange only heat. Thus, they do not exchange particles, and there is no potential coupling between the systems. This type of interaction is illustrated in

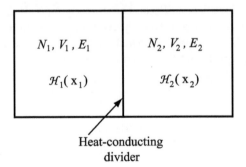

Heat-conducting
divider

Fig. 3.1 Two systems in thermal contact. System 1 (left) has N_1 particles in a volume V_1; system 2 (right) has N_2 particles in a volume V_2.

Fig. 3.1, which shows two systems (system 1 and system 2), each with fixed particle number and volume, separated by a heat-conducting divider. If system 1 has a phase-space vector x_1 and system 2 has a phase-space vector x_2, then the total Hamiltonian can be written as

$$\mathcal{H}(x) = \mathcal{H}_1(x_1) + \mathcal{H}_2(x_2). \tag{3.4.1}$$

Additionally, we let system 1 have N_1 particles in a volume V_1 and system 2 have N_2 particles in a volume V_2. The total particle number N and volume V are $N = N_1 + N_2$

and $V = V_1 + V_2$, respectively. The entropy of each system is given in terms of the partition function for each system as

$$S_1(N_1, V_1, E_1) = k \ln \Omega_1(N_1, V_1, E_1)$$
$$S_2(N_2, V_2, E_2) = k \ln \Omega_2(N_2, V_2, E_2), \tag{3.4.2}$$

where the partition functions are given by

$$\Omega_1(N_1, V_1, E_1) = M_{N_1} \int dx_1 \, \delta(\mathcal{H}_1(x_1) - E_1)$$

$$\Omega_2(N_2, V_2, E_2) = M_{N_2} \int dx_2 \, \delta(\mathcal{H}_2(x_2) - E_2). \tag{3.4.3}$$

Of course, if we solved Hamilton's equations for $\mathcal{H}(x)$ in eqn. (3.4.1), $\mathcal{H}_1(x_1)$ and $\mathcal{H}_2(x_2)$ would be separately conserved because $\mathcal{H}(x)$ is separable. However, the framework of the microcanonical ensemble allows us to consider the full set of microstates for which only $\mathcal{H}(x) = \mathcal{H}_1(x_1) + \mathcal{H}_2(x_2) = E$ without $\mathcal{H}_1(x_1)$ and $\mathcal{H}_2(x_2)$ being independently conserved. Indeed, since the systems can exchange heat, we do not expect $\mathcal{H}_1(x_1)$ and $\mathcal{H}_2(x_2)$ to be individually conserved. The total partition function is then

$$\Omega(N, V, E) = M_N \int dx \, \delta(\mathcal{H}_1(x_1) + \mathcal{H}_2(x_2) - E)$$

$$\neq \Omega_1(N_1, V_1, E_1)\Omega_2(N_2, V_2, E_2). \tag{3.4.4}$$

Because Ω_1 and Ω_2 both involve δ-functions, it can be shown that the total partition function is given by

$$\Omega(N, V, E) = C' \int_0^E dE_1 \, \Omega_1(N_1, V_1, E_1)\Omega_2(N_2, V_2, E - E_1), \tag{3.4.5}$$

where C' is an overall constant independent of the energy. In order to realize eqn. (3.4.5) by solving Hamilton's equations, we would need to solve the equations for all values of E_1 between 0 and E with $E_2 = E - E_1$. We can imagine accomplishing this by choosing a set of P closely spaced values for E_1, $E_1^{(1)}, ..., E_1^{(P)}$ and solving the equations of motion for each of these P values. In this case, eqn. (3.4.5) would be replaced by a Riemann sum expression:

$$\Omega(E) = C' \Delta \sum_{i=1}^P \Omega_1(E_1^{(i)})\Omega_2(E - E_1^{(i)}), \tag{3.4.6}$$

where Δ is the small energy interval $\Delta = E_1^{(i+1)} - E_1^{(i)}$. The integral is exact when $\Delta \to 0$ and $P \to \infty$. When the integral is written in this way, we can make use of a powerful theorem on sums with large numbers of terms.

Consider a sum of the form

$$\sigma = \sum_{i=1}^{P} a_i, \tag{3.4.7}$$

where $a_i > 0$ for all a_i. Let $a_{\max}$ be the largest of all the a_i's. Clearly, then

$$a_{\max} \leq \sum_{i=1}^{P} a_i$$

$$P a_{\max} \geq \sum_{i=1}^{P} a_i. \tag{3.4.8}$$

Thus, we have the inequality

$$a_{\max} \leq \sigma \leq P a_{\max}, \tag{3.4.9}$$

or

$$\ln a_{\max} \leq \ln \sigma \leq \ln a_{\max} + \ln P. \tag{3.4.10}$$

This gives upper and lower bounds on the value of $\ln \sigma$. Now suppose that $\ln a_{\max} >> \ln P$. Then the above inequality implies that

$$\ln \sigma \approx \ln a_{\max}. \tag{3.4.11}$$

This would be the case, for example, if $a_{\max} \sim e^P$. In this case, the value of the sum is given to a very good approximation by the value of its maximum term (McQuarrie, 2000).

Why should this theorem apply to the sum expression for $\Omega(N, V, E)$ in eqn. (3.4.6)? In the next section, we will see that the partition function of an ideal gas, that is, a collection of N free particles, varies as $\Omega \sim [g(E)V]^N$, where $g(E)$ is some function of the energy. This fact motivates the more general result that the terms in the sum vary exponentially with N. But the number of terms in the sum P also varies like N since $P = E/\Delta$ and $E \sim N$, since E is extensive. Thus, the terms in the sum under consideration obey the conditions for the application of the theorem.

Let the maximum term in the sum be defined by energies $\bar{E}_1$ and $\bar{E}_2 = E - \bar{E}_1$. Then, according to the above analysis,

$$
\begin{aligned}
S(N, V, E) &= k \ln \Omega(N, V, E) \\
&= k \ln \Delta + k \ln \left[\Omega_1(N_1, V_1, \bar{E}_1) \Omega_2(N_2, V_2, E - \bar{E}_1) \right] \\
&\quad + k \ln P + k \ln C'.
\end{aligned}
\tag{3.4.12}
$$

Note that since $\ln P$ is considered negligible, we can add it on the second line of eqn. (3.4.12) without affecting the result, and since $P = E/\Delta$, it follows that $\ln \Delta + \ln P = \ln \Delta + \ln E - \ln \Delta = \ln E$. But $E \sim N$, while $\ln \Omega_1 \sim N$. Since $N >> \ln N$, the above expression becomes, to a good approximation,

$$S(N, V, E) \approx k \ln \left[\Omega_1(N_1, V_1, \bar{E}_1) \Omega_2(N_2, V_2, E - \bar{E}_1) \right] + \mathcal{O}(\ln N) + \text{const.} \tag{3.4.13}$$

Thus, apart from constants, the entropy is approximately additive:

$$S(N, V, E) = k \ln \Omega_1(N_1, V_1, \bar{E}_1) + k \ln \Omega_2(N_2, V_2, \bar{E}_2)$$
$$= S_1(N_1, V_1, \bar{E}_1) + S_2(N_2, V_2, \bar{E}_2) + \mathcal{O}(\ln N) + \text{const.} \quad (3.4.14)$$

Finally, in order to compute the temperature of each system, we vary the energy $\bar{E}_1$ by a small amount $d\bar{E}_1$. But since $\bar{E}_1 + \bar{E}_2 = E$, $d\bar{E}_1 = -d\bar{E}_2$. Also, this variation is made such that the total entropy S and energy E remain constant. Thus, we obtain

$$0 = \frac{\partial S_1}{\partial \bar{E}_1} + \frac{\partial S_2}{\partial \bar{E}_1}$$

$$0 = \frac{\partial S_1}{\partial \bar{E}_1} - \frac{\partial S_2}{\partial \bar{E}_2}$$

$$0 = \frac{1}{T_1} - \frac{1}{T_2}, \quad (3.4.15)$$

from which it is clear that $T_1 = T_2$. Thus, when two systems in thermal contact reach equilibrium, their temperatures become equal. Note that this result holds independent of the relative sizes of systems 1 and 2, a fact we will need in the next chapter where we will consider systems in contact with large heat baths.

3.5 The free particle and the ideal gas

Our first example of the microcanonical ensemble is a single free particle in one spatial dimension and its extension to an ideal gas of N particles in three dimensions. The microcanonical ensemble for a single particle in one dimension is not especially interesting (it has only one degree of freedom). Nevertheless, it is instructive to go through the pedagogical exercise in order to show how the partition function is computed so that the subsequent calculation for the ideal gas becomes more transparent.

The particle is described by a single coordinate x, confined to a region of the real line between $x = 0$ and $x = L$, and a single momentum p. The free particle Hamiltonian is just

$$\mathcal{H} = \frac{p^2}{2m}. \quad (3.5.1)$$

The phase space is two-dimensional, and the microcanonical partition function is

$$\Omega(L, E) = \frac{E_0}{h} \int_0^L dx \int_{-\infty}^\infty dp \; \delta\left(\frac{p^2}{2m} - E\right). \quad (3.5.2)$$

Note that since $N = 1$, Ω only depends on E and the one-dimensional "volume" L. We see immediately that the integrand is independent of x so that the integral over x can be done immediately, yielding

$$\Omega(L, E) = \frac{E_0 L}{h} \int_{-\infty}^\infty dp \; \delta\left(\frac{p^2}{2m} - E\right). \quad (3.5.3)$$

In order to perform the momentum integral, we start by introducing a change of variables, $y = p/\sqrt{2m}$ so that the integral becomes

$$\Omega(L, E) = \frac{E_0 L \sqrt{2m}}{h} \int_{-\infty}^{\infty} \mathrm{d}y \, \delta \left(y^2 - E \right). \tag{3.5.4}$$

Then, using the properties of the Dirac δ-function in Appendix A, in particular eqn. (A.15), we obtain

$$\Omega(L, E) = \frac{E_0 L \sqrt{2m}}{h} \frac{1}{2\sqrt{E}} \int_{-\infty}^{\infty} \mathrm{d}y \left[\delta \left(y - \sqrt{E} \right) + \delta \left(y + \sqrt{E} \right) \right]. \tag{3.5.5}$$

Therefore, integrating over the δ-function using eqn. (A.2), we obtain

$$\Omega(L, E) = \frac{E_0 L \sqrt{2m}}{h \sqrt{E}}. \tag{3.5.6}$$

It is easily checked that eqn. (3.5.6) is a dimensionless number.

The free particle example leads naturally into a discussion of the classical ideal gas. An ideal gas is defined to be a system of particles that do not interact. An ideal gas of N particles is, therefore, simply a collection of N free particles. Therefore, in order to treat an ideal gas, we need to consider N free particles in three dimensions for which the Hamiltonian is

$$\mathcal{H} = \sum_{i=1}^{N} \frac{\mathbf{p}_i^2}{2m}. \tag{3.5.7}$$

A primary motivation for studying the ideal gas is that all real systems approach ideal gas behavior in the limit of low density and pressure.

In the present discussion, we shall consider an ideal gas of N classical particles in a cubic container of volume V with a total internal energy E. The partition function in this case is given by

$$\Omega(N, V, E) = \frac{E_0}{N! h^{3N}} \int \mathrm{d}^N \mathbf{p} \int_{D(V)} \mathrm{d}^N \mathbf{r} \, \delta \left(\sum_{i=1}^{N} \frac{\mathbf{p}_i^2}{2m} - E \right). \tag{3.5.8}$$

As in the one-dimensional case, the integrand is independent of the coordinates, hence the position-dependent part of the integral can be evaluated immediately as

$$\int_{D(V)} \mathrm{d}^N \mathbf{r} = \int_0^L \mathrm{d}x_1 \int_0^L \mathrm{d}y_1 \int_0^L \mathrm{d}z_1 \cdots \int_0^L \mathrm{d}x_N \int_0^L \mathrm{d}y_N \int_0^L \mathrm{d}z_N = L^{3N}. \tag{3.5.9}$$

Since $L^3 = V$, $L^{3N} = V^N$.

For the momentum part of the integral, we change integration variables to $\mathbf{y}_i = \mathbf{p}_i / \sqrt{2m}$ so that the partition function becomes

$$\Omega(N, V, E) = \frac{E_0 (2m)^{3N/2} V^N}{N! h^{3N}} \int \mathrm{d}^N \mathbf{y} \, \delta \left(\sum_{i=1}^{N} \mathbf{y}_i^2 - E \right). \tag{3.5.10}$$

Note that the condition required by the δ-function is

$$\sum_{i=1}^{N} \mathbf{y}_i^2 = E. \tag{3.5.11}$$

Equation (3.5.11) is the equation of a $(3N - 1)$-dimensional spherical surface of radius $\sqrt{E}$. Thus, the natural thing to do is transform to spherical coordinates; what we need, however, is a set of spherical coordinates for a $3N$-dimensional sphere. Recall that ordinary spherical coordinates in three dimensions consist of one radial coordinate, r, and two angular coordinates, θ and ϕ, with a volume element given by $r^2 dr d^2\omega$, where $d^2\omega = \sin\theta d\theta d\phi$ is the solid angle element. In $3N$ dimensions, spherical coordinates consist of one radial coordinate and $3N - 1$ angular coordinates, $\theta_1, ..., \theta_{3N-1}$ with a volume element $d^N \mathbf{y} = r^{3N-1} dr d^{3N-1}\omega$. The angles have ranges $\theta_1 \in [0, \pi], \theta_2 \in [0, \pi], ..., \theta_{3N-2} \in [0, \pi]$, and $\theta_{3N-1} \in [0, 2\pi]$. The radial coordinate is given simply by

$$r^2 = \sum_{i=1}^{N} \mathbf{y}_i^2. \tag{3.5.12}$$

After transforming to these coordinates, the partition function becomes

$$\Omega(N, V, E) = \frac{E_0 (2m)^{3N/2} V^N}{N! h^{3N}} \int d^{3N-1}\omega \int_0^\infty dr\, r^{3N-1} \delta\left(r^2 - E\right). \tag{3.5.13}$$

The solid angle element $d^n\omega$ is $\sin^{n-1}\theta_1 \sin^{n-2}\theta_2 \cdots \sin\theta_{n-1} d\theta_1 \cdots d\theta_n$, and the general formula for the solid angle integral is

$$\int d^n\omega = \frac{(n+1)\pi^{(n+1)/2}}{\Gamma\left(\frac{n+1}{2} + 1\right)}, \tag{3.5.14}$$

where $\Gamma(x)$ is the Gamma function defined by

$$\Gamma(x) = \int_0^\infty dt\, t^{x-1} e^{-t}. \tag{3.5.15}$$

According to eqn. (3.5.15), for any integer n, we have

$$\Gamma(n) = (n-1)!$$

$$\Gamma\left(n + \frac{1}{2}\right) = \frac{(2n-1)!!}{2^n} \pi^{1/2}. \tag{3.5.16}$$

Finally, expanding the δ-function using eqn. (A.15), we obtain

$$\Omega(N, V, E) = \frac{E_0 (2m)^{3N/2} V^N}{N! h^{3N}} \frac{3N\pi^{3N/2}}{\Gamma(1 + 3N/2)}$$

$$\times \int_0^\infty dr\, r^{3N-1} \frac{1}{2\sqrt{E}} \left[\delta(r - \sqrt{E}) + \delta(r + \sqrt{E})\right]. \tag{3.5.17}$$

Because the integration range on r is $[0, \infty)$, only the first δ-function gives a nonzero contribution, and, by eqn. (A.13), the result of the integration is

$$\Omega(N, V, E) = \frac{E_0(2m)^{3N/2}V^N}{N!h^{3N}} \frac{3N\pi^{3N/2}}{\Gamma(1+3N/2)} \frac{1}{2\sqrt{E}} E^{(3N-1)/2}$$

$$= \frac{E_0}{E} \frac{1}{N!} \frac{3N/2}{\Gamma(1+3N/2)} \left[V \left(\frac{2\pi mE}{h^2} \right)^{3/2} \right]^N. \tag{3.5.18}$$

Assuming $\Gamma(1+3N/2) = (3N/2)!$, we can replace $(3N/2)/\Gamma(1+3N/2)$ with $1/\Gamma(3N/2)$. The prefactor of $1/E$ causes the actual dependence of $\Omega(N, V, E)$ on E to be $E^{3N/2-1}$. In the thermodynamic limit, we may approximate $3N/2 - 1 \approx 3N/2$, in which case, we may simply neglect the E_0/E prefactor altogether. We may further simplify this expression by introducing *Stirling's approximation* for the factorial of a large number:

$$N! \approx e^{-N}N^N, \tag{3.5.19}$$

so that the Gamma function can be written as

$$\Gamma\left(\frac{3N}{2}\right) = \left(\frac{3N}{2} - 1\right)! \approx \left(\frac{3N}{2}\right)! \approx e^{-3N/2} \left(\frac{3N}{2}\right)^{3N/2}. \tag{3.5.20}$$

Substituting eqn. (3.5.20) into eqn. (3.5.18) gives the partition function expression

$$\Omega(N, V, E) = \frac{1}{N!} \left[\frac{V}{h^3} \left(\frac{4\pi mE}{3N} \right)^{3/2} \right]^N e^{3N/2}. \tag{3.5.21}$$

We have intentionally not applied Stirling's approximation to the prefactor $1/N!$ because, as was discussed earlier, this factor is appended *a posteriori* in order to account for the indistinguishability of the particles not treated in a classical description. Leaving this factor as is will allow us to assess its effects on the thermodynamics of the ideal gas, which we will do shortly.

Let us now use the machinery of statistical mechanics to compute the temperature of the ideal gas. From eqn. (3.2.21), we obtain

$$\frac{1}{kT} = \left(\frac{\partial \ln \Omega}{\partial E} \right)_{N,V}. \tag{3.5.22}$$

Since $\Omega \sim E^{3N/2}$, $\ln \Omega \sim (3N/2) \ln E$ so that the derivative yields

$$\frac{1}{kT} = \frac{3N}{2E}, \tag{3.5.23}$$

or

$$kT = \frac{2E}{3N} \qquad E = \frac{3}{2}NkT, \tag{3.5.24}$$

which expresses the familiar relationship between temperature and internal energy from kinetic theory. Similar, the pressure of the ideal gas is given by

$$\frac{P}{kT} = \left(\frac{\partial \ln \Omega}{\partial V}\right)_{N,E}. \tag{3.5.25}$$

Since $\Omega \sim V^N$, $\ln \Omega \sim N \ln V$ so that the derivative yields

$$\frac{P}{kT} = \frac{N}{V}, \tag{3.5.26}$$

or

$$P = \frac{NkT}{V} = \rho kT, \tag{3.5.27}$$

where we have introduced the *number density*, $\rho = N/V$, i.e., the number of particles per unit volume. Equation (3.5.27) is the familiar ideal gas equation of state or ideal gas law (cf. eqn. (2.2.1)), which can be expressed in terms of the number of moles by multiplying and dividing by Avogadro's number, N_0:

$$PV = \frac{N}{N_0} N_0 kT = nRT. \tag{3.5.28}$$

The product $N_0 k$ yields the *gas constant* R, whose value is 8.314472 J·mol^{-1} · K^{-1}.

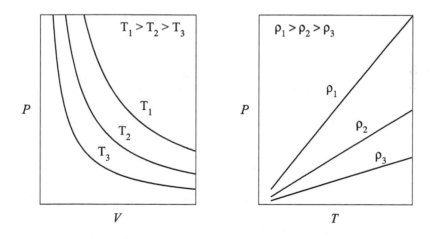

Fig. 3.2 (*Left*) Pressure vs. volume for different temperatures (isotherms of the equation of state (2.2.1)). (*Right*) Pressure vs. temperature for different densities $\rho = N/V$.

Figure 3.2 (left) is a plot of P vs. V for different values of T. The curves are known as the *isotherms* of the ideal gas. From the figure, the inverse relationship between pressure and volume can be clearly seen. Similarly, Fig. 3.2 (right) shows a plot of P vs. T for different densities. The lines are the *isochores* of the ideal gas. Because of the absence of interactions, the ideal gas can only exist as a gas under *all* thermodynamic conditions.

3.5.1 The Gibbs Paradox

According to eqn. (3.5.21), the entropy of an ideal gas is

$$S(N, V, E) = k \ln \Omega(N, V, E)$$

$$= Nk \ln \left[\frac{V}{h^3} \left(\frac{4\pi m E}{3N} \right)^{3/2} \right] + \frac{3}{2} Nk - k \ln N!, \qquad (3.5.29)$$

or, using eqn. (3.5.24), we obtain S as a function of N, V, and T:

$$S(N, V, T) = Nk \ln \left[\frac{V}{h^3} (2\pi m k T)^{3/2} \right] + \frac{3}{2} Nk - k \ln N!, \qquad (3.5.30)$$

Recall, however, that the $1/N!$ factor in eqn. (3.5.21) was added a *posteriori* to correct for overcounting the number of microstates due to the identical nature of the gas particles. If this factor is not included, then the entropy, known as the *classical entropy*, becomes

$$S^{(\text{cl})}(N, V, T) = Nk \ln \left[\frac{V}{h^3} (2\pi m k T)^{3/2} \right] + \frac{3}{2} Nk. \qquad (3.5.31)$$

Let us now work through a thought experiment that reveals the importance of the $1/N!$ correction. Consider an ideal gas of N indistinguishable particles in a container with a volume V and uniform temperature T. An impermeable partition separates the container into two sections with volumes V_1 and V_2, respectively, such that $V_1 + V_2 = V$. There are are N_1 particles in the volume V_1, and N_2 particles in the volume V_2, with $N = N_1 + N_2$ It is assumed that the number density $\rho = N/V$ is the same throughout the system so that $N_1/V_1 = N_2/V_2$. If the partition is now removed, will the total entropy increase or remain the same? Since the particles are identical, exchanges of particles before and after the partition is removed will yield identical microstates. Therefore, the entropy should remain the same. We will now analyze this thought experiment more carefully using eqns. (3.5.29) and (3.5.31) above.

Using eqn. (3.5.31), the classical entropy expressions for each of the two sections of the container are (apart from additive constants) are

$$S_1^{(\text{cl})} \sim N_1 k \ln V_1 + \frac{3}{2} N_1 k$$

$$S_2^{(\text{cl})} \sim N_2 k \ln V_2 + \frac{3}{2} N_2 k, \qquad (3.5.32)$$

and, since entropy is additive, the total entropy is $S^{(\text{cl})} = S_1^{(\text{cl})} + S_2^{(\text{cl})}$. After the partition is removed, the total classical entropy is

$$S^{(\text{cl})} \sim (N_1 + N_2) k \ln(V_1 + V_2) + \frac{3}{2} (N_1 + N_2) k. \qquad (3.5.33)$$

Therefore, the difference $\Delta S^{(\text{cl})}$ is

$$\Delta S^{(\text{cl})} = (N_1 + N_2) k \ln(V_1 + V_2) - N_1 k \ln V_1 - N_2 k \ln V_2$$
$$= N_1 k \ln(V/V_1) + N_2 k \ln(V/V_2) > 0, \qquad (3.5.34)$$

which contradicts the anticipated result that $\Delta S = 0$. Without the $1/N!$ correction factor, a paradoxical result is obtained, which is known as the *Gibbs paradox*.

Let us now repeat the analysis using eqn. (3.5.29). Introducing Stirling's approximation as a logarithm of eqn. (3.5.19), $\ln N! \approx N \ln N - N$, eqn. (3.5.29) can be rewritten as

$$S = Nk \ln \left[\frac{V}{Nh^3} \left(\frac{2\pi m}{\beta} \right)^{3/2} \right] + \frac{5}{2} Nk, \tag{3.5.35}$$

which is known as the *Sackur-Tetrode* equation. Using eqn. (3.5.35), the entropy difference ΔS becomes

$$\Delta S = (N_1 + N_2)k \ln \left(\frac{V_1 + V_2}{N_1 + N_2} \right) - N_1 k \ln(V_1/N_1) - N_2 k \ln(V_2/N_2)$$

$$= N_1 k \ln(V/V_1) + N_2 k \ln(V/V_2) - N_1 k \ln(N/N_1) - N_2 k \ln(N/N_2)$$

$$= N_1 k \ln \left(\frac{V}{N} \frac{N_1}{V_1} \right) + N_2 k \ln \left(\frac{V}{N} \frac{N_2}{V_2} \right). \tag{3.5.36}$$

However, since the density $\rho = N_1/V_1 = N_2/V_2 = N/V$ is constant, the logarithms all vanish, which leads to the expected results $\Delta S = 0$. A purely classical treatment of the particles is, therefore, unable to resolve the paradox. Only by accounting for the identical nature of the particles *a posteriori* or via a proper quantum treatment of the ideal gases (see Chapter 11) can a consistent thermodynamic picture be obtained.

3.6 The harmonic oscillator and harmonic baths

The second example we will study is a single harmonic oscillator in one dimension and its extension to a system of N oscillators in three dimensions (also known as a "harmonic bath"). We are returning to this problem because harmonic oscillators lie at the heart of a wide variety of important problems. They are often used to describe intramolecular bond and bend vibrations in biological force fields, they are used to describe ideal solids, they form the basis of normal mode analysis (see Section 1.7), and they turn up repeatedly in quantum mechanics.

Consider first a single particle in one dimension with coordinate x and momentum p moving in a harmonic potential

$$U(x) = \frac{1}{2} kx^2, \tag{3.6.1}$$

where k is the force constant. The Hamiltonian is given by

$$\mathcal{H} = \frac{p^2}{2m} + \frac{1}{2} kx^2. \tag{3.6.2}$$

In Section 1.3, we saw that the harmonic oscillator is an example of a bound phase space. We shall consider that the one-dimensional "container" is larger than the maximum value of x (as determined by the energy E), so that the integration can be taken over all space.

The partition function is

$$\Omega(E) = \frac{E_0}{h} \int_{-\infty}^{\infty} \mathrm{d}p \int_{-\infty}^{\infty} \mathrm{d}x \, \delta \left(\frac{p^2}{2m} + \frac{1}{2}kx^2 - E \right). \tag{3.6.3}$$

In order to evaluate the integral in eqn. (3.6.3), we first introduce a change of variables

$$\tilde{p} = \frac{p}{\sqrt{2m}} \qquad \tilde{x} = \sqrt{\frac{k}{2}}x, \tag{3.6.4}$$

so that $\mathrm{d}p\,\mathrm{d}x = (2\sqrt{m/k})\mathrm{d}\tilde{p}\,\mathrm{d}\tilde{x}$, and the partition function can be written as

$$\Omega(E) = \frac{2E_0}{h} \sqrt{\frac{m}{k}} \int_{-\infty}^{\infty} \mathrm{d}\tilde{p} \int_{-\infty}^{\infty} \mathrm{d}\tilde{x} \, \delta \left(\tilde{p}^2 + \tilde{x}^2 - E \right). \tag{3.6.5}$$

Recall from Section 1.3, however, that $\sqrt{k/m} = \omega$ is just the fundamental frequency of the oscillator. The partition function then becomes

$$\Omega(E) = \frac{2E_0}{h\omega} \int_{-\infty}^{\infty} \mathrm{d}\tilde{p} \int_{-\infty}^{\infty} \mathrm{d}\tilde{x} \, \delta \left(\tilde{p}^2 + \tilde{x}^2 - E \right). \tag{3.6.6}$$

The δ-function requires that $\tilde{p}^2 + \tilde{x}^2 = E$, which defines a circle in the scaled $(\tilde{p}, \tilde{x})$ phase space. Therefore, it is natural to introduce polar coordinates in the form

$$\tilde{p} = \sqrt{I\omega} \cos \theta$$

$$\tilde{x} = \sqrt{I\omega} \sin \theta. \tag{3.6.7}$$

Here, the usual "radial" coordinate has been expressed as $\sqrt{I\omega}$. The new coordinates (I, θ) are known as *action-angle* variables. They are chosen such that the Jacobian is simply a constant, $\omega/2$, so that the partition function becomes

$$\Omega(E) = \frac{E_0}{h} \int_0^{2\pi} \mathrm{d}\theta \int_0^{\infty} \mathrm{d}I \, \delta(I\omega - E). \tag{3.6.8}$$

In action-angle variables, the harmonic Hamiltonian has the rather simple form $\mathcal{H} = I\omega$. If one were to derive Hamilton's equations in terms of action-angle variables, the result would be simply, $\dot{\theta} = \partial\mathcal{H}/\partial I = \omega$ and $\dot{I} = -\partial\mathcal{H}/\partial\theta = 0$ so that the action I is a constant $I(0)$ for all time, and $\theta = \omega t + \theta(0)$. The constancy of the action is consistent with energy conservation; $I \propto E$. The angle then gives the oscillatory time dependence of x and p. In eqn. (3.6.8), the angular integration can be performed directly to yield

$$\Omega(E) = \frac{2\pi E_0}{h} \int_0^{\infty} \mathrm{d}I \, \delta(I\omega - E). \tag{3.6.9}$$

Changing the action variable to $I' = I\omega$, we obtain

$$\Omega(E) = \frac{E_0}{\hbar\omega} \int_0^{\infty} \mathrm{d}I' \, \delta(I' - E), \tag{3.6.10}$$

where $\hbar = h/2\pi$. The integration over I' now proceeds using eqn. (A.2) and yields unity, so that

$$\Omega(E) = \frac{E_0}{\hbar\omega}. \tag{3.6.11}$$

Interestingly, $\Omega(E)$ is a constant independent of E. All one-dimensional harmonic oscillators of frequency ω have the same number of accessible microstates! Thus, no interesting thermodynamic properties can be derived from this partition function, and the entropy S is simply a constant $k\ln(E_0/\hbar\omega)$.

Consider next a collection of N independent harmonic oscillators with different masses and force constants, for which the Hamiltonian is

$$\mathcal{H} = \sum_{i=1}^{N} \left[\frac{\mathbf{p}_i^2}{2m_i} + \frac{1}{2}k_i\mathbf{r}_i^2 \right]. \tag{3.6.12}$$

For this system, the microcanonical partition function is

$$\Omega(N, E) = \frac{E_0}{h^{3N}} \int \mathrm{d}^N\mathbf{p}\,\mathrm{d}^N\mathbf{r}\,\delta\left(\sum_{i=1}^{N}\left[\frac{\mathbf{p}_i^2}{2m_i} + \frac{1}{2}k_i\mathbf{r}_i^2 \right] - E \right). \tag{3.6.13}$$

Since the oscillators are all different, the $N!$ factor is not needed. Let us first introduce scaled variables according to

$$\mathbf{y}_i = \frac{\mathbf{p}_i}{\sqrt{2m_i}} \qquad \mathbf{u}_i = \sqrt{\frac{k_i}{2}}\mathbf{r}_i, \tag{3.6.14}$$

so that the partition function becomes

$$\Omega(N, E) = \frac{2^{3N}E_0}{h^{3N}} \prod_{i=1}^{N}\frac{1}{\omega_i^3} \int \mathrm{d}^N\mathbf{y}\,\mathrm{d}^N\mathbf{u}\,\delta\left(\sum_{i=1}^{N}(\mathbf{y}_i^2 + \mathbf{u}_i^2) - E \right), \tag{3.6.15}$$

where $\omega_i = \sqrt{k_i/m_i}$ is the natural frequency for each oscillator. As in the ideal gas example, we recognize that the condition $\sum_{i=1}^{N}(\mathbf{y}_i^2 + \mathbf{u}_i^2) = E$ defines a $(6N-1)$-dimensional spherical surface, and we may introduce $6N$-dimensional spherical coordinates to yield

$$\Omega(N, E) = \frac{8E_0}{h^{3N}} \prod_{i=1}^{N}\frac{1}{\omega_i^3} \int \mathrm{d}^{6N-1}\tilde{\omega} \int_0^{\infty} \mathrm{d}R\,R^{6N-1}\,\delta\left(R^2 - E\right). \tag{3.6.16}$$

Using eqn. (3.5.14) and eqn. (A.15) allows the integration to be carried out in full with the result:

$$\Omega(N, E) = \frac{2^{3N}E_0\pi^{3N}}{Eh^{3N}}\frac{E^N}{\Gamma(3N)} \prod_{i=1}^{N}\frac{1}{\omega_i^3}. \tag{3.6.17}$$

In the thermodynamic limit, $3N - 1 \approx 3N$, and we can neglect the prefactor E_0/E, leaving

$$\Omega(N, E) = \left(\frac{2\pi E}{h}\right)^{3N} \frac{1}{\Gamma(3N)} \prod_{i=1}^{N} \frac{1}{\omega_i^3},$$

(3.6.18)

or, using Stirling's approximation $\Gamma(3N) \approx (3N)! \approx (3N)^{3N} e^{-3N}$,

$$\Omega(N, E) = \left(\frac{2\pi E}{3Nh}\right)^{3N} e^{3N} \prod_{i=1}^{N} \frac{1}{\omega_i^3}.$$

(3.6.19)

We can now calculate the temperature of the collection of oscillators via

$$\frac{1}{kT} = \left(\frac{\partial \ln \Omega(N, E)}{\partial E}\right)_N = \frac{3N}{E},$$

(3.6.20)

which leads to the familiar relation $E = 3NkT$. Note that this result is readily evident from the virial theorem eqn. (3.3.1), which also dictates that the average of the potential and kinetic energies each be $3NkT/2$.

The harmonic bath and ideal gas systems illustrate that the microcanonical ensemble is not a particularly convenient ensemble in which to carry out equilibrium calculations due to the integrations that must be performed over the Dirac δ-function. In the next three chapters, three different statistical ensembles will be considered that employ different sets of thermodynamic control variables other than N, V, and E. It will be shown that all statistical ensembles become equivalent in the thermodynamic limit, and therefore one has the freedom to choose the most convenient statistical ensemble for a given problem (although some care is needed when applying this notion to finite systems). The importance of the microcanonical ensemble lies not so much in its utility for equilibrium calculations but rather in that it is the only ensemble in which the dynamics of a system can be rigorously generated. In the remainder of this chapter, therefore, we will begin our foray into the numerical simulation technique of molecular dynamics, which we will show is capable both of sampling an equilibrium distribution and producing true dynamical observables.

3.7 Introduction to molecular dynamics

Calculating the partition function and associated thermodynamic and equilibrium properties for a general many-body potential that includes nonlinear interactions becomes an insurmountable task if only analytical techniques are employed. Unless a clever transformation can be devised, it is very unlikely that the integrals in eqns. (3.2.16) and (3.2.22) can be performed analytically. In this case, the only recourses are to introduce simplifying approximations, replace a given system by a simpler model system, or employ numerical methods. In the remainder of this chapter, our discussion will focus on such a numerical approach, namely, the methodology of molecular dynamics.

Molecular dynamics is a technique that allows a numerical "thought experiment" to be carried out using a model that, to a limited extent, approximates a real physical or chemical system. Such a "virtual laboratory" approach has the advantage that many such "experiments" can be easily set up and carried out in succession by simply varying the control parameters. Moreover, extreme conditions, such as high temperature and pressure, can be created in a straightforward (and considerably safer) manner than they can in a physical laboratory. The obvious downside is that the results are only as good as the numerical model. In addition, the results can be artificially biased if the molecular dynamics calculation is unable to sample an adequate number of microstates over the time it is allowed to run.

One of the earliest examples of such a numerical thought experiment was the Fermi-Pasta-Ulam calculation (1955), in which the equations of motion for a one-dimensional chain of nonlinear oscillators were integrated numerically in order to quantify the degree of ergodicity and energy equipartitioning in the system. Later, Alder and Wainwright carried out the first condensed-phase molecular dynamics calculation on a hard-sphere system (Alder and Wainwright, 1957; Alder and Wainwright, 1959), showing that a solid-liquid phase transition exists. Following this, Rahman (1964) and Verlet (1967) carried out the first simulations using a realistic continuous potential for systems of 864 argon atoms. The next major milestone came when Berne and coworkers (Harp and Berne, 1968; Berne *et al.*, 1968; Harp and Berne, 1970; Berne, 1971) carried out molecular dynamics simulations of diatomic liquids and characterized the time dependence of molecular reorientation in these systems. Following these studies, Stillinger and Rahman (1971, 1972, 1974) carried out the first molecular dynamics simulations of liquid water. Soon thereafter, Karplus and coworkers reported the first molecular dynamics calculations of proteins (McCammon *et al.*, 1976; McCammon *et al.*, 1977). Explicit treatment of molecular systems was enabled by the introduction of techniques for maintaining specific bonding patterns either by stiff intramolecular forces (Berne and Harp, 1970a) or by imposing holonomic constraints into the simulation (Ryckaert *et al.*, 1977).

The evolution of the field of molecular dynamics has benefitted substantially by advances in high-performance computing. The original Alder and Wainwright calculations required the use of a "supercomputer" at Lawrence Livermore National Laboratory in California, namely, the UNIVAC system. Nowadays, molecular dynamics calculations with force fields can be carried out on desktop computers. Another major milestone in molecular dynamics is the technique now known as *ab initio* or first-principles molecular dynamics (Car and Parrinello, 1985; Marx and Hutter, 2009). In an *ab initio* molecular dynamics calculation, the interatomic interactions are computed "on the fly" directly from the electronic structure as the simulation proceeds, thereby allowing chemical bonding breaking and forming events to be treated explicitly. However, even with current large-scale, high-performance computing resources, the computational overhead of solving the electronic Schrödinger equation using widely employed approximation schemes is considerable, and novel algorithmic developments that address the computational bottlenecks are needed. The field of molecular dynamics is an exciting and rapidly evolving one, and the immediate availability of free software packages capable of performing many different types of molecular dynamics calculations has dramatically increased the number of users of the methodology.

We begin our treatment of the subject of molecular dynamics by noting a few important properties of the microcanonical ensemble. The microcanonical ensemble consists of all microscopic states on the constant-energy hypersurface $\mathcal{H}(x) = E$.

This fact suggests an intimate connection between the microcanonical ensemble and classical Hamiltonian mechanics. In the latter, we have seen that the equations of motion conserve the total energy, $d\mathcal{H}/dt = 0 \Rightarrow \mathcal{H}(x) = $ const. Imagine that we have a system evolving according to Hamilton's equations:

$$\dot{q}_\alpha = \frac{\partial \mathcal{H}}{\partial p_\alpha}, \qquad \dot{p}_\alpha = -\frac{\partial \mathcal{H}}{\partial q_\alpha}. \qquad (3.7.1)$$

Since the equations of motion conserve the Hamiltonian $\mathcal{H}(x)$, a trajectory computed via eqns. (3.7.1) will generate microscopic configurations belonging to a microcanonical ensemble with energy E. Suppose, further that given an infinite amount of time, the system with energy E is able to visit all configurations on the constant-energy hypersurface. A system with this property is said to be *ergodic* and can be used to generate a microcanonical ensemble. In general, dynamical systems provide a powerful approach for generating an ensemble and its associated averages, and they form the basis of the molecular dynamics methodology, which has evolved into one of the most widely used techniques for solving statistical mechanical problems.

Given an ergodic trajectory generated by a Hamiltonian $\mathcal{H}(x)$, microcanonical phase-space averages can be replaced by time averages over the trajectory according to

$$\langle a \rangle = \frac{\int dx \, a(x) \delta(\mathcal{H}(x) - E)}{\int dx \, \delta(\mathcal{H}(x) - E)} = \lim_{\mathcal{T} \to \infty} \frac{1}{\mathcal{T}} \int_0^{\mathcal{T}} dt \, a(x_t) \equiv \bar{a}. \qquad (3.7.2)$$

In a molecular dynamics calculation, eqns. (3.7.1) are solved numerically subject to a given set of initial conditions. Doing so requires the use of a particular *numerical integrator* or *solver* for the equations of motion, a topic we shall take up in the next section. An integrator generates phase-space vectors at discrete times that are multiples of a fundamental time discretization parameter, Δt, known as the *time step*. Starting with the initial condition x_0, phase-space vectors $x_{n\Delta t}$ where $n = 0, ..., M$ are generated by applying the integrator or solver iteratively. The ensemble average of a property $a(x)$ is then related to the discretized time average by

$$A = \langle a \rangle = \frac{1}{M} \sum_{n=1}^M a(x_{n\Delta t}) \equiv \bar{a}. \qquad (3.7.3)$$

The molecular dynamics method has the particular advantage of yielding equilibrium averages and dynamical information simultaneously. This is an aspect of molecular dynamics that is not shared by other equilibrium methods such as Monte Carlo (see Chapter 7). Although we will develop the foundations of molecular dynamics in detail over the next few chapters, we will restrict our usage of the technique, for now, to the calculation of equilibrium averages only. We will not see how to use the dynamical information available from molecular dynamics calculations until Chapter 13.

In the preceding discussion, many readers will have greeted the assumption of ergodicity, which seems to underly the molecular dynamics approach, with a dose of skepticism. Indeed, this assumption is a rather strong one that clearly will not hold for a system whose potential energy $U(\mathbf{r})$ possesses high barriers—regions where $U(\mathbf{r}) > E$, leading to separatrices in the phase space. In general, it is not possible to prove the ergodicity or lack thereof in a system with many degrees of freedom. The ergodic hypothesis tends to break down locally rather than globally. The virial theorem tells us that the average energy in a given mode is kT at equilibrium if the system has been able to equipartition the energy. Instantaneously, however, the energy of a mode or degree of freedom fluctuates. Thus, if some particular mode has a high barrier to surmount, a very long time will be needed for a fluctuation to occur that amasses sufficient energy in this mode to promote barrier-crossing. Biological macromolecules such as proteins and polypeptides exemplify this problem, as important conformations are often separated by barriers in the space of the backbone dihedral angles or other collective variables in the system. Many other types of systems have severe ergodicity problems that render them challenging to treat via numerical simulation, and one must always bear such problems in mind when applying numerical methods such as molecular dynamics. Keeping such caveats in mind, we begin with a discussion of numerical integrators.

3.8 Integrating the equations of motion: Finite difference methods

3.8.1 The Verlet algorithm

There are three principal aspects to a molecular dynamics calculation: 1) the model describing the interparticle interactions; 2) the calculation of energies and forces from the model, which should be done accurately and efficiently; 3) the algorithm used to integrate the equations of motion. Each of these can strongly influence the quality of the calculation and its ability to sample a sufficient number of microstates to obtain reliable averages. We will start by considering the problem of devising a numerical integrator or solver for the equations of motion. Later in this chapter, we will consider different types of models for physical systems. Technical aspects of force calculations are provided in Appendix C.

By far the simplest way to obtain a numerical integration scheme is to use a Taylor series. In this approach, the position of a particle at a time $t + \Delta t$ is expressed in terms of its position, velocity, and acceleration at time t according to:

$$\mathbf{r}_i(t + \Delta t) \approx \mathbf{r}_i(t) + \Delta t \dot{\mathbf{r}}_i(t) + \frac{1}{2}\Delta t^2 \ddot{\mathbf{r}}_i(t), \tag{3.8.1}$$

where all terms higher than second-order in Δt have been dropped. Since $\dot{\mathbf{r}}_i(t) = \mathbf{v}_i(t)$ and $\ddot{\mathbf{r}}_i(t) = \mathbf{F}_i(t)/m_i$ by Newton's second law, eqn. (3.8.1) can be written as

$$\mathbf{r}_i(t + \Delta t) \approx \mathbf{r}_i(t) + \Delta t \mathbf{v}_i(t) + \frac{\Delta t^2}{2m_i}\mathbf{F}_i(t). \tag{3.8.2}$$

Note that the shorthand notation for the force $\mathbf{F}_i(t)$ is used in place of the full expression, $\mathbf{F}_i(\mathbf{r}_1(t), ..., \mathbf{r}_N(t))$. A velocity-independent scheme can be obtained by writing a similar expansion for $\mathbf{r}_i(t - \Delta t)$:

$$\mathbf{r}_i(t - \Delta t) = \mathbf{r}_i(t) - \Delta t \mathbf{v}_i(t) + \frac{\Delta t^2}{2m_i}\mathbf{F}_i(t). \tag{3.8.3}$$

Adding eqns. (3.8.2) and (3.8.3), one obtains

$$\mathbf{r}_i(t + \Delta t) + \mathbf{r}_i(t - \Delta t) = 2\mathbf{r}_i(t) + \frac{\Delta t^2}{m_i}\mathbf{F}_i(t), \tag{3.8.4}$$

which, after rearrangement, becomes

$$\mathbf{r}_i(t + \Delta t) = 2\mathbf{r}_i(t) - \mathbf{r}_i(t - \Delta t) + \frac{\Delta t^2}{m_i}\mathbf{F}_i(t). \tag{3.8.5}$$

Equation (3.8.5) is a numerical solver known as the *Verlet algorithm* (Verlet, 1967). Given a set of initial coordinates $\{\mathbf{r}_1(0), ..., \mathbf{r}_N(0)\}$ and initial velocities $\{\mathbf{v}_1(0), ..., \mathbf{v}_N(0)\}$, eqn. (3.8.2) can be used to obtain a set of coordinates, $\{\mathbf{r}_1(\Delta t), ..., \mathbf{r}_N(\Delta t)\}$, after which eqn. (3.8.5) can be used to generate a trajectory of arbitrary length. Note that the Verlet algorithm only generates positions. If needed, the velocities can be constructed at any point in the trajectory via the centered difference formula

$$\mathbf{v}_i(t) = \frac{\mathbf{r}_i(t + \Delta t) - \mathbf{r}_i(t - \Delta t)}{2\Delta t}. \tag{3.8.6}$$

3.8.2 The velocity Verlet algorithm

Although appealing in its simplicity, the Verlet algorithm does not explicitly evolve the velocities, and this is somewhat inelegant, as phase space is composed of both positions and velocities (or momenta). Here, we will derive a variant of the Verlet integrator, known as the *velocity Verlet* algorithm (Swope *et al.*, 1982), that explicitly evolves positions and velocities. Consider, again, the expansion of the coordinates up to second-order in Δt:

$$\mathbf{r}_i(t + \Delta t) \approx \mathbf{r}_i(t) + \Delta t\mathbf{v}_i(t) + \frac{\Delta t^2}{2m_i}\mathbf{F}_i(t). \tag{3.8.7}$$

Interestingly, we could also *start* from $\mathbf{r}_i(t + \Delta t)$ and $\mathbf{v}_i(t + \Delta t)$, compute $\mathbf{F}_i(t + \Delta t)$ and evolve backwards in time to $\mathbf{r}_i(t)$ according to

$$\mathbf{r}_i(t) = \mathbf{r}_i(t + \Delta t) - \Delta t\mathbf{v}_i(t + \Delta t) + \frac{\Delta t^2}{2m_i}\mathbf{F}_i(t + \Delta t). \tag{3.8.8}$$

Substituting eqn. (3.8.7) for $\mathbf{r}_i(t + \Delta t)$ into eqn. (3.8.8) and solving for $\mathbf{v}_i(t + \Delta t)$ yields

$$\mathbf{v}_i(t + \Delta t) = \mathbf{v}_i(t) + \frac{\Delta t}{2m_i}\left[\mathbf{F}_i(t) + \mathbf{F}_i(t + \Delta t)\right]. \tag{3.8.9}$$

Thus, the velocity Verlet algorithm uses both eqns. (3.8.7) and (3.8.9) to evolve the positions and velocities simultaneously. The Verlet and velocity Verlet algorithms satisfy two properties that are crucial for the long-time stability of numerical solvers. The first is time-reversibility, which means that if we take as initial conditions $\mathbf{r}_1(t + \Delta t), ..., \mathbf{r}_N(t + \Delta t), \mathbf{v}_1(t + \Delta t), ..., \mathbf{v}_N(t + \Delta t)$ and step backward in time using a time step $-\Delta t$, we will arrive at the state $\mathbf{r}_1(t), ..., \mathbf{r}_N(t), \mathbf{v}_1(t), ..., \mathbf{v}_N(t)$. Time-reversibility is a fundamental symmetry of Hamilton's equations that should be preserved by a numerical integrator. The second is the symplectic property of eqn. (1.6.29); we will discuss the importance of the symplectic property for numerical stability in Section 3.13.

While there are classes of integrators that purport to be more accurate than the simple second-order Verlet and velocity Verlet algorithms, for example predictor-corrector methods, we note here that many of these methods are neither symplectic nor time-reversible and, therefore, lead to significant drifts in the total energy when used. In choosing a numerical integration method, one should *always* examine the properties of the integrator and verify its suitability for a given problem.

3.8.3 Choosing the initial conditions

At this point, it is worth saying a few words about how the initial conditions for a molecular dynamics calculation are chosen. Indeed, setting up an initial condition can, depending on the complexity of the system, be a nontrivial problem. For a simple liquid, one might start with initial coordinates corresponding to the solid phase of the substance and then simply melt the solid structure under thermodynamic conditions appropriate to the liquid. Alternatively, one can begin with random initial coordinates, requiring only that the distance between particles be large enough to avoid strong repulsive forces initially. For a molecular liquid, initial bond lengths and bend angles may be dictated by holonomic constraints or may simply be chosen to be equilibrium values. For more complex systems such as molecular crystals or biological macromolecules, it is usually necessary to obtain initial coordinates from an experimental X-ray crystal structure. Many such crystal structures are deposited into structure databases such as the Cambridge Structure Database, the Inorganic Crystal Structure Database, or the Protein Data Bank. When using experimental structures, it might be necessary to supply missing information, such as the coordinates of hydrogen atoms, that cannot be experimentally resolved. For biological systems, it is often necessary to solvate the macromolecule in a bath of water molecules. For this purpose, one might take coordinates from a large, well-equilibrated pure water simulation, place the macromolecule into the water bath, and then remove waters that are within a certain distance (e.g. 1.8 Å) of any atom in the macromolecule, being careful to retain crystallographic waters bound within the molecule. After such a procedure, it is necessary to re-establish equilibrium, which typically involves adjusting the energy to give a certain temperature and the volume (Chapter 4) to give a certain pressure (Chapter 5).

Once initial coordinates are specified, it remains to set the initial velocities. This is generally done by "sampling" the velocities from a Maxwell-Boltzmann distribution, taking care to ensure that the sampled velocities are consistent with any constraints imposed on the system. We will treat the problem of sampling a distribution more generally in Chapter 7; however, here we provide a simple algorithm for obtaining an initial set of velocities.

The Maxwell-Boltzmann distribution for the velocity v of a particle of mass m at temperature T is

$$f(v) = \left(\frac{m}{2\pi kT}\right)^{1/2} e^{-mv^2/2kT}. \tag{3.8.10}$$

The distribution $f(v)$ is an example of a *Gaussian distribution*. More generally, if x

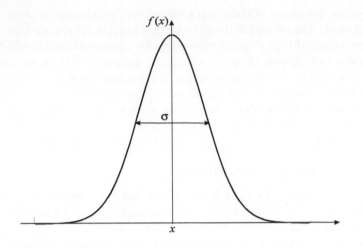

Fig. 3.3 Gaussian distribution given in eqn. (3.8.11).

is a Gaussian random variable with zero mean, its probability distribution is

$$f(x) = \left(\frac{1}{2\pi\sigma^2}\right)^{1/2} e^{-x^2/2\sigma^2}, \tag{3.8.11}$$

where σ is the width of the Gaussian (see Fig. 3.3). Here, $f(x)\mathrm{d}x$ is the probability that a given value of the variable, x, will lie in an interval between x and $x + \mathrm{d}x$. Note that $f(x)$ satisfies the requirements of a probability distribution function:

$$f(x) \geq 0$$

$$\int_{-\infty}^{\infty} \mathrm{d}x\, f(x) = 1. \tag{3.8.12}$$

The cumulative probability that a randomly chosen value of x lies in the interval $x \in (-\infty, X)$ for some upper limit X is

$$P(X) = \int_{-\infty}^{X} \mathrm{d}x\, f(x) = \left(\frac{1}{2\pi\sigma^2}\right)^{1/2} \int_{-\infty}^{X} \mathrm{d}x\, e^{-x^2/2\sigma^2}. \tag{3.8.13}$$

Since $P(X)$ is a number between 0 and 1, the problem of sampling $f(x)$ consists,

therefore, in choosing a probability $\xi \in [0,1]$ and solving the equation $P(X) = \xi$, the probability that $x \in (-\infty, X]$ for X. The resulting value of X is known as a *Gaussian random number*. If the equation is solved for M values $\xi_1, ..., \xi_M$ to yield values $X_1, ..., X_M$, then we simply set $x_i = X_i$, and we have a sampling of $f(x)$ (see Chapter 7 for a more detailed discussion).

Unfortunately, we do not have a simple closed form expression for $P(X)$ that allows us to solve the equation $P(X) = \xi$ easily for X. The trick we need comes from recognizing that if we square eqn. (3.8.13), we obtain a probability distribution for which a simple closed form does exist. Note that squaring the cumulative probability requires introduction of another variable, Y, yielding a two-dimensional Gaussian cumulative probability

$$P(X, Y) = \left(\frac{1}{2\pi\sigma^2}\right) \int_{-\infty}^{X} \int_{-\infty}^{Y} \mathrm{d}x\, \mathrm{d}y\, \mathrm{e}^{-(x^2 + y^2)/2\sigma^2}. \tag{3.8.14}$$

The integral in eqn. (3.8.14) can be carried out analytically by introducing polar coordinates:

$$x = r\cos\phi, \qquad y = r\sin\phi$$

$$X = R\cos\Phi, \qquad Y = R\sin\Phi. \tag{3.8.15}$$

Using this transformation, we obtain the cumulative probability of R and Φ as

$$P(R, \Phi) = \frac{1}{2\pi} \int_0^{\Phi} \mathrm{d}\phi\, \frac{1}{\sigma^2} \int_0^{R} \mathrm{d}r\, r\, \mathrm{e}^{-r^2/2\sigma^2}. \tag{3.8.16}$$

These are now elementary integrals, which can be performed to yield

$$P(R, \Phi) = \left(\frac{\Phi}{2\pi}\right) \left(1 - \mathrm{e}^{-R^2/2\sigma^2}\right). \tag{3.8.17}$$

Note that eqn. (3.8.17) is in the form of a product of two independent probabilities. One is a uniform probability that $\phi \leq \Phi$ and the other is the nonuniform radial probability that $r \leq R$. We may, therefore, set each of these equal to two different random numbers, ξ_1 and ξ_2, drawn from $[0, 1]$:

$$\frac{\Phi}{2\pi} = \xi_1$$

$$1 - \mathrm{e}^{-R^2/2\sigma^2} = \xi_2. \tag{3.8.18}$$

Introducing $\xi_2' = 1 - \xi_2$ (which is also a random number uniformly distributed on $[0, 1]$) and solving for R and Φ yields

$$\Phi = 2\pi\xi_1$$

$$R = \sigma\sqrt{-2\ln\xi_2'}. \tag{3.8.19}$$

Therefore, the values of X and Y are

$$X = \sigma\sqrt{-2\ln\xi_2'}\,\cos 2\pi\xi_1$$

$$Y = \sigma\sqrt{-2\ln\xi_2'}\,\sin 2\pi\xi_1. \tag{3.8.20}$$

Thus, we obtain two Gaussian random numbers X and Y. This algorithm for generating Gaussian random numbers is known as *Box-Muller sampling*. By applying the algorithm to the Maxwell-Boltzmann distribution in eqn. (3.8.10), $3N$ initial velocity vector components can be generated. Note, however, that if there are any constraints on the system, the velocities must be projected back to the surface of constraint after the sampling is complete in order to ensure that the first time derivatives of the constraint conditions are also satisfied. Moreover, for systems in which the total force $\sum_{i=1}^{N} \mathbf{F}_i = 0$, the center-of-mass velocity

$$\mathbf{v}_{\rm cm} = \frac{\sum_{i=1}^{N} m_i \mathbf{v}_i}{\sum_{i=1}^{N} m_i} \tag{3.8.21}$$

is a constant of the motion. Therefore, it is often useful to choose the initial velocities in such a way that $\mathbf{v}_{\rm cm} = 0$ in order to avoid an overall drift of the system in space.

Once the initial conditions are specified, all information needed to start a simulation is available, and an algorithm such as the Verlet or velocity Verlet algorithm can be used to integrate the equations of motion.

3.9 Systems subject to holonomic constraints

In Section 1.9, we discussed the formulation of classical mechanics for a system subject to a set of holonomic constraints, that is, constraints which depend only on the positions of the particles and possibly time:

$$\sigma_k(\mathbf{r}_1, ..., \mathbf{r}_N, t) = 0 \qquad k = 1, ..., N_c. \tag{3.9.1}$$

For the present discussion, we shall consider only time-independent constraints. In this case, according to eqn. (1.9.11), the equations of motion can be expressed as

$$m_i \ddot{\mathbf{r}}_i = \mathbf{F}_i + \sum_{k=1}^{N_c} \lambda_k \nabla_i \sigma_k, \tag{3.9.2}$$

where λ_k is a set of Lagrange multipliers for enforcing the constraints. Although it is possible to obtain an exact expression for the Lagrange multipliers using Gauss's principle of least constraint, the numerical integration of the equations of motion obtained by substituting the exact expression for λ_k into eqn. (3.9.2) would not exactly preserve the constraint condition due to numerical errors, which would lead to unwanted instabilities and artifacts in a simulation. In addition, Gauss's equations of motion are complicated non-Hamiltonian equations that cannot be treated using simple methods such as the Verlet and velocity Verlet algorithms. These problems can be circumvented by introducing a scheme for computing the multipliers "on the fly" in a simulation in such a way that the constraint conditions are exactly satisfied within a particular chosen numerical integration scheme. This is the approach we will now describe.

3.9.1 The SHAKE and RATTLE algorithms

For time-independent holonomic constraints, the Euler-Lagrange equations of motion, eqns. (1.9.11) and (1.9.12), in Cartesian coordinates are

$$\frac{\mathrm{d}}{\mathrm{d}t}\left(\frac{\partial \mathcal{L}}{\partial \dot{\mathbf{r}}_i}\right) - \frac{\partial \mathcal{L}}{\partial \mathbf{r}_i} = \sum_{k=1}^{N_c} \lambda_k \mathbf{a}_{ki}$$

$$\sum_{i=1}^{N} \mathbf{a}_{ki} \cdot \dot{\mathbf{r}}_i = 0, \tag{3.9.3}$$

where

$$\mathbf{a}_{ki} = \nabla_i \sigma_k(\mathbf{r}_1, ..., \mathbf{r}_N). \tag{3.9.4}$$

Note that these are equivalent to

$$m_i \ddot{\mathbf{r}}_i = \mathbf{F}_i + \sum_{k=1}^{N_c} \lambda_k \nabla_i \sigma_k$$

$$\frac{\mathrm{d}}{\mathrm{d}t} \sigma_k(\mathbf{r}_1, ..., \mathbf{r}_N) = 0. \tag{3.9.5}$$

The constraint problem amounts to integrating eqn. (3.9.3) subject to the conditions that $\sigma_k(\mathbf{r}_1, ..., \mathbf{r}_N) = 0$ and $\dot{\sigma}_k(\mathbf{r}_1, ..., \mathbf{r}_N, \dot{\mathbf{r}}_1, ..., \dot{\mathbf{r}}_N) = \sum_i \nabla_i \sigma_k(\mathbf{r}_1, ..., \mathbf{r}_N) \cdot \dot{\mathbf{r}}_i = 0$. We wish to develop a numerical scheme in which the constraint conditions are satisfied exactly as part of the integration algorithm.

Starting from the velocity Verlet approach, for example, we begin with the position update, which, when holonomic constraints are imposed, reads

$$\mathbf{r}_i(\Delta t) = \mathbf{r}_i(0) + \Delta t \mathbf{v}_i(0) + \frac{\Delta t^2}{2m_i} \mathbf{F}_i(0) + \frac{\Delta t^2}{2m_i} \sum_k \lambda_k \nabla_i \sigma_k(0), \tag{3.9.6}$$

where $\sigma_k(0) \equiv \sigma_k(\mathbf{r}_1(0), ..., \mathbf{r}_N(0))$. In order to ensure that the constraint is satisfied exactly at time Δt, we impose the constraint condition directly on the numerically obtained positions $\mathbf{r}_i(\Delta t)$ and determine, on the fly, the multipliers needed to enforce the constraint. Let us define

$$\mathbf{r}'_i = \mathbf{r}_i(0) + \Delta t \mathbf{v}_i(0) + \frac{\Delta t^2}{2m_i} \mathbf{F}_i(0) \tag{3.9.7}$$

so that

$$\mathbf{r}_i(\Delta t) = \mathbf{r}'_i + \frac{1}{m_i} \sum_k \tilde{\lambda}_k \nabla_i \sigma_k(0), \tag{3.9.8}$$

where $\tilde{\lambda}_k = (\Delta t^2/2)\lambda_k$. Then, for each constraint condition $\sigma_l(\mathbf{r}_1, ..., \mathbf{r}_N) = 0$, we impose

$$\sigma_l(\mathbf{r}_1(\Delta t), ..., \mathbf{r}_N(\Delta t)) = 0, \qquad l = 1, ..., N_c. \tag{3.9.9}$$

Substituting in for $\mathbf{r}_i(\Delta t)$, we obtain a set of N_c nonlinear equations for the N_c unknown multipliers $\tilde{\lambda}_1, ..., \tilde{\lambda}_{N_c}$:

$$\sigma_l \left(\mathbf{r}'_1 + \frac{1}{m_1} \sum_k \tilde{\lambda}_k \nabla_1 \sigma_k(0), ..., \mathbf{r}'_N + \frac{1}{m_N} \sum_k \tilde{\lambda}_k \nabla_N \sigma_k(0) \right) = 0. \tag{3.9.10}$$

Unless the constraints are of a particularly simple form, eqns. (3.9.10) will need to be solved iteratively. A simple procedure for doing this, known as the SHAKE algorithm (Ryckaert *et al.*, 1977), proceeds as follows. First, if a good initial guess of the solution, $\{\tilde{\lambda}_k^{(1)}\}$, is available (for example, the multipliers from the previous molecular dynamics time step), then the coordinates can be updated according to

$$\mathbf{r}_i^{(1)} = \mathbf{r}' + \frac{1}{m_i} \sum_k \tilde{\lambda}_k^{(1)} \nabla_i \sigma_k(0). \tag{3.9.11}$$

The exact solution for the multipliers is now written as $\tilde{\lambda}_k = \tilde{\lambda}_k^{(1)} + \delta\tilde{\lambda}_k^{(1)}$, and $\mathbf{r}_i(\Delta t) = \mathbf{r}_i^{(1)} + (1/m_i) \sum_k \delta\tilde{\lambda}_k^{(1)} \nabla_i \sigma_k(0)$, so that eqn. (3.9.10) becomes

$$\sigma_l\left(\mathbf{r}_1^{(1)} + \frac{1}{m_1}\sum_k \delta\tilde{\lambda}_k \nabla_1 \sigma_k(0), ..., \mathbf{r}_N^{(1)} + \frac{1}{m_N}\sum_k \delta\tilde{\lambda}_k^{(1)} \nabla_N \sigma_k(0)\right) = 0. \tag{3.9.12}$$

Next, eqn. (3.9.12) is expanded to first order in a Taylor series about $\delta\tilde{\lambda}_k^{(1)} = 0$:

$$\sigma_l(\mathbf{r}_1^{(1)}, ..., \mathbf{r}_N^{(1)}) +$$

$$\sum_{i=1}^{N} \sum_{k=1}^{N_c} \frac{1}{m_i} \nabla_i \sigma_l(\mathbf{r}_1^{(1)}, ..., \mathbf{r}_N^{(1)}) \cdot \nabla_i \sigma_k(\mathbf{r}_1(0), ..., \mathbf{r}_N(0)) \delta\tilde{\lambda}_k^{(1)} \approx 0. \tag{3.9.13}$$

Equation (3.9.13) is a matrix equation for the changes $\delta\tilde{\lambda}_k^{(1)}$ in the multipliers. If the dimensionality of this equation is not too large, then it can be inverted directly to yield the full set of $\delta\tilde{\lambda}_k^{(1)}$ simultaneously. This procedure is known as matrix-SHAKE or M-SHAKE (Kraeutler *et al.*, 2001). Because eqn. (3.9.12) was approximated by linearization, however, adding the correction $\sum_k \delta\tilde{\lambda}_k^{(1)} \nabla_i \sigma_k(0)$ to $\mathbf{r}_i^{(1)}$ does not yield a fully converged $\mathbf{r}_i(\Delta t)$. We, therefore, define $\mathbf{r}_i^{(2)} = \mathbf{r}_i^{(1)} + (1/m_i) \sum_k \delta\tilde{\lambda}_k^{(1)} \nabla_i \sigma_k(0)$ and write $\mathbf{r}_i(\Delta t) = \mathbf{r}_i^{(2)} + (1/m_i) \sum_k \delta\tilde{\lambda}_k^{(2)} \nabla_i \sigma_k(0)$ and use eqn. (3.9.13) with the "(1)" superscript replaced by "(2)" for another iteration. The procedure is repeated until the constraint conditions are satisfied to a given small tolerance.

If the dimensionality of eqn. (3.9.13) is high due to a large number of constraints, then a further time-saving approximation can be made. We replace the full matrix $A_{lk} = \sum_{i=1}^{N}(1/m_i)\nabla_i \sigma_l(\mathbf{r}_1^{(1)}, ..., \mathbf{r}_N^{(1)}) \cdot \nabla_i \sigma_k(\mathbf{r}_1(0), ..., \mathbf{r}_N(0))\delta\tilde{\lambda}_k^{(1)}$ by its diagonal elements only, leading to

$$\sigma_l(\mathbf{r}_1^{(1)}, ..., \mathbf{r}_N^{(1)}) + \sum_{i=1}^{N} \frac{1}{m_i} \nabla_i \sigma_l(\mathbf{r}_1^{(1)}, ..., \mathbf{r}_N^{(1)}) \cdot \nabla_i \sigma_l(\mathbf{r}_1(0), ..., \mathbf{r}_N(0))\delta\tilde{\lambda}_l^{(1)} \approx 0. \tag{3.9.14}$$

Equation (3.9.14) has a simple solution

$$\delta\tilde{\lambda}_l^{(1)} = -\frac{\sigma_l(\mathbf{r}_1^{(1)}, ..., \mathbf{r}_N^{(1)})}{\sum_{i=1}^{N}(1/m_i)\nabla_i \sigma_l(\mathbf{r}_1^{(1)}, ..., \mathbf{r}_N^{(1)}) \cdot \nabla_i \sigma_l(\mathbf{r}_1(0), ..., \mathbf{r}_N(0))}. \tag{3.9.15}$$

Equation (3.9.15) could be used, for example, to obtain $\delta\tilde{\lambda}_1^{(1)}$ followed immediately by an update of all $\mathbf{r}_i^{(1)}$ involved in the $k = 1$ constraint to obtain positions $\mathbf{r}_i^{(2)}$

for this constraint. Given the updated position, eqn. (3.9.15) is used to obtain $\delta\tilde{\lambda}_1^{(2)}$ immediately followed by an update of all $\mathbf{r}_i^{(2)}$ involved in the $k = 1$ constraint, and we iterate until the $k = 1$ constraint is satisfied. We then proceed to the $k = 2$ constraint and iterate until it is satisfied, which will cause a slight violation of the $k = 1$ constraint. Note that satisfying the kth constraint via this procedure causes all $l < k$ constraints to be slightly violated. Thus, after cycling through all of the constraints in this manner, the procedure must be repeated until the full set of constraints is converged to within a given tolerance. Once the converged multipliers are obtained, and the coordinates fully updated, the velocities must be updated as well according to

$$\mathbf{v}_i(\Delta t/2) = \mathbf{v}_i(0) + \frac{\Delta t}{2m_i}\mathbf{F}_i(0) + \frac{1}{m_i\Delta t}\sum_k \tilde{\lambda}_k \nabla_i \sigma_k(0). \qquad (3.9.16)$$

After eqn. (3.9.16) is applied, we can proceed to the next step of updating the velocities, which requires that the condition $\dot{\sigma}_k(\mathbf{r}_1, ..., \mathbf{r}_N) = 0$ be satisfied. Once the new forces are obtained from the updated positions, the final velocities are written as

$$\mathbf{v}_i(\Delta t) = \mathbf{v}_i(\Delta t/2) + \frac{\Delta t}{2m_i}\mathbf{F}_i(\Delta t) + \frac{\Delta t}{2m_i}\sum_k \mu_k \nabla_i \sigma_k(\Delta t)$$

$$= \mathbf{v}_i' + \frac{1}{m_i}\sum_k \tilde{\mu}_k \nabla_i \sigma_k(\Delta t), \qquad (3.9.17)$$

where μ_k has been used to denote the multipliers for the velocity step to indicate that they are different from those used for the position step, and $\tilde{\mu}_k = (\Delta t/2)\mu_k$. The multipliers μ_k are now obtained by enforcing the condition

$$\sum_{i=1}^N \nabla_i \sigma_k(\Delta t) \cdot \mathbf{v}_i(\Delta t) = 0 \qquad (3.9.18)$$

on the velocities. Substituting in for $\mathbf{v}_i(\Delta t)$, we obtain a set of N_c linear equations

$$\sum_{i=1}^N \nabla_i \sigma_k(\Delta t) \cdot \left(\mathbf{v}_i' + \frac{1}{m_i}\sum_l \tilde{\mu}_l \nabla_i \sigma_l(\Delta t) \right) = 0 \qquad (3.9.19)$$

for the multipliers $\tilde{\mu}_l$. These can be solved straightforwardly by matrix inversion or, for large systems, iteratively by satisfying the condition for each constraint in turn and then cycling through the constraints again to compute a new increment to the multiplier until convergence is reached as was proposed for the position update step. The latter iterative procedure is known as the RATTLE algorithm (Andersen, 1983). Once converged multipliers are obtained, the final velocity update is performed by substituting into eqn. (3.9.17). The SHAKE algorithm can be used in conjunction with the Verlet and velocity Verlet algorithms while RATTLE is particular to velocity Verlet. For other numerical solvers, constraint algorithms need to be adapted or tailored for consistency with the particulars of the solver.

3.10   The classical time evolution operator and numerical integrators

Thus far, we have discussed numerical integration in a somewhat simplistic way, relying on Taylor series expansions to generate update procedures. However, because there are certain formal properties of Hamiltonian systems that should be preserved by numerical integration methods, it is important to develop a formal structure that allows numerical solvers to be generated more rigorously. The framework we seek is based on the classical time evolution operator approach, and we will return to this framework repeatedly throughout the book.

We begin by considering the time evolution of any function $a(\mathbf{x})$ of the phase-space vector. If $a(\mathbf{x})$ is evaluated along a trajectory $\mathbf{x}_t$, then in generalized coordinates, the time derivative of $a(\mathbf{x}_t)$ is given by the chain rule

$$\frac{\mathrm{d}a}{\mathrm{d}t} = \sum_{\alpha=1}^{3N} \left[\frac{\partial a}{\partial q_\alpha} \dot{q}_\alpha + \frac{\partial a}{\partial p_\alpha} \dot{p}_\alpha \right]. \tag{3.10.1}$$

Hamilton's equations

$$\dot{q}_\alpha = \frac{\partial \mathcal{H}}{\partial p_\alpha}, \qquad \dot{p}_\alpha = -\frac{\partial \mathcal{H}}{\partial q_\alpha} \tag{3.10.2}$$

are now used for the time derivatives appearing in eqn. (3.10.1), which yields

$$\frac{\mathrm{d}a}{\mathrm{d}t} = \sum_{\alpha=1}^{3N} \left[\frac{\partial a}{\partial q_\alpha} \frac{\partial \mathcal{H}}{\partial p_\alpha} - \frac{\partial a}{\partial p_\alpha} \frac{\partial \mathcal{H}}{\partial q_\alpha} \right]$$

$$= \{a, \mathcal{H}\}. \tag{3.10.3}$$

The bracket $\{a, \mathcal{H}\}$ appearing in eqn. (3.10.3) is the Poisson bracket from eqns. (1.6.19) and (1.6.20). Equation (3.10.3) indicates that the Poisson bracket between $a(\mathbf{x})$ and $\mathcal{H}(\mathbf{x})$ is a generator of the time evolution of $a(\mathbf{x}_t)$.

The Poisson bracket allows us to introduce an operator on the phase space that acts on any phase-space function. Define an operator, iL, where $i = \sqrt{-1}$, by

$$iLa = \{a, \mathcal{H}\}, \tag{3.10.4}$$

where L is known as the *Liouville operator*. Note that iL can be expressed abstractly as $iL = \{..., \mathcal{H}\}$, which means "take whatever function iL acts on and substitute it for the ... in the Poisson bracket expression." It can also be written as a differential operator

$$iL = \sum_{\alpha=1}^{3N} \left[\frac{\partial \mathcal{H}}{\partial p_\alpha} \frac{\partial}{\partial q_\alpha} - \frac{\partial \mathcal{H}}{\partial q_\alpha} \frac{\partial}{\partial p_\alpha} \right]. \tag{3.10.5}$$

The equation $\mathrm{d}a/\mathrm{d}t = iLa$ can be solved formally for $a(\mathbf{x}_t)$ as

$$a(\mathbf{x}_t) = e^{iLt} a(\mathbf{x}_0). \tag{3.10.6}$$

In eqn. (3.10.6), the derivatives appearing in eqn. (3.10.5) must be taken to act on the components of the initial phase-space vector $\mathbf{x}_0$. The operator $\exp(iLt)$ appearing in eqn. (3.10.6) is known as the *classical propagator*. With the i appearing in the definition of iL, $\exp(iLt)$ strongly resembles the quantum propagator $\exp(-i\hat{\mathcal{H}}t/\hbar)$ in terms of the Hamiltonian operator $\hat{\mathcal{H}}$, which is why the i is formally included in eqn. (3.10.4). Indeed, the operator L can be shown to be a Hermitian operator so that the classical propagator $\exp(iLt)$ is a unitary operator on the phase space.

By applying eqn. (3.10.6) to the vector function $a(x) = x$, we have a formal solution to Hamilton's equations

$$x_t = e^{iLt}x_0. \tag{3.10.7}$$

Although elegant in its compactness, eqn. (3.10.7) amounts to little more than a formal device since we cannot evaluate the action of the operator $\exp(iLt)$ on x_0 exactly. If we could, then any and every problem in classical mechanics could be solved exactly analytically and we would not be in the business of developing numerical methods or buying expensive computers in the first place! What eqn. (3.10.7) does do is provide us with a very useful starting point for developing approximate solutions to Hamilton's equations. As eqn. (3.10.5) suggests, the Liouville operator can be written as a sum of two contributions

$$iL = iL_1 + iL_2, \tag{3.10.8}$$

where

$$iL_1 = \sum_{\alpha=1}^{3N} \frac{\partial \mathcal{H}}{\partial p_\alpha} \frac{\partial}{\partial q_\alpha}$$

$$iL_2 = -\sum_{\alpha=1}^{3N} \frac{\partial \mathcal{H}}{\partial q_\alpha} \frac{\partial}{\partial p_\alpha}. \tag{3.10.9}$$

The operators in eqn. (3.10.9) are examples of *noncommuting* operators. This means that, given any function $\phi(x)$ on the phase space,

$$iL_1 iL_2 \phi(x) \neq iL_2 iL_1 \phi(x). \tag{3.10.10}$$

That is, the order in which the operators is applied matters. The operator difference $iL_1 iL_2 - iL_2 iL_1$ is an object that arises frequently both in classical and quantum mechanics and is known as the *commutator* between the operators:

$$iL_1 iL_2 - iL_2 iL_1 \equiv [iL_1, iL_2]. \tag{3.10.11}$$

If $[iL_1, iL_2] = 0$, then the operators iL_1 and iL_2 are said to *commute*.

That iL_1 and iL_2 do *not* generally commute can be seen in a simple one-dimensional example. Consider the Hamiltonian

$$\mathcal{H} = \frac{p^2}{2m} + U(x). \tag{3.10.12}$$

According to eqn. (3.10.9),

$$iL_1 = \frac{p}{m} \frac{\partial}{\partial x}, \qquad iL_2 = F(x) \frac{\partial}{\partial p}, \tag{3.10.13}$$

where $F(x) = -dU/dx$. The action of $iL_1 iL_2$ on a function $\phi(x, p)$ is

$$\frac{p}{m}\frac{\partial}{\partial x}F(x)\frac{\partial}{\partial p}\phi(x, p) = \frac{p}{m}F(x)\frac{\partial^2\phi}{\partial p\partial x} + \frac{p}{m}F'(x)\frac{\partial\phi}{\partial p}, \tag{3.10.14}$$

whereas the action of $iL_2 iL_1$ on $\phi(x, p)$ is

$$F(x)\frac{\partial}{\partial p}\frac{p}{m}\frac{\partial}{\partial x}\phi(x, p) = F(x)\frac{p}{m}\frac{\partial^2\phi}{\partial p\partial x} + F(x)\frac{1}{m}\frac{\partial\phi}{\partial x}, \tag{3.10.15}$$

so that $[iL_1, iL_2]\phi(x, p)$ is

$$[iL_1, iL_2]\phi(x, p) = \frac{p}{m}F'(x)\frac{\partial\phi}{\partial p} - \frac{F(x)}{m}\frac{\partial\phi}{\partial x}. \tag{3.10.16}$$

Since the function $\phi(x, p)$ is arbitrary, we can conclude that the operator

$$[iL_1, iL_2] = \frac{p}{m}F'(x)\frac{\partial}{\partial p} - \frac{F(x)}{m}\frac{\partial}{\partial x}, \tag{3.10.17}$$

from which it can be seen that $[iL_1, iL_2] \neq 0$.

Since iL_1 and iL_2 generally do not commute, the classical propagator $\exp(iLt) = \exp[(iL_1+iL_2)t]$ *cannot* be separated into a simple product $\exp(iL_1 t)\exp(iL_2 t)$. This is unfortunate because in many instances, the action of the individual operators $\exp(iL_1 t)$ and $\exp(iL_2 t)$ on the phase-space vector can be evaluated exactly. Thus, it would be useful if the propagator could somehow be expressed in terms of these two factors. In fact, there is a way to do this using an important theorem known as the *Trotter theorem* (Trotter, 1959). This theorem states that given two operators A and B for which $[A, B] \neq 0$,

$$e^{A+B} = \lim_{P\to\infty}\left[e^{B/2P}e^{A/P}e^{B/2P}\right]^P, \tag{3.10.18}$$

where P is an integer. In fact, eqn. (3.10.18) is commonly referred to as the *symmetric Trotter theorem* or *Strang splitting formula* (Strang, 1968). The proof of the Trotter theorem is somewhat involved and is, therefore, presented in Appendix D for interested readers. Applying the symmetric Trotter theorem to the classical propagator yields

$$e^{iLt} = e^{(iL_1+iL_2)t} = \lim_{P\to\infty}\left[e^{iL_2 t/2P}e^{iL_1 t/P}e^{iL_2 t/2P}\right]^P. \tag{3.10.19}$$

Equation (3.10.19) can be expressed more suggestively by defining a time step $\Delta t = t/P$. Introducing Δt into eqn. (3.10.19) yields

$$e^{iLt} = \lim_{P\to\infty, \Delta t\to 0}\left[e^{iL_2\Delta t/2}e^{iL_1\Delta t}e^{iL_2\Delta t/2}\right]^P. \tag{3.10.20}$$

Equation (3.10.20) states that we can propagate a classical system using the separate factor $\exp(iL_2\Delta t/2)$ and $\exp(iL_1\Delta t)$ exactly for a finite time t in the limit that we let the number of steps we take go to infinity and the time step go to zero! Of course, this is not practical, but if we do not take these limits, then eqn. (3.10.20) leads to a useful approximation for classical propagation.

Note that for finite P, eqn. (3.10.20) implies an approximation to $\exp(iLt)$:

$$e^{iLt} \approx \left[e^{iL_2 \Delta t/2} e^{iL_1 \Delta t} e^{iL_2 \Delta t/2} \right]^P + \mathcal{O}\left(P \Delta t^3 \right), \qquad (3.10.21)$$

where the leading order error is proportional to $P \Delta t^3$. Since $P = t/\Delta t$, the error is actually proportional to Δt^2. According to eqn. (3.10.21), an approximate time propagation can be generated by performing P steps of *finite* length Δt using the factorized propagator

$$e^{iL \Delta t} \approx e^{iL_2 \Delta t/2} e^{iL_1 \Delta t} e^{iL_2 \Delta t/2} + \mathcal{O}\left(\Delta t^3 \right) \qquad (3.10.22)$$

for each step. Equation (3.10.22) results from taking the $1/P$ power of both sides of eqn. (3.10.21). An important difference between eqns. (3.10.22) and (3.10.21) should be noted. While the error in a single step of length Δt is proportional to Δt^3, the error in a trajectory of P steps is proportional to Δt^2. This distinguishes the *local* error in one step from the *global* error in a full trajectory of P steps. The utility of eqn. (3.10.22) is that if the contributions iL_1 and iL_2 to the Liouville operator are chosen such the action of the operators $\exp(iL_1 \Delta t)$ and $\exp(iL_2 \Delta t/2)$ can be evaluated analytically, then eqn. (3.10.22) can be used as a numerical propagation scheme for a single time step.

In order to see how this works, consider again the example of a single particle moving in one dimension with a Hamiltonian $\mathcal{H} = p^2/2m + U(x)$ and the two contributions to the overall Liouville operator given by eqn. (3.10.13). Using these operators in eqn. (3.10.22) gives the approximate single-step propagator:

$$\exp(iL \Delta t) \approx \exp\left(\frac{\Delta t}{2} F(x) \frac{\partial}{\partial p} \right) \exp\left(\Delta t \frac{p}{m} \frac{\partial}{\partial x} \right) \exp\left(\frac{\Delta t}{2} F(x) \frac{\partial}{\partial p} \right). \qquad (3.10.23)$$

The exact evolution specified by eqn. (3.10.7) is now replaced by the approximation evolution of eqn. (3.10.23). Thus, starting from an initial condition $(x(0), p(0))$, the approximation evolution can be expressed as

$$\begin{pmatrix} x(\Delta t) \\ p(\Delta t) \end{pmatrix} \approx \exp\left(\frac{\Delta t}{2} F(x(0)) \frac{\partial}{\partial p(0)} \right)$$

$$\times \exp\left(\Delta t \frac{p(0)}{m} \frac{\partial}{\partial x(0)} \right)$$

$$\times \exp\left(\frac{\Delta t}{2} F(x(0)) \frac{\partial}{\partial p(0)} \right) \begin{pmatrix} x(0) \\ p(0) \end{pmatrix}. \qquad (3.10.24)$$

In order to make the notation less cumbersome in the proceeding analysis, we will drop the "(0)" label and write eqn. (3.10.24) as

$$\begin{pmatrix} x(\Delta t) \\ p(\Delta t) \end{pmatrix} \approx \exp\left(\frac{\Delta t}{2} F(x) \frac{\partial}{\partial p} \right) \exp\left(\Delta t \frac{p}{m} \frac{\partial}{\partial x} \right) \exp\left(\frac{\Delta t}{2} F(x) \frac{\partial}{\partial p} \right) \begin{pmatrix} x \\ p \end{pmatrix}. \qquad (3.10.25)$$

The "(0)" label will be replaced at the end.

The propagation is determined by acting with each of the three operators in succession on x and p. But how do we apply exponential operators? Let us start by asking how the operator $\exp(c\partial/\partial x)$, where c is independent of x, acts on an arbitrary function $g(x)$. The action of the operator can be worked out by expanding the exponential in a Taylor series

$$\exp\left(c\frac{\partial}{\partial x}\right)g(x) = \sum_{k=0}^{\infty}\frac{1}{k!}\left(c\frac{\partial}{\partial x}\right)^k g(x)$$

$$= \sum_{k=0}^{\infty}\frac{1}{k!}c^k g^{(k)}(x), \tag{3.10.26}$$

where $g^{(k)}(x) = \mathrm{d}^k g/\mathrm{d}x^k$. The second line of eqn. (3.10.26) is just the Taylor expansion of $g(x+c)$ about $c=0$. Thus, we have the general result

$$\exp\left(c\frac{\partial}{\partial x}\right)g(x) = g(x+c), \tag{3.10.27}$$

which we can use to evaluate the action of the first operator in eqn. (3.10.25):

$$\exp\left(\frac{\Delta t}{2}F(x)\frac{\partial}{\partial p}\right)\begin{pmatrix} x \\ p \end{pmatrix} = \begin{pmatrix} x \\ p + \frac{\Delta t}{2}F(x) \end{pmatrix}. \tag{3.10.28}$$

The second operator, which involves a derivative with respect to position, acts on the x appearing in both components of the vector on the right side of eqn. (3.10.28):

$$\exp\left(\Delta t\frac{p}{m}\frac{\partial}{\partial x}\right)\begin{pmatrix} x \\ p + \frac{\Delta t}{2}F(x) \end{pmatrix} = \begin{pmatrix} x + \Delta t\frac{p}{m} \\ p + \frac{\Delta t}{2}F\left(x + \Delta t\frac{p}{m}\right) \end{pmatrix}. \tag{3.10.29}$$

In the same way, the third operator, which involves another derivative with respect to momentum, yields:

$$\exp\left(\frac{\Delta t}{2}F(x)\frac{\partial}{\partial p}\right)\begin{pmatrix} x + \Delta t\frac{p}{m} \\ p + \frac{\Delta t}{2}F\left(x + \Delta t\frac{p}{m}\right) \end{pmatrix}$$

$$= \begin{pmatrix} x + \frac{\Delta t}{m}\left(p + \frac{\Delta t}{2}F(x)\right) \\ p + \frac{\Delta t}{2}F(x) + \frac{\Delta t}{2}F\left(x + \frac{\Delta t}{m}\left(p + \frac{\Delta t}{2}F(x)\right)\right) \end{pmatrix}. \tag{3.10.30}$$

Using the fact that $v = p/m$, the final position $x(\Delta t)$ can be written as (replacing the "(0)" label on the initial conditions):

$$x(\Delta t) = x(0) + \Delta t v(0) + \frac{\Delta t^2}{2m}F(x(0)). \tag{3.10.31}$$

Equation (3.10.31) is equivalent to a second-order Taylor expansion of $x(\Delta t)$ up to second-order in Δt and is also the position update part of the velocity Verlet algorithm.

From eqn. (3.10.31), the momentum update step can be written compactly (using $v = p/m$ and replacing the "(0)" label) as

$$v(\Delta t) = v(0) + \frac{\Delta t}{2m} \left[F(x(0)) + F(x(\Delta t)) \right], \qquad (3.10.32)$$

which is the velocity update part of the velocity Verlet algorithm.

The above analysis demonstrates how we can obtain the velocity Verlet algorithm via the powerful formalism provided by the Trotter factorization scheme. Moreover, it is now manifestly clear that the velocity Verlet algorithm constitutes a symplectic, unitary, time-reversible solver that preserves the important symmetries of classical mechanics (see also Problem 1.5 in Chapter 1). Since eqns. (3.10.31) and (3.10.32) together constitute a symplectic algorithm, each term of the product implied by eqn. (3.10.20) is symplectic. Finally, in the limit $\Delta t \to 0$ and $P \to \infty$, exact classical mechanics is recovered, which allows us to conclude that time evolution under Hamilton's equations is, indeed, symplectic, as claimed in Section 1.6. While it may seem as though we arrived at eqns. (3.10.31) and (3.10.32) using an overly complicated formalism, the power of the Liouville operator approach will be readily apparent as we encounter increasingly complex numerical integration problems in the upcoming chapters.

Extending the analysis to N-particle systems in d dimensions is straightforward. For the standard Hamiltonian in Cartesian coordinates

$$\mathcal{H} = \sum_{i=1}^{N} \frac{\mathbf{p}_i^2}{2m_i} + U(\mathbf{r}_1, ..., \mathbf{r}_N), \qquad (3.10.33)$$

the Liouville operator is given by

$$iL = \sum_{i=1}^{N} \frac{\mathbf{p}_i}{m_i} \cdot \frac{\partial}{\partial \mathbf{r}_i} + \sum_{i=1}^{N} \mathbf{F}_i \cdot \frac{\partial}{\partial \mathbf{p}_i}. \qquad (3.10.34)$$

If we write $iL = iL_1 + iL_2$ with iL_1 and iL_2 defined in a manner analogous to eqn. (3.10.13), then it can be easily shown that the Trotter factorization of eqn. (3.10.22) yields the velocity Verlet algorithm of eqns. (3.8.7) and (3.8.9) because all of the terms in iL_1 commute with each other, as do all of the terms in iL_2. It should be noted that if a system is subject to a set of holonomic constraints imposed via the SHAKE and RATTLE algorithms, the symplectic and time-reversibility properties of the Trotter-factorized integrators are lost unless the iterative solutions for the Lagrange multipliers are iterated to *full* convergence.

Before concluding this section, one final point should be made. It is not necessary to grind out explicit finite difference equations by applying the operators in a Trotter factorization analytically. Note that the velocity Verlet algorithm can be expressed as a three-step procedure:

$$p(\Delta t/2) = p(0) + \frac{\Delta t}{2} F(x(0))$$

$$x(\Delta t) = x(0) + \frac{\Delta t}{m} p(\Delta t/2)$$

$$p(\Delta t) = p(\Delta t/2) + \frac{\Delta t}{2} F(x(\Delta t)). \qquad (3.10.35)$$

This three-step procedure can also be rewritten to resemble actual lines of computer code

$$p = p + 0.5 * \Delta t * F$$
$$x = x + \Delta t * p/m$$
Recalculate the force
$$p = p + 0.5 * \Delta t * F. \qquad (3.10.36)$$

Equations (3.10.35) and (3.10.36) could also be cast in terms of $v = p/m$ as in eqn. (3.10.32). The third line in (3.10.36) involves a call to some function or subroutine that updates the force from the new positions generated in the second line. When written this way, the specific instructions are: i) perform a momentum translation; ii) follow this by a position translation; iii) recalculate the force using the new position; iv) use the new force to perform a momentum translation. Note, however, that these are just the steps required by the operator factorization scheme of eqn. (3.10.23): The first operator that acts on the phase-space vector is $\exp[(\Delta t/2)F(x)\partial/\partial p]$, which produces the momentum translation; the next operator $\exp[\Delta t(p/m)\partial/\partial p]$ takes the output of the preceding step and performs the position translation; since this step changes the positions, the force must be recalculated; the last operator $\exp[(\Delta t/2)F(x)\partial/\partial p]$ produces the final momentum translation using the new force. The fact that instructions in computer code can be written directly from the operator factorization scheme, bypassing the lengthy algebra needed to derive explicit finite-difference equations, is an immensely powerful technique that we term the *direct translation* method (Martyna *et al.*, 1996). Because direct translation is possible, we can simply let a factorization of the classical propagator denote a particular integration algorithm; we will employ the direct translation technique in many of our subsequent discussions of numerical solvers.

3.11 Multiple time-scale integration

One of the most ubiquitous aspects of complex systems in classical mechanics is the presence of forces that generate motion with different time scales. Examples include long biological macromolecules such as proteins and other types of polymers. In fact, virtually any chemical system will span a wide range of time scales from very fast bond and bend vibrations to global conformational changes in macromolecules or slow diffusion/transport molecular liquids, to illustrate just a few cases. To make the discussion more concrete, consider a simple potential energy model commonly used for biological macromolecules:

$$U(\mathbf{r}_1,...,\mathbf{r}_N) = \sum_{\text{bonds}} \frac{1}{2} K_{\text{bond}} (r - r_0)^2 + \sum_{\text{bends}} \frac{1}{2} K_{\text{bend}} (\theta - \theta_0)^2$$

$$+ \sum_{\text{tors}} \sum_{n=0}^{6} A_n \left[1 + \cos(C_n \phi + \delta_n) \right]$$

$$+ \sum_{i,j \in nb} \left\{ 4\epsilon_{ij} \left[\left(\frac{\sigma_{ij}}{r_{ij}} \right)^{12} - \left(\frac{\sigma_{ij}}{r_{ij}} \right)^6 \right] + \frac{q_i q_j}{r_{ij}} \right\}. \qquad (3.11.1)$$

The first term is the energy for all covalently bonded pairs, which are treated as harmonic oscillators in the bond length r, each with their own force constant K_{bond}. The second term is the bend energy of all neighboring covalent bonds, and again the bending motion is treated as harmonic on the bend angle θ, each bend having a force constant K_{bend}. The third term is the conformational energy of dihedral angles ϕ, which generally involves multiple minima separated by energy barriers of various heights. The first three terms constitute the intramolecular energy due to bonding and connectivity. The last term describes the so-called *nonbonded* (nb) interactions, which include van der Waals forces between spheres of radius σ_i and σ_j ($\sigma_{ij} = (\sigma_i + \sigma_j)/2$) separated by a distance r_{ij} with well-depth ϵ_{ij}, expressed as a Lennard-Jones potential (Lennard-Jones, 1924), and Coulomb forces between particles with charges q_i and q_j separated by a distance r_{ij}. If the molecule is in a solvent such as water, then eqn. (3.11.1) also describes the solvent-solute and solvent-solvent interactions as well. The forces $\mathbf{F}_i = -\partial U / \partial \mathbf{r}_i$ derived from this potential will have large and rapidly varying components due to the intramolecular terms and smaller, slowly varying components due to long-range contributions to the nonbonded interactions. Moreover, the simple functional forms of the intramolecular terms renders the fast forces computationally inexpensive to evaluate while the slower forces, which involve sums over many pairs of particles, will be much more time-consuming to compute. On the time scale over which the fast forces vary naturally, the slow forces change very little. In the simple velocity Verlet scheme, one time step Δt is employed whose magnitude is limited by the fast forces, yet all force components must be computed at each step, including those that change very little over a time Δt. Ideally, it would be advantageous to develop a numerical solver capable of exploiting this separation of time scales for a gain in computational efficiency. Such an integrator should allow the slow forces to be recomputed less frequently than the fast forces, thereby saving the computational overhead lost by updating the slow forces at every step. The Liouville operator formalism allows this to be done in a rigorous manner, leading to a symplectic, time-reversible multiple time-scale solver.

We will show how the algorithm is developed using, once again, the example of a single particle in one dimension. Suppose the particle is subject to a force $F(x)$ that has two components $F_{\text{fast}}(x)$ and $F_{\text{slow}}(x)$. The equations of motion are

$$\dot{x} = \frac{p}{m}$$
$$\dot{p} = F_{\text{fast}}(x) + F_{\text{slow}}(x). \qquad (3.11.2)$$

Since the system is Hamiltonian, the equations of motion can be integrated using a symplectic solver. The Liouville operator is given by

$$iL = \frac{p}{m}\frac{\partial}{\partial x} + [F_{\text{fast}}(x) + F_{\text{slow}}(x)]\frac{\partial}{\partial p} \qquad (3.11.3)$$

and can be separated into pure kinetic and force components as was done in Section 3.10:

$$iL = iL_1 + iL_2$$

$$iL_1 = \frac{p}{m}\frac{\partial}{\partial x}$$

$$iL_2 = [F_{\text{fast}}(x) + F_{\text{slow}}(x)]\frac{\partial}{\partial p}. \qquad (3.11.4)$$

Using this separation in a Trotter factorization of the propagator would lead to the standard velocity Verlet algorithm. Consider instead separating the Liouville operator as follows:

$$iL = iL_{\text{fast}} + iL_{\text{slow}}$$

$$iL_{\text{fast}} = \frac{p}{m}\frac{\partial}{\partial x} + F_{\text{fast}}(x)\frac{\partial}{\partial p}$$

$$iL_{\text{slow}} = F_{\text{slow}}(x)\frac{\partial}{\partial p}. \qquad (3.11.5)$$

We now define a reference Hamiltonian system $\mathcal{H}_{\text{ref}}(x, p) = p^2/2m + U_{\text{fast}}(x)$, where $F_{\text{fast}}(x) = -dU_{\text{fast}}/dx$. The reference system obeys the equations of motion $\dot{x} = p/m$, $\dot{p} = F_{\text{fast}}(x)$ and has the associated single-time-step propagator $\exp(iL_{\text{fast}}\Delta t)$. The full propagator is then factorized by applying the Trotter scheme as follows:

$$\exp(iL\Delta t) = \exp\left(iL_{\text{slow}}\frac{\Delta t}{2}\right)\exp(iL_{\text{fast}}\Delta t)\exp\left(iL_{\text{slow}}\frac{\Delta t}{2}\right). \qquad (3.11.6)$$

This factorization leads to the *reference system propagator algorithm* (RESPA) (Tuckerman *et al.*, 1992). The idea behind the RESPA algorithm is that the time step Δt appearing in eqn. (3.11.6) is chosen according to the time scale of the slow forces. There are two ways to achieve this: Either the propagator $\exp(iL_{\text{fast}}\Delta t)$ is applied exactly analytically, or $\exp(iL_{\text{fast}}\Delta t)$ is further factorized with a smaller time step δt that is appropriate for the fast motion. We will discuss these two possibilities below.

First, suppose an analytical solution for the reference system is available. As a concrete example, consider a harmonic fast force $F_{\text{fast}}(x) = -m\omega^2 x$. Acting with the operators directly yields an algorithm of the form

$$x(\Delta t) = x(0)\cos\omega\Delta t + \frac{1}{\omega}\left[\frac{p(0)}{m} + \frac{\Delta t}{2m}F_{\text{slow}}(x(0))\right]\sin\omega\Delta t$$

$$p(\Delta t) = \left[p(0) + \frac{\Delta t}{2} F_{\text{slow}}(x(0)) \right] \cos \omega \Delta t - m\omega x(0) \sin \omega \Delta t$$

$$+ \frac{\Delta t}{2} F_{\text{slow}}(x(\Delta t)), \tag{3.11.7}$$

which can also be written as a step-wise set of instructions using the direct translation technique:

$$p = p + 0.5 * \mathrm{dt} * F_{\text{slow}}$$

$$x\_\text{temp} = x * \cos(\text{arg}) + p/(m * \omega) * \sin(\text{arg})$$

$$p\_\text{temp} = p * \cos(\text{arg}) - m * \omega * x * \sin(\text{arg})$$

$$x = x\_\text{temp}$$

$$p = p\_\text{temp}$$

Recalculate slow force

$$p = p + 0.5 * \mathrm{dt} * F_{\text{slow}}, \tag{3.11.8}$$

where $\text{arg} = \omega * \mathrm{dt}$. Generalizations of eqn. (3.11.8) for complex molecular systems described by potential energy models like eqn. (3.11.1) were recently presented by Janežič and coworkers (Janežič *et al.*, 2005). Similar schemes can be worked out for other analytically solvable systems.

When the reference system cannot be solved analytically, the RESPA concept can still be applied by introducing a second time step $\delta t = \Delta t/n$ and writing

$$\exp(iL_{\text{fast}}\Delta t) = \left[\exp\left(\frac{\delta t}{2} F_{\text{fast}} \frac{\partial}{\partial p} \right) \exp\left(\delta t \frac{p}{m} \frac{\partial}{\partial x} \right) \exp\left(\frac{\delta t}{2} F_{\text{fast}} \frac{\partial}{\partial p} \right) \right]^n. \tag{3.11.9}$$

Substitution of eqn. (3.11.9) into eqn. (3.11.6) yields a purely numerical RESPA propagator given by

$$\exp(iL\Delta t) = \exp\left(\frac{\Delta t}{2} F_{\text{slow}} \frac{\partial}{\partial p} \right)$$

$$\times \left[\exp\left(\frac{\delta t}{2} F_{\text{fast}} \frac{\partial}{\partial p} \right) \exp\left(\delta t \frac{p}{m} \frac{\partial}{\partial x} \right) \exp\left(\frac{\delta t}{2} F_{\text{fast}} \frac{\partial}{\partial p} \right) \right]^n$$

$$\times \exp\left(\frac{\Delta t}{2} F_{\text{slow}} \frac{\partial}{\partial p} \right). \tag{3.11.10}$$

In eqn. (3.11.10), two time steps appear. The large time step Δt is chosen according to the natural time scale of evolution of F_{slow}, while the small time step δt is chosen according to the natural time scale of F_{fast}. Translating eqn. (3.11.10) into a set of instructions yields the following pseudocode:

$$p = p + 0.5 * \Delta t * F_{\text{slow}}$$
for i = 1 to n
$$p = p + 0.5 * \delta t * F_{\text{fast}}$$
$$x = x + \delta t * p/m$$
Recalculate fast force
$$p = p + 0.5 * \delta t * F_{\text{fast}}$$
endfor
Recalculate slow force
$$p = p + 0.5 * \Delta t * F_{\text{slow}}. \qquad (3.11.11)$$

Note that the algorithm in eqn. (3.11.11) can be rewritten using a single time step δt as

$$F_{\text{now}} = F_{\text{fast}} + n * F_{\text{slow}}$$
for i = 1 to n
$$p = p + 0.5 * \delta t * F_{\text{now}}$$
$$x = x + \delta t * p/m$$
Recalculate fast force
$$F_{\text{now}} = F_{\text{fast}}$$
if(i = n) then
Recalculate slow force
$$F_{\text{now}} = F_{\text{fast}} + n * F_{\text{slow}}$$
endif
$$p = p + 0.5 * \delta t * F_{\text{now}}$$
endfor $(3.11.12)$

RESPA factorizations involving more than two time steps can be generated in the same manner. As an illustrative example, suppose the force $F(x)$ is composed of three contributions, $F(x) = F_{\text{fast}}(x) + F_{\text{intermed}}(x) + F_{\text{slow}}(x)$, with three different time scales. We can introduce three time steps δt, $\Delta t = n\delta t$, and $\Delta T = N\Delta t = nN\delta t$ and factorize the propagator as follows:

$$\exp(iL\Delta T) = \exp\left(\frac{\Delta T}{2} F_{\text{slow}} \frac{\partial}{\partial p}\right) \left\{ \exp\left(\frac{\Delta t}{2} F_{\text{intermed}} \frac{\partial}{\partial p}\right) \right.$$

$$\left[\exp\left(\frac{\delta t}{2} F_{\text{fast}} \frac{\partial}{\partial p}\right) \exp\left(\delta t \frac{p}{m} \frac{\partial}{\partial x}\right) \exp\left(\frac{\delta t}{2} F_{\text{fast}} \frac{\partial}{\partial p}\right) \right]^{n}$$

$$\left. \exp\left(\frac{\Delta t}{2} F_{\text{intermed}} \frac{\partial}{\partial p}\right) \right\}^{N} \exp\left(\frac{\Delta T}{2} F_{\text{slow}} \frac{\partial}{\partial p}\right). \qquad (3.11.13)$$

It is left as an exercise to the reader to translate eqn. (3.11.13) into a sequence of instructions in pseudocode. As we can see, an arbitrary number of RESPA levels can be generated for an arbitrary number of force components, each with different associated time scales.

In Appendix C, we discuss specific schemes for subdividing forces for force fields such as the simple fixed-charge model presented in eqn. (3.11.1). A scheme for subdividing forces in a more complex, polarizable model can be found in Margul and Tuckerman (2016).

3.11.1 The problem of resonances

Shortly after multiple time-step algorithms for molecular dynamics were introduced, it was determined (Schlick *et al.*, 1998; Sandu and Schlick, 1999; Ma *et al.*, 2003) that there is a limit on the size of the time step that can be associated with the slow forces in a given system. Furthermore, this limit is not necessarily set by the time scale on which these slow forces vary but, rather, by that of the fast forces, a result that seems counterintuitive. However, by performing a bit of mathematical analysis, this observation can be understood and rationalized (underscoring why one should not shy away from mathematics(!)). We will outline the analysis of a simple harmonic system that reveals the origin of resonances. Details of the derivation are left to the reader to be worked out in Problem 3.6.

Consider a simple harmonic system in one dimension with coordinate x, momentum p, and unit mass $(m = 1)$ for which the potential is given by

$$U(x) = \frac{1}{2}\left(\omega^2 + \Omega^2\right)x^2, \tag{3.11.14}$$

where ω and Ω are two frequencies with $\Omega \ll \omega$. Equation (3.11.14) is tantamount to taking an oscillator with a single frequency f and dividing it into two frequencies ω and Ω such that $\omega^2 + \Omega^2 = f^2$. The rationale for this somewhat artificial construct is to introduce a harmonic fast force $F_{\text{fast}}(x) = -\omega^2 x$ and a harmonic slow force $F_{\text{slow}}(x) = -\Omega^2 x$ so that a fully linear multiple time-scale integrator can be employed. In this case, we will use eqn. (3.11.7) with the given $F_{\text{slow}}(x)$; when this is done, the approximate evolution can be written in the form of a matrix equation:

$$\begin{pmatrix} x(\Delta t) \\ p(\Delta t) \end{pmatrix} = A(\omega, \Omega, \Delta t) \begin{pmatrix} x(0) \\ p(0) \end{pmatrix}. \tag{3.11.15}$$

Derivation of the precise form of this matrix is an objective in Problem 3.6. The two eigenvalues of $A(\omega, \Omega, \Delta t)$, denoted $\lambda_1(\omega, \Omega, \Delta t)$ and $\lambda_2(\omega, \Omega, \Delta t)$, are either complex conjugate pairs, in which case the motion is oscillatory, or they are both real, in which case the motion is hyperbolic and, therefore, unstable (see Problem 3.6). Which of the two conditions these eigenvalues obey depends on the value of Δt: If $\Delta t < \pi/\omega$, the eigenvalues are complex conjugate pairs, and if $\Delta t \geq \pi/\omega$, the motion is hyperbolic. The onset of unstable, hyperbolic motion occurs when $\Delta t = n\pi/\omega$, for $n = 1, 2, 3, \dots$. These are referred to as *resonant* time steps, with $n = 1$ corresponding to the first occurrence of such resonant behavior. This places a fundamental limit on the magnitude of Δt that can be employed, and while we might expect that this time step would be determined by the slow frequency Ω, we see that, in fact, the limit is set by the fast frequency ω. The higher the value of ω, the smaller Δt can be chosen, which ultimately limits the computational savings afforded by the algorithm, independent of how "slow" the slow force is.

At the end of this chapter, we explore the practical limit resonances place on the large time step in illustrative atomistic molecular dynamics simulation examples. Later, in Chapters 4 and 15, we will introduce a number of approaches for circumventing resonances, which permit the time step Δt to be chosen based on the variation in the slow force, thus allowing the full potential of multiple time-scale integrators to be realized.

3.12   Symplectic integration for quaternions

In Section 1.11, we showed that the rigid body equations of motion could be expressed in terms of quaternions (cf. eqns. (1.11.44) to (1.11.47)). Equation (1.11.40) relates the four-component quaternion vector $\mathbf{q} = (q_1, q_2, q_3, q_4)$ to the Euler angles. We also saw that when the rigid body equations of motion are expressed in terms of quaternions, we must impose an additional constraint that the vector $\mathbf{q}$ be a unit vector:

$$\sum_{i=1}^{4} q_i^2 = 1. \tag{3.12.1}$$

As was noted in Section 3.10, the iterations associated with the imposition of holonomic constraints affect the symplectic and time-reversibility properties of otherwise symplectic solvers. However, the simplicity of the constraint in eqn. (3.12.1) allows the quaternion equations of motion to be reformulated in such a way that the constraint is satisfied automatically by the dynamics, thereby allowing symplectic integrators to be developed using the Liouville operator. In this section, we will present such a scheme following the formulation of Miller *et al.* (2002).

Recall that the angular vector $\boldsymbol{\omega}$ was defined by $\boldsymbol{\omega} = (0, \omega_x, \omega_y, \omega_z)$, which has one trivial component that is always defined to be zero. Thus, there seems to be an unnatural asymmetry between the quaternion $\mathbf{q}$ and the angular $\boldsymbol{\omega}$. The idea of Miller *et al.* is to restore the symmetry between $\mathbf{q}$ and $\boldsymbol{\omega}$ by introducing a fourth angular velocity component ω_1 and redefining the angular velocity according to

$$\boldsymbol{\omega}^{(4)} = 2\mathbf{S}^{\mathsf{T}}(\mathbf{q})\dot{\mathbf{q}} \equiv (\omega_1, \omega_x, \omega_y, \omega_z). \tag{3.12.2}$$

The idea of extending phase spaces is a common trick in molecular dynamics, and we will examine numerous examples throughout the book. This new angular velocity component can be incorporated into the Lagrangian for one rigid body according to

$$\mathcal{L} = \frac{1}{2}\left[I_{11}\omega_1^2 + I_{xx}\omega_x^2 + I_{yy}\omega_y^2 + I_{zz}\omega_z^2\right] - U(\mathbf{q}). \tag{3.12.3}$$

Here I_{11} is an arbitrary moment of inertia associated with the new angular velocity component. We will see that the choice of I_{11} has no influence on the dynamics.

From eqns. (3.12.2) and (3.12.3), the momentum conjugate to the quaternion and corresponding Hamiltonian can be worked out to yield

$$\mathbf{p} = \frac{2}{|\mathbf{q}|^4}\mathbf{S}(\mathbf{q})\mathbf{D}^{-1}\boldsymbol{\omega}^{(4)}$$

$$\mathcal{H} = \frac{1}{8}\mathbf{p}^{\mathsf{T}}\mathbf{S}(\mathbf{q})\mathbf{D}^{-1}\mathbf{S}^{\mathsf{T}}(\mathbf{q})\mathbf{p} + U(\mathbf{q}), \tag{3.12.4}$$

where the matrix $\mathbf{D}$ is given by

$$\mathbf{D} = \begin{pmatrix} I_{11} & 0 & 0 & 0 \\ 0 & I_{xx} & 0 & 0 \\ 0 & 0 & I_{yy} & 0 \\ 0 & 0 & 0 & I_{zz} \end{pmatrix}. \tag{3.12.5}$$

Equation (3.12.4) leads to a slightly modified set of equations for the angular velocity components. Instead of those in eqn. (1.11.44), one obtains

$$\dot{\omega}_1 = \frac{\omega_1^2}{|\mathbf{q}|^2}$$

$$\dot{\omega}_x = \frac{\omega_1\omega_x}{|\mathbf{q}|^2} + \frac{\tau_x}{I_{xx}} + \frac{(I_{yy} - I_{zz})}{I_{xx}}\omega_y\omega_z$$

$$\dot{\omega}_y = \frac{\omega_1\omega_y}{|\mathbf{q}|^2} + \frac{\tau_y}{I_{yy}} + \frac{(I_{zz} - I_{xx})}{I_{yy}}\omega_z\omega_x$$

$$\dot{\omega}_z = \frac{\omega_1\omega_z}{|\mathbf{q}|^2} + \frac{\tau_z}{I_{zz}} + \frac{(I_{xx} - I_{yy})}{I_{zz}}\omega_x\omega_y. \tag{3.12.6}$$

If $\omega_1(0) = 0$ and eqn. (3.12.1) is satisfied by the initial quaternion $\mathbf{q}(0)$, then the new equations of motion will yield rigid body dynamics in which eqn. (3.12.1) is satisfied implicitly, thereby eliminating the need for an explicit constraint. This is accomplished through the extra terms in the angular velocity equations. Recall that implicit treatment of constraints can also be achieved via Gauss's principle of least constraint discussed in Section 1.10, which also leads to extra terms in the equations of motion. The difference here is that, unlike in Gauss's equations of motion, the extra terms are derived directly from a Hamiltonian and, therefore, the modified equations of motion are symplectic.

All that is needed now is an integrator for the new equations of motion. Miller *et al.* showed that the Hamiltonian could be decomposed into five contributions that are particularly convenient for the development of a symplectic solver. Defining four vectors $\mathbf{c}_1, .., \mathbf{c}_4$ as the columns of the matrix $\mathbf{S}(q)$, $\mathbf{c}_1 = (q_1, q_2, q_3, q_4)$, $\mathbf{c}_2 = (-q_2, q_1, q_4, -q_3)$, $\mathbf{c}_3 = (-q_3, -q_4, q_1, q_2)$, and $\mathbf{c}_4 = (-q_4, q_3, -q_2, q_1)$, the Hamiltonian can be written as

$$\mathcal{H}(\mathbf{q}, \mathbf{p}) = \sum_{k=1}^{4} h_k(\mathbf{q}, \mathbf{p}) + U(\mathbf{q})$$

$$h_k(\mathbf{q},\mathbf{p}) = \frac{1}{8I_k}(\mathbf{p}\cdot\mathbf{c}_k)^2, \tag{3.12.7}$$

where $I_1 = I_{11}$, $I_2 = I_{xx}$, $I_3 = I_{yy}$, and $I_4 = I_{zz}$. Note that if $\omega_1(0) = 0$, then $h_1(\mathbf{q},\mathbf{p}) = 0$ for all time. In terms of the Hamiltonian contributions $h_k(\mathbf{q},\mathbf{p})$, Liouville operator contributions $iL_k = \{...,h_k\}$ are introduced, with an additional Liouville operator $iL_5 = -(\partial U/\partial \mathbf{q})\cdot(\partial/\partial\mathbf{p})$, and a RESPA factorization scheme is introduced for the propagator

$$e^{iL\Delta t} \approx e^{iL_5\Delta t/2}$$

$$\times \left[e^{iL_4\delta t/2} e^{iL_3\delta t/2} e^{iL_2\delta t} e^{iL_3\delta t/2} e^{iL_4\delta t/2} \right]^n$$

$$\times e^{iL_5\Delta t/2}. \tag{3.12.8}$$

Note that since $h_1(\mathbf{q},\mathbf{p}) = 0$, the operator iL_1 does not appear in the integrator.

What is particularly convenient about this decomposition is that the operators $\exp(iL_k\delta t/2)$ for $k = 2, 3, 4$, can be applied analytically according to

$$e^{iL_k t}\mathbf{q} = \mathbf{q}\cos(\zeta_k t) + \sin(\zeta_k t)\mathbf{c}_k$$

$$e^{iL_k t}\mathbf{p} = \mathbf{p}\cos(\zeta_k t) + \sin(\zeta_k t)\mathbf{d}_k, \tag{3.12.9}$$

where

$$\zeta_k = \frac{1}{4I_k}\mathbf{p}\cdot\mathbf{c}_k, \tag{3.12.10}$$

and $\mathbf{d}_k$ is defined analogously to $\mathbf{c}_k$ but with the components of $\mathbf{p}$ replacing the components of $\mathbf{q}$. The action of $\exp(iL_5\Delta t)$ is just a translation $\mathbf{p} \leftarrow \mathbf{p} + (\Delta t/2)\mathbf{F}$, where $\mathbf{F} = -\partial U/\partial\mathbf{q}$. Miller *et al.* present several numerical examples that exhibit the performance of eqn. (3.12.8) on realistic systems, and the interested reader is referred to the aforementioned paper (Miller *et al.*, 2002) for more details.

3.13 Exactly conserved time-step dependent Hamiltonians

As already noted, the velocity Verlet algorithm is an example of a symplectic algorithm or *symplectic map*, the latter indicating that the algorithm maps the initial phase-space point x_0 into $x_{\Delta t}$ without destroying the symplectic property of classical mechanics. Although numerical solvers do not exactly conserve the Hamiltonian $\mathcal{H}(x)$, a symplectic solver has the important property that there exists a Hamiltonian $\tilde{\mathcal{H}}(x,\Delta t)$ such that, along a trajectory, $\tilde{\mathcal{H}}(x,\Delta t)$ remains *close* to the true Hamiltonian and is *exactly* conserved by the map. By close, we mean that $\tilde{\mathcal{H}}(x,\Delta t)$ approaches the true Hamiltonian $\mathcal{H}(x)$ as $\Delta t \to 0$. Because the auxiliary Hamiltonian $\tilde{\mathcal{H}}(x,\Delta t)$ is a close approximation to the true Hamiltonian, it is referred to as a "shadow" Hamiltonian (Yoshida, 1990; Toxvaerd, 1994; Gans and Shalloway, 2000; Skeel and Hardy, 2001). The existence of $\tilde{\mathcal{H}}(x,\Delta t)$ ensures the error in a symplectic map will be bounded. After presenting an illustrative example of a shadow Hamiltonian, we will indicate how to prove the existence of $\tilde{\mathcal{H}}(x, \Delta t)$.

The existence of a shadow Hamiltonian does not mean that this Hamiltonian can be constructed exactly for a general system. In fact, the general form of the shadow Hamiltonian is not known. Skeel and coworkers have described how approximate shadow Hamiltonians can be constructed practically (Skeel and Hardy, 2001) and have provided formulas for shadow Hamiltonians up to 24th order in the time step (Engle *et al.*, 2005). The one example for which the shadow Hamiltonian is known is, not surprisingly, the harmonic oscillator. Recall that the Hamiltonian for a harmonic oscillator of mass m and frequency ω is

$$\mathcal{H}(x, p) = \frac{p^2}{2m} + \frac{1}{2}m\omega^2 x^2. \tag{3.13.1}$$

If the equations of motion $\dot{x} = p/m$ $\dot{p} = -m\omega^2 x$ are integrated via the velocity Verlet algorithm

$$x(\Delta t) = x(0) + \Delta t \frac{p(0)}{m} - \frac{1}{2}\Delta t^2 \omega^2 x(0)$$

$$p(\Delta t) = p(0) - \frac{m\omega^2 \Delta t}{2}\left[x(0) + x(\Delta t)\right], \tag{3.13.2}$$

then it can be shown that the Hamiltonian

$$\tilde{\mathcal{H}}(x, p; \Delta t) = \frac{p^2}{2m(1 - \omega^2 \Delta t^2/4)} + \frac{1}{2}m\omega^2 x^2 \tag{3.13.3}$$

is exactly preserved by eqns. (3.13.2). Of course, the form of the shadow Hamiltonian will depend on the particular symplectic solver used to integrate the equations of motion. A phase-space plot of $\tilde{\mathcal{H}}$ vs. that of $\mathcal{H}$ is provided in Fig. 3.4. In this case, the eccentricity of the ellipse increases as the time step increases. The curves in Fig. 3.4 are exaggerated for illustrative purposes. For any reasonable (small) time step, the difference between the true phase space and that of the shadow Hamiltonian would be almost indistinguishable. As eqn. (3.13.3) indicates, if $\Delta t = 2/\omega$, $\tilde{\mathcal{H}}$ becomes ill-defined, and for $\Delta t > 2/\omega$, the trajectories are no longer bounded. Thus, the existence of $\tilde{\mathcal{H}}$ can only guarantee long-time stability for small Δt.

Although it is possible to envision developing novel simulation techniques based on a knowledge of $\tilde{\mathcal{H}}$ (Izaguirre and Hampton, 2004), the mere existence of $\tilde{\mathcal{H}}$ is sufficient to guarantee that the error in a symplectic map is bounded. That is, given that we have generated a trajectory $\tilde{x}_{n\Delta t}$, $n = 0, 1, 2, \ldots$ using a symplectic integrator, if we then evaluate $\mathcal{H}(\tilde{x}_{n\Delta t})$ at each point along the trajectory, it should not drift away from the true conserved value of $\mathcal{H}(x_t)$ evaluated along the exact (but, generally, unknown) trajectory x_t. Note, this does not mean that the numerical and true trajectories will follow each other. It simply means that $\tilde{x}_{n\Delta t}$ will remain on a constant-energy hypersurface that is close to the true constant-energy hypersurface. This is an important fact in developing molecular dynamics codes. If one uses a symplectic integrator and finds that the total energy exhibits a dramatic drift, the integrator cannot be blamed, and one should search for other causes!

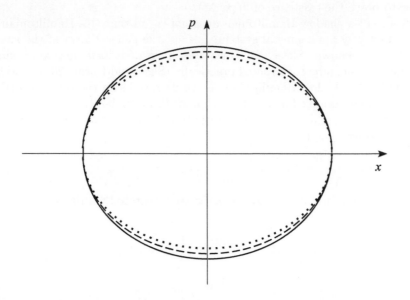

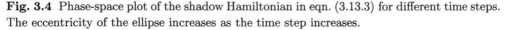

Fig. 3.4 Phase-space plot of the shadow Hamiltonian in eqn. (3.13.3) for different time steps. The eccentricity of the ellipse increases as the time step increases.

In order to understand why the shadow Hamiltonian exists, let us consider the Trotter factorization in eqn. (3.10.22). The factorization scheme is not an exact representation of the propagator $\exp(iL\Delta t)$; however, a formally exact relation

$$\exp\left[iL_2\frac{\Delta t}{2}\right]\exp\left[iL_1\Delta t\right]\exp\left[iL_2\frac{\Delta t}{2}\right] = \exp\left[\Delta t\left(iL + \sum_{k=1}^{\infty}\Delta t^{2k}C_k\right)\right], \quad (3.13.4)$$

which is known as the Baker-Campbell-Hausdorff formula, connects the two propagators (see, for example, Yoshida, 1990). Here, the operators C_k are nested commutators of the operators iL_1 and iL_2. For example, the operator C_1 is

$$C_1 = -\frac{1}{24}\left[iL_2 + 2iL_1, [iL_2, iL_1]\right]. \quad (3.13.5)$$

Now, an important property of the Liouville operator for Hamiltonian systems is that such commutators as those in eqn. (3.13.5) yield new Liouville operators that correspond to Hamiltonians derived from analogous expressions involving the Poisson bracket. Consider, for example, the simple commutator $[iL_1, iL_2] \equiv -iL_3$. It is possible to show that iL_3 is derived from the Hamiltonians $\mathcal{H}_1(\mathbf{x})$ and $\mathcal{H}_2(\mathbf{x})$, which define iL_1 and iL_2, respectively, via $iL_3 = \{..., \mathcal{H}_3\}$, where $\mathcal{H}_3(\mathbf{x}) = \{\mathcal{H}_1(\mathbf{x}), \mathcal{H}_2(\mathbf{x})\}$. The proof of this is straightforward and relies on an important identity satisfied by the Poisson bracket known as the *Jacobi identity*: If $P(\mathbf{x})$, $Q(\mathbf{x})$, and $R(\mathbf{x})$ are three functions on the phase space, then the Jacobi identity states

$$\{P, \{Q, R\}\} + \{R, \{P, Q\}\} + \{Q, \{R, P\}\} = 0. \quad (3.13.6)$$

Note that the second and third terms are generated from the first by moving the functions around in a cyclic manner. Thus, consider the action of $[iL_1, iL_2]$ on an arbitrary phase-space function $F(\mathrm{x})$. Since $iL_1 = \{..., \mathcal{H}_1(\mathrm{x})\}$ and $iL_2 = \{..., \mathcal{H}_2(\mathrm{x})\}$, we have

$$[iL_1, iL_2]F(\mathrm{x}) = \{\{F(\mathrm{x}), \mathcal{H}_2(\mathrm{x})\}, \mathcal{H}_1(\mathrm{x})\} - \{\{F(\mathrm{x}), \mathcal{H}_1(\mathrm{x})\}, \mathcal{H}_2(\mathrm{x})\}. \qquad (3.13.7)$$

From the Jacobi identity, it follows that

$$\{\{F(\mathrm{x}), \mathcal{H}_2(\mathrm{x})\}, \mathcal{H}_1(\mathrm{x})\} =$$

$$-\{\{\mathcal{H}_1(\mathrm{x}), F(\mathrm{x})\}, \mathcal{H}_2(\mathrm{x})\} - \{\{\mathcal{H}_2(\mathrm{x}), \mathcal{H}_1(\mathrm{x})\}, F(\mathrm{x})\}. \qquad (3.13.8)$$

Substituting eqn. (3.13.8) into eqn. (3.13.7) yields, after some algebra,

$$[iL_1, iL_2]F(\mathrm{x}) = -\{F(\mathrm{x}), \{\mathcal{H}_1(\mathrm{x}), \mathcal{H}_2(\mathrm{x})\}\} = -\{F(\mathrm{x}), \mathcal{H}_3(\mathrm{x})\}, \qquad (3.13.9)$$

from which we see that $iL_3 = \{..., \mathcal{H}_3(\mathrm{x})\}$.

A similar analysis can be carried out for each of the terms C_k in eqn. (3.13.4). Thus, for example, the operator C_1 corresponds to a Hamiltonian $\tilde{\mathcal{H}}_1(\mathrm{x})$ given by

$$\tilde{\mathcal{H}}_1(\mathrm{x}) = \frac{1}{24}\{\mathcal{H}_2 + 2\mathcal{H}_1, \{\mathcal{H}_2, \mathcal{H}_1\}\}. \qquad (3.13.10)$$

Consequently, each operator C_k corresponds to a Hamiltonian $\tilde{\mathcal{H}}_k(\mathrm{x})$, and it follows that the operator $iL + \sum_{k=1}^{\infty} \Delta t^{2k} C_k$ is generated by a Hamiltonian $\tilde{\mathcal{H}}(\mathrm{x}; \Delta t)$ of the form

$$\tilde{\mathcal{H}}(\mathrm{x}; \Delta t) = \mathcal{H}(\mathrm{x}) + \sum_{k=1}^{\infty} \Delta t^{2k} \tilde{\mathcal{H}}_k(\mathrm{x}). \qquad (3.13.11)$$

This Hamiltonian, which appears as a power series in Δt, is exactly conserved by the factorized operator appearing on the left side of eqn. (3.13.4). Note that $\tilde{\mathcal{H}}(\mathrm{x}; \Delta t) \to \mathcal{H}(\mathrm{x})$ as $\Delta t \to 0$. The existence of $\tilde{\mathcal{H}}(\mathrm{x}; \Delta t)$ guarantees the long-time stability of trajectories generated by the factorized propagator provided Δt is small enough that $\tilde{\mathcal{H}}$ and $\mathcal{H}$ are not that different, as the example of the harmonic oscillator above makes clear. Thus, care must be exercised in the choice of Δt, as the radius of convergence of eqn. (3.13.11) is generally unknown.

3.14 Illustrative examples of molecular dynamics calculations

In this section, we present a few illustrative examples of molecular dynamics calculations (in the microcanonical ensemble) employing symplectic numerical integration algorithms. We will focus primarily on investigating the properties of the numerical solvers, including accuracy and long-time stability, rather than on the direct calculation of observables (we will begin discussing observables in the next chapter). Throughout the section, energy conservation will be measured via the quantity

$$\Delta E(\delta t, \Delta t, \Delta T, ...) = \frac{1}{N_{\text{step}}} \sum_{k=1}^{N_{\text{step}}} \left| \frac{E_k(\delta t, \Delta t, \Delta T, ...) - E(0)}{E(0)} \right|, \qquad (3.14.1)$$

where the quantity $\Delta E(\delta t, \Delta t, \Delta T, ...)$ depends on however many time steps are employed. N_{step} is the total number of complete time steps taken (a "complete" time step is defined as an application of the full factorized classical propagator), $E_k(\delta t, \Delta t, \Delta T, ...)$ is the energy obtained at the kth step, and $E(0)$ is the initial energy. Equation (3.14.1) measures the average absolute relative deviation of the energy from its initial value (which determines the energy of the ensemble). Thus, it is a stringent measure of energy conservation that is sensitive to drifts in the total energy over time.

3.14.1 The harmonic oscillator

The phase space of a harmonic oscillator with frequency ω and mass m was shown in Fig. 1.3. Figure 3.5 (left) shows the comparison between the numerical solution of the equations of motion $\dot{x} = p/m$, $\dot{p} = -m\omega^2 x$ using the velocity Verlet algorithm analytical solutions. Here, we choose $\omega = 1$, $m = 1$, $\Delta t = 0.01$, $x(0) = 0$, and $p(0) = 1$. The figure shows that, over the few oscillation periods displayed, the numerical trajectory follows the analytical trajectory nearly perfectly. However, if the difference $|x_{\text{num}}(t) - x_{\text{analyt}}(t)|$ between the numerical and analytical solutions for the position is plotted over many periods, it can be seen that the solutions eventually diverge over time (Fig. 3.5 (middle)) (the error cannot grow indefinitely since the motion is bounded). Nevertheless, the numerical trajectory conserves energy to within approximately 10^{-5} as measured by eqn. (3.14.1), and shown in Fig. 3.5(right). It is also instructive to consider the time step dependence of ΔE depicted in Fig. 3.6. The figure shows $\log(\Delta E)$ vs. Δt and demonstrates that the time step dependence is a line with slope 2. This result confirms the fact that the global error in a long trajectory is

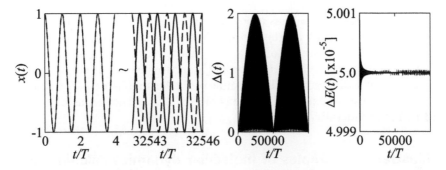

Fig. 3.5 (*Left*) Numerical (solid line) and analytical (dashed line) solutions for a harmonic oscillator of unit mass and frequency $\omega = 2$. The solutions are shown for $0 \leq t/T \leq 4$, where T is the period, and for $32542 \leq t/T \leq 32546$. (*Middle*) Deviation $\Delta(t) \equiv |x_{\text{num}}(t) - x_{\text{analyt}}(t)|$. (*Right*) Energy conservation measure over time as defined by eqn. (3.14.1)

proportional to Δt^2 as expected. Note that if a fourth-order integration scheme had been employed, then a similar plot would be expected to yield a line of slope 4.

Next, suppose we express the harmonic force $F(x) = -m\omega^2 x$ as

$$F(x) = -\lambda m\omega^2 x - (1-\lambda)m\omega^2 x = F_\lambda(x) + F_{1-\lambda}(x). \qquad (3.14.2)$$

It is clear that if λ is chosen very close to 1, then $F_\lambda(x)$ will generate motion on a time scale much faster than $F_{1-\lambda}(x)$. Thus, we have a simple example of a multiple time-scale problem to which the RESPA algorithm can be applied with $F_{\text{fast}}(x) = F_\lambda(x)$ and $F_{\text{slow}}(x) = F_{1-\lambda}(x)$. Note that this examples only serves to illustrate the use of the RESPA method; it is *not* recommended to separate harmonic forces in this manner! For the choice $\lambda = 0.9$, Fig. 3.6 shows how the energy conservation for fixed δt varies as Δt is increased. For comparison, the pure velocity Verlet result is also shown. Figure 3.6 demonstrates that the RESPA method yields a second-order integrator with significantly better energy conservation than the single time step case. A similar plot for $\lambda = 0.99$ is also shown in Fig. 3.6. These examples illustrate the fact that the RESPA method becomes more effective as the separation in time scales increases.

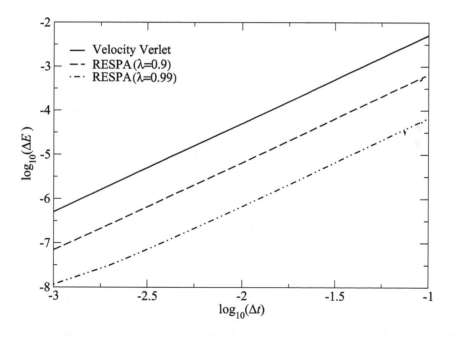

Fig. 3.6 Logarithm of the energy conservation measure in eqn. (3.14.1) vs. logarithm of the time step for a harmonic oscillator with $m = 1$, $\omega = 2$ using the velocity Verlet algorithm (solid line), RESPA with $\lambda = 0.9$ and a fixed small time step of $\delta t = 10^{-4}$ and variable large time step (dashed line), and RESPA with $\lambda = 0.99$ and the same fixed small time step (dotted-dashed line).

3.14.2 The Lennard-Jones fluid

We next consider a system of N identical particles interacting via a pair-wise additive potential of the form

$$U(\mathbf{r}_1, ..., \mathbf{r}_N) = \sum_{i=1}^{N-1} \sum_{j=i+1}^{N} 4\epsilon \left[\left(\frac{\sigma}{|\mathbf{r}_i - \mathbf{r}_j|} \right)^{12} - \left(\frac{\sigma}{|\mathbf{r}_i - \mathbf{r}_j|} \right)^{6} \right]. \qquad (3.14.3)$$

This potential, known as the Lennard-Jones potential (Lennard-Jones, 1924) is often used to describe the Van der Waals forces between simple rare-gas atom systems as well as in more complex systems. The numerical integration of Hamilton's equations $\dot{\mathbf{r}}_i = \mathbf{p}_i/m$, $\dot{\mathbf{p}}_i = -\nabla_i U$ requires both the specification of initial conditions, which was discussed in Section 3.8.3, as well as boundary conditions on the simulation cell. In this case, periodic boundary conditions are employed as a means of reducing the influence of the walls of the box. When periodic boundary conditions are employed, a particle that leaves the box through a particular face reenters the system at the same point of the face directly opposite. The correct handling of periodic boundary conditions within the force calculation is described in Appendix C. Numerical calculations in periodic boxes rarely make use of the Lennard-Jones potential in the form given in eqn. (3.14.3) but rather exploit the short-range nature of the function $u(r) = 4\epsilon[(\sigma/r)^{12} - (\sigma/r)^6]$ by introducing a truncated interaction $\tilde{u}(r) = u(r)S(r)$, where $S(r)$ is a *switching function* that smoothly truncates the Lennard-Jones potential to 0 at a value $r = r_c$, where r_c is typically chosen to be between 2.5σ and 3.0σ. A useful choice for $S(r)$ is

$$S(r) = \begin{cases} 1 & r < r_c - \lambda \\ 1 + R^2(2R - 3) & r_c - \lambda < r \leq r_c \\ 0 & r > r_c \end{cases} \qquad (3.14.4)$$

(Watanabe and Reinhardt, 1990), where $R = [r - (r_c - \lambda)]/\lambda$. The parameter λ is called the *healing length* of the switching function. This switching function has two continuous derivatives, thus ensuring that the forces, which require $\tilde{u}'(r) = u'(r)S(r) + u(r)S'(r)$, are continuous.

It is important to note several crucial differences between a simple system such as the harmonic oscillator and a highly complex system such as the Lennard-Jones (LJ) fluid. First, the LJ fluid is an example of a system that is highly *chaotic*. A key characteristic of a chaotic system is known as *sensitive dependence on initial conditions*. That is, two trajectories in phase space with only a minute difference in their initial conditions will diverge exponentially in time. In order to illustrate this fact, consider two trajectories for the Lennard-Jones potential whose initial conditions differ in just a single particle. In one of the trajectories, the initial position of a randomly chosen particle is different from that in the other by only $10^{-10}\%$. In this simulation, the Lennard-Jones parameters corresponding to fluid argon ($\epsilon = 119.8$ Kelvin, $\sigma = 3.405$ Å, $m = 39.948$ a.u.). Each system contains $N = 864$ particles in a cubic box of volume $V = 42811.0867$ Å^3, corresponding to a density of 1.34 g/cm^3. The equations of motion are integrated with a time step of 5.0 fs using a cutoff of $r_c = 2.5\sigma$. The value of the total Hamiltonian is approximately 0.65 Hartrees or

7.5×10^{-4} Hartrees/atom. The average temperature over each run is approximately 227 K. The thermodynamic parameters such as temperature and density, as well as the time step, can also be expressed in terms of the so-called Lennard-Jones *reduced units*, in which combinations of m, σ, and ϵ are multiplied by quantities such as number density ($\rho = N/V$), temperature and time step to yield dimensionless versions of these. Thus, the reduced density, denoted ρ^*, is given in terms of ρ by $\rho^* = \rho\sigma^3$. The reduced temperature, T^*, is $T^* = T/\epsilon$, and the reduced time step $\Delta t^* = \Delta t\sqrt{\epsilon/m\sigma^2}$. For the fluid argon parameters above, we find $\rho = 0.02\text{Å}^{-3}$ and $\rho^* = 0.8$, $T^* = 1.9$, and $\Delta t^* = 2.3 \times 10^{-3}$.

Figure 3.7 shows the y position of this particle (particle 1 in this case) in both trajectories as functions of time when integrated numerically using the velocity Verlet algorithm. Note that the trajectories follow each other closely for an initial period but then begin to diverge. Soon, the trajectories do not resemble each other at all. The implication of this exercise is that a single dynamical trajectory conveys very little information because a slight change in initial conditions changes the trajectory completely. In the spirit of the ensemble concept, dynamical observables do not rely on single trajectories. Rather, as we will explore further in Chapter 13, observables require averaging over an ensemble of trajectories each with different initial conditions. Thus, no single initial condition can be given special significance. Despite the sensitive dependence on initial conditions of the LJ fluid, Fig. 3.7 shows that the energy is well conserved over a single trajectory. The average value of the energy conservation based on eqn. (3.14.1), which is around 10^{-4}, is typical in molecular dynamics simulations.

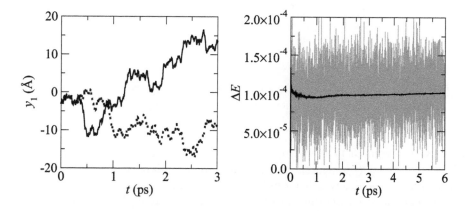

Fig. 3.7 (*Left*) The y coordinate for particle 1 as a function of time for two identical Lennard-Jones systems whose initial conditions differ by only $10^{-10}\%$ in the position of a single particle. (*Right*) The energy conservation as measured by eqn. (3.14.1) for one of the two systems. The light-grey background shows the instantaneous fluctuations of the summand in eqn. (3.14.1).

3.14.3 A realistic example: The alanine dipeptide in water

As a realistic example of a molecular dynamics calculation, we consider the alanine dipeptide in water. An isolated alanine dipeptide is depicted in Fig. 3.8. The solvated alanine dipeptide is one of the most studied simple peptide systems, both theoretically and experimentally, as it provides important clues about the conformational variability and thermodynamics of more complex polypeptides and biological macromolecules. At the same time, the system is simple enough that its conformational equilibria can be mapped out in great detail, which is important for benchmarking new models for the interactions. Figure 1.11 shows a schematic of the alanine dipeptide, which has been capped at both ends by methyl groups. In the present simulation, a force field

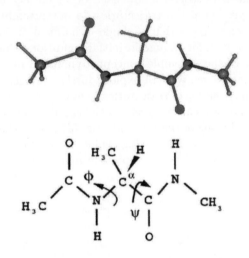

Fig. 3.8 (*Top*) Ball-and-stick model of the alanine dipeptide. Color code: red = O, blue = N, grey = C. (*Bottom*) Schematic representation of the alanine dipeptide, showing the angles ϕ and ψ.

of the type given in eqn. (3.11.1) is employed with the parameters corresponding to the CHARMM22 model (MacKerell *et al.*, 1998). In addition, water is treated as a completely rigid molecule, which requires three internal distance constraints (see Problem 3.3). The three constraints in each molecule are treated with the explicit matrix version of the constraint algorithms described in Section 3.9.

In this simulation, the dipeptide is solvated in a cubic box of length 25.64 Å on a side containing 558 water molecules for a total of 1696 atoms. A simulation of 1.5 ns is run using the RESPA integrator of Section 3.11 with a small time step of 1.0 fs and a large time step of 6.0 fs The reference system includes all intramolecular bonding and bending forces of the solute. Figure 3.9 shows energy conservation of eqn. (3.14.1) and its instantaneous fluctuations as well as the cumulative average and instantaneous temperature fluctuations produced by the total kinetic energy divided by $(3/2)Nk$. It can be seen that the temperature exhibits regular fluctuations, leading to a well-defined

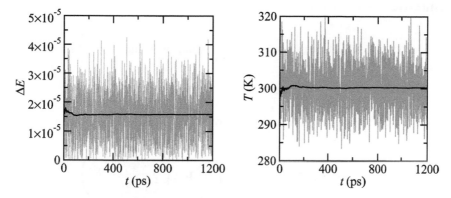

Fig. 3.9 (*Left*) Instantaneous and cumulative energy conservation measures for the alanine dipeptide in water. (*Right*) Instantaneous and cumulative temperature.

thermodynamic temperature of 300 K. The figure also shows that the energy is well conserved over this run. The CPU time needed for one step of molecular dynamics using the RESPA integrator is nearly the same as the time that would be required for a single time step method such as velocity Verlet with a 1.0 fs time step because of the low computational overhead of the bonding and bending forces compared to that of a full force calculation. Hence, the gain in efficiency using the RESPA solver is very close to a factor of 6. In order to examine the conformational changes taking place in the dipeptide over the run, we plot the dihedral angles ϕ and ψ in Fig. 3.10 as functions of time. Different values of these angles correspond to different stable conformations of the alanine dipeptide. We see from the figure that the motion of these angles is characterized by local fluctuations about these stable conformations with occasional abrupt transitions to a different stable conformation. The fact that such transitions are rare indicates that the full conformational space is not adequately sampled on the 1.5 ns time scale of the run. The problem of enhancing conformational sampling and treating rare events in molecular dynamics will be treated in detail in Chapter 8.

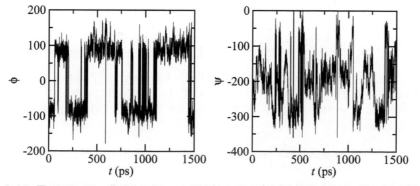

Fig. 3.10 Trajectories of the angles ϕ (*left*) and ψ (*right*) over a 1.5 ns run of the alanine dipeptide in water.

3.15   Problems

3.1. Consider the standard Hamiltonian for a system of N identical particles

$$\mathcal{H} = \sum_{i=1}^{N} \frac{\mathbf{p}_i^2}{2m} + U(\mathbf{r}_1, ..., \mathbf{r}_N).$$

a. Show that the microcanonical partition function can be expressed in the form

$$\Omega(N, V, E) = M_N \int \mathrm{d}E' \int \mathrm{d}^N\mathbf{p}\, \delta \left(\sum_{i=1}^{N} \frac{\mathbf{p}_i^2}{2m} - E' \right)$$

$$\times \int_{D(V)} \mathrm{d}^N\mathbf{r}\, \delta \left(U(\mathbf{r}_1, ..., \mathbf{r}_N) - E + E' \right),$$

which provides a way to separate the kinetic and potential contributions to the partition function.

b. Based on the result of part a, show that the partition function can, therefore, be expressed as

$$\Omega(N, V, E) = \frac{E_0}{N! \, \Gamma\left(\frac{3N}{2}\right)} \left[\left(\frac{2\pi m}{h^2} \right)^{3/2} \right]^N$$

$$\times \int_{D(V)} \mathrm{d}^N\mathbf{r}\, [E - U(\mathbf{r}_1, ..., \mathbf{r}_N)]^{3N/2-1} \, \theta \left(E - U(\mathbf{r}_1, ..., \mathbf{r}_N) \right),$$

where $\theta(x)$ is the Heaviside step function.

*3.2. Figure 1.7 illustrates the harmonic polymer model introduced in Section 1.7. If we take the equilibrium bond lengths all to be zero, then the potential energy takes the simple form

$$U(\mathbf{r}_1, ..., \mathbf{r}_N) = \frac{1}{2}m\omega^2 \sum_{k=0}^{N} (\mathbf{r}_k - \mathbf{r}_{k+1})^2,$$

where m is the mass of each particle, and ω is the frequency of the harmonic couplings. Let $\mathbf{r}$ and $\mathbf{r}'$ be the positions of the endpoints, with the definition that $\mathbf{r}_0 \equiv \mathbf{r}$ and $\mathbf{r}_{N+1} \equiv \mathbf{r}'$. Consider making the following change of coordinates:

$$\mathbf{r}_k = \mathbf{u}_k + \frac{k}{k+1}\mathbf{r}_{k+1} + \frac{1}{k+1}\mathbf{r}, \qquad k = 1, ..., N.$$

Using this change of coordinates, calculate the microcanonical partition function $\Omega(N, V, E)$ for this system. Assume the polymer to be in a cubic box of volume V.

Hint: Note that the transformation is defined recursively. How should you start the recursion? It might help to investigate how it works for a small number of particles, e.g. two or three.

3.3. A water molecule H_2O is subject to an external potential. Let the positions of the three atoms be denoted $\mathbf{r}_O$, $\mathbf{r}_{H_1}$, $\mathbf{r}_{H_2}$, so that the forces on the three atoms can be denoted $\mathbf{F}_O$, $\mathbf{F}_{H_1}$, and $\mathbf{F}_{H_2}$. Consider treating the molecule as completely rigid, with internal bond lengths d_{OH} and d_{HH}, so that the constraints are:

$$|\mathbf{r}_O - \mathbf{r}_{H_1}|^2 - d_{OH}^2 = 0$$
$$|\mathbf{r}_O - \mathbf{r}_{H_2}|^2 - d_{OH}^2 = 0$$
$$|\mathbf{r}_{H_1} - \mathbf{r}_{H_2}|^2 - d_{HH}^2 = 0.$$

a. Derive the constrained equations of motion for the three atoms in the molecule in terms of undetermined Lagrange multipliers.

b. Assume that the equations of motion are integrated numerically using the velocity Verlet algorithm. Derive a 3×3 matrix equation that can be used to solve for the multipliers in the SHAKE step.

c. Devise an iterative procedure for solving your matrix equation based on a linearization of the equation.

d. Derive a 3×3 matrix equation that can be used to solve for the multipliers in RATTLE step. Show that this equation can be solved analytically without iteration.

3.4. A one-dimensional harmonic oscillator of mass m and frequency ω is described by the Hamiltonian

$$\mathcal{H} = \frac{p^2}{2m} + \frac{1}{2}m\omega^2 x^2.$$

For the phase-space function $a(x, p) = p^2$, prove that the microcanonical ensemble average $\langle a \rangle$ and the time average

$$\bar{a} = \frac{1}{T}\int_0^T dt\, a(x(t), p(t))$$

are equal. Here, $T = 2\pi/\omega$ is one period of the motion.

3.5. Consider a single particle moving in one dimension with a Hamiltonian of the form $\mathcal{H} = p^2/2m + U(x)$, and consider factorizing the propagator $\exp(iL\Delta t)$ according to the following Trotter scheme:

$$\exp(iL\Delta t) \approx \exp\left(\frac{\Delta t}{2}\frac{p}{m}\frac{\partial}{\partial x}\right)\exp\left(\Delta t F(x)\frac{\partial}{\partial p}\right)\exp\left(\frac{\Delta t}{2}\frac{p}{m}\frac{\partial}{\partial x}\right).$$

a. Derive the finite-difference equations determining $x(\Delta t)$ and $p(\Delta t)$ for this factorization. This algorithm is known as the *position* Verlet algorithm (Tuckerman *et al.*, 1992).

b. From the matrix of partial derivatives

$$J = \begin{pmatrix} \frac{\partial x(\Delta t)}{\partial x(0)} & \frac{\partial x(\Delta t)}{\partial p(0)} \\ \frac{\partial p(\Delta t)}{\partial x(0)} & \frac{\partial p(\Delta t)}{\partial p(0)} \end{pmatrix},$$

show that the algorithm is measure-preserving and symplectic.

c. If $U(x) = m\omega^2 x^2/2$, find the exactly conserved Hamiltonian.

Hint: Assume the exactly conserved Hamiltonian takes the form

$$\tilde{\mathcal{H}}(x, p; \Delta t) = a(\Delta t)p^2 + b(\Delta t)x^2,$$

and determine a specific choice for the unknown coefficients a and b.

d. Write a program that implements this algorithm and verify that it exactly conserves your Hamiltonian for part c and that the true Hamiltonian remains stable for a suitably chosen small time step.

3.6. This problem illustrates the phenomenon of resonances in multiple time-stepping algorithms. Consider a single particle moving in one dimension that is subject to a potential of the form

$$U(x) = \frac{1}{2}m\left(\omega^2 + \Omega^2\right)x^2,$$

where $\Omega \ll \omega$. The forces associated with this potential have two time scales, $F_{\text{fast}} = -m\omega^2 x$ and $F_{\text{slow}} = -m\Omega^2 x$. Consider integrating this system for one time step Δt using the propagator factorization scheme in eqn. (3.11.6), where iL_{fast} is the full Liouville operator for the fast oscillator.

a. The action of the operator $\exp(iL_{\text{fast}}\Delta t)$ on the phase-space vector (x, p) can be evaluated analytically as in eqn. (3.11.8). Using this fact, show that the phase-space evolution can be written in the form

$$\begin{pmatrix} x(\Delta t) \\ p(\Delta t) \end{pmatrix} = A(\omega, \Omega, \Delta t) \begin{pmatrix} x(0) \\ p(0) \end{pmatrix},$$

where $A(\omega, \Omega, \Delta t)$ is a 2×2 matrix. Derive the explicit form of this matrix.

b. Show that $\det(A) = 1$.

c. Show that, depending on Δt, the eigenvalues of A are either complex conjugate pairs such that $-2 < \text{Tr}(A) < 2$, or both real, such that $|\text{Tr}(A)| \geq 2$.

d. Discuss the numerical implication of the choice $\Delta t = \pi/\omega$. This time step is known as a *resonant* time step (Schlick *et al.*, 1998; Ma *et al.*, 2003) and indicates that large step in the RESPA algorithm is fundamentally limited.

3.7. Consider again the harmonic system in Problem 3.6. However, suppose the propagator $\exp(iL_{\text{fast}}\Delta t)$ is replaced by the following approximation:

$$e^{iL_{\text{fast}}\Delta t} \approx (I - (\Delta t/2)K)^{-1} (I + (\Delta t/2)K),$$

where K is the matrix

$$K = \begin{pmatrix} 0 & 1 \\ -\omega^2 & 0 \end{pmatrix}.$$

This representation of $\exp(iL_{\text{fast}}\Delta t)$ is called the *Cayley* matrix representation (Korol *et al.*, 2019). Determine the matrix $A(\omega, \Omega, \Delta t)$ within this algorithm, find its eigenvalues, and determine if resonances exist in this approximation.

3.8. A single particle moving in one dimension is subject to a potential of the form

$$U(x) = \frac{1}{2}m\omega^2 x^2 + \frac{g}{4}x^4.$$

Choosing $m = 1$, $\omega = 1$, $g = 0.1$, $x(0) = 0$, and $p(0) = 1$, write a program that implements the RESPA algorithm for this problem. If the small time step δt is chosen to be 0.01, how large can the big time step Δt be chosen for accurate integration? Compare the RESPA trajectory to a single time step trajectory using a very small time step. Use your program to verify that the RESPA algorithm is globally second-order.

3.9. Use the direct translation technique to produce a pseudocode for the algorithm in eqn. (3.11.13).

3.10. Use the Legendre transform to determine the energy that results by transforming from volume V to pressure P in the microcanonical ensemble. What thermodynamic function does this energy represent?

3.11. A harmonic oscillator of unit mass and unit frequency with coordinate x and momentum p has a Hamiltonian $\mathcal{H}(x,p) = (p^2 + x^2)/2$. The Liouville operator is then given by

$$iL = p\frac{\partial}{\partial x} - x\frac{\partial}{\partial p}$$

The exact classical propagator $\exp(iLt)$ can be expanded in a Taylor series according to

$$e^{iLt} = \sum_{k=0}^{\infty} \frac{1}{k!} (iLt)^k$$

Using this expanion, show that the action of the exact propagator $\exp(iLt)$ on an initial condition (x, p) yields the expected solution

$$x(t) = x\cos(t) + p\sin(t)$$
$$p(t) = p\cos(t) - x\sin(t)$$

4

The canonical ensemble

4.1 Introduction: A different set of experimental conditions

The microcanonical ensemble is composed of a collection of systems isolated from any surroundings. Each system in the ensemble is characterized by fixed values of the particle number N, volume V, and total energy E. Moreover, since all members of the ensemble have the same underlying Hamiltonian $\mathcal{H}(\mathrm{x})$, the phase-space distribution of the system is uniform over the constant-energy hypersurface $\mathcal{H}(\mathrm{x}) = E$ and zero off the hypersurface. Therefore, the entire ensemble can be generated by a dynamical system evolving according to Hamilton's equations of motion $\dot{q}_\alpha = \partial \mathcal{H}/\partial p_\alpha$ and $\dot{p}_\alpha = -\partial \mathcal{H}/\partial q_\alpha$ under the assumption that the dynamical system is ergodic, *i.e.*, that in an infinite time, it visits all points on the constant-energy hypersurface. Under this assumption, a molecular dynamics calculation can be used to generate a microcanonical distribution.

The main disadvantage of the microcanonical ensemble is that conditions of constant total energy are not those under which experiments are performed. It is, therefore, important to develop ensembles that have different sets of thermodynamic control variables in order to reflect more common experimental setups. The *canonical ensemble* is an example. Its thermodynamic control variables are particle number N, volume V, and temperature T, which characterize a system in thermal contact with an infinite heat source. Although experiments are more commonly performed at conditions of constant pressure P, rather than constant volume, or they may fix the chemical potential μ rather than constant particle number, the canonical ensemble nevertheless forms the basis for the NPT (isothermal-isobaric) and μVT (grand canonical) ensembles, which will be discussed in the subsequent two chapters. Moreover, for large systems, the canonical distribution is often a good approximation to the isothermal-isobaric and grand canonical distributions, and when this is true, results from the canonical ensemble will not deviate much from results of the other ensembles.

In this chapter, we will formulate the basic thermodynamics of the canonical ensemble. Recall that thermodynamics always divides the universe into a system and its surroundings. When a system is in thermal contact with an infinite external heat source, its energy will fluctuate in such a way that its temperature remains fixed, leading to the conditions of the canonical ensemble. This thermodynamic paradigm will be used in a microcanonical formulation of the universe (system + surroundings) to derive the partition function and phase-space distribution of the system under these conditions. It will be shown that the Hamiltonian $\mathcal{H}(\mathrm{x})$ of the system, which is not conserved, obeys a *Boltzmann distribution* $\exp[-\beta \mathcal{H}(\mathrm{x})]$. Once we have laid out the the underlying statistical mechanics, we will work through a number of examples employing the canonical ensemble. In addition, we will examine how physical observables of experimental interest are obtained in this ensemble, including both thermodynamic and structural properties of real systems. Finally, we will show how molecular dynamics methods capable of generating a sampling of the canonical distribution can be devised.

4.2 Thermodynamics of the canonical ensemble

The Legendre transformation technique introduced in Section 1.5 is the method by which thermodynamic potentials are transformed between ensembles. Recall that in the microcanonical ensemble, the control variables are particle number N, volume V, and total energy E. The state function that depends on these is the entropy $S(N, V, E)$, and the thermodynamic variables obtained from the partial derivatives of the entropy are:

$$\frac{1}{T} = \left(\frac{\partial S}{\partial E}\right)_{N,V}, \qquad \frac{P}{T} = \left(\frac{\partial S}{\partial V}\right)_{N,E}, \qquad \frac{\mu}{T} = -\left(\frac{\partial S}{\partial N}\right)_{V,E}. \qquad (4.2.1)$$

Note that the entropy $S = S(N, V, E)$ can also be inverted to give E as a function, $E(N, V, S)$. In terms of E, using eqn. (3.2.7), the above thermodynamic relations become

$$T = \left(\frac{\partial E}{\partial S}\right)_{N,V}, \qquad P = -\left(\frac{\partial E}{\partial V}\right)_{N,S}, \qquad \mu = \left(\frac{\partial E}{\partial N}\right)_{V,S}. \qquad (4.2.2)$$

For transforming from the microcanonical to the canonical ensemble, eqn. (4.2.2) is preferable, as it gives the temperature directly, rather than $1/T$. Thus, we seek to transform the function $E(N, V, S)$ from a function of N, V, and S to a function of N, V, and T. Since $T = \partial E/\partial S$, the Legendre transform method can be applied. According to eqn. (1.5.5), the new function, which we will denote as $A(N, V, T)$, is given by

$$A(N, V, T) = E(N, V, S(N, V, T)) - \frac{\partial E}{\partial S} S(N, V, T)$$

$$= E(N, V, S(N, V, T)) - TS(N, V, T). \qquad (4.2.3)$$

The function $A(N, V, T)$ is a new state function known as the *Helmholtz free energy*. Physically, when a thermodynamic transformation of a system from state 1 to state 2 is carried out on a system along a reversible path, then the work needed to effect this transformation is equal to the change in the Helmholtz free energy ΔA. From eqn. (4.2.3), it is clear that A has both energetic and entropic contributions, and the delicate balance between these two contributions can sometimes have a sizeable effect on the free energy. Free energy is a particularly useful concept as it determines whether a process is thermodynamically favorable, indicated by a decrease in free energy, or unfavorable, indicated by an increase in free energy. It is important to note that although thermodynamics can determine if a process is favorable, it has nothing to say about the time scale on which the process occurs.

A process in which N, V, and T change by small amounts dN, dV, and dT leads to a change dA in the Helmholtz free energy of

$$dA = \left(\frac{\partial A}{\partial N}\right)_{V,T} dN + \left(\frac{\partial A}{\partial V}\right)_{N,T} dV + \left(\frac{\partial A}{\partial T}\right)_{N,V} dT \tag{4.2.4}$$

via the chain rule. However, since $A = E - TS$, the change in A can also be expressed as

$$dA = dE - SdT - TdS$$

$$= TdS - PdV + \mu dN - SdT - TdS$$

$$= -PdV + \mu dN - SdT, \tag{4.2.5}$$

where the second line follows from the first law of thermodynamics. By comparing the last line of eqn. (4.2.5) with eqn. (4.2.4), we see that the thermodynamic variables obtained from the partial derivatives of A are:

$$\mu = \left(\frac{\partial A}{\partial N}\right)_{V,T}, \qquad P = -\left(\frac{\partial A}{\partial V}\right)_{N,T}, \qquad S = -\left(\frac{\partial A}{\partial T}\right)_{N,V}. \tag{4.2.6}$$

These relations define the basic thermodynamics of the canonical ensemble. We must now establish the link between these thermodynamic relations and the microscopic description of the system in terms of its Hamiltonian $\mathcal{H}(\mathbf{x})$.

4.3 The canonical phase-space distribution and partition function

In the canonical ensemble, we assume that a system can only exchange heat with its surroundings. As was done in Section 3.2, we consider two systems in thermal contact. Let us denote the physical system as "system 1" and the surroundings as "system 2" (see Fig. 4.1). System 1 is assumed to contain N_1 particles in a volume V_1, while system 2 contains N_2 particles in a volume V_2. In addition, system 1 has an energy E_1, and system 2 has an energy E_2, such that the total energy $E = E_1 + E_2$. System 2 is taken to be much larger than system 1 so that $N_2 \gg N_1$, $V_2 \gg V_1$, $E_2 \gg E_1$. System 2 is often referred to as a *thermal reservoir*, which can exchange energy with system 1 without changing its energy appreciably. The thermodynamic "universe," composed of system 1 + system 2, is treated within the microcanonical ensemble. Thus, the total Hamiltonian $\mathcal{H}(\mathbf{x})$ of the universe is expressed as a sum of contributions, $\mathcal{H}_1(\mathbf{x}_1) + \mathcal{H}_2(\mathbf{x}_2)$ of system 1 and system 2, where $\mathbf{x}_1$ is the phase-space vector of system 1, and $\mathbf{x}_2$ is the phase-space vector of system 2.

As was argued in Section 3.4, if we simply solved Hamilton's equations for the total Hamiltonian $\mathcal{H}(\mathbf{x}) = \mathcal{H}_1(\mathbf{x}_1) + \mathcal{H}_2(\mathbf{x}_2)$, $\mathcal{H}_1(\mathbf{x}_1)$ and $\mathcal{H}_2(\mathbf{x}_2)$ would be separately conserved because the Hamiltonian is separable. However, the microcanonical distribution, which is proportional to $\delta(\mathcal{H}(\mathbf{x}) - E)$ allows us to consider all possible energies E_1 and E_2 for which $E_1 + E_2 = E$ without explicitly requiring a potential coupling

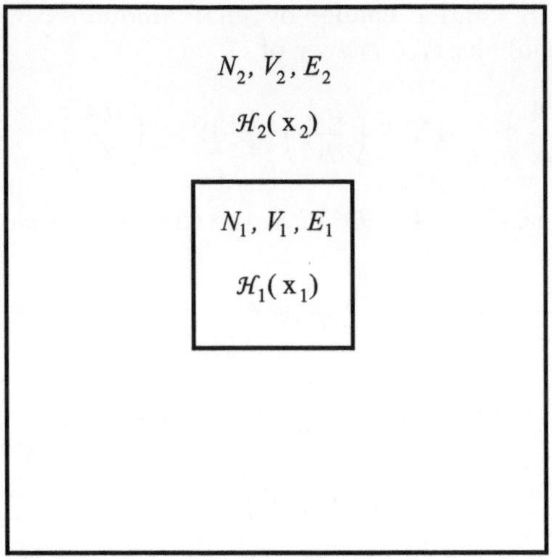

Fig. 4.1 A system (system 1) in contact with a thermal reservoir (system 2). System 1 has N_1 particles in a volume V_1; system 2 has N_2 particles in a volume V_2.

between the two systems. Since the two systems can exchange energy, we do not expect $\mathcal{H}(x_1)$ and $\mathcal{H}(x_2)$ to be separately conserved.

The microcanonical partition function of this thermodynamic universe is

$$\Omega(N, V, E) = M_N \int dx \, \delta(\mathcal{H}(x) - E)$$

$$= M_N \int dx_1 \, dx_2 \, \delta\left(\mathcal{H}_1(x_1) + \mathcal{H}_2(x_2) - E\right). \qquad (4.3.1)$$

The corresponding phase-space distribution function $f(x_1)$ is obtained by integrating only over the phase-space variables of system 2, yielding

$$f(x_1) = \int dx_2 \, \delta\left(\mathcal{H}_1(x_1) + \mathcal{H}_2(x_2) - E\right), \qquad (4.3.2)$$

which is unnormalized. Because thermodynamic quantities are obtained from derivatives of the logarithm of the partition function, it is preferable to work with the logarithm of the distribution:

$$\ln f(x_1) = \ln \int dx_2 \, \delta\left(\mathcal{H}_1(x_1) + \mathcal{H}_2(x_2) - E\right). \qquad (4.3.3)$$

We now exploit the fact that system 1 is small compared to system 2. Since $E_2 \gg E_1$, it follows that $\mathcal{H}_2(x_2) \gg \mathcal{H}_1(x_1)$. Thus, we expand eqn. (4.3.3) about $\mathcal{H}_1(x_1) = 0$ at an arbitrary phase-space point x_1. Carrying out the expansion to first order in $\mathcal{H}_1$ gives

$$\ln f(\mathbf{x}_1) \approx \ln \int d\mathbf{x}_2 \, \delta \left(\mathcal{H}_2(\mathbf{x}_2) - E \right)$$

$$+ \frac{\partial}{\partial \mathcal{H}(\mathbf{x}_1)} \ln \int d\mathbf{x}_2 \, \delta \left(\mathcal{H}_1(\mathbf{x}_1) + \mathcal{H}_2(\mathbf{x}_2) - E \right) \Bigg|_{\mathcal{H}_1(\mathbf{x}_1)=0} \mathcal{H}_1(\mathbf{x}_1). \quad (4.3.4)$$

Since the δ-function requires $\mathcal{H}_1(\mathbf{x}_1) + \mathcal{H}_2(\mathbf{x}_1) - E = 0$, we may differentiate with respect to E instead, using the fact that

$$\frac{\partial}{\partial \mathcal{H}_1(\mathbf{x}_1)} \delta(\mathcal{H}_1(\mathbf{x}_1) + \mathcal{H}_2(\mathbf{x}_2) - E) = -\frac{\partial}{\partial E} \delta(\mathcal{H}_1(\mathbf{x}_1) + \mathcal{H}_2(\mathbf{x}_2) - E). \quad (4.3.5)$$

Then, eqn. (4.3.4) becomes

$$\ln f(\mathbf{x}_1) \approx \ln \int d\mathbf{x}_2 \, \delta \left(\mathcal{H}_2(\mathbf{x}_2) - E \right)$$

$$- \frac{\partial}{\partial E} \ln \int d\mathbf{x}_2 \, \delta \left(\mathcal{H}_1(\mathbf{x}_1) + \mathcal{H}_2(\mathbf{x}_2) - E \right) \Bigg|_{\mathcal{H}_1(\mathbf{x}_1)=0} \mathcal{H}_1(\mathbf{x}_1). \quad (4.3.6)$$

Now, $\mathcal{H}_1(\mathbf{x}_1)$ can be set to 0 in the second term of eqn. (4.3.6) yielding

$$\ln f(\mathbf{x}_1) \approx \ln \int d\mathbf{x}_2 \, \delta \left(\mathcal{H}_2(\mathbf{x}_2) - E \right) - \frac{\partial}{\partial E} \ln \int d\mathbf{x}_2 \, \delta \left(\mathcal{H}_2(\mathbf{x}_2) - E \right) \mathcal{H}_1(\mathbf{x}_1). \quad (4.3.7)$$

Recognizing that

$$\int d\mathbf{x}_2 \, \delta \left(\mathcal{H}_2(\mathbf{x}_2) - E \right) \propto \Omega_2(N_2, V_2, E), \quad (4.3.8)$$

where $\Omega_2(N_2, V_2, E)$ is the microcanonical partition function of system 2 at energy E. Since $\ln \Omega_2(N_2, V_2, E) = S_2(N_2, V_2, E)/k$, eqn. (4.3.7) can be written (apart from overall normalization) as

$$\ln f(\mathbf{x}_1) = \frac{S_2(N_2, V_2, E)}{k} - \mathcal{H}_1(\mathbf{x}_1) \frac{\partial}{\partial E} \frac{S_2(N_2, V_2, E)}{k}. \quad (4.3.9)$$

Moreover, because $\partial S_2/\partial E = 1/T$, where T is the common temperature of systems 1 and 2, it follows that

$$\ln f(\mathbf{x}_1) = \frac{S_2(N_2, V_2, E)}{k} - \frac{\mathcal{H}_1(\mathbf{x}_1)}{kT}. \quad (4.3.10)$$

Exponentiating both sides, and recognizing that $\exp(S_2/k)$ is just an overall constant, we obtain

$$f(\mathbf{x}_1) \propto e^{-\mathcal{H}_1(\mathbf{x}_1)/kT}. \quad (4.3.11)$$

At this point, the "1" subscript is no longer necessary. In other words, we can conclude that the phase-space distribution of a system with Hamiltonian $\mathcal{H}(\mathbf{x})$ in equilibrium with a thermal reservoir at temperature T is

$$f(\mathbf{x}) \propto e^{-\mathcal{H}(\mathbf{x})/kT}.$$ \hfill (4.3.12)

The overall normalization of eqn. (4.3.12) must be proportional to

$$\int d\mathbf{x}\, e^{-\mathcal{H}(\mathbf{x})/kT}.$$

As was the case for the microcanonical ensemble, the integral is accompanied by an N-dependent factor that accounts for the identical nature of the particles and yields an overall dimensionless quantity. This factor is denoted C_N and is given by

$$C_N = \frac{1}{N!h^{3N}},$$ \hfill (4.3.13)

so that the phase-space distribution function becomes

$$f(\mathbf{x}) = \frac{C_N e^{-\beta\mathcal{H}(\mathbf{x})}}{Q(N,V,T)}.$$ \hfill (4.3.14)

(As we noted in Section 3.2, in a multicomponent system with N_A particles of type A, N_B particles of type B,..., and N total particles, C_N would be replaced by $C_{\{N\}} = 1/[h^{3N}(N_A!N_B!\cdots)]$.) The parameter $\beta = 1/kT$ has been introduced, and the denominator in eqn. (4.3.14) is

$$Q(N,V,T) = C_N \int d\mathbf{x}\, e^{-\beta\mathcal{H}(\mathbf{x})}.$$ \hfill (4.3.15)

The quantity $Q(N,V,T)$ (or, equivalently, $Q(N,V,\beta)$) is the partition function of the canonical ensemble, and, as with the microcanonical ensemble, it is a measure of the total number of accessible microscopic states. In contrast to the microcanonical ensemble, however, the Hamiltonian is not conserved. Rather, it obeys the Boltzmann distribution as a consequence of the fact that the system can exchange energy with its surroundings. This energy exchange changes the number of accessible microscopic states. Note that the canonical partition function $Q(N,V,T)$ can be directly related to the microcanonical partition function $\Omega(N,V,E)$ as follows:

$$Q(N,V,T) = \frac{1}{E_0} \int_0^\infty dE\, e^{-\beta E} M_N \int d\mathbf{x}\, \delta(\mathcal{H}(\mathbf{x}) - E)$$

$$= \frac{1}{E_0} \int_0^\infty dE\, e^{-\beta E} \Omega(N,V,E).$$ \hfill (4.3.16)

In the first line, if the integration over energy E is performed first, then the δ-function allows E to be replaced by the Hamiltonian $\mathcal{H}(\mathbf{x})$ in the exponential, leading to eqn. (4.3.15). The second line shows that the canonical partition function is simply the Laplace transform of the microcanonical partition function. Readers unfamiliar with Laplace transforms are referred to Appendix E.

The link between the macroscopic thermodynamic properties in eqn. (4.2.6) and the microscopic states contained in $Q(N, V, T)$ is provided through the relation

$$A(N, V, T) = -kT \ln Q(N, V, T) = -\frac{1}{\beta} \ln Q(N, V, \beta). \qquad (4.3.17)$$

In order to see that eqn. (4.3.17) provides the connection between the thermodynamic state function $A(N, V, T)$ and the partition function $Q(N, V, T)$, we note that $A = E - TS$ and that $S = -\partial A/\partial T$, from which we obtain

$$A = E + T \frac{\partial A}{\partial T}. \qquad (4.3.18)$$

We also recognize that $E = \langle \mathcal{H}(\mathbf{x}) \rangle$, the ensemble average of the Hamiltonian. By definition, this ensemble average is

$$\langle \mathcal{H} \rangle = \frac{C_N \int d\mathbf{x}\, \mathcal{H}(\mathbf{x}) e^{-\beta \mathcal{H}(\mathbf{x})}}{C_N \int d\mathbf{x}\, e^{-\beta \mathcal{H}(\mathbf{x})}}$$

$$= -\frac{1}{Q(N, V, T)} \frac{\partial}{\partial \beta} Q(N, V, T)$$

$$= -\frac{\partial}{\partial \beta} \ln Q(N, V, T). \qquad (4.3.19)$$

Thus, eqn. (4.3.18) becomes

$$A + \frac{\partial}{\partial \beta} \ln Q(N, V, \beta) + \beta \frac{\partial A}{\partial \beta} = 0, \qquad (4.3.20)$$

where the fact that

$$T \frac{\partial A}{\partial T} = T \frac{\partial A}{\partial \beta} \frac{\partial \beta}{\partial T} = -T \frac{\partial A}{\partial \beta} \frac{1}{kT^2} = -\beta \frac{\partial A}{\partial \beta} \qquad (4.3.21)$$

has been used. We just need to show that eqn. (4.3.17) is the solution to eqn. (4.3.20), which is a first-order differential equation for A. Differentiating eqn. (4.3.17) with respect to β gives

$$\beta \frac{\partial A}{\partial \beta} = \frac{1}{\beta} \ln Q(N, V, \beta) - \frac{\partial}{\partial \beta} \ln Q(N, V, \beta). \qquad (4.3.22)$$

Substituting eqns. (4.3.17) and (4.3.22) into eqn. (4.3.20) yields

$$-\frac{1}{\beta} \ln Q(N, V, \beta) + \frac{\partial}{\partial \beta} \ln Q(N, V, \beta) + \frac{1}{\beta} \ln Q(N, V, \beta) - \frac{\partial}{\partial \beta} \ln Q(N, V, \beta) = 0,$$

which verifies that $A = -kT \ln Q$ is the solution. Therefore, from eqn. (4.2.6), it is clear that the macroscopic thermodynamic observables are given in terms of the partition function by

$$\mu = -kT \left(\frac{\partial \ln Q}{\partial N} \right)_{N,V}$$

$$P = kT \left(\frac{\partial \ln Q}{\partial V} \right)_{N,T}$$

$$S = k \ln Q + kT \left(\frac{\partial \ln Q}{\partial T} \right)_{N,V}$$

$$E = - \left(\frac{\partial}{\partial \beta} \ln Q \right)_{N,V}. \tag{4.3.23}$$

Noting that

$$kT \frac{\partial \ln Q}{\partial T} = kT \frac{\partial \ln Q}{\partial \beta} \frac{\partial \beta}{\partial T} = -kT \frac{\partial \ln Q}{\partial \beta} \frac{1}{kT^2} = -\frac{1}{T} \frac{\partial \ln Q}{\partial \beta} = \frac{E}{T}, \tag{4.3.24}$$

one finds that the entropy is given by

$$S(N,V,T) = k \ln Q(N,V,T) + \frac{E(N,V,T)}{T}, \tag{4.3.25}$$

which is equivalent to $S = (-A + E)/T$. Other thermodynamic relations can be obtained as well. For example, the heat capacity C_V at constant volume is defined to be

$$C_V = \left(\frac{\partial E}{\partial T} \right)_{N,V}. \tag{4.3.26}$$

Differentiating the last line of eqn. (4.3.19) using $\partial / \partial T = -(k\beta^2) \partial / \partial \beta$ gives

$$C_V = k\beta^2 \frac{\partial^2}{\partial \beta^2} \ln Q(N,V,\beta). \tag{4.3.27}$$

Interestingly, the heat capacity in eqn. (4.3.27) is an extensive quantity. The corresponding intensive *molar heat capacity* is obtained from eqn. (4.3.27) by dividing by the number of moles in the system.

4.4 Canonical ensemble via entropy maximization

In this section, we will show that the equilibrium canonical phase-space probability distribution function $f(\mathbf{x}) \propto \exp(-\beta \mathcal{H}(\mathbf{x}))$ is the distribution that maximizes the entropy of the system. Just as a gas released from a small container into a larger volume will expand until it fills the available space, an ensemble reaches equilibrium once a maximal number of available microscopic states is achieved. Since the entropy is the logarithm of the number of microstates, equilibrium is equivalent to a state of maximum entropy. Thus, it should be possible to derive an equilibrium ensemble distribution function from an entropy maximization principle, provided we can formulate the entropy in a way that depends on a distribution function.

Since $f(x)dx$ is the fraction of microstates in the phase-space volume dx, the quantity $-k \ln f(x)$ represents a kind of "local entropy" at the phase-space point x. If we average this local entropy over the ensemble, we then obtain an expression for the entropy in terms of the distribution function (see also Problem 2.7)

$$S = -k \langle \ln f(x) \rangle = -k \int dx \, f(x) \ln f(x). \tag{4.4.1}$$

This entropy expression is often referred to as the *Shannon entropy* after the American mathematician and engineer Claude E. Shannon (1916–2001). When written in the form shown in eqn. (4.4.1), we see that the entropy depends on *all* values of $f(x)$ and, thus, is another example of a local functional, similar to the type of functional we encountered in Chapter 1 (see Section 1.8). Thus, we denote the integral in eqn. (4.4.1) as $S[f]$ to indicate its functional dependence on $f(x)$.

Maximization of $S[f]$ over all distribution functions $f(x)$ should then produce the correct equilibrium distribution for the canonical ensemble. However, as we search over all $f(x)$ for the distribution that maximizes $S[f]$, we need to account for two important conditions that $f(x)$ must obey. First, because $f(x)$ is a probability distribution, it must be normalized to 1:

$$\int dx \, f(x) = 1. \tag{4.4.2}$$

Second, the ensemble average of the Hamiltonian $\mathcal{H}(x)$ must yield a well-defined energy E, as the dependence of S on E is critical for obtaining the temperature via the thermodynamic derivative $\partial S / \partial E = 1/T$. Thus, the second condition we require is

$$\int dx \, \mathcal{H}(x) f(x) = E. \tag{4.4.3}$$

This condition, along with the normalization condition, will be accounted for in the maximization procedure using two Lagrange multipliers, μ, and λ. We thus define a modified entropy functional that includes the two constraint conditions

$$\tilde{S}[f] = -k \int dx \, f(x) \ln f(x) - \mu \left(\int dx \, f(x) - 1 \right)$$

$$- \lambda \left(\int dx \, \mathcal{H}(x) f(x) - E \right). \tag{4.4.4}$$

(See Appendix B for an introduction to functionals and their properties.) The condition for entropy maximization is obtained by setting the functional derivative of $\tilde{S}[f]$ with respect to $f(x)$ equal to 0, *i.e.*,

$$\frac{\delta \tilde{S}}{\delta f(x)} = 0. \tag{4.4.5}$$

Taking the functional derivative of eqn. (4.4.4) and setting the result equal to 0, we obtain

$$-k \left[\ln f(x) + 1 \right] - \mu - \lambda \mathcal{H}(x) = 0. \tag{4.4.6}$$

Solving this for $f(x)$, we obtain

$$f(x) = e^{-\mu/k}e^{-1}e^{-\lambda\mathcal{H}(x)/k}. \tag{4.4.7}$$

If we now require that $\partial\tilde{S}/\partial E = 1/T$, we can obtain an expression for the multiplier λ. First, note that $\partial\tilde{S}/\partial E = \lambda$. But since the derivative must also be $1/T$, it must follow that $\lambda = 1/T$. Similarly, since $f(x)$ must be normalized to 1, we can obtain an expression for the multiplier μ as

$$e^{-\mu/k}e^{-1} = \left[\int dx\, e^{-\lambda\mathcal{H}(x)/k}\right]^{-1} = \left[\int dx\, e^{-\mathcal{H}(x)/kT}\right]^{-1} = \frac{C_N}{Q(N,V,T)}. \tag{4.4.8}$$

Thus, the final expression for the distribution, obtained by maximizing the entropy, is

$$f(x) = \frac{C_N e^{-\mathcal{H}(x)/kT}}{Q(N,V,T)}, \tag{4.4.9}$$

which is the expected canonical distribution of eqn. (4.3.14). As we continue to develop the equilibrium ensembles, we will see that each distribution can be obtained from entropy maximization with additional conditions appropriate for each ensemble.

4.5   Energy fluctuations in the canonical ensemble

Since the Hamiltonian $\mathcal{H}(x)$ is not conserved in the canonical ensemble, it is natural to ask how the energy fluctuates. Energy fluctuations can be quantified using the standard statistical measure of variance. The variance of the Hamiltonian is given by

$$\Delta E = \sqrt{\langle(\mathcal{H}(x) - \langle\mathcal{H}(x)\rangle)^2\rangle}, \tag{4.5.1}$$

which measures the width of the energy distribution, *i.e.*, the root-mean-square deviation of $\mathcal{H}(x)$ from its average value. The quantity under the square root can also be expressed as

$$\langle(\mathcal{H}(x) - \langle\mathcal{H}(x)\rangle)^2\rangle = \langle(\mathcal{H}^2(x) - 2\mathcal{H}(x)\langle\mathcal{H}(x)\rangle + \langle\mathcal{H}(x)\rangle^2)\rangle$$

$$= \langle\mathcal{H}^2(x)\rangle - 2\langle\mathcal{H}(x)\rangle\langle\mathcal{H}(x)\rangle + \langle\mathcal{H}(x)\rangle^2$$

$$= \langle\mathcal{H}^2(x)\rangle - \langle\mathcal{H}(x)\rangle^2. \tag{4.5.2}$$

The first term in the last line of eqn. (4.5.2) is, by definition, given by

$$\langle\mathcal{H}^2(x)\rangle = \frac{C_N \int dx\, \mathcal{H}^2(x)e^{-\beta\mathcal{H}(x)}}{C_N \int dx\, e^{-\beta\mathcal{H}(x)}} = \frac{1}{Q(N,V,\beta)}\frac{\partial^2}{\partial\beta^2}Q(N,V,\beta). \tag{4.5.3}$$

Now, consider the quantity

$$\frac{\partial^2}{\partial\beta^2}\ln Q(N,V,\beta) = \frac{\partial}{\partial\beta}\left[\frac{1}{Q(N,V,\beta)}\frac{\partial Q(N,V,\beta)}{\partial\beta}\right]$$

$$= -\frac{1}{Q^2(N,V,\beta)} \left[\frac{\partial Q(N,V,\beta)}{\partial \beta}\right]^2 + \frac{1}{Q(N,V,\beta)} \frac{\partial^2 Q(N,V,\beta)}{\partial \beta^2}. \quad (4.5.4)$$

The first term in this expression is just the square of eqn. (4.3.19) or $\langle \mathcal{H}(\mathrm{x})\rangle^2$, while the second term is the average $\langle \mathcal{H}^2(\mathrm{x})\rangle$. Thus, we see that

$$\frac{\partial^2}{\partial \beta^2} \ln Q(N,V,\beta) = -\langle \mathcal{H}(\mathrm{x})\rangle^2 + \langle \mathcal{H}^2(\mathrm{x})\rangle = (\Delta E)^2. \quad (4.5.5)$$

However, from eqn. (4.3.27),

$$\frac{\partial^2}{\partial \beta^2} \ln Q(N,V,\beta) = kT^2 C_V = (\Delta E)^2. \quad (4.5.6)$$

Thus, the variance in the energy is directly related to the heat capacity at constant volume. If we now consider the energy fluctuations *relative* to the total energy, $\Delta E/E$, we find

$$\frac{\Delta E}{E} = \frac{\sqrt{kT^2 C_V}}{E}. \quad (4.5.7)$$

Since C_V is an extensive quantity, $C_V \sim N$. The same is true for the energy, $E \sim N$, as it is also extensive. Therefore, according to eqn. (4.5.7), the relative energy fluctuations should behave as

$$\frac{\Delta E}{E} \sim \frac{\sqrt{N}}{N} \sim \frac{1}{\sqrt{N}}. \quad (4.5.8)$$

In the thermodynamic limit, when $N \to \infty$, the relative energy fluctuations tend to zero. For very large systems, the magnitude of ΔE relative to the total average energy E becomes negligible. The implication of this result is that in the thermodynamic limit, the canonical ensemble becomes equivalent to the microcanonical ensemble, where, in the latter, the Hamiltonian is explicitly fixed. In the next two chapters, we will analyze fluctuations associated with other ensembles, and we will see that the tendency of these fluctuations to become negligible in the thermodynamic limit is a general result. The consequence of this fact is that all ensembles become equivalent in the thermodynamic limit. Thus, we are always at liberty to choose the ensemble that is most convenient for a particular problem and still obtain the same macroscopic observables. It must be stressed, however, that this freedom only exists in the thermodynamic limit. In numerical simulations, for example, systems are finite, and fluctuations, which decay slowly as $1/\sqrt{N}$, might be large, depending on the system size chosen. Thus, the choice of ensemble can influence the results of the calculation, and one should choose the ensemble that best reflects the experimental conditions of the problem.

Now that we have the fundamental principles of the classical canonical ensemble at hand, we proceed next to consider a few simple analytical solvable examples of this ensemble in order to demonstrate how it is used.

4.6   Simple examples in the canonical ensemble

4.6.1   The free particle and the ideal gas

Consider a free particle of mass m moving in a one-dimensional "box" of length L. The Hamiltonian is simply $\mathcal{H} = p^2/2m$. The partition function for an ensemble of such systems at temperature T is

$$Q(L,T) = \frac{1}{h} \int_0^L \mathrm{d}x \int_{-\infty}^{\infty} \mathrm{d}p \, e^{-\beta p^2/2m}. \tag{4.6.1}$$

The position x can be integrated trivially, yielding a factor of L. The momentum integral is an example of a *Gaussian integral*, for which the general formula is

$$\int_{-\infty}^{\infty} \mathrm{d}y \, e^{-\alpha y^2} = \sqrt{\frac{\pi}{\alpha}} \tag{4.6.2}$$

(see Section 3.8.3, where a method for performing Gaussian integrals is discussed). Applying eqn. (4.6.2) to the partition function gives the final result

$$Q(L,T) = L\sqrt{\frac{2\pi m}{\beta h^2}}. \tag{4.6.3}$$

The quantity $\sqrt{\beta h^2/2\pi m}$ appearing in eqn. (4.6.3) can be easily seen to have units of a length. For reasons that will become clear when we consider quantum statistical mechanics in Chapter 10, this quantity, denoted λ, is often referred to as the *thermal wavelength* of the particle. Thus, the partition function is simply the ratio of the box length L to the thermal wavelength of the particle:

$$Q(L,T) = \frac{L}{\lambda}. \tag{4.6.4}$$

We now extend this derivation to the case of N particles in three dimensions, *i.e.*, an ideal gas of N particles in a cubic box of side L (volume $V = L^3$), for which the Hamiltonian is

$$\mathcal{H} = \sum_{i=1}^{N} \frac{\mathbf{p}_i^2}{2m}. \tag{4.6.5}$$

Since each momentum vector $\mathbf{p}_i$ has three components, we may also write the Hamiltonian as

$$\mathcal{H} = \sum_{i=1}^{N} \sum_{\alpha=1}^{3} \frac{p_{\alpha i}^2}{2m}, \tag{4.6.6}$$

where $\alpha = (x,y,z)$ indexes the Cartesian components of $\mathbf{p}_i$. The sum in eqn. (4.6.6) contains $3N$ terms. Thus, the partition function is given by

$$Q(N,V,T) = \frac{1}{N!h^{3N}} \int_{D(V)} \mathrm{d}^N\mathbf{r} \int \mathrm{d}^N\mathbf{p} \, \exp\left[-\beta \sum_{i=1}^{N} \frac{\mathbf{p}_i^2}{2m}\right]. \tag{4.6.7}$$

Since the Hamiltonian is separable in the each of the N coordinates and momenta, the partition function can be simplified according to

$$Q(N,V,T) = \frac{1}{N!} \left[\frac{1}{h^3} \int_{D(V)} d\mathbf{r}_1 \int d\mathbf{p}_1 \, e^{-\beta \mathbf{p}_1^2/2m} \right] \left[\frac{1}{h^3} \int_{D(V)} d\mathbf{r}_2 \int d\mathbf{p}_2 \, e^{-\beta \mathbf{p}_2^2/2m} \right]$$

$$\cdots \left[\frac{1}{h^3} \int_{D(V)} d\mathbf{r}_N \int d\mathbf{p}_N \, e^{-\beta \mathbf{p}_N^2/2m} \right]. \tag{4.6.8}$$

Since each integral in brackets is the same, we can write eqn. (4.6.8) as

$$Q(N,V,T) = \frac{1}{N!} \left[\frac{1}{h^3} \int_{D(V)} d\mathbf{r} \int d\mathbf{p} \, e^{-\beta \mathbf{p}^2/2m} \right]^N. \tag{4.6.9}$$

The six-dimensional integral in brackets is just

$$\frac{1}{h^3} \int_{D(V)} d\mathbf{r} \int d\mathbf{p} \, e^{-\beta \mathbf{p}^2/2m} =$$

$$\frac{1}{h^3} \int_0^L dx \int_0^L dy \int_0^L dz$$

$$\times \int_{-\infty}^{\infty} dp_x e^{-\beta p_x^2/2m} \int_{-\infty}^{\infty} dp_y e^{-\beta p_y^2/2m} \int_{-\infty}^{\infty} dp_z e^{-\beta p_z^2/2m}. \tag{4.6.10}$$

Equation (4.6.10) can also be written as

$$\frac{1}{h^3} \int_{D(V)} d\mathbf{r} \int d\mathbf{p} \, e^{-\beta \mathbf{p}^2/2m} = \left[\frac{1}{h} \int_0^L dx \int_{-\infty}^{\infty} dp e^{-\beta p^2/2m} \right]^3, \tag{4.6.11}$$

which is just the cube of eqn. (4.6.1). Using eqn. (4.6.4), we obtain the partition function as

$$Q(N,V,T) = \frac{1}{N!} \left(\frac{L}{\lambda} \right)^{3N} = \frac{V^N}{N! \lambda^{3N}}. \tag{4.6.12}$$

From eqn. (4.6.12), the thermodynamics can now be derived. Using eqn. (4.3.23) to obtain the pressure yields

$$P = kT \frac{\partial}{\partial V} \ln \left[\frac{V^N}{N! \lambda^{3N}} \right] = NkT \frac{\partial \ln V}{\partial V} = \frac{NkT}{V}, \tag{4.6.13}$$

which we recognize as the ideal gas equation of state. Similarly, the energy is given by

$$E = -\frac{\partial}{\partial \beta} \ln \left[\frac{V^N}{N! \lambda^{3N}} \right] = 3N \frac{\partial \ln \lambda}{\partial \beta} = \frac{3N}{\beta} = \frac{3N}{2\beta} = \frac{3}{2} NkT, \tag{4.6.14}$$

which follows from the fact that $\lambda = \sqrt{\beta h^2/2\pi m}$ and is the expected result from the Virial theorem. From eqn. (4.6.14), it follows that the heat capacity at constant volume is

$$C_V = \left(\frac{\partial E}{\partial T}\right) = \frac{3}{2}Nk. \tag{4.6.15}$$

Note that if we multiply and divide by N_0, Avogadro's number, we obtain

$$C_V = \frac{3}{2}\frac{N}{N_0}N_0k = \frac{3}{2}nR, \tag{4.6.16}$$

where n is the number of moles of gas and R is the gas constant. Dividing by the number of moles yields the expected result for the *molar* heat capacity $c_V = 3R/2$.

4.6.2   The harmonic oscillator and the harmonic bath

We begin by considering a one-dimensional harmonic oscillator of mass m and frequency ω for which the Hamiltonian is

$$\mathcal{H} = \frac{p^2}{2m} + \frac{1}{2}m\omega^2 x^2. \tag{4.6.17}$$

The canonical partition function becomes

$$Q(\beta) = \frac{1}{h}\int dp\ dx\ e^{-\beta(p^2/2m + m\omega^2 x^2/2)}$$

$$= \frac{1}{h}\int_{-\infty}^{\infty} dp\ e^{-\beta p^2/2m}\int_{0}^{L} dx\ e^{-\beta m\omega^2 x^2/2}. \tag{4.6.18}$$

Although the coordinate integration is restricted to the physical box containing the oscillator, we will assume that the width of the distribution $\exp(-m\omega^2 x^2/2)$ is very small compared to the size of the (macroscopic) container so that we can perform the integration of x over all space with no significant loss of accuracy. Therefore, the partition function becomes

$$Q(\beta) = \frac{1}{h}\int_{-\infty}^{\infty} dp\ e^{-\beta p^2/2m}\int_{-\infty}^{\infty} dx\ e^{-\beta m\omega^2 x^2/2}$$

$$= \frac{1}{h}\left(\frac{2\pi m}{\beta}\right)^{1/2}\left(\frac{2\pi}{m\omega^2}\right)^{1/2}$$

$$= \frac{2\pi}{\beta h\omega} = \frac{1}{\beta\hbar\omega}, \tag{4.6.19}$$

where $\hbar = h/2\pi$. From eqn. (4.6.19), it follows that the energy is $E = kT$, the pressure is $P = 0$ (which is expected for a bound system), and the heat capacity is $C_V = k$.

If we now consider a collection of N uncoupled harmonic oscillators with different masses and frequencies with a Hamiltonian

$$\mathcal{H} = \sum_{i=1}^{N} \left[\frac{p_i^2}{2m_i} + \frac{1}{2} m_i \omega_i^2 x_i^2 \right]. \tag{4.6.20}$$

Since the oscillators are not identical, the $1/N!$ factor is not needed, and the partition function is just a product of single particle partition functions for the N oscillators:

$$Q(N, \beta) = \prod_{i=1}^{N} \frac{1}{\beta \hbar \omega_i}. \tag{4.6.21}$$

For this system, the energy is $E = NkT$, and the heat capacity is simply $C_v = Nk$.

4.6.3   The harmonic bead-spring model

Another important class of harmonic models is a simple model of a polymer chain based on harmonic nearest-neighbor interactions. Consider a polymer with endpoints at positions $\mathbf{r}$ and $\mathbf{r}'$ having N repeat units in between, each of which will be treated as a single 'particle'. The particles are indexed from 0 to $N+1$, and the Hamiltonian takes the form

$$\mathcal{H} = \sum_{i=0}^{N+1} \frac{\mathbf{p}_i^2}{2m} + \frac{1}{2} m \omega^2 \sum_{i=0}^{N} (\mathbf{r}_i - \mathbf{r}_{i+1})^2, \tag{4.6.22}$$

where $\mathbf{r}_0, ..., \mathbf{r}_{N+1}$ and $\mathbf{p}_0, ..., \mathbf{p}_{N+1}$ are the positions and momenta of the particles with the additional identification $\mathbf{r}_0 = \mathbf{r}$ and $\mathbf{r}_{N+1} = \mathbf{r}'$ and $\mathbf{p}_0 = \mathbf{p}$ and $\mathbf{p}_{N+1} = \mathbf{p}'$ as the positions and momenta of the endpoint particles, and $m\omega^2$ is the force constant. The polymer is placed in a cubic container of volume $V = L^3$ such that L is much larger than the average distance between neighboring particles $|\mathbf{r}_k - \mathbf{r}_{k+1}|$.

Let us first consider the case in which the endpoints are fixed at given positions $\mathbf{r}$ and $\mathbf{r}'$ so that $\mathbf{p} = \mathbf{p}' = 0$. We seek to calculate the partition function $Q(N, V, T, \mathbf{r}, \mathbf{r}')$ given by

$$Q(N, V, T, \mathbf{r}, \mathbf{r}') =$$

$$\frac{1}{h^{3N}} \int d^N \mathbf{p} \, d^N \mathbf{r} \exp \left\{ -\beta \left[\sum_{i=1}^{N} \frac{\mathbf{p}_i^2}{2m} + \frac{1}{2} m \omega^2 \sum_{i=0}^{N} (\mathbf{r}_i - \mathbf{r}_{i+1})^2 \right] \right\}. \tag{4.6.23}$$

We will regard the particles as truly distinguishable so that no $1/N!$ is needed. The Gaussian integrals over the N momenta can be performed immediately, yielding

$$Q(N, V, T, \mathbf{r}, \mathbf{r}') =$$

$$\frac{1}{h^{3N}} \left(\frac{2\pi m}{\beta} \right)^{3N/2} \int d^N \mathbf{r} \exp \left[-\frac{1}{2} \beta m \omega^2 \sum_{i=0}^{N} (\mathbf{r}_i - \mathbf{r}_{i+1})^2 \right]. \tag{4.6.24}$$

The coordinate integrations can be performed straightforwardly, if tediously, by simply integrating first over $\mathbf{r}_1$, then over $\mathbf{r}_2,...$ and recognizing the pattern that results after

$n < N$ such integrations have been performed. We will first follow this procedure, and then we will show how a simple change of integration variables can be used to simplify the integrations by uncoupling the harmonic interaction term.

Consider, first, the integration over $\mathbf{r}_1$. Defining $\alpha = \beta m \omega^2 / 2$, and using the fact that V is much larger than the average nearest-neighbor particle distance to extend the integration over all space, the integral that must be performed is

$$I_1 = \int_{\text{all space}} d\mathbf{r}_1 \, e^{-\alpha[(\mathbf{r}_1 - \mathbf{r})^2 + (\mathbf{r}_2 - \mathbf{r}_1)^2]}. \tag{4.6.25}$$

Expanding the squares gives

$$I_1 = e^{-\alpha(\mathbf{r}^2 + \mathbf{r}_2^2)} \int_{\text{all space}} d\mathbf{r}_1 \, e^{-2\alpha[\mathbf{r}_1^2 - \mathbf{r}_1 \cdot (\mathbf{r} + \mathbf{r}_2)]}. \tag{4.6.26}$$

Now, we can complete the square to give

$$I_1 = e^{-\alpha(\mathbf{r}^2 + \mathbf{r}_2^2)} e^{\alpha(\mathbf{r} + \mathbf{r}_2)^2 / 2} \int_{\text{all space}} d\mathbf{r}_1 \, e^{-2\alpha[\mathbf{r}_1 - (\mathbf{r} + \mathbf{r}_2)/2]^2}$$

$$= e^{-\alpha(\mathbf{r}_2 - \mathbf{r})^2 / 2} \left(\frac{\pi}{2\alpha} \right)^{3/2}. \tag{4.6.27}$$

We can now proceed to the $\mathbf{r}_2$ integration, which is of the form

$$I_2 = \left(\frac{\pi}{2\alpha} \right)^{3/2} \int_{\text{all space}} d\mathbf{r}_2 \, e^{-\alpha(\mathbf{r}_2 - \mathbf{r})^2 / 2} e^{-\alpha(\mathbf{r}_3 - \mathbf{r}_2)^2}. \tag{4.6.28}$$

Again, we begin by expanding the squares to yield

$$I_2 = \left(\frac{\pi}{2\alpha} \right)^{3/2} e^{-\alpha(\mathbf{r}^2 + 2\mathbf{r}_3^2)/2} \int_{\text{all space}} d\mathbf{r}_2 \, e^{-3\alpha[\mathbf{r}_2^2 - 2\mathbf{r}_2 \cdot (\mathbf{r} + 2\mathbf{r}_3)/3]/2}. \tag{4.6.29}$$

Completing the square gives

$$I_2 = \left(\frac{\pi}{2\alpha} \right)^{3/2} e^{-\alpha(\mathbf{r}^2 + 2\mathbf{r}_3^2)/2} e^{\alpha(\mathbf{r} + 2\mathbf{r}_3)^2 / 6} \int_{\text{all space}} d\mathbf{r}_2 \, e^{-3\alpha[\mathbf{r}_2 - (\mathbf{r} + 2\mathbf{r}_3)]^2 / 2}$$

$$= \left(\frac{\pi}{2\alpha} \right)^{3/2} \left(\frac{2\pi}{3\alpha} \right)^{3/2} e^{-\alpha(\mathbf{r} - \mathbf{r}_3)^2 / 3}$$

$$= \left(\frac{\pi^2}{3\alpha^2} \right)^{3/2} e^{-\alpha(\mathbf{r} - \mathbf{r}_3)^2 / 3}. \tag{4.6.30}$$

From the calculation of I_1 and I_2, a pattern can be discerned from which the result of performing all N integrations can be predicted. Specifically, after performing $n < N$ integrations, we find

$$I_n = \left(\frac{\pi^n}{(n+1)\alpha^n} \right)^{3/2} e^{-\alpha(\mathbf{r} - \mathbf{r}_{n+1})^2 / (n+1)}. \tag{4.6.31}$$

Thus, setting $n = N$, we obtain

$$I_N = \left(\frac{\pi^N}{(N+1)\alpha^N} \right)^{3/2} e^{-\alpha(\mathbf{r}-\mathbf{r}_{N+1})^2/(N+1)}. \tag{4.6.32}$$

Identifying $\mathbf{r}_{N+1} = \mathbf{r}'$ and attaching the prefactor $(2\pi m/\beta h^2)^{3N/2}$, we obtain the partition function for fixed $\mathbf{r}$ and $\mathbf{r}'$ as

$$Q(N, T, \mathbf{r}, \mathbf{r}') = \left(\frac{2\pi}{\beta h\omega} \right)^{3N} \frac{1}{(N+1)^{3/2}} e^{-\beta m\omega^2(\mathbf{r}-\mathbf{r}')^2/(N+1)}. \tag{4.6.33}$$

The volume dependence has dropped out because the integrations were extended over all space. Equation (4.6.33) can be regarded as a probability distribution function for the distance $|\mathbf{r}-\mathbf{r}'|^2$ between the endpoints of the polymer. Note that this distribution is Gaussian in the end-to-end distance $|\mathbf{r} - \mathbf{r}'|$.

If we now allow the endpoints to move, then the full partition function can be calculated by introducing the momenta $\mathbf{p}_0$ and $\mathbf{p}_{N+1}$ of the endpoints and performing the integration

$$Q(N, V, T) = \frac{1}{h^6} \left(\frac{2\pi}{\beta h\omega} \right)^{3N} \frac{1}{(N+1)^{3/2}} \int d\mathbf{p}_0 \, d\mathbf{p}_{N+1} \, e^{-\beta(\mathbf{p}_0^2+\mathbf{p}_{N+1}^2)/2m}$$

$$\times \int d\mathbf{r}_0 \, d\mathbf{r}_{N+1} \, e^{-\beta m\omega^2(\mathbf{r}_0-\mathbf{r}_{N+1})^2/(N+1)}. \tag{4.6.34}$$

Here, the extra factor of $1/h^6$ has been introduced along with the kinetic energy of the endpoints. Performing the momentum integrations gives

$$Q(N, V, T) = \frac{1}{h^6} \left(\frac{2\pi}{\beta h\omega} \right)^{3N} \left(\frac{2\pi m}{\beta} \right)^3 \frac{1}{(N+1)^{3/2}}$$

$$\times \int d\mathbf{r}_0 \, d\mathbf{r}_{N+1} \, e^{-\beta m\omega^2(\mathbf{r}_0-\mathbf{r}_{N+1})^2/(N+1)}. \tag{4.6.35}$$

We now introduce a change of variables to the center of mass $\mathbf{R} = (\mathbf{r}_0 + \mathbf{r}_{N+1})/2$ of the endpoint particles and their corresponding relative coordinate $\mathbf{s} = \mathbf{r}_0 - \mathbf{r}_{N+1}$. The Jacobian of the transformation is 1. With this transformation, we have

$$Q(N, V, T) = \frac{1}{h^6} \left(\frac{2\pi}{\beta h\omega} \right)^{3N} \left(\frac{2\pi m}{\beta} \right)^3 \frac{1}{(N+1)^{3/2}} \int d\mathbf{R} \, d\mathbf{s} \, e^{-\beta m\omega^2 \mathbf{s}^2/(N+1)}. \tag{4.6.36}$$

The integration over $\mathbf{s}$ can be performed over all space because the Gaussian rapidly decays to 0. However, the integration over the center of mass $\mathbf{R}$ is completely free and must be restricted to the containing volume V. The result of performing the last two coordinate integrations is

$$Q(N, V, T) = \left(\frac{V}{\lambda^3} \right) \left(\frac{2\pi}{\beta h\omega} \right)^{3(N+1)}, \tag{4.6.37}$$

where $\lambda = \sqrt{\beta h^2/2\pi m}$.

Now that we have seen how to perform the coordinate integrations directly, let us demonstrate how a change of integration variables in the partition function can simplify the problem considerably. The use of variable transformations in a partition function is a powerful technique that can lead to novel computational algorithms for solving complex problems (Tuckerman *et al.*, 1993; Zhu *et al.*, 2002; Minary *et al.*, 2007). Consider, once again, the polymer chain with fixed endpoints, so that the partition function is given by eqn. (4.6.23), and consider a change of integration variables from $\mathbf{r}_k$ to $\mathbf{u}_k$ given by

$$\mathbf{u}_k = \mathbf{r}_k - \frac{k\mathbf{r}_{k+1} + \mathbf{r}}{(k+1)}, \tag{4.6.38}$$

where, again, the condition $\mathbf{r}_{N+1} = \mathbf{r}'$ is implied. In order to express the harmonic coupling in terms of the new variables $\mathbf{u}_1, ..., \mathbf{u}_N$, we need the inverse of this transformation. Interestingly, if we simply solve eqn. (4.6.38) for $\mathbf{r}_k$, we obtain

$$\mathbf{r}_k = \mathbf{u}_k + \frac{k}{k+1}\mathbf{r}_{k+1} + \frac{1}{k+1}\mathbf{r}. \tag{4.6.39}$$

Note that eqn. (4.6.39) defines the inverse transformation *recursively*, since knowledge of how $\mathbf{r}_{k+1}$ depends on $\mathbf{u}_1, ..., \mathbf{u}_N$ allows the dependence of $\mathbf{r}_k$ on $\mathbf{u}_1, ..., \mathbf{u}_N$ to be determined. Consequently, the inversion process is "seeded" by starting with the $k = N$ term and working backwards to $k = 1$.

In order to illustrate how the recursive inverse works, consider the special case of $N = 3$. If we set $k = 3$ in eqn. (4.6.39), we find

$$\mathbf{r}_3 = \mathbf{u}_3 + \frac{3}{4}\mathbf{r}' + \frac{1}{4}\mathbf{r}, \tag{4.6.40}$$

where the fact that $\mathbf{r}_4 = \mathbf{r}'$ has been used. Next, setting $k = 2$,

$$\mathbf{r}_2 = \mathbf{u}_2 + \frac{2}{3}\mathbf{r}_3 + \frac{1}{3}\mathbf{r}$$

$$= \mathbf{u}_2 + \frac{2}{3}\left[\mathbf{u}_3 + \frac{3}{4}\mathbf{r}' + \frac{1}{4}\mathbf{r}\right] + \frac{1}{3}\mathbf{r}$$

$$= \mathbf{u}_2 + \frac{2}{3}\mathbf{u}_3 + \frac{1}{2}\mathbf{r}' + \frac{1}{2}\mathbf{r} \tag{4.6.41}$$

and similarly, we find that

$$\mathbf{r}_1 = \mathbf{u}_1 + \frac{1}{2}\mathbf{u}_2 + \frac{1}{3}\mathbf{u}_3 + \frac{1}{4}\mathbf{r}' + \frac{3}{4}\mathbf{r}. \tag{4.6.42}$$

Thus, if we now use these relations to evaluate $(\mathbf{r}'-\mathbf{r}_3)^2+(\mathbf{r}_3-\mathbf{r}_2)^2+(\mathbf{r}_2-\mathbf{r}_1)^2+(\mathbf{r}_1-\mathbf{r})^2$, after some algebra, we find

$$(\mathbf{r}'-\mathbf{r}_3)^2+(\mathbf{r}_3-\mathbf{r}_2)^2+(\mathbf{r}_2-\mathbf{r}_1)^2+(\mathbf{r}_1-\mathbf{r})^2 = 2\mathbf{u}_1^2+\frac{3}{2}\mathbf{u}_2^2+\frac{4}{3}\mathbf{u}_3^2+\frac{1}{4}(\mathbf{r}-\mathbf{r}')^2. \tag{4.6.43}$$

Extrapolating to arbitrary N, we have

$$\sum_{i=0}^{N}(\mathbf{r}_i - \mathbf{r}_{i+1})^2 = \sum_{i=1}^{N}\frac{i+1}{i}\mathbf{u}_i^2 + \frac{1}{N+1}(\mathbf{r}-\mathbf{r}')^2. \tag{4.6.44}$$

Finally, since the variable transformation must be applied to a multidimensional integral, we need to compute the Jacobian of the transformation. Consider, again, the special case of $N = 3$. For any of the spatial directions $\alpha = x, y, z$, the Jacobian matrix $J_{ij} = \partial r_{\alpha,i}/\partial u_{\alpha,j}$ is

$$J = \begin{pmatrix} 1 & 1/2 & 1/3 \\ 0 & 1 & 2/3 \\ 0 & 0 & 1 \end{pmatrix}. \tag{4.6.45}$$

This matrix, being both upper triangular and having 1s on the diagonal, has unit determinant, a fact that generalizes to arbitrary N, where the Jacobian matrix takes the form

$$J = \begin{pmatrix} 1 & 1/2 & 1/3 & 1/4 & \cdots & 1/N \\ 0 & 1 & 2/3 & 2/4 & \cdots & 2/N \\ 0 & 0 & 1 & 3/4 & \cdots & 3/N \\ \vdots & \vdots & \vdots & \vdots & \cdots & \vdots \\ 0 & 0 & 0 & 0 & \cdots & 1 \end{pmatrix}. \tag{4.6.46}$$

Thus, substituting the transformation into eqn. (4.6.24), we obtain

$$Q(N,V,T,\mathbf{r},\mathbf{r}') = \frac{1}{h^{3N}}\left(\frac{2\pi m}{\beta}\right)^{3N/2} e^{-\beta m\omega^2(\mathbf{r}-\mathbf{r}')^2/N+1}$$

$$\times \int d^N\mathbf{u} \exp\left[-\frac{1}{2}\beta m\omega^2 \sum_{i=1}^{N}\frac{i+1}{i}\mathbf{u}^2\right]. \tag{4.6.47}$$

Now, each of the integrals over $\mathbf{u}_1, ..., \mathbf{u}_N$ can be performed independently and straightforwardly to give

$$Q(N,V,T,\mathbf{r},\mathbf{r}') =$$

$$\frac{1}{h^{3N}}\left(\frac{2\pi m}{\beta}\right)^{3N/2}\left(\frac{2\pi}{\beta m\omega^2}\right)^{3N/2} e^{-\beta m\omega^2(\mathbf{r}-\mathbf{r}')^2/N+1}\prod_{i=1}^{N}\left(\frac{i}{i+1}\right)^{3/2}. \tag{4.6.48}$$

Expanding the product, we find

$$\prod_{i=1}^{N}\left(\frac{i}{i+1}\right)^{3/2} = \left(\prod_{i=1}^{N}\frac{i}{i+1}\right)^{3/2}$$

$$= \left(\frac{1}{2}\frac{2}{3}\frac{3}{4}\cdots\frac{N-1}{N}\frac{N}{N+1}\right)^{3/2}$$

$$= \left(\frac{1}{N+1}\right)^{3/2}. \tag{4.6.49}$$

Thus, substituting this result into eqn. (4.6.48) yields eqn. (4.6.33).

Finally, let us use the partition function expressions in eqns. (4.6.37) and (4.6.33) to compute an observable, specifically, the expectation value $\langle |\mathbf{r} - \mathbf{r}'|^2 \rangle$, known as the *mean-square end-to-end distance* of the polymer. From eqn. (4.6.35), we can set up the expectation value as

$$\langle |\mathbf{r} - \mathbf{r}'|^2 \rangle = \frac{1}{Q(N,V,T)} \frac{1}{h^6} \left(\frac{2\pi}{\beta h \omega} \right)^{3N} \left(\frac{2\pi m}{\beta} \right)^3$$

$$\times \frac{1}{(N+1)^{3/2}} \int \mathrm{d}\mathbf{r}\,\mathrm{d}\mathbf{r}'\, |\mathbf{r} - \mathbf{r}'|^2 \mathrm{e}^{-\beta m \omega^2 (\mathbf{r} - \mathbf{r}')^2 / (N+1)}. \qquad (4.6.50)$$

Using the fact that $1/Q(N,V,T) = (\lambda^3/V)(\beta h \omega/2\pi)^{3(N+1)}$, and transforming to center-of-mass ($\mathbf{R}$) and relative ($\mathbf{s}$) coordinates yields

$$\langle |\mathbf{r} - \mathbf{r}'|^2 \rangle = \left(\frac{\lambda^3}{V} \right) \left(\frac{\beta h \omega}{2\pi} \right)^{3N+3} \frac{1}{h^6} \left(\frac{2\pi}{\beta h \omega} \right)^{3N} \left(\frac{2\pi m}{\beta} \right)^3$$

$$\times \frac{1}{(N+1)^{3/2}} \int \mathrm{d}\mathbf{R}\,\mathrm{d}\mathbf{s}\, s^2 \mathrm{e}^{-\beta m \omega^2 s^2 / (N+1)}. \qquad (4.6.51)$$

The integration over $\mathbf{R}$ yields, again, a factor of V, which cancels the V factor in the denominator. For the $\mathbf{s}$ integration, we change to spherical polar coordinates, which yields

$$\langle |\mathbf{r} - \mathbf{r}'|^2 \rangle = \lambda^3 \left(\frac{\beta h \omega}{2\pi} \right)^{3N+3} \frac{1}{h^6} \left(\frac{2\pi}{\beta h \omega} \right)^{3N} \left(\frac{2\pi m}{\beta} \right)^3$$

$$\times \frac{4\pi}{(N+1)^{3/2}} \int_0^\infty \mathrm{d}s\, s^4 \mathrm{e}^{-\beta m \omega^2 s^2 / (N+1)}. \qquad (4.6.52)$$

A useful trick for performing integrals of the form $\int_0^\infty \mathrm{d}x\, x^{2n} \exp(-\alpha x^2)$ is to express them as

$$\int_0^\infty \mathrm{d}x\, x^{2n} \mathrm{e}^{-\alpha x^2} = (-1)^n \frac{\partial^n}{\partial \alpha^n} \int_0^\infty \mathrm{d}x\, \mathrm{e}^{-\alpha x^2}$$

$$= (-1)^n \frac{\partial^n}{\partial \alpha^n} \frac{1}{2} \sqrt{\frac{\pi}{\alpha}}$$

$$= \frac{(n+1)!!}{2 \cdot 2^n \alpha^n} \sqrt{\frac{\pi}{\alpha}}. \qquad (4.6.53)$$

Applying eqn. (4.6.53) yields, after some algebra,

$$\langle |\mathbf{r} - \mathbf{r}'|^2 \rangle = \frac{3}{4\sqrt{2}} \frac{(N+1)}{\beta m \omega^2}. \qquad (4.6.54)$$

The mean-square end-to-end distance increases both with temperature and with the number of repeat units in the polymer. Because of the latter, mean-square end-to-end distances are often reported in dimensionless form as $\langle |\mathbf{r} - \mathbf{r}'|^2 \rangle / (N d_0^2)$, where d_0 is some reference distance that is characteristic of the system. In an alkane chain, for example, d_0 might be the equilibrium carbon–carbon distance.

4.7   Structure and thermodynamics in real gases and liquids from spatial distribution functions

Characterizing the equilibrium properties of real systems is a significant challenge due to the rich variety of behavior arising from the particle interactions. In real gases and liquids, among the most useful properties that can be described statistically are the spatial distribution functions. That spatial correlations exist between the individual components of a system can be seen most dramatically in the example of liquid water at room temperature. Because a water molecule is capable of forming hydrogen bonds with other water molecules, liquid water is best described as a complex network of hydrogen bonds. Within this network, there is a well-defined local structure that arises from the fact that each water molecule can both donate and accept hydrogen bonds. Although it might seem natural to try to characterize this coordination shell in terms of a set of distances between the molecules, such an attempt misses something fundamental about the system. In a liquid at finite temperature, the individual atoms are constantly in motion, and distances are constantly fluctuating, as is the coordination pattern. Hence, a more appropriate measure of the solvation structure is a distribution function of distances in the coordination structure. In such a distribution, we would expect peaks at particular values characteristic of the average structural motifs present in the system, with the peak widths largely determined by the temperature, density, *etc.* This argument suggests that the spatial distribution functions in a system contain a considerable amount of information about the local structure and the fluctuations. In this section, we will discuss the formulation of such distribution functions as ensemble averages and relate these functions to the thermodynamics of the system.

We begin the discussion with the canonical partition function for a system of N identical particles interacting via a potential $U(\mathbf{r}_1, ..., \mathbf{r}_N)$.

$$Q(N, V, T) =$$

$$\frac{1}{N! h^{3N}} \int d^N \mathbf{p} \int_{D(V)} d^N \mathbf{r} \, \exp\left\{ -\beta \left[\sum_{i=1}^{N} \frac{\mathbf{p}_i^2}{2m} + U(\mathbf{r}_1, ..., \mathbf{r}_N) \right] \right\}. \quad (4.7.1)$$

Since the momentum integrations can be evaluated independently, the partition function can also be expressed as

$$Q(N, V, T) = \frac{1}{N! \lambda^{3N}} \int_{D(V)} d^N \mathbf{r} \, e^{-\beta U(\mathbf{r}_1, ..., \mathbf{r}_N)}. \quad (4.7.2)$$

Note that in the Hamiltonian

$$\mathcal{H} = \sum_{i=1}^{N} \frac{\mathbf{p}_i^2}{2m} + U(\mathbf{r}_1, ..., \mathbf{r}_N), \quad (4.7.3)$$

the kinetic energy term is a universal term that appears in all such Hamiltonians. It is only the potential $U(\mathbf{r}_1, ..., \mathbf{r}_N)$ that determines the particular properties of the system. In order to make this fact manifest in the partition function, we introduce the *configurational partition function*

$$Z(N, V, T) = \int_{D(V)} d\mathbf{r}_1 \cdots d\mathbf{r}_N \ e^{-\beta U(\mathbf{r}_1,...,\mathbf{r}_N)}, \tag{4.7.4}$$

in terms of which, $Q(N, V, T) = Z(N, V, T)/(N!\lambda^{3N})$. Note that the ensemble average of any coordinate-dependent function $a(\mathbf{r}_1, ..., \mathbf{r}_N)$ can be expressed as

$$\langle a \rangle = \frac{1}{Z} \int_{D(V)} d\mathbf{r}_1 \cdots d\mathbf{r}_N \ a(\mathbf{r}_1, ..., \mathbf{r}_N) e^{-\beta U(\mathbf{r}_1,...,\mathbf{r}_N)}. \tag{4.7.5}$$

(Throughout the discussion, the arguments of the configurational partition function $Z(N, V, T)$ will be left off for notational simplicity.) From the form of eqn. (4.7.4), we see that the probability of finding particle 1 in a small volume element $d\mathbf{r}_1$ about the point $\mathbf{r}_1$ and particle 2 in a small volume element $d\mathbf{r}_2$ about the point $\mathbf{r}_2$,..., and particle N in a small volume element $d\mathbf{r}_N$ about the point $\mathbf{r}_N$ is

$$P^{(N)}(\mathbf{r}_1, ..., \mathbf{r}_N) d\mathbf{r}_1 \cdots d\mathbf{r}_N = \frac{1}{Z} e^{-\beta U(\mathbf{r}_1,...,\mathbf{r}_N)} d\mathbf{r}_1 \cdots d\mathbf{r}_N. \tag{4.7.6}$$

Now suppose that we are interested in the probability of finding only the first $n < N$ particles in small volume elements about the points $\mathbf{r}_1, ..., \mathbf{r}_n$, respectively, independent of the locations of the remaining $n+1, ..., N$ particles. This probability can be obtained by simply integrating eqn. (4.7.6) over the last $N - n$ particles:

$$P^{(n)}(\mathbf{r}_1, ..., \mathbf{r}_n) d\mathbf{r}_1 \cdots d\mathbf{r}_n =$$

$$\frac{1}{Z} \left[\int_{D(V)} d\mathbf{r}_{n+1} \cdots d\mathbf{r}_N \ e^{-\beta U(\mathbf{r}_1,...,\mathbf{r}_N)} \right] d\mathbf{r}_1 \cdots d\mathbf{r}_n. \tag{4.7.7}$$

Since the particles are indistinguishable, we are actually interested in the probability of finding *any* particle in a volume element $d\mathbf{r}_1$ about the point $\mathbf{r}_1$ and *any* particle in $d\mathbf{r}_2$ about the point $\mathbf{r}_2$, *etc.*, which is given by the distribution

$$\rho^{(n)}(\mathbf{r}_1, ..., \mathbf{r}_n) d\mathbf{r}_1 \cdots d\mathbf{r}_n =$$

$$\frac{N!}{(N - n)! Z} \left[\int_{D(V)} d\mathbf{r}_{n+1} \cdots d\mathbf{r}_N \ e^{-\beta U(\mathbf{r}_1,...,\mathbf{r}_N)} \right] d\mathbf{r}_1 \cdots d\mathbf{r}_n. \tag{4.7.8}$$

The prefactor $N!/(N - n)! = N(N - 1)(N - 2) \cdots (N - n + 1)$ comes from the fact that the first particle can be chosen in N ways, the second particle in $N - 1$ ways, the third particle in $N - 2$ ways and so forth.

Equation (4.7.8) is really a measure of the spatial correlations among particles. If, for example, the potential $U(\mathbf{r}_1, ..., \mathbf{r}_N)$ is attractive at long and intermediate range, then the presence of a particle at $\mathbf{r}_1$ will increase the probability that another particle will be in its vicinity. If the potential contains strong repulsive regions, then a particle at $\mathbf{r}_1$ will increase the probability of a void in its vicinity. More formally, an n-particle correlation function is defined in terms of $\rho^{(n)}(\mathbf{r}_1, ..., \mathbf{r}_n)$ via

$$g^{(n)}(\mathbf{r}_1, ..., \mathbf{r}_n) = \frac{1}{\rho^n} \rho^{(n)}(\mathbf{r}_1, ..., \mathbf{r}_n)$$

$$= \frac{N!}{(N-n)!\rho^n Z} \int_{D(V)} d\mathbf{r}_{n+1} \cdots d\mathbf{r}_N \, e^{-\beta U(\mathbf{r}_1, ..., \mathbf{r}_N)}, \qquad (4.7.9)$$

where $\rho = N/V$ is the number density. The n-particle correlation function eqn. (4.7.9) can also be formulated in an equivalent manner by introducing δ-functions for the first n particles and integrating over all N particles:

$$g^{(n)}(\mathbf{r}_1, ..., \mathbf{r}_n) = \frac{N!}{(N-n)!\rho^n Z} \int_{D(V)} d\mathbf{r}'_1 \cdots d\mathbf{r}'_N e^{-\beta U(\mathbf{r}'_1, ..., \mathbf{r}'_N)} \prod_{i=1}^{n} \delta(\mathbf{r}_i - \mathbf{r}'_i). \quad (4.7.10)$$

Note that eqn. (4.7.10) is equivalent to an ensemble average of the quantity

$$\prod_{i=1}^{n} \delta(\mathbf{r}_i - \mathbf{r}'_i)$$

using $\mathbf{r}'_1, ..., \mathbf{r}'_N$ as integration variables:

$$g^{(n)}(\mathbf{r}_1, ..., \mathbf{r}_n) = \frac{N!}{(N-n)!\rho^n} \left\langle \prod_{i=1}^{n} \delta(\mathbf{r}_i - \mathbf{r}'_i) \right\rangle_{\mathbf{r}'_1, ..., \mathbf{r}'_N}. \qquad (4.7.11)$$

Of course, the most important cases of eqn. (4.7.11) are the first few integers for n. If $n = 1$, for example, $g^{(1)}(\mathbf{r}) = V\rho^{(1)}(\mathbf{r})$, where $\rho^{(1)}(\mathbf{r})d\mathbf{r}$ is the probability of finding a particle in $d\mathbf{r}$. For a perfect crystal, $\rho^{(1)}(\mathbf{r})$ is a periodic function, but in a liquid, due to isotropy, $\rho^{(1)}(\mathbf{r})$ is independent of $\mathbf{r}$. Thus, since $\rho^{(1)}(\mathbf{r}) = (1/V)g^{(1)}(\mathbf{r})$, and $\rho^{(1)}(\mathbf{r})$ is a probability

$$\frac{1}{N} \int d\mathbf{r}\, \rho^{(1)}(\mathbf{r}) = 1 = \frac{1}{V} \int d\mathbf{r}\, g^{(1)}(\mathbf{r}). \qquad (4.7.12)$$

However, $g^{(1)}(\mathbf{r})$ is also independent of $\mathbf{r}$ for an isotropic system, in which case, eqn. (4.7.12) implies that $g^{(1)}(\mathbf{r}) = 1$.

4.7.1 The pair correlation function and the radial distribution function

The case $n = 2$ is of particular interest. The function $g^{(2)}(\mathbf{r}_1, \mathbf{r}_2)$ that results when $n = 2$ is used in eqn. (4.7.11) is called the *pair correlation function*.

$$g^{(2)}(\mathbf{r}_1, \mathbf{r}_2) = \frac{1}{Z} \frac{N(N-1)}{\rho^2} \int_{D(V)} d\mathbf{r}_3 \cdots d\mathbf{r}_N \, e^{-\beta U(\mathbf{r}_1, \mathbf{r}_2, \mathbf{r}_3 ..., \mathbf{r}_N)}$$

$$= \frac{N(N-1)}{\rho^2} \langle \delta(\mathbf{r}_1 - \mathbf{r}'_1)\delta(\mathbf{r}_2 - \mathbf{r}'_2) \rangle_{\mathbf{r}'_1, ..., \mathbf{r}'_N}. \qquad (4.7.13)$$

Although eqn. (4.7.13) suggests that $g^{(2)}$ depends on $\mathbf{r}_1$ and $\mathbf{r}_2$ individually, in a homogeneous system such as a liquid, we anticipate that $g^{(2)}$ actually depends only on the relative position between two particles. Thus, it is useful to introduce a change of variables to center-of-mass and relative coordinates of particles 1 and 2:

$$\mathbf{R} = \frac{1}{2}(\mathbf{r}_1 + \mathbf{r}_2) \qquad \mathbf{r} = \mathbf{r}_1 - \mathbf{r}_2. \qquad (4.7.14)$$

The inverse of this transformation is

$$\mathbf{r}_1 = \mathbf{R} + \frac{1}{2}\mathbf{r} \qquad \mathbf{r}_2 = \mathbf{R} - \frac{1}{2}\mathbf{r}, \qquad (4.7.15)$$

and its Jacobian is unity: $d\mathbf{R}d\mathbf{r} = d\mathbf{r}_1 d\mathbf{r}_2$. Defining $\tilde{g}^{(2)}(\mathbf{r}, \mathbf{R}) = g^{(2)}(\mathbf{R}+\mathbf{r}/2, \mathbf{R}-\mathbf{r}/2)$, we find that

$$\tilde{g}^{(2)}(\mathbf{r}, \mathbf{R}) = \frac{N(N-1)}{\rho^2 Z} \int_{D(V)} d\mathbf{r}_3 \cdots d\mathbf{r}_N \, e^{-\beta U(\mathbf{R}+\frac{1}{2}\mathbf{r}, \mathbf{R}-\frac{1}{2}\mathbf{r}, \mathbf{r}_3, \dots, \mathbf{r}_N)}$$

$$= \frac{N(N-1)}{\rho^2} \left\langle \delta\left(\mathbf{R} + \frac{1}{2}\mathbf{r} - \mathbf{r}_1'\right) \delta\left(\mathbf{R} - \frac{1}{2}\mathbf{r} - \mathbf{r}_2'\right) \right\rangle_{\mathbf{r}_1', \dots, \mathbf{r}_N'}. \quad (4.7.16)$$

In a homogeneous system, the location of the particle pair, determined by the center-of-mass coordinate $\mathbf{R}$, is of little interest since, on average, the distribution of particles around a given pair does not depend on where the pair is in the system. Thus, we integrate over $\mathbf{R}$, yielding a new function $\tilde{g}(\mathbf{r})$ defined as

$$\tilde{g}(\mathbf{r}) \equiv \frac{1}{V} \int_{D(V)} d\mathbf{R} \, \tilde{g}^{(2)}(\mathbf{r}, \mathbf{R})$$

$$= \frac{(N-1)}{\rho Z} \int_{D(V)} d\mathbf{R} \, d\mathbf{r}_3 \cdots d\mathbf{r}_N e^{-\beta U(\mathbf{R}+\frac{1}{2}\mathbf{r}, \mathbf{R}-\frac{1}{2}\mathbf{r}, \mathbf{r}_3, \dots, \mathbf{r}_N)}$$

$$= \frac{(N-1)}{\rho} \langle \delta(\mathbf{r} - \mathbf{r}') \rangle_{\mathbf{R}', \mathbf{r}', \mathbf{r}_3', \dots, \mathbf{r}_N'}, \qquad (4.7.17)$$

where the last line follows from eqn. (4.7.16) by integrating one of the δ-functions over $\mathbf{R}$ and renaming $\mathbf{r}_1' - \mathbf{r}_2' = \mathbf{r}'$. Next, we recognize that a system such as a liquid is spatially isotropic, so that there are no preferred directions in space. Thus, the correlation function should only depend on the distance between the two particles, that is, on the magnitude $|\mathbf{r}|$. Thus, we introduce the spherical polar resolution of the vector $\mathbf{r} = (x, y, z)$

$$x = r \sin\theta \cos\phi$$
$$y = r \sin\theta \sin\phi$$
$$z = r \cos\theta, \qquad (4.7.18)$$

where θ is the polar angle and ϕ is the azimuthal angle. Defining the unit vector $\mathbf{n} = (\sin\theta \cos\phi, \sin\theta \sin\phi, \cos\theta)$, it is clear that $\mathbf{r} = r\mathbf{n}$. Also, the Jacobian is $dxdydz = r^2 \sin\theta dr d\theta d\phi$. Thus, integrating $\tilde{g}(\mathbf{r})$ over angles gives a new function

$$g(r) = \frac{1}{4\pi} \int_0^{2\pi} d\phi \int_0^{\pi} d\theta \sin\theta \tilde{g}(\mathbf{r})$$

$$= \frac{(N-1)}{4\pi\rho Z} \int_0^{2\pi} d\phi \int_0^{\pi} d\theta \sin\theta \int_{D(V)} d\mathbf{R} d\mathbf{r}_3 \cdots d\mathbf{r}_N$$

$$\times e^{-\beta U(\mathbf{R}+\frac{1}{2}r\mathbf{n}, \mathbf{R}-\frac{1}{2}r\mathbf{n}, \mathbf{r}_3, \dots, \mathbf{r}_N)}$$

$$= \frac{(N-1)}{4\pi\rho r^2} \langle \delta(r-r') \rangle_{r',\theta',\phi',\mathbf{R}',\mathbf{r}_3',\dots,\mathbf{r}_N'}, \tag{4.7.19}$$

known as the *radial distribution function*. The last line follows from the identity

$$\delta(\mathbf{r}-\mathbf{r}') = \frac{\delta(r-r')}{rr'} \delta(\cos\theta - \cos\theta')\delta(\phi-\phi'). \tag{4.7.20}$$

From the foregoing analysis, we see that the radial distribution function is a measure of the probability of finding two particles a distance r apart under the conditions of the canonical ensemble.

As an example of a radial distribution function, consider a system of N identical particles interacting via the pairwise additive Lennard-Jones potential of eqn. (3.14.3). The potential between any two particles is shown in Fig. 4.2(a), where we can clearly see an attractive well at $r = 2^{1/6}\sigma$ of depth ϵ. The radial distribution function for such a system with $\sigma = 3.405$ Å, $\epsilon = 119.8$ K, $m = 39.948$ amu, $\rho = 0.02$ Å^{-3} ($\rho^* = 0.8$) and a range of temperatures corresponding to liquid conditions is shown in Fig. 4.2(b). In all cases, the radial distribution functions show a pronounced peak in the range $r = 3.6$-3.7 Å depending on temperature, compared to the location of the potential energy minimum $r = 3.82$ Å. The presence of such a peak in the radial distribution function

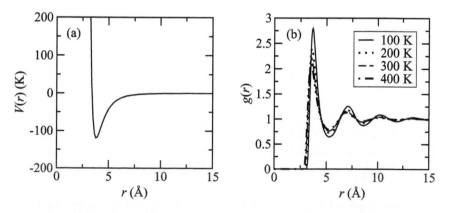

Fig. 4.2 (a) Potential as a function of the distance r between two particles with $\sigma = 3.405$ Å and $\epsilon = 119.8$ K. (b) Radial distribution functions at four temperatures.

indicates a well-defined coordination structure in the liquid. Figure 4.2 also shows clear secondary peaks at larger distances, indicating second and third solvation shell structures around each particle. We see, therefore, that spatial correlations survive out to at least two solvation shells at the higher temperatures and three (or nearly 11 Å) at the lower temperatures.

Note that the integral of $g(r)$ over all distances gives

$$4\pi\rho \int_0^\infty r^2 g(r)\, \mathrm{d}r = N - 1 \approx N, \tag{4.7.21}$$

indicating that if we integrate over the correlation function, we must find all of the particles. Equation (4.7.21) further suggests that the integration of the radial distribution function under the first peak should yield the number of particles coordinating a given particle in its first solvation shell. This number, known as the *coordination number*, can be written as

$$N_1 = 4\pi\rho \int_0^{r_{\min}} r^2 g(r)\, \mathrm{d}r, \tag{4.7.22}$$

where $r_{\min}$ is the location of the first minimum of $g(r)$. In fact, a more general "running" coordination number, defined as the average number of particles coordinating a given particle out to a distance r, can be calculated via according to

$$N(r) = 4\pi\rho \int_0^r \tilde{r}^2 g(\tilde{r})\, \mathrm{d}\tilde{r}. \tag{4.7.23}$$

It is clear that $N_1 = N(r_{\min})$. As an illustration of the running coordination number, we show a plot of the oxygen–oxygen and oxygen–hydrogen radial distribution func-

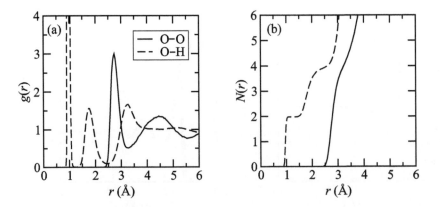

Fig. 4.3 (a) Oxygen–oxygen (O–O) and oxygen–hydrogen (O–H) radial distribution functions for a particular model of water (Lee and Tuckerman, 2006; Marx *et al.*, 2010). (b) The corresponding running coordination numbers computed from eqn. (4.7.23).

tions for a particular model of water (Lee and Tuckerman, 2006; Marx *et al.*, 2010) in Fig. 4.3(a) and the corresponding running coordination numbers in Fig. 4.3(b). For the oxygen–oxygen running coordination number, the plot is nearly linear except for a slight deviation in this trend around $N(r) = 4$. The r value of this deviation corresponds to the first minimum in the oxygen–oxygen radial distribution function of Fig. 4.3(a) and indicates a solvation shell with a coordination number close to 4 for this model. By contrast, the oxygen–hydrogen running coordination number shows more clearly defined plateaus at $N(r) = 2$ and $N(r) = 4$. The first plateau corresponds to the first minimum in the O–H radial distribution function and counts the two covalently bonded hydrogens to an oxygen. The second plateau counts two additional hydrogens that are donated in hydrogen bonds to the oxygen in the first solvation shell. The plateaus in the O–H running coordination number plot are more pronounced than in the O–O plot because the peaks in the O–H radial distribution function are sharper with correspondingly deeper minima due to the directionality of water's hydrogen-bonding pattern.

4.7.2 Scattering intensities and the radial distribution function

An important property of the radial distribution function is that many useful observables can be expressed in terms of $g(r)$. These include neutron or X-ray scattering intensities and various thermodynamic quantities. In this and the next subsections, we will analyze this aspect of radial distribution functions.

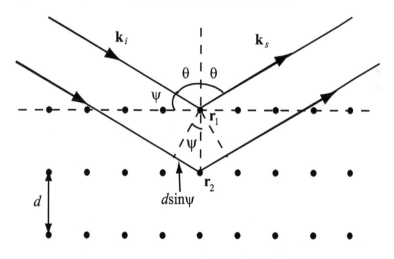

Fig. 4.4 Illustration of Bragg scattering from neighboring planes in a crystal

Let us first review the simple Bragg scattering experiment from ordered planes in a crystal, illustrated in Fig. 4.4. Recall that the condition for constructive interference is that the total path difference between radiation scattered from two different planes is an integral number of wavelengths. Since the path difference (see Fig. 4.4) is $2d \sin \psi$, where d is the distance between the planes, the condition can be expressed as

$$2d \sin \psi = n\lambda, \tag{4.7.24}$$

where λ is the wavelength of the radiation used. However, we can look at the scattering experiment in another way. Consider two atoms in the crystal at points $\mathbf{r}_1$ and $\mathbf{r}_2$ (see figure), with $\mathbf{r}_1 - \mathbf{r}_2$ the relative vector between them. Let $\mathbf{k}_i$ and $\mathbf{k}_s$ be the wave vectors of the incident and scattered radiation, respectively. Since the form of a free wave is $\exp(\pm i\mathbf{k} \cdot \mathbf{r})$, the phase of the incident wave at the point $\mathbf{r}_2$ is just $-\mathbf{k}_i \cdot \mathbf{r}_2$ (the negative sign arising from the fact that the wave is incoming), while the phase at $\mathbf{r}_1$ is $-\mathbf{k}_i \cdot \mathbf{r}_1$. Thus, the phase difference of the incident wave between the two points is $-\mathbf{k}_i \cdot (\mathbf{r}_1 - \mathbf{r}_2)$. If θ is the angle between $-\mathbf{k}_i$ and $\mathbf{r}_1 - \mathbf{r}_2$, then this phase difference can be written as

$$\delta\phi_i = -|\mathbf{k}_i||\mathbf{r}_1 - \mathbf{r}_2| \cos{(\pi - \theta)} = |\mathbf{k}_i||\mathbf{r}_1 - \mathbf{r}_2| \cos\theta = \frac{2\pi}{\lambda} d \cos\theta. \tag{4.7.25}$$

By a similar analysis, the phase difference of the scattered radiation between points $\mathbf{r}_1$ and $\mathbf{r}_2$ is

$$\delta\phi_s = |\mathbf{k}_s||\mathbf{r}_1 - \mathbf{r}_2| \cos\theta = \frac{2\pi}{\lambda} d \cos\theta. \tag{4.7.26}$$

The total phase difference is just the sum

$$\delta\phi = \delta\phi_i + \delta\phi_s = \mathbf{q} \cdot (\mathbf{r}_1 - \mathbf{r}_2) = \frac{4\pi}{\lambda} d \cos\theta, \tag{4.7.27}$$

where $\mathbf{q} = \mathbf{k}_s - \mathbf{k}_i$ is the *momentum transfer*. For constructive interference, the total phase difference must be an 2π times an integer, giving the equivalent Bragg condition

$$\frac{4\pi}{\lambda} d \cos\theta = 2\pi n$$

$$2d \cos\theta = n\lambda. \tag{4.7.28}$$

Since $\theta = \pi/2 - \psi$, $\cos\theta = \sin\psi$, and the original Bragg condition is recovered.

This simple analysis suggests that a similar scattering experiment performed in a liquid could reveal the presence of ordered structures, *i.e.*, significant probability that two atoms will be found a particular distance r apart, leading to a peak in the radial distribution function. If two atoms in a well-defined structure are at positions $\mathbf{r}_1$ and $\mathbf{r}_2$, then the function $\exp[i\mathbf{q} \cdot (\mathbf{r}_1 - \mathbf{r}_2)]$ will peak when the phase difference is an integer multiple of 2π. Of course, we need to consider all possible pairs of atoms, and we need to average over an ensemble because the atoms are constantly in motion. We, therefore, introduce a scattering function

$$S(\mathbf{q}) \propto \frac{1}{N} \left\langle \sum_{i,j} \exp{(i\mathbf{q} \cdot (\mathbf{r}_i - \mathbf{r}_j))} \right\rangle. \tag{4.7.29}$$

Note that eqn. (4.7.29) also contains terms involving the interference of incident and scattered radiation from the same atom. Moreover, the quantity inside the angle brackets is purely real, which becomes evident by writing the double sum as the square of a single sum:

$$S(\mathbf{q}) \propto \frac{1}{N} \left\langle \left| \sum_i \exp\left(i\mathbf{q}\cdot\mathbf{r}_i\right) \right|^2 \right\rangle. \qquad (4.7.30)$$

The function $S(\mathbf{q})$ is called the *structure factor*. Its precise shape will depend on certain details of the apparatus and type of radiation used. Indeed, $S(\mathbf{q})$ could also include **q**-dependent form factors, which is why eqns. (4.7.29) and (4.7.30) are written as proportionalities. For isotropic systems, $S(\mathbf{q})$ should only depend on the magnitude $|\mathbf{q}|$ of **q**, since there are no preferred directions in space. In this case, it is straightforward to show (see Problem 4.12) that $S(q)$ is related to the radial distribution function by

$$S(q) = 4\pi\rho \int_0^\infty dr\ r^2\left(g(r) - 1\right)\frac{\sin qr}{qr}. \qquad (4.7.31)$$

If a system contains several chemical species, then radial distribution functions $g_{\alpha\beta}(r)$ among the different species, known as *partial radial distribution functions*, can be introduced (see Fig. 4.3). Here, α and β range over the different species, with $g_{\alpha\beta}(r) = g_{\beta\alpha}(r)$. Equation (4.7.31) then generalizes to

$$S_{\alpha\beta}(q) = 4\pi\rho \int_0^\infty dr\ r^2\left(g_{\alpha\beta}(r) - 1\right)\frac{\sin qr}{qr}. \qquad (4.7.32)$$

$S_{\alpha\beta}(r)$ are called the *partial structure factors*, and ρ is the full atomic number density. Figure 4.5(a) shows the structure factor, $S(q)$, for the Lennard-Jones system studied in Fig. 4.2. Figure 4.5(b) shows a more realistic example of the N–N partial structure factor for liquid ammonia measured via neutron scattering (Ricci *et al.*, 1995). In both cases, the peaks occur at wavelengths where constructive interference occurs. Although it is not straightforward to read the structural features of a system off a plot of $S(q)$,

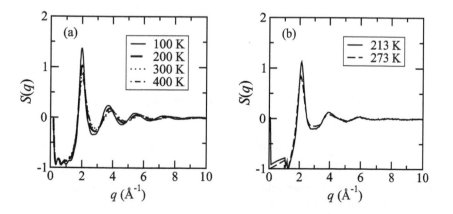

Fig. 4.5 (a) Structure factors corresponding to the radial distribution functions in Fig. 4.2. (b) N–N partial structure factors for liquid ammonia at 213 K and 273 K from Ricci *et al.* (1995).

examination of eqn. (4.7.31) shows that at values of r where $g(r)$ peaks, there will be corresponding peaks in $S(q)$ for those values of q for which $\sin(qr)/qr$ is maximal. The similarity between the structure factors of Fig. 4.5(a) and 4.5(b) indicate that, in both systems, London dispersion forces play an important role in their structural features.

4.7.3   Thermodynamic quantities from the radial distribution function

The spatial distribution functions discussed previously can be used to express a number of important thermodynamic properties of a system. Consider first the total internal energy. In the canonical ensemble, this is given by the thermodynamic derivative

$$E = -\frac{\partial}{\partial \beta} \ln Q(N, V, T). \tag{4.7.33}$$

Since $Q(N, V, T) = Z(N, V, T)/(N!\lambda^{3N})$, it follows that

$$E = -\frac{\partial}{\partial \beta} \left[\ln Z(N, V, T) - \ln N! - 3N \ln \lambda \right]. \tag{4.7.34}$$

Recall that λ is temperature dependent, so that $\partial \lambda/\partial \beta = \lambda/(2\beta)$. Thus, the energy is given by

$$\begin{aligned} E &= \frac{3N}{\lambda} \frac{\partial \lambda}{\partial \beta} - \frac{\partial \ln Z}{\partial \beta} \\ &= \frac{3N}{2} kT - \frac{1}{Z} \frac{\partial Z}{\partial \beta}. \end{aligned} \tag{4.7.35}$$

From eqn. (4.7.4), we obtain

$$-\frac{1}{Z} \frac{\partial Z}{\partial \beta} = \frac{1}{Z} \int d\mathbf{r}_1 \cdots d\mathbf{r}_N \, U(\mathbf{r}_1, ..., \mathbf{r}_N) e^{-\beta U(\mathbf{r}_1, ..., \mathbf{r}_N)} = \langle U \rangle, \tag{4.7.36}$$

and the total energy becomes

$$E = \frac{3}{2} NkT + \langle U \rangle. \tag{4.7.37}$$

Moreover, since $3NkT/2 = \langle \sum_{i=1}^{N} \mathbf{p}_i^2/2m_i \rangle$, we can write eqn. (4.7.37) as

$$E = \left\langle \sum_{i=1}^{N} \frac{\mathbf{p}_i^2}{2m_i} + U(\mathbf{r}_1, ..., \mathbf{r}_N) \right\rangle = \langle \mathcal{H}(\mathbf{r}, \mathbf{p}) \rangle, \tag{4.7.38}$$

which is just the sum of the average kinetic and average potential energies over the canonical ensemble. In eqns. (4.7.37) and (4.7.38), we have expressed a thermodynamic quantity as an ensemble average of a phase-space function. Such a phase-space function is referred to as an instantaneous *estimator* for the corresponding thermodynamic quantity. For the internal energy E, it should come as no surprise that the corresponding estimator is just the Hamiltonian $\mathcal{H}(\mathbf{r}, \mathbf{p})$.

Let us now apply eqn. (4.7.37) to a pair potential, such as that of eqn. (3.14.3). Taking the general form of the potential to be

$$U_{\text{pair}}(\mathbf{r}_1, ..., \mathbf{r}_N) = \sum_{i=1}^{N-1} \sum_{j=i+1}^{N} u(|\mathbf{r}_i - \mathbf{r}_j|),$$

(4.7.39)

the ensemble average of U_{pair} becomes

$$\langle U_{\text{pair}} \rangle = \frac{1}{Z} \sum_{i=1}^{N-1} \sum_{j=i+1}^{N} \int d\mathbf{r}_1 \cdots d\mathbf{r}_N \, u(|\mathbf{r}_i - \mathbf{r}_j|) e^{-\beta U_{\text{pair}}(\mathbf{r}_1, ..., \mathbf{r}_N)}.$$

(4.7.40)

Note, however, that every term in the sum over i and j in the above expression can be transformed into

$$\int d\mathbf{r}_1 \cdots d\mathbf{r}_N \, u(|\mathbf{r}_1 - \mathbf{r}_2|) e^{-\beta U_{\text{pair}}(\mathbf{r}_1, ..., \mathbf{r}_N)}$$

by simply relabeling the integration variables. Since there are $N(N-1)/2$ such terms, the average potential energy becomes

$$\langle U_{\text{pair}} \rangle = \frac{N(N-1)}{2Z} \int d\mathbf{r}_1 \cdots d\mathbf{r}_N \, u(|\mathbf{r}_1 - \mathbf{r}_2|) e^{-\beta U_{\text{pair}}(\mathbf{r}_1, ..., \mathbf{r}_N)}$$

$$= \frac{1}{2} \int d\mathbf{r}_1 \, d\mathbf{r}_2 \, u(|\mathbf{r}_1 - \mathbf{r}_2|)$$

$$\times \left[\frac{N(N-1)}{Z} \int d\mathbf{r}_3 \cdots d\mathbf{r}_N \, e^{-\beta U_{\text{pair}}(\mathbf{r}_1, ..., \mathbf{r}_N)} \right].$$

(4.7.41)

However, the quantity in the square brackets is nothing more than the pair correlation function $g^{(2)}(\mathbf{r}_1, \mathbf{r}_2)$. Thus,

$$\langle U_{\text{pair}} \rangle = \frac{\rho^2}{2} \int d\mathbf{r}_1 \, d\mathbf{r}_2 \, u(|\mathbf{r}_1 - \mathbf{r}_2|) g^{(2)}(\mathbf{r}_1, \mathbf{r}_2).$$

(4.7.42)

Proceeding as we did in deriving $g(r)$ (Section 4.7), we introduce the change of variables in eqn. (4.7.14), which gives

$$\langle U_{\text{pair}} \rangle = \frac{N^2}{2V^2} \int d\mathbf{r} \, d\mathbf{R} \, u(r) \tilde{g}^{(2)}(\mathbf{r}, \mathbf{R}).$$

(4.7.43)

Next, assuming $g^{(2)}$ is independent of $\mathbf{R}$, then integrating over this variable simply cancels a factor of volume in the denominator, yielding

$$\langle U_{\text{pair}} \rangle = \frac{N^2}{2V} \int d\mathbf{r} \, u(r) \tilde{g}(\mathbf{r}).$$

(4.7.44)

Introducing spherical polar coordinates and assuming $\tilde{g}(\mathbf{r})$ is independent of θ and ϕ, integrating over the angular variables leads to

$$\langle U_{\text{pair}} \rangle = \frac{N^2}{2V} \int_0^\infty dr 4\pi r^2 u(r)g(r). \tag{4.7.45}$$

Finally, inserting eqn. (4.7.45) into eqn. (4.7.37) gives the energy expression

$$E = \frac{3}{2}NkT + 2\pi N\rho \int_0^\infty dr \, r^2 \, u(r)g(r), \tag{4.7.46}$$

which involves only the functional form of the pair potential and the radial distribution function. Note that extending the integral over r from 0 to ∞ rather than limiting it to the physical domain is justified if the potential is short-ranged. Interestingly, if the potential energy $U(\mathbf{r}_1, ..., \mathbf{r}_N)$ includes additional N-body terms, such as 3-body or 4-body terms, then by extension of the above analysis, an expression analogous to eqn. (4.7.46) for the average energy would include additional terms involving general N-point correlation functions, e.g. $g^{(3)}$ and $g^{(4)}$, etc.

Let us next consider the pressure, which is given by the thermodynamic derivative

$$P = kT\frac{\partial}{\partial V} \ln Q(N, V, T) = \frac{kT}{Z(N, V, T)}\frac{\partial Z(N, V, T)}{\partial V}. \tag{4.7.47}$$

This derivative can be performed only if we have an explicit volume dependence in the expression for $Z(N, V, T)$. For a Hamiltonian of the standard form

$$\mathcal{H} = \sum_{i=1}^N \frac{\mathbf{p}_i^2}{2m_i} + U(\mathbf{r}_1, ..., \mathbf{r}_N), \tag{4.7.48}$$

the configurational partition function is

$$Z(N, V, T) = \int_{D(V)} d\mathbf{r}_1 \cdots \int_{D(V)} d\mathbf{r}_N \, e^{-\beta U(\mathbf{r}_1, ..., \mathbf{r}_N)}, \tag{4.7.49}$$

where $D(V)$ is spatial domain defined by the physical container. It can be seen immediately that the volume dependence is contained implicitly in the integration limits, so that the volume differentiation cannot be easily performed. The task would be made considerably simpler if the volume dependence could be moved into the integrand by some means. In fact, we can achieve this by a simple change of variables in the integral. The change of variables we seek should render the limits independent of the box size. In a cubic box of length L, for example, the range of all of the integrals is $[0, L]$, which suggests that if we introduce new Cartesian coordinates

$$\mathbf{s}_i = \frac{1}{L}\mathbf{r}_i \qquad i = 1, ..., N, \tag{4.7.50}$$

all of the integrals range from 0 to 1. The coordinates $\mathbf{s}_1, ..., \mathbf{s}_N$ are called *scaled coordinates*. For orthorhombic boxes, the transformation can be generalized to

$$\mathbf{s}_i = \frac{1}{V^{1/3}}\mathbf{r}_i, \tag{4.7.51}$$

where V is the volume of the box. Performing this change of variables in $Z(N, V, T)$ yields

$$Z(N, V, T) = V^N \int d\mathbf{s}_1 \cdots d\mathbf{s}_N \, \exp\left[-\beta U\left(V^{1/3}\mathbf{s}_1, ..., V^{1/3}\mathbf{s}_N\right)\right]. \qquad (4.7.52)$$

The volume derivative of $Z(N, V, T)$ may now be easily computed as

$$\frac{\partial Z}{\partial V} = \frac{N}{V} Z(N, V, T)$$

$$- \beta V^N \int d\mathbf{s}_1 \cdots d\mathbf{s}_N \frac{1}{3} V^{-2/3} \left[\sum_{i=1}^{N} \mathbf{s}_i \cdot \frac{\partial U}{\partial (V^{1/3}\mathbf{s}_i)}\right] e^{-\beta U(V^{1/3}\mathbf{s}_1, ..., V^{1/3}\mathbf{s}_N)}$$

$$= \frac{N}{V} Z(N, V, T) + \frac{\beta}{3V} \int d\mathbf{r}_1 \cdots d\mathbf{r}_N \left[\sum_{i=1}^{N} \mathbf{r}_i \cdot \mathbf{F}_i\right] e^{-\beta U(\mathbf{r}_1, ..., \mathbf{r}_N)}, \qquad (4.7.53)$$

where, in the last line, we have transformed back to $\mathbf{r}_1, ..., \mathbf{r}_N$. Thus,

$$\frac{1}{Z}\frac{\partial Z}{\partial V} = \frac{N}{V} + \frac{\beta}{3V} \left\langle \sum_{i=1}^{N} \mathbf{r}_i \cdot \mathbf{F}_i \right\rangle, \qquad (4.7.54)$$

so that the pressure becomes

$$P = \frac{NkT}{V} + \frac{1}{3V} \left\langle \sum_{i=1}^{N} \mathbf{r}_i \cdot \mathbf{F}_i \right\rangle. \qquad (4.7.55)$$

Again, using the fact that $NkT/V = (1/3V)\langle\sum_{i=1}^{N} \mathbf{p}_i^2/m_i\rangle$, eqn. (4.7.55) becomes

$$P = \frac{1}{3V} \left\langle \sum_{i=1}^{N} \left[\frac{\mathbf{p}_i^2}{m_i} + \mathbf{r}_i \cdot \mathbf{F}_i\right] \right\rangle. \qquad (4.7.56)$$

The quantity in the angle brackets in eqn. (4.7.56) is an instantaneous estimator $\mathcal{P}(\mathbf{r}, \mathbf{p})$ for the pressure

$$\mathcal{P}(\mathbf{r}, \mathbf{p}) = \frac{1}{3V} \sum_{i=1}^{N} \left[\frac{\mathbf{p}_i^2}{m_i} + \mathbf{r}_i \cdot \mathbf{F}_i\right]. \qquad (4.7.57)$$

Note the presence of the virial in eqns. (4.7.55) and (4.7.56). When, $\mathbf{F}_i = 0$, the pressure reduces to the usual ideal gas law. In addition, because of the virial theorem, the two terms in eqn. (4.7.57) largely cancel, so that this estimator essentially measures boundary effects. One final note concerns potentials that have an explicit volume dependence. Volume dependence in the potential arises, for example, in molecular dynamics calculations in systems with long-range forces. For such potentials, eqn. (4.7.57) is modified to read

$$\mathcal{P}(\mathbf{r}, \mathbf{p}) = \frac{1}{3V} \left[\sum_{i=1}^{N} \frac{\mathbf{p}_i^2}{m_i} + \sum_{i=1}^{N} \mathbf{r}_i \cdot \mathbf{F}_i - 3V \frac{\partial U}{\partial V}\right]. \qquad (4.7.58)$$

In d dimensions, the "3" eqns. (4.7.57) and (4.7.58) is replaced by d.

We now consider eqn. (4.7.55) for the case of a pair-wise additive potential (with no explicit volume dependence). For such a potential, it is useful to introduce the vector, $\mathbf{f}_{ij}$, which is the force on particle i due to particle j with

$$\mathbf{F}_i = \sum_{j\neq i} \mathbf{f}_{ij}. \tag{4.7.59}$$

From Newton's third law

$$\mathbf{f}_{ij} = -\mathbf{f}_{ji}. \tag{4.7.60}$$

In terms of $\mathbf{f}_{ij}$, the virial can be written as

$$\sum_{i=1}^{N} \mathbf{r}_i \cdot \mathbf{F}_i = \sum_{i=1}^{N}\sum_{j=1,j\neq i}^{N} \mathbf{r}_i \cdot \mathbf{f}_{ij} \equiv \sum_{i,j,i\neq j} \mathbf{r}_i \cdot \mathbf{f}_{ij}. \tag{4.7.61}$$

By interchanging the i and j summations in the above expression, we obtain

$$\sum_{i=1}^{N} \mathbf{r}_i \cdot \mathbf{F}_i = \frac{1}{2}\left[\sum_{i,j,i\neq j} \mathbf{r}_i \cdot \mathbf{f}_{ij} + \sum_{i,j,i\neq j} \mathbf{r}_j \cdot \mathbf{f}_{ji} \right], \tag{4.7.62}$$

so that, using Newton's third law, the virial can be expressed as

$$\sum_{i=1}^{N} \mathbf{r}_i \cdot \mathbf{F}_i = \frac{1}{2}\left[\sum_{i,j,i\neq j} \mathbf{r}_i \cdot \mathbf{f}_{ij} - \sum_{i,j,i\neq j} \mathbf{r}_j \cdot \mathbf{f}_{ij} \right]$$

$$= \frac{1}{2}\sum_{i,j,i\neq j} (\mathbf{r}_i - \mathbf{r}_j) \cdot \mathbf{f}_{ij} \equiv \frac{1}{2}\sum_{i,j,i\neq j} \mathbf{r}_{ij} \cdot \mathbf{f}_{ij}, \tag{4.7.63}$$

where $\mathbf{r}_{ij} = \mathbf{r}_i - \mathbf{r}_j$. The ensemble average of this quantity is

$$\frac{\beta}{3V}\left\langle \sum_{i=1}^{N} \mathbf{r}_i \cdot \mathbf{F}_i \right\rangle = \frac{\beta}{6V}\left\langle \sum_{i,j,i\neq j} \mathbf{r}_{ij} \cdot \mathbf{f}_{ij} \right\rangle$$

$$= \frac{\beta}{6VZ}\int d\mathbf{r}_1 \cdots d\mathbf{r}_N \left[\sum_{i,j,i\neq j} \mathbf{r}_{ij} \cdot \mathbf{f}_{ij} \right] e^{-\beta U_{\mathrm{pair}}(\mathbf{r}_1,\dots,\mathbf{r}_N)}. \tag{4.7.64}$$

As we saw in the derivation of the internal energy, all of the integrals can be made identical by changing the particle labels. Hence,

$$\frac{\beta}{3V}\left\langle \sum_{i=1}^{N} \mathbf{r}_i \cdot \mathbf{F}_i \right\rangle = \frac{\beta N(N-1)}{6VZ}\int d\mathbf{r}_1 \cdots d\mathbf{r}_N\, \mathbf{r}_{12} \cdot \mathbf{f}_{12}\, e^{-\beta U_{\mathrm{pair}}(\mathbf{r}_1,\dots,\mathbf{r}_N)}$$

$$= \frac{\beta}{6V}\int d\mathbf{r}_1 d\mathbf{r}_2\, \mathbf{r}_{12} \cdot \mathbf{f}_{12}\left[\frac{N(N-1)}{Z}\int d\mathbf{r}_3 \cdots d\mathbf{r}_N e^{-\beta U_{\mathrm{pair}}(\mathbf{r}_1,\dots,\mathbf{r}_N)} \right]$$

$$= \frac{\beta}{6V} \int d\mathbf{r}_1 d\mathbf{r}_2 \, \mathbf{r}_{12} \cdot \mathbf{f}_{12} \, \rho^{(2)}(\mathbf{r}_1, \mathbf{r}_2)$$

$$= \frac{\beta N^2}{6V^3} \int d\mathbf{r}_1 d\mathbf{r}_2 \, \mathbf{r}_{12} \cdot \mathbf{f}_{12} \, g^{(2)}(\mathbf{r}_1, \mathbf{r}_2). \tag{4.7.65}$$

Moreover,

$$\mathbf{f}_{12} = -\frac{\partial U_{\text{pair}}}{\partial \mathbf{r}_{12}} = -u'(|\mathbf{r}_1 - \mathbf{r}_2|)\frac{(\mathbf{r}_1 - \mathbf{r}_2)}{|\mathbf{r}_1 - \mathbf{r}_2|} = -u'(r_{12})\frac{\mathbf{r}_{12}}{r_{12}}, \tag{4.7.66}$$

where $u'(r) = du/dr$, and $r_{12} = |\mathbf{r}_{12}|$. Substituting eqn. (4.7.66) into the ensemble average gives

$$\frac{\beta}{3V}\left\langle \sum_{i=1}^{N} \mathbf{r}_i \cdot \mathbf{F}_i \right\rangle = -\frac{\beta N^2}{6V^3} \int d\mathbf{r}_1 d\mathbf{r}_2 \, u'(r_{12}) r_{12} g^{(2)}(\mathbf{r}_1, \mathbf{r}_2). \tag{4.7.67}$$

As was done for the average energy, we change variables using eqn. (4.7.14), which yields

$$\frac{\beta}{3V}\left\langle \sum_{i=1}^{N} \mathbf{r}_i \cdot \mathbf{F}_i \right\rangle = -\frac{\beta N^2}{6V^3} \int d\mathbf{r} \, d\mathbf{R} \, u'(r) r \tilde{g}^{(2)}(\mathbf{r}, \mathbf{R})$$

$$= -\frac{\beta N^2}{6V^2} \int d\mathbf{r} \, u'(r) r \tilde{g}(\mathbf{r})$$

$$= -\frac{\beta N^2}{6V^2} \int_0^{\infty} dr \, 4\pi r^3 u'(r) g(r). \tag{4.7.68}$$

Therefore, the pressure becomes

$$\frac{P}{kT} = \rho - \frac{2\pi \rho^2}{3kT} \int_0^{\infty} dr \, r^3 u'(r) g(r), \tag{4.7.69}$$

which is a simple expression for the pressure in terms of the derivative of the pair potential form and the radial distribution function.

Equation (4.7.69) is in the form of an equation of state and is exact for pair-wise potentials. The dependence of the second term on ρ and T is more complicated than it appears because $g(r)$ depends on both ρ and T: $g(r) = g(r; \rho, T)$. At low density, however, where the thermodynamic properties of a system should be dominated by those of an ideal gas, the second term, which has a leading ρ^2 dependence, should be small. This fact suggests that the low density limit can be accurately approximated by expanding the ρ dependence of $g(r)$ in a power series in ρ:

$$g(r; \rho, T) = \sum_{j=0}^{\infty} \rho^j g_j(r; T). \tag{4.7.70}$$

Substituting eqn. (4.7.70) into eqn. (4.7.69) gives the equation of state in the form

$$\frac{P}{kT} = \rho + \sum_{j=0}^{\infty} B_{j+2}(T)\rho^{j+2}. \tag{4.7.71}$$

Equation (4.7.71) is known as the *virial equation of state*. The coefficients $B_{j+2}(T)$ are given by

$$B_{j+2}(T) = -\frac{2\pi}{3kT} \int_0^{\infty} dr \; r^3 u'(r) g_j(r; T) \tag{4.7.72}$$

and are known as the *virial coefficients*. Equation (4.7.71) is still exact. However, in the low density limit, the expansion can be truncated after the first few terms. If we stop after the second-order term, for example, then the equation of state reads

$$\frac{P}{kT} \approx \rho + B_2(T)\rho^2 \tag{4.7.73}$$

with

$$B_2(T) \approx -\frac{2\pi}{3kT} \int_0^{\infty} dr \; r^3 u'(r) g(r) \tag{4.7.74}$$

since $g_0(r; T) \approx g(r)$. Thus, the *second virial coefficient* $B_2(T)$ gives the leading order deviation from ideal gas behavior. In this limit, the radial distribution function, itself, can be approximated by (see Problem 4.6)

$$g(r) \approx e^{-\beta u(r)}, \tag{4.7.75}$$

and the second virial coefficient is given approximately by

$$B_2(T) \approx -2\pi \int_0^{\infty} dr \; r^2 \left(e^{-\beta u(r)} - 1 \right). \tag{4.7.76}$$

These concepts will be important for our development of perturbation theory and the derivation of the van der Waals equation of state, to be treated in the next section.

4.7.4 Uniqueness of the radial distribution function

Let us consider two systems having the same number N of particles and the same average density but described by two distinct pair potentials $U_{\text{pair}}^{(1)}(\mathbf{r}_1, ..., \mathbf{r}_N)$ and $U_{\text{pair}}^{(2)}(\mathbf{r}_1, ..., \mathbf{r}_N)$ given by

$$U_{\text{pair}}^{(1)}(\mathbf{r}_1, ..., \mathbf{r}_N) = \sum_{i=1}^{N-1} \sum_{j=i+1}^{N} u_1(|\mathbf{r}_i - \mathbf{r}_j|)$$

$$U_{\text{pair}}^{(2)}(\mathbf{r}_1, ..., \mathbf{r}_N) = \sum_{i=1}^{N-1} \sum_{j=i+1}^{N} u_2(|\mathbf{r}_i - \mathbf{r}_j|), \tag{4.7.77}$$

where $u_1(r)$ and $u_2(r)$ are distinct in the sense that they differ by more than a trivial constant: $u_1(r) \neq u_2(r) + \text{const}$. We will show, in this subsection, that if each system is at the same temperature T, then the corresponding radial distribution functions $g_1(r)$ and $g_2(r)$ are uniquely determined by the pair potential of each system, a result known as the *Henderson theorem* (Henderson, 1974). The proof of this theorem stems from the well-known *Gibbs-Bogoliubov inequality*, which states that if two systems have the same number of particles but different Hamiltonians $\mathcal{H}_1(\mathbf{x})$ and $\mathcal{H}_2(\mathbf{x})$, then corresponding Helmholtz free energies A_1 and A_2 at a common temperature T satisfy

$$A_2 \leq A_1 + \langle \mathcal{H}_2(\mathbf{x}) - \mathcal{H}_1(\mathbf{x}) \rangle_1, \tag{4.7.78}$$

where $\langle \cdots \rangle_1$ indicates an average over the canonical ensemble of system 1. Equality holds if and only if $\mathcal{H}_1(\mathbf{x}) = \mathcal{H}_2(\mathbf{x})$. We will defer the proof of the Gibbs-Bogoliubov inequality until Chapter 8.

In order to prove Henderson's theorem, we start with eqn. (4.7.46) for pair potentials, which allows us to express $\langle \mathcal{H}_2(\mathbf{x}) - \mathcal{H}_1(\mathbf{x}) \rangle_1$ in terms of the radial distribution function $g_1(r)$ for system 1 and the two pair potentials as

$$\langle \mathcal{H}_2(\mathbf{x}) - \mathcal{H}_1(\mathbf{x}) \rangle_1 = 2\pi N \rho \int_0^\infty dr \, r^2 \left(u_2(r) - u_1(r) \right) g_1(r). \tag{4.7.79}$$

In deriving eqn. (4.7.79), we have used the fact that the ideal-gas contribution, being universal, is the same for both systems and, therefore, cancels out. Equation (4.7.78) applied to systems described by distinct pair potentials now becomes

$$A_2 < A_1 + 2\pi N \rho \int_0^\infty dr \, r^2 \left(u_2(r) - u_1(r) \right) g_1(r). \tag{4.7.80}$$

Note that eqn. (4.7.80) is expressed as a strict inequality because the potentials are assumed to be different. Equation (4.7.78) can be turned around and expressed in terms of an ensemble average over system 2 instead of system 1, i.e., $A_1 \leq A_2 + \langle \mathcal{H}_1(\mathbf{x}) - \mathcal{H}_2(\mathbf{x}) \rangle_2$ or $A_1 \leq A_2 - \langle \mathcal{H}_2(\mathbf{x}) - \mathcal{H}_1(\mathbf{x}) \rangle_2$. For systems described by distinct pair potentials, this inequality reads

$$A_1 < A_2 - 2\pi N \rho \int_0^\infty dr \, r^2 \left(u_2(r) - u_1(r) \right) g_2(r). \tag{4.7.81}$$

As a method of proof, let us assume the opposite of what we are trying to prove and show that this assumption leads to a logical contradiction. That is, we assume that $g_1(r) = g_2(r) \equiv g(r)$. Then eqns. (4.7.80) and (4.7.81) become

$$A_2 < A_1 + 2\pi N \rho \int_0^\infty dr \, r^2 \left(u_2(r) - u_1(r) \right) g(r)$$

$$A_1 < A_2 - 2\pi N \rho \int_0^\infty dr \, r^2 \left(u_2(r) - u_1(r) \right) g(r). \tag{4.7.82}$$

Adding these two inequalities gives

$$A_2 + A_1 < A_1 + A_2, \tag{4.7.83}$$

which is a nonsensical result (*reductio ad absurdum*). Therefore, it must be true that $g_1(r) \neq g_2(r)$, thus proving the theorem. It is also possible to establish that choosing

$u_1(r) = u_2(r)$ implies $g_1(r) = g_2(r)$ since, if the two pair potentials are the same, the two inequalities become, simultaneously, $A_1 < A_2$ and $A_2 < A_1$, which is equally nonsensical. In fact, if $u_1(r) = u_2(r)$, then it must follow that $A_1 = A_2$ because $\mathcal{H}_1(\mathrm{x}) = \mathcal{H}_2(\mathrm{x})$, in which case, the only possibility is that $g_1(r) = g_2(r) \equiv g(r)$. Hence, it must follow that if $g_1(r) \neq g_2(r)$, $u_1(r) \neq u_2(r)$.

The Henderson theorem not only establishes the uniqueness of $g(r)$, it also establishes the uniqueness of thermodynamic properties derived from $g(r)$. Moreover, since the theorem establishes that $g(r)$ is uniquely determined by the interaction potential, it is possible, via an approach known as reverse Monte Carlo (McGreevy and Pusztai, 1988), to take a measured structure factor from an x-ray or neutron diffraction experiment and determine corresponding partial radial distribution functions (Steinczinger and Pusztai, 2012), which is demonstrated by these authors for liquid water. Additional techniques (Soper, 1996) further allow an interaction potential of the system to be reverse engineered. Since partial radial distribution functions cannot be measured directly from experiment, the aforementioned approaches are critical for extracting these distributions from the measured structure data.

4.8 Perturbation theory and the van der Waals equation

Up to this point, the example systems we have considered (ideal gas, harmonic bead-spring model, *etc.*) have been simple enough to permit an analytical treatment but lack the complexity needed for truly interesting behavior. The theory of distributions presented in Section 4.7 is useful for characterizing structural and thermodynamic properties of real gases and liquids, and as Figs. 4.2, 4.3, and 4.5 suggest, these properties reflect the richness that arises from even mildly complex interparticle interactions. In particular, complex systems can exist in different phases (e.g. solid, liquid, or gas) and can undergo *phase transitions* between these different states. By contrast, the ideal gas, in which the molecular constituents do not interact, cannot exist as anything but a gas.

In this section, we will consider a model system sufficiently complex to exhibit a gas–liquid phase transition but simple enough to permit an approximate analytical treatment. We will see how the phase transition manifests itself in the equation of state, and we will introduce some of the basic concepts of critical phenomena (to be discussed in greater detail in Chapter 16).

Before introducing our real-gas model, we first need to develop some important machinery, specifically, a statistical mechanical perturbation theory for calculating partition functions. To this end, consider a system whose potential energy can be written in the form

$$U(\mathbf{r}_1, ..., \mathbf{r}_N) = U_0(\mathbf{r}_1, ..., \mathbf{r}_N) + U_1(\mathbf{r}_1, ..., \mathbf{r}_N). \tag{4.8.1}$$

Here, $U_1(\mathbf{r}_1, ..., \mathbf{r}_N)$ is assumed to be a small perturbation to the potential $U_0(\mathbf{r}_1, ..., \mathbf{r}_N)$. We define the configurational partition function for the unperturbed system, described by $U_0(\mathbf{r}_1, ..., \mathbf{r}_N)$, as

$$Z^{(0)}(N, V, T) = \int d\mathbf{r}_1 \cdots d\mathbf{r}_N \, e^{-\beta U_0(\mathbf{r}_1, ..., \mathbf{r}_N)}. \tag{4.8.2}$$

Then, the total configurational partition function

$$Z(N, V, T) = \int d\mathbf{r}_1 \cdots d\mathbf{r}_N \, e^{-\beta U(\mathbf{r}_1,...,\mathbf{r}_N)} \tag{4.8.3}$$

can be expressed as

$$Z(N, V, T) = \int d\mathbf{r}_1 \cdots d\mathbf{r}_N \, e^{-\beta U_0(\mathbf{r}_1,...,\mathbf{r}_N)} e^{-\beta U_1(\mathbf{r}_1,...,\mathbf{r}_N)}$$

$$Z(N, V, T) = \frac{Z^{(0)}(N, V, T)}{Z^{(0)}(N, V, T)} \int d\mathbf{r}_1 \cdots d\mathbf{r}_N \, e^{-\beta U_0(\mathbf{r}_1,...,\mathbf{r}_N)} e^{-\beta U_1(\mathbf{r}_1,...,\mathbf{r}_N)}$$

$$= Z^{(0)}(N, V, T) \langle e^{-\beta U_1} \rangle_0, \tag{4.8.4}$$

where an average over the unperturbed ensemble has been introduced. In general, an unperturbed average $\langle a \rangle_0$ is defined to be

$$\langle a \rangle_0 = \frac{1}{Z^{(0)}(N, V, T)} \int d\mathbf{r}_1 \cdots d\mathbf{r}_N \, a(\mathbf{r}_1, ..., \mathbf{r}_N) \, e^{-\beta U_0(\mathbf{r}_1,...,\mathbf{r}_N)}. \tag{4.8.5}$$

If U_1 is a small perturbation to U_0, then the average $\langle \exp(-\beta U_1) \rangle_0$ can be expanded in powers of U_1:

$$\langle e^{-\beta U_1} \rangle_0 = 1 - \beta \langle U_1 \rangle_0 + \frac{\beta^2}{2!} \langle U_1^2 \rangle_0 - \frac{\beta^3}{3!} \langle U_1^3 \rangle_0 + \cdots = \sum_{l=0}^{\infty} \frac{(-\beta)^l}{l!} \langle U_1^l \rangle_0. \tag{4.8.6}$$

Since the total partition function is given by

$$Q(N, V, T) = \frac{Z(N, V, T)}{N! \lambda^{3N}}, \tag{4.8.7}$$

the Helmholtz free energy becomes

$$A(N, V, T) = -\frac{1}{\beta} \ln \left(\frac{Z(N, V, T)}{N! \lambda^{3N}} \right)$$

$$= -\frac{1}{\beta} \ln \left(\frac{Z^{(0)}(N, V, T)}{N! \lambda^{3N}} \right) - \frac{1}{\beta} \ln \langle e^{-\beta U_1} \rangle_0. \tag{4.8.8}$$

The free energy naturally separates into two contributions

$$A(N, V, T) = A^{(0)}(N, V, T) + A^{(1)}(N, V, T), \tag{4.8.9}$$

where

$$A^{(0)}(N, V, T) = -\frac{1}{\beta} \ln \left(\frac{Z^{(0)}(N, V, T)}{N! \lambda^{3N}} \right) \tag{4.8.10}$$

is independent of U_1 and

$$A^{(1)}(N, V, T) = -\frac{1}{\beta} \ln \langle e^{-\beta U_1} \rangle_0 = -\frac{1}{\beta} \ln \sum_{l=0}^{\infty} \frac{(-\beta)^l}{l!} \langle U_1^l \rangle_0, \tag{4.8.11}$$

where, in the second expression, we have expanded the exponential in a power series. We easily see that $A^{(0)}$ is the free energy of the unperturbed system, and $A^{(1)}$ is a correction to be determined perturbatively. To this end, we propose an expansion for $A^{(1)}$ of the general form

$$A^{(1)} = \sum_{k=1}^{\infty} \frac{(-\beta)^{k-1}}{k!} \omega_k, \tag{4.8.12}$$

where $\{\omega_k\}$ is a set of (as yet) unknown expansion coefficients. These coefficients are determined by the condition that eqn. (4.8.12) be consistent with eqn. (4.8.11) at each order in the two expansions.

We can equate the two expressions for $A^{(1)}$ by further expanding the natural log in eqn. (4.8.11) using

$$\ln(1 + x) = \sum_{k=1}^{\infty} (-1)^{k-1} \frac{x^k}{k}. \tag{4.8.13}$$

Substituting eqn. (4.8.13) into eqn. (4.8.11) gives

$$
\begin{aligned}
A^{(1)}(N, V, T) &= -\frac{1}{\beta} \ln \langle e^{-\beta U_1} \rangle_0 \\
&= -\frac{1}{\beta} \ln \left(1 + \sum_{l=1}^{\infty} \frac{(-\beta)^l}{l!} \langle U_1^l \rangle_0 \right) \\
&= -\frac{1}{\beta} \sum_{k=1}^{\infty} (-1)^{k-1} \frac{1}{k} \left(\sum_{l=1}^{\infty} \frac{(-\beta)^l}{l!} \langle U_1^l \rangle_0 \right)^k.
\end{aligned}
\tag{4.8.14}
$$

Equating eqn. (4.8.14) to eqn. (4.8.12) and canceling an overall factor of $1/\beta$ gives

$$\sum_{k=1}^{\infty} (-1)^{k-1} \frac{1}{k} \left(\sum_{l=1}^{\infty} \frac{(-\beta)^l}{l!} \langle U_1^l \rangle_0 \right)^k = \sum_{k=1}^{\infty} (-\beta)^k \frac{\omega_k}{k!}. \tag{4.8.15}$$

In order to determine the unknown coefficients ω_k, we equate like powers of β on both sides. Note that this will yield an expansion in powers such as $\langle U_1 \rangle^k$ and $\langle U_1^k \rangle$, consistent with the perturbative approach we have been following. To see how the expansion develops, consider working to first order only and equating the β^1 terms on both sides. On the right side, the β^1 term is simply $-\beta \omega_1/1!$. On the left side, the term with $l = 1$, $k = 1$ is of order β^1 and is $-\beta \langle U_1 \rangle_0/1!$. Thus, equating these two expressions allows us to determine ω_1:

$$\omega_1 = \langle U_1 \rangle_0. \tag{4.8.16}$$

The coefficient ω_2 can be determined by equating terms on both sides proportional to β^2. On the right side, this term is $\beta^2 \omega_2/2!$. On the left side, the $l = 1$, $k = 2$ and $l = 2$, $k = 1$ terms both contribute, giving

$$\frac{\beta^2}{2} \left(\langle U_1^2 \rangle_0 - \langle U_1 \rangle_0^2 \right).$$

By equating the two expressions, we find that

$$\omega_2 = \langle U_1 \rangle_0^2 - \langle U_1^2 \rangle_0 = \left\langle (U_1 - \langle U_1 \rangle_0)^2 \right\rangle_0. \tag{4.8.17}$$

Interestingly, ω_2 is related to the fluctuation in U_1 in the unperturbed ensemble. This procedure can be repeated to generate as many orders in the expansion as desired. At third order, for example, the reader should verify that ω_3 is given by

$$\omega_3 = \langle U_1^3 \rangle_0 - 3\langle U_1 \rangle_0 \langle U_1^2 \rangle_0 + 2\langle U_1 \rangle_0^3. \tag{4.8.18}$$

The expressions for ω_1, ω_2, and ω_3 are known as the first, second, and third *cumulants* of $U_1(\mathbf{r}_1, ..., \mathbf{r}_N)$, respectively. The expansion in eqn. (4.8.12) is, therefore, known as a *cumulant expansion*, generally given by

$$A^{(1)} = \sum_{k=1}^{\infty} \frac{(-\beta)^{k-1}}{k!} \langle U_1^k \rangle_{c,0}, \tag{4.8.19}$$

where $\langle U_1^k \rangle_{c,0}$ denotes the kth cumulant of U_1 with respect to the unperturbed system.

In general, suppose a random variable y has a probability distribution function $P(y)$. The cumulants of y can all be obtained by the use of a *cumulant generating function*. Let λ be an arbitrary parameter. Then, the cumulant generating function $R(\lambda)$ is defined to be

$$R(\lambda) = \ln \left\langle e^{\lambda y} \right\rangle. \tag{4.8.20}$$

Note that a general power-series expansion of $R(\lambda)$ about $\lambda = 0$ takes the form

$$R(\lambda) = \sum_{n=0}^{\infty} \frac{1}{n!} \left. \frac{\mathrm{d}^n}{\mathrm{d}\lambda^n} R(\lambda) \right|_{\lambda=0} \lambda^n. \tag{4.8.21}$$

The nth cumulant of y, denoted $\langle y \rangle_c$ is then obtained from the derivative terms in this expansion as

$$\langle y^n \rangle_c = \left. \frac{\mathrm{d}^n}{\mathrm{d}\lambda^n} R(\lambda) \right|_{\lambda=0}. \tag{4.8.22}$$

Equation (4.8.22) can be generalized to N random variables $y_1, ..., y_N$ with a probability distribution function $P(y_1, ..., y_N)$. The cumulant generating function now depends on N parameters $\lambda_1, ..., \lambda_N$ and is defined to be

$$R(\lambda_1, ..., \lambda_N) = \ln \left\langle \exp \left(\sum_{i=1}^{N} \lambda_i y_i \right) \right\rangle. \tag{4.8.23}$$

The definition of the cumulants is now generalized as

$$\langle y_1^{\nu_1} y_2^{\nu_2} \cdots y_N^{\nu_N} \rangle_c = \left[\frac{\partial^{\nu_1}}{\partial \lambda_1^{\nu_1}} \frac{\partial^{\nu_2}}{\partial \lambda_2^{\nu_2}} \cdots \frac{\partial^{\nu_N}}{\partial \lambda_N^{\nu_N}} \right] R(\lambda_1, ..., \lambda_N) \Bigg|_{\lambda_1 = \cdots = \lambda_N = 0}. \tag{4.8.24}$$

More detailed discussions about cumulants and their application in quantum chemistry and quantum dynamics are provided by Kladko and Fulde (1998) and Causo *et al.* (2006), respectively, for the interested reader.

Substituting eqns. (4.8.16), (4.8.17), and (4.8.18) into eqn. (4.8.19) and adding $A^{(0)}$ gives the free energy up to third order in U_1:

$$A = A^{(0)} + \omega_1 - \frac{\beta}{2}\omega_2 + \frac{\beta^2}{6}\omega_3 \cdots$$

$$= -\frac{1}{\beta}\ln\left(\frac{Z^{(0)}(N,V,T)}{N!\lambda^{3N}}\right) + \langle U_1 \rangle_0$$

$$- \frac{\beta}{2}\left(\langle U_1^2 \rangle_0 - \langle U_1 \rangle_0^2\right) + \frac{\beta^2}{6}\left(\langle U_1^3 \rangle_0 - 3\langle U_1 \rangle_0\langle U_1^2 \rangle_0 + 2\langle U_1 \rangle_0^3\right) + \cdots . \qquad (4.8.25)$$

It is evident that each term in eqn. (4.8.25) involves increasingly higher powers of U_1 and its averages.

Suppose next that U_0 and U_1 are both pair-wise additive potentials of the form

$$U_0(\mathbf{r}_1, ..., \mathbf{r}_N) = \sum_{i=1}^{N-1}\sum_{j=i+1}^{N} u_0(|\mathbf{r}_i - \mathbf{r}_j|)$$

$$U_1(\mathbf{r}_1, ..., \mathbf{r}_N) = \sum_{i=1}^{N-1}\sum_{j=i+1}^{N} u_1(|\mathbf{r}_i - \mathbf{r}_j|). \qquad (4.8.26)$$

By the same analysis that led to eqn. (4.7.45), the unperturbed average of U_1 is

$$\langle U_1 \rangle_0 = 2\pi N\rho \int_0^\infty dr \, r^2 u_1(r)g_0(r), \qquad (4.8.27)$$

where $g_0(r)$ is the radial distribution function of the unperturbed system at a given density and temperature. In this case, the Helmholtz free energy, to first order in U_1, is

$$A(N,V,T) \approx -\frac{1}{\beta}\ln\left(\frac{Z^{(0)}(N,V,T)}{N!\lambda^{3N}}\right) + 2\pi\rho N \int_0^\infty dr \, r^2 u_1(r)g_0(r). \qquad (4.8.28)$$

We now wish to use the framework of perturbation theory to formulate a statistical mechanical model capable of describing real gases and a gas–liquid phase transition. In Fig. 4.2(a), we depicted a pair-wise potential energy capable of describing both gas and liquid phases. However, the form of this potential, eqn. (3.14.3), is too complicated for an analytical treatment. Thus, we seek a crude representation of such a potential that can be treated within perturbation theory. Consider replacing the $4\epsilon(\sigma/r)^{12}$ repulsive wall by a simpler *hard sphere* potential,

$$u_0(r) = \begin{cases} 0 & r > \sigma \\ \infty & r \leq \sigma \end{cases}, \qquad (4.8.29)$$

which we will use to define the unperturbed ensemble. Since we are interested in the gas–liquid phase transition, we will work in the low density limit appropriate for the

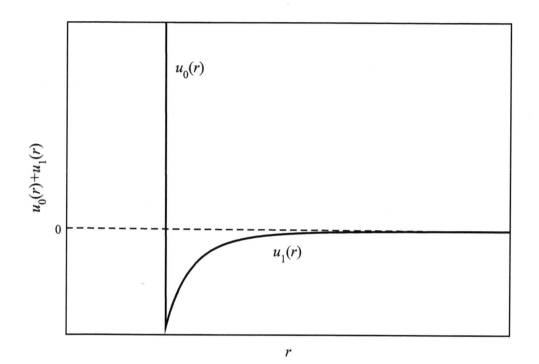

Fig. 4.6 Plot of the potential $u_0(r) + u_1(r)$. The dashed line corresponds to $r = 0$.

gas phase. In this limit, we can apply eqn. (4.7.75) and write the unperturbed radial distribution function as

$$g_0(r) \approx e^{-\beta u_0(r)} = \begin{cases} 1 & r > \sigma \\ 0 & r \leq \sigma \end{cases} = \theta(r - \sigma), \tag{4.8.30}$$

where $\theta(x)$ is the Heaviside step function. For the perturbation $u_1(r)$, we need to mimic the attractive part of Fig. 4.2(a), which is determined by the $-4\epsilon(\sigma/r)^6$ term. In fact, the particular form of $u_1(r)$ is not particularly important as long as $u_1(r) < 0$ for all r and $u_1(r)$ is short-ranged. Thus, our crude representation of Fig. 4.2(a) is shown in Fig. 4.6. Despite the simplicity of this model, some very interesting physics can be extracted.

Consider the perturbative correction $A^{(1)}$, which is given to first order in U_1 by

$$A^{(1)} \approx 2\pi N \rho \int_0^\infty r^2 u_1(r) g_0(r) \, dr$$

$$= 2\pi N \rho \int_0^\infty r^2 u_1(r) \theta(r - \sigma) \, dr$$

$$= 2\pi N \rho \int_\sigma^\infty r^2 u_1(r) \, dr \equiv -aN\rho,$$

where

$$a = -2\pi \int_{\sigma}^{\infty} r^2 u_1(r) \, dr > 0. \tag{4.8.31}$$

Since $u_1(r) < 0$, a must be positive. Next, in order to determine $A^{(0)}$, it is necessary to determine $Z^{(0)}(N, V, T)$. Note that if σ were equal to 0, the potential $u_0(r)$ would vanish, and $Z^{(0)}(N, V, T)$ would just be the ideal gas configurational partition function $Z^{(0)}(N, V, T) = V^N$. Thus, in the low density limit, we might expect that the unperturbed configurational partition function, to a good approximation, would be given by

$$Z^{(0)}(N, V, T) \approx V_{\text{available}}^N, \tag{4.8.32}$$

where $V_{\text{available}}$ is the total available volume to the system. For a hard sphere gas, $V_{\text{available}} < V$ since there is a distance of closest approach between any pair of particles. Smaller interparticle separations are forbidden, as the potential $u_0(r)$ suddenly increases to ∞. Thus, there is an *excluded* volume V_{excluded} that is not accessible to the system, and the available volume can be reexpressed as $V_{\text{available}} = V - V_{\text{excluded}}$. The excluded volume, itself, can be written as $V_{\text{excluded}} = Nb$ where b is the excluded volume per particle. In order to see what this excluded volume is, consider Fig. 4.7, which shows two spheres at their minimum separation, where the distance between

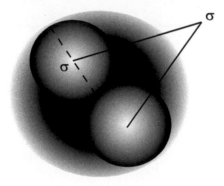

Fig. 4.7 Two hard spheres of diameter σ at closest contact. The distance between their centers is also σ. A sphere of radius σ just containing the two particles is shown in cross-section.

their centers is σ. If we now consider a larger sphere that encloses the two particles when they are at closest contact (shown as a dashed line), then the radius of this sphere is exactly σ, and the its volume is $4\pi\sigma^3/3$. This is the total excluded volume for two particles. Hence, the excluded volume *per particle* is just half of this, or $b = 2\pi\sigma^3/3$, and the unperturbed configurational partition function is given approximately by

$$Z^{(0)}(N, V, T) = \left(V - \frac{2N\pi\sigma^3}{3}\right)^N = (V - Nb)^N. \tag{4.8.33}$$

Therefore, the free energy, to first order, becomes

$$A(N, V, T) \approx -\frac{1}{\beta} \ln \left[\frac{(V - Nb)^N}{N! \lambda^{3N}}\right] - \frac{aN^2}{V}. \tag{4.8.34}$$

We now use this free energy to compute the pressure from

$$P = -\left(\frac{\partial A}{\partial V}\right),$$

(4.8.35)

which gives

$$P = \frac{NkT}{V - Nb} - \frac{aN^2}{V^2}$$

$$\frac{P}{kT} = \frac{\rho}{1 - \rho b} - \frac{a\rho^2}{kT}.$$

(4.8.36)

Equation (4.8.36) is known as the *van der Waals equation of state*. Specifically, it is an equation of state for a system described by the pair potential $u(r) = u_0(r) + u_1(r)$ to first order in perturbation theory in the low density limit. Given the many approximations made in the derivation of eqn. (4.8.36) and the crudeness of the underlying model, we cannot expect it to be applicable over a wide range of P, V, and T values. Nevertheless, if we plot the isotherms of the van der Waals equation, something quite interesting emerges (see Fig. 4.8). For temperatures larger than a certain temperature T_c, the isotherms resemble those of an ideal gas. At T_c, however, we see that the isotherm is flat in a small region. That is, at this point, the "flatness" of the isotherm is characterized by the conditions

$$\frac{\partial P}{\partial V} = 0, \qquad \frac{\partial^2 P}{\partial V^2} = 0.$$

(4.8.37)

The first and second conditions imply that the slope of the isotherm and its curvature, respectively, vanish at the point of "flatness." For temperatures below T_c, the isotherms take on an unphysical character: They all possess a region in which P and V simultaneously increase. As already noted, considering the many approximations made, regions of unphysical behavior should come as no surprise. A physically realistic isotherm for $T < T_c$ should have the unphysical region replaced by the thin solid line in Fig. 4.8. From the placement of this thin line, we see that the isotherm exhibits a discontinuous change in the volume for a very small change in pressure, signifying a gas–liquid phase transition. The isotherm at $T = T_c$ is a kind of "boundary" between isotherms along which V is continuous ($T > T_c$) and those that exhibit discontinuous volume changes ($T < T_c$). For this reason, the $T = T_c$ isotherm is called the *critical isotherm*. The point at which the isotherm is flat is known as the *critical point*. On a phase diagram, this would be the point at which the gas–liquid coexistence curve terminates. The conditions in eqn. (4.8.37) define the temperature, volume, and pressure at the critical point. The first and second derivatives of eqn. (4.8.36) with respect to V yield two equations in the two unknowns V and T:

$$-\frac{NkT}{(V - Nb)^2} + \frac{2aN^2}{V^3} = 0$$

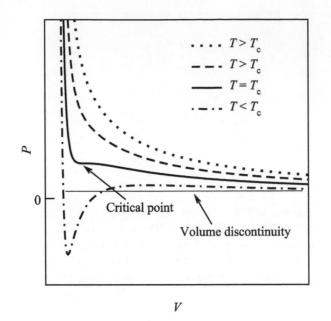

$$\begin{array}{ll}
\cdots\cdots & T > T_c \\
-\,\cdot\,- & T > T_c \\
\rule[0.5ex]{1.5em}{0.4pt} & T = T_c \\
-\,\cdot\,\cdot & T < T_c
\end{array}$$

Critical point

Volume discontinuity

Fig. 4.8 Isotherms of the van der Waals equation of state for four different temperatures.

$$\frac{2NkT}{(V - Nb)^3} - \frac{6aN^2}{V^4} = 0. \tag{4.8.38}$$

Solving these equations leads to the critical volume V_c and critical temperature T_c:

$$V_c = 3Nb, \qquad kT_c = \frac{8a}{27b}. \tag{4.8.39}$$

Substitution of the critical volume and temperature into the van der Waals equation gives the critical pressure P_c:

$$P_c = \frac{a}{27b^2}. \tag{4.8.40}$$

Let us now consider the behavior of a particular thermodynamic quantity as the critical point is approached. Because we are interested in the relationship between pressure and volume as the critical point is approached, it is useful to study the isothermal compressibility, defined to be

$$\kappa_T = -\frac{1}{V}\left(\frac{\partial V}{\partial P}\right)_T = -\frac{1}{V(\partial P/\partial V)}. \tag{4.8.41}$$

At $V = V_c$, the pressure derivative gives

$$\left.\frac{\partial P}{\partial V}\right|_{V=V_c} = -\frac{NkT}{2N^2b^2} + \frac{2aN^2}{27N^3b^3}$$

$$= \frac{1}{4Nb^2}\left(\frac{8a}{27b} - kT\right)$$

$$\sim (T_c - T), \tag{4.8.42}$$

so that

$$\kappa_T \sim (T - T_c)^{-1}. \tag{4.8.43}$$

This shows that at $V = V_c$, as T approaches T_c from above, the isothermal compressibility diverges according to a power law. That κ_T diverges is also confirmed experimentally. The experimentally observed power-law divergence of κ_T can be expressed generally in the form

$$\kappa_T \sim |T - T_c|^{-\gamma}, \tag{4.8.44}$$

where γ is an example of what is termed a *critical exponent*. The van der Waals theory clearly predicts that the value of $\gamma = 1$.

Briefly, critical exponents describe the behavior of systems near their *critical points*. A critical point is a point in the phase diagram where a coexistence curve terminates. For example, a simple molecular system that can exist as a solid, liquid, or gas has a critical point on the gas–liquid coexistence curve. Similarly, a ferromagnetic material has a critical point on the coexistence curve between its two ordered phases. As a critical point is approached, certain thermodynamic properties are observed to diverge according to power laws that are characterized by the critical exponents. These will be explored in more detail in Chapter 16. What is particularly fascinating about these exponents is that they are the same across large classes of systems that are otherwise very different physically. These classes are known as *universality classes*, and their existence suggests that the local detailed interactions among particles become swamped by long-range cooperative effects that dominate the behavior of a system at its critical point.

Other critical exponents are defined as follows: The heat capacity C_V at $V = V_c$ is observed to diverge as T approaches T_c, according to

$$C_V \sim |T - T_c|^{-\alpha}. \tag{4.8.45}$$

Near the critical point, the equation of state is observed to behave as

$$P - P_c \sim |\rho - \rho_c|^\delta \text{sign}(\rho - \rho_c). \tag{4.8.46}$$

Finally, the shape of the gas–liquid coexistence curve (in the ρ-T plane) near the critical point for $T < T_c$ behaves as

$$\rho_L - \rho_G \sim (T_c - T)^\beta, \tag{4.8.47}$$

where ρ_L and ρ_G are the liquid and gas densities, respectively. The four exponents, α, β, γ, δ comprise the four principal critical exponents.

In order to calculate α, we first compute the energy according to

$$E = -\frac{\partial}{\partial \beta} \ln Q(N, V, T) = \frac{\partial}{\partial \beta}[\beta A(N, V, T)]$$

$$= -\frac{\partial}{\partial \beta}\left\{\ln\left[\frac{(V - Nb)^N}{N!\lambda^{3N}}\right] + \frac{\beta a N^2}{V}\right\}. \tag{4.8.48}$$

Computing the derivative, we obtain $E = 3NkT/2 - aN^2/V$, so that the heat capacity $C_V = (\partial E/\partial T)$ is independent of T or simply $C_V \sim |T - T_c|^0$. From this, it follows that the van der Waals theory predicts $\alpha = 0$. The value of δ can be easily deduced as follows: In the van der Waals theory, the equation of state has the analytical form in eqn. (4.8.36). Thus, we may expand P in a power series in ρ about the critical values according to

$$P = P_c + \frac{\partial P}{\partial \rho}\bigg|_{\rho_c, T_c}(\rho - \rho_c) + \frac{1}{2}\frac{\partial^2 P}{\partial \rho^2}\bigg|_{\rho_c, T_c}(\rho - \rho_c)^2 + \frac{1}{6}\frac{\partial^3 P}{\partial \rho^3}\bigg|_{\rho_c, T_c}(\rho - \rho_c)^3 + \cdots. \tag{4.8.49}$$

The density derivatives of the pressure can be computed from the volume derivatives as

$$\frac{\partial P}{\partial \rho} = \frac{\partial P}{\partial V}\frac{\partial V}{\partial \rho}$$

$$\frac{\partial^2 P}{\partial \rho^2} = \left[\frac{\partial^2 P}{\partial V^2}\left(\frac{\partial V}{\partial \rho}\right)^2 + \frac{\partial P}{\partial V}\frac{\partial^2 V}{\partial \rho^2}\right]. \tag{4.8.50}$$

Both derivatives vanish at the critical point because of the conditions in eqn. (4.8.37). It can be easily verified, however, that the third derivative is not zero, so that the first nonvanishing term in eqn. (4.8.49) (apart from the constant term) is

$$P - P_c \sim (\rho - \rho_c)^3, \tag{4.8.51}$$

which leads to the prediction that $\delta = 3$. The calculation of β is somewhat more involved, so for now, we simply quote the result, namely, that the van der Waals theory predicts $\beta = 1/2$. We will discuss this exponent in more detail in Chapter 16. In summary, the van der Waals theory predicts the four principal exponents to be $\alpha = 0$, $\beta = 1/2$, $\gamma = 1$, and $\delta = 3$. Experimental determination of these exponents gives $\alpha = 0.1$, $\beta = 0.34$, $\gamma = 1.35$, and $\delta = 4.2$, and we can conclude that the van der Waals theory is only a qualitative theory: It predicts the existence of a gas-liquid phase transition but cannot yield quantitative agreement with experiment for the critical exponents of the transition.

4.9 Molecular dynamics in the canonical ensemble: Hamiltonian formulation in an extended phase space

Our treatment of the canonical ensemble naturally raises the question of how molecular dynamics simulations can be performed under the external conditions of this ensemble.

After all, as noted in the previous chapter, simply integrating Hamilton's equations of motion generates a microcanonical ensemble as a consequence of the conservation of the total Hamiltonian. By contrast, in a canonical ensemble, energy is not conserved but fluctuates so as to generate the Boltzmann distribution $\exp[-\beta \mathcal{H}(x)]$ due to exchange of energy between the system and the thermal reservoir to which it is coupled. Although we argued that these energy fluctuations vanish in the thermodynamic limit, most simulations are performed far enough from this limit that the fluctuations cannot be neglected.

In order to generate these fluctuations in a molecular dynamics simulation, we need to mimic the effect of the thermal reservoir. Various methods to achieve this have been proposed (Andersen, 1980; Nosé and Klein, 1983; Berendsen *et al.*, 1984; Nosé, 1984; Evans and Morriss, 1984; Hoover, 1985; Bulgac and Kusnezov, 1990; Martyna *et al.*, 1992; Liu and Tuckerman, 2000). We will discuss several of these approaches in the remainder of this chapter. It must be mentioned at the outset, however, that most canonical "dynamics" methods do not actually yield any kind of realistic dynamics for a system coupled to a thermal bath. Rather, the trajectories generated by these schemes comprise a set of microstates consistent with the canonical distribution. In other words, they produce a *sampling* of the canonical phase-space distribution from which equilibrium observables can be computed. The problem of generating dynamical properties consistent with a canonical distribution will be treated in Chapters 13 through 15.

The most straightforward approach to kinetic control is a simple periodic rescaling of the velocities such that the instantaneous kinetic energy corresponds to a desired temperature. While easy to implement, this approach does not guarantee that a canonical phase-space distribution is obtained. We can improve upon this approach by replacing the velocity scaling by a periodic resampling of the velocities from the Maxwell-Boltzmann distribution. Such a scheme only guarantees that a canonical momentum-space distribution is obtained. Nevertheless, it can be useful in the initial stages of a molecular dynamics calculation as a means of relaxing unfavorable contacts arising from poorly chosen initial positions. This method can be further refined (Andersen, 1980) by selecting a subset of velocities to be resampled at each time step according to a preset collision frequency ν. The probability that any particle will suffer a "collision" (a resampling event) in a time Δt is $\nu \Delta t$. Thus, if a random number in the interval $[0, 1]$ is less than $\nu \Delta t$, the particle's velocity is resampled.

Of all the canonical dynamics methods, by far the most popular are the "extended phase space" approaches (Andersen, 1980; Nosé and Klein, 1983; Nosé, 1984; Hoover, 1985; Bulgac and Kusnezov, 1990; Martyna *et al.*, 1992; Bond *et al.*, 1999; Liu and Tuckerman, 2000). These techniques supplement the physical phase space with additional variables that serve to mimic the effect of a heat bath within a continuous, deterministic dynamical scheme. The extended phase-space methodology allows the greatest amount of flexibility and creativity in devising canonical dynamics algorithms. Moreover, the idea of extending the phase space has lead to other important algorithmic advances including the Car-Parrinello molecular dynamics approach for marrying electronic structure with finite temperature dynamics (Car and Parrinello, 1985; Tuckerman, 2002; Marx and Hutter, 2009) and techniques for computing free energies (see Chapter 8).

4.9.1 The Nosé Hamiltonian

Extended phase-space methods can be either Hamiltonian or non-Hamiltonian in their formulation. Here, we begin with a Hamiltonian approach originally introduced by S. Nosé (1983, 1984). Nosé's approach can be viewed as a kind of Maxwell daemon. An additional "agent" is introduced into a system that "checks" whether the instantaneous kinetic energy is higher or lower than that prescribed by the desired temperature and then scales the velocities accordingly. Denoting this variable as s and its conjugate momentum as p_s, the Nosé Hamiltonian for a system with physical coordinates $\mathbf{r}_1, ..., \mathbf{r}_N \equiv \mathbf{r}$ and momenta $\mathbf{p}_1, ..., \mathbf{p}_N \equiv \mathbf{p}$, takes the form

$$\mathcal{H}_{\mathrm{N}} = \sum_{i=1}^{N} \frac{\mathbf{p}_i^2}{2m_i s^2} + U(\mathbf{r}_1, ..., \mathbf{r}_N) + \frac{p_s^2}{2Q} + gkT \ln s, \qquad (4.9.1)$$

where Q is a parameter that determines the time scale on which the daemon acts. Q is not a mass! In fact, it has units of energy $\times$ time^2. T is the desired temperature of the canonical distribution. If d is the number of spatial dimensions, then the phase space now has a total of $2dN + 2$ dimensions with the addition of s and p_s. The parameter g appearing in eqn. (4.9.1) will be determined by the condition that a *microcanonical* distribution of $2dN + 2$-dimensional phase space of $\mathcal{H}_{\mathrm{N}}$ yields a *canonical* distribution in the $2dN$-dimensional physical phase space. The presence of s in the kinetic energy is essentially what we would expect for an agent that must scale the kinetic energy in order to control its fluctuations. The choice $gkT \ln s$ as the potential in s, while seemingly mysterious, is carefully chosen to ensure that a canonical distribution in the physical phase space is obtained.

In order to see how the canonical distribution emerges from $\mathcal{H}_{\mathrm{N}}$, consider the microcanonical partition function of the full $2dN + 2$-dimensional phase space:

$$\Omega = \int \mathrm{d}^N \mathbf{r} \, \mathrm{d}^N \mathbf{p} \, \mathrm{d}s \, \mathrm{d}p_s$$

$$\times \delta \left(\sum_{i=1}^{N} \frac{\mathbf{p}_i^2}{2m_i s^2} + U(\mathbf{r}_1, ..., \mathbf{r}_N) + \frac{p_s^2}{2Q} + gkT \ln s - E \right), \qquad (4.9.2)$$

where E is the energy of the ensemble. (For clarity, prefactors preceding the integral have been left out.) The distribution of the physical phase space is obtained by integrating over s and p_s. We first introduce a change of momentum variables:

$$\tilde{\mathbf{p}}_i = \frac{\mathbf{p}_i}{s}, \qquad (4.9.3)$$

which gives

$$\Omega = \int \mathrm{d}^N \mathbf{r} \, \mathrm{d}^N \tilde{\mathbf{p}} \, \mathrm{d}s \, \mathrm{d}p_s \, s^{dN} \delta \left(\sum_{i=1}^{N} \frac{\tilde{\mathbf{p}}_i^2}{2m_i} + U(\mathbf{r}_1, ..., \mathbf{r}_N) + \frac{p_s^2}{2Q} + gkT \ln s - E \right)$$

$$= \int d^N \mathbf{r} \, d^N \mathbf{p} \, ds \, dp_s \, s^{dN} \delta \left(\mathcal{H}(\mathbf{r}, \mathbf{p}) + \frac{p_s^2}{2Q} + gkT \ln s - E \right), \tag{4.9.4}$$

where $\mathcal{H}(\mathbf{r}, \mathbf{p})$ is the physical Hamiltonian

$$\mathcal{H}(\mathbf{r}, \mathbf{p}) = \sum_{i=1}^{N} \frac{\mathbf{p}_i^2}{2m_i} + U(\mathbf{r}_1, ..., \mathbf{r}_N). \tag{4.9.5}$$

In the last line of eqn. (4.9.4), we have renamed $\tilde{\mathbf{p}}_i$ as $\mathbf{p}_i$. We can now integrate over s using the δ-function by making use of the following identity: Given a function $f(s)$ that has a single zero at s_0, $\delta(f(s))$ can be replaced by

$$\delta(f(s)) = \frac{\delta(s - s_0)}{|f'(s_0)|}. \tag{4.9.6}$$

Taking $f(s) = \mathcal{H}(\mathbf{r}, \mathbf{p}) + p_s^2/2Q + gkT \ln s - E$, the solution of $f(s_0) = 0$ is

$$s_0 = e^{(E - \mathcal{H}(\mathbf{r}, \mathbf{p}) - p_s^2/2Q)/gkT}$$

$$\frac{1}{|f'(s_0)|} = \frac{1}{gkT} e^{(E - \mathcal{H}(\mathbf{r}, \mathbf{p}) - p_s^2/2Q)/gkT}. \tag{4.9.7}$$

Substituting eqn. (4.9.7) into eqn. (4.9.4) yields

$$\Omega = \frac{1}{gkT} \int d^N \mathbf{p} \, d^N \mathbf{r} \, dp_s \, e^{(dN+1)(E - \mathcal{H}(\mathbf{r}, \mathbf{p}) - p_s^2/2Q)/gkT}. \tag{4.9.8}$$

Thus, if the parameter g is chosen to be $dN + 1$, then, after performing the p_s integration, eqn. (4.9.8) becomes

$$\Omega = \frac{e^{E/kT} \sqrt{2\pi QkT}}{(dN+1)kT} \int d^N \mathbf{p} \, d^N \mathbf{r} \, e^{-\mathcal{H}(\mathbf{r}, \mathbf{p})/kT}, \tag{4.9.9}$$

which is the canonical partition function, apart from the prefactors. Our analysis shows how a microcanonical distribution of the Nosé Hamiltonian $\mathcal{H}_N$ is equivalent to a canonical distribution in the physical Hamiltonian. This suggests that a molecular dynamics calculation performed using $\mathcal{H}_N$ should generate sampling of the canonical distribution $\exp[-\beta\mathcal{H}(\mathbf{r}, \mathbf{p})]$ under the usual assumptions of ergodicity. Because the Nosé Hamiltonian mimics the effect of a heat bath by controlling the fluctuations in the kinetic energy, the mechanism of the Nosé Hamiltonian is also known as a *thermostatting* mechanism.

The equations of motion generated by $\mathcal{H}_N$ are

$$\dot{\mathbf{r}}_i = \frac{\partial \mathcal{H}_N}{\partial \mathbf{p}_i} = \frac{\mathbf{p}_i}{m_i s^2}$$

$$\dot{\mathbf{p}}_i = -\frac{\partial \mathcal{H}_N}{\partial \mathbf{r}_i} = \mathbf{F}_i$$

$$\dot{s} = \frac{\partial \mathcal{H}_N}{\partial p_s} = \frac{p_s}{Q}$$

$$\dot{p}_s = -\frac{\partial \mathcal{H}_N}{\partial s} = \sum_{i=1}^{N} \frac{\mathbf{p}_i^2}{m_i s^3} - \frac{gkT}{s} = \frac{1}{s}\left[\sum_{i=1}^{N} \frac{\mathbf{p}_i^2}{m_i s^2} - gkT \right]. \tag{4.9.10}$$

The $\dot{\mathbf{r}}_i$ and $\dot{p}_s$ equations reveal that the thermostatting mechanism works on an unconventional kinetic energy $\sum_i \mathbf{p}_i^2/(2m_i s^2)$. This form suggests that the more familiar kinetic energy can be recovered by introducing the following (noncanonical) change of variables:

$$\mathbf{p}_i' = \frac{\mathbf{p}_i}{s}, \qquad p_s' = \frac{p_s}{s}, \qquad \mathrm{d}t' = \frac{\mathrm{d}t}{s}. \tag{4.9.11}$$

When eqn. (4.9.11) is substituted into eqns. (4.9.10), the equations of motion become

$$\frac{\mathrm{d}\mathbf{r}_i}{\mathrm{d}t'} = \frac{\mathbf{p}_i'}{m_i}$$

$$\frac{\mathrm{d}\mathbf{p}_i'}{\mathrm{d}t'} = \mathbf{F}_i - \frac{sp_s'}{Q}\mathbf{p}_i'$$

$$\frac{\mathrm{d}s}{\mathrm{d}t'} = \frac{s^2 p_s'}{Q}$$

$$\frac{\mathrm{d}p_s'}{\mathrm{d}t'} = \frac{1}{s}\left[\sum_{i=1}^{N} \frac{(\mathbf{p}_i')^2}{m_i} - gkT \right] - \frac{s(p_s')^2}{Q}. \tag{4.9.12}$$

Because of the noncanonical transformation, these equations lose their symplectic structure, meaning that they are no longer incompressible. In addition, they involve an unconventional definition of time due to the scaling by the variable s. This scaling makes the equations somewhat cumbersome to use directly in the form of (4.9.12).

   The disadvantage of adhering to a strictly Hamiltonian structure is that a measure of flexibility in the design of molecular dynamics algorithms for specific purposes is lost. In fact, there is no particular reason, apart from the purely mathematical, that a Hamiltonian structure must be preserved when seeking to develop molecular dynamics methods whose purpose is to sample an ensemble. Therefore, in the remainder of this chapter, we will focus on techniques that employ non-Hamiltonian equations of motion. We will illustrate how the freedom to stray outside the tight Hamiltonian framework allows a wider variety of algorithms to be created.

4.9.2   The Nosé-Hoover equations

In 1985, W. G. Hoover introduced a reformulation of the Nosé dynamics that has become one of the staples of molecular dynamics. Starting from the Nosé equations of motion, one introduces a noncanonical change of variables

$$\mathbf{p}_i' = \frac{\mathbf{p}_i}{s}, \qquad \mathrm{d}t' = \frac{\mathrm{d}t}{s}, \qquad \frac{1}{s}\frac{\mathrm{d}s}{\mathrm{d}t'} = \frac{\mathrm{d}\eta}{\mathrm{d}t'}, \qquad p_s = p_\eta \tag{4.9.13}$$

and a redefinition $g = dN$, which leads to new equations of motion of the form

$$\dot{\mathbf{r}}_i = \frac{\mathbf{p}_i}{m_i}$$

$$\dot{\mathbf{p}}_i = \mathbf{F}_i - \frac{p_\eta}{Q}\mathbf{p}_i$$

$$\dot{\eta} = \frac{p_\eta}{Q}$$

$$\dot{p}_\eta = \sum_{i=1}^{N} \frac{\mathbf{p}_i^2}{m_i} - dNkT. \tag{4.9.14}$$

(The introduction of the η variable was actually not in the original Hoover formulation but was later recognized by Martyna *et al.* (1992) as essential for the analysis of the phase-space distribution.) The additional term in the momentum equation acts as a kind of friction term, which, however, can be either negative or positive. In fact, the evolution of the "friction" variable p_η is driven by the difference in the instantaneous value of the kinetic energy (multiplied by 2) and its canonical average $dNkT$.

Equations (4.9.14) constitute an example of a *non-Hamiltonian* system. In this case, they are, in a sense, trivially non-Hamiltonian because they are derived from a Hamiltonian system using a noncanonical choice of variables. As we proceed through the remainder of this chapter, however, we will encounter examples of systems that are intrinsically non-Hamiltonian, meaning that there is no set of canonical variables that transforms the equations of motion into a Hamiltonian structure. In order to analyze any non-Hamiltonian system, whether trivial or not, we need to generalize some of the concepts from Chapter 2 for non-Hamiltonian phase spaces. Thus, before we can proceed to analyze the Nosé-Hoover equations, we must first visit this subject.

4.10 Classical non-Hamiltonian statistical mechanics

Generally, Hamiltonian mechanics describes a system in isolation from its surroundings. We have also seen that, with certain tricks, a Hamiltonian system can be used to generate a canonical distribution. But let us examine the problem of a system interacting with its surroundings more closely. If we are willing to treat the system plus surroundings together as an isolated system, then the use of Hamiltonian mechanics to describe the whole is appropriate within a classical description. The distribution of the system alone can be determined by integrating over the variables that represent the surroundings in the microcanonical partition function, as was done above. In most situations, when the surroundings are integrated out in this way, the microscopic equations of motion obeyed by the system are no longer Hamiltonian. In fact, it is often possible to model the effect of the surroundings by simply positing a set of non-Hamiltonian equations of motion and then proving that the equations of motion generate the desired ensemble distribution. Under such a protocol, it is possible to treat systems interacting with heat and particle reservoirs or systems subject to external driving forces. Consequently, it is important to develop an approach that allows us to predict the phase-space distribution function for a given set of non-Hamiltonian equations of motion.

Let us begin by assuming that a system interacting with its surroundings and possibly subject to driving forces is described by non-Hamiltonian microscopic equations of the form

$$\dot{x} = \xi(x, t). \tag{4.10.1}$$

We do not restrict the vector function $\xi(x, t)$ except to assume that it is smooth and at least once differentiable. In particular, the phase-space compressibility $\nabla_x \cdot \dot{x} = \nabla_x \cdot \xi(x, t)$ need not vanish for a non-Hamiltonian system. If it does not vanish, then the system is non-Hamiltonian. Note, however, that the converse is not necessarily true. That is, there are dynamical systems for which the phase-space compressibility is zero but which cannot be derived from a Hamiltonian. Recall that the vanishing of the phase-space compressibility was central to the derivation of the Liouville theorem and Liouville's equation in Sections 2.4 and 2.5. Thus, in order to understand how these results change when the dynamics is compressible, we need to revisit these derivations.

4.10.1 The phase-space metric

Recall from Section 2.4 that a collection of trajectories initially in a volume element dx_0 about the point x_0 will evolve to dx_t about the point x_t, and the transformation $x_0 \rightarrow x_t$ is a unique one with a Jacobian $J(x_t; x_0)$ satisfying the equation of motion

$$\frac{d}{dt} J(x_t; x_0) = J(x_t; x_0) \nabla_{x_t} \cdot \dot{x}_t. \tag{4.10.2}$$

Since the compressibility will occur many times in our discussion of non-Hamiltonian systems, we introduce the notation $\kappa(x_t, t)$, to represent this quantity

$$\kappa(x_t, t) = \nabla_{x_t} \cdot \dot{x}_t = \nabla_{x_t} \cdot \xi(x_t, t). \tag{4.10.3}$$

Since $\kappa(x_t, t)$ cannot be assumed to be zero, the Jacobian is not unity for all time, and the Liouville theorem $dx_t = dx_0$ no longer holds.

The Jacobian can be determined by solving eqn. (4.10.2) using the method of characteristics subject to the initial condition $J(x_0; x_0) = 1$ yielding

$$J(x_t; x_0) = \exp\left[\int_0^t ds\, \kappa(x_s, s)\right]. \tag{4.10.4}$$

However, eqn. (4.10.2) implies that there exists a function $w(x_t, t)$ such that

$$\kappa(x_t, t) = \frac{d}{dt} w(x_t, t) \tag{4.10.5}$$

or that there exists a function whose derivative yields the compressibility. Substitution of eqn. (4.10.5) into eqn. (4.10.4) yields

$$J(x_t; x_0) = \exp\left[w(x_t, t) - w(x_0, 0)\right]. \tag{4.10.6}$$

Since the phase-space volume element evolves according to

$$dx_t = J(x_t; x_0)dx_0, \qquad (4.10.7)$$

we have

$$dx_t = \exp\left[w(x_t, t) - w(x_0, 0)\right] dx_0$$
$$\exp\left[-w(x_t, t)\right] dx_t = \exp\left[-w(x_0, 0)\right] dx_0 \qquad (4.10.8)$$

(Tuckerman *et al.*, 1999; Tuckerman *et al.*, 2001). Equation (4.10.8) constitutes a generalization of Liouville's theorem; it implies that a *weighted* phase space volume $\exp[-w(x_t, t)]dx_t$ is conserved rather than simply dx_t.

Equation (4.10.8) implies that a conservation law exists on a phase space that does not follow the usual laws of Euclidean geometry. We therefore need to view the phase space of a non-Hamiltonian system in a more general way as a non-Euclidean or Riemannian space or *manifold*. Riemannian spaces are locally curved spaces and, therefore, it is necessary to consider local coordinates in each neighborhood of the space. The coordinate transformations needed to move from one neighborhood to another give rise to a nontrivial metric and a corresponding volume element denoted $\sqrt{g(x)}dx$, where $g(x)$ is the determinant of a second-rank tensor $g_{ij}(x)$, known as the *metric tensor*. Given a coordinate transformation from coordinates x to coordinates y, the Jacobian is simply the ratio of the metric determinant factors:

$$J(x; y) = \frac{\sqrt{g(y)}}{\sqrt{g(x)}}. \qquad (4.10.9)$$

It is clear, then, that eqn. (4.10.6) is nothing more than a statement of this fact for a coordinate transformation $x_0 \rightarrow x_t$

$$J(x_t; x_0) = \frac{\sqrt{g(x_0, 0)}}{\sqrt{g(x_t, t)}}, \qquad (4.10.10)$$

where

$$\sqrt{g(x_t, t)} = e^{-w(x_t, t)} \qquad (4.10.11)$$

when the metric $\sqrt{g(x_t, t)}$ is allowed to have an explicit time dependence. Although such coordinate and parameter-dependent metrics are not standard features in the theory of Riemannian spaces, they do occasionally arise (Sardanashvily, 2002*a*; Sardanashvily, 2002*b*). Most of the metric factors we will encounter in our treatments of non-Hamiltonian systems will not involve explicit time-dependence and will therefore obey eqn. (4.10.9). The implication of eqn. (4.10.8) is that any phase-space integral that represents an equilibrium ensemble average should be performed using $\sqrt{g(x)}dx$ as the volume element, when $\sqrt{g}$ has no explicit time dependence, so that the average can be performed at any instant in time.

Imbuing phase space with a metric is not as strange as it might at first seem. After all, phase space is a fictitious mathematical construction, a background space on which a dynamical system evolves. There is no particular reason that we need to attach the same fixed, Euclidean space to *every* dynamical system. In fact, it is more natural to allow the properties of a given dynamical system dictate the geometry of the phase space on which it lives. Thus, if imbuing a phase space with a metric that is particular to a given dynamical system leads to a volume conservation law, then such a phase space is the most natural choice for that dynamical system. Once the geometry of the phase space is chosen, the form of the Liouville equation and its equilibrium solution are determined, as we will now show.

4.10.2 Generalizing the Liouville equation

In order to generalize the Liouville equation for the phase-space distribution $f(\mathbf{x}_t, t)$ for a non-Hamiltonian system, it is necessary to recast the derivation of Section 2.5 on a space with a nontrivial metric. The mathematics required to do this are beyond the scope of the general discussion we wish to present here but are discussed elsewhere by Tuckerman *et al.* (1999, 2001), and we simply quote the final result,

$$\frac{\partial}{\partial t}\left(f(\mathbf{x},t)\sqrt{g(\mathbf{x},t)}\right) + \nabla_{\mathbf{x}} \cdot \left(\dot{\mathbf{x}}\sqrt{g(\mathbf{x},t)}f(\mathbf{x},t)\right) = 0. \tag{4.10.12}$$

Now, combining eqns. (4.10.10) and (4.10.2), we find that the phase-space metric factor $\sqrt{g(\mathbf{x},t)}$ satisfies

$$\frac{\mathrm{d}}{\mathrm{d}t}\sqrt{g(\mathbf{x}_t,t)} = -\kappa(\mathbf{x}_t,t)\sqrt{g(\mathbf{x}_t,t)}, \tag{4.10.13}$$

which, by virtue of eqn. (4.10.12), leads to an equation for $f(\mathbf{x},t)$ alone:

$$\frac{\partial}{\partial t}f(\mathbf{x},t) + \xi(\mathbf{x},t) \cdot \nabla_{\mathbf{x}}f(\mathbf{x},t) = 0 \tag{4.10.14}$$

or simply

$$\frac{\mathrm{d}}{\mathrm{d}t}f(\mathbf{x}_t,t) = 0. \tag{4.10.15}$$

That is, when the non-Euclidean nature of the non-Hamiltonian phase space is properly accounted for, the ensemble distribution function $f(\mathbf{x}_t, t)$ is conserved just as it is in the Hamiltonian case, but it is conserved on a different phase space, namely, one with a nontrivial metric. Consequently, eqn. (2.5.11) generalizes to

$$f(\mathbf{x}_t,t)\sqrt{g(\mathbf{x}_t,t)}\mathrm{d}\mathbf{x}_t = f(\mathbf{x}_0,0)\sqrt{g(\mathbf{x}_0,0)}\mathrm{d}\mathbf{x}_0. \tag{4.10.16}$$

Equation (4.10.12) assumes smoothness both of the metric factor $\sqrt{g(\mathbf{x},t)}$ and of the distribution function $f(\mathbf{x},t)$, which places some restrictions on the class of non-Hamiltonian systems for which it is valid. This and related issues have been discussed by others (Ramshaw, 2002; Ezra, 2004) and are beyond the scope of this book.

4.10.3 Equilibrium solutions

In equilibrium, both $f(\mathbf{x}_t, t)$ and $\sqrt{g(\mathbf{x}_t, t)}$ have no explicit time dependence, and eqn. (4.10.16) reduces to

$$f(\mathbf{x}_t)\sqrt{g(\mathbf{x}_t)}\mathrm{d}\mathbf{x}_t = f(\mathbf{x}_0)\sqrt{g(\mathbf{x}_0)}\mathrm{d}\mathbf{x}_0, \tag{4.10.17}$$

which means that equilibrium averages can be performed at any instant in time, just as in the Hamiltonian case.

Although the equilibrium Liouville equation takes the same form as it does in the Hamiltonian case

$$\xi(\mathbf{x}) \cdot \nabla_{\mathbf{x}} f(\mathbf{x}) = 0, \tag{4.10.18}$$

we cannot express this in terms of a Poisson bracket with the Hamiltonian because there is no Hamiltonian to generate the equations of motion $\dot{\mathbf{x}} = \xi(\mathbf{x})$. In cases for which we can determine the full metric tensor $g_{ij}(\mathbf{x})$, a non-Hamiltonian generalization of the Poisson bracket is possible (Sergi, 2003; Tarasov, 2004; Ezra, 2004); however, no general theory of this metric tensor yet exists. Nevertheless, the fact that $\mathrm{d}f/\mathrm{d}t = 0$ allows us to construct a general equilibrium solution that is suitable for our purposes in this book. The non-Hamiltonian systems we will be studying in subsequent chapters are assumed to be complete in the sense that they represent the physical system plus some additional variables that grossly represent the surroundings. Thus, in order to construct a distribution function $f(\mathbf{x})$ that satisfies $\mathrm{d}f/\mathrm{d}t = 0$, it is sufficient to know all of the conservation laws satisfied by the equations of motion. Let there be N_c conservation laws of the form

$$\Lambda_k(\mathbf{x}_t) - C_k = 0, \qquad \frac{\mathrm{d}}{\mathrm{d}t} \Lambda_k(\mathbf{x}_t) = 0, \tag{4.10.19}$$

where $k = 1, ..., N_c$. If we can identify these, then a general "microcanonical" solution for $f(\mathbf{x})$ can be constructed from these conservation laws in the form

$$f(\mathbf{x}) = \prod_{k=1}^{N_c} \delta(\Lambda_k(\mathbf{x}) - C_k). \tag{4.10.20}$$

This solution simply states that the distribution generated by the dynamics is one that samples the intersection of the hypersurfaces represented by all of the conservation laws in eqn. (4.10.19). Under the usual assumptions of ergodicity, the system will sample all of the points on this intersection surface in an infinite time. Consequently, the non-Hamiltonian system has an associated "microcanonical" partition function obtained by integrating the distribution in eqn. (4.10.20):

$$\mathcal{Z} = \int \mathrm{d}\mathbf{x} \, \sqrt{g(\mathbf{x})} f(\mathbf{x}) = \int \mathrm{d}\mathbf{x} \, \sqrt{g(\mathbf{x})} \prod_{k=1}^{N_c} \delta\left(\Lambda_k(\mathbf{x}) - C_k\right). \tag{4.10.21}$$

The appearance of the metric determinant in the phase-space integral conforms to the requirement of eqn. (4.10.17), which states that the number of microstates available to the system is determined by $f(\mathbf{x})$ when it is integrated with respect to the *conserved* volume element $\sqrt{g(\mathbf{x})}\mathrm{d}\mathbf{x}$. Equations (4.10.20) and (4.10.21) lie at the heart of our theory of non-Hamiltonian phase spaces and will be used to analyze a variety of non-Hamiltonian systems in this and subsequent chapters.

4.10.4 Analysis of the Nosé-Hoover equations

We now turn to the analysis of eqns. (4.9.14). Our goal is to determine the physical phase-space distribution generated by the equations of motion. We begin by identifying the conservation laws associated with the equations. First, there is a conserved energy of the form

$$\mathcal{H}'(\mathbf{r}, \eta, \mathbf{p}, p_\eta) = \mathcal{H}(\mathbf{r}, \mathbf{p}) + \frac{p_\eta^2}{2Q} + dNkT\eta, \tag{4.10.22}$$

where $\mathcal{H}(\mathbf{r}, \mathbf{p})$ is the physical Hamiltonian. If $\sum_{i=1}^{N} \mathbf{F}_i \neq 0$, then except for very simple systems, eqn. (4.10.22) is the only conservation law. Next, we compute the compressibility as

$$\kappa = \sum_{i=1}^{N} [\nabla_{\mathbf{p}_i} \cdot \dot{\mathbf{p}}_i + \nabla_{\mathbf{r}_i} \cdot \dot{\mathbf{r}}_i] + \frac{\partial \dot{\eta}}{\partial \eta} + \frac{\partial \dot{p}_\eta}{\partial p_\eta}$$

$$= -\sum_{i=1}^{N} d\frac{p_\eta}{Q}$$

$$= -dN\dot{\eta}, \tag{4.10.23}$$

from which it is clear that the metric $\sqrt{g} = \exp(-w) = \exp(dN\eta)$. The microcanonical partition function at a given temperature T can be constructed using $\sqrt{g}$ and the energy conservation condition,

$$\mathcal{Z}_T(N, V, C_1) = \int d^N\mathbf{p} \int_{D(V)} d^N\mathbf{r} \int dp_\eta \, d\eta \, e^{dN\eta}$$

$$\times \delta \left(\mathcal{H}(\mathbf{r}, \mathbf{p}) + \frac{p_\eta^2}{2Q} + dNkT\eta - C_1 \right), \tag{4.10.24}$$

where the T subscript indicates that the microcanonical partition function depends parametrically on the temperature T.

The distribution function of the physical phase space can now be obtained by integrating over η and p_η. Using the δ-function to perform the integration over η requires that

$$\eta = \frac{1}{dNkT} \left(C_1 - \mathcal{H}(\mathbf{r}, \mathbf{p}) - \frac{p_\eta^2}{2Q} \right). \tag{4.10.25}$$

Substitution of this result into eqn. (4.10.24) and using eqn. (4.9.6) yields

$$\mathcal{Z}_T(N, V, C_1) = \frac{e^{\beta C_1}}{dNkT} \int dp_\eta e^{-\beta p_\eta^2/2Q} \int d^N\mathbf{p} \int_{D(V)} d^N\mathbf{r} \, e^{-\beta \mathcal{H}(\mathbf{r}, \mathbf{p})}, \tag{4.10.26}$$

which is the canonical distribution function apart from constant prefactors. This demonstrates that the Nosé-Hoover equations are capable of generating a canonical distribution in the physical subsystem variables when $\mathcal{H}'$ is the *only* conserved quantity. Unfortunately, this is not the typical situation. In the absence of external forces, Newton's third law requires that $\sum_{i=1}^{N} \mathbf{F}_i = 0$, which leads to an additional conservation law

$$\mathbf{P}e^{\eta} = \mathbf{K}, \qquad (4.10.27)$$

where $\mathbf{P} = \sum_{i=1}^{N} \mathbf{p}_i$ is the center-of-mass momentum of the system and $\mathbf{K}$ is an arbitrary constant vector in d dimensions. When this additional conservation law is present, the Nosé-Hoover equations do not generate the correct distribution (see Problem 4.3). Figure 4.9 illustrates the failure of the Nosé-Hoover equations for a single free particle in one dimension. The distribution $f(p)$ should be a Gaussian $f(p) = \exp(-p^2/2mkT)/\sqrt{2\pi mkT}$, which it clearly is not. Finally, Fig. 4.10 shows that the Nosé-Hoover equations also fail for a simple harmonic oscillator, for which eqn. (4.10.27) does not hold. Problem 4.4 suggests that an additional conservation law different from eqn. (4.10.27) is a possible culprit in the failure of the Nosé-Hoover equations for the harmonic oscillator.

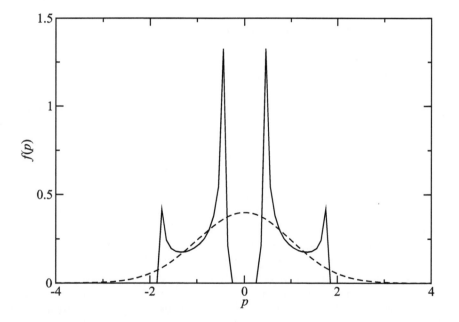

Fig. 4.9 Momentum distribution obtained by integrating the Nosé-Hoover equations $\dot{p} = -(p_\eta/Q)p, \dot{\eta} = p_\eta/Q, \dot{p}_\eta = p^2/m - kT$ for a free particle with $m = 1$, $Q = 1$, $kT = 1$, $p(0) = 1$, $\eta(0) = 0$, $p_\eta(0) = 1$, $\Delta t = 0.01$. The solid line is the distribution obtained from the simulation (see Problem 4.3), and the dashed line is the expected analytical distribution $f(p) = \exp(-p^2/2mkT)/\sqrt{2\pi mkT}$.

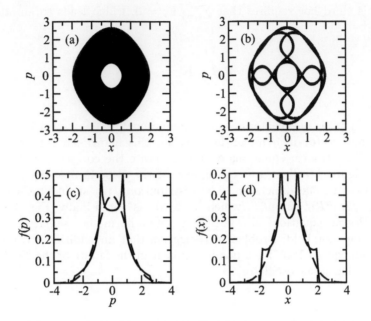

Fig. 4.10 Phase space and distribution functions obtained by integrating the Nosé-Hoover equations $\dot{x} = p/m$ $\dot{p} = -m\omega^2 x - (p_\eta/Q)p, \dot{\eta} = p_\eta/Q, \dot{p}_\eta = p^2/m - kT$ for a harmonic oscillator with $m = 1$, $\omega = 1$, $Q = 1$, $kT = 1$, $x(0) = 0$, $p(0) = 1$, $\eta(0) = 0$, $p_\eta(0) = 1$, $\Delta t = 0.01$. (a) shows the phase space p vs. x independent of η and p_η, (b) shows the phase space for $p_\eta \in [-\epsilon, \epsilon]$, where $\epsilon = 0.001$, and (c) and (d) show distributions $f(p)$ and $f(x)$ obtained from the simulation (solid line) compared with the correct canonical distributions (dashed line).

4.11 Nosé-Hoover chains

The reason for the failure of the Nosé-Hoover equations when more than one conservation law is obeyed by the system is that the equations of motion do not contain a sufficient number of variables in the extended phase space to offset the restrictions placed on the accessible phase space caused by multiple conservation laws. Each conservation law restricts the accessible phase space by one dimension. In order to counteract this effect, more phase space dimensions must be introduced, which can be accomplished by introducing additional variables. But how should these variables be added so as to give the correct distribution in the physical phase space? The answer can be gleaned from the fact that the momentum variable p_η in the Nosé-Hoover equations must have a Maxwell-Boltzmann distribution, just as the physical momenta do. In order to ensure that such a distribution is generated, p_η itself can be coupled to a Nosé-Hoover-type thermostat, which will bring in a new set of variables, $\tilde{\eta}$ and $p_{\tilde{\eta}}$. But once this is done, we have the problem that $p_{\tilde{\eta}}$ must also have a Maxwell-Boltzmann distribution, which requires introducing a thermostat for this variable. We could continue in this way *ad infinitum*, but the procedure must terminate at some point. If we terminate it after the addition of M new thermostat variable pairs η_j and p_{η_j}, $j = 1, ..., M$, then the equations of motion can be expressed as

$$\dot{\mathbf{r}}_i = \frac{\mathbf{p}_i}{m_i}$$

$$\dot{\mathbf{p}}_i = \mathbf{F}_i - \frac{p_{\eta_1}}{Q_1}\mathbf{p}_i$$

$$\dot{\eta}_j = \frac{p_{\eta_j}}{Q_j} \qquad\qquad j = 1, ..., M$$

$$\dot{p}_{\eta_1} = \left[\sum_{i=1}^{N}\frac{\mathbf{p}_i^2}{m_i} - dNkT\right] - \frac{p_{\eta_2}}{Q_2}p_{\eta_1}$$

$$\dot{p}_{\eta_j} = \left[\frac{p_{\eta_{j-1}}^2}{Q_{j-1}} - kT\right] - \frac{p_{\eta_{j+1}}}{Q_{j+1}}p_{\eta_j} \qquad\qquad j = 2, ..., M-1$$

$$\dot{p}_{\eta_M} = \left[\frac{p_{\eta_{M-1}}^2}{Q_{M-1}} - kT\right] \qquad\qquad\qquad (4.11.1)$$

(Martyna *et al.*, 1992). Equations (4.11.1) are known as the *Nosé-Hoover chain equations*. These equations ensure that the first $M-1$ thermostat momenta $p_{\eta_1}, ..., p_{\eta_{M-1}}$ have the correct Maxwell-Boltzmann distribution. Note that for $M=1$, the equations reduce to the simpler Nosé-Hoover equations. However, unlike the Nosé-Hoover equations, which are essentially Hamiltonian equations in noncanonical variables, the Nosé-Hoover chain equations have no underlying Hamiltonian structure, meaning no canonical variables exist that transform eqns. (4.11.1) into a Hamiltonian system.

Concerning the parameters $Q_1, ..., Q_M$, Martyna *et al.* (1992) suggested that an optimal choice for these is

$$Q_1 = dNkT\tau^2$$
$$Q_j = kT\tau^2, \qquad\qquad j = 2, ..., M, \qquad\qquad (4.11.2)$$

where τ is a characteristic time scale in the system. Since this time scale might not be known explicitly, in practical molecular dynamics calculations, a reasonable choice is $\tau \geq 20\Delta t$, where Δt is the time step.

In order to analyze the distribution of the physical phase space generated by eqns. (4.11.1), we first identify the conservation laws. If $\sum_{i=1}^{N}\mathbf{F}_i \neq 0$, then the equations of motion conserve

$$\mathcal{H}' = \mathcal{H}(\mathbf{r}, \mathbf{p}) + \sum_{j=1}^{M}\frac{p_{\eta_j}^2}{2Q_j} + dNkT\eta_1 + kT\sum_{j=2}^{M}\eta_j, \qquad\qquad (4.11.3)$$

which, in general, will be the only conservation law satisfied by the system. Next, the compressibility of eqns. (4.11.1) is

$$\nabla_{\mathbf{x}} \cdot \dot{\mathbf{x}} = -dN\frac{p_{\eta_1}}{Q_1} - \sum_{j=2}^{M}\frac{p_{\eta_j}}{Q_j} = -dN\dot{\eta}_1 - \dot{\eta}_c. \qquad\qquad (4.11.4)$$

Here, we have introduced the variable $\eta_c = \sum_{j=2}^{M} \eta_j$ as a convenience since this particular combination of the η variables comes up frequently. From the compressibility, we see that the phase-space metric is

$$\sqrt{g} = \exp\left[dN\eta_1 + \eta_c\right]. \tag{4.11.5}$$

Using eqns. (4.11.3) and (4.11.5), proving that the Nosé-Hoover chain equations generate a canonical ensemble is analogous to eqns. (4.10.24) through (4.10.26) for the Nosé-Hoover equations and, therefore, will not be repeated here but left as an exercise at the end of the chapter (see Problem 4.3).

An important property of the Nosé-Hoover chain equations is the fact that when $\sum_{i=1}^{N} \mathbf{F}_i = 0$, the equations of motion still generate a correct canonical distribution in all variables except the magnitude of the center-of-mass momentum $\mathbf{P}$ (see Problem 4.3). When there are no external forces, eqn. (4.10.27) becomes

$$\mathbf{K} = \mathbf{P}e^{\eta_1}. \tag{4.11.6}$$

In order to illustrate this for the simple cases considered in Figs. 4.9 and 4.10, Fig. 4.11 shows the momentum distribution of the one-dimensional free particle coupled to a Nosé-Hoover chain, together with the correct canonical distribution. The figure shows

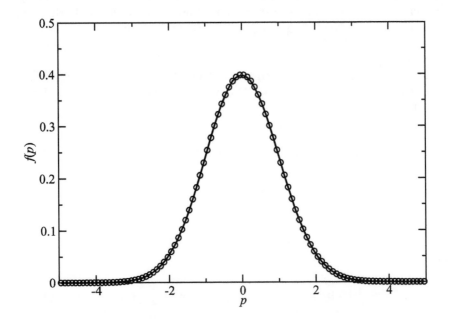

Fig. 4.11 Momentum distribution obtained by integrating the Nosé-Hoover chain equations for a free particle with $m = 1$, $Q = 1$, $kT = 1$, $\Delta t = 0.01$. Here, the distribution is generated only for $p > 0$ and has been symmetrized about $p = 0$. In addition, we set $\eta_k(0) = 0$, $p_{\eta_k}(0) = 1$. The solid line is the distribution obtained from the simulation (see Problem 4.3), and the circles are the correct distribution $f(p) = \exp(-p^2/2mkT)/\sqrt{2\pi mkT}$.

that the correct distribution is, indeed, obtained. Note that this distribution has been symmetrized about $p = 0$. As we will see in Section 4.12, the integrator we employ for the Nosé-Hoover chain equations scales the momentum by a positive-definite constant, which means that, for a free particle, the sign of the momentum cannot change over the course of a simulation and only half of the momentum distribution $f(p) = \exp(-p^2/2mkT)/\sqrt{2\pi mkT}$ can be generated. Therefore, in order to include the full domain of momentum values and the full range of $f(p)$, the distribution needs to be symmetrized or two simulations starting from two different initial conditions, $p(0) = 1$ and $p(0) = -1$, must be combined. This pathology only exists for free particles and for systems whose forces and/or dynamics cannot change the sign of the momentum. In addition, Fig. 4.12 also shows the physical phase space and position and momentum distributions for the harmonic oscillator coupled to a Nosé-Hoover chain. Again, it can be seen that the correct canonical distribution is generated, thereby solving the failure of the Nosé-Hoover equations. By working through Problem 4.3, it will become clear what mechanism is at work in the Nosé-Hoover chain equations that leads to the correct canonical distributions and why, therefore, these equations are recommended over the Nosé-Hoover equations.

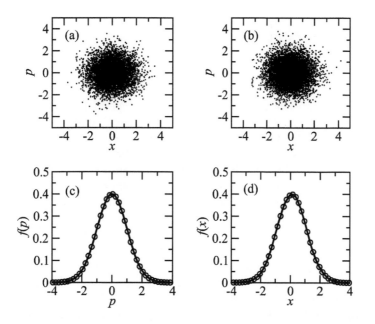

Fig. 4.12 Phase space and distribution functions obtained by integrating the Nosé-Hoover chain equations for a harmonic oscillator with $m = 1$, $\omega = 1$, $Q = 1$, $kT = 1$, $x(0) = 0$, $p(0) = 1$, $\eta_k(0) = 0$, $p_{\eta_1}(0) = p_{\eta_3}(0) = 1$, $p_{\eta_2}(0) = p_{\eta_4}(0) = -1$, $\Delta t = 0.01$. (a) shows the phase space p vs. x independent of η and p_η, (b) shows the phase space for $p_{\eta_1}, p_{\eta_2} \in [-\epsilon, \epsilon]$, where $\epsilon = 0.001$, (c) and (d) show distributions $f(p)$ and $f(x)$ obtained from the simulation (solid line) together with the correct canonical distributions (circles).

As one final yet important note, consider rewriting eqns. (4.11.1) such that each particle has its *own* Nosé-Hoover chain thermostat. This would be expressed in the equations by adding an additional index to the thermostat variables:

$$\dot{\mathbf{r}}_i = \frac{\mathbf{p}_i}{m_i}$$

$$\dot{\mathbf{p}}_i = \mathbf{F}_i - \frac{p_{\eta_1,i}}{Q_1}\mathbf{p}_i$$

$$\dot{\eta}_{j,i} = \frac{p_{\eta_j,i}}{Q_j} \qquad\qquad j = 1, ..., M$$

$$\dot{p}_{\eta_1,i} = \left[\frac{\mathbf{p}_i^2}{m_i} - dkT\right] - \frac{p_{\eta_2,i}}{Q_2}p_{\eta_1,i}$$

$$\dot{p}_{\eta_j,i} = \left[\frac{p_{\eta_{j-1},i}^2}{Q_{j-1}} - kT\right] - \frac{p_{\eta_{j+1},i}}{Q_{j+1}}p_{\eta_j} \qquad\qquad j = 2, ..., M-1$$

$$\dot{p}_{\eta_M,i} = \left[\frac{p_{\eta_{M-1},i}^2}{Q_{M-1}} - kT\right]. \tag{4.11.7}$$

The introduction of a separate thermostat for each particle has the immediate practical advantage of yielding a molecular dynamics scheme capable of rapidly equilibrating a system by ensuring that each particle satisfies the virial theorem. Even in a large homogeneous system such as the Lennard-Jones liquid studied in Section 3.14.2, where rapid energy transfer between particles usually leads to rapid equilibration, eqns. (4.11.7) provide a noticeable improvement in the convergence of the kinetic energy fluctuations as shown in Fig. 4.13. In complex, inhomogeneous systems such as a protein in aqueous solution, polymeric materials, or even "simple" molecular liquids such as water and methanol, there will be a wide range of time scales. Some of these time scales are only weakly coupled so that equipartition of the energy in accordance with the virial theorem happens only *very* slowly. In such systems, the use of separate thermostats as in eqns. (4.11.7) can be very effective. Unlike the global thermostat of eqns. (4.11.1), which can actually allow "hot" and "cold" spots to develop in a system while only ensuring that the average total kinetic energy is $dNkT$, eqns. (4.11.7) avoid this problem by allowing each particle to exchange energy with its own heat bath. Moreover, it can be easily seen that even if $\sum_i \mathbf{F}_i = 0$, conservation laws such as eqn. (4.11.6) no longer exist in the system, a fact which leads to a simplification of the proof that the canonical distribution is generated. In fact, it is possible to take this idea one step further and couple a Nosé-Hoover chain to *each Cartesian degree of freedom* in the system, for a total of dN heat baths. Such a scheme is known colloquially as "massive" thermostatting and was shown by Tobias *et al.* (1993) to lead to very rapid thermalization of a protein in aqueous solution.

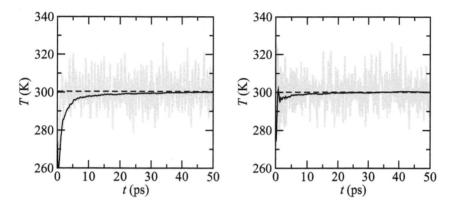

Fig. 4.13 Convergence of kinetic energy fluctuations (in Kelvin) normalized by the number of degrees of freedom for the argon system of Section 3.14.2 at a temperature of 300 K for a global Nosé-Hoover chain thermostat (*left*) and individual Nosé-Hoover chain thermostats ("massive" thermostatting) (*right*) attached to Cartesian degree of freedom of each particle.

4.12 Integrating the Nosé-Hoover chain equations

Numerical integrators for non-Hamiltonian systems such as the Nosé-Hoover chain (NHC) equations can be derived using the Liouville operator formalism developed in Section 3.10 (Martyna *et al.*, 1996). However, certain subtleties arise due to the generalized Liouville theorem in eqn. (4.10.8) and, therefore, the subject merits some discussion. Recall that for a Hamiltonian system, any numerical integration algorithm must preserve the symplectic property, in which case, it will also conserve the phase-space volume. For non-Hamiltonian systems, there is no clear analog of the symplectic property. Nevertheless, the existence of a generalized Liouville theorem, eqn. (4.10.8), provides us with a minimal requirement that numerical solvers for non-Hamiltonian systems should satisfy, specifically, the preservation of the measure $\sqrt{g(\mathbf{x})}d\mathbf{x}$. Integrators that fail to obey the generalized Liouville theorem cannot be guaranteed to generate correct distributions. Therefore, in devising numerical solvers for non-Hamiltonian systems, care must be taken to ensure that they are measure-preserving (Ezra, 2007).

Keeping in mind the generalized Liouville theorem, let us now develop an integrator for the Nosé-Hoover chain equations. Despite the fact that eqns. (4.11.1) are non-Hamiltonian, they can be expressed as an operator equation just as in the Hamiltonian case. Indeed, a general non-Hamiltonian system

$$\dot{\mathbf{x}} = \xi(\mathbf{x}) \tag{4.12.1}$$

can always be expressed as

$$\dot{\mathbf{x}} = iL\mathbf{x}, \tag{4.12.2}$$

where

$$iL = \xi(\mathbf{x}) \cdot \nabla_{\mathbf{x}}. \tag{4.12.3}$$

Note that we are considering systems with no explicit time dependence, although the Liouville operator formalism can be extended to systems with explicit time dependence (Suzuki, 1992). The Liouville operator corresponding to eqns. (4.11.1) can be written as

$$iL = iL_{\text{NHC}} + iL_1 + iL_2, \tag{4.12.4}$$

where

$$iL_1 = \sum_{i=1}^{N} \frac{\mathbf{p}_i}{m_i} \cdot \frac{\partial}{\partial \mathbf{r}_i}$$

$$iL_2 = \sum_{i=1}^{N} \mathbf{F}_i \cdot \frac{\partial}{\partial \mathbf{p}_i}$$

$$iL_{\text{NHC}} = -\sum_{i=1}^{N} \frac{p_{\eta_1}}{Q_1} \mathbf{p}_i \cdot \frac{\partial}{\partial \mathbf{p}_i} + \sum_{j=1}^{M} \frac{p_{\eta_j}}{Q_j} \frac{\partial}{\partial \eta_j}$$

$$+ \sum_{j=1}^{M-1} \left(G_j - p_{\eta_j} \frac{p_{\eta_{j+1}}}{Q_{j+1}} \right) \frac{\partial}{\partial p_{\eta_j}} + G_M \frac{\partial}{\partial p_{\eta_M}}. \tag{4.12.5}$$

Here, the thermostat "forces" are represented as

$$G_1 = \sum_{i=1}^{N} \frac{\mathbf{p}_i^2}{m_i} - dNkT$$

$$G_j = \frac{p_{\eta_{j-1}}^2}{Q_{j-1}} - kT. \tag{4.12.6}$$

Note that the sum $iL_1 + iL_2$ in eqn. (4.12.4) constitutes a purely Hamiltonian subsystem. The evolution of the full phase-space vector

$$\mathbf{x} = (\mathbf{r}_1, ..., \mathbf{r}_N, \eta_1, ..., \eta_M, \mathbf{p}_1, ..., \mathbf{p}_N, p_{\eta_1}, ..., p_{\eta_M}) \tag{4.12.7}$$

is given by the usual relation $\mathbf{x}_t = \exp(iLt)\mathbf{x}_0$. As was done in the Hamiltonian case, we will employ the Trotter theorem to factorize the propagator $\exp(iL\Delta t)$ for a single time step Δt. Consider a particular factorization of the form

$$e^{iL\Delta t} = e^{iL_{\text{NHC}}\Delta t/2} e^{iL_2\Delta t/2} e^{iL_1\Delta t} e^{iL_2\Delta t/2} e^{iL_{\text{NHC}}\Delta t/2} + \mathcal{O}\left(\Delta t^3\right). \tag{4.12.8}$$

Note that the three operators in the middle are identical to those in eqn. (3.10.22). By the analysis of Section 3.10, this factorization, on its own, would generate the velocity Verlet algorithm. However, in eqn. (4.12.8), it is sandwiched between the thermostat propagators. The separation between the Hamiltonian and non-Hamiltonian parts of the Liouville operator is both intuitively appealing and, as will be seen below, allows for easy implementation of both multiple time-scale (RESPA) schemes and constraints. When the NHC operator is placed on the ends of the Trotter factorization, this type of algorithm is termed an "end" scheme. An alternative factorization, known as a "middle" scheme, places the NHC operator in the middle, as follows (Zhang *et al.*, 2017; Zhang *et al.*, 2019):

$$e^{iL\Delta t} = e^{iL_2\Delta t/2}\,e^{iL_1\Delta t/2}e^{iL_{\mathrm{NHC}}\Delta t}e^{iL_1\Delta t/2}\,e^{iL_2\Delta t/2} + \mathcal{O}\left(\Delta t^3\right). \qquad (4.12.9)$$

A detailed error analysis, which lies beyond the scope of this book, shows that the end scheme has a small error in the configurational distribution of a harmonic system while the middle scheme does not. This can be easily demonstrated numerically for a thermostatted harmonic oscillator, as illustrated in Fig. 4.14, which shows the error in the position distribution for both the end and middle schemes as a function of the time step. The figure demonstrates that the error in the distribution is comparable up to time steps that would lead to acceptable conservation of eqn. (4.11.3). However, when the time step is increased, the error in the end scheme grows, indicating that the distribution is more accurately generated by the middle scheme.

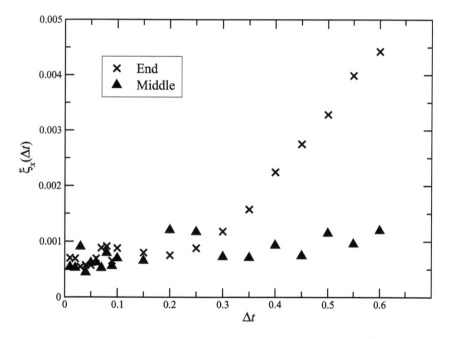

Fig. 4.14 Error in the position distribution as a function of time step Δt for a harmonic oscillator with $\omega = 1$, $m = 1$, $x(0) = 0$, $p(0) = 1$ coupled to a Nosé-Hoover chain of length 4. The "end" factorization (eqn. (4.12.8)) and "middle" factorization (eqn. (4.12.9)) are compared. The error measure is given by $\xi_x(\Delta t) = (1/N_{\mathrm{bins}})\sum_{i=1}^{N_{\mathrm{bins}}} |P_{\mathrm{calc}}(x_i) - P_{\mathrm{ex}}(x_i)|$, where N_{bins} is the number of bins in the distribution and $P_{\mathrm{calc}}(x)$ and $P_{\mathrm{ex}}(x)$ are the computed and exact distributions, respectively. The number M of simulation steps is chosen such that $M\Delta t = 2 \times 10^5$.

The operator iL_{NHC} contains many terms, so we still need to break down the operator $\exp(iL_{\mathrm{NHC}}\Delta t/2)$ further. Experience has shown, unfortunately, that a simple factorization of the operator based on the separate terms in iL_{NHC} is insufficient to achieve a robust integration scheme. The reason is that the thermostat forces in eqn. (4.12.6) vary rapidly, thereby limiting the time step. To alleviate this problem, we can apply the RESPA methodology of Section 3.11 to this part of the propagator. Once again, experience shows that several hundred RESPA steps are needed to resolve the thermostat part of the propagator accurately, so RESPA alone cannot easily handle the rapidly varying thermostat forces. Consider, however, employing a higher-order (than Δt^3) factorization together with RESPA to $\exp(iL_{\mathrm{NHC}}\Delta t/2)$ or $\exp(iL_{\mathrm{NHC}}\Delta t)$. A judiciously chosen algorithm could improve the accuracy of RESPA without adding significantly to the computational overhead. Fortunately, high order methods suitable for our purposes exist and are straightforward to apply. One scheme in particular, due to Suzuki (1991a, 1991b) and Yoshida (1990), has proved particularly useful for the Nosé-Hoover chain system.

The Suzuki-Yoshida scheme works as follows: Let $S(\lambda)$ be a *primitive factorization* of the operator $\exp[\lambda(A_1 + A_2)]$. For example, a primitive factorization could be the simple Trotter scheme $S(\lambda) = \exp(\lambda A_2/2)\exp(\lambda A_1)\exp(\lambda A_2/2)$. Next, introduce a set of n_{sy} weights w_α such that

$$\sum_{\alpha=1}^{n_{\mathrm{sy}}} w_\alpha = 1. \tag{4.12.10}$$

These weights are chosen in such a way that error terms up to a certain order $2s$ are eliminated in a general factorization of $\exp[\lambda(A_1 + A_2)]$, yielding a high order scheme. In the original Suzuki scheme, it was shown that

$$n_{\mathrm{sy}} = 5^{s-1}, \tag{4.12.11}$$

so that a fourth-order scheme would require 5 weights, a sixth-order scheme would require 25 weights, *etc.*, with all weights having a simple analytical form. For example, for $2s = 4$, the five weights are

$$w_1 = w_2 = w_4 = w_5 = \frac{1}{4 - 4^{1/3}}$$

$$w_3 = 1 - (w_1 + w_2 + w_4 + w_5).$$

Since the number of weights grows exponentially with the order, an alternative set of weights, introduced by Yoshida, proves beneficial. In the Yoshida scheme, a numerical procedure for obtaining the weights is introduced, leading to a much smaller number of weights. For example, only three weights are needed for a fourth-order scheme, and these are given by

$$w_1 = w_3 = \frac{1}{2 - 2^{1/3}} \qquad w_2 = 1 - w_1 - w_3. \tag{4.12.12}$$

For a sixth-order scheme, seven weights are needed, and these are obtained numerically with the result

$$w_1 = w_7 = 0.784513610477560$$
$$w_2 = w_6 = 0.235573213359357$$
$$w_3 = w_5 = -1.17767998417887$$
$$w_4 = 1 - w_1 - w_2 - w_3 - w_5 - w_6 - w_7. \tag{4.12.13}$$

Once a set of weights is chosen, the factorization of the operator is then expressed as

$$e^{\lambda(A_1+A_2)} \approx \prod_{\alpha=1}^{n_{\mathrm{sy}}} S(w_\alpha\lambda). \tag{4.12.14}$$

In the present discussion, we will let $S(\Delta t/\nu)$ be a primitive factorization of the operator $\exp(iL_{\mathrm{NHC}}\Delta t/\nu)$, where $\nu = 2$ for the end scheme [eqn. (4.12.8)] and $\nu = 1$ for the middle scheme [eqn. (4.12.9)]. Applying eqn. (4.12.14) to $\exp(iL_{\mathrm{NHC}}\Delta t/\nu)$, we obtain

$$e^{iL_{\mathrm{NHC}}\Delta t/\nu} \approx \prod_{\alpha=1}^{n_{\mathrm{sy}}} S(w_\alpha\Delta t/\nu). \tag{4.12.15}$$

Finally, RESPA is introduced very simply by applying the operator S n times with a time step $w_\alpha\Delta t/\nu n$, i.e.,

$$e^{iL_{\mathrm{NHC}}\Delta t/\nu} \approx \prod_{\alpha=1}^{n_{\mathrm{sy}}} [S(w_\alpha\Delta t/\nu n)]^n. \tag{4.12.16}$$

Using the Suzuki-Yoshida scheme allows the propagator in eqn. (4.12.8) to be written as

$$e^{iL\Delta t} \approx \prod_{\alpha=1}^{n_{\mathrm{sy}}} [S(w_\alpha\Delta t/2n)]^n \; e^{iL_2\Delta t/2} \, e^{iL_1\Delta t} \, e^{iL_2\Delta t/2} \prod_{i=1}^{n_{\mathrm{sy}}} [S(w_\alpha\Delta t/2n)]^n, \tag{4.12.17}$$

while the propagator in eqn. (4.12.9) becomes

$$e^{iL\Delta t} \approx \prod_{\alpha=1}^{n_{\mathrm{sy}}} e^{iL_2\Delta t/2} \, e^{iL_1\Delta t/2} \, [S(w_\alpha\Delta t/n)]^n \; e^{iL_1\Delta t/2} \, e^{iL_2\Delta t/2} \tag{4.12.18}$$

Finally, we need to choose a primitive factorization $S(w_\alpha\Delta t/\nu n)$ for the operator $\exp(iL_{\mathrm{NHC}}\Delta t/\nu)$. Although this choice is not unique, we must nevertheless ensure that our factorization scheme preserves the generalized Liouville theorem. Defining $\delta_\alpha = w_\alpha\Delta t/n$, one such possibility is the following:

$$S(\delta_\alpha/\nu) = \exp\left[\frac{\delta_\alpha}{2\nu}G_M\frac{\partial}{\partial p_{\eta_M}}\right]$$

$$\times \prod_{j=M'}^{1} \left\{\exp\left[-\frac{\delta_\alpha}{4\nu}\frac{p_{\eta_{j+1}}}{Q_{j+1}}p_{\eta_j}\frac{\partial}{\partial p_{\eta_j}}\right] \exp\left[\frac{\delta_\alpha}{2\nu}G_j\frac{\partial}{\partial p_{\eta_j}}\right] \exp\left[-\frac{\delta_\alpha}{4\nu}\frac{p_{\eta_{j+1}}}{Q_{j+1}}p_{\eta_j}\frac{\partial}{\partial p_{\eta_j}}\right]\right\}$$

$$
\times \prod_{i=1}^{N} \exp \left[\frac{\delta_\alpha}{\nu} \frac{p_{\eta_1}}{Q_1} \mathbf{p}_i \cdot \frac{\partial}{\partial \mathbf{p}_i} \right] \quad \prod_{j=1}^{M} \exp \left[-\frac{\delta_\alpha}{\nu} \frac{p_{\eta_j}}{Q_j} \frac{\partial}{\partial \eta_j} \right]
$$

$$
\times \prod_{j=1}^{M'} \left\{ \exp \left[-\frac{\delta_\alpha}{4\nu} \frac{p_{\eta_{j+1}}}{Q_{j+1}} p_{\eta_j} \frac{\partial}{\partial p_{\eta_j}} \right] \exp \left[\frac{\delta_\alpha}{2\nu} G_j \frac{\partial}{\partial p_{\eta_j}} \right] \exp \left[-\frac{\delta_\alpha}{4\nu} \frac{p_{\eta_{j+1}}}{Q_{j+1}} p_{\eta_j} \frac{\partial}{\partial p_{\eta_j}} \right] \right\}
$$

$$
\times \exp \left[\frac{\delta_\alpha}{2\nu} G_M \frac{\partial}{\partial p_{\eta_M}} \right], \tag{4.12.19}
$$

where $M' = M - 1$. In eqn. (4.12.19), the symbol $\prod_{j=M'}^{1}$ indicates a backward product in which j starts at M' and is decremented in each term of the product until $j = 1$ is reached. Equation (4.12.19) may look imposing, but each of the operators appearing in the primitive factorization has a straightforward effect on the phase space. In fact, one sees that most of the operators are just the translation operators introduced in Section 3.10. The only exceptions are operators of the general form $\exp(cx\partial/\partial x)$, which also appear in the factorization. What is the effect of this type of operator? We start by considering its action on x. We can work this out using a Taylor series:

$$
\exp \left[cx \frac{\partial}{\partial x} \right] x = \left[\sum_{k=0}^{\infty} \frac{c^k}{k!} \left(x \frac{\partial}{\partial x} \right)^k \right] x
$$

$$
= x \sum_{k=0}^{\infty} \frac{c^k}{k!}
$$

$$
= xe^c. \tag{4.12.20}
$$

We see that the operator scales x by the constant e^c. Thus, when such a "scaling" operator acts on a momentum, the sign of the momentum cannot be changed. This is why a simulation of a free particle coupled to a Nosé-Hoover chain thermostat, performed using the algorithms discussed in this section, will generate only half of the momentum distribution, as noted in Section 4.11. We further note that the action of the operator $\exp(cx\partial/\partial x)$ on a function $f(x)$ produces $f(xe^c)$. Using this general result, each of the operators in eqn. (4.12.19) can be turned into a simple instruction in code (either translation or scaling) via the direct translation technique from Section 3.10.

At this point, several comments are in order. First, the separation of the non-Hamiltonian component of the equations of motion from the Hamiltonian component in eqn. (4.12.8) makes implementation of RESPA integration with Nosé-Hoover chains relatively straightforward. For example, suppose a system has fast and slow forces as discussed in Section 3.11. Instead of decomposing the Liouville operator as was done in eqn. (4.12.8), we could express iL as

$$
iL = iL_{\text{fast}} + iL_{\text{slow}} + iL_{\text{NHC}} \tag{4.12.21}
$$

and further decompose iL_{fast} into kinetic and force terms $iL_{\text{fast}}^{(1)} + iL_{\text{fast}}^{(2)}$, respectively. Then, the "end" propagator can be factorized according to

$$e^{iL\Delta t} = e^{iL_{\mathrm{NHC}}\Delta t/2} \, e^{iL_{\mathrm{slow}}\Delta t/2} \left[e^{iL_{\mathrm{fast}}^{(2)}\delta t/2} \, e^{iL_{\mathrm{fast}}^{(1)}\delta t} e^{iL_{\mathrm{fast}}^{(2)}\delta t/2} \right]^n e^{iL_{\mathrm{slow}}\Delta t/2} e^{iL_{\mathrm{NHC}}\Delta t/2},$$

$$(4.12.22)$$

where $\delta t = \Delta t/n$. Similarly, the "middle" propagator becomes

$$e^{iL\Delta t} = e^{iL_{\mathrm{slow}}\Delta t/2} \left[e^{iL_{\mathrm{fast}}^{(2)}\delta t/2} \, e^{iL_{\mathrm{fast}}^{(1)}\delta t/2} \right]^n e^{iL_{\mathrm{NHC}}\Delta t} \left[e^{iL_{\mathrm{fast}}^{(1)}\delta t/2} \, e^{iL_{\mathrm{fast}}^{(2)}\delta t/2} \right]^n e^{iL_{\mathrm{slow}}\Delta t/2}.$$

$$(4.12.23)$$

In such a factorization, the τ parameter in eqn. (4.11.2) should be chosen according to the time scale of the slow forces. On the other hand, if we need the thermostats to act on a faster time scale, then they can be pulled into the reference system by writing the end propagator as:

$$e^{iL\Delta t} = e^{iL_{\mathrm{NHC}}\delta t/2} \, e^{iL_{\mathrm{slow}}\Delta t/2} e^{-iL_{\mathrm{NHC}}\delta t/2}$$
$$\times \left[e^{iL_{\mathrm{NHC}}\delta t/2} \, e^{iL_{\mathrm{fast}}^{(2)}\delta t/2} \, e^{iL_{\mathrm{fast}}^{(1)}\delta t} e^{iL_{\mathrm{fast}}^{(2)}\delta t/2} e^{iL_{\mathrm{NHC}}\delta t/2} \right]^n$$
$$\times e^{-iL_{\mathrm{NHC}}\delta t/2} \, e^{iL_{\mathrm{slow}}\Delta t/2} e^{iL_{\mathrm{NHC}}\delta t/2}.$$

$$(4.12.24)$$

Here, the operator $\exp(-iL_{\mathrm{NHC}}\delta t/2)$ is never actually applied; its presence in eqn. (4.12.24) indicates that in the first and last RESPA steps, the Nosé-Hoover chain part of the propagator acts on the ends but with the small time step. In a comparable manner, the middle propagator becomes

$$e^{iL\Delta t} = e^{iL_{\mathrm{slow}}\Delta t/2} \left[e^{iL_{\mathrm{fast}}^{(2)}\delta t/2} \, e^{iL_{\mathrm{fast}}^{(1)}\delta t/2} e^{iL_{\mathrm{NHC}}\delta t} e^{iL_{\mathrm{fast}}^{(1)}\delta t/2} \, e^{iL_{\mathrm{fast}}^{(2)}\delta t/2} \right]^n e^{iL_{\mathrm{slow}}\Delta t/2}.$$

$$(4.12.25)$$

We denote the schemes in eqn. (4.12.22) and eqn. (4.12.23) as XO-RESPA (eXtended-system Outer RESPA) and the schemes in eqn. (4.12.24) and eqn. (4.12.25) as XI-RESPA (eXtended-system Inner RESPA) (Martyna et al., 1996; Zhang et al., 2017; Zhang et al., 2019).

The next point we address concerns the use of Nosé-Hoover chains with holonomic constraints. Constraints were discussed in Section 1.9 in the context of Lagrangian mechanics, and numerical procedures for imposing them within a given integration algorithm were presented in Section 3.9. Recall from Section 3.9 that the numerical procedure requires imposing the constraint conditions

$$\sigma_k(\mathbf{r}_1, ..., \mathbf{r}_N) = 0 \qquad k = 1, ..., N_c \tag{4.12.26}$$

and their first time derivatives

$$\sum_{i=1}^{N} \nabla_i \sigma_k \cdot \dot{\mathbf{r}}_i = \sum_{i=1}^{N} \nabla_i \sigma_k \cdot \frac{\mathbf{p}_i}{m_i} = 0. \tag{4.12.27}$$

Note that the time derivatives above are linear in the velocities or the momenta. This is important for the present discussion, as each application of the factorization in eqn.

(4.12.19) scales the particle momenta by a constant. Therefore, provided all particles involved in a common constraint are coupled to the same thermostat, their momenta will be scaled by a common constant due to the operation

$$
\exp\left[-\frac{\delta_\alpha}{2}\frac{p_{\eta_1}}{Q_1}\mathbf{p}_i \cdot \frac{\partial}{\partial \mathbf{p}_i}\right]\mathbf{p}_i = \mathbf{p}_i \exp\left[-\frac{\delta_\alpha}{2}\frac{p_{\eta_1}}{Q_1}\right].
$$

However, since eqn. (4.12.27) can be multiplied by an arbitrary constant without violating the condition it prescribes, the operator $\exp(iL_{\text{NHC}}\Delta t/2)$ in eqn. (4.12.22) can be applied after the RATTLE step (see Section 3.9.1) without affecting the constraint. Unfortunately, this is only true for the end factorization scheme. The middle factorization is more subtle, as the scaling occurs prior to the imposition of the constraint. As the two operations do not commute, the propagation of the scaling factor through the constraint can change the temperature achieved by the thermostat. This problem can be cured by iterating between the constraint and the thermostat until the two reach self consistency. Other schemes are discussed by Zhang *et al.* (2019).

4.13 The isokinetic ensemble: A variant of the canonical ensemble

Extended phase-space methods are not unique in their ability to generate canonical distributions in molecular dynamics calculations. In this section, we will discuss an alternative approach known as the *isokinetic ensemble*. As the name implies, the isokinetic ensemble is one in which the total kinetic energy of a system is maintained at a constant value. It is, therefore, described by a partition function of the form

$$
\mathcal{Q}(N,V,T,K) = \frac{K_0}{N!h^{3N}} \int \mathrm{d}^N\mathbf{p} \int_{D(V)} \mathrm{d}^N\mathbf{r}\, \delta\left(\sum_{i=1}^{N}\frac{\mathbf{p}_i^2}{2m_i} - K\right) e^{-\beta U(\mathbf{r}_1,\dots,\mathbf{r}_N)},
$$

(4.13.1)

where K is preset value of the kinetic energy, and K_0 is an arbitrary constant having units of energy. Equation (4.13.1) indicates that while the momenta are constrained to a spherical hypersurface of constant kinetic energy, the position-dependent part of the distribution is canonical. Since this is the most important part of the distribution for the calculation of equilibrium properties, the fact that the momentum distribution is not canonical is of little consequence. Nevertheless, since the momentum- and position-dependent parts of the distribution are separable, the isokinetic partition function can be trivially related to the true canonical partition function by

$$
Q(N,V,T) = \frac{(1/N!)V^N(2\pi mkT/h^2)^{3N/2}}{(1/N!)(K_0/K)(1/\Gamma(3N/2))V^N(2\pi mK/h^2)^{3N/2}}\mathcal{Q}(N,V,T,K)
$$

$$
= \frac{Q_{\text{ideal}}(N,V,T)}{\Omega_{\text{ideal}}(N,V,K)}\mathcal{Q}(N,V,T,K),
$$

(4.13.2)

where Ω_{ideal} and Q_{ideal} are the ideal gas partition functions in the microcanonical and canonical ensembles, respectively.

Equations of motion for the isokinetic ensemble were first written down by D. J. Evans and G. P. Morriss (1980) by applying Gauss's principle of least constraint. The equations of motion are obtained by imposing a kinetic-energy constraint

$$\sum_{i=1}^{N} m_i \dot{\mathbf{r}}_i^2 = \sum_{i=1}^{N} \frac{\mathbf{p}_i^2}{m_i} = 2K \tag{4.13.3}$$

on the Hamiltonian dynamics of the system. According to the discussion in Section 1.9, eqn. (4.13.3) is a nonholonomic constraint, but one that can be expressed in differential form. Thus, the Lagrangian form of the equations of motion is

$$\frac{\mathrm{d}}{\mathrm{d}t} \left(\frac{\partial \mathcal{L}}{\partial \dot{\mathbf{r}}_i} \right) - \frac{\partial \mathcal{L}}{\partial \mathbf{r}_i} = \alpha m_i \dot{\mathbf{r}}_i, \tag{4.13.4}$$

which can also be put into Hamiltonian form

$$\dot{\mathbf{r}}_i = \frac{\mathbf{p}_i}{m_i}$$

$$\dot{\mathbf{p}}_i = \mathbf{F}_i - \alpha \mathbf{p}_i. \tag{4.13.5}$$

Here, α is the single Lagrange multiplier needed to impose the constraint. Using Gauss's principle of least constraint gives a closed-form expression for α. We first differentiate eqn. (4.13.3) once with respect to time, which yields

$$\sum_{i=1}^{N} \frac{\mathbf{p}_i}{m_i} \cdot \dot{\mathbf{p}}_i = 0. \tag{4.13.6}$$

Thus, substituting the second of eqns. (4.13.5) into eqn. (4.13.6) gives

$$\sum_{i=1}^{N} \frac{\mathbf{p}_i}{m_i} \cdot [\mathbf{F}_i - \alpha \mathbf{p}_i], \tag{4.13.7}$$

which can be solved for α giving

$$\alpha = \frac{\sum_{i=1}^{N} \mathbf{F}_i \cdot \mathbf{p}_i / m_i}{\sum_{i=1}^{N} \mathbf{p}_i^2 / m_i}. \tag{4.13.8}$$

When eqn. (4.13.8) is substituted into eqn. (4.13.5), the equations of motion for the isokinetic ensemble become

$$\dot{\mathbf{r}}_i = \frac{\mathbf{p}_i}{m_i}$$

$$\dot{\mathbf{p}}_i = \mathbf{F}_i - \left[\frac{\sum_{j=1}^{N} \mathbf{F}_j \cdot \mathbf{p}_j / m_j}{\sum_{j=1}^{N} \mathbf{p}_j^2 / m_j} \right] \mathbf{p}_i. \tag{4.13.9}$$

Because eqns. (4.13.9) were constructed to preserve eqn. (4.13.3), they manifestly *conserve* the kinetic energy; however, that eqn. (4.13.3) is a conservation law of the isokinetic equations of motion can also be verified by direct substitution. Equations (4.13.9) are non-Hamiltonian and can, therefore, be analyzed via the techniques Section 4.10.

In order to carry out the analysis, we first need to calculate the phase-space compressibility:

$$
\kappa = \sum_{i=1}^{N} [\nabla_{\mathbf{r}_i} \cdot \dot{\mathbf{r}}_i + \nabla_{\mathbf{p}_i} \cdot \dot{\mathbf{p}}_i]
$$

$$
= \sum_{i=1}^{N} \nabla_{\mathbf{p}_i} \cdot \left\{ \mathbf{F}_i - \left[\frac{\sum_{j=1}^{N} \mathbf{F}_j \cdot \mathbf{p}_j / m_j}{\sum_{j=1}^{N} \mathbf{p}_j^2 / m_j} \right] \mathbf{p}_i \right\}
$$

$$
= -\frac{(dN-1) \sum_{i=1}^{N} \mathbf{F}_i \cdot \mathbf{p}_i / m_i}{2K}
$$

$$
= \frac{(dN-1)}{2K} \frac{dU(\mathbf{r}_1, ..., \mathbf{r}_N)}{dt}. \tag{4.13.10}
$$

Thus, the function $w(\mathbf{x})$ is just $(dN-1)U(\mathbf{r}_1, ..., \mathbf{r}_N)/2K$, and the phase-space metric becomes

$$
\sqrt{g} = e^{-(dN-1)U(\mathbf{r}_1, ..., \mathbf{r}_N)/2K}. \tag{4.13.11}
$$

Since the equations of motion explicitly conserve the total kinetic energy $\sum_{i=1}^{N} \mathbf{p}_i^2 / m_i$, we can immediately write down the partition function generated by the equations of motion:

$$
\Omega = \int d^N \mathbf{p} \, d^N \mathbf{r} \, e^{-(dN-1)U(\mathbf{r}_1, ..., \mathbf{r}_N)/K} \delta \left(\sum_{i=1}^{N} \frac{\mathbf{p}_i^2}{m_i} - (dN-1)kT \right). \tag{4.13.12}
$$

The analysis shows that if the constant parameter K is chosen to be $(dN-1)kT$, then the partition function becomes

$$
\Omega = \int d^N \mathbf{p} \, d^N \mathbf{r} \, e^{-\beta U(\mathbf{r}_1, ..., \mathbf{r}_N)} \delta \left(\sum_{i=1}^{N} \frac{\mathbf{p}_i^2}{m_i} - (dN-1)kT \right), \tag{4.13.13}
$$

which is the partition function of the isokinetic ensemble. Indeed, the constraint condition $\sum_{i=1}^{N} \mathbf{p}_i^2 / m_i = (dN-1)kT$ is exactly what we would expect for a system with a single kinetic-energy constraint based on the virial theorem, since the number of degrees of freedom is $dN-1$ rather than dN.

A simple yet effective integrator for the isokinetic equations can be obtained by applying the Liouville operator approach. As usual, we begin by writing the total Liouville operator

$$
iL = \sum_{i=1}^{N} \left[\frac{\mathbf{p}_i}{m_i} \cdot \nabla_{\mathbf{r}_i} + \left(\mathbf{F}_i - \left[\frac{\sum_{j=1}^{N} (\mathbf{F}_j \cdot \mathbf{p}_j)/m_j}{2K} \right] \mathbf{p}_i \right) \cdot \nabla_{\mathbf{p}_i} \right] \tag{4.13.14}
$$

as the sum of two contributions $iL = iL_1 + iL_2$, where

$$iL_1 = \sum_{i=1}^{N} \frac{\mathbf{p}_i}{m_i} \cdot \nabla_{\mathbf{r}_i}$$

$$iL_2 = \sum_{i=1}^{N} \left(\mathbf{F}_i - \left[\frac{\sum_{j=1}^{N}(\mathbf{F}_j \cdot \mathbf{p}_j)/m_j}{2K} \right] \mathbf{p}_i \right) \cdot \nabla_{\mathbf{p}_i}. \tag{4.13.15}$$

The approximate evolution of an isokinetic system over a time Δt is obtained by acting with a Trotter factorized operator $\exp(iL\Delta t) = \exp(iL_2\Delta t/2)\exp(iL_1\Delta t)\exp(iL_2\Delta t/2)$ on an initial condition $\{\mathbf{p}(0), \mathbf{r}(0)\}$. The action of each of the operators in this factorization can be evaluated analytically (Zhang, 1997; Minary *et al.*, 2003). The action of $\exp(iL_2\Delta t/2)$ can be determined by first solving the coupled first-order differential equations

$$\frac{dp_{i,\alpha}}{dt} = F_{i,\alpha} - \left[\frac{\sum_{j=1}^{N}(\mathbf{F}_j \cdot \mathbf{p}_j)/m_j}{2K} \right] p_{i,\alpha}$$

$$= F_{i,\alpha} - \dot{h}(t)p_{i,\alpha} \tag{4.13.16}$$

with $\mathbf{r}_1, ..., \mathbf{r}_N$ (and hence $F_{i,\alpha}$) held fixed. Here, we explicitly index both the spatial components ($\alpha = 1, ..., d$) and particle numbers $i = 1, ..., N$. The solution to eqn. (4.13.16) can be expressed as

$$p_{i,\alpha}(t) = \frac{p_{i,\alpha}(0) + F_{i,\alpha}s(t)}{\dot{s}(t)}, \tag{4.13.17}$$

where $s(t)$ is a general integrating factor:

$$s(t) = \int_0^t dt' \exp[h(t')]. \tag{4.13.18}$$

By substituting into the time derivative of the constraint condition $\sum_{i=1}^{N} \mathbf{p}_i \cdot \dot{\mathbf{p}}_i/m_i = 0$, we find that $s(t)$ satisfies a differential equation of the form

$$\ddot{s}(t) = \dot{s}(t)\dot{h}(t)$$

$$= \left[\frac{\sum_{j=1}^{N}(\mathbf{F}_j \cdot \mathbf{p}_j(t))/m_j}{2K} \right] \dot{s}(t)$$

$$= \left[\frac{\sum_{j=1}^{N}(\mathbf{F}_j \cdot \mathbf{p}_j(0))/m_j}{2K} \right] + \left[\frac{\sum_{j=1}^{N}(\mathbf{F}_j \cdot \mathbf{F}_j)/m_j}{2K} \right] s(t)$$

whose solution is

$$s(t) = \frac{a}{b} \left(\cosh(t\sqrt{b}) - 1 \right) + \frac{1}{\sqrt{b}} \sinh(t\sqrt{b}), \tag{4.13.19}$$

where

$$a = \frac{\sum_{j=1}^{N}(\mathbf{F}_j \cdot \mathbf{p}_j(0))/m_j}{2K}$$

$$b = \frac{\sum_{j=1}^{N}(\mathbf{F}_j \cdot \mathbf{F}_j)/m_j}{2K}. \tag{4.13.20}$$

The operator is applied by simply evaluating eqn. (4.13.19) and the associated eqn. (4.13.20) at $t = \Delta t/2$. The action of the operator $\exp(iL_1 \Delta t)$ on a state $\{\mathbf{p}, \mathbf{r}\}$ yields

$$\exp(iL_1 \Delta t)\mathbf{p}_i = \mathbf{p}_i$$
$$\exp(iL_1 \Delta t)\mathbf{r}_i = \mathbf{r}_i + \Delta t \mathbf{p}_i, \tag{4.13.21}$$

which has no effect on the momenta.

The combined action of the three operators in the Trotter factorization leads to the following reversible, kinetic energy conserving algorithm for integrating the isokinetic equations:

1. Evaluate new $\{s(\Delta t/2), \dot{s}(\Delta t/2)\}$ and update the momenta according to

$$\mathbf{p}_i \longleftarrow \frac{\mathbf{p}_i + \mathbf{F}_i s(\Delta t/2)}{\dot{s}(\Delta t/2)}. \tag{4.13.22}$$

2. Using the new momenta, update the positions according to

$$\mathbf{r}_i \longleftarrow \mathbf{r}_i + \Delta t \mathbf{p}_i/m_i. \tag{4.13.23}$$

3. Calculate new forces using the new positions.
4. Evaluate new $\{s(\Delta t/2), \dot{s}(\Delta t/2)\}$ and update the momenta according to

$$\mathbf{p}_i \longleftarrow \frac{\mathbf{p}_i + \mathbf{F}_i s(\Delta t/2)}{\dot{s}(\Delta t/2)}. \tag{4.13.24}$$

Note $\{s(\Delta t/2), \dot{s}(\Delta t/2)\}$ are evaluated by substituting the present momenta and forces into eqns. (4.13.20) with $t = \Delta t/2$. The symbol, "$\longleftarrow$," indicates that on the computer, the values on the left-hand side are overwritten in memory by the values on the right-hand side.

The isokinetic ensemble method has recently been shown to be a useful method for generating a canonical coordinate distribution. First, it is a remarkably stable method, allowing very long time steps to be used, particularly when combined with the RESPA scheme. Unfortunately, the isokinetic approach suffers from some of the pathologies of the Nosé-Hoover approach so some care is needed when applying it. However, the isokinetic ensemble can be particularly useful when combined with Nosé-Hoover chains to circumvent resonance artifacts (see Problem 3.6) in multiple time-step algorithms (Minary *et al.*, 2004*b*), as we discuss next.

4.14   Isokinetic Nosé-Hoover chains: Achieving very large time steps

Problem 3.6 illustrates a phenomenon known as *resonances* in algorithms such as RESPA (Tuckerman *et al.*, 1992) that employ multiple time steps. Although it lies outside the scope of a statistical mechanics book to analyze resonances in great detail, readers are encouraged to work through Problem 3.6 in order to understand resonances and how these numerical artifacts limit the time step gain that can be achieved in multiple time-step integration algorithms. For example, in atomistic simulations of biomolecular systems, the largest time step possible in a multiple time-step algorithm is ~5–10 fs before the algorithm becomes numerically unstable. In this section, we will show how the isokinetic ensemble can be combined with the Nosé-Hoover chain thermostat to circumvent resonances in simulations within the isokinetic ensemble.

Resonance artifacts manifest as a breakdown in energy conservation (Schlick *et al.*, 1998; Ma *et al.*, 2003) traceable to the buildup of energy in one or more resonant modes in the system. The isokinetic ensemble offers a way to prevent energy buildup as it places a hard constraint on the kinetic energy. If isokinetic constraints could be imposed on individual modes, these modes would not become resonant. Unfortunately, applying an isokinetic constraint on a single degree of freedom produces motion with a constant momentum. (Try applying an isokinetic constraint $p^2/m = C$ to the equations of motion for a single variable $\dot{x} = p/m$, $\dot{p} = F(x)$, modifying the momentum equation to read $\dot{p} = F(x) - \alpha p$, where α is the Lagrange multiplier. You should find $\dot{p} = 0$.) However, if each mode were coupled via an isokinetic constraint to an external system of some kind, then the constant-momentum problem disappears. Minary *et al.* (2003) proposed the use of simple nonlinear systems such as double-well potentials as useful external systems. Subsequently, it was recognized that using a heat bath as the external system could be particularly effective (Minary *et al.*, 2004b). Combining what we have introduced so far, one natural choice for this heat bath is a Nosé-Hoover chain thermostat.

In order to see how isokinetic coupling between a system and a Nosé-Hoover chain can be established, consider a particle with coordinate x with mass m and corresponding momentum p subject to a potential $U(x)$ and force $F(x) = -dU/dx$. The particle is to be coupled through an isokinetic constraint to a Nosé-Hoover chain with two elements having momenta p_1 and p_2 (the additional "η" index is no longer needed since η no longer appears in this coupling scheme) having mass-like parameters Q_1 and Q_2, respectively. The equations of motion read

$$\dot{x} = \frac{p}{m}$$

$$\dot{p} = F - \alpha p$$

$$\dot{p}_1 = -\frac{p_2}{Q_2}p_1 - \alpha p_1$$

$$\dot{p}_2 = G(p_1), \tag{4.14.1}$$

where $G(p_1) = p_1^2/Q_1 - kT$. Here, α is the Lagrange multiplier introduced to enforce an isokinetic constraint. We might expect the constraint to take the form $p^2/m + p_1^2/Q_1 = kT$. However, the fact that the Nosé-Hoover chain thermostat is inherently non-Hamiltonian involves some unexpected subtleties that render this constraint incorrect. In fact, the two kinetic energy terms in the constraint cannot enter the constraint with the same weight. If we insist that the physical kinetic energy term p^2/m have a weight of 1, then an unknown weight should be placed in front of the thermostat kinetic energy term, so that the constraint becomes

$$\frac{p^2}{m} + \gamma \frac{p_1^2}{Q_1} = kT, \tag{4.14.2}$$

where the constant γ is to be determined via analysis of the non-Hamiltonian system in eqns. (4.14.1). Using Gauss's principle of least constraint, we compute the time derivative of eqn. (4.14.2) and use eqns. (4.14.1) to obtain an explicit expression for the multiplier α. This procedure yields:

$$\frac{p}{m}\dot{p} + \gamma \frac{p_1}{Q_1}\dot{p}_1 = 0$$

$$\frac{p}{m}(F - \alpha p) + \gamma \frac{p_1}{Q_1}\left(-\frac{p_2}{Q_2}p_1 - \alpha p_1\right) = 0$$

$$\alpha = \frac{Fp/m - \gamma p_2 p_1^2/(Q_1 Q_2)}{p^2/m + \gamma p_1^2/Q_1}. \tag{4.14.3}$$

When the multiplier is substituted back into eqns. (4.14.1), the resulting non-Hamiltonian equations of motion become

$$\dot{x} = \frac{p}{m}$$

$$\dot{p} = F - \left[\frac{Fp^2/m - \gamma p p_2 p_1^2/(Q_1 Q_2)}{p^2/m + \gamma p_1^2/Q_1}\right]$$

$$\dot{p}_1 = -\frac{p_2}{Q_2}p_1 - \left[\frac{Fpp_1/m - \gamma p_2 p_1^3/(Q_1 Q_2)}{p^2/m + \gamma p_1^2/Q_1}\right]$$

$$\dot{p}_2 = G(p_1). \tag{4.14.4}$$

In order to determine the equilibrium distribution generated by eqns. (4.14.4), we first need to compute the phase-space compressibility $\kappa = \partial\dot{x}/\partial x + \partial\dot{p}/\partial p + \partial\dot{p}_1/\partial p_1 + \partial\dot{p}_2/\partial p_2$ as we have done previously in this chapter. The algebra involved in this calculation is tedious but straightforward and yields the expression

$$\kappa = \beta\left[-\frac{Fp}{m} + 2\gamma p_2\frac{p_1^2}{Q_1^2}\right] - \frac{p_2}{Q_2}. \tag{4.14.5}$$

From this, we see that if we choose $\gamma = 1/2$, then κ becomes

$$\kappa = \beta\left[\frac{dU(x)}{dt} + \frac{d}{dt}\left(\frac{p_2^2}{2Q_2}\right)\right]. \tag{4.14.6}$$

The function $w(x) = \beta[U(x) + p_2^2/2Q_2]$, and the phase-space metric factor is $\sqrt{g}(x) = \exp(-\beta U(x) - \beta p_2^2/2Q_2)$. This gives the desired partition function

$$\Omega = \int dx \, dp \, dp_1 \, dp_2 \, e^{-\beta p_2^2/2Q_2} e^{-\beta U(x)} \delta \left(\frac{p^2}{m} + \frac{p_1^2}{2Q_1} - kT \right). \qquad (4.14.7)$$

The mechanism of the isokinetic/Nosé-Hoover chain coupling is that the physical degree of freedom can exchange energy with the thermostat, allowing for kinetic energy fluctuations. However, the total kinetic energy of the system plus the bath can never leave the perimeter of the ellipse defined by the constraint equation $p^2/m + p_1^2/2Q_1 = kT$. In this way, energy buildup in a degree of freedom is prevented, which effectively eliminates resonance artifacts.

We can enhance the efficacy of the isokinetic/Nosé-Hoover chain coupling mechanism in eqns. (4.14.1) by extending the chains in these equations to include M elements. In addition, in order to promote greater kinetic energy fluctuations in the system, the isokinetic coupling can be made to L Nosé-Hoover chains rather than to just one. Incorporating these additional enhancements, the equations of motion become

$$\dot{x} = \frac{p}{m}$$

$$\dot{p} = F - \alpha p$$

$$\dot{p}_{1,j} = -\frac{p_{2,j}}{Q_2} p_{1,j} - \alpha p_{1,j}, \qquad j = 1, ..., L$$

$$\dot{p}_{i,j} = G(p_{i-1,j}) - \frac{p_{i,j+1}}{Q_{j+1}} p_{i,j}, \qquad i = 2, ..., M-1$$

$$\dot{p}_{M,j} = G(p_{M-1,j}), \qquad (4.14.8)$$

where $G(p_j) = p_j^2/Q_j - kT$. The isokinetic constraint condition that defines the coupling takes the form

$$\frac{p^2}{m} + \frac{L}{L+1} \sum_{j=1}^{L} \frac{p_{1,j}^2}{Q_1} = LkT, \qquad (4.14.9)$$

which means that the Lagrange multiplier is

$$\alpha = \frac{Fp/m - L/(L+1) \sum_{j=1}^{L} p_{1,j}^2 p_{2,j}/Q_1 Q_1}{p^2/m + L/(L+1) \sum_{j=1}^{L} p_{1,j}^2/Q_1}, \qquad (4.14.10)$$

and the corresponding partition function becomes

$$\Omega = \int dx \, dp \left[\prod_{j=1}^{L} dp_{1,j} \, dp_{2,j} \, e^{-\beta p_{2,j}^2/2Q_2} \right] e^{-\beta U(x)} \delta \left(\frac{p^2}{m} + \frac{L}{L+1} \sum_{j=1}^{L} \frac{p_{1,j}^2}{Q_1} - LkT \right). \qquad (4.14.11)$$

The optimal manner for employing eqns. (4.14.8) is in "massive" mode wherein each degree of freedom is coupled to L Nosé-Hoover chains of length M. However, the method can be applied in other ways as well. An integrator for eqns. (4.14.8) is a combination of the algorithms already discussed for the Nosé-Hoover chain and the isokinetic schemes. In Chapter 15, we present a stochastic version of eqns. (4.14.8), and, hence, we will defer a discussion of an integrator for equations of motion of this type, including a multiple time-step version of the integrator, until then.

As an example of the application of eqns. (4.14.8), we consider a weakly perturbed harmonic oscillator with potential $U(x) = \omega^2 x^2/2 + gx^4/4$ with $L = 1$, $M = 3$, $\omega = 3$, $g = 0.1$, $Q_1 = Q_2 = 1$, and $kT = 1$. When eqns. (4.14.8) are integrated with a RESPA factorization (see, also, Section 15.5.3), it is denoted Isok-NHC-RESPA. In this example, we choose $\Delta t = \pi/\omega$, which is the time step where the first resonance occurs (see Problem 3.6), and $\delta t = \Delta t/100$. The equations are integrated for 5×10^7 steps. Figure 4.15 compares the numerical position probability distribution $P(x)$ generated using RESPA with Nosé-Hoover chains (NHC-RESPA) and Isok-NHC-RESPA with the analytical result for $P(x)$. The figure shows the effect of the resonance, which causes a significant distortion of the distribution. By contrast, the Isok-NHC-RESPA reproduces the analytical result, unaffected by resonances, clearly demonstrating the ability of the Isok-NHC-RESPA approach to avoid resonance artifacts.

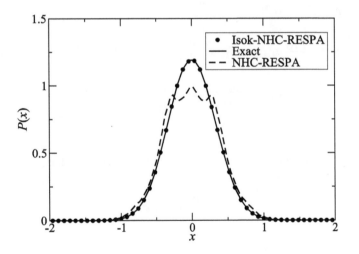

Fig. 4.15 Distribution function of the quartic oscillator integrated using Isok-NHC-RESPA and NHC-RESPA compared to the exact result.

4.15 Applying canonical molecular dynamics: Liquid structure

Figures 4.2 and 4.3 showed radial distributions functions for liquid argon and water, respectively. The importance of the radial distribution function in understanding the structure of liquids and approximating their thermodynamic properties was discussed in Section 4.7. In this section, we will describe how these plots can be extracted from a molecular dynamics trajectory. Since the radial distribution function is an equilibrium property, it is appropriate to employ a canonical sampling method such as Nosé-Hoover chains or the isokinetic ensemble for this purpose.

The argon system represented in Fig. 4.2 was simulated using the "massive" Nosé-Hoover chain approach on the argon system described in Section 3.14.2. The thermostats maintained the system at a temperature of 300 K by controlling the kinetic energy fluctuations. The system was integrated for a total of 10^5 steps using a time step of 10.0 fs. Each Cartesian degree of freedom of each particle was coupled to its own Nosé-Hoover chain thermostat with $M = 4$, using $n_{sy} = 7$ and $n = 4$ in the Suzuki-Yoshida integration scheme of eqn. (4.12.16). The parameter τ used to determine the value of $Q_1, ..., Q_4$ was taken as 200.0 fs.

The water system represented in Fig. 4.3 was simulated using, once again, the "massive" Nosé-Hoover chain approach on a system of 64 water molecules in a cubic box of length 12.4164 Å subject to periodic boundary conditions. The forces were obtained directly from density functional theory electronic structure calculations performed at each molecular dynamics step via the Car-Parrinello approach (Car and Parrinello, 1985; Tuckerman, 2002; Marx and Hutter, 2009). The system was maintained at a temperature of 300 K using a time step of 0.1 fs. For the thermostats, the following parameters were used: $n_{sy} = 7$, $n = 4$, and $\tau = 20$ fs. The system was run for a total of 60 ps.

After the molecular dynamics calculation has been performed, the saved trajectory is used to compute the radial distribution function using the following algorithm:

1. Divide the radial interval between $r = 0$ and $r = r_{max}$, where r_{max} is some radial value beyond which no significant structure exists, into N_r intervals of length Δr. It is important to note that the largest value r_{max} can have is half the length of the box edge. Let these intervals be indexed by an integer $i = 0, ..., N_r - 1$ with radial values $r_1, ..., r_{N_r}$.

2. Generate a histogram $h_{ab}(i)$ by counting the number of times the distance between two atoms of type a and b lies between r_i and $r_i + \Delta r$. For this histogram, all atoms of the desired types in the system can be used and all configurations generated in the simulation should be considered. Thus, if we are interested in the oxygen–oxygen histogram of water, we would use the oxygens of all waters in the system and all configurations generated in the simulation. For each distance r calculated, the index into the histogram is given by

$$i = \text{int}(r/\Delta r). \tag{4.15.1}$$

3. Once the histogram is generated, the radial distribution function is obtained by

$$g_{ab}(r_i) = \frac{h_{ab}(i)}{4\pi\rho_b r_i^2 \Delta r N_{conf} N_a}, \tag{4.15.2}$$

where N_{conf} is the number of configurations in the simulation, N_a is the number of atoms of type a, and ρ_b is the number density of the atom type b.

This procedure was employed to produce the plots in Figs. 4.2 and 4.3.

In order to illustrate the performance of the Isok-NHC-RESPA method of Section 4.14 in a simulation of the structure of liquid water, we show, in Fig. 4.16, radial distribution functions for a flexible model of water (Paesani *et al.*, 2006) generated at $T = 300$ K using Nosé-Hoover chains with a time step of 0.5 fs and the Isok-NHC-RESPA method with $M = 4$, $L = 4$, $\delta t = 0.5$ fs used for OH bonding forces and

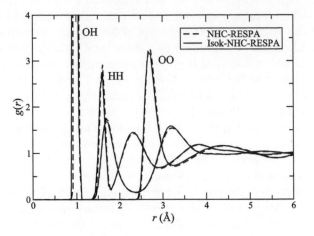

Fig. 4.16 Radial distribution functions of a flexible model of liquid water simulated at 300 K using the Nosé-Hoover chain RESPA method and the Isok-NHC-RESPA scheme of Section 4.14 using an outer time step of 99 fs.

HOH bending forces, $\Delta t = 3.0$ fs for short-range intermolecular forces cut off at 6.0 Å, and $\Delta T = 99$ fs used for long-range intermolecular forces cut off at 12.0 Å using the RESPA1 force decomposition scheme described in Appendix C. The length of each run is 600 ps. The figure shows that accurate distribution functions can be obtained with a quite large outer time step of 99 fs. If the same set of time steps were to be used in a Nosé-Hoover chain/RESPA run, the simulation would eventually become numerically unstable.

4.16   Problems

4.1. Consider a one-dimensional system with momentum p and coordinate q coupled to an extended-system thermostat for which the equations of motion take the form

$$\dot{q} = \frac{p}{m}$$

$$\dot{p} = F(q) - \frac{p_{\eta_1}}{Q_1}p - \frac{p_{\eta_2}}{Q_2}\left[(kT)p + \frac{p^3}{3m}\right]$$

$$\dot{\eta}_1 = \frac{p_{\eta_1}}{Q_1}$$

$$\dot{\eta}_2 = \left[(kT) + \frac{p^2}{m}\right]\frac{p_{\eta_2}}{Q_2}$$

$$\dot{p}_{\eta_1} = \frac{p^2}{m} - kT$$

$$\dot{p}_{\eta_2} = \frac{p^4}{3m^2} - (kT)^2.$$

a. Show that these equations of motion are non-Hamiltonian.

b. Show that the equations of motion conserve the following energy:

$$\mathcal{H}' = \frac{p^2}{2m} + U(q) + \frac{p_{\eta_1}^2}{2Q_1} + \frac{p_{\eta_2}^2}{2Q_2} + kT(\eta_1 + \eta_2).$$

c. Use the non-Hamiltonian formalism of Section 4.10 to show that these equations of motion generate the canonical distribution in the physical Hamiltonian $\mathcal{H} = p^2/2m + U(q)$.

*d. These equations of motion are designed to control the fluctuations in the first two moments of the Maxwell-Boltzmann distribution $P(p) \propto \exp(-\beta p^2/2m)$. A set of equations of motion designed to fix an arbitrary number M of these moments is

$$\dot{q} = \frac{p}{m}$$

$$\dot{p} = F(q) - \sum_{n=1}^{M} \sum_{k=1}^{n} \frac{p_{\eta_n}}{Q_n} \frac{(kT)^{n-k}}{C_{k-1}} \frac{p^{2k-1}}{m^{k-1}}$$

$$\dot{\eta}_n = \left[(kT)^{n-1} + \sum_{k=2}^{n} \frac{(kT)^{n-k}}{C_{k-2}} \left(\frac{p^2}{m} \right)^{k-1} \right] \frac{p_{\eta_n}}{Q_n}$$

$$\dot{p}_{\eta_n} = \frac{1}{C_{n-1}} \left(\frac{p^2}{m} \right)^n - (kT)^n,$$

where $C_n = \prod_{k=1}^{n}(1 + 2k)$ and $C_0 \equiv 1$. These equations were first introduced by Liu and Tuckerman (2000) (who also introduced versions of these for N-particle systems) and constitute what is known as the *generalized Gaussian moment thermostat* approach because they control the fluctuations in n moments of canonical momentum distribution. Show that the equations conserve the energy

$$\mathcal{H}' = \frac{p^2}{2m} + U(q) + \sum_{n=1}^{M} \frac{p_{\eta_n}^2}{2Q_n} + kT \sum_{n=1}^{M} \eta_n$$

and therefore that they generate a canonical distribution in the Hamiltonian $\mathcal{H} = p^2/2m + U(q)$.

4.2. The following algorithm is introduced to generate a canonical distribution in a physical Hamiltonian $\mathcal{H}(q, p)$ for a one-dimensional particle with coordinate q and momentum p. The algorithm is an extended phase-space approach for which the equations of motion are

$$\dot{q} = \frac{\partial \mathcal{H}}{\partial p} - p_\xi f(q, p)$$

$$\dot{p} = -\frac{\partial \mathcal{H}}{\partial q} - p_\eta g(q, p)$$

$$\dot{\xi} = p_\xi \frac{\partial f}{\partial q}$$

$$\dot{\eta} = p_\eta \frac{\partial g}{\partial p}$$

$$\dot{p}_\xi = \frac{\partial \mathcal{H}}{\partial q} f(q, p) - \frac{1}{\beta} \frac{\partial f}{\partial q}$$

$$\dot{p}_\eta = \frac{\partial \mathcal{H}}{\partial p} g(q, p) - \frac{1}{\beta} \frac{\partial g}{\partial p}.$$

Here, $f(q, p)$ and $g(q, p)$ are arbitrary differentiable functions of the coordinate and momentum, $\beta = 1/kT$, and ξ and η are heat-bath variables with corresponding momenta p_ξ and p_η, respectively. The equations of motion also conserve the following energy function:

$$\mathcal{H}'(q, \xi, \eta, p, p_\xi, p_\eta) = \mathcal{H}(q, p) + \frac{p_\eta^2}{2} + \frac{p_\xi^2}{2} + \frac{1}{\beta}(\eta + \xi).$$

a. Provide a detailed analysis of the equations of motion showing that they conserve $\mathcal{H}'$ and generate a canonical distribution.

b. Write down the Liouville operator for these equations of motion.

4.3. a. Consider the Nosé-Hoover equations for a single free particle of mass m moving in one spatial dimension. The equations of motion are

$$\dot{p} = -\frac{p_\eta}{Q} p, \qquad \dot{\eta} = \frac{p_\eta}{Q}, \qquad \dot{p}_\eta = \frac{p^2}{m} - kT.$$

Show that these equations obey the following two conservation laws:

$$C = \frac{p^2}{2m} + \frac{p_\eta^2}{2Q} + kT\eta \equiv \mathcal{H}'$$

$$K = p e^\eta.$$

b. Show, therefore, that the distribution function in the physical momentum p generated by the equations is

$$f(p) = \frac{\sqrt{2Q}}{\sqrt{p^2 \left(C - (p^2/2m) + kT \ln(p/K)\right)}}$$

rather than the expected Maxwell-Boltzmann distribution

$$f(p) = \frac{1}{\sqrt{2\pi mkT}} \exp(-p^2/2mkT).$$

c. Plot the distribution $f(p)$ and show that it matches the distribution shown in Fig. 4.9.

d. Write a program that integrates the equations of motion using the algorithm of Section 4.12 and verify that the numerical distribution matches that of part c.

e. Next, consider the Nosé-Hoover chain equations with $M = 2$ for the same free particle:

$$\dot{p} = -\frac{p_{\eta_1}}{Q}p, \qquad \dot{\eta}_k = \frac{p_{\eta_k}}{Q}, \qquad \dot{p}_{\eta_1} = \frac{p^2}{m} - kT - \frac{p_{\eta_2}}{Q}p_{\eta_1}, \qquad \dot{p}_{\eta_2} = \frac{p_{\eta_1}^2}{Q} - kT.$$

Here, $k = 1, 2$. Show that these equations of motion generate the correct Maxwell-Boltzmann distribution in p.

*f. Will these equations yield the correct Maxwell-Boltzmann distribution in practice if implemented using the Liouville-based integrator of Section 4.12?

Hint: Consider how an initial momentum $p(0) > 0$ evolves under the action of the integrator. What happens if $p(0) < 0$?

*g. Derive the general distribution generated by eqns. (4.9.14) when no external forces are present and the conservation law in eqn. (4.10.27) is obeyed.

Hint: Since the conservation law involves the center-of-mass momentum $\mathbf{P}$, it is useful to introduce a canonical transformation to center-of-mass momentum and position $(\mathbf{R}, \mathbf{P})$ and the $d(N-1)$ corresponding relative coordinates $\mathbf{r}_1', \mathbf{r}_2', \ldots$ and momenta $\mathbf{p}_1', \mathbf{p}_2', \ldots$.

*h. Show that when $\sum_{i=1}^{N} \mathbf{F}_i = 0$, the Nosé-Hoover chain equations generate the correct canonical distribution in all variables except the center-of-mass momentum.

i. Finally, show the conservation $\sum_{i=1}^{N} \mathbf{F}_i = 0$ is *not* obeyed by eqns. (4.11.7) and, therefore, that they also generate a correct canonical distribution in *all* variables.

4.4. Consider a modified version of the Nosé-Hoover equations for a harmonic oscillator with unit mass, unit frequency, and $kT = 1$:

$$\dot{x} = p - p_\eta x, \quad \dot{p} = -x - p_\eta p \quad \dot{\eta} = p_\eta, \quad \dot{p}_\eta = p^2 + x^2 - 2.$$

a. Show that these equations have the two conservation laws:

$$C = \frac{1}{2}\left(p^2 + x^2 + p_\eta^2\right) + 2\eta$$

$$K = \frac{1}{2}\left(p^2 + x^2\right)e^{2\eta}.$$

b. Determine the distribution $f(H)$ of the physical Hamiltonian $\mathcal{H}(x, p) = (p^2 + x^2)/2$. Is the distribution the expected canonical distribution $f(H) \propto \exp(-H)$?

Hint: Try using the two conservation laws to eliminate the variables η and p_η.

*c. Show that a plot of the physical phase space p vs. x necessarily must have a hole centered at $(x, p) = (0, 0)$, and find a condition that determines the size of the hole.

4.5. In Section 4.5, it was shown that the energy fluctuations in the canonical ensemble satisfy

$$(\Delta E)^2 = k_{\mathrm{B}} T^2 C_V$$

where C_V is the constant-volume heat capacity. Define the third-order fluctuation $(\Delta E)^3$ as

$$(\Delta E)^3 = \left\langle (\mathcal{H}(\mathbf{x}) - \langle \mathcal{H}(\mathbf{x}) \rangle)^3 \right\rangle.$$

Derive an expression for $(\Delta E)^3$ in terms of T, C_V, and any derivatives of C_V you think are necessary.

4.6. Suppose the interactions in an N-particle system are described by a pair potential of the form

$$U(\mathbf{r}_1, ..., \mathbf{r}_N) = \sum_{i=1}^{N-1} \sum_{j=i+1}^{N} u(|\mathbf{r}_i - \mathbf{r}_j|).$$

In the low density limit, we can assume that each particle interacts with *at most* one other particle.

a. Show that the canonical partition function in this limit can be expressed as

$$Q(N,V,T) = \frac{(N-1)!!V^{N/2}}{N!\lambda^{3N}} \left[4\pi \int_0^\infty dr\, r^2 e^{-\beta u(r)} \right]^{N/2}.$$

b. Show that the radial distribution function is proportional to $\exp[-\beta u(r)]$ in this limit.

c. Show that the second virial coefficient in the low density limit becomes

$$B_2(T) = -2\pi \int_0^\infty dr\, r^2 f(r)$$

where $f(r) = e^{-\beta u(r)} - 1$.

d. Suppose the interaction potential between a pair of particles in a gas is given by

$$u(r) = \begin{cases} \infty & r \leq \sigma \\ \frac{\epsilon}{\sigma} \frac{r-\lambda\sigma}{\lambda-1} & \sigma \leq r \leq \lambda\sigma \\ 0 & r > \lambda\sigma, \end{cases}$$

where ϵ, λ, and σ are constants. Sketch a plot of this potential as a function of r and calculate the second virial coefficient $B_2(T)$ for this potential. Plot $B_2(T)$ as a function of T.

4.7. An ideal gas of N particles of mass m at temperature T is in a cylindrical container with radius a and length L. The container rotates about its cylindrical axis (taken to be the z axis) with angular velocity w. In addition, the gas is subject to a uniform gravitational field of strength g. Therefore, the Hamiltonian for the gas is

$$\mathcal{H} = \sum_{i=1}^{N} h(\mathbf{r}_i, \mathbf{p}_i),$$

where $h(\mathbf{r}, \mathbf{p})$ is the Hamiltonian for a single particle

$$h(\mathbf{r}, \mathbf{p}) = \frac{\mathbf{p}^2}{2m} - w(\mathbf{r} \times \mathbf{p})_z + mgz.$$

Here, $(\mathbf{r} \times \mathbf{p})_z$ is the z-component of the cross product between $\mathbf{r}$ and $\mathbf{p}$.

a. Show, in general, that when the Hamiltonian is separable in this manner, the canonical partition function $Q(N,V,T)$ is expressible as

$$Q(N,V,T) = \frac{1}{N!} [q(V,T)]^N,$$

where

$$q(V,T) = \frac{1}{h^3} \int d\mathbf{p} \int_{D(V)} d\mathbf{r} \, e^{-\beta h(\mathbf{r},\mathbf{p})}.$$

b. Show, in general, that the chemical potential $\mu(N,V,T)$ is given by

$$\mu(N,V,T) = kT \ln \left[\frac{Q(N-1,V,T)}{Q(N,V,T)} \right],$$

where $Q(N-1,V,T)$ is the partition function for an $(N-1)$-particle system.

c. Calculate the partition function for this ideal gas.

d. Calculate the Helmholtz free energy of the gas.

e. Calculate the total internal energy of the gas.

f. Calculate the heat capacity of the gas.

*g. What is the equation of state of the gas?

4.8. A classical system of N noninteracting diatomic molecules enclosed in a cubic box of length L and volume $V = L^3$ is held at a fixed temperature T. The Hamiltonian for a single molecule is

$$h(\mathbf{r}_1, \mathbf{r}_2, \mathbf{p}_1, \mathbf{p}_2) = \frac{\mathbf{p}_1^2}{2m_1} + \frac{\mathbf{p}_2^2}{2m_2} + \epsilon |r_{12} - r_0|,$$

where $r_{12} = |\mathbf{r}_1 - \mathbf{r}_2|$ is the distance between the atoms in the diatomic.

a. Calculate the canonical partition function.

b. Calculate the Helmholtz free energy.

c. Calculate the total internal energy.

d. Calculate the heat capacity.

e. Calculate the mean-square molecular bond length $\langle |\mathbf{r}_1 - \mathbf{r}_2|^2 \rangle$.

4.9. Write a program to integrate the Nosé-Hoover chain equations for a harmonic oscillator with mass $m = 1$, frequency $\omega = 1$, and temperature $kT = 1$ using the integrator of Section 4.12. Verify that the correct momentum and position distributions are obtained by comparison with the analytical results

$$f(p) = \frac{1}{\sqrt{2\pi mkT}} e^{-p^2/2mkT}, \qquad f(x) = \sqrt{\frac{m\omega^2}{2\pi kT}} e^{-m\omega^2 x^2/2kT}.$$

*4.10. Consider a system of N particles subject to a single holonomic constraint

$$\sigma(\mathbf{r}_1, ..., \mathbf{r}_N) \equiv \sigma(\mathbf{r}) = 0.$$

Recall that the equations of motion derived using Gauss's principle of least constraint are

$$\dot{\mathbf{r}}_i = \frac{\mathbf{p}_i}{m_i}$$

$$\dot{p}_i = \mathbf{F}_i - \left[\frac{\sum_j \mathbf{F}_j \cdot \nabla_j \sigma / m_j + \sum_{j,k} \nabla_j \nabla_k \sigma \cdot \cdot \mathbf{p}_j \mathbf{p}_k / (m_j m_k)}{\sum_j (\nabla_j \sigma)^2 / m_j} \right] \nabla_i \sigma.$$

Show using the techniques of Section 4.10 that these equations of motion generate the partition function

$$\Omega = \int d^N \mathbf{p} \, d^N \mathbf{r} \, Z(\mathbf{r}) \delta(\mathcal{H}(\mathbf{r}, \mathbf{p}) - E) \delta(\sigma(\mathbf{r})) \delta(\dot{\sigma}(\mathbf{r}, \mathbf{p})),$$

where

$$Z(\mathbf{r}) = \sum_{i=1}^{N} \frac{1}{m_i} \left(\frac{\partial \sigma}{\partial \mathbf{r}_i} \right)^2.$$

This result was first derived by Ryckaert and Ciccotti (1983).

4.11. The canonical ensemble version of the classical virial theorem is credited to Richard C. Tolman (1918). Prove that the canonical average

$$\left\langle x_i \frac{\partial \mathcal{H}}{\partial x_j} \right\rangle = \frac{1}{N! h^{3N} Q(N, V, T)} \int dx \, x_i \frac{\partial \mathcal{H}}{\partial x_j} e^{-\beta \mathcal{H}(x)} = kT \delta_{ij}$$

holds. What assumptions must be made in the derivation of this result?

4.12. Prove that the structure factor $S(q)$ of a one-component isotropic liquid or gas is related to the radial distribution function $g(r)$ via eqn. (4.7.31).

4.13. Consider a system of N identical noninteracting molecules, each molecule being comprised of n atoms with some chemical bonding pattern within the molecule. The atoms in each molecule are held together by a potential $u(\mathbf{r}_1^{(i)}, ..., \mathbf{r}_n^{(i)})$, $i = 1, ..., N$, which rapidly increases as the distance between any two pairs of atoms increases, and becomes infinite as the distance between any two atoms in the molecule becomes infinite. Assume the atoms in each molecule have masses m_k, where $k = 1, ..., n$.

a. Write down the Hamiltonian and the canonical partition function for this system and show that the partition function can be reduced to a product of single-molecule partition functions.

b. Make the following change of coordinates in your single-molecule partition function:

$$\mathbf{s}_1 = \frac{1}{M} \sum_{k=1}^{n} m_k \mathbf{r}_k$$

$$\mathbf{s}_k = \mathbf{r}_k - \frac{1}{m_k'} \sum_{l=1}^{k-1} m_l \mathbf{r}_l \qquad k = 2, ..., n,$$

where

$$m_k' \equiv \sum_{l=1}^{k-1} m_l,$$

M is the total mass of a molecule, and the i superscript has been dropped for simplicity. What is the meaning of the coordinate $\mathbf{s}_1$? Show that if $u(\mathbf{r}_1, ..., \mathbf{r}_n)$ only depends on the *relative* coordinates between pairs of atoms in the molecule, then single molecule partition function is of the general form

$$Q(N, V, T) = \frac{(Vf(n, T))^N}{N!},$$

where $f(n, T)$ is a pure function of n and T.

c. Show, therefore, that the equation of state is *always* that of an ideal gas, independent of the type of molecule in the system.

d. Denote the single-molecule partition function as $q(n, V, T) = Vf(n, T)$. Now suppose that the system is composed of *different types of molecules* (see comment following eqn. (4.3.14)). Specifically, suppose the system contains N_A molecules of type A, N_B molecules of type B, N_C molecules of type C and N_D molecules of type D. Suppose further that the molecules can undergo the following chemical reaction:

$$aA + bB \rightleftharpoons cC + dD,$$

which is a chemical equilibrium. The Helmholtz free energy A must now be a function of V, T, N_A, N_B, N_C, and N_D. When chemical equilibrium is reached, the free energy is a minimum, so that $dA = 0$. Assume that the volume and temperature of the system are kept constant. Let λ be a variable such that $dN_A = ad\lambda$, $dN_B = bd\lambda$, $dN_C = -cd\lambda$, and $dN_D = -dd\lambda$. λ is called the *reaction extent*. Show that, at equilibrium,

$$a\mu_A + b\mu_B - c\mu_C - d\mu_D = 0, \qquad (4.16.3)$$

where μ_A is the chemical potential of species A:

$$\mu_A = -kT \frac{\partial \ln Q(V, T, N_A, N_B, N_C, N_D)}{\partial N_A}$$

with similar definitions for μ_B, μ_C, and μ_D.

e. Finally, show that eqn. (4.16.3) implies

$$\frac{\rho_C^c \rho_D^d}{\rho_A^a \rho_B^b} = \frac{(q_C/V)^c (q_D/V)^d}{(q_A/V)^a (q_B/V)^b}$$

and that both sides are pure functions of temperature. Here, q_A is the one-molecule partition function for a molecule of type A, q_B, the one-molecule partition function for a molecule of type B, *etc.*, and ρ_A is the number density of type A molecules, *etc.* How is the quantity on the right related to the usual equilibrium constant

$$K = \frac{P_C^c P_D^d}{P_A^a P_B^b}$$

for the reaction? Here, P_A, P_B,... are the partial pressures of species A, species, B,..., respectively.

4.14. Consider a system of N identical particles interacting via a pair potential

$$u(\mathbf{r}_1, ..., \mathbf{r}_N) = \frac{1}{2} \sum_{i,j,i \neq j} u(|\mathbf{r}_i - \mathbf{r}_j|),$$

where $u(r)$ is a general repulsive potential of the form

$$u(r) = \frac{A}{r^n},$$

where n is an integer and $A > 0$. In the low density limit, compute the pressure of such a system as a function of n. Explain why a system described by such a potential cannot exist stably for certain values of n.

Hint: You may express the answer in terms of the Γ-function

$$\Gamma(x) = \int_0^\infty dt\ t^{x-1} e^{-t}.$$

Also, the following properties of the Γ-function may be useful:

$$\Gamma(x) > 0 \qquad x > 0,$$
$$\Gamma(0) = \infty,$$
$$\Gamma(n) = \infty \qquad \text{for integer } n < 0$$
$$\Gamma(-1/2) = -2\sqrt{\pi}.$$

4.15. Often a pair potential is insufficient to describe accurately the behavior of many real liquids and gases. One then often includes *three-body* terms in the potential, which appear as follows:

$$U(\mathbf{r}_1, ..., \mathbf{r}_N) = \sum_{i>j} u(|\mathbf{r}_i - \mathbf{r}_j|) + \sum_{i>j>k} v(|\mathbf{r}_i - \mathbf{r}_j|, |\mathbf{r}_j - \mathbf{r}_k|, |\mathbf{r}_i - \mathbf{r}_k|),$$

where the first term is the usual pair interaction term and the second contains the three-body contributions.

a. Derive an expression for the average energy in terms of $g^{(2)}(\mathbf{r}_1, \mathbf{r}_2)$ and $g^{(3)}(\mathbf{r}_1, \mathbf{r}_2, \mathbf{r}_3)$. What is the expression for $g^{(3)}$?

b. Explain why $g^{(3)}(\mathbf{r}_1, \mathbf{r}_2, \mathbf{r}_3)$ should only depend on $\mathbf{r}_1 - \mathbf{r}_2, \mathbf{r}_3 - \mathbf{r}_2$ and $\mathbf{r}_3 - \mathbf{r}_1$. By making the following coordinate transformation:

$$\mathbf{R} = \frac{1}{3}(\mathbf{r}_1 + \mathbf{r}_2 + \mathbf{r}_3)$$
$$\mathbf{r} = \mathbf{r}_1 - \mathbf{r}_2$$
$$\mathbf{s} = \mathbf{r}_3 - \frac{1}{2}(\mathbf{r}_1 + \mathbf{r}_2),$$

show that $g^{(3)}$ really only depends on $\mathbf{r}$ and $\mathbf{s}$. By integrating over the variable $\mathbf{R}$, obtain a new distribution function $\tilde{g}^{(3)}(\mathbf{r}, \mathbf{s})$.

c. Express the average energy in terms of the radial distribution function $g(r)$ and the new function $\tilde{g}^{(3)}(\mathbf{r}, \mathbf{s})$.

d. For isotropic systems, explain why $g^{(3)}$ should only depend on the three distances $|\mathbf{r}_1 - \mathbf{r}_2|$, $|\mathbf{r}_3 - \mathbf{r}_2|$ and $|\mathbf{r}_3 - \mathbf{r}_1|$. Show, therefore, that in terms of the variables $\mathbf{r}$ and $\mathbf{s}$, the simplest distribution function can be expressed as $\bar{g}^{(3)}(r, \mu, \nu)$, i.e., as a function of the three variables r, μ, and ν, where r, μ and ν are given by

$$r = |\mathbf{r}|$$
$$\mu = \left|\mathbf{s} + \frac{1}{2}\mathbf{r}\right| + \left|\mathbf{s} - \frac{1}{2}\mathbf{r}\right|$$
$$\nu = \left|\mathbf{s} + \frac{1}{2}\mathbf{r}\right| - \left|\mathbf{s} - \frac{1}{2}\mathbf{r}\right|.$$

(You do not need to obtain an explicit expression for $\bar{g}^{(3)}$.)

4.16. Derive eqn. (4.14.5).

4.17. A classical ideal gas of N massless particles with coordinates $\mathbf{r}_1, ..., \mathbf{r}_N \equiv \mathbf{r}$ and momenta $\mathbf{p}_1, ..., \mathbf{p}_N \equiv \mathbf{p}$ is held at a constant volume V and external temperature T. The Hamiltonian of the gas is

$$\mathcal{H}(\mathbf{r}, \mathbf{p}) = \sum_{i=1}^{N} c|\mathbf{p}_i|$$

where c is a constant. The number of particles N is also kept fixed. Determine the equation of state, the total energy, the heat capacity, and the entropy of the gas under these conditions.

5

The isobaric ensembles

5.1 Why constant pressure?

Standard handbooks of thermodynamic data report numerical values of physical properties, including standard enthalpies, entropies and free energies of formation, redox potentials, equilibrium constants (such as acid ionization constants, solubility products, inhibition constants) and other such data, under conditions of constant temperature and pressure. This makes the isothermal-isobaric ensemble one of the most important ensembles since it most closely reflects the conditions under which many condensed-phase experiments are performed.

In order to maintain a fixed internal pressure, the volume of a system must be allowed to fluctuate. We may therefore view an isobaric system as coupled to an isotropic "piston" that compresses or expands the system uniformly in response to instantaneous internal pressure fluctuations such that the average internal pressure is equal to an external applied pressure. Remember that an instantaneous pressure estimator is the total force exerted by the particles on the walls of their container, and the average of this quantity gives the observable internal pressure. Coupling a system to the piston leads to an ensemble known as the *isoenthalpic-isobaric* ensemble, since the enthalpy remains fixed as well as the pressure. Recall that the enthalpy is $H = E + PV$. If the system also exchanges heat with a thermal reservoir, which maintains a fixed temperature T, then the system is described by the *isothermal-isobaric* ensemble.

In this chapter, the basic thermodynamics of isobaric ensembles will be derived by performing a Legendre transformation on the volume starting with the microcanonical and canonical ensembles, respectively. The condition of a fluctuating volume will be seen to affect the ensemble distribution function, which must be viewed as a function of both the phase-space vector x and the volume V. Indeed, when considering how the volume fluctuates in an isobaric ensemble, it is important to note that both isotropic and anisotropic fluctuations are possible. Bulk liquids and gases in equilibrium only support isotropic fluctuations. However, in any system that is not isotropic by nature, anisotropic volume fluctuations are possible even if the applied external pressure is isotropic. For solids, if one is interested in structural phase transitions under an external applied pressure or in mapping out the space of crystal structures of complex molecular systems that exhibit polymorphism, it is often critical to include anisotropic shape changes of the containing volume or supercell. Other examples that support anisotropic volume changes include biological membranes, amorphous materials, and interfaces, to name a few.

After developing the basic statistical mechanics of the isobaric ensembles, we will see how the extended phase-space techniques of the previous chapter can be adapted for molecular dynamics calculations in these ensembles. We will show how the volume and density distributions can be generated by treating the volume as an additional dynamical variable with a corresponding momentum, the latter serving as a barostatic control of the fluctuations in the internal pressure. This idea will be extended to anisotropic volume shape-changes by treating the cell vectors as dynamical variables.

5.2 Thermodynamics of isobaric ensembles

We begin by considering the isoenthalpic-isobaric ensemble, which derives from a Legendre transformation performed on the microcanonical ensemble. In the microcanonical ensemble, the energy E is constant and is expressed as a function of the number of particles N, the volume V, and the entropy S: $E = E(N, V, S)$. Since we seek to use an external applied pressure P as the control variable in place of the volume V, it is necessary to perform a Legendre transform of E with respect to the volume V. Denoting the new energy as $\tilde{E}$, we find

$$\tilde{E}(N, P, S) = E(N, V(P), S) - \frac{\partial E}{\partial V} V(P). \qquad (5.2.1)$$

However, since $P = -\partial E/\partial V$, the new energy is just $\tilde{E} = E + PV$, which we recognize as the enthalpy H:

$$H(N, P, S) = E(N, V(P), S) + PV(P). \qquad (5.2.2)$$

The enthalpy is naturally a function of N, P, and S. Thus, for a process in which these variables change by small amounts, dN, dP, and dS, respectively, the change in the enthalpy is

$$dH = \left(\frac{\partial H}{\partial N}\right)_{P,S} dN + \left(\frac{\partial H}{\partial P}\right)_{N,S} dP + \left(\frac{\partial H}{\partial S}\right)_{N,P} dS. \qquad (5.2.3)$$

Since $H = E + PV$, it also follows that

$$dH = dE + PdV + VdP$$

$$= TdS - PdV + \mu dN + PdV + VdP$$

$$= TdS + VdP + \mu dN, \qquad (5.2.4)$$

where the second line follows from the first law of thermodynamics. Comparing eqns. (5.2.3) and (5.2.4) leads to the thermodynamic relations

$$\mu = \left(\frac{\partial H}{\partial N}\right)_{P,S}, \qquad \langle V \rangle = \left(\frac{\partial H}{\partial P}\right)_{N,S}, \qquad T = \left(\frac{\partial H}{\partial S}\right)_{N,P}. \qquad (5.2.5)$$

The notation $\langle V \rangle$ for the volume appearing in eqn. (5.2.5) serves to remind us that the observable volume results from a sampling of instantaneous volume fluctuations. Equations (5.2.5) constitute the basic thermodynamic relations in the isoenthalpic-isobaric ensemble. The reason that enthalpy is designated as a control variable rather than the entropy is the same as for the microcanonical ensemble: It is not possible to "dial up" a desired entropy, whereas, in principle, the enthalpy can be set by the external conditions, even if it is never done in practice (except in computer simulations).

The isothermal-isobaric ensemble results from performing the same Legendre transform on the canonical ensemble. The volume in the Helmholtz free energy $A(N, V, T)$ is transformed into the external pressure P yielding a new free energy denoted $G(N, P, T)$:

$$G(N, P, T) = A(N, V(P), T) - V(P)\frac{\partial A}{\partial V}. \tag{5.2.6}$$

Using the fact that $P = -\partial A/\partial V$, we obtain

$$G(N, P, T) = A(N, V(P), T) + PV(P). \tag{5.2.7}$$

The function $G(N, P, T)$ is known as the *Gibbs free energy*. Since G is a function of N, P, and T, a small change in each of these control variables yields a change in G given by

$$dG = \left(\frac{\partial G}{\partial N}\right)_{P,T} dN + \left(\frac{\partial G}{\partial P}\right)_{N,T} dP + \left(\frac{\partial G}{\partial T}\right)_{N,P} dT. \tag{5.2.8}$$

However, since $G = A + PV$, the differential change dG can also be expressed as

$$dG = dA + PdV + VdP$$

$$= -PdV + \mu dN - SdT + PdV + VdP$$

$$= \mu dN + VdP - SdT, \tag{5.2.9}$$

where the second line follows from eqn. (4.2.5). Thus, equating eqn. (5.2.9) with eqn. (5.2.8), the thermodynamic relations of the isothermal-isobaric ensemble follow:

$$\mu = \left(\frac{\partial G}{\partial N}\right)_{P,T}, \qquad \langle V \rangle = \left(\frac{\partial G}{\partial P}\right)_{N,T}, \qquad S = -\left(\frac{\partial G}{\partial T}\right)_{N,P}. \tag{5.2.10}$$

As before, the volume in eqn. (5.2.10) must be regarded as an average over instantaneous volume fluctuations.

5.3 Isobaric phase-space distributions and partition functions

The relationship between the isoenthalpic-isobaric and isothermal-isobaric ensembles is similar to that between the microcanonical and canonical ensembles. In the isoenthalpic-isobaric ensemble, the *instantaneous* enthalpy is given by $\mathcal{H}(x) + PV$, where V is the instantaneous volume and $\mathcal{H}(x)$ is the Hamiltonian. Note that $\mathcal{H}(x) + PV$ is strictly conserved under isoenthalpic conditions. Thus, the ensemble is defined by a collection of systems evolving according to Hamilton's equations in a containing volume; in turn, the volume of the container adjusts to keep the internal pressure equal to the external applied pressure such that $\mathcal{H}(\mathbf{x}) + PV$ is constant. The term PV in the instantaneous enthalpy represents the work done by the system against the external pressure.

The fact that $\mathcal{H}(\mathbf{x}) + PV$ is conserved implies that the ensemble is the collection of all microstates on the constant enthalpy hypersurface defined by the condition

$$\mathcal{H}(\mathbf{x}) + PV = H, \tag{5.3.1}$$

analogous to the constant-energy hypersurface in the microcanonical ensemble. Since the ensemble distribution function must satisfy the equilibrium Liouville equation and therefore be a function $F(\mathcal{H}(\mathbf{x}))$ of the Hamiltonian, the appropriate solution for the isoenthalpic-isobaric ensemble is simply a δ-function expressing the conservation of the instantaneous enthalpy,

$$f(\mathbf{x}) = F(\mathcal{H}(\mathbf{x})) = \mathcal{M}\delta(\mathcal{H}(\mathbf{x}) + PV - H), \tag{5.3.2}$$

where $\mathcal{M}$ is an overall normalization constant.

As in the microcanonical ensemble, the partition function (the number of accessible microstates) is obtained by integrating over the constant enthalpy hypersurface. However, as the volume is not fixed in this ensemble, each volume accessible to the system has an associated manifold of accessible phase-space points because the size of the configuration is determined by the volume. The partition function must, therefore, contain an integration over both the phase space *and* the volume. Denoting the partition function as $\Gamma(N, P, H)$, we have

$$\Gamma(N, P, H) = \mathcal{M} \int_0^\infty dV \int d\mathbf{p}_1 \cdots \int d\mathbf{p}_N$$

$$\times \int_{D(V)} d\mathbf{r}_1 \cdots \int_{D(V)} d\mathbf{r}_N \; \delta(\mathcal{H}(\mathbf{r}, \mathbf{p}) + PV - H), \tag{5.3.3}$$

where the volume can, in principle, be any positive number. It is important to note that the volume and position integrations cannot be interchanged, since the position integration is restricted to the domain defined by each volume. For this reason, the volume integration cannot be used to integrate over the δ-function. The definition of the normalization constant $\mathcal{M}$ is similar to the microcanonical ensemble except that an additional reference volume V_0 is needed to make the partition function dimensionless:

$$\mathcal{M} \equiv \mathcal{M}_N = \frac{H_0}{V_0 N! h^{3N}}. \tag{5.3.4}$$

Although we can write eqn. (5.3.3) more compactly as

$$\Gamma(N, P, H) = \mathcal{M} \int_0^\infty dV \int d\mathbf{x} \; \delta(\mathcal{H}(\mathbf{x}) + PV - H), \tag{5.3.5}$$

where the volume dependence of the phase-space integration is implicit, this volume dependence must be determined before the integration over V can be performed.

Noting that the thermodynamic relations in eqn. (5.2.5) can also be written in terms of the entropy $S = S(N, P, H)$ as

$$\frac{1}{T} = \left(\frac{\partial S}{\partial H}\right)_{N,P}, \qquad \frac{\langle V \rangle}{T} = \left(\frac{\partial S}{\partial P}\right)_{N,H}, \qquad \frac{\mu}{T} = \left(\frac{\partial S}{\partial N}\right)_{V,H}, \qquad (5.3.6)$$

the thermodynamics can be related to the number of microscopic states by the analogous Boltzmann relation

$$S(N, P, H) = k \ln \Gamma(N, P, H), \qquad (5.3.7)$$

so that eqns. (5.3.6) can be expressed in terms of the partition function as

$$\frac{1}{kT} = \left(\frac{\partial \ln \Gamma}{\partial H}\right)_{N,P}, \qquad \frac{\langle V \rangle}{kT} = \left(\frac{\partial \ln \Gamma}{\partial P}\right)_{N,H}, \qquad \frac{\mu}{kT} = \left(\frac{\partial \ln \Gamma}{\partial N}\right)_{V,H}. \qquad (5.3.8)$$

The partition function for the isothermal-isobaric ensemble can be derived in much the same way as the canonical ensemble is derived from the microcanonical ensemble. The proof is similar to that in Section 4.3 and is left as an exercise (see Problem 5.1). As an alternative, we present a derivation of the partition function that parallels the development of the thermodynamics: We will make explicit use of the canonical ensemble.

Consider two systems coupled to a common thermal reservoir so that each system is described by a canonical distribution at temperature T. Systems 1 and 2 have N_1 and N_2 particles respectively with $N_2 \gg N_1$ and volumes V_1 and V_2 with $V_2 \gg V_1$. System 2 is coupled to system 1 as a "barostat," allowing the volume to fluctuate such that the internal pressure P of system 2 functions as an external applied pressure to system 1 while keeping its internal pressure equal to P (see Fig. 5.1). The total particle number and volume are $N = N_1 + N_2$ and $V = V_1 + V_2$, respectively. Let $\mathcal{H}_1(x_1)$ be the Hamiltonian of system 1 and $\mathcal{H}_2(x_2)$ be the Hamiltonians of system 2. The total Hamiltonian is $\mathcal{H}(x) = \mathcal{H}_1(x_1) + \mathcal{H}_2(x_2)$.

If the volume of each system were fixed, the total canonical partition function $Q(N, V, T)$ would be

$$Q(N, V, T) = C_N \int dx_1 \, dx_2 \, e^{-\beta \mathcal{H}_1(x_1) + \mathcal{H}_2(x_2)}$$

$$= g(N, N_1, N_2) C_{N_1} \int dx_1 \, e^{-\beta \mathcal{H}_1(x_1)} C_{N_2} \int dx_2 \, e^{-\beta \mathcal{H}_2(x_2)}$$

$$\propto Q_1(N_1, V_1, T) Q_2(N_2, V_2, T), \qquad (5.3.9)$$

where $g(N, N_1, N_2)$ is an overall normalization constant. Equation (5.3.9) does not produce a proper counting of all possible microstates, as it involves only one specific choice of V_1 and V_2, and these volumes need to be varied over *all* possible values. A proper counting, therefore, requires that we integrate over all V_1 and V_2, subject to the condition that $V_1 + V_2 = V$. Since $V_2 = V - V_1$, we only need to integrate explicitly

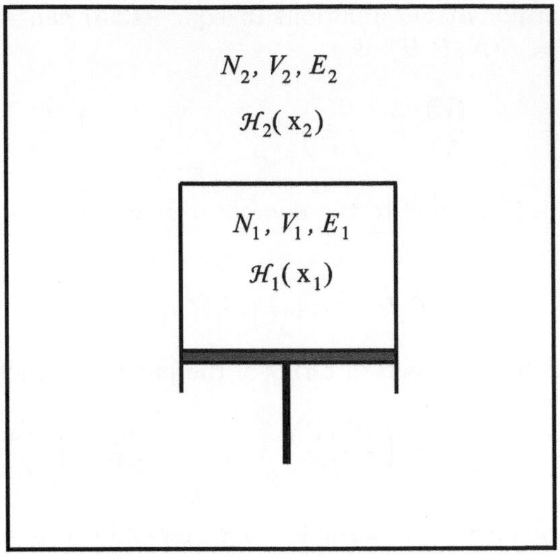

Fig. 5.1 Two systems in contact with a common thermal reservoir at temperature T. System 1 has N_1 particles in a volume V_1; system 2 has N_2 particles in a volume V_2. Both V_1 and V_2 can vary.

over one of the volumes, say V_1. Thus, we write the correct canonical partition function for the total system as

$$Q(N, V, T) = g(N, N_1, N_2) \int_0^V dV_1 \, Q_1(N_1, V_1, T) Q_2(N_2, V - V_1, T). \qquad (5.3.10)$$

The canonical phase-space distribution function $f(\mathbf{x})$ of the combined system 1 and 2 is

$$f(\mathbf{x}) = \frac{C_N e^{-\beta \mathcal{H}(\mathbf{x})}}{Q(N, V, T)}. \qquad (5.3.11)$$

In order to determine the distribution function $f_1(\mathbf{x}_1, V_1)$ of system 1, we need to integrate over the phase space of system 2:

$$f_1(\mathbf{x}_1, V_1) = \frac{g(N, N_1, N_2)}{Q(N, V, T)} C_{N_1} e^{-\beta \mathcal{H}_1(\mathbf{x}_1)} C_{N_2} \int d\mathbf{x}_2 \, e^{-\beta \mathcal{H}_2(\mathbf{x}_2)}$$

$$= \frac{Q_2(N_2, V - V_1, T)}{Q(N, V, T)} g(N, N_1, N_2) C_{N_1} e^{-\beta \mathcal{H}_1(\mathbf{x}_1)}. \qquad (5.3.12)$$

The distribution in eqn. (5.3.12) satisfies the normalization condition:

$$\int_0^V dV_1 \int d\mathbf{x}_1 \, f_1(\mathbf{x}_1, V_1) = 1. \qquad (5.3.13)$$

The ratio of partition functions can be expressed in terms of Helmholtz free energies according to

$$Q_2(N_2, V - V_1, T) = e^{-\beta A(N_2, V - V_1, T)}$$

$$Q(N, V, T) = e^{-\beta A(N, V, T)}$$

$$\frac{Q_2(N_2, V - V_1, T)}{Q(N, V, T)} = e^{-\beta[A(N - N_1, V - V_1, T) - A(N, V, T)]}. \tag{5.3.14}$$

Recalling that $N \gg N_1$ and $V \gg V_1$, the free energy $A(N - N_1, V - V_1, T)$ can be expanded to first order about $N_1 = 0$ and $V_1 = 0$, which yields

$$A(N - N_1, V - V_1, T) \approx A(N, V, T) - N_1 \left(\frac{\partial A}{\partial N} \right) \bigg|_{N_1 = 0, V_1 = 0}$$

$$- V_1 \left(\frac{\partial A}{\partial V} \right) \bigg|_{N_1 = 0, V_1 = 0}. \tag{5.3.15}$$

The minus signs arise from differentiating with respect to N and V instead of N_1 and V_1. Using the relations $\mu = \partial A / \partial N$ and $P = -\partial A / \partial V$, eqn. (5.3.15) becomes

$$A(N - N_1, V - V_1, T) \approx A(N, V, T) - \mu N_1 + P V_1. \tag{5.3.16}$$

Substituting eqn. (5.3.16) into eqn. (5.3.12) yields the distribution

$$f_1(\mathbf{x}_1, V_1) = g(N, N_1, N_2) e^{\beta \mu N_1} e^{-\beta P V_1} e^{-\beta \mathcal{H}_1(\mathbf{x}_1)}. \tag{5.3.17}$$

System 2 has now been eliminated, and we can drop the extraneous "1" subscript. Rearranging eqn. (5.3.17), integrating both sides, and taking the thermodynamic limit, we obtain

$$e^{-\beta \mu N} \int_0^\infty dV \int d\mathbf{x}\, f(\mathbf{x}, V) = I_N \int_0^\infty dV \int d\mathbf{x}\, e^{-\beta(\mathcal{H}(\mathbf{x}) + PV)}. \tag{5.3.18}$$

Equation (5.3.18) defines the partition function of the isothermal-isobaric ensemble as

$$\Delta(N, P, T) = I_N \int_0^\infty dV \int d\mathbf{x}\, e^{-\beta(\mathcal{H}(\mathbf{x}) + PV)}, \tag{5.3.19}$$

where the definition of the prefactor I_N is analogous to the microcanonical and canonical ensembles but with an additional reference volume to make the overall expression dimensionless:

$$I_N = \frac{1}{V_0 N! h^{3N}}. \tag{5.3.20}$$

As noted in Sections 3.2 and 4.3, the factors I_N and $\mathcal{M}_N$ (see eqn. (5.3.4)) should be generalized to $I_{\{N\}}$ and $\mathcal{M}_{\{N\}}$ for multicomponent systems.

Equation (5.3.18) illustrates an important point. Since eqn. (5.3.13) is true in the limit $V \to \infty$, and $\Delta(N, P, T) = \exp(-\beta G(N, P, T))$ (we will prove this shortly), it follows that

$$e^{-\beta \mu N} = e^{-\beta G(N, P, T)}, \tag{5.3.21}$$

or $G(N, P, T) = \mu N$. This relation is a special case of a more general result known as *Euler's theorem* (see Section 6.2). Euler's theorem implies that if a thermodynamic function depends on extensive variables such as N and V, it can be reexpressed as a sum of these variables multiplied by their thermodynamic conjugates. Since $G(N, P, T)$ depends only one extensive variable N, and μ is conjugate to N, $G(N, P, T)$ is a simple product μN.

The partition function of the isothermal-isobaric ensemble is essentially a canonical partition function in which the Hamiltonian $\mathcal{H}(\mathbf{x})$ is replaced by the "instantaneous enthalpy" $\mathcal{H}(\mathbf{x}) + PV$ and an additional volume integration is included. Since $I_N = C_N/V_0$, it is readily seen that eqn. (5.3.19) is

$$\Delta(N, P, T) = \frac{1}{V_0} \int_0^\infty dV \, e^{-\beta PV} Q(N, V, T). \tag{5.3.22}$$

According to eqn. (5.3.22), the isothermal-isobaric partition function is the Laplace transform (see Appendix E) of the canonical partition function with respect to volume, just as the canonical partition function is the Laplace transform of the microcanonical partition function with respect to energy. In both cases, the variable used to form the Laplace transform between partition functions is the same variable used to form the Legendre transform between thermodynamic functions.

We now show that the Gibbs free energy is given by the relation

$$G(N, P, T) = -\frac{1}{\beta} \ln \Delta(N, P, T). \tag{5.3.23}$$

Recall that $G = A + P\langle V \rangle = E + P\langle V \rangle - TS$, which can be expressed as

$$G = \langle \mathcal{H}(\mathbf{x}) + PV \rangle + T \frac{\partial G}{\partial T} \tag{5.3.24}$$

with the help of eqn. (5.2.10). Note that the average of the instantaneous enthalpy is

$$H = \langle \mathcal{H}(\mathbf{x}) + PV \rangle = \frac{I_N \int_0^\infty dV \int dx \, (\mathcal{H}(\mathbf{x}) + PV) e^{-\beta(\mathcal{H}(\mathbf{x}) + PV)}}{I_N \int_0^\infty dV \int dx \, e^{-\beta(\mathcal{H}(\mathbf{x}) + PV)}}$$

$$= -\frac{1}{\Delta(N, P, T)} \frac{\partial}{\partial \beta} \Delta(N, P, T)$$

$$= -\frac{\partial}{\partial \beta} \ln \Delta(N, P, T). \tag{5.3.25}$$

Therefore, eqn. (5.3.24) becomes

$$G + \frac{\partial}{\partial \beta} \ln \Delta(N, P, T) + \beta \frac{\partial G}{\partial \beta} = 0, \tag{5.3.26}$$

which is analogous to eqn. (4.3.20). Thus, following the procedure in Section 4.3 used to prove that $A = -(1/\beta) \ln Q$, we can easily show that $G = -(1/\beta) \ln \Delta$ is a solution to eqn. (5.3.26). other thermodynamic quantities follow in a manner similar to the canonical ensemble. The average volume is

$$\langle V \rangle = -kT \left(\frac{\partial \ln \Delta(N, P, T)}{\partial P} \right)_{N,T}, \tag{5.3.27}$$

the chemical potential is given by

$$\mu = -kT \left(\frac{\partial \ln \Delta(N, P, T)}{\partial N} \right)_{N,P}, \tag{5.3.28}$$

the heat capacity at constant pressure C_P is

$$C_P = \left(\frac{\partial H}{\partial T} \right)_{N,P}$$

$$= k\beta^2 \frac{\partial^2}{\partial \beta^2} \ln \Delta(N, P, T), \tag{5.3.29}$$

and the entropy is obtained from

$$S(N, P, T) = k \ln \Delta(N, P, T) + \frac{H(N, P, T)}{T}. \tag{5.3.30}$$

The equivalence of the isobaric ensembles to the canonical and microcanonical ensembles is explored in Problem 5.2 which examines the behavior of volume fluctuations as a functions of system size.

5.4   Isothermal-isobaric ensemble via entropy maximization

In Section 4.4, we demonstrated that the canonical ensemble could be derived via entropy maximization by defining an entropy functional of the phase-space distribution function based on the Shannon entropy definition. The maximization was performed subject to two constraints: (i) the distribution must be properly normalized and (ii) $\langle \mathcal{H}(\mathbf{x}) \rangle = E$, i.e., the canonical average of the Hamiltonian must yield a well-defined energy. The constrained maximization revealed that the only distribution function possible is $f(\mathbf{x}) = C_N \exp[-\beta \mathcal{H}(\mathbf{x})]/Q(N, V, T)$. We will now show that the distribution function $f(\mathbf{x}, V)$ of the isothermal-isobaric ensemble can be derived in a similar manner. When the volume integration is included, the analog to eqn. (4.4.1) for an isobaric ensemble is

$$S = -k\langle \ln f(\mathbf{x}) \rangle = -k \int_0^\infty dV \int d\mathbf{x} \, f(\mathbf{x}, V) \ln f(\mathbf{x}, V). \tag{5.4.1}$$

We need to maximize this functional subject to two constraints. As in Section 4.4, the first constraint is the normalization of the distribution function

$$\int_0^\infty dV \int d\mathbf{x} \, f(\mathbf{x}, V) = 1, \tag{5.4.2}$$

while the second is dictated by eqn. (5.3.25), which requires that $\langle \mathcal{H}(x) + PV \rangle$ yield a well-defined enthalpy H. These two constraints are added with Lagrange multipliers μ and λ to yield the modified entropy functional

$$\tilde{S}[f] = -k \int_0^\infty dV \int dx \, f(x, V) \ln f(x, V) - \mu \left(\int_0^\infty dV \int dx \, f(x, V) - 1 \right)$$

$$- \lambda \left(\int_0^\infty dV \int dx \, (\mathcal{H}(x) + PV) \, f(x, V) - H \right). \tag{5.4.3}$$

As we did in Section 4.4, we take the functional derivative of $\tilde{S}[f]$ with respect to $f(x, V)$ and set it equal to 0, which yields

$$\frac{\delta \tilde{S}}{\delta f(x, V)} = -k \left[\ln f(x, V) + 1 \right] - \mu - \lambda \left(\mathcal{H}(x) + PV \right) = 0. \tag{5.4.4}$$

Solving for $f(x, V)$, we obtain

$$f(x, V) = e^{-\mu/k} e^{-1} e^{-\lambda(\mathcal{H}(x) + PV)/k}. \tag{5.4.5}$$

Referring back to eqn. (5.2.5), we can obtain an expression for the Lagrange multiplier λ by requiring that

$$\frac{\partial \tilde{S}}{\partial H} = \frac{1}{T}. \tag{5.4.6}$$

From eqn. (5.4.3), we see that $(\partial \tilde{S}/\partial H) = \lambda$, from which it follows that $\lambda = 1/T$. Substituting this into eqn. (5.4.5), we find that

$$f(x, V) = e^{-(1+\mu/k)} e^{-(\mathcal{H}(x) + PV)/kT}, \tag{5.4.7}$$

and once the distribution is properly normalized, we obtain

$$f(x, V) = \frac{I_N e^{-(\mathcal{H}(x) + PV)/kT}}{\Delta(N, P, T)}, \tag{5.4.8}$$

which is the expected result for the isothermal-isobaric ensemble.

5.5 Pressure and work virial theorems

In this chapter, we must distinguish between the external applied pressure, here denoted P, and the internal pressure of the system, denoted $P^{(\text{int})}$, obtained by averaging the estimator $\mathcal{P}(\mathbf{r}, \mathbf{p})$ in eqns. (4.7.57) and (4.7.58) over a canonical ensemble. In the isobaric ensembles, the volume adjusts so that the volume-averaged internal pressure $\langle P^{(\text{int})} \rangle$ is equal to the external applied pressure P. Recall that the internal pressure $P^{(\text{int})}$ at a particular volume V is given in terms of the canonical partition function by

$$P^{(\text{int})} = \langle \mathcal{P}(\mathbf{r}, \mathbf{p}) \rangle = kT \frac{\partial \ln Q}{\partial V} = \frac{kT}{Q} \frac{\partial Q}{\partial V}. \tag{5.5.1}$$

In order to determine the volume-averaged internal pressure, we need to average eqn. (5.5.1) over an isothermal-isobaric distribution according to

$$\langle P^{(\text{int})} \rangle = \frac{1}{\Delta(N, P, T)} \int_0^\infty dV \, e^{-\beta PV} Q(N, V, T) \frac{kT}{Q(N, V, T)} \frac{\partial}{\partial V} Q(N, V, T)$$

$$= \frac{1}{\Delta(N,P,T)} \int_0^\infty \mathrm{d}V \; e^{-\beta PV} kT \frac{\partial}{\partial V} Q(N,V,T). \qquad (5.5.2)$$

Integrating by parts in eqn. (5.5.2), we obtain

$$\langle P^{(\mathrm{int})} \rangle = \frac{1}{\Delta} \left[e^{-\beta PV} kT Q(N,V,T) \right] \Big|_0^\infty - \frac{1}{\Delta} \int_0^\infty \mathrm{d}V kT \left(\frac{\partial}{\partial V} e^{-\beta PV} \right) Q(N,V,T)$$

$$= P \frac{1}{\Delta} \int_0^\infty \mathrm{d}V e^{-\beta PV} Q(N,V,T) = P. \qquad (5.5.3)$$

The boundary term in the first line of eqn. (5.5.3) vanishes at both endpoints: At $V = 0$, the configurational integrals in $Q(N,V,T)$ over a box of zero volume must vanish, and at $V = \infty$, the exponential $\exp(-\beta PV)$ decays faster than $Q(N,V,T)$ increases with V.[1] Recognizing that the integral in the last line of eqn. (5.5.3) is just the partition function $\Delta(N,P,T)$, it follows that

$$\langle P^{(\mathrm{int})} \rangle = P. \qquad (5.5.4)$$

Equation (5.5.4) expresses the expected result that the volume-averaged internal pressure is equal to the external pressure. This result is known as the *pressure virial theorem*. Any computational approach that seeks to generate the isothermal-isobaric ensemble must obey this theorem.

We next consider the average of the pressure-volume product $\langle P^{(\mathrm{int})} V \rangle$. At a fixed volume V, the product $P^{(\mathrm{int})} V$ is given in terms of the canonical partition function by

$$P^{(\mathrm{int})} V = kTV \frac{\partial \ln Q}{\partial V} = \frac{kTV}{Q} \frac{\partial Q}{\partial V}. \qquad (5.5.5)$$

Averaging eqn. (5.5.5) over an isothermal-isobaric ensemble yields

$$\langle P^{(\mathrm{int})} V \rangle = \frac{1}{\Delta} \int_0^\infty \mathrm{d}V e^{-\beta PV} kTV \frac{\partial}{\partial V} Q(N,V,T). \qquad (5.5.6)$$

As was done for eqn. (5.5.3), we integrate eqn. (5.5.6) by parts, which gives

$$\langle P^{(\mathrm{int})} V \rangle = \frac{1}{\Delta} \left[e^{-\beta PV} kTV Q(N,V,T) \right] \Big|_0^\infty - \frac{1}{\Delta} \int_0^\infty \mathrm{d}V kT \left[\frac{\partial}{\partial V} V e^{-\beta PV} \right] Q(N,V,T)$$

$$= \frac{1}{\Delta} \left[-kT \int_0^\infty \mathrm{d}V e^{-\beta PV} Q(V) + P \int_0^\infty \mathrm{d}V e^{-\beta PV} V Q(V) \right]$$

$$= -kT + P\langle V \rangle, \qquad (5.5.7)$$

or

$$\langle P^{(\mathrm{int})} V \rangle + kT = P\langle V \rangle. \qquad (5.5.8)$$

Equation (5.5.8) is known as the *work virial theorem*. Note the presence of the extra kT term on the left side. Since $P\langle V \rangle$ and $\langle P^{(\mathrm{int})} V \rangle$ are both extensive quantities and hence

---

[1] Recall that as $V \to \infty$, $Q(N,V,T)$ approaches the ideal-gas and grows as V^N.

proportional to N, the extra kT term can be neglected in the thermodynamic limit, and eqn. (5.5.8) becomes $\langle P^{(\text{int})}V \rangle \approx P\langle V \rangle$. Nevertheless, eqn. (5.5.8) is rigorously correct, and it is interesting to consider the origin of the extra kT term, since it will arise again in Section 5.10, where we discuss molecular dynamics algorithms for the isothermal-isobaric ensemble.

The quantity $\langle P^{(\text{int})}V \rangle$ can be defined for any ensemble. However, because the volume can fluctuate in an isobaric ensemble, we can think of the volume as an additional degree of freedom that is not present in the microcanonical and canonical ensembles. If energy is equipartitioned, there should be an additional kT of energy in the volume motion, giving rise to a difference of kT between $\langle P^{(\text{int})}V \rangle$ and $P\langle V \rangle$. Since the motion of the volume is driven by an imaginary "piston" that acts to adjust the internal pressure to the external pressure, this piston also adds an amount of energy kT to the system so that eqn. (5.5.8) is satisfied.

5.6 An ideal gas in the isothermal-isobaric ensemble

As an example application of the isothermal-isobaric ensemble, we compute the partition function and thermodynamic properties of an ideal gas. Recall from Section 4.6 that canonical partition function for the ideal gas is

$$Q(N, V, T) = \frac{V^N}{N!\lambda^{3N}}, \tag{5.6.1}$$

where $\lambda = \sqrt{\beta h^2 / 2\pi m}$. Substituting eqn. (5.6.1) into eqn. (5.3.22) gives the isothermal-isobaric partition function

$$\Delta(N, P, T) = \frac{1}{V_0} \int_0^\infty dV \ e^{-\beta PV} \frac{V^N}{N!\lambda^{3N}} = \frac{1}{V_0 N!\lambda^{3N}} \int_0^\infty dV \ e^{-\beta PV} V^N. \tag{5.6.2}$$

The volume integral can be rendered dimensionless by letting $x = \beta PV$, leading to

$$\Delta(N, P, T) = \frac{1}{V_0 N!\lambda^{3N}} \frac{1}{(\beta P)^{N+1}} \int_0^\infty dx \ x^N e^{-x}. \tag{5.6.3}$$

The value of the integral is just $N!$. Hence, the isothermal-isobaric partition function for an ideal gas is

$$\Delta(N, P, T) = \frac{1}{V_0 \lambda^{3N} (\beta P)^{N+1}}. \tag{5.6.4}$$

The thermodynamics of the ideal gas follow from the relations derived in Section 5.3. For the equation of state, we obtain the average volume from

$$\langle V \rangle = -kT \left(\frac{\partial \ln \Delta}{\partial P} \right) = \frac{(N+1)kT}{P}, \tag{5.6.5}$$

or

$$P\langle V \rangle = (N+1)kT \approx NkT, \tag{5.6.6}$$

where the last expression follows from the thermodynamic limit. Using eqn. (5.5.8), we can express eqn. (5.6.6) in terms of the average $P^{(\text{int})}V$ product:

$$\langle P^{(\text{int})} V \rangle = NkT. \tag{5.6.7}$$

Equation (5.6.7) is generally true even away from the thermodynamic limit. The average enthalpy of the ideal gas is given by

$$H = -\frac{\partial}{\partial \beta} \ln \Delta = (N+1)kT + \frac{3}{2}NkT \approx \frac{5}{2}NkT, \tag{5.6.8}$$

from which the constant pressure heat capacity is given by

$$C_P = \left(\frac{\partial H}{\partial T} \right) = \frac{5}{2}Nk. \tag{5.6.9}$$

Equations (5.6.8) and (5.6.9) are usually first encountered in elementary physics and chemistry textbooks with no microscopic justification. This derivation shows the microscopic origin of eqn. (5.6.9). Note that the difference between the constant volume and constant pressure heat capacities is

$$C_P = C_V + Nk = C_V + nR, \tag{5.6.10}$$

where the product Nk has been replaced by nR, with n the number of moles of gas and R the gas constant. (This relation is obtained by multiplying and dividing by N_0, Avogadro's number, $Nk = (N/N_0)N_0 k = nR$.) Dividing eqn. (5.6.10) by the number of moles leads to the familiar relation for the molar heat capacities:

$$c_P = c_V + R. \tag{5.6.11}$$

5.7 Extending the isothermal-isobaric ensemble: Anisotropic cell fluctuations

In this section, we will show how to account for anisotropic volume fluctuations within the isothermal-isobaric ensemble. Anisotropic volume fluctuations can occur under a wide variety of external conditions; however, we will limit ourselves to those that develop under an applied isotropic external pressure. Other external conditions, such as an applied pressure in two dimensions, would generate a constant surface tension ensemble. The formalism developed in this chapter will provide the reader with the tools to understand and develop computational approaches for different external conditions.

When the volume of a system can undergo anisotropic fluctuations, it is necessary to allow the containing volume to change its basic shape. Consider a system contained within a general parallelepiped. The parallelepiped represents the most general "box" shape and is appropriate for describing, for example, solids whose unit cells are generally triclinic. As shown in Fig. 5.2, any parallelepiped can be specified by the three vectors $\mathbf{a}$, $\mathbf{b}$, and $\mathbf{c}$ that lie along three edges originating from a vertex. Simple geometry tells us that the volume V of the parallelepiped is given by

$$V = \mathbf{a} \cdot \mathbf{b} \times \mathbf{c}. \tag{5.7.1}$$

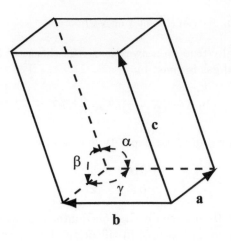

Fig. 5.2 A general parallelepiped showing the convention for the cell vectors and angles.

Since each edge vector contains three components, nine numbers can be used to characterize the parallelepiped; these are often collected in the columns of a 3×3 matrix **h** called the *box matrix* or *cell matrix*:

$$\mathbf{h} = \begin{pmatrix} a_x & b_x & c_x \\ a_y & b_y & c_y \\ a_z & b_z & c_z \end{pmatrix}. \tag{5.7.2}$$

In terms of the cell matrix, the volume V is easily seen to be

$$V = \det(\mathbf{h}). \tag{5.7.3}$$

On the other hand, a little reflection shows that, in fact, only six numbers are needed to specify the cell: the lengths of the edges $a = |\mathbf{a}|$, $b = |\mathbf{b}|$, and $c = |\mathbf{c}|$ and the angles α, β, and γ between them. By convention, these three angles are defined such that α is the angle between vectors **b** and **c**, β is the angle between vectors **a** and **c**, and γ is the angle between vectors **a** and **b**. It is clear, therefore, that the full cell matrix contains redundant information—in addition to providing information about the cell lengths and angles, it also describes overall rotations of the cell in space, as specified by the three Euler angles (see Section 1.11), which accounts for the three extra degrees of freedom.

In order to separate isotropic from anisotropic cell fluctuations, we introduce a *unit box matrix* $\mathbf{h}_0$ related to **h** by $\mathbf{h} = V^{1/3}\mathbf{h}_0$ such that $\det(\mathbf{h}_0) = 1$. Focusing on the isothermal-isobaric ensemble, the changing cell shape under the influence of an isotropic applied pressure P can be incorporated into the partition function by writing $\Delta(N, P, T)$ as

$$\Delta(N, P, T) = \frac{1}{V_0} \int_0^\infty dV \int d\mathbf{h}_0 \, e^{-\beta PV} Q(N, V, \mathbf{h}_0, T) \, \delta\left(\det(\mathbf{h}_0) - 1\right), \tag{5.7.4}$$

where $\int d\mathbf{h}_0$ is an integral over all nine components of $\mathbf{h}_0$ and the δ-function restricts the integration to unit box matrices satisfying $\det(\mathbf{h}_0) = 1$. In eqn. (5.7.4), the explicit dependence of the canonical partition function Q on both the volume V and the shape of the cell described by $\mathbf{h}_0$ is shown.

Rather than integrate over V and $\mathbf{h}_0$ with the constraint of $\det(\mathbf{h}_0) = 1$, it is preferable to perform an unconstrained integration over $\mathbf{h}$. This can be accomplished by a change of variables from $\mathbf{h}_0$ to $\mathbf{h}$. Since each element of $\mathbf{h}_0$ is multiplied by $V^{1/3}$ to obtain $\mathbf{h}$, the integration measure, which is a nine-dimensional integration, transforms as $d\mathbf{h}_0 = V^{-3}d\mathbf{h}$. In addition, $\det(\mathbf{h}_0) = \det(\mathbf{h})/V$. Thus, substituting the cell-matrix transformation into eqn. (5.7.4) yields

$$\Delta(N, P, T) = \frac{1}{V_0} \int_0^\infty dV \int d\mathbf{h}\, V^{-3} e^{-\beta PV} Q(N, \mathbf{h}, T)\, \delta\left(\frac{1}{V}\det(\mathbf{h}) - 1\right)$$

$$= \frac{1}{V_0} \int_0^\infty dV \int d\mathbf{h}\, V^{-3} e^{-\beta PV} Q(N, \mathbf{h}, T) V\, \delta\left(\det(\mathbf{h}) - V\right)$$

$$= \frac{1}{V_0} \int_0^\infty dV \int d\mathbf{h}\, V^{-2} e^{-\beta PV} Q(N, \mathbf{h}, T)\, \delta\left(\det(\mathbf{h}) - V\right), \quad (5.7.5)$$

where the dependence of Q on V and $\mathbf{h}_0$ has been expressed as an equivalent dependence only on $\mathbf{h}$. Performing the integration over the volume using the δ-function, we obtain for the partition function

$$\Delta(N, P, T) = \frac{1}{V_0} \int d\mathbf{h}\, [\det(\mathbf{h})]^{-2}\, e^{-\beta P \det(\mathbf{h})} Q(N, \mathbf{h}, T). \quad (5.7.6)$$

In an arbitrary number d of spatial dimensions, the transformation is $\mathbf{h} = V^{1/d}\mathbf{h}_0$, and the partition function becomes

$$\Delta(N, P, T) = \frac{1}{V_0} \int d\mathbf{h}\, [\det(\mathbf{h})]^{1-d}\, e^{-\beta P \det(\mathbf{h})} Q(N, \mathbf{h}, T). \quad (5.7.7)$$

Before describing the generalization of the virial theorems of Section 5.5, we note that the internal pressure of a canonical ensemble with a fixed cell matrix $\mathbf{h}$ describing an anisotropic system cannot be described by a single scalar quantity as is possible for an isotropic system. Rather, a *tensor* is needed; this tensor is known as the *pressure tensor*, $\mathbf{P}^{(\text{int})}$. Since the Helmholtz free energy $A = A(N, \mathbf{h}, T)$ depends on the full cell matrix, the pressure tensor, which is a 3×3 (or rank 2) tensor, has components given by

$$P_{\alpha\beta}^{(\text{int})} = -\frac{1}{\det(\mathbf{h})} \sum_{\gamma=1}^{3} h_{\beta\gamma} \left(\frac{\partial A}{\partial h_{\alpha\gamma}}\right)_{N,T}, \quad (5.7.8)$$

which can be expressed in terms of the canonical partition function as

$$P_{\alpha\beta}^{(\text{int})} = \frac{kT}{\det(\mathbf{h})} \sum_{\gamma=1}^{3} h_{\beta\gamma} \left(\frac{\partial \ln Q}{\partial h_{\alpha\gamma}}\right)_{N,T}. \quad (5.7.9)$$

In Section 5.8, an appropriate microscopic estimator for the pressure tensor will be derived.

If we now consider the average of the pressure tensor in the isothermal-isobaric ensemble, a *tensorial* version of virial theorem can be proved for an applied isotropic external pressure P. The average of the internal pressure tensor is

$$\langle P_{\alpha\beta}^{(\text{int})} \rangle = \frac{1}{\Delta(N,P,T)} \int d\mathbf{h} \, [\det(\mathbf{h})]^{-2} e^{-\beta P \det(\mathbf{h})} \frac{kTQ(N,\mathbf{h},T)}{\det(\mathbf{h})} \sum_{\gamma=1}^{3} h_{\beta\gamma} \left(\frac{\partial \ln Q}{\partial h_{\alpha\gamma}} \right)_{N,T}$$

$$= \frac{1}{\Delta(N,P,T)} \int d\mathbf{h} \, [\det(\mathbf{h})]^{-2} e^{-\beta P \det(\mathbf{h})} \frac{kT}{\det(\mathbf{h})} \sum_{\gamma=1}^{3} h_{\beta\gamma} \left(\frac{\partial Q}{\partial h_{\alpha\gamma}} \right)_{N,T} . \quad (5.7.10)$$

An integration by parts can be performed as was done in Section 5.5, and, recognizing that the boundary term vanishes, we obtain

$$\langle P_{\alpha\beta}^{(\text{int})} \rangle = -\frac{kT}{\Delta(N,P,T)} \int d\mathbf{h} \sum_{\gamma=1}^{3} \frac{\partial}{\partial h_{\alpha\gamma}} \left\{ [\det(\mathbf{h})]^{-2} e^{-\beta P \det(\mathbf{h})} \frac{kT}{\det(\mathbf{h})} h_{\beta\gamma} \right\} Q(N,\mathbf{h},T)$$

$$= -\frac{kT}{\Delta(N,P,T)} \int d\mathbf{h} \sum_{\gamma=1}^{3} \left\{ -3[\det(\mathbf{h})]^{-4} \frac{\partial \det(\mathbf{h})}{\partial h_{\alpha\gamma}} h_{\beta\gamma} \right.$$

$$- \beta P [\det(\mathbf{h})]^{-3} \frac{\partial \det(\mathbf{h})}{\partial h_{\alpha\gamma}} h_{\beta\gamma}$$

$$\left. + [\det(\mathbf{h})]^{-3} \frac{\partial h_{\beta\gamma}}{\partial h_{\alpha\gamma}} \right\} e^{-\beta P \det(\mathbf{h})} Q(N,\mathbf{h},T). \quad (5.7.11)$$

In order to proceed, we need to know how to calculate the derivative of the determinant of a matrix with respect to one of its elements. The determinant of a matrix M can be written as $\det(M) = \exp[\text{Tr} \ln(M)]$. Taking the derivative of this expression with respect to an element M_{ij}, we obtain

$$\frac{\partial[\det(M)]}{\partial M_{ij}} = e^{\text{Tr} \ln(M)} \text{Tr} \left[M^{-1} \frac{\partial M}{\partial M_{ij}} \right]$$

$$= \det(M) \sum_{k,l} M_{kl}^{-1} \frac{\partial M_{lk}}{\partial M_{ij}}, \quad (5.7.12)$$

where the trace has been written out explicitly. The derivative $\partial M_{lk}/\partial M_{ij} = \delta_{il}\delta_{kj}$. Thus, performing the sums over k and l leaves

$$\frac{\partial[\det(M)]}{\partial M_{ij}} = \det(M) M_{ji}^{-1}. \quad (5.7.13)$$

Applying eqn. (5.7.13) to eqn. (5.7.11), and using the fact that $\sum_{\gamma} \partial h_{\beta\gamma}/\partial h_{\alpha\gamma} = \sum_{\gamma} \delta_{\beta\alpha}\delta_{\gamma\gamma} = 3\delta_{\beta\alpha}$, it can be seen that the first and last terms in the curly brackets of eqn. (5.7.11) cancel, leaving

$$\langle P_{\alpha\beta}^{(\text{int})} \rangle = \frac{kT}{\Delta(N, P, T)} \int d\mathbf{h} \, \beta P \delta_{\alpha\beta} e^{-\beta P \det(\mathbf{h})} Q(N, \mathbf{h}, T) = P \delta_{\alpha\beta}, \quad (5.7.14)$$

which states that, on the average, the pressure tensor should be diagonal with each diagonal element equal to the external applied pressure P. This is the generalization of the pressure virial theorem of eqn. (5.5.4). In a similar manner, the generalization of the work virial in eqn. (5.5.8) can be shown to be

$$\langle P_{\alpha\beta}^{(\text{int})} \det(\mathbf{h}) \rangle + kT \delta_{\alpha\beta} = P \langle \det(\mathbf{h}) \rangle \delta_{\alpha\beta}, \quad (5.7.15)$$

according to which the average $\langle P_{\alpha\beta}^{(\text{int})} \det(\mathbf{h}) \rangle$ is diagonal.

5.8 Derivation of the pressure tensor estimator from the canonical partition function

Molecular dynamics calculations in isobaric ensembles require explicit microscopic estimators for the pressure. In Section 4.7.3, we derived an estimator for the isotropic internal pressure (see eqn. (4.7.57)). In this section, we generalize the derivation and obtain an estimator for the pressure tensor. For readers wishing to skip over the mathematical details of this derivation, we present the final result:

$$\mathcal{P}_{\alpha\beta}^{(\text{int})}(\mathbf{r}, \mathbf{p}) = \frac{1}{\det(\mathbf{h})} \sum_{i=1}^{N} \left[\frac{(\mathbf{p}_i \cdot \hat{\mathbf{e}}_\alpha)(\mathbf{p}_i \cdot \hat{\mathbf{e}}_\beta)}{m_i} + (\mathbf{F}_i \cdot \hat{\mathbf{e}}_\alpha)(\mathbf{r}_i \cdot \hat{\mathbf{e}}_\beta) \right], \quad (5.8.1)$$

where $\hat{\mathbf{e}}_\alpha$ and $\hat{\mathbf{e}}_\beta$ are unit vectors along the α and β spatial directions, respectively. Thus, $(\mathbf{p}_i \cdot \hat{\mathbf{e}}_\alpha)$ is just the αth component of the momentum vector $\mathbf{p}_i$, with $\alpha = x, y, z$. The internal pressure tensor $P_{\alpha\beta}^{(\text{int})}$ at fixed $\mathbf{h}$ is simply a canonical ensemble average of the estimator in eqn. (5.8.1).

The derivation of the pressure tensor requires a transformation from the primitive Cartesian variables $\mathbf{r}_1, ..., \mathbf{r}_N, \mathbf{p}_1, ..., \mathbf{p}_N$ to scaled variables, as was done in Section 4.7.3 for the isotropic pressure estimator. In order to make the dependence of the Hamiltonian and the partition function on the box matrix $\mathbf{h}$ explicit, we introduce scaled variables $\mathbf{s}_1, ..., \mathbf{s}_N$ related to the primitive Cartesian positions by

$$\mathbf{r}_i = \mathbf{h} \mathbf{s}_i. \quad (5.8.2)$$

The right side of eqn. (5.8.2) is a matrix-vector product, which, in component form, appears as

$$\mathbf{r}_i \cdot \hat{\mathbf{e}}_\alpha = \sum_\beta h_{\alpha\beta} (\mathbf{s}_i \cdot \hat{\mathbf{e}}_\beta), \quad (5.8.3)$$

or in more compact notation,

$$r_{i,\alpha} = \sum_\beta h_{\alpha\beta} s_{i,\beta}, \quad (5.8.4)$$

where $r_{i,\alpha} = \mathbf{r}_i \cdot \hat{\mathbf{e}}_\alpha$ and $s_{i,\beta} = \mathbf{s}_i \cdot \hat{\mathbf{e}}_\beta$.

Not unexpectedly, the corresponding transformation for the momenta requires multiplication by the inverse box matrix $\mathbf{h}^{-1}$. However, since $\mathbf{h}$ and $\mathbf{h}^{-1}$ are not symmetric, should the matrix be multiplied on the right or on the left. The Lagrangian formulation of classical mechanics of Section 1.4 provides us with a direct route for answering this question. Recall that the Lagrangian is given by

$$\mathcal{L}(\mathbf{r}, \dot{\mathbf{r}}) = \frac{1}{2} \sum_i m_i \dot{\mathbf{r}}_i^2 - U(\mathbf{r}_1, ..., \mathbf{r}_N). \tag{5.8.5}$$

The Lagrangian can be transformed into the scaled coordinates by substituting eqn. (5.8.4) and the corresponding velocity transformation

$$\dot{r}_{i,\alpha} = \sum_\beta h_{\alpha\beta} \dot{s}_{i,\beta} \tag{5.8.6}$$

into eqn. (5.8.5) to yield

$$\mathcal{L}(\mathbf{s}, \dot{\mathbf{s}}) = \frac{1}{2} \sum_i m_i \sum_{\alpha,\beta,\gamma} h_{\alpha\beta} \dot{s}_{i,\beta} h_{\alpha\gamma} \dot{s}_{i,\gamma} - U(\mathbf{hs}_1, ..., \mathbf{hs}_N)$$

$$= \frac{1}{2} \sum_{\alpha,\beta,\gamma} h_{\alpha\beta} h_{\alpha\gamma} \sum_i m_i \dot{s}_{i,\beta} \dot{s}_{i,\gamma} - U(\mathbf{hs}_1, ..., \mathbf{hs}_N). \tag{5.8.7}$$

A component of the momentum $\boldsymbol{\pi}_j$ conjugate to $\mathbf{s}_j$ is computed according to

$$\pi_{j,\lambda} = \frac{\partial \mathcal{L}}{\partial \dot{s}_{j,\lambda}}. \tag{5.8.8}$$

The trickiest part of this derivative is keeping track of the indices. Since all of the indices in eqn. (5.8.7) are summed over or *contracted*, eqn. (5.8.7) contains many terms. The only terms that contribute to the momentum in eqn. (5.8.8) are those for which $i = j$ and $\beta = \lambda$ or $\gamma = \lambda$. The easiest way to keep track of the bookkeeping is to replace factors of $\dot{s}_{i,\beta}$ or $\dot{s}_{i,\gamma}$ with $\delta_{ij}\delta_{\beta\lambda}$ and $\delta_{ij}\delta_{\gamma\lambda}$, respectively, when computing the derivative, and then perform the sums with the aid of the Kroenecker deltas:

$$\pi_{j,\lambda} = \frac{1}{2} \sum_{\alpha,\beta,\gamma} h_{\alpha\beta} h_{\alpha\gamma} \sum_i m_i \left[\delta_{ij}\delta_{\beta\lambda}\dot{s}_{i,\gamma} + \dot{s}_{i,\beta}\delta_{ij}\delta_{\gamma\lambda} \right]$$

$$= \frac{1}{2} m_j \left[\sum_{\alpha,\gamma} h_{\alpha\lambda} h_{\alpha\gamma} \dot{s}_{j,\gamma} + \sum_{\alpha,\beta} h_{\alpha\beta} h_{\alpha\lambda} \dot{s}_{j,\beta} \right]. \tag{5.8.9}$$

Since the two sums appearing in the last line of eqn. (5.8.9) are the same, the factor of $1/2$ can be cancelled, yielding

$$\pi_{j,\lambda} = m_j \sum_{\alpha,\gamma} h_{\alpha\lambda} h_{\alpha\gamma} \dot{s}_{j,\gamma} = m_j \sum_\alpha \dot{r}_{j,\alpha} h_{\alpha\lambda}. \tag{5.8.10}$$

Writing this in vector notation, we find

$$\boldsymbol{\pi}_j = m_j \dot{\mathbf{r}}_j \mathbf{h} = \mathbf{p}_j \mathbf{h} \tag{5.8.11}$$

or

$$\mathbf{p}_j = \boldsymbol{\pi}_j \mathbf{h}^{-1}. \tag{5.8.12}$$

Thus, we see that $\boldsymbol{\pi}_j$ must be multiplied *on the right* by $\mathbf{h}^{-1}$.

Having obtained the Lagrangian in scaled coordinates and the momentum transformation from the Lagrangian, we must now derive the Hamiltonian in order to determine the canonical partition function. The Hamiltonian is given by the Legendre transform rule:

$$\mathcal{H} = \sum_i \boldsymbol{\pi}_i \cdot \dot{\mathbf{s}}_i - \mathcal{L} = \sum_i \sum_\alpha \pi_{i,\alpha} \dot{s}_{i,\alpha} - \mathcal{L}. \tag{5.8.13}$$

Using the fact that $\dot{\mathbf{s}}_i = \mathbf{h}^{-1}\dot{\mathbf{r}}_i = \mathbf{h}^{-1}\mathbf{p}_i/m_i$ together with eqn. (5.8.12) to substitute $\mathbf{p}_i$ in terms of $\boldsymbol{\pi}_i$, the Hamiltonian becomes

$$\mathcal{H} = \sum_i \sum_{\alpha,\beta} \frac{1}{m_i} \pi_{i,\alpha} h_{\alpha\beta}^{-1} p_{i,\beta} - \mathcal{L} = \sum_i \sum_{\alpha,\beta,\gamma} \frac{1}{m_i} \pi_{i,\alpha} h_{\alpha\beta}^{-1} \pi_{i,\gamma} h_{\gamma\beta}^{-1} - \mathcal{L}. \tag{5.8.14}$$

Since the kinetic energy term in L is just $1/2$ of the first term in eqn. (5.8.14), the Hamiltonian becomes

$$\mathcal{H} = \sum_i \sum_{\alpha,\beta,\gamma} \frac{\pi_{i,\alpha} \pi_{i,\gamma} h_{\alpha\beta}^{-1} h_{\gamma\beta}^{-1}}{2m_i} + U(\mathbf{hs}_1, ..., \mathbf{hs}_N). \tag{5.8.15}$$

The pressure tensor in the canonical ensemble is given by

$$
\begin{aligned}
P_{\alpha\beta}^{(\mathrm{int})} &= \frac{1}{Q(N,\mathbf{h},T)} \frac{kT}{\det(\mathbf{h})} \sum_\gamma h_{\beta\gamma} \frac{\partial Q(N,\mathbf{h},T)}{\partial h_{\alpha\gamma}} \\
&= \frac{kT}{\det(\mathbf{h})} \frac{1}{Q(N,\mathbf{h},T)} \int d^N \boldsymbol{\pi} d^N \mathbf{s} \sum_\gamma h_{\beta\gamma} \left(-\beta \frac{\partial \mathcal{H}}{\partial h_{\alpha\gamma}} \right) e^{-\beta\mathcal{H}} \\
&= -\left\langle \frac{1}{\det(\mathbf{h})} \sum_\gamma h_{\beta\gamma} \frac{\partial \mathcal{H}}{\partial h_{\alpha\gamma}} \right\rangle.
\end{aligned}
\tag{5.8.16}
$$

Equation (5.8.16) requires the derivative of the Hamiltonian with respect to an arbitrary element of $\mathbf{h}$. This derivative must be obtained from eqn. (5.8.15), which requires more index bookkeeping. Let us first rewrite the Hamiltonian using a different set of summation indices:

$$\mathcal{H} = \sum_i \sum_{\mu,\nu,\lambda} \frac{\pi_{i,\mu} \pi_{i,\nu} h_{\mu\lambda}^{-1} h_{\nu\lambda}^{-1}}{2m_i} + U(\mathbf{hs}_1, ..., \mathbf{hs}_N). \tag{5.8.17}$$

Computing the derivative with respect to $h_{\alpha\gamma}$, we obtain

$$\frac{\partial \mathcal{H}}{\partial h_{\alpha\gamma}} = \sum_i \sum_{\mu,\nu,\lambda} \frac{\pi_{i,\mu} \pi_{i,\nu}}{2m_i} \left(\frac{\partial h_{\mu\lambda}^{-1}}{\partial h_{\alpha\gamma}} h_{\nu\lambda}^{-1} + h_{\mu\lambda}^{-1} \frac{\partial h_{\nu\lambda}^{-1}}{\partial h_{\alpha\gamma}} \right) + \frac{\partial}{\partial h_{\alpha\gamma}} U(\mathbf{hs}_1, ..., \mathbf{hs}_N). \tag{5.8.18}$$

In order to proceed, we will derive an identity for the derivative of the inverse of a matrix $M(\lambda)$ with respect to an arbitrary parameter λ. We start by differentiating the matrix identity

$$M(\lambda)M^{-1}(\lambda) = I \tag{5.8.19}$$

with respect to λ to obtain

$$\frac{dM}{d\lambda}M^{-1} + M\frac{dM^{-1}}{d\lambda} = 0. \tag{5.8.20}$$

Solving eqn. (5.8.20) for $dM^{-1}/d\lambda$ yields

$$\frac{dM^{-1}}{d\lambda} = -M^{-1}\frac{dM}{d\lambda}M^{-1}. \tag{5.8.21}$$

Applying eqn. (5.8.21) to eqn. (5.8.18), we obtain

$$\frac{\partial \mathcal{H}}{\partial h_{\alpha\gamma}} = -\sum_i \sum_{\mu,\nu,\lambda} \frac{\pi_{i,\mu}\pi_{i,\nu}}{2m_i} \sum_{\rho,\sigma} \left(h_{\mu\rho}^{-1}\frac{\partial h_{\rho\sigma}}{\partial h_{\alpha\gamma}}h_{\sigma\lambda}^{-1}h_{\nu\lambda}^{-1} + h_{\mu\lambda}^{-1}h_{\nu\rho}^{-1}\frac{\partial h_{\rho\sigma}}{\partial h_{\alpha\gamma}} \right) h_{\sigma\lambda}^{-1}$$

$$+ \frac{\partial}{\partial h_{\alpha\gamma}}U(\mathbf{hs}_1, ..., \mathbf{hs}_N). \tag{5.8.22}$$

Using $\partial h_{\rho\sigma}/\partial h_{\alpha\gamma} = \delta_{\alpha\rho}\delta_{\sigma\gamma}$ and performing the sums over ρ and σ, we find

$$\frac{\partial \mathcal{H}}{\partial h_{\alpha\gamma}} = -\sum_i \sum_{\mu,\nu,\lambda} \frac{\pi_{i,\mu}\pi_{i,\nu}}{2m_i} \left(h_{\mu\alpha}^{-1}h_{\gamma\lambda}^{-1}h_{\nu\lambda}^{-1} + h_{\mu\lambda}^{-1}h_{\nu\alpha}^{-1}h_{\gamma\lambda}^{-1} \right)$$

$$+ \frac{\partial}{\partial h_{\alpha\gamma}}U(\mathbf{hs}_1, ..., \mathbf{hs}_N). \tag{5.8.23}$$

Since

$$\frac{\partial}{\partial h_{\alpha\gamma}}U(\mathbf{hs}_1, ..., \mathbf{hs}_N) = \sum_i \sum_{\mu,\nu} \frac{\partial U}{\partial(\mathbf{hs}_i)_\mu}\frac{\partial h_{\mu\nu}}{\partial h_{\alpha\gamma}}s_{i,\nu}$$

$$= \sum_i \sum_{\mu,\nu} \frac{\partial U}{\partial(\mathbf{hs}_i)_\mu}\delta_{\alpha\mu}\delta_{\gamma\nu}s_{i,\nu}$$

$$= \sum_i \frac{\partial U}{\partial(\mathbf{hs}_i)_\alpha}s_{i,\gamma}, \tag{5.8.24}$$

we arrive at the result

$$\frac{\partial \mathcal{H}}{\partial h_{\alpha\gamma}} = -\sum_i \sum_{\mu,\nu,\lambda} \frac{\pi_{i,\mu}\pi_{i,\nu}}{2m_i} \left(h_{\mu\alpha}^{-1}h_{\gamma\lambda}^{-1}h_{\nu\lambda}^{-1} + h_{\mu\lambda}^{-1}h_{\nu\alpha}^{-1}h_{\gamma\lambda}^{-1} \right) + \sum_i \frac{\partial U}{\partial(\mathbf{hs}_i)_\alpha}s_{i,\gamma}. \tag{5.8.25}$$

To obtain the pressure tensor estimator, we must multiply by $h_{\beta\gamma}$ and sum over γ. When this is done and the sum over γ is performed according to $\sum_\gamma h_{\beta\gamma}h_{\gamma\lambda}^{-1} = \delta_{\beta\lambda}$, then the sum over λ can be performed as well, yielding

$$\sum_{\gamma} h_{\beta\gamma} \frac{\partial \mathcal{H}}{\partial h_{\alpha\gamma}} = -\sum_{i} \sum_{\mu,\nu} \frac{\pi_{i,\mu}\pi_{i,\nu}}{2m_i} \left(h_{\mu\alpha}^{-1} h_{\nu\beta}^{-1} + h_{\mu\beta}^{-1} h_{\nu\alpha}^{-1} \right)$$

$$+ \sum_{i} \sum_{\gamma} \frac{\partial U}{\partial (\mathbf{hs}_i)_{\alpha}} h_{\beta\gamma} s_{i,\gamma}. \tag{5.8.26}$$

We now recognize that $\sum_{\alpha} \pi_{i,\mu} h_{\mu\alpha}^{-1} = p_{i,\alpha}$, $\sum_{\nu} \pi_{i,\nu} h_{\nu\beta}^{-1} = p_{i,\beta}$, $\partial U/\partial(\mathbf{hs}_i) = \partial U/\partial \mathbf{r}_i$ and $\sum_{\gamma} h_{\beta\gamma} s_{i,\gamma} = r_{i,\beta}$. Substituting these results into eqn. (5.8.26) and multiplying by $-1/\det(\mathbf{h})$ gives

$$\mathcal{P}_{\alpha\beta}^{(\text{int})}(\mathbf{r}_1, ..., \mathbf{r}_N, \mathbf{p}_1, ..., \mathbf{p}_N) = \frac{1}{\det(\mathbf{h})} \sum_{i=1}^{N} \left[\frac{p_{i,\alpha} p_{i,\beta}}{m_i} + F_{i,\alpha} r_{i,\beta} \right], \tag{5.8.27}$$

which is equivalent to eqn. (5.8.1), thus completing the derivation. The isotropic pressure estimator for $P^{(\text{int})}$ in eqn. (4.7.57) can be obtained directly from the pressure tensor estimator by tracing:

$$\mathcal{P}^{(\text{int})}(\mathbf{r}, \mathbf{p}) = \frac{1}{3} \sum_{\alpha} \mathcal{P}_{\alpha\alpha}^{(\text{int})}(\mathbf{r}, \mathbf{p}) = \frac{1}{3} \text{Tr} \left[\mathbf{P}^{(\text{int})}(\mathbf{r}, \mathbf{p}) \right], \tag{5.8.28}$$

where $\mathbf{P}^{(\text{int})}(\mathbf{r}, \mathbf{p})$ is the tensorial representation of eqn. (5.8.27). Finally, note that if the potential has an explicit dependence on the cell matrix $\mathbf{h}$, then the estimator is modified to read

$$\mathcal{P}_{\alpha\beta}^{(\text{int})}(\mathbf{r}, \mathbf{p}) = \frac{1}{\det(\mathbf{h})} \sum_{i=1}^{N} \left[\frac{(\mathbf{p}_i \cdot \hat{\mathbf{e}}_{\alpha})(\mathbf{p}_i \cdot \hat{\mathbf{e}}_{\beta})}{m_i} + (\mathbf{F}_i \cdot \hat{\mathbf{e}}_{\alpha})(\mathbf{r}_i \cdot \hat{\mathbf{e}}_{\beta}) \right]$$

$$- \frac{1}{\det(\mathbf{h})} \sum_{\gamma=1}^{3} \frac{\partial U}{\partial h_{\alpha\gamma}} h_{\gamma\beta}. \tag{5.8.29}$$

5.9 Molecular dynamics in the isoenthalpic-isobaric ensemble

The derivation of the isobaric ensembles requires that the volume be allowed to vary in order to keep the internal pressure equal, on average, to the applied external pressure. This suggests that if we wish to develop a molecular dynamics technique for generating isobaric ensembles, we could introduce the volume as an independent dynamical variable in the phase space. Indeed, the work-virial theorem of eqn. (5.5.8) strongly supports such a notion, since it effectively assigns an energy of kT to a "volume mode." The idea of incorporating the volume into the phase space as an additional dynamical degree of freedom, together with its conjugate momentum, as a means of generating an isobaric ensemble was first introduced by Andersen (1980) and later generalized for anisotropic volume fluctuations by Parrinello and Rahman (1980). This idea inspired numerous other powerful techniques based on extended phase spaces, including the canonical molecular dynamics methods from Chapter 4, the Car-Parrinello approach (Car and Parrinello, 1985; Tuckerman, 2002; Marx and Hutter, 2009) for performing molecular dynamics with forces obtained from "on the fly" electronic structure calculations, and schemes for including nuclear quantum effects in molecular dynamics (see Chapter 12). In this section, we present Andersen's original method for the isoenthalpic-isobaric ensemble and then use this idea as the basis for a non-Hamiltonian isothermal-isobaric molecular dynamics approach in Section 5.10.

Andersen's method is based on the remarkably simple yet very elegant idea that the scaling transformation used to derive the pressure

$$\mathbf{s}_i = V^{-1/3}\mathbf{r}_i, \qquad \boldsymbol{\pi}_i = V^{1/3}\mathbf{p}_i \tag{5.9.1}$$

is all we need to derive an isobaric molecular dynamics method. This transformation is used not only to make the volume dependence of the coordinates and momenta explicit but also to promote the volume to a dynamical variable. Moreover, it leads to a force that is used to propagate the volume. In order to make the volume dynamical, we need to define a momentum p_V conjugate to the volume and introduce a kinetic energy $p_V^2/2W$ term into the Hamiltonian. Here, W is a mass-like parameter that determines the time scale of volume motion. Since we already know that the instantaneous pressure estimator is $-\partial\mathcal{H}/\partial V$, we seek a Hamiltonian and associated equations of motion that drive the volume according to the difference between the instantaneous pressure and the external applied pressure P. The Hamiltonian postulated by Andersen is obtained from the standard Hamiltonian for an N-particle system by substituting eqn. (5.9.1) for the coordinates and momenta into the Hamiltonian, adding the volume kinetic energy and an additional term PV for the action of the imaginary "piston" driving the volume fluctuations. Andersen's Hamiltonian is

$$\mathcal{H}_\mathrm{A} = \sum_{i=1}^{N} \frac{V^{-2/3}\boldsymbol{\pi}_i^2}{2m_i} + U(V^{1/3}\mathbf{s}_1, ..., V^{1/3}\mathbf{s}_N) + \frac{p_V^2}{2W} + PV. \tag{5.9.2}$$

The parameter W is determined by a relation similar to eqn. (4.11.2)

$$W = (3N+1)kT\tau_\mathrm{b}^2, \tag{5.9.3}$$

where τ_b is a time scale for the volume motion. The factor of $3N+1$ arises because the barostat scales all N particles and the volume. Equation (5.9.2) is now used to derive equations of motion for generating the isoenthalpic-isobaric ensemble. Applying Hamilton's equations, we obtain

$$\dot{\mathbf{s}}_i = \frac{\partial\mathcal{H}_\mathrm{A}}{\partial\boldsymbol{\pi}_i} = \frac{V^{-2/3}\boldsymbol{\pi}_i}{m_i}$$

$$\dot{\boldsymbol{\pi}}_i = -\frac{\partial\mathcal{H}_\mathrm{A}}{\partial\mathbf{s}_i} = -\frac{\partial U}{\partial(V^{1/3}\mathbf{s}_i)}V^{1/3}$$

$$\dot{V} = \frac{\partial\mathcal{H}_\mathrm{A}}{\partial p_V} = \frac{p_V}{W}$$

$$\dot{p}_V = -\frac{\partial \mathcal{H}_A}{\partial V} = \frac{2}{3}V^{-5/3}\sum_i \frac{\pi_i^2}{m_i} - \frac{1}{3}V^{-2/3}\sum_i \frac{\partial U}{\partial(V^{1/3}\mathbf{s}_i)}\cdot\mathbf{s}_i - P. \qquad (5.9.4)$$

These equations of motion could be integrated numerically using the techniques introduced in Section 3.10 to yield a trajectory in the scaled coordinates. However, it is not always convenient to work in these coordinates, as they do not correspond to the physical coordinates. Fortunately, eqns. (5.9.4) can be easily transformed back into the original Cartesian coordinates by inverting the transformation as follows:

$$\mathbf{s}_i = V^{-1/3}\mathbf{r}_i$$

$$\dot{\mathbf{s}}_i = V^{-1/3}\dot{\mathbf{r}}_i - \frac{1}{3}V^{-4/3}\dot{V}\mathbf{r}_i$$

$$\boldsymbol{\pi}_i = V^{1/3}\mathbf{p}_i$$

$$\dot{\boldsymbol{\pi}}_i = V^{1/3}\dot{\mathbf{p}}_i + \frac{1}{3}V^{-2/3}\dot{V}\mathbf{p}_i. \qquad (5.9.5)$$

Substituting eqns. (5.9.5) into eqns. (5.9.4) yields

$$\dot{\mathbf{r}}_i = \frac{\mathbf{p}_i}{m_i} + \frac{1}{3}\frac{\dot{V}}{V}\mathbf{r}_i$$

$$\dot{\mathbf{p}}_i = -\frac{\partial U}{\partial \mathbf{r}_i} - \frac{1}{3}\frac{\dot{V}}{V}\mathbf{p}_i$$

$$\dot{V} = \frac{p_V}{W}$$

$$\dot{p}_V = \frac{1}{3V}\sum_i \left[\frac{\mathbf{p}_i^2}{m_i} - \frac{\partial U}{\partial \mathbf{r}_i}\cdot\mathbf{r}_i\right] - P. \qquad (5.9.6)$$

Note that the right side of the equation of motion for p_V is simply the difference between the instantaneous pressure estimator of eqn. (4.7.57) or (4.7.58) and the external pressure P. Although eqns. (5.9.6) cannot be derived from a Hamiltonian, they nevertheless possess the important conservation law

$$\mathcal{H}' = \sum_{i=1}^N \frac{\mathbf{p}_i^2}{2m_i} + U(\mathbf{r}_1,...,\mathbf{r}_N) + \frac{p_V^2}{2W} + PV$$

$$= \mathcal{H}(\mathbf{r},\mathbf{p}) + \frac{p_V^2}{2W} + PV, \qquad (5.9.7)$$

and they are incompressible. Here, $\mathcal{H}$ is the physical Hamiltonian of the system. Equations (5.9.6) therefore generate a partition function of the form

$$\Omega_P = \int \mathrm{d}p_V \int_0^\infty \mathrm{d}V \int \mathrm{d}^N\mathbf{p} \int_{D(V)} \mathrm{d}^N\mathbf{r}\, \delta\left(\mathcal{H}(\mathbf{r},\mathbf{p}) + \frac{p_V^2}{2W} + PV - H\right) \qquad (5.9.8)$$

at a pressure P.[2] Equation (5.9.8) is not precisely equivalent to the true isoenthalpic-isobaric partition function given in eqn. (5.3.3) because the conserved energy in eqn. (5.9.7) differs from the true enthalpy by $p_V^2/2W$. However, when the system is equipartitioned, then according to the classical virial theorem, $\langle p_V^2/W \rangle = kT$, and for N very large, this constitutes only a small deviation from the true enthalpy. In fact, this kT is related to the extra kT appearing in the work-virial theorem of eqn. (5.5.8). If the fluctuations in $p_V^2/2W$ are small, then the instantaneous enthalpy $\mathcal{H}(\mathbf{r},\mathbf{p}) + PV$ is confined to a thin shell between H and $H + \Delta$.

In most molecular dynamics calculations, the isoenthalpic-isobaric ensemble is employed only seldom: the most common experimental conditions are constant pressure and temperature. Nevertheless, eqns. (5.9.6) provide the foundation for molecular dynamics algorithms capable of generating an isothermal-isobaric ensemble, which we discuss next.

5.10 Molecular dynamics in the isothermal-isobaric ensemble I: Isotropic volume fluctuations

Since most condensed-phase experiments are carried out under the conditions of constant temperature and pressure (e.g. thermochemistry), the majority of isobaric molecular dynamics calculations are performed in the isothermal-isobaric ensemble. Because N, P, and T are the control variables, we often refer to the NPT ensemble for short. Calculations in the NPT ensemble require one of the canonical methods of Chapter 4 to be grafted onto an isoenthalpic method in order to induce fluctuations in the enthalpy. In this section, we will develop molecular dynamics techniques for isotropic volume fluctuations under isothermal conditions. Following this, we will proceed to generalize the method for anisotropic cell fluctuations.

Although several algorithms have been proposed in the literature for generating an NPT ensemble, they do not all give the correct ensemble distribution function (Martyna *et al.*, 1994; Tuckerman *et al.*, 2001). Therefore, we will restrict ourselves to the approach of Martyna, Tobias, and Klein (1994) (MTK), which has been proved to yield the correct volume distribution. The failure of other schemes is the subject of Problem 5.9.

The starting point for developing the MTK algorithm is eqns. (5.9.6). In order to avoid having to write $\dot{V}/3V$ repeatedly, we introduce, as a convenience, the variable $\epsilon = (1/3)\ln(V/V_0)$, where V_0 is the reference volume appearing in the isothermal-isobaric partition function of eqn. (5.3.22). A momentum p_ϵ corresponding to ϵ can be defined according to $\dot{\epsilon} = p_\epsilon/W = \dot{V}/3V$. Note that in d dimensions, $\epsilon = (1/d)\ln(V/V_0)$ and $p_\epsilon = \dot{V}/dV$. In terms of these variables, eqns. (5.9.6) become, in d dimensions,

$$\dot{\mathbf{r}}_i = \frac{\mathbf{p}_i}{m_i} + \frac{p_\epsilon}{W}\mathbf{r}_i$$

---

[2]If $\sum_i \mathbf{F}_i = -\sum_i \partial U/\partial \mathbf{r}_i = 0$, then an additional conservation law of the form $\mathbf{K} = \mathbf{P}\exp[(1/3)\ln V]$ exists, and the equations will not generate eqn. (5.9.8). Note that the equations of motion in scaled variables, eqns. (5.9.4), do not suffer from this pathology.

$$\dot{\mathbf{p}}_i = -\frac{\partial U}{\partial \mathbf{r}_i} - \frac{p_\epsilon}{W}\mathbf{p}_i$$

$$\dot{V} = \frac{dV\,p_\epsilon}{W}$$

$$\dot{p}_\epsilon = dV(\mathcal{P}^{(\text{int})} - P), \qquad (5.10.1)$$

where $\mathcal{P}^{(\text{int})}$ is the internal pressure estimator of eqn. (4.7.57) or eqn. (4.7.58). Although eqns. (5.10.1) are isobaric, they still lack a proper isothermal coupling and therefore, they do not generate an NPT ensemble. However, we know from Section 4.11 that temperature control can be achieved by coupling eqns. (5.10.1) to a thermostat.

Before we discuss the thermostat coupling, however, we need to analyze eqns. (5.10.1) in greater detail, for in introducing the "convenient" variables ϵ and p_ϵ, we have transformed the incompressible equations (5.9.6) into compressible ones; the compressibility of eqns. (5.10.1) now leads to an incorrect volume dependence in the phase-space measure. Applying the rules of Section 4.10 for analyzing non-Hamiltonian systems, we find that the compressibility of eqns. (5.10.1) is

$$\kappa = \sum_{i=1}^{N}[\nabla_{\mathbf{r}_i} \cdot \dot{\mathbf{r}}_i + \nabla_{\mathbf{p}_i} \cdot \dot{\mathbf{p}}_i] + \frac{\partial \dot{V}}{\partial V}$$

$$= dN\frac{p_\epsilon}{W} - dN\frac{p_\epsilon}{W} + d\frac{p_\epsilon}{W}$$

$$= d\frac{p_\epsilon}{W}$$

$$= \frac{\dot{V}}{V}$$

$$= \frac{d}{dt}\ln\left(\frac{V}{V_0}\right). \qquad (5.10.2)$$

Thus, the function $w(\mathbf{x}) = \ln(V/V_0)$ and the phase-space metric becomes $\sqrt{g} = \exp(-w) = V_0/V$. The inverse volume dependence in the phase-space measure leads to an incorrect volume distribution. The origin of this problem is the volume dependence of the transformation leading to eqns. (5.10.1).

We can make the compressibility vanish, however, by a minor modification of eqns. (5.10.1). All we need is to add a term that yields an extra $-dp_\epsilon/W$ in the compressibility. One way to proceed is to modify the momentum equation and add a term to the p_ϵ equation to ensure conservation of energy. If the momentum equation is modified to read

$$\dot{\mathbf{p}}_i = \tilde{\mathbf{F}}_i - \left(1 + \frac{d}{N_f}\right)\frac{p_\epsilon}{W}\mathbf{p}_i, \qquad (5.10.3)$$

where N_f is the number of degrees of freedom ($dN - N_c$) with N_c the number of constraints, then the compressibility κ will be zero, as required for a proper isobaric ensemble. Here, $\tilde{\mathbf{F}}_i$ is the total force on atom i including any forces of constraint. If $N_c = 0$, then $\tilde{\mathbf{F}}_i = \mathbf{F}_i = -\partial U/\partial \mathbf{r}_i$. In addition, if the p_ϵ equation is modified to read:

$$\dot{p}_\epsilon = dV(\mathcal{P}^{(\text{int})} - P) + \frac{d}{N_f} \sum_{i=1}^{N} \frac{\mathbf{p}_i^2}{m_i}, \tag{5.10.4}$$

then eqns. (5.10.1), together with these two modifications, will conserve eqn. (5.9.7). Since, eqns. (5.10.1), together with eqns. (5.10.3) and (5.10.4), possess the correct phase-space metric and conserved energy, they can now be coupled to a thermostat in order to generate a true isothermal-isobaric ensemble. Choosing the Nosé-Hoover chain approach of Section 4.11, we obtain the equations of motion

$$\dot{\mathbf{r}}_i = \frac{\mathbf{p}_i}{m_i} + \frac{p_\epsilon}{W} \mathbf{r}_i$$

$$\dot{\mathbf{p}}_i = \tilde{\mathbf{F}}_i - \left(1 + \frac{d}{N_f}\right) \frac{p_\epsilon}{W} \mathbf{p}_i - \frac{p_{\eta_1}}{Q_1} \mathbf{p}_i$$

$$\dot{V} = \frac{dVp_\epsilon}{W}$$

$$\dot{p}_\epsilon = dV(\mathcal{P}^{(\text{int})} - P) + \frac{d}{N_f} \sum_{i=1}^{N} \frac{\mathbf{p}_i^2}{m_i} - \frac{p_{\xi_1}}{Q_1'} p_\epsilon$$

$$\dot{\eta}_j = \frac{p_{\eta_j}}{Q_j}, \qquad \dot{\xi}_j = \frac{p_{\xi_j}}{Q_j'}$$

$$\dot{p}_{\eta_j} = G_j - \frac{p_{\eta_{j+1}}}{Q_{j+1}} p_{\eta_j}$$

$$\dot{p}_{\eta_M} = G_M$$

$$\dot{p}_{\xi_j} = G_j' - \frac{p_{\xi_{j+1}}}{Q_{j+1}'} p_{\xi_j}$$

$$\dot{p}_{\xi_M} = G_M', \tag{5.10.5}$$

where the G_j are defined in eqn. (4.12.6). Note that eqns. (5.10.5) possess two Nosé-Hoover chains. One is coupled to the particles and the other to the volume. The reason for this seemingly baroque scheme is that the particle positions and momenta move on a considerably faster time scale than the volume. Thus, for practical applications, they need to be thermalized independently. The volume thermostat forces G_j' are defined in a manner analogous to the particle thermostat forces:

$$G_1' = \frac{p_\epsilon^2}{W} - kT$$

$$G_j' = \frac{p_{\xi_{j-1}}^2}{Q_{j-1}'} - kT. \tag{5.10.6}$$

Equations (5.10.5) are the MTK equations, which have the conserved energy

$$\mathcal{H}' = \sum_{i=1}^{N} \frac{\mathbf{p}_i^2}{2m_i} + U(\mathbf{r}_1, ..., \mathbf{r}_N) + \frac{p_\epsilon^2}{2W} + PV$$

$$+ \sum_{j=1}^{M} \left[\frac{p_{\eta_j}^2}{2Q_j} + \frac{p_{\xi_j}^2}{2Q_j'} + kT\xi_j \right] + N_f kT\eta_1 + kT \sum_{j=2}^{M} \eta_j. \tag{5.10.7}$$

The metric factor associated with these equations is $\sqrt{g} = \exp(dN\eta_1 + \eta_2 + \cdots + \eta_M + \xi_1 + \cdots \xi_M)$. With this metric and eqn. (5.10.7), it is straightforward to prove, using the techniques of Section 4.10, that these equations do, indeed, generate the correct isothermal-isobaric phase-space distribution (see Problem 5.7). Moreover, they can be modified to include a thermostat on each particle or on each degree of freedom ("massive" thermostatting), as discussed in Section 4.11.

To illustrate the use of eqns. (5.10.5), consider the simple example of a particle of mass m moving in a one-dimensional box with length L subject to periodic potential. Let p and q be the momentum and coordinate of the particle, respectively. The potential is given by

$$U(q, L) = \frac{m\omega^2 L^2}{4\pi^2} \left[1 - \cos\left(\frac{2\pi q}{L}\right) \right], \tag{5.10.8}$$

where ω is a parameter having units of inverse time. Such a potential could be used, for example, as a simple model for the motion of particles through a nanowire. We will use eqns. (5.10.5) to determine the position and box-length distributions for a given pressure P and temperature T. These distributions are given by

$$P(q) \propto \int_q^\infty dL \exp\left[-\beta PL\right] \exp\left\{ -\beta \frac{m\omega^2 L^2}{4\pi^2} \left[1 - \cos\left(\frac{2\pi q}{L}\right) \right] \right\}$$

$$P(L) \propto \exp\left[-\beta PL\right] \int_0^L dq \, \exp\left\{ -\beta \frac{m\omega^2 L^2}{4\pi^2} \left[1 - \cos\left(\frac{2\pi q}{L}\right) \right] \right\}$$

$$\propto L \exp\left[-\beta PL\right] \int_0^1 ds \, \exp\left\{ -\beta \frac{m\omega^2 L^2}{4\pi^2} \left[1 - \cos\left(2\pi s\right) \right] \right\}, \tag{5.10.9}$$

where the last line is obtained by introducing the scaled coordinate $s = q/L$. The one-dimensional integrals can be performed using a standard numerical quadrature scheme, yielding "analytical" distributions that can be compared to the simulated ones. The simulations are carried out using a numerical integration scheme that we will present in Section 5.13. Figure 5.3 shows the comparison for the specific case that $\omega = 1$, $m = 1$, $kT = 1$, and $P = 1$. The parameters of the simulation are: $W = 18$, $M = 4$, $Q_j = 1$, $Q_j' = 1$, and $\Delta t = 0.05$. It can be seen that the simulated and analytical distributions match extremely well, indicating that eqns. (5.10.5) generate the correct phase-space distribution.

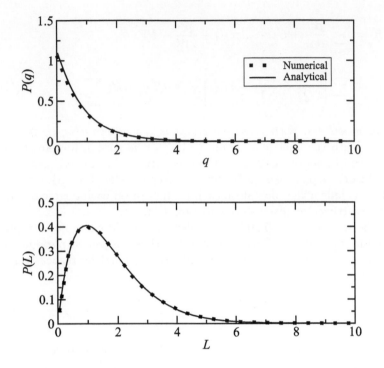

Fig. 5.3 Position and box-length distributions for a particle moving in the one-dimensional potential of eqn. (5.10.8).

5.11 Molecular dynamics in the isothermal-isobaric ensemble II: Anisotropic cell fluctuations

Suppose we wish to map out the space of stable crystal structures for a given substance. We can only do this within a molecular dynamics framework if we can sample different cell shapes. For this reason, the development of molecular dynamics approaches with a fully flexible cell or box is an extremely important problem. We have already laid the groundwork in the isotropic scheme developed above and in our derivation of the pressure tensor estimator in eqn. (5.8.1). The key modification we need here is that the nine components of the box matrix $\mathbf{h}$ must be treated as dynamical variables with nine corresponding momenta. Moreover, we must devise a set of equations of motion whose compressibility leads to the metric factor

$$\sqrt{g} = [\det(\mathbf{h})]^{1-d} \exp\left[dN\eta_1 + \eta_c + d^2\xi_1 + \xi_c\right], \tag{5.11.1}$$

where $\eta_c = \sum_{j=2}^{M} \eta_j$ and $\xi_c = \sum_{j=2}^{M} \xi_j$, as required by the partition function in eqn. (5.7.6).

We begin by defining the 3×3 matrix of box momenta, denoted $\mathbf{p}_g$. $\mathbf{p}_g$ is analogous to p_ϵ in that we let $\mathbf{p}_g/W_g = \dot{\mathbf{h}}\mathbf{h}^{-1}$ where W_g is the time-scale parameter analogous to W in the isotropic case. Rather than repeat the full development presented for the isotropic case, here we will simply propose a set of equations of motion that represent a generalization of eqns. (5.10.5) for fully flexible cells and then prove that they generate the correct distribution. A proposed set of equations of motion is (Martyna *et al.*, 1994)

$$\dot{\mathbf{r}}_i = \frac{\mathbf{p}_i}{m_i} + \frac{\mathbf{p_g}}{W_g}\mathbf{r}_i$$

$$\dot{\mathbf{p}}_i = \tilde{\mathbf{F}}_i - \frac{\mathbf{p_g}}{W_g}\mathbf{p}_i - \frac{1}{N_f}\frac{\mathrm{Tr}\,[\mathbf{p_g}]}{W_g}\mathbf{p}_i - \frac{p_{\eta_1}}{Q_1}\mathbf{p}_i$$

$$\dot{\mathbf{h}} = \frac{\mathbf{p_g}\mathbf{h}}{W_g}$$

$$\dot{\mathbf{p}}_g = \det[\mathbf{h}](\mathbf{P}^{(\mathrm{int})} - \mathbf{I}P) + \frac{1}{N_f}\sum_{i=1}^{N}\frac{\mathbf{p}_i^2}{m_i}\mathbf{I} - \frac{p_{\xi_1}}{Q_1'}\mathbf{p_g}$$

$$\dot{\eta}_j = \frac{p_{\eta_j}}{Q_j}, \qquad\qquad \dot{\xi}_j = \frac{p_{\xi_j}}{Q_j'}$$

$$\dot{p}_{\eta_j} = G_j - \frac{p_{\eta_{j+1}}}{Q_{j+1}}p_{\eta_j}$$

$$\dot{p}_{\eta_M} = G_M$$

$$\dot{p}_{\xi_j} = G_j' - \frac{p_{\xi_{j+1}}}{Q_{j+1}'}p_{\xi_j}$$

$$\dot{p}_{\xi_M} = G_M', \tag{5.11.2}$$

where $\mathbf{P}^{(\mathrm{int})}$ is the internal pressure tensor, whose components are given by eqn. (5.8.1) or (5.8.29), $\mathbf{I}$ is the 3×3 identity matrix, the thermostat forces G_j are given by eqns. (4.12.6), and

$$G_1' = \frac{\mathrm{Tr}\,[\mathbf{p}_g^T\mathbf{p}_g]}{W_g} - d^2kT$$

$$G_j' = \frac{p_{\xi_{j-1}}^2}{Q_{j-1}'} - kT. \tag{5.11.3}$$

The matrix $\mathbf{p}_g^T$ is the transpose of $\mathbf{p_g}$. Equations (5.11.2) have the conserved energy

$$\mathcal{H}' = \sum_{i=1}^{N}\frac{\mathbf{p}_i^2}{2m_i} + U(\mathbf{r}_1, ..., \mathbf{r}_N) + \frac{\mathrm{Tr}\,[\mathbf{p}_g^T\mathbf{p}_g]}{2W_g} + P\det[\mathbf{h}]$$

$$+ \sum_{j=1}^{M}\left[\frac{p_{\eta_j}^2}{2Q_j} + \frac{p_{\xi_j}^2}{2Q_j'}\right] + N_fkT\eta_1 + d^2kT\xi_1 + kT\,(\eta_c + \xi_c). \tag{5.11.4}$$

Furthermore, if $\sum_i \tilde{\mathbf{F}}_i = 0$, *i.e.*, there are no external forces on the system, then when a global thermostat is used on the particles, there is an additional vector conservation law of the form

$$\mathbf{K} = \mathbf{h}\mathbf{P}\,\{\det[\mathbf{h}]\}^{1/N_f}\,e^{\eta_1}, \tag{5.11.5}$$

where $\mathbf{P} = \sum_i \mathbf{p}_i$ is the center-of-mass momentum.

We will now proceed to show that eqns. (5.11.2) generate the ensemble described by eqn. (5.7.7). For the purpose of this analysis, we will assume that there are no constraints on the system, so that $N_f = dN$ and that $\sum_i \mathbf{F}_i \neq 0$. The slightly more complex case that arises when $\sum_i \mathbf{F}_i = 0$ will be left for the reader to ponder in Problem 5.8.

We start by calculating the compressibility of eqns. (5.11.2). Since the matrix multiplications give rise to a mixing among the components of the position and momentum vectors, it is useful to write the equations of motion for $\mathbf{r}_i$, $\mathbf{p}_i$, $\mathbf{h}$, and $\mathbf{p}_\mathrm{g}$ explicitly in terms of their Cartesian components:

$$\dot{r}_{i,\alpha} = \frac{p_{i,\alpha}}{m_i} + \sum_\beta \frac{p_{\mathrm{g},\alpha\beta}}{W_\mathrm{g}} r_{i,\beta}$$

$$\dot{p}_{i,\alpha} = F_{i,\alpha} - \sum_\beta \frac{p_{\mathrm{g},\alpha\beta}}{W_\mathrm{g}} p_{i,\beta} - \frac{1}{dN} \frac{\mathrm{Tr}\,[\mathbf{p}_\mathrm{g}]}{W_\mathrm{g}} p_{i,\alpha} - \frac{p_{\eta_1}}{Q_1} p_{i,\alpha}$$

$$\dot{h}_{\alpha\beta} = \sum_\gamma \frac{p_{\mathrm{g},\alpha\gamma} h_{\gamma\beta}}{W_\mathrm{g}}$$

$$\dot{p}_{\mathrm{g},\alpha\beta} = \det(\mathbf{h}) \left[\mathcal{P}_{\alpha\beta}^{(\mathrm{int})} - P\delta_{\alpha\beta} \right] + \frac{1}{dN} \sum_i \frac{\mathbf{p}_i^2}{m_i} \delta_{\alpha\beta} - \frac{p_{\xi_i}}{Q_1'} p_{\mathrm{g},\alpha\beta}. \qquad (5.11.6)$$

Now, the compressibility is given by

$$\kappa = \sum_{i,\alpha} \left[\frac{\partial \dot{r}_{i,\alpha}}{\partial r_{i,\alpha}} + \frac{\partial \dot{p}_{i,\alpha}}{\partial p_{i,\alpha}} \right] + \sum_{\alpha,\beta} \left[\frac{\partial \dot{h}_{\alpha\beta}}{\partial h_{\alpha\beta}} + \frac{\partial \dot{p}_{\mathrm{g},\alpha\beta}}{\partial p_{\mathrm{g},\alpha\beta}} \right]$$

$$+ \sum_{j=1}^M \left[\frac{\partial \dot{\eta}_j}{\partial \eta_j} + \frac{\partial \dot{p}_{\eta_j}}{\partial p_{\eta_j}} + \frac{\partial \dot{\xi}_j}{\partial \xi_j} + \frac{\partial \dot{p}_{\xi_j}}{\partial p_{\xi_j}} \right]. \qquad (5.11.7)$$

Carrying out the differentiation using eqns. (5.11.7) and (5.11.2), we find that

$$\kappa = N \sum_{\alpha,\beta} \frac{p_{\mathrm{g},\alpha\beta}}{W_\mathrm{g}} \delta_{\alpha\beta} - N \sum_{\alpha,\beta} \frac{p_{\mathrm{g},\alpha\beta}}{W_\mathrm{g}} \delta_{\alpha\beta} - \frac{1}{dN} \frac{\mathrm{Tr}\,[\mathbf{p}_\mathrm{g}]}{W_\mathrm{g}} dN$$

$$- dN \frac{p_{\eta_1}}{Q_1} - \sum_{j=2}^M \frac{p_{\eta_j}}{Q_j} + d \frac{p_{\mathrm{g},\alpha\beta}}{W_\mathrm{g}} \delta_{\alpha\beta} - d^2 \frac{p_{\xi_1}}{Q_1'} - \sum_{j=2}^M \frac{p_{\xi_j}}{Q_j'}$$

$$= -(1-d) \frac{\mathrm{Tr}\,[\mathbf{p}_\mathrm{g}]}{W_\mathrm{g}} - dN \frac{p_{\eta_1}}{Q_1} - d^2 \frac{p_{\xi_1}}{Q_1'} - \sum_{j=2}^M \left[\frac{p_{\eta_j}}{Q_j} + \frac{p_{\xi_j}}{Q_j'} \right]. \qquad (5.11.8)$$

Since $\mathbf{p}_\mathrm{g}/W_\mathrm{g} = \dot{\mathbf{h}}\mathbf{h}^{-1}$,

$$\frac{\mathrm{Tr}\,[\mathbf{p}_\mathrm{g}]}{W_\mathrm{g}} = \mathrm{Tr}\left[\dot{\mathbf{h}}\mathbf{h}^{-1} \right]. \qquad (5.11.9)$$

Using the identity $\det[\mathbf{h}] = \exp\left[\mathrm{Tr}(\ln \mathbf{h}) \right]$, we have

$$\frac{\mathrm{d}}{\mathrm{d}t} \det[\mathbf{h}] = e^{\mathrm{Tr}[\ln \mathbf{h}]} \mathrm{Tr}\left[\dot{\mathbf{h}}\mathbf{h}^{-1} \right] = \det[\mathbf{h}] \mathrm{Tr}\left[\dot{\mathbf{h}}\mathbf{h}^{-1} \right]$$

$$\mathrm{Tr}\left[\dot{\mathbf{h}}\mathbf{h}^{-1}\right] = \frac{1}{\det[\mathbf{h}]}\frac{\mathrm{d}}{\mathrm{d}t}\det[\mathbf{h}] = \frac{\mathrm{d}}{\mathrm{d}t}\ln\left[\det(\mathbf{h})\right]. \tag{5.11.10}$$

Thus, the compressibility becomes

$$\kappa = -(1-d)\frac{\mathrm{d}}{\mathrm{d}t}\ln\left[\det(\mathbf{h})\right] - dN\dot{\eta}_1 - d^2\dot{\xi}_1 - \left[\dot{\eta}_c + \dot{\xi}_c\right], \tag{5.11.11}$$

which leads to the metric in eqn. (5.11.1). Assuming that eqn. (5.11.4) is the only conservation law, then by combining the metric in eqn. (5.11.1) with eqn. (5.11.4) and inserting these into eqn. (4.10.21) we obtain

$$\mathcal{Z} = \int \mathrm{d}^N\mathbf{p}\,\mathrm{d}^N\mathbf{r}\,\mathrm{d}\mathbf{h}\,\mathrm{d}\mathbf{p_g}\,\mathrm{d}\eta_1\,\mathrm{d}\eta_c\,\mathrm{d}\xi_1\,\mathrm{d}\xi_c\,\mathrm{d}^M p_\eta \mathrm{d}^M p_\xi\,[\det(\mathbf{h})]^{1-d}\,e^{dN\eta_1 + \eta_c}e^{d^2\xi_1 + \xi_c}$$

$$\times\,\delta\left(\mathcal{H}(\mathbf{r},\mathbf{p}) + \sum_{j=1}^{M}\left[\frac{p_{\eta_j}^2}{2Q_j} + \frac{p_{\xi_j}^2}{2Q_j'}\right] + N_f kT\eta_1 + d^2 kT\xi_1 + kT\left[\eta_c + \xi_c\right]\right.$$

$$\left. + \frac{\mathrm{Tr}\left[\mathbf{p_g}^T\mathbf{p_g}\right]}{2W_\mathrm{g}} + P\det[\mathbf{h}] - H\right). \tag{5.11.12}$$

If we now integrate over η_1 using the δ-function, we find

$$\mathcal{Z} \propto \int \mathrm{d}\mathbf{h}\,[\det(\mathbf{h})]^{1-d}\,e^{-\beta P\det(\mathbf{h})}\,\mathrm{d}^N\mathbf{p}\,\mathrm{d}^N\mathbf{r}\,e^{-\beta\mathcal{H}(\mathbf{r},\mathbf{p})}, \tag{5.11.13}$$

where the constant of proportionality includes uncoupled integrations over the remaining thermostat/barostat variables. Thus, the correct isothermal-isobaric partition function for fully flexible cells is recovered.

5.12 Atomic and molecular virials

The isotropic pressure estimator in eqn. (5.8.28) and pressure tensor estimator in eqn. (5.8.1) were derived assuming a scaling or matrix multiplication of all atomic positions. The resulting virial term in the estimator

$$\sum_{i=1}^{N}\mathbf{r}_i \cdot \mathbf{F}_i$$

is, therefore, known as an *atomic virial*. Although mathematically correct and physically sensible for purely atomic systems, the atomic virial might seem to be an overkill for molecular systems. In a collection of molecules, assuming no constraints, the force $\mathbf{F}_i$ appearing in the atomic virial contains both intramolecular and intermolecular components. If the size of the molecule is small compared to its container, it is more intuitive to think of the coordinate scaling (or multiplication by the cell matrix) as acting only on the centers of mass of the molecules rather than on each atom individually. That is, the scaling should only affect the relative positions of the molecules rather than the bond lengths and angles within each molecule. In fact, an alternative pressure estimator can be derived by scaling only the positions of the molecular centers of mass rather than individual atomic positions.

Consider a system of N molecules with centers of mass at positions $\mathbf{R}_1, ..., \mathbf{R}_N$. For isotropic volume fluctuations, we would define the scaled coordinates $\mathbf{S}_1, ..., \mathbf{S}_N$ of the centers of mass by

$$\mathbf{S}_i = V^{-1/d}\mathbf{R}_i. \tag{5.12.1}$$

If each molecule has n atoms with masses $m_{i,1}, ..., m_{i,n}$ and atomic positions $\mathbf{r}_{i,1}, ..., \mathbf{r}_{i,n}$, then the center-of-mass position is

$$\mathbf{R}_i = \frac{\sum_{\alpha=1}^{n} m_{i,\alpha}\mathbf{r}_{i,\alpha}}{\sum_{\alpha=1}^{n} m_{i,\alpha}}. \tag{5.12.2}$$

We saw in Section 1.11 that the center-of-mass motion of each molecule can be separated from internal motion relative to a body-fixed frame. Thus, if the derivation leading up to eqn. (4.7.57) is repeated using the transformation in eqn. (5.12.1), the following pressure estimator is obtained:

$$\mathcal{P}_{\text{mol}}(\mathbf{P}, \mathbf{R}) = \frac{1}{dV} \sum_{i=1}^{N} \left[\frac{\mathbf{P}_i^2}{M_i} + \mathbf{R}_i \cdot \mathcal{F}_i \right], \tag{5.12.3}$$

where M_i is the mass of the ith molecule, and $\mathbf{P}_i$ is the momentum of its center of mass:

$$\mathbf{P}_i = \sum_{\alpha=1}^{n} \mathbf{p}_{i,\alpha}, \tag{5.12.4}$$

and $\mathcal{F}_i$ is the force on the center of mass

$$\mathcal{F}_i = \sum_{\alpha=1}^{n} \mathbf{F}_{i,\alpha}. \tag{5.12.5}$$

The virial term appearing in eqn. (5.12.3)

$$\sum_{i=1}^{N} \mathbf{R}_i \cdot \mathcal{F}_i$$

is known as the *molecular virial*.

Given the molecular virial, it is straightforward to derive a molecular dynamics algorithm for the isoenthalpic-isobaric ensemble that uses a molecular virial. The key feature of this algorithm is that the barostat coupling acts only on the center-of-mass positions and momenta. Assuming three spatial dimensions and no constraints between the molecules, the equations of motion take the form

$$\dot{\mathbf{r}}_{i,\alpha} = \frac{\mathbf{p}_{i,\alpha}}{m_{i,\alpha}} + \frac{p_\epsilon}{W}\mathbf{R}_i$$

$$\dot{\mathbf{p}}_{i,\alpha} = \mathbf{F}_{i,\alpha} - \left(1 + \frac{1}{N}\right) \frac{p_\epsilon}{W} \frac{m_{i,\alpha}}{M_i}\mathbf{P}_i$$

$$\dot{V} = \frac{dVp_\epsilon}{W}$$

$$\dot{p}_\epsilon = dV\left(\mathcal{P}_{\mathrm{mol}} - P\right) + \frac{1}{N}\sum_{i=1}^{N}\frac{\mathbf{P}_i^2}{M_i}. \tag{5.12.6}$$

These equations have the conserved energy

$$\mathcal{H}' = \sum_{i,\alpha}\frac{\mathbf{p}_{i,\alpha}^2}{2m_{i,\alpha}} + U(\mathbf{r}) + \frac{p_\epsilon^2}{2W} + PV, \tag{5.12.7}$$

where the $\mathbf{r}$ in $U(\mathbf{r})$ denotes the full set of atomic positions. The proof that these equations generate the correct isobaric-isoenthalpic ensemble is left as an exercise in Problem 5.10. These equations can easily be generalized for the isothermal-isobaric ensemble with a molecular virial by coupling Nosé-Hoover chain thermostats as in eqns. (5.10.5). Moreover, starting from the transformation for anisotropic cell fluctuations

$$\mathbf{S}_i = \mathbf{h}^{-1}\mathbf{R}_i, \tag{5.12.8}$$

the algorithm in eqn. (5.12.6) can be turned into an algorithm capable of handling anisotropic volume fluctuations with a molecular virial.

5.13 Integrating the Martyna-Tobias-Klein equations of motion

Integrating the MTK equations is only slightly more difficult than integrating the NHC equations and builds on the methodology we have already developed. We begin with the isotropic case, and for the present, we consider a system in which no constraints are imposed so that $N_f = dN$ and $\tilde{\mathbf{F}}_i = \mathbf{F}_i = -\partial U/\partial \mathbf{r}_i$. In Section 5.14, we will see how to account for forces of constraint. We first write the total Liouville operator as

$$iL = iL_1 + iL_2 + iL_{\epsilon,1} + iL_{\epsilon,2} + iL_{\mathrm{NHC-baro}} + iL_{\mathrm{NHC-part}}, \tag{5.13.1}$$

where

$$iL_1 = \sum_{i=1}^{N}\left[\frac{\mathbf{p}_i}{m_i} + \frac{p_\epsilon}{W}\mathbf{r}_i\right]\cdot\frac{\partial}{\partial\mathbf{r}_i}$$

$$iL_2 = \sum_{i=1}^{N}\left[\mathbf{F}_i - \alpha\frac{p_\epsilon}{W}\mathbf{p}_i\right]\cdot\frac{\partial}{\partial\mathbf{p}_i}$$

$$iL_{\epsilon,1} = \frac{p_\epsilon}{W}\frac{\partial}{\partial\epsilon}$$

$$iL_{\epsilon,2} = G_\epsilon\frac{\partial}{\partial p_\epsilon}, \tag{5.13.2}$$

and the operators $iL_{\mathrm{NHC-part}}$ and $iL_{\mathrm{NHC-baro}}$ are the particle and barostat Nosé-Hoover chain Liouville operators, respectively, which are defined in the last two lines of eqn. (4.12.5). In eqn. (5.13.2), $\alpha = 1 + d/N_f = 1 + 1/N$, and

$$G_\epsilon = \alpha \sum_{i=1}^{N} \frac{\mathbf{p}_i^2}{m_i} + \sum_{i=1}^{N} \mathbf{r}_i \cdot \mathbf{F}_i - dV \frac{\partial U}{\partial V} - dPV. \tag{5.13.3}$$

The propagator is factorized following the scheme of eqn. (4.12.8) as

$$\exp(iL\Delta t) = \exp\left(iL_{\mathrm{NHC-baro}}\frac{\Delta t}{2}\right) \exp\left(iL_{\mathrm{NHC-part}}\frac{\Delta t}{2}\right)$$

$$\times \exp\left(iL_{\epsilon,2}\frac{\Delta t}{2}\right) \exp\left(iL_2\frac{\Delta t}{2}\right)$$

$$\times \exp\left(iL_{\epsilon,1}\Delta t\right) \exp\left(iL_1\Delta t\right)$$

$$\times \exp\left(iL_2\frac{\Delta t}{2}\right) \exp\left(iL_{\epsilon,2}\frac{\Delta t}{2}\right)$$

$$\times \exp\left(iL_{\mathrm{NHC-part}}\frac{\Delta t}{2}\right) \exp\left(iL_{\mathrm{NHC-baro}}\frac{\Delta t}{2}\right) + \mathcal{O}(\Delta t^3) \tag{5.13.4}$$

(Tuckerman *et al.*, 2006). In evaluating the action of this propagator, the Suzuki-Yoshida decomposition developed in eqns. (4.12.17) and (4.12.19) is applied to the operators $\exp(iL_{\mathrm{NHC-baro}}\Delta t/2)$ and $\exp(iL_{\mathrm{NHC-part}}\Delta t/2)$. The operators $\exp(iL_{\epsilon,1}\Delta t)$ and $\exp(iL_{\epsilon,2}\Delta t/2)$ are simple translation operators. The operators $\exp(iL_1\Delta t)$ and $\exp(iL_2\Delta t/2)$ are somewhat more complicated than their microcanonical or canonical ensemble counterparts due to the barostat coupling and need further explication. The action of the operator $\exp(iL_1\Delta t)$ can be determined by solving the first-order differential equation

$$\dot{\mathbf{r}}_i = \boldsymbol{v}_i + v_\epsilon \mathbf{r}_i, \tag{5.13.5}$$

keeping $\boldsymbol{v}_i = \mathbf{p}_i/m_i$ and $v_\epsilon = p_\epsilon/W$ constant with an arbitrary initial condition $\mathbf{r}_i(0)$ and then evaluating the solution at $t = \Delta t$. Note that $\boldsymbol{v}_i$ must not be confused with the atomic velocity $\mathbf{v}_i = \dot{\mathbf{r}}_i = \boldsymbol{v}_i + v_\epsilon \mathbf{r}_i$. $\boldsymbol{v}_i = \mathbf{p}_i/m_i$, introduced here for notational convenience to avoid having to write $\mathbf{p}_i/m_i$ explicitly everywhere. Solving eqn. (5.13.5) yields the finite-difference expression

$$\mathbf{r}_i(\Delta t) = \mathbf{r}_i(0)e^{v_\epsilon \Delta t} + \Delta t \boldsymbol{v}_i e^{v_\epsilon \Delta t/2} \frac{\sinh(v_\epsilon \Delta t/2)}{v_\epsilon \Delta t/2}. \tag{5.13.6}$$

Similarly, the action of $\exp(iL_2\Delta t/2)$ can be determined by solving the differential equation

$$\dot{\boldsymbol{v}}_i = \frac{\mathbf{F}_i}{m_i} - \alpha v_\epsilon \boldsymbol{v}_i, \tag{5.13.7}$$

keeping $\mathbf{F}_i$ and v_ϵ constant with an arbitrary initial condition $\boldsymbol{v}_i(0)$ and then evaluating the solution at $t = \Delta t/2$. This yields the evolution

$$\boldsymbol{v}_i(\Delta t/2) = \boldsymbol{v}_i(0)e^{-\alpha v_\epsilon \Delta t/2} + \frac{\Delta t}{2m_i}\mathbf{F}_i e^{-\alpha v_\epsilon \Delta t/4} \frac{\sinh(\alpha v_\epsilon \Delta t/4)}{\alpha v_\epsilon \Delta t/4}. \tag{5.13.8}$$

In practice, the factor $\sinh(x)/x$ should be evaluated by a power series for small x to avoid numerical instabilities.[3]

Equations (5.13.4), (5.13.6) and (5.13.8), together with the Suzuki-Yoshida factorization of the thermostat operators, completely define an integrator for eqns. (5.10.5). The integrator can be easily coded using the direct translation technique.

Integrating eqns. (5.11.2) for the fully flexible case employs the same basic factorization scheme as in eqn. (5.13.4). First, we decompose the total Liouville operator as

$$iL = iL_1 + iL_2 + iL_{\text{g},1} + iL_{\text{g},2} + iL_{\text{NHC}-\text{baro}} + iL_{\text{NHC}-\text{part}}, \tag{5.13.10}$$

where

$$iL_1 = \sum_{i=1}^{N} \left[\frac{\mathbf{p}_i}{m_i} + \frac{\mathbf{p}_\text{g}}{W_\text{g}} \mathbf{r}_i \right] \cdot \frac{\partial}{\partial \mathbf{r}_i}$$

$$iL_2 = \sum_{i=1}^{N} \left[\mathbf{F}_i - \left(\frac{\mathbf{p}_\text{g}}{W_\text{g}} + \frac{1}{N_f} \frac{\text{Tr}\,[\mathbf{p}_\text{g}]}{W_\text{g}} \mathbf{I} \right) \mathbf{p}_i \right] \cdot \frac{\partial}{\partial \mathbf{p}_i}$$

$$iL_{\text{g},1} = \frac{\mathbf{p}_\text{g}\mathbf{h}}{W_\text{g}} \cdot \frac{\partial}{\partial \mathbf{h}}$$

$$iL_{\text{g},2} = \mathbf{G}_\text{g} \frac{\partial}{\partial \mathbf{p}_\text{g}}, \tag{5.13.11}$$

with

$$G_\text{g} = \det[\mathbf{h}](\mathbf{P}^{(\text{int})} - \mathbf{I}P) + \frac{1}{N_f} \sum_{i=1}^{N} \frac{\mathbf{p}_i^2}{m_i} \mathbf{I}. \tag{5.13.12}$$

The propagator is factorized exactly as in eqn. (5.13.4) with the contributions to iL_ϵ replaced by the contributions to iL_g. In the flexible case, the application of the operators $\exp(iL_1 \Delta t)$ and $\exp(iL_2 \Delta t/2)$ requires solution of the following matrix-vector equations:

$$\dot{\mathbf{r}}_i = \mathbf{v}_i + \mathbf{v}_\text{g} \mathbf{r}_i \tag{5.13.13}$$

$$\dot{\mathbf{v}}_i = \frac{\mathbf{F}_i}{m_i} - \mathbf{v}_\text{g} \mathbf{v}_i - b \text{Tr}\,[\mathbf{v}_\text{g}]\, \mathbf{v}_i, \tag{5.13.14}$$

where $\mathbf{v}_\text{g} = \mathbf{p}_\text{g}/W_\text{g}$, and $b = 1/N_f$. In order to solve eqn. (5.13.13), we introduce a transformation

$$\mathbf{x}_i = \mathbf{O}\mathbf{r}_i, \tag{5.13.15}$$

---

[3]The power series expansion of $\sinh(x)/x$ up to tenth order is

$$\frac{\sinh(x)}{x} \approx \sum_{n=0}^{5} a_{2n} x^{2n}, \tag{5.13.9}$$

where $a_0 = 1$, $a_2 = 1/6$, $a_4 = 1/120$, $a_6 = 1/5040$, $a_8 = 1/362880$, $a_{10} = 1/39916800$.

where $\mathbf{O}$ is a constant orthogonal matrix. We also let $\mathbf{u}_i = \mathbf{O}\mathbf{v}_i$. Since $\mathbf{O}$ is orthogonal, it satisfies $\mathbf{O}^\mathsf{T}\mathbf{O} = \mathbf{I}$. Introducing this transformation into eqn. (5.13.13) yields

$$\mathbf{O}\dot{\mathbf{r}}_i = \mathbf{O}\mathbf{v}_i + \mathbf{O}\mathbf{v}_\mathrm{g}\mathbf{r}_i$$

$$\dot{\mathbf{x}}_i = \mathbf{u}_i + \mathbf{O}\mathbf{v}_\mathrm{g}\mathbf{O}^\mathsf{T}\mathbf{O}\mathbf{r}_i = \mathbf{u}_i + \mathbf{O}\mathbf{v}_\mathrm{g}\mathbf{O}^\mathsf{T}\mathbf{x}_i, \tag{5.13.16}$$

where the second line follows from the orthogonality of $\mathbf{O}$. Now, since the pressure tensor is symmetric, $\mathbf{v}_\mathrm{g}$ is also symmetric. Therefore, it is possible to choose $\mathbf{O}$ to be the orthogonal matrix that diagonalizes $\mathbf{v}_\mathrm{g}$ according to

$$\mathbf{v}_\mathrm{g}^{(\mathrm{d})} = \mathbf{O}\mathbf{v}_\mathrm{g}\mathbf{O}^\mathsf{T}, \tag{5.13.17}$$

where $\mathbf{v}_\mathrm{g}^{(\mathrm{d})}$ is a diagonal matrix with the eigenvalues of $\mathbf{v}_\mathrm{g}$ on the diagonal. The columns of $\mathbf{O}$ are just the eigenvectors of $\mathbf{v}_\mathrm{g}$. Let λ_α, $\alpha = 1, 2, 3$ be the eigenvectors of $\mathbf{v}_\mathrm{g}$. Since $\mathbf{v}_\mathrm{g}$ is symmetric, its eigenvalues are real. In this representation, the three components of $\mathbf{x}_i$ are uncoupled in eqn. (5.13.16) and can be solved independently using eqn. (5.13.6). The solution at $t = \Delta t$ for each component of $\mathbf{x}_i$ is

$$x_{i,\alpha}(\Delta t) = x_{i,\alpha}(0)\mathrm{e}^{\lambda_\alpha \Delta t} + \Delta t v_{i,\alpha}\mathrm{e}^{\lambda_\alpha \Delta t/2}\frac{\sinh(\lambda_\alpha \Delta t/2)}{\lambda_\alpha \Delta t/2}. \tag{5.13.18}$$

Transforming back to $\mathbf{r}_i$, we find that

$$\mathbf{r}_i(\Delta t) = \mathbf{O}^\mathsf{T}\mathbf{D}\mathbf{O}\mathbf{r}_i(0) + \Delta t \mathbf{O}^\mathsf{T}\tilde{\mathbf{D}}\mathbf{O}\mathbf{v}_i, \tag{5.13.19}$$

where the matrices $\mathbf{D}$ and $\tilde{\mathbf{D}}$ have the elements

$$D_{\alpha\beta} = \mathrm{e}^{\lambda_\alpha \Delta t}\delta_{\alpha\beta}$$

$$\tilde{D}_{\alpha\beta} = \mathrm{e}^{\lambda_\alpha \Delta t/2}\frac{\sinh(\lambda_\alpha \Delta t/2)}{\lambda_\alpha \Delta t/2}\delta_{\alpha\beta}. \tag{5.13.20}$$

In a similar manner, eqn. (5.13.14) can be solved for $\mathbf{v}_i(t)$ and the solution evaluated at $t = \Delta t/2$ with the result

$$\mathbf{v}_i(\Delta t/2) = \mathbf{O}^\mathsf{T}\boldsymbol{\Delta}\mathbf{O}\mathbf{v}_i(0) + \frac{\Delta t}{2m_i}\mathbf{O}^\mathsf{T}\tilde{\boldsymbol{\Delta}}\mathbf{O}\mathbf{F}_i, \tag{5.13.21}$$

where the matrices $\boldsymbol{\Delta}$ and $\tilde{\boldsymbol{\Delta}}$ are given by their elements

$$\Delta_{\alpha\beta} = \mathrm{e}^{-(\lambda_\alpha + b\mathrm{Tr}[\mathbf{v}_\mathrm{g}])\Delta t/2}\delta_{\alpha\beta}$$

$$\tilde{\Delta}_{\alpha\beta} = \mathrm{e}^{-(\lambda_\alpha + b\mathrm{Tr}[\mathbf{v}_\mathrm{g}])\Delta t/4}\frac{\sinh[(\lambda_\alpha + b\mathrm{Tr}[\mathbf{v}_\mathrm{g}])\Delta t/4]}{(\lambda_\alpha + b\mathrm{Tr}[\mathbf{v}_\mathrm{g}])\Delta t/4}\delta_{\alpha\beta}. \tag{5.13.22}$$

A technical comment is in order at this point. As noted in Section 5.7, if all nine elements of the box matrix $\mathbf{h}$ are allowed to vary independently, then the simulation box could execute overall rotational motion, which makes analysis of molecular dynamics trajectories difficult. Overall cell rotations can be eliminated straightforwardly, however (Tobias *et al.*, 1993). One scheme for accomplishing this is to restrict the box matrix to be upper (or lower) triangular only. Consider, for example, what an upper triangular box matrix represents. According to eqn. (5.7.2), if **h** is upper triangular, then the vector **a** has only one nonzero component, which is its x-component. Hence, **h** lies entirely along the x direction. Similarly, **b** lies entirely in the x-y plane. Only **c** has complete freedom. With the base of the box firmly rooted in the x-y plane with its **a** vector pinned to the x-axis, overall rotations of the cell are eliminated. The other option, which is preferable when the system is subject to holonomic constraints, is explicit symmetrization of the pressure tensor $P_{\alpha\beta}^{(\mathrm{int})}$. That is, we can simply replace occurrences of $P_{\alpha\beta}^{(\mathrm{int})}$ in eqns. (5.11.2) with $\tilde{P}_{\alpha\beta}^{(\mathrm{int})} = (P_{\alpha\beta}^{(\mathrm{int})} + P_{\beta\alpha}^{(\mathrm{int})})/2$. This has the effect of ensuring that $\mathbf{p}_\mathrm{g}$ and $\mathbf{v}_\mathrm{g}$ are symmetric matrices. If the initial conditions are chosen such that the angular momentum of the cell is initially zero, then the cell should not rotate. Both techniques can actually be derived using simple holonomic constraints and Lagrange undetermined multipliers (see Problem 5.15). When the number of degrees of freedom in the cell matrix is restricted, factors of d^2 in eqns. (5.11.3) and (5.11.4) must be replaced by the correct number of degrees of freedom. If overall cell rotations are eliminated, then this number is $d^2 - d$.

The new NPT integrator can also be applied within the multiple time-step RESPA framework of Section 3.11. For two time steps, δt and $\Delta t = n\delta t$, the following contributions to the total Liouville operator are defined as

$$iL_1 = \sum_{i=1}^{N} \left[\frac{\mathbf{p}_i}{m_i} + \frac{p_\epsilon}{W}\mathbf{r}_i \right] \cdot \frac{\partial}{\partial \mathbf{r}_i}$$

$$iL_2^{(\mathrm{fast})} = \sum_{i=1}^{N} \left[\mathbf{F}_i^{(\mathrm{fast})} - \alpha\frac{p_\epsilon}{W}\mathbf{p}_i \right] \cdot \frac{\partial}{\partial \mathbf{p}_i}$$

$$iL_2^{(\mathrm{slow})} = \sum_{i=1}^{N} \mathbf{F}_i^{(\mathrm{slow})} \cdot \frac{\partial}{\partial \mathbf{p}_i}$$

$$iL_{\epsilon,1} = \frac{p_\epsilon}{W}\frac{\partial}{\partial \epsilon}$$

$$iL_{\epsilon,2}^{(\mathrm{fast})} = G_\epsilon^{(\mathrm{fast})}\frac{\partial}{\partial p_\epsilon}$$

$$iL_{\epsilon,2}^{(\mathrm{slow})} = G_\epsilon^{(\mathrm{slow})}\frac{\partial}{\partial p_\epsilon}, \tag{5.13.23}$$

where fast and slow components are designated with superscripts with

$$G_\epsilon^{(\mathrm{fast})} = \alpha\sum_i \frac{\mathbf{p}_i^2}{m_i} + \sum_{i=1}^{N}\mathbf{r}_i \cdot \mathbf{F}_i^{(\mathrm{fast})} - dV\frac{\partial U^{(\mathrm{fast})}}{\partial V} - dP^{(\mathrm{fast})}V \tag{5.13.24}$$

$$G_\epsilon^{(\text{slow})} = \sum_{i=1}^{N} \mathbf{r}_i \cdot \mathbf{F}_i^{(\text{slow})} - dV \frac{\partial U^{(\text{slow})}}{\partial V} - dP^{(\text{slow})} V. \tag{5.13.25}$$

The variables $P^{(\text{fast})}$ and $P^{(\text{slow})}$ are external pressure components corresponding to the fast and slow virial contributions and must be chosen such that $P = P^{(\text{fast})} + P^{(\text{slow})}$. Although the subdivision of the pressure is arbitrary, a physically meaningful choice can be made. One possibility is to perform a short calculation with a single time step and compute the contributions to the pressure from

$$P^{(\text{fast})} = \left\langle \frac{1}{dV} \left\{ \sum_{i=1}^{N} \left[\frac{\mathbf{p}_i^2}{m_i} + \mathbf{r}_i \cdot \mathbf{F}_i^{(\text{fast})} \right] - dV \frac{\partial U^{(\text{fast})}}{\partial V} \right\} \right\rangle$$

$$P^{(\text{slow})} = \left\langle \frac{1}{dV} \left\{ \sum_{i=1}^{N} \mathbf{r}_i \cdot \mathbf{F}_i^{(\text{slow})} - dV \frac{\partial U^{(\text{slow})}}{\partial V} \right\} \right\rangle, \tag{5.13.26}$$

that is, using the definitions of the reference system and correction contributions to the internal pressure. Another simple choice is

$$P^{(\text{fast})} = \frac{n}{n+1} P$$

$$P^{(\text{slow})} = \frac{1}{n+1} P. \tag{5.13.27}$$

The factorized propagator then takes the form

$$\exp(iL\Delta t) = \exp\left(iL_{\text{NHC-baro}} \frac{\Delta t}{2} \right) \exp\left(iL_{\text{NHC-part}} \frac{\Delta t}{2} \right)$$

$$\times \exp\left(iL_{\epsilon,2}^{(\text{slow})} \frac{\Delta t}{2} \right) \exp\left(iL_2^{(\text{slow})} \frac{\Delta t}{2} \right)$$

$$\times \left[\exp\left(iL_2^{(\text{fast})} \frac{\delta t}{2} \right) \exp\left(iL_{\epsilon,2}^{(\text{fast})} \frac{\delta t}{2} \right) \right.$$

$$\times \exp\left(iL_{\epsilon,1} \delta t \right) \exp\left(iL_1 \delta t \right)$$

$$\times \left. \exp\left(iL_{\epsilon,2}^{(\text{fast})} \frac{\delta t}{2} \right) \exp\left(iL_2^{(\text{fast})} \frac{\delta t}{2} \right) \right]^n$$

$$\times \exp\left(iL_2^{(\text{slow})} \frac{\Delta t}{2} \right) \exp\left(iL_{\epsilon,2}^{(\text{slow})} \frac{\Delta t}{2} \right)$$

$$\times \exp\left(iL_{\text{NHC-part}} \frac{\Delta t}{2} \right) \exp\left(iL_{\text{NHC-baro}} \frac{\Delta t}{2} \right) + \mathcal{O}(\Delta t^3). \tag{5.13.28}$$

Note that because G_ϵ depends on the forces $\mathbf{F}_i$, it is necessary to update both the particles and the barostat in the reference system.

The integrators presented in this section can be generalized to handle systems with constraints under constant pressure. It is not entirely straightforward, however, because self-consistency conditions arise from the nonlinearity of some of the operators. A detailed discussion of the implementation of constraints under conditions of constant pressure can be found in Section 5.14. Note that these integrators can also be formulated using the "middle" scheme discussed in Section 4.12.

5.13.1 Example: Liquid argon at constant pressure

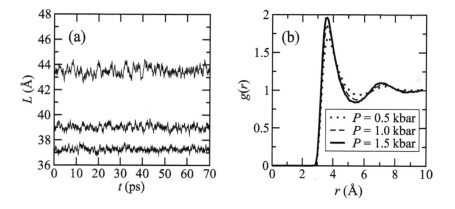

Fig. 5.4 (a) Box-length fluctuations at pressures of $P = 0.5$ kbar (top curve), $P = 1.0$ kbar (middle curve), and $P = 1.5$ kbar (bottom curve), respectively. (b) Radial distribution functions at each of the three pressures.

As an illustrative example of molecular dynamics in the isothermal-isobaric ensemble, we consider first the argon system of Section 3.14.2. Three simulations at applied external pressures of 0.5 kbar, 1.0 kbar, and 1.5 kbar and a temperature of 300 K are carried out, and the radial distribution functions computed at each pressure. The parameters of the Lennard-Jones potential are described in Section 3.14.2, together with the integration time step used in eqn. (5.13.4). Temperature control is achieved using the "massive" Nosé-Hoover chain scheme of Section 4.11. The values of τ for the particle and barostat Nosé-Hoover chains are 100.0 fs and 1000.0 fs, respectively, while $\tau_b = 500.0$ fs. Nosé-Hoover chains of length $M = 4$ are employed using $n_{sy} = 7$ and $n = 4$ in eqn. (4.12.17). Each simulation is 75 ps in length and carried out in a cubic box with periodic boundary conditions subject only to isotropic volume fluctuations. In Fig. 5.4(a), we show the fluctuations in the box length at each pressure, while in Fig. 5.4(b), we show the radial distribution functions obtained at each pressure. Both panels exemplify the expected behavior of the system. As the pressure increases, the box length decreases. Similarly, as the pressure increases, the liquid becomes more structured, and the first and second peaks in the radial distribution function become sharper. Figure 5.5 shows the density distribution (in reduced units) obtained form

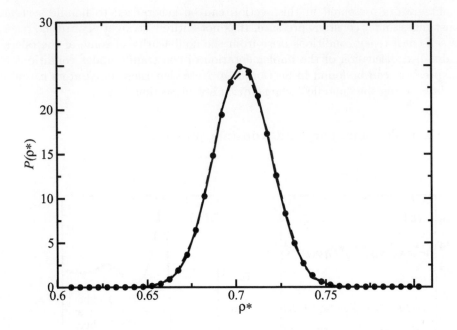

Fig. 5.5 Density distribution for the argon system at $P = 0.5$ kbar for two different values of τ_b (Tuckerman *et al.*, 2006). The solid curve with filled circles represents the fit to the Gaussian form in eqn. (5.13.29).

the simulation at $P = 0.5$ kbar ($P^* = P\sigma^3/\epsilon = 1.279$). The solid and dashed curves correspond to τ_b values of 500.0 fs and 5000.0 fs, respectively. It can be seen that the distribution is fairly sharply peaked in both cases around a density value $\rho^* \approx 0.704$, and that the distribution is only sensitive to the value of τ_b near the peak. Interestingly, the distribution can be fit very accurately to a Gaussian form,

$$P_{\mathrm{G}}(\rho^*) = \frac{1}{\sqrt{2\pi\sigma^2}} e^{-(\rho^* - \rho_0)^2/2\sigma^2} \tag{5.13.29}$$

with a width $\sigma = 0.01596$ and average $\rho_0 = 0.7038$. Such a fit is shown in circles on the solid curve in Fig. 5.5.

5.13.2 Example: Structural transition of paracetamol crystal as a function of temperature

Structural diversity abounds in nature, and an area where this structural diversity has a significant impact is the study of organic molecular crystals. An important issue in the study of these crystals is *polymorphism*, which refers to the ability of a compound to crystallize into multiple forms. In some systems under certain conditions, the land-scape of polymorphs may contain multiple thermodynamically close-lying structures. When the most useful form is not the lowest free-energy structure but a metastable

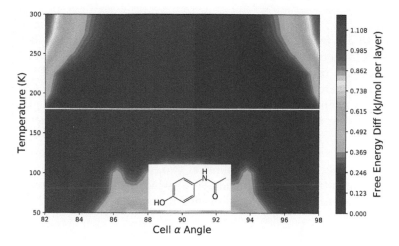

Fig. 5.6 Contours of the Gibbs free energy of paracetamol form III as a function of temperature T and cell angle α (see Fig. 5.2); the molecular diagram of paracetamol is shown in the box in the lower portion of the figure. The white horizontal line indicates the temperature at which the transition from monoclinic to orthorhombic occurs (reprinted with permission from Hong *et al.*, *Cryst. Growth & Design*, **2**, 886 (2021), copyright American Chemical Society).

one, then an unforeseen transition to a more stable crystal structure can cause pharmaceuticals to fail, affect the performance of organic semiconductors, or change the lethality of a contact insecticide (Yang *et al.*, 2017; Yang *et al.*, 2020; Zhu *et al.*, 2021).

When pharmaceutical formulations are created from compressed powders of molecular crystals (the process of "tabletting"), the cost of manufacturing these formulations correlates with the ease of crystallization and compressibility of the resulting crystal. Paracetamol (see Fig. 5.6), also known sometimes as acetaminophen, is an active pharmaceutical compound used as an analgesic. In total, nine different polymorphs of paracetamol have been identified, as discussed by Shtukenberg *et al.* (2019), with forms I, II, and IIII existing stably under ambient conditions. Of these, paracetamol I is used to create pharmaceutical formulations. Paracetamol I possesses a herringbone-like structure, which has a low compressibility and, therefore, is not ideal for tabletting. Forms II and III, on the other hand, are composed of stacked layers, which are more compressible and, therefore, much better for tabletting. Unfortunately, forms II and III are difficult to crystallize. Form III is particularly interesting in that the stacked layers are arranged in pairs. Intermolecular interactions within each pair are stronger than those between the layer pairs, which means that these layer pairs can slip past each other, thus creating a difference between low- and high-temperature structures of form III. At low temperatures (up to around 175 K), form III appears as a monoclinic structure with the α angle (see Fig. 5.2) taking on values around 82–84°, while at temperatures above these, it is thought that the structure can be characterized as orthorhombic with $\alpha = \beta = \gamma = 90°$. By mapping out the Gibbs free energy as a function of temperature at $P = 1$ atm using molecular dynamics simulations in the isothermal-isobaric ensemble, the location of the transition between the monoclinic and orthorhombic structures can be identified (Hong *et al.*, 2021). If, in addition, one integrates over all cell parameters except α and plot G as a function of both T and α, these different structures can be characterized in greater detail. Note that G is not a natural function of α. However, we can think of α as a structural variable for distinguishing the different structures (an idea we will expand upon in Chapter 8). The free energy as a function of T and α is shown in Fig. 5.6. Interestingly, the free energy indicates that at high temperature (above the white line), the structure is best characterized as an ensemble of monoclinic structures with different values of α, the distribution of which peaks at 90° and ranges between 82° and 98°. Below the white line, the free energy clearly shows two minima at two equivalent monoclinic structures. The transition temperature occurs around 180 K, which agrees with an experimental prediction of the transition temperature, placing it between 170 K and 220 K (Reiss *et al.*, 2018).

5.14 The isothermal-isobaric ensemble with constraints: The ROLL algorithm

Incorporating holonomic constraints into molecular dynamics calculations in the isobaric ensembles introduces new technical difficulties. The forces in the virial contributions to the pressure and pressure tensor estimators must also include the forces of constraint. According to eqn. (3.9.5), the force on atom i in an N-particle system is $\tilde{\mathbf{F}}_i = \mathbf{F}_i + \sum_k \lambda_k \mathbf{F}_{c,i}^{(k)}$, where $\mathbf{F}_{c,i}^{(k)} = \nabla_i \sigma_k(\mathbf{r}_1, ..., \mathbf{r}_N)$, where $\mathbf{F}_i = -\partial U/\partial \mathbf{r}_i$, and the virial part of the pressure is

$$\mathcal{P}^{(\text{vir})} = \frac{1}{dV} \sum_{i=1}^{N} \left[\mathbf{r}_i \cdot \mathbf{F}_i + \mathbf{r}_i \cdot \sum_k \lambda_k \mathbf{F}_{c,i}^{(k)} \right]. \tag{5.14.1}$$

The integration algorithm for eqns. (5.10.5) encoded in the factorization of eqn. (5.13.4) generates a nonlinear dependence of the coordinates and velocities on the barostat variables v_ϵ or $\mathbf{v}_g$, while these variables, in turn, depend linearly on the pressure or pressure tensor. The consequence is that the coordinates and velocities acquire a complicated dependence on the Lagrange multipliers, and solving for multipliers is much less straightforward than in the constant-volume ensembles (see Section 3.9).

In order to tackle this problem, we need to modify the SHAKE and RATTLE algorithms of Section 3.9. We refer to the modified algorithm as the "ROLL" algorithm.[4] It is worth noting that a version of the ROLL algorithm was developed by Martyna, *et al.* (1996), however, the version that will be described here based on eqn. (5.13.4) is somewhat simpler.

Here, we will only consider the problem of isotropic cell fluctuations; the extension to fully flexible cells is straightforward, though tedious (Yu *et al.*, 2010). Because of the highly nonlinear dependence of eqn. (5.13.6) on the Lagrange multipliers, the operators $\exp(iL_{\epsilon,1}\Delta t) \exp(iL_1\Delta t) \exp(iL_2\Delta t/2) \exp(iL_{\epsilon,2}\Delta t/2)$ must be applied in an iterative fashion until a self-consistent solution that satisfies the constraints is obtained. The

---

[4]Yes, the "ROLL" moniker does fit well with "SHAKE" and "RATTLE"; however, there is an actual "rolling" procedure in the ROLL algorithm when used in fully flexible cell calculations.

full evolution of the coordinates $\mathbf{r}_i$ is obtained by combining eqns. (5.13.6) and (5.13.8) to give

$$\mathbf{r}_i(\Delta t) = \mathbf{r}_i(0)e^{v_\epsilon \Delta t} + \Delta t \boldsymbol{v}_i(\Delta t/2)e^{v_\epsilon \Delta t/2}\frac{\sinh(v_\epsilon \Delta t/2)}{v_\epsilon \Delta t/2}$$

$$= \mathbf{r}_i(0)e^{v_\epsilon \Delta t} + \Delta t e^{v_\epsilon \Delta t/2}\frac{\sinh(v_\epsilon \Delta t/2)}{v_\epsilon \Delta t/2} \tag{5.14.2}$$

$$\times \left[\boldsymbol{v}_i^{(\text{NHC})} e^{-\alpha v_\epsilon \Delta t/2} + \frac{\Delta t}{2m_i}\left(\mathbf{F}_i(0) + \sum_k \lambda_k \mathbf{F}_{c,i}^{(k)}(0) \right) \frac{\sinh(\alpha v_\epsilon \Delta t/4)}{\alpha v_\epsilon \Delta t/4} \right],$$

or

$$\mathbf{r}_i(\Delta t) = \mathbf{r}_i(0)e^{v_\epsilon \Delta t} + \Delta t \boldsymbol{v}_i^{(\text{NHC})} e^{-v_\epsilon(\alpha-1)\Delta t/2}\frac{\sinh(v_\epsilon \Delta t/2)}{v_\epsilon \Delta t/2} \tag{5.14.3}$$

$$+ \frac{\Delta t^2}{2m_i}\left[\mathbf{F}_i(0) + \sum_k \lambda_k \mathbf{F}_{c,i}^{(k)}(0) \right] e^{-v_\epsilon(\alpha-2)\Delta t/4}\frac{\sinh(v_\epsilon \Delta t/2)}{v_\epsilon \Delta t/2}\frac{\sinh(\alpha v_\epsilon \Delta t/4)}{\alpha v_\epsilon \Delta t/4},$$

where $\alpha = 1 + d/N_f$. Here, $\boldsymbol{v}_i^{(\text{NHC})}$ is the "velocity" generated by the thermostat operator, $\exp(iL_{\text{NHC}-\text{part}}\Delta t/2)$. Because the evolution of v_ϵ is determined by the pressure, many of the factors in eqn. (5.14.4) depend on the Lagrange multipliers. Thus, let us write eqn. (5.14.4) in the suggestive shorthand form

$$\mathbf{r}_i(\Delta t) = R_{xx}(\lambda, 0)\mathbf{r}_i(0) + R_{vx}(\lambda, 0)\Delta t \boldsymbol{v}_i^{(\text{NHC})}$$

$$+ \frac{\Delta t^2}{2m_i}R_{Fx}(\lambda, 0)\left[\mathbf{F}_i(0) + \sum_k \lambda_k \mathbf{F}_{c,i}^{(k)}(0) \right], \tag{5.14.4}$$

where λ denotes the full set of Lagrange multipliers. The factors $R_{xx}(\lambda, 0)$, $R_{vx}(\lambda, 0)$ and $R_{Fx}(\lambda, 0)$ denote the v_ϵ-dependent factors in eqn. (5.14.4); we refer to them as the "ROLL scalars." (In the fully flexible cell case, these scalars are replaced by 3×3 matrices.) Note the three operators $\exp(iL_t\Delta t)\exp(iL_2\Delta t/2)\exp(iL_{\epsilon,2}\Delta t/2)$ also generate the following half-step velocities:

$$\boldsymbol{v}_i(\Delta t/2) = \boldsymbol{v}_i^{(\text{NHC})} e^{-\alpha v_\epsilon \Delta t/2} + \frac{\Delta t}{2m_i}\left[\mathbf{F}_i(0) + \sum_k \lambda_k \mathbf{F}_{c,i}^{(k)}(0) \right] e^{-\alpha v_\epsilon \Delta t/4}\frac{\sinh(\alpha v_\epsilon \Delta t/4)}{\alpha v_\epsilon \Delta t/4}$$

$$\equiv R_{vv}(\lambda, 0)\boldsymbol{v}_i^{(\text{NHC})} + \frac{\Delta t}{2m_i}R_{Fv}(\lambda, 0)\left[\mathbf{F}_i(0) + \sum_k \lambda_k \mathbf{F}_{c,i}^{(k)}(0) \right], \tag{5.14.5}$$

where we have introduce the ROLL scalars $R_{vv}(\lambda, 0)$ and $R_{Fv}(\lambda, 0)$.

The first half of the ROLL algorithm is derived by requiring that the coordinates in eqn. (5.14.4) satisfy the constraint conditions $\sigma_k(\mathbf{r}_1(\Delta t), ..., \mathbf{r}_N(\Delta t)) = 0$. That is, eqns. (5.14.4) are inserted into the conditions $\sigma_k(\mathbf{r}_1(\Delta t), ..., \mathbf{r}_N(\Delta t)) = 0$, which are then solved for the Lagrange multipliers λ. Once the multipliers are determined, they are substituted into eqns. (5.14.4), (5.14.5), and (5.14.1) to generate final coordinates, half-step velocities, and the virial contribution to the pressure. Unfortunately, unlike the NVE and NVT cases, where the coordinates and velocities depend linearly on the Lagrange multipliers, the highly nonlinear dependence of eqn. (5.14.4) on λ complicates the task of solving for the multipliers. To see how we can solve this problem, we begin by letting $\tilde{\lambda}_k = (\Delta t^2/2)\lambda_k$. We now seed the ROLL algorithm with a guess $\{\tilde{\lambda}_k^{(1)}\}$ for the multipliers and write the exact multipliers as $\tilde{\lambda}_k = \tilde{\lambda}_k^{(1)} + \delta\tilde{\lambda}_k^{(1)}$. We also assume, at first, that the ROLL scalars are independent of the multipliers. Thus, when this ansatz for the multipliers is substituted into eqn. (5.14.4), the coordinates can be expressed as

$$\mathbf{r}_i(\Delta t) = \mathbf{r}_i^{(1)} + \frac{1}{m_i} R_{Fx}(\lambda, 0) \sum_k \delta\tilde{\lambda}_k^{(1)} \mathbf{F}_{c,i}^{(k)}(0), \tag{5.14.6}$$

where $\mathbf{r}_i^{(1)}$ contains everything except the $\delta\tilde{\lambda}_k^{(1)}$-dependent term. Since we are ignoring the dependence of the ROLL scalars on the multipliers, $\mathbf{r}_i^{(1)}$ has no dependence on $\delta\tilde{\lambda}_k^{(1)}$. The constraint conditions now become

$$\sigma_l \left(\mathbf{r}_1^{(1)} + \frac{1}{m_1} R_{Fx}(\lambda, 0) \sum_k \delta\tilde{\lambda}_k^{(1)} \mathbf{F}_{c,1}^{(k)}(0), ..., \right.$$

$$\left. \mathbf{r}_N^{(1)} + \frac{1}{m_N} R_{Fx}(\lambda, 0) \sum_k \delta\tilde{\lambda}_k^{(1)} \mathbf{F}_{c,N}^{(k)}(0) \right) = 0. \tag{5.14.7}$$

As we did in eqn. (3.9.13), we linearize these conditions using a first-order Taylor expansion:

$$\sigma_l(\mathbf{r}_1^{(1)}, ..., \mathbf{r}_N^{(1)}) + \sum_{i=1}^{N} \sum_{k=1}^{N_c} \mathbf{F}_{c,i}^{(k)}(1) \cdot \frac{1}{m_i} R_{Fx}(\lambda, 0) \delta\tilde{\lambda}_k^{(1)} \mathbf{F}_{c,i}^{(k)}(0) \approx 0, \tag{5.14.8}$$

where $\mathbf{F}_{c,i}^{(k)}(1) = \nabla_i \sigma_k(\mathbf{r}_1^{(1)}, ..., \mathbf{r}_N^{(1)})$ are the constraint forces evaluated at the positions $\mathbf{r}_i^{(1)}$. As noted in Section 3.9, we can either solve the full matrix equation in eqn. (5.14.8) if the dimensionality is not too large, or as a time-saving measure, we neglect the dependence of eqn. (5.14.8) on $l \neq k$ terms, write the condition as

$$\sigma_l(\mathbf{r}_1^{(1)}, ..., \mathbf{r}_N^{(1)}) + \sum_{i=1}^{N} \mathbf{F}_{c,i}^{(l)}(1) \cdot \frac{1}{m_i} R_{Fx}(\lambda, 0) \delta\tilde{\lambda}_l^{(1)} \mathbf{F}_{c,i}^{(l)}(0) \approx 0, \tag{5.14.9}$$

and iterate the corrections $\delta\tilde{\lambda}_l^{(1)}$ to convergence as in Section 3.9. Equation (5.14.9) can be solved easily for the multiplier corrections $\delta\tilde{\lambda}_l^{(1)}$ to yield

$$\delta\tilde{\lambda}_l^{(1)} = -\frac{\sigma_l(\mathbf{r}_1^{(1)}, ..., \mathbf{r}_N^{(1)})}{\sum_{i=1}^{N} (1/m_i) R_{Fx}(\lambda, 0) \mathbf{F}_{c,i}^{(l)}(1) \cdot \mathbf{F}_{c,i}^{(l)}(0)}. \tag{5.14.10}$$

Whichever procedure is used to obtain the corrections $\delta \tilde{\lambda}_l^{(1)}$, once we have them, we substitute them into eqn. (5.14.1) to obtain a new update to the pressure virial. Using this new pressure virial, we now cycle again through the operators

$$\exp(iL_{\epsilon,1}\Delta t) \exp(iL_1\Delta t) \exp(iL_2\Delta t/2) \exp(iL_{\epsilon,2}\Delta t/2),$$

which are applied on the original coordinates $\mathbf{r}_i(0)$ and $\mathbf{v}_i^{(\text{NHC})}$. This will generate a new set of ROLL scalars, which we use to generate a new set of corrections $\delta \tilde{\lambda}_l^{(1)}$ using the above procedure. This cycle is now iterated, each producing successively smaller corrections $\delta \tilde{\lambda}_l^{(n)}$ to the multipliers, until the ROLL scalars stop changing. Once this happens, the constraints will be satisfied and the pressure virial will be fully converged. Using the final multipliers, the half-step velocities are obtained from eqn. (5.14.5). It is important to note that, unlike the algorithm proposed by Martyna *et al.* (1996), this version of the first half of the ROLL algorithm requires no iteration through the thermostat operators.

The second half of the ROLL algorithm requires an iteration through the operators $\exp(iL_{\epsilon,2}\Delta t/2)\exp(iL_2\Delta t/2)$. However, it is also necessary to apply the operators $\exp(iL_{\text{NHC}-\text{part}}\Delta t/2)$ and $\exp(iL_{\text{NHC}-\text{baro}}\Delta t/2)$ in order to obtain the overall scaling factors on the velocities $v_i(\Delta t)$ and $v_\epsilon(\Delta t)$, which we will denote $S_i(\Delta t)$ and $S_\epsilon(\Delta t)$. Thus, the entire operator whose application must be iterated is

$$\hat{O} = \exp(iL_{\text{NHC}-\text{baro}}\Delta t/2) \exp(iL_{\text{NHC}-\text{part}}\Delta t/2)$$

$$\times \exp(iL_{\epsilon,2}\Delta t/2) \exp(iL_2\Delta t/2). \tag{5.14.11}$$

The evolution of v_i can now be expressed as

$$v_i(\Delta t) = \left\{ v_i(\Delta t/2)e^{-\alpha v_\epsilon \Delta t/2} + \frac{\Delta t}{2m_i} \left(\mathbf{F}_i(\Delta t) + \sum_k \mu_k \mathbf{F}_{c,i}^{(k)}(\Delta t) \right) \right.$$

$$\left. \times e^{-\alpha v_\epsilon \Delta t/4} \left[\frac{\sinh(\alpha v_\epsilon \Delta t/4)}{\alpha v_\epsilon \Delta t/4} \right] \right\} S_i(\Delta t), \tag{5.14.12}$$

which we can also write as

$$v_i(\Delta t) = \left\{ R_{vv}(\mu, \Delta t)v_i(\Delta t/2) \right.$$

$$\left. + \frac{\Delta t}{2m_i} R_{Fv}(\mu, \Delta t) \left[\mathbf{F}_i(\Delta t) + \sum_k \mu_k \mathbf{F}_{c,i}^{(k)}(\Delta t) \right] \right\} S_i(\Delta t), \tag{5.14.13}$$

and for v_ϵ, we obtain

$$v_\epsilon(\Delta t) = \left[v_\epsilon(\Delta t/2) + \frac{\Delta t}{2W} G_\epsilon(\mu, \Delta t) \right] S_\epsilon(\Delta t). \tag{5.14.14}$$

In eqns. (5.14.13) and (5.14.14), the use of μ_k and μ for the Lagrange multipliers indicates that these multipliers are used to enforce the first time derivative of the constraint conditions as described in Section 3.9. Let $\tilde{\mu}_k = (\Delta t/2)\mu_k$, and suppose we have a good initial guess to the multipliers $\tilde{\mu}_k^{(1)}$. Then, $\tilde{\mu}_k = \tilde{\mu}_k^{(1)} + \delta\tilde{\mu}_k^{(1)}$, and we can write eqns. (5.14.13) and (5.14.14) in shorthand as

$$\boldsymbol{v}_i(\Delta t) = \boldsymbol{v}_i^{(1)} + \frac{1}{m_i}R_{Fv}(\lambda,\Delta t)\sum_k \delta\tilde{\mu}_k^{(1)}\mathbf{F}_{c,i}^{(k)}(\Delta t)S_i(\Delta t)$$

$$v_\epsilon(\Delta t) = v_\epsilon^{(1)} + \frac{1}{W}\tilde{S}_\epsilon(\Delta t)\sum_i\sum_k \delta\tilde{\mu}_k^{(1)}\mathbf{r}_i(\Delta t)\cdot\mathbf{F}_{c,i}^{(k)}(\Delta t), \qquad (5.14.15)$$

where $\tilde{S}_\epsilon(\Delta t) = (\Delta t/2)S_\epsilon(\Delta t)$.

As in the first half of the ROLL algorithm, we assume that the ROLL scalars and scaling factors are independent of the multipliers and use eqns. (5.14.15) to determine the corrections $\delta\tilde{\mu}_k^{(1)}$ such that the first time derivative of each constraint condition vanishes. This requires

$$\dot{\sigma}_k = \sum_{i=1}^N F_{c,i}^{(k)}\cdot\dot{\mathbf{r}}_i = 0. \qquad (5.14.16)$$

However, a slight subtlety arises because according to eqns. (5.10.5), $\dot{\mathbf{r}}_i \neq \boldsymbol{v}_i$ but rather $\dot{\mathbf{r}}_i = \boldsymbol{v}_i + v_\epsilon\mathbf{r}_i$. Thus, eqn. (5.14.16) becomes a condition involving both $\boldsymbol{v}_i(\Delta t)$ and $v_\epsilon(\Delta t)$ at $t = \Delta t$:

$$\sum_i \mathbf{F}_{c,i}^{(k)}(\Delta t)\cdot[\boldsymbol{v}_i(\Delta t) + v_\epsilon(\Delta t)\mathbf{r}_i(\Delta t)] = 0. \qquad (5.14.17)$$

Substituting eqns. (5.14.15) into eqn. (5.14.17) yields

$$\sum_i \mathbf{F}_{c,i}^{(k)}(\Delta t)\cdot\left[\boldsymbol{v}_i^{(1)} + \frac{1}{m_i}R_{Fv}(\mu,\Delta t)S_i(\Delta t)\sum_l \delta\tilde{\mu}_l^{(1)}\mathbf{F}_{c,i}^{(l)}(\Delta t)\right.$$

$$\left. + \mathbf{r}_i(\Delta t)\left(v_\epsilon^{(1)} + \frac{1}{W}\tilde{S}_\epsilon(\Delta t)\sum_j\sum_l \delta\tilde{\mu}_l^{(1)}\mathbf{r}_j(\Delta t)\cdot\mathbf{F}_{c,j}^{(l)}(\Delta t)\right)\right] = 0. \qquad (5.14.18)$$

As we did with eqn. (5.14.8), we can solve eqn. (5.14.18) as a full matrix equation, or we can make the approximation of independent constraints and iterate to convergence as in Section 3.9. When the latter procedure is used, eqn. (5.14.18) becomes

$$\sum_i \mathbf{F}_{c,i}^{(l)}(\Delta t)\cdot\left[\boldsymbol{v}_i^{(1)} + \frac{1}{m_i}R_{Fv}(\mu,\Delta t)S_i(\Delta t)\delta\tilde{\mu}_l^{(1)}\mathbf{F}_{c,i}^{(l)}(\Delta t)\right.$$

$$\left. + \mathbf{r}_i(\Delta t)\left(v_\epsilon^{(1)} + \frac{1}{W}\tilde{S}_\epsilon(\Delta t)\sum_j \delta\tilde{\mu}_l^{(1)}\mathbf{r}_j(\Delta t)\cdot\mathbf{F}_{c,j}^{(l)}(\Delta t)\right)\right] = 0. \qquad (5.14.19)$$

Denoting $\mathbf{F}_{c,i}^{(l)}\cdot[\boldsymbol{v}_i^{(1)} + v_\epsilon^{(1)}\mathbf{r}_i(\Delta t)]$ as $\dot{\sigma}_l(\Delta t)$, eqn. (5.14.19) can be solved for the multiplier corrections $\delta\tilde{\mu}_l^{(1)}$ to yield $\delta\tilde{\mu}_l^{(1)} = -\dot{\sigma}_l(\Delta t)/D$, where D is given by

$$D = \sum_i \frac{1}{m_i}R_{Fv}(\lambda,\Delta t)S_i(\Delta t)\mathbf{F}_{c,i}^{(l)}(\Delta t)\cdot\mathbf{F}_{c,i}^{(l)}(\Delta t)$$

$$+ \frac{1}{W} S_\epsilon(\Delta t) \left[\sum_i \mathbf{r}_i(\Delta t) \cdot \mathbf{F}_{c,i}^{(l)}(\Delta t) \right]^2 . \tag{5.14.20}$$

As in the first part of the ROLL algorithm, once a fully converged set of correction multipliers $\delta \tilde{\mu}_l^{(1)}$ is obtained, we update the pressure virial according to

$$\mathcal{P}^{(\text{vir})} = \frac{1}{dV} \sum_{i=1}^N \left[\mathbf{r}_i \cdot \mathbf{F}_i + \mathbf{r}_i \cdot \sum_k \left(\tilde{\mu}_k^{(1)} + \delta \tilde{\mu}_k^{(1)} \right) \mathbf{F}_{c,i}^{(k)} \right] . \tag{5.14.21}$$

We then apply the operators in eqn. (5.14.11) again on the velocities and the v_ϵ that emerged from the first part of the ROLL procedure in order to obtain a new set of ROLL scalars and scaling factors. We cycle through this procedure, obtaining successively smaller corrections $\delta \tilde{\mu}_l^{(n)}$, until the ROLL scalars stop changing.

5.15 Problems

5.1. Show that the distribution function for the isothermal-isobaric ensemble can be derived starting from a microcanonical description of a system coupled to both a thermal reservoir and a mechanical piston.

5.2. Calculate the volume fluctuations ΔV given by

$$\Delta V = \sqrt{\langle V^2 \rangle - \langle V \rangle^2}$$

in the isothermal-isobaric ensemble. Express the answer in terms of the isothermal compressibility κ defined to be

$$\kappa = -\frac{1}{\langle V \rangle} \left(\frac{\partial \langle V \rangle}{\partial P} \right)_{N,T} .$$

Show that $\Delta V / \langle V \rangle \sim 1/\sqrt{N}$ and hence vanish in the thermodynamic limit.

5.3. Prove the tensorial version of the work virial theorem in eqn. (5.7.15).

5.4. The thermodynamic relation for the pressure tensor of a system in d dimensions undergoing full cell fluctuations at temperature T is

$$P_{\alpha\beta}^{(\text{int})} = \frac{k_\text{B} T}{\det(\text{h})} \sum_{\gamma=1}^d \text{h}_{\beta\gamma} \left(\frac{\partial \ln Q}{\partial \text{h}_{\alpha\gamma}} \right)_{N,T} ,$$

where h is the cell matrix and Q is the canonical partition function, while the thermodynamic relation for the isotropic pressure is

$$P^{(\text{int})} = k_{\text{B}}T \left(\frac{\partial \ln Q}{\partial V} \right)_{N,T},$$

where $V = \det(\text{h})$ is the volume of the system. Using only these thermodynamic relations, prove that $P^{(\text{int})}$ is related to the trace of the pressure tensor by

$$P^{(\text{int})} = \frac{1}{3} \sum_{\alpha=1}^{d} P_{\alpha\alpha}^{(\text{int})}.$$

N.B.: You cannot use expressions for estimators of the pressure or pressure tensor, as these expressions *assume* the validity of the thermodynamic relations. Thus, you need to use the aforementioned thermodynamic relations and show that they are consistent.

5.5. Repeat the entropy maximization calculation of Section 5.4 using three separate constraints of normalization and

$$\int_0^\infty dV \int dx\, \mathcal{H}(x) f(x, V) = E, \qquad \int_0^\infty dV \int dx\, V f(x, V) = \langle V \rangle,$$

using three Lagrange multipliers. Show that the isothermal-isobaric distribution results as the only solution to the maximization problem.

*5.6. a. For the ideal gas in Problem 4.7 of Chapter 4, calculate the isothermal-isobaric partition function assuming that only the length of the cylinder can vary.

   Hint: You might find the binomial theorem helpful in this problem.

   b. Derive an expression for the average length of the cylinder.

5.7. Prove that the isotropic NPT equations of motion in eqns. (5.10.5) generate the correct ensemble distribution function using the techniques of Section 4.10 for the following cases:

   a. $\sum_{i=1}^{N} \mathbf{F}_i \neq 0$,

   b. $\sum_{i=1}^{N} \mathbf{F}_i = 0$, for which there is an additional conservation law

   $$\mathbf{K} = \mathbf{P} \exp\left[\left(1 + \frac{d}{N_f} \right) \epsilon + \eta_1 \right],$$

   where $\mathbf{P} = \sum_{i=1}^{N} \mathbf{p}_i$ is the center-of-mass momentum;

   c. "Massive" thermostatting is used on the particles.

5.8. Prove that eqns. (5.11.2) for generating anisotropic volume fluctuations generate the correct ensemble distribution when $\sum_{i=1}^{N} \mathbf{F}_i = 0$.

5.9. One of the first algorithms proposed for generating the isotropic NPT ensemble via molecular dynamics is given by the equations of motion

$$\dot{\mathbf{r}}_i = \frac{\mathbf{p}_i}{m_i} + \frac{p_\epsilon}{W}\mathbf{r}_i$$

$$\dot{\mathbf{p}}_i = -\frac{\partial U}{\partial \mathbf{r}_i} - \frac{p_\epsilon}{W}\mathbf{p}_i - \frac{p_\eta}{Q}\mathbf{p}_i$$

$$\dot{V} = \frac{dV p_\epsilon}{W}$$

$$\dot{p}_\epsilon = dV(\mathcal{P}^{(\text{int})} - P) - \frac{p_\eta}{Q}p_\epsilon$$

$$\dot{\eta} = \frac{p_\eta}{Q}$$

$$\dot{p}_\eta = \sum_{i=1}^{N} \frac{\mathbf{p}_i^2}{m_i} + \frac{p_\epsilon^2}{W} - (N_f + 1)kT,$$

(Hoover, 1985), where $\mathcal{P}^{(\text{int})}$ is the pressure estimator of eqn. (5.8.28). These equations have the conserved energy

$$\mathcal{H}' = \mathcal{H}(\mathbf{r}, \mathbf{p}) + \frac{p_\epsilon^2}{2W} + \frac{p_\eta^2}{2Q} + (N_f + 1)kT\eta + PV.$$

Determine the ensemble distribution function $f(\mathbf{r}, \mathbf{p}, V)$ generated by these equations for

a. $\sum_{i=1}^{N} \mathbf{F}_i \neq 0$. Would the distribution be expected to approach the correct isothermal-isobaric ensemble distribution in the thermodynamic limit?

*b. $\sum_{i=1}^{N} \mathbf{F} = 0$, in which case, there is an additional conservation law

$$\mathbf{K} = \mathbf{P}e^{\epsilon + \eta},$$

where $\mathbf{P}$ is the center-of-mass momentum. Be sure to integrate over *all* nonphysical variables.

5.10. Prove that eqns. (5.12.6) generate the correct isobaric-isoenthalpic ensemble distribution when the pressure is determined using a molecular virial.

5.11. A simple model for the motion of particles through a nanowire consists of a one-dimensional ideal gas of N particles moving in a periodic potential. Let the Hamiltonian for one particle with coordinate q and momentum p be

$$h(q,p) = \frac{p^2}{2m} + \frac{kL^2}{4\pi^2}\left[1 - \cos\left(\frac{2\pi q}{L}\right)\right],$$

where m is the mass of the particle, k is a constant, and L is the length of the one-dimensional "box" or unit cell.

a. Calculate the change in the Helmholtz free energy *per particle* required to change the length of the "box" from L_1 to L_2. Express your answer in terms of the zeroth-order modified Bessel function

$$I_0(x) = \frac{1}{\pi}\int_0^\pi d\theta e^{\pm x \cos\theta}.$$

b. Calculate the equation of state by determining the one-dimensional "pressure" P. Do you obtain an ideal-gas equation of state? Why or why not? You might find the following properties of modified Bessel functions useful:

$$\frac{dI_\nu(x)}{dx} = \frac{1}{2}\left[I_{\nu+1}(x) + I_{\nu-1}(x)\right], \qquad I_\nu(x) = I_{-\nu}(x).$$

c. Write down integral expressions for the position and length distribution functions in the isothermal-isobaric ensemble.

5.12. Write a program to integrate the isotropic NPT equations of motion (5.10.5) for the one-dimensional periodic potential in eqn. (5.10.8) using the integrator in eqn. (5.13.4). The program should be able to generate the distributions in Fig. 5.3.

5.13. How should the algorithm in Section 4.15 for calculating the radial distribution function be modified for the isotropic NPT ensemble?

*5.14. Generalize the ROLL algorithm of Section 5.14 to the case of anisotropic cell fluctuations based on eqns. (5.11.2) and the integrator defined by eqns. (5.13.10) and (5.13.11).

5.15. a. Using the constraint condition on the box matrix $h_{\alpha\beta} = 0$ for $\alpha > \beta$, show using Lagrange undetermined multipliers, that overall cell rotations in eqns. (5.11.2) can be eliminated simply by working with an upper triangular box matrix.

b. Using the constraint condition that $\mathbf{p_g} - \mathbf{p_g^T} = 0$, show using Lagrange undetermined multipliers, that overall cell rotations in eqns. (5.11.2) can be eliminated by explicitly symmetrizing the pressure tensor $P_{\alpha\beta}^{(int)}$. Is this scheme easier to implement within the ROLL algorithm of Section 5.14?

6
The grand canonical ensemble

6.1 Introduction: The need for yet another ensemble

The ensembles discussed thus far all have the common feature that the particle number N is kept fixed as one of the control variables. The fourth ensemble to be discussed, the grand canonical ensemble, differs in that it permits fluctuations in the particle number at *constant chemical potential* μ. Why is such an ensemble necessary? As useful as the isothermal-isobaric and canonical ensembles are, numerous physical situations correspond to a system in which the particle number varies. These include liquid–vapor equilibria, capillary condensation, and, notably, molecular electronics and batteries, in which a device is assumed to be coupled to an electron source. In computational molecular design, one seeks to sample a complete "chemical space" of compounds in order to optimize a particular property (e.g. binding energy to a target), which requires varying both the number and chemical identity of the constituent atoms. Finally, in certain cases, it simply proves easier to work in the grand canonical ensemble, and given that all ensembles become equivalent in the thermodynamic limit, we are free to choose the ensemble that proves most convenient for the problem at hand.

In this chapter, we introduce the basic thermodynamics and classical statistical mechanics of the grand canonical ensemble. We will begin with a discussion of Euler's theorem and a derivation of the free energy. Following this, we will consider the partition function of a physical system coupled to both thermal and particle reservoirs. Finally, we will discuss the procedure for obtaining an equation of state within the framework of the grand canonical ensemble.

Because of the inherently discrete nature of particle fluctuations, the grand canonical ensemble does not easily fit into the continuous molecular dynamics framework we have discussed so far for kinetic-energy and volume fluctuations. Therefore, a discussion of computational approaches to the grand canonical ensemble will be deferred until Chapters 7 and 8. These chapters will develop the machinery needed to design computational approaches suitable for the grand canonical ensemble.

6.2 Euler's theorem

Euler's theorem is a general statement about a certain class of functions known as *homogeneous functions of degree* n. Consider a function $f(x_1, ..., x_N)$ of N variables that satisfies

$$f(\lambda x_1, ..., \lambda x_k, x_{k+1}, ..., x_N) = \lambda^n f(x_1, ..., x_k, x_{k+1}, ...x_N) \tag{6.2.1}$$

for an arbitrary parameter, λ. We call such a function a *homogeneous function of degree* n in the variables $x_1, ..., x_k$. The function $f(x) = x^2$, for example, is a homogeneous function of degree 2. The function $f(x, y, z) = xy^2 + z^3$ is a homogeneous function of degree 3 in all three variables x, y, and z. The function $f(x, y, z) = x^2(y^2 + z)$ is a homogeneous function of degree 2 in x but not in y and z. The function $f(x, y) = e^{xy} - xy$ is not a homogeneous function in either x or y.

Euler's theorem states the following: Let $f(x_1, ..., x_N)$ be a homogeneous function of degree n in $x_1, ..., x_k$. Then,

$$nf(x_1, ..., x_N) = \sum_{i=1}^{k} x_i \frac{\partial f}{\partial x_i}. \tag{6.2.2}$$

The proof of Euler's theorem is straightforward. Beginning with eqn. (6.2.1), we differentiate both sides with respect to λ to yield:

$$\frac{\mathrm{d}}{\mathrm{d}\lambda} f(\lambda x_1, ..., \lambda x_k, x_{k+1}, ..., x_N) = \frac{\mathrm{d}}{\mathrm{d}\lambda} \lambda^n f(x_1, ..., x_k, x_{k+1}, ..., x_N)$$

$$\sum_{i=1}^{k} x_i \frac{\partial f}{\partial(\lambda x_i)} = n\lambda^{n-1} f(x_1, ..., x_k, x_{k+1}, ..., x_N). \tag{6.2.3}$$

Since λ is arbitrary, we may freely choose $\lambda = 1$, which yields

$$\sum_{i=1}^{k} x_i \frac{\partial f}{\partial x_i} = nf(x_1, ..., x_k, x_{k+1}, ..., x_N) \tag{6.2.4}$$

and proves the theorem.

What does Euler's theorem have to do with thermodynamics? Consider, for example, the Helmholtz free energy $A(N, V, T)$, which depends on two extensive variables, N and V. Since A is, itself, extensive, $A \sim N$, and since $V \sim N$, A must be a homogeneous function of degree 1 in N and V, i.e., $A(\lambda N, \lambda V, T) = \lambda A(N, V, T)$. Applying Euler's theorem, it follows that

$$A(N, V, T) = V \frac{\partial A}{\partial V} + N \frac{\partial A}{\partial N}. \tag{6.2.5}$$

From the thermodynamic relations of the canonical ensemble for pressure and chemical potential, we have $P = -(\partial A/\partial V)$ and $\mu = (\partial A/\partial N)$. Thus,

$$A = -PV + \mu N. \tag{6.2.6}$$

We can verify this result by recalling that

$$A(N, V, T) = E - TS. \tag{6.2.7}$$

From the first law of thermodynamics,

$$E - TS = -PV + \mu N, \tag{6.2.8}$$

so that

$$A(N, V, T) = -PV + \mu N, \tag{6.2.9}$$

which agrees with Euler's theorem. Similarly, the Gibbs free energy $G(N, P, T)$ is a homogeneous function of degree 1 in N only, *i.e.*, $G(\lambda N, P, T) = \lambda G(N, P, T)$. Thus, from Euler's theorem,

$$G(N, P, T) = N \frac{\partial G}{\partial N} = \mu N, \tag{6.2.10}$$

which agrees with the definition $G = E - TS + PV = \mu N$. From these two examples, we see that Euler's theorem allows us to derive alternative expressions for extensive thermodynamic functions such as the Gibbs and Helmholtz free energies. As will be shown in the next section, Euler's theorem simplifies the derivation of the thermodynamic relations of the grand canonical ensemble.

6.3 Thermodynamics of the grand canonical ensemble

In the grand canonical ensemble, the control variables are the chemical potential μ, the volume V, and the temperature T. The free energy of the ensemble can be obtained by performing a Legendre transformation of the Helmholtz free energy $A(N, V, T)$. Let $\tilde{A}(\mu, V, T)$ be the transformed free energy, which we obtain as

$$\tilde{A}(\mu, V, T) = A(N(\mu), V, T) - N \left(\frac{\partial A}{\partial N} \right)_{V,T}$$

$$\tilde{A}(\mu, V, T) = A(N(\mu), V, T) - N(\mu)\mu. \tag{6.3.1}$$

Since $\tilde{A}$ is a function of μ, V, and T, a small change in each of these variables leads to a change in $\tilde{A}$ given by

$$d\tilde{A} = \left(\frac{\partial \tilde{A}}{\partial \mu} \right)_{V,T} d\mu + \left(\frac{\partial \tilde{A}}{\partial V} \right)_{\mu,T} dV + \left(\frac{\partial \tilde{A}}{\partial T} \right)_{\mu,V} dT. \tag{6.3.2}$$

However, from the first law of thermodynamics,

$$d\tilde{A} = dA - N d\mu - \mu dN$$

$$= -P dV - S dT + \mu dN - N d\mu - \mu dN$$

$$= -P dV - S dT - N d\mu, \tag{6.3.3}$$

and we obtain the thermodynamic relations

$$\langle N \rangle = -\left(\frac{\partial \tilde{A}}{\partial \mu} \right)_{V,T}, \qquad P = -\left(\frac{\partial \tilde{A}}{\partial V} \right)_{\mu,T}, \qquad S = -\left(\frac{\partial \tilde{A}}{\partial T} \right)_{V,\mu}. \tag{6.3.4}$$

In the above relations, $\langle N \rangle$ denotes the average particle number. Euler's theorem can be used to determine a relation for $\tilde{A}$ in terms of other thermodynamic variables. Since

$\tilde{A}$ depends on a single extensive variable, V, it is a homogeneous function of degree 1 in V, i.e., $\tilde{A}(\mu, \lambda V, T) = \lambda \tilde{A}(\mu, V, T)$. From Euler's theorem,

$$\tilde{A} = V \frac{\partial \tilde{A}}{\partial V}, \tag{6.3.5}$$

which, according to eqn. (6.3.4), becomes

$$\tilde{A} = -PV. \tag{6.3.6}$$

Thus, $-PV$ is the natural free energy of the grand canonical ensemble. Unlike other ensembles, $\tilde{A} = -PV$ is not given a unique symbol. Rather, because it leads directly to the equation of state, the free energy is simply denoted $-PV$.

6.4 Grand canonical phase space and the partition function

Since the grand canonical ensemble uses μ, V, and T as its control variables, it is convenient to think of this ensemble as a canonical ensemble coupled to a *particle reservoir*, which drives the fluctuations in the particle number. As the name implies, a particle reservoir is a system that can gain or lose particles without appreciably changing its own particle number. Thus, we imagine two systems coupled to a common thermal reservoir at temperature T, such that system 1 has N_1 particles and volume V_1 and system 2 has N_2 particles and a volume V_2. The two systems can exchange particles, with system 2 acting as a particle reservoir (see Fig. 6.1). Hence, $N_2 \gg N_1$. The total particle number and volume are

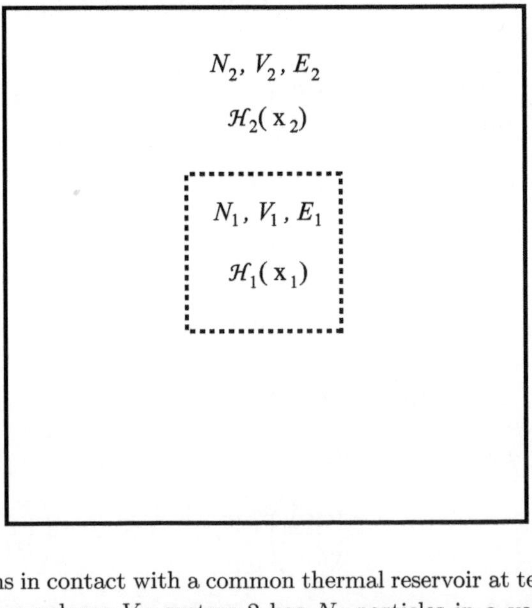

Fig. 6.1 Two systems in contact with a common thermal reservoir at temperature T. System 1 has N_1 particles in a volume V_1; system 2 has N_2 particles in a volume V_2. The dashed lines indicate that systems 1 and 2 can exchange particles.

$$N = N_1 + N_2, \qquad V = V_1 + V_2. \tag{6.4.1}$$

In order to carry out the derivation of the ensemble distribution function, we will need to consider explicitly the dependence of the Hamiltonian on particle number, usually appearing as the upper limit of sums in the kinetic and potential energies. Therefore, let $\mathcal{H}_1(x_1, N_1)$ be the Hamiltonian of system 1 and $\mathcal{H}(x_2, N_2)$ be the Hamiltonian of system 2. As usual, we will take the total Hamiltonian to be

$$\mathcal{H}(x, N) = \mathcal{H}_1(x_1, N_1) + \mathcal{H}_2(x_2, N_2). \tag{6.4.2}$$

Consider first the simpler case in which systems 1 and 2 do not exchange particles. The overall canonical partition function in this limit is

$$
\begin{aligned}
Q(N, V, T) &= \frac{1}{N! h^{3N}} \int dx_1 \int dx_2 \; e^{-\beta[\mathcal{H}_1(x_1, N_1) + \mathcal{H}_2(x_2, N_2)]} \\
&= \frac{N_1! N_2!}{N!} \frac{1}{N_1! h^{3N_1}} \int dx_1 \; e^{-\beta \mathcal{H}_1(x_1, N_1)} \frac{1}{N_2! h^{3N_2}} \int dx_2 \; e^{-\beta \mathcal{H}_2(x_2, N_2)} \\
&= \frac{N_1! N_2!}{N!} Q_1(N_1, V_1, T) Q_2(N_2, V_2, T), \tag{6.4.3}
\end{aligned}
$$

where $Q_1(N_1, V_1, T)$ and $Q_2(N_2, V_2, T)$ are the canonical partition functions of systems 1 and 2, respectively, at the common temperature T.

When the systems are allowed to exchange particles, the right side of eqn. (6.4.3) represents one specific choice of N_1 particles for system 1 and $N_2 = N - N_1$ particles for systems 2. In order to account for particle number variations in systems 1 and 2, the true partition function must contain a sum over all possible values of N_1 and N_2 on the right side of eqn. (6.4.3) subject to the restriction that $N_1 + N_2 = N$. This restriction is accounted for by summing only N_1 or N_2 over the range $[0, N]$. For concreteness, we will carry out the sum over N_1 and set $N_2 = N - N_1$. Additionally, we need to weight each term in the sum by a degeneracy factor $g(N_1, N_2) = g(N_1, N - N_1)$ that accounts for the number of distinct configurations that exist for particular values of N_1 and N_2. Thus, the partition function for varying particle numbers is

$$Q(N, V, T) = \sum_{N_1=0}^{N} g(N_1, N - N_1) \frac{N_1!(N - N_1)!}{N!}$$

$$\times Q_1(N_1, V_1, T) Q_2(N - N_1, V - V_1, T), \tag{6.4.4}$$

where we have used the fact that $V_1 + V_2 = V$.

We now determine the degeneracy factor $g(N_1, N - N_1)$. For the $N_1 = 0$ term, $g(0, N)$ represents the number of ways in which system 1 can have 0 particles and system 2 can have all N particles. There is only one way to create such a configuration, hence $g(0, N) = 1$. For $N_1 = 1$, $g(1, N - 1)$ represents the number of ways in which system 1 can have one particle and system 2 can have $(N-1)$ particles. Since there are N ways to choose that one particle to place in system 1, it follows that $g(1, N-1) = N$.

When $N_1 = 2$, we need to place two particles in system 1. The first particle can be chosen in N ways, while the second can be chosen in $(N-1)$ ways, which seems to lead to a product $N(N-1)$ ways that this configuration can be created. However, choosing particle 1, for example, as the first particle to put into system 1 and particle 2 as the second particle leads to the same physical configuration as choosing particle 2 as the first particle and particle 1 as the second. Thus, the degeneracy factor $g(2, N-2)$ is actually $N(N-1)/2$. In general, $g(N_1, N_1 - 1)$ is nothing more than the number of ways of placing N "labeled" objects into 2 containers, which is just the well-known binomial coefficient

$$g(N_1, N_1 - N) = \frac{N!}{N_1!(N - N_1)!}. \tag{6.4.5}$$

We can check eqn. (6.4.5) against the specific examples we analyzed:

$$g(0, N) = \frac{N!}{0!N!} = 1$$

$$g(1, N - 1) = \frac{N!}{1!(N - 1)!} = N$$

$$g(2, N - 2) = \frac{N!}{2!(N - 2)!} = \frac{N(N - 1)}{2}. \tag{6.4.6}$$

Interestingly, the degeneracy factor exactly cancels the $N_1!(N - N_1)!/N!$ appearing in eqn. (6.4.4). This cancellation is not unexpected since, as we recall, the latter factor was included as a "fudge factor" to correct for the fact that classical particles are always distinguishable, and we need our results to be consistent with the indistinguishable nature of the particles (recall Section 3.5.1). Thus, all N configurations in which one particle is in system 1 are physically the same, and so forth. Inserting eqn. (6.4.5) into eqn. (6.4.4) gives

$$Q(N, V, T) = \sum_{N_1=0}^{N} Q_1(N_1, V_1, T)Q_2(N - N_1, V - V_1, T). \tag{6.4.7}$$

Now the total phase-space distribution function

$$f(\mathbf{x}, N) = \frac{e^{-\beta \mathcal{H}(\mathbf{x}, N)}}{N!h^{3N}Q(N, V, T)} \tag{6.4.8}$$

satisfies the normalization condition

$$\int d\mathbf{x}\, f(\mathbf{x}, N) = 1, \tag{6.4.9}$$

since it is just a canonical distribution. However, the phase-space distribution of system 1, obtained by integrating over $\mathbf{x}_2$ according to

$$f_1(\mathbf{x}_1, N_1) = \left(\frac{e^{-\beta \mathcal{H}_1(\mathbf{x}_1, N_1)}}{Q(N, V, T) N_1! h^{3N_1}} \right) \frac{1}{(N - N_1)! h^{3(N-N_1)}} \int d\mathbf{x}_2 \, e^{-\beta \mathcal{H}_2(\mathbf{x}_2, N-N_1)}$$

$$= \frac{Q_2(N - N_1, V - V_1, T)}{Q(N, V, T)} \frac{1}{N_1! h^{3N_1}} e^{-\beta \mathcal{H}_1(\mathbf{x}_1, N_1)}, \tag{6.4.10}$$

satisfies the normalization condition

$$\sum_{N_1=0}^{N} \int d\mathbf{x}_1 \, f(\mathbf{x}_1, N_1) = 1. \tag{6.4.11}$$

Since the total partition function is canonical, $Q(N, V, T) = \exp[-\beta A(N, V, T)]$ where $A(N, V, T)$ is the Helmholtz free energy, and it follows that

$$\frac{Q_2(N - N_1, V - V_1, T)}{Q(N, V, T)} = e^{-\beta [A(N-N_1, V-V_1, T) - A(N, V, T)]}, \tag{6.4.12}$$

where we have assumed that system 1 and system 2 are described by the same set of physical interactions, so that the functional form of the free energy is the same for both systems and for the total system. Since $N \gg N_1$ and $V \gg V_1$, we may expand $A(N - N_1, V - V_1, T)$ about $N_1 = 0$ and $V_1 = 0$. To first order, the expansion yields

$$A(N - N_1, V - V_1, T) \approx A(N, V, T) - \frac{\partial A}{\partial N} N_1 - \frac{\partial A}{\partial V} V_1$$

$$= A(N, V, T) - \mu N_1 + P V_1. \tag{6.4.13}$$

Thus, the phase-space distribution of system 1 becomes

$$f(\mathbf{x}_1, N_1) = \frac{1}{N_1! h^{3N_1}} e^{\beta \mu N_1} e^{-\beta P V_1} e^{-\beta \mathcal{H}_1(\mathbf{x}_1, N_1)}$$

$$= \frac{1}{N_1! h^{3N_1}} e^{\beta \mu N_1} \frac{1}{e^{\beta P V_1}} e^{-\beta \mathcal{H}_1(\mathbf{x}_1, N_1)}. \tag{6.4.14}$$

Since system 2 quantities no longer appear in eqn. (6.4.14), we may drop the "1" subscript and write the phase-space distribution for the grand canonical ensemble as

$$f(\mathbf{x}, N) = \frac{1}{N! h^{3N}} e^{\beta \mu N} \frac{1}{e^{\beta P V}} e^{-\beta \mathcal{H}(\mathbf{x}, N)}. \tag{6.4.15}$$

Moreover, taking the thermodynamic limit, the summation over N is now unrestricted ($N \in [0, \infty)$), so the normalization condition becomes

$$\sum_{N=0}^{\infty} \int d\mathbf{x} \, f(\mathbf{x}, N) = 1, \tag{6.4.16}$$

which implies that

$$\frac{1}{e^{\beta PV}} \sum_{N=0}^{\infty} e^{\beta \mu N} \frac{1}{N! h^{3N}} \int dx\, e^{-\beta \mathcal{H}(x,N)} = 1. \tag{6.4.17}$$

Taking the $\exp(\beta PV)$ factor to the right side, we obtain

$$\sum_{N=0}^{\infty} e^{\beta \mu N} \frac{1}{N! h^{3N}} \int dx\, e^{-\beta \mathcal{H}(x,N)} = e^{\beta PV}. \tag{6.4.18}$$

However, recall that $-PV = \tilde{A}(\mu, V, T)$ is the free energy of the grand canonical ensemble. Thus, $\exp(\beta PV) = \exp[-\beta(-PV)]$ is equal to the partition function. In the grand canonical ensemble, we denote the partition function as $\mathcal{Z}(\mu, V, T)$, and it is given by

$$\mathcal{Z}(\mu, V, T) = \sum_{N=0}^{\infty} e^{\beta \mu N} \frac{1}{N! h^{3N}} \int dx\, e^{-\beta \mathcal{H}(x,N)}$$

$$= \sum_{N=0}^{\infty} e^{\beta \mu N} Q(N, V, T). \tag{6.4.19}$$

The product PV is thus related to $\mathcal{Z}(\mu, V, T)$ by

$$\frac{PV}{kT} = \ln \mathcal{Z}(\mu, V, T). \tag{6.4.20}$$

According to eqn. (6.4.20), the equation of state can be obtained directly from the partition function in the grand canonical ensemble. Recall, however, that the equation of state is of the general form (cf. eqn. (2.2.1))

$$g(\langle N \rangle, P, V, T) = 0, \tag{6.4.21}$$

which is a function of $\langle N \rangle$ rather than μ. This suggests that a second equation for the average particle number $\langle N \rangle$ is needed. By definition,

$$\langle N \rangle = \frac{1}{\mathcal{Z}(\mu, V, T)} \sum_{N=0}^{\infty} N e^{\beta \mu N} Q(N, V, T), \tag{6.4.22}$$

which can be expressed as a derivative of $\mathcal{Z}$ with respect to μ as

$$\langle N \rangle = kT \left(\frac{\partial}{\partial \mu} \ln \mathcal{Z}(\mu, V, T) \right)_{V,T}. \tag{6.4.23}$$

Equations (6.4.23) and (6.4.20) give a prescription for finding the equation of state in the grand canonical ensemble. Equation (6.4.23) must be solved for μ in terms of $\langle N \rangle$ and then substituted back into eqn. (6.4.20) in order to obtain an equation in the proper form.

For other thermodynamic quantities, it is convenient to introduce a new variable

$$\zeta = e^{\beta\mu} \tag{6.4.24}$$

known as the *fugacity*. Since ζ and μ are directly related, the fugacity can be viewed as an alternative external control variable for the grand canonical ensemble, and the partition function can be expressed in terms of ζ as

$$\mathcal{Z}(\zeta, V, T) = \sum_{N=0}^{\infty} \zeta^N Q(N, V, T), \tag{6.4.25}$$

so that

$$\frac{PV}{kT} = \ln \mathcal{Z}(\zeta, V, T). \tag{6.4.26}$$

Since

$$\frac{\partial}{\partial\mu} = \frac{\partial\zeta}{\partial\mu}\frac{\partial}{\partial\zeta} = \beta\zeta\frac{\partial}{\partial\zeta}, \tag{6.4.27}$$

the average particle number can be computed from $\mathcal{Z}(\zeta, V, T)$ by

$$\langle N \rangle = \zeta\frac{\partial}{\partial\zeta} \ln \mathcal{Z}(\zeta, V, T). \tag{6.4.28}$$

Thus, the equation of state results when eqn. (6.4.28) is solved for ζ in terms of $\langle N \rangle$ and substituted back into eqn. (6.4.26). Other thermodynamic quantities can be obtained as well. The average energy, $E = \langle \mathcal{H}(\mathbf{x}, N) \rangle$, is given by

$$E = \langle \mathcal{H}(\mathbf{x}, N) \rangle = \frac{1}{\mathcal{Z}} \sum_{N=0}^{\infty} \zeta^N \frac{1}{N!h^{3N}} \int d\mathbf{x}\, \mathcal{H}(\mathbf{x}, N) e^{-\beta\mathcal{H}(\mathbf{x},N)}$$

$$= -\left(\frac{\partial}{\partial\beta} \ln \mathcal{Z}(\zeta, V, T)\right)_{\zeta, V}. \tag{6.4.29}$$

In eqn. (6.4.29), it must be emphasized that the average energy is computed as the derivative with respect to β of $\ln \mathcal{Z}$ at fixed V and ζ *rather* than at fixed V and μ. The entropy is given in terms of the derivative of the free energy with respect to T:

$$S(\mu, V, T) = -\left(\frac{\partial(-PV)}{\partial T}\right)_{\mu, V}$$

$$= k \ln \mathcal{Z}(\mu, V, T) - k\beta\left(\frac{\partial}{\partial\beta} \ln \mathcal{Z}(\mu, V, T)\right)_{\mu, V}. \tag{6.4.30}$$

For the entropy, the temperature derivative must be taken at fixed μ rather than at fixed ζ.

6.5   Grand canonical ensemble via entropy maximization

As we did in Chapters 4 and 5, we will use entropy maximization to derive the phase-space distribution function of the grand canonical ensemble. Following the procedure of Section 4.4 and Section 5.4, we define the Shannon entropy functional in terms of the distribution function $f(\mathbf{x}, N)$:

$$S[f] = -k \sum_{N=0}^{\infty} \int d\mathbf{x}\, f(\mathbf{x}, N) \ln f(\mathbf{x}, N). \tag{6.5.1}$$

The correct phase-space distribution of the grand canonical ensemble is the one that maximizes the entropy subject to a constraint of normalization (cf. eqn. (6.4.16)) and the following constraints on the ensemble average of the Hamiltonian $\mathcal{H}(\mathbf{x})$ and particle number:

$$\sum_{N=0}^{\infty} \int d\mathbf{x}\, \mathcal{H}(\mathbf{x}) f(\mathbf{x}, N) = E$$

$$\sum_{N=0}^{\infty} \int d\mathbf{x}\, N f(\mathbf{x}, N) = \langle N \rangle. \tag{6.5.2}$$

Here, E and $\langle N \rangle$ are the correct grand canonical ensemble averages. The three constraints will be incorporated with Lagrange multipliers μ (not to be confused with the chemical potential), λ, and γ via the modified entropy functional

$$\tilde{S}[f] = -k \sum_{N=0}^{\infty} \int d\mathbf{x}\, f(\mathbf{x}, N) \ln f(\mathbf{x}, N) - \mu \left(\sum_{N=0}^{\infty} \int d\mathbf{x}\, f(\mathbf{x}, N) - 1 \right) \tag{6.5.3}$$

$$- \lambda \left(\sum_{N=0}^{\infty} \int d\mathbf{x}\, \mathcal{H}(\mathbf{x}) f(\mathbf{x}, N) - E \right) - \gamma \left(\sum_{N=0}^{\infty} \int d\mathbf{x}\, N f(\mathbf{x}, N) - \langle N \rangle \right).$$

Taking the functional derivative and setting it equal to 0, we obtain

$$\frac{\delta \tilde{S}}{\delta f(\mathbf{x}, N)} = -k \left[\ln f(\mathbf{x}, N) + 1 \right] - \mu - \lambda \mathcal{H}(\mathbf{x}) - \gamma N = 0. \tag{6.5.4}$$

Solving for $f(\mathbf{x}, N)$, we find

$$f(\mathbf{x}, N) = e^{-(1+\mu/k)} e^{-\lambda \mathcal{H}(\mathbf{x})/k} e^{-\gamma N/k}. \tag{6.5.5}$$

The Lagrange multipliers are determined by requiring that $\tilde{S}$ satisfy the conditions

$$\frac{\partial \tilde{S}}{\partial E} = \frac{1}{T}, \qquad \frac{\partial \tilde{S}}{\partial \langle N \rangle} = -\frac{\mu}{T}. \tag{6.5.6}$$

However, from eqn. (6.5.4), we see that $\partial \tilde{S}/\partial E = \lambda$ and $\partial \tilde{S}/\partial \langle N \rangle = \gamma$, from which it follows that $\lambda = 1/T$ and $\gamma = -\mu/T$. Substituting these into eqn. (6.5.5), we obtain

$$f(\mathbf{x}, N) = e^{-(1+\mu/k)} e^{-\mathcal{H}(\mathbf{x})/kT} e^{\mu N/kT}, \tag{6.5.7}$$

and after normalization, we find

$$f(\mathbf{x}, N) = \frac{C_N e^{-(\mathcal{H}(\mathbf{x}) - \mu N)/kT}}{\mathcal{Z}(\mu, V, T)}, \tag{6.5.8}$$

which is the expected result for the grand canonical phase-space distribution.

6.6 Illustration of the grand canonical ensemble: The ideal gas

In Chapter 11, the grand canonical ensemble will be used to derive the properties of the quantum ideal gases. It will be seen that the use of the grand canonical ensemble greatly simplifies the treatment over the canonical ensemble. Thus, in order to prepare for this analysis, it is instructive to illustrate the grand canonical procedure for deriving the equation of state with a simple example, namely, the classical ideal gas. Since the partition function of the grand canonical ensemble is given by eqn. (6.4.25), we can start by recalling the expression of the canonical partition function of the classical ideal gas

$$Q(N, V, T) = \frac{1}{N!} \left[V \left(\frac{2\pi m}{\beta h^2} \right)^{3/2} \right]^N = \frac{1}{N!} \left(\frac{V}{\lambda^3} \right)^N. \tag{6.6.1}$$

Substituting this expression into eqn. (6.4.25) gives

$$\mathcal{Z}(\zeta, V, T) = \sum_{N=0}^{\infty} \zeta^N \frac{1}{N!} \left(\frac{V}{\lambda^3} \right)^N$$

$$= \sum_{N=0}^{\infty} \frac{1}{N!} \left(\frac{V\zeta}{\lambda^3} \right)^N. \tag{6.6.2}$$

Equation (6.6.2) is in the form of a Taylor series expansion for the exponential:

$$e^x = \sum_{k=0}^{\infty} \frac{x^k}{k!}. \tag{6.6.3}$$

Equation (6.6.2) can, therefore, be summed over N to yield

$$\mathcal{Z}(\zeta, V, T) = e^{V\zeta/\lambda^3}. \tag{6.6.4}$$

The procedure embodied in eqns. (6.4.28) and (6.4.26) requires first the calculation of ζ as a function of $\langle N \rangle$. From eqn. (6.4.28),

$$\langle N \rangle = \zeta \frac{\partial}{\partial \zeta} \ln \mathcal{Z}(\zeta, V, T) = \frac{V\zeta}{\lambda^3}. \tag{6.6.5}$$

Thus,

$$\zeta(\langle N \rangle) = \frac{\langle N \rangle \lambda^3}{V}. \tag{6.6.6}$$

From eqn. (6.6.4), we have

$$\frac{PV}{kT} = \ln \mathcal{Z}(\zeta, V, T) = \frac{V\zeta}{\lambda^3}. \tag{6.6.7}$$

By substituting $\zeta(\langle N \rangle)$ into eqn. (6.6.7), the expected equation of state results:

$$\frac{PV}{kT} = \langle N \rangle, \tag{6.6.8}$$

which contains the average particle number $\langle N \rangle$ instead of N as would appear in the canonical ensemble. Similarly, the average energy is given by

$$E = -\frac{\partial}{\partial \beta} \ln \mathcal{Z}(\zeta, V, T) = -\frac{\partial}{\partial \beta} \frac{V\zeta}{\lambda^3} = \frac{3V\zeta}{\lambda^4} \frac{\partial \lambda}{\partial \beta} = \frac{3}{2} \langle N \rangle kT. \tag{6.6.9}$$

Finally, in order to compute the entropy, $\mathcal{Z}$ must be expressed in terms of μ rather than ζ, *i.e.*,

$$\ln \mathcal{Z}(\mu, V, T) = \frac{V e^{\beta\mu}}{\lambda^3}. \tag{6.6.10}$$

Then,

$$S(\mu, V, T) = k \ln \mathcal{Z}(\mu, V, T) - k\beta \left(\frac{\partial \ln \mathcal{Z}(\mu, V, T)}{\partial \beta} \right)_{\mu, V}$$

$$= k \frac{V e^{\beta\mu}}{\lambda^3} - k\beta \left[\frac{V \mu e^{\beta\mu}}{\lambda^3} - \frac{3V e^{\beta\mu}}{\lambda^4} \frac{\partial \lambda}{\partial \beta} \right]. \tag{6.6.11}$$

Using the facts that

$$\frac{V e^{\beta\mu}}{\lambda^3} = \frac{V\zeta}{\lambda^3} = \langle N \rangle, \qquad \frac{\partial \lambda}{\partial \beta} = \frac{\lambda}{2\beta}, \tag{6.6.12}$$

we obtain

$$S = k\langle N \rangle - k\beta\langle N \rangle kT \ln \zeta + k\beta \frac{3}{2} \frac{\langle N \rangle}{\beta}$$

$$= \frac{5}{2} \langle N \rangle k - \langle N \rangle k \ln \left(\frac{\langle N \rangle \lambda^3}{V} \right)$$

$$= \frac{5}{2} \langle N \rangle k + \langle N \rangle k \ln \left(\frac{V}{\langle N \rangle \lambda^3} \right), \tag{6.6.13}$$

which is the Sackur-Tetrode equation derived in Section 3.5.1. Note that because the $1/N!$ is included *a posteriori* in the expression for $Q(N, V, T)$, the correct quantum mechanical entropy expression results.

6.7 Particle number fluctuations in the grand canonical ensemble

In the grand canonical ensemble, the total particle number fluctuates at constant chemical potential. It is, therefore, instructive to analyze these fluctuations, as was done for the energy fluctuations in the canonical ensemble (Section 4.5) and volume fluctuations in the isothermal-isobaric ensemble (see Problem 5.2 in Chapter 5). Particle number fluctuations in the grand canonical ensemble can be studied by considering the variance

$$\Delta N = \sqrt{\langle N^2 \rangle - \langle N \rangle^2}. \tag{6.7.1}$$

In order to compute this quantity, we start by examining the operation

$$\zeta \frac{\partial}{\partial \zeta} \zeta \frac{\partial}{\partial \zeta} \ln \mathcal{Z}(\zeta, V, T). \tag{6.7.2}$$

Using eqn. (6.4.25), this becomes

$$\zeta \frac{\partial}{\partial \zeta} \zeta \frac{\partial}{\partial \zeta} \ln \mathcal{Z}(\zeta, V, T) = \zeta \frac{\partial}{\partial \zeta} \frac{1}{\mathcal{Z}} \sum_{N=0}^{\infty} N \zeta^N Q(N, V, T)$$

$$= \frac{1}{\mathcal{Z}} \sum_{N=0}^{\infty} N^2 \zeta^N Q(N, V, T) - \frac{1}{\mathcal{Z}^2} \left[\sum_{N=0}^{\infty} N \zeta^N Q(N, V, T) \right]^2$$

$$= \langle N^2 \rangle - \langle N \rangle^2. \tag{6.7.3}$$

Thus, we have

$$(\Delta N)^2 = \zeta \frac{\partial}{\partial \zeta} \zeta \frac{\partial}{\partial \zeta} \ln \mathcal{Z}(\zeta, V, T). \tag{6.7.4}$$

Expressing eqn. (6.7.4) as derivatives of $\mathcal{Z}(\mu, V, T)$ with respect to μ, we obtain

$$(\Delta N)^2 = (kT)^2 \frac{\partial^2}{\partial \mu^2} \ln \mathcal{Z}(\mu, V, T) = (kT)^2 \frac{\partial^2}{\partial \mu^2} \frac{PV}{kT}. \tag{6.7.5}$$

Since μ, V, and T are the independent variables in the ensemble, only the pressure in the above expression depends on μ, and we can write

$$(\Delta N)^2 = kTV \frac{\partial^2 P}{\partial \mu^2}. \tag{6.7.6}$$

Therefore, computing the particle number fluctuations amounts to computing the second derivative of the pressure with respect to chemical potential. This is a rather nontrivial bit of thermodynamics, which can be carried out in a variety of ways. One approach is the following: Let $A(N, V, T)$ be the canonical Helmholtz free energy at a particular value of N. Recall that the pressure can be obtained from $A(N, V, T)$ via

$$P = - \left(\frac{\partial A}{\partial V} \right). \tag{6.7.7}$$

Since $A(N, V, T)$ is an extensive quantity, and we want to make the N dependence in the analysis as explicit as possible, we define an intensive Helmholtz free energy $a(v, T)$ by

$$a(v, T) = \frac{1}{N} A\left(N, \frac{V}{N}, T\right), \tag{6.7.8}$$

where $v = V/N$ is the volume per particle and $a(v, T)$ is clearly the Helmholtz free energy per particle. Then,

$$P = -N\frac{\partial a}{\partial v}\frac{\partial v}{\partial V} = -N\frac{\partial a}{\partial v}\frac{1}{N} = -\frac{\partial a}{\partial v}. \tag{6.7.9}$$

From eqn. (6.7.9), it follows that

$$\frac{\partial P}{\partial \mu} = \frac{\partial P}{\partial v}\frac{\partial v}{\partial \mu} = -\frac{\partial^2 a}{\partial v^2}\frac{\partial v}{\partial \mu}. \tag{6.7.10}$$

We can obtain an expression for $\partial\mu/\partial v$ by

$$\mu = \frac{\partial A}{\partial N}$$

$$= a(v, T) + N\frac{\partial a}{\partial v}\frac{\partial v}{\partial N}$$

$$= a(v, T) - v\frac{\partial a}{\partial v}, \tag{6.7.11}$$

so that

$$\frac{\partial \mu}{\partial v} = \frac{\partial a}{\partial v} - \frac{\partial a}{\partial v} - v\frac{\partial^2 a}{\partial v^2}$$

$$= -v\frac{\partial^2 a}{\partial v^2}. \tag{6.7.12}$$

Substituting this result into eqn. (6.7.10) gives

$$\frac{\partial P}{\partial \mu} = -\frac{\partial^2 a}{\partial v^2}\left[\frac{\partial \mu}{\partial v}\right]^{-1} = \frac{\partial^2 a}{\partial v^2}\left[v\frac{\partial^2 a}{\partial v^2}\right]^{-1} = \frac{1}{v}. \tag{6.7.13}$$

Differentiating eqn. (6.7.13) once again with respect to μ gives

$$\frac{\partial^2 P}{\partial \mu^2} = -\frac{1}{v^2}\frac{\partial v}{\partial \mu} = \frac{1}{v^2}\left[v\frac{\partial^2 a}{\partial v^2}\right]^{-1} = -\frac{1}{v^3\partial P/\partial v}. \tag{6.7.14}$$

Now, recall that the isothermal compressibility is given by

$$\kappa_T = -\frac{1}{V}\frac{\partial V}{\partial P} = -\frac{1}{v}\frac{\partial v}{\partial P} = -\frac{1}{v\partial P/\partial v} \tag{6.7.15}$$

and is an intensive quantity. It is clear from eqn. (6.7.14) that $\partial^2 P/\partial \mu^2$ can be expressed in terms of κ_T as

$$\frac{\partial^2 P}{\partial \mu^2} = \frac{1}{v^2}\kappa_T, \tag{6.7.16}$$

so that

$$(\Delta N)^2 = kT\langle N\rangle v \frac{1}{v^2}\kappa_T = \frac{\langle N\rangle kT\kappa_T}{v}, \tag{6.7.17}$$

where the specific value of N has been replaced by its average value $\langle N\rangle$ in the grand canonical ensemble. The relative fluctuations in particle number can now be computed from

$$\frac{\Delta N}{\langle N\rangle} = \frac{1}{\langle N\rangle}\sqrt{\frac{\langle N\rangle kT\kappa_T}{v}} = \sqrt{\frac{kT\kappa_T}{\langle N\rangle v}} \sim \frac{1}{\sqrt{\langle N\rangle}}. \tag{6.7.18}$$

Thus, as $\langle N\rangle \longrightarrow \infty$ in the thermodynamic limit, the particle fluctuations vanish and the grand canonical ensemble is seen to be equivalent to the other ensembles in this limit.

6.8 Potential distribution theorem

Up to this point, our treatment of the grand canonical ensemble has been restricted to one-component systems of N particles. Extending the ensemble formalism to mixed systems, such as solutions or co-crystals, that contain multiple components, is straightforward in all ensembles except the grand canonical ensemble, since for mixtures we need to do the bookkeeping of the extra particle-number sums in constructing the partition function and ensemble averages. As a vehicle for illustrating the treatment of multiple species within the grand canonical ensemble, we will present a derivation of an important result in the theory of solutions, namely the *potential distribution theorem*, which was originally derived by Widom (1963, 1982). This theorem provides a statistical mechanical framework for computing the excess chemical potential associated with the addition of an extra solute molecule to a solution, a topic we will revisit in Chapter 8.

Let us consider a solution containing N_l solvent particles and N_s solute particles. Even though $N_l + N_s = N$, the total particle number, N, varies from 0 to ∞, which means we can vary N_l and N_s separately. Let $\mathbf{r}_1^{(l)}, ..., \mathbf{r}_{N_l}^{(l)} \equiv \mathbf{r}^{(l)}$ denote the solvent particle coordinates, and let $\mathbf{r}_1^{(s)}, ..., \mathbf{r}_{N_s}^{(s)} \equiv \mathbf{r}^{(s)}$ denote the solute particle coordinates. The grand canonical partition function of the solution is

$$\mathcal{Z}(\zeta_l, \zeta_s, V, T) = e^{\beta PV} = \sum_{N_l=0}^{\infty}\sum_{N_s=0}^{\infty} \zeta_l^{N_l}\zeta_s^{N_s} Q(N_l, N_s, V, T). \tag{6.8.1}$$

Equation (6.8.1) reflects the fact that each species in the mixture has its own chemical potential and corresponding fugacity: $\beta\mu_l = \ln\zeta_l$, $\beta\mu_s = \ln\zeta_s$. The potential distribution theorem starts by considering the average number of solute particles in the solution, which can be expressed as a grand canonical ensemble average of the form

$$\langle N_s \rangle = e^{-\beta PV} \sum_{N_l=0}^{\infty} \sum_{N_s=0}^{\infty} N_s \zeta_l^{N_l} \zeta_s^{N_s} Q(N_l, N_s, V, T). \tag{6.8.2}$$

The presence of N_s in the summand means that the N_s sum begins at 1, since the first term is 0. Thus,

$$\langle N_s \rangle = e^{-\beta PV} \sum_{N_l=0}^{\infty} \sum_{N_s=1}^{\infty} N_s \zeta_l^{N_l} \zeta_s^{N_s} Q(N_l, N_s, V, T). \tag{6.8.3}$$

If we require that the N_s sum start at 0, we need to replace N_s by $N_s + 1$, in which case, eqn. (6.8.3) becomes

$$\langle N_s \rangle = e^{-\beta PV} \sum_{N_l=0}^{\infty} \sum_{N_s=0}^{\infty} (N_s + 1) \zeta_l^{N_l} \zeta_s^{N_s+1} Q(N_l, N_s + 1, V, T). \tag{6.8.4}$$

We now factor our ζ_s out of eqn. (6.8.4) and multiply and divide by a partition function factor $Q(0, 1, V, T)$, which is simply the partition function of one solute particle. This gives

$$\langle N_s \rangle = e^{-\beta PV} \zeta_s Q(0, 1, V, T) \sum_{N_l=0}^{\infty} \sum_{N_s=0}^{\infty} (N_s + 1) \zeta_l^{N_l} \zeta_s^{N_s+1} \frac{Q(N_l, N_s + 1, V, T)}{Q(0, 1, V, T)}. \tag{6.8.5}$$

The factor $Q(0, 1, V, T)$ is the partition function of a single solute particle in the gas phase and is, therefore, given by the canonical ideal gas relation with $N_s = 1$:

$$Q(0, 1, V, T) = \frac{q_s^{(\text{int})} V}{\Lambda_s^3}. \tag{6.8.6}$$

If a solute "particle" is a molecule of n atoms, then $q_s^{(\text{int})}$ results from the internal potential of the molecule and V is the volume factor associated with the center of mass (see Problem 4.13). The factor Λ_s is simply a product of the individual thermal wavelengths of each of the n atoms, $\Lambda_s = \lambda_{s,1} \cdots \lambda_{s,n}$. Obviously, if the solute particle is a single atom, then $q_s = V/\lambda_s^3$. Introducing the solute density $\rho_s = \langle N_s \rangle / V$, eqn. (6.8.5) can be written as

$$\frac{\rho_s \Lambda_s^3}{\zeta_s q_s^{(\text{int})}} = e^{-\beta PV} \sum_{N_l=0}^{\infty} \sum_{N_s=0}^{\infty} (N_s + 1) \zeta_l^{N_l} \zeta_s^{N_s+1} \frac{Q(N_l, N_s + 1, V, T)}{Q(0, 1, V, T)}. \tag{6.8.7}$$

The partition function $Q(N_l, N_s + 1, V, T)$ requires specifying the potential for N_l solvent particles and $N_s + 1$ solute particles, which we write as

$$U(\mathbf{r}^{(l)}, \mathbf{r}^{(s)}, \mathbf{r}_{N_s+1}^{(s)}) = U(\mathbf{r}^{(l)}, \mathbf{r}^{(s)}) + \Delta U(\mathbf{r}^{(l)}, \mathbf{r}^{(s)}, \mathbf{r}_{N_s+1}^{(s)}), \tag{6.8.8}$$

where $U(\mathbf{r}^{(l)}, \mathbf{r}^{(s)})$ is the potential of the original solution and $\Delta U(\mathbf{r}^{(l)}, \mathbf{r}^{(s)}, \mathbf{r}_{N_s+1}^{(s)})$ is the interaction between the $(N_s + 1)^{\text{st}}$ particle and the solution. With the definitions of the terms in eqn. (6.8.8), we can write

$$(N_s + 1)Q(N_l, N_s + 1, V, T) = \frac{(N_s + 1)}{(N_s + 1)! N_l! \Lambda^{3N_l} \Lambda^{3(N_s + 1)}}$$

$$\times \int \mathrm{d}\mathbf{r}^{(l)} \mathrm{d}\mathbf{r}^{(s)} \mathrm{d}\mathbf{r}^{(s)}_{N_s + 1} \mathrm{e}^{-\beta(U(\mathbf{r}^{(l)}, \mathbf{r}^{(s)}) + \Delta U(\mathbf{r}^{(l)}, \mathbf{r}^{(s)}, \mathbf{r}^{(s)}_{N_s + 1}))}. \qquad (6.8.9)$$

Since $(N_s + 1)/(N_s + 1)! = 1/N_s!$, if we now multiply and divide by $Q(N_l, N_s, V, T)$, which is the canonical partition function of the original solution, we obtain

$$\frac{(N_s + 1)Q(N_l, N_s + 1, V, T)}{Q(0, 1, V, T)} = Q(N_l, N_s, V, T) \left[\frac{1}{Q(N_l, N_s, V, T)Q(0, 1, V, T)} \right.$$

$$\left. \times \frac{1}{N_l! N_s! \Lambda_l^{3N_l} \Lambda_s^{3(N_s + 1)}} \int \mathrm{d}\mathbf{r}^{(l)} \mathrm{d}\mathbf{r}^{(s)} \mathrm{d}\mathbf{r}^{(s)}_{N_s + 1} \mathrm{e}^{-\beta(U(\mathbf{r}^{(l)}, \mathbf{r}^{(s)}) + \Delta U(\mathbf{r}^{(l)}, \mathbf{r}^{(s)}, \mathbf{r}^{(s)}_{N_s + 1}))} \right]. \quad (6.8.10)$$

The term in square brackets is an average of $\exp(-\beta \Delta U(\mathbf{r}^{(l)}, \mathbf{r}^{(s)}, \mathbf{r}^{(s)}_{N_s + 1}))$ over a canonical ensemble of N_l solvent particles and $N_s + 1$ solute particles but with no interaction between the $(N_s + 1)^{\text{st}}$ particle and the original solution, an average we denote as $\langle \exp(-\beta \Delta U) \rangle_0 (N_l, N_s, V, T)$. With this notation, eqn. (6.8.10) becomes

$$\frac{(N_s + 1)Q(N_l, N_s + 1, V, T)}{Q(0, 1, V, T)} = Q(N_l, N_s, V, T) \left\langle \mathrm{e}^{-\beta \Delta U} \right\rangle_0 (N_l, N_s, V, T). \qquad (6.8.11)$$

When eqn. (6.8.11) is substituted into eqn. (6.8.7), we obtain

$$\frac{\rho_s \Lambda_s^3}{\zeta_s q_s^{(\text{int})}} = \mathrm{e}^{-\beta PV} \sum_{N_l = 0}^{\infty} \sum_{N_s = 0}^{\infty} \zeta_l^{N_l} \zeta_s^{N_s} \left\langle \mathrm{e}^{-\beta \Delta U} \right\rangle_0 (N_l, N_s, V, T) Q(N_l, N_s, V, T)$$

$$= \left\langle \mathrm{e}^{-\beta \Delta U} \right\rangle = \mathrm{e}^{-\beta \mu_{\text{ex}}}. \qquad (6.8.12)$$

The first line of eqn. (6.8.12) is a grand canonical average of $\langle \mathrm{e}^{-\beta \Delta U} \rangle_0 (N_l, N_s, V, T)$, which leads to our final result in the second line of eqn. (6.8.12). The last line expresses the work needed to add the $(N_s + 1)^{\text{st}}$ solute particle to the solution, which is referred to as the *excess chemical potential*, μ_{ex}. The fact that μ_{ex} can be obtained from this grand canonical average $\langle \mathrm{e}^{-\beta \Delta U} \rangle$ is the *potential distribution theorem*, which is an important starting point for theories of solutions and solvation (Pratt and Laviolette, 1998; Beck, 2011).

6.9 Molecular dynamics in the grand canonical ensemble

Our previous discussions of molecular dynamics simulation in the canonical and isobaric ensembles suggest that equations of motion are rather natural for treating fluctuations in ensemble control variables, such as temperature and pressure, that can be expressed as averages of continuous functions of the phase-space vector x. The key element proves to be extending the phase space with variables designed to control the fluctuations in the phase-space estimators of these control variables. The same is not true for particle-number fluctuations, which are discrete by nature; for this reason, molecular dynamics is an unnatural choice for generating a grand canonical ensemble distribution. In principle, one could use a very large canonical simulation box and select a small subcell within it to compute grand canonical averages. Such a calculation would closely match the setup in Fig. 6.1, and the number of particles within this subcell will exhibit fluctuations consistent with a grand canonical distribution.

There are, however, a number of disadvantages to performing simulations in the manner described above. First, it can only be applied in spatially homogeneous systems where we expect the same particle-number fluctuations throughout the system; many applications are not of this type. Second, since particles can only enter a subcell through its boundary, convergence of particle-number fluctuations could be quite slow, which is problematic given the high computational overhead of running a large canonical simulation box in the first place. This problem becomes especially severe if a subcell within the canonical box must, itself, also be large, since the surface-area-to-volume ratio decreases as the size of the subcell increases.

Ideally, we would like to be able to perform grand canonical simulations using a single simulation box of any desired size, and allow particles to appear and disappear anywhere within the box; this would lead to maximum efficiency in the convergence of particle-number fluctuations. However, as noted above, simulating discrete changes is not easily achieved within the molecular dynamics framework. We will see in Chapter 7 that particle-number fluctuations can be handled via Monte Carlo simulation methods, which are more flexible in the types of changes they allow. Nevertheless, there is an approach to grand canonical molecular dynamics that avoids the overhead of large simulation boxes through multiple scales of spatial resolution. We close out this chapter with a discussion of this approach.

The method is known as Hamiltonian Adaptive Resolution Simulation (H-AdResS) (Potestio *et al.*, 2013; Heidari *et al.*, 2020) and is close in spirit to the scheme in Fig. 6.1. However, H-AdResS exploits the fact that we can design a particle reservoir whose spatial resolution is coarser than that of the system and within which the interparticle interactions are highly simplified compared to those in the system. With these simplifications, the computational demand is significantly lower than using a small subcell within a large canonical box. A full understanding of the details of the H-AdResS approach requires techniques to be discussed in Chapters 7 and 8; however, we can proceed to outline the basic ideas here.

Figure 6.2 illustrates the setup for an H-AdResS simulation of a molecular fluid. The simulation cell is divided into three regions: (i) an atomistic region, whose grand canonical properties we seek; (ii) a coarse-grained region containing a set of reservoir particles, described at a low level of spatial resolution, that can flow in and out of the atomistic region; and (iii) a hybrid region, in which coarse-grained reservoir particles are transformed into molecules at full atomistic resolution. If the coarse-grained particles are the same molecules as in the atomistic region but with simplified interactions, *e.g.*, occurring only between their centers of mass, then the transformation between the coarse and atomistic regions is simple, and the Hamiltonian can be written as

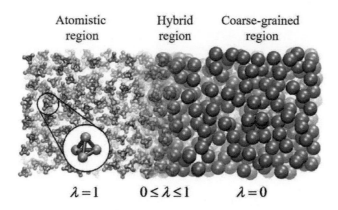

Atomistic region   Hybrid region   Coarse-grained region

$\lambda = 1$ $0 \leq \lambda \leq 1$ $\lambda = 0$

Fig. 6.2 Illustration of the H-AdResS simulation box setup (reproduced with permission from Potestio *et al.*, *Phys. Rev. Lett.* **110**, 108301 (2013), copyright American Physical Society).

$$\mathcal{H} = \sum_{\alpha=1}^{N} \sum_{i=1}^{n} \frac{\mathbf{p}_{\alpha,i}^2}{2m_i} + \sum_{\alpha=1}^{N} \left[\lambda_\alpha U_\alpha^{(\mathrm{AA})} + (1 - \lambda_\alpha) U_\alpha^{(\mathrm{CG})} \right] + U_{\mathrm{intra}}, \qquad (6.9.1)$$

where the system is assumed to have N total molecules, each containing n atoms with mass m_i, $i = 1, ..., n$. The potential $U_\alpha^{(\mathrm{AA})}$ is felt by the αth molecule in the atomistic region, $U_\alpha^{(\mathrm{CG})}$ is the potential of the αth molecular mass center in the coarse-grained region, λ_α is a function of the molecular mass centers that switches smoothly from 0 to 1 through the hybrid region, and U_{intra} is the intramolecular potential. Note that the molecules retain their structure in the coarse-grained region, but since the interaction is only between centers of mass, individual atoms are carried along as "excess baggage". A technical issue with the Hamiltonian in eqn. (6.9.1) concerns differences between the atomistic and coarse-grained interaction potentials. Because of these differences, the density and chemical potentials in the atomistic and coarse-grained regions are not the same, which can cause unphysical gradients between the two regions. It is, therefore, necessary to add a compensating term into eqn. (6.9.1) derived from a free-energy calculation, which we will discuss in Chapter 8. Finally, in order to maintain a constant chemical potential in the coarse-grained region, which, in turn, fixes the chemical potential of the atomistic region, a Monte Carlo particle insertion/deletion algorithm is employed (see Section 7.3.3). In practice, it helps if the potential in the coarse-grained region is one for which the chemical potential can be derived analytically; an obvious choice is to take the coarse-grained region to be a reservoir of ideal gas particles (Heidari *et al.*, 2020). A feature of the H-AdResS scheme is that it can be employed at and away from equilibrium (Delle Site and Praprotnik, 2017; Ciccotti and Delle Site, 2019; Cortes-Huerto *et al.*, 2021).

6.10   Problems

6.1. Using a Legendre transform, determine if it is possible to define an ensemble in which μ, P, and T are the control variables. Can you rationalize your result based on Euler's theorem?

6.2.   a. Derive the thermodynamic relations for an ensemble in which μ, V, and S are the control variables.

   b. Determine the partition function for this ensemble.

6.3. For the ideal gas in Problem 4.7 of Chapter 4, imagine dividing the cylinder into rings of radius r, thickness Δr, and height Δz. Within each ring, assume that r and z are constant.

   a. Within each ring, explain why it is possible to work within the grand canonical ensemble.

   b. Show that the grand canonical partition function within each ring satisfies

$$\mathcal{Z}(\mu, V_{\text{ring}}, r, z, T) = \mathcal{Z}^{(0)}(\mu_{\text{eff}}(r, z), V_{\text{ring}}, T),$$

   where $\mathcal{Z}^{(0)}$ is the grand canonical partition function for $\omega = 0$ and $g = 0$, V_{ring} is the volume of each ring, and $\mu_{\text{eff}}(r, z)$ is an effective local chemical potential that varies from ring to ring. Derive an expression for $\mu_{\text{eff}}(r, z)$.

   c. Is this result true even if there are interactions among the particles? Why or why not?

6.4. Consider an equilibrium chemical reaction involving K molecular species denoted $X_1, ..., X_K$, where some of the species are reactants and some are products. Denote the chemical equation governing the reaction as

$$\sum_{i=1}^{K} \nu_i X_i = 0,$$

where ν_i are the stoichiometric coefficients in the reaction. Using this notation, the coefficients of the products are, by definition, negative. As the reaction proceeds, there will be a change δN_i in the number N_i of each species such that the law of mass balance is

$$\frac{\delta N_1}{\nu_1} = \frac{\delta N_2}{\nu_2} = \cdots \frac{\delta N_K}{\nu_K}.$$

In order to find a condition describing the chemical equilibrium, we can make use of the Helmholtz free energy $A(N_1, N_2, ..., N_K, V, T)$. At equilibrium, the changes δN_i should not change the free energy to first order. That is, $\delta A = 0$.

a. Show that this assumption leads to the equilibrium condition

$$\sum_{i=1}^{K} \mu_i \nu_i = 0.$$

b. Now consider the reaction

$$2H_2(g) + O_2(g) \rightleftharpoons 2H_2O(g).$$

Let ρ_0 be the initial density of H_2 molecules and $\rho_0/2$ be the initial density of O_2 molecules, and let the initial amount of H_2O be zero. Calculate the equilibrium densities of the three components as a function of temperature and ρ_0.

*6.5. Prove the following fluctuation theorems for the grand canonical ensemble:

a.

$$\langle N\mathcal{H}(\mathbf{x}) \rangle - \langle N \rangle \langle \mathcal{H}(\mathbf{x}) \rangle = \left(\frac{\partial E}{\partial N} \right)_{V,T} (\Delta N)^2.$$

b.

$$\Delta \mathcal{F}^2 = kT^2 C_V + \left[\left(\frac{\partial E}{\partial N} \right)_{V,T} - \mu \right]^2 (\Delta N)^2,$$

where C_V is the constant-volume heat capacity, $\mathcal{F} = E - N\mu = TS - PV$, and $\Delta \mathcal{F} = \sqrt{\langle \mathcal{F}^2 \rangle - \langle \mathcal{F} \rangle^2}$.

6.6. In a multicomponent system with K components, show that the fluctuations in the particle numbers of each component are related by

$$\Delta N_i \Delta N_j = kT \left(\frac{\partial \langle N_i \rangle}{\partial \mu_j} \right)_{V,T,\mu_i} = kT \left(\frac{\partial \langle N_j \rangle}{\partial \mu_i} \right)_{V,T,\mu_j},$$

where $\Delta N_i = \sqrt{\langle N_i^2 \rangle - \langle N_i \rangle^2}$, with a similar definition for ΔN_j.

6.7. Derive eqns. (6.8.6) and (6.8.11).

6.8. The Potential Distribution Theorem has a useful inverse form:

$$\langle e^{\beta \Delta U} \rangle = e^{\beta \mu_{ex}}.$$

Using the approach in Section 6.8, prove this result and give a physical interpretation of it.

6.9. For the gas in Problem 4.17 of Chapter 4, suppose, instead of a fixed volume V, the external pressure P is held fixed. Calculate the partition function, the equation of state, the enthalpy, the heat capacity, and the entropy under these conditions.

7
Monte Carlo

7.1   Introduction to the Monte Carlo method

In our treatment of the equilibrium ensembles, we have, thus far, exclusively developed and employed dynamical techniques for sampling the phase-space distributions. This choice was motivated by the natural connection between the statistical ensembles and classical (Hamiltonian or non-Hamiltonian) mechanics. The dynamical aspect of these approaches is, however, irrelevant for equilibrium statistical mechanics, as we are interested only in sampling the accessible microscopic states of the ensemble.

In this chapter, we will introduce another class of sampling techniques known as *Monte Carlo* methods. As the name implies, Monte Carlo techniques are based on games of chance (driven by sequences of random numbers) which, when played many times, yield outcomes that are the solutions to particular problems. The first use of random methods to solve a physical problem dates back to 1930 when Enrico Fermi (1901-1954) employed such an approach to study the properties of neutrons. Monte Carlo simulations also played a central role in the Manhattan Project. It was not until computers could be leveraged that the power of Monte Carlo methods would be realized. In the 1950s, for example, Monte Carlo methods were carried out on the MANIAC at Los Alamos National Laboratory in New Mexico for research on the hydrogen bomb. Eventually, it was determined that Monte Carlo techniques constitute a powerful suite of tools for solving statistical mechanical problems involving integrals of very high dimension.

As a simple illustrative example, consider the evaluation of the definite integral

$$I = \int_0^1 \mathrm{d}x \int_0^{\sqrt{1-x^2}} \mathrm{d}y = \frac{\pi}{4}. \tag{7.1.1}$$

The result $\pi/4$ can be obtained straightforwardly, since this is an elementary integral. Note that the answer $\pi/4$ is also the ratio of the area of a circle of arbitrary radius to the area of its circumscribed square. This fact suggests that the following game could be used to solve the integral: Draw a square and an inscribed circle on a piece of paper, tape the paper to a dart board, and throw darts randomly at the board. The ratio of the number of darts that land in the circle to the number of darts that land anywhere in the square will, in the limit of a very large number of dart throws, yield a good estimate of the area ratio and hence of the integral in eqn. (7.1.1).[1] In

---

[1]Kalos and Whitlock (1986) suggested putting a round cake pan in a square one, placing the combination in a rain storm, and measuring the ratio of raindrops that fall in the round cake pan to those that fall in the square one.

practice, it would take about 10^6 such dart throws to achieve a reasonable estimate of $\pi/4$, which would try the patience of even the most avid dart player. For this reason, it is more efficient to have the computer throw the darts. Nevertheless, this example shows that a simple random process can be used to produce a numerical estimate of a two-dimensional integral; no fancy sets of dynamical differential equations are needed.

In this chapter, we will discuss an important underpinning of the Monte Carlo technique, namely the central limit theorem, and then proceed to describe a number of commonly used Monte Carlo algorithms for evaluating high-dimensional integrals of the type that are ubiquitous in classical equilibrium statistical mechanics.

7.2   The Central Limit theorem

The integrals that must be evaluated in equilibrium statistical mechanics are generally of the form

$$I = \int \mathrm{dx}\ \phi(\mathrm{x})f(\mathrm{x}), \tag{7.2.1}$$

where x is an n-dimensional vector, $\phi(\mathrm{x})$ is an arbitrary function, and $f(\mathrm{x})$ is a function satisfying the properties of a probability distribution function, namely $f(\mathrm{x}) \geq 0$ and

$$f(\mathrm{x}) \geq 0$$

$$\int \mathrm{dx}\ f(\mathrm{x}) = 1. \tag{7.2.2}$$

The integral in eqn. (7.2.1) represents the ensemble average of a physical observable in equilibrium statistical mechanics. Let $\mathrm{x}_1, ..., \mathrm{x}_M$ be a set of M n-dimensional vectors that are *sampled* from $f(\mathrm{x})$. That is, the vectors $\mathrm{x}_1, ..., \mathrm{x}_M$ are distributed according to $f(\mathrm{x})$, so that the probability that the vector x_i is in a small region dx of the n-dimensional space on which the vectors $\mathrm{x}_1, ..., \mathrm{x}_M$ are defined is $f(\mathrm{x}_i)\mathrm{dx}$. Recall that in Section 3.8.3, we described an algorithm for sampling the Maxwell-Boltzmann distribution, which is a particularly simple case. In general, the problem of sampling a distribution $f(\mathrm{x})$ is a nontrivial one that we will address in this chapter. For now, however, let us assume that an algorithm exists for carrying out the sampling of $f(\mathrm{x})$ and generating the vectors $\mathrm{x}_1, ..., \mathrm{x}_M$. We will establish that the simple arithmetic average

$$\tilde{I}_M = \frac{1}{M}\sum_{i=1}^{M}\phi(\mathrm{x}_i) \tag{7.2.3}$$

is an *estimator* for the integral I, meaning that

$$\lim_{M\to\infty}\tilde{I}_M = I. \tag{7.2.4}$$

This result is guaranteed by a theorem known as the *central limit theorem*, which we will now prove. Readers wishing to proceed immediately to the specifics of Monte Carlo methodology can take the results in eqns. (7.2.3) and (7.2.4) as given and skip to the next section.

For simplicity, we introduce the notation

$$\int dx\, \phi(x) f(x) = \langle \phi \rangle_f, \tag{7.2.5}$$

where $\langle \cdots \rangle_f$ indicates an average of $\phi(x)$ with respect to the distribution $f(x)$. We wish to compute the probability $\mathcal{P}(y)$ that the estimator $\tilde{I}_M$ will have a value y. This probability is given formally by

$$\mathcal{P}(y) = \int dx_1 \cdots dx_M \left[\prod_{i=1}^{M} f(x_i) \right] \delta\left(\frac{1}{M} \sum_{i=1}^{M} \phi(x_i) - y \right), \tag{7.2.6}$$

where the Dirac δ-function restricts the integral to those sets of vectors $x_1, ..., x_M$ for which the estimator is equal to y. Equation (7.2.6) can be simplified by introducing the integral representation of the δ-function (see Appendix A)

$$\delta(z) = \frac{1}{2\pi} \int_{-\infty}^{\infty} d\sigma\, e^{iz\sigma}. \tag{7.2.7}$$

Substituting eqn. (7.2.7) into eqn. (7.2.6) and using the general property of δ-functions that $\delta(ax) = (1/|a|)\delta(x)$ yields

$$\mathcal{P}(y) = M \int dx_1 \cdots dx_M \left[\prod_{i=1}^{M} f(x_i) \right] \delta\left(\sum_{i=1}^{M} \phi(x_i) - My \right)$$

$$= \frac{M}{2\pi} \int dx_1 \cdots dx_M \left[\prod_{i=1}^{M} f(x_i) \right] \int_{-\infty}^{\infty} d\sigma\, e^{i\sigma\left(\sum_{i=1}^{M} \phi(x_i) - My \right)}. \tag{7.2.8}$$

Interchanging the order of integrations gives

$$\mathcal{P}(y) = \frac{M}{2\pi} \int_{-\infty}^{\infty} d\sigma\, e^{-iM\sigma y} \int dx_1 \cdots dx_M \left[\prod_{i=1}^{M} f(x_i) \right] e^{i\sigma \sum_{i=1}^{M} \phi(x_i)}$$

$$= \frac{M}{2\pi} \int_{-\infty}^{\infty} d\sigma\, e^{-iM\sigma y} \left[\int dx\, f(x) e^{i\sigma\phi(x)} \right]^M$$

$$= \frac{M}{2\pi} \int_{-\infty}^{\infty} d\sigma\, e^{-iM\sigma y} e^{M \ln \int dx\, f(x) e^{i\sigma\phi(x)}}$$

$$= \frac{M}{2\pi} \int_{-\infty}^{\infty} d\sigma\, e^{MF(\sigma, y)}, \tag{7.2.9}$$

where in the second line, we have used the fact that the integrals over $x_1, ..., x_M$ in the product are all identical. In the last line of eqn. (7.2.9), the function $F(\sigma, y)$ is defined to be

$$F(\sigma, y) = -i\sigma y + g(\sigma) \tag{7.2.10}$$

with

$$g(\sigma) = \ln \int \mathrm{d}x\, f(\mathrm{x}) e^{i\sigma\phi(\mathrm{x})}. \qquad (7.2.11)$$

Although we cannot evaluate the integral over σ in eqn. (7.2.9) exactly, we can approximate it by a technique known as the *stationary phase method*. This technique applies to integrals of functions $F(\sigma, y)$ that are sharply peaked about a global maximum at $\sigma = \tilde{\sigma}(y)$ where the integral is expected to have its dominant contribution. For $\sigma = \tilde{\sigma}(y)$ to be a maximum, the following conditions must hold:

$$\left.\frac{\partial F}{\partial \sigma}\right|_{\sigma = \tilde{\sigma}(y)} = 0, \qquad \left.\frac{\partial^2 F}{\partial \sigma^2}\right|_{\sigma = \tilde{\sigma}(y)} < 0 \qquad (7.2.12)$$

when $\sigma = \tilde{\sigma}(y)$. Thus, the function of $\tilde{\sigma}(y)$ is derived from the solution of the condition on the left in eqn. (7.2.12), which depends on the value of y. We will return to this point shortly. Expanding $F(\sigma, y)$ in a Taylor series about $\sigma = \tilde{\sigma}(y)$ up to second-order and taking into account that $\partial F/\partial \sigma = 0$ at $\sigma = \tilde{\sigma}(y)$, gives

$$F(\sigma, y) = F(\tilde{\sigma}(y), y) + \frac{1}{2}\left.\frac{\partial^2 F}{\partial \sigma^2}\right|_{\sigma = \tilde{\sigma}(y)} (\sigma - \tilde{\sigma}(y))^2 + \cdots. \qquad (7.2.13)$$

Substituting eqn. (7.2.13) into eqn. (7.2.9) yields

$$\mathcal{P}(y) \approx \frac{M}{2\pi} e^{MF(\tilde{\sigma}(y),y)} \int_{-\infty}^{\infty} \mathrm{d}\sigma \exp\left[\frac{M}{2}\left.\frac{\partial^2 F}{\partial \sigma^2}\right|_{\sigma = \tilde{\sigma}(y)} (\sigma - \tilde{\sigma}(y))^2\right]. \qquad (7.2.14)$$

Since the integral in eqn. (7.2.14) is now just a Gaussian integral over σ, it can be performed straightforwardly to give

$$\mathcal{P}(y) \approx \sqrt{\frac{M}{-2\pi\,(\partial^2 F/\partial \sigma^2)|_{\sigma = \tilde{\sigma}(y)}}}\; e^{MF(\tilde{\sigma}(y),y)}. \qquad (7.2.15)$$

Thus, in order to specify the distribution, we need to find $\tilde{\sigma}(y)$ and $(\partial^2 F/\partial \sigma^2)|_{\sigma = \tilde{\sigma}(y)}$. The condition $\partial F/\partial \sigma = 0$ leads to $-iy + g'(\sigma) = 0$ or

$$y = -ig'(\sigma) = \frac{\int \mathrm{d}x\, \phi(\mathrm{x}) f(\mathrm{x}) e^{i\sigma\phi(\mathrm{x})}}{\int \mathrm{d}x\, f(\mathrm{x}) e^{i\sigma\phi(\mathrm{x})}}. \qquad (7.2.16)$$

The last line in eqn. (7.2.16) can, in principle, be inverted to give the solution $\sigma = \tilde{\sigma}(y)$. Moreover,

$$\left.\frac{\partial^2 F}{\partial \sigma^2}\right|_{\sigma = \tilde{\sigma}(y)} = g''(\tilde{\sigma}(y)). \qquad (7.2.17)$$

Therefore,

$$\mathcal{P}(y) = \sqrt{\frac{M}{-2\pi g''(\tilde{\sigma}(y))}}\; e^{MF(\tilde{\sigma}(y),y)}. \qquad (7.2.18)$$

Equation (7.2.18) is a function of y alone, and we can now analyze its y dependence in greater detail. First, the extrema of $F(\tilde{\sigma}(y), y)$ are given by the solution of

$$0 = \frac{\mathrm{d}F}{\mathrm{d}y} = \frac{\partial F}{\partial y} + \frac{\partial F}{\partial \tilde{\sigma}} \frac{\partial \tilde{\sigma}}{\partial y} = -i\tilde{\sigma}(y). \tag{7.2.19}$$

Since $\partial F / \partial \tilde{\sigma}(y) = 0$ by the definition of $\tilde{\sigma}(y)$, the extrema of F occur where $\tilde{\sigma}(y) = 0$. According to eqn. (7.2.16), this implies

$$y = \frac{\int \mathrm{d}x\, \phi(x) f(x)}{\int \mathrm{d}x\, f(x)} = \langle \phi \rangle_f. \tag{7.2.20}$$

Because this solution is unique, we can expand $F(\tilde{\sigma}(y), y)$ to second-order about $y = \langle \phi \rangle_f$. For this, we need

$$\frac{\mathrm{d}^2 F}{\mathrm{d}y^2}\bigg|_{\tilde{\sigma}=0} = -i \frac{\mathrm{d}\tilde{\sigma}}{\mathrm{d}y}\bigg|_{\tilde{\sigma}=0}. \tag{7.2.21}$$

Differentiating eqn. (7.2.16) at $\sigma = \tilde{\sigma}(y)$ with respect to y, we obtain

$$1 = -ig''(\tilde{\sigma}) \frac{\mathrm{d}\tilde{\sigma}}{\mathrm{d}y}, \tag{7.2.22}$$

so that

$$\frac{\mathrm{d}\tilde{\sigma}}{\mathrm{d}y}\bigg|_{\tilde{\sigma}=0} = \frac{i}{g''(0)}. \tag{7.2.23}$$

Note, however, that

$$g''(0) = i\left[\frac{\int \mathrm{d}x\, \phi^2(x) f(x)}{\int \mathrm{d}x\, f(x)} - \frac{\left(\int \mathrm{d}x\, \phi(x) f(x)\right)^2}{\left(\int \mathrm{d}x\, f(x)\right)^2} \right] = i\left[\langle \phi^2 \rangle_f - \langle \phi \rangle_f^2 \right], \tag{7.2.24}$$

which (apart from the factor of i) is just the square of the fluctuation $\delta\phi$ in $\phi(x)$ with respect to the distribution $f(x)$. From this analysis, we see that $\mathcal{P}(y)$ has a single maximum at $y = \langle \phi \rangle_f$ and decreases monotonically in either direction from this point. In the limit that M becomes very large, all higher-order contributions, which are simply higher-order moments of f with respect to $\mathcal{P}(y)$, vanish, so that $\mathcal{P}(y)$ becomes just a Gaussian normal distribution

$$\mathcal{P}(y) \longrightarrow \sqrt{\frac{M}{2\pi\delta\phi^2}} \exp\left[-\frac{M(y - \langle \phi \rangle_f)^2}{2\delta\phi^2} \right]. \tag{7.2.25}$$

We conclude, finally, that for large M, eqn. (7.2.5) can be approximated via eqn. (7.2.3) with a variance consistent with a normal distribution in the limit of large M, i.e.,

$$\int \mathrm{d}x\, \phi(x) f(x) = \frac{1}{M} \sum_{i=1}^{M} \phi(x_i) \pm \frac{1}{\sqrt{M}} \left[\langle \phi^2 \rangle_f - \langle \phi \rangle_f^2 \right]^{1/2} = \sum_{i=1}^{M} \phi(x_i) \pm \delta\phi, \tag{7.2.26}$$

thus guaranteeing convergence in the limit $M \to \infty$. Since the variance (second) term in eqn. (7.2.26), decreases as $1/\sqrt{M}$, efficient convergence relies on making this variance as small as possible, which is one of the challenges in designing Monte Carlo algorithms. Otherwise, a very large sample will be needed before $\tilde{I}_M$ becomes a good estimator for the integral.

7.3 Sampling distributions

7.3.1 Sampling simple distributions

In Section 3.8.3, we showed how to sample a Gaussian distribution using the Box-Muller method. In that context, we introduced some of the basic principles underlying sampling schemes. Here, we review and generalize the discussion for arbitrary distribution functions.

Consider a simple one-dimensional distribution function $f(x)$, $x \in [a, b]$ satisfying $f(x) \geq 0$ on the interval $[a, b]$ and normalized such that

$$\int_a^b f(x)\mathrm{d}x = 1. \tag{7.3.1}$$

Since $f(x)$ is normalized, the value of an integral of the form

$$P(X) = \int_a^X f(x)\mathrm{d}x, \tag{7.3.2}$$

with $X \in [a, b]$, lies in the interval $[0, 1]$. $P(X)$ measures the probability that a particular x randomly chosen from the distribution $f(x)$ lies in the interval $[a, X]$. Because $f(x) \geq 0$, $P(X)$ is a monotonically increasing function of X. Note also that $f(X) = \mathrm{d}P/\mathrm{d}X$.

Now, suppose we perform a variable transformation from x to y, where $y = g(x)$ and $g(x)$ is a nondecreasing function of x. In this case, if $X \geq x$, then $g(X) \geq g(x)$, and the probability $\tilde{P}(Y)$ that $g(X) = Y \geq y = g(x)$ must be equal to the probability that $X \geq x$, since the function g uniquely maps each value of x onto a value of y. Thus, the cumulative probabilities $\tilde{P}(Y)$ and $P(X)$ are equal:

$$\tilde{P}(Y) = P(X). \tag{7.3.3}$$

Conventional random number generators produce random sequences that are uniformly distributed on the interval $[0, 1]$. (In actuality, the numbers are not truly random since they are generated by a deterministic algorithm. Thus, they are more accurately called *pseudo* random numbers, although it is conventional to refer to them as "random.") If r is a random number, it will have a probability distribution $w(r)$ given by

$$w(r) = \begin{cases} 1 & 0 \leq r \leq 1 \\ 0 & \text{otherwise} \end{cases}. \tag{7.3.4}$$

The cumulative probability $W(\xi)$ is then

$$W(\xi) = \int_0^\xi w(r)\,\mathrm{d}r = \begin{cases} 0 & \xi < 0 \\ \xi & 0 \leq \xi \leq 1 \\ 1 & \xi > 1 \end{cases}. \tag{7.3.5}$$

The function $W(\xi) = \xi$ for $\xi \in [0, 1]$ is the probability that a random number r chosen by a random number generator lies in the interval $[0, \xi]$. Let us assume that there exists a variable transformation $r = g(x)$ where $g(x)$ is a nondecreasing function of x. This function maps the interval $x \in [a, b]$ onto the interval $r \in [0, 1]$. In this case, by eqn. (7.3.3), the cumulative probabilities can be equated in the interval $\xi \in [0, 1]$, yielding the relation

$$P(X) = \xi. \tag{7.3.6}$$

Thus, a sampling of the distribution function $f(x)$ can be achieved by randomly choosing a probability between 0 and 1 and then solving eqn. (7.3.6) for the corresponding probability that x chosen from $f(x)$ lies in the interval $x \in [a, X]$. The invertibility of eqn. (7.3.6) to yield X as a function of ξ guarantees the existence of the transformation $r = g(x)$. Therefore, for a set of M random numbers $\xi_1, ..., \xi_M$, eqn. (7.3.6) yields M values $X_1, ..., X_M$. We simply set $x_i = X_i$, and we have a sampling of M values from the distribution $f(x)$.

As an example, consider the distribution function $f(x) = ce^{-cx}$ on the interval $x \in [0, \infty)$. Clearly, $f(x)$ satisfies the conditions of a properly normalized probability distribution function. In order to sample $f(x)$, we first need $P(X)$:

$$P(X) = \int_0^X ce^{-cx} \, dx = 1 - e^{-cX}. \tag{7.3.7}$$

Next, we equate $1 - \exp(-cX)$ to the random number ξ, *i.e.*,

$$1 - e^{-cX} = \xi, \tag{7.3.8}$$

and solve for X, which can be done straightforwardly to yield

$$X = -\frac{1}{c} \ln(1 - \xi). \tag{7.3.9}$$

The example from Section 3.8.3 using the Box-Muller method to sample the Gaussian distribution illustrates that a single-variable distribution can be sampled by turning it into a two-variable distribution $F(x, y)$ that could be factorized into a product of two identical single-variable distributions $f(x)f(y)$. A simple change of variables to polar coordinates yielded another separable distribution $g(r)h(\theta)$, each factor of which could be sampled straightforwardly using eqn. (7.3.6) (see eqns. (3.8.14) to (3.8.20)). In general, the problem of sampling a multi-variable distribution $f(\mathbf{x})$, where $\mathbf{x} = (x_1, ..., x_n)$ is an n-dimensional vector, is nontrivial and will be discussed below. However, let us consider the special case that $f(\mathbf{x})$ is separable into a product of n single-variable distributions,

$$f(\mathbf{x}) = \prod_{\alpha=1}^{n} f_\alpha(x_\alpha). \tag{7.3.10}$$

If there exists a general transformation $y_\alpha = g_\alpha(\mathbf{x})$ such that the new distribution $\tilde{f}(\mathbf{y})$, $\mathbf{y} = (y_1, ..., y_n)$, is separable in the transformed variables

$$\tilde{f}(\mathbf{y}) = \prod_{\alpha=1}^{n} \tilde{f}_\alpha(y_\alpha), \tag{7.3.11}$$

then eqn. (7.3.6) can be applied to each individual distribution $f_\alpha(x_\alpha)$ or $\tilde{f}_\alpha(y_\alpha)$ to yield variables $X_1, ..., X_n$ or $Y_1, ..., Y_n$ and hence a complete sampling of the multi-variable distribution.

7.3.2 Importance sampling

Let us return to the problem of calculating the multidimensional integral

$$I = \int \mathrm{dx}\, \phi(\mathrm{x}) f(\mathrm{x}). \tag{7.3.12}$$

Instead of sampling the distribution $f(\mathrm{x})$, we could sample a different distribution $h(\mathrm{x})$ by rewriting the integral as

$$I = \int \mathrm{dx}\, \left[\frac{\phi(\mathrm{x}) f(\mathrm{x})}{h(\mathrm{x})} \right] h(\mathrm{x}) \tag{7.3.13}$$

and introducing $\psi(\mathrm{x}) = \phi(\mathrm{x}) f(\mathrm{x})/h(\mathrm{x})$. When this is done, eqn. (7.2.3) leads to

$$I = \int \mathrm{dx}\psi(\mathrm{x}) h(\mathrm{x}) = \frac{1}{M} \sum_{i=1}^{M} \psi(\mathrm{x}_i) \pm \frac{1}{\sqrt{M}} \left[\langle \psi^2 \rangle_h - \langle \psi \rangle_h^2 \right]^{1/2}, \tag{7.3.14}$$

where the vectors x_i are sampled from the distribution $h(\mathrm{x})$. The use of the distribution $h(\mathrm{x})$ in lieu of $f(\mathrm{x})$ is known as *importance sampling*.

There are several reasons to employ an importance function $h(\mathrm{x})$ in a Monte Carlo calculation. First, the function $h(\mathrm{x})$ might be easier to sample than $f(\mathrm{x})$. If $h(\mathrm{x})$ retains some of the most important features of $f(\mathrm{x})$, then $h(\mathrm{x})$ will be a good choice for an importance function. In this sense, employing importance sampling is akin to using a reference potential in molecular dynamics, which we discussed in Section 3.11. A second reason concerns the behavior of the integrand $\phi(\mathrm{x})$ itself. If $\phi(\mathrm{x})$ is a highly oscillatory function, then positive and negative contributions will tend to cancel in the Monte Carlo evaluation of eqn. (7.3.12), rendering the convergence of the sampling algorithm extremely slow and inefficient because of the large variance. A judiciously chosen importance function can help tame such oscillatory behavior, leading to a smaller variance and better convergence.

We now ask if there is an optimal choice for an importance function $h(\mathrm{x})$. The best choice is one that leads to the smallest possible variance. According to eqn. (7.3.14), the variance, which is a functional of $h(\mathrm{x})$, is given by

$$\sigma^2[h] = \left[\int \mathrm{dx}\psi^2(\mathrm{x}) h(\mathrm{x}) - \left(\int \mathrm{dx}\, \psi(\mathrm{x}) h(\mathrm{x}) \right)^2 \right]$$

$$= \left[\int \mathrm{dx} \frac{\phi^2(\mathrm{x}) f^2(\mathrm{x})}{h^2(\mathrm{x})} h(\mathrm{x}) - \left(\int \mathrm{dx}\, \phi(\mathrm{x}) f(\mathrm{x}) \right)^2 \right]. \tag{7.3.15}$$

We seek to minimize this variance with respect to the choice of $h(\mathrm{x})$ subject to the constraint that $h(\mathrm{x})$ be properly normalized:

$$\int \mathrm{dx}\, h(\mathrm{x}) = 1. \tag{7.3.16}$$

This can be done by introducing a Lagrange multiplier and minimizing the functional

$$F[h] = \sigma^2[h] - \lambda \int dx\, h(x). \tag{7.3.17}$$

Computing the functional derivative $\delta F/\delta h(x)$, we obtain the condition

$$\frac{\phi^2(x)f^2(x)}{h^2(x)} + \lambda = 0 \tag{7.3.18}$$

or

$$h(x) = \frac{1}{\sqrt{-\lambda}}\phi(x)f(x). \tag{7.3.19}$$

The Lagrange multiplier can be determined by requiring that $h(x)$ be normalized so that

$$\int dx\, h(x) = \frac{1}{\sqrt{-\lambda}} \int dx\, \phi(x)f(x) = 1 \tag{7.3.20}$$

or $\sqrt{-\lambda} = -\int dx \phi(x)f(x) = -I$. Thus, the optimal choice for $h(x)$ is

$$h(x) = \frac{\phi(x)f(x)}{I}. \tag{7.3.21}$$

In fact, with this choice of $h(x)$, the variance is identically zero, meaning that a perfect Monte Carlo algorithm can be constructed based on $h(x)$. Of course, this choice of $h(x)$ is only of academic interest because if we knew I, we would not need to perform the calculation in the first place! However, eqn. (7.3.21) provides a guideline for choosing $h(x)$ so as to keep the variance low.

As an example, consider the Monte Carlo evaluation of the integral

$$I = \int_0^1 dx\, e^{-x} = 1 - \frac{1}{e} = 0.632120558829. \tag{7.3.22}$$

The simplest Monte Carlo sampling scheme for this problem consists in sampling x uniformly on the interval $(0,1)$ ($f(x) = 1$ for $x \in (0,1)$ and 0 otherwise) and then evaluating the function $\phi(x) = \exp(-x)$. The integrand $\exp(-x)$ is shown as the solid line in Fig. 7.1(a), and the instantaneous value of the estimator $\phi(x) = \exp(-x)$ is shown in Fig. 7.1(b). After 10^6 steps, uniform sampling gives the answer as $I \approx 0.6322 \pm 0.000181$. Now let us attempt to devise an importance function $h(x)$ capable of reducing the variance. We might be tempted to try a first-order Taylor expansion $\exp(-x) \approx 1 - x$, which is shown as the dotted line in Fig. 7.1(a). After normalization, $h(x)$ becomes $h(x) = 2(1-x)$, and the use of this importance function gives, after 10^6 steps of sampling, $I \approx 0.6318 \pm 0.000592$. Interestingly, this importance function makes things worse, yielding a larger variance than simple uniform sampling! The reason for the failure of this importance function is that $1 - x$ is only a good representation of $\exp(-x)$ for x very close to 0, as Fig. 7.1(a) clearly shows. Over the full interval $x \in (0,1)$, however, $1 - x$ does not accurately represent $\exp(-x)$ and, therefore, biases the sampling toward regions of x that are more unfavorable than uniform sampling. Consider, next, using a more general linear function $h(x) = 1 - ax$, where the parameter a is chosen to give a better representation of $\exp(-x)$ over the

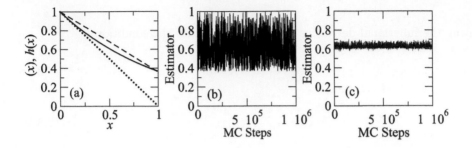

Fig. 7.1 (a) The integrand $\exp(-x)$ (solid line) and two possible importance functions $1 - x$ (dotted line) and $1 - 0.64x$ (dashed line). (b) Instantaneous value of the estimator $\phi(x) = \exp(-x)$ for a Monte Carlo calculation with $f(x) = 1$ on the interval $x \in (0, 1)$. (c) Instantaneous value of the estimator $\phi(x) = \exp(-x)/h(x)$ when the importance function $h(x) = (1 - ax)/(1 - a/2)$ is used with $a = 0.64$ to improve sampling.

full interval. The dashed line in Fig. 7.1(a) shows this linear function for $a = 0.64$, which, although a poorer representation of $\exp(-x)$ near $x = 0$ than the Taylor series, is still a better representation of $\exp(-x)$ overall for the full interval. The general normalization of $h(x)$ is $h(x) = (1 - ax)/(1 - a/2)$. After 10^6 Monte Carlo sampling steps with $a = 0.64$, we obtain the approximate answer $I \approx 0.63212 \pm 0.0000234$, which is nearly a full order of magnitude better in the variance than simple uniform sampling. The lesson from this example is that importance functions must be chosen carefully. Unless the importance function is a good representation of the integrand, the sampling efficiency can actually degrade over simple uniform sampling.

7.3.3   The M(RT)^2 algorithm: Acceptance and rejection

In the application of eqn. (7.2.3) to any Monte Carlo calculation, the vectors $x_1, ..., x_M$ can be generated independently and randomly. However, better convergence can often be achieved if the vectors are generated sequentially $x_1 \to x_2 \to \cdots \to x_M$ with a rule that specifies how to generate x_{i+1} given x_i. Such a sequence of vectors, in which x_{i+1} is generated based only knowledge of x_i is called a *Markov chain*. Markov chains are the core of many Monte Carlo algorithms.

Let $R(x|y)$ be a probability distribution function for obtaining a vector x from a given vector y. If x and y are accessible microstates of a physical system, then $R(x|y)$ is the probability for moving to a state x given that the system is currently in the state y. Thus, $R(x|y)$ potentially constitutes a rule for generating a Markov chain. However, in order for $R(x|y)$ to be valid as such a rule, it must satisfy the condition of *detailed balance*, which states that

$$R(x|y)f(y) = R(y|x)f(x). \tag{7.3.23}$$

Here, $R(x|y)f(y)$ is the *a priori* probability of a move from y to x, which is simply the probability $R(x|y)$ for the move from y to x times the probability $f(y)$ that the system is at y when the move is initiated. The detailed balance condition ensures that the Markov process is microscopically reversible and hence guarantees unbiased sampling of the state space. It has been argued that the detailed balance condition is a sufficient but not strictly necessary condition to ensure proper sampling of the state space (Manousiouthakis and Deam, 1999). In the present discussion, however, we will assume that eqn. (7.3.23) is satisfied by the process $R(\mathrm{x}|\mathrm{y})$.

The sampling technique to be discussed in this section, which was proposed by Metropolis *et al.* in 1953, which we refer to as the $\mathrm{M(RT)}^2$ algorithm ($\mathrm{M(RT)}^2$ stands in for the last names of the five authors), belongs to a class of Monte Carlo schemes known as *rejection methods*. The $\mathrm{M(RT)}^2$ method starts with a rule for generating trial or proposed moves from y to x, denoted $T(\mathrm{x}|\mathrm{y})$. The normalization condition on $T(\mathrm{x}|\mathrm{y})$ is

$$\int \mathrm{dx}\, T(\mathrm{x}|\mathrm{y}) = 1. \tag{7.3.24}$$

Once a trial move is generated, a decision is made either to accept the move, in which case the system is advanced to x, or to reject it, in which case it is returned to y by setting $\mathrm{x} = \mathrm{y}$. Let $A(\mathrm{x}|\mathrm{y})$ be the probability that the move is accepted. Then, the transition probability $R(\mathrm{x}|\mathrm{y})$ can be expressed as

$$R(\mathrm{x}|\mathrm{y}) = A(\mathrm{x}|\mathrm{y})T(\mathrm{x}|\mathrm{y}). \tag{7.3.25}$$

When eqn. (7.3.25) is substituted into eqn. (7.3.23), we find

$$A(\mathrm{x}|\mathrm{y})T(\mathrm{x}|\mathrm{y})f(\mathrm{y}) = A(\mathrm{y}|\mathrm{x})T(\mathrm{y}|\mathrm{x})f(\mathrm{x}), \tag{7.3.26}$$

so that the acceptance probabilities are related by

$$A(\mathrm{x}|\mathrm{y}) = \frac{T(\mathrm{y}|\mathrm{x})f(\mathrm{x})}{T(\mathrm{x}|\mathrm{y})f(\mathrm{y})}A(\mathrm{y}|\mathrm{x}) = r(\mathrm{x}|\mathrm{y})A(\mathrm{y}|\mathrm{x}). \tag{7.3.27}$$

The implication of eqn. (7.3.27) is an interesting one. Suppose $A(\mathrm{x}|\mathrm{y}) = 1$ so that the move from y to x is favorable. In this case, we expect that the reverse move from x to y is less favorable so that $A(\mathrm{y}|\mathrm{x}) < 1$, implying that $r(\mathrm{x}|\mathrm{y}) > 1$. On the other hand, if $A(\mathrm{x}|\mathrm{y}) < 1$ so that the move x to y is not entirely favorable, then we expect $A(\mathrm{y}|\mathrm{x}) = 1$ so that $r(\mathrm{x}|\mathrm{y}) < 1$. Combining these facts, we see that the acceptance probability is given conveniently by

$$A(\mathrm{x}|\mathrm{y}) = \min[1, r(\mathrm{x}|\mathrm{y})], \tag{7.3.28}$$

where function $\min[a, b]$ chooses the smaller of a and b.

Given the ideas developed above, the $\mathrm{M(RT)}^2$ algorithm can now be stated succinctly. At the kth step of a Markov chain, the trial distribution $T(\mathrm{x}_{k+1}|\mathrm{x}_k)$ is used to generate a proposed move from x_k to x_{k+1}. Using x_k and x_{k+1}, we compute the ratio

$$r(\mathrm{x}_{k+1}|\mathrm{x}_k) = \frac{T(\mathrm{x}_k|\mathrm{x}_{k+1})f(\mathrm{x}_{k+1})}{T(\mathrm{x}_{k+1}|\mathrm{x}_k)f(\mathrm{x}_k)}. \tag{7.3.29}$$

Next, the decision must be made whether this trial move is to be accepted or rejected. If $r(\mathrm{x}_{k+1}|\mathrm{x}_k) \geq 1$, then the move is accepted with probability 1 according to eqn. (7.3.28). If, however, $r(\mathrm{x}_{k+1}|\mathrm{x}_k) < 1$, then a random number $\xi \in [0, 1]$ is generated. If

$r(\mathrm{x}_{k+1}|\mathrm{x}_k) > \xi$, the move is accepted, and the value of x_{k+1} is retained. Otherwise, if $r(\mathrm{x}_{k+1}|\mathrm{x}_k) < \xi$, the move is rejected, and x_{k+1} is set equal to x_k. Remember, however, that even when a move is rejected, the value x_k to which x_{k+1} is "reset" must be considered as the next point in the Markov chain and must be used in the calculation of estimators. Thus, in any Markov chain, there will be points that are repeated and used more than once in the calculation of averages.

In the Markov chain generated by the $\mathrm{M(RT)}^2$ algorithm, each of the points $\mathrm{x}_1, \mathrm{x}_2, ..., \mathrm{x}_n$ will have an associated probability $\pi_1(\mathrm{x}), \pi_2(\mathrm{x}), ..., \pi_n(\mathrm{x})$. We wish to prove that in the limit of an infinite chain, $\lim_{n \to \infty} \pi_n(\mathrm{x}) = f(\mathrm{x})$, for which we will use inductive reasoning. That is, we will show that if $\pi_n(\mathrm{x}) = f(\mathrm{x})$ for some n, then it follows that $\pi_{n+1}(\mathrm{x}) = f(\mathrm{x})$. The proof proceeds first by finding a recursive relation between for $\pi_{n+1}(\mathrm{x})$ in terms of $\pi_n(\mathrm{x})$, and then by showing that $f(\mathrm{x})$ is a *fixed-point* of the recursion.

The recursion is derived as follows: $\pi_{n+1}(\mathrm{x})$ has a contribution from attempted moves that start at y, lead to x, and are accepted. It also has contributions from attempted moves that start at x, lead to y, and are rejected. Let us begin with the former. The quantity $\pi_n(\mathrm{y})\mathrm{dy}$ is the probability that a given point is in a neighborhood dy about the value y in the nth step of the Markov chain. Thus, the probability that a trial move yields a new point x starting from y in dy is $A(\mathrm{x}|\mathrm{y})T(\mathrm{x}|\mathrm{y})\pi_n(\mathrm{y})\mathrm{dy}$, which involves a product of the trial and acceptance probabilities, as expected. Therefore, the probability that a point x is reached from any starting point is obtained by integrating the product over all y:

$$\int A(\mathrm{x}|\mathrm{y})T(\mathrm{x}|\mathrm{y})\pi_n(\mathrm{y})\mathrm{dy}.$$

Similarly, the probability that an attempted move to any y starting from a particular x is rejected is

$$\int [1 - A(\mathrm{y}|\mathrm{x})]\, T(\mathrm{y}|\mathrm{x})\mathrm{dy},$$

which needs to be multiplied by $\pi_n(\mathrm{x})$, the probability density for being at x to begin with. When these two expressions are combined, the total probability density $\pi_{n+1}(\mathrm{x})$ becomes

$$\pi_{n+1}(\mathrm{x}) = \int A(\mathrm{x}|\mathrm{y})T(\mathrm{x}|\mathrm{y})\pi_n(\mathrm{y})\mathrm{dy} + \pi_n(\mathrm{x}) \int [1 - A(\mathrm{y}|\mathrm{x})]\, T(\mathrm{y}|\mathrm{x})\mathrm{dy}. \qquad (7.3.30)$$

From eqn. (7.3.30), it is possible to show that the distribution $f(\mathrm{x})$ is a fixed point of the recursion. We substitute the assumed condition of the induction, $\pi_n(\mathrm{x}) = f(\mathrm{x})$, into the recursion relation, which yields

$$\pi_{n+1}(\mathrm{x}) = \int A(\mathrm{x}|\mathrm{y})T(\mathrm{x}|\mathrm{y})f(\mathrm{y})\mathrm{dy} + f(\mathrm{x}) \int [1 - A(\mathrm{y}|\mathrm{x})]\, T(\mathrm{y}|\mathrm{x})\mathrm{dy}. \qquad (7.3.31)$$

However, due to the detailed balance condition in eqn. (7.3.23), eqn. (7.3.31) reduces to

$$\pi_{n+1}(\mathrm{x}) = f(\mathrm{x}) \int T(\mathrm{y}|\mathrm{x})\mathrm{dy} = f(\mathrm{x}), \qquad (7.3.32)$$

which follows from eqn. (7.3.24) and completes the proof.

A simple yet fairly standard way to apply the M(RT)^2 algorithm is based on choosing the trial probability $T(\text{x}|\text{y})$ to be uniform for x in some domain of radius Δ about y. For this choice, $T(\text{x}|\text{y})$ takes the form

$$T(\text{x}|\text{y}) = \begin{cases} 1/\Delta & |\text{x} - \text{y}| < \Delta/2 \\ 0 & \text{otherwise} \end{cases}. \tag{7.3.33}$$

This $T(\text{x}|\text{y})$ clearly satisfies eqn. (7.3.24) and also has the property that $T(\text{x}|\text{y}) = T(\text{y}|\text{x})$. Consequently, the acceptance probability becomes simply

$$A(\text{x}|\text{y}) = \min\left[1, \frac{f(\text{x})}{f(\text{y})}\right]. \tag{7.3.34}$$

Sampling the canonical distribution

Equations (7.3.33) and (7.3.34) can be straightforwardly applied to the problem of calculating the canonical configurational partition function for a system of monatomic particles such as a Lennard-Jones liquid (see Section 3.14.2). Recall that the partition function for a system of N particles with coordinates $\mathbf{r}_1, ..., \mathbf{r}_N$ and potential energy $U(\mathbf{r}_1, ..., \mathbf{r}_N)$ is given by

$$Q(N, V, T) = \frac{1}{N!\lambda^{3N}} \int d\mathbf{r}_1 \cdots d\mathbf{r}_N e^{-\beta U(\mathbf{r}_1, ..., \mathbf{r}_N)}, \tag{7.3.35}$$

where $\lambda = \sqrt{\beta h^2/2\pi m}$ and the integral is the configurational partition function. Introducing the usual notation $\mathbf{r} \equiv \mathbf{r}_1, ..., \mathbf{r}_N$ as the complete set of coordinates, we wish to devise a trial move from $\mathbf{r}$ to $\mathbf{r}'$ and determine the corresponding acceptance probability. If the move is based on eqn. (7.3.33), then, since $f(\mathbf{r}) \propto \exp[-\beta U(\mathbf{r})]$, the acceptance probability is simply

$$A(\mathbf{r}'|\mathbf{r}) = \min\left[1, e^{-\beta[U(\mathbf{r}') - U(\mathbf{r})]}\right]. \tag{7.3.36}$$

In other words, the acceptance probability is determined solely by the change in the potential energy that results from the move. If the potential energy decreases in a trial move, the move will be accepted with probability 1; if the energy increases, the move will be accepted with a probability that decreases exponentially with the change in energy.

An immediate problem arises when attempting to apply eqn. (7.3.33) as written. As a general rule, it is not possible to move *all* of the particles simultaneously! Remember that the potential energy U is an extensive quantity, meaning that $U \sim N$. Thus, if we attempt to move all of the particles at once, the change in the potential energy can be quite large unless the particle positions are varied only minimally. This problem becomes increasingly severe as the number of particles grows. According to eqn. (7.3.36), a large change in the potential energy leads to a very small probability that the move is accepted, and the M(RT)^2 algorithm becomes inefficient as a means of generating canonical configurations.

A simple remedy for this problem lies at the other extreme, in which we attempt to only move one particle at a time. Of course, eqn. (7.3.33) includes this possibility as well. Thus, we begin by choosing a particle at random from among the N particles in the system. Suppose the randomly chosen particle has an index i. Each of the three components of $\mathbf{r}_i$ is displaced at random using three uniform random numbers ξ_x, ξ_y, and ξ_z, with $\xi_\alpha \in [0,1]$, $\alpha = x, y, z$. The displacements are then given by

$$x_i' = x_i + \frac{1}{\sqrt{3}}\left(\xi_x - 0.5\right)\Delta$$

$$y_i' = y_i + \frac{1}{\sqrt{3}}\left(\xi_y - 0.5\right)\Delta$$

$$z_i' = z_i + \frac{1}{\sqrt{3}}\left(\xi_z - 0.5\right)\Delta. \tag{7.3.37}$$

All other particle coordinates remain unchanged. The random numbers are shifted to the interval $[-0.5, 0.5]$ to ensure that the sphere of radius Δ is centered on $\mathbf{r}_i$, and the $\sqrt{3}$ factor ensures that $|\mathbf{r}_i' - \mathbf{r}| < \Delta/2$ and consequently that $|\mathbf{r}' - \mathbf{r}| < \Delta/2$ as required by eqn. (7.3.33). A *Monte Carlo pass* through the system is a collection of N such trial moves which, in principle, is a sufficient number to attempt a move on each of the particles, although in any Monte Carlo pass, attempts will be made on some particles more than once while others will have no attempted moves.

At this point, several comments are in order. First, for a large majority of systems, it is not necessary to recompute the potential energy $U(\mathbf{r}')$ in full in order to determine the acceptance probability when a single particle is moved. We only need to recompute the terms that involve $\mathbf{r}_i$. Thus, for a short-ranged pair potential, only the interaction of particle i with other particles that lie within the cutoff radius of i need to be recomputed, which is a relatively inexpensive operation (see Appendix C). Second, it is natural to ask if maximal efficiency could be achieved via an optimal target for the average number of accepted moves. While this question can be answered in the affirmative, it is impossible to give a particular number for the target fraction of accepted moves. Note that the acceptance probability depends on the choice of Δ, hence the efficiency of the algorithm depends on this critical parameter. A large value for Δ generates large displacements for each particle with possibly significant increases in the potential energy and consequently, a low acceptance probability. A small value for Δ generates small displacements and a correspondingly high acceptance probability. Thus, choosing Δ is a compromise between large displacements and a reasonable number of accepted moves. One occasionally reads in the literature that a good target is between 20% and 50% acceptance of trial moves, with an optimal value around 30% (Allen and Tildesley, 1989; Frenkel and Smit, 2002). This range can be a useful rule of thumb or serve as a starting point for refining the target acceptance rate. However, the optimal value depends on the system, the thermodynamic control variables N, V, and T, and even how the computer program is written. In general, tests should be performed for each system to determine the optimal acceptance probability and displacement Δ.

As a final observation, it is interesting to compare Monte Carlo with molecular dynamics as methods for sampling ensemble distributions. A key difference, readily apparent from the preceding discussion, is that in a Monte Carlo calculation, particles are moved one at a time (or, at most, a few at a time); in molecular dynamics, all of the particles are moved simultaneously. It might seem, therefore, that molecular dynamics calculations are more efficient than Monte Carlo, but if we recall that molecular dynamics moves are limited by the size of the time step, we find that, in many instances, the methods are comparable in their efficiency when molecular dynamics is used with appropriate thermostatting and/or barostatting schemes. An advantage of molecular dynamics over Monte Carlo is that it is straightforward to couple and uncouple thermostats and barostats in order to switch between sampling and dynamics calculations, which makes writing an elegant, object-oriented code that encompasses both types of calculations conceptually seamless. Monte Carlo, as described here, is only useful as a sampling technique, and therefore a separate molecular dynamics module would be needed to study the dynamics of a system. Because molecular dynamics moves all particles simultaneously, it is also easier to devise and implement algorithms suitable for parallel computing architectures in order to tackle very large-scale applications. On the other hand, Monte Carlo allows for considerable flexibility to invent new types of moves since one need worry only about satisfying detailed balance. It is, of course, likewise possible to devise clever molecular dynamics methods, as we have seen in Chapters 4 and 5. However, in molecular dynamics, "cleverness" appears in the equations of motion and the demonstration that the algorithm achieves its objective. Finally, due to inherent randomness, Monte Carlo calculations are, by construction, ergodic, even if a large number of Monte Carlo passes is required to achieve converged results. In molecular dynamics, because of its deterministic nature, achieving ergodicity is a significant challenge.

Before considering other ensembles, it is worth mentioning how the algorithm in eqn. (7.3.37) is modified for systems consisting of rigid molecules. Since a rigid body has both translational and rotational degrees of freedom (see Section 1.11), two types of uniform moves are needed. Suppose a rigid body has a center-of-mass position $\mathbf{R}$ and n constituent particles with coordinates $\mathbf{r}_1, ..., \mathbf{r}_n$ *relative* to the center of mass. In a system consisting of N such rigid bodies, we first choose one of them at random. Then, eqn. (7.3.37) is applied to the center of mass in order to generate a move from $\mathbf{R}$ to $\mathbf{R}'$ according to:

$$X' = X + \frac{1}{\sqrt{3}} \left(\xi_x - 0.5 \right) \Delta$$

$$Y' = Y + \frac{1}{\sqrt{3}} \left(\xi_y - 0.5 \right) \Delta$$

$$Z' = Z + \frac{1}{\sqrt{3}} \left(\xi_z - 0.5 \right) \Delta. \tag{7.3.38}$$

Equation (7.3.38) generates a translation of the rigid body. Next, a unit vector $\mathbf{n}$ is randomly chosen to define an axis through the center of mass. This can be accomplished by choosing three additional random numbers ζ_x, ζ_y, and ζ_z to give the components of a random vector and normalizing the vector by the length

$$l = \sqrt{\zeta_x^2 + \zeta_y^2 + \zeta_z^2} \tag{7.3.39}$$

to give $\mathbf{n} = (\zeta_x/l, \zeta_y/l, \zeta_z/l)$. One final random number η is used to determine a random rotation angle $\theta = 2\pi\eta$, and the rotation formula (see eqn. (1.11.3))

$$\mathbf{r}_i' = \mathbf{r}_i \cos\theta + \mathbf{n}(\mathbf{n} \cdot \mathbf{r}_i)(1 - \cos\theta) + (\mathbf{r}_i \times \mathbf{n}) \sin\theta \tag{7.3.40}$$

is applied to each particle in the rigid body. Once the trial move is generated, eqn. (7.3.36) is used to determine if the move is accepted or rejected. Such rotational moves can also be generated using the three Euler angles or quaternions described in Section 1.11.

Sampling the isothermal-isobaric distribution

The isothermal-isobaric partition function for a system of N particles at constant external pressure P and temperature T is

$$\Delta(N, P, T) = \frac{1}{V_0} \int_0^\infty dV \, e^{-\beta PV} Q(N, V, T)$$

$$= \frac{1}{V_0 N! \lambda^{3N}} \int_0^\infty dV \, e^{-\beta PV} \int_{D(V)} d\mathbf{r}_1 \cdots d\mathbf{r}_N \, e^{-\beta U(\mathbf{r}_1, \dots, \mathbf{r}_N)}, \tag{7.3.41}$$

where the coordinate integrations are limited to the spatial domain $D(V)$ defined by the containing volume.

Sampling the isothermal-isobaric distribution requires sampling both the particle coordinates and the volume V. The former can be done using the uniform sampling schemes of the previous subsection. A trial volume move from V to V' can also be generated from a uniform distribution. A random number ξ_V is generated and the trial volume move is given by

$$V' = V + (\xi_V - 0.5)\,\delta, \tag{7.3.42}$$

where δ determines the size of the volume displacement. Volume moves need to be handled with some care because each time the volume changes, the particle coordinates must be scaled by $\mathbf{r}_i' = (V'/V)^{1/3}\mathbf{r}_i$ and the total potential energy recalculated before the decision to accept or reject the move can be made. Also, the dependence of the integration limits in eqn. (7.3.42) on the volume presents an additional complication. As we saw in eqn. (4.7.51), we can make the volume dependence explicit by transforming the spatial integrals to scaled coordinates $\mathbf{s}_i = \mathbf{r}_i/V^{1/3}$, which gives

$$\Delta(N, P, T) = \frac{1}{V_0 N! \lambda^{3N}} \int_0^\infty dV \, e^{-\beta PV} V^N$$

$$\times \int ds_1 \cdots ds_N \, e^{-\beta U(V^{1/3} s_1, \ldots, V^{1/3} s_N)}. \tag{7.3.43}$$

From eqn. (7.3.43), we see that the acceptance probability for volume moves is, therefore, given by

$$A(V'|V) = \min \left[1, e^{-\beta P(V'-V)} e^{-\beta [U(r') - U(r)]} e^{N \ln(V'/V)} \right]. \tag{7.3.44}$$

Since a volume move leads to a change in potential energy that increases with N, volume moves have a low probability of acceptance unless δ is small. Moreover, the fact that all terms in the potential energy must be updated when the volume changes means that volume moves are computationally demanding. For this reason, volume moves are usually made less frequently and have a slightly higher target average acceptance rate than particle moves.

Sampling the grand canonical distribution

The grand canonical partition function for a system of particles maintained at constant chemical potential μ in a volume V at temperature T is

$$\mathcal{Z}(\mu, V, T) = \sum_{N=0}^\infty e^{\beta \mu N} Q(N, V, T)$$

$$= \sum_{N=0}^\infty e^{\beta \mu N} \frac{1}{N! \lambda^{3N}} \int dr_1 \cdots dr_N e^{-\beta U(r_1, \ldots, r_N)}. \tag{7.3.45}$$

The relative ease with which this ensemble can be sampled in Monte Carlo is an interesting advantage over molecular dynamics.

Sampling the grand canonical ensemble requires sampling the particle coordinates and the particle number N. Particle moves can, once again, be generated using the scheme of eqn. (7.3.37). Sampling the particle number N is achieved via attempted *particle insertion* and *particle deletion* moves. As these names imply, periodic attempts are made to insert a particle at a randomly chosen spatial location or to delete a randomly chosen particle from the system as a means of generating particle-number fluctuations. From eqn. (7.3.45), the acceptance probability of a trial insertion move can be seen to be

$$A(N+1|N) = \min \left[1, \frac{V}{\lambda^3(N+1)} e^{\beta \mu} e^{-\beta [U(r') - U(r)]} \right], \tag{7.3.46}$$

where r' is the configuration of an $(N+1)$-particle system generated by the insertion and r is the original N-particle configuration. The volume factor in eqn. (7.3.46) arises from the use of scaled coordinates in the configurational partition function in eqn. (7.3.45). Similarly, the acceptance probability of a trial deletion move is

$$A(N-1|N) = \min \left[1, \frac{\lambda^3 N}{V} e^{-\beta\mu} e^{-\beta[U(\mathbf{r}') - U(\mathbf{r})]} \right], \tag{7.3.47}$$

where $\mathbf{r}'$ is the configuration of an $(N-1)$-particle system generated by the deletion and $\mathbf{r}$ is, again, the original N-particle configuration. Particle insertion and deletion moves require calculation of only the change in potential energy due to the addition or removal of one particle, which is computationally no more expensive than the calculation associated with a particle displacement. Thus, particle insertion and deletion moves can be performed with greater frequency than can volume moves in the isothermal-isobaric ensemble. In general, one Monte Carlo pass consists of N particle displacements and $N_{i/d}$ insertion/deletion attempts. Note that N_i and N_d are new parameters that need to be optimized for each particular system.

7.4 Hybrid Monte Carlo

In the remaining sections of this chapter, we will discuss several algorithms that build on the basic ideas developed so far. Although we will focus on sampling the canonical distribution, the techniques we will introduce can be easily generalized to other ensembles.

In Section 7.3, we showed how the canonical distribution can be generated using uniform trial particle displacements. We argued that we can only attempt to move one or just a few particles at a time in order to maintain a reasonable average acceptance probability. We also noted that a key difference between Monte Carlo and molecular dynamics calculations is the ability of the latter to generate moves of the entire system (global moves) with acceptance probability 1. In molecular dynamics, however, such moves are deterministic and fundamentally limited by the time step Δt, which needs to be sufficiently small to yield reasonable energy conservation. In this section, we describe the *hybrid Monte Carlo* approach (Duane *et al.*, 1987), which is a synthesis of $M(RT)^2$ Monte Carlo and molecular dynamics and, therefore, derives advantages from each of these methods.

Hybrid Monte Carlo seeks to relax the restriction on the size of Δt in a molecular dynamics calculation by introducing an acceptance criterion for molecular dynamics moves with a large Δt. Of course, when Δt is too large, the numerical integration algorithm for Hamilton's equations leads to large changes in the Hamiltonian $\mathcal{H}$, which should normally be approximately conserved. In principle, this is not a problem since in the canonical ensemble, $\mathcal{H}$ is not constant but is allowed to fluctuate as the system exchanges energy with a surrounding thermal bath. By using molecular dynamics as an engine for generating moves, the system naturally tends to move toward regions of configuration space that are energetically favored, and hence the moves are more "intelligent" than simple uniform displacements. However, when we use a time step that is too large, we are simply performing a "bad" molecular dynamics calculation, and the changes in $\mathcal{H}$ caused by inaccurate integration of the equations of motion will not be consistent with the canonical distribution. Thus, we need a device for ensuring this consistency. We can achieve this by accompanying each "bad" molecular dynamics move by an acceptance step based on the change $\Delta\mathcal{H}$ in the Hamiltonian. Since hybrid Monte Carlo uses the full phase space, the acceptance probability is expressed in terms of the change from one phase-space point $(\mathbf{r}, \mathbf{p})$ to another $(\mathbf{r}', \mathbf{p}')$, where $\mathbf{p}$ and $\mathbf{r}$ are sets of N momenta and coordinates, respectively, and similarly for $\mathbf{p}'$ and $\mathbf{r}'$. Thus, the acceptance probability is given by

$$A(\mathbf{r}', \mathbf{p}'|\mathbf{r}, \mathbf{p}) = \min\left[1, e^{-\beta\{\mathcal{H}(\mathbf{r}', \mathbf{p}') - \mathcal{H}(\mathbf{r}, \mathbf{p})\}}\right] = \min\left[1, e^{-\beta\Delta\mathcal{H}}\right]. \qquad (7.4.1)$$

Equation (7.4.1) ensures that the acceptance is based on the correct canonical distribution $f(\mathbf{r}, \mathbf{p}) \propto \exp[-\beta\mathcal{H}(\mathbf{r}, \mathbf{p})]$.

There is a subtlety in eqn. (7.4.1). As with the other $M(RT)^2$ schemes we have considered, eqn. (7.4.1) assumes that the probability distribution $T(\mathbf{r}', \mathbf{p}'|\mathbf{r}, \mathbf{p})$ for trial moves satisfies

$$T(\mathbf{r}', \mathbf{p}'|\mathbf{r}, \mathbf{p}) = T(\mathbf{r}, -\mathbf{p}|\mathbf{r}', -\mathbf{p}'), \qquad (7.4.2)$$

i.e., that the original configuration $\mathbf{r}$ can be reached from $\mathbf{r}'$ by simply reversing the momenta and running the integrator backwards. In addition, detailed balance requires that phase-space volume be preserved. These conditions will only be met if the molecular dynamics move is carried out using a symplectic, reversible integration algorithm such as the velocity Verlet integrator of eqns. (3.8.7) and (3.8.9). If $\mathcal{H}$ is given by

$$\mathcal{H}(\mathbf{r}, \mathbf{p}) = \sum_{i=1}^{N} \frac{\mathbf{p}_i^2}{2m_i} + U(\mathbf{r}_1, ..., \mathbf{r}_N), \qquad (7.4.3)$$

then the molecular dynamics move can be expressed in terms of the Trotter-factorized classical propagator scheme of Section 3.10. If $\mathbf{F}_i = -\partial U/\partial \mathbf{r}_i$ is the force on particle i, then the Liouville operator is

$$iL = \sum_{i=1}^{N} \frac{\mathbf{p}_i}{m_i} \cdot \frac{\partial}{\partial \mathbf{r}_i} + \sum_{i=1}^{N} \mathbf{F}_i \cdot \frac{\partial}{\partial \mathbf{p}_i} = iL_1 + iL_2, \qquad (7.4.4)$$

and the trial molecular dynamics move can be expressed as

$$\begin{pmatrix} \mathbf{p}' \\ \mathbf{r}' \end{pmatrix} = \left[e^{iL_2 \Delta t/2} e^{iL_1 \Delta t} e^{iL_2 \Delta t/2} \right]^m \begin{pmatrix} \mathbf{p} \\ \mathbf{r} \end{pmatrix}, \qquad (7.4.5)$$

which in general, allows the point $(\mathbf{r}', \mathbf{p}')$ to be generated from the initial condition $(\mathbf{r}, \mathbf{p})$ from m iterations of the velocity Verlet integrator using the time step Δt before eqn. (7.4.1) is applied.

When implementing hybrid Monte Carlo, the values of m, Δt, and the target average acceptance probability need to be optimized. As with the $M(RT)^2$ algorithms previously presented, it is difficult to provide guidelines for choosing m and Δt, as optimal values are strongly dependent on the system and efficiency of the computer program. It is important to note, however, that a single trial move generated via eqn. (7.4.5) requires m full force evaluations, in comparison to the relatively inexpensive uniform moves of Section 7.3. Typically, in a hybrid Monte Carlo calculation, one aims for a higher average acceptance probability (40% to 70%) than was recommended in Section 7.3; indeed, larger m and smaller Δt yield better trial moves that are more likely to be accepted. If force calculations are very expensive and/or the code is inefficient, then smaller m and larger Δt are preferable. Finally, if a move is rejected, the positions $\mathbf{r}'$ are set to their original values $\mathbf{r}$. However, if we similarly reset the momenta to their original values and reapply eqn. (7.4.5), we will end up at the same point $(\mathbf{r}', \mathbf{p}')$ and reject the move again. Thus, the momenta should be resampled from a Maxwell-Boltzmann distribution before initiating the next trial move via eqn. (7.4.5).

We conclude this section with a short proof that the detailed balance condition in eqn. (7.3.23) is satisfied when a time-reversible integrator is used. Since we are only interested in the configurational distribution, the detailed balance condition for hybrid Monte Carlo can be stated as

$$\int \mathrm{d}^N \mathbf{p}\, \mathrm{d}^N \mathbf{p}'\, T(\mathbf{r}', \mathbf{p}' | \mathbf{r}, \mathbf{p}) A(\mathbf{r}', \mathbf{p}' | \mathbf{r}, \mathbf{p}) f(\mathbf{r}, \mathbf{p})$$

$$= \int \mathrm{d}^N \mathbf{p}\, \mathrm{d}^N \mathbf{p}'\, T(\mathbf{r}, \mathbf{p} | \mathbf{r}', \mathbf{p}') A(\mathbf{r}, \mathbf{p} | \mathbf{r}', \mathbf{p}') f(\mathbf{r}', \mathbf{p}'). \quad (7.4.6)$$

In order to prove this, we first note that

$$A(\mathbf{r}', \mathbf{p}' | \mathbf{r}, \mathbf{p}) f(\mathbf{r}, \mathbf{p}) = \frac{1}{Q(N, V, T)} \min \left[1, \mathrm{e}^{-\beta[\mathcal{H}(\mathbf{r}', \mathbf{p}') - \mathcal{H}(\mathbf{r}, \mathbf{p})]} \right] \mathrm{e}^{-\beta \mathcal{H}(\mathbf{r}, \mathbf{p})}$$

$$= \frac{1}{Q(N, V, T)} \min \left[\mathrm{e}^{-\beta \mathcal{H}(\mathbf{r}, \mathbf{p})}, \mathrm{e}^{-\beta \mathcal{H}(\mathbf{r}', \mathbf{p}')} \right]. \quad (7.4.7)$$

Similarly,

$$A(\mathbf{r}, \mathbf{p} | \mathbf{r}', \mathbf{p}') f(\mathbf{r}', \mathbf{p}') = \frac{1}{Q(N, V, T)} \min \left[1, \mathrm{e}^{-\beta[\mathcal{H}(\mathbf{r}, \mathbf{p}) - \mathcal{H}(\mathbf{r}', \mathbf{p}')]} \right] \mathrm{e}^{-\beta \mathcal{H}(\mathbf{r}', \mathbf{p}')}$$

$$= \frac{1}{Q(N, V, T)} \min \left[\mathrm{e}^{-\beta \mathcal{H}(\mathbf{r}', \mathbf{p}')}, \mathrm{e}^{-\beta \mathcal{H}(\mathbf{r}, \mathbf{p})} \right]. \quad (7.4.8)$$

Therefore

$$A(\mathbf{r}', \mathbf{p}' | \mathbf{r}, \mathbf{p}) f(\mathbf{r}, \mathbf{p}) = A(\mathbf{r}, \mathbf{p} | \mathbf{r}', \mathbf{p}') f(\mathbf{r}', \mathbf{p}'). \quad (7.4.9)$$

Multiplying both sides of eqn. (7.4.9) by $T(\mathbf{r}', \mathbf{p}' | \mathbf{r}, \mathbf{p})$ and integrating over momenta gives

$$\int \mathrm{d}^N \mathbf{p}\, \mathrm{d}^N \mathbf{p}'\, T(\mathbf{r}', \mathbf{p}' | \mathbf{r}, \mathbf{p}) A(\mathbf{r}', \mathbf{p}' | \mathbf{r}, \mathbf{p}) f(\mathbf{r}, \mathbf{p})$$

$$= \int \mathrm{d}^N \mathbf{p}\, \mathrm{d}^N \mathbf{p}'\, T(\mathbf{r}', \mathbf{p}' | \mathbf{r}, \mathbf{p}) A(\mathbf{r}, \mathbf{p} | \mathbf{r}', \mathbf{p}) f(\mathbf{r}', \mathbf{p}'). \quad (7.4.10)$$

By using the property of the integrator that $T(\mathbf{r}', \mathbf{p}' | \mathbf{r}, \mathbf{p}) = T(\mathbf{r}, -\mathbf{p} | \mathbf{r}', -\mathbf{p}')$, changing the integration variables on the right side from $\mathbf{p}$ and $\mathbf{p}'$ to $-\mathbf{p}$ and $-\mathbf{p}'$, and noting that $d(-\mathbf{p})\,d(-\mathbf{p}') = d\mathbf{p}\,d\mathbf{p}'$, we obtain eqn. (7.4.6), which emphasizes the importance of using a reversible, measure-preserving integration algorithm such as velocity Verlet.

As a final comment, we note that the trial moves in eqn. (7.4.5) need not be purely Hamiltonian in nature. We could have used thermostatted equations of motion, for example, as described in Chapter 4 and still generated a proper canonical sampling using the acceptance criterion in eqn. (7.4.1) or a modified acceptance criterion based on a conserved extended energy (see eqn. (4.11.3), for example).

7.5 Replica exchange Monte Carlo

One of the most challenging computational problems met by researchers in statistical mechanics is the development of methods capable of sampling a canonical distribution when the potential energy $U(\mathbf{r}_1, ..., \mathbf{r}_N)$ is characterized by a large number of local minima separated by high barriers. Such potential energy functions describe many physical systems including proteins, glasses, polymer membranes, and polymer blends, to name just a few. An illustration of such a surface is shown in Fig. 7.2. Potential energy surfaces that resemble Fig. 7.2 (but in $3N$ dimensions) are referred to as *rough energy landscapes*. The ongoing development of general and robust techniques capable of adequately sampling statistically relevant configurations on such a surface continues to impact computational biology and materials science in important ways as newer and more sophisticated methods become available.

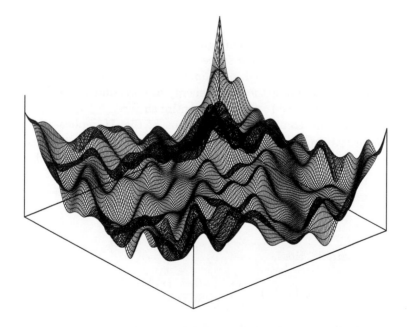

Fig. 7.2 A two-dimensional rough potential energy surface.

Clearly, a straightforward molecular dynamics or Monte Carlo calculation carried out on a rough potential energy surface exhibits hopelessly slow convergence of equilibrium properties because the probability of crossing a barrier of height $U^{\ddagger}$ is proportional to $\exp(-\beta U^{\ddagger})$. Thus, the system tends to become trapped in a single local minimum and requires an enormously long time to escape the minimum. As an illustration, the Boltzmann factor for a barrier of height 15 kJ/mol at a temperature of 300 K is roughly 3×10^{-3}, and for a barrier height of 30 kJ/mol, it is roughly 6×10^{-6}. As a result, barrier crossing becomes a "rare event." In Chapter 8, we will discuss a number of methods for addressing the rare-event problem. Here, we begin studying this problem by introducing a powerful and popular method, replica exchange Monte Carlo, designed to accelerate barrier crossing.

The term "replica exchange" refers to a class of Monte Carlo methods in which simultaneous calculations are performed on a set of M independent copies or replicas of a physical system. Each replica is assigned a different value of some physical control parameter, and Monte Carlo moves in the form of exchanges of the coordinates between different replicas are attempted. In this section, we will describe a replica exchange approach called *parallel tempering* (Marinari and Parisi, 1992; Tesi *et al.*, 1996), in which temperature is used as the control variable, and different temperatures are assigned to the replicas.

In the parallel tempering scheme, a set of temperatures $T_1, ..., T_M$, with $T_M > T_{M-1} > \cdots > T_1$ is selected and assigned to the M replicas. The lowest temperature T_1 is taken to be the temperature T of the canonical distribution to be sampled. The motivation for this scheme is that the high-temperature replicas can easily cross barriers on the potential energy surface if the temperatures are high enough. The attempted exchanges between the replicas cause the coordinates of the high-temperature copies to "percolate" down to the low-temperature copies, allowing the latter to sample larger portions of the configuration space at the correct temperature. The idea is illustrated in Fig. 7.3. On the rough one-dimensional surface shown in the figure, the low-temperature copies sample the lowest energy minima on the surface, while the high-temperature copies are able to "scan" the entire surface.

Let $\mathbf{r}^{(1)}, ..., \mathbf{r}^{(M)}$ be the complete configurations of the M replicas, i.e., $\mathbf{r}^{(K)} \equiv \mathbf{r}_1^{(K)}, ..., \mathbf{r}_N^{(K)}$. Since the replicas are independent, the total probability distribution $F(\mathbf{r}^{(1)}, ..., \mathbf{r}^{(K)})$ for the full set of replicas is just a product of the individual distribution functions of the replicas

$$F(\mathbf{r}^{(1)}, ..., \mathbf{r}^{(K)}) = \prod_{K=1}^{M} f_K\left(\mathbf{r}^{(K)}\right) \tag{7.5.1}$$

and is, therefore, separable. Here,

$$f_K\left(\mathbf{r}^{(K)}\right) = \frac{\exp\left[-\beta_K U\left(\mathbf{r}^{(K)}\right)\right]}{Q(N, V, T_K)}. \tag{7.5.2}$$

A replica exchange calculation proceeds by performing either a molecular dynamics or simple M(RT)^2 Monte Carlo calculation on each individual replica. Periodically, a neighboring pair of replicas K and $K + 1$ is selected, and an attempted move

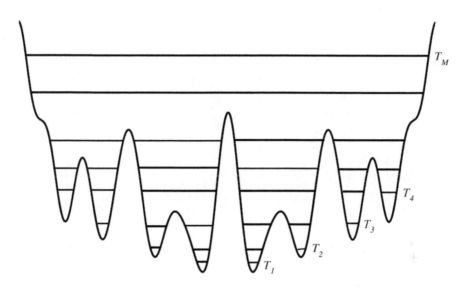

Fig. 7.3 Schematic of the parallel-tempering replica exchange Monte Carlo.

$(\mathbf{r}^{(K)}, \mathbf{r}^{(K+1)}) \rightarrow (\tilde{\mathbf{r}}^{(K)}, \tilde{\mathbf{r}}^{(K+1)})$ is made, where $\tilde{\mathbf{r}}^{(K)} = \mathbf{r}^{(K+1)}$ and $\tilde{\mathbf{r}}^{(K+1)} = \mathbf{r}^{(K)}$; this move is simply an exchange of coordinates between the systems. Since the co-ordinates are not actually *changed* (they are not displaced, rotated, *etc.*) but merely *exchanged*, the probability distribution for such trial moves satisfies

$$T\left(\tilde{\mathbf{r}}^{(K)}, \tilde{\mathbf{r}}^{(K+1)} | \mathbf{r}^{(K)}, \mathbf{r}^{(K+1)}\right) = T\left(\mathbf{r}^{(K)}, \mathbf{r}^{(K+1)} | \tilde{\mathbf{r}}^{(K)}, \tilde{\mathbf{r}}^{(K+1)}\right), \qquad (7.5.3)$$

so that the acceptance probability becomes

$$A\left(\tilde{\mathbf{r}}^{(K)}, \tilde{\mathbf{r}}^{(K+1)} | \mathbf{r}^{(K)}, \mathbf{r}^{(K+1)}\right) = A\left(\mathbf{r}^{(K+1)}, \mathbf{r}^{(K)} | \mathbf{r}^{(K)}, \mathbf{r}^{(K+1)}\right)$$

$$= \min\left[1, \frac{f_K\left(\mathbf{r}^{(K+1)}\right) f_{K+1}\left(\mathbf{r}^{(K)}\right)}{f_K\left(\mathbf{r}^{(K)}\right) f_{K+1}\left(\mathbf{r}^{(K+1)}\right)}\right]$$

$$= \min\left[1, e^{-\Delta_{K,K+1}}\right], \qquad (7.5.4)$$

where

$$\Delta_{K,K+1} = (\beta_K - \beta_{K+1})\left[U\left(\mathbf{r}^{(K)}\right) - U\left(\mathbf{r}^{(K+1)}\right)\right]. \qquad (7.5.5)$$

The improvement in conformational sampling efficiency gained by employing a parallel-tempering replica exchange Monte Carlo approach is illustrated using the sim-ple example of a 50-mer alkane system $C_{50}H_{102}$ in the gas phase using the CHARMM22 force field (MacKerell *et al.*, 1998). Here, the conformational preferences of the molecule can be characterized using the set of backbone dihedral angles. Since each dihedral angle has three attractive basins corresponding to two *gauche* and a *trans* conforma-tion, the number of local minima is $3^{50} \approx 7 \times 10^{23}$. Figure 7.4 displays a histogram

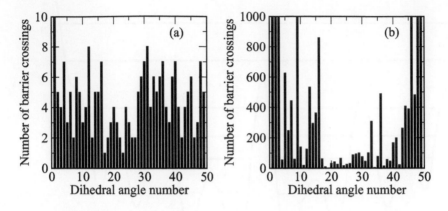

Fig. 7.4 Comparison of hybrid Monte Carlo (a) and parallel tempering replica exchange (b) for $C_{50}H_{102}$.

of the number of times each of the backbone dihedral angles crosses an energy barrier(Minary *et al.*, 2007). The replica exchange calculations are carried out using hybrid Monte Carlo to evolve each of ten individual replicas. The temperatures of the replicas all lie in the range 300 K to 1000 K with a distribution chosen to give an average acceptance probability of 20%, as recommended by Rathore *et al.* (2005) and Kone and Kofke (2005). The replica exchange Monte Carlo calculations are compared to a straight hybrid Monte Carlo calculation on a single system at a temperature of $T = 300$ K. The figure shows a significant improvement in sampling efficiency with replica exchange compared to simple hybrid Monte Carlo. Although the replica exchange calculation is ten times more expensive than straight hybrid Monte Carlo, the gain in efficiency more than offsets this cost. It is interesting to note, however, that even for this simple system, replica exchange does not improve the sampling uniformly over the entire chain. Achieving more uniform sampling requires algorithms of considerable sophistication. An example of such an approach was introduced by Zhu *et al.* (2002) and by Minary *et al.* (2007).

It is important to note that a direct correlation exists between the number of replicas and the acceptance probability. If sufficient computational resources are available, then a replica exchange calculation can be set up that contains a large number of replicas. In principle, this facilitates exchanges between neighboring replicas, thereby increasing the average acceptance probability. However, more attempted exchanges (and hence more computational time) are needed for high-temperature copies to percolate down to the low-temperature copies. Thus, the number of replicas and average acceptance probability need to be optimized for each system, computer program, and available computational platform. Indeed, since $U \sim N$, as the system size increases, one is forced to use a finer temperature "grid" in order to have a reasonable average acceptance probability, thereby increasing the overhead of the method considerably. Several improvements to the algorithm have been suggested to alleviate this problem. For example, for biomolecules in aqueous solution, Berne and coworkers introduced a modification of the algorithm in which attempted exchanges are made between coordinates of the solute only rather than between the complete set of coordinates (Liu *et al.*, 2005).

7.6 Wang-Landau sampling

In this section, we will consider a rather different approach to Monte Carlo calculations pioneered by Wang and Landau (2001). Until now, we have focused on Monte Carlo methods aimed at generating the canonical distribution of the coordinates, motivated by the fact that the canonical partition function can be written in the form of eqn. (7.3.35). However, let us recall that the canonical partition function can also be expressed in terms of the microcanonical partition function $\Omega(N, V, E)$ as

$$Q(N, V, T) = \frac{1}{E_0} \int_0^\infty dE \, e^{-\beta E} \Omega(N, V, E) \tag{7.6.1}$$

(see eqn. (4.3.16)), where E_0 is an arbitrary reference energy. Since we know that N and V are fixed, let us simplify the notation by dropping N and V in this section, set $E_0 = 1$, and write eqn. (7.6.1) as

$$Q(\beta) = \int_0^\infty dE \, e^{-\beta E} \Omega(E). \tag{7.6.2}$$

Equation (7.6.2) suggests that we can calculate the partition function, and hence all thermodynamic quantities derivable from it, if we can devise a method to generate the unknown function $\Omega(E)$ for a wide range of energies. $\Omega(E)$, in addition to being the microcanonical partition function, is also referred to as the *density of states*, since it is a measure of the number of microscopic states available to a system at a given energy E.[2] The probability, therefore, that a microscopic state with energy E will be visited is proportional to $1/\Omega(E)$.

The approach of Wang and Landau is to sample the function $\Omega(E)$ directly and, once known, calculate the partition function via eqn. (7.6.2). The rub, of course, is that we do not actually know $\Omega(E)$ *a priori*, and trying to generate it using ordinary M(RT)^2 sampling and eqn. (7.3.36) is extremely inefficient. The Wang-Landau algorithm is a simple yet elegant approach that can generate $\Omega(E)$ with impressive efficiency. We begin by assuming that $\Omega(E) = 1$ for all values of E. Numerical implementation requires that $\Omega(E)$ be discretized into a number of energy bins. Now imagine that a trial move, such as a uniform displacement as in eqn. (7.3.37), is attempted. The move changes the energy of the system from E_1 to E_2. For such a move, the energy is determined entirely by the potential energy U. For any trial move, the acceptance probability is taken to be

$$A(E_2|E_1) = \min\left[1, \frac{\Omega(E_1)}{\Omega(E_2)}\right]. \tag{7.6.3}$$

Of course, the move is initially accepted with probability 1. After such a move, the system will have an energy E that is either E_1 or E_2, depending on whether the

---

[2]In fact, if the E_0 is left off the prefactor in eqn. (3.2.20), then $\Omega(N, V, E)$ has units of inverse energy.

move is accepted or rejected. The key step in the Wang-Landau algorithm is that the density of states $\Omega(E)$ is modified after each move according to $\Omega(E) \rightarrow \Omega(E)f$, where f is a scaling factor with $f > 1$. Note that the scaling is applied *only* to the energy bin in which E happens to fall. All other bins remain unchanged. In addition, we accumulate a histogram $h(E)$ of each energy visited as a result of such moves. After many iterations of this procedure $h(E)$ starts to "flatten out," meaning that it has roughly the same value in each energy bin. Once we decide that $h(E)$ is "flat enough" for the given value of f, we start refining the procedure. We choose a new value of f, for example, $f_{\text{new}} = \sqrt{f_{\text{old}}}$ (an arbitrary formula suggested by Wang and Landau) and begin a new cycle. As before, we wait until $h(E)$ is flat enough and then switch to a new value of f (using the square-root formula again, for example). After many refinement cycles, we will find that $f \rightarrow 1$ and $h(E)$ becomes smoothly flat. When this happens, $\Omega(E)$ is a converged density of states. The Wang-Landau approach can be used for both discrete lattice-based models as well as continuous systems such as simple fluids (Yan *et al.*, 2002) and proteins (Rathore *et al.*, 2003).

An interesting point concerning the Wang-Landau algorithm is that it does not satisfy detailed balance due to the application of the scaling factor to the density of states $\Omega(E)$, causing the latter to change continually throughout the calculation. As f approaches unity, the algorithm just starts to satisfy detailed balance. Thus, the Wang-Landau approach represents a Monte Carlo method that can work without strict adherence to detailed balance throughout the sampling procedure. Note that it is also possible to use molecular dynamics to generate trial moves before application of eqn. (7.6.3). If a large time step is used, for example, then after m steps, energy will not be conserved (which also occurs in the hybrid Monte Carlo scheme), meaning that there will be an energy change from E_1 to E_2. Thus, the potential energy will change from U_1 to U_2, and the kinetic energy will also change from K_1 to K_2, with $E_1 = K_1 + U_1$ and $E_2 = K_2 + U_2$. In this case, the acceptance criterion should be modified, as suggested by Rathmore *et al.* (2003), according to

$$A(E_2|E_1) = \min \left[1, e^{-\beta \Delta K} \frac{\Omega(U_1)}{\Omega(U_2)} \right], \tag{7.6.4}$$

where $\Delta K = K_2 - K_1$.

7.7 Transition path sampling and the transition path ensemble

The last technique we will describe in this chapter is something of a departure from the methods we have discussed thus far. Up to now, we have discussed approaches for sampling configurations from a specified probability distribution, and because our aim has *only* been to sample, there is no dynamical information in the sequence of configurations we have produced. Consequently, these methods can only be used to calculate equilibrium and thermodynamic properties. But suppose we could devise a Monte Carlo scheme capable of producing actual dynamical trajectories from one part of phase space to another. The framework of Monte Carlo is flexible enough that such a thing ought to be possible. In fact, this is something that can be achieved, but doing so requires a small shift in the way we think of statistical ensembles and the sampling problem.

The technique we will discuss, known as *transition path sampling*, was pioneered by Chandler and coworkers (Dellago *et al.*, 1998; Bolhuis *et al.*, 2002; Dellago *et al.*, 2002; Frenkel and Smit, 2002). This approach is a particularly powerful one for generating dynamical trajectories between two regions of phase space when the passage of the system from one to the other is exceedingly rare. When such *rare events* (which we will discuss in more detail in Section 8.5) cannot be accessed using ordinary molecular dynamics, transition path sampling provides a means of producing the desired trajectories. A classic example of a rare event is the dissociation of a water molecule according to $2H_2O(l) \longrightarrow H_3O^+(aq) + OH^-(aq)$ (Geissler *et al.*, 2001). The reaction ostensibly only requires transferring a proton from one water molecule to another. However, if we attach ourselves to a particular water molecule and wait for the chemical reaction to occur, the average time we would have to wait is ten hours for a single event, a time scale that is well beyond that which can be accessed in an ordinary molecular dynamics calculation. Generally, a process is termed a rare-event process when the system must cross one or more high energy barriers (see Fig. 7.2). The actual passage time over the barrier can be quite short, while most of the time is spent waiting for a sufficient amount of energy to amass in a small number of modes in the system to allow the barrier to be surmounted. When a system is in equilibrium, where equipartitioning holds, such a fluctuation is, indeed, a rare event. This is illustrated in Fig. 7.5, which shows a thermostatted trajectory of a particle in a one-dimensional double-well potential with minima at $x = \pm 1$ and a barrier height of $8kT$. The figure shows that actual crossing events are rapid but times between such events are quite long.

In the transition path sampling method, we approach the problem in a way that is qualitatively different from what we have done until now. Let us assume that we know *a priori* the regions of phase space in which the trajectory initiates and in which it finally ends up. We denote these regions generically as $\mathcal{A}$ and $\mathcal{B}$, respectively (see

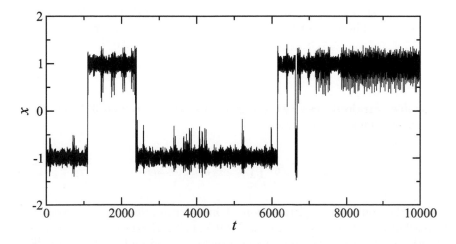

Fig. 7.5 Rare-event trajectory in a double-well potential with barrier height $8kT$.

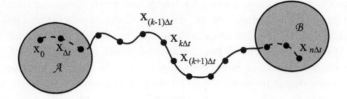

Fig. 7.6 A representative transition path from $\mathcal{A}$ to $\mathcal{B}$. Also shown are discrete phase-space points along the path: Initial (x_0), final ($x_{n\Delta t} = x_t$), and several intermediate ($x_{(k-1)\Delta t}$, $x_{k\Delta t}$, $x_{(k+1)\Delta t}$) points.

Fig. 7.6). Let us also assume that a time $\mathcal{T}$ is needed for the system to pass from $\mathcal{A}$ to $\mathcal{B}$. If we could generate a molecular dynamics trajectory from $\mathcal{A}$ to $\mathcal{B}$, then this trajectory would consist of a sequence of discrete phase-space points $x_0, x_{\Delta t}, x_{2\Delta t}, ..., x_{n\Delta t}$, where $n\Delta t = \mathcal{T}$, such that $x_0 \in \mathcal{A}$ and $x_{n\Delta t} \in \mathcal{B}$ (see Fig. 7.6). Let us denote this set of phase-space points as $X(\mathcal{T})$. That is, $X(\mathcal{T})$ is a time-ordered sequence of microscopic states visited by the system as it passes from $\mathcal{A}$ to $\mathcal{B}$. If we view $X(\mathcal{T})$ as belonging to an ensemble of trajectories from $\mathcal{A}$ to $\mathcal{B}$, then we can derive a probability $\mathcal{P}_{A\mathcal{B}}[X(\mathcal{T})]$ associated with a given trajectory. Note that we are using functional notation because $\mathcal{P}_{A\mathcal{B}}$ depends on the entire trajectory. If we regard the sequence of phase-space points $x_0, x_{\Delta t}, x_{2\Delta t}, ..., x_{n\Delta t}$ as a Markov chain, then there exists a rule $T(x_{(k+1)\Delta t}|x_{k\Delta t})$ for generating $x_{(k+1)\Delta t}$ given $x_{k\Delta t}$. For example, suppose we posit that the trajectory is to be generated deterministically via a symplectic integrator such as the velocity Verlet method of Section 3.10. Then, $x_{(k+1)\Delta t}$ would be generated from $x_{k\Delta t}$ using a Trotter factorization of the propagator

$$x_{(k+1)\Delta t} = e^{iL_2\Delta t/2}e^{iL_1\Delta t}e^{iL_2\Delta t/2}x_{k\Delta t} \equiv \phi_{\Delta t}(x_{k\Delta t}). \qquad (7.7.1)$$

Here, $\phi_{\Delta t}(x_{k\Delta t})$ is a shorthand notation for the Trotter factorized single-step propagator acting on $x_{k\Delta t}$. The rule $T(x_{(k+1)\Delta t}|x_{k\Delta t})$ must specify that there is only one possible choice for $x_{(k+1)\Delta t}$ given $x_{k\Delta t}$, which means we must take the rule as

$$T(x_{(k+1)\Delta t}|x_{k\Delta t}) = \delta\left(x_{(k+1)\Delta t} - \phi_{\Delta t}(x_{k\Delta t})\right). \qquad (7.7.2)$$

In an ensemble of trajectories $X(\mathcal{T})$, the general statistical weight $\mathcal{P}[X(\mathcal{T})]$ that we would assign to any single trajectory is

$$\mathcal{P}[X(\mathcal{T})] = f(x_0) \prod_{k=0}^{n-1} T(x_{(k+1)\Delta t}|x_{k\Delta t}), \qquad (7.7.3)$$

where $f(x_0)$ is the equilibrium distribution of initial conditions x_0, for example, a canonical distribution $f(x_0) = \exp(-\beta\mathcal{H}(x_0))/Q(N, V, T)$. However, since our interest is in trajectories that start in the phase-space region $\mathcal{A}$ and end in the region $\mathcal{B}$, we need to restrict the trajectory distribution in eqn. (7.7.3) to this subset of trajectories. We do this by multiplying eqn. (7.7.3) by functions $h_A(x_0)$ and $h_\mathcal{B}(x_{n\Delta t})$, where

$h_A(\mathrm{x}) = 1$ if $\mathrm{x} \in \mathcal{A}$ and $h_A(\mathrm{x}) = 0$ otherwise, with a similar definition for $h_{\mathcal{B}}(\mathrm{x})$. This gives the transition path probability $\mathcal{P}_{A\mathcal{B}}[X(\mathcal{T})]$ as

$$\mathcal{P}_{A\mathcal{B}}[X(\mathcal{T})] = \frac{1}{\mathcal{F}_{A\mathcal{B}}(\mathcal{T})} h_A(\mathrm{x}_0)\mathcal{P}[X(\mathcal{T})]h_{\mathcal{B}}(\mathrm{x}_{n\Delta t}), \tag{7.7.4}$$

where $\mathcal{F}_{A\mathcal{B}}$ is a normalization constant given by

$$\mathcal{F}_{A\mathcal{B}}(\mathcal{T}) = \int d\mathrm{x}_0 \cdots d\mathrm{x}_{n\Delta t} h_A(\mathrm{x}_0)\mathcal{P}[X(\mathcal{T})]h_{\mathcal{B}}(\mathrm{x}_{n\Delta t}). \tag{7.7.5}$$

Equations (7.7.4) and (7.7.5) can be regarded as the probability distribution and partition function for an ensemble of trajectories that begin in $\mathcal{A}$ and end in $\mathcal{B}$ and thus, they can be regarded as defining an ensemble called the *transition path ensemble* (Dellago *et al.*, 1998; Bolhuis *et al.*, 2002; Dellago *et al.*, 2002). Although eqns. (7.7.4) and (7.7.5) are valid for any trajectory rule $T(\mathrm{x}_{(k+1)\Delta t}|\mathrm{x}_{k\Delta t})$, if we take the specific example of deterministic molecular dynamics in eqn. (7.7.2), then eqn. (7.7.5) becomes

$$\mathcal{F}_{A\mathcal{B}}(\mathcal{T}) = \int d\mathrm{x}_0 \cdots d\mathrm{x}_{n\Delta t} h_A(\mathrm{x}_0)f(\mathrm{x}_0) \prod_{k=0}^{n-1} \delta\left(\mathrm{x}_{(k+1)\Delta t} - \phi_{\Delta t}(\mathrm{x}_{k\Delta t})\right) h_{\mathcal{B}}(\mathrm{x}_{n\Delta t})$$

$$= \int d\mathrm{x}_0 f(\mathrm{x}_0)h_A(\mathrm{x}_0)h_{\mathcal{B}}(\mathrm{x}_{n\Delta t}(\mathrm{x}_0)), \tag{7.7.6}$$

where we have used the Dirac δ-functions to integrate over all points $\mathrm{x}_{k\Delta t}$ except $k = 0$. When this is done the $h_{\mathcal{B}}$ factor looks like $h_{\mathcal{B}}(\phi_{\Delta t}(\phi_{\Delta t}(\cdots\phi_{\Delta t}(\mathrm{x}_0))))$, where the integrator $\phi_{\Delta t}$ acts n times on the initial condition x_0 to give the unique numerical solution $\mathrm{x}_{n\Delta t}(\mathrm{x}_0)$ appearing in eqn. (7.7.6). Thus, for deterministic molecular dynamics, eqn. (7.7.6) simply counts those microstates belonging to the equilibrium ensemble $f(\mathrm{x}_0)$ that are contained in $\mathcal{A}$ and, when integrated for n steps, end in $\mathcal{B}$. In Chapter 15, we will show how to define and generate the transition path ensemble for trajectories obeying stochastic rather than deterministic dynamics.

Having now defined the ensemble of transition paths, we need a Monte Carlo algorithm for sampling this ensemble. The method we will describe here is an adaptation of the M(RT)^2 algorithm for an ensemble of paths rather than one of configurations. Accordingly, we seek to generate a Markov chain of M trajectories $X_1(\mathcal{T}), ..., X_M(\mathcal{T})$, and to accomplish this, we begin, as we did in Section 7.3.3, with a generalization of the detailed balance condition. Let $\mathcal{R}_{A\mathcal{B}}[X(\mathcal{T})|Y(\mathcal{T})]$ be the conditional probability to generate a trajectory $X(\mathcal{T})$ starting from $Y(\mathcal{T})$. Both $X(\mathcal{T})$ and $Y(\mathcal{T})$ must be transition paths from $\mathcal{A}$ to $\mathcal{B}$. The detailed balance condition appropriate for the transition path ensemble is

$$\mathcal{R}_{A\mathcal{B}}[X(\mathcal{T})|Y(\mathcal{T})]\mathcal{P}_{A\mathcal{B}}[Y(\mathcal{T})] = \mathcal{R}_{A\mathcal{B}}[Y(\mathcal{T})|X(\mathcal{T})]\mathcal{P}_{A\mathcal{B}}[X(\mathcal{T})]. \tag{7.7.7}$$

As we did in eqn. (7.3.25), we decompose $\mathcal{R}_{A\mathcal{B}}[X(\mathcal{T})|Y(\mathcal{T})]$ into a product

$$\mathcal{R}_{A\mathcal{B}}[X(\mathcal{T})|Y(\mathcal{T})] = \Lambda_{A\mathcal{B}}[X(\mathcal{T})|Y(\mathcal{T})]\mathcal{T}_{A\mathcal{B}}[X(\mathcal{T})|Y(\mathcal{T})] \tag{7.7.8}$$

of a trial probability $\mathcal{T}_{A\mathcal{B}}$ and an acceptance probability $\Lambda_{A\mathcal{B}}$. The same logic used to obtain eqn. (7.3.28) leads to the acceptance rule for transition path sampling

$$\Lambda[X(\mathcal{T})|Y(\mathcal{T})] = \min\left[1, \frac{\mathcal{P}_{A\mathcal{B}}[X(\mathcal{T})]\mathcal{T}_{A\mathcal{B}}[Y(\mathcal{T})|X(\mathcal{T})]}{\mathcal{P}_{A\mathcal{B}}[Y(\mathcal{T})]\mathcal{T}_{A\mathcal{B}}[X(\mathcal{T})|Y(\mathcal{T})]}\right]. \tag{7.7.9}$$

Since the trajectory $Y(\mathcal{T})$ is assumed to be a proper transition path from A to $\mathcal{B}$, $h_A(y_0) = 1$ and $h_\mathcal{B}(y_{n\Delta t}) = 1$. Thus, we can write eqn. (7.7.9) as

$$\Lambda[X(\mathcal{T})|Y(\mathcal{T})] = h_A(x_0)h_\mathcal{B}(x_{n\Delta t})\min\left[1, \frac{\mathcal{P}[X(\mathcal{T})]\mathcal{T}_{A\mathcal{B}}[Y(\mathcal{T})|X(\mathcal{T})]}{\mathcal{P}[Y(\mathcal{T})]\mathcal{T}_{A\mathcal{B}}[X\mathcal{T})|Y(\mathcal{T})]}\right], \tag{7.7.10}$$

which is zero unless the new trajectory $X(\mathcal{T})$ is also a proper transition path.

As with any Monte Carlo algorithm, the key to efficient sampling of transition paths is the design of the rule $\mathcal{T}_{A\mathcal{B}}[X(\mathcal{T})|Y(\mathcal{T})]$ for generating trial moves from one path to another. Here, we will discuss a particular type of trial move known as a "shooting move." Shooting moves are conceptually simple. We randomly select a point $y_{j\Delta t}$ (by randomly choosing the integer j) from the starting trajectory $Y(\mathcal{T})$ and modify it in some way to give a point $x_{j\Delta t}$ on the trial trajectory $X(\mathcal{T})$, referred to as a "shooting point." Starting from this point, trajectories are launched forward and backward in time. If the new trajectory $X(\mathcal{T})$ thus generated is a transition path from A to $\mathcal{B}$, it is accepted with some probability; otherwise, it is rejected. The idea of shooting moves is illustrated in Fig. 7.7.

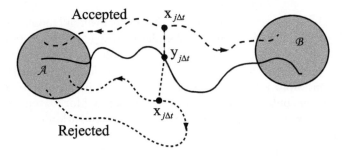

Fig. 7.7 The shooting algorithm. The original path $Y(\mathcal{T})$ is shown as the solid line. A point $y_{j\Delta t}$ randomly chosen from this path is used to determine the shooting point $x_{j\Delta t}$. Two example shooting paths are shown as dashed lines. The long dashed line is a successful path that is accepted, while the short dashed line shows an unsuccessful path that is rejected.

Let $\tau(x_{j\Delta t}|y_{j\Delta t})$ denote the rule for generating a trial shooting point $x_{j\Delta t}$ from $y_{j\Delta t}$. Then, we can express $\mathcal{T}_{A\mathcal{B}}[X(\mathcal{T})|Y(\mathcal{T})]$ as

$$\mathcal{T}_{A\mathcal{B}}[X(\mathcal{T})|Y(\mathcal{T})] =$$

$$\tau(x_{j\Delta t}|y_{j\Delta t})\left[\prod_{k=j}^{n-1} T(x_{(k+1)\Delta t}|x_{k\Delta t})\right]\left[\prod_{k=1}^{j} T(x_{(k-1)\Delta t}|x_{k\Delta t})\right]. \tag{7.7.11}$$

The first product in eqn. (7.7.11) is the probability for the forward trajectory, and the second product, which requires the rule to generate $x_{(k-1)\Delta t}$ from $x_{k\Delta t}$ via time-reversed dynamics, is the weight for the backward trajectory. For molecular dynamics, this part of the trajectory is just obtained via eqn. (7.7.1) using $-\Delta t$ instead of Δt, i.e., $x_{(k-1)\Delta t} = \phi_{-\Delta t}(x_{k\Delta t})$, which can be generated by integrating forward in time with time step Δt but with velocities reversed at the shooting point. Combining eqn. (7.7.11) with (7.7.10), we obtain for the acceptance probability

$$
\Lambda[X(\mathcal{T})|Y(\mathcal{T})] = h_{\mathcal{A}}(x_0)h_{\mathcal{B}}(x_{n\Delta t})\min\left[1, \frac{f(x_0)}{f(y_0)}\left(\prod_{k=0}^{n-1}\frac{T(x_{(k+1)\Delta t}|x_{k\Delta t})}{T(y_{(k+1)\Delta t}|y_{k\Delta t})}\right)\right.
$$

$$
\left. \times \left(\frac{\tau(y_{j\Delta t}|x_{j\Delta t})}{\tau(x_{j\Delta t}|y_{j\Delta t})}\prod_{k=j}^{n-1}\frac{T(y_{(k+1)\Delta t}|y_{k\Delta t})}{T(x_{(k+1)\Delta t}|x_{k\Delta t})}\prod_{k=0}^{j-1}\frac{T(y_{k\Delta t}|y_{(k+1)\Delta t})}{T(x_{k\Delta t}|x_{(k+1)\Delta t})}\right)\right]
$$

$$
= h_{\mathcal{A}}(x_0)h_{\mathcal{B}}(x_{n\Delta t})
$$

$$
\times\min\left[1, \frac{f(x_0)}{f(y_0)}\frac{\tau(y_{j\Delta t}|x_{j\Delta t})}{\tau(x_{j\Delta t}|y_{j\Delta t})}\prod_{k=0}^{j-1}\frac{T(y_{k\Delta t}|y_{(k+1)\Delta t})T(x_{(k+1)\Delta t}|x_{k\Delta t})}{T(x_{k\Delta t}|x_{(k+1)\Delta t})T(y_{(k+1)\Delta t}|y_{k\Delta t})}\right]. \quad (7.7.12)
$$

Although eqn. (7.7.12) might seem rather involved, consider what happens when the trajectories are generated by molecular dynamics, with the trial probability given by eqn. (7.7.2). Since a symmetric Trotter factorization of the classical propagator is time reversible, as discussed in Section 3.10, the ratio $T(y_{k\Delta t}|ry_{(k+1)\Delta t})/T(y_{(k+1)\Delta t}|y_{k\Delta t})$ is unity, as is the ratio $T(x_{(k+1)\Delta t}|x_{k\Delta t})/T(x_{k\Delta t}|x_{(k+1)\Delta t})$, and the acceptance criterion simplifies to

$$
\Lambda[X(\mathcal{T})|Y(\mathcal{T})] = h_{\mathcal{A}}(x_0)h_{\mathcal{B}}(x_{n\Delta t})\min\left[1, \frac{f(x_0)}{f(y_0)}\frac{\tau(y_{j\Delta t}|x_{j\Delta t})}{\tau(x_{j\Delta t}|y_{j\Delta t})}\right]. \quad (7.7.13)
$$

Finally, suppose the new shooting point $x_{j\Delta t}$ is generated from the old point $y_{j\Delta t}$ using the following rule;

$$
x_{j\Delta t} = y_{j\Delta t} + \Delta, \quad (7.7.14)
$$

where the phase-space displacement Δ is chosen randomly from a symmetric distribution $\pi(\Delta)$ satisfying $\pi(-\Delta) = \pi(\Delta)$. In this case, the ratio $\tau(y_{j\Delta t}|x_{j\Delta t})/\tau(x_{j\Delta t}|y_{j\Delta t})$ in eqn. (7.7.13) is unity, and the acceptance rule becomes simply

$$
\Lambda[X(\mathcal{T})|Y(\mathcal{T})] = h_{\mathcal{A}}(x_0)h_{\mathcal{B}}(x_{n\Delta t})\min\left[1, \frac{f(x_0)}{f(y_0)}\right], \quad (7.7.15)
$$

which is determined just by the initial conditions and whether the new trajectory is a proper transition path from $\mathcal{A}$ to $\mathcal{B}$.

At this point, several comments on the shooting algorithm are in order. First, a very common and simple choice for the phase-space displacement is $\Delta = (0, \delta p)$, meaning that only the momenta are altered, while the configuration is left unchanged (Dellago *et al.*, 2002). If δp is chosen from a Maxwell-Boltzmann distribution, then the symmetry condition is satisfied and will continue to be satisfied if the new momenta are projected onto a surface of constraint or modified to give zero total linear or angular momentum in the system. As a rule of thumb, the displacement Δ should be chosen to give roughly 40% acceptance probability (Dellago *et al.*, 2002).

The basic steps of the shooting algorithm can be summarized as follows:

1. Choose an index j randomly on the old trajectory $Y(\mathcal{T})$.
2. Generate a random phase-space displacement Δ in order to generate the new shooting point $x_{j\Delta t}$ from the old point $y_{j\Delta t}$.
3. Integrate the equations of motion backwards in time from the shooting point to the initial condition x_0.
4. If the initial condition x_0 is not in the phase-space region $\mathcal{A}$, reject the trial move.
5. If $x_0 \in \mathcal{A}$, accept the move with probability $\min[1, f(x_0)/f(y_0)]$. Note that if the distribution of initial conditions is microcanonical rather than canonical (or isothermal-isobaric), this step can be skipped.
6. Integrate the equations of motion forward in time to generate the final point $x_{n\Delta t}$.
7. If $x_{n\Delta t} \in \mathcal{B}$, accept the trial move, and reject it otherwise.
8. If the path is rejected at steps 4, 5, or 7, then the old trajectory $Y(\mathcal{T})$ is counted again in the calculation of averages over the transition path ensemble. Otherwise, invert the momenta along the backward path of the path to yield a forward moving transition path $X(\mathcal{T})$ and replace the old trajectory $Y(\mathcal{T})$ by the new trajectory $X(\mathcal{T})$.

A useful variant of the shooting algorithm is known as "aimless shooting" (Peters and Trout, 2006). In this approach, rather than displacing a point $y_{j\Delta t}$ to obtain a new shooting point $x_{j\Delta t}$, a shooting point $y_{j\Delta t}$ from the old trajectory is selected, and the momenta are resampled from a Boltzmann distribution at that point. The resulting new phase-space point is denoted $x_{j\Delta t}$. As in the shooting algorithm above, the equations of motion for the new trajectory originating at the new shooting point $x_{j\Delta t}$ are integrated backward to $t = 0$ to determine whether the trajectory will return to the phase-space region $\mathcal{A}$. If it does, then the equations of motion are integrated forward to $t = n\Delta t = \mathcal{T}$ to determine whether the trajectory will terminate in the phase-space region $\mathcal{B}$. If it does, then the trajectory is accepted; otherwise, the trajectory is rejected. In aimless shooting, the shooting point does not correspond to an arbitrary choice of j, but rather, j is chosen to be $(n/2) - 1$, $n/2$, or $(n/2) + 1$ with equal probability, so that the corresponding point on the new trajectory is $x_{((n/2)-1)\Delta t}$, $x_{(n/2)\Delta t}$, or $x_{((n/2)+1)\Delta t}$, respectively. We note, in passing, that aimless shooting simplifies the condition for accepting a trajectory. Importantly, aimless shooting is designed to generate most of its shooting points near a region in the phase space for which there is roughly equal probability that trajectories initiated at these points will end in $\mathcal{A}$ or $\mathcal{B}$, which we identify as a transition state. Further discussion of this point will be deferred to Section 8.12, where the concept of a committor and its distribution is introduced.

Another important point to note about the transition path sampling approach is that an initial transition path $X_0(\mathcal{T})$ is needed in order to seed the algorithm. Generating such a path can be difficult, particularly for extremely rare event processes. However, a few tricks can be employed, for example running a system at high temperature to accelerate a process or possibly starting a path in $\mathcal{B}$ and letting it evolve to $\mathcal{A}$. The latter could be employed, for example, in protein folding, where it is generally easier to induce a protein to unfold than fold. Although initial paths generated via such tricks are not likely to have a large weight in the transition path ensemble, they should quickly relax to more probable paths under the shooting algorithm. However, as with any Monte Carlo scheme, this fast relaxation cannot be guaranteed if the initial path choice is a particularly poor one. Just as configurational Monte Carlo methods can become trapped in local regions of configuration space, so transition path sampling can become trapped in local regions of path space where substantial barriers keep the system from accessing regions of the space containing paths of higher probability. The design of path-generation algorithms capable of enhancing sampling of the transition path ensemble is still an open and interesting question with room for novel improvements.

7.8 Problems

7.1. Write a Monte Carlo program to calculate the integral

$$I = \int_0^1 e^{-x^2} \mathrm{d}x$$

using
a. uniform sampling of x on the interval $[0, 1]$, and

b. an importance function $h(x)$, where

$$h(x) = \frac{3}{2}\left(1 - x^2\right)$$

constitutes the first two terms in the Taylor series expansion of $\exp(-x^2)$. In both cases, compare the converged result you obtain to the value of I generated using a simple numerical integration algorithm such as Simpson's rule.

7.2. Devise an importance function for performing the integral

$$I = \int_0^1 \cos\left(\frac{\pi x}{2}\right) \mathrm{d}x,$$

and show using a Monte Carlo program that your importance function leads to a smaller variance than uniform sampling for the same number of Monte Carlo moves. How many steps are required with your importance function to converge the Monte Carlo estimator to within 10^{-6} of the analytical value of I?

7.3. The following example (Kalos and Whitlock, 1986) illustrates the recursion associated with the $M(RT)^2$ algorithm. Consider the $M(RT)^2$ algorithm for sampling the one-dimensional probability distribution $f(x) = 2x$ for $x \in (0, 1)$. Let the probability for trial moves from y to x be

$$T(x|y) = \begin{cases} 1 & x \in (0,1) \\ 0 & \text{otherwise} \end{cases}.$$

In this case, $r(x|y) = x/y$ for in $y \in (0, 1)$.

a. Show that the sequence of distributions $\pi_n(x)$ satisfies the recursion

$$\pi_{n+1}(x) = \int_x^1 \frac{x}{y} \pi_n(y) \mathrm{d}y + \int_0^x \pi_n(y) \mathrm{d}y + \pi_n(x) \int_0^x \left(1 - \frac{y}{x}\right) \mathrm{d}y.$$

b. Show, therefore, that $\pi_n(x) = cx$ is a fixed point of the recursion, where c is an arbitrary constant. By normalization, c must be equal to 2.

c. Now suppose that we start the recursion with $\pi_0(x) = 3x^2$ and that at the nth step of the iteration

$$\pi_n(x) = a_n x + c_n x^{n+2},$$

where a_n and c_n are constants. Show that as $n \to \infty$, c_n goes asymptotically to 0, leaving a distribution that is purely linear.

*7.4. In this problem, we will compare some of the Monte Carlo schemes introduced in this chapter to thermostatted molecular dynamics for a one-dimensional harmonic oscillator for which

$$U(x) = \frac{1}{2}m\omega^2 x^2$$

in the canonical ensemble. For this problem, you can take the mass m and frequency ω both equal to 1.

a. Write a Monte Carlo program that uses uniform sampling of x within the $M(RT)^2$ algorithm to calculate the canonical ensemble average $\langle x^4 \rangle$. Try to optimize the step size Δ and average acceptance probability to obtain the lowest possible variance.

b. Write a hybrid Monte Carlo program that uses the velocity Verlet integrator to generate trial moves of x. Use your program to calculate the same average $\langle x^4 \rangle$. Try to optimize the time step and average acceptance probability to obtain the lowest possible variance.

c. Write a thermostatted molecular dynamics program using the Nosé-Hoover chain equations together with the integrator described by eqns. (4.12.8)-(4.12.19). Try using $n_{\text{sy}} = 7$ with the weights in eqn. (4.12.13) and $n = 4$. Adjust the time step so that the energy in eqn. (4.11.3) is conserved to 10^{-4} as measured by eqn. (3.14.1).

d. Compare the number of steps of each algorithm needed to converge the average $\langle x^4 \rangle$ to within the same error as measured by the variance. What are your conclusions about the efficiency of Monte Carlo versus molecular dynamics for this problem?

*7.5. Consider Hamilton's equations in the form $\dot{x} = \eta(x)$. Suppose $x = (q, p)$ is a two-dimensional phase-space vector, and let the equations of motion be integrated using the following numerical solver:

$$x(\Delta t) = x(0) + \Delta t \eta \left(x(0) + \frac{\Delta t}{2} \eta(x(0)) \right).$$

Can this algorithm be used in conjunction with hybrid Monte Carlo? Why or why not?

7.6. Write a replica exchange Monte Carlo algorithm to sample the Boltzmann distribution corresponding to a double-well potential $U(x)$ of the form

$$U(x) = D_0 \left(x^2 - a^2 \right)^2$$

for $a = 1$ and D_0 values of 5 and 10. For each case, optimize the temperature ladder $T_1, ..., T_M$, the number of replicas M, and the frequency of exchange attempts. Use separate Nosé-Hoover chains to control the temperatures on each of the M replicas. For each simulation, plot the following measure of convergence:

$$\zeta_k = \frac{1}{N_{\text{bins}}} \sum_{i=1}^{N_{\text{bins}}} |P_k(x_i) - P_{\text{exact}}(x_i)|,$$

where $P_{\text{exact}}(x)$ is the exact probability distribution, $P_k(x)$ is the probability at the kth step of the simulation, and N_{bins} is the number of bins used in the calculation of the histogram of the system with temperature T_1.

*7.7 Suppose a non-Hamiltonian molecular dynamics algorithm, such as the Nosé-Hoover chain method of Section 4.11, is used to generate paths in the transition path sampling algorithm. Assuming symmetry of the rule for generating shooting points, i.e., $\tau(y_{j\Delta t}|x_{j\Delta t})/\tau(x_{j\Delta t}|y_{j\Delta t})$, show that the acceptance rule in eqn. (7.7.15) must be modified to read

$$\Lambda[X(\mathcal{T})|Y(\mathcal{T})] = h_A(x_0) h_B(x_{n\Delta t}) \min \left[1, \frac{f(x_0)}{f(y_0)} \frac{J(y_{j\Delta t}; y_0)}{J(x_{j\Delta t}; x_0)} \right],$$

where $J(x_t; x_0)$ is the Jacobian of the transformation from x_0 to $x_t(x_0)$ determined by eqn. (4.10.2).

7.8. The probability distribution function

$$f(x) = \frac{C}{x^2 + a^2},$$

where $x \in (-\infty, \infty)$, is known as a *Lorentzian distribution*. Here, a is a constant.

a. Find an expression for the constant C so that $f(x)$ is properly normalized.

b. Design an algorithm for generating random samples from the Lorentzian distribution function.

c. Repeat parts a and b for the distribution

$$f(x) = \frac{Cx}{x^2 + a^2},$$

where $x \in [0, a]$.

8
Free-energy calculations

Our treatment of the classical equilibrium ensembles makes clear that the free energy is a quantity of particular importance in statistical mechanics. Being related to the logarithm of the partition function, the free energy is the generator through which other thermodynamic quantities are obtained via differentiation. Often, however, we are less interested in the absolute free energy than we are in the free-energy *difference* between two thermodynamic states. Free-energy differences tell us, for example, whether a chemical reaction occurs spontaneously or requires input of work. They tell us whether a given solute is hydrophobic or hydrophilic, and they are directly related to equilibrium constants for chemical processes. Thus, from free-energy differences, we can compute acid or base ionization constants. We can also quantify the therapeutic viability of a candidate drug compound by calculating its inhibition constant or IC50 value from the binding free energy. Another type of free energy often sought is the free energy as a function of one or more generalized or collective coordinates in a system, *e.g.*, a set of backbone dihedral angles in an oligopeptide. This surface provides a map of the stable conformations of the molecule, the relative stability of these conformations, and the barrier heights that must be crossed for a change in conformation. It should, therefore, be clear that free-energy surfaces encode the conformational preferences of complex molecules such as proteins and nucleic acids.

In this chapter, we describe a variety of widely used techniques that have been developed for calculating free energies and discuss the relative merits and disadvantages of the methods. The fact that the free energy is a state function, ensuring that the system can be transformed from one state to another along physical or unphysical paths without affecting the free-energy difference, allows for considerable flexibility in the design of novel techniques and will be frequently exploited in the developments we will present.

The techniques described in this chapter are constructed within the framework of the canonical ensemble with the aim of obtaining Helmholtz free-energy differences ΔA. Generalization to the isothermal-isobaric ensemble and the Gibbs free-energy difference ΔG is, in all cases, straightforward. (For a useful compendium of free-energy calculation methods, readers are referred to *Free-Energy Calculations*, C. Chipot and A. Pohorille, eds. (2007) or *Rugged Free-Energy Landscapes*, W. Janke, ed. (2010) .

8.1 Free-energy perturbation theory

We begin our treatment of free-energy differences by considering the problem of transforming a system from one thermodynamic state to another. Let these states be

denoted generically as $\mathcal{A}$ and $\mathcal{B}$. At the microscopic level, these two states are charac-
terized by potential energy functions $U_\mathcal{A}(\mathbf{r}_1, ..., \mathbf{r}_N)$ and $U_\mathcal{B}(\mathbf{r}_1, ..., \mathbf{r}_N)$. For example,
in a drug-binding study, the state $\mathcal{A}$ might correspond to the unbound ligand and en-
zyme while $\mathcal{B}$ would correspond to the bound complex. In this case, the potential $U_\mathcal{A}$
would exclude all interactions between the ligand and the enzyme, but the potential
$U_\mathcal{B}$ would include them.

The Helmholtz free-energy difference between the states $\mathcal{A}$ and $\mathcal{B}$ is simply $\Delta A_{\mathcal{A}\mathcal{B}} = A_\mathcal{B} - A_\mathcal{A}$. The two free energies $A_\mathcal{A}$ and $A_\mathcal{B}$ are given in terms of their respective canonical partition functions $Q_\mathcal{A}$ and $Q_\mathcal{B}$, respectively by $A_\mathcal{A} = -kT \ln Q_\mathcal{A}$ and
$A_\mathcal{B} = -kT \ln Q_\mathcal{B}$, where

$$Q_\mathcal{A}(N, V, T) = C_N \int d^N\mathbf{p}\, d^N\mathbf{r}\, \exp\left\{ -\beta \left[\sum_{i=1}^N \frac{\mathbf{p}_i^2}{2m_i} + U_\mathcal{A}(\mathbf{r}_1, ..., \mathbf{r}_N) \right] \right\}$$

$$= \frac{Z_\mathcal{A}(N, V, T)}{N! \lambda^{3N}}$$

$$Q_\mathcal{B}(N, V, T) = C_N \int d^N\mathbf{p}\, d^N\mathbf{r}\, \exp\left\{ -\beta \left[\sum_{i=1}^N \frac{\mathbf{p}_i^2}{2m_i} + U_\mathcal{B}(\mathbf{r}_1, ..., \mathbf{r}_N) \right] \right\}$$

$$= \frac{Z_\mathcal{B}(N, V, T)}{N! \lambda^{3N}}. \tag{8.1.1}$$

The $1/N!$ combinatorial factor should be adjusted if multiple particle species are
present in the system (see footnote below eqn. (3.2.22)). The free-energy difference
is, therefore,

$$\Delta A_{\mathcal{A}\mathcal{B}} = A_\mathcal{B} - A_\mathcal{A} = -kT \ln\left(\frac{Q_\mathcal{B}}{Q_\mathcal{A}} \right) = -kT \ln\left(\frac{Z_\mathcal{B}}{Z_\mathcal{A}} \right), \tag{8.1.2}$$

where $Z_\mathcal{A}$ and $Z_\mathcal{B}$ are the configurational partition functions for states $\mathcal{A}$ and $\mathcal{B}$,
respectively:

$$Z_\mathcal{A} = \int d^N\mathbf{r}\, e^{-\beta U_\mathcal{A}(\mathbf{r}_1, ..., \mathbf{r}_N)}$$

$$Z_\mathcal{B} = \int d^N\mathbf{r}\, e^{-\beta U_\mathcal{B}(\mathbf{r}_1, ..., \mathbf{r}_N)}. \tag{8.1.3}$$

The ratio of full partition functions $Q_\mathcal{B}/Q_\mathcal{A}$ reduces to the ratio of configurational
partition functions $Z_\mathcal{B}/Z_\mathcal{A}$ because the momentum integrations in the former cancel
out of the ratio.

Equation (8.1.2) is difficult to implement in practice because in any numerical
calculation based either on molecular dynamics or Monte Carlo, we can compute averages of phase-space functions, but we do not have direct access to the partition
function.[1] Thus, eqn. (8.1.2) can be computed directly if it can be expressed in terms

---

[1]The Wang-Landau method of Section 7.6 is an exception.

of a phase-space average. To this end, consider inserting unity into the expression for $Z_{\mathcal{B}}$ as follows:

$$Z_{\mathcal{B}} = \int d^N \mathbf{r} \, e^{-\beta U_{\mathcal{B}}(\mathbf{r}_1,...,\mathbf{r}_N)}$$

$$= \int d^N \mathbf{r} \, e^{-\beta U_{\mathcal{B}}(\mathbf{r}_1,...,\mathbf{r}_N)} e^{-\beta U_{\mathcal{A}}(\mathbf{r}_1,...,\mathbf{r}_N)} e^{\beta U_{\mathcal{A}}(\mathbf{r}_1,...,\mathbf{r}_N)}$$

$$= \int d^N \mathbf{r} \, e^{-\beta U_{\mathcal{A}}(\mathbf{r}_1,...,\mathbf{r}_N)} e^{-\beta(U_{\mathcal{B}}(\mathbf{r}_1,...,\mathbf{r}_N) - U_{\mathcal{A}}(\mathbf{r}_1,...,\mathbf{r}_N))}. \qquad (8.1.4)$$

If we now take the ratio $Z_{\mathcal{B}}/Z_{\mathcal{A}}$, we find

$$\frac{Z_{\mathcal{B}}}{Z_{\mathcal{A}}} = \frac{1}{Z_{\mathcal{A}}} \int d^N \mathbf{r} \, e^{-\beta U_{\mathcal{A}}(\mathbf{r}_1,...,\mathbf{r}_N)} e^{-\beta(U_{\mathcal{B}}(\mathbf{r}_1,...,\mathbf{r}_N) - U_{\mathcal{A}}(\mathbf{r}_1,...,\mathbf{r}_N))}$$

$$= \left\langle e^{-\beta(U_{\mathcal{B}}(\mathbf{r}_1,...,\mathbf{r}_N) - U_{\mathcal{A}}(\mathbf{r}_1,...,\mathbf{r}_N))} \right\rangle_{\mathcal{A}}, \qquad (8.1.5)$$

where the notation $\langle \cdots \rangle_{\mathcal{A}}$ indicates an average taken with respect to the canonical configurational distribution of the state $\mathcal{A}$. Substituting eqn. (8.1.5) into eqn. (8.1.2) gives

$$\Delta A_{\mathcal{A}\mathcal{B}} = -kT \ln \left\langle e^{-\beta(U_{\mathcal{B}} - U_{\mathcal{A}})} \right\rangle_{\mathcal{A}}. \qquad (8.1.6)$$

Equation (8.1.6) is known as the *free-energy perturbation* formula (Zwanzig, 1954); it should be reminiscent of the thermodynamic perturbation formula in eqn. (4.8.8). Recalling the potential distribution theorem (Widom, 1963; Beck *et al.*, 2006) for calculating the excess chemical potential of a solute molecule in solution (Section 6.8), we see that it is essentially a variant of eqn. (8.1.6) applied to the Gibbs free energy $G = \mu N$.

Equation (8.1.6) can be interpreted as follows: We sample a set of configurations $\{\mathbf{r}_1,...,\mathbf{r}_N\}$ from the canonical distribution of state $\mathcal{A}$ and simply use these configurations, unchanged, to sample the canonical distribution of state $\mathcal{B}$ with potential $U_{\mathcal{B}}$. However, because these configurations are not sampled from $\exp(-\beta U_{\mathcal{B}})/Z_{\mathcal{B}}$ directly, we need to "unbias" our sampling by removing the factor $\exp(-\beta U_{\mathcal{A}})$ and reweighting with $\exp(-\beta U_{\mathcal{B}})$, which leads to eqn. (8.1.6). The difficulty with this approach is that the configuration spaces of states $\mathcal{A}$ and $\mathcal{B}$ might not have significant overlap. By this, we mean that configurations sampled from the canonical distribution of state $\mathcal{A}$ may not be states of high probability in the canonical distribution of state $\mathcal{B}$. When this is the case, the potential energy difference $U_{\mathcal{B}} - U_{\mathcal{A}}$ becomes large, and hence the exponential factor $\exp[-\beta(U_{\mathcal{B}} - U_{\mathcal{A}})]$ becomes negligibly small. Thus, most of the configurations have very low weight in the ensemble average, and the free-energy difference converges slowly. For this reason, it is clear that the free-energy perturbation formula is only useful when the two states $\mathcal{A}$ and $\mathcal{B}$ do not differ significantly. In other words, the state $\mathcal{B}$ must be a small perturbation to the state $\mathcal{A}$.

Even if $\mathcal{B}$ is not a small perturbation to $\mathcal{A}$, the free-energy perturbation idea can still be salvaged by introducing a set of $M - 2$ intermediate states with potentials

$U_\alpha(\mathbf{r}_1, ..., \mathbf{r}_N)$, where $\alpha = 1, ..., M$, $\alpha = 1$ corresponds to the state $\mathcal{A}$, and $\alpha = M$ corresponds to the state $\mathcal{B}$. Let $\Delta U_{\alpha, \alpha+1} = U_{\alpha+1} - U_\alpha$. We now transform the system from state $\mathcal{A}$ to state $\mathcal{B}$ along a path through each of the intermediate states and compute the average of $\Delta U_{\alpha, \alpha+1}$ in each state α. The free-energy difference $\Delta A_{\mathcal{AB}}$ is the sum of contributions obtained using the free-energy perturbation formula from each intermediate state along the path:

$$\Delta A_{\mathcal{AB}} = -kT \sum_{\alpha=1}^{M-1} \ln \left\langle e^{-\beta \Delta U_{\alpha, \alpha+1}} \right\rangle_\alpha, \tag{8.1.7}$$

where $\langle \cdots \rangle_\alpha$ represents an average taken over the distribution $\exp(-\beta U_\alpha)$. The key to applying eqn. (8.1.7) is to choose the thermodynamic path between $\mathcal{A}$ and $\mathcal{B}$ so as to achieve sufficient overlap between successive intermediate states without requiring a large number of them.

8.2 Adiabatic switching and thermodynamic integration

The free-energy perturbation approach evokes a physical picture in which configurations sampled from the canonical distribution of state $\mathcal{A}$ are immediately "switched" to the state $\mathcal{B}$ by simply changing the potential from $U_\mathcal{A}$ to $U_\mathcal{B}$. When there is insufficient overlap between the states $\mathcal{A}$ and $\mathcal{B}$, a set of intermediate states can be employed to define an optimal transformation path. The use of intermediate states conjures a picture in which the system is slowly switched from $\mathcal{A}$ to $\mathcal{B}$. In this section, we will discuss an alternative approach in which the system is continuously, *adiabatically* switched from $\mathcal{A}$ to $\mathcal{B}$. An adiabatic path is one along which the system is fully relaxed at each point of the path. In order to effect the switching from one state to the other, we employ a common trick of introducing an "external" switching variable λ in order to parameterize the adiabatic path. This parameter is used to define a new potential energy function, sometimes called a "metapotential," defined as

$$U(\mathbf{r}_1, ..., \mathbf{r}_N, \lambda) \equiv f(\lambda)U_\mathcal{A}(\mathbf{r}_1, ..., \mathbf{r}_N) + g(\lambda)U_\mathcal{B}(\mathbf{r}_1, ..., \mathbf{r}_N). \tag{8.2.1}$$

The functions $f(\lambda)$ and $g(\lambda)$ are *switching functions* that must satisfy the conditions $f(0) = 1$, $f(1) = 0$, $g(0) = 0$, $g(1) = 1$. Thus $U(\mathbf{r}_1, ..., \mathbf{r}_N, 0) = U_\mathcal{A}(\mathbf{r}_1, ..., \mathbf{r}_N)$ and $U(\mathbf{r}_1, ..., \mathbf{r}_N, 1) = U_\mathcal{B}(\mathbf{r}_1, ..., \mathbf{r}_N)$. Apart from these conditions, $f(\lambda)$ and $g(\lambda)$ are completely arbitrary. The mechanism of eqn. (8.2.1) is one in which an imaginary external controlling influence ("hand of God" in the form of the λ parameter) starts the system off in state $\mathcal{A}$ ($\lambda = 0$) and slowly switches off the potential $U_\mathcal{A}$ while simultaneously switching on the potential $U_\mathcal{B}$. The process is complete when $\lambda = 1$. A simple choice for the functions $f(\lambda)$ and $g(\lambda)$ is $f(\lambda) = 1 - \lambda$ and $g(\lambda) = \lambda$.

   In order to see how eqn. (8.2.1) is used to compute the free-energy difference $\Delta A_{\mathcal{AB}}$, consider the canonical partition function of a system described by the potential of eqn. (8.2.1) for a particular choice of λ:

$$Q(N, V, T, \lambda) = C_N \int d^N\mathbf{p}\, d^N\mathbf{r}\, \exp\left\{-\beta \left[\sum_{i=1}^N \frac{\mathbf{p}_i^2}{2m_i} + U(\mathbf{r}_1, ..., \mathbf{r}_N, \lambda) \right] \right\}. \tag{8.2.2}$$

This partition function leads to a free energy $A(N, V, T, \lambda)$ via

$$A(N, V, T, \lambda) = -kT \ln Q(N, V, T, \lambda). \tag{8.2.3}$$

In Section 4.2, we showed that derivatives of A with respect to N, V, or T give the chemical potential, pressure, or entropy, respectively. For $A(N, V, T, \lambda)$, what does the derivative with respect to λ represent? According to eqn. (8.2.3),

$$\frac{\partial A}{\partial \lambda} = -\frac{kT}{Q} \frac{\partial Q}{\partial \lambda} = -\frac{kT}{Z} \frac{\partial Z}{\partial \lambda}. \tag{8.2.4}$$

Computing the derivative of Z with respect to λ, we find

$$\frac{kT}{Z} \frac{\partial Z}{\partial \lambda} = \frac{kT}{Z} \frac{\partial}{\partial \lambda} \int d^N \mathbf{r}\, e^{-\beta U(\mathbf{r}_1, \dots, \mathbf{r}_N, \lambda)}$$

$$= \frac{kT}{Z} \int d^N \mathbf{r} \left(-\beta \frac{\partial U}{\partial \lambda} \right) e^{-\beta U(\mathbf{r}_1, \dots, \mathbf{r}_N, \lambda)}$$

$$= -\left\langle \frac{\partial U}{\partial \lambda} \right\rangle. \tag{8.2.5}$$

Note that the free-energy difference $\Delta A_{\mathcal{A}\mathcal{B}}$ can be obtained trivially from the relation

$$\Delta A_{\mathcal{A}\mathcal{B}} = \int_0^1 \frac{\partial A}{\partial \lambda} d\lambda. \tag{8.2.6}$$

Substituting eqns. (8.2.4) and (8.2.5) into eqn. (8.2.6) yields the free-energy difference as

$$\Delta A_{\mathcal{A}\mathcal{B}} = \int_0^1 \left\langle \frac{\partial U}{\partial \lambda} \right\rangle_\lambda d\lambda, \tag{8.2.7}$$

where $\langle \cdots \rangle_\lambda$ denotes an average over the canonical ensemble described by the distribution $\exp[-\beta U(\mathbf{r}_1, \dots, \mathbf{r}_N, \lambda)]$ with λ fixed at a particular value. The special choice of $f(\lambda) = 1 - \lambda$ and $g(\lambda) = \lambda$ has a simple interpretation: With these functions, eqn. (8.2.7) becomes

$$\Delta A_{\mathcal{A}\mathcal{B}} = \int_0^1 \langle U_{\mathcal{B}} - U_{\mathcal{A}} \rangle_\lambda d\lambda. \tag{8.2.8}$$

Equation (8.2.8) recalls the relationship between work and free energy from the second law of thermodynamics. If, in transforming the system from state $\mathcal{A}$ to state $\mathcal{B}$, an amount of work W is performed on the system, then

$$W \geq \Delta A_{\mathcal{A}\mathcal{B}}, \tag{8.2.9}$$

where equality holds only if the transformation is carried out along a reversible path. We will refer to this inequality as the "work-free-energy inequality." Since reversible work is related to a change in potential energy (see Section 1.6), eqn. (8.2.8) is actually a statistical version of eqn. (8.2.9) for the special case of equality. Equation (8.2.8) tells us that the free-energy difference is the ensemble average of the microscopic reversible work needed to change the potential energy of each configuration from $U_\mathcal{A}$ to $U_\mathcal{B}$ along the chosen λ-path. Note, however, that eqn. (8.2.7), which is known as the *thermodynamic integration* formula (Kirkwood, 1935), is independent of the choice of $f(\lambda)$ and $g(\lambda)$, which means that eqn. (8.2.7) always yields the reversible work via the free-energy difference. The flexibility in the choice of the λ-path, however, can be exploited to design adiabatic switching algorithms of greater efficiency than can be achieved with the simple choice $f(\lambda) = 1 - \lambda$, $g(\lambda) = \lambda$.

In practice, the thermodynamic integration formula is implemented as follows: A set of M values of λ is chosen from the interval $[0, 1]$, and at each chosen value λ_k a full molecular dynamics or Monte Carlo calculation is carried out in order to generate the average $\langle \partial U / \partial \lambda_k \rangle_{\lambda_k}$. The resulting values of $\langle \partial U / \partial \lambda_k \rangle_{\lambda_k}$, $k = 1, ..., M$, are then substituted into eqn. (8.2.7), and the result is integrated numerically to produce the free-energy difference $\Delta A_{\mathcal{AB}}$. The selected values $\{\lambda_k\}$ can be evenly spaced, for example, or they could be a set of Gaussian quadrature nodes, depending on the anticipated variation of $A(N, V, T, \lambda)$ with λ for particular $f(\lambda)$ and $g(\lambda)$.

Let us now consider an example of a particular type of free-energy calculation of particular relevance, specifically, the binding of a lead drug candidate to the active site of an enzyme E. The purpose of the drug candidate is to inhibit the catalytic mechanism of an enzyme used, for instance, by a virus to attack a host cell, hijack its cellular machinery, or replicate itself. We will refer to the candidate drug compound as "I" (inhibitor). The efficacy of the compound as an inhibitor of the enzyme is measured by an equilibrium constant known as the *inhibition constant* $K_i = [E][I]/[EI]$, where [E], [I], and [EI] refer to the concentrations in aqueous solution of the uncomplexed enzyme, uncomplexed inhibitor, and enzyme-inhibitor complex EI, respectively. Since K_i is an equilibrium constant, it is also related to the Gibbs free energy of binding ΔG_b, which is the free energy of the reaction $E(aq) + I(aq) \rightleftharpoons EI(aq)$. That is, $K_i = \exp(\Delta G_b / kT)$. For the purposes of this discussion, we will assume that the binding Helmholtz free energy ΔA_b is approximately equal to the Gibbs free energy, so that the former can be reasonably used to estimate the inhibition constant. If we wish to determine K_i for a given drug candidate by calculating the binding free energy, a technical complication immediately arises. In principle, we can let the potential $U_\mathcal{A}$ contain all interactions except that between the enzyme and the inhibitor and then let this excluded interaction be included in $U_\mathcal{B}$. Now consider placing an enzyme and inhibitor in a bath of water molecules in order to perform the calculation. First, in order to sample the unbound state, the enzyme and inhibitor need to be separated by a distance large enough that both are fully solvated. If we then attempt to let them bind by turning on the enzyme-inhibitor interaction in stages, the probability that they will "find" each other and bind properly under this interaction is small, and the calculation will be inefficient. For this reason, a more efficient thermodynamic path for this problem is a three-stage one in which the enzyme and the inhibitor are first desolvated by transferring them from solution to vacuum. Following this, the enzyme and inhibitor are allowed to bind in vacuum, and finally, the complex EI is solvated by transferring it back to solution. Figure 8.1 illustrates the direct and indirect paths. Since free energy is a state function, the final result is independent of the path taken.

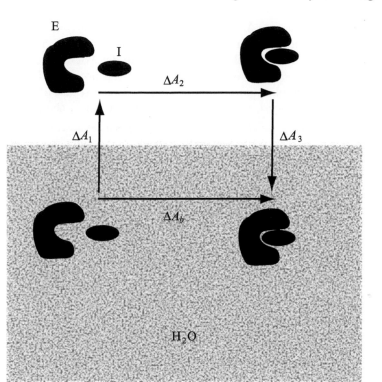

Fig. 8.1 Two thermodynamic pathways for the calculation of the binding free energy of an enzyme E and inhibitor I. According the figure, $\Delta A_b = \Delta A_1 + \Delta A_2 + \Delta A_3$.

The advantage of the indirect path, however, is that once desolvated, the enzyme and inhibitor no longer need to be at such a large separation. Hence, we can start them much closer to each other in order to obtain the vacuum binding free energy. Moreover, this part of the calculation will have a low computational overhead because the expensive water–water interactions have been removed. Although the desolvation and solvation parts of the cycle are still expensive to carry out, they are considerably more straightforward than direct binding in solution. The method shown in Fig. 8.1 is known as the "double decoupling method" (Gilson *et al.*, 1997; Deng and Roux, 2009).

As with free-energy perturbation theory, the thermodynamic integration approach can be implemented easily. An immediate disadvantage of the method, however, is the same as applies to eqn. (8.1.7): in order to perform the numerical integration, it is necessary to perform many simulations of a system at physically uninteresting intermediate values of λ where the potential $U(\mathbf{r}_1, ..., \mathbf{r}_N, \lambda)$ is, itself, unphysical. Only $\lambda = 0, 1$ correspond to actual physical states, and ultimately we can only attach physical meaning to the free-energy difference $\Delta A_{AB} = A(N, V, T, 1) - A(N, V, T, 0)$. Nevertheless, the intermediate averages must be accurately calculated in order for the integration to yield a correct result. The approach in the next section attempts to reduce the time spent in such unphysical intermediate states, thereby focusing the sampling in the important regions $\lambda = 0, 1$.

In carrying out calculations that involve coupling and decoupling from a solvent, it is often useful to couple/decouple electrostatic and van der Waals interactions separately, which requires two λ pathways. For example, if the potential $U_{\mathcal{B}}(\mathbf{r}_1, ..., \mathbf{r}_N)$ contains a van der Waals coupling potential $u_{\text{vdw}}(r_{ij})$ between solute atom i and solvent atom j, where $r_{ij} = |\mathbf{r}_i - \mathbf{r}_j|$, then the corresponding term in the metapotential would be $u_{\text{vdw}}^{(\text{meta})}(r_{ij}, \lambda) = g(\lambda)u_{\text{vdw}}(r_{ij})$. When $u_{\text{vdw}}(r_{ij})$ is a Lennard-Jones potential $u_{\text{vdw}}(r_{ij}) = 4\epsilon[(\sigma/r_{ij})^{12} - (\sigma/r_{ij})^6]$, as $\lambda \to 0$, particles can easily penetrate the repulsive core, which can become problematic when λ increases, and the Lennard-Jones interaction once again becomes active. Greater numerical stability can be achieved by modifying the form of the potential in such a way as to "soften" the hard core of the potential with the full Lennard-Jones potential being recovered when $\lambda = 1$. One such form (Beutler *et al.*, 1994) is

$$u_{\text{vdw}}^{(\text{meta,soft})}(r_{ij}, \lambda) = 4\epsilon g(\lambda) \left[\left(\frac{\sigma^6}{r_{ij}^6 + \frac{(1-\lambda)\sigma^6}{2}} \right)^2 - \left(\frac{\sigma^6}{r_{ij}^6 + \frac{(1-\lambda)\sigma^6}{2}} \right) \right]. \qquad (8.2.10)$$

Clearly, when $\lambda = 1$, $u_{\text{vdw}}^{(\text{meta,soft})}(r_{ij}, 1) = u_{\text{vdw}}(r_{ij})$. Once the van der Waals potential is switched on ($\lambda = 1$), a second pathway for the Coulomb potential $u_{\text{Coul}}(r_{ij})$ can be defined in terms of a second switching parameter λ' with a contribution to the metapotential $u_{\text{Coul}}^{(\text{meta})}(r_{ij}, \lambda') = g(\lambda')q_i q_j / r_{ij}$, where q_i and q_j are the charges on atoms i and j, respectively. Once λ' reaches 1, full coupling is achieved. It is worth noting that solvation free-energy calculations can be carried out with very large time steps using the isokinetic methods introduced in Section 4.14 and will be discussed further in Chapter 15 (see, also, Fig. 15.4) (Abreu and Tuckerman, 2020).

8.3 Adiabatic free-energy dynamics

Although we cannot entirely eliminate the need to visit the unphysical states between $\lambda = 0$ and $\lambda = 1$ in adiabatic switching methods, we can substantially alleviate it. If, instead of preselecting a set of λ values, we are willing to allow λ to vary continuously in a molecular dynamics calculation as an additional dynamical degree of freedom, albeit a fictitious one, then we can exploit the flexibility in our choice of the switching functions $f(\lambda)$ and $g(\lambda)$ to make the region between $\lambda = 0$ and $\lambda = 1$ energetically unfavorable. When such a choice is used within a molecular dynamics calculation, λ will spend most of its time in the physically relevant regions close to $\lambda = 0$ and $\lambda = 1$.

To understand how to devise such a scheme, let us consider a Hamiltonian that includes a kinetic energy term $p_\lambda^2/2m_\lambda$, where p_λ is a "momentum" conjugate to λ and m_λ is a mass-like parameter needed to define the kinetic energy. The parameter m_λ also determines the time scale on which λ evolves dynamically. The total Hamiltonian is then

$$\mathcal{H}_\lambda(\mathbf{r}, \lambda, \mathbf{p}, p_\lambda) = \frac{p_\lambda^2}{2m_\lambda} + \sum_{i=1}^{N} \frac{\mathbf{p}_i^2}{2m_i} + U(\mathbf{r}_1, ..., \mathbf{r}_N, \lambda). \qquad (8.3.1)$$

In eqn. (8.3.1), λ and its conjugate momentum p_λ are now part of an extended phase space (see, for example, the discussion Section 4.9). We can now define a canonical partition function for the Hamiltonian in eqn. (8.3.1),

$$Q(N, V, T) = \int dp_\lambda \int d^N \mathbf{p} \int_0^1 d\lambda \int_{D(V)} d^N \mathbf{r} \, e^{-\beta \mathcal{H}_\lambda(p_\lambda, \lambda, \mathbf{p}, \mathbf{r})}, \qquad (8.3.2)$$

and therefore compute any ensemble average with respect to the corresponding canonical distribution. In particular, the probability distribution $P(\lambda') = \langle \delta(\lambda - \lambda') \rangle_\lambda$, leads directly to a λ-dependent free-energy function $A(\lambda)$ through the relation

$$A(\lambda') = -kT \ln P(\lambda'). \qquad (8.3.3)$$

Equation (8.3.3) defines an important quantity known as a *free-energy profile*. We will have more to say about free-energy profiles starting in Section 8.6. Note that the free-energy difference

$$A(1) - A(0) = -kT \ln \left[\frac{P(1)}{P(0)} \right] = -kT \ln \left[\frac{Q_\mathcal{B}}{Q_\mathcal{A}} \right] = \Delta A_{\mathcal{A}\mathcal{B}}, \qquad (8.3.4)$$

since $P(0)$ and $P(1)$ are the partition functions $Q_\mathcal{A}$ and $Q_\mathcal{B}$, respectively. The distribution function $\langle \delta(\lambda - \lambda') \rangle_\lambda$ can be generated straightforwardly in a molecular dynamics calculation by accumulating a histogram of λ values visited over the course of the trajectory.

We still need to answer the question of how to maximize the time λ spends near the endpoints $\lambda = 0$ and $\lambda = 1$. Equation (8.3.3) tells us that free energy is a direct measure of probability. Thus, consider choosing the functions $f(\lambda)$ and $g(\lambda)$ such that $A(\lambda)$ has a significant barrier separating the regions near $\lambda = 0$ and $\lambda = 1$. According to eqn. (8.3.3), where the free energy is high, the associated phase-space probability is low. Thus, if the region between $\lambda = 0$ and $\lambda = 1$ is transformed into a low-probability region, λ will spend a very small fraction of its time there in a molecular dynamics run, and sampling near the endpoints $\lambda = 0$ and $\lambda = 1$ will be enhanced.

To illustrate how we can achieve a barrier in the free-energy profile $A(\lambda)$, let us examine a simple example. Consider taking two uncoupled harmonic oscillators and using λ-switching to "grow in" a bilinear coupling between them. If x and y represent the coordinates of the two oscillators with masses m_x and m_y and frequencies ω_x and ω_y, respectively, then the two potential energy functions U_A and U_B take the form

$$U_A(x, y) = \frac{1}{2} m_x \omega_x^2 x^2 + \frac{1}{2} m_y \omega_y^2 y^2$$

$$U_B(x, y) = \frac{1}{2} m_x \omega_x^2 x^2 + \frac{1}{2} m_y \omega_y^2 y^2 + \kappa xy, \qquad (8.3.5)$$

where κ determines the strength of the coupling between the oscillators. For this problem, the integration over x and y can be performed analytically (see Problem 8.1), leading to the exact probability distribution function in λ

$$P(\lambda) = \frac{C}{\sqrt{m_x\omega_x^2 m_y\omega_y^2 \left(f(\lambda) + g(\lambda)\right)^2 - \kappa^2 g^2(\lambda)}}, \tag{8.3.6}$$

where C is a constant, from which the free-energy profile $A(\lambda) = -kT \ln P(\lambda)$ becomes

$$A(\lambda) = \frac{kT}{2} \ln \left[m_x\omega_x^2 m_y\omega_y^2 \left(f(\lambda) + g(\lambda)\right)^2 - \kappa^2 g^2(\lambda)\right]. \tag{8.3.7}$$

The reader can easily verify that the free-energy difference $\Delta A = A(1) - A(0)$ does not depend on the choice of $f(\lambda)$ and $g(\lambda)$. For concreteness, let us set the parameters $m_x = m_y = 1$, $\omega_x = 1$, $\omega_y = 2$, $kT = 1$, and $\kappa = 1$. First, consider switches of the form $f(\lambda) = (\lambda^2 - 1)^2$ and $g(\lambda) = ((\lambda - 1)^2 - 1)^2$. The solid line in Fig. 8.2(a) shows the free-energy profile obtained from eqn. (8.3.7). The free-energy profile clearly contains a barrier between $\lambda = 0$ and $\lambda = 1$. If, on the other hand, we choose $f(\lambda) = (\lambda^2 - 1)^4$ and $g(\lambda) = ((\lambda - 1)^2 - 1)^4$, the free-energy profile appears as the dashed line in Fig. 8.2(a), and we see that the profile exhibits a deep well. A well indicates a region

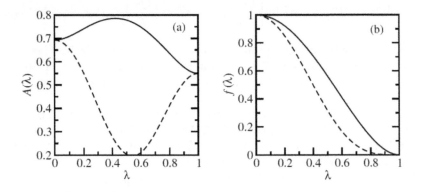

Fig. 8.2 (a) Free-energy profiles from eqn. (8.3.7). The solid line indicates switches $f(\lambda) = (\lambda^2 - 1)^2$ and $g(\lambda) = ((\lambda - 1)^2 - 1)^2$, and the dashed line indicates $f(\lambda) = (\lambda^2 - 1)^4$ and $g(\lambda) = ((\lambda - 1)^2 - 1)^4$. (b) Corresponding switch $f(\lambda)$.

of high probability and suggests that in a molecular dynamics calculation, λ will spend considerably more time in this irrelevant region than it will near the endpoints. In this case, therefore, the quartic switches are preferable. For comparison, these two choices for $f(\lambda)$ ($g(\lambda)$ is just the mirror image of $f(\lambda)$) are shown in Fig. 8.2(b). It can be seen that small differences in the shape of the switches lead to considerable differences in the free-energy profiles.

Suppose we now try to use a molecular dynamics calculation based on the Hamiltonian in eqn. (8.3.1) to generate a free-energy profile $A(\lambda)$ with a substantial barrier between $\lambda = 0$ and $\lambda = 1$. We immediately encounter a problem: The probability for λ to cross this barrier becomes exponentially small! So have we actually accomplished anything by introducing the barrier? After all, what good is enhancing the sampling in the endpoint regions if the barrier between them cannot be easily crossed? It seems that we have simply traded the problem of inefficient sampling at the endpoints of the λ-path for the problem of crossing a high barrier. That is, we have created what is commonly referred to as a *rare-event* problem (see also Section 8.5).

In order to overcome the rare-event problem, we introduce an approach in this section known as *adiabatic free-energy dynamics* (AFED) (Rosso *et al.*, 2002; Rosso *et al.*, 2005; Abrams *et al.*, 2006). A related method, called canonical adiabatic free-energy sampling, was proposed by VandeVondele and Röthlisberger (2002). Let $U^{\ddagger}$ be the value of the potential energy at the top of the barrier. In the canonical ensemble, the probability that the system will visit a configuration whose potential energy is $U^{\ddagger}$ at temperature T is proportional to $\exp[-U^{\ddagger}/kT]$, which is exceedingly small when $kT \ll U^{\ddagger}$. The exponential form of the probability suggests that we could promote barrier crossing by simply raising the temperature. If we do this naïvely, however, we broaden the ensemble distribution and change the thermodynamics. On the other hand, suppose we raise the "temperature" of just the λ degree of freedom. We can achieve this by coupling λ to a thermostat designed to keep the average $\langle p_\lambda^2/2m_\lambda \rangle = kT_\lambda$, where $T_\lambda > T$. In general, the thermodynamics would still be affected. However, under certain conditions, we can still recover the correct free energy. In particular, we must also increase the "mass" parameter m_λ to a value high enough that λ is *adiabatically* decoupled from all other degrees of freedom. When this is done, it can be shown that the adiabatically decoupled dynamics generates the correct free-energy profile even though the phase-space distribution is not the true canonical one. Here, we will give a heuristic argument showing how the modified ensemble distribution and free energy can be predicted and corrected. (Later, in Section 8.10, we will analyze the adiabatic dynamics more thoroughly and derive the phase-space distribution rigorously.)

Under the assumption of adiabatic decoupling between λ and the physical degrees of freedom, λ evolves very slowly, thereby allowing the physical degrees of freedom to sample large portions of their available phase space while λ samples only a very localized region of its part of phase space. Since the physical degrees of freedom are coupled to a thermostat at the physical temperature T, we expect the adiabatic dynamics to generate a distribution $Z(\lambda, \beta)$ as a function of λ in which the physical coordinates sample essentially all of their configuration space at temperature T at each λ, leading to

$$Z(\lambda, \beta) = \int d^N \mathbf{r} \ e^{-\beta U(\mathbf{r}_1, \ldots, \mathbf{r}_N, \lambda)}. \tag{8.3.8}$$

In the extreme limit, where λ is fixed at each value, this is certainly the correct distribution function if the physical degrees of freedom are properly thermostatted. The adiabatic decoupling approximates the more extreme situation of fixed λ. Equation (8.3.8) leads to an important quantity known as the *potential of mean force* in λ, obtained from $-(1/\beta) \ln Z(\lambda, \beta)$. In the limit of adiabatic decoupling, the potential of mean force becomes an effective potential on which λ can be assumed to move quasi-independently from the physical degrees of freedom. Note that $-(1/\beta) \ln Z(\lambda, \beta)$ is also equal to the free-energy profile $A(\lambda)$ we originally sought to determine. Using the potential of mean force, we can construct an effective Hamiltonian for λ:

$$\mathcal{H}_{\text{eff}}(\lambda, p_\lambda) = \frac{p_\lambda^2}{2m_\lambda} - \frac{1}{\beta} \ln Z(\lambda, \beta). \tag{8.3.9}$$

Now, if λ is thermostatted to a temperature T_λ, then a canonical distribution in eqn. (8.3.9) at temperature T_λ will be generated. This distribution takes the form

$$P_{\mathrm{adb}}(\lambda, p_\lambda, \beta, \beta_\lambda) \propto \mathrm{e}^{-\beta_\lambda \mathcal{H}_{\mathrm{eff}}(\lambda, p_\lambda)}, \qquad (8.3.10)$$

where $\beta_\lambda = 1/kT_\lambda$, and the "adb" subscript indicates that the distribution is valid in the limit of adiabatic decoupling of λ. Integrating over p_λ yields a distribution $\tilde{P}_{\mathrm{adb}}(\lambda, \beta_\lambda, \beta) \propto [Z(\lambda, \beta)]^{\beta_\lambda/\beta}$, from which the free-energy profile can be computed as

$$A(\lambda) = -kT_\lambda \ln \tilde{P}_{\mathrm{adb}}(\lambda, \beta, \beta_\lambda) = -kT \ln Z(\lambda, \beta) + \mathrm{const}. \qquad (8.3.11)$$

Thus, apart from a trivial additive constant, the free-energy profile can be computed from the distribution $\tilde{P}_{\mathrm{adb}}(\lambda)$ generated by the adiabatic dynamics. Note that eqn. (8.3.11) closely resembles eqn. (8.3.3), the only difference being the prefactor of kT_λ rather than kT. Despite the fact that $\ln \tilde{P}_{\mathrm{adb}}(\lambda, \beta, \beta_\lambda)$ is multiplied by kT_λ, the free-energy profile is obtained at the correct ensemble temperature T. As an example, Abrams and Tuckerman (2006) employed the AFED approach to calculate the hydration free energies of alanine and serine side-chain analogs using the CHARMM22 force field (MacKerell *et al.*, 1998) in a bath of 256 water molecule. These simulations required $kT_\lambda = 40kT = 12,000$ K, m_λ one-thousand times the mass of an oxygen atom and could obtain the desired free energies, 1.98 kcal/mol (for the alanine analog) and -4.31/kcal/mol (for the serine analog), to within an error of 0.25 kcal/mol in 1-2 ns using a time step of 1.0 fs. In order to keep the λ degree of freedom in the range $[0, 1]$, reflecting boundaries were placed at $\lambda = -\epsilon$ and $1 + \epsilon$ where $\epsilon = 0.01$.

8.4 Jarzynski's equality and nonequilibrium methods

In this section, we investigate the connection between free energy and nonequilibrium work. We have already introduced the work-free-energy inequality in eqn. (8.2.9), which states that if an amount of work $W_{A\mathcal{B}}$ takes a system from state $\mathcal{A}$ to state $\mathcal{B}$, then $W_{A\mathcal{B}} \geq \Delta A_{A\mathcal{B}}$, where equality holds only if the work is performed reversibly. $W_{A\mathcal{B}}$ is a thermodynamic quantity, which means that it can be expressed as an ensemble average of a phase-space function. Specifically, $W_{A\mathcal{B}}$ must be an average of the mechanical $\mathcal{W}_{A\mathcal{B}}(\mathrm{x})$ performed on a single member of the ensemble to drive it from a microstate of $\mathcal{A}$ to a microstate of $\mathcal{B}$. However, we need to be careful about how we define this ensemble average because, as we saw in Chapter 1 (eqn. (1.2.6)), the work $\mathcal{W}_{A\mathcal{B}}(\mathrm{x})$ is defined along a particular path or trajectory, while equilibrium averages are performed over the microstates that describe a particular thermodynamic state. This distinction is emphasized by the fact that the work could be carried out irreversibly, such that the system is driven out of equilibrium.

To illustrate the use of the microscopic function $\mathcal{W}_{A\mathcal{B}}(\mathrm{x})$, suppose we prepare an initial distribution of microstates x_0 belonging to $\mathcal{A}$ and then initiate a trajectory from each of these initial states. The ensemble average that determines the thermodynamic work $W_{A\mathcal{B}}$ is an average of $\mathcal{W}_{A\mathcal{B}}(\mathrm{x}_0)$ over this initial ensemble, which we will take to be a canonical ensemble. The trajectory x_t along which the work is computed is a unique function of the initial condition x_0, *i.e.*, $\mathrm{x}_t = \mathrm{x}_t(\mathrm{x}_0)$. Thus, the work $\mathcal{W}_{A\mathcal{B}}(\mathrm{x}_0)$ is actually a functional of the path $\mathcal{W}_{\mathcal{AB}}[\mathbf{x}_t]$. However, since the trajectory $\mathbf{x}_t$ is uniquely determined by the initial condition $\mathbf{x}_0$, $\mathcal{W}_{\mathcal{AB}}$ is also determined by $\mathbf{x}_0$, and we have

$$W_{\mathcal{AB}} = \langle \mathcal{W}_{\mathcal{AB}}(\mathbf{x}_0) \rangle_{\mathcal{A}} = \frac{C_N}{Q_{\mathcal{A}}(N,V,T)} \int d\mathbf{x}_0 \; e^{-\beta \mathcal{H}_{\mathcal{A}}(\mathbf{x}_0)} \mathcal{W}_{\mathcal{AB}}(\mathbf{x}_0). \qquad (8.4.1)$$

Thus, the work-free-energy inequality can be stated as $\langle \mathcal{W}_{\mathcal{AB}}(\mathbf{x}_0) \rangle_{\mathcal{A}} \geq \Delta A_{\mathcal{AB}}$.

From this inequality, it would seem that performing work on a system as a method to compute free energy leads, at best, to an upper bound on the free energy. It turns out, however, that irreversible work can be used to calculate free-energy differences by virtue of a connection between the two quantities first established by C. Jarzynski (1997) that is now referred to as the *Jarzynski equality*. The equality states that if instead of averaging $\mathcal{W}_{\mathcal{AB}}(\mathbf{x}_0)$ over the initial canonical distribution (that of state $\mathcal{A}$), an average of $\exp[-\beta \mathcal{W}_{\mathcal{AB}}(\mathbf{x}_0)]$ is performed over the same distribution, the result is $\exp[-\beta \Delta A_{\mathcal{AB}}]$, that is,

$$e^{-\beta \Delta A_{\mathcal{AB}}} = \left\langle e^{-\beta \mathcal{W}_{\mathcal{AB}}(\mathbf{x}_0)} \right\rangle_{\mathcal{A}} = \frac{C_N}{Q_{\mathcal{A}}(N,V,T)} \int d\mathbf{x}_0 \; e^{-\beta \mathcal{H}_{\mathcal{A}}(\mathbf{x}_0)} e^{-\beta \mathcal{W}_{\mathcal{AB}}(\mathbf{x}_0)}. \qquad (8.4.2)$$

Hence, the free-energy difference $\Delta A_{\mathcal{AB}} = -kT \ln \langle \exp(-\beta \mathcal{W}_{\mathcal{AB}}(\mathbf{x}_0)) \rangle_{\mathcal{A}}$. This remarkable result not only provides a foundation for the development of nonequilibrium free-energy methods but has important implications for thermodynamics in general.[2] Since its introduction, the Jarzynski equality has been the subject of both theoretical and experimental investigation (Park *et al.*, 2003; Liphardt *et al.*, 2002).

The Jarzynski equality can be derived using different strategies, as we will now show. Consider first a time-dependent Hamiltonian of the form

$$\mathcal{H}(\mathbf{r},\mathbf{p},t) = \sum_{i=1}^{N} \frac{\mathbf{p}_i^2}{2m_i} + U(\mathbf{r}_1,...,\mathbf{r}_N,t). \qquad (8.4.3)$$

For time-dependent Hamiltonians, the usual conservation law $d\mathcal{H}/dt = 0$ does not hold, which can be seen by computing

$$\frac{d\mathcal{H}}{dt} = \nabla_{\mathbf{x}_t} \mathcal{H} \cdot \dot{\mathbf{x}}_t + \frac{\partial \mathcal{H}}{\partial t}, \qquad (8.4.4)$$

where the phase-space vector $\mathbf{x} = (\mathbf{r}_1,...,\mathbf{r}_N,\mathbf{p}_1,...,\mathbf{p}_N) \equiv (\mathbf{r},\mathbf{p})$ has been introduced. Integrating both sides over time from $t = 0$ to an endpoint $t = \tau$, we find

$$\int_0^\tau dt \, \frac{d\mathcal{H}}{dt} = \int_0^\tau dt \, \nabla_{\mathbf{x}_t} H \cdot \dot{\mathbf{x}}_t + \int_0^\tau dt \, \frac{\partial \mathcal{H}}{\partial t}. \qquad (8.4.5)$$

Equation (8.4.5) can be regarded as a microscopic version of the first law of thermodynamics, where the first and second terms represent the heat absorbed by the

---

[2] Jarzynski's equality is actually implied by a more general theorem known as the *Crooks fluctuation theorem* (Crooks, 1998,1999).

system and the work done on the system over the trajectory, respectively.[3] That the work depends on the initial phase-space vector x_0 can be seen by defining the work associated with the trajectory $x_t(x_0)$ obtained up to time $t = t'$ as

$$W_{t'}(x_0) = \int_0^{t'} dt \, \frac{\partial}{\partial t} \mathcal{H}(x_t(x_0), t). \tag{8.4.6}$$

Note that $\mathcal{W}_{AB}(x_0) = W_\tau(x_0)$.

The derivation of the Jarzynski equality requires the calculation of the ensemble average of $\exp[-\beta \mathcal{W}_{AB}(x_0)] = \exp[-\beta W_\tau(x_0)]$ over a canonical distribution in the initial conditions x_0. Before we examine how this average might be performed along a molecular dynamics trajectory for a finite system, let us consider the simpler case where each initial condition x_0 evolves in isolation according to Hamilton's equations as derived from eqn. (8.4.3). If Hamilton's equations are obeyed, then the first (heat) term on the right in eqn. (8.4.4) vanishes, $\nabla_{x_t} \mathcal{H} \cdot \dot{x}_t = 0$, and we can express the work as

$$W_{t'} = \int_0^{t'} dt \, \frac{d}{dt} \mathcal{H}(x_t(x_0), t) = \mathcal{H}(x_{t'}(x_0), t') - \mathcal{H}(x_0, 0). \tag{8.4.7}$$

Taking $t' = \tau$, and recognizing that $\mathcal{H}(x_0, 0) = \mathcal{H}_A(x_0)$, we can write the ensemble average of $\exp[-\beta \mathcal{W}_{AB}(x_0)]$ as

$$\left\langle e^{-\beta \mathcal{W}_{AB}} \right\rangle_A = \frac{C_N}{Q_A(N,V,T)} \int dx_0 \, e^{-\beta \mathcal{H}_A(x_0)} e^{-\beta[\mathcal{H}(x_\tau(x_0), \tau) - \mathcal{H}_A(x_0)]}$$

$$= \frac{C_N}{Q_A(N,V,T)} \int dx_0 \, e^{-\beta \mathcal{H}(x_\tau(x_0), \tau)}. \tag{8.4.8}$$

Now let us change variables from x_0 to $x_\tau(x_0)$ in the integral in eqn. (8.4.8). Since the trajectory $x_t(x_0)$ is generated from Hamilton's equations of motion, the mapping of x_0 to $x_\tau(x_0)$ is unique. Moreover, by Liouville's theorem, the phase-space measure satisfies $dx_\tau = dx_0$. Therefore, we find that

$$\left\langle e^{-\beta \mathcal{W}_{AB}} \right\rangle_A = \frac{C_N}{Q_A(N,V,T)} \int dx_\tau \, e^{-\beta \mathcal{H}_B(x_\tau)}$$

$$= \frac{Q_B(N,V,T)}{Q_A(N,V,T)}$$

$$= e^{-\beta \Delta A_{AB}}, \tag{8.4.9}$$

which is Jarzynski's equality. Note that by Jensen's inequality,

---

[3]To see this, consider, for example, a mechanical piston slowly compressing a gas. Such a device can be viewed as an explicitly time-dependent external agent acting on the system, which can be incorporated into the potential. For this reason, the second term on the right in eqn. (8.4.5) represents the work performed on the system. However, even in the absence of an external agent, if the system interacts with a thermal reservoir (see Section 4.3) then the Hamiltonian is not conserved, and the first term on the right in eqn. (8.4.5) will be nonzero. Hence, this term represents the heat absorbed.

$$\left\langle \mathrm{e}^{-\beta \mathcal{W}_{A\mathcal{B}}} \right\rangle_A \geq \mathrm{e}^{-\beta \langle \mathcal{W}_{A\mathcal{B}} \rangle_A}, \tag{8.4.10}$$

which implies that

$$\mathrm{e}^{-\beta \Delta A_{A\mathcal{B}}} \geq \mathrm{e}^{-\beta \langle \mathcal{W}_{A\mathcal{B}} \rangle_A}. \tag{8.4.11}$$

Taking the natural log of both sides of eqn. (8.4.11) leads to the work-free-energy inequality. In fact, from eqn. (8.4.11), it becomes clear how to prove the Gibbs-Bogoliubov inequality in eqn. (4.7.78). If we set $t' = \tau$ and $\mathrm{x}_{\mathcal{B}} = \mathrm{x}_A = \mathrm{x}$, then $\langle \mathcal{W}_{A\mathcal{B}} \rangle = \langle \mathcal{H}_{\mathcal{B}}(\mathrm{x}) - \mathcal{H}_A(\mathrm{x}) \rangle_A$, so that we obtain the expected result

$$A_{\mathcal{B}} \leq A_A + \langle \mathcal{H}_{\mathcal{B}}(\mathrm{x}) - \mathcal{H}_A(\mathrm{x}) \rangle_A. \tag{8.4.12}$$

We now present a proof of Jarzynski's equality that is relevant for finite systems coupled to thermostats typically employed in molecular dynamics. The original version of this derivation is due to Cuendet (2006) and was subsequently generalized by Schöll-Paschinger and Dellago (2006). The proof does not depend critically on the particular thermostatting mechanism as long as the scheme rigorously generates a canonical distribution. For notational simplicity, we will employ the Nosé-Hoover scheme in eqns. (4.9.14). Although we already know that the Nosé-Hoover equations have many weaknesses which can be fixed using, for example, Nosé-Hoover chains, the former is sufficient for our purpose here and keeps the notation simpler. We start, therefore, with the equations of motion

$$\dot{\mathbf{r}}_i = \frac{\mathbf{p}_i}{m_i}$$

$$\dot{\mathbf{p}}_i = -\frac{\partial}{\partial \mathbf{r}_i} U(\mathbf{r}, t) - \frac{p_\eta}{Q} \mathbf{p}_i$$

$$\dot{\eta} = \frac{p_\eta}{Q}$$

$$\dot{p}_\eta = \sum_i \frac{\mathbf{p}_i^2}{m_i} - 3NkT, \tag{8.4.13}$$

which are identical to eqns. (4.9.14) except for the time-dependent potential $U(\mathbf{r}, t)$ and the choice of $d = 3$ for the dimensionality. As was discussed in Section 4.10, these equations have an associated phase-space metric $\sqrt{g} = \mathrm{e}^{3N\eta}$. Note that the heat term in eqn. (8.4.4) does not vanish for the Nosé-Hoover equations because energy is exchanged between the physical system and the thermostat. Nevertheless, the work in eqn. (8.4.6) allows us to construct a conserved energy according to

$$\tilde{\mathcal{H}}(\mathrm{x}_t, t) = \sum_{i=1}^{N} \frac{\mathbf{p}_i^2}{2m_i} + U(\mathbf{r}, t) + \frac{p_\eta^2}{2Q} + 3NkT\eta - W_\tau(\mathrm{x}_0)$$

$$\equiv \mathcal{H}'(\mathbf{x}_t, t) - W_\tau(\mathbf{x}_0). \tag{8.4.14}$$

Here x is the extended phase-space vector $\mathbf{x} = (\mathbf{r}_1, ..., \mathbf{r}_N, \eta, \mathbf{p}_1, ..., \mathbf{p}_N, p_\eta)$. According to the procedure outlined described in Section 4.10, the average of $\exp[-\beta W_\tau(\mathbf{x}_0)]$ is computed as

$$\left\langle \mathrm{e}^{-\beta W_{\mathcal{A}\mathcal{B}}} \right\rangle_{\mathcal{A}} = \frac{1}{\mathcal{Z}_T(0)} \int \mathrm{dx}_0 \, \mathrm{e}^{3N\eta} \mathrm{e}^{-\beta W_\tau(\mathbf{x}_0)} \delta \left(\mathcal{H}'(\mathbf{x}_0, 0) - C \right), \tag{8.4.15}$$

where $\mathrm{dx}_0 = \mathrm{d}^N \mathbf{p}_0 \mathrm{d}^N \mathbf{r}_0 \mathrm{d} p_{\eta,0} \mathrm{d} \eta_0$, C is a constant and $\mathcal{Z}_T(0)$ is the "microcanonical" partition function generated by the Nosé–Hoover equations for the $t = 0$ Hamiltonian

$$\mathcal{Z}_T(0) = \int \mathrm{dx}_0 \, \mathrm{e}^{3N\eta_0} \delta \left(\mathcal{H}'(\mathbf{x}_0, 0) - C \right). \tag{8.4.16}$$

In eqns. (8.4.15) and (8.4.16), the ensemble distribution must be the distribution of the *initial* state, which is the state $\mathcal{A}$. In Section 4.10, we showed that when eqn. (8.4.16) is integrated over η, the canonical partition function is obtained in the form

$$\mathcal{Z}_T(0) = \frac{\mathrm{e}^{\beta C}}{3NkT} \int \mathrm{d} p_{\eta,0} \, \mathrm{e}^{-\beta p_{\eta,0}^2/2Q} \int \mathrm{d}^N \mathbf{p}_0 \, \mathrm{d}^N \mathbf{r}_0 \, \mathrm{e}^{-\beta \mathcal{H}(\mathbf{r}_0, \mathbf{p}_0)}$$

$$\propto Q_{\mathcal{A}}(N, V, T). \tag{8.4.17}$$

In order to complete the proof, we need to carry out the integration over η in the numerator of eqn. (8.4.15). As was done above, we change variables from $\mathbf{x}_0$ to $\mathbf{x}_\tau(\mathbf{x}_0)$. Recalling from Section 4.10 that the measure $\exp(3N\eta)\mathrm{dx}$ is conserved, it follows that $\exp(3N\eta_0)\mathrm{dx}_0 = \exp(3N\eta_\tau)\mathrm{dx}_\tau$. Therefore, the variable transformation leads to

$$\left\langle \mathrm{e}^{-\beta W_{\mathcal{A}\mathcal{B}}} \right\rangle_{\mathcal{A}} = \frac{1}{\mathcal{Z}_T(0)} \int \mathrm{dx}_\tau \, \mathrm{e}^{3N\eta_\tau} \mathrm{e}^{-\beta W_\tau(\mathbf{x}_0(\mathbf{x}_\tau))} \delta \left(\mathcal{H}'(\mathbf{x}_0(\mathbf{x}_\tau), 0) - C \right). \tag{8.4.18}$$

From eqn. (8.4.14), it follows that $\mathcal{H}'(\mathbf{x}_0(\mathbf{x}_\tau), 0) = \mathcal{H}'(\mathbf{x}_\tau, \tau) - W_\tau(\mathbf{x}_0(\mathbf{x}_\tau))$. Inserting this into eqn. (8.4.18), we obtain

$$\left\langle \mathrm{e}^{-\beta W_{\mathcal{A}\mathcal{B}}} \right\rangle_{\mathcal{A}} = \frac{1}{\mathcal{Z}_T(0)} \int \mathrm{dx}_\tau \, \mathrm{e}^{3N\eta_\tau} \mathrm{e}^{-\beta W_\tau(\mathbf{x}_0(\mathbf{x}_\tau))}$$

$$\times \delta \left(\mathcal{H}'(\mathbf{x}_\tau, \tau) - W_\tau(\mathbf{x}_0(\mathbf{x}_\tau)) - C \right). \tag{8.4.19}$$

The integration over η_τ can now be performed by noting that the δ-function requires

$$\eta_\tau = \frac{1}{3NkT} \left[C - \mathcal{H}(\mathbf{r}_\tau, \mathbf{p}_\tau) - \frac{p_{\eta,\tau}^2}{2Q} + W_\tau(\mathbf{x}_0(\mathbf{x}_\tau)) \right]. \tag{8.4.20}$$

Using eqn. (8.4.20), together with eqn. (4.9.6), to perform the integration over η_τ causes the exponential factors of $W_\tau(\mathbf{x}_0(\mathbf{x}_\tau))$ to cancel, yielding

$$\left\langle \mathrm{e}^{-\beta W_{\mathcal{A}\mathcal{B}}} \right\rangle_{\mathcal{A}} = \frac{1}{\mathcal{Z}_T(0)} \frac{\mathrm{e}^{\beta C}}{3NkT} \int \mathrm{d} p_{\eta,\tau} \, \mathrm{e}^{-\beta p_{\eta,\tau}^2/2Q} \int \mathrm{d}^N \mathbf{p}_\tau \, \mathrm{d}^N \mathbf{r}_\tau \, \mathrm{e}^{-\beta \mathcal{H}(\mathbf{r}_\tau, \mathbf{p}_\tau)}$$

$$\propto \frac{1}{\mathcal{Z}_T(0)} Q_{\mathcal{B}}(N, V, T), \tag{8.4.21}$$

where, since the integration is performed using the phase-space vector and Hamiltonian at $t = \tau$, the result is proportional to the canonical partition function in state $\mathcal{B}$. In fact, when eqn. (8.4.17) for $\mathcal{Z}_T(0)$ is substituted into eqn. (8.4.21), the prefactors cancel, yielding simply

$$\left\langle e^{-\beta W_{\mathcal{AB}}} \right\rangle_{\mathcal{A}} = \frac{Q_{\mathcal{B}}}{Q_{\mathcal{A}}} = e^{-\beta \Delta A_{\mathcal{AB}}}, \tag{8.4.22}$$

which, again, is Jarzynski's equality.

Jarzynski's equality has an intriguing connection with mechanical pulling experiments involving laser trapping (Liphardt *et al.*, 2002) as suggested by Hummer and Szabo (2001), or atomic force microscopy (Binnig *et al.*, 1986) applied, for example, to biomolecules (Fernandez and Li, 2004). Experiments such as these can be mimicked in molecular dynamics calculations (Park *et al.*, 2003; Park and Schulten, 2004). For instance, suppose we wish to unfold a protein or polypeptide by "pulling" on the ends as illustrated in Fig. 8.3. Within the Jarzynski framework, we could perform nonequilib-

Fig. 8.3 Extension of deca-alanine by pulling via eqn. (8.4.23) (reprinted with permission from Park *et al.*, *J. Chem. Phys.* **119**, 3559 (2003), copyright American Institute of Physics).

rium calculations and obtain the free-energy change ΔA associated with the unfolding process. This could be accomplished by adding a time-dependent term to the potential that "drives" the distance between the two ends of the molecule from its (small) value in the folded state to a final (large) value in the unfolded state. For concreteness, let us designate the atomic coordinates at the ends of the molecule by $\mathbf{r}_1$ and $\mathbf{r}_N$. (In practice, $\mathbf{r}_1$ and $\mathbf{r}_N$ could be nitrogen atoms at the N- and C-termini of a protein or polypeptide.) The time-dependent potential would then take the form

$$U(\mathbf{r}_1, ..., \mathbf{r}_N, t) = U_0(\mathbf{r}_1, ..., \mathbf{r}_N) + \frac{1}{2}\kappa \left(|\mathbf{r}_1 - \mathbf{r}_N| - r_{\text{eq}} - vt\right)^2, \qquad (8.4.23)$$

where U_0 is the internal potential described, for example, by a force field. The second term in eqn. (8.4.23) is a harmonic potential with force constant κ that drives the end-to-end distance $|\mathbf{r}_1 - \mathbf{r}_N|$ away from its equilibrium value at the folded state r_{eq} by means of the time-dependent term vt, where v is the pulling rate. In practice, applying Jarzynski's formula to such a problem requires generating an ensemble of initial conditions $\mathbf{x}_0$ and then performing the pulling "experiment" in order to obtain a work value $W_\tau(\mathbf{x}_0)$ for each chosen $\mathbf{x}_0$. The final average of $\exp(-\beta W_\tau(\mathbf{x}_0))$ then leads to the free-energy difference.

Several challenges arise in the use of Jarzynski's formula. First, while it is an elegant approach to free-energy calculations, a potential bottleneck needs to be considered. The work values that are generated from an ensemble of nonequilibrium processes have a distribution $P(W_\tau)$. Thus, we could imagine calculating the average $\exp(-\beta W_\tau)$ using this distribution according to

$$\langle e^{-\beta W_\tau} \rangle = \int dW_\tau \, P(W_\tau) e^{-\beta W_\tau}. \qquad (8.4.24)$$

However, as illustrated in Fig. 8.4, $P(W_\tau)$ and $P(W_\tau)\exp(-\beta W_\tau)$ can peak at very different locations, depending on the value of β. In Fig. 8.4, the average is dominated by small values of W_τ, which lie predominantly in the left tail of the work probability distribution $P(W_\tau)$ and occur only rarely. Consequently, many "fast-switching" trajectories with high pulling rates v are needed in order to sample the tail of the work distribution adequately. Alternatively, trajectories with very slow pulling rates ("slow-switching" trajectories) could be used, in which case a smaller ensemble can be employed; however, the method will then have an efficiency comparable to equilibrium methods (Oberhofer *et al.*, 2005). It is worth noting that if the work distribution $P(W_\tau)$ is Gaussian or nearly Gaussian, then the exponential average can be computed reliably using a truncated cumulant expansion (see eqns. (4.8.22) and (4.8.24)):

$$\ln\langle e^{-\beta W_\tau} \rangle \approx -\beta\langle W_\tau \rangle + \frac{\beta^2}{2}\left(\langle W_\tau^2 \rangle - \langle W_\tau \rangle^2\right), \qquad (8.4.25)$$

which eliminates the problem of poor overlap between $P(W_\tau)$ and $P(W_\tau)\exp(-\beta W_\tau)$ in the exponential average.

A second challenge raises a more fundamental question concerning the proof of the equality. In the proof we have presented, it is tacitly assumed that after the transformation $\mathbf{x}_0$ to $\mathbf{x}_\tau$ is made, the integral that results constitutes an actual equilibrium

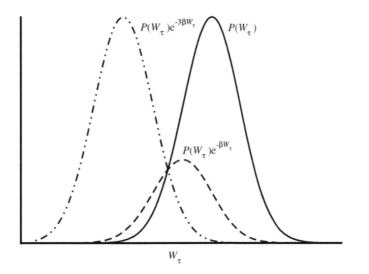

$P(W_\tau)e^{-3\beta W_\tau}$

$P(W_\tau)$

$P(W_\tau)e^{-\beta W_\tau}$

W_τ

Fig. 8.4 Shift in a Gaussian work distribution as a result of the multiplication by $\exp(-\beta W_\tau)$ for various values of β.

partition function (see eqn. (8.4.9)). That is, we assume an equilibrium distribution of phase-space points x_τ, but if the the driving force is too strong, this might not be valid in a finite system. Certainly, if the driving force is sufficiently mild to maintain the system close to equilibrium along the time-dependent path between states $\mathcal{A}$ and $\mathcal{B}$, then the assumption is valid, and indeed, in this limit, the Jarzynski equality seems to be most effective in actual applications (Oberhofer *et al.*, 2005). Away from this limit, it might be necessary to allow the final states $x_\tau(x_0)$ to relax to an equilibrium distribution by performing a short run with a thermostat coupling, even if such a relaxation process has been shown to be formally unnecessary (Jarzynski, 2004).

8.5 The problem of rare events

In Section 8.3, we alluded to the rare-event problem associated with large barriers separating important minima on a potential energy surface. Such energy surfaces are known as *rough energy landscapes* and characterize, for example, proteins, glasses, and polymers. As we noted in Section 8.3, when high barriers separate the minima, the probability that a fluctuation capable of driving the system from one minimum to another over such a barrier will occur becomes exponentially small with the ratio of the barrier height to kT. Consequently, such an event is described as a *rare event* (see Fig. 7.5). In the remainder of this chapter, we will discuss this problem at length together with methods for enhancing conformational sampling on rough potential energy surfaces.

In order to illustrate the concept of "roughness," consider the alanine dipeptide, shown in Fig. 3.8. This small dipeptide can exist in a number of stable conformations, which can be characterized by two backbone dihedral angles ϕ and ψ known as the

Ramachandran angles (see Fig. 3.8). Figure 8.5 shows the two-dimensional free-energy surface in these two angles obtained by "integrating out" the remaining degrees of freedom (we will discuss methods for generating such surfaces later in this chapter). The figure shows that there are four pronounced minima on the surface for the particular model employed, indicating four stable conformations with different relative free energies. These relative free energies can be used to rank the minima in terms of their thermodynamic importance. The full free-energy surface also contains the locations and heights of barriers, from which information about transition states and rates of conformational changes can be obtained. In a study by Chen *et al.* (2012) of the free-energy surface of the alanine tripeptide using the four backbone Ramachandran angles, 14 minima were found, which is close to $4^2 = 16$. If we now consider

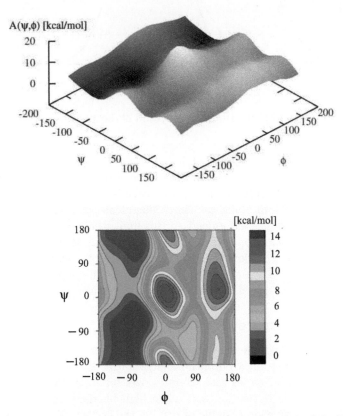

Fig. 8.5 Free-energy surface (*top*) and contours (*bottom*) as a function of the Ramachandran angles for an alanine dipeptide in solution.

that a free-energy surface of just two dihedral angles contains four minima, and one containing four dihedral angles contains approximately 16 minima, then in a protein of modest size containing 50 such pairs, the number of possible conformations would be roughly $4^{50} = 2^{100}$. From this simple exercise, we have a crude measure of the roughness of an energy landscape.

The large number of free-energy minima exhibited by our modest protein is far more than can be sampled in a typical computer simulation, and in fact, many of these minima tend to be high enough in energy as to contribute little to an ensemble average. How, then, can we limit sampling to the most important regions of an energy landscape? There is no definitive answer at present, and the question remains an active area of research. In many cases, however, it is possible to identify a subset of generalized coordinates that is particularly well suited for characterizing a given process. In the next few sections, we will discuss the selection of collective variables and describe how to make use of such variables in enhanced sampling schemes for mapping out their corresponding free-energy surfaces.

8.6 Collective variables

It is frequently the case that the progress of some chemical, mechanical, or thermo-dynamics process can be characterized using a small set of generalized coordinates in a system. When generalized coordinates are used in this manner, they are typically referred to as *collective variables*. They may also be referred to as *reaction coordinates* or *order parameters* depending on the context, type of system, and their information content. The term "reaction coordinate" or "order parameter" generally refers to a single coordinate capable of describing the full pathway of a particular process such as a chemical reaction, conformational change, change in phase, *etc.* In this case, we think of two basins separated by a saddle point or transition state, all of which is accurately described by the chosen coordinate. By contrast, a collective variable refers to one of a possibly large set of functions of the particle coordinates in a system that charac-terizes the conformational landscape, which might include many minima and saddle points separating them. Collective variables are typically chosen for their ability to distinguish the basins of attraction in the configuration space with less regard for how well they characterize transition states. These functions can certainly be considered as generalized coordinates; however, it is rarely necessary to obtain a full generalized coordinate transformation that contains the chosen collective variables as a subset of the set of generalized coordinates. Usually, in choosing a set of collective variables, one seeks to identify these minima and obtain free-energy differences between them. If, by chance, a collective variable is able to characterize a transition state between two basins accurately, it could be identified as a candidate reaction coordinate for a process that takes the system from one basin to the other through the transition state. Because of the distinction between the terms "collective variables" and "reaction coor-dinates" or "order parameters", we will mostly use the former to refer to the variables used in the enhanced sampling methods to be discussed in the next few sections and employ the latter two terms only when when referring to coordinates describing a specific transition between two basins and passing through a proper transition state (see Section 8.12).

As an example of a collective variable, consider a simple gas-phase diatomic disso-ciation process AB $\longrightarrow$ A+B. If $\mathbf{r}_A$ and $\mathbf{r}_B$ denote the Cartesian coordinates of atoms A and B, respectively, then a useful collective generalized coordinate for describing a dissociation reaction is simply the distance $r = |\mathbf{r}_B - \mathbf{r}_A|$. As we saw in Section 1.4.2, a set of generalized coordinates that contains r as one of the coordinates is the center of mass $\mathbf{R} = (m_A\mathbf{r}_A + m_B\mathbf{r}_B)/(m_A + m_B)$, the magnitude of the relative coordinate $r = |\mathbf{r}_B - \mathbf{r}_A|$, and the two angles $\phi = \tan^{-1}(y/x)$ and $\theta = \tan^{-1}(\sqrt{x^2 + y^2}/z)$, where x, y, and z are the components of $\mathbf{r} = \mathbf{r}_B - \mathbf{r}_A$. Of course, in the gas phase, r is the most relevant coordinate when the potential depends only on r, and r is likely to be a proper reaction coordinate for the gas-phase dissociation process as well. If the dissociation reaction takes place in solution, however, then some thought is needed as to whether a simple coordinate like r is sufficient to serve as a reaction coordinate. If the solvent drives or hinders the reaction by some mechanism, then a more complex coordinate that involves solvent degrees of freedom is likely needed. If the role of the solvent is a more "passive" one, then the free energy $A(r)$, obtained by "integrating out" all other degrees of freedom, will yield important information about the thermodynamics and kinetics of the entire dissociation process.

Another example is the gas-phase proton transfer reaction A–H$\cdots$B$\longrightarrow$A$\cdots$H–B. Here, although the two distances $|\mathbf{r}_H - \mathbf{r}_A|$ and $|\mathbf{r}_H - \mathbf{r}_B|$ can be used to monitor the progress of the proton away from A and toward B, respectively, neither distance alone is sufficient to follow the progress of the reaction. However, a collective variable equal to the difference $\delta = |\mathbf{r}_H - \mathbf{r}_B| - |\mathbf{r}_H - \mathbf{r}_A|$, which is positive (negative) when the proton is near A (B) and zero when the proton is equidistant between A and B, is a reasonable choice for describing the process. Therefore, δ is likely also a proper reaction coordinate in the gas phase. As it happens, a complete set of generalized coordinates involving δ can be constructed as follows. If $\mathbf{r}_A$, $\mathbf{r}_B$, and $\mathbf{r}_H$ denote the Cartesian coordinates of the three atoms, then first introduce the center of mass $\mathbf{R} = (m_A\mathbf{r}_A + m_B\mathbf{r}_B + m_H\mathbf{r}_H)/(m_A + m_B + m_H)$, the relative coordinate between A and B, $\mathbf{r} = \mathbf{r}_B - \mathbf{r}_A$, and a third relative coordinate $\mathbf{s}$ between H and the center of mass of A and B, $\mathbf{s} = \mathbf{r} - (m_A\mathbf{r}_A + m_B\mathbf{r}_B)/(m_A + m_B)$. Finally, $\mathbf{r}$ is transformed into spherical polar coordinates (r, θ, ϕ), and from $\mathbf{r}$ and $\mathbf{s}$, three more coordinates are formed as

$$\sigma = \left|\mathbf{s} + \frac{m_B}{m_A + m_B}\mathbf{r}\right| + \left|\mathbf{s} - \frac{m_A}{m_A + m_B}\mathbf{r}\right|,$$

$$\delta = \left|\mathbf{s} + \frac{m_B}{m_A + m_B}\mathbf{r}\right| - \left|\mathbf{s} - \frac{m_A}{m_A + m_B}\mathbf{r}\right|, \tag{8.6.1}$$

and the angle α, which measures the "tilt" of the plane containing the three atoms from the vertical. These coordinates could also be useful if the reaction takes place in solution, but as noted above, some careful thought about relevant solvent coordinates is needed.

As a third example, conformational changes in small peptides, such as di- and tripeptides, can be described in terms of the Ramachandran backbone dihedral angles ϕ and ψ (see Fig. 3.8, for example). For longer oligopeptides that can fold into protein secondary structure elements, such as helices and β-sheets, other coordinates could be used in addition to the Ramachandran angles such as the radius of gyration of N_b heavy backbone atoms, given by

$$R_G = \sqrt{\frac{1}{N_b}\sum_{i=1}^{N_b}\left(\mathbf{r}_i - \frac{1}{N_b}\sum_{j=1}^{N_b}\mathbf{r}_j\right)^2}, \tag{8.6.2}$$

or the number of hydrogen bonds of length approximately d_0 between n_O oxygens and n_H hydrogens, which can be expressed as

$$N_H = \sum_{i=1}^{n_O} \sum_{j=1}^{n_H} \frac{1 - [(\mathbf{r}_i - \mathbf{r}_j)/d_0]^6}{1 - [(\mathbf{r}_i - \mathbf{r}_j)/d_0]^{12}}. \tag{8.6.3}$$

These coordinates have been shown to be useful in characterizing both the folded and unfolded states of polypeptides (Bussi *et al.*, 2006). Given some knowledge of secondary structural elements, collective variables capable of distinguishing α-helix patterns from β-strands are also useful in this context (Bonomi *et al.*, 2009). In all of these examples, the collective variables are functions of the primitive Cartesian coordinates of some or all of the atoms in the system. If we are interested in the packing of domains of secondary structural elements in proteins, then Cartesian centroids of these domains can be employed (Abrams and Vanden-Eijnden, 2010).

As a final example, collective variables for exploring the polymorph landscape of a molecular crystal might include the cell matrix $\mathbf{h}$ of eqn. (5.7.2), as different crystal structures in different space groups will often be characterized by different values of the cell parameters (Yu and Tuckerman, 2011). In addition to the cell matrix, collective variables measuring the internal ordering of molecules are also important in this context. For example, the Shannon entropy can be formulated as a collective variable given the full three-dimensional spatial distribution $\tilde{g}(\mathbf{r})$ of eqn. (4.7.17) as

$$S(\mathbf{r}_1, ..., \mathbf{r}_N) = -N\rho k \int d\mathbf{r} \, [\tilde{g}(\mathbf{r}, \mathbf{r}_1, ..., \mathbf{r}_N) \ln \tilde{g}(\mathbf{r}, \mathbf{r}_1, ..., \mathbf{r}_N)$$
$$- (\tilde{g}(\mathbf{r}, \mathbf{r}_1, ..., \mathbf{r}_N) - 1)], \tag{8.6.4}$$

(Piaggi and Parrinello, 2018), where ρ is the number density. If $\tilde{g}(\mathbf{r}, \mathbf{r}_1, ..., \mathbf{r}_N)$ is modeled as a sum of Gaussians in distances r_{ij} and relative orientations $\cos\theta_{ij}$ of pairs (ij) of molecules according to

$$\tilde{g}(\mathbf{r}, \mathbf{r}_1, ..., \mathbf{r}_N) = \frac{1}{4\pi r^2 N\rho\sigma_r^2\sigma_\theta^2} \sum_{i,j} e^{-(r-r_{ij})^2/2\sigma_r^2} e^{-(\cos\theta - \cos\theta_{ij})^2/2\sigma_\theta^2}, \tag{8.6.5}$$

where σ_r and σ_θ are the Gaussian widths, then $S(\mathbf{r}_1, ..., \mathbf{r}_N)$ along with $\mathbf{h}$ can be particularly effective in searching for polymorphs of crystals of small rigid molecules such as benzene, resorcinol, urea, 5-fluorouracil, or co-crystals containing more than one of these components (Song *et al.*, 2020).

While collective variables are potentially powerful constructs and intuitively appealing, they must be used with care. Enhanced sampling approaches applied to poorly chosen collective variables can bias the system in misleading ways and generate erroneous predictions of free-energy barriers, transition states, and mechanisms. A dramatic illustration of this can be seen with the autodissociation of liquid water according to the classic reaction $2H_2O(l) \longrightarrow H_3O^+(aq) + OH^-(aq)$, as discussed in Section 7.7. The reaction ostensibly only requires transferring a proton from one water molecule to another. If we pursue this simple picture, seemingly sensible collective variables might be the distance between the oxygen and the transferring proton or the number of hydrogens covalently bonded to one of the oxygens. As it turns out, these coordinates do not capture the mechanism of the reaction and consequently fail to yield an accurate free energy or autoionization constant K_w (Trout and Parrinello, 1998). A clearer picture of this reaction was provided by Geissler *et al.* (2001) using the transition path sampling (Dellago *et al.*, 1998; Bolhuis *et al.*, 2002; Dellago *et al.*, 2002) discussed in Section 7.7. From their calculations, these authors posited that the dissociation reaction is complete only when the H_3O^+ and OH^- ions separate such that no contiguous path of hydrogen bonds exists between them that would allow them to recombine through a series of proton transfer steps. In order to describe such a process correctly, a collective variable that involves many solvent degrees of freedom would be needed and could ultimately be employed as a reaction coordinate. Later, in Section 8.12, we will describe a technique for judging the fitness of a collective variable to serve as a reaction coordinate.

The examples presented in this section suggest that there is neither an automated procedure for selecting collective variables *a priori*, nor is there a criterion for determining the number of collective variables needed to characterize a particular configurational landscape of a complex system. In Chapter 17, where machine learning techniques are introduced, we will discuss how machine learning can be leveraged to assist in the selection of collective variables and reaction coordinates.

Keeping in mind all of our caveats about choosing collective variables, we now describe a number of popular methods designed to enhance sampling along preselected collective variables. These methods are designed to generate, either directly or indirectly, the probability distribution function of a subset of n collective variables of interest in a system. When these collective variables are expressed as functions of the N primitive particle coordinates $r_1, ..., r_N$ as $q_\alpha = f_\alpha(r_1, ..., r_N)$, $\alpha = 1, ..., n$, then the probability density that these n coordinates will have values $q_\alpha = s_\alpha$ in the canonical ensemble is

$$P(s_1, ..., s_n) = \frac{C_N}{Q(N, V, T)} \int d^N\mathbf{p}\, d^N\mathbf{r}\, e^{-\beta\mathcal{H}(\mathbf{r},\mathbf{p})} \prod_{\alpha=1}^{n} \delta(f_\alpha(\mathbf{r}_1, ..., \mathbf{r}_N) - s_\alpha), \quad (8.6.6)$$

where the δ-functions are introduced to fix the collective variables $q_1, ..., q_n$ at $s_1, ..., s_n$. The function $P(s_1, ..., s_n)$ is known as the *marginal probability distribution function*. Once the marginal distribution $P(s_1, ..., s_n)$ is known, the free-energy surface in these coordinates is defined in terms of the marginal distribution as

$$A(s_1, ..., s_n) = -kT \ln P(s_1, ..., s_n). \quad (8.6.7)$$

Generating a free-energy surface is far from simple, and the remainder of this chapter will be devoted to advanced techniques for obtaining and understanding these often complicated functions. We will revisit the problem of generating and representing free-energy surfaces in Section 17.7.

8.7 The blue moon ensemble approach

The term "blue moon," as the name implies, colloquially describes a rare event.[4] The blue moon ensemble approach was introduced by Carter *et al.* (1989) and Sprik and Ciccotti (1998) for computing the free-energy profile of a collective variable or reaction coordinate when one or more high barriers along this coordinate direction lead to a rare-event problem in an ordinary thermostatted molecular dynamics calculation.

Suppose a process of interest can be monitored by a single collective coordinate $q_1 = f_1(\mathbf{r}_1, ..., \mathbf{r}_N)$. Then according to eqns. (8.6.6) and (8.6.7), the probability that $f_1(\mathbf{r}_1, ..., \mathbf{r}_N)$ has the value s and the associated free-energy profile, are given by

$$P(s) = \frac{C_N}{Q(N,V,T)} \int d^N\mathbf{p}\, d^N\mathbf{r}\, e^{-\beta \mathcal{H}(\mathbf{r},\mathbf{p})} \delta(f_1(\mathbf{r}_1, ..., \mathbf{r}_N) - s)$$

$$= \frac{1}{N!\lambda^{3N} Q(N,V,T)} \int d^N\mathbf{r}\, e^{-\beta U(\mathbf{r})} \delta(f_1(\mathbf{r}_1, ..., \mathbf{r}_N) - s)$$

$$A(s) = -kT \ln P(s). \tag{8.7.1}$$

In the second line, the integration over momenta has been performed giving the thermal prefactor factor λ^{3N}. In the blue moon ensemble approach, a holonomic constraint (see Section 1.9 and Section 3.9) $\sigma(\mathbf{r}_1, ..., \mathbf{r}_N) = f_1(\mathbf{r}_1, ..., \mathbf{r}_N) - s$ is introduced in a molecular dynamics calculation as a means of "driving" the coordinate q_1 from an initial value $s^{(i)}$ of the parameter s to a final value $s^{(f)}$ via a set of intermediate points $s^{(1)}, ..., s^{(n)}$ between $s^{(i)}$ and $s^{(f)}$. As we saw in Section 3.9, the introduction of a holonomic constraint does not yield the single δ-function condition $\delta(\sigma(\mathbf{r})) = \delta(f_1(\mathbf{r}) - s)$, where $\mathbf{r} \equiv \mathbf{r}_1, ..., \mathbf{r}_N$, as required by eqn. (8.7.1), but rather the product $\delta(\sigma(\mathbf{r}))\delta(\dot{\sigma}(\mathbf{r}, \mathbf{p}))$, since both the constraint and its first time derivative are imposed in a constrained dynamics calculation. We will return to this point shortly. In addition, the blue moon ensemble approach does not yield $A(s)$ directly but rather the derivative

$$\frac{dA}{ds} = -\frac{kT}{P(s)} \frac{dP}{ds}, \tag{8.7.2}$$

from which the free-energy profile $A(s)$ along the chosen collective variable or reaction coordinate and the free-energy difference $\Delta A = A(s_f) - A(s_i)$ are given by the integrals

$$A(s) = A(s^{(i)}) + \int_{s^{(i)}}^{s} \frac{dA}{ds'} ds', \qquad \Delta A = \int_{s^{(i)}}^{s^{(f)}} \frac{dA}{ds} ds. \tag{8.7.3}$$

In the free-energy profile expression, $A(s^{(i)})$ is just an additive constant that can be left off or adjusted so the minimum value of the profile at $s_{\min}$ corresponds to $A(s_{\min}) = 0$. In practice, these integrals are evaluated numerically using the integration points $s^{(1)}, ..., s^{(n)}$. These points can be chosen equally-spaced between $s^{(i)}$ and $s^{(f)}$, so that the integrals can be evaluated using a standard numerical quadrature, or they can

---

[4]A "blue moon" usually refers to the occurrence of a second full moon in a calendar month. The extra full moon occurs roughly every 2.7 years and is, therefore, a rare event.

be chosen according to a more sophisticated quadrature scheme. If the full profile $A(s)$ is desired, however, the number of quadrature points should be sufficient to capture the detailed shape of the profile.

We next show how to evaluate the derivative in eqn. (8.7.2). Noting that $P(s) = \langle \delta(f_1(\mathbf{r}) - s) \rangle$, the derivative can be written as

$$\frac{1}{P(s)} \frac{\mathrm{d}P}{\mathrm{d}s} = \frac{C_N}{Q(N,V,T)} \frac{\int \mathrm{d}^N \mathbf{p}\, \mathrm{d}^N \mathbf{r}\, \mathrm{e}^{-\beta \mathcal{H}(\mathbf{r},\mathbf{p})} \frac{\partial}{\partial s} \delta(f_1(\mathbf{r}) - s)}{\langle \delta(f_1(\mathbf{r}) - s) \rangle}. \tag{8.7.4}$$

In order to avoid the derivative of the δ-function, an integration by parts is performed. We first introduce a complete set of $3N$ generalized coordinates $q_\alpha = f_\alpha(\mathbf{r}_1, ..., \mathbf{r}_N)$ and their conjugate momenta p_α. This transformation, being canonical, has a unit Jacobian so that $\mathrm{d}^N \mathbf{p}\, \mathrm{d}^N \mathbf{r} = \mathrm{d}^{3N} p\, \mathrm{d}^{3N} q$. Denoting the transformed Hamiltonian as $\tilde{\mathcal{H}}(q,p)$, eqn. (8.7.4) becomes

$$\frac{1}{P(s)} \frac{\mathrm{d}P}{\mathrm{d}s} = \frac{C_N}{Q(N,V,T)} \frac{\int \mathrm{d}^{3N} p\, \mathrm{d}^{3N} q\, \mathrm{e}^{-\beta \tilde{\mathcal{H}}(q,p)} \frac{\partial}{\partial s} \delta(q_1 - s)}{\langle \delta(q_1 - s) \rangle}. \tag{8.7.5}$$

Next, we change the derivative in front of the δ-function from $\partial/\partial s$ to $\partial/\partial q_1$, using the fact that

$$\frac{\partial}{\partial s} \delta(q_1 - s) = -\frac{\partial}{\partial q_1} \delta(q_1 - s). \tag{8.7.6}$$

Finally, we integrate by parts to obtain

$$\frac{1}{P(s)} \frac{\mathrm{d}P}{\mathrm{d}s} = \frac{C_N}{Q(N,V,T)} \frac{\int \mathrm{d}^{3N} p\, \mathrm{d}^{3N} q\, \left[\frac{\partial}{\partial q_1} \mathrm{e}^{-\beta \tilde{\mathcal{H}}(q,p)} \right] \delta(q_1 - s)}{\langle \delta(q_1 - s) \rangle}$$

$$= -\frac{\beta C_N}{Q(N,V,T)} \frac{\int \mathrm{d}^{3N} p\, \mathrm{d}^{3N} q\, \frac{\partial \tilde{\mathcal{H}}}{\partial q_1} \mathrm{e}^{-\beta \tilde{\mathcal{H}}(q,p)} \delta(q_1 - s)}{\langle \delta(q_1 - s) \rangle}$$

$$= -\frac{\beta}{\langle \delta(q_1 - s) \rangle} \left\langle \left(\frac{\partial \tilde{\mathcal{H}}}{\partial q_1} \right) \delta(q_1 - s) \right\rangle. \tag{8.7.7}$$

The last line defines a new ensemble average. Specifically, the average must be performed subject to the condition $q_1 = s$. Note, however, that this is not equivalent to a mechanical constraint since the additional condition $\dot{q}_1 = 0$ is not imposed. This new ensemble average will be denoted $\langle \cdots \rangle_s^{\mathrm{cond}}$. Hence, the derivative $\mathrm{d}P/\mathrm{d}s$ can be expressed as

$$\frac{1}{P(s)} \frac{\mathrm{d}P}{\mathrm{d}s} = -\beta \left\langle \frac{\partial \tilde{\mathcal{H}}}{\partial q_1} \right\rangle_s^{\mathrm{cond}}. \tag{8.7.8}$$

Substituting eqn. (8.7.8) yields the free-energy profile

$$A(s) = A(s_i) + \int_{s^{(i)}}^{s} \mathrm{d}s' \left\langle \frac{\partial \tilde{\mathcal{H}}}{\partial q_1} \right\rangle_{s'}^{\mathrm{cond}}. \tag{8.7.9}$$

Noting that $-\langle \partial \tilde{\mathcal{H}} / \partial q_1 \rangle_{s'}^{\mathrm{cond}}$ is the expression for the average of the generalized force on q_1 when $q_1 = s'$, the integral represents the work done on the system in moving from $s^{(i)}$ to an arbitrary final point q. Since the conditional average implies a full simulation at each fixed value of q_1, the thermodynamic transformation is carried out reversibly, and eqn. (8.7.9) is consistent with the work-free-energy inequality.

Equation (8.7.9) provides insight into the underlying statistical mechanical expression for the free energy. Technically, however, the need for a full canonical transformation to generalized coordinates and conjugate momenta is inconvenient (see eqn. (1.4.16)). A more useful expression results when we perform the momentum integrations before introducing the transformation to generalized coordinates. Starting again with eqn. (8.7.4), we integrate out the momenta to yield

$$\frac{1}{P(s)}\frac{dP}{ds} = \frac{1}{N!\lambda^{3N}Q(N,V,T)}\frac{\int d^N\mathbf{r}\, e^{-\beta U(\mathbf{r})}\frac{\partial}{\partial s}\delta(f_1(\mathbf{r}) - s)}{\langle\delta(f_1(\mathbf{r}) - s)\rangle}. \tag{8.7.10}$$

Next, we transform just the coordinates to generalized coordinates $q_\alpha = f_\alpha(\mathbf{r}_1, ..., \mathbf{r}_N)$. However, because there is no corresponding momentum transformation, the Jacobian of the transformation is not unity. Let $J(q) = |\partial(\mathbf{r}_1, ..., \mathbf{r}_N)/\partial(q_1, ..., q_{3N})|$ denote this Jacobian. Equation (8.7.10) then becomes

$$\frac{1}{P(s)}\frac{dP}{ds} = \frac{1}{N!\lambda^{3N}Q(N,V,T)}\frac{\int d^{3N}q\, J(q)e^{-\beta\tilde{U}(q)}\frac{\partial}{\partial s}\delta(q_1 - s)}{\langle\delta(q_1 - s)\rangle}$$

$$= \frac{1}{N!\lambda^{3N}Q(N,V,T)}\frac{\int d^{3N}q\, e^{-\beta\left(\tilde{U}(q)-kT\ln J(q)\right)}\frac{\partial}{\partial s}\delta(q_1 - s)}{\langle\delta(q_1 - s)\rangle}, \tag{8.7.11}$$

where, in the last line, the Jacobian has been exponentiated. Changing the derivative $\partial/\partial s$ to $\partial/\partial q_1$ and performing the integration by parts as was done in eqn. (8.7.7), we obtain

$$\frac{1}{P(s)}\frac{dP}{ds} = \frac{1}{N!\lambda^{3N}Q(N,V,T)}\frac{\int d^{3N}q\, \frac{\partial}{\partial q_1}e^{-\beta\left(\tilde{U}(q)-kT\ln J(q)\right)}\delta(q_1 - s)}{\langle\delta(q_1 - s)\rangle}$$

$$= -\frac{\beta}{N!\lambda^{3N}Q(N,V,T)}$$

$$\times\frac{\int d^{3N}q\,\left[\frac{\partial\tilde{U}}{\partial q_1} - kT\frac{\partial}{\partial q_1}\ln J(q)\right]e^{-\beta\left(\tilde{U}(q)-kT\ln J(q)\right)}\delta(q_1 - s)}{\langle\delta(q_1 - s)\rangle}$$

$$= -\beta\left\langle\left[\frac{\partial\tilde{U}}{\partial q_1} - kT\frac{\partial}{\partial q_1}\ln J(q)\right]\right\rangle_s^{\text{cond}}. \tag{8.7.12}$$

Therefore, the free-energy profile becomes

$$A(s) = A(s^{(i)}) + \int_{s^{(i)}}^{s} ds'\left\langle\left[\frac{\partial\tilde{U}}{\partial q_1} - kT\frac{\partial}{\partial q_1}\ln J(q)\right]\right\rangle_{s'}^{\text{cond}}. \tag{8.7.13}$$

The derivative of $\tilde{U}$, the transformed potential, can be computed form the original potential U using the chain rule

$$\frac{\partial \tilde{U}}{\partial q_1} = \sum_{i=1}^{N} \frac{\partial U}{\partial \mathbf{r}_i} \cdot \frac{\partial \mathbf{r}_i}{\partial q_1}. \tag{8.7.14}$$

Equation (8.7.13) can be applied straightforwardly to simple collective variables or reaction coordinates for which the full transformation to generalized coordinates is known. Let us return to the problem of computing conditional ensemble averages from constrained molecular dynamics. We will use this discussion as a vehicle for introducing yet another expression for $A(q)$ that does not require a coordinate transformation at all.

Recall from Section 1.9 that the equations of motion for a system subject to a single holonomic constraint $\sigma(\mathbf{r}_1, ..., \mathbf{r}_N) = 0$ are

$$\dot{\mathbf{r}}_i = \frac{\mathbf{p}_i}{m_i}$$

$$\dot{\mathbf{p}}_i = \mathbf{F}_i + \lambda \frac{\partial \sigma}{\partial \mathbf{r}_i}, \tag{8.7.15}$$

where λ is the Lagrange multiplier needed to impose the constraint. In order to carry out the statistical mechanical analysis of the constrained dynamics, we shall make use of Gauss's principle of least constraint introduced in Section 1.10. There we showed that the Lagrange multiplier is given by

$$\lambda = -\frac{\sum_j \mathbf{F}_j \cdot \nabla_j \sigma / m_j + \sum_{j,k} \nabla_j \nabla_k \sigma \cdot \cdot \mathbf{p}_j \mathbf{p}_k / (m_j m_k)}{\sum_j (\nabla_j \sigma)^2 / m_j}. \tag{8.7.16}$$

When eqn. (8.7.16) is substituted into eqn. (8.7.15), we obtain a set of non-Hamiltonian equations of motion that explicitly satisfy the two conservation laws $\sigma(\mathbf{r}) = 0$ and $\dot{\sigma}(\mathbf{r}, \mathbf{p}) = 0$. In addition, the equations of motion conserve the Hamiltonian $\mathcal{H}(\mathbf{r}, \mathbf{p})$, since the forces of constraint do no work on the system. The methods of classical non-Hamiltonian statistical mechanics introduced in Section 4.10 allow us to derive the phase-space distribution sampled by eqns. (8.7.15). According to Section 4.10, we need to determine the conservation laws and the phase-space metric in order to construct the "microcanonical" partition function in eqn. (4.10.21). The partition function corresponding to eqns. (8.7.15) is given by

$$\mathcal{Z} = \int \mathrm{d}^N \mathbf{r} \, \mathrm{d}^N \mathbf{p} \, \sqrt{g(\mathbf{r}, \mathbf{p})} \delta \left(\mathcal{H}(\mathbf{r}, \mathbf{p}) - E \right) \delta \left(\sigma(\mathbf{r}) \right) \delta \left(\dot{\sigma}(\mathbf{r}, \mathbf{p}) \right). \tag{8.7.17}$$

From eqn. (4.10.11), the metric factor is $\sqrt{g} = \exp(-w)$, where $\mathrm{d}w/\mathrm{d}t = \kappa$, and κ is the compressibility of the system,

$$\kappa = \sum_{i=1}^{N} [\nabla_{\mathbf{r}_i} \cdot \dot{\mathbf{r}}_i + \nabla_{\mathbf{p}_i} \cdot \dot{\mathbf{p}}_i]. \tag{8.7.18}$$

Note that $\nabla_{\mathbf{r}_i} \cdot \dot{\mathbf{r}}_i = 0$ so that

$$\kappa = \sum_i \nabla_{\mathbf{p}_i} \cdot \dot{\mathbf{p}}_i = -\frac{2 \sum_i (\nabla_i \sigma / m_i) \cdot \nabla_i \sum_j \nabla_j \sigma \cdot \mathbf{p}_j / m_j}{\sum_i (\nabla_i \sigma)^2 / m_i}$$

$$= -\frac{2 \sum_i \nabla_i \sigma \cdot \nabla_i \dot{\sigma} / m_i}{\sum_i (\nabla_i \sigma)^2 / m_i}$$

$$= -\frac{\mathrm{d}}{\mathrm{d}t} \ln \left[\sum_i (\nabla_i \sigma)^2 / m_i \right]$$

$$= \frac{\mathrm{d}w}{\mathrm{d}t}. \tag{8.7.19}$$

The metric, therefore, becomes

$$\sqrt{g} = \mathrm{e}^{-w} = \sum_i \frac{1}{m_i} \left(\frac{\partial \sigma}{\partial \mathbf{r}_i} \right)^2 \equiv z(\mathbf{r}). \tag{8.7.20}$$

This metric factor arises frequently when holonomic constraints are imposed on a system and is, therefore, given the special symbol $z(\mathbf{r})$. The partition function generated by eqns. (8.7.15) now becomes

$$\mathcal{Z} = \int \mathrm{d}^N \mathbf{r} \, \mathrm{d}^N \mathbf{p} \, z(\mathbf{r}) \delta \left(\mathcal{H}(\mathbf{r}, \mathbf{p}) - E \right) \delta \left(\sigma(\mathbf{r}) \right) \delta \left(\dot{\sigma}(\mathbf{r}, \mathbf{p}) \right). \tag{8.7.21}$$

The energy-conserving δ-function, $\delta(\mathcal{H}(\mathbf{p}, \mathbf{r}) - E)$ can be replaced by a canonical distribution $\exp[-\beta \mathcal{H}(\mathbf{r}, \mathbf{p})]$ simply by coupling eqns. (8.7.15) to a thermostat such as the Nosé-Hoover chain thermostat (see Section 4.11), and eqn. (8.7.21) is replaced by

$$\mathcal{Z} = \int \mathrm{d}^N \mathbf{r} \, \mathrm{d}^N \mathbf{p} \, z(\mathbf{r}) \mathrm{e}^{-\beta \mathcal{H}(\mathbf{r}, \mathbf{p})} \delta \left(\sigma(\mathbf{r}) \right) \delta \left(\dot{\sigma}(\mathbf{r}, \mathbf{p}) \right). \tag{8.7.22}$$

In order to compare eqn. (8.7.21) to eqn. (8.7.1), we need to perform the integration over the momenta in order to clear the second δ-function. Noting that

$$\dot{\sigma}(\mathbf{r}, \mathbf{p}) = \sum_i \frac{\partial \sigma}{\partial \mathbf{r}_i} \cdot \dot{\mathbf{r}}_i = \sum_i \frac{\partial \sigma}{\partial \mathbf{r}_i} \cdot \frac{\mathbf{p}_i}{m_i}, \tag{8.7.23}$$

the partition function becomes

$$\mathcal{Z} = \int \mathrm{d}^N \mathbf{r} \, \mathrm{d}^N \mathbf{p} \, z(\mathbf{r}) \mathrm{e}^{-\beta \mathcal{H}(\mathbf{r}, \mathbf{p})} \delta \left(\sigma(\mathbf{r}) \right) \delta \left(\sum_i \frac{\partial \sigma}{\partial \mathbf{r}_i} \cdot \frac{\mathbf{p}_i}{m_i} \right). \tag{8.7.24}$$

Fortunately, the second δ-function is linear in the momenta and can be integrated over relatively easily (Problem 8.4), yielding

$$\mathcal{Z} \propto \int \mathrm{d}^N \mathbf{r} \, z^{1/2}(\mathbf{r}) \mathrm{e}^{-\beta U(\mathbf{r})} \delta \left(\sigma(\mathbf{r}) \right). \tag{8.7.25}$$

Apart from prefactors irrelevant to the free energy, eqns. (8.7.25) and eqn. (8.7.1), differ only by the factor of $z^{1/2}(\mathbf{r})$. We have already seen that the conditional average of any function $\mathcal{O}(\mathbf{r})$ of the positions is

$$\langle \mathcal{O}(\mathbf{r}) \rangle_s^{\text{cond}} = \frac{\int d\mathbf{r} \, e^{-\beta U(\mathbf{r})} \mathcal{O}(\mathbf{r}) \delta(f_1(\mathbf{r}) - s)}{\langle \delta(f_1(\mathbf{r}) - s) \rangle}. \tag{8.7.26}$$

The above analysis suggests that the average of $\mathcal{O}(\mathbf{r})$ in the ensemble generated by the constrained dynamics is

$$\langle \mathcal{O}(\mathbf{r}) \rangle_s^{\text{constr}} = \frac{\int d\mathbf{r} \, e^{-\beta U(\mathbf{r})} z^{1/2}(\mathbf{r}) \mathcal{O}(\mathbf{r}) \delta(f_1(\mathbf{r}) - s)}{\langle z^{1/2}(\mathbf{r}) \delta(f_1(\mathbf{r}) - s) \rangle}, \tag{8.7.27}$$

since $\sigma(\mathbf{r}) = f_1(\mathbf{r}) - s$. Thus, the conditional average of $\mathcal{O}(\mathbf{r})$ can be generated using the constrained ensemble if, instead of computing the average of $\mathcal{O}(\mathbf{r})$ in this ensemble, we compute the average of $z^{-1/2}(\mathbf{r})\mathcal{O}(\mathbf{r})$ and normalize by the average of $z^{-1/2}(\mathbf{r})$. That is,

$$\langle \mathcal{O}(\mathbf{r}) \rangle_s^{\text{cond}} = \frac{\langle z^{-1/2}(\mathbf{r}) \mathcal{O}(\mathbf{r}) \rangle_s^{\text{constr}}}{\langle z^{-1/2}(\mathbf{r}) \rangle_s^{\text{constr}}}. \tag{8.7.28}$$

Given the connection between the conditional and constrained averages, eqn. (8.7.13) for the free-energy profile can be written as

$$A(s) = A(s^{(i)}) + \int_{s^{(i)}}^s ds' \, \frac{\left\langle z^{-1/2}(\mathbf{r}) \left[\frac{\partial \tilde{U}}{\partial q_1} - kT \frac{\partial}{\partial q_1} \ln J(q) \right] \right\rangle_{s'}^{\text{constr}}}{\langle z^{-1/2}(\mathbf{r}) \rangle_{s'}^{\text{constr}}}. \tag{8.7.29}$$

Having now demonstrated how to compute a conditional average from constrained dynamics, we quote an important result of Sprik and Ciccotti (1998) who showed that a transformation to a complete set of generalized coordinates is not required. Rather, all we need is the form of the function $f_1(\mathbf{r}_1, ..., \mathbf{r}_N)$ associated with the collective variable or reaction coordinate q_1. Then, when eqns. (8.7.15) are used to constrain $f_1(\mathbf{r}_1, ..., \mathbf{r}_N)$, the free-energy profile can be expressed as

$$A(s) = A(s^{(i)}) + \int_{s^{(i)}}^s ds' \, \frac{\langle z^{-1/2}(\mathbf{r}) [\lambda + kTG] \rangle_{s'}^{\text{constr}}}{\langle z^{-1/2}(\mathbf{r}) \rangle_{s'}^{\text{constr}}}, \tag{8.7.30}$$

where λ is the Lagrange multiplier for the constraint and

$$G = \frac{1}{z^2(\mathbf{r})} \sum_{i,j} \frac{1}{m_i m_j} \frac{\partial f_1}{\partial \mathbf{r}_i} \cdot \frac{\partial^2 f_1}{\partial \mathbf{r}_i \partial \mathbf{r}_j} \cdot \frac{\partial f_1}{\partial \mathbf{r}_j}. \tag{8.7.31}$$

In a constrained molecular dynamics calculation, the Lagrange multiplier λ is calculated "on the fly" at every step and can be used, together with the calculation of G from eqn. (8.7.31), to construct the average in eqn. (8.7.30) at each value of the constraint. An interesting twist on the blue moon method was introduced by Darve and Pohorille (2001, 2007, 2008) who suggested that the free-energy derivative could

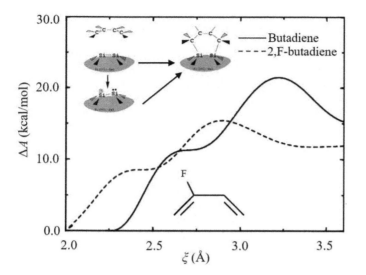

Fig. 8.6 Free-energy profiles for the addition of 1,3-butadiene and 2-F-1,3-butadiene to a Si(100)-2×1 surface. The inset at the upper left shows two possible reaction paths, concerted and non-concerted; the figure in the lower part of the plot is the molecule 2-F-butadiene.

also be computed using unconstrained dynamics by connecting the former to the instantaneous force acting on the collective variable or reaction coordinate.

As an illustration of the use of the blue moon ensemble method, we show Helmholtz free-energy profiles for the addition of two different organic molecules, 1,3-butadiene, and 2-F-1,3-butadiene to a silicon (100)-2×1 reconstructed surface (see Fig. 8.6). The surface contains rows of silicon dimers with a strong double-bond character that can form [4+2] Diels-Alder type adducts with these two molecules (see inset to Fig. 8.6). An important challenge is the design of molecules that can chemisorb to the surface but can also be selectively removed for surface patterning, thus requiring a relatively low free energy for the retro-Diels-Alder reaction. By computing the free-energy profile using a reaction coordinate $\xi = (1/2)|(\mathbf{r}_{Si_1} + \mathbf{r}_{Si_2}) - (\mathbf{r}_{C_1} + \mathbf{r}_{C_4})|$, where Si_1 and Si_2 are the two silicon atoms in a surface dimer, and C_1 and C_4 are the outer carbons in the organic molecule, it was possible to show that a simple modification of the molecule substantially lowers the free energy of the retro-Diels-Alder reaction. Specifically, if the hydrogen at the 2 position is replaced by a fluorine, the free energy is lowered by 8-10 kcal/mol (Minary and Tuckerman, 2004; Iftimie *et al.*, 2005). These calculations are performed using a molecular dynamics protocol in which forces are obtained "on the fly" from an electronic structure calculation (this is called *ab initio* molecular dynamics (Car and Parrinello, 1985; Tuckerman, 2002; Marx and Hutter, 2009)). The calculations are performed at 300 K using 13 different values for ξ separated by a distance of 0.15 Å.

8.8   Umbrella sampling and weighted histogram methods

In this section, we will discuss another free-energy technique known as *umbrella sampling* (Torrie and Valleau, 1974; Torrie and Valleau, 1977). This method bears some similarity to the blue moon ensemble approach; however, rather than constraining the collective variable or reaction coordinate, the latter is *restrained* with a biasing potential. The bias is usually taken to be a harmonic potential of the form

$$W(f_1(\mathbf{r}_1, ..., \mathbf{r}_N), s) = \frac{1}{2}\kappa\left(f_1(\mathbf{r}_1, ..., \mathbf{r}_N) - s\right)^2, \tag{8.8.1}$$

which is also known as an *umbrella potential*. Equation (8.8.1) is added to $U(\mathbf{r})$ so that the total potential is $U(\mathbf{r}) + W(f_1(\mathbf{r}), s)$. This potential is then used in a molecular dynamics or Monte Carlo calculation. As is done in the blue moon ensemble approach, the equilibrium value of the harmonic restraining potential is chosen to be a set of intermediate points $s^{(1)}, ..., s^{(n)}$ between $s^{(i)}$ and $s^{(f)}$, so that the chosen coordinate $q_1 = f_1(\mathbf{r})$ is "driven" between the two endpoint values. Each molecular dynamics or Monte Carlo simulation will then yield a biased probability distribution $\tilde{P}(s, s^{(k)})$ $k = 1, ..., n$ with $s^{(1)} = s^{(i)}$ and $s^{(n)} = s^{(f)}$ about each point $s^{(k)}$. From this set of distribution functions, the true free-energy profile across the entire range of the collective variable or reaction coordinate must be reconstructed.

When a biasing potential is used, reconstructing the full distribution requires an unbiasing procedure. This parallels the use of a constraint in the blue moon ensemble approach, which also requires an unbiasing factor in ensemble averages. The technique we will present in this section is known as the *weighted histogram analysis method*, or WHAM (Kumar *et al.*, 1992). WHAM begins by defining the biased probability distribution generated from each molecular dynamics or Monte Carlo simulation

$$\tilde{P}(s, s^{(k)}) = e^{\beta A_k} \int d^N\mathbf{r}\, e^{-\beta U(\mathbf{r})} e^{-\beta W(f_1(\mathbf{r}), s^{(k)})} \delta(f_1(\mathbf{r}) - s), \tag{8.8.2}$$

where A_k is (apart from additive constants) the free energy associated with the biased potential

$$e^{-\beta A_k} = \int d^N\mathbf{r}\, e^{-\beta U(\mathbf{r})} e^{-\beta W(f_1(\mathbf{r}), s^{(k)})} = e^{-\beta A_0} \left\langle e^{-\beta W(f_1(\mathbf{r}), s^{(k)})} \right\rangle. \tag{8.8.3}$$

In eqn. (8.8.3), the average is taken with respect to the unbiased potential $U(\mathbf{r})$, and

$$e^{-\beta A_0} = \int d^N\mathbf{r}\, e^{-\beta U(\mathbf{r})} \tag{8.8.4}$$

is just the unbiased configurational partition function. The factor $\exp(\beta A_k)$ is, therefore, the correct normalization constant for eqn. (8.8.2).

Next, we define the unbiased probability distribution function $P_k(s)$ corresponding to the distribution $\tilde{P}(s, s^{(k)})$ as

$$P_k(s) = e^{-\beta(A_k - A_0)} e^{\beta W(s, s^{(k)})} \tilde{P}(s, s^{(k)}). \tag{8.8.5}$$

We now "glue" these distributions together to give the full probability distribution function $P(s)$ by expressing $P(s)$ as a linear combination of the distributions $P_k(s)$:

$$P(s) = \sum_{k=1}^{n} C_k(s) P_k(s)$$

$$= \sum_{k=1}^{n} C_k(s) \left[e^{-\beta(A_k - A_0)} e^{\beta W(s, s^{(k)})} \tilde{P}(s, s^{(k)}) \right], \qquad (8.8.6)$$

where $\{C_k(s)\}$ is a set of coefficients that must be optimized to give the best representation of the true distribution $P(s)$. The coefficients must satisfy the constraint

$$\sum_{k=1}^{n} C_k(s) = 1. \qquad (8.8.7)$$

In order to determine the coefficients, we seek to minimize the statistical error in the distribution generated by the WHAM procedure. Let $\tilde{H}_k(s)$ be the (biased) histogram obtained from each molecular dynamics or Monte Carlo simulation. Then, the biased distribution is estimated by

$$\tilde{P}(s, s^{(k)}) \approx \frac{1}{n_k \Delta s} \tilde{H}_k(s), \qquad (8.8.8)$$

where n_k is the number of configurations sampled in the kth simulation and Δs is the bin width used to compute the histogram. The statistical error in the biased distribution for the kth umbrella window is then $\tilde{\sigma}_k^2 = \epsilon_k(s) \tilde{H}_k(s) / (n_k \Delta s)$, where $\epsilon_k(s)$ measures the deviation as a function of q between the numerically sampled distribution and the true distribution $P(s)$ in the kth umbrella window. The error in $P_k(s)$ is then given by the square of the unbiasing factor

$$\sigma_k^2 = e^{-2\beta(A_k - A_0)} e^{2\beta W(s, s^{(k)})} \tilde{\sigma}_k^2. \qquad (8.8.9)$$

We aim to minimize the total error

$$\sigma^2 = \sum_{k=1}^{n} C_k^2(s) \sigma_k^2 \qquad (8.8.10)$$

subject to eqn. (8.8.7). The constraint can be imposed by means of a Lagrange multiplier λ; the error function to be minimized is

$$\Sigma^2 = \sum_{k=1}^{n} C_k^2(s) e^{-2\beta(A_k - A_0)} e^{2\beta W(s, s^{(k)})} \frac{\epsilon_k(s) \tilde{H}_k(s)}{(n_k \Delta s)} - \lambda \left(\sum_{k=1}^{n} C_k(s) - 1 \right). \qquad (8.8.11)$$

Thus, setting the derivative $\partial \Sigma^2 / \partial C_k(s) = 0$ and solving for $C_k(s)$ in terms of the Lagrange multiplier, we find

$$C_k(s) = \frac{\lambda n_k \Delta s}{2\epsilon_k(s) \tilde{H}_k(s) e^{-2\beta(A_k - A_0)} e^{2\beta W(s, s^{(k)})}}. \qquad (8.8.12)$$

The Lagrange multiplier is now determined by substituting eqn. (8.8.12) into eqn. (8.8.7). This yields

$$\lambda \sum_{k=1}^{n} \frac{n_k \Delta s}{2\epsilon_k(s)\tilde{H}_k(s)e^{-2\beta(A_k-A_0)}e^{2\beta W(s,s^{(k)})}} = 1 \tag{8.8.13}$$

so that

$$\lambda = \frac{1}{\sum_{k=1}^{n} n_k \Delta s / [2\epsilon_k(s)\tilde{H}_k(s)e^{-2\beta(A_k-A_0)}e^{2\beta W(s,s^{(k)})}]}. \tag{8.8.14}$$

Substituting the Lagrange multiplier back into eqn. (8.8.12) gives the coefficients as

$$C_k(s) = \frac{n_k / [\epsilon_k(s)\tilde{H}_k(s)e^{-2\beta(A_k-A_0)}e^{2\beta W(s,s^{(k)})}]}{\sum_{j=1}^{n} n_j / [\epsilon_j(s)\tilde{H}_j(s)e^{-2\beta(A_j-A_0)}e^{2\beta W(s,s^{(j)})}]}. \tag{8.8.15}$$

At this point, we make two vital assumptions. First, we assume that the error function $\epsilon_k(s)$ is the same in all n umbrella windows, which is tantamount to assuming that the sampling is of equal quality for each simulation. Note that this does not necessarily mean that all simulations should be of equal length, as the relaxation of the system along directions in configuration space orthogonal to q might be longer or shorter depending on the value of q. The second assumption is that the biased histogram in each umbrella window $\tilde{H}_k(s)$ is well estimated by simply applying the biasing factor directly to the target distribution $P(s)$, i.e.,

$$\tilde{H}_k(s) \propto e^{\beta(A_k-A_0)}e^{-\beta W(s,s^{(k)})}P(s), \tag{8.8.16}$$

which, again, will be approximately true if there is adequate sampling. Once these assumptions are introduced into eqn. (8.8.15), the coefficients are finally given by

$$C_k(s) = \frac{n_k e^{\beta A_k}e^{-\beta W(s,s^{(k)})}}{\sum_{j=1}^{n} n_j e^{\beta A_j}e^{-\beta W(s,s^{(j)})}}. \tag{8.8.17}$$

Therefore, the distribution becomes

$$P(s) = \frac{\sum_{k=1}^{n} n_k P_k(s)}{\sum_{k=1}^{n} n_k e^{\beta(A_k-A_0)}e^{-\beta W(s,s^{(k)})}}. \tag{8.8.18}$$

Although the WHAM procedure might seem straightforward, eqn. (8.8.18) only defines $P(s)$ implicitly because the free-energy factors in eqn. (8.8.18) are directly related to $P(s)$ by

$$e^{-\beta(A_k-A_0)} = \int dq\, P(s)e^{-\beta W(s,s^{(k)})}. \tag{8.8.19}$$

Equations (8.8.18) and (8.8.19), therefore, constitute a set of self-consistent equations, and the solution for the coefficients and the free-energy factors must be iterated to self-consistency. The iteration is usually started with an initial guess for the free-energy factors A_k. Note that the WHAM procedure only yields A_k up to an overall additive constant A_0. When applying the WHAM procedure, care must be taken that the assumption of equal quality sampling in each umbrella window is approximately satisfied. If this is not the case, the WHAM iteration can yield unphysical results which might, for example, appear as holes in the final distribution. Once $P(s)$ is known, the free-energy profile is given by eqn. (8.6.7).

We note, finally, that Kästner and Thiel (2005) showed how to combine the umbrella sampling and thermodynamic integration techniques. Their approach makes use of the bias in eqn. (8.8.1) as a means of obtaining the free-energy derivative $\mathrm{d}A/\mathrm{d}q$ in each umbrella window. The method assumes that if the bias is sufficiently strong to keep the chosen coordinate $q_1 = f_1(\mathbf{r})$ very close to $s^{(k)}$ in the kth window, then the probability distribution $P(s)$ can be well represented by a Gaussian distribution:

$$P(s) \approx \frac{1}{\sqrt{2\pi\sigma_k^2}} e^{-(s-\bar{s}_k)^2/2\sigma_k^2}. \tag{8.8.20}$$

Here, $\bar{s}_k$ is the average value of the target coordinate in the kth window and σ_k^2 is the variance, both of which are computed via a molecular dynamics or Monte Carlo simulation in the window. Because we have assumed a specific form for the distribution, we can calculate the unbiased approximation to the derivative $\mathrm{d}A_k/\mathrm{d}q$ in the kth window

$$\frac{\mathrm{d}A_k}{\mathrm{d}q} = \frac{1}{\beta\sigma_k^2}(s - \bar{s}_k) - \kappa(s - s^{(k)}). \tag{8.8.21}$$

Let n_w and n represent the number of umbrella windows and thermodynamic integration points, respectively. We now obtain the full derivative profile $\mathrm{d}A/\mathrm{d}q$ at each integration point $q^{(i)}$ by "gluing" the windows together as in the original WHAM procedure:

$$\frac{\mathrm{d}A}{\mathrm{d}q^{(i)}} = \sum_{k=1}^{n_w} C_k(s^{(i)}) \left.\frac{\mathrm{d}A_k}{\mathrm{d}q}\right|_{q=q^{(i)}}. \tag{8.8.22}$$

The coefficients $C_k(s)$ satisfy eqn. (8.8.7). Equations (8.8.20), (8.8.21), and (8.8.22) are the starting points for the development of a weighted histogram method that is considerably simpler than the one developed previously as it eliminates the global constant A_0. Once the values of $\mathrm{d}A/\mathrm{d}q^{(i)}$ are obtained (see Problem 8.11), a numerical integration is used to obtain the full free-energy profile $A(s)$.

8.9 Wang-Landau sampling

In Section 7.6 of Chapter 7, we introduced the Wang-Landau approach for obtaining a flat density of energy states $g(E)$. There we showed that in addition to a move in configuration space from $\mathbf{r}_0$ to $\mathbf{r}$ leading to an energy change from E_0 to E with a Metropolis acceptance probability

$$\mathrm{acc}(\mathbf{r}_0 \to \mathbf{r}) = \min\left[1, \frac{g(E_0)}{g(E)}\right], \tag{8.9.1}$$

the density of states $g(E)$ is scaled at each energy E visited by a factor f: $g(E) \to fg(E)$. Here $\mathbf{r}_0$ is a complete set of initial Cartesian coordinates, and $\mathbf{r}$ is the complete set of trial Cartesian coordinates. Initially, we take $g(E)$ to be 1 for all possible energies.

The Wang-Landau sampling scheme has been extended to collective variable by Calvo (2002). The idea is to use the Wang-Landau scaling f (see Section 7.6) to generate a function that approaches the probability $P(s)$ in eqn. (8.7.1) over many Monte Carlo passes. Let $g(s)$ be a function that we initially take to be 1 over the entire range of s, *i.e.*, over the entire range of the collective variable $q_1 = f_1(\mathbf{r}_1, ..., \mathbf{r}_N) \equiv f(\mathbf{r})$. Let $h(s) = \ln g(s)$ so that $h(s)$ is initially zero everywhere. A Monte Carlo simulation is performed with the Metropolis acceptance rule:

$$\mathrm{acc}(\mathbf{r}_0 \to \mathbf{r}) = \min \left[1, \frac{\exp\left(-\beta U(\mathbf{r})\right)}{\exp\left(-\beta U(\mathbf{r}_0)\right)} \frac{g(s_0)}{g(s)} \right]$$

$$= \min \left[1, \frac{\exp\left(-\beta U(\mathbf{r})\right)}{\exp\left(-\beta U(\mathbf{r}_0)\right)} \frac{\exp\left(-h(s)\right)}{\exp\left(-h(s_0)\right)} \right]. \tag{8.9.2}$$

Here $s_0 = f_1(\mathbf{r}_0)$ and $s = q_1 = f_1(\mathbf{r})$. In addition to this acceptance rule, for each value s of the chosen coordinate $q_1 = f_1(\mathbf{r})$ visited, the function $h(s)$ is updated according to $h(s) \to h(s) + \alpha$, where $\alpha = \ln f$. This is equivalent to scaling $g(s) \to fg(s)$. As the simulation proceeds, $g(s)$ approaches the true probability $P(s)$, and the histogram $H(s)$ will become flat. Typically, f is initially chosen large, e.g. $f = \mathrm{e}^1$, and is gradually reduced to 1, which means that α is initially chosen to near 1 and is reduced to 0. Note that the Metropolis acceptance rule in eqn. (8.9.2) is equivalent to the usual acceptance rule using a modified potential $\tilde{U}(\mathbf{r}_1, ..., \mathbf{r}_N) = U(\mathbf{r}_1, ..., \mathbf{r}_N) - kTh(f_1(\mathbf{r}_1, ..., \mathbf{r}_N))$.

8.10 Driven adiabatic free-energy dynamics

In this section, we show how the adiabatic free-energy dynamics approach introduced in Section 8.3 can be extended to treat a set of collective variables. We will provide a detailed analysis of the approach, demonstrating how it leads to the free-energy profile directly from the adiabatic probability distribution function.

Let us begin by considering what type of calculation eqn. (8.6.6), as written, would require. The presence of the Dirac δ-functions in the expression dictates that we impose a restriction on any simulation we use to compute the integral such that each function $f_\alpha(\mathbf{r}_1, ..., \mathbf{r}_N)$ has a constant value s_α. This amounts to n restrictions, and a simulation that includes these n restrictions will then generate just one point of the marginal probability distribution function $P(s_1, ..., s_n)$. Now imagine trying to generate the entire marginal distribution numerically this way by laying down a grid of M points along each of the n directions in the space of s variables. In this case, the total number of points needed to characterize the marginal is M^n. When $n = 1$, this is essentially what is done in the blue moon and umbrella sampling methods, where it is computationally tractable. Recall that in the former, a constraint is used to enforce the restriction, while in the latter, a simple harmonic restraint is applied as a bias potential. Even if $n = 2$, it might still be possible to compute $P(s_1, s_2)$ on a grid of M^2 points in a *tour de force* calculation, but for $n > 2$, the number of points needed grows exponentially with n, and the problem rapidly becomes intractable. In this section, we will apply the idea of adiabatic free-energy dynamics to devise an approach to sample $P(s_1, ..., s_n)$, thus generating the statistically relevant regions of the marginal and the corresponding free-enrgy hypersurface $A(s_1, ..., s_n)$.

The notion of sampling the marginal distribution $P(s_1, ..., s_n)$ is no different in principle from sampling the high-dimensional canonical Boltzmann configuration distribution $\exp(-\beta U(\mathbf{r}_1, ..., \mathbf{r}_N))/Z(N, V, T)$. In each case, we are primarily interested in sampling the statistically important regions. Because we know its explicit functional form *a priori*, we can sample the Boltzmann distribution directly from molecular dynamics or Monte Carlo simulations. However, we do not have such an explicit functional form for the marginal distribution $P(s_1, ..., s_n)$. Rather, it emerges numerically from a simulation by applying eqn. (8.6.6). The obvious question thus arises of how we can sample a distribution function whose form we do not know *a priori*. In fact, this can be done "on the fly" in a simulation if we are able to generate sufficient information about $P(s_1, ..., s_n)$ in a *local* neighborhood of each n-dimensional point $(s_1, ..., s_n) \equiv \mathbf{s}$ as the simulation progresses. As a matter of nomenclature, we refer to the variables $s_1, ..., s_n$ as *coarse-grained variables*.

The first step in realizing this kind of "on the fly" sampling is to replace each Dirac δ-function in eqn. (8.6.6) with a Gaussian using the identity in eqn. (A.5) in Appendix A:

$$\delta(x - a) = \lim_{\sigma \to 0} \frac{1}{\sqrt{2\pi\sigma^2}} e^{-(x-a)^2/2\sigma^2}. \tag{8.10.1}$$

Applying this identity to the δ-functions in eqn. (8.6.6), we obtain

$$\prod_{\alpha=1}^{n} \delta\left(f_\alpha(\mathbf{r}) - s_\alpha\right) = \lim_{\sigma \to 0} \left(\frac{1}{2\pi\sigma^2}\right)^{n/2} \prod_{\alpha=1}^{n} e^{-(f_\alpha(\mathbf{r})-s_\alpha)^2/2\sigma^2}$$

$$= \lim_{\kappa \to \infty} \left(\frac{\beta\kappa}{2\pi}\right)^{n/2} \prod_{\alpha=1}^{n} e^{-\beta\kappa(f_\alpha(\mathbf{r})-s_\alpha)^2/2}, \tag{8.10.2}$$

where $\mathbf{r} \equiv (\mathbf{r}_1, ..., \mathbf{r}_N)$ denotes the full set of N coordinate vectors. In the last line of eqn. (8.10.2), we have replaced the width parameter σ with an inverse width measure κ with the definition $\sigma^2 = 1/(\beta\kappa)$. Here, β is an arbitrary constant; however, we will identify it with the physical temperature in the usual way as $\beta = 1/kT$ going forward. With this identity, eqn. (8.6.6) becomes

$$P(s_1, ..., s_n) = \lim_{\kappa \to \infty} \frac{C_N (\beta\kappa/2\pi)^{n/2}}{Q(N, V, T)} \int d^N\mathbf{p}\, d^N\mathbf{r}\, e^{-\beta\mathcal{H}(\mathbf{r},\mathbf{p})} \prod_{\alpha=1}^{n} e^{-\beta\kappa(f_\alpha(\mathbf{r})-s_\alpha)^2/2}. \tag{8.10.3}$$

Equation (8.10.3) is exact and, therefore, no different from eqn. (8.6.6). As written, it is not practical because of the limit sending $\kappa \to \infty$. Obviously, in a computer simulation, we must choose a finite value of κ sufficiently large that the Gaussian becomes a reasonable approximation of the δ-function. In fact, discretization of δ-functions is exactly what is done when generating a histogram, and therefore, we have a criterion for choosing κ at a given temperature, such that the width parameter $\sigma^2 = kT/\kappa$ of the Gaussian matches that of the bin width one might use to generate the histogram. Following this line of reasoning, if we allow κ to be finite, then there is no reason to assume one value of κ for all n components of **s**. Rather, for greater generality, we can choose a set of n values $\kappa_1, ..., \kappa_n \equiv \boldsymbol{\kappa}$ selected to match the histogram that would be needed to resolve each of the directions of **s**, which is helpful if each of these components is a different type of collective variable with a different associated scale. Within the approximation of finite $\boldsymbol{\kappa}$, the marginal distribution becomes

$$P_{\boldsymbol{\kappa}}(s_1, ..., s_n) = \frac{N C_N}{Q(N, V, T)} \int \mathrm{d}^N \mathbf{p} \, \mathrm{d}^N \mathbf{r} \, e^{-\beta\left(\mathcal{H}(\mathbf{r}, \mathbf{p}) + \frac{1}{2} \sum_{\alpha=1}^{n} \kappa_\alpha (f_\alpha(\mathbf{r}) - s_\alpha)^2\right)}, \quad (8.10.4)$$

where $N = \prod_{\alpha=1}^{n} (\beta \kappa_\alpha / 2\pi)^{1/2}$. Clearly, in the limit that all components of $\boldsymbol{\kappa}$ go to infinity, $P_{\boldsymbol{\kappa}}(s_1, ..., s_n) \rightarrow P(s_1, ..., s_n)$. However, within the approximation of eqn. (8.10.4), we see that when the δ-functions are converted to Gaussians of the form given in eqn. (8.10.2), the term $\sum_{\alpha=1}^{n} \kappa_\alpha (f_\alpha(\mathbf{r}) - s_\alpha)^2 / 2$ adds to the physical Hamiltonian $\mathcal{H}(\mathbf{r}, \mathbf{p})$ as a kind of harmonic coupling between the collective variables $f_\alpha(\mathbf{r})$ and the points s_α on which the marginal distribution depends. This harmonic coupling can be considered as an additional term in the potential. We now see that the unknown approximate marginal distribution $P_{\boldsymbol{\kappa}}(s_1, ..., s_n) \equiv P_{\boldsymbol{\kappa}}(\mathbf{s})$ can be sampled using molecular dynamics if we consider **s** as a set of dynamical variables. Of course, in order to do this, we must extend the phase space to include a set of momenta conjugate to **s**, which we will denote as $\boldsymbol{\pi} \equiv (\pi_1, ..., \pi_n)$. This construct leads to an extended phase-space Hamiltonian of the form

$$\mathcal{H}_{\boldsymbol{\kappa}}(\mathbf{r}, \mathbf{s}, \mathbf{p}, \boldsymbol{\pi}) = \mathcal{H}(\mathbf{r}, \mathbf{p}) + \sum_{\alpha=1}^{n} \frac{\pi_\alpha^2}{2\mu_\alpha} + \sum_{\alpha=1}^{n} \frac{1}{2}\kappa_\alpha (f_\alpha(\mathbf{r}) - s_\alpha)^2 \quad (8.10.5)$$

$$= \sum_{i=1}^{N} \frac{\mathbf{p}_i^2}{2m_i} + \sum_{\alpha=1}^{n} \frac{\pi_\alpha^2}{2\mu_\alpha} + U(\mathbf{r}) + \sum_{\alpha=1}^{n} \frac{1}{2}\kappa_\alpha (f_\alpha(\mathbf{r}) - s_\alpha)^2,$$

where $\mu_1, ..., \mu_n$ is a set of fictitious "mass-like" parameters that define the time scale on which the extended phase-space variables $(\mathbf{s}, \boldsymbol{\pi})$ evolve. The equations of motion obtained from this Hamiltonian are

$$\dot{\mathbf{r}}_i = \frac{\partial \mathcal{H}_{\boldsymbol{\kappa}}}{\partial \mathbf{p}_i} = \frac{\mathbf{p}_i}{m_i}$$

$$\dot{\mathbf{p}}_i = -\frac{\partial \mathcal{H}_{\boldsymbol{\kappa}}}{\partial \mathbf{r}_i} = \mathbf{F}_i - \sum_{\alpha=1}^{n} \kappa_\alpha (f_\alpha(\mathbf{r}) - s_\alpha) \frac{\partial f_\alpha}{\partial \mathbf{r}_i}$$

$$\dot{s}_\alpha = \frac{\partial \mathcal{H}_{\boldsymbol{\kappa}}}{\partial \pi_\alpha} = \frac{\pi_\alpha}{\mu_\alpha}$$

$$\dot{\pi}_\alpha = -\frac{\partial \mathcal{H}_{\boldsymbol{\kappa}}}{\partial s_\alpha} = \kappa_\alpha (f_\alpha(\mathbf{r}) - s_\alpha). \quad (8.10.6)$$

Unfortunately, as we saw in Section 8.3, if we simply integrate these numerically as written, we will not properly sample the marginal in eqn. (8.10.4). On-the-fly sampling of the marginal distribution requires that we effect an adiabatic separation between the physical phase space $(\mathbf{r}, \mathbf{p})$ and the extended phase space $(\mathbf{s}, \boldsymbol{\pi})$ such that the latter moves on a time scale much shorter than the former. This adiabatic separation is realized through the choice of the fictitious parameters $\mu_1, ..., \mu_n$: these must be chosen sufficiently large to ensure the time-scale separation of the motion of the two sets of variables. At the end of this section, we provide a detailed proof showing how adiabatic separation allows the marginal distribution to be correctly generated. However, as was discussed in Section 8.3, we can use a simple argument to illustrate this point. Let us first write the marginal distribution as an average of the product of Gaussians over the physical phase space $(\mathbf{r}, \mathbf{p})$:

$$P_{\boldsymbol{\kappa}}(s_1, ..., s_n) = \mathcal{N} \left\langle e^{-\frac{\beta}{2} \sum_{\alpha=1}^{n} \kappa_\alpha (f_\alpha(\mathbf{r}) - s_\alpha)^2} \right\rangle. \tag{8.10.7}$$

In the limit of perfect adiabatic separation, eqns. (8.10.6), when coupled to a canonical sampling approach such as a Nosé–Hoover chain, isokinetic Nosé–Hoover chain, or generalized Gaussian moment (see Problem 4.1) thermostat, can be seen to generate the average in eqn. (8.10.7). If we define a $\boldsymbol{\kappa}$-dependent free-energy surface

$$A_{\boldsymbol{\kappa}}(s_1, ..., s_n) = -kT \ln P_{\boldsymbol{\kappa}}(s_1, ..., s_n), \tag{8.10.8}$$

then we can compute free-energy derivatives and express them as averages according to

$$\frac{\partial A_{\boldsymbol{\kappa}}}{\partial s_\gamma} = -\frac{kT}{P(s_1, ..., s_n)} \frac{\partial P_{\boldsymbol{\kappa}}}{\partial s_\gamma}$$

$$= -\frac{\left\langle \kappa_\gamma \left(f_\gamma(\mathbf{r}) - s_\gamma\right) e^{-\frac{\beta}{2} \sum_{\alpha=1}^{n} \kappa_\alpha (f_\alpha(\mathbf{r}) - s_\alpha)^2} \right\rangle}{\left\langle e^{-\frac{\beta}{2} \sum_{\alpha=1}^{n} \kappa_\alpha (f_\alpha(\mathbf{r}) - s_\alpha)^2} \right\rangle}. \tag{8.10.9}$$

From eqn. (8.10.9), we see that under perfect adiabatic separation, the limiting form of the equations of motion for $(\mathbf{s}, \boldsymbol{\pi})$ from eqns. (8.10.6) becomes

$$\dot{s}_\alpha = \frac{\pi_\alpha}{\mu_\alpha}$$

$$\dot{\pi}_\alpha = -\frac{\partial A_{\boldsymbol{\kappa}}}{\partial s_\alpha} + \text{Therm}(T), \tag{8.10.10}$$

which means that these variables will explore the free-energy landscape via a dynamics driven by the free-energy derivative. Here, the notation $\text{Therm}(T)$ is a shorthand indicating a coupling to one of the aforementioned thermostats, conveniently allowing us to express this without having to write out explicit equations of motion for all of the thermostat variables! This result will be proved more rigorously at the end of this section.

While adiabatic separation is a necessary condition to ensure proper sampling of the marginal distribution, it will lead to inefficient sampling, particularly if the potential energy surface is a rough surface, containing numerous barriers that are high compared to kT. Therefore, in order to ensure that these barriers can be surmounted and that sampling is efficient, we can choose to sample $(\mathbf{s}, \boldsymbol{\pi})$ variables at a temperature higher than T. Since these variables constitute the extended phase space, then so long as adiabatic separation is maintained, the only effect this higher temperature will have is to change the definition of the marginal distribution in eqn. (8.10.8). If we define $T_s \gg T$ as the (higher) temperature assigned to the extended phase-space variables, then the effect of introducing this temperature boost is to change the marginal distribution expressed as a Boltzmann factor of the free-energy surface $A_\kappa(s_1, ..., s_n)$ according to

$$P_{\mathrm{adb}}(s_1, ..., s_n, T_s) \propto e^{-\beta_s A_\kappa(s_1, ..., s_n)}$$

$$\propto \left[e^{-\beta A_\kappa(s_1, ..., s_n)} \right]^{\beta_s / \beta}$$

$$\propto \left[P(s_1, ..., s_n) \right]^{T/T_s}, \tag{8.10.11}$$

where $\beta_s = 1/kT_s$ and $P_{\mathrm{adb}}(s_1, ..., s_n, T_s)$ denotes the marginal distribution generated under adiabatic separation and at temperature T_s. Equation (8.10.11) was first derived by Marx *et al.* (1999*b*) and was later adapted for direct collective-variable-based adiabatic free-energy dynamics (Rosso *et al.*, 2002). Given eqn. (8.10.11), we see that the free-energy surface at the physical temperature T can be obtained from the simple relation

$$A_\kappa(s_1, ..., s_n) = -kT_s \ln P_{\mathrm{adb}}(s_1, ..., s_n, T_s). \tag{8.10.12}$$

Importantly, as $\kappa \to \infty$, the exact free energy $A(s_1, ..., s_n)$ is recovered. The introduction of a high temperature T_s for the $(\mathbf{s}, \boldsymbol{\pi})$ phase space is tantamount to replacing eqns. (8.10.10) with

$$\dot{s}_\alpha = \frac{\pi_\alpha}{\mu_\alpha}$$

$$\dot{\pi}_\alpha = -\frac{\partial A_\kappa}{\partial s_\alpha} + \mathrm{Therm}(T_s). \tag{8.10.13}$$

This, then, brings us to the practical scheme employed to generate the adiabatic marginal distribution, which amounts to rewriting eqns. (8.10.6) as

$$\dot{\mathbf{r}}_i = \frac{\mathbf{p}_i}{m_i}$$

$$\dot{\mathbf{p}}_i = \mathbf{F}_i - \sum_{\alpha=1}^{n} \kappa_\alpha \left(f_\alpha(\mathbf{r}) - s_\alpha \right) \frac{\partial f_\alpha}{\partial \mathbf{r}_i} + \mathrm{Therm}(T)$$

$$\dot{s}_\alpha = \frac{\pi_\alpha}{\mu_\alpha}$$

$$\dot{\pi}_\alpha = \kappa_\alpha \left(f_\alpha(\mathbf{r}) - s_\alpha \right) + \mathrm{Therm}(T_s). \tag{8.10.14}$$

The procedure for employing eqns. (8.10.14) to compute an n-dimensional free-energy landscape is to accumulate an n-dimensional histogram in the variables $s_1, ..., s_n$ over the course of a simulation based on these equations of motion, and after normalization, the histogram is fed into eqn. (8.10.12) to obtain the free-energy surface. In practice, the force constants κ need to be large enough to reproduce the $\kappa_\alpha \to \infty$ limit within error bars. However, as these large force constants lead to high-frequency motion, use of multiple time-stepping methods in the integration of eqns. (8.10.14) allows larger time steps to be used than would be needed for the resulting fast forces. Resonance phenomena, discussed in Section 3.11.1, can be handled using isokinetic thermostatting methods; resonance-avoiding methods allow calculations involving eqn. (8.10.14) to be performed with very large time steps (Chen and Tuckerman, 2018).

Equations (8.10.14) comprise the method known as *driven adiabatic free-energy dynamics* (d-AFED) (Abrams and Tuckerman, 2008) or *temperature-accelerated molecular dynamics* (TAMD) (Maragliano and Vanden-Eijnden, 2006), and they are an extended phase-space version of a scheme originally proposed by Rosso *et al.* (2002). An important feature of these equations is that they can be used to explore and generate free-energy surfaces of high dimension, by which we usually mean a value of n ranging anywhere from 4 to several 10s of collective variables (Abrams and Vanden-Eijnden, 2010). Not surprisingly, as the dimension n grows, the associated free-energy surfaces become increasingly difficult to visualize, represent mathematically, and analyze for key features such as minima and saddle points, which are also called *landmark points* (Chen *et al.*, 2015). It is important to point out that d-AFED/TAMD generates full sweeps over the free-energy surface, generating a scattering of points in each sweep. Over long simulation times, the points fill in to provide a global surface. However, as will be discussed in Chapter 17, techniques of machine learning can serve as regression models to fill in details of a free-energy surface even if a d-AFED/TAMD run is short and the scattering of points is sparse (Maragliano and Vanden-Eijnden, 2008; Chen *et al.*, 2012; Schneider *et al.*, 2017).

As we have already noted, there is no general protocol for selecting optimal collective variables *a priori*, which also means that the optimal number, n, of collective variables cannot be known at the outset. Nevertheless, an important advantage of d-AFED/TAMD over other sampling methods is that it can generate a free-energy surface as a function of a large number of possibly redundant collective variables. This surface can then be analyzed *a posteriori* to determine a lower dimensional set of variables that captures slow motions of a system and, consequently, the true rare events. This lower dimensional set of collective variables should correspond to a putative set of optimal collective variables. This type of analysis will be described in Chapter 17. Additional approaches discussed in Chapter 17 will provide a route to constructing the free-energy surface using the sampled gradients in eqn. (8.10.9), an approach that often converges to a final free-energy surface more rapidly than histograms in eqn. (8.10.12) (Chen *et al.*, 2012). Finally, Chapter 17 will explore machine learning methods for constructing optimal collective variables that can be targeted in enhanced sampling methods like d-AFED/TAMD or to construct meaningful reaction coordinates for specific processes.

We conclude this section with a physically motivated proof that d-AFED/TAMD generates the correct free-energy surface under the assumption of ergodic sampling of the extended phase space. (This analysis is rather technical and not critical for implementing d-AFED/TAMD; readers may proceed to the discussion and examples at the end of this section without loss of continuity.) The analysis is based on the Liouville operator formalism we have been using to derive numerical integrators for molecular dynamics simulations. In order to simplify the notation, let $\mathbf{r} \equiv \mathbf{r}_1, ..., \mathbf{r}_N$ denote the full set of physical coordinates, $\mathbf{p} \equiv \mathbf{p}_1, ..., \mathbf{p}_N$ the full set of physical momenta, and $\mathbf{F} \equiv \mathbf{F}_1, ..., \mathbf{F}_N$ the full set of physical forces. We will continue to use $(\mathbf{s}, \boldsymbol{\pi})$ to denote the $2n$-dimensional extended phase space, and we will use $\boldsymbol{\phi} \equiv \phi_1, ..., \phi_n$ to denote the forces on $\mathbf{s}$, i.e., $\phi_\alpha = \kappa_\alpha(q_\alpha(\mathbf{r}) - s_\alpha)$. Finally, let M be a $3N \times 3N$ diagonal matrix of physical particle masses and μ be an $n \times n$ diagonal matrix of extended phase-space mass-like parameters. The full Liouville operator for d-AFED/TAMD can be written as

$$iL = \mathrm{M}^{-1}\mathbf{p} \cdot \frac{\partial}{\partial \mathbf{r}} + \mu^{-1}\boldsymbol{\pi} \cdot \frac{\partial}{\partial \mathbf{s}} + \mathbf{F} \cdot \frac{\partial}{\partial \mathbf{p}} + \boldsymbol{\phi} \cdot \frac{\partial}{\partial \boldsymbol{\pi}} + iL_{\mathrm{therm}}^{(\mathrm{r})}(T). \tag{8.10.15}$$

Here, $iL_{\mathrm{therm}}^{(\mathrm{r})}$ is a thermostat operator for the physical phase space. For ease of notation, we will not couple a thermostat to the extended phase space in the present analysis $(\mathbf{s}, \boldsymbol{\pi})$. The thermostat on $(\mathbf{r}, \mathbf{p})$ could be one of the deterministic methods we discussed in Chapter 4 or one of the stochastic methods we will introduce in Chapter 15. The particular form of this thermostat is not important so long as the corresponding propagator generates canonical sampling in the physical subspace. For this analysis, we will introduce a factorization of the classical propagator $\exp(iL\Delta t)$, which will lead to a formal numerical scheme equivalent to eqns. (8.10.10) or (8.10.13).

The factorization we seek starts with the following decomposition of the Liouville operator in eqn. (8.10.15):

$$iL = iL^{(\mathrm{r})} + iL_{\mathrm{ref}}^{(\mathrm{s})} \tag{8.10.16}$$

where

$$iL_{\mathrm{ref}}^{(\mathrm{s})} = \mu^{-1}\boldsymbol{\pi} \cdot \frac{\partial}{\partial \mathbf{s}} \tag{8.10.17}$$

and

$$iL^{(\mathrm{r})} = iL_{\mathrm{ref}}^{(\mathrm{r})} + \mathbf{F} \cdot \frac{\partial}{\partial \mathbf{p}} = \boldsymbol{\phi} \cdot \frac{\partial}{\partial \boldsymbol{\pi}} \tag{8.10.18}$$

with

$$iL_{\mathrm{ref}}^{(\mathrm{r})} = \mathrm{M}^{-1}\mathbf{p} \cdot \frac{\partial}{\partial \mathbf{r}} + iL_{\mathrm{therm}}^{(\mathrm{r})}(T). \tag{8.10.19}$$

Equation (8.10.16) is now used to generate a factorization for $\exp(iL\Delta t)$ in which the time step Δt is chosen to integrate the slow motion accurately. This factorization is

$$e^{iL\Delta t} = e^{iL^{(\mathrm{r})}\Delta t/2} e^{iL_{\mathrm{ref}}^{(\mathrm{s})}\Delta t} e^{iL^{(\mathrm{r})}\Delta t/2}. \tag{8.10.20}$$

Note that the propagator $\exp(iL^{(\mathrm{r})}\Delta t/2)$ is left intact and, therefore, generates formally exact motion for the (fast) physical variables. Without loss of generality, we can also represent the operator exactly using the Trotter theorem:

$$e^{iL^{(\mathbf{r})}\Delta t/2} = \lim_{M\to\infty} \left[\exp\left(\frac{\Delta t}{4M}\boldsymbol{\phi}\cdot\frac{\partial}{\partial\boldsymbol{\pi}}\right)\exp\left(\frac{\Delta t}{4M}\mathbf{F}\cdot\frac{\partial}{\partial\mathbf{p}}\right)\exp\left(iL_{\text{ref}}^{(\mathbf{r})}\frac{\Delta t}{2M}\right)\right.$$

$$\left. \times \exp\left(\frac{\Delta t}{4M}\mathbf{F}\cdot\frac{\partial}{\partial\mathbf{p}}\right)\exp\left(\frac{\Delta t}{4M}\boldsymbol{\phi}\cdot\frac{\partial}{\partial\boldsymbol{\pi}}\right)\right]^{M}. \tag{8.10.21}$$

When eqn. (8.10.21) is substituted into eqn. (8.10.20), the result is a formally exact analog of a multiple time-scale factorization (see Section 3.11), which, nevertheless, can be applied analytically. When the operator $\exp(iL^{(\mathbf{r})}\Delta t/2)$ is applied in the exactly factorized form given in eqn. (8.10.21), the $\mathbf{s}$ variables do not change through the first application of this operator; however, the momenta $\boldsymbol{\pi}$ do. In each of the M applications of the terms in the factorization, both $\mathbf{r}$ and $\mathbf{p}$ change, producing their exact evolution up to $\Delta t/2$, and since $\boldsymbol{\phi}$ depends on both $\mathbf{s}$ and $\mathbf{r}$, the $\boldsymbol{\phi}$ forces change as $\mathbf{r}$ changes. Let this evolution of $\mathbf{r}$ be denoted $\mathbf{r}_{\text{adb}}(t;\mathbf{r}(0),\mathbf{p}(0),\gamma(0)) \equiv \mathbf{r}_{\text{adb}}^{(1)}(t)$, where γ denotes a set of thermostat variables acting on the physical space $(\mathbf{r},\mathbf{p})$ and the "adb" subscript indicates evolution under the conditions of adiabatic separation between the physical $(\mathbf{r},\mathbf{p})$ and extended $(\mathbf{s},\boldsymbol{\pi})$ phase spaces. The evolution of $\boldsymbol{\pi}$ under $\exp(iL^{(\mathbf{r})}\Delta t/2)$ in its factorized form can be written as

$$\boldsymbol{\pi}(\Delta t/2) = \boldsymbol{\pi}(0) + \lim_{M\to\infty}\frac{\Delta t}{2M}\left[\frac{1}{2}\boldsymbol{\phi}(\mathbf{r}_{\text{adb}}^{(1)}(0),\mathbf{s}(0)) + \frac{1}{2}\boldsymbol{\phi}(\mathbf{r}_{\text{adb}}^{(1)}(\Delta t/2),\mathbf{s}(0))\right.$$

$$\left. + \sum_{k=1}^{M-1}\boldsymbol{\phi}(\mathbf{r}_{\text{adb}}^{(1)}(k\Delta t/2M),\mathbf{s}(0))\right]. \tag{8.10.22}$$

The term in brackets in the limit that $M\to\infty$ is the trapezoidal rule representation of a time integral, which means that the evolution in eqn. (8.10.22) can be written exactly as

$$\boldsymbol{\pi}(\Delta t/2) = \boldsymbol{\pi}(0) + \int_{0}^{\Delta t/2}dt\,\boldsymbol{\phi}(\mathbf{r}_{\text{adb}}^{(1)}(t),\mathbf{s}(0)). \tag{8.10.23}$$

Application of the remaining operators completes the evolution of $\boldsymbol{\pi}$ as

$$\boldsymbol{\pi}(\Delta t) = \boldsymbol{\pi}(\Delta t/2) + \int_{\Delta t/2}^{\Delta t}dt\,\boldsymbol{\phi}(\mathbf{r}_{\text{adb}}^{(2)}(t),\mathbf{s}(\Delta t)), \tag{8.10.24}$$

where $\mathbf{r}_{\text{adb}}^{(2)}(t) = \mathbf{r}_{\text{adb}}(t;\mathbf{r}(\Delta t/2),\mathbf{p}(\Delta t/2),\gamma(\Delta t/2))$, indicating the second half of the evolution of the physical coordinates. Of course, the factorized propagator prescribes the evolution of all other coordinates and momenta in the system, but we do not need to concern ourselves with these for the purposes of this analysis.

The next step of the proof is to invoke the conditions of adiabaticity and the assumption of ergodic evolution of the physical variables. Under these conditions, we assume that the physical variables $(\mathbf{r},\mathbf{p})$ sample enough of the local phase space at $\mathbf{s}(0)$

(or at $\mathbf{s}(\Delta t/2)$) that we can replace the time integrals in eqns. (8.10.23) and (8.10.24) with integrals over the physical coordinates, *i.e.*,

$$\frac{2}{\Delta t} \int_{\tau}^{\tau+\Delta t/2} dt\, \phi(\mathbf{r}_{\mathrm{adb}}(t;\mathbf{r}(\tau),\mathbf{p}(\tau),\gamma(\tau),\mathbf{s}) = \frac{\left\langle \phi(\mathbf{r},\mathbf{s})\mathrm{e}^{-\frac{\beta}{2}\sum_{\alpha=1}^{n}\kappa_{\alpha}(f_{\alpha}(\mathbf{r})-s_{\alpha})^2} \right\rangle_{\mathbf{r}}}{\left\langle \mathrm{e}^{-\frac{\beta}{2}\sum_{\alpha=1}^{n}\kappa_{\alpha}(f_{\alpha}(\mathbf{r})-s_{\alpha})^2} \right\rangle_{\mathbf{r}}}$$

$$= -\frac{\partial A_{\kappa}}{\partial \mathbf{s}}, \tag{8.10.25}$$

where $\langle\cdots\rangle_{\mathbf{r}}$ indicates that the integrals are performed over $\mathbf{r}$, keeping $\mathbf{s}$ fixed. Thus, if we denote the exact free-energy gradient as $\bar{\phi}(\mathbf{s})$, then eqn. (8.10.25) shows that the numerical evolution of $\boldsymbol{\pi}$ will be a velocity Verlet evolution step on the exact free-energy surface:

$$\boldsymbol{\pi}(\Delta t) = \boldsymbol{\pi}(0) + \frac{\Delta t}{2}\left[\bar{\phi}(\mathbf{s}(0)) + \bar{\phi}(\mathbf{s}(\Delta t))\right]. \tag{8.10.26}$$

Note that the use of a high temperature $T_s \gg T$ on the $(\mathbf{s},\boldsymbol{\pi})$ extended space is not necessary for the analysis we have performed; only the adiabatic decoupling is required to generate motion on the free-energy surface. The effect of adding a second thermostat on the extended phase space would be that the free-energy surface is sampled under canonical evolution at high temperature, but this would not change the underlying surface on which $(\mathbf{s},\boldsymbol{\pi})$ evolves. Thus, the canonical distribution that would be sampled at temperature T_s would be $\exp(-\beta_s A_{\kappa}(\mathbf{s})) = [\exp(-\beta A_{\kappa}(\mathbf{s}))]^{\beta_s/\beta} = [\exp(-\beta A_{\kappa}(\mathbf{r}))]^{T/T_s}$, from which the free energy can be easily recovered via eqn. (8.10.12). This completes our proof.

As an illustration of adiabatic dynamics, consider a simple problem with two degrees of freedom x and y. The potential is chosen to be a double well in x coupled linearly to a harmonic oscillator in y:

$$U(x,y) = D_0\left(x^2 - a^2\right)^2 + \frac{1}{2}ky^2 + \lambda xy. \tag{8.10.27}$$

The free-energy profile $A(s)$ for this system can be derived analytically with the result

$$A(s) = D_0\left(s^2 - a^2\right)^2 - \frac{\lambda^2}{2k}s^2. \tag{8.10.28}$$

The bare double well has minima at $x = \pm a$ and a barrier height of $D_0 a^4$ while the free energy in eqn. (8.10.28) has minima at $s = \pm\sqrt{a^2 + \lambda^2/(4D_0 k)}$ and a barrier height of $D_0 a^4 + \lambda^2 a^2/(2k) + \lambda^4/(16D_0 k^2)$. Thus, in order to ensure sufficient barrier-crossing in an adiabatic dynamics simulation, the temperature of the extended variable, s, should satisfy $kT_s > D_0 a^4$. In Fig. 8.7, we examine how the free-energy profile generated by eqns. (8.10.14) depends on T_s, μ, and κ for the potential in eqn. (8.10.27) and the corresponding free-energy profile in eqn. (8.10.28) for a simulation of 10^9 steps of length $\Delta t = 0.0025$ using a generalized Gaussian moment thermostat (Liu and Tuckerman, 2000). The figure shows that if T_s is too low, the barrier is inadequately

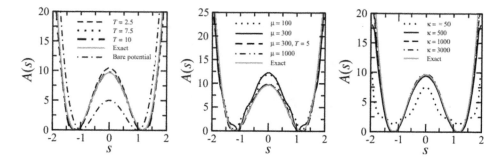

Fig. 8.7 (*Left*) Free-energy profiles for the potential in eqn. (8.10.27) generated by d-AFED/TAMD for different values of T_s with $\kappa = 3000$ and $\mu = 1000$. (*Middle*) Free-energy profiles for different values of μ with $T_s = 7.5$ (except where $T_s = 5$ as indicated) and $\kappa = 3000$. (*Right*) Free-energy profiles for different values of κ with $T_s = 7.5$ and $\mu = 1000$.

sampled due to insufficient crossing. If the mass μ is too low, the free-energy profile is distorted due to insufficient adiabatic decoupling. Finally, if the spring constant κ is too low, the profile is distorted due to poor approximation of the δ-function.

As a second example of a realistic system, d-AFED/TAMD is applied to explore the landscape of crystal structures of solid xenon at 2700 K and 25 GPa pressure. In this application, an isothermal-isobaric version of the d-AFED/TAMD algorithm is employed (Yu *et al.*, 2014) to a system of 4000 xenon atoms in an orthorhombic box. Atoms interact via a Buckingham potential from Ross and McMahan (1980). The simulation employs five collective variables, specifically, the three diagonal elements of the cell matrix $\mathbf{h}$ and two crystalline order parameters known as continuous Steinhardt parameters (Steinhardt *et al.*, 1983; Quigley and Rodger, 2008; Quigley and Rodger, 2009), defined as $Q_l(\mathbf{r}) = [(4\pi/(2l+1)) \sum_{m=-l}^{l} |Q_{lm}(\mathbf{r})|^2]^{1/2}$, where

$$Q_{lm}(\mathbf{r}) = \frac{1}{N N_{\text{coord}}} \sum_{b=1}^{N_b} f_c(r_b) Y_{lm}(\hat{\mathbf{r}}_b) \qquad (8.10.29)$$

and where N_b is the total number of atom pairs separated by a distance $r_{\max}$, N_{coord} is the first-shell coordination number of each atom, $Y_{lm}(\hat{\mathbf{r}}_b)$ is a spherical harmonic, $\hat{\mathbf{r}}_b$ is the unit vector along the direction $\mathbf{r}_b$, $r_b = |\mathbf{r}_b|$, and $f_c(r)$ is a smooth switching function that switches the order parameter off at $r = r_{\max}$ (here, set to 4.5 Å). The two parameters employed correspond to $l = 4$ and $l = 6$. In these simulations, T_s is set to 10^5 K and κ is set to 10^{10} K. A simulation of length 200 ns with a time step of 5 fs is performed from which 5×10^5 snapshots are extracted for free-energy surface reconstruction (Schneider *et al.*, 2017). A projection of the five-dimensional free-energy surface onto the Q_4-Q_6 plane is shown in Fig. 8.8. The figure clearly indicates three polymorphs corresponding to the face-centered cubic, body-centered cubic, and hexagonal close-packed (FCC, BCC, and HCP) Bravais lattices. Moreover, we see that the FCC and BCC structures are the most stable with the HCP structure appearing as a metastable state between the two stable structures, in accordance with experiments that have reported pressure-induced FCC-to-HCP transitions (Cynn

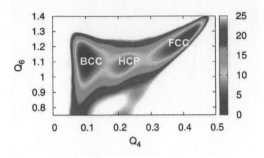

Fig. 8.8 Projection onto the $Q_4 - Q_6$ plane of the five-dimensional free-energy surface of crystalline xenon at 2700 K and 25 GPa from a d-AFED/TAMD simulation.

et al., 2001). It is worth noting that use of Q_4 and Q_6 alone as collective variables is insufficient to converge the free energy and reveal the three polymorphs (Yu *et al.*, 2014). Thus, it is often essential to work in a high-dimensional collective-variable space in order to explore fully the structural diversity of a system.

8.11 Metadynamics

In this section, we will describe a popular method for computing a free-energy hypersurface in terms of a set of collective variables that is akin to a dynamical version of the Wang-Landau approach in Section 7.6. The *metadynamics* method (Laio and Parrinello, 2002) is a dynamical scheme wherein energy basins are "filled in" using a time-dependent potential that depends on the history of the system's trajectory. Once a basin is filled in, the system is driven into the next basin, which is subsequently filled in, and so forth until the entire landscape is "flat." When this state is achieved, the accumulated time-dependent potential is used to construct the free-energy profile. Note that, in contrast to d-AFED/TAMD, in which basins are visited repeatedly as the system sweeps over the free-energy landscape, in metadynamics, once a basin has been filled in, the system is effectively repelled from visiting it again for any significant length of time.

In order to see how such a dynamics can be constructed, consider once again the probability distribution function in eqn. (8.6.6). Since $P(s_1, ..., s_n)$ is an ensemble average

$$P(s_1, ..., s_n) = \left\langle \prod_{\alpha=1}^{n} \delta(f_\alpha(\mathbf{r}_1, ..., \mathbf{r}_N) - s_\alpha) \right\rangle, \tag{8.11.1}$$

we can replace the phase-space average with a time average over a trajectory as

$$P(s_1, ..., s_n) = \lim_{\mathcal{T} \to \infty} \frac{1}{\mathcal{T}} \int_0^{\mathcal{T}} dt \prod_{\alpha=1}^{n} \delta(f_\alpha(\mathbf{r}_1(t), ..., \mathbf{r}_N(t)) - s_\alpha), \tag{8.11.2}$$

under the assumption of ergodic dynamics. In the metadynamics approach, we express the δ-function as the limit of a Gaussian function as the width goes to 0 and the height is goes to infinity:

$$\delta(x - a) = \lim_{\sigma \to 0} \frac{1}{\sqrt{2\pi\sigma^2}} e^{-(x-a)^2/2\sigma^2}. \tag{8.11.3}$$

Using eqn. (8.11.3), eqn. (8.11.2) can be rewritten as

$$P(s_1, ..., s_n) =$$

$$\lim_{\mathcal{T} \to \infty} \lim_{\Delta s \to 0} \frac{1}{\sqrt{2\pi\Delta s^2 \mathcal{T}}} \int_0^{\mathcal{T}} dt \prod_{\alpha=1}^{n} \exp\left(-\frac{(s_\alpha - f_\alpha(\mathbf{r}_1(t), ..., \mathbf{r}_N(t)))^2}{2\Delta s^2} \right). \tag{8.11.4}$$

Thus, for finite $\mathcal{T}$ and Δs, eqn. (8.11.4) represents an approximation to $P(s_1, ..., s_n)$, which becomes increasingly accurate as $\mathcal{T}$ increases and the Gaussian width Δs decreases. To see how eqn. (8.11.4) builds up the probability distribution at the point $s_1, ..., s_n$, consider a dynamical trajectory initiated such that $f_\alpha(\mathbf{r}_1(0), ..., \mathbf{r}_N(0)) = s_\alpha$ for all α. If $s_1, ..., s_n$ is a point of low free energy, this is also a point of high probability, and the trajectory will sample more points for which $f_\alpha(\mathbf{r}_1(t), ..., \mathbf{r}_N(t))$ is in the neighborhood of s_α, where the Gaussian has its peak, than it would if $s_1, ..., s_n$ were a point of high free energy or low probability. In the latter case, the trajectory will tend to drift away from such points of low probability, $f_\alpha(\mathbf{r}_1(t), ..., \mathbf{r}_N(t))$ will no longer be close to $s_1, ..., s_n$, and the Gaussian will contribute negligibly to the overall probability.

For numerical evaluation, the integral in eqn. (8.11.4) is written as a discrete sum so that the approximation becomes

$$P(s_1, ..., s_n) \approx$$

$$\frac{1}{\sqrt{2\pi\Delta s^2 \mathcal{T}}} \sum_{k=0}^{N-1} \exp\left(-\sum_{\alpha=1}^{n} \frac{(s_\alpha - f_\alpha(\mathbf{r}_1(k\Delta t), ..., \mathbf{r}_N(k\Delta t)))^2}{2\Delta s^2} \right). \tag{8.11.5}$$

Again, assuming that $f_\alpha(\mathbf{r}_1(0), ..., \mathbf{r}_N(0)) = s_\alpha$, we can write eqn. (8.11.5) as

$$P(s_1, ..., s_n) \approx$$

$$\frac{1}{\sqrt{2\pi\Delta s^2 \mathcal{T}}} \left[1 + \sum_{k=1}^{N-1} \exp\left(-\sum_{\alpha=1}^{n} \frac{(s_\alpha - f_\alpha(\mathbf{r}_1(k\Delta t), ..., \mathbf{r}_N(k\Delta t)))^2}{2\Delta s^2} \right) \right]. \tag{8.11.6}$$

Now, since $A(s_1, ..., s_n) = -kT \ln P(s_1, ..., s_n)$, if we assume the Gaussians are generally small in amplitude, then using $\ln(1 + x) \approx x$, we can approximate the free energy (apart from an arbitrary offset) as

$$A(s_1, ..., s_n) \approx -kT \sum_{k=1}^{N-1} \exp\left(-\sum_{\alpha=1}^{n} \frac{(s_\alpha - f_\alpha(\mathbf{r}_1(k\Delta t), ..., \mathbf{r}_N(k\Delta t)))^2}{2\Delta s^2} \right) + \text{const.}$$

$$\tag{8.11.7}$$

Equation (8.11.7) suggests an intriguing bias potential that can be added to the original potential $U(\mathbf{r}_1, ..., \mathbf{r}_N)$ to help the system sample the free-energy hypersurface while allowing for a straightforward reconstruction of this surface directly from the dynamics. Consider a bias potential of the form (Laio and Parrinello, 2002; Laio *et al.*, 2005)

$$U_{\mathrm{G}}(\mathbf{r}, t) = W \sum_{t'=\tau_{\mathrm{G}}, 2\tau_{\mathrm{G}}, \cdots,}^{t} \exp\left[-\sum_{\alpha=1}^{n} \frac{(f_\alpha(\mathbf{r}) - f_\alpha(\mathbf{r}_{\mathrm{G}}(t')))^2}{2\Delta s^2}\right], \qquad (8.11.8)$$

where $\mathbf{r} \equiv \mathbf{r}_1, ..., \mathbf{r}_N$, as usual, and $\mathbf{r}_{\mathrm{G}}(t)$ is the time evolution of the complete set of Cartesian coordinates up to time t under the action of the potential $U(\mathbf{r}) + U_{\mathrm{G}}(\mathbf{r}, t)$, and τ_{G} is a time interval. The purpose of this bias potential is to add Gaussians of height W and width Δs at intervals τ_{G} to the potential energy so that as time increases, these Gaussians accumulate. If the system starts in a deep basin on the

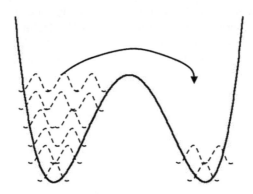

Fig. 8.9 Illustration of the metadynamics procedure for a symmetric double well.

potential energy surface, then this basin will be "filled in" by the Gaussians, thereby lifting the system up toward the barrier until it is able to cross into the next basin, which is subsequently filled by Gaussians until the system can escape into the next basin, and so forth. The process is illustrated in Fig. 8.9.

Our analysis of the d-AFED/TAMD approach shows that if the reaction coordinates move relatively slowly, then they move instantaneously not on the bare potential energy surface but on the potential of mean force surface. Thus, if small Gaussians (adjusted through the parameter W) are added slowly enough, then as time increases, U_{G} takes on the shape of $-A(s_1, ..., s_n)$, since it has maxima where A has minima, and vice versa. Thus, given a long trajectory $\mathbf{r}_{\mathrm{G}}(t)$ generated using the bias potential, the free-energy hypersurface is constructed using

$$A(s_1, ..., s_n) \approx -W \sum_{t'=\tau_{\mathrm{G}}, 2\tau_{\mathrm{G}}, \cdots,} \exp\left[-\sum_{\alpha=1}^{n} \frac{(s_\alpha - f_\alpha(\mathbf{r}_{\mathrm{G}}(t')))^2}{2\Delta s^2}\right]. \qquad (8.11.9)$$

Given the points $\mathbf{r}_G(t)$ generated during a metadynamics run, the times at which Gaussians were deposited, and the parameters W and Δs, eqn. (8.11.9) can be used to generate the free-energy surface by evaluating it at $s_1, ..., s_n \equiv \mathbf{s}$. Of course, the free-energy surface can only be reconstructed in this way if $\mathbf{s}$ lies in regions sampled during the metadynamics run, for otherwise the Gaussians will vanish if $\mathbf{s}$ lies in a region that was never visited during the run. Moreover, if free-energy surfaces of high dimensionality are sought, *e.g.*, $n > 3$, the high dimensionality of the Gaussians makes it difficult for eqn. (8.11.9) to achieve sufficient coverage that a smooth surface is generated. For this reason, metadynamics typically fails in generating free-energy surfaces of dimensionality beyond $n = 3$, in contrast to d-AFED/TAMD, which can be employed with larger values of n (Abrams and Vanden-Eijnden, 2010).

A difficulty with metadynamics concerns the question of when to terminate a run. Once all basins are filled, the metadynamics procedure prescribes the continued addition of Gaussians, which means that the free energy will never truly converge but, rather, will simply fluctuate about the correct result, leading to an error in the free energy that is proportional to the square root of the deposition rate of the Gaussians (Barducci *et al.*, 2008). However, we can use the history of the bias potential itself to create an *adaptive* bias that converges to the correct free-energy surface. A scheme that implements this idea is *well-tempered metadynamics* (Barducci *et al.*, 2008). In a well-tempered metadynamics run, a bias potential similar to that in eqn. (8.11.8) is constructed as follows:

$$U_{\mathrm{WT}}(\mathbf{r}, t) = W \sum_{t'=\tau_G, 2\tau_G, ...,}^{t} e^{-U_{\mathrm{WT}}(\mathbf{r}, t')/\Delta T}\, e^{-\sum_{\alpha=1}^{n}\left(f_\alpha(\mathbf{r}) - f_\alpha(\mathbf{r}_G(t'))\right)^2 / 2\Delta s^2}, \quad (8.11.10)$$

where ΔT is a parameter that defines an elevated "pseudotemperature" $T + \Delta T > T$, which helps accelerate barrier crossing. The key feature of well-tempered metadynamics is that the rate at which the bias potential is added decreases by an exponential factor given by $\exp(-U_{\mathrm{WT}}(\mathbf{r}_1, ..., \mathbf{r}_N)/\Delta T)$ such that in the limit that $t \to \infty$, the rate drops to zero and a probability distribution $P_{\mathrm{WT}}(s_1, ..., s_n)$ of the form (Barducci *et al.*, 2008; Dama *et al.*, 2014)

$$P_{\mathrm{WT}}(s_1, ..., s_n) \propto e^{-A(s_1, ..., s_n)/k(T + \Delta T)}$$

$$\propto \left[e^{-A(s_1, ..., s_n)/kT} \right]^{T/(T + \Delta T)} \quad (8.11.11)$$

is generated. Note the similarity between eqn. (8.11.11) and eqn. (8.10.11), indicating the advantage of using temperature boosting in both d-AFED/TAMD and in metadynamics. It is worth noting other forms of metadynamics, including the so-called parallel-replica/bias-exchange metadynamics (Piana and Laio, 2007), in which metadynamics is performed on a set of n replicas, each of which accumulates a bias on just one of the n collective variables, thereby permitting larger n, and variational metadynamics (Valsson and Parrinello, 2014), which uses a variational procedure to find an optimal metadynamics bias potential.

8.11.1 Unifying metadynamics and d-AFED/TAMD

Both metadynamics and well-tempered metadynamics can be formulated in the extended phase-space framework used in d-AFED/TAMD (Cuendet and Tuckerman, 2014). This feature makes combining metadynamics with d-AFED/TAMD straightforward and leads to a method that leverages the advantages of both methodologies. Consider incorporating a bias potential of the form

$$U_{\text{bias}}(\mathbf{s}, t) = W \sum_{t'=\tau_G, 2\tau_G, \ldots,}^{t} \exp\left[-\sum_{\alpha=1}^{n} \frac{(s_\alpha - s_\alpha(t'))^2}{2\Delta s^2} \right] \tag{8.11.12}$$

into eqns. (8.10.14), such that the modified equations of motion read

$$\dot{\mathbf{r}}_i = \frac{\mathbf{p}_i}{m_i}$$

$$\dot{\mathbf{p}}_i = \mathbf{F}_i - \sum_{\alpha=1}^{n} \kappa_\alpha \left(f_\alpha(\mathbf{r}) - s_\alpha \right) \frac{\partial f_\alpha}{\partial \mathbf{r}_i} + \text{Therm}(T)$$

$$\dot{s}_\alpha = \frac{\pi_\alpha}{\mu_\alpha}$$

$$\dot{\pi}_\alpha = \kappa_\alpha \left(f_\alpha(\mathbf{r}) - s_\alpha \right) - \frac{\partial U_{\text{bias}}(\mathbf{s}, t)}{\partial s_\alpha} + \text{Therm}(T_s). \tag{8.11.13}$$

Equations (8.11.13) constitute the *unified free-energy dynamics* (UFED) method (Chen *et al.*, 2012). The mechanism of eqns. (8.11.13) is similar to d-AFED/TAMD except that as the system sweeps over the free-energy landscape, Gaussians are deposited at a specified rate, leading to a more uniform filling of basins, which, in turn, accelerates convergence of d-AFED/TAMD calculations. In a UFED calculation, the bias potential, $U_{\text{bias}}(\mathbf{s}, t)$, serves only to enhance the temperature-accelerated adiabatic sampling but is not used for free-energy reconstruction. Although it is beyond the scope of this book, a detailed analysis by Chen *et al.* (2012) shows that in the limit of adiabatic decoupling, the mean force average becomes independent of T_s and $U_{\text{bias}}(\mathbf{s}, t)$, which means that the average in eqn. (8.10.9) can be used directly in the same way it is used in a d-AFED/TAMD simulation.

In order to illustrate how adding $U_{\text{bias}}(\mathbf{s}, t)$ can accelerate a d-AFED/TAMD simulation, we show, in Fig. 8.10, the convergence of the four-dimensional free-energy surface (FES) of the gas-phase alanine tripeptide using the two pairs of Ramachandran angles as collective variables. The free-energy surface is generated using both UFED and d-AFED/TAMD for the CHARMM22 force field (MacKerell *et al.*, 1998). The d-AFED/TAMD parameters are $T_s = 1500$ K, $\kappa = 168.0$ amu·Å^2/rad^2, and $\mu = 1008$ kcal/mol·rad^2. For the bias potential, the Gaussian height W is 2 kcal/mol, the width Δs is 0.63 radians, and the deposition rate τ_G is 100 fs. The total simulation time is 200 ns for d-AFED/TAMD and 300 ns for UFED. The convergence of the simulation is measured by computing an L_1 norm between the four-dimensional surface generated from either d-AFED/TAMD or UFED and a reference surface generated from a 1 μs

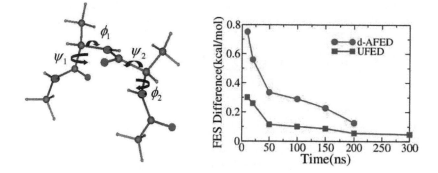

Fig. 8.10 (*Left*) Snapshot of the alanine tripeptide showing the two sets of Ramachandran angles $(\phi_1, \psi_1, \phi_2, \psi_2)$. Color code: red = O, blue = N, grey = C. (*Right*) Comparison of the L_1 norm of the corresponding four-dimensional free-energy surface difference between UFED or d-AFED/TAMD and a reference free-energy surface for the alanine tripeptide in the gas phase space (Chen *et al.*, 2012).

run using parallel tempering (see Section 7.5). The figure shows that the use of the bias potential $U_{\text{bias}}(\mathbf{s}, t)$ causes the L_1 error measure to drop below 0.1 kcal/mol by 50 ns, which is roughly a factor of 4 improvement over use of d-AFED/TAMD alone.

As we noted above for metadynamics, it is important to remember that the efficacy of the bias potential $U_{\text{bias}}(\mathbf{s}, t)$ will degrade as the dimensionality increases, which means that d-AFED/TAMD would need to be employed on its own in order to treat large numbers of collective variables. The next section describes a scheme for applying different sampling methods to different subsets of collective variables.

8.11.2 Unifying metadynamics, d-AFED/TAMD, and umbrella sampling

According to eqns. (8.11.13), the UFED method adds the bias potential $U_{\text{bias}}(\mathbf{s}, t)$ to all of the collective variables equally. This is reasonable if the number of collective variables is not too large, *i.e.*, $n \lesssim 4\text{–}6$, and if all collective variables are of the same type. However, it might happen that different types of collective variables, *e.g.*, distances, dihedral angles, and energies, are targeted simultaneously in one simulation. One might then expect that optimal treatment of different types of collective variables could be achieved by tailoring the enhanced sampling approach to each type of variable. For example, local collective variables, such as dihedral angles, having finite domains and ranges characterized by large barriers, might best be treated using d-AFED/TAMD, while collective variables such as distances, radii of gyration, or energies, having wider domains and more global character, could be treated with metadynamics, UFED, the blue moon ensemble, or umbrella sampling. The *temperature-accelerated sliced sampling* (TASS) method, introduced by Awasthi and Nair (2017), combines d-AFED/TAMD, metadynamics or UFED, and umbrella sampling. As an illustration of TASS, suppose, of the n collective variables employed in an enhanced-sampling simulation, we wish to treat s_1 with umbrella sampling, s_2 with UFED, and the rest with d-AFED/TAMD. In such a simulation, all n components of **s** are kept at the high temperature T_s and are adiabatically decoupled from the physical variables. For this example, the Hamiltonian in eqn. (8.10.6) would be replaced with

$$\mathcal{H}_{\kappa}^{(\text{TASS})}(\mathbf{r}, \mathbf{s}, \mathbf{p}, \boldsymbol{\pi}) = \mathcal{H}(\mathbf{r}, \mathbf{p}) + \sum_{\alpha=1}^{n} \frac{\pi_{\alpha}^2}{2\mu_{\alpha}} + \sum_{\alpha=1}^{n} \frac{1}{2} \kappa_{\alpha} \left(f_{\alpha}(\mathbf{r}) - s_{\alpha} \right)^2$$

$$+ U_{\text{ub}}(s_1) + U_{\text{bias}}(s_2, t). \tag{8.11.14}$$

Here, $U_{\text{bias}}(s_2, t)$ is the metadynamics bias in eqn. (8.11.12) evaluated only as a function of s_2, and $U_{\text{ub}}(s_1)$ is an umbrella potential (cf. eqn. (8.8.1)) $U_{\text{sb}}(s_1) = \kappa_{\text{ub}}(s_1 - \sigma)^2/2$. The parameter σ is the fixed value around which s_1 is restrained with force constant κ_{ub}. In an application of TASS from Paul *et al.* (2019), a set of ten collective variables, encompassing both distances and dihedral angles, was chosen to study a product-release reaction in a sugar nucleotidyltransferase enzyme. The simulations revealed low free-energy pathways for the reaction that could not have been uncovered using umbrella sampling alone.

8.12   The committor distribution and the histogram test

We conclude this chapter with a discussion of the following question: How do we know if a given reaction coordinate is a good choice for representing a particular process of interest? After all, reaction coordinates are often chosen based on some intuitive mental picture we might have of the process, and intuition can be misleading. Therefore, it is important to have a test capable of revealing the quality of a chosen reaction coordinate. To this end, we introduce the concept of a *committor* and its associated probability distribution function (Geissler *et al.*, 1999).

Let us consider a process that takes a system from state $\mathcal{A}$ to state $\mathcal{B}$. We define the *committor* as the probability $p_{\mathcal{B}}(\mathbf{r}_1, ..., \mathbf{r}_N) \equiv p_{\mathcal{B}}(\mathbf{r})$ that a trajectory initiated from a configuration $\mathbf{r}_1, ..., \mathbf{r}_N \equiv \mathbf{r}$ with velocities sampled from a Maxwell-Boltzmann distribution will arrive in state $\mathcal{B}$ before state $\mathcal{A}$. If the configuration $\mathbf{r}$ corresponds to a true transition state, then $p_{\mathcal{B}}(\mathbf{r}) = 1/2$. Inherent in the definition of the committor is the assumption that the trajectory is stopped as soon as it ends up in either state $\mathcal{A}$ or $\mathcal{B}$. Therefore, $p_{\mathcal{B}}(\mathbf{r}) = 1$ if $\mathbf{r}$ belongs to the state $\mathcal{B}$ and $p_{\mathcal{B}}(\mathbf{r}) = 0$ if $\mathbf{r}$ belongs to $\mathcal{A}$. It can be seen that, In principle, $p_{\mathcal{B}}(\mathbf{r})$ is an exact and universal reaction coordinate for any system. The idea of the committor is illustrated in Fig. 8.11.

Unfortunately, we do not have an analytical expression for the committor, and mapping out $p_{\mathcal{B}}(\mathbf{r})$ numerically is intractable for large systems. Nevertheless, the committor forms the basis of a useful test that is able to determine the quality of a chosen reaction coordinate. This test, referred to as the *histogram test* (Geissler *et al.*, 1999; Bolhuis *et al.*, 2002; Dellago *et al.*, 2002; Peters, 2006), applies the committor concept to a reaction coordinate $q(\mathbf{r})$. If $q(\mathbf{r})$ is a good reaction coordinate, then the isosurfaces $q(\mathbf{r}) = \text{const}$ should approximate the isosurfaces $p_{\mathcal{B}}(\mathbf{r}) = \text{const}$ of the committor. Thus, we can test the quality of $q(\mathbf{r})$ by calculating an approximation to the *committor distribution* on an isosurface of $q(\mathbf{r})$. The committor distribution is defined to be the probability that $p_{\mathcal{B}}(\mathbf{r})$ has the value p when $q(\mathbf{r}) = q^{\ddagger}$, the value of $q(\mathbf{r})$ at a maximum

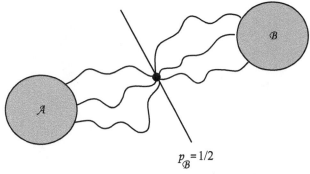

$$p_{\mathcal{B}} = 1/2$$

Isocommittor surface

Fig. 8.11 Schematic of the committor concept. In the figure, trajectories are initiated from the isocommittor surface $p_{\mathcal{B}}(\mathbf{r}) = 1/2$, which is also the transition state surface, so that an equal number of trajectories "commit" to basins $\mathcal{A}$ and $\mathcal{B}$.

between $\mathcal{A}$ and $\mathcal{B}$, which is a presumptive transition state or separatrix between the two basins. This probability distribution is given by

$$P(p) = \frac{C_N}{Q(N,V,T)} \int \mathrm{d}^N \mathbf{p} \int_{q(\mathbf{r})=q^{\ddagger}} \mathrm{d}^N \mathbf{r} \, e^{-\beta \mathcal{H}(\mathbf{r},\mathbf{p})} \delta(p_{\mathcal{B}}(\mathbf{r}_1,...,\mathbf{r}_N) - p). \quad (8.12.1)$$

In discussing the histogram test, we will assume that $q(\mathbf{r})$ is the generalized coordinate $q_1(\mathbf{r})$. The histogram test is then performed as follows: 1) Fix the value of $q_1(\mathbf{r})$ at $q^{\ddagger}$. 2) Sample an ensemble of M configurations $q_2(\mathbf{r}),...,q_{3N}(\mathbf{r})$ corresponding to the remaining degrees of freedom. This will lead to many values of each remaining coordinate. Denote this set of orthogonal coordinates as $q_2^{(k)}(\mathbf{r}),...,q_{3N}^{(k)}(\mathbf{r})$, where $k = 1,...,M$. 3) For each of these sampled configurations, sample a set of initial velocities from a Maxwell-Boltzmann distribution. 4) For the configuration $q^{\ddagger}, q_2^{(k)},...,q_{3N}^{(k)}$, use each set of sampled initial velocities to initiate a trajectory and run the trajectory until the system ends up in $\mathcal{A}$ or $\mathcal{B}$, at which point, the trajectory is stopped. Assign the trajectory a value of 1 if it ends up in state $\mathcal{B}$ and a value of 0 if it ends up in state $\mathcal{A}$. When the complete set of sampled initial velocities is exhausted for this particular orthogonal configuration, average the 1s and 0s, and record the average value as $p^{(k)}$. 5) Repeat for all of the configurations sampled in step 2 until the full set of averaged probabilities $p^{(1)},...,p^{(M)}$ is generated. 6) Plot a histogram of the probabilities $p^{(1)},...,p^{(M)}$. If the histogram from step 6 peaks sharply at $1/2$, then $q(\mathbf{r})$ is a good reaction coordinate. However, if the histogram is broad over the entire range $(0,1)$, then $q(\mathbf{r})$ is a poor reaction coordinate. Illustrations of good and poor reaction coordinates obtained from the histogram test are shown in Fig. 8.12. Although the histogram test can be expensive to carry out, it is, nevertheless, an important evaluation of the quality of a reaction coordinate and its associated free-energy profile. Once the investment in the histogram test is made, the payoff can be considerable, regardless of whether the reaction coordinate passes the test. If it does pass the test, then the same coordinate can be used in subsequent studies of similar systems. If it does not pass

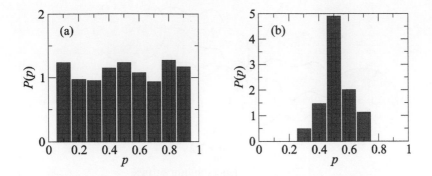

Fig. 8.12 Example histogram tests for evaluating the quality of a reaction coordinate. (a) An example of a poor reaction coordinate; (b) An example of a good reaction coordinate.

the test, then it is clear that the coordinate $q(\mathbf{r})$ should be avoided for the present and similar systems. In Chapter 17, we demonstrate how machine learning can be leveraged to estimate a committor from an enhanced sampling simulation in which a large and possibly redundant set of collective variables is targeted, thereby showing how to reduce such collective-variable sets to a single, optimal reaction coordinate.

8.13   Problems

8.1. Derive eqn. (8.3.6).

8.2. Write a program to compute the free-energy profile in eqn. (8.3.6) using thermodynamic integration. How many λ points do you need to compute the integral accurately enough to obtain the correct free-energy difference $A(1) - A(0)$?

8.3. Write a program to compute the free-energy difference $A(1) - A(0)$ from eqn. (8.3.6) using the free-energy perturbation approach. Can you obtain an accurate answer using a one-step perturbation, or do you need intermediate states?

8.4. Derive eqn. (8.7.25).

8.5. Derive eqn. (8.10.28).

8.6. Consider a classical system with two degrees of freedom x and y described by a potential energy

$$U(x,y) = \frac{U_0}{a^4}\left(x^2 - a^2\right)^2 + \frac{1}{2}ky^2 + \lambda xy,$$

and consider a process in which x is moved from the position $x = -a$ to the position $x = 0$.

a. Calculate the Helmholtz free-energy difference ΔA for this process in a canonical ensemble.

b. Consider now an irreversible process in which the ensemble is frozen in time and, in each member of the ensemble, x is moved instantaneously from $x = -a$ to $x = 0$, i.e., the value of y remains fixed in each ensemble member during this process. The work performed on each system in the ensemble is related to the change in potential energy in this process by

$$W = U(0, y) - U(-a, y)$$

(see eqn. (1.4.2)). By performing the average of W over the initial ensemble, that is, an ensemble in which $x = -a$ for each member of the ensemble, show that $\langle W \rangle > \Delta A$.

c. Now perform the average of $\exp(-\beta W)$ for the work in part b using the same initial ensemble and show that the Jarzynski equality $\langle \exp(-\beta W) \rangle = \exp(-\beta \Delta A)$ holds.

8.7. Calculate the unbiasing $(Z(\mathbf{r}))$ and curvature $(G(\mathbf{r}))$ factors (see eqns. (8.7.20) and (8.7.31)) in the blue moon ensemble method for the following constraints:

a. a distance between two positions $\mathbf{r}_1$ and $\mathbf{r}_2$;

b. the difference of distances between $\mathbf{r}_1$ and $\mathbf{r}_2$ and $\mathbf{r}_1$ and $\mathbf{r}_3$, i.e., $\sigma = |\mathbf{r}_1 - \mathbf{r}_2| - |\mathbf{r}_1 - \mathbf{r}_3|$;

c. the bend angle between the three positions $\mathbf{r}_1$, $\mathbf{r}_2$, and $\mathbf{r}_3$. Treat $\mathbf{r}_1$ as the central position;

*d. the dihedral angle involving the four positions $\mathbf{r}_1$, $\mathbf{r}_2$, $\mathbf{r}_3$, and $\mathbf{r}_4$.

8.8. For the enzyme-inhibitor binding free-energy calculation illustrated in Fig. 8.1, describe, in detail, the algorithm that would be needed to perform the calculation along the indirect path. What are the potential energy functions that would be needed to describe each endpoint?

*8.9. a. Write a program to perform an adiabatic free-energy dynamics calculation of the free-energy profile $A(x)$ corresponding to the potential in Problem 8.6. Using the following values in your program: $a = 1$, $U_0 = 5$, $kT_y = 1$, $kT_x = 5$, $m_y = 1$, $m_x = 1000$, $\lambda = 2.878$. Use separate Nosé-Hoover chains to control the x and y temperatures.

b. Use your program to perform the histogram test of Section 8.12. Does your histogram peak at $p = 1/2$?

8.10. Write adiabatic dynamics and thermodynamic integration codes to generate the λ free-energy profile of Fig. 8.2 using the switches $f(\lambda) = (\lambda^2 - 1)^4$ and $g(\lambda) = ((\lambda - 1)^2 - 1)^4$. In your adiabatic dynamics code, use $kT_\lambda = 0.3$, $kT = 1$, $m_\lambda = 250$, $m = 1$. For the remaining parameters, take $\omega_x = 1$, $\omega_y = 2$, and $\kappa = 1$.

*8.11. Develop a weighted histogram procedure to obtain the free-energy derivative dA/dq_i at a set of integration points q_i starting with eqn. (8.8.22). Describe the difference between your algorithm and that corresponding to the original WHAM procedure for obtaining A_k.

*8.12. In this problem, we will illustrate how a simple change of integration variables in the partition function can be used to create an enhanced sampling method. The approach was originally introduced by Zhu *et al.* (2002) and later enhanced by Minary *et al.* (2007). Consider the double-well potential

$$U(x) = \frac{U_0}{a^4} \left(x^2 - a^2 \right)^2 .$$

The configurational partition function is

$$Z(\beta) = \int dx \, e^{-\beta U(x)}.$$

a. Consider the change of variables $q = f(x)$. Assume that the inverse $x = f^{-1}(q) \equiv g(q)$ exists. Show that the partition function can be expressed as an integral of the form

$$Z(\beta) = \int dq \, e^{-\beta \phi(q)},$$

and give an explicit form for the potential $\phi(q)$.

b. Now consider the transformation

$$q = f(x) = \int_{-a}^{x} dy \, e^{-\beta \tilde{U}(y)}$$

for $-a \le x \le a$ and $q = x$ for $|x| > a$ and $\tilde{U}(x)$ a continuous potential energy function. This transformation is known as a *spatial-warping transformation* (Zhu *et al.*, 2002; Minary *et al.*, 2007). Show that $f(x)$ is a monotonically increasing function of x and, therefore, that $f^{-1}(q)$ exists. Write down the partition function that results from this transformation.

c. If the function $\tilde{U}(x)$ is chosen to be $\tilde{U}(x) = U(x)$ for $-a \le x \le a$ and $\tilde{U}(x) = 0$ for $|x| > a$, then the function $\phi(x)$ is a single-well potential energy function. Sketch a plot of q vs. x, and compare the shape of $\phi(x)$ as a function of x to $\phi(q)$ as a function of q.

d. Argue, therefore, that a Monte Carlo calculation carried out based on $\phi(q)$, or molecular dynamics calculation performed using the Hamiltonian $H(q, p) = p^2/2m + \phi(q)$, leads to an enhanced sampling algorithm for high barriers, and show that the same equilibrium and thermodynamic properties will result.

Hint: From the plot of q vs. x, argue that a small change in q leads to a change in x large enough to move it from one well of $U(x)$ to the other.

e. Develop a Monte Carlo approach for sampling the distribution function

$$P(q) = \frac{1}{Z}e^{-\beta\phi(q)}$$

from part d.

f. Derive molecular dynamics equations of motion, including full expressions for the force on q using the chain rule on the derivatives $(dU/dx)(dx/dq)$ and $(d\tilde{U}/dx)(dx/dq)$, and develop a numerical procedure for obtaining these forces.

Hint: Consider expanding $\exp[-\beta\tilde{U}(x)]$ in a set of orthogonal polynomials such as Legendre polynomials $P_l(\alpha(x))$ with $\alpha(x) \in [-1, 1]$. What should the function $\alpha(x)$ be? [5]

8.13. It has been suggested (Peters *et al.*, 2007) that the committor probability $p_B(\mathbf{r})$ for a single reaction coordinate $q(\mathbf{r})$ can be approximated by a function $\pi_B(q(\mathbf{r}))$ that depends on $\mathbf{r}$ only through $q(\mathbf{r})$.

a. What are the advantages and disadvantages of such an approximation?

b. Suppose that $\pi_B(q(\mathbf{r}))$ can be accurately fit to the following functional form

$$\pi_B(q(\mathbf{r})) = \frac{1 + \tanh(q(\mathbf{r}))}{2}.$$

Is $q(\mathbf{r})$ a good reaction coordinate? Why or why not?

8.14. Starting from eqn. (8.10.7), derive an expression for the Hessian of the free energy

$$H_{\alpha\beta} = \frac{\partial^2 A}{\partial s_\alpha \partial s_\beta}$$

and, in particular, show that it can be expressed in terms of a covariance matrix involving $\kappa_\alpha(f_\alpha(\mathbf{r}) - s_\alpha)$, $\kappa_\beta(f_\beta(\mathbf{r}) - s_\beta)$, and averages of these quantities and their products.

---

[5]It is worth noting that machine learning techniques (cf. Chapter 17) exist for generating transformations $x \to q$ in multiple dimensions such that $P(q)$ is smoother and more easily sampled than $P(x)$. These are known as *Boltzmann generators* (Noe et al., 2019; Tuckerman, 2019), generated from a type of model known as a *normalizing flow*. They rely on similar ideas as addressed in this problem.

8.15. Suppose we have generated a multidimensional free-energy surface $A(s_1, ..., s_n)$ at temperature T as a function of n variables but we wish to project it down to a function $\tilde{A}(s_1, s_2)$ for presentation purposes. Derive a formula for obtaining the projected function $\tilde{A}(s_1, s_2)$ from the full free-energy function $A(s_1, ..., s_n)$.

9
Quantum mechanics

9.1   Introduction: Waves and particles

The first half of the twentieth century witnessed a revolution in physics. Classical mechanics, with its deterministic world view, was shown not to provide a correct description of nature. New experiments were looking deeper into the microscopic world than had been hitherto possible, and the results could not be rationalized using classical concepts. Consequently, a paradigm shift occurred: classical determinism needed to be overthrown, and a new perspective on the physical world emerged.

One of the earliest breakthroughs concerned the blackbody radiation problem. The classical theory of electromagnetism predicts that the intensity of electromagnetic radiation from a blackbody at wavelength λ is proportional to $1/\lambda^2$, which diverges as $\lambda \to 0$. Experimentally, the intensity is observed to vanish as $\lambda \to 0$. In 1901, the German physicist Max Planck postulated that the radiated energy cannot take on any value but is quantized according to the formula $E = nh\nu$, where ν is the frequency of the radiation, n is an integer, and h is a constant. With this simple hypothesis, Planck correctly predicted the shape of the intensity versus wavelength curves and determined the value of h. The constant h is now known as *Planck's constant* and has the value of $h = 6.6208 \times 10^{-34}$ J·s.

A second breakthrough concerned the so-called photoelectric effect. It is observed that when light of sufficiently high frequency impinges on a metallic surface, electrons are ejected from the surface with a residual kinetic energy that depends on the light's frequency. According to classical mechanics, the energy carried by an electromagnetic wave is proportional to its amplitude, independent of its frequency, which contradicts the observation. Planck's hypothesis, however, implies that the impinging light carries energy proportional to its frequency. Applying Planck's idea, Albert Einstein was able to provide a correct explanation of the photoelectric effect in 1905 and was awarded the Nobel prize for this work in 1921. The photoelectric effect also suggests that, in the context of the experiment, the impinging light behaves less like a wave and more like a massless "particle" that is able to transfer energy to the electrons.

Finally, a fascinating experiment carried out by Davisson and Germer in 1927 investigated the interference patterns registered by a photosensitive detector when electrons impinge on a diffraction grating. This experiment revealed an interference pattern very similar to that produced when coherent light impinges on a diffraction grating, suggesting that the electrons in such a diffraction experiment behave less like particles and more like waves. Moreover, where an individual electron strikes the detector cannot be predicted. All that can be predicted is the *probability* that the electron will strike the detector in some small region. This observation suggests an object that exhibits "wave-like" behavior over one that follows a precise particle-like trajectory predictable from a deterministic equation of motion.

The notions of energy quantization, unpredictability of individual experimental outcomes, and particle-wave duality are aspects of the modern theory of the microscopic world known as *quantum mechanics*. Yet even this particle/wave description is incomplete. For what exactly does it mean for a particle to behave like a wave and a wave to behave like a particle? To answer this, we need to specify more precisely what we mean by "wave" and "particle." In general, a wave is a type of field describing something that can vary over an extended region of space as a function of time. Examples are the displacement of a plucked string over its length or the air pressure inside of an organ pipe. Mathematically, a wave is described by an *amplitude*, $A(x,t)$ (in one dimension) that depends on both space and time. In classical wave mechanics, the form of $A(x,t)$ is determined by solving the (classical) wave equation. Quantum theory posits that the probability of an experimental outcome is determined from a particular "wave" that assigns to each possible outcome a (generally complex) *probability amplitude* Ψ. If, for example, we are interested in the probability that a particle will strike a detector at a location x at time t, then there is an amplitude $\Psi(x,t)$ for this outcome, which is also referred to as the *wave function*. From the amplitude, the probability that the particle will strike the detector in a small region $\mathrm{d}x$ about the point x at time t is given by $P(x,t)\mathrm{d}x = |\Psi(x,t)|^2\mathrm{d}x$. Here,

$$P(x,t) = |\Psi(x,t)|^2 \tag{9.1.1}$$

is known as the *probability density* or *probability distribution*. Such probability amplitudes are fundamental in quantum mechanics because they directly relate to the possible outcomes of experiments and lead to predictions of average quantities obtained over many trials of an experiment. These averages are known as *expectation values*. The spatial probability amplitude, $\Psi(x,t)$, is determined by a particular type of wave equation known as the Schrödinger equation. As we will see shortly, the framework of quantum mechanics describes how to compute the probabilities and associated expectation values of any type of physical observable beyond the spatial probability distribution.

We now seek to understand what is meant by "particle" in quantum mechanics. A particularly elegant description was provided by Richard Feynman in the context of his *path integral* formalism (to be discussed in detail in Chapter 12). As we noted above, the classical notion that particles follow precise, deterministic trajectories breaks down in the microscopic realm. Indeed, if an experiment can have many possible outcomes with different associated probabilities, then it should follow that a particle can follow many different possible paths between the initiation and detection points of an experimental setup. Moreover, it must trace all of these paths *simultaneously*! In order to build up a probability distribution $P(x,t)$, the different paths that a particle can follow will have different associated weights or amplitudes. Since the particle evolves unobserved between initiation and detection, it is impossible to conclude that a particle follows a particular path from one point to the other, and according to Feynman's concept, physical predictions can only be made by summing over *all* possible paths that lead from the initiation point to the detection point. This sum over paths is referred to as the *Feynman path integral*. As we will see in Chapter 12, the classical path, *i.e.*, the path predicted by extremizing the classical action, is the most probable path, thereby indicating that classical mechanics naturally emerges as an approximation to quantum mechanics.

Proceeding as we did for classical statistical mechanics, this chapter will review the basic principles of quantum mechanics. In the next chapter, we will lay out the statistical mechanical rules for connecting the quantum description of the microscopic world to macroscopic observables. These chapters are by no means meant to be an overview of the entire field of quantum mechanics, which could (and does) fill entire books. Here, we seek only to develop the quantum-mechanical concepts that we will use in our treatment of quantum statistical mechanics.

9.2 Review of the fundamental postulates of quantum mechanics

The fundamental postulates and definitions of quantum mechanics address the following questions:

1. How is the physical state of a system described?
2. How are physical observables represented?
3. What are the possible outcomes of a given experiment?
4. What is the expected result when an average over a very large number of observations is performed?
5. How does the physical state of a system evolve in time?
6. What types of measurements are compatible with each other?

Let us begin by detailing how we describe the physical state of a system.

9.2.1 The state vector

In quantum theory, it is not possible to determine the precise outcome of a given single experimental measurement. Thus, unlike in classical mechanics, where the microscopic state of a system is specified by providing a definite set of coordinates and velocities of the particles at any time t, the microscopic state of a system in quantum mechanics is specified in terms of the probability amplitudes for the possible outcomes of different measurements made on the system. Since we must be able to describe any type of measurement, the specification of the amplitudes remains abstract until a particular measurement is explicitly considered. The procedure for converting a set of abstract amplitudes to probabilities associated with the outcomes of particular measurements will be given shortly. For now, let us choose a mathematically useful construct for listing these amplitudes. Such a list is conveniently represented as a vector of complex numbers, which we can specify as a column vector:

$$|\Psi\rangle = \begin{pmatrix} \alpha_1 \\ \alpha_2 \\ \alpha_3 \\ \cdot \\ \cdot \\ \cdot \end{pmatrix}. \tag{9.2.1}$$

We have introduced a special type of notation for this column vector, "$|\Psi\rangle$" with half of an angle bracket, which is called a *Dirac ket vector*, after its inventor, the English physicist P. A. M. Dirac (1902–1984). This notation is now standard in quantum mechanics. The components of $|\Psi\rangle$ are complex probability amplitudes α_k that are related to the corresponding probabilities by

$$P_k = |\alpha_k|^2. \tag{9.2.2}$$

The vector $|\Psi\rangle$ is called the *state vector* (conceptually, it bears some similarity to the phase-space vector used to hold the physical state in classical mechanics). The dimension of $|\Psi\rangle$ must be equal to the number of possible states in which the system might be observed. For example, if the physical system were a coin, then we might observe the coin in a "heads-up" or a "tails-up" state, and a coin-toss experiment is needed to realize one of these states. In this example, the dimension of $|\Psi\rangle$ is 2, and $|\Psi\rangle$ could be represented as follows:

$$|\Psi\rangle = \begin{pmatrix} \alpha_{\mathrm{H}} \\ \alpha_{\mathrm{T}} \end{pmatrix}, \tag{9.2.3}$$

where α_{H} and α_{T} are the (complex) amplitudes for heads-up and tails-up states, respectively. Since the sum of all the probabilities must be unity

$$\sum_k P_k = 1, \tag{9.2.4}$$

it follows that

$$\sum_k |\alpha_k|^2 = 1. \tag{9.2.5}$$

In the coin-toss example, an unbiased coin would have amplitudes $\alpha_{\mathrm{H}} = \alpha_{\mathrm{T}} = 1/\sqrt{2}$.

Dirac ket vectors live in a vector space known as the *Hilbert space*, which we will denote as **H**. A complementary or *dual* space to **H** can also be defined in terms of vectors of the form

$$\langle\Psi| = (\, \alpha_1^* \ \alpha_2^* \ \alpha_3^* \ \cdots\,), \tag{9.2.6}$$

which is known as a *Dirac bra vector*. Hilbert spaces have numerous interesting properties; however, the most important one for our present purposes is the inner or scalar product between $\langle\Psi|$ and $|\Psi\rangle$. This product is defined to be

$$\langle\Psi|\Psi\rangle = \sum_k \alpha_k^* \alpha_k = \sum_k |\alpha_k|^2. \tag{9.2.7}$$

Note that the inner product requires both a bra vector and a ket vector. The terms "bra" and "ket" are meant to denote two halves of a "bracket" ($\langle\cdots|\cdots\rangle$), which is formed when an inner product is constructed. Combining eqn. (9.2.7) with (9.2.5), we see that $|\Psi\rangle$ is a unit vector since $\langle\Psi|\Psi\rangle = 1$.

A more general inner product between two Hilbert-space vectors

$$|\phi\rangle = \begin{pmatrix} \phi_1 \\ \phi_2 \\ \phi_3 \\ \cdot \\ \cdot \\ \cdot \end{pmatrix} \qquad |\psi\rangle = \begin{pmatrix} \psi_1 \\ \psi_2 \\ \psi_3 \\ \cdot \\ \cdot \\ \cdot \end{pmatrix} \qquad (9.2.8)$$

is defined to be

$$\langle\psi|\phi\rangle = \sum_k \psi_k^* \phi_k. \qquad (9.2.9)$$

Note that $\langle\phi|\psi\rangle = \langle\psi|\phi\rangle^*$.

9.2.2 Representation of physical observables

In quantum mechanics, physical observables are represented by linear Hermitian *operators*, which act on the vectors of the Hilbert space (we will see shortly why the operators must be Hermitian). When the vectors of **H** and its dual space are represented as ket and bra vectors, respectively, such operators are represented by matrices. Thus, if $\hat{A}$ is an operator corresponding to a physical observable, we can represent it as

$$\hat{A} = \begin{pmatrix} A_{11} & A_{12} & A_{13} & \cdots \\ A_{21} & A_{22} & A_{13} & \cdots \\ A_{31} & A_{32} & A_{33} & \cdots \\ \vdots & \vdots & \vdots & \ddots \end{pmatrix}. \qquad (9.2.10)$$

(The overhat notation is commonly used in quantum mechanics to denote Hilbert-space operators.) That $\hat{A}$ must be a Hermitian operator means that its matrix elements satisfy

$$A_{ji}^* = A_{ij}. \qquad (9.2.11)$$

The Hermitian conjugate of $\hat{A}$ is defined as

$$\hat{A}^\dagger = \begin{pmatrix} A_{11}^* & A_{21}^* & A_{31}^* & \cdots \\ A_{12}^* & A_{22}^* & A_{31}^* & \cdots \\ A_{13}^* & A_{23}^* & A_{33}^* & \cdots \\ \vdots & \vdots & \vdots & \ddots \end{pmatrix}, \qquad (9.2.12)$$

and the requirement that $\hat{A}$ be Hermitian means $\hat{A}^\dagger = \hat{A}$. Since the vectors of **H** are column vectors, it is clear that an operator $\hat{A}$ can act on a vector $|\phi\rangle$ to yield a new vector $|\phi'\rangle$ via $\hat{A}|\phi\rangle = |\phi'\rangle$, which is a simple matrix-vector product.

9.2.3 Possible outcomes of a physical measurement

Quantum mechanics postulates that if a measurement is performed on a physical observable represented by an operator $\hat{A}$, the result must be one of the eigenvalues of $\hat{A}$. From this postulate, we now see why observables must be represented by Hermitian operators: A physical measurement must yield a *real* number, and Hermitian operators have strictly real eigenvalues. In order to prove this, consider the eigenvalue problem for $\hat{A}$ cast in Dirac notation:

$$\hat{A}|a_k\rangle = a_k|a_k\rangle, \tag{9.2.13}$$

where $|a_k\rangle$ denotes an eigenvector of $\hat{A}$ with eigenvalue a_k. For a general $\hat{A}$, the corresponding equation cast in Dirac bra form would be

$$\langle a_k|\hat{A}^\dagger = \langle a_k|a_k^*. \tag{9.2.14}$$

However, since $\hat{A}^\dagger = \hat{A}$, this reduces to

$$\langle a_k|\hat{A} = \langle a_k|a_k^*. \tag{9.2.15}$$

Thus, if we multiply eqn. (9.2.13) by the bra vector $\langle a_k|$ and eqn. (9.2.15) by the ket vector $|a_k\rangle$, we obtain the following two equations:

$$\langle a_k|\hat{A}|a_k\rangle = a_k\langle a_k|a_k\rangle$$

$$\langle a_k|\hat{A}|a_k\rangle = a_k^*\langle a_k|a_k\rangle. \tag{9.2.16}$$

Consistency between these two relations requires that $a_k = a_k^*$, which proves that the eigenvalues are real. Note that the operator $\hat{A}$ can be expressed in terms of its eigenvalues and eigenvectors as

$$\hat{A} = \sum_k a_k|a_k\rangle\langle a_k|. \tag{9.2.17}$$

The product $|a_k\rangle\langle a_k|$ is known as the *outer* or *tensor* product between the ket and bra vectors and could be used to project an arbitrary vector $|\phi\rangle$ along the direction of $|a_k\rangle$.

Another important property of Hermitian operators is that their eigenvectors form a complete orthonormal set of vectors that span the Hilbert space. In order to prove orthonormality of the eigenvectors, we multiply eqn. (9.2.13) by the bra vector $\langle a_j|$, which gives

$$\langle a_j|\hat{A}|a_k\rangle = a_k\langle a_j|a_k\rangle. \tag{9.2.18}$$

On the other hand, if we start with the bra equation (remembering that $\hat{A} = \hat{A}^\dagger$ and $a_j = a_j^*$)

$$\langle a_j|\hat{A} = a_j\langle a_j| \tag{9.2.19}$$

and multiply by the ket vector $|a_k\rangle$, we obtain

$$\langle a_j|\hat{A}|a_k\rangle = a_j\langle a_j|a_k\rangle. \tag{9.2.20}$$

Subtracting eqn. (9.2.18) from (9.2.20) gives

$$0 = (a_k - a_j)\langle a_j|a_k\rangle. \tag{9.2.21}$$

If the eigenvalues of $\hat{A}$ are not degenerate, then for $k \neq j$, $a_k \neq a_j$, and it is clear that $\langle a_j|a_k\rangle = 0$. If $k = j$, then $(a_j - a_j) = 0$, and $\langle a_j|a_j\rangle$ can take on any value. This arbitrariness reflects the arbitrariness of the overall normalization of the eigenvectors of $\hat{A}$. The natural choice for this normalization is $\langle a_j|a_j\rangle = 1$, so that the eigenvectors of $\hat{A}$ are unit vectors. Therefore, the eigenvectors are orthogonal and have unit length, hence, they are orthonormal. If some of the eigenvalues of $\hat{A}$ are degenerate, we can choose the eigenvectors to be orthogonal by taking appropriate linear combinations of the degenerate eigenvectors using a procedure such as Gram-Schmidt orthogonalization to produce an orthogonal set (see Problem 9.1). The last property we need to prove is completeness of the eigenvectors of $\hat{A}$. Since a rigorous proof is considerably more involved, we will simply sketch out the main points of the proof. Let **G** be the orthogonal complement space to **H**. By this, we mean that any vector that lies entirely in **G** has no components along the axes of **H**. Let $|b_j\rangle$ be a vector in **G**. Since $\hat{A}$ is defined entirely in **H**, matrix elements of the form $\langle b_j|\hat{A}|a_k\rangle$ and $\langle a_k|\hat{A}|b_j\rangle$ vanish. Thus, $\hat{A}|b_j\rangle$ has no components along any of the directions $|a_k\rangle$. As a consequence, the operator $\hat{A}$ maps vectors of **G** back into **G**. This implies that $\hat{A}$ must have at least one eigenvector in **G**. However, this conclusion contradicts our original assumption that **G** is the orthogonal complement to **H**. Consequently, **G** must be a null space, which means that the eigenvectors of $\hat{A}$ span **H**.[1]

The most important consequence of the completeness relation is that an arbitrary vector $|\phi\rangle$ on the Hilbert space can be expanded in terms of the eigenvectors of any Hermitian operator. For the operator $\hat{A}$, we have

$$|\phi\rangle = \hat{I}|\phi\rangle = \sum_k |a_k\rangle\langle a_k|\phi\rangle = \sum_k C_k|a_k\rangle, \qquad (9.2.22)$$

where the expansion coefficient C_k is given by

$$C_k = \langle a_k|\phi\rangle, \qquad (9.2.23)$$

and $|a_k\rangle\langle a_k|\phi\rangle$ is the projection of $|\phi\rangle$ along $|a_k\rangle$. Thus, to obtain the expansion coefficient C_k, we simply compute the inner product of the vector to be expanded with the eigenvector $|a_k\rangle$. Finally, we note that any function $g(\hat{A})$ will have the same eigenvectors of $\hat{A}$ with eigenvalues $g(a_k)$ satisfying

$$g(\hat{A})|a_k\rangle = g(a_k)|a_k\rangle. \qquad (9.2.24)$$

Now that we have derived the properties of Hermitian operators and their eigenvector/eigenvalue spectra, we next consider several other aspects of the measurement process in quantum mechanics. We stated that the result of a measurement of an observable associated with a Hermitian operator $\hat{A}$ must yield one of its eigenvalues. If the state vector of a system is $|\Psi\rangle$, then the probability amplitude that a specific eigenvalue a_k will be obtained in a measurement of $\hat{A}$ is determined by taking the inner product of the corresponding eigenvector $|a_k\rangle$ with the state vector,

$$\alpha_k = \langle a_k|\Psi\rangle, \qquad (9.2.25)$$

---

[1]Note that the argument pertains to finite-dimensional discrete vector spaces. In Section 9.2.5, continuous vector spaces will be introduced, for which such proofs are considerably more subtle.

and the corresponding probability is $P_k = |\alpha_k|^2$. Interestingly, $\{\alpha_k\}$ are just the coefficients of an expansion of $|\Psi\rangle$ in the eigenvectors of $\hat{A}$:

$$|\Psi\rangle = \sum_k \alpha_k |a_k\rangle. \tag{9.2.26}$$

Thus, the more aligned the state vector is with a given eigenvector of $\hat{A}$, the greater is the probability of obtaining the corresponding eigenvalue in a given measurement. Clearly, if $|\Psi\rangle$ is one of the eigenvectors of $\hat{A}$, then the corresponding eigenvalue must be obtained with 100% probability, since no other result is possible in this state.

Although we have not yet discussed the time evolution of the state vector, one aspect of this evolution can be established immediately. According to our discussion, when a measurement is made and yields a particular eigenvalue of $\hat{A}$, then immediately following the measurement, the state vector must somehow "collapse" onto the corresponding eigenvector, for at that moment, we know with 100% certainty that a particular eigenvalue was obtained as the result. Therefore, the act of measurement changes the state of the system and its subsequent time development. Moreover, this change is abrupt and discontinuous.[2]

Finally, suppose a measurement of $\hat{A}$ is performed many times, with each repetition carried out on the same state $|\Psi\rangle$. If we average over the outcomes of these measurements, what is the result? We know that each measurement yields a result a_k with probability $|\alpha_k|^2$. The average over these trials yields the *expectation value* of $\hat{A}$ defined by

$$\langle \hat{A} \rangle = \langle \Psi | \hat{A} | \Psi \rangle. \tag{9.2.27}$$

In order to verify this definition, consider, again, the expansion in eqn. (9.2.26). Substituting eqn. (9.2.26) into eqn. (9.2.27) gives

$$\langle \hat{A} \rangle = \sum_{j,k} \alpha_j^* \alpha_k \langle a_j | \hat{A} | a_k \rangle$$

$$= \sum_{j,k} \alpha_j^* \alpha_k a_k \langle a_j | a_k \rangle$$

$$= \sum_{j,k} \alpha_j^* \alpha_k a_k \delta_{jk}$$

$$= \sum_j a_j |\alpha_j|^2. \tag{9.2.28}$$

The last line shows that the expectation value is determined by summing the possible outcomes of a measurement of $\hat{A}$ (the eigenvalues a_j) times the probability $|\alpha_j|^2$ that

---

[2]It is important to note that the notion of a "collapsing" wave function belongs to one of several interpretations of quantum mechanics and the measurement process known as the *Copenhagen Interpretation*. Another interpretation, the so-called "many-worlds" interpretation, states that our universe is part of an essentially infinite "multiverse"; when $\hat{A}$ is measured, a different outcome is obtained in each member of the multiverse. It has been suggested that a more fundamental theory of the universe's origin (e.g. string theory or loop quantum gravity) will encode a more fundamental interpretation. Many interesting articles and books exist on this subject for curious readers who wish to explore the subject further.

each of these results is obtained. This is precisely what we would expect the average over many trials to yield as the number of trials goes to infinity, so that every possible outcome is ultimately obtained, including those with very low probabilities. The quantity $|\alpha_j|^2$ gives the fraction of all the measurements that yield a_j as a result.

We noted above that the act of measuring an operator $\hat{A}$ causes a "collapse" of the state vector onto one of the eigenvectors of $\hat{A}$. Given this, it follows that no experiment can be designed that can measure two observables simultaneously *unless the two observables have a common set of eigenvectors*. This is simply a consequence of the fact that the state vector cannot simultaneously collapse onto two different eigenvectors. Suppose two observables represented by Hermitian operators $\hat{A}$ and $\hat{B}$ have a common set of eigenvectors $\{|a_k\rangle\}$ so that the two eigenvalue equations

$$\hat{A}|a_k\rangle = a_k|a_k\rangle, \qquad \hat{B}|a_k\rangle = b_k|a_k\rangle \qquad (9.2.29)$$

are satisfied. It is then clear that

$$\hat{A}\hat{B}|a_k\rangle = a_k b_k|a_k\rangle$$
$$\hat{B}\hat{A}|a_k\rangle = b_k a_k|a_k\rangle$$
$$\hat{A}\hat{B}|a_k\rangle = \hat{B}\hat{A}|a_k\rangle$$
$$(\hat{A}\hat{B} - \hat{B}\hat{A})|a_k\rangle = 0. \qquad (9.2.30)$$

Since $|a_k\rangle$ is not a null vector, $\hat{A}\hat{B} - \hat{B}\hat{A}$ must vanish as an operator. The operator

$$\hat{A}\hat{B} - \hat{B}\hat{A} \equiv [\hat{A}, \hat{B}] \qquad (9.2.31)$$

is known as the *commutator* between $\hat{A}$ and $\hat{B}$. If the commutator between two operators vanishes, then the two operators have a common set of eigenvectors and hence can be simultaneously measured. Conversely, two operators $\hat{A}$ and $\hat{B}$ that do not commute ($[\hat{A}, \hat{B}] \neq 0$) are said to be *incompatible observables* and cannot be simultaneously measured.

9.2.4 Time evolution of the state vector

So far, we have referred to the state vector $|\Psi\rangle$ as a static object. In actuality, the state vector is dynamic, and one of the postulates of quantum mechanics specifies how the time evolution is determined. Suppose the system is characterized by a Hamiltonian operator $\hat{\mathcal{H}}$. The eigenvalues of $\hat{\mathcal{H}}$ give the allowed energy levels of the system. (How the Hamiltonian is obtained for a quantum mechanical system when the classical Hamiltonian is known will be described in the next subsection.) As in classical mechanics, the quantum Hamiltonian plays the special role of determining the time evolution of the physical state. Quantum mechanics postulates that in the absence of a measurement, the time dependence of the state vector $|\Psi(t)\rangle$ is determined by

$$i\hbar\frac{\partial}{\partial t}|\Psi(t)\rangle = \hat{\mathcal{H}}|\Psi(t)\rangle, \qquad (9.2.32)$$

which is known as the *time-dependent Schrödinger equation* after the Austrian physicist Erwin Schrödinger (1887–1961) (for which he was awarded the Nobel Prize in

1933). Here $\hbar$ is related to Planck's constant by $\hbar = h/2\pi$ and is also referred to as Planck's constant. Since eqn. (9.2.32) is a first-order differential equation, it must be solved subject to an initial condition $|\Psi(0)\rangle$. Interestingly, eqn. (9.2.32) bears a marked mathematical similarity to the classical equation that determines the evolution of the phase-space vector $\dot{x} = iLx$. The Schrödinger equation can be formally solved to yield the evolution

$$|\Psi(t)\rangle = e^{-i\hat{\mathcal{H}}t/\hbar}|\Psi(0)\rangle. \tag{9.2.33}$$

Again, note the formal similarity to the classical relation $x(t) = \exp(iLt)x(0)$. The unitary operator

$$\hat{U}(t) = e^{-i\hat{\mathcal{H}}t/\hbar} \tag{9.2.34}$$

is known as the *time evolution operator* or the *quantum propagator*. The term *unitary* means that $\hat{U}^\dagger(t)\hat{U}(t) = \hat{I}$. Consequently, the action of $\hat{U}(t)$ on the state vector cannot change the magnitude of the vector, only its direction. This is crucial, as $|\Psi(t)\rangle$ must always be normalized to 1 in order that it generate proper probabilities. Suppose the eigenvectors $|E_k\rangle$ and energy levels or eigenvalues E_k of the Hamiltonian are known from the eigenvalue equation

$$\hat{\mathcal{H}}|E_k\rangle = E_k|E_k\rangle. \tag{9.2.35}$$

It is then straightforward to show that

$$|\Psi(0)\rangle = \sum_k |E_k\rangle\langle E_k|\Psi(0)\rangle$$

$$|\Psi(t)\rangle = \sum_k e^{-iE_kt/\hbar}|E_k\rangle\langle E_k|\Psi(0)\rangle. \tag{9.2.36}$$

If we know the initial amplitudes for obtaining the various eigenvalues of $\hat{\mathcal{H}}$ in an experiment designed to measure the energy, the time evolution of the state vector can be determined. In general, the calculation of the eigenvectors and eigenvalues of $\hat{\mathcal{H}}$ is an extremely difficult problem that can only be solved for systems with a very small number of degrees of freedom, and alternative methods for calculating observables are typically needed.

9.2.5   Position and momentum operators

Up to now, we have formulated the theory of measurement in quantum mechanics for observables with discrete eigenvalue spectra. While there certainly are observables that satisfy this condition, we must also consider operators whose spectra are possibly continuous. The most notable examples are the position and momentum operators corresponding to the classical position and momentum variables.[3] In infinite space, the classical position and momentum variables are continuous, so that in a quantum description, we require operators with continuous eigenvalue spectra. If $\hat{x}$ and $\hat{p}$ denote

---

[3]Note, however, that there are important cases in which the momentum eigenvalues are discrete. An example is a free particle confined to a finite spatial domain, where the discrete momentum eigenvalues are related to the properties of standing waves. This case will be discussed in Section 9.3.

the quantum mechanical position and momentum operators, respectively, then these will satisfy eigenvalue equations of the form

$$\hat{x}|x\rangle = x|x\rangle, \qquad \hat{p}|p\rangle = p|p\rangle, \tag{9.2.37}$$

where x and p are the continuous eigenvalues. In place of the discrete orthonormality and completeness relations, we have continuous analogs, which take the form

$$\langle x|x'\rangle = \delta(x - x'), \qquad \langle p|p'\rangle = \delta(p - p')$$

$$\int dx\, |x\rangle\langle x| = \hat{I}, \qquad \int dp\, |p\rangle\langle p| = \hat{I}$$

$$|\phi\rangle = \int dx\, |x\rangle\langle x|\phi\rangle, \qquad |\phi\rangle = \int dp\, |p\rangle\langle p|\phi\rangle. \tag{9.2.38}$$

The last line shows how to expand an arbitrary vector $|\phi\rangle$ in terms of the position or momentum eigenvectors.

Quantum mechanics postulates that the position and momentum of a particle are not compatible observables. That is, no experiment can measure both properties simultaneously. This postulate is known as the *Heisenberg uncertainty principle* after the German physicist Werner Heisenberg (1901–1976) and is expressed as a relation between the statistical uncertainties $\Delta x \equiv \sqrt{\langle \hat{x}^2 \rangle - \langle \hat{x} \rangle^2}$ and $\Delta p \equiv \sqrt{\langle \hat{p}^2 \rangle - \langle \hat{p} \rangle^2}$, namely

$$\Delta x \Delta p \geq \frac{\hbar}{2}. \tag{9.2.39}$$

Since Δx and Δp are inversely proportional, the more certainty we have about a particle's position, the less certain we are about its momentum, and vice versa. Thus, any experiment designed to measure a particle's position with a small uncertainty must cause a large uncertainty in the particle's momentum. The uncertainty principle also tells us that the concepts of classical microstates and phase spaces are fictions, as these require a specification of a particle's position and momentum simultaneously. Thus, a point in phase space cannot correspond to anything physical. The uncertainty principle, therefore, supports the idea of a "coarse-graining" of phase space, which was considered in Problem 2.7 and in Section 3.2. A two-dimensional phase space should be represented as a tiling with squares of minimum area $\hbar/2$. These squares would represent the smallest area into which the particle's position and momentum can be localized. Similarly, the phase space of an N-particle system should be coarse-grained into hypervolumes of size $(\hbar/2)^{3N}$. In the classical limit, which involves letting $\hbar \to 0$, we recover the notion of a continuous phase space as an approximation.

The action of the operators $\hat{x}$ and $\hat{p}$ on an arbitrary Hilbert-space vector $|\phi\rangle$ can be expressed in terms of a projection of the resulting vector onto the basis of either position or momentum eigenvectors. Consider the vector $\hat{x}|\phi\rangle$ and multiply on the left by $\langle x|$, which yields $\langle x|\hat{x}|\phi\rangle$. Since $\langle x|\hat{x} = \langle x|x$, this becomes $x\langle x|\phi\rangle$. Remembering that the eigenvalue x is continuous, the vectors $|x\rangle$ form a continuous set of vectors, and hence, the inner product $\langle x|\phi\rangle$ is a continuous function of x, which we can denote as $\phi(x)$. Similarly, the inner project $\langle p|\phi \rangle$ is a continuous function of p, which we can denote as $\phi(p)$.

The uncertainty principle tells us that $\hat{x}$ and $\hat{p}$ do not commute. Can we, nevertheless, determine what $[\hat{x}, \hat{p}]$ is? If we take the particle-wave duality as our starting point, then we can, indeed, derive this commutator. Consider a free particle, for which the classical Hamiltonian is $\mathcal{H} = p^2/2m$. The corresponding quantum operator is obtained by *promoting* the classical momentum p to the quantum operator $\hat{p}$ to give the quantum Hamiltonian $\hat{\mathcal{H}} = \hat{p}^2/2m$. Since this Hamiltonian is a function of $\hat{p}$ alone, it follows that $[\hat{\mathcal{H}}, \hat{p}] = 0$, so that $\hat{\mathcal{H}}$ and $\hat{p}$ have simultaneous eigenvectors. Consider, therefore, the eigenvalue equation for $\hat{p}$

$$\hat{p}|p\rangle = p|p\rangle. \tag{9.2.40}$$

When this equation is projected into the coordinate basis, we obtain

$$\langle x|\hat{p}|p\rangle = p\langle x|p\rangle. \tag{9.2.41}$$

The quantity $\langle x|p \rangle$ is a continuous function of the eigenvalues x and p. We can write eqn. (9.2.41) as

$$\hat{p}\langle x|p\rangle = p\langle x|p\rangle \tag{9.2.42}$$

if we specify how $\hat{p}$ acts on the continuous function $\langle x|p \rangle$. Equation (9.2.42) is actually an equation for a continuous *eigenfunction* of $\hat{p}$ with eigenvalue p. This eigenfunction must be a continuous function of x. According to the particle-wave duality, a free particle should behave as if it were a wave with amplitude $\psi(x) = \exp(\pm ikx)$, where k is the wave vector $k = 2\pi/\lambda$. Indeed, the *de Broglie hypothesis* assigns a wavelength $\lambda = h/p$ to a particle of momentum p, so that $k = p/\hbar$. We now posit that the function $\exp(\pm ipx/\hbar)$ is an eigenfunction of $\hat{p}$ and, therefore, a solution to eqn. (9.2.42) with eigenvalue p. This means that, with proper normalization,

$$\langle x|p\rangle = \frac{1}{\sqrt{2\pi\hbar}}e^{ipx/\hbar}. \tag{9.2.43}$$

However, eqn. (9.2.42) will only be true if $\hat{p}$ acts on $\langle x|p \rangle$ as the derivative

$$\hat{p} \to \frac{\hbar}{i}\frac{\partial}{\partial x}. \tag{9.2.44}$$

Now, consider the commutator $\hat{x}\hat{p} - \hat{p}\hat{x}$. If we sandwich this between the vectors $\langle x|$ and $|p\rangle$, we obtain

$$\langle x|\hat{x}\hat{p} - \hat{p}\hat{x}|p\rangle = \langle x|\hat{x}\hat{p}|p\rangle - \langle x|\hat{p}\hat{x}|p\rangle$$

$$= xp\langle x|p\rangle - \hat{p}\langle x|\hat{x}|p\rangle$$

$$= xp\langle x|p\rangle - \frac{\hbar}{i}\frac{\partial}{\partial x}\left(x\langle x|p\rangle\right)$$

$$= xp\langle x|p\rangle + i\hbar\langle x|p\rangle - x\hat{p}\langle x|p\rangle$$

$$= i\hbar\langle x|p\rangle, \tag{9.2.45}$$

where the penultimate line follows from eqns. (9.2.40) and (9.2.44). Since $|x\rangle$ and $|p\rangle$ are not null vectors, eqn. (9.2.45) implies that the operator

$$\hat{x}\hat{p} - \hat{p}\hat{x} = [\hat{x}, \hat{p}] = i\hbar\hat{I}. \tag{9.2.46}$$

Next, consider a classical particle of mass m moving in one dimension with a Hamiltonian

$$\mathcal{H}(x, p) = \frac{p^2}{2m} + U(x). \tag{9.2.47}$$

The quantum Hamiltonian operator $\hat{\mathcal{H}}$ is obtained by promoting both $\hat{p}$ and $\hat{x}$ to operator, which yields

$$\hat{\mathcal{H}}(\hat{x}, \hat{p}) = \frac{\hat{p}^2}{2m} + U(\hat{x}). \tag{9.2.48}$$

The promotion of a classical phase-space function to a quantum operator via the substitution $x \to \hat{x}$ and $p \to \hat{p}$ is known as the *quantum-classical correspondence principle*. Using eqn. (9.2.44), we can now project the Schrödinger equation onto the basis of position eigenvectors:

$$\langle x|\hat{\mathcal{H}}(\hat{x}, \hat{p})|\Psi(t)\rangle = i\hbar\frac{\partial}{\partial t}\langle x|\Psi(t)\rangle$$

$$-\frac{\hbar^2}{2m}\frac{\partial^2}{\partial x^2}\Psi(x, t) + U(x)\Psi(x, t) = i\hbar\frac{\partial}{\partial t}\Psi(x, t), \tag{9.2.49}$$

where $\Psi(x, t) \equiv \langle x|\Psi(t)\rangle$. Equation (9.2.49) is a partial differential equation that is often referred to as the *Schrödinger wave equation*, and the function $\Psi(x, t)$ is referred to as the *wave function*. Despite the nomenclature, eqn. (9.2.49) differs from a classical wave equation in that it is complex and only first-order in time, and it includes a multiplicative potential energy term $U(x)\Psi(x, t)$. A solution $\Psi(x, t)$ is then used to compute expectation values at time t of any operator. In general, the promotion of classical phase-space functions $a(x)$ or $b(p)$, which depend only on position or momentum, to quantum operators follows by simply replacing x by the operator $\hat{x}$ and p by the operator $\hat{p}$. In this case, the expectation values $\hat{A}(\hat{x})$ or $\hat{B}(\hat{p})$ are defined by

$$\langle\hat{A}\rangle_t = \langle\Psi(t)|\hat{A}(\hat{x})|\Psi(t)\rangle = \int dx\ \Psi^*(x, t)\Psi(x, t)a(x)$$

$$\langle\hat{B}\rangle_t = \langle\Psi(t)|\hat{B}(\hat{p})|\Psi(t)\rangle = \int dx\ \Psi^*(x, t)b\left(\frac{\hbar}{i}\frac{\partial}{\partial x}\right)\Psi(x, t). \tag{9.2.50}$$

For phase-space functions $a(x, p)$ that depend on both position and momentum, promotion to a quantum operator is less straightforward for the reason that in a classical function, how the variables x and p are arranged is irrelevant, but the order matters considerably in quantum mechanics! Therefore, a rule is needed as to how the operators $\hat{x}$ and $\hat{p}$ are ordered when the operator $\hat{A}(\hat{x}, \hat{p})$ is constructed. Since we will not encounter such operators in this book, we will not belabor the point except to refer to one rule for such an ordering due to H. Weyl (1927) (see also Hillery *et al.* (1984)). If a classical phase-space function has the form $a(x, p) = x^n p^m$, its Weyl ordering is

$$x^n p^m \longrightarrow \frac{1}{2^n} \sum_{r=0}^{n} \binom{n}{r} \hat{x}^{n-r} \hat{p}^m \hat{x}^r \tag{9.2.51}$$

for $n < m$.

Using the analysis leading up to eqn. (9.2.50), the eigenvalue equation for the Hamiltonian can also be expressed as a differential equation:

$$\left[-\frac{\hbar^2}{2m} \frac{d^2}{dx^2} + U(x) \right] \psi_k(x) = E_k \psi_k(x), \tag{9.2.52}$$

where $\psi_k(x) \equiv \langle x | E_k \rangle$. The functions, $\psi_k(x)$ are the *eigenfunctions* of the Hamiltonian. Because eqns. (9.2.49) and (9.2.52) differ only in their right-hand sides, the former and latter are often referred to as the "time-dependent" and "time-independent" Schrödinger equations, respectively.

Equation (9.2.52) yields the well-known quantum-mechanical phenomenon of energy quantization. Even in one dimension, the number of potential functions $U(x)$ for which eqns. (9.2.49) or (9.2.52) can be solved analytically is remarkably small.[4] In solving eqn. (9.2.52), if for any given eigenvalue E_k there exist M independent eigenfunctions, then that energy level is said to be M-fold *degenerate*.

Finally, let us extend this framework to three spatial dimensions. The position and momentum operators are now vectors $\hat{\mathbf{r}} = (\hat{x}, \hat{y}, \hat{z})$ and $\hat{\mathbf{p}} = (\hat{p}_x, \hat{p}_y, \hat{p}_z)$. The components of vectors satisfy the commutation relations

$$[\hat{x}, \hat{y}] = [\hat{x}, \hat{z}] = [\hat{y}, \hat{z}] = 0$$

$$[\hat{p}_x, \hat{p}_y] = [\hat{p}_x, \hat{p}_z] = [\hat{p}_y, \hat{p}_z] = 0$$

$$[\hat{x}, \hat{p}_x] = [\hat{y}, \hat{p}_y] = [\hat{z}, \hat{p}_z] = i\hbar \hat{I}. \tag{9.2.53}$$

All other commutators between position and momentum components are 0. Therefore, given a Hamiltonian of the form

$$\hat{\mathcal{H}} = \frac{\hat{\mathbf{p}}^2}{2m} + U(\hat{\mathbf{r}}), \tag{9.2.54}$$

the eigenvalue problem can be expressed as a partial differential equation using the momentum operator substitutions $\hat{p}_x \rightarrow -i\hbar(\partial/\partial x)$, $\hat{p}_y \rightarrow -i\hbar(\partial/\partial y)$, $\hat{p}_z \rightarrow -i\hbar(\partial/\partial z)$. This leads to an equation of the form

---

[4]An excellent treatise on such problems can be found in the book by S. Flügge, *Practical Quantum Mechanics* (1994).

$$\left[-\frac{\hbar^2}{2m}\nabla^2 + U(\mathbf{r})\right]\psi_{\mathbf{k}}(\mathbf{r}) = E_{\mathbf{k}}\psi_{\mathbf{k}}(\mathbf{r}), \tag{9.2.55}$$

where the label $\mathbf{k} = (k_x, k_y, k_z)$ indicates that three quantum numbers are needed to characterize the states.

9.2.6 The Heisenberg picture

An important fact about quantum mechanics is that it supports multiple equivalent formulations, which allows us to choose the formulation that is most convenient for the problem at hand. The picture of quantum mechanics we have been describing postulates that the state vector $|\Psi(t)\rangle$ evolves in time according to the Schrödinger equation and the operators corresponding to physical observables are static. This formulation is known as the *Schrödinger picture* of quantum mechanics. In fact, there exists a perfectly equivalent alternative formulation in which the state vector is taken to be static and the operators evolve in time. This formulation is known as the *Heisenberg picture*.

In the Heisenberg picture, an operator $\hat{A}$ corresponding to an observable evolves in time according to the *Heisenberg equation of motion*:

$$\frac{\mathrm{d}\hat{A}}{\mathrm{d}t} = \frac{1}{i\hbar}[\hat{A}, \hat{\mathcal{H}}]. \tag{9.2.56}$$

Note the mathematical similarity to the evolution of a classical phase-space function:

$$\frac{\mathrm{d}A}{\mathrm{d}t} = \{A, \mathcal{H}\}. \tag{9.2.57}$$

This similarity suggests that the commutator $[\hat{A}, \hat{\mathcal{H}}]/i\hbar$ becomes the Poisson bracket $\{A, \mathcal{H}\}$ in the classical limit. Like the Schrödinger equation, the Heisenberg equation can be solved formally to yield

$$\hat{A}(t) = e^{i\hat{\mathcal{H}}t/\hbar}\hat{A}(0)e^{-i\hat{\mathcal{H}}t/\hbar} = \hat{U}^\dagger(t)\hat{A}(0)\hat{U}(t). \tag{9.2.58}$$

The initial value $\hat{A}(0)$ that appears in eqn. (9.2.58) is the operator $\hat{A}$ in the Schrödinger picture. Thus, given a state vector $|\Psi\rangle$, the expectation value of the operator $\hat{A}(t)$ in the Heisenberg picture is simply

$$\langle\hat{A}(t)\rangle = \langle\Psi|\hat{A}(t)|\Psi\rangle. \tag{9.2.59}$$

The Heisenberg picture makes clear that any operator $\hat{A}$ that commutes with the Hamiltonian satisfies $\mathrm{d}\hat{A}/\mathrm{d}t = 0$ and, hence, does not evolve in time. Such an operator is referred to as a *constant of the motion*. In the Schrödinger picture, if an operator is a constant of the motion, the probabilities associated with the eigenvalues of the operator do not evolve in time. To see this, consider the evolution of the state vector in the Schrödinger picture:

$$|\Psi(t)\rangle = e^{-i\hat{\mathcal{H}}t/\hbar}|\Psi(0)\rangle. \tag{9.2.60}$$

The probability of obtaining an eigenvalue a_k of $\hat{A}$ at time t is given by $|\langle a_k | \Psi(t) \rangle|^2$. Thus, taking the inner product on both sides with $\langle a_k |$, we find

$$\langle a_k | \Psi(t) \rangle = \langle a_k | e^{-i\hat{\mathcal{H}}t/\hbar} | \Psi(0) \rangle. \tag{9.2.61}$$

If $[\hat{A}, \hat{\mathcal{H}}] = 0$, then $|a_k\rangle$ is an eigenvector of $\hat{\mathcal{H}}$ with an eigenvalue, say E_k. Hence, the amplitude for obtaining a_k at time t is

$$\langle a_k | \Psi(t) \rangle = e^{-iE_k t/\hbar} \langle a_k | \Psi(0) \rangle. \tag{9.2.62}$$

Taking the absolute squares of both sides, the complex exponential disappears, and we obtain

$$|\langle a_k | \Psi(t) \rangle|^2 = |\langle a_k | \Psi(0) \rangle|^2, \tag{9.2.63}$$

which implies that the probability at time t is the same as at $t = 0$. Any operator that is a constant of the motion can be simultaneously diagonalized with the Hamiltonian, and the eigenvalues of the operator can be used to characterize the physical states along with those of the Hamiltonian. As these eigenvalues are often expressed in terms of integers, these integers are referred to as the *quantum numbers* of the state.

9.3 Simple examples

In this section, we will consider two simple examples, the free particle and the harmonic oscillator, which illustrate how energy quantization arises and how the eigenstates of the Hamiltonian can be determined and manipulated.

9.3.1 The free particle

The first example is a single free particle in one dimension. In a sense, we solved this problem in Section 9.2.5 using an argument based on the particle-wave duality. Here, we will work backwards, assuming eqn. (9.2.44) is true, and solve the eigenvalue problem explicitly. The Hamiltonian is

$$\hat{\mathcal{H}} = \frac{\hat{p}^2}{2m}. \tag{9.3.1}$$

The eigenvalue problem for $\hat{\mathcal{H}}$ can be expressed as

$$\frac{\hat{p}^2}{2m} | E_k \rangle = E_k | E_k \rangle, \tag{9.3.2}$$

which, from eqn. (9.2.52), is equivalent to the differential equation

$$-\frac{\hbar^2}{2m} \frac{\mathrm{d}^2}{\mathrm{d}x^2} \psi_k(x) = E_k \psi_k(x). \tag{9.3.3}$$

Solution of eqn. (9.3.3) requires determining the functions $\psi_k(x)$, the eigenvalues E_k, and the appropriate quantum number k. The problem can be simplified considerably by noting that $\hat{\mathcal{H}}$ commutes with $\hat{p}$. Therefore, $\psi_k(x)$ are also eigenfunctions of $\hat{p}$, which means we can determine these by solving the simpler equation $\hat{p}|p\rangle = p|p\rangle$. In the coordinate basis, this is a simple differential equation

$$\frac{\hbar}{i}\frac{d}{dx}\phi_p(x) = p\phi_p(x), \tag{9.3.4}$$

which has the solution $\phi_p(x) = C\exp(ipx/\hbar)$. Here, C is a normalization constant to be determined by the requirement of orthonormality. First, let us note that these eigenfunctions are characterized by the eigenvalue p of momentum, hence the p subscript. We can verify that the functions are also eigenfunctions of $\hat{\mathcal{H}}$ by substituting them into eqn. (9.3.3). When this is done, we find that the energy eigenvalues are also characterized by p and are given by $E_p = p^2/2m$. We can, therefore, write the energy eigenfunctions as

$$\psi_p(x) = \phi_p(x) = Ce^{ipx/\hbar}, \tag{9.3.5}$$

and it is clear that different eigenvalues and eigenfunctions are distinguished by their value of p.

The requirement that the momentum eigenfunctions be orthonormal is expressed via eqn. (9.2.38), *i.e.*, $\langle p|p'\rangle = \delta(p - p')$. By inserting the identity operator in the form of $\hat{I} = \int dx |x\rangle\langle x|$ between the bra and ket vectors, we can express this condition as

$$\langle p|p'\rangle = \int_{-\infty}^{\infty} dx \langle p|x\rangle\langle x|p'\rangle = \delta(p - p'). \tag{9.3.6}$$

Since, by definition, $\langle p|x\rangle = \phi_p(x) = \psi_p(x)$, we have

$$|C|^2 \int_{-\infty}^{\infty} dx\, e^{-ipx/\hbar}e^{ip'x/\hbar} = |C|^2 2\pi\hbar\delta(p - p'), \tag{9.3.7}$$

and it follows that $C = 1/\sqrt{2\pi\hbar}$. Hence, the normalized energy and momentum eigenfunctions are $\psi_p(x) = \exp(ipx/\hbar)/\sqrt{2\pi\hbar}$. These eigenfunctions are known as *plane waves*. Note that they are oscillating functions of x defined over the entire spatial range $x \in (-\infty, \infty)$. Moreover, the corresponding probability distribution function $P_p(x) = |\psi_p(x)|^2$ is spatially uniform. If we consider the time dependence of the eigenfunctions

$$\psi_p(x, t) \sim \exp\left[\frac{ipx}{\hbar} - \frac{iE_p t}{\hbar}\right] \tag{9.3.8}$$

(which can be easily shown to satisfy the time-dependent Schrödinger equation), then this represents a free wave moving to the right for $p > 0$ and to the left for $p < 0$ with frequency $\omega = E_p/\hbar$.

As noted in Section 9.2.5, the momentum and energy eigenvalues are continuous because p is a continuous parameter that can range from $-\infty$ to ∞. This results from the fact that the system is unbounded. Let us now consider placing our free particle in a one-dimensional box of length L, which is more in keeping with the paradigm of statistical mechanics. If x is restricted to the interval $[0, L]$, then we need to impose boundary conditions at $x = 0$ and $x = L$. We first analyze the case of periodic boundary conditions, for which we require that $\psi_p(0) = \psi_p(L)$. Imposing this on the eigenfunctions leads to

$$Ce^{ip\cdot 0/\hbar} = 1 = C^{ipL/\hbar}.\qquad(9.3.9)$$

Since $e^{i\theta} = \cos\theta + i\sin\theta$, the only way to satisfy this condition is to require that $pL/\hbar$ is an integer multiple of 2π. Denoting this integer as n, we have the requirement

$$\frac{pL}{\hbar} = 2\pi n \quad \Rightarrow \quad p = \frac{2\pi\hbar}{L}n \equiv p_n,\qquad(9.3.10)$$

and we see immediately that the momentum eigenvalues are no longer continuous but are quantized. Similarly, the energy eigenvalues are now also quantized as

$$E_n = \frac{p_n^2}{2m} = \frac{2\pi^2\hbar^2}{mL^2}n^2.\qquad(9.3.11)$$

In eqns. (9.3.10) and (9.3.11), n can be any integer.

This example illustrates the important concept that the quantized energy eigenvalues are determined by the boundary conditions. In this case, the fact that the energies are discrete leads to a discrete set of eigenfunctions distinguished by the value of n and given by

$$\psi_n(x) = Ce^{ip_n x/\hbar} = Ce^{2\pi inx/L}.\qquad(9.3.12)$$

These functions are orthogonal but not normalized. The normalization condition determines the constant C:

$$\int_0^L |\psi_n(x)|^2 \mathrm{d}x = 1$$

$$|C|^2 \int_0^L e^{-2\pi inx/L}e^{2\pi inx/L} \mathrm{d}x = 1$$

$$|C|^2 \int_0^L \mathrm{d}x = 1$$

$$|C|^2 L = 1$$

$$C = \frac{1}{\sqrt{L}}.\qquad(9.3.13)$$

Hence, the normalized functions for a particle in a periodic box are

$$\psi_n(x) = \frac{1}{\sqrt{L}}\exp(2\pi inx/L).\qquad(9.3.14)$$

Another interesting boundary condition is $\psi_p(0) = \psi_p(L) = 0$, which corresponds to hard walls at $x = 0$ and $x = L$. We can no longer satisfy the boundary condition with a right- or left-propagating plane wave. Rather, we need to take a linear combination of right- and left-propagating waves to form a sin wave, which is also a free standing wave in the box. This is possible because the Schrödinger equation is linear, hence any linear combination of eigenfunctions with the same eigenvalue is also an eigenfunction. In this case, we need to take

$$\psi_p(x) = C \sin\left(\frac{px}{\hbar}\right) = \frac{C}{2i}\left[e^{ipx/\hbar} - e^{-ipx/\hbar}\right], \tag{9.3.15}$$

which manifestly satisfies the boundary condition at $x = 0$. This function satisfies the boundary condition at $x = L$ only if $pL/\hbar = n\pi$, where n is a positive integer. This leads to the momentum quantization condition $p = n\pi L/\hbar \equiv p_n$ and the energy eigenvalues

$$E_n = \frac{p_n^2}{2m} = \frac{\hbar^2\pi^2}{2mL^2}n^2. \tag{9.3.16}$$

The eigenfunctions become

$$\psi_n(x) = C \sin\left(\frac{n\pi x}{L}\right). \tag{9.3.17}$$

Normalizing yields $C = \sqrt{2/L}$ for the constant. From eqn. (9.3.17), it is clear why n must be strictly positive. If $n = 0$, then $\psi_n(x) = 0$ everywhere, which would imply that the particle exists nowhere. Finally, since the eigenfunctions are already constructed from combinations of right- and left-propagating waves, to form standing waves in the box, allowing $n < 0$ only changes the sign of the eigenfunctions (which is a trivial phase factor) but not the physical content of the eigenfunctions (probabilities and expectation values are not affected by an overall sign). Note that the probability distribution $P_n(x) = (2/L)\sin^2(n\pi x/L)$ is no longer uniform.

9.3.2 The harmonic oscillator

The second example we will consider is a single one-dimensional particle moving in a harmonic potential $U(x) = m\omega^2 x^2/2$, so that the Hamiltonian becomes

$$\hat{\mathcal{H}} = \frac{\hat{p}^2}{2m} + \frac{1}{2}m\omega^2\hat{x}^2. \tag{9.3.18}$$

The eigenvalue equation for $\hat{\mathcal{H}}$ becomes, according to eqn. (9.2.52),

$$\left[-\frac{\hbar^2}{2m}\frac{d^2}{dx^2} + \frac{1}{2}m\omega^2 x^2\right]\psi_n(x) = E_n\psi_n(x). \tag{9.3.19}$$

Here, we have anticipated that because the particle is asymptotically bound ($U(x) \to \infty$ as $x \to \pm\infty$), the energy eigenvalues will be discrete and characterized by an integer n. Since the potential becomes infinitely large as $x \to \pm\infty$, we have the boundary conditions $\psi_n(\infty) = \psi_n(-\infty) = 0$.

The solution of this second-order differential equation is not trivial and, therefore, we will not carry out its solution in detail. However, the interested reader is referred to the excellent treatment in *Principles of Quantum Mechanics* by R. Shankar (1994).

The solution does, indeed, lead to a discrete set of energy eigenvalues given by the familiar formula

$$E_n = \left(n + \frac{1}{2} \right) \hbar\omega \qquad n = 0, 1, 2, ... \tag{9.3.20}$$

and a set of normalized eigenfunctions

$$\psi_n(x) = \left(\frac{m\omega}{2^{2n}(n!)^2 \pi \hbar} \right)^{1/4} e^{-m\omega x^2 / 2\hbar} H_n\left(\sqrt{\frac{m\omega}{\hbar}} x \right), \tag{9.3.21}$$

where $\{H_n(y)\}$ are the Hermite polynomials

$$H_n(y) = (-1)^n e^{y^2} \frac{d^n}{dy^n} e^{-y^2}. \tag{9.3.22}$$

The first few of these eigenfunctions are

$$\psi_0(x) = \left(\frac{\alpha}{\pi} \right)^{1/4} e^{-\alpha x^2 / 2}$$

$$\psi_1(x) = \left(\frac{4\alpha^3}{\pi} \right)^{1/4} x e^{-\alpha x^2 / 2}$$

$$\psi_2(x) = \left(\frac{\alpha}{4\pi} \right)^{1/4} \left(2\alpha x^2 - 1 \right) e^{-\alpha x^2 / 2}$$

$$\psi_3(x) = \left(\frac{\alpha^3}{9\pi} \right)^{1/4} \left(2\alpha x^3 - 3x \right) e^{-\alpha x^2 / 2}, \tag{9.3.23}$$

where $\alpha = m\omega/\hbar$. These are plotted in Fig. 9.1. Note that the number of nodes in each eigenfunction is equal to n. Doing actual calculations with these eigenfunctions is mathematically cumbersome. It turns out, however, that there is a simple and convenient framework for the harmonic oscillator in terms of the abstract set of ket vectors $|n\rangle$ that define the eigenfunctions through $\langle x|n\rangle = \psi_n(x)$.

If we exploit the symmetry between $\hat{p}$ and $\hat{x}$ in the harmonic-oscillator Hamiltonian, we can factorize the sum of squares to give

$$\hat{\mathcal{H}} = \left[\frac{\hat{p}^2}{2m\hbar\omega} + \frac{m\omega}{2\hbar} \hat{x}^2 \right] \hbar\omega$$

$$= \left[\left(\sqrt{\frac{m\omega}{2\hbar}} \hat{x} - \frac{i}{\sqrt{2m\hbar\omega}} \hat{p} \right) \left(\sqrt{\frac{m\omega}{2\hbar}} \hat{x} + \frac{i}{\sqrt{2m\hbar\omega}} \hat{p} \right) + \frac{1}{2} \right] \hbar\omega. \tag{9.3.24}$$

The extra $1/2$ appearing in eqn. (9.3.24) arises from the nonzero commutator between $\hat{x}$ and $\hat{p}$, $[\hat{x}, \hat{p}] = i\hbar \hat{I}$. Let us now define two operators

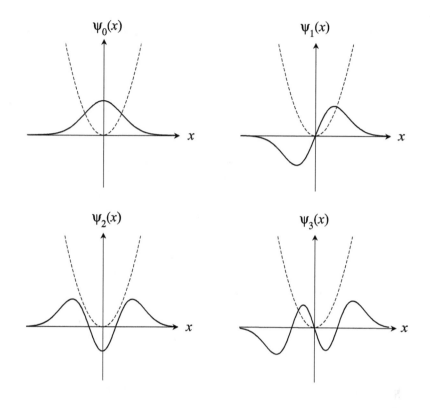

Fig. 9.1 The first four eigenfunctions of a harmonic oscillator.

$$\hat{a} = \sqrt{\frac{m\omega}{2\hbar}}\,\hat{x} + \frac{i}{\sqrt{2m\hbar\omega}}\hat{p}$$

$$\hat{a}^\dagger = \sqrt{\frac{m\omega}{2\hbar}}\,\hat{x} - \frac{i}{\sqrt{2m\hbar\omega}}\hat{p}\,, \tag{9.3.25}$$

which can be shown to satisfy the commutation relation

$$[\hat{a}, \hat{a}^\dagger] = 1. \tag{9.3.26}$$

In terms of these operators, the Hamiltonian can be easily derived with the result

$$\hat{\mathcal{H}} = \left(\hat{a}^\dagger \hat{a} + \frac{1}{2}\right)\hbar\omega. \tag{9.3.27}$$

The action of $\hat{a}$ and $\hat{a}^\dagger$ on the eigenfunctions of $\hat{\mathcal{H}}$ can be worked out using the fact that

$$\hat{a}\psi_n(x) = \left[\sqrt{\frac{\alpha}{2}}x + \frac{1}{\sqrt{2\alpha}}\frac{\mathrm{d}}{\mathrm{d}x}\right]\psi_n(x) \tag{9.3.28}$$

together with the recursion relation for $H_n(y)$: $H_n'(y) = 2nH_{n-1}(y)$. Here, we have used the fact that $p = (\hbar/i)(\mathrm{d}/\mathrm{d}x)$. After some algebra, we find that

$$\hat{a}\psi_n(x) = \sqrt{n}\psi_{n-1}(x). \tag{9.3.29}$$

Similarly, it can be shown that

$$\hat{a}^\dagger\psi_n(x) = \sqrt{n+1}\psi_{n+1}(x). \tag{9.3.30}$$

These relations make it possible to bypass the eigenfunctions and work in an abstract ket representation of the energy eigenvectors, which we denote simply as $|n\rangle$. The above relations can be expressed compactly as

$$\hat{a}|n\rangle = \sqrt{n}|n-1\rangle, \qquad \hat{a}^\dagger|n\rangle = \sqrt{n+1}|n+1\rangle. \tag{9.3.31}$$

Because the operator $\hat{a}^\dagger$ changes an eigenvector of $\hat{\mathcal{H}}$ into the eigenvector corresponding to the next highest energy, it is called a *raising* operator or *creation operator*. Similarly, the operator $\hat{a}$ changes an eigenvector of $\hat{\mathcal{H}}$ into the eigenvector corresponding to the next lowest energy, and hence it is called a *lowering* operator or *annihilation operator*. Note that $\hat{a}|0\rangle = 0$ by definition.

The raising and lowering operators simplify calculations for the harmonic oscillator considerably. Suppose, for example, we wish to compute the expectation value of the operator $\hat{x}^2$ for a system prepared in one of the eigenstates $\psi_n(x)$ of $\hat{\mathcal{H}}$. In principle, one could work out the scary-looking integral

$$\langle n|\hat{x}^2|n\rangle = \left(\frac{\alpha}{\pi 2^{2n}(n!)^2}\right)^{1/2}\int_{-\infty}^{\infty} x^2 e^{-\alpha x^2} H_n^2(\sqrt{\alpha}x)\mathrm{d}x. \tag{9.3.32}$$

However, since $\hat{x}$ has a simple expression in terms of the $\hat{a}$ and $\hat{a}^\dagger$,

$$\hat{x} = \sqrt{\frac{\hbar}{2m\omega}}\left(\hat{a} + \hat{a}^\dagger\right), \tag{9.3.33}$$

the expectation value can be evaluated in a few lines. Note that $\langle n|n'\rangle = \delta_{nn'}$ by orthogonality. Thus,

$$\langle n|\hat{x}^2|n\rangle = \frac{\hbar}{2m\omega}\langle n|\left(\hat{a}^2 + \hat{a}\hat{a}^\dagger + \hat{a}^\dagger\hat{a} + (\hat{a}^\dagger)^2\right)|n\rangle$$

$$= \frac{\hbar}{2m\omega}\left[\sqrt{n(n-1)}\langle n|n-2\rangle + (n+1)\langle n|n\rangle\right.$$

$$\left. + n\langle n|n\rangle + \sqrt{(n+1)(n+2)}\langle n|n+2\rangle\right]$$

$$= \frac{\hbar}{2m\omega}(2n+1). \tag{9.3.34}$$

Thus, by expressing $\hat{x}$ and $\hat{p}$ in terms of $\hat{a}$ and $\hat{a}^\dagger$, we can easily calculate expectation values and arbitrary matrix elements, such as $\langle n|\hat{x}^2|n'\rangle$.

9.4 Identical particles in quantum mechanics: Spin statistics

In 1922, an experiment carried out by Otto Stern and Walter Gerlach showed that quantum particles possess an intrinsic property that, unlike charge and mass, has no classical analog. When a beam of silver atoms was sent through an inhomogeneous magnetic field with a field increasing from the south to north poles of the magnet, the beam split into two distinct beams. The experiment was repeated in 1927 by T. E. Phipps and J. B. Taylor with hydrogen atoms in their ground state in order to ensure that the effect truly revealed an electronic property. The result of the experiment suggests that the particles comprising the beam possess an intrinsic property that couples to the magnetic field and takes on discrete values. This property is known as the magnetic moment $\hat{\mu}_M$ of the particle, which is defined in terms of a more fundamental property called *spin* denoted by the vector operator $\hat{\mathbf{S}}$. These two quantities are related by $\hat{\mu}_M = \gamma \hat{\mathbf{S}}$, where the constant of proportionality γ is the *spin gyromagnetic ratio*, $\gamma = -e/m_e c$. The energy of a particle of spin $\hat{\mathbf{S}}$ fixed in space but interacting with a magnetic field $\mathbf{B}$ is $E = -\hat{\mu}_M \cdot \mathbf{B} = -\gamma \hat{\mathbf{S}} \cdot \mathbf{B}$. Unlike charge and mass, which are simple scalar quantities, spin is expressed as a vector operator and can take on multiple values for a given particle. When the beam in a Stern-Gerlach experiment splits in the magnetic field, for example, this indicates that there are two possible spin states. Since a particle with a magnetic moment resembles a tiny bar magnet, the spin state that has the south pole of the bar magnet pointing toward the north pole of the external magnetic field will be attracted to the stronger field region, and the opposite spin state will be attracted toward the weaker field region.

The three components of the spin operator vector $\hat{\mathbf{S}} = (\hat{S}_x, \hat{S}_y, \hat{S}_z)$ satisfy the commutation relations

$$[\hat{S}_x, \hat{S}_y] = i\hbar \hat{S}_z, \qquad [\hat{S}_y, \hat{S}_z] = i\hbar \hat{S}_x, \qquad [\hat{S}_z, \hat{S}_x] = i\hbar \hat{S}_y. \qquad (9.4.1)$$

These commutation relations are similar to those satisfied by the three components of the angular momentum operator $\hat{\mathbf{L}} = \hat{\mathbf{r}} \times \hat{\mathbf{p}}$. A convenient way to remember the commutation relations is to note that they can be expressed compactly as $\hat{\mathbf{S}} \times \hat{\mathbf{S}} = i\hbar \hat{\mathbf{S}}$. Since spin is an intrinsic property, a particle is said to be a spin-s particle, where s can be either an integer or a half-integer. A spin-s particle can exist in $2s+1$ possible spin states, which, by convention, are taken to be the eigenvectors of the operator $\hat{S}_z$. The eigenvalues of $\hat{S}_z$ then range from $-s\hbar, (-s+1)\hbar, ..., (s-1)\hbar, s\hbar$. For example, the spin operators for a spin-1/2 particle can be represented by 2×2 matrices of the form

$$\hat{S}_x = \frac{\hbar}{2}\begin{pmatrix} 0 & 1 \\ 1 & 0 \end{pmatrix}, \qquad \hat{S}_y = \frac{\hbar}{2}\begin{pmatrix} 0 & -i \\ i & 0 \end{pmatrix}, \qquad \hat{S}_z = \frac{\hbar}{2}\begin{pmatrix} 1 & 0 \\ 0 & -1 \end{pmatrix}, \qquad (9.4.2)$$

and the spin-1/2 Hilbert space is a two-dimensional space. The two spin states have associated spin eigenvalues $s_z = m\hbar/2$, where $m = -1/2$ and $m = 1/2$, and the corresponding eigenvectors are given by

$$|m = 1/2\rangle \equiv |\chi_{1/2}\rangle = \begin{pmatrix} 1 \\ 0 \end{pmatrix}, \qquad |m = -1/2\rangle \equiv |\chi_{-1/2}\rangle = \begin{pmatrix} 0 \\ 1 \end{pmatrix}, \qquad (9.4.3)$$

which are (arbitrarily) referred to as "spin-up" and "spin-down," respectively. The spin-up and spin-down states are also sometimes denoted $|\alpha\rangle$ and $|\beta\rangle$, though we will not make use of this nomenclature. Note that the operator $\hat{S}^2 = \mathbf{\hat{S}} \cdot \mathbf{\hat{S}} = \hat{S}_x^2 + \hat{S}_y^2 + \hat{S}_z^2$ is diagonal and, therefore, shares common eigenvectors with $\hat{S}_z$. These eigenvectors are degenerate, however, having the eigenvalue $s(s+1)\hbar^2$. Finally, if the Hamiltonian is independent of the spin operator, then eigenvectors of $\hat{\mathcal{H}}$ are also eigenvectors of $\hat{S}^2$ and $\hat{S}_z$, since all three can be simultaneously diagonalized.

How the physical states of identical particles are constructed depends on the spin of the particles. Consider the example of two identical spin-s particles. Suppose a measurement is performed that can determine that one of the particles has an $\hat{S}_z$ eigenvalue of $m_a\hbar$ and the other $m_b\hbar$ such that $m_a \neq m_b$. Is the state vector of the total system just after this measurement $|m_a; m_b\rangle \equiv |m_a\rangle \otimes |m_b\rangle$ or $|m_b; m_a\rangle \equiv |m_b\rangle \otimes |m_a\rangle$? Note that, in the first state, particle 1 has an $\hat{S}_z$ eigenvalues $m_a\hbar$, and particle 2 has $m_b\hbar$ as the $\hat{S}_z$ eigenvalue. In the second state, the labeling is reversed. The answer is that neither state is correct. Since the particles are identical, the measurement is not able to assign the particular spin states of each particle. In fact, the two states $|m_a; m_b\rangle$ and $|m_b; m_a\rangle$ are not physically equivalent states. Two states $|\Psi\rangle$ and $|\Psi'\rangle$ can only be physically equivalent if there is a complex scalar α such that

$$|\Psi\rangle = \alpha|\Psi'\rangle, \qquad (9.4.4)$$

and there is no such number relating $|m_a; m_b\rangle$ to $|m_b; m_a\rangle$. Therefore, we need to construct a new state vector $|\Psi(m_a, m_b)\rangle$ such that $|\Psi(m_b, m_a)\rangle$ is physically equivalent to $|\Psi(m_a, m_b)\rangle$. Such a state is the only possibility for correctly representing the physical state of the system immediately after the measurement. Let us take as an ansatz

$$|\Psi(m_a, m_b)\rangle = C|m_a; m_b\rangle + C'|m_b; m_a\rangle. \qquad (9.4.5)$$

If we require that

$$|\Psi(m_a, m_b)\rangle = \alpha|\Psi(m_b, m_a)\rangle, \qquad (9.4.6)$$

then

$$C|m_a; m_b\rangle + C'|m_b; m_a\rangle = \alpha\left(C|m_b; m_a\rangle + C'|m_a; m_b\rangle\right), \qquad (9.4.7)$$

from which it can be seen that

$$C = \alpha C' \qquad C' = \alpha C \qquad (9.4.8)$$

or

$$C' = \alpha^2 C'. \qquad (9.4.9)$$

The only solution to these equations is $\alpha = \pm 1$ and $C = \pm C'$. This gives us two possible physical states of the system: a state that is symmetric (S) under an exchange of $\hat{S}_z$ eigenvalues and one that is antisymmetric (A) under such an exchange. These states are given by

$$|\Psi_S(m_a, m_b)\rangle \propto |m_a; m_b\rangle + |m_b; m_a\rangle$$

$$|\Psi_A(m_a, m_b)\rangle \propto |m_a; m_b\rangle - |m_b; m_a\rangle. \tag{9.4.10}$$

Similarly, suppose we have two identical particles in one dimension, and we perform an experiment capable of determining the position of each particle. If the measurement determines that one particle is at position $x = a$ and the other is at $x = b$, then the state of the system after the measurement would be one of the two following possibilities:

$$|\Psi_S(a, b)\rangle \propto |a\ b\rangle + |b\ a\rangle$$

$$|\Psi_A(a, b)\rangle \propto |a\ b\rangle - |b\ a\rangle. \tag{9.4.11}$$

How do we know whether a given pair of identical particles will opt for the symmetric or antisymmetric state? In order to resolve this ambiguity, the standard postulates of quantum mechanics need to be supplemented by an additional postulate that specifies which of the two possible physical states the particle pair will assume. The new postulate states the following: In nature, particles are of two possible types—those that are *always* found in symmetric (S) multiparticle states and those that are *always* found in antisymmetric (A) multiparticle states. The former are known as bosons (named for the Indian physicist Satyendra Nath Bose (1894–1974)) and the latter as fermions (named for the Italian physicist Enrico Fermi (1901-1954)). Fermions are half-integer-spin particles ($s = 1/2, 3/2, 5/2,...$), while bosons are integer-spin particles ($s = 0, 1, 2,...$). Examples of fermions are electrons, protons, neutrons, and ^{3}He nuclei, all of which are spin-1/2 particles. Examples of bosons are ^{4}He nuclei, which are spin-0, and photons, which are spin-1. Note that the antisymmetric state has the important property that if $m_a = m_b$, $|\Psi_A(m_a, m_a)\rangle = |\Psi_A(m_b, m_b)\rangle = 0$. Since identical fermions are found in antisymmetric states, it follows that *no two identical fermions can be found in nature in exactly the same quantum state.* Put another way, no two identical fermions can have the same set of quantum numbers. This statement is known as the *Pauli exclusion principle* after the Austrian physicist Wolfgang Pauli (1900–1958).

Suppose a system is composed of N identical fermions or bosons with coordinate labels $\mathbf{r}_1, ..., \mathbf{r}_N$ and spin labels $s_1, ..., s_N$. The spin labels designate the eigenvalue of $\hat{S}_z$ for each particle. Let us define, for each particle, a combined label $\mathbf{x}_i \equiv \mathbf{r}_i, s_i$. Then, for a given permutation $P(1), ..., P(N)$ of the particle indices $1,..,N$, the wave function will be totally symmetric if the particles are bosons:

$$\Psi_B(\mathbf{x}_1, ..., \mathbf{x}_N) = \Psi_B(\mathbf{x}_{P(1)}, ..., \mathbf{x}_{P(N)}). \tag{9.4.12}$$

For fermions, as a result of the Pauli exclusion principle, the wave function is antisymmetric with respect to an exchange of any two particles in the system. Therefore, in creating the given permutation, the wave function will pick up a factor of -1 for *each* exchange of two particles that is performed:

$$\Psi_F(\mathbf{x}_1, ..., \mathbf{x}_N) = (-1)^{N_{\text{ex}}} \Psi_F(\mathbf{x}_{P(1)}, ..., \mathbf{x}_{P(N)}), \tag{9.4.13}$$

where N_{ex} is the total number of exchanges of two particles required in order to achieve the permutation $P(1), ..., P(N)$. An N-particle bosonic or fermionic state can be created from a state $\Phi(\mathbf{x}_1, ..., \mathbf{x}_N)$ which is not properly symmetrized but which, nevertheless, is an eigenfunction of the Hamiltonian

$$\hat{\mathcal{H}}\Phi = E\Phi. \tag{9.4.14}$$

Since there are $N!$ possible permutations of the N particle labels in an N-particle state, the bosonic state $\Psi_B(\mathbf{x}_1, ..., \mathbf{x}_N)$ is created from $\Phi(\mathbf{x}_1, ..., \mathbf{x}_N)$ according to

$$\Psi_B(\mathbf{x}_1, ..., \mathbf{x}_N) = \frac{1}{\sqrt{N!}} \sum_{\alpha=1}^{N!} \hat{P}_\alpha \Phi(\mathbf{x}_1, ..., \mathbf{x}_N), \tag{9.4.15}$$

where $\hat{P}_\alpha$ creates 1 of the $N!$ possible permutations of the indices. The fermionic state is created from

$$\Psi_F(\mathbf{x}_1, ..., \mathbf{x}_N) = \frac{1}{\sqrt{N!}} \sum_{\alpha=1}^{N!} (-1)^{N_{\text{ex}}(\alpha)} \hat{P}_\alpha \Phi(\mathbf{x}_1, ..., \mathbf{x}_N), \tag{9.4.16}$$

where $N_{\text{ex}}(\alpha)$ is the number of exchanges needed to create permutation α. The $N!$ that appears in the physical states is exactly the $N!$ introduced *ad hoc* in the expressions for the classical partition functions to account for the identical nature of the particles not explicitly treated in classical mechanics.

9.5 Problems

9.1. Generalize the proof in Section 9.2.3 of orthogonality of the eigenvectors of a Hermitian operator to the case that some of the eigenvalues of the operator are degenerate. Start by considering two degenerate eigenvectors: If $|a_j\rangle$ and $|a_k\rangle$ are two eigenvectors of $\hat{A}$ with eigenvalue a_j, show that two new eigenvectors $|a_j'\rangle$ and $|a_k'\rangle$ can be constructed such that $|a_j'\rangle = |a_j\rangle$ and $|a_k'\rangle = |a_k\rangle + c|a_j\rangle$, where c is a constant, and determine c such that $\langle a_j'|a_k'\rangle = 0$. Generalize the procedure to an arbitrary degeneracy.

9.2. A spin-1/2 particle that is fixed in space interacts with a uniform magnetic field $\mathbf{B}$. The magnetic field lies entirely along the z-axis, so that $\mathbf{B} = (0, 0, B)$. The Hamiltonian for this system is therefore

$$\hat{\mathcal{H}} = -\gamma B \hat{S}_z.$$

The dimensionality of the Hilbert space for this problem is 2.

a. Determine the eigenvalues and eigenvectors of $\hat{\mathcal{H}}$.

b. Suppose the system is prepared with an initial state vector

$$|\Psi(0)\rangle = \begin{pmatrix} 1 \\ 0 \end{pmatrix}.$$

Determine the state vector $|\Psi(t)\rangle$ at time t.

c. Determine the expectation values of the operators $\hat{S}_x$, $\hat{S}_y$, and $\hat{S}_z$ in the time-dependent state computed in part b.

d. Suppose, instead, the system is prepared with an initial state vector

$$|\Psi(0)\rangle = \frac{1}{\sqrt{2}} \begin{pmatrix} 1 \\ 1 \end{pmatrix}.$$

Determine the state vector $|\Psi(t)\rangle$ at time t.

e. Using the time-dependent state computed in part d, determine the following expectation values: $\langle\Psi(t)|\hat{S}_x|\Psi(t)\rangle$, $\langle\Psi(t)|\hat{S}_y|\Psi(t)\rangle$, $\langle\Psi(t)|\hat{S}_z|\Psi(t)\rangle$.

f. For the time-dependent state in part d, determine the uncertainties ΔS_x, ΔS_y, and ΔS_z, where $\Delta S_\alpha = \sqrt{\langle\Psi(t)|\hat{S}_\alpha^2|\Psi(t)\rangle - \langle\Psi(t)|\hat{S}_\alpha|\Psi(t)\rangle^2}$, for $\alpha = x, y, z$.

9.3. Consider a free particle in a one-dimensional box that extends from $x = -L/2$ to $x = L/2$. Assuming periodic boundary conditions, determine the eigenvalues and eigenfunctions of the Hamiltonian for this problem. Repeat for infinite walls at $x = -L/2$ and $x = L/2$.

9.4. A rigid homonuclear diatomic molecule rotates in the xy plane about an axis through its center of mass. Let m be the mass of each atom in the molecule, and let R be its bond length. Show that the molecule has a discrete set of energy levels (energy eigenvalues) and determine the corresponding eigenfunctions.

9.5. Given only the commutator relation between $\hat{x}$ and $\hat{p}$, $[\hat{x}, \hat{p}] = i\hbar\hat{I}$ and the fact that $\hat{p} \to -i\hbar(\mathrm{d}/\mathrm{d}x)$ when projected into the coordinate basis, show that the inner product relation

$$\langle x|p\rangle = \frac{1}{\sqrt{2\pi\hbar}} e^{ipx/\hbar}$$

follows.

9.6. Using raising and lowering operators, calculate the expectation value $\langle n|\hat{x}^4|n\rangle$ and general matrix element $\langle n'|\hat{x}^4|n\rangle$ for a one-dimensional harmonic oscillator.

9.7. Consider an unbound free particle in one dimension such that $x \in (-\infty, \infty)$. An initial wave function $\Psi(x, 0)$, where

$$\Psi(x, 0) = \left[\frac{1}{2\pi\sigma^2} \right]^{1/4} e^{-x^2/4\sigma^2}$$

is prepared.

    a. Determine the time evolution of the initial wave function and the corresponding time-dependent probability density.

    b. Calculate the uncertainties in $\hat{x}$ and $\hat{p}$ at time t. What is the product $\Delta x \Delta p$?

*9.8. A charged particle with charge q and mass m moves in an external magnetic field $\mathbf{B} = (0, 0, B)$. Let $\hat{\mathbf{r}}$ and $\hat{\mathbf{p}}$ be the position and momentum operators for the particle, respectively. The Hamiltonian for the system is

$$\hat{\mathcal{H}} = \frac{1}{2m} \left(\hat{\mathbf{p}} - \frac{q}{c} \mathbf{A}(\hat{\mathbf{r}}) \right)^2,$$

where c is the speed of light and $\mathbf{A}(\mathbf{r})$ is called the *vector potential*. $\mathbf{A}$ is related to the magnetic field $\mathbf{B}$ by

$$\mathbf{B} = \nabla \times \mathbf{A}(\mathbf{r}).$$

One possible choice for $\mathbf{A}$ is

$$\mathbf{A}(\mathbf{r}) = (-By, 0, 0).$$

The particles occupy a cubic box of side L that extends from 0 to L in each spatial direction subject to periodic boundary conditions. Find the energy eigenvalues and eigenfunctions for this problem. Are any of the energy levels degenerate?

Hint: Try a solution of the form

$$\psi(x, y, z) = Ce^{i(p_x x + p_z z)/\hbar} \phi(y)$$

and show that $\phi(y)$ satisfies a harmonic oscillator equation with frequency $\omega = qB/mc$ and equilibrium position $y_0 = -(cp_x/qB)$. You may assume L is much larger than the range of $y - y_0$.

*9.9. Consider a system of N identical particles moving in one spatial dimension. Suppose the Hamiltonian for the system is *separable*, meaning that it can be expressed as a sum

$$\hat{\mathcal{H}} = \sum_{i=1}^{N} \hat{h}(\hat{x}_i, \hat{p}_i),$$

where $\hat{x}_i$ and $\hat{p}_i$ are the coordinate and momentum operators for particle i. These operators satisfy the commutation relations

$$[\hat{x}_i, \hat{x}_j] = 0, \qquad [\hat{p}_i, \hat{p}_j] = 0, \qquad [\hat{x}_i, \hat{p}_j] = i\hbar \delta_{ij}.$$

a. If the Hamiltonian $\hat{h}(\hat{x}, \hat{p})$ is of the form

$$\hat{h}(\hat{x}, \hat{p}) = \frac{\hat{p}^2}{2m} + U(\hat{x}),$$

show that the eigenvalue problem for $\hat{\mathcal{H}}$ can be expressed as N *single-particle* eigenvalue problems of the form

$$\left[-\frac{\hbar^2}{2m} \frac{\partial^2}{\partial x^2} + U(x) \right] \psi_{k_i}(x) = \varepsilon_{k_i} \psi_{k_i}(x),$$

such that the N-particle eigenvalues $E_{k_1,\dots,k_N}$, which are characterized by N quantum numbers, are given by

$$E_{k_1,\dots,k_N} = \sum_{i=1}^{N} \varepsilon_{k_i}.$$

b. Show that if the particles could be treated as distinguishable, then the eigenfunctions of $\hat{\mathcal{H}}$ could be expressed as a product

$$\Phi_{k_1,\dots k_N}(x_1, \dots, x_N) = \prod_{i=1}^{N} \psi_{k_i}(x_i).$$

c. Show that if the particles are identical fermions, then the application of eqn. (9.4.16) leads to a set of eigenfunctions $\Psi_{k_1,\dots,k_N}(x_1, \dots, x_N)$ that is expressible as the determinant of a matrix whose rows are of the form

$$\psi_{k_1}(x_{P(1)}) \quad \psi_{k_2}(x_{P(2)}) \cdots \psi_{k_N}(x_{P(N)}).$$

Recall that $P(1), \dots, P(N)$ is one of the $N!$ permutations of the indices $1,\dots,N$. Give the general form of this determinant. (This determinant is called a *Slater determinant* after its inventor John C. Slater (1900–1976).)

Hint: Try it first for $N = 2$ and $N = 3$.

d. Show that if the particles are bosons rather than fermions, then the eigenfunctions are exactly the same as those of part c except for a replacement of the determinant by a permanent.

Hint: The permanent of a matrix can be generated from the determinant by replacing all of the minus signs with plus signs. Thus, for a 2×2 matrix

$$M = \begin{pmatrix} a & b \\ c & d \end{pmatrix},$$

$\text{perm}(M) = ad + bc.$

9.10 A single particle in one dimension is subject to a potential $U(x)$. Another particle in one dimension is subject to a potential $V(x)$. Suppose $U(x) \neq V(x) + C$, where C is a constant. Prove that the ground-state wave functions $\psi_0(x)$ and $\phi_0(x)$ for each problem must be different.

Hint: Try using proof by contradiction. That is, assume $\psi_0(x) = \phi_0(x)$. What relation between $U(x)$ and $V(x)$ is obtained?

9.11. In a certain nuclear magnetic resonance experiment, a spin-1/2 nucleus (*e.g.*, a proton) is subject to a magnetic field $\mathbf{B} = (0, 0, B)$ in an environment in which its shielding is σ. The spin-1/2 nucleus is well separated from other nuclei, allowing spin-spin interactions to be neglected. The Hamiltonian in this case is then

$$\hat{H}_0 = -\gamma B(1 - \sigma)\hat{S}_z$$

where γ is the spin gyromagnetic ratio and $\hat{S}_z$ is the spin operator in the z-direction. Assume this Hamiltonian describes the system for time $t < 0$. At $t = 0$, a time-dependent magnetic field of strength B_1 aligned along the x-direction is applied, so as to induce transitions among the energy levels of $\hat{H}_0$. The total Hamiltonian for $t \geq 0$ is

$$\hat{H}(t) = \hat{H}_0 + \hat{P}(t) + \hat{P}^\dagger(t)$$

where the operator $\hat{P}(t)$ is

$$\hat{P}(t) = \gamma B_1 e^{i\omega t}\hat{S}_+$$

in which ω is the frequency of the time-dependent field, and

$$\hat{S}_+ = \hbar \begin{pmatrix} 0 & 1 \\ 0 & 0 \end{pmatrix}.$$

This Hamiltonian will cause the nuclear spin to acquire a time-dependent component off the z-axis, which is determined by solving the time-dependent Schrödinger equation for $t \geq 0$:

$$\hat{H}(t)|\Psi(t)\rangle = i\hbar\frac{\mathrm{d}}{\mathrm{dt}}|\Psi(t)\rangle.$$

This problem will walk you through a procedure for solving the time-dependent Schrödinger equation for this system.

a. Assume that for $t < 0$, the nucleus is in its lowest energy state. Based on this, determine the state vector $|\Psi(0)\rangle$ at $t = 0$ immediately before the time-dependent field is switched on. This state vector will be the initial condition on the time-dependent Schrödinger equation.

b. Derive a 2×2 matrix representation of the time-dependent Hamiltonian $\hat{H}(t)$ and show that $\hat{H}(t)$ is Hermitian.

c. In order to solve the equation, we will introduce a transformation of the state vector

$$|\Psi(t)\rangle = R^\dagger(t)|\Phi(t)\rangle$$

and derive an equation for $|\Phi(t)\rangle$. Here,

$$R(t) = \begin{pmatrix} e^{-i\omega t/2} & 0 \\ 0 & e^{i\omega t/2} \end{pmatrix}.$$

For this transformation, show that $|\Psi(0)\rangle = |\Phi(0)\rangle$.

d. Substitute the transformation $|\Psi(t)\rangle = R^\dagger(t)|\Phi(t)\rangle$ into the time-dependent Schrödinger equation and show that the resulting equation for $|\Phi(t)\rangle$ is

$$\hat{H}(t)R^\dagger(t)|\Phi(t)\rangle = i\hbar R^\dagger(t)\frac{d}{dt}|\Phi(t)\rangle + i\hbar\left(\frac{d}{dt}R^\dagger(t)\right)|\Phi(t)\rangle.$$

e. Multiply the equation in part d through on the left by $R(t)$ and, after simplifying, show that the resulting equation for $|\Phi(t)\rangle$ is

$$R(t)\hat{H}(t)R^\dagger(t)|\Phi(t)\rangle + \omega\hat{S}_z|\Phi(t)\rangle = i\hbar\frac{d}{dt}|\Phi(t)\rangle.$$

f. Show that the operator on the right can be written as a 2×2 matrix

$$\hat{H}_T \equiv R(t)\hat{H}(t)R^\dagger(t) + \omega\hat{S}_z = \begin{pmatrix} -H_{11} & H_{12} \\ H_{12} & H_{11} \end{pmatrix},$$

where

$$H_{11} = \frac{\gamma B(1-\sigma)\hbar - \hbar\omega}{2}, \qquad H_{12} = \gamma B_1\hbar.$$

g. Given that $\hat{H}_T$ is time-independent, we can express the solution for $|\Phi(t)\rangle$ in the familiar form

$$|\Phi(t)\rangle = |E_1\rangle\langle E_1|\Phi(0)\rangle e^{-iE_1 t/\hbar} + |E_2\rangle\langle E_2|\Phi(0)\rangle e^{-iE_2 t/\hbar}$$

where E_1 and E_2 are the eigenvalues of $\hat{H}_T$, and $|E_1\rangle$ and $|E_2\rangle$ are the corresponding eigenvectors. Determine the eigenvalues of $\hat{H}_T$ in terms of H_{11} and H_{12}, and show that the eigenvectors are

$$C\begin{pmatrix} -\frac{H_{12}}{D} \\ 1 \end{pmatrix}, \qquad C\begin{pmatrix} 1 \\ \frac{H_{12}}{D} \end{pmatrix}$$

where

$$D = H_{11} - \sqrt{H_{11}^2 + H_{12}^2}, \qquad C = \frac{1}{\sqrt{1 + (H_{12}^2/D^2)}}.$$

Finally, show that these eigenvectors form an orthonormal set.

h. Use the eigenvalues and eigenvectors from part g and the initial condition you obtained in part a to determine the time dependence of $|\Phi(t)\rangle$ and then transform $|\Phi(t)\rangle$ back to $|\Psi(t)\rangle$ using the matrix $R^\dagger(t)$ to obtain the solution to the time-dependent Schrödinger equation. You may express your answer in terms of H_{11}, H_{12}, and D.

i. In the state $|\Psi(t)\rangle$, find the probability as a function of time t that a measurement of $\hat{S}_z$ yields the value $\hbar/2$. You may express your answer in terms of H_{11}, H_{12}, and D.

j. Finally, show that if, at some point in time $\tau > 0$, the time-dependent field is switched off ($B_1 = 0$, $\omega = 0$), $|\Psi(t)\rangle$ reverts back to its initial form from Part a.

9.12. In quantum computing, a qubit is a linear combination of two basic states $|0\rangle$ and $|1\rangle$ whose definitions are the same as those of a spin-1/2 particle, *i.e.*,

$$|1\rangle = \begin{pmatrix} 1 \\ 0 \end{pmatrix}, \qquad |0\rangle = \begin{pmatrix} 0 \\ 1 \end{pmatrix}.$$

One of the most useful single-qubit logic gates in quantum computing is one that takes each input $|0\rangle$ and $|1\rangle$ and creates the following superposition states:

$$|0\rangle \longrightarrow \frac{|0\rangle + |1\rangle}{\sqrt{2}}, \qquad |1\rangle \longrightarrow \frac{|0\rangle - |1\rangle}{\sqrt{2}}.$$

a. Find a combination of the Pauli matrices

$$\sigma_x = \begin{pmatrix} 0 & 1 \\ 1 & 0 \end{pmatrix}, \qquad \sigma_y = \begin{pmatrix} 0 & -i \\ i & 0 \end{pmatrix}, \qquad \sigma_z = \begin{pmatrix} 1 & 0 \\ 0 & -1 \end{pmatrix}$$

that transforms each input this way, and show that your combination leads to a *unitary* matrix capable of performing the action of this logic gate.

b. Now consider the truth table on the next page for a quantum XOR gate: The two-input gate shown in the figure on the next page is a classical XOR gate ('XOR' = exclusive OR). It produces the output $(A \oplus B) \equiv (A + B) \times (\text{NOT}(A) + \text{NOT}(B))$. The quantum analog of the XOR gate would require two entangled qubits that produce a single-qubit output according to $|0\ 0\rangle \longrightarrow |0\rangle$, $|0\ 1\rangle \longrightarrow |1\rangle$, $|1\ 0\rangle \longrightarrow |1\rangle$, $|1\ 1\rangle \longrightarrow |0\rangle$. Can such a quantum gate be designed using only a unitary transformation? If so, specify what that transformation needs to be. If not, explain what other operations would be needed, and write down the full set of operations you propose.

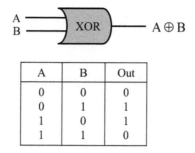

Fig. 9.2 Truth table of the quantum XOR gate.

c. Show that the two-qubit state

$$\frac{1}{2\sqrt{2}} \left(\sqrt{3}|0\,0\rangle - \sqrt{3}|0\,1\rangle + |1\,0\rangle - |1\,1\rangle \right)$$

is not an entangled state, meaning that it can be written as a product of two states $|\psi_0\rangle|\psi_1\rangle$. Find the two states $|\psi_0\rangle$ and $|\psi_1\rangle$ whose product yields this state.

d. Show that the two-qubit state

$$\frac{1}{\sqrt{2}} \left(|0\,0\rangle + |1\,1\rangle \right)$$

is an entangled state, meaning that no two states $|\psi_0\rangle$ and $|\psi_1\rangle$ can be found whose product yields this state.

10

Quantum ensembles and the density matrix

10.1   The difficulty of many-body quantum mechanics

We begin our discussion of the quantum equilibrium ensembles by considering a system of N identical particles in a container of volume V. This is the same setup we studied in Section 3.1 in developing the classical ensembles. In principle, the physical properties of such a large quantum system can be obtained by solving the full time-dependent Schrödinger equation. Suppose the Hamiltonian of the system is

$$\hat{\mathcal{H}} = \sum_{i=1}^{N} \frac{\hat{\mathbf{p}}_i^2}{2m} + U(\hat{\mathbf{r}}_1, ..., \hat{\mathbf{r}}_N). \tag{10.1.1}$$

In d dimensions, there will be dN position and momentum operators. All the position operators commute with each other as do all of the momentum operators. The commutation rule between position and momentum operators is

$$[\hat{r}_{i\alpha}, \hat{p}_{j\beta}] = i\hbar \delta_{ij} \delta_{\alpha\beta}, \tag{10.1.2}$$

where α and β index the d spatial directions and i and j index the particle number. Given the commutation rules, the many-particle coordinate and momentum eigenvectors are direct products (also called *tensor products*) of the eigenvectors of the individual operators. For example, a many-particle coordinate eigenvector in three dimensions is

$$|x_1\,y_1\,z_1 \cdots x_N\,y_N\,z_N\rangle = |x_1\rangle \otimes |y_1\rangle \otimes |z_1\rangle \cdots |x_N\rangle \otimes |y_N\rangle \otimes |z_N\rangle. \tag{10.1.3}$$

Thus, projecting the Schrödinger equation onto the coordinate basis, the N-particle Schrödinger equation in three dimensions becomes

$$\left[-\frac{\hbar^2}{2m} \sum_{i=1}^{N} \nabla_i^2 + U(\mathbf{r}_1, ..., \mathbf{r}_N) \right] \Psi(\mathbf{x}_1, ..., \mathbf{x}_N, t) = i\hbar \frac{\partial}{\partial t} \Psi(\mathbf{x}_1, ..., \mathbf{x}_N, t), \tag{10.1.4}$$

where $\mathbf{x}_i = \mathbf{r}_i, s_i$, and the expectation value of a Hermitian operator $\hat{A}$ corresponding to an observable is $\langle \hat{A} \rangle_t = \langle \Psi(t)|\hat{A}|\Psi(t)\rangle$. The problem inherent in solving eqn. (10.1.4) and evaluating the expectation value (which is a dN-dimensional integral) is that, unless an analytical solution is available, the computational overhead for a numerical solution grows exponentially with the number of degrees of freedom. If eqn. (10.1.4) were to be solved on a spatial grid with M points along each spatial direction, then the total number of points needed would be M^{3N}. Thus, even on a very coarse grid with just $M = 10$ points, for $N \sim 10^{23}$ particles, the total number of grid points would be on the order of $10^{10^{23}}$ points! But even for a small molecule of just $N = 10$ atoms in the gas phase, after we subtract out translations and rotations, Ψ is still a function of 24 coordinates and time. The size of the grid needed to solve eqn. (10.1.4) is large enough that the calculation is beyond the capability of current computing resources. The same is true for the N-particle eigenvalue equation

$$\left[-\frac{\hbar^2}{2m} \sum_{i=1}^{N} \nabla_i^2 + U(\mathbf{r}_1, ..., \mathbf{r}_N) \right] \psi_{\{\mathbf{k}, m_s\}}(\mathbf{x}_1, ..., \mathbf{x}_N) = E_{\{\mathbf{k}, m_s\}} \psi_{\{\mathbf{k}, m_s\}}(\mathbf{x}_1, ..., \mathbf{x}_N)$$

(10.1.5)

(see Problem 9.9). Here $\{\mathbf{k}, m_s\} \equiv \mathbf{k}_1, ..., \mathbf{k}_N, m_{s_1}, ..., m_{s_N}$ are the $4N$ quantum numbers, including $\hat{S}_z$ values, needed to characterize the eigenfunctions and eigenvalues. In fact, explicit solution of the eigenvalue equation for just 4 or 5 particles is considered a *tour de force* calculation. While such calculations yield a wealth of highly accurate dynamical information about small systems, if one wishes to move beyond the limits of the Schrödinger equation and the explicit calculation of the eigenvalues and eigenfunctions of $\hat{\mathcal{H}}$, statistical methods are needed. Now that we have a handle on the magnitude of the many-body quantum mechanical problem, we proceed to introduce the basic principles of quantum equilibrium ensemble theory.

10.2 The ensemble density matrix

Quantum ensembles are conceptually very much like their classical counterparts. Our treatment here will follow somewhat the development presented by Feynman (1998). We begin by considering a collection of $\mathcal{Z}$ quantum systems, each with a unique state vector $|\Psi^{(\lambda)}\rangle$, $\lambda = 1, ..., \mathcal{Z}$, corresponding to a unique microscopic state. At this stage, we imagine that our quantum ensemble is frozen in time, so that the state vectors are fixed. (In Section 10.3 below, we will see how the ensemble develops in time.) As in the classical case, it is assumed that the microscopic states of the ensemble are consistent with a set of macroscopic thermodynamic observables, such as temperature, pressure, chemical potential, *etc.* The principal goal is to predict observables in the form of expectation values. Therefore, we define the expectation value of an operator $\hat{A}$ as the ensemble average of expectation values with respect to each microscopic state in the ensemble. That is,

$$\langle \hat{A} \rangle = \frac{1}{\mathcal{Z}} \sum_{\lambda=1}^{\mathcal{Z}} \langle \Psi^{(\lambda)} | \hat{A} | \Psi^{(\lambda)} \rangle.$$

(10.2.1)

Since each state vector is an abstract object, it proves useful to work in a particular basis. Thus, we introduce a complete set of orthonormal vectors $|\phi_k\rangle$ on the Hilbert space and expand each state of the ensemble in this basis according to

$$|\Psi^{(\lambda)}\rangle = \sum_{k} C_k^{(\lambda)} |\phi_k\rangle,$$

(10.2.2)

where $C_k^{(\lambda)} = \langle \phi_k | \Psi^{(\lambda)} \rangle$. Substituting eqn. (10.2.2) into eqn. (10.2.1) yields

$$\langle \hat{A} \rangle = \frac{1}{\mathcal{Z}} \sum_{\lambda=1}^{\mathcal{Z}} \sum_{k,l} C_k^{(\lambda)*} C_l^{(\lambda)} \langle \phi_k | \hat{A} | \phi_l \rangle$$

$$= \sum_{k,l} \left(\frac{1}{\mathcal{Z}} \sum_{\lambda=1}^{\mathcal{Z}} C_l^{(\lambda)} C_k^{(\lambda)*} \right) \langle \phi_k | \hat{A} | \phi_l \rangle. \tag{10.2.3}$$

Equation (10.2.3) is in the form of the trace of a matrix product. Hence, let us introduce a matrix

$$\rho_{lk} = \sum_{\lambda=1}^{\mathcal{Z}} C_l^{(\lambda)} C_k^{(\lambda)*} \tag{10.2.4}$$

and a normalized matrix $\tilde{\rho}_{lk} = \rho_{lk}/\mathcal{Z}$. The matrix ρ_{lk} (or, equivalently, $\tilde{\rho}_{lk}$) is known as the *ensemble density matrix*. Introducing ρ_{lk} into eqn. (10.2.3), we obtain

$$\langle \hat{A} \rangle = \frac{1}{\mathcal{Z}} \sum_{k,l} \rho_{lk} A_{kl} = \frac{1}{\mathcal{Z}} \sum_{l} (\hat{\rho} \hat{A})_{ll} = \frac{1}{\mathcal{Z}} \text{Tr}(\hat{\rho} \hat{A}) = \text{Tr}(\tilde{\rho} \hat{A}). \tag{10.2.5}$$

Here, $A_{kl} = \langle \phi_k | \hat{A} | \phi_l \rangle$ and $\hat{\rho}$ is the operator whose matrix elements in the basis are ρ_{lk}. Thus, we see that the expectation value of $\hat{A}$ is expressible as a trace of the product of $\hat{A}$ with the ensemble density matrix. According to eqn. (10.2.2), the operator $\hat{\rho}$ can be written formally using the microscopic state vectors:

$$\hat{\rho} = \sum_{\lambda=1}^{\mathcal{Z}} |\Psi^{(\lambda)} \rangle \langle \Psi^{(\lambda)}|. \tag{10.2.6}$$

It is straightforward to show that this operator has the matrix elements given in eqn. (10.2.4).

According to eqn. (10.2.6), $\hat{\rho}$ is a Hermitian operator, so that $\hat{\rho}^\dagger = \hat{\rho}$ and $\tilde{\rho}^\dagger = \tilde{\rho}$. Therefore, its eigenvectors, which satisfy the eigenvalue equation

$$\tilde{\rho} |w_k \rangle = w_k |w_k \rangle, \tag{10.2.7}$$

form a complete orthonormal basis on the Hilbert space. Here, we have defined w_k as being an eigenvalue of $\tilde{\rho}$. In order to see what the eigenvalues of $\tilde{\rho}$ mean physically, let us consider eqn. (10.2.5) for the choice $\hat{A} = \hat{I}$. Since $\langle \hat{I} \rangle = 1$, it follows that

$$1 = \frac{1}{\mathcal{Z}} \text{Tr}(\hat{\rho}) = \text{Tr}(\tilde{\rho}) = \sum_{k} w_k. \tag{10.2.8}$$

Thus, the eigenvalues of $\tilde{\rho}$ must sum to 1. Next, let $\hat{A}$ be a projector onto an eigenstate of $\tilde{\rho}$, $\hat{A} = |w_k \rangle \langle w_k| \equiv \hat{P}_k$. Then

$$\langle \hat{P}_k \rangle = \mathrm{Tr}(\tilde{\rho}|w_k\rangle\langle w_k|) = \sum_l \langle w_l|\tilde{\rho}|w_k\rangle\langle w_k|w_l\rangle$$

$$= \sum_l w_k \delta_{kl}$$

$$= w_k, \tag{10.2.9}$$

where we have used eqn. (10.2.7) and the orthogonality of the eigenvectors of $\tilde{\rho}$. Note, however, that

$$\langle \hat{P}_k \rangle = \frac{1}{\mathcal{Z}} \sum_{\lambda=1}^{\mathcal{Z}} \langle \Psi^{(\lambda)}|w_k\rangle\langle w_k|\Psi^{(\lambda)}\rangle = \frac{1}{\mathcal{Z}} \sum_{\lambda=1}^{\mathcal{Z}} |\langle \Psi^{(\lambda)}|w_k\rangle|^2 \geq 0. \tag{10.2.10}$$

Equations (10.2.9) and (10.2.10) imply that $w_k \geq 0$. Combining the facts that $w_k \geq 0$ and $\sum_k w_k = 1$, we see that $0 \leq w_k \leq 1$. Thus, the w_k satisfy the properties of probabilities.

With this key property of w_k in mind, we can now assign a physical meaning to the density matrix. Let us now consider the expectation value of a projector $|a_k\rangle\langle a_k| \equiv \hat{\mathcal{P}}_{a_k}$ onto one of the eigenstates of the operator $\hat{A}$. The expectation value of this operator is given by

$$\langle \hat{\mathcal{P}}_{a_k} \rangle = \frac{1}{\mathcal{Z}} \sum_{\lambda=1}^{\mathcal{Z}} \langle \Psi^{(\lambda)}|\hat{\mathcal{P}}_{a_k}|\Psi^{(\lambda)}\rangle = \frac{1}{\mathcal{Z}} \sum_{\lambda=1}^{\mathcal{Z}} \langle \Psi^{(\lambda)}|a_k\rangle\langle a_k|\Psi^{(\lambda)}\rangle$$

$$= \frac{1}{\mathcal{Z}} \sum_{\lambda=1}^{\mathcal{Z}} |\langle a_k|\Psi^{(\lambda)}\rangle|^2. \tag{10.2.11}$$

However, $|\langle a_k|\Psi^{(\lambda)}\rangle|^2 \equiv P_{a_k}^{(\lambda)}$ is just the probability that a measurement of the operator $\hat{A}$ in the λth member of the ensemble will yield the eigenvalue a_k. Similarly,

$$\langle \hat{\mathcal{P}}_{a_k} \rangle = \frac{1}{\mathcal{Z}} \sum_{\lambda=1}^{\mathcal{Z}} P_{a_k}^{(\lambda)} \tag{10.2.12}$$

is just the ensemble average of the probability of obtaining the value a_l in each member of the ensemble. However, the expectation value of $\hat{\mathcal{P}}_{a_k}$ can also be written as

$$\langle \hat{\mathcal{P}}_{a_k} \rangle = \mathrm{Tr}(\tilde{\rho}\hat{\mathcal{P}}_{a_k}) = \sum_l \langle w_l|\tilde{\rho}\hat{\mathcal{P}}_{a_k}|w_l\rangle$$

$$= \sum_l w_l \langle w_l|a_k\rangle\langle a_k|w_l\rangle$$

$$= \sum_l w_l |\langle a_k|w_l\rangle|^2. \tag{10.2.13}$$

Equating the results of eqns. (10.2.12) and (10.2.13) gives

$$\frac{1}{Z}\sum_{\lambda=1}^{Z}\langle P_{a_k}^{(\lambda)}\rangle = \sum_{l} w_l |\langle a_k | w_l \rangle|^2. \tag{10.2.14}$$

We now interpret $\{|w_l\rangle\}$ as a complete set of microscopic states appropriate for the ensemble, with w_l the probability that a randomly selected member of the ensemble is in the state $|w_l\rangle$. Hence, the quantity on the right is the sum of probabilities that a measurement of $\hat{A}$ in a state $|w_l\rangle$ yields the result a_k weighted by the probability that an ensemble member is in the state $|w_l\rangle$. This is equal to the ensemble averaged probability on the left. Thus, the density operator $\tilde{\rho}$ gives the probability w_l for an ensemble member to be in a particular microscopic state $|w_l\rangle$ consistent with a set of macroscopic observables, and therefore, it plays the same role in quantum statistical mechanics as the phase-space distribution function $f(\mathbf{x})$ plays in classical statistical mechanics.

10.3   Time evolution of the density matrix

The evolution in time of the density matrix is determined by the time evolution of each of the state vectors $|\Psi^{(\lambda)}\rangle$. The latter are determined by the time-dependent Schrödinger equation. Starting from eqn. (10.2.6), we write the time-dependent density operator as

$$\hat{\rho}(t) = \sum_{\lambda=1}^{Z} |\Psi^{(\lambda)}(t)\rangle\langle\Psi^{(\lambda)}(t)|. \tag{10.3.1}$$

An equation of motion for $\hat{\rho}(t)$ can be determined by taking the time derivative of both sides of eqn. (10.3.1):

$$\frac{\partial\hat{\rho}}{\partial t} = \sum_{\lambda=1}^{Z}\left[\left(\frac{\partial}{\partial t}|\Psi^{(\lambda)}(t)\rangle\right)\langle\Psi^{(\lambda)}(t)| + |\Psi^{(\lambda)}(t)\rangle\left(\frac{\partial}{\partial t}\langle\Psi^{(\lambda)}(t)|\right)\right]. \tag{10.3.2}$$

However, since $\partial|\Psi^{(\lambda)}(t)\rangle/\partial t = (1/i\hbar)\hat{\mathcal{H}}|\Psi^{(\lambda)}(t)\rangle$ from the Schrödinger equation, eqn. (10.3.2) becomes

$$\frac{\partial\hat{\rho}}{\partial t} = \frac{1}{i\hbar}\sum_{\lambda=1}^{Z}\left[\left(\hat{\mathcal{H}}|\Psi^{(\lambda)}(t)\rangle\right)\langle\Psi^{(\lambda)}(t)| - |\Psi^{(\lambda)}(t)\rangle\left(\langle\Psi^{(\lambda)}(t)|\hat{\mathcal{H}}\right)\right]$$

$$= \frac{1}{i\hbar}(\hat{\mathcal{H}}\hat{\rho} - \hat{\rho}\hat{\mathcal{H}}) \tag{10.3.3}$$

or

$$\frac{\partial\hat{\rho}}{\partial t} = \frac{1}{i\hbar}[\hat{\mathcal{H}}, \hat{\rho}]. \tag{10.3.4}$$

Equation (10.3.4) is known as the *quantum Liouville equation*, and it forms the basis of quantum statistical mechanics just as the classical Liouville equation derived in Section 2.5 forms the basis of classical statistical mechanics.

Recall that the time evolution of a Hermitian operator representing a physical observable in the Heisenberg picture is given by eqn. (9.2.56). Although $\hat{\rho}$ is a Hermitian operator, its evolution equation differs from eqn. (9.2.56), as eqn. (10.3.4) makes clear. This difference underscores the fact that $\hat{\rho}$ does not actually represent a physical observable.

The quantum Liouville equation can be solved formally as

$$\hat{\rho}(t) = e^{-i\hat{\mathcal{H}}t/\hbar}\hat{\rho}(0)e^{i\hat{\mathcal{H}}t/\hbar} = \hat{U}(t)\hat{\rho}(0)\hat{U}^{\dagger}(t). \qquad (10.3.5)$$

Equation (10.3.4) is often cast into a form that closely resembles the classical Liouville equation by defining a quantum Liouville operator

$$iL = \frac{1}{i\hbar}[..., \hat{\mathcal{H}}]. \qquad (10.3.6)$$

In terms of this operator, the quantum Liouville equation becomes

$$\frac{\partial \hat{\rho}}{\partial t} = -iL\hat{\rho}, \qquad (10.3.7)$$

which has the formal solution

$$\hat{\rho}(t) = e^{-iLt}\hat{\rho}(0). \qquad (10.3.8)$$

There is a subtlety associated with the quantum Liouville operator iL. As eqn. (10.3.6) implies, iL is not an operator in the sense described in Section 9.2. The operators we have encountered so far act on the vectors of the Hilbert space to yield new vectors. By contrast, iL acts on an operator and returns a new operator. For this reason, it is often called a "superoperator" or "tetradic" operator.[1]

10.4 Quantum equilibrium ensembles

As in the classical case, quantum equilibrium ensembles are defined by a density matrix with no explicit time dependence, *i.e.*, $\partial\hat{\rho}/\partial t = 0$. Thus, the equilibrium Liouville equation becomes $[\hat{\mathcal{H}}, \hat{\rho}] = 0$. This is precisely the condition required for a quantity to be a constant of the motion. The general solution to the equilibrium Liouville equation is any function $F(\hat{\mathcal{H}})$ of the Hamiltonian. Consequently, $\hat{\mathcal{H}}$ and $\hat{\rho}$ have simultaneous eigenvectors. If $|E_k\rangle$ are the eigenvectors of $\hat{\mathcal{H}}$ with eigenvalues E_k, then

$$\hat{\rho}|E_k\rangle = F(\hat{\mathcal{H}})|E_k\rangle = F(E_k)|E_k\rangle. \qquad (10.4.1)$$

Starting from eqn. (10.4.1), we could derive the quantum equilibrium ensembles in much the same manner as we did for the classical equilibrium ensembles. That is, we could begin by defining the microcanonical ensemble based on the conservation of $\hat{\mathcal{H}}$ and then derive the canonical, isothermal-isobaric, and grand canonical ensembles by coupling the system to a heat bath, mechanical piston, particle reservoir, *etc.* However, since we have already carried out this program for the classical ensembles, we can exploit the quantum-classical correspondence principle and simply promote the classical equilibrium phase-space distribution functions, all of which are functions of

---

[1]As an example from the literature of the use of the superoperator formalism, S. Mukamel, in his book *Principles of Nonlinear Optical Spectroscopy* (1995), uses the quantum Liouville operator approach to develop an elegant framework for analyzing various types of nonlinear spectroscopies.

the classical Hamiltonian, to quantum operators. Thus, for the canonical ensemble at temperature T, the normalized density operator becomes

$$\tilde{\rho}(\hat{\mathcal{H}}) = \frac{e^{-\beta\hat{\mathcal{H}}}}{Q(N,V,T)}. \tag{10.4.2}$$

Since $\tilde{\rho}$ must have unit trace, the partition function is given by

$$Q(N,V,T) = \mathrm{Tr}\left[e^{-\beta\hat{\mathcal{H}}}\right]. \tag{10.4.3}$$

Here, $Q(N,V,T)$ is identified with the number $\mathcal{Z}$, the total number of microscopic states in the ensemble. Thus, the unnormalized density matrix $\hat{\rho}$ is $\exp(-\beta\hat{\mathcal{H}})$. Casting eqns. (10.4.2) and (10.4.3) into the basis of the eigenvectors of $\hat{\mathcal{H}}$, we obtain

$$\langle E_k|\tilde{\rho}|E_k\rangle = \frac{e^{-\beta E_k}}{Q(N,V,T)}$$

$$Q(N,V,T) = \sum_k e^{-\beta E_k}. \tag{10.4.4}$$

Equation (10.4.4) indicates that the microscopic states corresponding to the canonical ensemble are eigenstates of $\hat{\mathcal{H}}$, i.e., $w_k = F(E_k) = \exp(-\beta E_k)/Q(N,V,T)$, and that the probability of any member of the ensemble being in a state $|E_k\rangle$ is $\exp(-\beta E_k)/Q(N,V,T)$. Once $Q(N,V,T)$ is known from eqn. (10.4.4), the thermodynamics of the canonical ensemble are determined as usual from eqn. (4.3.23). Finally, the expectation value of any operator $\hat{A}$ in the canonical ensemble is given by

$$\langle\hat{A}\rangle = \mathrm{Tr}\left(\tilde{\rho}\hat{A}\right) = \frac{1}{Q(N,V,T)}\sum_k e^{-\beta E_k}\langle E_k|\hat{A}|E_k\rangle. \tag{10.4.5}$$

(Feynman regarded eqns. (10.4.4) and (10.4.5) as the core of statistical mechanics, and they appear on the first page of his book *Statistical Mechanics: A Set of Lectures*.)[2]) If there are degeneracies among the eigenvalues, then a factor $g(E_k)$, which is the degeneracy of the energy level E_k, i.e., the number of independent eigenstates with this energy, must be introduced into the above sums over eigenstates. Thus, for example, the partition function becomes

$$Q(N,V,T) = \sum_k g(E_k)e^{-\beta E_k}, \tag{10.4.6}$$

and the expectation value of the operator $\hat{A}$ is given by

$$\langle\hat{A}\rangle = \frac{1}{Q(N,V,T)}\sum_k g(E_k)e^{-\beta E_k}\langle E_k|\hat{A}|E_k\rangle. \tag{10.4.7}$$

---

[2]In reference to eqn.(10.4.5), Feynman casually remarks, "This law is the summit of statistical mechanics, and the entire subject is either the slide-down from this summit, as the principle is applied to various cases, or the climb-up to where the fundamental law is derived and the concepts of thermal equilibrium and temperature T clarified" (Feynman, 1998). Our program in the next two chapters will be the former, as we apply the principle and develop analytical and computational tools for carrying out quantum statistical mechanical calculations for complex systems.

In an isothermal-isobaric ensemble at temperature T and pressure P, the density operator, partition function, and expectation value are given, respectively, by

$$\tilde{\rho}(\hat{\mathcal{H}}, V) = \frac{e^{-\beta(\hat{\mathcal{H}}+PV)}}{\Delta(N, P, T)},$$

$$\langle E_k|\tilde{\rho}(\hat{\mathcal{H}}, V)|E_k\rangle = \frac{e^{-\beta(E_k+PV)}}{\Delta(N, P, T)}, \tag{10.4.8}$$

$$\Delta(N, P, T) = \int_0^\infty dV \ \text{Tr}\left[e^{-\beta(\hat{\mathcal{H}}+PV)}\right]$$

$$= \int_0^\infty dV \ \sum_k e^{-\beta(E_k+PV)}, \tag{10.4.9}$$

$$\langle \hat{A} \rangle = \frac{1}{\Delta(N, P, T)} \int_0^\infty dV \ \text{Tr}\left[\hat{A}e^{-\beta(\hat{\mathcal{H}}+PV)}\right]$$

$$= \frac{1}{\Delta(N, P, T)} \int_0^\infty dV \ \sum_k e^{-\beta(E_k+PV)}\langle E_k|\hat{A}|E_k\rangle. \tag{10.4.10}$$

Again, if there are degeneracies, then a factor of $g(E_k)$ must be introduced into the sums:

$$\Delta(N, P, T) = \int_0^\infty dV \ \sum_k g(E_k)e^{-\beta(E_k+PV)}$$

$$\langle \hat{A} \rangle = \frac{1}{\Delta(N, P, T)} \int_0^\infty dV \ \sum_k g(E_k)e^{-\beta(E_k+PV)}\langle E_k|\hat{A}|E_k\rangle. \tag{10.4.11}$$

Finally, for the grand canonical ensemble at temperature T and chemical potential μ, the density operator, partition function, and expectation value are given by

$$\tilde{\rho}(\hat{\mathcal{H}}, N) = \frac{e^{-\beta(\hat{\mathcal{H}}-\mu N)}}{\mathcal{Z}(\mu, V, T)},$$

$$\langle E_k|\tilde{\rho}(\hat{\mathcal{H}}, N)|E_k\rangle = \frac{e^{-\beta(E_k-\mu N)}}{\mathcal{Z}(\mu, V, T)}, \tag{10.4.12}$$

$$\mathcal{Z}(\mu, V, T) = \sum_{N=0}^\infty \text{Tr}\left[e^{-\beta(\hat{\mathcal{H}}-\mu N)}\right]$$

$$= \sum_{N=0}^\infty \sum_k e^{-\beta(E_k-\mu N)}, \tag{10.4.13}$$

$$\langle \hat{A} \rangle = \frac{1}{\mathcal{Z}(\mu, V, T)} \sum_{N=0}^{\infty} \mathrm{Tr} \left[\hat{A} e^{-\beta(\hat{\mathcal{H}} - \mu N)} \right]$$

$$= \frac{1}{\mathcal{Z}(\mu, V, T)} \sum_{N=0}^{\infty} \sum_{k} e^{-\beta(E_k - \mu N)} \langle E_k | \hat{A} | E_k \rangle. \tag{10.4.14}$$

As before, if there are degeneracies, then a factor of $g(E_k)$ must be introduced into the above sums:

$$\mathcal{Z}(\mu, V, T) = \sum_{N=0}^{\infty} \sum_{k} g(E_k) e^{-\beta(E_k - \mu N)}$$

$$\langle \hat{A} \rangle = \frac{1}{\mathcal{Z}(\mu, V, T)} \sum_{N=0}^{\infty} \sum_{k} g(E_k) e^{-\beta(E_k - \mu N)} \langle E_k | \hat{A} | E_k \rangle. \tag{10.4.15}$$

The quantum grand canonical ensemble will prove particularly useful in our treatment of the quantum ideal gases to be discussed in Chapter 11.

In the above list of definitions, a definition of the quantum microcanonical ensemble is conspicuously missing for the reason that it is very rarely used for condensed-phase systems. Moreover, in order to define this ensemble, the quantum-classical correspondence must be applied carefully because the eigenvalues of $\hat{\mathcal{H}}$ are assumed to be discrete. Hence, the δ-function used in the classical microcanonical ensemble does not make sense for quantum systems because a given eigenvalue may or may not be equal to the energy E used to define the ensemble. However, if we define an energy shell between E and $E + \Delta E$, then we can certainly find a subset of energy eigenvalues in this shell. The partition function is then related to the number of energy levels E_k satisfying $E < E_k < E + \Delta E$. Typically, when we take the thermodynamic limit of a system, the energy levels become very closely spaced, and we can shrink the thickness ΔE of the shell to zero.

10.4.1 Canonical ensemble via entropy maximization

In this subsection, we show how to derive the quantum canonical ensemble using a quantum mechanical version of the entropy maximization derivation presented in Section 4.4. Derivations of the quantum isothermal-isobaric and grand-canonical ensembles via entropy maximization are left as exercises for the reader (see Problem 10.2).

The idea is to express the entropy S using the density matrix eigenvalues w_k and then show that the form that maximizes S is $w_k = \exp(-\beta E_k)/Q(N, V, T)$. Treating the eigenvalues as discrete, the quantum mechanical analog of the Shannon entropy expression in eqn. (4.4.1) is

$$S = -k \sum_{k} w_k \ln w_k \tag{10.4.16}$$

(be careful not to confuse the leading Boltzmann's constant k with the summation index k). We seek to find the form of w_k that maximizes S. However, as in Section 4.4, the maximization must be performed subject to certain constraints. First, since the density matrix $\tilde{\rho}$ has unit trace, we need $\sum_k w_k = 1$. In addition, we need the ensemble average of the Hamiltonian $\hat{\mathcal{H}}$ to yield a well-defined energy E: $\langle \hat{\mathcal{H}} \rangle = E$. This gives the second constraint $\sum_k w_k E_k = E$. These two constraints are introduced with Lagrange multipliers μ and λ, respectively, giving us the modified entropy expression

$$\tilde{S} = -k \sum_k w_k \ln w_k - \mu \left(\sum_k w_k - 1 \right) - \lambda \left(\sum_k w_k E_k - E \right). \tag{10.4.17}$$

In contrast to Section 4.4, which requires a functional derivative, in the quantum context, the discreteness of the eigenvalues allows us to express the entropy maximization condition as a simple partial derivative

$$\frac{\partial \tilde{S}}{\partial w_i} = 0, \tag{10.4.18}$$

which leads to the condition

$$-k \left[\ln w_i + 1 \right] - \mu - \lambda E_i = 0. \tag{10.4.19}$$

Solving for w_i yields

$$w_i = e^{-(1+\mu/k)} e^{-\lambda E_i/k}. \tag{10.4.20}$$

The multiplier λ is determined by the condition

$$\frac{\partial \tilde{S}}{\partial E} = \frac{1}{T}, \tag{10.4.21}$$

which gives $\lambda = 1/T$. The normalization condition $\sum_i w_i = 1$, together with the definition $\sum_i \exp(-\beta E_i) = Q(N, V, T)$, then gives the multiplier μ as

$$e^{-(1+\mu/k)} = \frac{1}{Q(N, V, T)}, \tag{10.4.22}$$

so that $w_i = \exp(-\beta E_i)/Q(N, V, T)$, as expected.

10.4.2 Example: The canonical harmonic oscillator

In order to illustrate the application of a quantum equilibrium ensemble, we consider the case of a simple one-dimensional harmonic oscillator of frequency ω. We will derive the properties of this system using the canonical ensemble. Recall from Section 9.3 that the energy eigenvalues are given by

$$E_n = \left(n + \frac{1}{2} \right) \hbar\omega \qquad n = 0, 1, 2, ... \tag{10.4.23}$$

The canonical partition function is, therefore,

$$Q(\beta) = \sum_{n=0}^{\infty} e^{-\beta E_n} = \sum_{n=0}^{\infty} e^{-\beta(n+1/2)\hbar\omega}. \tag{10.4.24}$$

Recalling that the sum of a geometric series is given by

$$\sum_{n=0}^{\infty} r^n = \frac{1}{1-r},$$

(10.4.25)

where $0 < r < 1$, the partition function becomes

$$Q(\beta) = e^{-\beta\hbar\omega/2} \sum_{n=0}^{\infty} e^{-n\beta\hbar\omega} = e^{-\beta\hbar\omega/2} \sum_{n=0}^{\infty} \left(e^{-\beta\hbar\omega}\right)^n = \frac{e^{-\beta\hbar\omega/2}}{1 - e^{-\beta\hbar\omega}}.$$

(10.4.26)

From the partition function, various thermodynamic quantities can be determined. First, the free energy is given by

$$A = -\frac{1}{\beta} \ln Q(\beta) = \frac{\hbar\omega}{2} + \frac{1}{\beta} \ln\left(1 - e^{-\beta\hbar\omega}\right),$$

(10.4.27)

while the total energy is

$$E = -\frac{\partial}{\partial\beta} \ln Q(\beta) = \frac{\hbar\omega}{2} + \frac{\hbar\omega e^{-\beta\hbar\omega}}{1 - e^{-\beta\hbar\omega}} = \left(\frac{1}{2} + \langle n \rangle\right)\hbar\omega.$$

(10.4.28)

Thus, even if $\langle n \rangle = 0$, there is still a finite amount of energy, $\hbar\omega/2$ in the system. This residual energy is known as the *zero-point energy*. Next, from the average energy, the heat capacity can be determined

$$\frac{C}{k} = \frac{(\beta\hbar\omega)^2 e^{-\beta\hbar\omega}}{(1 - e^{-\beta\hbar\omega})^2}.$$

(10.4.29)

Note that as $\hbar \to 0$, $C/k \to 1$ in agreement with the classical result. Finally, the entropy is given by

$$S = k \ln Q(\beta) + \frac{E}{T} = -k \ln\left(1 - e^{-\beta\hbar\omega}\right) + \frac{\hbar\omega}{T} \frac{e^{-\beta\hbar\omega}}{1 - e^{-\beta\hbar\omega}},$$

(10.4.30)

which is consistent with the third law of thermodynamics, as $S \to 0$ as $T \to 0$.

The expressions we have derived for the thermodynamic observables are often used to estimate thermodynamic quantities of molecular systems under the assumption that the system can be approximately decomposed into a set of uncoupled harmonic oscillators corresponding to the normal modes of Section 1.7. By summing the expressions in eqns. (10.4.27), (10.4.29), or (10.4.30) over a set of frequencies generated in a normal-mode calculation, estimates of the quantum thermodynamic properties such as free energy, heat capacity, and entropy, can be easily obtained.

As a concluding remark, we note that the formulation of the quantum equilibrium ensembles in terms of the eigenvalues and eigenvectors of $\hat{\mathcal{H}}$ suggests that the computational problems inherent in many-body quantum mechanics have not been alleviated. After all, one still needs to solve the eigenvalue problem for the Hamiltonian, which involves solution eqn. (10.1.5). In Section 10.1, we described the difficulty inherent in this approach. The eigenvalue equation can be solved explicitly only for systems with a very small number of degrees of freedom. Looking ahead, in Chapter 12 we will develop a framework, known as the Feynman path integral formulation of statistical mechanics, that allows the calculation of N-particle eigenvalues to be circumvented, thereby allowing quantum equilibrium properties of large condensed-phase systems to be evaluated using molecular dynamics and Monte Carlo methods. Before exploring this formalism, however, we will use the traditional eigenvalue approach to study the quantum ideal gases, the subject of the next chapter.

10.5   Problems

10.1. a. Prove that the trace of a matrix A is independent of the basis in which the trace is performed.
   b. Prove the cyclic property of the trace

$$\text{Tr}(\text{ABC}) = \text{Tr}(\text{CAB}) = \text{Tr}(\text{BCA}).$$

10.2. Use the quantum mechanical formulation of the Shannon entropy to derive density matrix eigenvalues for the quantum isothermal-isobaric and grand-canonical ensembles by maximizing the entropy subject to appropriate constraints. Use the derivations in Sections 5.4 and 6.5 as guides for how to set up the appropriate modified entropy expressions.

10.3. Recall from Problem 9.2 that the energy of a quantum particle with magnetic moment μ in a magnetic field $\mathbf{B}$ is $E = -\mu \cdot \mathbf{B}$. Consider a spin-1/2 particle such as an electron fixed in space in a uniform magnetic field in the z direction, so that $\mathbf{B} = (0, 0, B)$. The Hamiltonian for the particle is given by

$$\hat{\mathcal{H}} = -\gamma B \hat{S}_z.$$

The spin operators are given in eqn. (9.4.2).
   a. Suppose an ensemble of such systems is prepared such that the density matrix initially is

$$\tilde{\rho}(0) = \begin{pmatrix} 1/2 & 0 \\ 0 & 1/2 \end{pmatrix}.$$

Calculate $\tilde{\rho}(t)$.

   b. What are the expectation values of the operators $\hat{S}_x$, $\hat{S}_y$, and $\hat{S}_z$ at any time t?

   c. Suppose now that the initial density matrix is

$$\tilde{\rho}(0) = \begin{pmatrix} 1/2 & -i/2 \\ i/2 & 1/2 \end{pmatrix}.$$

For this case, calculate $\tilde{\rho}(t)$.

d. What are the expectation values of the operators $\hat{S}_x$, $\hat{S}_y$, and $\hat{S}_z$ at time t for this case?

e. What is the fluctuation or uncertainty in $\hat{S}_x$ at time t? Recall that

$$\Delta \hat{S}_x = \sqrt{\langle \hat{S}_x^2 \rangle - \langle \hat{S}_x \rangle^2}$$

f. Suppose finally that the density matrix is given initially by a canonical density matrix:

$$\tilde{\rho}(0) = \frac{e^{-\beta \hat{\mathcal{H}}}}{\mathrm{Tr}(e^{-\beta \hat{\mathcal{H}}})}$$

What is $\tilde{\rho}(t)$?

g. What are the expectation values of $\hat{S}_x$, $\hat{S}_y$, and $\hat{S}_z$ at time t?

10.4. Consider a one-dimensional quantum harmonic oscillator of frequency ω, for which the energy eigenvalues are

$$E_n = \left(n + \frac{1}{2} \right) \hbar\omega \qquad n = 0, 1, 2, \dots.$$

Using the canonical ensemble at temperature T, calculate $\langle \hat{x}^2 \rangle$, $\langle \hat{p}^2 \rangle$, and the uncertainties Δx and Δp.

Hint: Might the raising and lowering operators of Section 9.3 be useful?

*10.5. A weakly anharmonic oscillator of frequency ω has energy eigenvalues given by

$$E_n = \left(n + \frac{1}{2} \right) \hbar\omega - \kappa \left(n + \frac{1}{2} \right)^2 \hbar\omega \qquad n = 0, 1, 2, \dots.$$

Show that, to first order in κ and fourth order in $r = \beta\hbar\omega$, the heat capacity in the canonical ensemble is given by

$$\frac{C}{k} = \left[\left(1 - \frac{r^2}{12} + \frac{r^4}{240} \right) + 4\kappa \left(\frac{1}{r} + \frac{r^3}{80} \right) \right]$$

(Pathria, 1972).

10.6. Suppose a quantum system has degenerate eigenvalues.
   a. If $g(E_n)$ is the degeneracy of the energy level E_n, show that the expression for the canonical partition function must be modified to read

$$Q(N, V, T) = \sum_n g(E_n) e^{-\beta E_n}.$$

b. A harmonic oscillator of frequency ω in d dimensions has energy eigenvalues given by

$$E_n = \left(n + \frac{d}{2}\right)\hbar\omega,$$

but the energy levels become degenerate. The degeneracy of each level is

$$g(E_n) = \frac{(n+d-1)!}{n!(d-1)!}.$$

Calculate the canonical partition function, free energy, total energy, and heat capacity in this case.

10.7. The Hamiltonian for a free particle in one dimension is

$$\hat{\mathcal{H}} = \frac{\hat{p}^2}{2m}.$$

a. Using the free particle eigenfunctions, show that the canonical density matrix is given by

$$\langle x|e^{-\beta\hat{\mathcal{H}}}|x'\rangle = \left(\frac{m}{2\pi\beta\hbar^2}\right)^{1/2}\exp\left[-\frac{m}{2\beta\hbar^2}(x-x')^2\right]$$

b. Recall that an operator $\hat{A}$ in the Heisenberg picture evolves in time according to

$$\hat{A}(t) = e^{i\hat{\mathcal{H}}t/\hbar}\hat{A}e^{-i\hat{\mathcal{H}}t/\hbar}.$$

Now consider a transformation from real time t to an imaginary time variable τ via $t = -i\tau\hbar$. In imaginary time, the evolution of an operator becomes

$$\hat{A}(\tau) = e^{\tau\hat{\mathcal{H}}}\hat{A}e^{-\tau\hat{\mathcal{H}}}$$

Using this evolution, derive an expression for the imaginary-time mean-square displacement of a free particle defined to be

$$R^2(\tau) = \langle[\hat{x}(0) - \hat{x}(\tau)]^2\rangle.$$

Assume the particle is a one-dimensional box of length L. This function can be used to quantify the quantum delocalization of a particle at temperature T.

10.8. The following theorem is due to Peierls (1938): Let $\{|\phi_n\rangle\}$ be an arbitrary set of orthonormal functions on the Hilbert space of a quantum system whose Hamiltonian is $\hat{\mathcal{H}}$. The functions $\{|\phi_n\rangle\}$ are assumed to satisfy the same boundary and symmetry conditions of the physical system. It follows that the canonical partition function $Q(N,V,T)$ satisfies the inequality

$$Q(N,V,T) \geq \sum_n e^{-\beta\langle\phi_n|\hat{\mathcal{H}}|\phi_n\rangle},$$

where equality holds only if $\{|\phi_n\rangle\}$ are the eigenfunctions of $\hat{\mathcal{H}}$. Prove this theorem.

Hint: You might find the Ritz variational principle of quantum mechanics helpful. The Ritz principle states that for an arbitrary wave function $|\Psi\rangle$, the ground-state energy E_0 obeys the inequality

$$E_0 \leq \langle \Psi | \hat{\mathcal{H}} | \Psi \rangle$$

where equality only holds if $|\Psi\rangle$ is the ground state wave function of $\hat{\mathcal{H}}$.

10.9. Prove the following inequality: If A_1 and A_2 are the Helmholtz free energies for systems with Hamiltonians $\hat{\mathcal{H}}_1$ and $\hat{\mathcal{H}}_2$, respectively, then

$$A_1 \leq A_2 + \langle \hat{\mathcal{H}}_1 - \hat{\mathcal{H}}_2 \rangle_2$$

where $\langle \cdots \rangle_2$ indicates an ensemble average calculated with respect to the density matrix of system 2. This inequality is known as the *Gibbs-Bogliubov* inequality (Feynman, 1998).

10.10 The density matrix of an equilibrium ensemble can be thought of as arising from an imaginary time reformulation of the time-dependent Schrödinger equation for a system with a time-independent Hamiltonian $\hat{H}$.

a. Starting with the time-dependent Schrödinger equation for a state vector $|\Psi(t)\rangle$, change the time variable t to $\tau = it$, and show that the resulting imaginary-time Schrödinger equation is

$$\hat{H}|\Psi(\tau)\rangle = -\hbar \frac{\partial}{\partial \tau}|\Psi(\tau)\rangle.$$

b. Now suppose an ensemble of Z members is constructed of imaginary-time dependent states $|\Psi_\lambda(\tau)\rangle$ with a density operator

$$\tilde{\rho}(\tau) = \frac{1}{Z}\sum_{\lambda=1}^{Z} |\Psi(\tau)\rangle\langle\Psi(\tau)|.$$

Since this density operator describes an equilibrium ensemble, we will assume that $[\hat{H}, \tilde{\rho}(\tau)] = 0$, that is, $\tilde{\rho}$ commutes with the Hamiltonian. Under this condition, use the imaginary-time Schrödinger equation to show that

$$\hat{H}\tilde{\rho}(\tau) = -\frac{\hbar}{2}\frac{\partial}{\partial \tau}\tilde{\rho}(\tau).$$

c. Show that if $\tau = \beta\hbar/2$, where $\beta = 1/k_{\mathrm{B}}T$, then the equation for the density operator becomes the so-called *Bloch equation*

$$\hat{H}\tilde{\rho}(\beta) = -\frac{\partial}{\partial \beta}\tilde{\rho}(\beta).$$

d. Taking $\tilde{\rho}(\beta = 0) = \hat{I}$ to be the identity operator, show that the Bloch equation is formally solved by the canonical density matrix.

*10.11. A simple model of a one-dimensional classical polymer consists of assigning discrete energy states to different configurations of the polymer. Suppose the polymer consists of flat, elliptical disc-shaped molecules that can align either along their long axis (length $2a$) or short axis (length a). The energy of a monomer aligned along its short axis is higher by an amount ε so that the total energy of the molecule is $E = n\varepsilon$, where n is the number of monomers aligned along the short axis.

    a. Calculate the canonical partition function $Q(N,T)$ for such a polymer consisting of N monomers.

    b. What is the average length of the polymer?

11

Quantum ideal gases: Fermi-Dirac and Bose-Einstein statistics

11.1 Complexity without interactions

In Chapters 3 through 6, the classical ideal gas was used to illustrate how the tools of classical statistical mechanics are applied to a simple problem. The classical ideal gas was seen to be a relatively trivial system with an uninteresting phase diagram. The situation with the quantum ideal gas is dramatically different.

The symmetry conditions imposed on the wave function for a system of N non-interacting bosons or fermions lead to surprisingly rich and complex behavior that is purely quantum mechanical in origin. For bosonic systems, the ideal gas admits a fascinating effect known as *Bose-Einstein condensation*. From the fermionic ideal gas, we arrive at the notion of a *Fermi surface*. Moreover, many of the results derived for an ideal gas of fermions have been used to develop approximations to the electronic structure theory known as *density functional theory* (Hohenberg and Kohn, 1964; Kohn and Sham, 1965). Thus, a detailed treatment of the quantum ideal gases is instructive.

In this chapter, we will study the general problem of a quantum-mechanical ideal gas using the rules of quantum statistical mechanics developed in the previous chapter. Following this, we will specialize our treatment for the fermionic and bosonic cases, examine a number of important limits, and finally, derive the general concepts that emerge from these limits.

11.2 General formulation of the quantum-mechanical ideal gas

The Hamiltonian operator for an ideal gas of N identical particles is

$$\hat{\mathcal{H}} = \sum_{i=1}^{N} \frac{\hat{\mathbf{p}}_i^2}{2m}. \tag{11.2.1}$$

In order to compute the partition function, we must solve for the eigenvalues of this Hamiltonian. In so doing, we will also determine the N-particle eigenfunctions. The eigenvalue problem for the Hamiltonian in the coordinate basis reads

$$-\frac{\hbar^2}{2m} \sum_{i=1}^{N} \nabla_i^2 \Phi(\mathbf{x}_1, ..., \mathbf{x}_N) = E\Phi(\mathbf{x}_1, ..., \mathbf{x}_N), \tag{11.2.2}$$

where $\mathbf{x}_i$ is the combined coordinate and spin label $\mathbf{x}_i = (\mathbf{r}_i, s_i)$. The N-particle function $\Phi(\mathbf{x}_1, ..., \mathbf{x}_N)$ is the solution to eqn. (11.2.2) before any symmetry conditions are imposed. Since eqn. (11.2.2) is completely separable in the N-particle coordinate/spin labels $\mathbf{x}_1, ..., \mathbf{x}_N$, the Hamiltonian can be written as a sum of single-particle Hamiltonians:

$$\hat{\mathcal{H}} = \sum_{i=1}^{N} \hat{h}_i$$

$$\hat{h}_i = \frac{\hat{\mathbf{p}}_i^2}{2m}. \tag{11.2.3}$$

Moreover, since $\hat{\mathcal{H}}$ is independent of spin, the eigenfunctions must also be eigenfunctions of $\hat{S}^2$ and $\hat{S}_z$. Therefore, the unsymmetrized solution to eqn. (11.2.2) can be written as a product:

$$\Phi_{\alpha_1 m_1, ..., \alpha_N m_N}(\mathbf{x}_1, ..., \mathbf{x}_N) = \prod_{i=1}^{N} \phi_{\alpha_i m_i}(\mathbf{x}_i), \tag{11.2.4}$$

where $\phi_{\alpha_i m_i}(\mathbf{x}_i)$ is a single-particle wave function characterized by a set of spatial quantum numbers α_i and S_z eigenvalues m_i. The spatial quantum numbers α_i are chosen to characterize the spatial part of the eigenfunctions according to a set of observables that commute with the Hamiltonian. Each single-particle function $\phi_{\alpha_i m_i}(\mathbf{x}_i)$ can be further decomposed into a product of a spatial function $\psi_{\alpha_i}(\mathbf{r}_i)$ and a spin eigenfunction $\chi_{m_i}(s_i)$. The spin eigenfunctions are defined via components of the eigenvectors of $\hat{S}_z$ given in eqn. (9.4.3):

$$\chi_m(s) = \langle s | \chi_m \rangle = \delta_{ms}. \tag{11.2.5}$$

Thus, $\chi_{\hbar/2}(\hbar/2) = 1$, $\chi_{\hbar/2}(-\hbar/2) = 0$, and so forth. Substituting this ansatz into the wave equation yields a *single-particle wave equation*:

$$-\frac{\hbar^2}{2m} \nabla_i^2 \psi_{\alpha_i}(\mathbf{r}_i) = \varepsilon_{\alpha_i} \psi_{\alpha_i}(\mathbf{r}_i). \tag{11.2.6}$$

Here, ε_{α_i} is a single-particle energy eigenvalue, and the N-particle eigenvalues are just sums of these:

$$E_{\alpha_1, ..., \alpha_N} = \sum_{i=1}^{N} \varepsilon_{\alpha_i}. \tag{11.2.7}$$

Note that the single-particle wave equation is completely separable in x, y, and z. If we impose periodic boundary conditions in all three directions, then the solution of the wave equation is simply a product of one-dimensional wave functions of the form given in eqn. (9.3.14). The one-dimensional wave functions are characterized by integers $n_{x,i}$, $n_{y,i}$, and $n_{z,i}$ that arise from the quantization of momentum due to the periodicity of the box. These can be collected into a vector $\mathbf{n}_i = (n_{x,i}, n_{y,i}, n_{z,i})$ of integers, which leads to the following solution to eqn. (11.2.6):

$$\psi_{\mathbf{n}_i}(\mathbf{r}_i) = \left(\frac{1}{L}\right)^{3/2} \exp(2\pi i n_{x,i} x_i / L) \exp(2\pi i n_{y,i} y_i / L) \exp(2\pi i n_{z,i} z_i / L)$$

$$= \frac{1}{\sqrt{V}} \exp(2\pi i \mathbf{n}_i \cdot \mathbf{r}_i / L). \tag{11.2.8}$$

Similarly, each component of momentum is quantized, so that the momentum eigenvalues can be expressed as

$$\mathbf{p}_{\mathbf{n}_i} = \frac{2\pi \hbar}{L} \mathbf{n}_i, \tag{11.2.9}$$

and the energy eigenvalues in eqn. (11.2.6) are just sums of the energies in eqn. (9.3.11) over x, y, and z:

$$\varepsilon_{\mathbf{n}_i} = \frac{\mathbf{p}_{\mathbf{n}_i}^2}{2m} = \frac{2\pi^2 \hbar^2}{mL^2} |\mathbf{n}_i|^2. \tag{11.2.10}$$

Multiplying the functions in eqn. (11.2.8) by spin eigenfunctions, the complete single-particle eigenfunctions become

$$\langle \mathbf{x}_i | \mathbf{n}_i\, m_i \rangle = \phi_{\mathbf{n}_i m_i}(\mathbf{x}_i) = \frac{1}{\sqrt{V}} e^{2\pi i \mathbf{n}_i \cdot \mathbf{r}_i / L} \chi_{m_i}(s_i), \tag{11.2.11}$$

and the total energy eigenvalues are given by a sum over single-particle eigenvalues

$$E_{\mathbf{n}_1,\dots,\mathbf{n}_N} = \sum_{i=1}^{N} \frac{2\pi^2 \hbar^2}{mL^2} |\mathbf{n}_i|^2. \tag{11.2.12}$$

Finally, since the eigenvalue problem is separable, complete fermionic and bosonic wave functions can be constructed as follows. Begin by constructing a matrix

$$M = \begin{pmatrix} \phi_{\mathbf{n}_1,m_1}(\mathbf{x}_1) & \phi_{\mathbf{n}_2,m_2}(\mathbf{x}_1) & \cdots & \phi_{\mathbf{n}_N,m_N}(\mathbf{x}_1) \\ \phi_{\mathbf{n}_1,m_1}(\mathbf{x}_2) & \phi_{\mathbf{n}_2,m_2}(\mathbf{x}_2) & \cdots & \phi_{\mathbf{n}_N,m_N}(\mathbf{x}_2) \\ \cdot & \cdot & \cdots & \cdot \\ \cdot & \cdot & \cdots & \cdot \\ \cdot & \cdot & \cdots & \cdot \\ \phi_{\mathbf{n}_1,m_1}(\mathbf{x}_N) & \phi_{\mathbf{n}_2,m_2}(\mathbf{x}_N) & \cdots & \phi_{\mathbf{n}_N,m_N}(\mathbf{x}_N) \end{pmatrix}. \tag{11.2.13}$$

The properly symmetrized fermionic and bosonic wave functions are ultimately given by

$$\Psi^{(F)}_{\mathbf{n}_1,m_1,\dots,\mathbf{n}_N,m_N}(\mathbf{x}_1, \dots, \mathbf{x}_N) = \det(M)$$

$$\Psi^{(B)}_{\mathbf{n}_1,m_1,\dots,\mathbf{n}_N,m_N}(\mathbf{x}_1, \dots, \mathbf{x}_N) = \text{perm}(M), \tag{11.2.14}$$

where det and perm refer to the determinant and permanent of M, respectively. (The permanent of a matrix is just a determinant in which all the minus signs are changed to plus signs.[1]) In the fermion case, the determinant leads to a wave function that is

---

[1] The permanent of a 2×2 matrix

$$A = \begin{pmatrix} a & b \\ c & d \end{pmatrix}$$

is perm(A) = $ad + bc$.

completely antisymmetric with respect to an exchange of any two particle spin labels. Such an exchange is equivalent to interchanging two rows of the matrix M, which has the effect of changing the sign of the determinant. These determinants are known as *Slater determinants* after the physicist John C. Slater (1900–1976) who introduced the procedure.

In the preceding discussion, each individual particle was treated separately, with total energy eigenvalues expressed as sums of single-particle eigenvalues, and over-all wave functions given as determinants/permanents constructed from single-particle wave functions. We will now introduce an alternative framework for solving the quantum ideal-gas problem that proves more convenient for the quantum statistical mechanical treatment to follow. Let us consider again the single-particle eigenfunction and eigenvalue for a given vector of integers $\mathbf{n}$ and spin eigenvalue m:

$$\phi_{\mathbf{n},m}(\mathbf{x}) = \frac{1}{\sqrt{V}} e^{2\pi i \mathbf{n}\cdot\mathbf{r}/L} \chi_m(s)$$

$$\varepsilon_{\mathbf{n}} = \frac{2\pi^2\hbar^2}{mL^2} |\mathbf{n}|^2. \tag{11.2.15}$$

We now ask: How many particles in the N-particle system are described by this wave function and energy? Let this number be $f_{\mathbf{n}m}$, which is called an *occupation number*. The occupation number $f_{\mathbf{n}m}$ tells us how many particles have an energy $\varepsilon_{\mathbf{n}}$ and probability amplitude $\phi_{\mathbf{n},m}(\mathbf{x})$. Since there is an infinite number of accessible states $\phi_{\mathbf{n},m}(\mathbf{x})$ and associated energies $\varepsilon_{\mathbf{n}}$, there are infinitely many occupation numbers, and only a finite subset of these can be nonzero. Indeed, the occupation numbers are subject to the restriction that the sum over them yield the number of particles in the system:

$$\sum_m \sum_{\mathbf{n}} f_{\mathbf{n}m} = N, \tag{11.2.16}$$

where

$$\sum_{\mathbf{n}} \equiv \sum_{n_x=-\infty}^{\infty} \sum_{n_y=-\infty}^{\infty} \sum_{n_z=-\infty}^{\infty}, \tag{11.2.17}$$

and

$$\sum_m \equiv \sum_{m=-s}^{s} \tag{11.2.18}$$

runs over the $(2s+1)$ possible values of m for a spin-s particle. The occupation numbers can be used to characterize the total energy eigenvalues of the system. The total energy eigenvalue can be expressed as

$$E_{\{f_{\mathbf{n}m}\}} = \sum_m \sum_{\mathbf{n}} \varepsilon_{\mathbf{n}} f_{\mathbf{n}m}, \tag{11.2.19}$$

which is just a sum over all possible energies multiplied by the number of particles having each energy. The formulation of the eigenvalue problem in terms of accessible states $\phi_{\mathbf{n},m}(\mathbf{x})$, energy levels $\varepsilon_{\mathbf{n}}$, and occupation numbers for these states and energies is known as *second quantization*. The framework of second quantization leads to a simple and elegant procedure for constructing the partition function.

11.3 An ideal gas of distinguishable quantum particles

To illustrate the use of occupation numbers in the evaluation of the quantum partition function, let us suppose we can ignore the symmetry of the wave function under particle exchange. Neglect of spin statistics leads to an approximation known as *Boltzmann statistics*. Boltzmann statistics are equivalent to an assumption that the particles are distinguishable because the N-particle wave function for Boltzmann particles is just a simple product of the functions $\phi_{n_i m_i}(\mathbf{x}_i)$. In this case, spin can also be neglected. The canonical partition function $Q(N, V, T)$ can be expressed as a sum over the quantum numbers $\mathbf{n}_1, ..., \mathbf{n}_N$ for each particle:

$$Q(N, V, T) = \sum_{\mathbf{n}_1} \sum_{\mathbf{n}_2} \cdots \sum_{\mathbf{n}_N} e^{-\beta E_{\mathbf{n}_1, ..., \mathbf{n}_N}}$$

$$= \sum_{\mathbf{n}_1} \sum_{\mathbf{n}_2} \cdots \sum_{\mathbf{n}_N} e^{-\beta \varepsilon_{\mathbf{n}_1}} e^{-\beta \varepsilon_{\mathbf{n}_2}} \cdots e^{-\beta \varepsilon_{\mathbf{n}_N}}$$

$$= \left(\sum_{\mathbf{n}_1} e^{-\beta \varepsilon_{\mathbf{n}_1}} \right) \left(\sum_{\mathbf{n}_2} e^{-\beta \varepsilon_{\mathbf{n}_2}} \right) \cdots \left(\sum_{\mathbf{n}_N} e^{-\beta \varepsilon_{\mathbf{n}_N}} \right)$$

$$= \left(\sum_{\mathbf{n}} e^{-\beta \varepsilon_{\mathbf{n}_N}} \right)^N. \tag{11.3.1}$$

In terms of occupation numbers, the partition function is

$$Q(N, V, T) = \sum_{\{f\}} g(\{f_{\mathbf{n}}\}) e^{-\beta \sum_{\mathbf{n}} \varepsilon_{\mathbf{n}} f_{\mathbf{n}}}, \tag{11.3.2}$$

where $g(\{f_{\mathbf{n}}\})$ is a factor that tells how many distinct physical states can be represented by a given set of occupation numbers $\{f_{\mathbf{n}}\}$. For Boltzmann particles, exchanging the momentum labels $\mathbf{n}_i$ of two particles leads to a new physical state but leaves the occupation numbers unchanged. Thus, the counting problem becomes one of determining how many different ways N particles can be placed in the physical states. This means that $g(\{f_{\mathbf{n}}\})$ is given simply by the combinatorial factor

$$g(\{f_{\mathbf{n}}\}) = \frac{N!}{\prod_{\mathbf{n}} f_{\mathbf{n}}!}. \tag{11.3.3}$$

For example, if there were only two states, then the occupation numbers are f_1 and f_2 where $f_1 + f_2 = N$. The above formula gives

$$g(f_1, f_2) = \frac{N!}{f_1! f_2!} = \frac{N!}{f_1!(N - f_1)!}, \tag{11.3.4}$$

which is the expected binomial coefficient.

Substituting eqn. (11.3.3) into eqn. (11.3.2) gives

$$Q(N, V, T) = \sum_{\{f_{\mathbf{n}}\}} \frac{N!}{\prod_{\mathbf{n}} f_{\mathbf{n}}!} \prod_{\mathbf{n}} e^{-\beta f_{\mathbf{n}} \varepsilon_{\mathbf{n}}}, \tag{11.3.5}$$

which is just a multinomial expansion for

$$Q(N, V, T) = \left(\sum_{\mathbf{n}} e^{-\beta \varepsilon_{\mathbf{n}}} \right)^N. \tag{11.3.6}$$

Again, if there were two states, then the partition function would be

$$(e^{-\beta \varepsilon_1} + e^{-\beta \varepsilon_2})^N = \sum_{f_1, f_2, f_1 + f_2 = N} \frac{N!}{f_1! f_2!} e^{-f_1 \beta \varepsilon_1} e^{-f_2 \beta \varepsilon_2} \tag{11.3.7}$$

from the binomial theorem. Therefore, in order to evaluate the partition function, we just need to perform the sum

$$\sum_{\mathbf{n}} e^{-\beta \varepsilon_{\mathbf{n}}} = \sum_{\mathbf{n}} e^{-2\pi^2 \beta \hbar^2 |\mathbf{n}|^2 / mL^2}. \tag{11.3.8}$$

Ultimately, we are interested in the thermodynamic limit, where $L \to \infty$. In this limit, the spacing between the single-particle energy levels becomes quite small, and the discrete sum over $\mathbf{n}$ can, to a very good approximation, be replaced by an integral over a continuous variable (which we also denote as $\mathbf{n}$):

$$\sum_{\mathbf{n}} e^{-2\pi^2 \beta \hbar^2 |\mathbf{n}|^2 / mL^2} \longrightarrow \int d\mathbf{n} \, e^{-2\pi^2 \beta \hbar^2 |\mathbf{n}|^2 / mL^2}. \tag{11.3.9}$$

Since the single-particle eigenvalues only depend on the magnitude of $\mathbf{n}$, we can transform the integral over n_x, n_y, and n_z into spherical polar coordinates (n, θ, ϕ), where $n = |\mathbf{n}|$, and θ and ϕ retain their usual meaning. Thus, the integral becomes

$$4\pi \int_0^\infty dn \, n^2 e^{-2\pi^2 \beta \hbar^2 |\mathbf{n}|^2 / mL^2} = V \left(\frac{m}{2\pi \beta \hbar^2} \right)^{3/2} = \left(\frac{V}{\lambda^3} \right), \tag{11.3.10}$$

where λ is the thermal wavelength. The partition function now becomes

$$Q(N, V, T) = \left(\frac{V}{\lambda^3} \right)^N, \tag{11.3.11}$$

which is just the classical canonical partition function for an ideal gas. Therefore, we see that an ideal gas of distinguishable particles, even when treated quantum mechanically, has precisely the same properties as a classical ideal gas. Thus, we conclude that all the quantum effects are contained in the particle spin statistics, which we will now consider.

11.4 General formulation for fermions and bosons

For systems of identical fermions or identical bosons, an exchange of particles does not change the physical state. Therefore, the factor $g(\{f_{\mathbf{nm}}\})$ is simply 1 for either particle type. For fermions, the Pauli exclusion principle forbids two identical particles from having the same set of quantum numbers. Note that the Slater determinant vanishes if, for any two particles i and j, $\mathbf{n}_i = \mathbf{n}_j$ and $m_i = m_j$. In the second quantization formalism, this means that no two particles may occupy the same state $\phi_{\mathbf{n},m}(\mathbf{x})$. Consequently, the occupation numbers are restricted to be only 0 or 1:

$$f_{\mathbf{n}m} = 0, 1 \qquad \text{(Fermions)}. \tag{11.4.1}$$

By contrast, since a permanent does not vanish if $\mathbf{n}_i = \mathbf{n}_j$ and $m_i = m_j$, the occupation numbers $f_{\mathbf{n}m}$ for a system of identical bosons have no such restriction and can, therefore, take on any value between 0 and N:

$$f_{\mathbf{n}m} = 0, 1, 2, ..., N \qquad \text{(Bosons)}. \tag{11.4.2}$$

For either set of occupation numbers, the canonical partition function can be written generally as

$$Q(N, V, T) = \sum_{\{f_{nm}\}} e^{-\beta \sum_m \sum_{\mathbf{n}} f_{\mathbf{n}m} \varepsilon_{\mathbf{n}}} = \sum_{\{f_{nm}\}} \prod_{\mathbf{n}} \prod_m e^{-\beta f_{\mathbf{n}m} \varepsilon_{\mathbf{n}}}. \tag{11.4.3}$$

Note that the sum over occupation numbers in eqn. (11.4.3) must be performed subject to the restriction

$$\sum_m \sum_{\mathbf{n}} f_{\mathbf{n}m} = N. \tag{11.4.4}$$

This restriction makes performing the sum in eqn. (11.4.3) nontrivial when $g(\{f_{\mathbf{n}m}\}) = 1$. Evidently, the canonical ensemble is not the most convenient choice for deriving the thermodynamics of bosonic or fermionic ideal gases.

Fortunately, since all ensembles are equivalent in the thermodynamic limit, we may choose from any of the other remaining ensembles. Of these, we will see shortly that working in the grand canonical ensemble makes our task considerably easier. Recall that in the grand canonical ensemble, μ, V, and T are the control variables, and the partition function is given by

$$\mathcal{Z}(\mu, V, T) = \sum_{N=0}^{\infty} \zeta^N Q(N, V, T)$$

$$= \sum_{N=0}^{\infty} e^{\beta \mu N} \sum_{\{f_{nm}\}} \prod_m \prod_{\mathbf{n}} e^{-\beta f_{\mathbf{n}m} \varepsilon_{\mathbf{n}}}. \tag{11.4.5}$$

Note that the inner sum in eqn. (11.4.5) over occupation numbers is still subject to the restriction $\sum_m \sum_{\mathbf{n}} f_{\mathbf{n}m} = N$. However, in the grand canonical ensemble, there is a final sum over all possible values of N, and this sum allows us to lift the restriction on the inner sum. The final sum over N combined with the restricted sum over occupation numbers is mathematically equivalent to an unrestricted sum over occupation numbers. For if we simply perform an unrestricted sum over occupation numbers, then all possible values of N will be generated automatically. Thus, we can see why the grand canonical ensemble is preferable for fermions and bosons. The grand canonical partition function can be written compactly as

$$\mathcal{Z}(\mu, V, T) = \sum_{\{f_{nm}\}} \prod_m \prod_n e^{\beta(\mu - \varepsilon n)f_{nm}}. \tag{11.4.6}$$

A second simplification results from rewriting the sum of products as a product of sums:

$$\sum_{f_1} \sum_{f_2} \sum_{f_3} \cdots e^{\beta(\mu - \varepsilon_1)f_1} e^{\beta(\mu - \varepsilon_2)f_2} e^{\beta(\mu - \varepsilon_3)f_3} \cdots$$

$$= \left(\sum_{f_1} e^{\beta(\mu - \varepsilon_1)f_1} \right) \left(\sum_{f_2} e^{\beta(\mu - \varepsilon_1)f_2} \right) \left(\sum_{f_3} e^{\beta(\mu - \varepsilon_1)f_3} \right) \cdots$$

$$= \prod_m \prod_n \sum_{\{f_{nm}\}} e^{\beta(\mu - \varepsilon n)f_{nm}}. \tag{11.4.7}$$

For fermions, each occupation-number sum contains only two terms corresponding to $f_{nm} = 0$ and $f_{nm} = 1$, which yields

$$\mathcal{Z}(\mu, V, T) = \prod_m \prod_n \left(1 + e^{\beta(\mu - \varepsilon n)} \right) \qquad \text{(Fermions)}. \tag{11.4.8}$$

For bosons, each occupation-number sum ranges from 0 to ∞ and can be computed using the sum formula for a geometric series $\sum_{n=0}^{\infty} r^n = 1/(1-r)$ for $0 < r < 1$. Thus, eqn. (11.4.7) becomes

$$\mathcal{Z}(\mu, V, T) = \prod_m \prod_n \frac{1}{1 - e^{\beta(\mu - \varepsilon n)}} \qquad \text{(Bosons)}. \tag{11.4.9}$$

Note that, in each case, the summands are independent of the quantum number m so that we may perform the product over m values trivially with the result

$$\mathcal{Z}(\mu, V, T) = \left[\prod_n \left(1 + e^{\beta(\mu - \varepsilon n)} \right) \right]^g \tag{11.4.10}$$

for fermions, and

$$\mathcal{Z}(\mu, V, T) = \left[\prod_n \frac{1}{1 - e^{\beta(\mu - \varepsilon n)}} \right]^g \tag{11.4.11}$$

for bosons, where $g = (2s + 1)$ is the number of eigenstates of $\hat{S}_z$, which is also known as the *spin degeneracy*. For spin-1/2 particles such as electrons, $g = 2$.

At this point, let us recall the procedure for calculating the equation of state in the grand canonical ensemble. The free energy in this ensemble is $-PV$, and

$$\frac{PV}{kT} = \ln \mathcal{Z}(\zeta, V, T). \tag{11.4.12}$$

Moreover, the average particle number is the thermodynamic derivative with respect to the fugacity ζ:

$$\langle N \rangle = \zeta \frac{\partial}{\partial \zeta} \ln \mathcal{Z}(\zeta, V, T). \tag{11.4.13}$$

Next, the fugacity ζ must be eliminated in favor of $\langle N \rangle$ by solving for ζ in terms of $\langle N \rangle$ and substituting into eqn. (11.4.12). Thus, in order to obtain the equation of state in the grand canonical ensemble, we must carry out the products in eqn. (11.4.10) and then apply the above procedure. Although we saw in Section 6.6 that this is straightforward for the classical ideal gas, the procedure cannot be performed exactly analytically for the quantum ideal gases. For an ideal gas of identical fermions, the equations we must solve are

$$\frac{PV}{kT} = \ln \mathcal{Z}(\zeta, V, T) = \ln \left[\prod_{\mathbf{n}} \left(1 + \zeta e^{-\beta \varepsilon_{\mathbf{n}}} \right) \right]^g = g \sum_{\mathbf{n}} \ln \left(1 + \zeta e^{-\beta \varepsilon_{\mathbf{n}}} \right)$$

$$\langle N \rangle = \zeta \frac{\partial}{\partial \zeta} \ln \mathcal{Z} = g \sum_{\mathbf{n}} \frac{\zeta e^{-\beta \varepsilon_{\mathbf{n}}}}{1 + \zeta e^{-\beta \varepsilon_{\mathbf{n}}}}, \tag{11.4.14}$$

and for bosons, they become

$$\frac{PV}{kT} = \ln \mathcal{Z}(\zeta, V, T) = \ln \left[\prod_{\mathbf{n}} \frac{1}{1 - \zeta e^{-\beta \varepsilon_{\mathbf{n}}}} \right]^g = -g \sum_{\mathbf{n}} \ln \left(1 - \zeta e^{-\beta \varepsilon_{\mathbf{n}}} \right)$$

$$\langle N \rangle = \zeta \frac{\partial}{\partial \zeta} \ln \mathcal{Z} = g \sum_{\mathbf{n}} \frac{\zeta e^{-\beta \varepsilon_{\mathbf{n}}}}{1 - \zeta e^{-\beta \varepsilon_{\mathbf{n}}}}. \tag{11.4.15}$$

It is not difficult to see that the problem of solving for ζ in terms of $\langle N \rangle$ is nontrivial for both particle types. In the next two sections, we will analyze the ideal fermion and boson gases individually and investigate the limits and approximations that can be applied to compute their thermodynamic properties.

11.5 The ideal fermion gas

As we did for the ideal Boltzmann gas in Section 11.3, we will consider the thermodynamic limit $L \to \infty$ of the ideal fermion gas, so that the spacing between energy levels becomes small. Then the sums in eqns. (11.4.14) can be replaced by integrals over a continuous variable denoted $\mathbf{n}$. For the pressure, this replacement leads to

$$\frac{PV}{kT} = g \int d\mathbf{n} \, \ln\left(1 + \zeta e^{-\beta \varepsilon \mathbf{n}}\right)$$

$$= g \int d\mathbf{n} \, \ln\left(1 + \zeta e^{-2\pi^2 \beta \hbar^2 |\mathbf{n}|^2 / mL^2}\right)$$

$$= 4\pi g \int_0^\infty dn \, n^2 \ln\left(1 + \zeta e^{-2\pi^2 \beta \hbar^2 |\mathbf{n}|^2 / mL^2}\right), \tag{11.5.1}$$

where, in the last line, we have transformed to spherical polar coordinates. Next, we introduce a change of variables

$$x = \sqrt{\frac{2\pi^2 \beta \hbar^2}{mL^2}} n, \tag{11.5.2}$$

which gives

$$\frac{PV}{kT} = 4\pi g V \left(\frac{m}{2\pi^2 \beta \hbar^2}\right)^{3/2} \int_0^\infty dx \, x^2 \ln\left(1 + \zeta e^{-x^2}\right)$$

$$= \frac{4Vg}{\sqrt{\pi}\lambda^3} \int_0^\infty dx \, x^2 \ln\left(1 + \zeta e^{-x^2}\right). \tag{11.5.3}$$

The remaining integral can be evaluated by expanding the log in a power series and integrating the series term by term. Using the fact that

$$\ln(1 + y) = \sum_{l=1}^\infty (-1)^{l+1} \frac{y^l}{l}, \tag{11.5.4}$$

we obtain

$$\ln\left(1 + \zeta e^{-x^2}\right) = \sum_{l=1}^\infty \frac{(-1)^{l+1} \zeta^l}{l} e^{-lx^2}$$

$$\frac{PV}{kT} = \frac{4Vg}{\sqrt{\pi}\lambda^3} \sum_{l=1}^\infty \frac{(-1)^{l+1} \zeta^l}{l} \int_0^\infty dx \, x^2 e^{-lx^2}$$

$$= \frac{Vg}{\lambda^3} \sum_{l=1}^\infty \frac{(-1)^{l+1} \zeta^l}{l^{5/2}}. \tag{11.5.5}$$

In the same way, it can be shown that the average particle number $\langle N \rangle$ is given by the expression

$$\langle N \rangle = \frac{Vg}{\lambda^3} \sum_{l=1}^\infty \frac{(-1)^{l+1} \zeta^l}{l^{3/2}}. \tag{11.5.6}$$

Multiplying eqns. (11.5.5) and (11.5.6) by $1/V$, we obtain

$$\frac{P\lambda^3}{gkT} = \sum_{l=1}^{\infty} \frac{(-1)^{l+1}\zeta^l}{l^{5/2}}$$

$$\frac{\rho\lambda^3}{g} = \sum_{l=1}^{\infty} \frac{(-1)^{l+1}\zeta^l}{l^{3/2}}, \tag{11.5.7}$$

where $\rho = \langle N \rangle / V$ is the number density. Although we cannot solve these equations to obtain a closed form for the equation of state, two interesting limits can be worked out to a very good approximation, which we examine next.

11.5.1 The high-temperature, low-density limit

Solving for ζ as a function of $\langle N \rangle$ is equivalent to solving for ζ as a function of ρ. Hence, in the low-density limit, we can take an ansatz for $\zeta = \zeta(\rho)$ in the form of a power series:

$$\zeta(\rho) = a_1\rho + a_2\rho^2 + a_3\rho^3 + \cdots. \tag{11.5.8}$$

How rapidly this series converges depends on how low the density actually is. Writing out the first few terms in the pressure and density equations, we have

$$\frac{P\lambda^3}{gkT} = \zeta - \frac{\zeta^2}{2^{5/2}} + \frac{\zeta^3}{3^{5/2}} - \frac{\zeta^4}{4^{5/2}} + \cdots$$

$$\frac{\rho\lambda^3}{g} = \zeta - \frac{\zeta^2}{2^{3/2}} + \frac{\zeta^3}{3^{3/2}} - \frac{\zeta^4}{4^{3/2}} + \cdots. \tag{11.5.9}$$

Substituting eqn. (11.5.8) into eqns. (11.5.9) gives

$$\frac{\rho\lambda^3}{g} = (a_1\rho + a_2\rho^2 + a_3\rho^3 + \cdots) - \frac{1}{2^{3/2}}(a_1\rho + a_2\rho^2 + a_3\rho^3 + \cdots)^2$$

$$+ \frac{1}{3^{3/2}}(a_1\rho + a_2\rho^2 + a_3\rho^3 + \cdots)^3 + \cdots. \tag{11.5.10}$$

Equation (11.5.10) can be solved perturbatively, equating like powers of ρ on both sides. For example, if we work only to first order in ρ, then we have

$$\frac{\rho\lambda^3}{g} = a_1\rho \quad \Rightarrow \quad a_1 = \frac{\lambda^3}{g} \quad \Rightarrow \zeta \approx \frac{\lambda^3\rho}{g}. \tag{11.5.11}$$

When eqn. (11.5.11) is substituted into eqn. (11.5.9) for the pressure and only terms first order in the density are kept, we obtain

$$\frac{P\lambda^3}{gkT} = \frac{\rho\lambda^3}{g} \quad \Rightarrow \quad \frac{P}{kT} = \rho = \frac{\langle N \rangle}{V}, \tag{11.5.12}$$

which is just the classical ideal gas equation. If we now go out to second-order in ρ, eqn. (11.5.9) gives

$$\frac{\lambda^3 \rho}{g} = \frac{\lambda^3 \rho}{g} + a_2 \rho^2 - \frac{1}{2^{3/2}} \frac{\lambda^6 \rho^2}{g^2}, \tag{11.5.13}$$

or

$$a_2 = \frac{\lambda^6}{2^{3/2} g^2}, \tag{11.5.14}$$

from which

$$\zeta \approx \frac{\lambda^3 \rho}{g} + \frac{\lambda^6}{2^{3/2} g^2} \rho^2, \tag{11.5.15}$$

and the equation of state becomes

$$\frac{P}{kT} = \rho + \frac{\lambda^3}{2^{5/2} g} \rho^2. \tag{11.5.16}$$

From the equation of state, we can read off the second virial coefficient

$$B_2(T) = \frac{\lambda^3}{2^{5/2} g} \approx 0.1768 \frac{\lambda^3}{g} > 0. \tag{11.5.17}$$

Even at second-order, we observe a nontrivial quantum effect, in particular, a second virial coefficient with a nonzero value despite the absence of interactions among the particles. The implication of eqn. (11.5.17) is that there is an effective "interaction" among the particles as a result of the fermionic spin statistics. This "interaction" tends to increase the pressure above the classical ideal gas result $(B_2(T) > 0)$ and hence is repulsive in nature. This result is a consequence of the Pauli exclusion principle: If we imagine filling the energy levels, then since no two particles can occupy the same quantum state, once the ground state $\mathbf{n} = (0,0,0)$ is fully occupied by particles with different $\hat{S}_z$ eigenvalues, the next particle must go into a higher energy state. The result is an effective "repulsion" among the particles that pushes them into increasingly higher energy states so as not to violate the Pauli principle.

If the third-order contribution is worked out, one finds (see Problem 11.1) that

$$a_3 = \left(\frac{1}{4} - \frac{1}{3^{3/2}} \right) \frac{\lambda^9}{g^3}$$

$$\zeta = \frac{\lambda^3 \rho}{g} + \frac{\lambda^6}{2^{3/2} g^2} \rho^2 + \left(\frac{1}{4} - \frac{1}{3^{3/2}} \right) \frac{\lambda^9}{g^3} \rho^3$$

$$\frac{P}{kT} = \rho + \frac{\lambda^3}{2^{5/2} g} \rho^2 + \frac{\lambda^6}{g^2} \left(\frac{1}{8} - \frac{2}{3^{5/2}} \right) \rho^3, \tag{11.5.18}$$

so that $B_3(T) < 0$. Since the third-order term is a second-order correction to the ideal-gas equation of state, the fact that $B_3(T) < 0$ is consistent with time-independent perturbation theory, wherein the second-order correction lowers all of the energy levels.

11.5.2 The high-density, low-temperature limit

The high-density, low-temperature limit exhibits the largest departure from classical behavior. Operating with $\zeta(\partial/\partial\zeta)$ on eqn. (11.5.3), as indicated in eqn. (11.4.14), we obtain the following integral expression for the density:

$$\rho\lambda^3 = \frac{4g}{\sqrt{\pi}} \int_0^\infty \frac{x^2 dx}{\zeta^{-1} e^{x^2} + 1}. \tag{11.5.19}$$

Starting with this expression, we can derive an expansion in the inverse powers of $\ln\zeta \equiv \mu/kT$, as these inverse powers will become decreasingly small as $T \to 0$, allowing the leading order behavior to be deduced. We begin by introducing the variable

$$\nu = \ln\zeta = \frac{\mu}{kT} \tag{11.5.20}$$

and developing an expansion in its inverse powers. We will sketch out briefly how this is accomplished. We first introduce a change of variable $y = x^2$, from which $x = \sqrt{y}$ and $dx = dy/(2\sqrt{y})$. When this change is made in eqn. (11.5.19), we obtain

$$\rho\lambda^3 = \frac{2g}{\sqrt{\pi}} \int_0^\infty \frac{\sqrt{y}dy}{e^{y-\nu} + 1}. \tag{11.5.21}$$

The integral can be carried out by parts using

$$u = \frac{1}{e^{y-\nu} + 1}, \qquad du = -\frac{1}{(e^{y-\nu} + 1)^2} e^{y-\nu} dy$$

$$dv = y^{1/2} dy, \qquad v = \frac{2}{3} y^{3/2}, \tag{11.5.22}$$

which gives

$$\rho\lambda^3 = \frac{4g}{3\sqrt{\pi}} \int_0^\infty \frac{y^{3/2} e^{y-\nu} dy}{(e^{y-\nu} + 1)^2}. \tag{11.5.23}$$

Next, we expand $y^{3/2}$ about $y = \nu$:

$$y^{3/2} = \nu^{3/2} + \frac{3}{2}\nu^{1/2}(y - \nu) + \frac{3}{8}\nu^{-1/2}(y - \nu)^2 + \cdots. \tag{11.5.24}$$

This expansion is now substituted into eqn. (11.5.23) and the resulting integrals over y are performed, which yields

$$\rho\lambda^3 = \frac{4g}{3\sqrt{\pi}} \left[(\ln\zeta)^{3/2} + \frac{\pi^2}{8}(\ln\zeta)^{-1/2} + \cdots \right] + \mathcal{O}(1/\zeta), \tag{11.5.25}$$

where the fact that $\mu/kT \gg 1$ has been used for the low temperature limit. The high density limit implies a high chemical potential, which makes $\zeta(\rho) = e^{\beta\mu(\rho)}$ large as well. A large ζ also helps ensure the convergence of the series in eqn. (11.5.25), since the error falls off with powers of $1/\zeta$.

As $T \to 0$, $\zeta \to \infty$ and only the first term in the above expansion survives:

$$\rho \lambda^3 = \rho \left(\frac{2\pi\hbar^2}{mkT} \right)^{3/2} \approx \frac{4g}{3\sqrt{\pi}} \left(\ln \zeta \right)^{3/2} = \frac{4g}{3\sqrt{\pi}} \left(\frac{\mu}{kT} \right)^{3/2}. \tag{11.5.26}$$

According to the procedure of the grand canonical ensemble, we need to solve for ζ as a function of ρ or equivalently for μ as a function of ρ. From eqn. (11.5.26), we find

$$\mu = \frac{\hbar^2}{2m} \left(\frac{6\pi^2 \rho}{g} \right)^{2/3} \equiv \mu_0 = \varepsilon_F, \tag{11.5.27}$$

which is independent of T. The special value of the chemical potential $\mu_0 = \mu(T = 0)$ is known as the *Fermi energy*, ε_F. The Fermi energy plays an important role in characterizing free or quasi-free many-fermion systems such as metals and semiconductors. In order to shed more light on the physical significance of the Fermi energy, consider the expression for the average number of particles:

$$\langle N \rangle = \sum_m \sum_n \frac{\zeta e^{-\beta \varepsilon_n}}{1 + \zeta e^{-\beta \varepsilon_n}}. \tag{11.5.28}$$

However, recall that the occupation numbers must sum to the total number of particles in the system:

$$\sum_m \sum_n f_{nm} = N. \tag{11.5.29}$$

Thus, taking an average of both sides over the grand canonical ensemble, we obtain

$$\langle N \rangle = \sum_m \sum_n \langle f_{nm} \rangle. \tag{11.5.30}$$

Comparing eqns. (11.5.28) and (11.5.30), we can deduce that the average occupation number of a given state with quantum numbers $\mathbf{n}$ and m is

$$\langle f_{nm} \rangle = \frac{e^{-\beta(\varepsilon_n - \mu)}}{1 + e^{-\beta(\varepsilon_n - \mu)}} = \frac{1}{1 + e^{\beta(\varepsilon_n - \mu)}}. \tag{11.5.31}$$

Equation (11.5.31) gives the average occupancy of each quantum state in the ideal fermion gas and is known as the *Fermi-Dirac distribution function*. As $T \to 0$, $\beta \to \infty$, and $e^{\beta(\varepsilon_n - \mu_0)} \to \infty$ if $\varepsilon_n > \mu_0$, and $e^{\beta(\varepsilon_n - \mu_0)} \to 0$ if $\varepsilon_n < \mu_0$. Recognizing that $\mu_0 = \varepsilon_F$, we have the $T = 0$ result

$$\langle f_{nm} \rangle = \begin{cases} 0 & \text{for} \quad \varepsilon_n > \varepsilon_F \\ 1 & \text{for} \quad \varepsilon_n < \varepsilon_F \end{cases}. \tag{11.5.32}$$

That is, at zero temperature, the Fermi-Dirac distribution becomes a simple step function:

$$\langle f_{nm} \rangle = \theta(\varepsilon_F - \varepsilon_n). \tag{11.5.33}$$

A plot of the average occupation number versus ε_n at $T = 0$ is shown in Fig. 11.1.

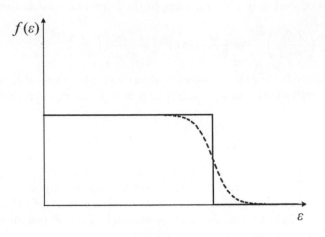

Fig. 11.1 The Fermi-Dirac distribution for $T = 0$ in eqn. (11.5.32) (solid line) and finite temperature using eqn. (11.5.31) (dashed line).

The implication of eqn. (11.5.33) is that at $T = 0$, the particles fill all of the available energy levels up to an energy value ε_F, above which all energy levels are unoccupied. Thus, ε_F represents a natural cutoff between occupied and unoccupied subspaces of energy levels. The highest occupied energy level must satisfy the condition $\varepsilon_\mathbf{n} = \varepsilon_F$, which implies

$$\frac{2\pi^2\hbar^2}{mL^2}|\mathbf{n}|^2 = \frac{2\pi^2\hbar^2}{mL^2}(n_x^2 + n_y^2 + n_z^2) = \varepsilon_F. \qquad (11.5.34)$$

Equation (11.5.34) defines a spherical surface in $\mathbf{n}$ space, which is known as the *Fermi surface*. Although the Fermi surface is a simple sphere for the ideal gas, for interacting systems the geometry of the Fermi surface will be considerably more complicated. In fact, characterizing the shape of a Fermi surface is an important component in the understanding of a wide variety of properties (thermal, electrical, optical, magnetic) of solid-state systems.

As T is increased, the probability of an excitation above the Fermi energy becomes nonzero, and on average, some of the energy levels above the Fermi energy will be occupied, leaving some of the energy levels below the Fermi energy vacant. This situation is represented with the dashed line in Fig. 11.1, which shows eqn. (11.5.31) for $T > 0$. The combination of a particle excitation to an energy level above ε_F and a depletion of an energy level below ε_F constitutes an "exciton-hole" pair. In real materials such as metals, an exciton-hole pair can also be created by bombarding the material with photons. The familiar concept of a *work function*—the energy needed to just remove an electron from one of the occupied energy levels—is closely related to the Fermi energy.

11.5.3 Zero-temperature thermodynamics

The fact that states of finite energy are occupied even at zero temperature in the fermion gas means that the thermodynamic properties at $T = 0$ are nontrivial. Consider, for example, the average particle number. In order to obtain an expression for this quantity, recall that

$$\langle N \rangle = \sum_m \sum_n \langle f_{nm} \rangle = \sum_m \sum_n \theta(\varepsilon_F - \varepsilon_n) = g \sum_n \theta(\varepsilon_F - \varepsilon_n). \qquad (11.5.35)$$

In the thermodynamic limit, the sum may be replaced by an integration in spherical polar coordinates

$$\langle N \rangle = g \int d\mathbf{n}\, \theta(\varepsilon_F - \varepsilon_n)$$

$$= 4\pi g \int_0^\infty dn\, n^2 \theta(\varepsilon_F - \varepsilon_n). \qquad (11.5.36)$$

However, since the energy eigenvalues are given by

$$\varepsilon_n = \frac{2\pi^2 \hbar^2}{mL^2} n^2, \qquad (11.5.37)$$

it proves useful to change variables of integration from n to ε using eqn. (11.5.37):

$$n = \left(\frac{mL^2}{2\pi^2 \hbar^2} \right)^{1/2} \varepsilon^{1/2}$$

$$dn = \frac{1}{2} \left(\frac{mL^2}{2\pi^2 \hbar^2} \right)^{1/2} \varepsilon^{-1/2} d\varepsilon. \qquad (11.5.38)$$

Inserting eqn. (11.5.38) into eqn. (11.5.36), we obtain

$$\langle N \rangle = 4\pi g \int_0^\infty dn\, n^2 \theta(\varepsilon_F - \varepsilon)$$

$$= 2\pi g \left(\frac{mL^2}{2\pi^2 \hbar^2} \right)^{3/2} \int_0^\infty d\varepsilon\, \varepsilon^{1/2} \theta(\varepsilon_F - \varepsilon)$$

$$= 2\pi g \left(\frac{mL^2}{2\pi^2 \hbar^2} \right)^{3/2} \int_0^{\varepsilon_F} d\varepsilon\, \varepsilon^{1/2}$$

$$\langle N \rangle = \frac{4\pi g}{3} \left(\frac{m}{2\pi \hbar^2} \right)^{3/2} V \varepsilon_F^{3/2}. \qquad (11.5.39)$$

By a similar procedure, we can obtain an expression for the average energy. Recall that the total energy for a given set of occupation numbers is given by

$$E_{\{f_n\}} = \sum_m \sum_n f_{nm} \varepsilon_n. \qquad (11.5.40)$$

Taking the ensemble average of both sides yields

$$\langle \hat{\mathcal{H}} \rangle = E = \sum_m \sum_n \langle f_{nm} \rangle \varepsilon_n. \tag{11.5.41}$$

At $T = 0$, this becomes

$$E = g \sum_n \theta(\varepsilon_F - \varepsilon_n) \varepsilon_n$$

$$\rightarrow g \int d\mathbf{n} \, \theta(\varepsilon_F - \varepsilon_n) \varepsilon_n$$

$$= 4\pi g \int_0^\infty dn \, n^2 \, \theta(\varepsilon_F - \varepsilon_n) \varepsilon_n, \tag{11.5.42}$$

where, as usual, we have replaced the sum by an integral and transformed to spherical polar coordinates. If the change of variables in Equation (11.5.38) is made, we find

$$E = 4\pi g \int_0^\infty d\varepsilon \, \frac{1}{2} \left(\frac{mL^2}{2\pi^2 \hbar^2} \right)^{3/2} \varepsilon^{3/2} \theta(\varepsilon_F - \varepsilon)$$

$$= 2\pi g \left(\frac{m}{2\pi^2 \hbar^2} \right)^{3/2} V \int_0^{\varepsilon_F} d\varepsilon \, \varepsilon^{3/2}$$

$$= \frac{4\pi g}{5} \left(\frac{m}{2\pi^2 \hbar^2} \right)^{3/2} V \varepsilon_F^{5/2}. \tag{11.5.43}$$

Combining eqns. (11.5.43) and (11.5.39), the following relation between E and $\langle N \rangle$ can be established:

$$E = \frac{3}{5} \langle N \rangle \varepsilon_F. \tag{11.5.44}$$

Moreover, since $\varepsilon_F \sim \rho^{2/3}$, we see that the total energy is related to the density ρ by

$$\frac{E}{V} = C_K \rho^{5/3}, \tag{11.5.45}$$

where C_K is an overall constant, $C_K = (3\hbar^2/10m)(6\pi^2/g)^{2/3}$. Note that if we perform a spatial integration on both sides of eqn. (11.5.45) over the containing volume, we obtain the total energy as

$$E = \int_{D(V)} d\mathbf{r} \, \frac{E}{V} = C_K \int_{D(V)} d\mathbf{r} \, \rho^{5/3} = V C_K \rho^{5/3}. \tag{11.5.46}$$

In one of the early theories of the electronic structure of multielectron atoms, the Thomas-Fermi theory, eqns. (11.5.45) and (11.5.46) were used to derive an expression for the electron kinetic energy. In a fermion ideal gas, the density ρ is constant, whereas in an interacting many-electron system, the density ρ varies in space and is, therefore, a function $\rho(\mathbf{r})$. A key assumption in the Thomas-Fermi theory is that in a multielectron atom, the spatial variation in $\rho(\mathbf{r})$ is mild enough that the kinetic energy can be approximated by replacing the constant ρ in eqn. (11.5.45) with $\rho(\mathbf{r})$ and then performing a spatial integration over both sides. The result is an approximate kinetic-energy functional given by

$$T[\rho] = C_K \int d\mathbf{r} \, \rho^{5/3}(\mathbf{r}). \tag{11.5.47}$$

Since the functional in eqn. (11.5.47) depends on the function $\rho(\mathbf{r})$, it is known as a *density functional*. In 1964, Hohenberg and Kohn proved that the total energy of a quantum multielectron system can be expressed as a unique functional $E[\rho]$ of the density $\rho(\mathbf{r})$ and that the minimum of this functional over the set of all densities $\rho(\mathbf{r})$ derivable from the set of all ground-state wave functions leads to the ground-state density of the particular system under consideration. The implication is that knowledge of the ground-state density $\rho_0(\mathbf{r})$ uniquely defines the quantum Hamiltonian of the system. This theorem has led to the development of the modern theory of electronic structure known as *density functional theory*, which has become one of the most widely used electronic structure methods. The Hohenberg-Kohn theorem amounts to an existence proof, since the exact form of the functional $E[\rho]$ is unknown. The kinetic energy functional in eqn. (11.5.47) is only an approximation to the exact kinetic-energy functional known as a *local density approximation* because the integrand of the functional depends only on one spatial point $\mathbf{r}$. Equation (11.5.47) is no longer used for actual applications because it, together with the rest of Thomas-Fermi theory, is unable to describe chemical bonding. In fact, Thomas-Fermi theory and its variants have been largely supplanted by the orbital-dependent version of density functional theory introduced by Kohn and Sham (1965); however, important recent work is showing that a nonlocal version of orbital-free density functional theory that builds on the Thomas-Fermi model can also yield accurate results in some cases (Huang and Carter, 2010). In Section 11.5.4 below, we will use our solution to the fermion ideal gas to derive another approximation commonly used in density functional theory, which is still used within the Kohn-Sham theory for certain classes of systems.

The pressure at $T = 0$ can now be obtained straightforwardly. We first recognize that the pressure is given by the sum in eqn. (11.5.5):

$$\frac{PV}{kT} = \frac{Vg}{\lambda^3} \sum_{l=1}^{\infty} \frac{(-1)^{l+1} \zeta^l}{l^{5/2}} = \ln \mathcal{Z}(\zeta, V, T). \tag{11.5.48}$$

Recall, however, the total energy can be obtained as a thermodynamic derivative of the partition function via

$$E = -\left(\frac{\partial}{\partial \beta} \ln \mathcal{Z}(\zeta, V, T)\right)_{\zeta, V}, \tag{11.5.49}$$

from which it follows that

$$E = \frac{3}{2\beta} \frac{Vg}{\lambda^3} \sum_{l=1}^{\infty} \frac{(-1)^{l+1} \zeta^l}{l^{5/2}}. \tag{11.5.50}$$

Comparing eqns. (11.5.48) and (11.5.50), we see that E and P are related by

$$E = \frac{3}{2}PV \quad \Rightarrow \quad P = \frac{2}{3}\frac{E}{V}.$$ (11.5.51)

Note that the energy and the pressure at $T = 0$ are not zero. The zero-temperature values of these quantities are

$$E = \frac{3}{5}\langle N \rangle \varepsilon_F$$

$$P = \frac{2}{5}\frac{\langle N \rangle}{V}\varepsilon_F.$$ (11.5.52)

These are referred to as the *zero-point* energy and pressure and are purely quantum mechanical in nature. The fact that the pressure does not vanish at $T = 0$ is again a consequence of the Pauli exclusion principle and the effective repulsive interaction that also appeared in the low density, high-temperature limit.

11.5.4   Derivation of the local density approximation

In Section 11.5.3, we referred to the local density approximation to density functional theory. In this section, we will derive the local density approximation to the exact exchange energy in density functional theory. The functional we will obtain is still used in many density functional calculations and serves as the basis for more sophisticated approximation schemes. The exact exchange energy is a component of the electronic structure method known as Hartree-Fock theory. It takes the form

$$E_x = -\frac{e^2}{4}\int \mathrm{d}\mathbf{r}\,\mathrm{d}\mathbf{r}'\,\frac{|\rho_1(\mathbf{r},\mathbf{r}')|^2}{|\mathbf{r}-\mathbf{r}'|},$$ (11.5.53)

where $\rho_1(\mathbf{r},\mathbf{r}')$ is known as the *one-particle density matrix*:

$$\rho_1(\mathbf{r},\mathbf{r}') = \sum_{s,s'}\sum_m\sum_n \langle f_{nm} \rangle \phi_{nm}(\mathbf{x})\phi_{nm}^*(\mathbf{x}').$$ (11.5.54)

Thus, for this calculation, we need both the energy levels and the corresponding eigenfunctions of the quantum ideal gas. We will show that for an ideal gas of electrons, the exchange energy is given exactly by

$$E_x = VC_x\rho^{4/3},$$ (11.5.55)

where

$$C_x = -\frac{3}{4}\left(\frac{3}{\pi}\right)^{1/3}e^2.$$ (11.5.56)

As we did for the kinetic energy, the volume factor in eqn. (11.5.55) can be written as an integral:

$$E_x = \int \mathrm{d}\mathbf{r}\,C_x\rho^{4/3}.$$ (11.5.57)

The local density approximation consists in replacing the constant density in eqn. (11.5.57) with the spatially varying density $\rho(\mathbf{r})$ of a system of interacting electrons. When this is done, we obtain the local density approximation to the exchange energy:

$$E_\mathrm{x} = \int \, d\mathbf{r} \, C_\mathrm{x} \rho^{4/3}(\mathbf{r}). \tag{11.5.58}$$

The remainder of this section will be devoted to the derivation of eqn. (11.5.55).

Since we are interested in the $T = 0$ limit, we will make use of the zero-temperature occupation numbers in eqn. (11.5.33), and we will assume that the fermions are electrons (spin-1/2) so that the spin degeneracy factor is $g = 2$. The first step in the derivation is to determine the one-particle density matrix using the eigenvalues and eigenfunctions in eqn. (11.2.15). Substituting these into eqn. (11.5.54) gives

$$\rho_1(\mathbf{r}, \mathbf{r}') = \frac{1}{V} \sum_{s,s'} \sum_m \sum_\mathbf{n} \chi_m(s) \chi_m(s') e^{2\pi i \mathbf{n} \cdot (\mathbf{r}-\mathbf{r}')/L} \theta(\varepsilon_\mathrm{F} - \varepsilon_\mathbf{n})$$

$$= \frac{1}{V} \sum_{s,s'} \sum_m \sum_\mathbf{n} \delta_{ms} \delta_{ms'} e^{2\pi i \mathbf{n} \cdot (\mathbf{r}-\mathbf{r}')/L} \theta(\varepsilon_\mathrm{F} - \varepsilon_\mathbf{n})$$

$$= \frac{2}{V} \sum_\mathbf{n} e^{2\pi i \mathbf{n} \cdot (\mathbf{r}-\mathbf{r}')/L} \theta(\varepsilon_\mathrm{F} - \varepsilon_\mathbf{n})$$

$$= \frac{2}{V} \int \, d\mathbf{n} \, e^{2\pi i \mathbf{n} \cdot (\mathbf{r}-\mathbf{r}')/L} \theta(\varepsilon_\mathrm{F} - \varepsilon_\mathbf{n}), \tag{11.5.59}$$

where in the last line, the summation has been replaced by integration, and the factor of 2 comes from the summation over spin states. At this point, notice that $\rho_1(\mathbf{r}, \mathbf{r}')$ does not depend on $\mathbf{r}$ and $\mathbf{r}'$ separately but only on the relative vector $\mathbf{s} = \mathbf{r} - \mathbf{r}'$. Thus, we can write the last line of eqn. (11.5.59) as

$$\rho_1(\mathbf{s}) = \frac{2}{V} \int \, d\mathbf{n} \, e^{2\pi \mathbf{n} \cdot \mathbf{s}/L} \theta(\varepsilon_\mathrm{F} - \varepsilon_\mathbf{n}). \tag{11.5.60}$$

The integral over $\mathbf{n}$ can be performed by orienting the $\mathbf{n}$ coordinate system such that the vector $\mathbf{s}$ lies along the n_z axis. Then, transforming to spherical polar coordinates in $\mathbf{n}$, we find that ρ_1 only depends on the magnitude $s = |\mathbf{s}|$ of $\mathbf{s}$ and the angle between $\mathbf{s}$ and $\mathbf{n}$:

$$\rho_1(s) = \frac{2}{V} \int_0^\infty \, dn \, n^2 \theta(\varepsilon_\mathrm{F} - \varepsilon_\mathbf{n}) \int_0^{2\pi} d\phi \int_0^\pi d\theta \, \sin\theta \, e^{2\pi n s \cos\theta/L}. \tag{11.5.61}$$

Performing the angular integrals, we obtain

$$\rho_1(s) = \frac{4\pi}{V} \int_0^\infty \, dn \, n^2 \, \theta(\varepsilon_\mathrm{F} - \varepsilon_\mathbf{n}) \frac{L}{2\pi i n s} \left(e^{2\pi i n s/L} - e^{-2\pi i n s/L} \right)$$

$$= \frac{4}{L^2 s} \int_0^\infty dn\, n\, \theta(\varepsilon_F - \varepsilon_n) \sin\left(\frac{2\pi n s}{L}\right). \tag{11.5.62}$$

For the remaining integral over n, transforming from n to ε_n is not convenient because of the sin function in the integrand. However, since $n > 0$, we recognize that the step function simply restricts the upper limit of the integral by the condition

$$\frac{2\pi^2 \hbar^2}{m L^2} n^2 < \varepsilon_F$$

$$n < \left(\frac{m L^2 \varepsilon_F}{2\pi^2 \hbar^2}\right)^{1/2} \equiv n_F. \tag{11.5.63}$$

Therefore,

$$\rho_1(s) = \frac{4}{L^2 s} \int_0^{n_F} dn\, n \sin\left(\frac{2\pi n s}{L}\right)$$

$$= \frac{1}{\pi^2 s^3} \left[\sin\left(\frac{2\pi n_F s}{L}\right) - \frac{s}{l_F} \cos\left(\frac{2\pi n_F s}{L}\right)\right], \tag{11.5.64}$$

where $l_F = \sqrt{\hbar^2 / 2m\varepsilon_F}$.

Given $\rho_1(s)$, we can now evaluate the exchange energy. First, we need to transform from integrations over $\mathbf{r}$ and $\mathbf{r}'$ to center-of-mass and relative coordinates

$$\mathbf{R} = \frac{1}{2}(\mathbf{r} + \mathbf{r}'), \qquad \mathbf{s} = \mathbf{r} - \mathbf{r}'. \tag{11.5.65}$$

This transformation yields for E_x:

$$E_x = -\frac{1}{4} \int d\mathbf{R}\, d\mathbf{s}\, \frac{\rho_1^2(s)}{s}. \tag{11.5.66}$$

Integrating over $\mathbf{R}$, transforming the $\mathbf{s}$ integral into spherical polar coordinates, and performing the angular part of the $\mathbf{s}$ integration gives

$$E_x = -\frac{V}{4} \int d\mathbf{s}\, \frac{\rho_1^2(s)}{s} = -\pi V \int_0^\infty ds\, s \rho_1^2(s)$$

$$= \frac{V}{\pi^3} \int_0^\infty \frac{ds}{s^5} \left[\sin(k_F s) - \frac{s}{l_F} \cos(k_F s)\right]^2, \tag{11.5.67}$$

where $k_F = 2\pi n_F / L$. If we now introduce the change of variables $x = k_F s$, we find that the expression separates into a density-dependent part and a purely numerical factor in the form of an integral:

$$E_x = -\frac{V}{\pi^3} k_F^4 \int_0^\infty dx\, \frac{(\sin x - x \cos x)^2}{x^5}. \tag{11.5.68}$$

Even without performing the remaining integral over x, we can see that $E_x \sim k_F^4$ and, therefore, $E_x \sim \rho^{4/3}$. However, the integral turns out to be straightforward to perform, despite its foreboding appearance. The trick (Parr and Yang, 1989) is to let $y = \sin x / x$. Then, it can be shown that

$$\frac{dy}{dx} = -\frac{\sin x - x \cos x}{x^2}$$

$$\frac{d^2 y}{dx^2} = -\frac{2}{x} \frac{dy}{dx} - y. \tag{11.5.69}$$

Finally,

$$\int_0^\infty dx \, \frac{(\sin x - x \cos x)^2}{x^5} = \int_0^\infty dx \frac{(\sin x - x \cos x)}{x^2} \frac{(\sin x - x \cos x)}{x^3}$$

$$= \int_0^\infty dx \left(\frac{dy}{dx}\right) \left(\frac{1}{x} \frac{dy}{dx}\right)$$

$$= -\frac{1}{2} \int_0^\infty dx \left(\frac{d^2 y}{dx^2} + y\right) \left(\frac{dy}{dx}\right)$$

$$= -\frac{1}{4} \int_0^\infty dx \, \frac{d}{dx} \left[y^2 + \left(\frac{dy}{dx}\right)^2 \right]$$

$$= -\frac{1}{4} \left[y^2 + \left(\frac{dy}{dx}\right)^2 \right] \Bigg|_0^\infty. \tag{11.5.70}$$

Both y and dy/dx vanish at $x = \infty$. In addition, by L'Hôpital's rule, dy/dx vanishes at $x = 0$. Thus, only the $\sin x / x$ term does not vanish at $x = 0$, and the result of the integral is simply $1/4$. Using the definitions of k_F and n_F, we ultimately find that

$$E_x = C_x V \rho^{4/3}, \tag{11.5.71}$$

which is the desired result.

11.5.5 Thermodynamics at finite temperature

At low but finite temperature, the Fermi-Dirac distribution appears as the dashed line in Fig. 11.1, which shows that small excitations above the Fermi surface are possible due to thermal fluctuations. These excitations will give rise to finite-temperature corrections to the thermodynamic quantities we derived above at $T = 0$. In order to see what is needed to compute thermodynamic properties at finite temperature, we consider the calculation of the average particle number using eqn. (11.5.30) but with the average occupation number given by eqn. (11.5.31). At finite temperature, the sum in eqn. (11.5.30) becomes

$$\langle N \rangle = \sum_m \sum_n \frac{1}{1 + e^{\beta(\varepsilon_n - \mu)}} = g \sum_n \frac{1}{1 + e^{\beta(\varepsilon_n - \mu)}}. \tag{11.5.72}$$

As was done in the zero-temperature case in Section 11.5.3, we assume the box is sufficiently large that we can replace the sum with an integral

$$\langle N \rangle = g \int d\mathbf{n} \, \frac{1}{1 + e^{\beta(\varepsilon_{\mathbf{n}} - \mu)}} = 4\pi g \int_0^\infty d\mathbf{n} \, n^2 \frac{1}{1 + e^{\beta(\varepsilon_{\mathbf{n}} - \mu)}}, \tag{11.5.73}$$

and we introduce the change of variables in eqn. (11.5.38) to give

$$\langle N \rangle = 2\pi g V \left(\frac{m}{2\pi^2 \hbar^2} \right)^{3/2} \int_0^\infty \frac{\varepsilon^{1/2} d\varepsilon}{1 + e^{\beta(\varepsilon - \mu)}}. \tag{11.5.74}$$

A similar derivation shows that the energy is given by

$$E = \sum_m \sum_n \frac{\varepsilon_{\mathbf{n}}}{1 + e^{\beta(\varepsilon - \mu)}}$$

$$= 2\pi g V \left(\frac{m}{2\pi^2 \hbar^2} \right)^{3/2} \int_0^\infty \frac{\varepsilon^{3/2} d\varepsilon}{1 + e^{\beta(\varepsilon - \mu)}}. \tag{11.5.75}$$

The integrals in eqns. (11.5.74) and (11.5.75) are nontrivial; however, they can be evaluated approximately in the low-temperature limit where we expect only small corrections to the zero-temperature results.

In order to proceed, we introduce a change of variables $\varepsilon = \mu + kTx$. The integrals for $\langle N \rangle$ and E then become

$$\langle N \rangle = 2\pi g V (kT)^{3/2} \left(\frac{m}{2\pi^2 \hbar^2} \right)^{3/2} \int_{-\beta\mu}^\infty \frac{(x + \beta\mu)^{1/2} \, dx}{1 + e^x}$$

$$E = 2\pi g V (kT)^{5/2} \left(\frac{m}{2\pi^2 \hbar^2} \right)^{3/2} \int_{-\beta\mu}^\infty \frac{(x + \beta\mu)^{3/2} \, dx}{1 + e^x}. \tag{11.5.76}$$

The integrals in these expressions are of the more general form

$$I = \int_{-\beta\mu}^\infty \frac{f(x + \beta\mu) \, dx}{1 + e^x}, \tag{11.5.77}$$

where $f(y) = y^{1/2}$ for $\langle N \rangle$ and $f(y) = y^{3/2}$ for E. The integral can be simplified by writing it as two integrals according to

$$I = \int_{-\beta\mu}^{0} \frac{f\left(x + \beta\mu\right) dx}{1 + e^{x}} + \int_{0}^{\infty} \frac{f\left(x + \beta\mu\right) dx}{1 + e^{x}}$$

$$= \int_{0}^{\beta\mu} \frac{f\left(-x + \beta\mu\right) dx}{1 + e^{-x}} + \int_{0}^{\infty} \frac{f\left(x + \beta\mu\right) dx}{1 + e^{x}}$$

$$= \int_{0}^{\beta\mu} f\left(-x + \beta\mu\right) dx - \int_{0}^{\beta\mu} \frac{f\left(-x + \beta\mu\right) dx}{1 + e^{x}} + \int_{0}^{\infty} \frac{f\left(x + \beta\mu\right) dx}{1 + e^{x}}$$

$$= \int_{0}^{\beta\mu} f(y) dy - \int_{0}^{\beta\mu} \frac{f\left(-x + \beta\mu\right) dx}{1 + e^{x}} + \int_{0}^{\infty} \frac{f\left(x + \beta\mu\right) dx}{1 + e^{x}}. \tag{11.5.78}$$

In the second line, the transformation $x \to -x$ was made; in the third line, the identity $1/(1 + e^{-x}) = 1 - 1/(1 + e^{x})$ was used; in the fourth line, we set $y = -x + \beta\mu$.

Until now, everything we have done is exact, and, in fact, the first term on the fourth line of eqn. (11.5.78) can be evaluated analytically to yield $(2/3)(\beta\mu)^{3/2}$ when $f(y) = y^{1/2}$ and $(2/5)(\beta\mu)^{5/2}$ when $f(y) = y^{3/2}$. However, the second two integrals on the fourth line of eqn. (11.5.78) can be simplified by working in the low-temperature limit. In this limit, we can let $\beta\mu \to \infty$, and we obtain

$$I \approx \int_{0}^{\beta\mu} f(y) dy + \int_{0}^{\infty} \frac{f\left(x + \beta\mu\right) - f\left(-x + \beta\mu\right)}{1 + e^{x}} dx. \tag{11.5.79}$$

Since, in the low-temperature limit, the most important contributions to the second integral are for $x \ll \beta\mu$, we can expand $f(x + \beta\mu)$ and $f(-x + \beta\mu)$ about $x = 0$:

$$f\left(x + \beta\mu\right) = f(\beta\mu) + f'(\beta\mu)x + \frac{1}{2}f''(\beta\mu)x^{2} + \frac{1}{6}f'''(\beta\mu)x^{3} + \cdots$$

$$f\left(-x + \beta\mu\right) = f(\beta\mu) - f'(\beta\mu)x + \frac{1}{2}f''(\beta\mu)x^{2} - \frac{1}{6}f'''(\beta\mu)x^{3} + \cdots$$

$$f\left(x + \beta\mu\right) - f\left(-x + \beta\mu\right) = 2f'(\beta\mu)x + \frac{1}{3}f'''(\beta\mu)x^{3} + \cdots \tag{11.5.80}$$

With this approximation, the integral I becomes

$$I = \int_{0}^{\beta\mu} f(y)\, dy + 2f'(\beta\mu) \int_{0}^{\infty} \frac{x\, dx}{1 + e^{x}} + \frac{1}{3}f'''(\beta\mu) \int_{0}^{\infty} \frac{x^{3}\, dx}{1 + e^{x}}. \tag{11.5.81}$$

Now, all of the integrals in I can be evaluated analytically. In the second and third terms, we use the known definite integrals

$$\int_{0}^{\infty} \frac{x\, dx}{1 + e^{x}} = \frac{\pi^{2}}{12}$$

$$\int_{0}^{\infty} \frac{x^{3}\, dx}{1 + e^{x}} = \frac{7\pi^{4}}{120}. \tag{11.5.82}$$

With this approximation for I, we obtain expressions for $\langle N \rangle$ (or, more precisely, $\rho\lambda^{3}$, by dividing by V and multiplying by λ^{3}):

$$
\rho\lambda^3 = \frac{4g}{3\sqrt{\pi}} \left[\left(\frac{\mu}{kT}\right)^{3/2} + \frac{\pi^2}{8} \left(\frac{\mu}{kT}\right)^{-1/2} + \cdots \right]
$$

$$
= \frac{4g}{3\sqrt{\pi}} \left[\left(\frac{\mu}{kT}\right)^{3/2} \left(1 + \frac{\pi^2}{8} \left(\frac{kT}{\mu}\right)^2 + \cdots \right) \right]. \tag{11.5.83}
$$

The term proportional to T^2 is a small thermal correction to the $T = 0$ limit. Working only to order T^2, we can replace the μ appearing in this term with $\mu_0 = \varepsilon_{\mathrm{F}}$, which yields

$$
\rho\lambda^3 = \frac{4g}{3\sqrt{\pi}} \left[\left(\frac{\mu}{kT}\right)^{3/2} \left(1 + \frac{\pi^2}{8} \left(\frac{kT}{\varepsilon_{\mathrm{F}}}\right)^2 + \cdots \right) \right]. \tag{11.5.84}
$$

Solving eqn. (11.5.84) for μ (which is equivalent to solving for ζ) gives

$$
\mu \approx kT \left[\frac{3\rho\lambda^3\sqrt{\pi}}{4g} \right]^{2/3} \left[1 + \frac{\pi^2}{8} \left(\frac{kT}{\varepsilon_{\mathrm{F}}}\right)^2 \right]^{-2/3}
$$

$$
\approx \varepsilon_{\mathrm{F}} \left[1 - \frac{\pi^2}{12} \left(\frac{kT}{\varepsilon_{\mathrm{F}}}\right)^2 + \cdots \right], \tag{11.5.85}
$$

where the second line in eqn. (11.5.85) is obtained by expanding $1/(1+x)^{2/3}$ about $x = 0$. Similarly, for the energy, we obtain

$$
E = 2\pi g V (kT)^{5/2} \left(\frac{m}{2\pi^2\hbar^2}\right)^{3/2} \left[\frac{2}{5} \left(\frac{\mu}{kT}\right)^{5/2} + \frac{\pi^2}{4} \left(\frac{\mu}{kT}\right)^{1/2} - \frac{7\pi^2}{960} \left(\frac{\mu}{kT}\right)^{-3/2} + \cdots \right] \tag{11.5.86}
$$

or, solving in a manner similar to that done for eqn. (11.5.85), we obtain

$$
E = \frac{3}{5} N\varepsilon_{\mathrm{F}} \left[1 + \frac{5}{12}\pi^2 \left(\frac{kT}{\varepsilon_{\mathrm{F}}}\right)^2 + \cdots \right]. \tag{11.5.87}
$$

This thermal correction is necessary in order to obtain the heat capacity at constant volume (which is zero at $T = 0$):

$$
C_V = \left(\frac{\partial E}{\partial T}\right)_V
$$

$$
\frac{C_V}{\langle N\rangle k} = \frac{\pi^2 kT}{2\varepsilon_{\mathrm{F}}}. \tag{11.5.88}
$$

From eqn. (11.5.87), the pressure can be obtained immediately as

$$
P = \frac{2}{5}\rho\varepsilon_{\mathrm{F}} \left[1 + \frac{5}{12}\pi^2 \left(\frac{kT}{\varepsilon_{\mathrm{F}}}\right)^2 + \cdots \right], \tag{11.5.89}
$$

which constitutes a low-temperature equation of state.

11.6   The ideal boson gas

The behavior of the ideal boson gas is dramatically different from that of the ideal fermion gas. Indeed, bosonic systems have received considerable attention in the literature because of a phenomenon known as *Bose-Einstein condensation*, which we will derive in this section.

As with the fermion case, the treatment of the ideal boson gas begins with the equations for the pressure and average particle number in terms of the fugacity:

$$\frac{PV}{kT} = -g \sum_{\mathbf{n}} \ln\left(1 - \zeta e^{-\beta \varepsilon_{\mathbf{n}}}\right) \tag{11.6.1}$$

$$\langle N \rangle = g \sum_{\mathbf{n}} \frac{\zeta e^{-\beta \varepsilon_{\mathbf{n}}}}{1 - \zeta e^{-\beta \varepsilon_{\mathbf{n}}}}. \tag{11.6.2}$$

Careful examination of eqns. (11.6.1) and (11.6.2) reveals an immediate problem: The term $\mathbf{n} = (0,0,0)$ diverges for both the pressure and the average particle number as $\zeta \to 1$. These terms need to be treated carefully, hence we split them off from the rest of the sums in eqns. (11.6.1) and (11.6.2), which gives

$$\frac{PV}{kT} = -g \sum_{\mathbf{n}}{}' \ln\left(1 - \zeta e^{-\beta \varepsilon_{\mathbf{n}}}\right) - g \ln(1 - \zeta)$$

$$\langle N \rangle = g \sum_{\mathbf{n}}{}' \frac{\zeta e^{-\beta \varepsilon_{\mathbf{n}}}}{1 - \zeta e^{-\beta \varepsilon_{\mathbf{n}}}} + g \frac{\zeta}{1 - \zeta}. \tag{11.6.3}$$

Here, $\sum'$ means that the $\mathbf{n} = (0,0,0)$ term is excluded. With these divergent terms written separately, we can take the thermodynamic limit straightforwardly and convert the remaining sums to integrals as was done in the fermion case. For the pressure, we obtain

$$\frac{PV}{kT} = -g \int d\mathbf{n} \ln\left(1 - \zeta e^{-\beta \varepsilon_{\mathbf{n}}}\right) - g \ln(1 - \zeta)$$

$$= -4\pi g \int_0^\infty dn \, n^2 \ln\left(1 - \zeta e^{-2\pi^2 \beta \hbar^2 n^2 / mL^2}\right) - g \ln(1 - \zeta)$$

$$= -\frac{4Vg}{\sqrt{\pi}\lambda^3} \int_0^\infty dx \, x^2 \ln(1 - \zeta e^{-x^2}) - g \ln(1 - \zeta), \tag{11.6.4}$$

where the change of variables in eqn. (11.5.2) has been made. Now, the function $\ln(1 - x)$ has the following power series expansion:

$$\ln(1 - y) = -\sum_{l=1}^{\infty} \frac{y^l}{l}. \tag{11.6.5}$$

Using eqn. (11.6.5) allows the pressure to be expressed as

$$\frac{P\lambda^3}{gkT} = \sum_{l=1}^{\infty} \frac{\zeta^l}{l^{5/2}} - \frac{\lambda^3}{V} \ln(1-\zeta), \tag{11.6.6}$$

and by a similar procedure, the average particle number becomes

$$\frac{\rho\lambda^3}{g} = \sum_{l=1}^{\infty} \frac{\zeta^l}{l^{3/2}} + \frac{\lambda^3}{V}\frac{\zeta}{1-\zeta}. \tag{11.6.7}$$

In eqn. (11.6.7), the term that has been split off represents the average occupation of the ground ($\mathbf{n} = (0,0,0)$) state:

$$\langle f_{0m} \rangle = \frac{\zeta}{1-\zeta}, \tag{11.6.8}$$

where $f_{0m} \equiv f_{\mathbf{n}=(0,0,0)m}$. Since $\langle f_{0m} \rangle$ must be greater than or equal to 0, it follows that there are restrictions on the allowed values of the fugacity ζ. First, since $\zeta = \exp(\beta\mu)$, ζ must be positive. However, in order that the average occupation of the ground state be positive, we must also have $\zeta < 1$. Therefore, $\zeta \in (0,1)$, so that $\mu < 0$. The fact that $\mu < 0$ suggests that adding particles to the ground state is favorable, which turns out to have fascinating consequences away from the classical limit. Before exploring these in Section 11.6.2, however, we first treat the low-density, high-temperature limit, where classical effects dominate.

11.6.1   Low-density, high-temperature limit

In a manner analogous to the fermion case, the low-density, high-temperature limit can be treated using a perturbative approach. At high temperature, the fugacity is sufficiently far from unity that the divergent terms in the pressure and density expressions can be safely neglected. Although it may not be obvious that $\zeta \ll 1$ at high temperature, recall that $\zeta = \exp(-|\mu|/kT)$. Moreover, μ decreases sharply in the low-density limit, and since $\mu < 0$, this means $|\mu|$ is large, and $\zeta \ll 1$. Thus, if ζ is very different from 1, the divergent terms in eqns. (11.6.6) and (11.6.7), which have a λ^3/V prefactor, vanish in the thermodynamic limit.

As in the fermion case, we assume that the fugacity can be expanded as

$$\zeta = a_1\rho + a_2\rho^2 + a_3\rho^3 + \cdots. \tag{11.6.9}$$

Then, from eqn. (11.6.7), the density becomes

$$\frac{\rho\lambda^3}{g} = (a_1\rho + a_2\rho^2 + a_3\rho^3 + \cdots) + \frac{1}{2^{3/2}}(a_1\rho + a_2\rho^2 + a_3\rho^3 + \cdots)^2$$

$$+ \frac{1}{3^{3/2}}(a_1\rho + a_2\rho^2 + a_3\rho^3 + \cdots)^3 + \cdots. \tag{11.6.10}$$

By equating like powers of ρ on both sides, the coefficients $a_1, a_2, a_3, \ldots$ can be determined as for the fermion gas. Working to first order in ρ gives

$$a_1 = \frac{\lambda^3}{g}, \qquad \zeta \approx \frac{\lambda^3 \rho}{g}, \qquad (11.6.11)$$

and the equation of state is the expected classical result

$$\frac{P}{kT} = \rho. \qquad (11.6.12)$$

Working to second-order, we find

$$a_2 = -\frac{\lambda^6}{2^{3/2} g^2} \qquad \zeta = \frac{\lambda^3 \rho}{g} - \frac{\lambda^6}{2^{3/2} g^2} \rho^2, \qquad (11.6.13)$$

and the second-order equation of state becomes

$$\frac{P}{kT} = \rho - \frac{\lambda^3}{2^{5/2} g} \rho^2. \qquad (11.6.14)$$

The second virial coefficient can be read off and is given by

$$B_2(T) = -\frac{1}{2^{5/2} g} \lambda^3 = -\frac{0.1768}{g} \lambda^3 < 0. \qquad (11.6.15)$$

In contrast to the fermion case, the bosonic pressure decreases from the classical value as a result of spin statistics. Thus, there appears to be an "effective attraction" between the particles. Unlike the fermion gas, where the occupation numbers of the available energy levels are restricted by the Pauli exclusion principle, any number of bosons can occupy a given energy state. Thus, at temperatures slightly lower than those at which a classical description is valid, particles can "condense" into lower energy states and cause small deviations from a strict Maxwell-Boltzmann distribution of kinetic energies.

11.6.2 The high-density, low-temperature limit

At high density, the work needed to insert an additional particle into the system becomes large. Since μ measures this work and $\mu < 0$, the high-density limit is equivalent to the $\mu \to 0$ or the $\zeta \to 1$ limit. In this limit, the full problem, including the divergent terms, must be solved:

$$\frac{P \lambda^3}{g k T} = \sum_{l=1}^{\infty} \frac{\zeta^l}{l^{5/2}} - \frac{\lambda^3}{V} \ln(1 - \zeta)$$

$$\frac{\rho \lambda^3}{g} = \sum_{l=1}^{\infty} \frac{\zeta^l}{l^{3/2}} + \frac{\lambda^3}{V} \frac{\zeta}{1 - \zeta}. \qquad (11.6.16)$$

We will need to refer to the two sums in eqns. (11.6.16) often in this section, so let us define them as follows:

$$g_{3/2}(\zeta) = \sum_{l=1}^{\infty} \frac{\zeta^l}{l^{3/2}}$$

$$g_{5/2}(\zeta) = \sum_{l=1}^{\infty} \frac{\zeta^l}{l^{5/2}}. \tag{11.6.17}$$

Thus, eqns. (11.6.16) can be expressed as

$$\frac{P\lambda^3}{gkT} = g_{5/2}(\zeta) - \frac{\lambda^3}{V}\ln(1-\zeta) \tag{11.6.18}$$

$$\frac{\rho\lambda^3}{g} = g_{3/2}(\zeta) + \frac{\lambda^3}{V}\frac{\zeta}{1-\zeta}. \tag{11.6.19}$$

First, consider eqn. (11.6.19) for the density. The term $\zeta/(1-\zeta)$ diverges at $\zeta = 1$. It is instructive to ask about the behavior of $g_{3/2}(\zeta)$ at $\zeta = 1$. In fact, $g_{3/2}(1)$, given by

$$g_{3/2}(1) = \sum_{l=1}^{\infty} \frac{1}{l^{3/2}}, \tag{11.6.20}$$

is a special type of a mathematical function known as a *Riemann zeta-function*. In general, the Riemann zeta-function $R(n)$ is defined to be

$$R(n) = \sum_{l=1}^{\infty} \frac{1}{l^n} \tag{11.6.21}$$

(values of $R(n)$ are provided in many standard math tables). The quantity $g_{3/2}(1) = R(3/2)$ is a pure number whose approximate value is 2.612. Moreover, from the form of $g_{3/2}(\zeta)$, it is clear that, since $\zeta < 1$, $g_{3/2}(1)$ is the maximum value of $g_{3/2}(\zeta)$. A plot of $g_{3/2}(\zeta)$ is given in Fig. 11.2.

The figure also indicates that the derivative $g'_{3/2}(\zeta)$ diverges at $\zeta = 1$, despite the value of the function being finite. Since $\zeta < 1$, it follows that

$$g_{3/2}(\zeta) < g_{5/2}(\zeta). \tag{11.6.22}$$

It is possible to solve eqn. (11.6.19) for ζ by noting that unless ζ is very close to 1, the divergent term must vanish in the thermodynamic limit as a result of the λ^3/V prefactor. It is, therefore, useful to ask precisely how close ζ must be to 1 for the divergent term to be important. Because of the λ^3/V prefactor, ζ can only be different from 1 by an amount on the order of $1/V$. In order to see this, let us assume that ζ can be written in the form

$$\zeta = 1 - \frac{a}{V}, \tag{11.6.23}$$

where a is a positive constant to be determined. The magnitude of a is a measure of the amount by which ζ deviates from 1 at a given volume. Substituting this ansatz into eqn. (11.6.19) gives

$$\frac{\rho\lambda^3}{g} = g_{3/2}(1 - a/V) + \frac{\lambda^3}{V}\frac{1-a/V}{a/V}. \tag{11.6.24}$$

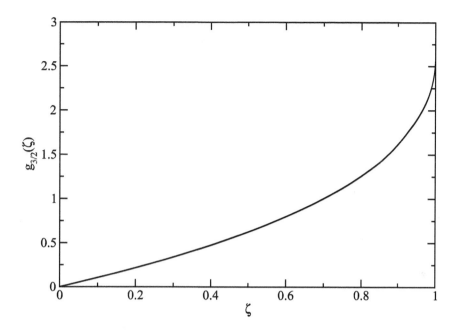

Fig. 11.2 The function $g_{3/2}(\zeta)$.

Since $g_{3/2}(\zeta)$ does not change its value much if ζ is displaced slightly from 1, we can replace the first term to a very good approximation by $R(3/2)$, which yields

$$\frac{\rho\lambda^3}{g} \approx g_{3/2}(1) + \frac{\lambda^3}{V}\frac{1 - a/V}{a/V}. \tag{11.6.25}$$

Equation (11.6.25) can be solved for the unknown parameter a to give

$$a = \frac{\lambda^3}{\frac{\rho\lambda^3}{g} - R(3/2)}, \tag{11.6.26}$$

where we have neglected a term proportional to λ^3/V, which vanishes in the thermodynamic limit. Since a must be positive, this solution is only valid for $\rho\lambda^3/g > R(3/2)$. For $\rho\lambda^3/g < R(3/2)$, ζ will be different from 1 by more than $1/V$, and the divergent term proportional to $\zeta/(1 - \zeta)$ can, therefore, be safely neglected. Thus, for $\rho\lambda^3/g < R(3/2)$, we only need to solve $\rho\lambda^3/g = g_{3/2}(\zeta)$ for ζ. Combining these results, the general solution for ζ valid at high density and low temperature can be expressed as

$$\zeta = \begin{cases} 1 - \frac{\lambda^3/V}{(\rho\lambda^3/g) - R(3/2)} & \frac{\rho\lambda^3}{g} > R(3/2) \\ \text{root of } g_{3/2}(\zeta) = \frac{\rho\lambda^3}{g} & \frac{\rho\lambda^3}{g} < R(3/2) \end{cases}, \tag{11.6.27}$$

which in the thermodynamic limit becomes

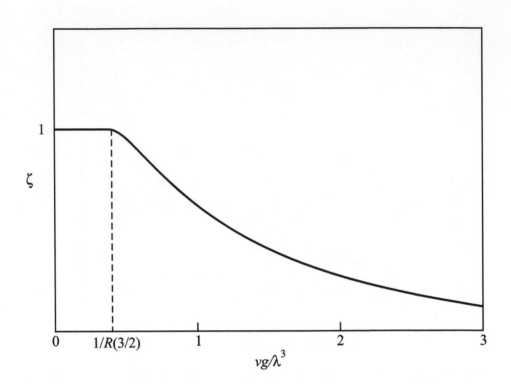

Fig. 11.3 Plot of eqn. (11.6.28).

$$\zeta = \begin{cases} 1 & \dfrac{\rho\lambda^3}{g} > R(3/2) \\ \text{root of } g_{3/2}(\zeta) = \dfrac{\rho\lambda^3}{g} & \dfrac{\rho\lambda^3}{g} < R(3/2) \end{cases}. \qquad (11.6.28)$$

A plot of ζ vs. $vg/\lambda^3 = Vg/\langle N\rangle\lambda^3$ is shown in Fig. 11.3. According to the figure, the point $R(3/2)$ is special, as ζ undergoes a transition there to the (approximately) constant value of 1.

In order to see what the effect of this transition has on the average occupation numbers, recall that the latter can be determined using

$$\langle N\rangle = \sum_{\mathbf{n},m} \frac{\zeta e^{-\beta\varepsilon\mathbf{n}}}{1 - \zeta e^{-\beta\varepsilon\mathbf{n}}} = \sum_{\mathbf{n},m}\langle f_{\mathbf{n}m}\rangle, \qquad (11.6.29)$$

from which it can be seen that the average occupation of each energy level is given by

$$\langle f_{\mathbf{n}m}\rangle = \frac{\zeta e^{-\beta\varepsilon\mathbf{n}}}{1 - \zeta e^{-\beta\varepsilon\mathbf{n}}} = \frac{1}{e^{\beta(\varepsilon\mathbf{n}-\mu)} - 1}. \qquad (11.6.30)$$

Equation (11.6.30) is known as the *Bose-Einstein distribution function*. For the ground state $(\mathbf{n} = (0,0,0))$, the occupation number expression is

$$\langle f_{0m} \rangle = \frac{\zeta}{1 - \zeta}. \tag{11.6.31}$$

Substituting the ansatz in eqn. (11.6.23) for ζ into eqn. (11.6.31) gives

$$\langle f_{0m} \rangle \approx \frac{V}{a} = \frac{V}{\lambda^3} \left(\frac{\rho \lambda^3}{g} - R(3/2) \right) \tag{11.6.32}$$

for $\rho \lambda^3 / g > R(3/2)$. At $\rho \lambda^3 / g = R(3/2)$, $\zeta \to 0$, and the occupation of the ground state becomes 0. The temperature at which the $\mathbf{n} = (0,0,0)$ level starts to become occupied can be computed by solving

$$\frac{\rho \lambda^3}{g} = R(3/2)$$

$$\frac{\rho}{g} \left(\frac{2\pi \hbar^2}{mkT_0} \right)^{3/2} = R(3/2)$$

$$kT_0 = \left(\frac{\rho}{gR(3/2)} \right)^{2/3} \frac{2\pi \hbar^2}{m}. \tag{11.6.33}$$

For temperatures less than T_0, the occupation of the ground state becomes

$$\langle f_{0m} \rangle = \frac{\rho V}{g} \left[1 - \frac{g}{\rho \lambda^3} R(3/2) \right]$$

$$= \frac{\langle N \rangle}{g} \left[1 - \frac{g}{\rho \lambda^3} R(3/2) \right]$$

$$= \frac{\langle N \rangle}{g} \left[1 - \frac{gR(3/2)}{\rho} \left(\frac{mkT}{2\pi \hbar^2} \right)^{3/2} \left(\frac{kT_0}{kT_0} \right)^{3/2} \right]$$

$$= \frac{\langle N \rangle}{g} \left[1 - \left(\frac{T}{T_0} \right)^{3/2} \right]$$

$$\frac{\langle f_{0m} \rangle}{\langle N \rangle} = \frac{1}{g} \left[1 - \left(\frac{T}{T_0} \right)^{3/2} \right]. \tag{11.6.34}$$

At $T = 0$,

$$\langle f_{0m} \rangle = \frac{\langle N \rangle}{g}. \tag{11.6.35}$$

Summing both sides of eqn. (11.6.35) over m cancels the degeneracy factor g on the right, yielding

$$\sum_m \langle f_{0m} \rangle = \sum_m \frac{\langle N \rangle}{g}$$

$$\langle \bar{f}_0 \rangle = \langle N \rangle, \tag{11.6.36}$$

where $\langle \bar{f}_0 \rangle$ indicates that the spin degeneracy has been summed over. For $T > T_0$, $\rho \lambda^3 / g < R(3/2)$ and ζ is not within $1/V$ of 1, implying that $\zeta/(1 - \zeta)$ is finite and

$$\frac{\langle \bar{f}_0 \rangle}{\langle N \rangle} = \frac{1}{\langle N \rangle} \frac{\zeta}{1 - \zeta} \longrightarrow 0 \tag{11.6.37}$$

as $\langle N \rangle \to \infty$. Thus, for the occupation of the ground state, we obtain

$$\frac{\langle \bar{f}_0 \rangle}{\langle N \rangle} = \begin{cases} 1 - (T/T_0)^{3/2} & T < T_0 \\ 0 & T > T_0 \end{cases}. \tag{11.6.38}$$

A plot of eqn. (11.6.38) is given in Fig. 11.4.

The occupation of the ground state undergoes a transition from a finite value at $T = 0$ to zero at $T = T_0$, and for all higher temperatures, it remains zero. Now, $\langle \bar{f}_0 \rangle / \langle N \rangle$ represents the probability that a particle will be found in the ground state and is, therefore, the fraction of the total number of particles occupying the ground state on average. For $T \ll T_0$, this number is very close to 1, and at $T = 0$, it becomes exactly 1, implying that at $T = 0$, all particles are in the ground state. This phenomenon, in which all particles "condense" into the ground state, is known as *Bose-Einstein* condensation. The temperature, T_0, at which "condensation" begins is known as the *Bose-Einstein condensation temperature*.

Bose-Einstein condensates were first realized experimentally using low-temperature (170 nano-Kelvin) magnetically confined rubidium atoms (Anderson *et al.*, 1995). These and other experiments have sparked considerable interest in the problem of creating Bose-Einstein condensates for technological applications, including superfluidity. Indeed, Bose-Einstein condensation is a striking example of a large-scale cooperative quantum phenomenon.

Note that there is also a critical density corresponding to the Bose-Einstein condensation temperature, which is given by the solution of

$$\frac{\rho \lambda^3}{g} = R(3/2). \tag{11.6.39}$$

Equation (11.6.39) can be solved to yield

$$\rho = \frac{g R(3/2)}{\lambda^3} = g R(3/2) \left(\frac{m k T_0}{2 \pi \hbar^2} \right)^{3/2} \equiv \rho_0, \tag{11.6.40}$$

and the occupation number, expressed in terms of the density is

$$\frac{\langle \bar{f}_0 \rangle}{\langle N \rangle} = \begin{cases} 1 - (\rho_0/\rho) & \rho > \rho_0 \\ 0 & \rho < \rho_0 \end{cases}. \tag{11.6.41}$$

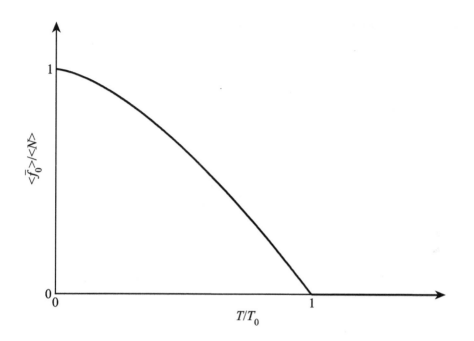

Fig. 11.4 Plot of eqn. (11.6.38).

The divergent term in eqn. (11.6.18), $-(\lambda^3/V)\ln(1-\zeta)$, becomes, for ζ very close to 1,

$$\frac{\lambda^3}{V}\ln(V/a) \sim \frac{\ln V}{V},\tag{11.6.42}$$

which clearly vanishes in the thermodynamic limit, since $V \sim \langle N \rangle$. Thus, the pressure simplifies even for ζ very close to 1, and the equation of state can be written as

$$\frac{P}{gkT} = \begin{cases} g_{5/2}(1)/\lambda^3 & \rho > \rho_0 \\ g_{5/2}(\zeta)/\lambda^3 & \rho < \rho_0 \end{cases},\tag{11.6.43}$$

where ζ is obtained by solving $\rho\lambda^3/g = g_{3/2}(\zeta)$. It is interesting to note that the pressure is approximately independent of the density for $\rho > \rho_0$. Isotherms of the ideal boson gas are shown in Fig. 11.5. Here, v_0 is the volume corresponding to the critical density ρ_0. The figure shows that $P \sim T^{5/2}$, which is quite different from the classical ideal gas. This is likewise in contrast to the fermion ideal gas, where as $T \to 0$, the pressure remains finite. For the boson gas, as $T \to 0$, the pressure vanishes, in keeping with the notion of an effective "attraction" between the particles that causes them to condense into the ground state, which is a state of zero energy.

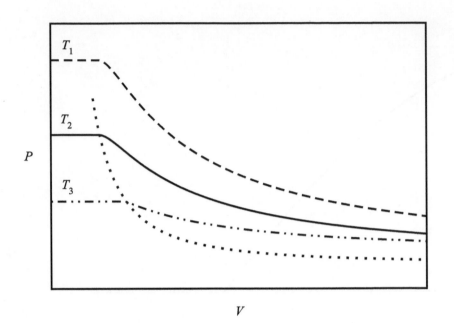

Fig. 11.5 Plot of the isotherms of the equation of state in eqn. (11.6.43). Here $T_1 > T_2 > T_3$. The dotted line connects the transition points from constant to decreasing pressure and is of the form $P \sim V^{-5/3}$.

Other thermodynamic quantities follow from the equation of state. The energy can be obtained from $E = 3PV/2$, yielding

$$E = \begin{cases} \frac{3}{2} \frac{kTV}{\lambda^3} g_{5/2}(1) & \rho > \rho_0, T < T_0 \\ \frac{3}{2} \frac{kTV}{\lambda^3} g_{5/2}(\zeta) & \rho < \rho_0, T > T_0 \end{cases}, \qquad (11.6.44)$$

and the heat capacity at constant volume is obtained from $C_V = (\partial E/\partial T)_v$, which gives

$$\frac{C_V}{\langle N \rangle k} = \begin{cases} \frac{15}{4} \frac{g_{5/2}(1)}{\rho \lambda^3} & T < T_0 \\ \frac{15}{4} \frac{g_{5/2}(\zeta)}{\rho \lambda^3} - \frac{9}{4} \frac{g_{3/2}(\zeta)}{g_{1/2}(\zeta)} & T > T_0 \end{cases}. \qquad (11.6.45)$$

The plot of the heat capacity in Fig. 11.6 exhibits a cusp at $T = T_0$. Experiments carried out on liquid ^{4}He, which has been observed to undergo Bose-Einstein condensation at around $T=2.18$ K, have measured an actual discontinuity in the heat capacity at the transition temperature, suggesting that Bose-Einstein condensation is a phase transition known as a λ *transition*. By contrast, the heat capacity of the ideal boson gas exhibits a discontinuous change at the transition temperature, signifying a first-order phase transition (see also Section 16.1). However, using the mass and density of liquid He^4 in the expression for T_0 in eqn. (11.6.33), we obtain $T_0 \approx 3.14$ K from the

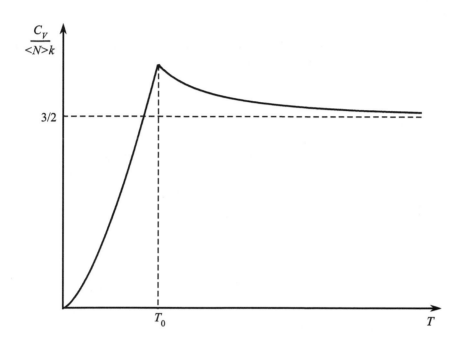

Fig. 11.6 C_V as a function of T from eqn. (11.6.45). For $T < T_0$, the curve increases as $T^{3/2}$.

ideal gas, which is not far off the experimental transition temperature of 2.18 K for real liquid helium.

11.7 Problems

11.1. Derive eqn. (11.5.18). What is the analogous term for bosons?

11.2.   a. Can Bose-Einstein condensation occur for an ideal gas of bosons in one dimension? If so, determine the temperature T_0. If not, prove that it is not possible.

      b. Can Bose-Einstein condensation occur for an ideal gas of bosons in two dimensions? If so, determine the temperature T_0. If not, prove that it is not possible.

11.3. Determine how the average energy of an ideal gas of identical fermions in one dimension at zero temperature depends on density. Repeat for a gas in two dimensions.

11.4. Consider an ideal gas of massless spin-1/2 fermions in a cubic periodic box of side L. The Hamiltonian for the system is

$$\hat{\mathcal{H}} = \sum_{i=1}^{N} c|\hat{\mathbf{p}}_i|,$$

where c is the speed of light.

a. Calculate the equation of state in the high-temperature, low-density limit up to second-order in the density. What is the second virial coefficient? What is the classical limit of the equation of state?

b. Calculate the Fermi energy, ε_F, of the gas.

c. Determine how the total energy depends on the density.

*11.5. Problem 9.7 of Chapter 9 considers the case of a charged fermion in a uniform magnetic field. In that problem, the eigenfunctions and eigenvalues of the Hamiltonian were determined. This problem uses your solution for these eigenvalues and eigenfunctions.

a. For N non-interacting charged fermions in a uniform magnetic field, calculate the grand canonical partition function $\mathcal{Z}(\zeta, V, T)$ in the high-temperature ($\hbar\omega/kT \ll 1$) and thermodynamic limits. In this limit, it is sufficient to work to *first order* in the fugacity, ζ.

Hint: Beware of degeneracies in the energy levels besides the spin degeneracy.

b. The magnetic susceptibility per unit volume is defined by

$$\chi = \frac{\partial \mathcal{M}}{\partial B},$$

where $\mathcal{M}$ is the average induced magnetization per unit volume along the direction of the magnetic field and is given by

$$\mathcal{M} = \frac{kT}{V} \left(\frac{\partial \ln \mathcal{Z}}{\partial B} \right)_{\zeta, V, T}.$$

Calculate $\mathcal{M}$ and χ for this system. Curie's Law for the magnetic susceptibility states that $|\chi| \propto 1/T$. Is your result in accordance with Curie's Law? If not, explain why it should not be.

c. If the fermions are replaced by Boltzmann particles, does the resulting susceptibility still accord with Curie's Law?

Hint: Consider using the canonical ensemble in this case.

11.6. Consider a system of N identical bosons. Each particle can occupy one of two single-particle energy levels with energies $\varepsilon_1 = 0$ and $\varepsilon_2 = \varepsilon$. Determine a condition in terms of N, β, and ε that must be obeyed if the thermally averaged occupation of the lower energy level is twice that of the upper level.

11.7. Derive expressions for the isothermal compressibility at zero temperature for ideal boson and fermion gases. Recall that the isothermal compressibility is given by

$$\kappa_T = -\frac{1}{V}\left(\frac{\partial V}{\partial P}\right)_T.$$

11.8. A cylinder is separated into two compartments by a piston that can slide freely along the length of the cylinder. In one of the compartments is an ideal gas of spin-1/2 particles, and in the other is an ideal gas of spin-3/2 particles. All particles have the same mass. At equilibrium, calculate the relative density of the two gases at $T = 0$ and at high temperature (Huang, 1963).

11.9. a. Two identical, noninteracting fermions of mass m are in a harmonic oscillator potential $U(x) = m\omega x^2/2$, where ω is the oscillator frequency. Calculate the canonical partition function of the system at temperature T.

   b. Repeat for two identical, noninteracting bosons.

11.10. Consider a system with a Hamiltonian $\hat{\mathcal{H}}_0$ that has two eigenstates $|\psi_1\rangle$ and $|\psi_2\rangle$ with the same energy eigenvalue E:

$$\hat{\mathcal{H}}_0|\psi_1\rangle = E|\psi_1\rangle$$

$$\hat{\mathcal{H}}_0|\psi_2\rangle = E|\psi_2\rangle$$

with the orthonormality condition

$$\langle\psi_i|\psi_j\rangle = \delta_{ij}.$$

Let a perturbation $\hat{\mathcal{H}}'$ be applied that breaks the degeneracy such that the new eigenstates of $\hat{\mathcal{H}} = \hat{\mathcal{H}}_0 + \hat{\mathcal{H}}'$ are

$$|\psi_+\rangle = \frac{1}{\sqrt{2}}\left[|\psi_1\rangle + |\psi_2\rangle\right]$$

$$|\psi_-\rangle = \frac{1}{\sqrt{2}}\left[|\psi_1\rangle - |\psi_2\rangle\right]$$

with corresponding energies E_+ and E_-:

$$\hat{\mathcal{H}}|\psi_+\rangle = E_+|\psi_+\rangle$$

$$\hat{\mathcal{H}}|\psi_-\rangle = E_-|\psi_-\rangle.$$

Here, $E_+ < E_-$, and

$$E_\pm = E \mp \frac{1}{2}\Delta.$$

a. Show that if $\Delta/kT \ll 1$ for an ensemble of such systems at temperature T, then Δ is given approximately by

$$\Delta \approx 2kT\frac{\langle\psi_1|\hat{\rho}|\psi_2\rangle}{\langle\psi_1|\hat{\rho}|\psi_1\rangle},$$

where $\hat{\rho}$ is the canonical density matrix of the full Hamiltonian $\hat{\mathcal{H}}$.

b. For a system of N noninteracting Boltzmann particles with allowed energies E_+ and E_-, calculate the canonical partition function, average energy, and chemical potential.

c. For a system of N noninteracting bosons with allowed energy E_+ and E_-, calculate the canonical partition function, average energy, and chemical potential.

11.11. Consider an ideal boson gas and let $\nu = -\ln\zeta$. Near $\zeta = 1$, it can be shown that that following expansion is valid (Huang, 1963):

$$g_{5/2}(\zeta) = a\nu^{3/2} + b + c\nu + d\nu^2 + \cdots,$$

where $a = 2.36$, $b = 1.342$, $c = -2.612$, $d = -0.730$. Using the recursion formula $g_{n-1} = -dg_n/d\nu$, show that the heat capacity exhibits a discontinuity given by

$$\lim_{T\to T_c^+}\left(\frac{\partial}{\partial T}\frac{C_V}{Nk}\right) - \lim_{T\to T_c^-}\left(\frac{\partial}{\partial T}\frac{C_V}{Nk}\right) = \frac{\lambda}{T_c}$$

and derive an approximate numerical value for λ.

11.12. Reproduce the plots in Figs. 11.3, 11.5, and 11.6.

11.13. The isothermal compressibility of a system is defined to be

$$\kappa_T = -\frac{1}{V}\left(\frac{\partial V}{\partial P}\right)_T.$$

Derive an expression for κ_T of an ideal electron gas at $T = 0$ in terms of the density and the Fermi energy.

11.14. An ideal electron gas at finite temperature T has the zero kinetic energy state half filled (for each of the two spin orientations). Determine the chemical potential μ and the temperature T of the gas.

11.15. The energy of an ideal electron gas at $T = 0$ can be computed using the kinetic energy density, defined as

$$\tau(\mathbf{r}) = \frac{\hbar^2}{2m} \sum_s \sum_m \sum_\mathbf{n} \langle f_{\mathbf{nm}} \rangle |\nabla \phi_{\mathbf{nm}}(\mathbf{x})|^2.$$

In terms of $\tau(\mathbf{r})$, the energy is computed from

$$E = \int_{D(V)} d\mathbf{r} \, \tau(\mathbf{r}).$$

For the ideal electron gas, compute $\tau(\mathbf{r})$ and E and compare to the result in eqn. (11.5.43).

12
The Feynman path integral

12.1 Quantum mechanics as a sum over paths

The strangeness of the quantum world is evident in systems as simple as ideal gases of the previous chapter, where nothing more than the spin statistics leads to remarkably complex behavior. We still have not yet included interactions in our treatment of quantum systems, and as noted in Chapter 10, the eigenvalue problem for the Hamiltonian when interactions are included can only be solved for very small systems. For large systems, we need a statistical approach, and this brings us to the formulation of quantum mechanics proposed by Feynman (Feynman, 1948; Feynman and Hibbs, 1965). Not surprisingly, quantum strangeness is no less apparent in Feynman's formulation of quantum mechanics than it is in the pictures we have studied thus far. Although mathematically equivalent to the Heisenberg and Schrödinger pictures of quantum mechanics, Feynman's view represents a qualitative departure from these formulations.

In order to introduce Feynman's picture, consider a particle prepared in a state initially localized at a point x that evolves unobserved to a point x'. Invoking the quantum wave-like nature of the particle, the wave packet representing the initial state evolves under the action of the propagator $U(t) = \exp(-i\hat{\mathcal{H}}t/\hbar)$, and the wave packet spreads in time, causing the state to become increasingly delocalized spatially until it is finally observed at the point x' through a measurement of position, where it again localizes due to the collapse of the wave function. In contrast, Feynman's view vaguely resembles a classical particle picture, in which the particle evolves unobserved from x to x'. There is a key difference, however, from the classical view. Classically, if we do not observe the particle, we will not know what path it will take, but we do know that it will follow a definite path. In quantum mechanics, by contrast, it is not our ignorance that prevents us from specifying a particle's path (which we could, but it is the very quantum nature of the particle itself that makes specifying a path impossible. Instead of following a unique path between x and x', the particle follows a myriad of paths, specifically all possible paths, simultaneously. These paths represent *interfering alternatives*, meaning that the total amplitude for the particle to be observed at x' at time t is the sum of the amplitudes associated with all possible paths between x and x'. Thus, according to the Feynman picture, we must calculate the amplitude for each of the infinitely many paths the particle follows and then sum them to obtain the complete amplitude for the process. Recall that the probability $P(x', t)$ for the particle to be observed at x' is the square modulus of the amplitude; since the latter is a sum of amplitudes for individual paths, the cross terms constitute the interference between individual amplitudes. This picture of quantum mechanics is known as the *sum over paths* or *path integral* formulation. How we calculate the amplitude for a path will be made explicit in the next section.

Before delving into the mathematics of path integrals, let us apply the Feynman picture heuristically to a concrete example. Consider an experiment in which electrons from a source S impinge on a double-slit system and are registered on a detection screen placed at D (see Fig. 12.1). Invoking a wave-like picture, the interference of electron

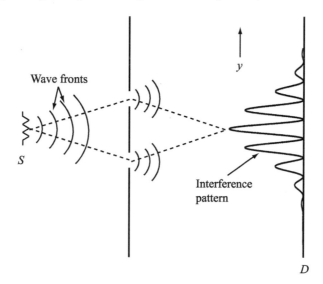

Fig. 12.1 Interference pattern observed in the electron double-slit experiment.

waves emanating from the two slits leads to the well-known interference pattern at the detector, as shown in the figure. This pattern is actually observed when the experiment is performed (Merli *et al.*, 1976). Next, consider Feynman's sum over paths formulation. If the electrons followed definite paths, as they would if they were classical particles, each path would have a separate probability and no interference pattern would be seen between the paths. We would, therefore, expect two bright spots on the screen directly opposite the slits as shown in Fig. 12.2. However, the quantum sum over paths requires that we consider all possible paths from the source S through the double-slit apparatus and finally to the detector D. Several of these paths are shown in Fig. 12.3. Let each path have a corresponding amplitude $A_i(y)$. Specifically, $A_i(y)$ is the amplitude that an electron following path i is detected at a point y on the screen at time t. The total amplitude $A(y)$ for observing an electron at y is, therefore, the sum $A(y) = A_1(y) + A_2(y) + A_3(y) + \cdots$, and the corresponding probability $P(y)$ is given by $P(y) = |A(y)|^2 = |A_1(y) + A_2(y) + A_3(y) + \cdots|^2$. Suppose there were only two such paths. Then $P(y) = |A_1(y) + A_2(y)|^2$. Since each amplitude is complex, we can write

$$A_1(y) = |A_1(y)|e^{i\phi_1(y)}, \qquad A_2(y) = |A_2(y)|e^{i\phi_2(y)}, \qquad (12.1.1)$$

and it can be easily shown that

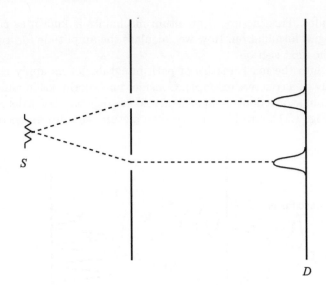

Fig. 12.2 Interference pattern expected for classical electrons impinging on the double-slit apparatus.

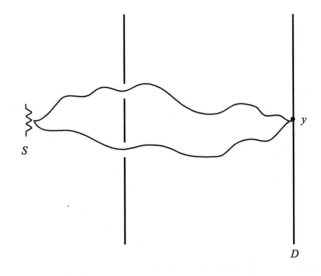

Fig. 12.3 Illustration of possible paths of quantum electrons through a double-slit apparatus.

$$P(y) = |A_1(y)|^2 + |A_2(y)|^2 + 2|A_1(y)||A_2(y)|\cos\left(\phi_1(y) - \phi_2(y)\right). \tag{12.1.2}$$

The third term in eqn. (12.1.2) is the interference term between the two paths, and the expression for $P(y)$ gives us our first clue that this term, which contains an oscillating function of the phase difference multiplied by an amplitude $|A_1(y)||A_2(y)|$, is largely responsible for the overall shape of the observed interference pattern.

Ultimately, an infinite number of amplitudes must be summed in order to obtain the overall probability, which we can express suggestively as

$$P(y) = \left| \sum_{\text{paths}} A_{\text{path}}(y) \right|^2 . \qquad (12.1.3)$$

In such an expression, the number of interference terms is infinite. Nevertheless, if the sum over paths is applied to the double-slit experiment, the correct observed interference pattern, whose intensity $I(y)$ is proportional to $P(y)$, is obtained.

In their book *Quantum Mechanics and Path Integrals*, Feynman and Hibbs (1965) employ an interesting visual device to help understand the nature of the many paths. Imagine modifying the double-slit experiment by introducing a large number of intermediate gratings, each containing many slits, as shown in Fig. 12.4. The electrons

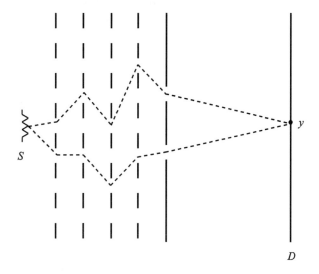

Fig. 12.4 Passage of electrons through a large number of intermediate gratings in the apparatus of the double-slit experiment.

may now pass through any sequence of slits before reaching the detector; the number of possible paths increases with both the number of intermediate gratings and the number of slits in each grating. If we now take the limit in which infinitely many gratings are placed between the source and the detector, each with an infinite number of slits, there will be an infinite number of possible paths the electrons can follow. However, when the number of slits in each intermediate grating becomes infinite, the space between the slits goes to zero, and the gratings disappear, reverting to empty space. The suggestion of this thought experiment is that empty space allows for an infinity of possible paths, and since the electrons are not observed until they reach the detector, we must sum over all of these possible paths. Indeed, this sum of path amplitudes should exactly recover or build up the amplitude pattern in a wave-like picture of the particles.

With this heuristic introduction to the sum over paths in mind, we now proceed to derive the Feynman path integral more rigorously and, in the process, learn how to determine the path amplitudes.

12.2   Derivation of path integrals for the canonical density matrix and the time evolution operator

In this and subsequent sections, the path integral concept will be given a more precise mathematical formulation and its computational advantages elucidated. For simplicity, the discussion will initially focus on a single particle moving in one spatial dimension with a Hamiltonian

$$\hat{\mathcal{H}} = \frac{\hat{p}^2}{2m} + U(\hat{x}) \equiv \hat{K} + \hat{U}. \qquad (12.2.1)$$

As noted in Section 12.1, the path integral describes a process in which a particle moves unobserved between an initiation point x and a detection point x'. That is, the particle is initially prepared in an eigenstate $|x\rangle$ of the position operator, which is subsequently allowed to evolve under the action of the propagator $\exp(-i\hat{\mathcal{H}}t/\hbar)$. After a time t, we ask what the amplitude will be for detection of a particle at a point x'. This amplitude A is given by

$$A = \langle x'|e^{-i\hat{\mathcal{H}}t/\hbar}|x\rangle \equiv U(x, x'; t). \qquad (12.2.2)$$

Therefore, what we seek are the coordinate-space matrix elements of the quantum propagator. More generally, if a system has an initial state vector $|\Psi(0)\rangle$, then from eqn. (9.2.33), at time t, the state vector is $|\Psi(t)\rangle = \exp(-i\hat{\mathcal{H}}t/\hbar)|\Psi(0)\rangle$. Projecting this into the coordinate basis gives

$$\langle x'|\Psi(t)\rangle = \Psi(x', t) = \langle x'|e^{-i\hat{\mathcal{H}}t/\hbar}|\Psi(0)\rangle$$

$$= \int dx\, \langle x'|e^{-i\hat{\mathcal{H}}t/\hbar}|x\rangle\langle x|\Psi(0)\rangle$$

$$= \int dx\, \langle x'|e^{-i\hat{\mathcal{H}}t/\hbar}|x\rangle\Psi(x, 0), \qquad (12.2.3)$$

which also requires the coordinate-space matrix elements of the propagator. The Feynman path integral provides a technique whereby these matrix elements can be computed via a sum over all possible paths leading from x to x' in time t.

Before presenting the detailed derivation of the path integral, it is worth noting an important connection between the propagator and the canonical density matrix. If we denote the latter by $\hat{\rho}(\beta) = \exp(-\beta\hat{\mathcal{H}})$, then it is clear that

$$\hat{\rho}(\beta) = \hat{U}(-i\beta\hbar), \qquad \hat{U}(t) = \hat{\rho}(it/\hbar). \qquad (12.2.4)$$

Equation (12.2.4) implies that the canonical density matrix can be obtained by evaluating the propagator at an imaginary time $t = -i\beta\hbar$. For this reason, the density matrix is often referred to as an *imaginary-time propagator*. (In Section 12.5, we will present a way to think about imaginary-time propagation.) Similarly, the real-time propagator can be obtained by evaluating the density matrix at an imaginary inverse temperature $\beta = it/\hbar$. In fact, if we allow time and temperature to be complex components of a general complex time parameter $\theta = t + i\beta\hbar$, then the transformations $t = -i\beta\hbar$ and $\beta = it/\hbar$ can be performed by rotations in the complex θ-plane from the real axis to the imaginary axis, as shown in Fig. 12.5. These rotations are known as

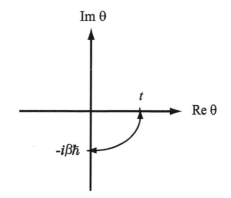

Fig. 12.5 Wick rotation in the complex time plane.

Wick rotations after the Italian physicist Giancarlo Wick (1909–1992), and they permit, in principle, the determination of the propagator given the density matrix, and vice versa. Since it is generally easier to work with a damped exponential rather than a complex one, we shall derive the Feynman path integral for the canonical density matrix and then exploit eqn. (12.2.4) to obtain a corresponding path integral expression for the quantum propagator.

Let us denote the coordinate-space matrix elements of $\hat{\rho}(\beta)$ as

$$\rho(x, x'; \beta) \equiv \langle x'|e^{-\beta\hat{\mathcal{H}}}|x\rangle. \tag{12.2.5}$$

Note that $\hat{\mathcal{H}}$ is the sum of two operators $K(\hat{p})$ and $U(\hat{x})$ that do not commute with each other ($[K(\hat{p}), U(\hat{x})] \neq 0$). Consequently, the operator $\exp(-\beta\hat{\mathcal{H}})$ cannot be evaluated straightforwardly. However, as we did in Section 3.10, we can exploit the Trotter theorem (see eqn. (3.10.18) and Appendix D) to express the operator as

$$e^{-\beta(\hat{K}+\hat{U})} = \lim_{P\to\infty} \left[e^{-\beta\hat{U}/2P}e^{-\beta\hat{K}/P}e^{-\beta\hat{U}/2P}\right]^P. \tag{12.2.6}$$

Substituting eqn. (12.2.6) into eqn. (12.2.5) yields

$$\rho(x, x'; \beta) = \lim_{P\to\infty} \langle x'| \left[e^{-\beta\hat{U}/2P}e^{-\beta\hat{K}/P}e^{-\beta\hat{U}/2P}\right]^P |x\rangle. \tag{12.2.7}$$

Let us now define an operator $\hat{\Omega}$ by

$$\hat{\Omega} = e^{-\beta\hat{U}/2P}e^{-\beta\hat{K}/P}e^{-\beta\hat{U}/2P}. \tag{12.2.8}$$

Substituting $\hat{\Omega}$ into eqn. (12.2.7) gives

$$\rho(x, x'; \beta) = \lim_{P \to \infty} \langle x' | \hat{\Omega}^P | x \rangle = \lim_{P \to \infty} \langle x' | \hat{\Omega}\hat{\Omega}\hat{\Omega} \cdots \hat{\Omega} | x \rangle. \tag{12.2.9}$$

In order to simplify the evaluation of eqn. (12.2.9), we introduce an identity operator in the form of

$$\hat{I} = \int dx \, |x\rangle\langle x| \tag{12.2.10}$$

(see also eqn. (9.2.38)) between each factor of $\hat{\Omega}$. Since there are P factors of $\hat{\Omega}$, $P-1$ insertions of the identity operator are needed. This will introduce $P-1$ integrations over coordinate labels giving the following expression for the density matrix:

$$\rho(x, x'; \beta) = \lim_{P \to \infty} \int dx_2 \cdots dx_P$$

$$\times \langle x' | \hat{\Omega} | x_P \rangle \langle x_P | \hat{\Omega} | x_{P-1} \rangle \langle x_{P-1} | \cdots | x_2 \rangle \langle x_2 | \hat{\Omega} | x \rangle. \tag{12.2.11}$$

Inserting the identity operator $P-1$ times is analogous to inserting $P-1$ gratings with many holes in Fig. 12.4. The integration over each x_i is analogous to summing over all possible ways a particle can pass through the infinitely many holes in each grating.

The advantage of employing the Trotter theorem is that the matrix elements in eqn. (12.2.11) can be evaluated in closed form. Consider the general matrix element

$$\langle x_{k+1} | \hat{\Omega} | x_k \rangle = \langle x_{k+1} | e^{-\beta \hat{U}/2P} e^{-\beta \hat{K}/P} e^{-\beta \hat{U}/2P} | x_k \rangle. \tag{12.2.12}$$

Note that $\hat{U} = U(\hat{x})$ is a function of the coordinate operator. Thus, $|x_k\rangle$ and $|x_{k+1}\rangle$, being coordinate eigenvectors, are eigenvectors of $\exp(-\beta U(\hat{x})/2P)$ with eigenvalues $\exp(-\beta U(x_k)/2P)$ and $\exp(-\beta U(x_{k+1})/2P)$, respectively. Hence, eqn. (12.2.12) simplifies to

$$\langle x_{k+1} | \hat{\Omega} | x_k \rangle = e^{-\beta U(x_{k+1})/2P} \langle x_{k+1} | e^{-\beta \hat{K}/P} | x_k \rangle e^{-\beta U(x_k)/2P}. \tag{12.2.13}$$

Since $\hat{K}$ is a function of the momentum operator, the matrix element of $\exp(-\beta \hat{K}/P)$ is less trivial to evaluate. However, if we insert another identity operator, this time expressed in terms of momentum eigenvectors as

$$\hat{I} = \int dp \, |p\rangle\langle p|, \tag{12.2.14}$$

into eqn. (12.2.13), we obtain

$$\langle x_{k+1} | e^{-\beta \hat{K}/P} | x_k \rangle = \int dp \, \langle x_{k+1} | e^{-\beta \hat{K}/P} | p \rangle \langle p | x_k \rangle. \tag{12.2.15}$$

Now the operator $\exp(-\beta \hat{K}/P)$ acts on one of its eigenvectors $|p\rangle$ to yield

$$\langle x_{k+1}|e^{-\beta\hat{K}/P}|x_k\rangle = \int dp\, \langle x_{k+1}|p\rangle\langle p|x_k\rangle e^{-\beta p^2/2mP}. \tag{12.2.16}$$

Finally, using eqn. (9.2.43), eqn. (12.2.16) becomes

$$\langle x_{k+1}|e^{-\beta\hat{K}/P}|x_k\rangle = \frac{1}{2\pi\hbar}\int dp\, e^{-\beta p^2/2mP}e^{ip(x_{k+1}-x_k)/\hbar}. \tag{12.2.17}$$

Since the range of the momentum integration is $p \in (-\infty, \infty)$, the above integral is a typical Gaussian integral that can be evaluated by completing the square. Thus, we write

$$\frac{\beta p^2}{2mP} - \frac{ip(x_{k+1}-x_k)}{\hbar}$$

$$= \frac{\beta}{2mP}\left[p^2 - \frac{2imPp(x_{k+1}-x_k)}{\beta\hbar}\right]$$

$$= \frac{\beta}{2mP}\left\{\left[p - \frac{imP(x_{k+1}-x_k)}{\beta\hbar}\right]^2 + \frac{m^2P^2(x_{k+1}-x_k)^2}{\beta^2\hbar^2}\right\}$$

$$= \frac{\beta}{2mP}\left[p - \frac{imP(x_{k+1}-x_k)}{\beta\hbar}\right]^2 + \frac{mP}{2\beta\hbar^2}(x_{k+1}-x_k)^2. \tag{12.2.18}$$

When the two last lines of eqn. (12.2.18) are substituted back into eqn. (12.2.17), and a change of variables

$$\tilde{p} = p - \frac{imP(x_{k+1}-x_k)}{\beta\hbar} \tag{12.2.19}$$

is made, we find

$$\langle x_{k+1}|e^{-\beta\hat{K}/P}|x_k\rangle = \frac{1}{2\pi\hbar}\exp\left[-\frac{mP}{2\beta\hbar^2}(x_{k+1}-x_k)^2\right]\int_{-\infty}^{\infty} d\tilde{p}\, e^{-\beta\tilde{p}^2/2mP}$$

$$= \left(\frac{mP}{2\pi\beta\hbar^2}\right)^{1/2}\exp\left[-\frac{mP}{2\beta\hbar^2}(x_{k+1}-x_k)^2\right]. \tag{12.2.20}$$

Now, eqn. (12.2.20) is combined with eqn. (12.2.13) to yield

$$\langle x_{k+1}|\hat{\Omega}|x_k\rangle = \left(\frac{mP}{2\pi\beta\hbar^2}\right)^{1/2}\exp\left[-\frac{\beta}{2P}(U(x_{k+1})+U(x_k))\right]$$

$$\times \exp\left[-\frac{mP}{2\beta\hbar^2}(x_{k+1}-x_k)^2\right]. \tag{12.2.21}$$

Finally, multiplying all P matrix elements together and integrating over the $P-1$ coordinate variables, we obtain for the density matrix:

$$\rho(x, x'; \beta) = \lim_{P \to \infty} \left(\frac{mP}{2\pi \beta \hbar^2} \right)^{P/2} \int dx_2 \cdots dx_P$$

$$\times \exp \left\{ -\frac{1}{\hbar} \sum_{k=1}^{P} \left[\frac{mP}{2\beta\hbar} (x_{k+1} - x_k)^2 + \frac{\beta\hbar}{2P} \left(U(x_{k+1}) + U(x_k) \right) \right] \right\} \Bigg|_{\substack{x_{P+1}=x' \\ x_1 = x}} . \quad (12.2.22)$$

In eqn. (12.2.22), the quantum kinetic energy is present in the form of a harmonic nearest-neighbor coupling term that acts between points along the path. The spring constant for this interaction is $mP/\beta^2 \hbar^2$.

Equation (12.2.22) is the limit $P \to \infty$ of a *discretized path integral* representation for the density matrix. As eqn. (12.2.22) indicates, the endpoints of the paths at points x_1 and x_{P+1} are fixed at the "initiation" and "detection" points, x and x', respectively. The intermediate integrations over $x_2, ..., x_P$ constitute the sum over all possible paths from x to x' in imaginary time $-i\beta\hbar$. For finite P, because the potential U only acts at the discrete points x_k, the paths are lines between successive imaginary-time points, as suggested by Fig. 12.4. Note that if the particle is confined to an interval $x \in [0, L]$, then all of the coordinate integrations must be restricted to this interval as well. The weight or amplitude assigned to each path is the value of the integrand in eqn. (12.2.22) evaluated along the discrete path.

A path integral representation for the real-time propagator can now be derived from eqn. (12.2.22) by applying eqn. (12.2.4) and setting $\beta = it/\hbar$. This yields a path integral expression for the coordinate-space matrix elements of the propagator, $\hat{U}(t)$:

$$U(x, x'; t) = \lim_{P \to \infty} \left(\frac{mP}{2\pi i t \hbar} \right)^{P/2} \int dx_2 \cdots dx_P$$

$$\times \exp \left\{ \frac{i}{\hbar} \sum_{k=1}^{P} \left[\frac{mP}{2t} (x_{k+1} - x_k)^2 - \frac{t}{2P} \left(U(x_{k+1}) + U(x_k) \right) \right] \right\} \Bigg|_{\substack{x_{P+1}=x' \\ x_1 = x}} . \quad (12.2.23)$$

Notice the change in relative sign between the kinetic and potential energy terms between eqns. (12.2.22) and (12.2.23) in the path integral expressions for the density matrix and the propagator. The path sums in eqns. (12.2.22) and (12.2.23) are represented pictorially in Fig. 12.6.

From eqn. (12.2.22), a path integral expression for the canonical partition function $Q(L, T)$ for a system confined to $x \in [0, L]$ can be derived. Recall that $Q(L, T) = \text{Tr}[\exp(-\beta \hat{\mathcal{H}})]$. Evaluating the trace in the coordinate basis gives

$$Q(L, T) = \int_0^L dx \, \langle x | e^{-\beta \hat{\mathcal{H}}} | x \rangle = \int_0^L dx \, \rho(x, x; \beta). \quad (12.2.24)$$

In order to evaluate eqn. (12.2.24), the diagonal elements of the density matrix in the coordinate basis are needed; these can be obtained by setting $x_1 = x_{P+1} = x$ in eqn. (12.2.22). Finally, an integration over the diagonal elements must be performed. Since

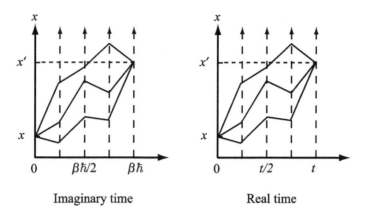

Imaginary time   Real time

Fig. 12.6 Representative paths in the path sums of eqns. (12.2.22) and (12.2.23).

$x_1 = x$, we may rename the integration variable in eqn. (12.2.24) x_1 and perform a P-dimensional integration

$$Q(L,T) = \lim_{P\to\infty} \left(\frac{mP}{2\pi\beta\hbar^2}\right)^{P/2} \int dx_1 \cdots dx_P$$

$$\times \exp\left\{-\frac{1}{\hbar}\sum_{k=1}^{P}\left[\frac{mP}{2\beta\hbar}(x_{k+1}-x_k)^2 + \frac{\beta\hbar}{2P}\left(U(x_{k+1}) + U(x_k)\right)\right]\right\}\Bigg|_{x_{P+1}=x_1}, \quad (12.2.25)$$

which is subject to the condition $x_{P+1} = x_1$. This condition restricts the integration to paths that begin and end at the same point. All of the coordinate integrations in eqn. (12.2.25), must be restricted to the spatial domain $x \in [0, L]$, which we will denote as $D(L)$. Finally, note that $\sum_{k=1}^{P}(1/2)[U(x_k) + U(x_{k+1})] = (1/2)[U(x_1) + U(x_2) + U(x_2) + U(x_3) + \cdots + U(x_{P-1}) + U(x_P) + U(x_P) + U(x_1)] = \sum_{k=1}^{P} U(x_k)$, where the condition $x_1 = x_{P+1}$ has been used. Thus, eqn. (12.2.25) simplifies to

$$Q(L,T) = \lim_{P\to\infty} \left(\frac{mP}{2\pi\beta\hbar^2}\right)^{P/2} \int_{D(L)} dx_1 \cdots dx_P$$

$$\times \exp\left\{-\frac{1}{\hbar}\sum_{k=1}^{P}\left[\frac{mP}{2\beta\hbar}(x_{k+1}-x_k)^2 + \frac{\beta\hbar}{P}U(x_k)\right]\right\}\Bigg|_{x_{P+1}=x_1}. \quad (12.2.26)$$

The integration over cyclic paths implied by eqn. (12.2.26) is illustrated in Fig. 12.7. Interestingly, as the temperature $T \to \infty$ and $\beta \to 0$, the harmonic spring constant connecting neighboring points along the paths becomes infinite, which causes the cyclic paths in the partition function to collapse onto a single point corresponding to a classical point particle. Thus, the path integral formalism shows that the high-temperature limit is equivalent to the classical limit. Finally, note that the partition function can be expressed compactly as the limit of an expression that resembles a classical configurational partition function

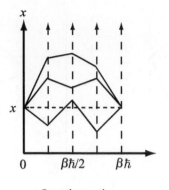

Imaginary time

Fig. 12.7 Representative paths in the discrete path sum for the canonical partition function.

$$Q(L, T) = \lim_{P \to \infty} \left(\frac{mP}{2\pi\beta\hbar^2} \right)^{P/2} \int_{D(L)} dx_1 \cdots dx_P \, e^{-\beta\phi(x_1, \ldots, x_P)}, \qquad (12.2.27)$$

an analogy we will revisit when we discuss numerical methods for evaluating path integrals in Section 12.8. Here,

$$\phi(x_1, \ldots, x_P) = \sum_{k=1}^{P} \left[\frac{1}{2} m\omega_P^2 (x_k - x_{k+1})^2 + \frac{1}{P} U(x_k) \right], \qquad (12.2.28)$$

where $\omega_P = \sqrt{P}/\beta\hbar$ and $x_{P+1} = x_1$. The function $\phi(x_1, \ldots, x_P)$, with the condition $x_{P+1} = x_1$, resembles a potential energy for a cyclic chain of classical-like pseudoparticles with harmonic springs of frequency ω_P connecting nearest neighbors on the chain (see Fig. 12.12 for an illustration of a cyclic polymer, or "ring polymer", snapshot described by eqn. (12.2.28) with $P = 8$). We will have more to say about this classical-like potential and how it can be effectively sampled in Section 12.8.

An analytical calculation of the density matrix, partition function, or propagator via path integration proceeds first by carrying out the P-dimensional integration and then taking the limit of the result as $P \to \infty$. As a simple example, consider the density matrix for a free particle ($U(x) = 0$). Assume $x \in (-\infty, \infty)$. The density matrix in this case is given by

$$\rho(x, x'; \beta) = \lim_{P \to \infty} \left(\frac{mP}{2\pi\beta\hbar^2} \right)^{P/2}$$

$$\times \int dx_2 \cdots dx_P \exp\left\{ -\sum_{k=1}^{P} \left[\frac{mP}{2\beta\hbar^2} (x_{k+1} - x_k)^2 \right] \right\} \Bigg|_{\substack{x_{P+1}=x' \\ x_1=x}}. \qquad (12.2.29)$$

In fact, we previously solved this problem in Section 4.6. Equation (4.6.33) is the partition function for a classical polymer with harmonic nearest-neighbor particle couplings and fixed endpoints. Applying the result of eqn. (4.6.33) to eqn. (12.2.29), recognizing the extra factor of P in the force constant and the fact that eqn. (12.2.29) has $P - 1$ integrations in one spatial dimension, we obtain

$$\rho(x, x'; \beta) = \left(\frac{m}{2\pi\beta\hbar^2}\right)^{1/2} \exp\left[-\frac{m}{2\beta\hbar^2}(x - x')^2\right]. \tag{12.2.30}$$

Interestingly, the P dependence completely disappears so that the limit can be taken trivially. Moreover, by substituting $\beta = it/\hbar$, the quantum propagator for a free particle can also be deduced from eqn. (12.2.30):

$$U(x, x'; t) = \left(\frac{m}{2\pi i\hbar t}\right)^{1/2} \exp\left[\frac{im}{2\hbar t}(x - x')^2\right]. \tag{12.2.31}$$

Analytical evaluation of the path integral is only possible for general quadratic potentials. Nevertheless, the path integral formalism renders quantum statistical mechanical calculations tenable with modern computers, even for large systems for which determination of the eigenvalues of $\hat{\mathcal{H}}$ is intractable. Of course, such computations can only be performed numerically for finite P, which leads to discrete path integral representations of the density matrix and partition function. P should be large enough that the difference between the discrete path integral and the formal limit $P \to \infty$ is negligible. Methodology for performing path integral calculations in imaginary time will be discussed in Sections 12.8.1 and 12.8.2. We will see that such calculations are a little more complicated than analogous calculations in the classical canonical ensemble (see Section 4.9) but straightforward, in principle. Moreover, they converge on time scales similar to those of classical calculations. Unfortunately, the same is not true for the quantum propagator in eqn. (12.2.23) due to the complex exponential in the integrand. The latter causes numerical calculations to oscillate wildly as different paths are sampled, leading to a severe convergence problem known as the *dynamical sign problem*. Thus, while computing quantum equilibrium properties via path integrals has become routine, the calculation of dynamical properties from path integrals remains one of the most challenging problems in computational physics and chemistry. As of the writing of this book, no truly satisfactory solution has been achieved.

12.3   Thermodynamics and expectation values from path integrals

Path integral expressions for expectation values of Hermitian operators follow from the basic relation

$$\langle\hat{A}\rangle = \frac{1}{Q(L, T)}\mathrm{Tr}\left[\hat{A}e^{-\beta\hat{\mathcal{H}}}\right]. \tag{12.3.1}$$

Performing the trace in the coordinate basis gives

$$\langle\hat{A}\rangle = \frac{1}{Q(L, T)}\int dx \, \langle x|\hat{A}e^{-\beta\hat{\mathcal{H}}}|x\rangle. \tag{12.3.2}$$

(We will not continue to include the spatial domain $D(L)$ in the expressions, but it must be remembered that the spatial integrals carry this restriction implicitly.) A common case for which we need to evaluate eqn. (12.3.2) is ultimately the simplest. If $\hat{A}$ is purely a function of $\hat{x}$, then $|x\rangle$ is an eigenvector of $\hat{A}(\hat{x})$ satisfying

$$\hat{A}(\hat{x})|x\rangle = a(x)|x\rangle, \tag{12.3.3}$$

where $a(x)$ is the corresponding eigenvalue, and eqn. (12.3.2) reduces to

$$\langle \hat{A} \rangle = \frac{1}{Q(L,T)} \int dx\, a(x)\langle x|e^{-\beta\hat{\mathcal{H}}}|x\rangle. \tag{12.3.4}$$

Thus, for operators that are functions only of position, eqn. (12.3.4) indicates that only the diagonal elements of the density matrix are needed. Substituting eqn. (12.2.22) for $x = x'$ into eqn. (12.3.4) leads to a path integral expression for the expectation value of $\hat{A}(\hat{x})$:

$$\langle \hat{A} \rangle = \frac{1}{Q(L,T)} \lim_{P\to\infty} \left(\frac{mP}{2\pi\beta\hbar^2} \right)^{P/2} \int dx_1 \cdots dx_P\, a(x_1)$$

$$\times \exp\left\{ -\frac{1}{\hbar} \sum_{k=1}^{P} \left[\frac{mP}{2\beta\hbar}(x_{k+1} - x_k)^2 + \frac{\beta\hbar}{P}U(x_k) \right] \right\}\Bigg|_{x_{P+1}=x_1}. \tag{12.3.5}$$

Although eqn. (12.3.5) is perfectly correct, it appears to favor one particular position variable (x_1) over the others, since $a(x)$ is evaluated only at this point. Equation (12.3.5) will consequently converge slowly and is not particularly useful for actual computations. Because the paths are cyclic, however, all points $x_1, ..., x_P$ of a path are equivalent. The equivalence can be proved by noting that the argument of the exponential is invariant under a cyclic relabeling of the coordinate variables

$$x_2' = x_1, \quad x_3' = x_2, \quad \cdots \quad x_P' = x_{P-1}, \quad x_1' = x_P. \tag{12.3.6}$$

If such a relabeling is introduced into eqn. (12.3.5), a completely equivalent expression for the expectation value results:

$$\langle \hat{A} \rangle = \frac{1}{Q(L,T)} \lim_{P\to\infty} \left(\frac{mP}{2\pi\beta\hbar^2} \right)^{P/2} \int dx_1' \cdots dx_P'\, a(x_2')$$

$$\times \exp\left\{ -\frac{1}{\hbar} \sum_{k=1}^{P} \left[\frac{mP}{2\beta\hbar}(x_{k+1}' - x_k')^2 + \frac{\beta\hbar}{P}U(x_k') \right] \right\}\Bigg|_{x_{P+1}'=x_1'}. \tag{12.3.7}$$

A second relabeling, $x_3'' = x_2', x_4'' = x_3',....,x_P'' = x_{P-1}', x_1'' = x_P', x_2'' = x_1'$, would yield a similar expression with $a(x)$ evaluated at x_3''. Since P such relabelings are possible, we can derive P equivalent expressions for the expectation value, each involving the evaluation of the $a(x)$ at the different coordinates $x_1, ..., x_P$. If these expressions are added together and divided by P, we find

$$\langle \hat{A} \rangle = \frac{1}{Q(L,T)} \lim_{P \to \infty} \left(\frac{mP}{2\pi\beta\hbar^2} \right)^{P/2} \int dx_1 \cdots dx_P \left[\frac{1}{P} \sum_{k=1}^{P} a(x_k) \right]$$

$$\times \exp\left\{ -\frac{1}{\hbar} \sum_{k=1}^{P} \left[\frac{mP}{2\beta\hbar} (x_{k+1} - x_k)^2 + \frac{\beta\hbar}{P} U(x_k) \right] \right\} \Bigg|_{x_{P+1}=x_1} , \qquad (12.3.8)$$

which treats the P coordinates $x_1, ..., x_P$ on an equal footing.

Equation (12.3.8) can be put into a compact form as follows: First, we define a probability distribution function $f(x_1, ..., x_P)$ by

$$f(x_1, ..., x_P) = \frac{1}{Q_P(L,T)} \left(\frac{mP}{2\pi\beta\hbar^2} \right)^{P/2}$$

$$\times \exp\left\{ -\frac{1}{\hbar} \sum_{k=1}^{P} \left[\frac{mP}{2\beta\hbar} (x_{k+1} - x_k)^2 + \frac{\beta\hbar}{P} U(x_k) \right] \right\} \Bigg|_{x_{P+1}=x_1} , \qquad (12.3.9)$$

where $Q_P(L,T)$ is the partition function for finite P, which is obtained by removing the limit as $P \to \infty$ from eqn. (12.2.26):

$$Q_P(L,T) = \left(\frac{mP}{2\pi\beta\hbar^2} \right)^{P/2} \int dx_1 \cdots dx_P$$

$$\times \exp\left\{ -\frac{1}{\hbar} \sum_{k=1}^{P} \left[\frac{mP}{2\beta\hbar} (x_{k+1} - x_k)^2 + \frac{\beta\hbar}{P} U(x_k) \right] \right\} \Bigg|_{x_{P+1}=x_1} . \qquad (12.3.10)$$

Clearly, $Q(L,T) = \lim_{P \to \infty} Q_P(L,T)$. The function $f(x_1, ..., x_P)$ satisfies the conditions of a probability distribution: $f(x_1, ..., x_P) \geq 0$ for all $x_1, ..., x_P$ and

$$\int dx_1 \cdots dx_P \, f(x_1, ..., x_P) = 1. \qquad (12.3.11)$$

In Section 7.2, we introduced the concept of an *estimator* for a multi-dimensional integral. In path integral calculations, equilibrium expectation values can be approximated using estimator functions that depend on the P coordinates $x_1, ..., x_P$. Thus, for eqn. (12.3.8), an appropriate estimator for $\langle \hat{A} \rangle$ is the function $a_P(x_1, ..., x_P)$ defined to be

$$a_P(x_1, ..., x_P) = \frac{1}{P} \sum_{k=1}^{P} a(x_k). \qquad (12.3.12)$$

The expectation value $\langle \hat{A} \rangle$ can be approximated for finite P as an average of the estimator in eqn. (12.3.12) with respect to the probability distribution function $f(x_1, ..., x_P)$. We write this approximation as

$$\langle \hat{A} \rangle_P = \langle a_P(x_1, ..., x_P) \rangle_f, \qquad (12.3.13)$$

where $\langle \cdots \rangle_f$ indicates an average over the probability distribution function $f(x_1, ..., x_P)$. It follows that $\langle \hat{A} \rangle = \lim_{P \to \infty} \langle \hat{A} \rangle_P$.

Suppose, next, that $\hat{A}$ is a function of just the momentum operator: $\hat{A} = \hat{A}(\hat{p})$. In this case, it is no longer possible to express $\hat{A}$ in terms of the diagonal elements of the density matrix. Hence, starting with eqn. (12.3.2), $|x\rangle$ is no longer an eigenvector of $\hat{A}(\hat{p})$ and cannot be brought outside the matrix element $\langle x|\hat{A}(\hat{p}) \exp(-\beta\hat{\mathcal{H}})|x\rangle$. However, if we insert an identity operator in the form of eqn. (12.2.10) between $\hat{A}$ and $\exp(-\beta\hat{\mathcal{H}})$, then we have a product of two matrix elements:

$$\langle\hat{A}\rangle = \frac{1}{Q(L,T)} \int \mathrm{d}x \, \mathrm{d}x' \, \langle x|\hat{A}|x'\rangle\langle x'|\mathrm{e}^{-\beta\hat{\mathcal{H}}}|x\rangle. \tag{12.3.14}$$

Equation (12.3.14) requires diagonal and off-diagonal elements of the density matrix. Substituting eqn. (12.2.22) into eqn. (12.3.14) gives a path integral expression for $\langle\hat{A}\rangle$:

$$\langle\hat{A}\rangle = \frac{1}{Q(L,T)} \lim_{P\to\infty} \left(\frac{mP}{2\pi\beta\hbar^2}\right)^{P/2} \int \mathrm{d}x_1 \cdots \mathrm{d}x_{P+1}\langle x_1|\hat{A}|x_{P+1}\rangle$$

$$\times \exp\left\{-\frac{1}{\hbar}\sum_{k=1}^{P}\left[\frac{mP}{2\beta\hbar}(x_{k+1} - x_k)^2 + \frac{\beta\hbar}{2P}(U(x_{k+1}) + U(x_k))\right]\right\}. \tag{12.3.15}$$

Note that the paths in eqn. (12.3.15) are no longer cyclic, and $x_1 \neq x_{P+1}$. In general, a sum over open paths is more difficult to evaluate than a sum over closed, cyclic paths because of the large fluctuations in the endpoints and quantities such as $\langle x_1|\hat{A}(\hat{p})|x_{P+1}\rangle$ that depend on them. According to eqn. (4.6.33), the distribution of the end-to-end distance for a free particle is a Gaussian whose width grows as $T \to 0$. An interesting example of a quantity that requires such off-diagonal elements is the momentum distribution $n(p)$, which is obtained by taking $\hat{A}(\hat{p}) = \delta(\hat{p} - p'\hat{I})$, where p' is a pure number, so that

$$n(p') = \langle\delta(\hat{p} - p'\hat{I})\rangle = \frac{1}{2\pi\hbar} \int \mathrm{d}x \, \mathrm{d}x' \mathrm{e}^{ip'(x-x')}\langle x'|\mathrm{e}^{-\beta\hat{\mathcal{H}}}|x\rangle. \tag{12.3.16}$$

This distribution can be measured in neutron Compton scattering experiments and can be computed using an algorithm introduced by Morrone *et al.* (2007).

Expectation values of operator functions that depend on both position and momentum can be equally difficult to evaluate depending on how the operators $\hat{x}$ and $\hat{p}$ appear in $\hat{A}(\hat{x}, \hat{p})$ (see eqn. (9.2.51)). The thermodynamic functions in the canonical ensemble are exceptional in that they can be evaluated using cyclic path integrals, as we will now demonstrate.

Consider first the evaluation of the average energy

$$E = \langle\hat{\mathcal{H}}\rangle = \left\langle\frac{\hat{p}^2}{2m} + U(\hat{x})\right\rangle. \tag{12.3.17}$$

Although the Hamiltonian is a function of both position and momentum, and it would, therefore, seem that both closed and open paths are needed to evaluate $\langle\hat{\mathcal{H}}\rangle$, we can evaluate E straightforwardly via the thermodynamic relation

$$E = -\frac{\partial}{\partial \beta} \ln Q(L,T) = -\frac{1}{Q(L,T)} \frac{\partial Q(L,T)}{\partial \beta}. \tag{12.3.18}$$

Since $Q(L,T)$ is expressible using only cyclic paths, these are all we need to calculate E. Taking the derivative of eqn. (12.2.26) with respect to β, we obtain the following expression for the energy:

$$E = \frac{1}{Q(L,T)} \lim_{P \to \infty} \left(\frac{mP}{2\pi\beta\hbar^2} \right)^{P/2} \int dx_1 \cdots dx_P \; \varepsilon_P(x_1, ..., x_P)$$

$$\times \exp\left\{ -\frac{1}{\hbar} \sum_{k=1}^{P} \left[\frac{mP}{2\beta\hbar}(x_{k+1} - x_k)^2 + \frac{\beta\hbar}{P}U(x_k) \right] \right\} \Bigg|_{x_{P+1}=x_1}$$

$$= \lim_{P \to \infty} \langle \varepsilon_P(x_1, ..., x_P) \rangle_f, \tag{12.3.19}$$

where

$$\varepsilon_P(x_1, ..., x_P) = \frac{P}{2\beta} - \sum_{k=1}^{P} \frac{mP}{2\beta^2\hbar^2}(x_{k+1} - x_k)^2 + \frac{1}{P} \sum_{k=1}^{P} U(x_k). \tag{12.3.20}$$

Therefore, $\varepsilon_P(x_1, ..., x_P)$ is an estimator for the energy, and the average $\langle \hat{\mathcal{H}} \rangle_P = \langle \varepsilon_P(x_1, ..., x_P) \rangle_f$ converges to the true thermodynamic energy E in the limit $P \to \infty$.

Similarly, we can obtain an estimator for the one-dimensional "pressure," which we will denote Π, from the thermodynamic relation

$$\Pi = kT \frac{\partial \ln Q}{\partial L} = \frac{kT}{Q} \frac{\partial Q}{\partial L}. \tag{12.3.21}$$

As was done in Section 4.7.3, the one-dimensional "volume" L is made explicit by introducing scaled variables $s_k = x_k/L$ into the path integral for the partition function, which yields

$$Q(L,T) = \lim_{P \to \infty} \left(\frac{mP}{2\pi\beta\hbar^2} \right)^{P/2} L^P \int ds_1 \cdots ds_P$$

$$\times \exp\left[-\frac{1}{\hbar} \sum_{k=1}^{P} \left(\frac{mP}{2\beta\hbar}L^2(s_{i+1} - s_i)^2 + \frac{\beta\hbar}{P}U(Ls_i) \right) \right] \Bigg|_{s_{P+1}=s_1}. \tag{12.3.22}$$

Equation (12.3.22) can now be differentiated with respect to L and transformed back to the original path variables $x_1, ..., x_P$ to yield

$$\Pi = \frac{1}{Q(L,T)} \lim_{P \to \infty} \left(\frac{mP}{2\pi\beta\hbar^2} \right)^{P/2} \int dx_1 \cdots dx_P \, \mathcal{P}_P(x_1, ..., x_P)$$

$$\times \exp \left\{ -\frac{1}{\hbar} \sum_{k=1}^{P} \left[\frac{mP}{2\beta\hbar} (x_{k+1} - x_k)^2 + \frac{\beta\hbar}{P} U(x_k) \right] \right\} \Bigg|_{x_{P+1}=x_1}$$

$$= \lim_{P \to \infty} \langle \mathcal{P}_P(\mathbf{x}_1, ..., \mathbf{x}_P) \rangle_f, \qquad (12.3.23)$$

where

$$\mathcal{P}_P(x_1, ..., x_P) = \frac{P}{\beta L} - \frac{1}{L} \sum_{k=1}^{P} \left[\frac{mP}{\beta^2\hbar^2} (x_{k+1} - x_k)^2 + \frac{1}{P} x_k \frac{\partial U}{\partial x_k} \right]. \qquad (12.3.24)$$

Thus, $\mathcal{P}_P(x_1, ..., x_P)$ is an estimator for the pressure (Martyna *et al.*, 1999) so that we can calculate the average pressure from $P = \lim_{P \to \infty} \langle \mathcal{P}_P(x_1, ..., x_P) \rangle_f$. As we discussed in Section 4.7.3, if the potential U has an explicit length (volume) dependence, then the estimator becomes

$$\mathcal{P}_P(x_1, ..., x_P) =$$

$$\frac{P}{\beta L} - \frac{1}{L} \sum_{k=1}^{P} \left[\frac{mP}{\beta^2\hbar^2} (x_{k+1} - x_k)^2 + \frac{1}{P} x_k \frac{\partial U}{\partial x_k} - \frac{L}{P} \frac{\partial U(x_k, L)}{\partial L} \right]. \qquad (12.3.25)$$

The basic thermodynamic relations of the canonical ensemble can be used to derive estimators for other thermodynamic quantities such as the constant-volume heat capacity (Glaesemann and Fried, 2002) (see Problem 12.3). We will explore the utility of expressions like eqns. (12.3.20) and (12.3.24) in practical calculations in Section 12.8.1.

12.4 The continuous limit: Functional integrals

Before we discuss the numerical implementation of path integrals, let us examine the physical content of the path integral in greater detail by formally analyzing the $P \to \infty$ limit. This limit gives rise to a mathematical construct known as a *functional integral*. Because the physical picture associated with the functional integral is clearer for real-time quantum mechanics, we will begin the discussion by analyzing the propagator of eqn. (12.2.23) and then perform the Wick rotation to imaginary time to obtain the canonical density matrix and partition function. For this analysis, it is convenient to introduce a parameter $\epsilon = t/P$, so that $P \to \infty$ implies $\epsilon \to 0$. In terms of ϵ, eqn. (12.2.23) can be written as

$$U(x, x'; t) = \lim_{\substack{P \to \infty \\ \epsilon \to 0}} \left(\frac{m}{2\pi i \epsilon \hbar} \right)^{P/2} \int dx_2 \cdots dx_P$$

$$\times \exp\left\{ \frac{i\epsilon}{\hbar} \sum_{k=1}^{P} \left[\frac{m}{2} \left(\frac{x_{k+1} - x_k}{\epsilon} \right)^2 - \frac{1}{2} \left(U(x_{k+1}) + U(x_k) \right) \right] \right\} \Bigg|_{\substack{x_{P+1}=x' \\ x_1=x}} . \quad (12.4.1)$$

In the limit $P \to \infty$ and $\epsilon \to 0$, the time interval between the points $x_1, x_2, ..., x_P, x_{P+1}$ become infinitely small, while the number of points becomes infinite. Thus, in this limit, $x_1, ..., x_{P+1}$ becomes the complete set of points needed to specify a continuous function $x(s)$ satisfying $x(0) = x$, $x(t) = x'$, with the identification

$$x_k = x(s = (k-1)\epsilon). \quad (12.4.2)$$

Moreover, in the limit $\epsilon \to 0$, the quantity $(x_{k+1} - x_k)/\epsilon$ becomes

$$\lim_{\epsilon \to 0} \left(\frac{x_{k+1} - x_k}{\epsilon} \right) = \frac{dx}{ds}. \quad (12.4.3)$$

Finally, in the limit $\epsilon \to 0$, the argument of the exponential

$$\epsilon \sum_{k=1}^{P} \left[\frac{m}{2} \left(\frac{x_{k+1} - x_k}{\epsilon} \right)^2 - \frac{1}{2} \left(U(x_{k+1}) + U(x_k) \right) \right]$$

is just a Riemann sum representation of an integral. Thus, we can write

$$\lim_{\epsilon \to 0} \epsilon \sum_{k=1}^{P} \left[\frac{m}{2} \left(\frac{x_{k+1} - x_k}{\epsilon} \right)^2 - \left(\frac{U(x_{k+1}) + U(x_k)}{2} \right) \right]$$

$$= \int_0^t ds \left[\frac{1}{2} m \dot{x}^2(s) - U(x(s)) \right], \quad (12.4.4)$$

where $\dot{x}(s) = dx/ds$. Interestingly, we see that the integrand of eqn. (12.4.4) is the classical Lagrangian $\mathcal{L}(x, \dot{x}) = (m/2)\dot{x}^2 - U(x)$ for the system (see Section 1.4). In eqn. (12.4.4), the integral of the Lagrangian is taken along the path $x(s)$, and this integral is just the action integral of action functional from Section 1.8:

$$A[x] = \int_0^t ds \left[\frac{1}{2} m \dot{x}^2(s) - U(x(s)) \right]. \quad (12.4.5)$$

Thus, the weight factor for a given path $x(s)$ that begins at x and ends at x' in time t is just the complex exponential $\exp(iA[x]/\hbar)$.

We turn next to the integration measure $dx_2 \cdots dx_P$. As noted previously, the points $x_1, ..., x_{P+1}$ comprise all of the points of the function $x(s)$ in the limit $P \to \infty$, with $x_1 = x$ and $x_{P+1} = x'$. Thus, the integration over $x_2, ..., x_P$ constitutes an integration over all possible functions $x(s)$ that satisfy the endpoint conditions $x(0) = x$, $x(t) = x'$. In other words, as $P \to \infty$, integrating over $x_2, ..., x_P$ varies all points of the function $x(s)$, which is equivalent to varying the function, itself keeping $x(0)$ and $x(t)$ fixed at x and x', respectively. This type of integration is referred to as *functional integration*. Symbolically, it is written as follows:

$$\lim_{P \to \infty, \epsilon \to 0} \left(\frac{m}{2\pi i \epsilon \hbar}\right)^{P/2} \mathrm{d}x_2 \cdots \mathrm{d}x_P \equiv \mathcal{D}x(s). \tag{12.4.6}$$

Thus, the functional integral representation of the real-time propagator is

$$U(x, x'; t) = \int_{x(0)=x}^{x(t)=x'} \mathcal{D}x(s) \, \exp\left\{\frac{i}{\hbar} \int_0^t \mathrm{d}s \, \left[\frac{1}{2}m\dot{x}^2(s) - U(x(s))\right]\right\}$$

$$= \int_{x(0)=x}^{x(t)=x'} \mathcal{D}x(s) \, \exp\left\{\frac{i}{\hbar} \int_0^t \mathrm{d}s \, \mathcal{L}(x(s), \dot{x}(s))\right\}$$

$$= \int_x^{x'} \mathcal{D}x \, e^{iA[x]/\hbar}. \tag{12.4.7}$$

At this point, several comments are in order. Equation (12.4.7) reveals that the functional integral is truly an integral over all paths $x(s)$ that begin at x at $s = 0$ and end at x' at $s = t$ with a weight $\exp(iA[x]/\hbar)$ assigned to each path. The integral over paths is illustrated in Fig. 12.8. The last line in eqn. (12.4.7) implies that the time

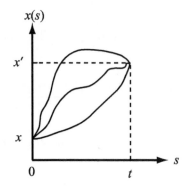

Fig. 12.8 Representative continuous paths in the path integral for the quantum propagator in eqn. (12.4.7).

label s is irrelevant, since the propagator $U(x, x'; t)$ only depends on the endpoints of the paths and the time t associated with paths. Similarly, the action $A[x]$ is a function only of x, x', and t, $A(x, x'; t)$. Consequently, the path integral sometimes appears in the literature as

$$U(x, x'; t) = \int_{x(0)=x}^{x(t)=x'} \mathcal{D}x(\cdot) \, e^{iA[x(\cdot)]/\hbar} \tag{12.4.8}$$

to indicate that the symbol used as the integration variable in the action integral is irrelevant. Finally, we point out that eqn. (12.4.7) is exactly equivalent to eqn.

(12.2.23); the former is only a symbolic representation of the latter. The functional integral notation provides a convenient and compact way of representing the more complicated discrete path integral expressions such as eqn. (12.2.23). Nevertheless, as we will see shortly, functional integrals can be directly manipulated and used for analytical calculations involving path integrals. Hence, the functional integral notation serves both a notational and a practical purpose.

Equation (12.4.7) contains some fascinating physical content. First, the weight factor $\exp(iA[x]/\hbar)$ implies that in the space of all possible paths $x(s)$, $x(0) = x$, $x(t) = x'$, the most important regions of the "path space" are those for which the action changes very little upon moving from one path to another. Indeed, when the variation in A is small, then the complex exponential oscillates very slowly in moving from path to path. Without frequent sign changes in $\exp(iA[x]/\hbar)$, the paths in this region of path space contribute significantly to the integral. On the other hand, there are other regions of path space for which A varies significantly in moving from one path to another. In this case, the exponential oscillates wildly, and the paths contribute negligibly to the path integral because contributions from closely spaced paths tend to cancel each other. More concretely, if we consider a path $x(s)$ and a slightly different path $\tilde{x}(s) = x(s) + \delta x(s)$, where $\delta x(s)$ is a small variation in $x(s)$, then these two paths will contribute significantly to the path integral if $\delta A \equiv A[x + \delta x] - A[x]$ is small. This condition is satisfied in a region where $A[x]$ is flat, that is where $\delta A/\delta x(s) \approx 0$. Note that since $x(0) = x$ and $x(t) = x'$, it follows that $\delta x(0) = \delta x(t) = 0$. Indeed, the most significant contribution occurs when $\delta A = 0$. However, recall from Section 1.8 that the condition $\delta A = 0$ is precisely the condition that leads to the Euler-Lagrange equation for the classical path:

$$\delta A = 0 \quad \Rightarrow \quad \frac{\mathrm{d}}{\mathrm{d}s}\left(\frac{\partial L}{\partial \dot{x}(s)}\right) - \frac{\partial L}{\partial x(s)} = 0. \tag{12.4.9}$$

For $L = (m/2)\dot{x}^2(s) - U(x(s))$, the equation of motion is the usual Newtonian form

$$m\frac{\mathrm{d}^2 x}{\mathrm{d}s^2} = -\frac{\partial U}{\partial x}. \tag{12.4.10}$$

We see, therefore, that the action integral and the principle of action extremization emerge naturally from the path integral formulation of quantum mechanics. This remarkable fact tells us that the most important contribution to the path integral is the region of path space around a classical path. The importance of paths that deviate from classical paths depends on the extent to which quantum effects dominate in a given system. For example, when a process occurs via quantum tunneling, paths that deviate considerably from classical paths have a significant contribution to the path integral since tunneling is a classically forbidden phenomenon. In other cases, where quantum effects are less important but not negligible, it may be possible to compute a path integral to a reasonable level of accuracy by performing an expansion about a classical path and working to a low order in the "quantum corrections." This popular approach is the basis of *semiclassical methods* for quantum dynamics. Finally, we note that the solution to the Euler-Lagrange equation with endpoint conditions $x(0) = x$ and $x(t) = x'$ may not be unique. We noted in Section 1.8 that the solution of the

Euler-Lagrange equation subject to initial values for x and $\dot{x}$ is unique; but as the path integral requires that the paths satisfy endpoint conditions, there are contribution from regions in path space around *each* classical path satisfying the endpoint conditions.

Equation (12.4.7) represents an integral over continuous real-time paths for the quantum mechanical propagator. However, due to eqn. (12.2.4), we may perform a Wick rotation and obtain functional integral expressions for the quantum mechanical density matrix and partition function in the canonical ensemble. This Wick rotation is performed by substituting $t = -i\beta\hbar$ into eqn. (12.4.7). Let the paths now be parameterized by a variable τ related to s by $\tau = is$. When $s = t = -i\beta\hbar$, $\tau = \beta\hbar$, and the action integral becomes

$$
\int_0^t ds \left[\frac{m}{2} \left(\frac{dx}{ds} \right)^2 - U(x(s)) \right] = \int_0^{-i\beta\hbar} d(-i\tau) \left[\frac{m}{2} \left(\frac{dx}{d(-i\tau)} \right)^2 - U(x(-i\tau)) \right]
$$

$$
= i \int_0^{\beta\hbar} d\tau \left[\frac{m}{2} \left(\frac{dx}{d\tau} \right)^2 + U(x(\tau)) \right]
$$

$$
\equiv iS[x]. \tag{12.4.11}
$$

Note that the action $S[x]$ is now the action for paths in *imaginary time* τ that start at $x(0) = x$ and end at $x(\tau) = x'$ in imaginary time $\tau = \beta\hbar$. The imaginary-time action $S[x]$ differs from the real-time action $A[x]$ by the sign of the potential. The action $S[x]$ is often called the *Euclidean action*. In terms of imaginary-time paths, we may write the density matrix elements as

$$
\rho(x, x'; \beta) = \int_{x(0)=x}^{x(\beta\hbar)=x'} \mathcal{D}x(\tau) \, \exp\left\{ -\frac{1}{\hbar} \int_0^{\beta\hbar} d\tau \left[\frac{1}{2} m \dot{x}^2(\tau) + U(x(\tau)) \right] \right\}
$$

$$
= \int_{x(0)=x}^{x(\beta\hbar)=x'} \mathcal{D}x(\tau) \, \exp\left\{ -\frac{1}{\hbar} \int_0^{\beta\hbar} d\tau \, \Lambda(x(\tau), \dot{x}(\tau)) \right\}
$$

$$
= \int_x^{x'} \mathcal{D}x \, e^{-S[x]/\hbar}. \tag{12.4.12}
$$

The quantity $\Lambda(x, \dot{x}) = (m/2)\dot{x}^2 + U(x)$ is called the *imaginary-time Lagrangian* or *Euclidean Lagrangian*. The density matrix is constructed by integrating over all paths $x(\tau)$ that satisfy $x(0) = x$, $x(\beta\hbar) = x'$ weighted by $\exp(-S[x]/\hbar)$. Since this weight factor is positive definite, we can find the most important contributions to the functional integral in eqn. (12.4.12) by minimizing the Euclidean action with respect to the path $x(\tau)$. As we did for the propagator, we consider a path $x(\tau)$ and a nearby path $\tilde{x}(\tau) = x(\tau) + \delta x(\tau)$. If $x(0) = x$ and $x(\beta\hbar) = x'$, then $\delta x(0) = \delta x(\beta\hbar) = 0$. We require that the variation $\delta S = S[x + \delta x] - S[x]$ vanish to first order in the path variation δx. Following the procedure in Section 1.8, the resulting equation of motion will be exactly the form of the Euler-Lagrange equation applied to Λ:

$$\frac{\mathrm{d}}{\mathrm{d}\tau} \left(\frac{\partial \Lambda}{\partial \dot{x}(\tau)} \right) - \frac{\partial \Lambda}{\partial x(\tau)} = 0. \tag{12.4.13}$$

However, when we apply the Euler-Lagrange equation to the Euclidean Lagrangian, we obtain an equation of motion of the form

$$m\frac{\mathrm{d}^2 x}{\mathrm{d}\tau^2} = \frac{\partial U}{\partial x}. \tag{12.4.14}$$

Equation (12.4.14) resembles Newton's second law except that the force is calculated using not the potential $U(x)$ but an inverted potential surface $-U(x)$. This result is not unexpected: If we transform Newton's second law in real time $m\mathrm{d}^2 x/\mathrm{d}s^2 = -\partial U/\partial x$ to imaginary time using $s = -i\tau$, then the equation of motion becomes $m\mathrm{d}^2 x/\mathrm{d}s^2 \longrightarrow m\mathrm{d}^2 x/\mathrm{d}(-i\tau)^2 = -m\mathrm{d}^2 x/\mathrm{d}\tau^2 = -\partial U/\partial x$, which is just eqn. (12.4.14). Thus, dominant paths are solutions to eqn. (12.4.14) subject to the endpoint conditions $x(0) = x$ and $x(\beta\hbar) = x'$.

We can now use eqn. (12.4.12) to construct a functional integral expression for the partition function $Q(\beta)$. Since

$$Q(\beta) = \int \mathrm{d}x \, \rho(x, x; \beta), \tag{12.4.15}$$

we begin by taking the diagonal element of $\hat{\rho}$ in eqn. (12.4.12):

$$\rho(x, x; \beta) = \int_{x(0)=x}^{x(\beta\hbar)=x} \mathcal{D}x(\tau) \, \exp\left\{ -\frac{1}{\hbar} \int_0^{\beta\hbar} \mathrm{d}\tau \left[\frac{1}{2}m\dot{x}^2(\tau) + U(x(\tau)) \right] \right\}$$

$$= \int_{x(0)=x}^{x(\beta\hbar)=x} \mathcal{D}x(\tau) \, \exp\left\{ -\frac{1}{\hbar} \int_0^{\beta\hbar} \mathrm{d}\tau \, \Lambda(x(\tau), \dot{x}(\tau)) \right\}$$

$$= \int_x^x \mathcal{D}x \, \mathrm{e}^{-S[x]/\hbar}. \tag{12.4.16}$$

Note that although the upper and lower limits of integration on the functional measure $\mathcal{D}x$ are the same, the integral does not vanish as would be the case for an ordinary integral. Rather, eqn. (12.4.16) denotes an integral over all paths that begin and end at the same point x ($x(0) = x(\beta\hbar) = x$). In order to construct the partition function, we must integrate over all x, which gives

$$Q(\beta) = \int \mathrm{d}x \int_{x(0)=x}^{x(\beta\hbar)=x} \mathcal{D}x(\tau) \, \exp\left\{ -\frac{1}{\hbar} \int_0^{\beta\hbar} \mathrm{d}\tau \left[\frac{1}{2}m\dot{x}^2(\tau) + U(x(\tau)) \right] \right\}$$

$$= \int \mathrm{d}x \int_{x(0)=x}^{x(\beta\hbar)=x} \mathcal{D}x(\tau) \, \exp\left\{ -\frac{1}{\hbar} \int_0^{\beta\hbar} \mathrm{d}\tau \Lambda(x(\tau), \dot{x}(\tau)) \right\}$$

$$= \int \mathrm{d}x \int_x^x \mathcal{D}x \, \mathrm{e}^{-S[x]/\hbar}$$

$$\equiv \oint \mathcal{D}x \, e^{-S[x]/\hbar}. \tag{12.4.17}$$

The $\oint$ symbol in last line of eqn. (12.4.17) indicates that the functional integral is to be taken over all paths that satisfy the condition $x(0) = x(\beta\hbar)$. These paths are periodic in imaginary time with period $\beta\hbar$. Consequently, the dominant contribution to the path integral for the partition function are paths near the solutions to eqn. (12.4.14) that satisfy $x(0) = x(\beta\hbar)$.

12.4.1   Example: The harmonic oscillator

In order to illustrate the use of the functional integral formalism, consider a simple harmonic oscillator of mass m and frequency ω described by the Hamiltonian of eqn. (9.3.18). The functional integral for the full density matrix is

$$\rho(x, x'; \beta) = \int_{x(0)=x}^{x(\beta\hbar)=x'} \mathcal{D}x(\tau) \exp\left[-\frac{1}{\hbar} \int_0^{\beta\hbar} d\tau \left(\frac{1}{2}m\dot{x}^2 + \frac{1}{2}m\omega^2 x^2\right)\right]. \tag{12.4.18}$$

As we have already seen, paths in the vicinity of the classical path on the inverted potential dominate the functional integral. Thus, in order to perform the functional integral, we utilize a technique known as *expansion about the classical path*. Suppose we are able to solve eqn. (12.4.14) for a classical path $x_{\text{cl}}(\tau)$ satisfying $x_{\text{cl}}(0) = x$ and $x_{\text{cl}}(\beta\hbar) = x'$. Given this path, we perform a "change of variables" in the functional integral; that is, we change the function of integration $x(\tau)$ to a new function $y(\tau)$ via the transformation $x(\tau) = x_{\text{cl}}(\tau) + y(\tau)$. This transformation is similar to a change of variables of the form $x = a + y$ in an ordinary integral $\int f(x)dx$, where a is a constant, so that $dx = dy$. Here, since $x_{\text{cl}}(\tau)$ is a single function, it is analogous to the constant a, and $\mathcal{D}x(\tau) = \mathcal{D}y(\tau)$. For the harmonic oscillator, $x_{\text{cl}}(\tau)$ satisfies the classical equation of motion $\ddot{x}_{\text{cl}} = \omega^2 x_{\text{cl}}$ on the inverted potential surface $-U(x) = -m\omega^2 x^2/2$, with $x_{\text{cl}}(0) = x$ and $x_{\text{cl}}(\beta\hbar) = x'$. Consequently, $y(0) = y(\beta\hbar) = 0$.

Substitution of this change of variables into the action integral yields

$$
\begin{aligned}
S &= \int_0^{\beta\hbar} d\tau \left[\frac{1}{2}m\dot{x}^2 + \frac{1}{2}m\omega^2 x^2\right] \\
&= \int_0^{\beta\hbar} d\tau \left[\frac{1}{2}m(\dot{x}_{\text{cl}} + \dot{y})^2 + \frac{1}{2}m\omega^2(x_{\text{cl}} + y)^2\right] \\
&= \int_0^{\beta\hbar} d\tau \left[\frac{1}{2}m\dot{x}_{\text{cl}}^2 + \frac{1}{2}m\omega^2 x_{\text{cl}}^2\right] + \int_0^{\beta\hbar} d\tau \left[\frac{1}{2}m\dot{y}^2 + \frac{1}{2}m\omega^2 y^2\right] \\
&\quad + \int_0^{\beta\hbar} d\tau \left[m\dot{x}_{\text{cl}}\dot{y} + m\omega^2 x_{\text{cl}}y\right].
\end{aligned} \tag{12.4.19}
$$

The last line of eqn. (12.4.19) contains cross terms between $x_{\text{cl}}(\tau)$ and $y(\tau)$, but these terms can be shown to vanish using an integration by parts:

$$\int_0^{\beta\hbar} d\tau \left[m\dot{x}_{\text{cl}}\dot{y} + m\omega^2 x_{\text{cl}}y\right] = m\dot{x}_{\text{cl}}y\Big|_0^{\beta\hbar} + \int_0^{\beta\hbar} d\tau \left[-m\ddot{x}_{\text{cl}} + m\omega^2 x_{\text{cl}}\right] y$$

$$= 0. \tag{12.4.20}$$

The boundary term vanishes because $y(0) = y(\beta\hbar) = 0$, and the second term vanishes because $x_{\text{cl}}(\tau)$ satisfies $\ddot{x}_{\text{cl}} = \omega^2 x_{\text{cl}}$.

The first term in the penultimate line of eqn. (12.4.19) is the classical Euclidean action integral. The solution of $\ddot{x}_{\text{cl}} = \omega^2 x_{\text{cl}}$ that satisfies the endpoint conditions is

$$x_{\text{cl}}(\tau) = \frac{x \left(e^{-\omega(\tau - \beta\hbar)} - e^{\omega(\tau - \beta\hbar)} \right) + x' \left(e^{\omega\tau} - e^{-\omega\tau} \right)}{e^{\beta\hbar\omega} - e^{-\beta\hbar\omega}}, \tag{12.4.21}$$

which we derive by assuming a solution of the form $x_{\text{cl}}(\tau) = A \exp(\omega\tau) + B \exp(-\omega\tau)$ and using the endpoint conditions to solve for the constants A and B. When this solution is substituted into the classical action integral, we obtain

$$\int_0^{\beta\hbar} d\tau \left[\frac{1}{2} m \dot{x}_{\text{cl}}^2(\tau) + \frac{1}{2} m\omega^2 x_{\text{cl}}^2(\tau) \right] =$$

$$\frac{m\omega}{2\sinh(\beta\hbar\omega)} \left[\left(x^2 + (x')^2 \right) \cosh(\beta\hbar\omega) - 2xx' \right]. \tag{12.4.22}$$

Inserting eqn. (12.4.22) into eqn. (12.4.18), we obtain the density matrix for the harmonic oscillator as

$$\rho(x, x'; \beta) = I_0 \exp \left\{ -\frac{m\omega}{2\hbar\sinh(\beta\hbar\omega)} \left[\left(x^2 + (x')^2 \right) \cosh(\beta\hbar\omega) - 2xx' \right] \right\}, \tag{12.4.23}$$

where I_0 is the path integral

$$I_0 = \int_{y(0)=0}^{y(\beta\hbar)=0} \mathcal{D}y(\tau) \exp \left[-\frac{1}{\hbar} \int_0^{\beta\hbar} d\tau \left(\frac{m}{2} \dot{y}^2 + \frac{m\omega^2}{2} y^2 \right) \right]. \tag{12.4.24}$$

Note that the remaining functional integral I_0 does not depend on the points x and x' and therefore can only contribute an overall (temperature-dependent) constant to the density matrix. This affects the thermodynamics but not any averages of physical observables.[1] Nevertheless, it is instructive to see how such a functional integral is performed.

We first note that I_0 is a functional integral over functions $y(\tau)$ satisfying $y(0) = y(\beta\hbar) = 0$. Because of these endpoint conditions, the paths $y(\tau)$ can be expanded in a Fourier sine series:

$$y(\tau) = \sum_{n=1}^{\infty} c_n \sin(\omega_n \tau), \tag{12.4.25}$$

where

$$\omega_n = \frac{n\pi}{\beta\hbar}. \tag{12.4.26}$$

Since a given $y(\tau)$ is uniquely determined by its expansion coefficients c_n, integrating over the functions $y(\tau)$ is equivalent to integrating over all possible values of the

---

[1]This is only the case for the harmonic oscillator. For anharmonic potentials, a stationary-phase approximation to the path integral, which also employs an expansion about classical paths, allows the dependence of I_0 on x and x' to be approximated.

expansion coefficients. Thus, we seek to change from an integral over the functions $y(\tau)$ to an integral over the coefficients c_n. Using eqn. (12.4.25), let us first determine the argument of the exponential. We first note that

$$\dot{y}(\tau) = \sum_{n=1}^{\infty} \omega_n c_n \cos(\omega_n \tau). \tag{12.4.27}$$

Thus, the terms in the action are:

$$\int_0^{\beta\hbar} d\tau \frac{m}{2}\dot{y}^2 = \frac{m}{2} \sum_{n=1}^{\infty} \sum_{n'=1}^{\infty} c_n c_{n'} \omega_n \omega_{n'} \int_0^{\beta\hbar} d\tau \cos(\omega_n \tau)\cos(\omega_{n'}\tau). \tag{12.4.28}$$

Since the cosines are orthogonal on the interval $\tau \in [0, \beta\hbar]$, the integral simplifies to

$$\int_0^{\beta\hbar} d\tau \frac{m}{2}\dot{y}^2 = \frac{m}{2} \sum_{n=1}^{\infty} c_n^2 \omega_n^2 \int_0^{\beta\hbar} d\tau \cos^2(\omega_n \tau)$$

$$= \frac{m}{2} \sum_{n=1}^{\infty} c_n^2 \omega_n^2 \int_0^{\beta\hbar} d\tau \left[\frac{1}{2} + \frac{1}{2}\cos(2\omega_n\tau)\right]$$

$$= \frac{m\beta\hbar}{4} \sum_{n=1}^{\infty} c_n^2 \omega_n^2. \tag{12.4.29}$$

In a similar manner, we can show that

$$\int_0^{\beta\hbar} d\tau \frac{1}{2}m\omega^2 y^2 = \frac{m\beta\hbar}{4}\omega^2 \sum_{n=1}^{\infty} c_n^2. \tag{12.4.30}$$

Next, we need to change the integration measure from $\mathcal{D}y(\tau)$ to an integration over the coefficients c_n. This is rather subtle since we are transforming from a continuous functional measure to a discrete one, and it is not immediately clear how the Jacobian is computed. The simplest way to transform the measure is to assume that

$$\mathcal{D}y(\tau) = g_0 \prod_{n=1}^{\infty} g_n dc_n, \tag{12.4.31}$$

where the g_0 and g_n are constants, and then adjust these parameters so that the final result yields the correct free particle limit $\omega = 0$. With this change of variables, I_0 becomes

$$I_0 = g_0 \prod_{n=1}^{\infty} \int_{-\infty}^{\infty} g_n dc_n \exp\left[-\frac{m\beta}{4}(\omega^2 + \omega_n^2)c_n^2\right]$$

$$= g_0 \prod_{n=1}^{\infty} g_n \left[\frac{4\pi}{m\beta(\omega^2 + \omega_n^2)} \right]^{1/2}. \tag{12.4.32}$$

From eqn. (12.4.32), we see that the free particle limit can be recovered by choosing

$$g_0 = \left[\frac{m}{2\pi\beta\hbar^2} \right]^{1/2}, \qquad g_n = \left[\frac{m\beta\omega_n^2}{2\pi} \right]^{1/2}. \tag{12.4.33}$$

When $\omega = 0$, the product is exactly 1 for this choice of g_0 and g_n, which leaves the overall free particle prefactor in eqn. (12.2.30).

For $\omega \neq 0$, the infinite product is

$$\prod_{n=1}^{\infty} \left[\frac{\omega_n^2}{\omega^2 + \omega_n^2} \right]^{1/2} = \prod_{n=1}^{\infty} \left[\frac{\pi^2 n^2 / \beta^2 \hbar^2}{\omega^2 + \pi^2 n^2 / \beta^2 \hbar^2} \right]^{1/2}$$

$$= \left[\prod_{n=1}^{\infty} \left(1 + \frac{\beta^2 \hbar^2 \omega^2}{\pi^2 n^2} \right) \right]^{-1/2}. \tag{12.4.34}$$

The product in the square brackets is one of many infinite product formulas for simple functions.[2] In this case, the product formula of interest is

$$\frac{\sinh(x)}{x} = \prod_{n=1}^{\infty} \left[1 + \frac{x^2}{\pi^2 n^2} \right]. \tag{12.4.35}$$

Using this formula, eqn. (12.4.34) becomes

$$I_0 = g_0 \left[\frac{\beta\hbar\omega}{\sinh(\beta\hbar\omega)} \right]^{1/2} = \left[\frac{m\omega}{2\pi\hbar\sinh(\beta\hbar\omega)} \right]^{1/2}. \tag{12.4.36}$$

Thus, the density matrix for a harmonic oscillator is finally given by

$$\rho(x, x'; \beta) = \left[\frac{m\omega}{2\pi\hbar\sinh(\beta\hbar\omega)} \right]^{1/2}$$

$$\times \exp\left\{ -\frac{m\omega}{2\hbar\sinh(\beta\hbar\omega)} \left[(x^2 + (x')^2) \cosh(\beta\hbar\omega) - 2xx' \right] \right\}. \tag{12.4.37}$$

Note that in the free-particle limit, we take the limit $\omega \to 0$, set $\sinh(\beta\hbar\omega) \approx \beta\hbar\omega$ and $\cosh(\beta\hbar\omega) \approx 1$, so that eqn. (12.4.37) reduces to eqn. (12.2.30).

12.5 How to think about imaginary time propagation

In the previous sections, we introduced the concept of imaginary time, suggesting that the quantum canonical density matrix $\hat{\rho}(\beta)$ is a quantum propagator in imaginary time

---

[2]See, for example, Weber and Arfken's *Methods of Mathematical Physics* (2005).

$-i\hbar$. We now introduce an imaginary-time parameter $\tau \in [0, \beta\hbar]$ and an imaginary time propagator $u(x, x'; \tau) = \langle x' | \exp(-\tau\hat{\mathcal{H}}/\hbar) | x \rangle$, such that $\rho(x, x'; \beta) = u(x, x'; \beta\hbar)$. The interpretation of $u(x, x'; \tau)$ is that a system is initiated in the state $|x\rangle$, the state is then propagated in imaginary time from 0 to $\beta\hbar$, and the inner product with $\langle x' |$ is taken in order to obtain an imaginary-time "amplitude" to find the system in the state $|x'\rangle$ at $\tau = \beta\hbar$. Since $\beta = 1/kT$, it is useful to introduce a variable temperature ϑ, such that $\tau = 1/k\vartheta$, with $\vartheta = (\infty, T]$. Thus, when $\tau = 0$, $\vartheta = \infty$, *i.e.*, at $\tau = 0$, the system is at infinite temperature. At $\tau = 0$ ($\vartheta = \infty$), $\exp(-\tau\hat{\mathcal{H}}/\hbar) = \hat{I}$ becomes the identity operator, and $u(x, x'; 0) \to \langle x' | x \rangle = \delta(x - x')$, which provides another way to view the classical (high-temperature) limit. Additionally, if we cast $u(x, x'; \tau)$ in an energy basis, where

$$u(x, x'; \tau) = \sum_n \langle x' | E_n \rangle \langle E_n | x \rangle e^{-\tau E_n}, \qquad (12.5.1)$$

then when $\tau \to 0$, this becomes

$$u(x, x'; 0) = \sum_n \langle x' | E_n \rangle \langle E_n | x \rangle = \delta(x - x'), \qquad (12.5.2)$$

which follows from the completeness of the energy eigenstates. Thus, we see that at infinite temperature, all energy states are occupied with equal probability (recall the discussion of equal *a priori* probabilities in Section 3.2). In addition, we see that the density matrix is diagonal and describes a localized particle with zero thermal de Broglie wavelength.

Now suppose we propagate a coordinate eigenstate $|x\rangle$ in imaginary time starting from $\tau = 0$ up to an imaginary time $\tau = \beta\hbar$. If we do this in small steps $\delta\tau$, then these correspond to small steps in temperature $\delta\vartheta$ related by $\delta\tau = -(1/k\vartheta^2)\delta\vartheta$. We see that uniform steps in τ are not uniform in temperature. At $\tau = 0$, the temperature is infinite, and since $\delta\tau > 0$ and $\delta\vartheta < 0$, initial steps in τ are steps *down* in temperature ϑ. If $\delta\tau$ is fixed, the prefactor $1/k\vartheta^2$, which is small at large ϑ, causes the initial steps $\delta\vartheta$ to be large so that the temperature decreases rapidly at first, ultimately slowing until the physical temperature T is reached. That is, propagation in imaginary time can be interpreted as cooling a system from infinite temperature down to physical temperature T. As the system cools, the probabilities of occupying different energy states decrease from 1 to a set of nonuniform weights $\exp(-\tau E_n)$, such that the highest energy states depopulate relative to lower energy states. As the system becomes increasingly localized in a few low-lying energy states, it becomes increasingly delocalized spatially, consistent with the uncertainty principle, causing off-diagonal elements in $u(x, x'; \tau) = \langle x' | \exp(-\tau\hat{\mathcal{H}}/\hbar) | x \rangle$ to appear. Changes in the diagonal elements $u(x, x; \tau)$ of the imaginary-time propagator correspond to the growth of the ring polymer as the temperature decreases.

The interpretation of imaginary-time propagation as a cooling down of a system from infinite temperature to physical temperature T is not meant to be entirely rigorous, but it does provide a perspective for thinking about imaginary-time paths. At $\tau = 0$, a matrix element such as $u(x, x'; 0) = \langle x' | \hat{I} | x \rangle = \delta(x' - x)$ tells us that at infinite temperature, the system is completely delocalized in energy space and, therefore, fully localized spatially. If we now divide the interval $[0, \beta\hbar]$ into P discrete steps of length

ϵ, then a single step in imaginary time is equivalent to a cooling step that slightly localizes the system in energy space and slightly delocalizes the system spatially. Thus, a matrix element of the form $u(x, x'; \epsilon) = \langle x' | \exp(-\epsilon \hat{\mathcal{H}}/\hbar) | x \rangle$ provides a measure of how far x' is from x as a result of this slight delocalization caused by cooling. If we now take another step in imaginary time, *i.e.*, another cooling step, which further localizes the system in energy space, then the sequence of matrix elements that occurs in the construction of an imaginary-time path, $\langle x' | \exp(-\epsilon \hat{\mathcal{H}}/\hbar) | x_1 \rangle \langle x_1 | \exp(-\epsilon \hat{\mathcal{H}}/\hbar) | x \rangle$, measures how far x is from an intermediate point x_1 as a result of the first cooling step and then how far x' is from x_1 as a result of the second cooling step. Of course, in order to construct $u(x, x'; 2\epsilon)$, we need to integrate over all x_1 and consider all possible intermediate points, but each individual x_1 is part of a single cooling or imaginary-time path. It now becomes clear that the sequence of P matrix elements occurring in the integrand of eqn. (12.2.11)

$$\langle x' | \exp(-\epsilon \hat{\mathcal{H}}/\hbar) | x_{P-1} \rangle \langle x_{P-1} | \cdots | x_2 \rangle \langle x_2 | \exp(-\epsilon \hat{\mathcal{H}}/\hbar) | x_1 \rangle \langle x_1 | \exp(-\epsilon \hat{\mathcal{H}}/\hbar) | x \rangle,$$

which represents a single imaginary time or cooling path, measures the accessibility of each coordinate eigenstate $|x_i\rangle$ from the previous state $|x_{i-1}\rangle$ after each individual cooling step. Thus, the set of states $x, x_1, x_2, ..., x_{P-1}, x'$ provides one realization of how a system sequentially delocalizes as it steps down (nonuniformly) in temperature and will tell us about the accessibility of the state $|x'\rangle$ when the system passes through this particular sequence of states as it cools. Taking the $\epsilon \to 0$ limit then gives us a continuous path $x(\tau)$ or a continuous set of states between x and x' through which the system can pass as it cools continuously and delocalizes spatially. Of course, it must be kept in mind that the operator $\exp(-\tau \hat{\mathcal{H}}/\hbar)$ is not unitary and, therefore, is neither norm-preserving nor "imaginary-time" reversible, which tells us that the "cooling" a system undergoes along an imaginary-time path is neither unitary nor reversible. However, we would not expect it to be either of these since the very process of cooling causes a simultaneous localization in energy space and delocalization in coordinate space. It might be helpful to keep this picture of imaginary-time propagation as a putative cooling process in mind as we now proceed to ramp up the complexity and extend the concepts developed thus far to many-body systems.

12.6 Many-body path integrals

Construction of a path integral for a system of N indistinguishable particles is nontrivial because we must take into account the symmetry of the physical states. Consider, for example, the case of two identical particles described by a Hamiltonian $\hat{\mathcal{H}}$. If we wish to compute the partition function $Q = \text{Tr}[\exp(-\beta \hat{\mathcal{H}})]$ by performing the trace in the coordinate basis, how we write down the proper coordinate eigenvectors depends on whether the overall state is symmetric or antisymmetric. If the coordinate labels are x_1 and x_2, then, as we saw in Section 9.4, the coordinate eigenvectors for bosons and fermions take the form

$$|\{x_1 \ x_2\}_B\rangle = \frac{1}{\sqrt{2}} [\ |x_1 \ x_2\rangle + |x_2 \ x_1\rangle] \qquad \text{(bosons)}$$

$$|\{x_1 \, x_2\}_{\mathrm{F}}\rangle = \frac{1}{\sqrt{2}} \left[\, |x_1 \, x_2\rangle - |x_2 \, x_1\rangle \right] \qquad \text{(fermions)},$$

respectively. The completeness relations for these coordinate eigenstates take the form

$$\int \mathrm{d}x_1 \, \mathrm{d}x_2 \, |\{x_1 \, x_2\}_{\mathrm{B}}\rangle \langle \{x_1 \, x_2\}_{\mathrm{B}}| = \hat{I}$$

$$\int \mathrm{d}x_1 \, \mathrm{d}x_2 \, |\{x_1 \, x_2\}_{\mathrm{F}}\rangle \langle \{x_1 \, x_2\}_{\mathrm{F}}| = \hat{I}. \qquad (12.6.1)$$

Thus, for bosons, the partition function is given by

$$Q = \frac{1}{2} \int \mathrm{d}x_1 \, \mathrm{d}x_2 \, \left[\langle x_1 \, x_2| + \langle x_2 \, x_1| \, \right] \mathrm{e}^{-\beta \hat{\mathcal{H}}} \left[\, |x_1 \, x_2\rangle + |x_2 \, x_1\rangle \right]$$

$$= \int \mathrm{d}x_1 \, \mathrm{d}x_2 \, \left[\langle x_1 \, x_2| \mathrm{e}^{-\beta \hat{\mathcal{H}}} |x_1 \, x_2\rangle + \langle x_1 \, x_2| \mathrm{e}^{-\beta \hat{\mathcal{H}}} |x_2 \, x_1\rangle \right], \qquad (12.6.2)$$

while for fermions, it is

$$Q = \int \mathrm{d}x_1 \, \mathrm{d}x_2 \, \left[\langle x_1 \, x_2| \mathrm{e}^{-\beta \hat{\mathcal{H}}} |x_1 \, x_2\rangle - \langle x_1 \, x_2| \mathrm{e}^{-\beta \hat{\mathcal{H}}} |x_2 \, x_1\rangle \right]. \qquad (12.6.3)$$

Functional integral expressions for each of the two matrix elements appearing in eqns. (12.6.2) and (12.6.3) can be derived using the techniques already developed in Section 12.4. These expressions are

$$\langle x_1 \, x_2| \mathrm{e}^{-\beta \hat{\mathcal{H}}} |x_1 \, x_2\rangle = \int_{x_1(0)=x_1, x_2(0)=x_2}^{x_1(\beta\hbar)=x_1, x_2(\beta\hbar)=x_2} \mathcal{D}x_1 \, \mathcal{D}x_2 \, \mathrm{e}^{-S[x_1, x_2]/\hbar}$$

$$\langle x_1 \, x_2| \mathrm{e}^{-\beta \hat{\mathcal{H}}} |x_2 \, x_1\rangle = \int_{x_1(0)=x_1, x_2(0)=x_2}^{x_1(\beta\hbar)=x_2, x_2(\beta\hbar)=x_1} \mathcal{D}x_1 \, \mathcal{D}x_2 \, \mathrm{e}^{-S[x_1, x_2]/\hbar}. \qquad (12.6.4)$$

These two terms are illustrated in Fig. 12.9. In particular, note that the first term involves two independent closed paths for particles 1 and 2, respectively, in which the paths $x_1(\tau)$ and $x_2(\tau)$ satisfy $x_1(0) = x_1(\beta\hbar) = x_1$ and $x_2(0) = x_2(\beta\hbar) = x_2$. This is exactly the term that would result if the physical state of the system had no particular symmetry and could be simply described as $|x_1 \, x_2\rangle$ or $|x_2 \, x_1\rangle$. The second term, which results from the symmetry conditions placed on the state vector, "ties" the paths together at the endpoints because of the endpoint conditions $x_1(0) = x_2(\beta\hbar) = x_1$ and $x_2(0) = x_1(\beta\hbar) = x_2$. This second term, called an *exchange term*, is a purely quantum mechanical effect arising from the symmetry of the state vector. Exchange effects involve long-range correlations of delocalized wave functions and can often be neglected for particles such as protons unless the system is at a very low temperature. For electrons, however, such effects are nearly always important and need to be included.

In order to underscore the difficulties associated with exchange effects, consider writing eqns. (12.6.2) and (12.6.3) as limits of discrete path integrals. For compactness, we use the notation of eqn. (12.2.28), with the discretized paths denoted as $x_1^{(1)}, ..., x_1^{(P)}$

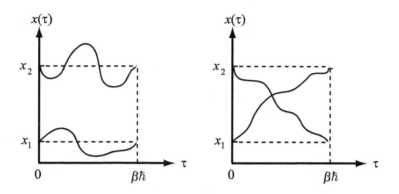

Fig. 12.9 Representative paths in the direct (*left*) and exchange (*right*) terms in the path integral of eqn. (12.6.4).

and $x_2^{(1)}, ..., x_2^{(P)}$ for particles 1 and 2, respectively. Note that the path index is now a superscript. The partition functions can be written as

$$Q(L,T) = \lim_{P\to\infty} \left(\frac{mP}{2\pi\beta\hbar^2}\right)^P \int dx_1^{(1)} \cdots dx_1^{(P)} dx_2^{(1)} \cdots dx_2^{(P)}$$

$$\times \left[e^{-\beta\phi\left(x_1^{(1)},...,x_1^{(P)},x_2^{(1)},...,x_2^{(P)}\right)} \pm e^{-\beta\tilde{\phi}\left(x_1^{(1)},...,x_1^{(P)},x_2^{(1)},...,x_2^{(P)}\right)} \right], \qquad (12.6.5)$$

where $+$ and $-$ are used for bosons and fermions, respectively, and

$$\phi\left(x_1^{(1)}, ..., x_1^{(P)}, x_2^{(1)}, ..., x_2^{(P)}\right) =$$

$$\sum_{k=1}^{P} \left\{ \frac{1}{2}m\omega_P^2 \left[\left(x_1^{(k)} - x_1^{(k+1)}\right)^2 + \left(x_2^{(k)} - x_2^{(k+1)}\right)^2 \right] + \frac{1}{P}U(x_1^{(k)}, x_2^{(k)}) \right\}, \qquad (12.6.6)$$

with $x_1^{(P+1)} = x_1^{(1)}$ and $x_2^{(P+1)} = x_2^{(1)}$. The definition of the function $\tilde{\phi}$ has the same mathematical form as eqn. (12.6.6) but with the endpoint conditions $x_1^{(P+1)} = x_2^{(1)}$ and $x_2^{(P+1)} = x_1^{(1)}$. If the term $\exp(-\beta\phi)$ is factored out of the brackets in eqn. (12.6.5), then the partition functions for fermions and bosons can be shown to be

$$Q(L,T) = \lim_{P\to\infty} \left(\frac{mP}{2\pi\beta\hbar^2}\right)^P \int dx_1^{(1)} \cdots dx_1^{(P)} dx_2^{(1)} \cdots dx_2^{(P)}$$

$$\times e^{-\beta\phi\left(x_1^{(1)},...,x_1^{(P)},x_2^{(1)},...,x_2^{(P)}\right)} \left[\det\left(\tilde{A}\right) \right] \qquad (12.6.7)$$

for fermions and

$$Q(L,T) = \lim_{P\to\infty} \left(\frac{mP}{2\pi\beta\hbar^2}\right)^P \int dx_1^{(1)} \cdots dx_1^{(P)} dx_2^{(1)} \cdots dx_2^{(P)}$$

$$\times \, e^{-\beta\phi\left(x_1^{(1)},...,x_1^{(P)},x_2^{(1)},...,x_2^{(P)}\right)} \left[\mathrm{perm}\left(\tilde{A}\right)\right] \tag{12.6.8}$$

for bosons. The matrix $\tilde{A}$ is defined to be $\tilde{A}_{ij} = A_{ij}/A_{ii}$ with

$$A_{ij} = \exp\left[-\frac{1}{2}\beta m\omega_P^2 \left(x_i^{(P)} - x_j^{(1)}\right)^2\right], \tag{12.6.9}$$

and det and perm denote the determinant and permanent of the matrix $\tilde{A}$, respectively. In this two-particle example, A and $\tilde{A}$ are 2×2 matrices. For N-particle systems, they are $N \times N$ matrices, giving rise to $N!$ terms from the determinant or permanent.

The presence of the permanent or determinant in eqns. (12.6.7) and (12.6.8), respectively, can be treated as objects to be averaged over direct paths; alternatively, they can be added as additional terms in ϕ in the form $-(1/\beta)\ln\det(\tilde{A})$ or $-(1/\beta)\ln\mathrm{perm}(\tilde{A})$. When exchange effects are important, the fermion case becomes particularly problematic, as the determinant is composed of the difference of two terms that are large and similar in magnitude. Hence, the determinant becomes a small difference of two large numbers, which is very difficult to converge. This problem is known as the *Fermi sign problem*, which is only exacerbated in a system of N fermions where $\det(\tilde{A})$ is the difference of two sums each containing $N!/2$ terms. Consequently, when $\det(\tilde{A})$ is absorbed into ϕ, it exhibits large fluctuations, which are numerically problematic. The sign problem does not exist for bosons, since $\tilde{A}$ and its permanent are positive definite, which means that numerical calculations for bosonic systems are tractable. Techniques for treating bosonic systems are discussed in detail, for example, in the review by Ceperley (1995), and a novel Monte Carlo scheme was suggested by Booth *et al.* (2009) for approaching the many-fermion problem, which involves sampling the space of fermion determinants. A decade later, Hirshberg *et al.* (2019, 2020) discovered a recursion relation satisfied by the permanents and determinants that show up in the treatment of N-body systems of bosons and fermions, respectively. This recursion formula yields a practical path integral molecular dynamics framework for treating bosons and fermions. This approach is an elegant one, which we will describe after the proceeding discussion of the N-body Boltzmann statistics problem.

Let us now suppose that exchange terms can be safely ignored, which is the case of Boltzmann statistics discussed in Section 11.3. In this limit, the path integral reduces to a sum over independent particle paths. Consider the Hamiltonian of an N-particle system in d dimensions of the standard form

$$\hat{\mathcal{H}} = \sum_{i=1}^{N} \frac{\hat{\mathbf{p}}_i^2}{2m_i} + U(\hat{\mathbf{r}}_1, ..., \hat{\mathbf{r}}_N). \tag{12.6.10}$$

As the limit of a discrete path integral, each particle will be characterized by a path in d dimensions specified by points $\mathbf{r}_i^{(1)}, ..., \mathbf{r}_i^{(P)}$, and the path integral for the partition function takes the form

$$Q(N,V,T) = \lim_{P\to\infty} \prod_{i=1}^{N} \left(\frac{m_i P}{2\pi\beta\hbar^2}\right)^{dP/2} \int \prod_{i=1}^{N} d\mathbf{r}_i^{(1)} \cdots d\mathbf{r}_i^{(P)}$$

$$\times \exp \left\{ -\sum_{k=1}^{P} \left[\sum_{i=1}^{N} \frac{m_i P}{2\beta \hbar^2} \left(\mathbf{r}_i^{(k+1)} - \mathbf{r}_i^{(k)} \right)^2 + \frac{\beta}{P} U \left(\mathbf{r}_1^{(k)}, ..., \mathbf{r}_N^{(k)} \right) \right] \right\}_{\mathbf{r}_i^{(P+1)} = \mathbf{r}_i^{(1)}} \quad (12.6.11)$$

where we now let i index the particles and k index the imaginary-time intervals. Eqn. (12.6.11) can also be written as a dN-dimensional functional integral:

$$Q(N, V, T) = \oint \mathcal{D}\mathbf{r}_1(\tau) \cdots \mathcal{D}\mathbf{r}_N(\tau)$$

$$\times \exp \left\{ -\frac{1}{\hbar} \int_0^{\beta \hbar} d\tau \sum_{i=1}^{N} \frac{1}{2} m_i \dot{\mathbf{r}}_i^2(\tau) + U(\mathbf{r}_1(\tau), ..., \mathbf{r}_N(\tau)) \right\}. \quad (12.6.12)$$

Using the techniques from Section 12.3, eqn. (12.6.11) yields the following estimators for the energy and pressure:

$$\epsilon_P \left(\{ \mathbf{r}^{(1)}, ..., \mathbf{r}^{(P)} \} \right) = \frac{dNP}{2\beta} - \sum_{k=1}^{P} \sum_{i=1}^{N} \frac{1}{2} m_i \omega_P^2 \left(\mathbf{r}_i^{(k)} - \mathbf{r}_i^{(k+1)} \right)^2 + \frac{1}{P} \sum_{k=1}^{P} U(\mathbf{r}_1^{(k)}, ..., \mathbf{r}_N^{(k)})$$

$$\mathcal{P}_P \left(\{ \mathbf{r}^{(1)}, ..., \mathbf{r}^{(P)} \} \right) = \frac{NP}{\beta V} - \frac{1}{dV} \sum_{k=1}^{P} \sum_{i=1}^{N} \left[m_i \omega_P^2 \left(\mathbf{r}_i^{(k)} - \mathbf{r}_i^{(k+1)} \right)^2 + \frac{1}{P} \mathbf{r}_i^{(k)} \cdot \nabla_{\mathbf{r}_i^{(k)}} U \right],$$

$$(12.6.13)$$

where $\{ \mathbf{r}^{(1)}, ..., \mathbf{r}^{(P)} \}$ represents the full set of N particle paths. If U has an explicit volume dependence, then an additional term

$$-\frac{1}{P} \sum_{k=1}^{P} \frac{\partial}{\partial V} U(\mathbf{r}_1^{(k)}, ..., \mathbf{r}_N^{(k)}, V)$$

must be added to eqn. (12.6.13) (Martyna *et al.*, 1999).

Unless one is interested in studying matter under extreme conditions or using path integrals to solve electronic structure problems, the approximation of Boltzmann statistics is generally sufficiently accurate to account for quantum effects in a calculation. However, when bosonic or fermionic statistics cannot be neglected, then the problem of including the requisite permanents or determinants must be faced. Although bosonic problems are a bit easier, as they do not involve a sign problem, the fluctuations in the permanent are still extremely large and difficult to converge, and taming these fluctuations is quite challenging. By contrast, fluctuations in fermion determinants arising from the sign problem are substantially more severe, making fermionic calculations significantly more difficult. Here, we will derive recursion relations based on the work of Hirshberg *et al.* for bosons and fermions that helps tame the violent fluctuations associated with bosonic permanents or fermionic determinants, rendering calculations involving identical bosons or fermions accessible to the numerical methods to be discussed in the remaining sections of this chapter.

In order to keep the notation simple, we introduce the following shorthand. Let $\mathbf{R}_i = (\mathbf{r}_i^{(1)}, ..., \mathbf{r}_i^{(P)})$ denote the set of beads on the ring polymer of particle i. Let

$$\mathcal{V}(\mathbf{R}_1, ..., \mathbf{R}_N) = \frac{1}{P} \sum_{k=1}^{P} U(\mathbf{r}_1^{(k)}, ..., \mathbf{r}_N^{(k)}) \qquad (12.6.14)$$

denote the interaction potential between the ring polymers. Let

$$\mathcal{K}(\mathbf{R}_1, ..., \mathbf{R}_N) = \frac{1}{2} m \omega_P^2 \sum_{i=1}^{N} \sum_{k=1}^{P} \left(\mathbf{r}_i^{(k+1)} - \mathbf{r}_i^{(k)} \right)^2, \qquad \mathbf{r}_i^{(P+1)} = \mathbf{r}_i^{(1)} \qquad (12.6.15)$$

denote the ring-polymer kinetic energy terms. Finally, let $\mathcal{N} = (mP/2\pi\beta\hbar^2)^{dNP/2}$ denote the overall normalization constant. Note that in these expressions, all of the masses are taken to be the same since the particles are assumed to be identical. Using the shorthand, we can write the canonical partition function of the N-particle system under the assumption of Boltzmann statistics as

$$Q(N, V, T) = \mathcal{N} \int d\mathbf{R}_1 \cdots d\mathbf{R}_N \ e^{-\beta(\mathcal{K}(\mathbf{R}_1, ..., \mathbf{R}_N) + \mathcal{V}(\mathbf{R}_1, ..., \mathbf{R}_N))}. \qquad (12.6.16)$$

In order to construct the partition function for an N-particle bosonic or fermionic system, we need to add in the exchange terms, the number of which grows as $N!$. However, if a simple recursion formula could be derived, then we could start with a small number of terms and build up the full N-particle partition function iteratively. In Fig. 12.10, we show ring-polymer configurations corresponding to the direct and exchange terms for $N = 2$ and $N = 3$. Terms having $\pm$ in front of them are added for bosons and subtracted for fermions. The two-particle partition functions were shown in eqns. (12.6.2) and (12.6.3). Let us start with the bosonic case. We can write the three-particle partition function as

$$Q_B(V, T) = \mathcal{N} \int d\mathbf{R}_1 \ d\mathbf{R}_2 \ d\mathbf{R}_3 \ e^{-\beta \left(\mathcal{K}_B^{(3)}(\mathbf{R}_1, \mathbf{R}_2, \mathbf{R}_3) + \mathcal{V}(\mathbf{R}_1, \mathbf{R}_2, \mathbf{R}_3) \right)}, \qquad (12.6.17)$$

where

$$e^{-\beta \mathcal{K}_B^{(3)}(\mathbf{R}_1, \mathbf{R}_2, \mathbf{R}_3)} = \frac{1}{6} \left(e^{-\beta \mathcal{K}_{ooo}(\mathbf{R}_1, \mathbf{R}_2, \mathbf{R}_3)} + 3 e^{-\beta \mathcal{K}_{oO}(\mathbf{R}_1, \mathbf{R}_2, \mathbf{R}_3)} \right.$$

$$\left. + 2 e^{-\beta \mathcal{K}_O(\mathbf{R}_1, \mathbf{R}_2, \mathbf{R}_3)} \right). \quad (12.6.18)$$

In this expression,

$$\mathcal{K}_{ooo}(\mathbf{R}_1, \mathbf{R}_2, \mathbf{R}_3) = \mathcal{K}(\mathbf{R}_1, \mathbf{R}_2, \mathbf{R}_3)$$
$$\mathcal{K}_{oO}(\mathbf{R}_1, \mathbf{R}_2, \mathbf{R}_3) = \mathcal{K}(\mathbf{R}_1) + \mathcal{K}_3^{(2)}(\mathbf{R}_2, \mathbf{R}_3)$$

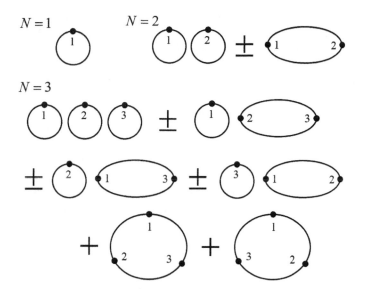

Fig. 12.10 Direct and exchange term ring-polymer configurations for $N = 1$, $N = 2$, and $N = 3$. Terms with $+/-$ are added/subtracted for bosons/fermions.

$$\mathcal{K}_\bigcirc(\mathbf{R}_1, \mathbf{R}_2, \mathbf{R}_3) = \mathcal{K}_3^{(3)}(\mathbf{R}_1, \mathbf{R}_2, \mathbf{R}_3) \tag{12.6.19}$$

with

$$\mathcal{K}_N^{(j)}(\mathbf{R}_{N-j+1}, ..., \mathbf{R}_N) = \frac{1}{2}m\omega_P^2 \sum_{i=N-j+1}^{N} \sum_{k=1}^{P} \left(\mathbf{r}_i^{(k+1)} - \mathbf{r}_i^{(k)}\right)^2, \tag{12.6.20}$$

subject to the endpoint conditions $\mathbf{r}_N^{(P+1)} = \mathbf{r}_{N-j+1}^{(1)}$, $\mathbf{r}_i^{(P+1)} = \mathbf{r}_{i+1}^{(1)}$. The small circle "o" indicates a one-particle ring polymer, "O" designates a two-particle ring polymer created by joining the endpoints, and "$\bigcirc$" signifies a three-particle ring polymer. The notation of "$\mathcal{K}$" for these energy terms indicates that they are all spring energies for the ring polymers in Fig. 12.10. Let us check eqn. (12.6.20) for $j = 2$ and $j = 3$ when $N = 3$. If $j = 2$ and $N = 3$, then the particle sum in eqn. (12.6.20) starts at $i = 2$ and, therefore, runs over two particles. The endpoint conditions are $\mathbf{r}_3^{(P+1)} = \mathbf{r}_2^{(1)}$ and $\mathbf{r}_2^{(P+1)} = \mathbf{r}_3^{(1)}$, which corresponds to a ring polymer composed of two individual polymers with their endpoints joined. If $j = 3$ and $N = 3$, then the particle sum in eqn. (12.6.20) starts at $i = 1$ and runs over all three particles. The endpoint conditions are $\mathbf{r}_3^{(P+1)} = \mathbf{r}_1^{(1)}$, $\mathbf{r}_1^{(P+1)} = \mathbf{r}_2^{(1)}$, and $\mathbf{r}_2^{(P+1)} = \mathbf{r}_3^{(1)}$, corresponding to a ring polymer composed of three individual polymers with polymer 1 joined to polymer 2, polymer 2 joined to polymer 3, and polymer 3 joined to polymer 1.

In order to generate a recursion relation for generating all of the distinct ring-polymer configurations, consider that eqn. (12.6.18) contains spring energies of topologically distinct terms for $N = 3$, "ooo", "oO", and "$\bigcirc$" with weights corresponding to the number of such terms appearing in Fig. 12.10. Based on this consideration, let us now see how to generate ring-polymer configurations for $N = 2$ and $N = 3$ starting from $N = 1$. First, to generate $N = 2$, we only need to add a single ring polymer corresponding to particle 2 and a double ring polymer consisting of particles 1 and 2 joined at the endpoints. We can now take the ring-polymer configurations for $N = 1$ and $N = 2$ as a starting point to generate the ring-polymer configurations we need for $N = 3$. In order to reach $N = 3$, we would need to add to the configuration for $N = 1$ a double ring polymer consisting of particles 2 and 3 joined at the endpoints and assign the new configuration a weight of 2. We would also need to add to the configurations for $N = 2$ a single ring polymer representing particle 3 and assign the two new configurations generated a weight of 1. Finally, we need to add a triple ring polymer consisting of all three particles and assign this configuration a weight of 2. If we now generalize to N particles, the following recursion formula emerges:

$$
e^{-\beta \mathcal{K}_B^{(N)}(\mathbf{R}_1,\ldots,\mathbf{R}_N)} = \frac{1}{N} \sum_{j=1}^{N} e^{-\beta \left(\mathcal{K}_B^{(N-j)}(\mathbf{R}_1,\ldots,\mathbf{R}_{N-j}) + \mathcal{K}_N^{(j)}(\mathbf{R}_{N-j+1},\ldots,\mathbf{R}_N) \right)} \quad (12.6.21)
$$

with the definition that $\exp(-\beta \mathcal{K}_B^{(0)}) = 1$. Let us now check the recursion relation for $N = 3$. According to eqn. (12.6.21), when $N = 3$,

$$
e^{-\beta \mathcal{K}_B^{(N)}} = \frac{1}{3} \left[e^{-\beta \left(\mathcal{K}_B^{(2)} + \mathcal{K}_3^{(1)} \right)} + e^{-\beta \left(\mathcal{K}_B^{(1)} + \mathcal{K}_3^{(2)} \right)} + e^{-\beta \left(\mathcal{K}_B^{(0)} + \mathcal{K}_3^{(3)} \right)} \right]. \quad (12.6.22)
$$

Then, since $\exp(-\beta \mathcal{K}_B^{(0)}) = 1$, $\exp(-\beta \mathcal{K}_B^{(1)}) = \exp(-\beta \mathcal{K}_\circ)$, and $\exp(-\beta \mathcal{K}_B^{(2)}) = [\exp(-\beta \mathcal{K}_{\circ\circ}) + \exp(-\beta \mathcal{K}_O)]/2$, where the "o" and "O" notation has been used, we obtain

$$
e^{-\beta \mathcal{K}_B^{(N)}} = \frac{1}{3} \left[\frac{1}{2} \left(e^{-\beta \mathcal{K}_{\circ\circ}} + e^{-\beta \mathcal{K}_O} \right) e^{-\beta \mathcal{K}_\circ} + e^{-\beta \mathcal{K}_\circ} e^{-\beta \mathcal{K}_O} + e^{-\beta \mathcal{K}_O} \right]
$$

$$
= \frac{1}{6} e^{-\beta \mathcal{K}_{\circ\circ\circ}} + \frac{1}{2} e^{-\beta \mathcal{K}_{\circ\circ O}} + \frac{1}{3} e^{-\beta \mathcal{K}_O}, \quad (12.6.23)
$$

which matches eqn. (12.6.18). Here, we have used the additive property of the spring energies, e.g., $\mathcal{K}_\circ + \mathcal{K}_{\circ\circ} = \mathcal{K}_{\circ\circ\circ}$, $\mathcal{K}_\circ O = \mathcal{K}_\circ + \mathcal{K}_O$, etc.

For fermions, a recursion relation very similar to eqn. (12.6.21) can be derived, differing only by a change of sign for certain terms. The resulting recursion formula is

$$
e^{-\beta \mathcal{K}_F^{(N)}(\mathbf{R}_1,\ldots,\mathbf{R}_N)} = \frac{1}{N} \sum_{j=1}^{N} (-1)^{j-1} e^{-\beta \left(\mathcal{K}_F^{(N-j)}(\mathbf{R}_1,\ldots,\mathbf{R}_{N-j}) + \mathcal{K}_N^{(j)}(\mathbf{R}_{N-j+1},\ldots,\mathbf{R}_N) \right)}. \quad (12.6.24)
$$

In practice, we write $\exp(-\beta \mathcal{K}_{B/F}^{(N)}(\mathbf{R}_1,\ldots,\mathbf{R}_N))$ as a single Boltzmann factor by taking the logarithm of both sides of eqns. (12.6.21) or (12.6.24) to obtain

$$\mathcal{K}^{(N)}_{B/F}(\mathbf{R}_1, ..., \mathbf{R}_N) =$$

$$-\frac{1}{\beta} \ln \left[\frac{1}{N} \sum_{j=1}^{N} \xi^{j-1} e^{-\beta \left(\mathcal{K}^{(N-j)}_{B/F}(\mathbf{R}_1,...,\mathbf{R}_{N-j}) + \mathcal{K}^{(j)}_{N}(\mathbf{R}_{N-j+1},...,\mathbf{R}_N) \right)} \right], \qquad (12.6.25)$$

where $\xi = 1$ for bosons and -1 for fermions. Then, the full potential energy used in path integral molecular dynamics for N-particle bosonic or fermionic systems is $\mathcal{K}^{(N)}_{B/F}(\mathbf{R}_1, ..., \mathbf{R}_N) + \mathcal{V}(\mathbf{R}_1, ..., \mathbf{R}_N)$. Evaluation of eqn. (12.6.25) requires evaluating $\mathcal{K}^{(0)}_{B/F}, ..., \mathcal{K}^{(N-1)}_{B/F}$ in order to generate all terms in the recursion relation.

Let us now discuss practical issues in the evaluation of $\mathcal{K}^{(N)}_{B/F}(\mathbf{R}_1, ..., \mathbf{R}_N)$ for bosons and fermions. For bosons, $\mathcal{K}^{(N)}_B$ is positive definite, making the use of molecular dynamics or Monte Carlo to evaluate bosonic path integrals possible. However, $\mathcal{K}^{(j)}_N \geq 0$ and grows with j, so that $\exp(-\beta \mathcal{K}^{(j)}_N)$ can become exponentially small. In order to stabilize the numerical calculation, we add and subtract a constant energy E_0 as

$$\mathcal{K}^{(N)}_B(\mathbf{R}_1, ..., \mathbf{R}_N) = E_0$$

$$-\frac{1}{\beta} \ln \left[\frac{1}{N} \sum_{j=1}^{N} \xi^{j-1} e^{-\beta \left(\mathcal{K}^{(N-j)}_{F}(\mathbf{R}_1,...,\mathbf{R}_{N-j}) + \mathcal{K}^{(j)}_{N}(\mathbf{R}_{N-j+1},...,\mathbf{R}_N) - E_0 \right)} \right]. \quad (12.6.26)$$

The results are independent of the choice of E_0. As of the writing of this book, interacting bosonic systems containing $N = 32$ particles have been treated using this formulation (Hirshberg *et al.*, 2019). For fermionic systems, $\exp(-\beta(\mathcal{K}^{(N)}_F(\mathbf{R}_1, ..., \mathbf{R}_N))$ is not positive definite, meaning that $\mathcal{K}^{(N)}_F(\mathbf{R}_1, ..., \mathbf{R}_N)$ is a complex potential and cannot be evaluated using standard simulation techniques. There are complex overdamped Langevin equation techniques (Fredrickson, 2006) that can be used for complex potential functions, but these have not been applied to the many fermion problem. (The Langevin and generalized Langevin equations are discussed in Chapter 15.) Because bosonic systems can be evaluated, an alternative approach is to find a related bosonic system and reweight averages of observables with the factor $\exp(-\beta(\mathcal{K}^{(N)}_F - \mathcal{K}^{(N)}_B)$ according to procedures discussed in Chapter 8. This fact allows systems containing small numbers ($N < 10$) to be simulated straightforwardly. For larger systems of interacting fermions, the Gibbs-Bogliubov inequality (see Problem 10.9) can be exploited to simulate a simpler system, *e.g.*, a system of strongly interacting fermions versus a system of weakly interacting fermions, in order to obtain an upper bound on the free energy of an N-fermion system (Hirshberg *et al.*, 2020).

When evaluating estimators of N-body quantum systems, whether Boltzmann, fermion, or boson statistics are used, the quantum (spring) kinetic energy terms grow linearly with P. From a numerical viewpoint, this is problematic, as these harmonic terms become quite stiff for systems with strong quantum effects and exhibit large, rapid fluctuations, making them difficult to converge. In the remainder of this chapter, we will discuss numerical techniques for the evaluation of path integrals that explicitly address how to handle these stiff harmonic interactions.

12.7 Quantum free-energy profiles

In Chapter 8, we discussed the calculation of classical free energies. In this section, we will discuss quantum free energies. For simplicity, we restrict the discussion to one-dimensional free-energy profiles whose features are used to calculate quantum rates using the approximate quantum dynamical techniques discussed in Chapter 14. Extension to multidimensional quantum free-energy surfaces is straightforward.

Suppose we have a quantum N-particle system obeying Boltzmann statistics. Let $\hat{\mathbf{r}}_1, ..., \hat{\mathbf{r}}_N$ be the quantum position operators for the N particles. Given a collective variable operator $\hat{q} = \mathcal{F}(\hat{\mathbf{r}}_1, ..., \hat{\mathbf{r}}_N)$, the analog of eqn. (8.7.1) for a quantum system is

$$P(s) = \frac{1}{Q(N, V, T)} \text{Tr} \left[e^{-\beta \hat{\mathcal{H}}} \delta \left(\mathcal{F}(\hat{\mathbf{r}}_1, ..., \hat{\mathbf{r}}_N) - s \right) \right] \tag{12.7.1}$$

which yields the marginal distribution $P(s)$. Equation (12.7.1) is an expectation value for a position-dependent operator expressed in terms of a δ-function. Because of eqns. (12.3.5) and (12.3.8), we can express the marginal as a path integral average using two possible estimators:

$$\pi_1 \left(\left\{ \mathbf{r}^{(1)} \right\}; s \right) = \delta \left(\mathcal{F}(\mathbf{r}_1^{(1)}, ..., \mathbf{r}_N^{(1)}) - s \right) \tag{12.7.2}$$

$$\pi_P \left(\left\{ \mathbf{r}^{(1)}, ..., \mathbf{r}^{(P)} \right\}; s \right) = \frac{1}{P} \sum_{k=1}^{P} \delta \left(\mathcal{F}(\mathbf{r}_1^{(k)}, ..., \mathbf{r}_N^{(k)}) - s \right). \tag{12.7.3}$$

If a path integral simulation (which will be described in Section 12.8) is able to sample the entire domain of $P(s)$, then the optimal estimator is that in eqn. (12.7.3) since the statistics of P beads leads to more rapid convergence than the single-bead estimator of eqn. (12.7.2). If the corresponding free-energy $A(s) = -kT \ln P(s)$ has one or more high barriers leading to rare events such that the enhanced sampling approaches in Chapter 8 are needed, then eqn. (12.7.3) would be of little utility. Suppose, for example, we were to employ the bluemoon method of Section 8.7. Recalling that the bluemoon method imposes a constraint on the system, if the constraint implied by the estimator in eqn. (12.7.2) were employed, the constraint would be imposed only among beads of index 1 on each particle, *i.e.*, the constraint would be $\mathcal{F}(\mathbf{r}_1^{(1)}, ..., \mathbf{r}_N^{(1)}) = s$. On the other hand, if we attempted to use the estimator in eqn. (12.7.3), then we would need to run P bluemoon simulations with the constraint imposed on each bead index separately and compute the average. However, given the cyclic invariance of the ring polymers, these P simulations would be expected to give exactly the same result within the statistical sampling error, and hence, the computational time needed to run the P simulations would be wasted. Yet since the convergence of a single-bead estimator would be expected to be quite slow, a common approximation made in the calculation of free energies consists in replacing the single-bead estimator in eqn. (12.7.2) with an estimator based on the N particle *path centroids*, defined as the average position of the P beads in a ring polymer:

$$\mathbf{r}_i^{(c)} = \frac{1}{P} \sum_{k=1}^{P} \mathbf{r}_i^{(k)}. \tag{12.7.4}$$

The centroid estimator then takes the form

$$\pi_c \left(\left\{ \mathbf{r}^{(c)} \right\}; s \right) = \delta \left(\mathcal{F}(\mathbf{r}_1^{(c)}, ..., \mathbf{r}_N^{(c)}) - s \right). \tag{12.7.5}$$

The estimator in eqn. (12.7.5) might be expected to be a good approximation for the estimators in eqns. (12.7.2) and (12.7.3) if the ring polymers are localized, so that the centroid accurately captures, in an average sense, the locations of the beads. However, if a ring polymer is highly delocalized, with beads divided between two sides of a potential energy barrier, then the centroid will be located in a central location near the top of the barrier, which would not accurately capture the fact that very few beads are located there. In this case, one might expect the free energy derived from the centroid

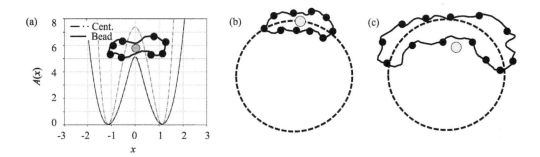

Fig. 12.11 (a) Free-energy profiles $A(x)$ at temperature $kT = 1$ for the potential $U(x,y) = 5\left(x^2 - 1\right)^2 + y^2/2 + 2.878xy$ using the centroid estimator in eqn. (12.7.5) and the all-bead estimator in eqn. (12.7.3). The division of beads among the two minima is illustrated, with the centroid shown as the grey sphere. (b) Snapshot of beads in a localized ring polymer on a spherical surface (shown in cross section), as would occur, for example, for a harmonic bond potential at high temperature. The centroid position is again shown as the grey sphere. (c) Same for a delocalized ring polymer on the spherical surface, as would occur, for example, for a harmonic bond potential at low temperature. The centroid is shown as the grey sphere and is located inside the sphere.

estimator to be a poor representation of the actual bead locations, as illustrated on the left in Fig. 12.11(a). Alternatively, the ring polymer could be delocalized over the surface of a sphere, as occurs, for example, in a radial potential such as a bond potential. In this situation, the centroid would likely be located *inside* the sphere, also providing a poor representation of the bead locations (Witt *et al.*, 2009), as illustrated in panels (b) and (c) of the figure. Nevertheless, imposing a constraint on the centroids is as straightforward as imposing a constraint on a single bead, and eqn. (12.7.5) has the advantage of rapid convergence. The corresponding approximate quantum free-energy profile $A_c(s) = -kT \ln \langle \pi_c(\{\mathbf{r}^{(c)}\}) \rangle_f$ is an essential component in path integral based quantum rate theories (Geva *et al.*, 2001; Craig and Manolopoulos, 2005).

It is important to note that dynamical enhanced sampling methods such as the metadynamics and d-AFED/TAMD methods described in Chapter 8 allow us to use the all-bead estimator in eqn. (12.7.3) to compute the quantum free energy (Cendagorta *et al.*, 2021). To see how this can be done, we write the average of the estimator in eqn. (12.7.3) as

$$\left\langle \pi_P \left(\left\{ \mathbf{r}^{(1)}, ..., \mathbf{r}^{(P)} \right\} ; s \right) \right\rangle_f = \frac{\int d\sigma \, \left\langle \pi_P \left(\left\{ \mathbf{r}^{(1)}, ..., \mathbf{r}^{(P)} \right\} ; s \right) \right\rangle_f(\sigma) e^{-\beta A_c(\sigma)}}{\int d\sigma \, e^{-\beta A_c(\sigma)}} = e^{-\beta A(s)},$$

(12.7.6)

where

$$\left\langle \pi_P \left(\left\{ \mathbf{r}^{(1)}, ..., \mathbf{r}^{(P)} \right\} ; s \right) \right\rangle_f(\sigma) = \frac{\left\langle \pi_c \left(\left\{ \mathbf{r}^{(c)} \right\} ; \sigma \right) \pi_P \left(\left\{ \mathbf{r}^{(1)}, ..., \mathbf{r}^{(P)} \right\} ; s \right) \right\rangle_f}{\left\langle \pi_c \left(\left\{ \mathbf{r}^{(c)} \right\} ; \sigma \right) \right\rangle_f}$$

$$= \frac{\left\langle \delta \left(\mathcal{F}(\mathbf{r}_1^{(c)}, ..., \mathbf{r}_N^{(c)}) - \sigma \right) \pi_P \left(\left\{ \mathbf{r}^{(1)}, ..., \mathbf{r}^{(P)} \right\} ; s \right) \right\rangle_f}{\left\langle \delta \left(\mathcal{F}(\mathbf{r}_1^{(c)}, ..., \mathbf{r}_N^{(c)}) - \sigma \right) \right\rangle_f}.$$

(12.7.7)

According to eqns. (12.7.6) and (12.7.7), we only need to perform an enhanced sampling simulation targeting $\mathcal{F}(\mathbf{r}_1^{(c)}, ..., \mathbf{r}_N^{(c)})$ and save the result as a function of σ and the corresponding probability distribution in the denominator of eqn. (12.7.7), i.e., $P_c(\sigma) = \langle \delta \left(\mathcal{F}(\mathbf{r}_1^{(c)}, ..., \mathbf{r}_N^{(c)}) - \sigma \right) \rangle_f$ and associated free energy. The average of the all-bead free-energy estimator is then computed from eqn. (12.7.6) with these saved quantities.

12.8   Numerical evaluation of path integrals

Early numerical studies using path integrals to study condensed-phase problems focused on one or a few quantum particles in simple liquids (Sprik *et al.*, 1985; Sprik *et al.*, 1986; Coker *et al.*, 1987; Wallqvist *et al.*, 1987; Martyna *et al.*, 1993; Liu and Berne, 1993). Somewhat later, path integrals were applied to study quantum effects in bulk fluids such as water (Delbuono *et al.*, 1991; Chen *et al.*, 2003; Fanourgakis *et al.*, 2006; Paesani *et al.*, 2007; Paesani and Voth, 2009; Morrone and Car, 2008) and aqueous solutions (Marx *et al.*, 1999*a*; Tuckerman *et al.*, 2002), and in biological processes such as enzyme catalysis (Hwang *et al.*, 1991; Hwang and Warshel, 1996). Numerical path integration is also central to studies of lattice gauge theories (Weingarten and Petcher, 1981; Weingarten, 1982; Brown and Christ, 1988; Brown *et al.*, 1991). The application of path integrals to many types problems using increasingly sophisticated models and computational algorithms is now becoming routine. In this section, we will discuss the use of molecular dynamics and Monte Carlo techniques to evaluate path integrals numerically, identifying several technical challenges that affect the construction of the numerical algorithms and the formulation of the thermodynamic estimators. While our focus here will be on the quantum canonical ensemble, it is possible to perform isothermal-isobaric and grand canonical path integral simulations, the latter using the adaptive resolution approach outlined in Section 6.9. These will be discussed further in Section 12.8.7.

12.8.1 Path integral molecular dynamics

We begin our discussion with the molecular dynamics approach. It must be mentioned at the outset that molecular dynamics is used here *only* as a means of sampling the quantum canonical distribution. No quantum dynamical properties can be generated using the techniques in this subsection. In Chapter 14, we will revisit the quantum dynamics problem and see how *approximate* dynamical quantities can be generated within a path integral molecular dynamics framework.

Let us start by considering, once again, a single particle moving in a one-dimensional potential $U(\hat{x})$. In eqn. (12.3.9), we introduced the notion of the discrete partition function $Q_P(L, T)$, which is explicitly defined to be

$$Q_P(L,T) = \left(\frac{mP}{2\pi\beta\hbar^2}\right)^{P/2} \int_{D(L)} \mathrm{d}x_1 \cdots \mathrm{d}x_P$$

$$\times \exp\left\{-\frac{1}{\hbar}\sum_{k=1}^{P}\left[\frac{mP}{2\beta\hbar}(x_{k+1} - x_k)^2 + \frac{\beta\hbar}{P}U(x_k)\right]\right\}\bigg|_{x_{P+1}=x_1}. \tag{12.8.1}$$

Equation (12.8.1) can be manipulated to resemble the classical canonical partition function of a cyclic ring polymer moving in a classical potential $U(x)/P$ by recasting the prefactor as a set of Gaussian integrals over variables we will call $p_1, ..., p_P$ so that they resemble momenta conjugate to $x_1, ..., x_P$:

$$Q_P(L,T) = \int \mathrm{d}p_1 \cdots \mathrm{d}p_P \int_{D(L)} \mathrm{d}x_1 \cdots \mathrm{d}x_P$$

$$\times \exp\left\{-\beta\sum_{k=1}^{P}\left[\frac{p_k^2}{2m'} + \frac{1}{2}m\omega_P^2(x_{k+1} - x_k)^2 + \frac{1}{P}U(x_k)\right]\right\}\bigg|_{x_{P+1}=x_1}. \tag{12.8.2}$$

The partition function represented in eqn. (12.8.2) resembles a classical phase-space integral of a system of P pseudoparticles with momenta $p_1, ..., p_P$ in an effective potential

$$\phi(x_1, ..., x_P) = \frac{1}{2}m\omega_P^2(x_{k+1} - x_k)^2 + \frac{1}{P}U(x_k). \tag{12.8.3}$$

In the exponential of eqn. (12.8.2), we have replaced the prefactor of $1/\hbar$ with a β prefactor. We have also introduced a frequency $\omega_P = \sqrt{P}/(\beta\hbar)$, which we call the *chain frequency*, since it is the frequency of the harmonic nearest-neighbor coupling of our cyclic chain. Finally, the parameter m' appearing in the Gaussian integrals is formally given by $m' = mP/(2\pi\hbar)^2$. However, since the prefactor does not affect any equilibrium averages, including those used to calculate thermodynamic estimators, we are free to choose m' as we like. The resemblance of the partition function in eqn. (12.8.2) to that of a classical cyclic ring polymer of P points led Chandler and Wolynes (1981) to coin the term "classical isomorphism" and to exploit the isomorphism between the classical and approximate (since P is finite) quantum partition functions. The classical isomorphism is illustrated in Fig. 12.12. The figure depicts a cyclic ring polymer

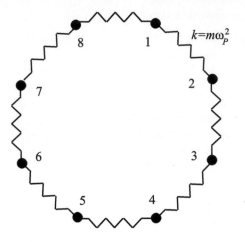

Fig. 12.12 Classical isomorphism: The figure shows a cyclic ring polymer having $P = 8$ described by the partition function in eqn. (12.8.2).

having $P = 8$ with a harmonic nearest-neighbor coupling constant $k = m\omega_P^2$. Because the ring polymer resembles a necklace, its P points are often referred to as "beads." According to the classical isomorphism, we can treat the cyclic ring polymer using all the techniques we have developed thus far for classical systems to obtain approximate quantum properties, and the latter can be systematically improved simply by increasing P.

The classical isomorphism allows us, in principle, to introduce a molecular dynamics scheme for eqn. (12.8.2), starting with the classical Hamiltonian

$$\mathcal{H}_{\text{cl}}(x,p) = \sum_{k=1}^{P} \left[\frac{p_k^2}{2m'} + \frac{1}{2} m\omega_P^2 (x_{k+1} - x_k)^2 + \frac{1}{P} U(x_k) \right] \Bigg|_{x_{P+1}=x_1} , \qquad (12.8.4)$$

which yields the following equations of motion:

$$\dot{x}_k = \frac{p_k}{m'}, \qquad \dot{p}_k = -m\omega_p^2 \left(2x_k - x_{k+1} - x_{k-1} \right) - \frac{1}{P} \frac{\partial U}{\partial x_k}. \qquad (12.8.5)$$

If eqns. (12.8.5) are coupled to a thermostat, as discussed in Section 4.11, then the dynamics will sample the canonical distribution in eqn. (12.8.2). Although one of the first path integral molecular dynamics calculations by Parrinello and Rahman (1984) employed a scheme of this type, it was simultaneously recognized by Hall and Berne (1984) that path integral molecular dynamics based on eqns. (12.8.5) can suffer from very slow convergence problems due to the wide range of time scales present in the dynamics. If one applies a normal mode transformation (see Section 1.7) to the harmonic coupling term in eqn. (12.8.4), the normal-mode frequencies range densely from 0 to $4P/(\beta\hbar)$. Thus, even for moderately large P, this constitutes a broad frequency spectrum. The time step that can be employed in a molecular dynamics algorithm is limited by the highest frequency, which means that the low-frequency modes, which are associated with large-scale changes in the shape of the cyclic chain, will be inadequately sampled unless very long runs are performed.

Because the normal-mode frequencies are closely spaced (approaching a continuum as $P \to \infty$), multiple time-scale integration algorithms such as RESPA (see Section 3.11) are insufficient to solve the problem. However, if we can devise a suitable coordinate transformation capable of uncoupling the harmonic term in eqn. (12.8.4), then we can replace the single fictitious mass m' in the classical Hamiltonian with a set of masses $m'_1, ..., m'_P$ such that only one harmonic frequency remains. Finally, we can then adjust the time step for stable integration of motion having that characteristic frequency and/or employ RESPA to the problem. This will ensure adequate sampling of all modes of the cyclic chain. In fact, we have already seen an example of such a transformation in Section 4.6. Equations (4.6.38) and (4.6.39) illustrate how a simple transformation uncouples the harmonic term for a model polymer that, when made cyclic, is identical to the discrete path integral. The one-dimensional analog of this transformation appropriate for eqn. (12.8.2) is

$$u_1 = x_1$$
$$u_k = x_k - \frac{(k-1)x_{k+1} + x_1}{k}, \qquad k = 2, ..., P, \qquad (12.8.6)$$

the inverse of which is

$$x_1 = u_1$$
$$x_k = u_k + \frac{k-1}{k}x_{k+1} + \frac{1}{k}u_1, \qquad k = 2, ..., P. \qquad (12.8.7)$$

Note that, as in eqn. (4.6.39), eqn. (12.8.7) is defined recursively. Because $x_1 = u_1$, the recursion can be seeded by starting with the $k = P$ term and working backwards to $k = 2$. The inverse can also be expressed in closed form as

$$x_1 = u_1$$
$$x_k = u_1 + \sum_{l=k}^{P} \frac{k-1}{l-1}u_l, \qquad k = 2, ..., P. \qquad (12.8.8)$$

The transformation defined in eqns. (12.8.6), (12.8.7), and (12.8.8) is known as a *staging transformation* because of its connection to a particular path integral Monte Carlo algorithm (Ceperley and Pollock, 1984), which we will discuss in the next section. The staging transformation was first introduced for path integral molecular dynamics by Tuckerman *et al.* (1993). The variables $u_1, ..., u_P$ are known as *staging variables*, as distinguished from the original variables, which are referred to as *primitive variables*.

We now proceed to develop a molecular dynamics scheme in terms of the staging variables. When the harmonic coupling term is evaluated using these variables, the result is

$$\sum_{k=1}^{P}(x_k - x_{k+1})^2 = \sum_{k=2}^{P}\frac{k}{k-1}u_k^2, \tag{12.8.9}$$

which is completely separable. Since the Jacobian of the transformation is 1, as we showed in eqn. (4.6.45), the discrete partition function becomes

$$Q_P(L,T) = \int \mathrm{d}p_1 \cdots \mathrm{d}p_P \int \mathrm{d}u_1 \cdots \mathrm{d}u_P$$

$$\times \exp\left\{-\beta \sum_{k=1}^{P}\left[\frac{p_k^2}{2m_k'} + \frac{1}{2}m_k\omega_P^2 u_k^2 + \frac{1}{P}U(x_k(u))\right]\right\}. \tag{12.8.10}$$

In eqn. (12.8.10), the parameters m_k are defined to be

$$m_1 = 0, \qquad m_k = \frac{k}{k-1}m, \qquad k = 2, ..., P, \tag{12.8.11}$$

and $m_1' = m$, $m_k' = m_k$. The notation $x_k(u)$ indicates the inverse transformation in eqn. (12.8.7) or (12.8.8). In order to evaluate eqn. (12.8.10), we can employ a classical Hamiltonian of the form

$$\tilde{\mathcal{H}}_{\mathrm{cl}}(u,p) = \sum_{k=1}^{P}\left[\frac{p_k^2}{2m_k'} + \frac{1}{2}m_k\omega_P^2 u_k^2 + \frac{1}{P}U(x_k(u))\right], \tag{12.8.12}$$

which leads to the equations of motion

$$\dot{u}_k = \frac{p_k}{m_k'}$$

$$\dot{p}_k = -m_k\omega_P^2 u_k - \frac{1}{P}\frac{\partial U}{\partial u_k}. \tag{12.8.13}$$

From the chain rule, the forces on the staging variables can be expressed recursively as

$$\frac{1}{P}\frac{\partial U}{\partial u_1} = \frac{1}{P}\sum_{l=1}^{P}\frac{\partial U}{\partial x_l}$$

$$\frac{1}{P}\frac{\partial U}{\partial u_k} = \frac{1}{P}\left[\frac{\partial U}{\partial x_k} + \frac{k-2}{k-1}\frac{\partial U}{\partial u_{k-1}}\right]. \tag{12.8.14}$$

The recursive staging force calculation is performed starting with $k = 2$ and using the first expression for $\partial U/\partial u_1$. Equations (12.8.13) need to be thermostatted to ensure that the canonical distribution is generated. The presence of the high-frequency force on each staging variable combined with the $1/P$ factor that attenuates the potential-energy derivatives leads to a weak coupling between these two forces. Therefore, it is important to have as much thermalization as possible in order to achieve equipartitioning of the energy. It is, therefore, strongly recommended (Tuckerman *et al.*, 1993) that path integral molecular dynamics calculations be carried out using the "massive" thermostatting mechanism described in Section 4.11. This protocol requires that a separate thermostat be attached to each Cartesian component of every staging variable. Thus, for the single-particle one-dimensional system described by eqns. (12.8.13), if Nosé-Hoover chain thermostats of length M are employed, the actual equations of motion would be

$$\dot{u}_k = \frac{p_k}{m'_k}$$

$$\dot{p}_k = -m_k \omega_P^2 u_k - \frac{1}{P}\frac{\partial U}{\partial u_k} - \frac{p_{\eta_{k,1}}}{Q_1} p_k$$

$$\dot{\eta}_{k,\gamma} = \frac{p_{\eta_{k,\gamma}}}{Q_k}$$

$$\dot{p}_{\eta_{k,1}} = \frac{p_k^2}{m'_k} - kT - \frac{p_{\eta_{k,2}}}{Q_k} p_{\eta_{k,1}}$$

$$\dot{p}_{\eta_{k,\gamma}} = \left[\frac{p_{\eta_{k,\gamma-1}}^2}{Q_k} - kT\right] - \frac{p_{\eta_{k,\gamma+1}}}{Q_k} p_{\eta_{k,\gamma}} \qquad \gamma = 2, ..., M-1$$

$$\dot{p}_{\eta_{k,M}} = \left[\frac{p_{\eta_{k,M-1}}^2}{Q_k} - kT\right], \tag{12.8.15}$$

where γ indexes the thermostat chain elements. When ω_P is the highest frequency in the system, the optimal choice for the parameters $Q_1, ..., Q_P$ are $Q_1 = kT\tau^2$ and $Q_k = kT/\omega_P^2$ for $k = 2, ..., P$. Here τ is a characteristic time scale of the corresponding classical system. Since each staging variable has its own thermostat of length M, the dimensionality of the thermostat phase space is $2MP$, which is considerably larger than the physical phase space! Luckily, with the exception of simple "toy" problems, the computational overhead of "massive" thermostatting is low relative to that of a force calculation in a complex system. Moreover, the massive thermostatting method is rapidly convergent, particularly when integrated using a multiple time scale algorithm such as the RESPA method of Section 3.11. Alternatively, Cayley matrix methods (see Problem 3.7) also allow increased efficiency by improving numerical stability and allowing large time steps to be employed (Korol *et al.*, 2019; Korol *et al.*, 2020; Rosa-Raíces *et al.*, 2021). The staging transformation is simple to implement and because of the recursive relations in eqns. (12.8.7) and (12.8.14), it scales linearly with P.

As an interesting alternative to the staging transformation, it is also possible to use the normal modes of the cyclic chain (Tuckerman *et al.*, 1993; Cao and Voth, 1994*b*). The normal mode transformation can be derived straightforwardly from a Fourier expansion of the periodic path

$$x_k = \sum_{l=1}^{P} a_l e^{2\pi i(k-1)(l-1)/P}. \tag{12.8.16}$$

The complex expansion coefficients a_l are then used to construct a transformation to a set of normal mode variables $u_1, ..., u_P$ via

$$u_1 = a_1, \qquad\qquad u_P = a_{(P+2)/2}$$

$$u_{2k-2} = \text{Re}(a_k), \qquad u_{2k-1} = \text{Im}(a_k). \tag{12.8.17}$$

The normal-mode transformation can also be constructed as follows: 1) Generate the matrix $A_{ij} = 2\delta_{ij} - \delta_{i,j-1} - \delta_{i,j+1}$, $i, j = 1, ..., P$, with the path periodicity conditions $A_{i0} = A_{iP}$, $A_{i,P+1} = A_{i1}$; 2) Diagonalize the matrix and save the eigenvalues and eigenvectors; 3) From the eigenvectors, construct the orthogonal matrix O_{ij} that diagonalizes A. The forward and inverse transformations are then given by

$$u_k = \frac{1}{\sqrt{P}} \sum_{l=1}^{P} O_{kl} x_l$$

$$x_k = \sqrt{P} \sum_{l=1}^{P} O_{kl}^{\mathsf{T}} u_l. \tag{12.8.18}$$

Although the eigenvalues emerge directly from the diagonalization procedure, they can also be constructed by hand according to

$$\lambda_{2k-1} = \lambda_{2k-2} = 2 \left[1 - \cos\left(\frac{2\pi(k-1)}{P} \right) \right] = 4 \sin^2\left(\frac{\pi(k-1)}{P} \right). \tag{12.8.19}$$

Note the twofold degeneracy. When evaluated in terms of the normal mode variables, the harmonic coupling term becomes

$$\sum_{k=1}^{P} (x_k - x_{k+1})^2 = \sum_{k=2}^{P} \lambda_k u_k^2. \tag{12.8.20}$$

As with the staging transformation, the harmonic term is now separable. The transformation also has unit Jacobian. Thus, the transformed partition function is identical to eqn. (12.8.10) if the masses m_k are defined as $m_k = m\lambda_k$, $m_1' = m$, and $m_k' = m_k$. With this identification, eqns. (12.8.12), (12.8.13), and (12.8.15) are applicable to the normal-mode case exactly as written. The only difference occurs when the chain rule is used to obtain the forces on the normal mode variables, whence we obtain

$$\frac{1}{P} \frac{\partial U}{\partial u_1} = \frac{1}{P} \sum_{l=1}^{P} \frac{\partial U}{\partial x_l}$$

$$\frac{1}{P} \frac{\partial U}{\partial u_k} = \frac{1}{\sqrt{P}} \sum_{l=1}^{P} \frac{\partial U}{\partial x_l} O_{lk}^{\mathsf{T}}. \tag{12.8.21}$$

In addition to its utility as a computational scheme, the normal-mode formulation of the path integral has several other interesting features. First, the variable u_1 can be shown to be equal to

$$u_1 = \frac{1}{P} \sum_{k=1}^{P} x_k \xrightarrow[P \to \infty]{} \frac{1}{\beta \hbar} \int_0^{\beta \hbar} d\tau \, x(\tau). \qquad (12.8.22)$$

That is, the sum on the left becomes the continuous integral on the right when $P \to \infty$. The variable u_1 is an average over all of the path variables and, therefore, corresponds, for finite P, to the center of mass of the ring polymer. This point is also path centroid defined previously. Note that the force on this mode is also just the average force $(1/P) \sum_{k=1}^{P} \partial U / \partial x_k$. It can be shown that, for the staging transformation, the force on the mode u_1 is also the average force; however this mode is not physically the same as the centroid.

In a seminal paper by Feynman and Kleinert (1986), it was shown that the path centroid could be used to capture approximate quantum effects in a system. Consider eqn. (12.8.10) with the variables $u_1, \dots, u_P$ representing the normal modes. If we integrate over the variables $u_2, \dots, u_P$ and the corresponding momenta $p_1, \dots, p_P$, then the result can, in the spirit of Section 8.10, be written as

$$Q_P(L,T) \propto \int dp_1 \, du_1 \exp \left\{ -\beta \left[\frac{p_1^2}{2m_1} + W(u_1) \right] \right\}, \qquad (12.8.23)$$

where $W(u_1)$ is the potential of mean force on the centroid given by

$$W(u_1) = -kT \ln \left\{ \int du_2 \cdots du_P \right.$$

$$\times \left. \exp \left[-\beta \sum_{k=1}^{P} \left(\frac{1}{2} m_k \omega_P^2 u_k^2 + \frac{1}{P} U(x_k(u)) \right) \right] \right\} \qquad (12.8.24)$$

up to an additive constant. Although we cannot determine $W(u_1)$ for an arbitrary potential $U(x)$, Feynman and Kleinert were able to derive an analytical expression for $W(u_1)$ for a harmonic oscillator using the functional integral techniques discussed in Section 12.4. In particular, they showed how to derive the parameters of a harmonic oscillator potential $U_{\text{ho}}(x; u_1)$ that minimize the expectation value $\langle U(x) - U_{\text{ho}}(x; u_1) \rangle$. The parameters of the potential depend on the position of the centroid so that the harmonic potential takes the general form $U_{\text{ho}}(x; u_1) = (1/2)\Omega^2(u_1)(x - u_1)^2 + L(u_1)$. That is, the frequency and vertical shift depend on the centroid u_1. The optimization procedure leads to a simple potential function $\tilde{W}(u_1)$ of the centroid that can then be used in eqn. (12.8.23) to obtain approximate quantum equilibrium and thermodynamic properties.

We close this section by showing how the path integral molecular dynamics protocol extends to N Boltzmann particles in d dimensions. Since most path integral calculations fall into this category, we will limit our discussion to these. Excellent descriptions of path integral algorithms for bosons and fermions can be found in the literature (Ceperley, 1995; Miura and Okazaki, 2000). The discrete N-particle partition function for Boltzmann particles follows directly from eqn. (12.6.11). After introducing dNP momentum integrations as in eqn. (12.8.2), the discrete partition function becomes

$$Q_P(N, V, T) = \prod_{i=1}^{N} \int \prod_{i=1}^{N} d\mathbf{r}_i^{(1)} \cdots d\mathbf{r}_i^{(P)} d\mathbf{p}_i^{(1)} \cdots d\mathbf{p}_i^{(P)} \tag{12.8.25}$$

$$\times \exp\left\{ -\beta \sum_{k=1}^{P} \left[\sum_{i=1}^{N} \frac{\mathbf{p}_i^{(k)2}}{2m_i'} + \sum_{i=1}^{N} \frac{1}{2} m_i \omega_P^2 \left(\mathbf{r}_i^{(k+1)} - \mathbf{r}_i^{(k)} \right)^2 + \frac{1}{P} U \left(\mathbf{r}_1^{(k)}, ..., \mathbf{r}_N^{(k)} \right) \right] \right\}$$

with the condition $\mathbf{r}_i^{(P+1)} = \mathbf{r}_i^{(1)}$. An important point to make about eqn. (12.8.25) is that the potential $U(\mathbf{r}_1^{(k)}, ..., \mathbf{r}_N^{(k)})$ only acts between beads with the same imaginary-time index k. This means that all beads with imaginary-time index 1 interact with each other, but these do not interact with beads having imaginary-time indices 2,3,....,P and so forth. This is illustrated for the case of two particles in Fig. 12.13.

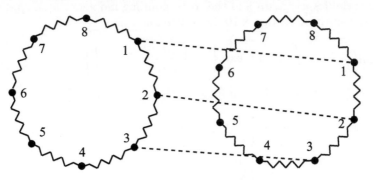

Fig. 12.13 Interaction pattern between two quantum particles represented as ring polymers within the discrete path integral framework. The cyclic chains obey the rule that only "beads" with the same imaginary-time index on different chains interact with each other.

The construction of a path integral molecular dynamics algorithm for N Boltzmann particles in d dimensions proceeds in the same manner as for a single particle in one dimension. First, a transformation from primitive to staging or normal-mode variables is performed for each quantum particle's cyclic path. In staging or normal-mode variables, the classical Hamiltonian from which the equations of motion are derived is

$$\mathcal{H} = \sum_{k=1}^{P} \left[\sum_{i=1}^{N} \frac{\mathbf{p}_i^{(k)2}}{2m_i^{(k)'}} + \sum_{i=1}^{N} \frac{1}{2} m_i^{(k)} \omega_P^2 \mathbf{u}_i^{(k)2} + \frac{1}{P} U \left(\mathbf{r}_1^{(k)}(\mathbf{u}_1), ..., \mathbf{r}_N^{(k)}(\mathbf{u}_N) \right) \right]. \tag{12.8.26}$$

Here, each primitive variable $\mathbf{r}_i^{(k)}$ depends on the staging or normal-mode variables with the same particle index i. In deriving the equations of motion from eqn. (12.8.26), the forces on the mode variables are obtained using eqns. (12.8.14) or (12.8.21). Importantly, if the equations of motion are coupled to Nosé-Hoover chains, it is critical to follow the protocol of coupling each component of each staging or normal-mode variable to its own thermostat, for a total of dNP thermostats. At first sight, this might seem like overkill because it adds $dNMP$ additional degrees of freedom to a system, where M is the length of each Nosé-Hoover chains. However, if we think back to Fig. 4.12 and note that a path integral, according to eqn. (12.8.26), is a collection of weakly coupled harmonic oscillators, then this protocol makes sense. Generally, the computational overhead of $dNMP$ thermostats is small compared to that associated with the calculation of the forces.

12.8.2 Path integral Monte Carlo

We saw in Chapter 7 that Monte Carlo methods are very effective for sampling an equilibrium distribution such as the canonical ensemble. Therefore, it is worth using a little space to discuss the calculation of path integrals using a Monte Carlo approach. The algorithm we will describe here uses many of the same ideas discussed in the previous subsection. In particular, for a quantum free particle in one dimension ($U(\hat{x}) = 0$), the discretized action expressed in terms of staging or normal-mode variables is just a sum of uncoupled harmonic oscillators, and as we saw in Section 3.8.3, these can be easily sampled using the Box-Muller method. The idea of path integral Monte Carlo, then, is to construct an $M(RT)^2$ algorithm (see Section 7.3.3), in which we sample the free particle distribution directly and use the change in the potential energy to build an acceptance probability. However, unlike path integral molecular dynamics, where a staging or normal-mode transformation can be applied to the entire ring polymer, the same cannot always be done in path integral Monte Carlo. The reason for this is that if P is sufficiently large, the complete set of staging or normal modes is simply too large to be sampled in its entirety in one Monte Carlo move: the average acceptance probability would, for most problems, simply be too low. Thus, in path integral Monte Carlo, staging or normal-mode transformations are applied to segments of the ring polymer of a certain length j that must be optimized to give a desired average acceptance probability.

In fact, we have already seen how to perform both normal-mode and staging transformations to a set of j particles with a harmonic nearest-neighbor coupling anchored to fixed endpoints in Sections 1.7 and 4.6, respectively. We first describe the staging transformation. The idea of staging was originally introduced by Ceperley and Pollock (1984) as a means of constructing an efficient Monte Carlo scheme. However, explicit variable transformations were not employed in the original work. Here we modify the original staging algorithm to incorporate explicit transformations. In order to sample a segment of length j of the free particle distribution for the ring polymer, we start by randomly choosing a starting bead. Suppose that the chosen bead has an imaginary-time index l. This bead forms one of the fixed endpoints of the segment and the other is $l+j+1$ beads away from this one. This leaves us with j intermediate beads having primitive coordinates $x_{l+1}, ..., x_{l+j}$. We now transform these primitive variables to staging variables using eqn. (4.6.38), which, for this case, appears as follows:

$$ u_{l+k} = x_{l+k} - \frac{k x_{l+k+1} - x_l}{(k+1)} \qquad k = 1, ..., j. \tag{12.8.27} $$

Equation (4.6.39) also allows a recursive inverse to be defined as

$$ x_{l+k} = u_{l+k} + \frac{k}{k+1} x_{l+k+1} + \frac{1}{k+1} x_l. \tag{12.8.28} $$

Applying the transformation to the following portion of the quantum kinetic energy

$$\frac{1}{2} m \omega_P^2 \sum_{k=0}^{j} (x_{l+k} - x_{l+k+1})^2 \,,$$

we obtain

$$\sum_{k=0}^{j} (x_{l+k} - x_{l+k+1})^2 = \sum_{k=1}^{j} \frac{k+1}{k} u_{l+k}^2 + \frac{1}{j+1} (x_{l+j+1} - x_l)^2 \qquad (12.8.29)$$

(see, also, eqn. (4.6.44)). Since the j staging variables have a distribution proportional to $\exp(-\beta m_k \omega_P^2 u_{l+k}^2 / 2)$, where $m_k = (k+1)m/k$, we sample each u_{l+k} randomly from its corresponding Gaussian distribution and then use eqn. (12.8.28) to generate the proposed move to new primitive variables x'_{l+k}, $k = 1, ..., j$. Let x denote the original coordinates x_{l+k}, with $k = 1, ..., j$, and let x′ denote the coordinates x'_{l+k} of the proposed move. The change in potential energy will be

$$\Delta U(\mathrm{x}, \mathrm{x}') = \frac{1}{P} \sum_{k=1}^{j} \left[U(x'_{l+k}) - U(x_{l+k}) \right], \qquad (12.8.30)$$

and the acceptance probability is then

$$A(\mathrm{x}'|\mathrm{x}) = \min \left[1, e^{-\beta \Delta U(\mathrm{x}, \mathrm{x}')} \right]. \qquad (12.8.31)$$

Thus, if the move lowers the potential energy, it will be accepted with probability 1; otherwise, it is accepted with probability $\exp[-\beta \Delta U(\mathrm{x}, \mathrm{x}')]$. On the average, P/j such moves will displace all the beads of the ring polymer, and hence the set of P/j staging moves is called a *Monte Carlo pass*.

The algorithm works equally well with a normal-mode transformation between fixed endpoints x_l and x_{l+j+1}, as in eqn. (1.7.12). In this case, the normal-mode variables are sampled from independent Gaussian distributions, and eqn. (12.8.31) is used to determine whether the move is accepted or rejected.

The staging and normal-mode schemes described above move only the internal modes of the cyclic chain. Therefore, in both algorithms, one additional move is needed, which is a displacement of the chain as a whole. This can be achieved by an attempted displacement of the uncoupled mode variable u_1 (recall that u_1 is the centroid variable in the normal-mode scheme) according to

$$u'_{1,\alpha} = u_{1,\alpha} + \frac{1}{\sqrt{d}} \left(\zeta_\alpha - 0.5 \right) \Delta. \qquad (12.8.32)$$

Here, α runs over the spatial dimensions and Δ is a displacement length (see, also, eqn. (7.3.33)). Since the positions of all the beads change under such a trial move, the potential energy changes as

$$\Delta U(\mathrm{x}, \mathrm{x}') = \frac{1}{P} \sum_{k=1}^{P} \left[U(x'_k) - U(x_k) \right]. \qquad (12.8.33)$$

Again, eqn. (12.8.31) determines whether the move is accepted or rejected. A complete Monte Carlo pass, therefore, requires P/j staging or normal-mode moves plus one move of the centroid. Typically, the parameter j is chosen such that the average acceptance probability is 40%. For a system of N particles, the algorithm extends straightforwardly. First a ring polymer is chosen at random, and a Monte Carlo pass is performed on that chain. On average, N such passes will move the entire system.

We conclude this subsection by noting an important difference between path integral Monte Carlo and path integral molecular dynamics. In molecular dynamics, a single time step generates a move of the entire system, while in Monte Carlo, each individual attempted move only changes a part of the system. This difference becomes important when implementing path integrals on parallel computing platforms. path integral molecular dynamics parallelizes much more readily than staging or normal-mode path integral Monte Carlo. Thus, if a molecular dynamics algorithm could be constructed with a convergence efficiency that rivals Monte Carlo, then the former becomes a competitive method. As part of our discussion of thermodynamic estimators in the next subsection, we will also present a comparison of the molecular dynamics and Monte Carlo approaches for a simple system.

12.8.3 Numerical aspects of thermodynamic estimators

As noted in Section 12.3, the estimators in eqns. (12.3.20), (12.3.24), and (12.6.13) suffer from large fluctuations in the kinetic energy due to their linear dependence on P. The consequence of this dependence is that in highly quantum systems, which require a large number of discretizations, it becomes increasingly difficult to converge such estimators. A solution to this dilemma was presented by Herman, Bruskin, and Berne (1982), who employed a path integral version of the virial theorem. For a single particle in one dimension, the theorem states

$$\frac{P}{2\beta} - \left\langle \frac{1}{2}m\omega_P^2 \sum_{k=1}^{P}(x_k - x_{k+1})^2 \right\rangle_f = \left\langle \frac{1}{2P}\sum_{k=1}^{P}x_k\frac{\partial U}{\partial x_k} \right\rangle_f. \tag{12.8.34}$$

Before we prove this theorem, we demonstrate its advantage in a simple application. First, note that when eqn. (12.8.34) is substituted into eqn. (12.3.19), a new energy estimator known as the *virial energy estimator* results:

$$\varepsilon_{\text{vir}}(x_1, ..., x_P) = \frac{1}{P}\sum_{k=1}^{P}\left[\frac{1}{2}x_k\frac{\partial U}{\partial x_k} + U(x_k)\right]. \tag{12.8.35}$$

The elimination of the kinetic energy yields an energy estimator with a much lower variance and, therefore, better convergence behavior than the primitive estimator of eqn. (12.3.20). In Fig. 12.14, we show a comparison between the instantaneous fluctuations and cumulative averages of the primitive and virial estimators for a harmonic oscillator with $m = 1$ and $\omega = 10$ computed using staging molecular dynamics with $P = 32$, $P = 64$, and $P = 128$ beads. The figure shows how the fluctuations, shown in grey, grow with P while the fluctuations in the virial estimator are insensitive to P. Despite the fact that the fluctuations of the primitive estimator grow with P, the cumulative averages between the two estimators agree for all P. This illustrates the idea that in any path integral simulation, one should monitor both estimators and

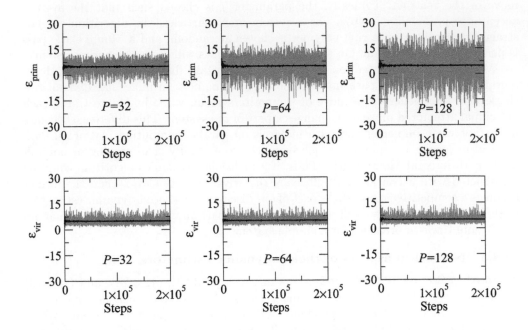

Fig. 12.14 Instantaneous fluctuations (grey) and cumulative averages (black) of the primitive (top row) and virial (bottom row) energy estimators for a harmonic oscillator simulated with staging path integral molecular dynamics using $P = 32$ (left column), $P = 64$ (middle column), and $P = 128$ (right column) beads.

ensure that they agree. If they do not, this should be taken as a sign of a problem in the simulation.

We now proceed to prove the theorem. First, we define a function $\alpha(x_1, ..., x_P)$ as

$$\alpha(x_1, ..., x_P) = \frac{1}{2} m\omega_P^2 \sum_{k=1}^{P} (x_k - x_{k+1})^2 . \qquad (12.8.36)$$

Note that the effective potential in eqn. (12.2.28) can now be written as

$$\phi(x_1, ..., x_P) = \alpha(x_1, ..., x_P) + \frac{1}{P} \sum_{k=1}^{P} U(x_k)$$

$$\equiv \alpha(x_1, ..., x_P) + \gamma(x_1, ..., x_P). \qquad (12.8.37)$$

Recalling the discussion of Euler's theorem in Section 6.2, the function $\alpha(x_1, ..., x_P)$ is a homogeneous function of degree 2. Hence, applying Euler's theorem, we can write $\alpha(x_1, ..., x_P)$ as

$$\alpha(x_1, ..., x_P) = \frac{1}{2} \sum_{k=1}^{P} x_k \frac{\partial \alpha}{\partial x_k}. \qquad (12.8.38)$$

Now consider the average $\langle \alpha \rangle_f$ over the finite-P path integral distribution f of eqn. (12.3.9), which we can write as

$$
\langle \alpha \rangle_f = \frac{C_P}{2Q_P} \int dx_1 \cdots dx_P \sum_{k=1}^{P} x_k \frac{\partial \alpha}{\partial x_k} e^{-\beta \alpha(x_1, \ldots, x_P)} e^{-\beta \gamma(x_1, \ldots, x_P)}
$$

$$
= -\frac{C_P}{2\beta Q_P} \int dx_1 \cdots dx_P \sum_{k=1}^{P} \left(x_k \frac{\partial}{\partial x_k} e^{-\beta \alpha(x_1, \ldots, x_P)} \right) e^{-\beta \gamma(x_1, \ldots, x_P)}. \tag{12.8.39}
$$

where $C_P = (mP/2\pi\beta\hbar^2)^{1/2}$. Integrating eqn. (12.8.39) by parts yields

$$
\langle \alpha \rangle_f = \frac{C_P}{2\beta Q_P} \int dx_1 \cdots dx_P \; e^{-\beta \alpha(x_1, \ldots, x_P)} \sum_{k=1}^{P} \frac{\partial}{\partial x_k} \left[x_k e^{-\beta \gamma(x_1, \ldots, x_P)} \right]
$$

$$
= \frac{C_P}{2\beta Q_P} \int dx_1 \cdots dx_P \left[P - \frac{\beta}{P} \sum_{k=1}^{P} x_k \frac{\partial U}{\partial x_k} \right] e^{-\beta \gamma(x_1, \ldots, x_P)} e^{-\beta \alpha(x_1, \ldots, x_P)}
$$

$$
= \frac{P}{2\beta} - \left\langle \frac{1}{2P} \sum_{k=1}^{P} x_k \frac{\partial U}{\partial x_k} \right\rangle_f, \tag{12.8.40}
$$

from which eqn. (12.8.34) follows.

Using the virial estimator, we now present a comparison (Fig. 12.15) of path integral molecular dynamics with no variable transformations (top row), path integral molecular dynamics with a staging transformation (middle row), and staging path integral Monte Carlo (bottom row). The system is a one-dimensional harmonic oscillator with $\beta\hbar\omega = 15.8$, $m\omega/\hbar = 0.03$, and $P = 400$. With these parameters, the thermodynamic energy is dominated by the ground-state value $\hbar\omega/2$. The figure shows the instantaneous value of the virial estimator (left column), the cumulative average of the virial estimator (middle column), and the error bar in the value of the estimator. The error bar is calculated by grouping individual samplings from molecular dynamics or Monte Carlo into blocks of size n, computing the average over each block, and then computing the error bar from these block averages with respect to the global average (Cao and Berne, 1989). The purpose of this type of "block averaging" is to remove unwanted correlations between successive samplings. As the right column indicates, the error bar starts off small and then reaches a plateau when the blocks are large enough that correlations are no longer present. This example demonstrates that without variable transformations, path integral molecular dynamics performs rather poorly, but when a staging transformation is employed, molecular dynamics and staging Monte Carlo are equally efficient, as evidenced by the fact that they converge to the same error bar after the same number of steps.

As written, the virial estimator in eqn. (12.8.35) is only valid for bound systems because the term $x_k(\partial U/\partial x_k)$ is not translationally invariant. This problem can be circumvented by defining the virial part of the estimator in terms of the path centroid. The generalization of eqn. (12.8.34) for N particles in d dimensions is

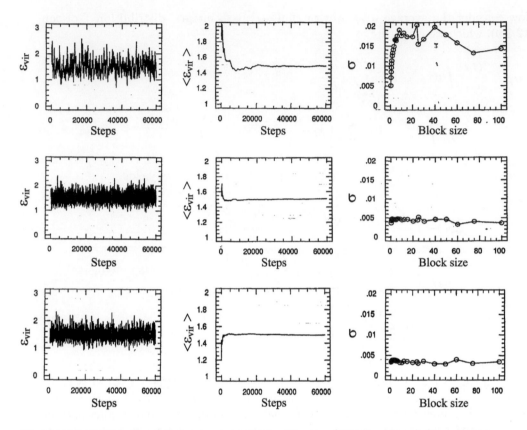

Fig. 12.15 (*Left column*) Instantaneous virial estimator. (*Middle column*) Cumulative average of virial estimator. (*Right column*) Error bar as a function of block size. (*Top row*) Path integral molecular dynamics with no variable transformations. (*Middle row*) path integral molecular dynamics with staging transformation. (*Bottom row*) Staging path integral Monte Carlo with $j = 80$. All energies are in units of $\hbar\omega$ (reprinted with permission from Tuckerman et al., *J. Chem. Phys.* **99**, 2796 (1993), copyright, American Institute of Physics).

$$\frac{dNP}{2\beta} - \left\langle \sum_{i=1}^{N} \sum_{k=1}^{P} \frac{m_i \omega_P^2}{2} \left(\mathbf{r}_i^{(k)} - \mathbf{r}_i^{(k+1)} \right)^2 \right\rangle_f = \left\langle \frac{1}{2P} \sum_{i=1}^{N} \sum_{k=1}^{P} \mathbf{r}_i^{(k)} \cdot \frac{\partial U}{\partial \mathbf{r}_i^{(k)}} \right\rangle_f, \quad (12.8.41)$$

and for N particles in d dimensions, the generalization of the virial estimator is

$$\varepsilon_{\text{vir}}(\{\mathbf{r}^{(1)}, ..., \mathbf{r}^{(P)}\}) = \frac{dN}{2\beta} + \sum_{k=1}^{P} \sum_{i=1}^{N} \frac{1}{2P} \left(\mathbf{r}_i^{(k)} - \mathbf{r}_i^{(c)} \right) \cdot \frac{\partial U}{\partial \mathbf{r}_i^{(k)}}$$

$$+ \frac{1}{P} \sum_{k=1}^{P} U \left(\mathbf{r}_1^{(k)}, ..., \mathbf{r}_N^{(k)} \right), \quad (12.8.42)$$

where $\mathbf{r}_i^{(c)}$ is the centroid of particle i. Similarly, by applying the path integral virial theorem to the pressure estimator in eqn. (12.6.13), one can derive a virial pressure estimator

$$\mathcal{P}_{\text{vir}}(\{\mathbf{r}^{(1)}, ..., \mathbf{r}^{(P)}\}) = \frac{dNkT}{V} - \frac{1}{V} \sum_{i=1}^{N} \frac{1}{P} \mathbf{r}_i^{(c)} \cdot \sum_{k=1}^{P} \frac{\partial}{\partial \mathbf{r}_i^{(c)}} U(\mathbf{r}_1^{(k)}, ..., \mathbf{r}_N^{(k)}, V)$$

$$- \frac{1}{P} \sum_{k=1}^{P} \frac{\partial}{\partial V} U(\mathbf{r}_1^{(k)}, ..., \mathbf{r}_N^{(k)}, V), \qquad (12.8.43)$$

which includes the possibility that the potential depends explicitly on the volume.

Path integral simulations should be applied whenever nuclear quantum effects are expected to be important, for example, when light nuclei such as hydrogen are present. Proton transfer reactions will often exhibit nontrivial quantum effects such as tunneling and zero-point motion. In malonaldehyde ($C_3H_4O_2$), a small, cyclic organic molecule with an internal $O-H\cdots O$ hydrogen bond, the hydrogen bond can reverse its polarity and become $O\cdots H-O$ via a proton transfer reaction (see Fig. 12.16, top). A free-energy profile for this reaction can be computed using the blue moon ensemble approach of Section 8.7 with a reaction coordinate $\delta = d_{O_1H} - d_{O_2H}$, where d_{O_1H} and d_{O_2H} are the distances between the two oxygens and the transferring proton. The free-energy profile in this reaction coordinate exhibits a typical double-well shaped. Interestingly, even at 300 K, there is a pronounced quantum effect on this free energy. The quantum free-energy profiles can be computed using the centroid of the reaction coordinate denoted δ^c (see eqn. (12.7.5)) in the figure (Voth *et al.*, 1989; Voth, 1993). Figure 12.16 shows that, at the level of theory utilized in the simulation (Tuckerman and Marx, 2001), the quantum free-energy barrier to the reaction decreases by approximately 2 kcal/mol from 3.6 kcal/mol to 1.6 kcal/mol due to the inclusion of nuclear quantum effects, which include both zero-point energy and tunneling contributions. Since enzymatic reaction barriers are in this energetic neighborhood, this simple example illustrates the important role that nuclear quantum effects, particularly quantum tunneling, can play in real biological proton transfer reaction. Another interesting point is that if only the transferring H is treated quantum mechanically, the reduction in the free energy is underestimated by roughly 0.4 kcal/mol, which shows that secondary nuclear quantum effects of the molecular skeleton are also important. These free-energy profiles are generated using *ab initio* molecular dynamics (Car and Parrinello, 1985; Marx and Hutter, 2009) and *ab initio* path integral techniques, in which a dynamical or path integral simulation is driven by forces generated from electronic structure calculations performed "on the fly" as the simulation is carried out (Marx and Parrinello, 1994; Marx and Parrinello, 1996; Tuckerman *et al.*, 1996; Marx and Hutter, 2009; Tuckerman, 2002).

In the malonaldehyde example of Fig. 12.16, nuclear quantum effects reduced the classical barrier by approximately 2 kcal/mol. Let us consider another example, in which nuclear quantum effects can actually lead to an increase of a barrier relative to the classical result. The example is a bulk host-guest system, structure-II hydrogen clathrate, which consists of cages of hydrogen-bonded water molecules that hold hydrogen molecules within the cages. More specifically, this material is composed of small cages made up of twelve pentagonal rings and large cages made up of twelve pentagonal

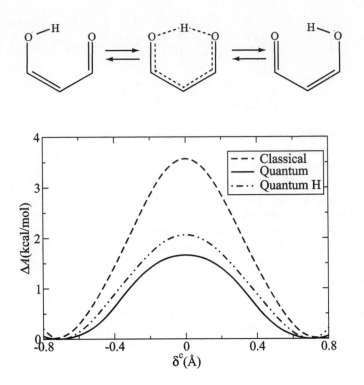

Fig. 12.16 (*Top*) Sketch of the internal proton transfer reaction in malonaldehyde. (*Bottom*) Classical, quantum, and quantum-H free-energy profiles at 300 K.

and four hexagonal rings. Hydrogen molecules can diffuse among the cages, typically moving from large cage to large cage through the hexagonal faces. The rate of diffusion is predominantly determined by the free-energy barrier for a hydrogen molecule to pass through a hexagonal face, a barrier that is strongly influenced by nuclear quantum effects at different temperatures. These barriers were computed by Cendagorta *et al.* (2016) using path integral based enhanced sampling, specifically a quantum version of d-AFED/TAMD (Cendagorta *et al.*, 2021). In their study, Cendagorta *et al.* employed a flexible quantum TIP4P potential (q-TIP4P/F) (Habershon *et al.*, 2009) to model water-water interactions, while H_2-H_2O interactions consisted of Coulomb and Lennard-Jones contributions. The point charges for the water molecules were taken from the q-TIP4P/F model, and charges on the H_2 atoms were set at $0.4238e$ on each H atom and $-0.9864e$ on the molecular center of mass. Lennard-Jones interactions were taken between the H_2 mass centers and between the H_2 mass center and the water oxygens. The parameters were $\sigma = 3.038$ Å and $\epsilon = 0.06816$ kcal/mol on the H_2 mass center and $\sigma = 3.166$ Å and $\epsilon = 0.1554$ kcal/mol on the water oxygen (Alavi *et al.*, 2005). Progress of the transfer process was measured by the reaction coordinate

$$q(\mathbf{r}) = \left[\frac{1}{2}\left(\mathbf{r}_{H_1} + \mathbf{r}_{H_2}\right) - \mathbf{R}_A\right] \cdot \mathbf{u}_{AB} - \frac{|\mathbf{R}_B - \mathbf{R}_A|}{2}. \qquad (12.8.44)$$

Here, $\mathbf{r}_{H_1}$ and $\mathbf{r}_{H_2}$ are the coordinates of the two H atoms in the H_2 molecule, $\mathbf{R}_A$ and $\mathbf{R}_B$ are the centers of the left and right cages, respectively, and $\mathbf{u}_{AB}$ is the unit vector along the direction between these two centers. Free-energy profiles using the centroid-based free-energy estimator of eqn. (12.7.5) for a single H_2 molecule transferring from one cage to another (all other cages are taken to be unoccupied) at temperatures of 50 K, 17 K, and 8 K are shown in Fig. 12.17 together with snapshots showing the path integral beads of the H_2 molecule during a transfer event. As with malonaldehyde, the free-energy profiles are constructed using the centroid estimator in eqn. (12.7.5). The figure shows that at 50 K, the quantum barrier is actually higher than the classical barrier. This is due to the delocalization of the H_2 ring polymer along the direction orthogonal to the line that joins the two cage centers. The increased barrier reflects the fact that it is more difficult for a delocalized object to pass through the hexagonal ring than it is for a classical point particle to do so. As the temperature is decreased to 17 K and 8 K, we see from the snapshots that the H_2 ring polymer spreads and is simultaneously located in both cages; this spreading is accompanied by a decrease in the bead density close to the plane of the hexagonal ring. This phenomenon is particularly pronounced at 17 K and 8 K: the quantum free-energy profiles at these temperatures flatten, and the quantum barrier is now below the classical barrier, which is a signature of deep tunneling of the H_2 molecule through the free-energy barrier.

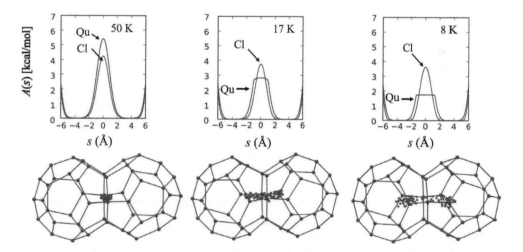

Fig. 12.17 Classical (Cl) and quantum (Qu) free-energy profiles for H_2 transfer between two large cages in structure-II hydrogen clathrate through a hexagonal face at temperatures 50 K, 17 K, and 8 K with snapshots of a path configuration of H_2 during transfer shown below the profiles (reprinted with permission from Cendagorta *et al.*, *Phys. Chem. Chem. Phys.* **18**, 32169 (2016), copyright Royal Society of Chemistry).

Before leaving this section, we note that care is needed when deciding whether to apply path integrals to a given problem. One must consider both the physical nature of the problem and the source of the potential-energy model used in each application before deciding to embark on a path integral investigation. Consider, for example, an empirical potential-energy function $U(\mathbf{r}_1, ..., \mathbf{r}_N)$ whose parameters are obtained by careful fits to experimental data. Since experiments are inherently quantum mechanical, the potential model U contains nuclear quantum effects by construction. Therefore, if such a model were used in conjunction with path integrals, quantum effects would be "double counted." By contrast, simulations performed with potential models whose parameters are fit to *ab initio* calculations do not contain quantum effects implicitly, and therefore, these models are strictly correct *only* when used in conjunction with path integral methods. Examples include the water model of Fanourgakis *et al.* (2006, 2006), Paesani *et al.* (2007), the many-body polarizable (MB-Pol) water model of Medders *et al.* (2014), and machine learning potentials for water and ice (Kapil *et al.*, 2016); these generally yield accurate results when simulated via path integrals. When nuclear quantum effects can be safely neglected, then simulations using classical molecular dynamics or Monte Carlo algorithms can be performed. Similarly, when potential energies and forces in a simulation are computed "on the fly" from the electronic structure via the *ab initio* molecular dynamics technique (Car and Parrinello, 1985; Marx and Hutter, 2009), these simulations should, strictly speaking, be performed within the path integral framework since nuclear quantum effects are not implicitly included in this approach. Simulations using this technique have yielded important insights, for example, into the solvation and transport of charge defects (in the form of hydronium and hydroxide ions) in aqueous solution (Marx *et al.*, 1999*a*; Tuckerman *et al.*, 2002; Marx *et al.*, 2010) and in acid crystal hydrates (Hayes *et al.*, 2009; Hayes *et al.*, 2011).

12.8.4   Open-chain path integral molecular dynamics

We saw in Section 12.3 that the calculation of momentum-dependent observables requires sampling off-diagonal elements of the density matrix, which can only be achieved by sampling thermal paths with different endpoints. For a single particle in one dimension, the integral that must be performed is given by eqn. (12.3.15). Molecular dynamics or Monte Carlo can also be used to sample configurations of an open-chain path integral, and, as was the case with cyclic or ring polymers, transforming to mode variables ensures efficient sampling. An open chain path integral is equivalent to the bead-spring model of Section 1.7 and Section 4.6.3, in which, respectively, normal mode and staging transformations were introduced. Here, let us consider the complete staging transformation for an open chain with beads at positions $\mathbf{r}_1, ..., \mathbf{r}_{P+1}$ for a single particle in three dimensions. The transformation takes the form

$$\mathbf{u}_1 = \frac{1}{2}\left(\mathbf{r}_1 + \mathbf{r}_{P+1}\right)$$

$$\mathbf{u}_k = \mathbf{r}_k - \frac{1}{k}\left[(k-1)\mathbf{r}_{k+1} + \mathbf{r}_1\right], \qquad k = 2, ..., P$$

$$\mathbf{u}_{P+1} = \left(\mathbf{r}_1 - \mathbf{r}_{P+1}\right) \tag{12.8.45}$$

(Morrone *et al.*, 2007; Pérez and Tuckerman, 2011). Note that $\mathbf{u}_1$ and $\mathbf{u}_{P+1}$ are the average and relative positions of the two endpoint beads, respectively. In open-chain staging variables, the potential $\phi(\mathbf{r}_1, ..., \mathbf{r}_{P+1})$, analogous to that in eqn. (12.2.28) but corresponding to the open-chain distribution in eqn. (12.3.15), becomes

$$\phi(\mathbf{r}_1, ..., \mathbf{r}_{P+1}) = \sum_{k=1}^{P} \left[\frac{1}{2} m\omega_P^2 (\mathbf{r}_{k+1} - \mathbf{r}_k)^2 + \frac{1}{2P} \left(U(\mathbf{r}_{k+1}) + U(\mathbf{r}_k) \right) \right]$$

$$\phi(\mathbf{u}_1, ..., \mathbf{u}_{P+1}) = \frac{1}{2P} m\omega_P^2 \mathbf{u}_{P+1}^2 + \sum_{k=2}^{P} \frac{1}{2} m_k \omega_P^2 \mathbf{u}_k^2 + \frac{1}{P} \sum_{k=2}^{P} U(\mathbf{r}_k(\mathbf{u}_1, ..., \mathbf{u}_{P+1}))$$

$$+ \frac{1}{2P} \left[U(\mathbf{r}_1(\mathbf{u}_1, \mathbf{u}_{P+1})) + U(\mathbf{r}_{P+1}(\mathbf{u}_1, \mathbf{u}_{P+1})) \right] \qquad (12.8.46)$$

where

$$\mathbf{r}_k(\mathbf{u}_1, ..., \mathbf{u}_{P+1}) = \mathbf{u}_1 + \frac{(P/2 - k + 1)}{P} \mathbf{u}_{P+1} + \sum_{l=k}^{P} \frac{k-1}{l-1} \mathbf{u}_l, \qquad (12.8.47)$$

$\mathbf{r}_1(\mathbf{u}_1, \mathbf{u}_{P+1}) = \mathbf{u}_1 + \mathbf{u}_{P+1}/2$, $\mathbf{r}_{P+1}(\mathbf{u}_1, \mathbf{u}_{P+1}) = \mathbf{u}_1 - \mathbf{u}_{P+1}/2$, and $m_k = [k/(k-1)]m$. In a normal mode transformation, the same assignment would be made to $\mathbf{u}_1$ and $\mathbf{u}_{P+1}$, and $\mathbf{u}_k$, $k = 2, ..., P$ would be obtained by a normal mode transformation. It is also straightforward to generalize these expressions for N particles in d dimensions obeying Boltzmann statistics. As an example application, Morrone *et al.* (2007) studied the momentum distribution of a proton distribution in bulk water, and in this study, the ring polymer of a single proton was opened for the calculation of the distribution, while the paths of the remaining particles were kept closed. We note, for completeness, that it is also possible to use closed paths within a perturbative approach to obtain a momentum distribution (Lin *et al.*, 2011).

In general, because they are less constrained by the cyclic wrapping condition of ring polymers, convergence of the distribution of open chains is slower than that of cyclic chains, which makes open path integral simulations more time consuming. In the last subsections of this chapter, we will discuss schemes for reducing the cost of both closed and open chain path integral calculations.

12.8.5 Reducing the number of beads for improved efficiency

Path integral Monte Carlo calculations and path integral molecular dynamics calculations require P evaluations of potential energies and forces for molecular dynamics. This means that such calculations are P times more costly in computational time than corresponding classical Monte Carlo and molecular dynamics calculations, for which $P = 1$. However, the index-wise interaction pattern between beads on different ring polymers (or between different sub-chains of joined ring polymers in systems of identical bosons or fermions) mean that path integral calculations exhibit nearly perfect parallel scaling with P and one can simply spread a path integral calculation out over P nodes on a parallel computer. As it happens, however, a variety of techniques exists for reducing the number of beads needed to converge properties of interest from a path integral calculation, and these can be leveraged to reduce the computational time. We will discuss two such strategies in this section.

The first approach harkens back to a concept introduced in Section 3.13 for the classical propagator. In that section, we introduced the Baker-Campbell-Hausdorff formula (cf. eqn. (3.13.4)) for expressing a low-order factorization of an operator of the form $\exp(\lambda(\hat{A} + \hat{B}))$ when $[\hat{A}, \hat{B}] \neq 0$. When we apply eqn. (3.13.4) to the high-temperature Boltzmann operator $\exp[-\beta\hat{\mathfrak{H}}/P]$, we obtain

$$e^{-\beta\hat{U}/2P}e^{-\beta\hat{K}/P}e^{-\beta\hat{U}/2P} = \exp\left[-\frac{\beta}{P}\left(\hat{\mathfrak{H}} + \sum_{k=1}^{\infty}\left(\frac{\beta}{P}\right)^{2k}\hat{C}_k\right)\right], \qquad (12.8.48)$$

where $\hat{\mathfrak{H}} = \hat{K} + \hat{U}$. As noted in Section 3.13, the operators $\hat{C}_k$ are multiply nested commutators of $\hat{K}$ and $\hat{U}$, e.g.,

$$\hat{C}_1 = -\frac{1}{24}[\hat{U} + 2\hat{K}, [\hat{U}, \hat{K}]]. \qquad (12.8.49)$$

If the sum over k in eqn. (12.8.48) is truncated after the $k = 1$ term, and a trace is taken of both sides, it can be shown that

$$\text{Tr}\left[e^{-\beta\hat{\mathfrak{H}}}\right] \approx \text{Tr}\left(\left[e^{-\beta\hat{K}/P}e^{-\beta\hat{C}/P}\right]^P\right) + \mathcal{O}\left(\beta^5 P^{-4}\right), \qquad (12.8.50)$$

where $\hat{C}$ is the operator

$$\hat{C} = \hat{U} + \frac{1}{24}\left(\frac{\beta}{P}\right)^2\left[\hat{U}, \left[\hat{K}, \hat{U}\right]\right] \qquad (12.8.51)$$

(Takahashi and Imada, 1984). The double commutator can be evaluated using the fact that $[\hat{p}, f(\mathbf{r})] = -i\hbar(\partial f/\partial\mathbf{r})$, so that when eqn. (12.8.50) is used to evaluate the partition function for N Boltzmann particles, the result is (see Problem 12.9)

$$Q_P^{(4)}(N, V, T) = \prod_{i=1}^{N}\left(\frac{m_i P}{2\pi\beta\hbar^2}\right)^{dP/2}\int\prod_{i=1}^{N}d\mathbf{r}_i^{(1)}\cdots d\mathbf{r}_i^{(P)} \qquad (12.8.52)$$

$$\times \exp\left\{-\beta\sum_{k=1}^{P}\left[\sum_{i=1}^{N}\frac{1}{2}m_i\omega_P^2\left(\mathbf{r}_i^{(k+1)} - \mathbf{r}_i^{(k)}\right)^2 + U^{(4)}\left(\mathbf{r}_1^{(k)}, ..., \mathbf{r}_N^{(k)}\right)\right]\right\},$$

where

$$U^{(4)}(\mathbf{r}_1, ..., \mathbf{r}_N) = \frac{1}{P}U(\mathbf{r}_1, ..., \mathbf{r}_N) + \sum_{i=1}^{N}\frac{1}{24m_i P^2\omega_P^2}\left[\frac{\partial U}{\partial\mathbf{r}_i}\right]^2$$

$$\equiv \frac{1}{P}U(\mathbf{r}_1, ..., \mathbf{r}_N) + U_{\text{TI}}(\mathbf{r}_1, ..., \mathbf{r}_N). \qquad (12.8.53)$$

Here, $U_{\text{TI}}(\mathbf{r}_1, ..., \mathbf{r}_N)$ is referred to as the *Takahashi-Imada* (TI) potential. It is evident from the form of eqn. (12.8.53) that use of the TI potential $U^{(4)}(\mathbf{r}_1, ..., \mathbf{r}_N)$ in path integral molecular dynamics would require calculation of second derivatives of the potential. We will shortly describe some ways in which this can be handled efficiently within path integral molecular dynamics; use of path integral Monte Carlo would clearly avoid the need for second derivatives. However, there is a more immediate technical issue concerning the calculation of observables within the TI framework. Because eqn. (12.8.52) can only be derived for the trace of an exponential operator, an expression for $\langle \hat{A} \rangle$ requires recasting the expectation value as

$$\langle \hat{A} \rangle = -\frac{1}{\beta Q(N,V,T)} \frac{\mathrm{d}}{\mathrm{d}\lambda} \mathrm{Tr} \left[e^{-\beta(\hat{\mathcal{H}} + \lambda \hat{A})} \right]_{\lambda=0} \tag{12.8.54}$$

so that the quantity $\hat{U} + \lambda \hat{A}$ can be treated as an effective "potential" for the purposes of applying the TI factorization. When this is done for a position-dependent operator, the resulting estimator for $\langle \hat{A} \rangle$ becomes

$$a_P^{(4)}(\mathbf{R}) = \frac{1}{P} \sum_{k=1}^{P} \left[a\left(\mathbf{r}_1^{(k)}, ..., \mathbf{r}_N^{(k)}\right) + \sum_{i=1}^{N} \frac{1}{12 m_i P \omega_P^2} \left(\frac{\partial U}{\partial \mathbf{r}_i^{(k)}} \right) \cdot \left(\frac{\partial a}{\partial \mathbf{r}_i^{(k)}} \right) \right], \tag{12.8.55}$$

where, for simplicity, we have introduced the notation $\mathbf{R}$ to denote the full set of N ring-polymer bead coordinates. A primitive energy estimator can also be derived using the standard relation $E = -(\partial/\partial\beta) \ln Q^{(4)}(N,V,T)$ following the procedure in Section 12.3. This yields

$$\varepsilon^{(4)}(\mathbf{R}) = \frac{3NP}{2\beta} - \frac{\omega_P^2}{2} \sum_{i=1}^{N} \sum_{k=1}^{P} m_i \left(\mathbf{r}_i^{(k)} - \mathbf{r}_i^{(k+1)} \right)^2 + \frac{1}{P} \sum_{k=1}^{P} U(\mathbf{r}_1^{(k)}, ..., \mathbf{r}_N^{(k)})$$

$$+ 3U_{\mathrm{TI}}(\mathbf{r}_1^{(k)}, ..., \mathbf{r}_N^{(k)}). \tag{12.8.56}$$

If we follow the procedure for deriving a virial estimator, the result is an expression that involves the second derivative of the potential. We will return to this issue later in the present section.

The most straightforward way to compute an average at fourth order within the TI scheme that also avoids the need to compute second derivatives of the potential is simply to reweight an average of $a_P^{(4)}$ using standard (second-order) path integral molecular dynamics. If done this way, a fourth-order average would be computed using

$$\langle \hat{A} \rangle^{(4)} = \frac{\langle a_P^{(4)}(\mathbf{R}) e^{-\beta U_{\mathrm{TI}}(\mathbf{R})} \rangle^{(2)}}{\langle e^{-\beta U_{\mathrm{TI}}(\mathbf{R})} \rangle^{(2)}} \tag{12.8.57}$$

(Jang *et al.*, 2001; Pérez and Tuckerman, 2011), where $\langle \cdots \rangle^{(2)}$ indicates an average over a standard (second-order) path integral sampling. The advantage of reweighting at fourth order is that it can be done *a posteriori*, with no changes to a path integral code, simply by saving all of the quantities needed to construct the average. However, in systems with many degrees of freedom, extensivity of $U_{\mathrm{TI}}(\mathbf{R})$ will cause slow convergence of reweighted averages as a function of simulation time if the system visits many configurations at second order for which $U_{\mathrm{TI}}(\mathbf{R})$ is large. Another technical difficulty with the TI scheme concerns the need to compute the derivative of the function $a_P(\mathbf{r}_1^{(k)}, ..., \mathbf{r}_N^{(k)})$ with respect to $\mathbf{r}_i^{(k)}$ in order to evaluate the second term in eqn. (12.8.55). For certain observables, the derivative can be performed straightforwardly; however, should we wish to compute, for example, a radial distribution function (not an uncommon task), the observable would involve δ-functions of the form $\delta(|\mathbf{r}_i^{(k)} - \mathbf{r}_j^{(k)}| - r)$ and its derivatives with respect to $\mathbf{r}_i^{(k)}$ and $\mathbf{r}_j^{(k)}$. Similar derivatives would be needed for other types of distributions. The presence of derivatives of a δ-function is troublesome at best (see Appendix A); if we attempt to integrate the second term in the average by parts, we again obtain terms involving second derivatives of the potential.

An approach that simplifies the calculation of higher-order path integrals is that of Suzuki and Chin (Suzuki, 1995; Chin, 1997) and is similar in spirit to the approach of Suzuki and Yoshida discussed in Section 4.12. Here, imaginary-time steps of different lengths are used to cancel error terms in an exponential operator factorization scheme. Unlike the Suzuki–Yoshida approach, which has positive and negative step lengths, for the Boltzmann operator it is critical that all steps have the same sign. The Suzuki–Chin (SC) factorization takes the form

$$
e^{-\beta(\hat{K}+\hat{U})/P} = e^{-\beta\hat{U}_e/6P} e^{-\beta\hat{K}/2P} e^{-2\beta\hat{U}_m/3P} e^{-\beta\hat{K}/2P} e^{-\beta\hat{U}_e/6P} + \mathcal{O}\left(\frac{\beta}{P}\right)^5, \quad (12.8.58)
$$

where the operators $\hat{U}_e$ and $\hat{U}_m$ are defined as

$$
\hat{U}_e(\mathbf{r}_1, ..., \mathbf{r}_N) = \hat{U}(\mathbf{r}_1, ..., \mathbf{r}_N) + \sum_{i=1}^{N} \frac{\alpha}{6m_i} \left(\frac{\beta\hbar}{P}\frac{\partial\hat{U}}{\partial\mathbf{r}_i}\right)^2
$$

$$
\hat{U}_m(\mathbf{r}_1, ..., \mathbf{r}_N) = \hat{U}(\mathbf{r}_1, ..., \mathbf{r}_N) + \sum_{i=1}^{N} \frac{1-\alpha}{12m_i} \left(\frac{\beta\hbar}{P}\frac{\partial\hat{U}}{\partial\mathbf{r}_i}\right)^2, \quad (12.8.59)
$$

with $\alpha \in [0, 1]$ being an adjustable parameter. Within the SC scheme, the canonical partition function is given by

$$
Q_P^{(4)}(N, V, T) = \prod_{i=1}^{N} \left(\frac{m_i P}{2\pi\beta\hbar^2}\right)^{dP/2} \int \prod_{i=1}^{N} d\mathbf{r}_i^{(1)} \cdots d\mathbf{r}_i^{(P)} \, w_{\mathrm{SC}}(\mathbf{R}) \quad (12.8.60)
$$

$$
\times \exp\left\{-\beta \sum_{k=1}^{P} \left[\sum_{i=1}^{N} \frac{1}{2} m_i \omega_P^2 \left(\mathbf{r}_i^{(k+1)} - \mathbf{r}_i^{(k)}\right)^2 + U\left(\mathbf{r}_1^{(k)}, ..., \mathbf{r}_N^{(k)}\right),\right]\right\}
$$

where

$$
w_{\mathrm{SC}}(\mathbf{R}) = w_{\mathrm{JJ}}(\mathbf{R}) \exp\left\{-\frac{\beta}{3P} \sum_{k=1}^{P/2} \left[U(\mathbf{r}_1^{(2k)}, ..., \mathbf{r}_N^{(2k)}) - U(\mathbf{r}_1^{(2k-1)}, ..., \mathbf{r}_N^{(2k-1)})\right]\right\}
$$

$$
(12.8.61)
$$

and

$$
w_{\mathrm{JJ}}(\mathbf{R}) = \exp\left\{ -\beta \sum_{i=1}^{N} \sum_{k=1}^{P/2} \frac{1}{9m_i \omega_P^2} \left[\frac{\alpha}{P^2} \left(\frac{\partial U(\mathbf{r}_1^{(2k-1)}, ..., \mathbf{r}_N^{(2k-1)})}{\partial \mathbf{r}_i^{(2k-1)}} \right)^2 \right. \right.
$$

$$
\left. \left. + \frac{1-\alpha}{P^2} \left(\frac{\partial U(\mathbf{r}_1^{(2k)}, ..., \mathbf{r}_N^{(2k)})}{\partial \mathbf{r}_i^{(2k)}} \right)^2 \right] \right\}. \tag{12.8.62}
$$

The advantage of the SC factorization is that it applies directly to the Boltzmann operator, as eqn. (12.8.58) makes clear, rather than to traces involving the Boltzmann operator. For this reason, the traces needed to compute observables take a more familiar form. For example, if we wish to compute the ensemble average of an operator $\hat{A}$ at fourth order within the SC approach, we could start with a weighted version of eqn. (12.3.5),

$$
\langle \hat{A} \rangle^{(4)} = \frac{\langle a(\mathbf{r}_1^{(1)}, ..., \mathbf{r}_N^{(1)}) w_{\mathrm{SC}}(\mathbf{R}) \rangle^{(2)}}{\langle w_{\mathrm{SC}}(\mathbf{R}) \rangle^{(2)}}. \tag{12.8.63}
$$

Ultimately, we would like to apply the cyclic invariance of the ring polymer to recast eqn. (12.8.63) in the form of eqn. (12.3.7), which has better convergence properties. However, in the SC formulation, even beads and odd beads have different effective potentials and are, thus, not equivalent. Consequently, cyclic invariance of the ring polymer in the SC framework only applies to cyclic relabelings of the form $k \to k+2$, rather than $k \to k+1$, as could be used to derive eqn. (12.3.7). Performing the cyclic relabeling $k \to k+2$, we obtain $P/2$ equivalent expressions and can, therefore, express eqn. (12.8.63) as

$$
\langle \hat{A} \rangle^{(4)} = \frac{2}{P} \sum_{k=1}^{P/2} \frac{\langle a(\mathbf{r}_1^{(2k-1)}, ..., \mathbf{r}_N^{(2k-1)}) w_{\mathrm{SC}}(\mathbf{R}) \rangle^{(2)}}{\langle w_{\mathrm{SC}}(\mathbf{R}) \rangle^{(2)}}, \tag{12.8.64}
$$

which allows us to compute ensemble averages within the SC formalism using the reweighting approach. (Problem 12.16 asks for a derivation of primitive and virial energy estimators within the SC scheme.) Another important point is that the SC approach can be easily extended to open chains (Pérez and Tuckerman, 2011; Cendagorta *et al.*, 2018) (see Section 12.8.4), which is not possible within the TI framework.

Let us now return to the question of how to avoid reweighting the averages within a fourth-order approximation. If we wished to generate the path integral molecular dynamics trajectories directly within either the TI or SC framework, we would need forces from $U_{\mathrm{TI}}(\mathbf{r}_1, ..., \mathbf{r}_N)$ or from terms in the exponential of $w_{\mathrm{JJ}}(\mathbf{R})$. Consider, for example, the derivative of $U_{\mathrm{TI}}(\mathbf{r}_1, ..., \mathbf{r}_N)$, given analytically by

$$
\frac{\partial U_{\mathrm{TI}}}{\partial \mathbf{r}_j} = C \sum_{i=1}^{N} \frac{1}{m_i} \frac{\partial^2 U}{\partial \mathbf{r}_j \partial \mathbf{r}_i} \cdot \frac{\partial U}{\partial \mathbf{r}_i} = -C \sum_{i=1}^{N} \frac{\partial^2 U}{\partial \mathbf{r}_j \partial \mathbf{r}_i} \cdot \frac{\mathbf{f}_i}{m_i}, \tag{12.8.65}
$$

where $C = 1/(12P^2\omega_P^2)$ and $\mathbf{f}_i = -\partial U/\partial \mathbf{r}_i$. Rather than calculate the second deriva-

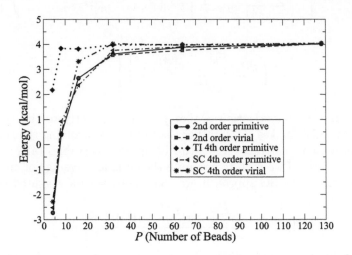

Fig. 12.18 Convergence as a function of P of second-order primitive and virial estimators, the TI primitive estimator, and the SC primitive and virial estimators for a system of 64 q-SPC/Fw water molecules.

tive matrix of the potential, we can use a finite-difference formula (Kapil *et al.*, 2016)

$$\frac{\partial U_{\mathrm{TI}}}{\partial \mathbf{r}_j} = C \lim_{\epsilon \to 0} \frac{1}{2\epsilon\delta} \left[\mathbf{f}_i(\mathbf{r} - \epsilon\delta\mathbf{u}) - \mathbf{f}_i(\mathbf{r} + \epsilon\delta\mathbf{u}) \right], \tag{12.8.66}$$

where $\mathbf{r} \equiv \mathbf{r}_1, ..., \mathbf{r}_N$, $\mathbf{u} \equiv \mathbf{u}_1, ..., \mathbf{u}_N$, with

$$\mathbf{u}_i = \frac{\mathbf{f}_i}{m_i}$$

$$\delta = \left[\frac{1}{NP} \sum_{k=1}^{P} \mathbf{f}_k \cdot \mathbf{f}_k \right]^{-1/2} . \tag{12.8.67}$$

(Problem 12.17 asks for a derivation of eqn. (12.8.66), which can be adapted for the SC approach as well.) Because the fourth-order corrections are small, the finite-difference approximation, although inexact, nevertheless yields stable path integral molecular dynamics trajectories (Kapil *et al.*, 2016).

A simple application to liquid water serves to demonstrate the convergence of the energy estimators for the different second-order and fourth-order methods as a function of P. In this example, the q-SPC/Fw model (Paesani *et al.*, 2006) is employed. Path integral molecular dynamics simulations of a system of 64 water molecules in a periodic box of length 12.416 Å are carried out at 300 K with $P = 2, 4, 8, 16, 32, 64,$ and 128 beads. Figure 12.18 compares second-order primitive and virial estimators, the TI fourth-order estimator, and primitive and virial fourth-order estimators (see Problem 12.16) for the SC scheme with $\alpha = 0$. All averages are computed using the reweighting scheme. The figure clearly shows the improvements in convergence with P for the fourth-order methods. Interestingly, the TI method, despite its technical difficulties, exhibits the best convergence of the different methods (although the convergence of the SC virial estimator is only slightly less efficient). From this example, it is clear that fourth-order methods offer an advantage. In the next section, we explore methods for reducing the computational overhead of path integral molecular dynamics within the second-order approach.

12.8.6 Ring polymer contraction

In Section 3.11, we introduced the idea of multiple time-stepping in classical molecular dynamics to avoid having to evaluate expensive and slowly varying forces at every time point in a simulation, thus increasing the efficiency of the simulation. Path inte-

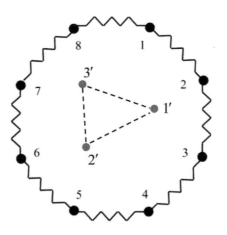

Fig. 12.19 Illustration of imaginary-time multiple time-stepping, in which expensive and slowly varying potential and force evaluations are performed on a smaller ring polymer with $n < P$ beads (here, $P = 8$, $n = 3$), connected by dashed lines.

gration in statistical mechanics involves stepping through imaginary time at intervals of $\epsilon = \beta\hbar/P$. As is clear from eqn. (12.8.25), the full potential $U(\mathbf{r}_1, ..., \mathbf{r}_N)$ must be evaluated at each step in imaginary time. It is possible to develop an analogous scheme in imaginary time for potentials that can be decomposed into a relatively cheap and rapidly varying term $U_f(\mathbf{r}_1, ..., \mathbf{r}_N)$, which is evaluated at every imaginary-time slice, and an expensive but slowly varying term $U_s(\mathbf{r}_1, ..., \mathbf{r}_N)$, which only needs to be evaluated at a small fraction of the imaginary-time slices. In this way, we could reduce the computational time needed to evaluate path integrals by reducing the number of expensive potential and force evaluations around the ring polymer. However, because of the strong coupling of the ring-polymer modes, rather than subdividing beads, it proves more optimal to subdivide modes and evaluate the slow potential on a ring poly-mer generated from low-frequency modes only. Such a ring polymer is "contracted" down to one having fewer beads, which achieves the goal of reducing the number of evaluations of the slow potential. This idea is illustrated in Fig. 12.19. The notion of a contracted ring polymer was first introduced by Markland and Manolopoulos (2008a, 2008b), who formulated a contraction scheme in normal-mode variables.

In order to construct the contracted ring polymer in Fig. 12.19 using normal modes, we start with the P modes with frequencies given in eqn. (12.8.19) and we discard $P-n$ of these modes, where $n < P$. We thus retain the n lowest-frequency modes, and we use these n modes to construct the contracted ring polymer. To this end, we simply need an orthogonal transformation $\tilde{O}_{kl}$ for transforming the n lowest-frequency mode variables $\mathbf{u}_i^{(k)}$ for particle i to n primitive variables $\mathbf{s}_i^{(k)}$ of the contracted ring polymer. The transformation matrix $\tilde{O}_{kl}$ is constructed in the same way as the original matrix O_{kl} of eqn. (12.8.18) except that we start with a matrix $\tilde{A}_{ij} = 2\delta_{ij} - \delta_{i,j-1} - \delta_{i,j+1}$, where $i,j = 1,...,n$ with the condition $\tilde{A}_{i0} = \tilde{A}_{in}$, $\tilde{A}_{i,n+1} = \tilde{A}_{i1}$. This smaller $n \times n$ matrix is then diagonalized, and the resulting orthogonal matrix that diagonalizes $\tilde{A}$ is used to construct the coordinates $\mathbf{s}_i^{(k)}$ of the contracted ring polymer via

$$\mathbf{s}_i^{(k)} = \sqrt{n} \sum_{l=1}^{n} \tilde{O}_{kl}^{\mathsf{T}} \mathbf{u}_i^{(l)}. \tag{12.8.68}$$

However, eqn. (12.8.18) allows us to express the coordinates of the contracted ring polymer in terms of the coordinates of the full P-bead ring polymer as

$$\mathbf{s}_i^{(k)} = \sqrt{\frac{n}{P}} \sum_{l=1}^{n} \sum_{m=1}^{P} \tilde{O}_{kl}^{\mathsf{T}} O_{lm} \mathbf{r}_i^{(m)}$$

$$\equiv \sum_{m=1}^{P} T_{km} \mathbf{r}_i^{(m)}, \tag{12.8.69}$$

where the matrix T is given by

$$T_{km} = \sqrt{\frac{n}{P}} \sum_{l=1}^{n} \tilde{O}_{kl}^{\mathsf{T}} O_{lm}. \tag{12.8.70}$$

Note that T is not a square matrix but is, rather, $n \times P$. Moreover, note that if $n = 1$, then the $1 \times P$ T would simply generate the centroid of the original P-bead ring polymer, *i.e.*,

$$\mathbf{s}_i^{(1)} = \mathbf{r}_i^{(c)} = \frac{1}{P} \sum_{m=1}^{P} \mathbf{r}_i^{(m)}, \tag{12.8.71}$$

indicating that the original ring polymer had contracted down to a single point. Equa-tion (12.8.71) also makes clear that the original and contracted ring polymers always have the same centroid position, even if $n > 1$. The computational efficiency of ring polymer contraction results from approximating U_s as

$$\frac{1}{P}\sum_{k=1}^{P} U_\mathrm{s}(\mathbf{r}_1^{(k)}, ..., \mathbf{r}_N^{(k)}) \approx \frac{1}{n}\sum_{k=1}^{n} U_\mathrm{s}(\mathbf{s}_1^{(k)}, ..., \mathbf{s}_N^{(k)}). \qquad (12.8.72)$$

Finally, the chain rule allows us to relate the forces on the contracted ring polymer from the slow potential U_s to those needed to evolve the full P-bead ring polymer. The relation connecting these forces is

$$\frac{1}{P}\frac{\partial U_\mathrm{s}}{\partial \mathbf{r}_i^{(m)}} = \frac{1}{n}\sum_{k=1}^{n} \frac{\partial U_\mathrm{s}}{\partial \mathbf{s}_i^{(k)}} T_{km}. \qquad (12.8.73)$$

Equation (12.8.73) allows us to evaluate the forces from the slow potential only on the n beads of the contracted ring polymer and then to distribute those forces throughout the full P-bead ring polymer. The generality of formulating the ring polymer contraction scheme in terms of transformation matrices allows the approach to be extended to staging modes with relative ease (see Problem 12.19).

As a simple example of the ring polymer contraction scheme, we consider the quantum analog of the problem laid out in eqn. (3.14.2). That is, we take a harmonic oscillator potential $U(\hat{x}) = m\omega^2\hat{x}^2/2$ and break it up into two contributions of different frequencies

$$U(\hat{x}) = \frac{1}{2}\lambda m\omega^2\hat{x}^2 + \frac{1}{2}(1 - \lambda)m\omega^2\hat{x}^2, \qquad (12.8.74)$$

where $\lambda \in [0, 1]$. If λ is close to 1, then the first term represents a high-frequency oscillator, which we denote $U_\mathrm{fast}(\hat{x})$, and the second term represents a low-frequency oscillator, which we denote $U_\mathrm{slow}(\hat{x})$. In Fig. 12.20, we show the instantaneous and cumulative average of the virial energy estimator for the harmonic oscillator potential, subdivided as in eqn. (12.8.74), with $\lambda = 0.93$ and the same parameters used to generate Fig. 12.15. The analytical result is $\varepsilon_\mathrm{vir} = 1.5$ in the units of the problem. The three panels show different choices of j and n: Larger values of n indicate more evaluations of the slow potential in the ring polymer contraction scheme. We see from the figure that there is very little difference in the behavior of the cumulative average and instantaneous fluctuations in the virial when 100 evaluations of the slow potential per time step are compared with only $n = 1$ evaluation per time step. In simulations for which the slow potential carries a high computational overhead, the gains in efficiency when using ring polymer contraction can be substantial. As an example, the reader is referred to the work of Marsalek and Markland (2016), who studied several small protonated or deprotonated water clusters and liquid water at an *ab initio* level. In their work, these authors found that a crude, semi-empirical potential could be used as U_fast, while the difference between the desired *ab initio* and semi-empirical potentials could serve as U_slow and could be evaluated on a few beads, or even just one bead, without compromising the accuracy of the results. In this way, full *ab initio* accuracy of path integral calculations could be achieved at a computational overhead similar to that of an *ab initio* calculation with classical nuclei, but using a crude, inexpensive

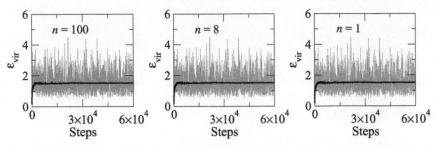

Fig. 12.20 Instantaneous and cumulative average of the virial energy estimator for the harmonic oscillator with potential divided into high- and low-frequency terms as given in eqn. (12.8.74). The parameters of the simulation are the same as those used to generate Fig. 12.15.

electronic structure level of theory as the "fast" potential and the difference between an accurate and the cruder level of theory as the "slow" potential, showing substantial gains in efficiency.

12.8.7 Path integrals in other ensembles

Throughout this chapter, we have focused on the representation of the quantum canonical partition function and associated equilibrium observables in terms of the Feynman path integral. However, other quantum ensembles can also be formulated in terms of a path integral. Before closing out this chapter, we will comment briefly on these formulations and point to specific references where details can be found. In a quantum isothermal-isobaric ensemble, one starts with the trace formula for the partition function $\Delta(N, P, T) = \text{Tr}\{\exp[-\beta(\hat{\mathcal{H}} + PV)]\}$, and in a quantum grand canonical ensemble, one would start with $\mathcal{Z}(\mu, V, T) = \text{Tr}\{\exp[-\beta(\hat{\mathcal{H}} - \mu N)]\}$. Recall, however, that these partition functions can be expressed in terms of the quantum canonical partition function with an additional volume integration (see Section 5.3) or particle sum (see Section 6.4). Consequently, performing path integral molecular dynamics or Monte Carlo calculations in these ensembles requires specifying how the volume fluctuations or particle number fluctuations can be sampled.

The literature on the quantum isothermal-isobaric ensemble contains descriptions of both Monte Carlo and molecular dynamics algorithms for path integral calculations (Marchi *et al.*, 1990; Scharf *et al.*, 1993; Martyna *et al.*, 1999). The procedure for deriving the path integral is exactly the same as described in this chapter; specifically, the trace is performed in the coordinate basis using a Trotter decomposition of the quantum canonical density matrix. Generating the associated volume fluctuations can be achieved using a virial estimator for the pressure involving all of the path integral ring-polymer modes; however, this tends to lead to large pressure fluctuations. If, instead, the path integral virial theorem is used, a centroid-based pressure estimator can be derived (Martyna *et al.*, 1999) (cf. eqn. (12.8.43)), which reduces the pressure fluctuations considerably and leads to more efficient path integral Monte Carlo and path integral algorithms. Although pressure is an intensive thermodynamic quantity, it is also macroscopic, and one would not expect quantum fluctuations within the ring polymers to play a significant role in determining the magnitude of the volume fluctuations, suggesting that the rapidly varying terms from the high-frequency ring-polymer modes largely cancel out over long simulation times but lead to large errors due to imperfect cancellation. Once it is determined that a centroid-based pressure estimator can be employed, the formulation of an isothermal-isobaric path integral molecular dynamics algorithm in terms of normal modes or staging variables becomes straightforward.

In a grand canonical ensemble, the challenge to varying the number of quantum particles is the insertion and deletion of entire ring polymers, and once again, both Monte Carlo (Wang *et al.*, 1997) and molecular dynamics algorithms (Kreis *et al.*, 2016; Agarwal and Delle Site, 2016; Kreis *et al.*, 2017) have been developed for this purpose. Because direct insertion of a ring polymer would generally be expected to be a low-probability operation due to their extended nature, an efficient grand canonical path integral Monte Carlo algorithm would require a database of pre-equilibrated configurations. These could be drawn from an ideal-gas distribution or from a previously equilibrated canonical path integral calculation on the exact same system for which one seeks to perform the grand canonical ensemble calculation. However, even the availability of such a database leads only to modest increases in the acceptance probability. The technical problems associated with the insertion of ring polymers in path integral grand canonical Monte Carlo simulations can be elegantly addressed within molecular dynamics using a quantum analog of the adaptive resolution approach of Section 6.9. The key to this type of algorithm is to "collapse" the ring polymers to roughly the size of point particles as they move from the fine-grained region, where observables are computed, to the coarse-grained or "classical" reservoir region, as illustrated in Fig. 12.21. The figure shows how ring polymers in the fine-grained quantum region have physical mass and are fully expanded at the desired temperature. As they pass through the hybrid region, the mass increases such that when they enter the coarse-grained or classical region, they collapse roughly to point particles. Such a procedure allows for seamless flow of particles between the two regions. Collapse of the ring polymers can be achieved by assigning a position-dependent mass to each quantum particle, the position dependence of which is controlled by the λ parameter discussed in Section 6.9.

Let us now analyze how a position-dependent mass in a quantum Hamiltonian influences the derivation of the path integral. Consider a single quantum particle in one dimension with position operator $\hat{x}$ and momentum operator $\hat{p}$; the Hamiltonian takes the form

$$\hat{\mathcal{H}} = \frac{1}{2}\hat{p}\mu^{-1}(\hat{x})\hat{p} + U(\hat{x}), \tag{12.8.75}$$

where $\mu(\hat{x})$ is an operator corresponding to the position-dependent mass and $\mu^{-1}(\hat{x})$ is its inverse. Kreis *et al.* (2016) discuss the path integral quantization of such a Hamiltonian, which involves modifying some of the coordinate-space matrix elements needed for the path integral. In particular, it can be shown that for the kinetic energy operator, the coordinate-space matrix elements become

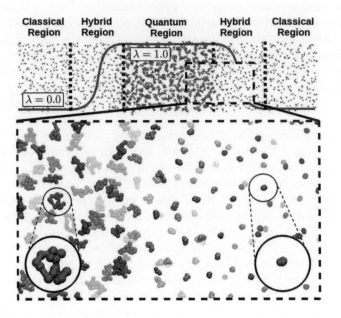

| Classical | Hybrid | Quantum | Hybrid | Classical |
| Region | Region | Region | Region | Region |

Fig. 12.21 Illustration of the quantum adaptive resolution scheme (reproduced with permission from Kreis *et al.*, *J. Chem. Theor. Comput.* **12**, 3030 (2016), copyright American Chemical Society).

$$\langle x_{k+1}|e^{-\beta\hat{p}\mu^{-1}(\hat{x})\hat{p}/2P}|x_k\rangle = \left(\frac{\mu(x_{k+1})P}{2\pi\beta\hbar^2}\right)^{1/2}$$

$$\times \exp\left\{-\frac{\beta\mu(x_{k+1})P}{2(\beta\hbar)^2}\left[(x_{k+1}-x_k)-\frac{\beta\hbar^2}{2P}\frac{\mathrm{d}\mu^{-1}}{\mathrm{d}x}\bigg|_{x_{k+1}}\right]^2\right\}. \quad (12.8.76)$$

The additional mass-derivative term can be neglected assuming that the average nearest-neighbor bead coupling is comparatively large, and this fact provides some guidance on the choice of the mass function $\mu(\hat{x})$: we should choose this mass function such that $\mu(\hat{x})$ has the value of the physical mass in the fine-grained region and is sufficiently large in the coarse-grained region so as to cause the desired collapse of the ring polymers (see Fig. 12.21). With this condition, particle number fluctuations consistent with a grand canonical ensemble can be generated.

12.9   Problems

12.1. Recall the definitions of the quantum propagator $U(t)$ and the canonical density matrix $\rho(\beta)$:

$$U(t) = e^{-i\hat{H}t/\hbar}, \qquad \rho(\beta) = e^{-\beta\hat{H}}.$$

Prove the following relations:

$$\rho(\beta) = C_1 \int_{-\infty}^{\infty} ds\, e^{-i\beta\hbar s} \int_0^{\infty} dt\, e^{-st} U(t)$$

$$U(t) = C_2 \int_{-\infty}^{\infty} d\omega\, e^{\omega t/\hbar} \int_0^{\infty} d\beta\, e^{i\omega\beta} \rho(\beta)$$

In the course of your proofs, find the (constant and possibly complex) coefficients C_1 and C_2 that make the relations true.

12.2. Derive primitive and virial estimators for the full pressure tensor $P_{\alpha\beta}$ defined by

$$P_{\alpha\beta} = \frac{kT}{\det(\mathbf{h})} \sum_{\gamma} \frac{\partial \ln Q}{\partial h_{\alpha\gamma}} h_{\beta\gamma},$$

where $h_{\mu\nu}$ is the cell matrix.

*12.3. Derive a virial form for the heat capacity at constant volume using the thermodynamic relation

$$C_V = k\beta^2 \frac{\partial^2 \ln Q}{\partial \beta^2}.$$

12.4. Derive eqns. (12.6.7) and (12.6.8) and generalize these equations to the case of N particles in three dimensions.

12.5. Derive eqn. (12.4.22).

12.6. The following problem considers the path integral theory for the tunneling of a particle through a barrier.

a. Show that the path integral expression for the density matrix can be written as:

$$\rho(x, x'; \beta) = \int_{x(-\beta\hbar/2)=x}^{x(\beta\hbar/2)=x'} \mathcal{D}[x] \exp\left[-\frac{1}{\hbar} \int_{-\beta\hbar/2}^{\beta\hbar/2} d\tau \left(\frac{1}{2} m\dot{x}^2(\tau) + U(x(\tau)) \right) \right].$$

b. Consider a double-well potential of the form

$$U(x) = \frac{\omega^2}{8a^2}(x^2 - a^2)^2.$$

Show that, for a particle of unit mass, the dominant path for the density matrix $\rho(-a, a; \beta)$ is given by

$$x(\tau) = a\tanh[(\tau - \tau_0)\omega/2]$$

in the low-temperature limit with negligible error in the endpoint conditions. This path is called an *instanton* or *kink* solution. Discuss the behavior of this trajectory in imaginary time τ.

   c. Calculate the classical imaginary-time action for the kink solution.

*12.7. Consider two distinguishable particles in one dimension with respective coordinates x and y and conjugate momenta p_x and p_y with a Hamiltonian

$$\hat{\mathcal{H}} = \frac{\hat{p}_x^2}{2m} + \frac{\hat{p}_y^2}{2M} + U(\hat{x}) + \frac{1}{2}M\omega^2\hat{y}^2 - \lambda\hat{x}\hat{y}.$$

   a. Show that the density matrix $\rho(x, y, x', y'; \beta)$ can be written in the form

$$\rho(x, y, x', y'; \beta)$$

$$= \int_{x(0)=x}^{x(\beta\hbar)=x'} \mathcal{D}x(\tau) \exp\left[-\frac{1}{\hbar}\int_0^{\beta\hbar} d\tau \left(\frac{1}{2}m\dot{x}^2(\tau) + U(x(\tau))\right)\right] T[x; y, y'],$$

where $T[x; y, y']$ is known as the *influence functional*. What is the functional integral expression for $T[x; y, y']$, and of what function is $T[x; y, y']$ a functional?

   b. Using the method of expansion about the classical path, derive a closed form expression for $T(x(\tau), y, y')$ by evaluating the functional integral.

12.8. In Section 12.6, we discussed a recursion relation for the path integral formulation of N-particle bosonic and fermionic systems. Consider a system of $N = 4$ bosons or fermions. Generate a set of ring-polymer configurations analogous to Fig. 12.10 for this four-particle system, and determine the number of topologically distinct configurations. Finally, use the recursion relations in eqns. (12.6.21) and (12.6.24) to generate $\exp(-\beta\mathcal{K}_{B/F}^{(4)})$ and show these relations reproduce the correct exchange terms with the corresponding topological weighting factors.

12.9. A fourth-order Trotter formula valid for traces is

$$\mathrm{Tr}\left[e^{-\lambda(\hat{A}+\hat{B})}\right] \approx \mathrm{Tr}\left\{\left[e^{-\lambda\hat{A}/P}e^{-\lambda\hat{C}/P}\right]^P\right\} + \mathcal{O}\left(\lambda^5 P^{-4}\right),$$

when $[\hat{A}, \hat{B}] \neq 0$.

$$\hat{C} = \hat{B} + \frac{1}{24}\left(\frac{\lambda}{P}\right)^2\left[\hat{B}, [\hat{A}, \hat{B}]\right].$$

Derive the discrete path integral expression for the canonical partition function $Q(N, V, T)$ for N Boltzmann particles in three dimensions that results from applying this approximation. In particular, show that the N-particle potential $U(\mathbf{r}_1, ..., \mathbf{r}_N)$ is replaced by a new effective potential $\tilde{U}(\mathbf{r}_1, ..., \mathbf{r}_N)$ and derive the expression for this new potential.

12.10. In path integral molecular dynamics, convergence to the true quantum limit as a function of P can be slow, requiring large values of P. It is claimed that this problem can be circumvented by *starting* from the continuous form of the path integral (in which the limit $P \to \infty$ is already taken) and deriving an algorithm based on it. In order to explore this claim, the goal of this problem is to derive a path integral molecular dynamics algorithm for sampling continuous paths and consider its convergence behavior.

Recall that the continuous form of the path integral for the canonical partition function of a single quantum particle of mass m subject to a one-dimensional potential $U(x)$ is

$$Q(\beta) = \oint \mathcal{D}x(\tau) \, \exp\left\{ -\frac{1}{\hbar} \int_0^{\beta\hbar} d\tau \, \left[\frac{1}{2}m\dot{x}^2(\tau) + U(x(\tau))\right] \right\},$$

where $\dot{x} = dx/d\tau$. Consider representing the paths $x(\tau)$ in a basis of sine and cosine functions as

$$x(\tau) = x_c + \sum_{k=1}^{\infty} \left[a_k \sin\left(\frac{2\pi k\tau}{\beta\hbar}\right) + b_k \cos\left(\frac{2\pi k\tau}{\beta\hbar}\right) \right],$$

where x_c is the path centroid

$$x_c = \frac{1}{\beta\hbar} \int_0^{\beta\hbar} x(\tau)d\tau.$$

Starting from this continuous (not discrete) path integral representation of the partition function and the above expansion of the paths, design a path integral molecular dynamics algorithm that propagates, and therefore samples the distribution of, the path centroid x_c as well as the expansion coefficients a_k and b_k. In the design of your algorithm, you should provide clear answers to the following questions:

a. What is the most simplified representation of the imaginary-time action in terms of x_c and the expansion coefficients a_k and b_k? This means explicitly performing as many integrals analytically as possible.

b. What is the classical Hamiltonian from which your algorithm is derived, and how did you derive it?

c. What are the equations of motion for your algorithm? In deriving these equations, be sure to obtain explicit expressions for the forces on the centroid and expansion coefficients.

    d. How will you ensure that the correct ensemble distribution is generated by your equations of motion? Give an example of how you would address this question.

    e. Since, on a computer, the number of sine and cosine basis functions must be finite, what parameter determines the convergence of the algorithm to the true quantum limit?

12.11. Consider a system of two distinguishable degrees of freedom with position operators $\hat{x}$ and $\hat{X}$ and corresponding momenta $\hat{p}$ and $\hat{P}$, respectively, with Hamiltonian

$$\hat{\mathcal{H}} = \frac{\hat{p}^2}{2m} + \frac{\hat{P}^2}{2M} + U(\hat{x}, \hat{X}).$$

Assume that the masses M and m are such that $M \gg m$, meaning that the two degrees of freedom are adiabatically decoupled.

    a. Show that the partition function of the system can be approximated as

$$Q(\beta) = \sum_n \oint \mathcal{D}X(\tau) \exp\left\{ -\frac{1}{\hbar} \int_0^{\beta\hbar} d\tau \left[\frac{1}{2} M\dot{X}^2(\tau) + \varepsilon_n\left(X(\tau)\right) \right] \right\},$$

where $\varepsilon_n(X)$ are the eigenvalues that result from the solution of the Schrödinger equation

$$\left[-\frac{\hbar^2}{2m} \frac{\partial^2}{\partial x^2} + U(x, X) \right] \psi_n(x; X) = \varepsilon_n(X)\psi_n(x; X)$$

for the light degree of freedom at a fixed value X of the heavy degree of freedom. This approximation is known as the *path integral Born-Oppenheimer approximation* (Cao and Berne, 1993). The eigenvalues $\varepsilon_n(X)$ are the Born-Oppenheimer surfaces.

    b. Under what conditions can the sum over n in the above expression be approximated by a single term involving only the ground-state surface $\varepsilon_0(X)$?

12.12.   a. Consider making the following transformation in eqn. (12.3.14):

$$r = \frac{1}{2}(x + x'), \qquad s = x - x'.$$

Show that the ensemble average of $\hat{A}(\hat{p})$ can be written as

$$\langle \hat{A} \rangle = \frac{1}{Q(L, T)} \int dp \, dr \, a(p)\rho_W(r, p),$$

where $\rho_W(r, p)$, known as the *Wigner distribution function* after Eugene P. Wigner (1902–1995), is defined by

$$\rho_W(r, p) = \int ds \, e^{ips/\hbar} \left\langle r - \frac{s}{2} \left| e^{-\beta\hat{\mathcal{H}}} \right| r + \frac{s}{2} \right\rangle.$$

b. Calculate $\rho_W(r,p)$ for a harmonic oscillator of mass m and frequency ω and show that in the classical limit, $\rho_W(r,p)$ becomes the classical canonical distribution.

12.13. Path integrals can be used to calculate isotope fractionation of an element X between two phases. If X has two isotopes I and I′ with masses m_I and $m_{I'}$, then the required factor we need in each phase is denoted $\ln \beta^{I'}$, given by

$$\ln \beta^{I'} = \frac{\Delta A_{I-I'}}{kT} - \frac{3}{2} \ln \frac{m_{I'}}{m_I},$$

where $\Delta A_{I-I'}$ is the free-energy difference associated with a change from isotope I to isotope I′. This free-energy difference can be determined via thermodynamic integration by letting the mass of X vary from m_I to $m_{I'}$ using a λ-path for the mass $m(\lambda)$, where $m(0) = m_I$ and $m(1) = m_{I'}$. Suppose there is only one atom of X in a system of N quantum particles obeying Boltzmann statistics. Derive a thermodynamic integration scheme for the free-energy difference $\Delta A_{I-I'}$ by determining an estimator for the free-energy derivative $dA/d\lambda$.

12.14. Show that the transition path ensemble of Section 7.7 can be formulated as a kind of path integral when the limit $n \to \infty$ and $\Delta t \to 0$ is taken. Give an explicit functional integral expression for partition function in eqn. (7.7.5).

12.15. Write down a complete set of path integral molecular dynamics equations of motion (using Nosé-Hoover chain thermostats of length M) for the numerical evaluation of an imaginary-time path integral for a system of N quantum particles in d dimensions obeying Boltzmann statistics at temperature T.

12.16. Using the partition function in eqn. (12.8.60), derive primitive and virial estimators for the energy within the Suzuki-Chin fourth-order scheme.

12.17. Derive eqn. (12.8.66).

12.18. Using the commutation relation

$$f(\hat{x})\hat{p} = \hat{p}f(\hat{x}) + [f(\hat{x}), \hat{p}] = \hat{p}f(\hat{x}) + i\hbar \frac{df}{dx}$$

for an arbitrary differentiable function $f(\hat{x})$ of the coordinate operator, derive eqn. (12.8.76).

12.19. Reformulate the ring polymer contraction scheme in Section 12.8.6 in terms of staging modes. In staging variables, do the original P-bead ring polymer and contracted ring polymer have any beads in common?

13

Classical time-dependent statistical mechanics

13.1 Ensembles of driven systems

Our discussion of both classical and quantum statistical mechanics has thus far been restricted to equilibrium ensembles. The most fundamental of these, the microcanonical ensemble, consists of a collection of systems and evolving according to Hamilton's equations of motion in isolation from the surroundings. Other ensembles are generated by coupling a physical system to a heat bath, a barostat, or a particle reservoir in order to control other equilibrium thermodynamic variables. The equilibrium ensembles allow a wide variety of thermodynamic and structural properties of systems to be computed.

However, there are many properties of interest that can only be measured by subjecting the system to an external perturbation of some kind. For example, if we wish to measure the coefficient of shear viscosity of a system, we could subject it to an external shear force by placing the system between two movable plates, pulling the plates in opposite directions (see Fig. 13.1), and measuring the response of the system to the force caused by the plate motion. Properties of this type are known as *transport properties*. The coefficient of shear viscosity is an example of a *transport coefficient*. Other examples of transport coefficients are the diffusion constant, the thermal conductivity, the coefficient of bulk viscosity, and the electrical conductivity.

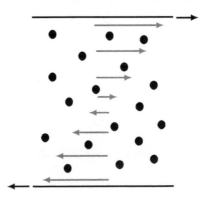

Fig. 13.1 A model shearing experiment: A fluid placed between two plates is subject to a shearing force by pulling the plates in opposite directions.

Similarly, to measure a vibrational spectrum, such as an infrared or Raman spectrum, it is necessary to induce transitions between different vibrational states by subjecting the system to an external electromagnetic field of a given frequency and measuring the frequencies at which excitations occur (see Chapter 14). In general, the perturbations needed to measure such dynamical properties are time-dependent and drive the system slightly away from equilibrium. Thus, in order to calculate dynamical properties, we need to develop a statistical mechanical framework for treating systems weakly perturbed from the equilibrium state by possibly time-dependent external perturbations. In this chapter, we will develop the classical theory of such weakly perturbed systems, and in the next chapter, the corresponding quantum theory will be developed.

To see how the effect of a driving force can change the nature of an ensemble in a simple and familiar example, consider the case of a harmonic oscillator of mass m and frequency ω. In the absence of any driving forces, the equations of motion for the oscillator are given by eqns. (1.6.30), and the motion conserves the Hamiltonian $\mathcal{H}(x,p) = p^2/2m + m\omega^2 x^2/2$. The trajectory traces out the phase-space curve shown in Fig. 1.3, which is simply the ellipse $\mathcal{H}(x,p) = E$. Suppose, now, that the oscillator is subject to an external driving force $F_e(t)$. Hamilton's equations of motion now read

$$\dot{x} = \frac{p}{m}, \qquad\qquad \dot{p} = -m\omega^2 x + F_e(t). \qquad\qquad (13.1.1)$$

Depending on the form of $F_e(t)$, the phase-space trajectory generated by eqns. (13.1.1) is considerably more complicated than that of the undriven oscillator. For example, suppose $F_e(t) = F_0 \cos\Omega t$, with $\Omega/\omega = \sqrt{2}$. The phase-space trajectory generated is shown in Fig. 13.2(a). Comparing Fig. 13.2(a) to Fig. 1.3, perhaps the most striking feature of Fig. 13.2(a) is the fact that many more phase-space points are visited, even on the short time scale of the simulation used to generate the figure, than in Fig. 1.3. Recalling that the equations of motion generate a distribution of accessible microscopic states, the number of accessible states is larger for the driven oscillator. Thus, the ensembles generated by the driven and undriven oscillators are not equiva-

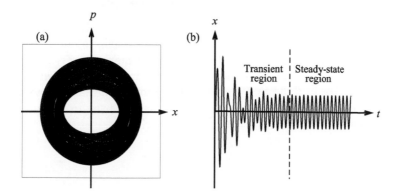

Fig. 13.2 (a) Phase space of a driven oscillator satisfying $m\ddot{x} = -m\omega^2 x + F_0 \cos\Omega t$ for $\Omega/\omega = \sqrt{2}$. (b) Trajectory of a damped-driven oscillator satisfying $m\ddot{x} = -m\omega^2 x - \gamma\dot{x} + F_0 \cos\Omega t$ for $\Omega/\omega = \sqrt{2}$.

lent. Indeed, the undriven oscillator generates a microcanonical ensemble, while eqns. (13.1.1) generate a more complex ensemble, which is an example of a *nonequilibrium* ensemble. Moreover, if $F_e(t)$ were more complicated than the simple periodic function considered, this complexity would be reflected in the phase-space distribution.

Suppose, next, that the oscillator is subject to a frictional force $-\gamma p/m$ in addition to the driving force $F_e(t)$. The equations of motion now read

$$\dot{x} = \frac{p}{m}, \qquad \dot{p} = -m\omega^2 x - \gamma\frac{p}{m} + F_e(t). \qquad (13.1.2)$$

A typical position trajectory $x(t)$, for $F_e(t) = F_0 \cos \Omega t$ with $\Omega/\omega = \sqrt{2}$, is shown in Fig. 13.2(b). The combination of damping and driving forces generates motion with two components. There is an initial *transient* phase that disappears after a certain length of time, leaving a regular component known as a *steady state*. The steady state persists in the long time limit. In general, the phase-space distribution of an ensemble that is allowed to reach a steady state is significantly different from that of the corresponding equilibrium distribution and, therefore, has different properties. Moreover, it is clear that the phase space distribution function $f(x, p, t)$ can have an explicit time dependence due to the presence of the time-dependent driving force and the transient component of the motion. From our analysis of this simple system, we can conclude that ensembles of driven systems are more complex than equilibrium ensembles. They generally contain many more accessible microscopic states due to the presence of the driving forces, they are described by time-dependent phase-space distribution functions, and they often exhibit steady-state behavior.

We now proceed to make these simple arguments more formal by deriving an approximation to the phase-space distribution of a system weakly driven away from equilibrium by a time-dependent perturbation. Interested readers are also referred to the detailed article by B. J. Berne (1971) on time-dependent properties in the condensed phase.

13.2   Driven systems and linear response theory

In this section, we will consider a general class of driven classical systems and their corresponding phase-space distributions. Consider a classical system described by $3N$ generalized coordinates $q_1, ..., q_{3N} \equiv q$, $3N$ generalized momenta $p_1, ..., p_{3N} \equiv p$, and a Hamiltonian $\mathcal{H}(q, p)$, which, in the absence of driving forces, satisfies Hamilton's equations of motion, eqns. (1.6.11). We now wish to include the effect of a weak driving force that is assumed to perturb the system only slightly away from equilibrium. To this end, we introduce the following equations of motion:

$$\dot{q}_i = \frac{\partial \mathcal{H}}{\partial p_i} + C_i(q, p) F_{\mathrm{e}}(t)$$

$$\dot{p}_i = -\frac{\partial \mathcal{H}}{\partial q_i} + D_i(q, p) F_{\mathrm{e}}(t), \tag{13.2.1}$$

where $F_{\mathrm{e}}(t)$ is a time-dependent driving function and $C_i(q, p)$ and $D_i(q, p)$ are phase-space functions whose forms are determined by the particular external perturbation. In our treatment, we will not consider forces that give rise to a nonzero phase-space compressibility, such as frictional forces. Rather, we will impose the requirement that eqns. (13.2.1), although possibly non-Hamiltonian, nevertheless satisfy an incompressibility condition:

$$\sum_{i=1}^{3N} \left[\frac{\partial \dot{q}_i}{\partial q_i} + \frac{\partial \dot{p}_i}{\partial p_i} \right] = 0. \tag{13.2.2}$$

We note, however, that if eqns. (13.2.1) can be generated by a time-dependent Hamiltonian $\mathcal{H}(\mathbf{x}, t)$, where $\mathbf{x} = (q, p)$ is the phase-space vector, then the usual conservation law $d\mathcal{H}/dt = 0$ is replaced by

$$\frac{d\mathcal{H}}{dt} = \frac{\partial \mathcal{H}}{\partial t}, \tag{13.2.3}$$

which states that a nonzero total time derivative of the Hamiltonian arises solely from the explicit time dependence of $\mathcal{H}$.

Substituting eqns. (13.2.1) into eqn. (13.2.2) leads to a restriction on the choice of the phase-space functions $C_i(q, p)$ and $D_i(q, p)$

$$\sum_{i=1}^{3N} \left[\frac{\partial^2 \mathcal{H}}{\partial p_i \partial q_i} + \frac{\partial C_i}{\partial q_i} F_{\mathrm{e}}(t) - \frac{\partial^2 \mathcal{H}}{\partial q_i \partial p_i} + \frac{\partial D_i}{\partial p_i} F_{\mathrm{e}}(t) \right] = 0 \tag{13.2.4}$$

or simply

$$\sum_{i=1}^{3N} \left[\frac{\partial C_i}{\partial q_i} + \frac{\partial D_i}{\partial p_i} \right] = 0. \tag{13.2.5}$$

As we showed in Section 2.5, when the equations of motion have zero phase space compressibility, the phase-space distribution function $f(\mathbf{x}, t)$ satisfies the Liouville equation

$$\frac{\partial}{\partial t} f(\mathbf{x}, t) + iLf(\mathbf{x}, t) = 0, \tag{13.2.6}$$

where $iL = \dot{\mathbf{x}} \cdot \nabla_{\mathbf{x}}$ is the Liouville operator.

Solving the Liouville equation for an ensemble of systems described by eqns. (13.2.1) is nontrivial, in general, especially for large N. However, if we assume that the external driving forces constitute only a small perturbation to Hamilton's equations, so that the ensemble remains relatively close to its equilibrium distribution, then we can use a perturbative approach to solve the Liouville equation. When the external perturbation is small, we assume that the solution $f(\mathbf{x}, t)$ can be written in the form

$$f(\mathbf{x}, t) = f_0(\mathcal{H}(\mathbf{x})) + \Delta f(\mathbf{x}, t), \tag{13.2.7}$$

where $\mathcal{H}(\mathbf{x}) = \mathcal{H}(q, p)$ is the Hamiltonian and $f_0(\mathcal{H}(\mathbf{x}))$ is the equilibrium phase-space distribution function generated by the corresponding unperturbed system ($C_i = D_i =$

0). Equations (13.2.1) could be coupled to a heat bath, barostat, or particle reservoir, so that $f_0(\mathcal{H}(\mathbf{x}))$ can be any of the equilibrium ensemble distributions introduced thus far. All that is required of $f_0(\mathcal{H}(\mathbf{x}))$ is that it satisfy the equilibrium Liouville equation

$$iL_0 f_0(\mathcal{H}(\mathbf{x})) = 0, \tag{13.2.8}$$

where iL_0 is the unperturbed Liouville operator $iL_0 = \{..., \mathcal{H}\}$. We will assume that $f(\mathbf{x}, t)$ is normalized, so that

$$\int d\mathbf{x}\, f(\mathbf{x}, t) = 1. \tag{13.2.9}$$

Using the ansatz in eqn. (13.2.7) gives the expression for the ensemble average of any function $a(\mathbf{x})$:

$$\langle a \rangle_t = \int d\mathbf{x}\, a(\mathbf{x}) f(\mathbf{x}, t)$$

$$= \int d\mathbf{x}\, a(\mathbf{x}) f_0(\mathcal{H}(\mathbf{x})) + \int d\mathbf{x}\, a(\mathbf{x}) \Delta f(\mathbf{x}, t)$$

$$= \langle a \rangle + \int d\mathbf{x}\, a(\mathbf{x}) \Delta f(\mathbf{x}, t)$$

$$= A(t), \tag{13.2.10}$$

where $\langle a \rangle$ is the average of $a(\mathbf{x})$ in the unperturbed ensemble described by $f_0(\mathcal{H}(\mathbf{x}))$, and the notation $A(t) = \langle a \rangle_t$ indicates an average in the nonequilibrium ensemble corresponding to the time-dependent property $A(t)$. Note that if we assume the system to be in equilibrium at $t = 0$, then $\langle a \rangle_0 = \langle a \rangle$.

Since eqns. (13.2.1) are of the form $\dot{\mathbf{x}} = \dot{\mathbf{x}}_0 + \Delta \dot{\mathbf{x}}(t)$, the Liouville operator can be written as

$$iL = \dot{\mathbf{x}} \cdot \nabla_{\mathbf{x}} = (\dot{\mathbf{x}}_0 + \Delta \dot{\mathbf{x}}(t)) \cdot \nabla_{\mathbf{x}} = iL_0 + i\Delta L(t). \tag{13.2.11}$$

Thus, the Liouville equation becomes

$$\frac{\partial}{\partial t}(f_0(\mathcal{H}(\mathbf{x})) + \Delta f(\mathbf{x}, t)) + (iL_0 + i\Delta L(t))(f_0(\mathcal{H}(\mathbf{x})) + \Delta f(\mathbf{x}, t)) = 0. \tag{13.2.12}$$

Assuming that the driving force terms in eqns. (13.2.1) constitute a small perturbation, we neglect the second-order term $i\Delta L \Delta f(\mathbf{x}, t)$. This approximation constitutes a linearization of the Liouville equation, which is the basis of *linear response theory*. Thus, the results we derive by neglecting second-order terms will only be valid within this approximation. Interestingly, however, the approximation of linear response theory proves to be remarkably robust (Bianucci *et al.*, 1996). Within linear response theory, eqn. (13.2.12) reduces to

$$\left(\frac{\partial}{\partial t} + iL_0 \right) \Delta f(\mathbf{x}, t) = -i\Delta L(t) f_0(\mathcal{H}(\mathbf{x})), \tag{13.2.13}$$

which follows from the facts that $\partial f_0/\partial t = 0$ and $iL_0 f_0(\mathcal{H}(\mathbf{x})) = 0$. In order to solve eqn. (13.2.13), we take the driving force to be 0 for $t < 0$, so that at $t = 0$, the ensemble is described by $f(\mathbf{x}, 0) = f_0(\mathcal{H}(\mathbf{x}))$, and $\Delta f(\mathbf{x}, 0) = 0$. Equation (13.2.13) is a simple first-order inhomogeneous differential equation that can be solved using the unperturbed classical propagator $\exp(iL_0 t)$ as an integrating factor. The solution that satisfies the initial condition is

$$\Delta f(\mathbf{x}, t) = -\int_0^t ds \, e^{-iL_0(t-s)} i\Delta L(s) f_0(\mathcal{H}(\mathbf{x})). \tag{13.2.14}$$

In order to simplify eqn. (13.2.14), we note that

$$i\Delta L(s) f_0(\mathcal{H}(\mathbf{x})) = (iL(s) - iL_0) f_0(\mathcal{H}(\mathbf{x}))$$

$$= iL(s) f_0(\mathcal{H}(\mathbf{x}))$$

$$= \dot{\mathbf{x}}(s) \cdot \nabla_{\mathbf{x}} f_0(\mathcal{H}(\mathbf{x})) \tag{13.2.15}$$

since $iL_0 f_0(\mathcal{H}(\mathbf{x})) = 0$. However,

$$\dot{\mathbf{x}}(s) \cdot \nabla_{\mathbf{x}} f_0(\mathcal{H}(\mathbf{x})) = \dot{\mathbf{x}}(s) \cdot \frac{\partial f_0}{\partial \mathcal{H}} \frac{\partial \mathcal{H}}{\partial \mathbf{x}}$$

$$= \frac{\partial f_0}{\partial \mathcal{H}} \sum_{i=1}^{3N} \left[\dot{p}_i(s) \frac{\partial \mathcal{H}}{\partial p_i} + \dot{q}_i(s) \frac{\partial \mathcal{H}}{\partial q_i} \right]$$

$$= \frac{\partial f_0}{\partial \mathcal{H}} \sum_{i=1}^{3N} \left[\frac{\partial \mathcal{H}}{\partial p_i} \left(-\frac{\partial \mathcal{H}}{\partial q_i} + D_i(\mathbf{x}) F_{\mathrm{e}}(s) \right) + \frac{\partial \mathcal{H}}{\partial q_i} \left(\frac{\partial \mathcal{H}}{\partial p_i} + C_i(\mathbf{x}) F_{\mathrm{e}}(s) \right) \right]$$

$$= \frac{\partial f_0}{\partial \mathcal{H}} \sum_{i=1}^{3N} \left[D_i(\mathbf{x}) \frac{\partial \mathcal{H}}{\partial p_i} + C_i(\mathbf{x}) \frac{\partial \mathcal{H}}{\partial q_i} \right] F_{\mathrm{e}}(s). \tag{13.2.16}$$

The quantity

$$j(\mathbf{x}) = -\sum_{i=1}^{3N} \left[D_i(\mathbf{x}) \frac{\partial \mathcal{H}}{\partial p_i} + C_i(\mathbf{x}) \frac{\partial \mathcal{H}}{\partial q_i} \right] \tag{13.2.17}$$

appearing in eqn. (13.2.16) is known as the *dissipative flux*. In terms of this quantity, we have

$$iL(s) f_0(\mathcal{H}(\mathbf{x})) = -\frac{\partial f_0}{\partial \mathcal{H}} j(\mathbf{x}) F_{\mathrm{e}}(s). \tag{13.2.18}$$

Now, suppose that $f_0(\mathcal{H}(\mathbf{x}))$ is given by a canonical distribution

$$f_0(\mathcal{H}(\mathbf{x})) = \frac{C_N \exp[-\beta \mathcal{H}(\mathbf{x})]}{Q(N, V, T)}. \tag{13.2.19}$$

Then, since $\partial f_0 / \partial \mathcal{H} = -\beta f_0$, eqn. (13.2.18) becomes

$$iL(s) f_0(\mathcal{H}(\mathbf{x})) = \beta f_0(\mathcal{H}(\mathbf{x})) j(\mathbf{x}) F_{\mathrm{e}}(s). \tag{13.2.20}$$

Note that eqn. (13.2.16) also holds when $f_0(\mathcal{H}(\mathbf{x}))$ is an isothermal-isobaric or grand canonical distribution. When we substitute eqn. (13.2.20) into eqn. (13.2.14), we obtain

$$\Delta f(\mathbf{x}, t) = -\beta \int_0^t ds\, e^{-iL_0(t-s)} f_0(\mathcal{H}(\mathbf{x})) j(\mathbf{x}) F_e(s). \qquad (13.2.21)$$

We now substitute eqn. (13.2.21) into eqn. (13.2.10) to express the average of $a(\mathbf{x})$ in the nonequilibrium ensemble as

$$A(t) = \langle a \rangle_t = \langle a \rangle - \beta \int d\mathbf{x}\, a(\mathbf{x}) \int_0^t ds\, e^{-iL_0(t-s)} f_0(\mathcal{H}(\mathbf{x})) j(\mathbf{x}) F_e(s)$$

$$= \langle a \rangle - \beta \int_0^t ds \int d\mathbf{x}\, a(\mathbf{x}) e^{-iL_0(t-s)} f_0(\mathcal{H}(\mathbf{x})) j(\mathbf{x}) F_e(s)$$

$$= \langle a \rangle - \beta \int_0^t ds \int d\mathbf{x}\, f_0(\mathcal{H}(\mathbf{x})) a(\mathbf{x}) e^{-iL_0(t-s)} j(\mathbf{x}) F_e(s). \qquad (13.2.22)$$

The second term in eqn. (13.2.22) involves a classical propagator $\exp[-iL_0(t-s)]$. However, since the propagator contains only iL_0, its action generates the evolution of the undriven system obtained by solving Hamilton's equations in the absence of any perturbation. If we write the propagator as $\exp[iL_0(-(t-s))]$, it is clear that this propagator evolves a system backwards in time. Recall from Section 3.10 that if the phase-space vector evolves in time from an initial condition $\mathbf{x}_0$ to $\mathbf{x}_t$ according to Hamilton's equations (under the action of $\exp[iL_0t]$), the evolution of any phase-space function $a(\mathbf{x})$ is determined by

$$\frac{da}{dt} = iL_0 a$$

$$a(\mathbf{x}_t) = e^{iL_0 t} a(\mathbf{x}_0). \qquad (13.2.23)$$

However, since it can be shown that L_0 is a Hermitian operator, if we take the adjoint of both sides, we obtain

$$a^*(\mathbf{x}_t) = a^*(\mathbf{x}_0) e^{-iL_0 t}$$

$$a(\mathbf{x}_t) = a(\mathbf{x}_0) e^{-iL_0 t}, \qquad (13.2.24)$$

where the last line follows from the fact that physical observables are real. In eqn. (13.2.24), the propagator acts to the left on $a(\mathbf{x}_0)$. Thus, we see that the action of $\exp(-iL_0 t)$ on the left evolves $a(\mathbf{x}_0)$ *forward* in time just as $\exp(iL_0 t)$ produces forward evolution when it acts to the right. According to this result, the propagator

$\exp[-iL_0(t-s)]$ in eqn. (13.2.22) can be taken to act to the left on $a(\mathrm{x})$ to produce $a(\mathrm{x}_{t-s})$, assuming x is an initial condition to the undriven equations of motion. Consequently, we could also write eqn. (13.2.22) as

$$A(t) = \langle a \rangle - \beta \int_0^t ds \int d\mathrm{x}\, f_0(\mathcal{H}(\mathrm{x})) j(\mathrm{x}) e^{iL_0(t-s)} a(\mathrm{x}) F_e(s). \qquad (13.2.25)$$

Both eqns. (13.2.25) and (13.2.22) indicate that the nonequilibrium ensemble average can be expressed as

$$A(t) = \langle a \rangle - \beta \int_0^t ds\, F_e(s) \int d\mathrm{x}\, f_0(\mathcal{H}(\mathrm{x})) a(\mathrm{x}_{t-s}) j(\mathrm{x}). \qquad (13.2.26)$$

Since every solution to Hamilton's equations is a unique function of the initial conditions, *i.e.*, $\mathrm{x}_{t-s} = \mathrm{x}_{t-s}(\mathrm{x})$ is a unique function of the initial condition x, we can write eqn. (13.2.26) more explicitly as

$$A(t) = \langle a \rangle - \beta \int_0^t ds\, F_e(s) \int d\mathrm{x}\, f_0(\mathcal{H}(\mathrm{x})) a(\mathrm{x}_{t-s}(\mathrm{x})) j(\mathrm{x}). \qquad (13.2.27)$$

Equation (13.2.27) is an expression for the ensemble average of a phase-space function $a(\mathrm{x})$ for the driven system described by eqns. (13.2.1) valid within linear response theory.

At this point, several comments are in order. First, we interpret the second term in eqn. (13.2.27) as follows: We take each point x in phase space and use it as an initial condition for Hamilton's equations of motion, evolving each initial condition up to time $t-s$. This evolution yields a new phase space point x_{t-s}, which depends uniquely on x. We evaluate the phase-space function $a(\mathrm{x})$ at the point x_{t-s}, giving $a(\mathrm{x}_{t-s}(\mathrm{x}))$. We then take an average of $a(\mathrm{x}_{t-s}(\mathrm{x})) j(\mathrm{x})$ over the phase space with respect to the unperturbed distribution function $f_0(\mathcal{H}(\mathrm{x}))$ of all possible initial conditions. Finally, we integrate the result multiplied by $F_e(s)$ over s from 0 to t. The quantity

$$\int d\mathrm{x}\, f_0(\mathcal{H}(\mathrm{x})) a(\mathrm{x}_{t-s}(\mathrm{x})) j(\mathrm{x}) \equiv \langle a(t-s) j(0) \rangle \qquad (13.2.28)$$

has a form we have not previously encountered. Specifically, it is known as an *equilibrium time correlation function*; these functions play a fundamental role in time-dependent statistical mechanics. The right side of eqn. (13.2.28) is a commonly used shorthand notation for a time correlation function that stands in for the left side of the equation. Using this notation, eqn. (13.2.27) can be written compactly as

$$A(t) = \langle a \rangle_t = \langle a \rangle - \beta \int_0^t ds\, F_e(s) \langle a(t-s) j(0) \rangle. \qquad (13.2.29)$$

Second, as eqn. (13.2.27) suggests, in linear response theory, the observable $A(t)$ obtained by averaging the phase-space function $a(\mathrm{x})$ over the nonequilibrium ensemble is expressible solely in terms of averages over the equilibrium ensemble characterized by $f_0(\mathcal{H}(\mathrm{x}))$. All information concerning the response of the system to the external perturbation is embodied in the equilibrium time correlation function. This remarkable result indicates that, within linear response theory, a nonequilibrium average can be generated entirely within an equilibrium calculation. We will study several applications of linear response theory in Section 13.3.

13.2.1 Properties of equilibrium time correlation functions

Before we apply the linear response theory to specific examples, we first explore some of the properties of time correlation functions. We define the equilibrium time correlation function $C_{AB}(t)$ between two observables A and B, corresponding to phase-space functions $a(\mathrm{x})$ and $b(\mathrm{x})$, with respect to a normalized equilibrium distribution function $f(\mathrm{x})$ and dynamics generated by a Liouville operator iL as

$$C_{AB}(t) = \langle a(0)b(t) \rangle = \int \mathrm{dx}\, f(\mathrm{x})a(\mathrm{x})e^{iLt}b(\mathrm{x})$$

$$= \int \mathrm{dx}\, f(\mathrm{x})a(\mathrm{x})b(\mathrm{x}_t(\mathrm{x})). \tag{13.2.30}$$

Since the propagator $\exp(iLt)$ can be taken to act either to the right as a forward propagator or to the left as a backward propagator, the time correlation function satisfies the property

$$\langle a(0)b(t) \rangle = \langle a(-t)b(0) \rangle. \tag{13.2.31}$$

At $t = 0$,

$$C_{AB}(0) = \langle ab \rangle = \int \mathrm{dx}\, f(\mathrm{x})a(\mathrm{x})b(\mathrm{x}), \tag{13.2.32}$$

which is a simple equilibrium average of $a(\mathrm{x})b(\mathrm{x})$. The long time $(t \to \infty)$ limit, by contrast, is a little more subtle. In complex many-body systems characterized by highly nonlinear forces, the influence of each initial condition on the resultant trajectories generated by $\exp(iLt)$ rapidly becomes negligible as time proceeds (recall the rapid decay of the transient component of the forced-damped harmonic oscillator in Section 13.1). This loss of memory of the initial condition means that there is a characteristic time, called the *correlation time*, over which the trajectory $\mathrm{x}_t(\mathrm{x})$ appears to be particular to a given choice of x and beyond which $\mathrm{x}_t(\mathrm{x})$ is essentially indistinguishable from any other trajectory. This is merely a conceptual device, as every trajectory is uniquely determined for all time by its initial conditions. However, in complex many-body systems, nearly all trajectories will exhibit the same type of chaotic behavior and will resemble each other as they traverse the phase space. In order to see what the existence of a correlation time implies for a correlation function, consider the special case $a(\mathrm{x}) = b(\mathrm{x})$. The time correlation function

$$C_{AA}(t) = \langle a(0)a(t) \rangle = \int \mathrm{dx}\, f(\mathrm{x})a(\mathrm{x})a(\mathrm{x}_t(\mathrm{x})) \tag{13.2.33}$$

is known as an *autocorrelation function*. For very short times, $a(\mathrm{x}_t(\mathrm{x}))$ and $a(\mathrm{x})$ are not very different, hence they are highly correlated. As long as t is small compared to the correlation time, the trajectory $\mathrm{x}_t(\mathrm{x})$ appears to be particular to the initial condition x, and $a(\mathbf{x}_t(\mathbf{x}))$ remains correlated with $a(\mathbf{x})$. However, for times longer than the correlation time, the trajectory loses memory of its initial condition and $a(\mathbf{x}_t(\mathbf{x}))$ and $a(\mathbf{x})$ become *uncorrelated*. (Recall Fig. 3.7 which shows how rapidly two very similar initial conditions diverge in time for a Lennard-Jones liquid.) Clearly, the length of a correlation time depends on the nature of the system and the property under consideration. However, the notion of two properties becoming uncorrelated after a sufficiently long time is the basis of the *Onsager regression hypothesis*. The latter states that in complex systems in which memory of initial condition is not retained, the long-time behavior of a time correlation function is given by

$$\lim_{t \to \infty} C_{AB}(t) = \langle a \rangle \langle b \rangle. \tag{13.2.34}$$

Clearly, this hypothesis does not apply to all systems and cannot be proved in general. In fact, a harmonic oscillator is an example of a pathological system that has an infinitely long memory of its initial condition and therefore violates the regression hypothesis. However, for sufficiently chaotic systems with finite correlation times, the regression hypothesis generally holds. Note that the initial value of an autocorrelation function $C_{AA}(0) = \langle a^2 \rangle$ is the equilibrium average of $a^2(\mathbf{x})$. If we define a phase-space function

$$\delta a(\mathbf{x}) = a(\mathbf{x}) - \langle a \rangle, \tag{13.2.35}$$

then the corresponding macroscopic observable $\delta A = \langle \delta a \rangle = 0$. However, the autocorrelation function of δA is

$$C_{\delta A \delta A}(t) = \langle \delta a(0) \delta a(t) \rangle = \langle (a(\mathbf{x}_0) - \langle a \rangle)(a(\mathbf{x}_t) - \langle a \rangle) \rangle, \tag{13.2.36}$$

whose initial value is

$$C_{\delta A \delta A}(0) = \langle (a(\mathbf{x}_0) - \langle a \rangle)(a(\mathbf{x}_0) - \langle a \rangle) \rangle = \langle a^2 \rangle - \langle a \rangle^2, \tag{13.2.37}$$

which is just the equilibrium fluctuation in A. In the remainder of this and in Chapters 14 and 15, we will see that time correlation functions play a central role in the theory of transport coefficients and vibrational spectra.

13.3 Applying linear response theory: Green-Kubo relations for transport coefficients

13.3.1 Shear viscosity

The coefficient of shear viscosity (denoted η) is an example of a transport property that characterizes the resistance of a system to flow under the action of a shearing force. As described in Section 13.1, a shearing force can be generated by placing the system between movable plates and pulling the plates apart in opposite directions (see Fig. 13.1). Thinking for a moment in terms of a hydrodynamic description of the system rather than an atomistic one, the shearing force sets up a flow field in the direction of the force as shown in Fig. 13.1. The speed of the flow is maximum at the top and bottom plates. Since the plates move in opposite directions, there must be a point at which the speed of the flow is zero. Let this point be $y = 0$ as shown in the figure, and let the plates be located at the points $y = \pm y_{\max}$. Note that, although the direction of the flow is in the x direction, the flow pattern has a gradient in the y direction: the flow velocity in the positive x direction increases from $y = 0$ to $y = y_{\max}$ and increases in the negative x direction from $y = 0$ to $y = -y_{\max}$. We will assume that the rate γ at which the plates are pulled is small enough that the gradient of the flow pattern is constant. The rate γ is known as the *shear rate* and has units of inverse time. In this case, the flow rate itself increases linearly with y. Such a flow pattern is known as *planar Couette flow* and can be expressed in terms of a flow field, which is an equation giving the velocity $\mathbf{u}(\mathbf{r})$ of the flow as a function of each point in space. For planar Couette flow, using the coordinate frame in Fig. 13.1, the flow field is given by

$$\mathbf{u}(\mathbf{r}) = u_{\mathrm{x}}(\mathbf{r})\hat{\mathbf{e}}_x + u_{\mathrm{y}}(\mathbf{r})\hat{\mathbf{e}}_y + u_{\mathrm{z}}(\mathbf{r})\hat{\mathbf{e}}_z = \gamma y \hat{\mathbf{e}}_x, \tag{13.3.1}$$

which is valid for $-y_{\max} \leq y \leq y_{\max}$. Here $\hat{\mathbf{e}}_\alpha$, where $\alpha = x, y, z$ is the unit vector along the α axis. Equation (13.3.1) expresses the fact that the flow is entirely in the x direction. Consequently, only the x-component of the velocity vector field is nonzero, and that the magnitude of the velocity in the x direction depends on how far from the center the flow is observed. Therefore, the only nonvanishing component of the gradient $\nabla \mathbf{u}(\mathbf{r})$ (which is generally a tensor quantity) is the y-derivative of the x component:

$$\frac{\partial u_{\mathrm{x}}}{\partial y} = \gamma. \tag{13.3.2}$$

Because the flow profile is linear, the gradient is constant.

The application of the external shearing force breaks the usual spatial isotropy of the system, causing numerous properties to develop a dependence on different spatial directions. The coefficient of shear viscosity is related to the anisotropy in the pressure, as expressed through the xy component of the pressure tensor P_{xy} (see Section 5.7 for a discussion of the pressure tensor, in particular eqn. (5.7.8)) via Newton's law of viscosity. The latter states that the coefficient of shear viscosity η is the constant of proportionality between the pressure anisotropy and the gradient of the flow field, which we express as

$$P_{xy} = -\eta \frac{\partial u_x}{\partial y} = -\eta \gamma. \tag{13.3.3}$$

(The minus sign arises because we choose to work with the pressure tensor $P_{\alpha\beta}$ rather than $-P_{\alpha\beta}$, which is the stress tensor $\sigma_{\alpha\beta}$.) Therefore,

$$\eta = -\frac{P_{xy}}{\gamma}. \tag{13.3.4}$$

Having provided a hydrodynamic picture of flow under the action of a shearing force, we now seek an atomistic description in terms of a microscopic dynamics in the form of eqns. (13.2.1) so that we can apply the linear response formula in eqn. (13.2.29). We consider a system with Cartesian coordinates $\mathbf{r}_1, ..., \mathbf{r}_N$ and conjugate momenta $\mathbf{p}_1, ..., \mathbf{p}_N$, described by a Hamiltonian of the usual form

$$\mathcal{H} = \sum_{i=1}^{N} \frac{\mathbf{p}_i^2}{2m_i} + U(\mathbf{r}_1, ..., \mathbf{r}_N), \tag{13.3.5}$$

where $U(\mathbf{r}_1, ..., \mathbf{r}_N)$ is the potential. The equation of motion for $\dot{\mathbf{r}}_i$ can be understood on simple physical grounds. Since the shearing force induces a flow field $\mathbf{u}(\mathbf{r}_i)$ at the position $\mathbf{r}_i$ of a particle, we expect each velocity $\dot{\mathbf{r}}_i$ to have a contribution from this flow field in addition to a contribution $\mathbf{p}_i/m_i$ from the usual mechanical kinetic energy. Thus, we can write the equation of motion for $\mathbf{r}_i$ as

$$\dot{\mathbf{r}}_i = \frac{\mathbf{p}_i}{m_i} + \mathbf{u}(\mathbf{r}_i) = \frac{\mathbf{p}_i}{m_i} + \gamma y_i \hat{\mathbf{e}}_x$$

$$= \frac{\mathbf{p}_i}{m_i} + \gamma (\mathbf{r}_i \cdot \hat{\mathbf{e}}_y) \hat{\mathbf{e}}_x. \tag{13.3.6}$$

If we average $\dot{\mathbf{r}}_i$ over an equilibrium distribution such as the canonical distribution, the quantity $\langle \mathbf{p}_i/m_i \rangle$ vanishes, leaving only the overall flow component $\gamma \langle y_i \rangle \hat{\mathbf{e}}_x$, indicating that, on average, the net flow of the system follows the flow field, as expected from hydrodynamics. The form of the momentum equation must now be chosen such that the overall phase-space compressibility is zero. A possible choice consistent with this requirement is

$$\dot{\mathbf{p}}_i = \mathbf{F}_i - \gamma p_{y_i} \hat{\mathbf{e}}_x$$

$$= \mathbf{F}_i - \gamma (\mathbf{p}_i \cdot \hat{\mathbf{e}}_y) \hat{\mathbf{e}}_x, \tag{13.3.7}$$

where $\mathbf{F}_i = -\partial U/\partial \mathbf{r}_i$.

Equations (13.3.6) and (13.3.7) constitute the microscopic equations of motion for a system subject to a shearing force and have the conserved energy

$$\mathcal{H}' = \sum_{i=1}^{N} \frac{1}{2m_i} [\mathbf{p}_i + m_i \gamma y_i \hat{\mathbf{e}}_x]^2 + U(\mathbf{r}_1, ..., \mathbf{r}_N).$$

$$= \sum_{i=1}^{N} \frac{1}{2m_i} [\mathbf{p}_i + m_i \gamma (\mathbf{r}_i \cdot \hat{\mathbf{e}}_y) \hat{\mathbf{e}}_x]^2 + U(\mathbf{r}_1, ..., \mathbf{r}_N). \tag{13.3.8}$$

Equations (13.3.6) and (13.3.7) could, in fact, be used in a molecular dynamics calculation to simulate the effect of a shearing force. In such a calculation, additional couplings to a thermostat and possibly a barostat would be included in order generate a canonical or isothermal-isobaric equilibrium distribution. We will describe such simulations in detail in Section 13.5. Note that eqn. (13.3.7) is not the only possibility; we could also choose $\dot{\mathbf{p}}_i = \mathbf{F}_i - \gamma (\mathbf{p}_i \cdot \hat{\mathbf{e}}_x) \hat{\mathbf{e}}_y$. Interestingly, this is what would be obtained for an equation of motion for $\mathbf{p}_i$ if eqn. (13.3.8) is used as a Hamiltonian to derive the driven equations of motion under shear flow.

For the purposes of the present analysis, we will assume an initial equilibrium distribution and focus on the influence of the external field terms. In order to apply eqn. (13.3.4), we need to cast it into a form useful for microscopic analysis. Recall from eqn. (5.8.1) that the microscopic estimator for the xy component of the pressure tensor is given by the phase-space function

$$\mathcal{P}_{xy}(\mathbf{x}) = p_{xy}(\mathbf{r}, \mathbf{p}) = \frac{1}{V} \sum_{i=1}^{N} \left[\frac{(\mathbf{p}_i \cdot \hat{\mathbf{e}}_x)(\mathbf{p}_i \cdot \hat{\mathbf{e}}_y)}{m_i} + (\mathbf{r}_i \cdot \hat{\mathbf{e}}_x)(\mathbf{F}_i \cdot \hat{\mathbf{e}}_y) \right]. \tag{13.3.9}$$

The pressure-tensor component P_{xy} appearing in eqn. (13.3.4) is the average of this estimator over the nonequilibrium ensemble once a steady state has been achieved, *i.e.*, after transient behavior has died away in the presence of the external field. Thus, we can rewrite eqn. (13.3.4) as

$$\eta = -\lim_{t \to \infty} \frac{\langle \mathcal{P}_{xy} \rangle_t}{\gamma}. \tag{13.3.10}$$

We will compute the nonequilibrium average $\langle \mathcal{P}_{xy} \rangle_t$ using eqn. (13.2.27). In order to use linear response theory, we must first compute the dissipative flux. From eqns. (13.3.6) and (13.3.7), we identify $\mathbf{C}_i(\mathbf{r}, \mathbf{p})$ and $\mathbf{D}_i(\mathbf{r}, \mathbf{p})$ as

$$\mathbf{C}_i(\mathbf{r}, \mathbf{p}) = \gamma(\mathbf{r}_i \cdot \hat{\mathbf{e}}_y)\hat{\mathbf{e}}_x$$

$$\mathbf{D}_i(\mathbf{r}, \mathbf{p}) = -\gamma(\mathbf{p}_i \cdot \hat{\mathbf{e}}_y)\hat{\mathbf{e}}_x. \tag{13.3.11}$$

Also, $F_e(t) = 1$. In Cartesian coordinates with a Hamiltonian given by eqn. (13.3.5), eqn. (13.2.17) for the dissipative flux becomes

$$j(\mathbf{r}, \mathbf{p}) = \sum_{i=1}^{N} \left[\mathbf{C}_i(\mathbf{r}, \mathbf{p}) \cdot \mathbf{F}_i - \mathbf{D}_i(\mathbf{r}, \mathbf{p}) \cdot \frac{\mathbf{p}_i}{m_i} \right]. \tag{13.3.12}$$

Substituting eqn. (13.3.11) into eqn. (13.3.12) gives

$$j(\mathbf{r}, \mathbf{p}) = \gamma \sum_{i=1}^{N} \left[(\mathbf{r}_i \cdot \hat{\mathbf{e}}_y)(\hat{\mathbf{e}}_x \cdot \mathbf{F}_i) + (\mathbf{p}_i \cdot \hat{\mathbf{e}}_y) \left(\hat{\mathbf{e}}_x \cdot \frac{\mathbf{p}_i}{m_i} \right) \right]$$

$$= \gamma V \mathcal{P}_{xy}. \tag{13.3.13}$$

Substituting eqn. (13.3.13) into eqn. (13.2.27) yields for $\langle \mathcal{P}_{xy} \rangle_t$

$$\langle \mathcal{P}_{xy} \rangle_t = \langle \mathcal{P}_{xy} \rangle - \beta \gamma V \int_0^t ds \, \langle \mathcal{P}_{xy}(0) \mathcal{P}_{xy}(t-s) \rangle. \tag{13.3.14}$$

Finally, introducing the change of variables $\tau = t - s$ into the integral gives

$$\langle \mathcal{P}_{xy} \rangle_t = \langle \mathcal{P}_{xy} \rangle - \beta \gamma V \int_0^t d\tau \, \langle \mathcal{P}_{xy}(0) \mathcal{P}_{xy}(\tau) \rangle. \tag{13.3.15}$$

Since the first term involves the average of $\mathcal{P}_{xy}$ over an equilibrium ensemble, which describes an isotropic system, it is not difficult to see, again on purely physical grounds, that $\langle \mathcal{P}_{xy} \rangle = 0$; however, this can also be proved analytically from the virial theorem (see Problem 13.9). Thus, taking the limit $t \to \infty$, an expression for the coefficient of shear viscosity is obtained:

$$\eta = \frac{V}{kT} \int_0^\infty d\tau \, \langle \mathcal{P}_{xy}(0)\mathcal{P}_{xy}(\tau) \rangle = \frac{V}{kT} \int_0^\infty dt \, \langle \mathcal{P}_{xy}(0)\mathcal{P}_{xy}(t) \rangle, \qquad (13.3.16)$$

where we have renamed the integration variable "t" instead of "τ" in the final expression to comply with standard notation. Equation (13.3.16) is an example of a *Green-Kubo relation*, which expresses a transport coefficient in terms of the integral of an equilibrium time correlation function. In this case, the coefficient of shear viscosity is given as the time integral of the autocorrelation function of the xy component of the pressure tensor. Interestingly, despite the fact that $\langle \mathcal{P}_{xy} \rangle = 0$, the equilibrium time correlation function of $\mathcal{P}_{xy}$ does not vanish, suggesting that, even in equilibrium, there are short-lived anisotropic fluctuations that cause $\mathcal{P}_{xy}$ to have a nonzero correlation time. The length of the correlation time depends entirely on the details of the potential and the external thermodynamic conditions of the equilibrium ensemble.

It is worth pointing out that the choice of the x-y plane for the anisotropy is arbitrary. In fact, we could have chosen flow along the x-direction and a gradient along z, or flow along the y-direction and a gradient along z. The corresponding elements of the pressure tensor would then be $\mathcal{P}_{xz}$ and $\mathcal{P}_{yz}$, respectively. We could ultimately use all three of these off-diagonal elements of the pressure tensor in the Green-Kubo relation and average over them. If this is done, then eqn. (13.3.16) is modified to read

$$\eta = \frac{V}{3kT} \sum_{\alpha=1}^2 \sum_{\beta=\alpha+1}^3 \int_0^\infty dt \, \langle \mathcal{P}_{\alpha\beta}(0)\mathcal{P}_{\alpha\beta}(t) \rangle. \qquad (13.3.17)$$

In practical applications, one also considers anisotropies within the diagonal elements and uses the following Green-Kubo formula for η:

$$\eta = \frac{V}{6kT} \sum_{\alpha=1}^3 \sum_{\beta=\alpha}^3 \int_0^\infty dt \, \langle \bar{\mathcal{P}}_{\alpha\beta}(0)\bar{\mathcal{P}}_{\alpha\beta}(t) \rangle, \qquad (13.3.18)$$

where $\bar{\mathcal{P}}_{xy} = \mathcal{P}_{xy}$, $\bar{\mathcal{P}}_{xz} = \mathcal{P}_{xz}$, $\bar{\mathcal{P}}_{yz} = \mathcal{P}_{yz}$, $\bar{\mathcal{P}}_{xx} = 0.5(\mathcal{P}_{xx} - \mathcal{P}_{yy})$, $\bar{\mathcal{P}}_{yy} = 0.5(\mathcal{P}_{yy} - \mathcal{P}_{zz})$, $\bar{\mathcal{P}}_{zz} = 0.5(\mathcal{P}_{xx} - \mathcal{P}_{zz})$.

An application of eqn. (13.3.18) is shown in Fig. 13.3(a), which shows the temperature dependence of the viscosity of a 2:1 molar mixture of ethylene glycol and choline chloride (Zhang *et al.*, 2020), commonly referred to as "ethaline" (Smith *et al.*, 2014). The ionic nature of this liquid, due to the presence of choline cations and chloride anions, causes its viscosity to be quite high compared, for example, to the viscosity of pure ethylene glycol (viscosities of ethaline at this molar ratio tend to be around four to five time higher than those of pure ethylene glycol). In the study by Zhang *et al.*, molecular dynamics simulations are performed using the generalized Amber force field (Wang *et al.*, 2004). From Fig. 13.3(a), one can see that at most temperatures, the agreement between viscosities computed from the Green-Kubo relation with this force field, the experimental results reported by Zhang *et al.*, and those of Mjalli and Naser (2015) is reasonably good. For illustrative purposes, Fig. 13.3(b) shows the time

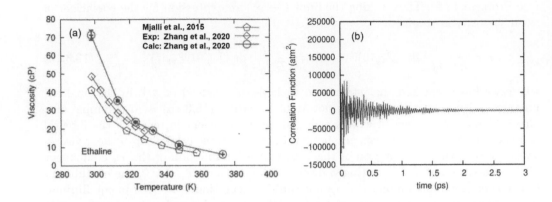

Fig. 13.3 (a) Viscosity of ethaline as a function of temperature from molecular dynamics simulations (Zhang *et al.*, 2020) (red), the experiments reported by Zhang *et al.*, and the experiments of Mjalli and Naser (2015). (b) Pressure autocorrelation function at 298 K from a molecular dynamics simulation (reprinted with permission from Zhang *et al.*, J. Phys. Chem. B, **124**, 5251 (2020), copyright American Chemical Society).

dependence of the pressure autocorrelation function at 298 K computed using a full atomic virial. As the figure shows, the fluctuations in the pressure autocorrelation function with an atomic virial can be both rapid and large in magnitude. Thus, an accurate integration of these fluctuations requires a well-converged correlation function. In this study, although the correlation function is shown out to 3 ps, where the magnitude of fluctuations becomes negligible, the total length of the trajectory is 20 ns, indicating the importance of using correlation times very short compared to the total simulation time.

13.3.2 The diffusion constant

The diffusion constant is a measure of the tendency of particles to drift through a system under the action of a constant external force (see Fig. 13.4). A microscopic

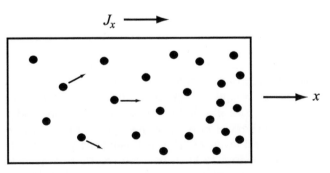

Fig. 13.4 A model diffusion experiment: A fluid is subject to an external force f in the positive x direction, which gives rise to a particle current J_x.

description of diffusion can be provided by the simple addition of a constant force of magnitude f to the negative gradient of the potential as

$$\dot{\mathbf{r}}_i = \frac{\mathbf{p}_i}{m_i}$$

$$\dot{\mathbf{p}}_i = \mathbf{F}_i + f\hat{\mathbf{e}}_x, \tag{13.3.19}$$

where we have arbitrarily chosen the constant force to act in the position x direction. Equations (13.3.19) conserve the total energy:

$$\mathcal{H}' = \sum_{i=1}^{N} \frac{\mathbf{p}_i^2}{2m_i} + U(\mathbf{r}_1, ..., \mathbf{r}_N) - f\sum_{i=1}^{N} x_i. \tag{13.3.20}$$

Equation (13.3.20) indicates that the external force arises from an external potential field of the form

$$\phi(x) = -fx \tag{13.3.21}$$

since the new term in the Hamiltonian is separable and of the form $\sum_{i=1}^{N} \phi(x_i)$. The potential field has a nonzero gradient given by

$$\nabla\phi = -f\hat{\mathbf{e}}_x. \tag{13.3.22}$$

This external potential field causes particles to drift "down" the potential gradient, which is in the positive x direction in this example. This drift causes a concentration gradient ∇c to develop. In general, the concentration $c(x)$ of particles in an external potential $\phi(x)$ follows a Boltzmann distribution $c(x) = c(x = 0)\exp(-\beta\phi(x))$, where we take $\phi(x = 0) = 0$. Assuming that $\phi(x)$ is a weak perturbation and that the concentration is defined such that $c(x = 0) = 1$ (the standard state), then this expression can be linearized to give $c(x) \approx 1 - \beta\phi(x)$, whose gradient is

$$\nabla c = -\frac{1}{kT}\nabla\phi = \frac{f}{kT}\hat{\mathbf{e}}_x$$

$$\frac{\partial c}{\partial x} = -\frac{1}{kT}\frac{\partial\phi}{\partial x} = \frac{f}{kT} \tag{13.3.23}$$

in a direction opposite to that of the potential gradient. Again, because the concentration is linear in x, the gradient is constant.

The drift of particles in a given direction can be quantified in terms of a drift velocity averaged over all the particles, which can be described by the following phase-space function:

$$u_x(\mathbf{r}, \mathbf{p}) = \frac{1}{N}\sum_{i=1}^{N}\dot{x}_i = \frac{1}{N}\sum_{i=1}^{N}\frac{\mathbf{p}_i}{m_i}\cdot\hat{\mathbf{e}}_x. \tag{13.3.24}$$

The average of $u_x(\mathbf{r}, \mathbf{p})$ over the nonequilibrium ensemble, once a steady state has been achieved, is denoted J_x, the average particle current. Thus, J_x is given by

$$J_x = \lim_{t\to\infty}\langle u_x\rangle_t. \tag{13.3.25}$$

The particle current J_x can be related to the concentration gradient $\partial c/\partial x$ using Fick's law of diffusion. The latter states that

$$J_x = D\frac{\partial c}{\partial x}. \tag{13.3.26}$$

The constant of proportionality is the *diffusion constant*, denoted D, which has units of $(\text{Length})^2/\text{Time}$. Substituting eqn. (13.3.23) into eqn. (13.3.26) gives

$$J_x = -\frac{D}{kT}\frac{\partial\phi}{\partial x} = \frac{D}{kT}f. \tag{13.3.27}$$

Thus, the diffusion constant can be written in terms of nonequilibrium ensemble average as

$$D = \frac{kT}{f}J_x = \frac{kT}{f}\lim_{t\to\infty}\langle u_x\rangle_t. \tag{13.3.28}$$

The average $\langle u_x\rangle_t$ can now be evaluated using linear response theory. From eqns. (13.3.19), we see that $F_{\text{e}}(t) = 1$, $\mathbf{D}_i(\mathbf{r},\mathbf{p}) = f\hat{\mathbf{e}}_x$, $\mathbf{C}_i(\mathbf{r},\mathbf{p}) = 0$, so that the dissipative flux becomes

$$j(\mathbf{r},\mathbf{p}) = -f\sum_{i=1}^{N}\frac{p_{x_i}}{m_i} = -f\sum_{i=1}^{N}\dot{x}_i = -Nfu_x. \tag{13.3.29}$$

Therefore, substituting eqn. (13.3.29) into eqn. (13.2.27), we obtain

$$\langle u_x\rangle_t = \langle u_x\rangle + \beta Nf\int_0^t ds\,\langle u_x(0)u_x(t-s)\rangle. \tag{13.3.30}$$

Once again, letting $\tau = t - s$ gives

$$\langle u_x\rangle_t = \langle u_x\rangle + \beta Nf\int_0^t d\tau\,\langle u_x(0)u_x(\tau)\rangle. \tag{13.3.31}$$

The average of u_x over an equilibrium canonical ensemble vanishes, since the average of any component of the velocity is zero. Letting the upper limit of the time integral go to infinity, and changing the integration variable from τ to t, the diffusion constant expression becomes

$$D = N\int_0^\infty dt\,\langle u_x(0)u_x(t)\rangle. \tag{13.3.32}$$

Substituting eqn. (13.3.24) into eqn. (13.3.32) gives

$$D = \int_0^\infty dt\,\frac{1}{N}\left\langle\left(\sum_{i=1}^{N}\dot{x}_i(0)\right)\left(\sum_{i=1}^{N}\dot{x}_i(t)\right)\right\rangle. \tag{13.3.33}$$

Recall that in equilibrium, the velocity (momentum) distribution is a product of independent Gaussian distributions. Hence, $\langle\dot{x}_i\dot{x}_j\rangle$ is 0, and moreover, all cross correlations

$\langle \dot{x}_i(0)\dot{x}_j(t) \rangle$ vanish when $i \neq j$. Thus, the Green-Kubo relation for the diffusion constant becomes

$$D = \int_0^\infty dt \, \frac{1}{N} \sum_{i=1}^N \langle \dot{x}_i(0)\dot{x}_i(t) \rangle. \tag{13.3.34}$$

The correlation function in eqn. (13.3.34) is known as the *velocity autocorrelation function*. Since we could have chosen any direction for the external force, we can compute D by averaging over the three spatial directions and obtain

$$D = \frac{1}{3} \int_0^\infty dt \, \frac{1}{N} \sum_{i=1}^N \langle \dot{\mathbf{r}}_i(0) \cdot \dot{\mathbf{r}}_i(t) \rangle. \tag{13.3.35}$$

An example of a velocity autocorrelation function for a particular model of heavy water (D_2O) at 300 K (Lee and Tuckerman, 2007) is shown in Fig. 13.5(b). We note that the velocity autocorrelation function exhibits a long-time algebraic decay in time that is ubiquitous in the velocity autocorrelation functions of diffusing particles and is of hydrodynamic origin.[1] In practice, convergence of the long-time tail of the velocity autocorrelation function is slow and is influenced by finite-size effects as well. The noise in this part of the correlation function makes the calculation of the diffusion constant via eqn. (13.3.35) numerically difficult. Note, however, that the diffusion constant can also be computed using the Einstein relation

$$D = \frac{1}{6} \lim_{t\to\infty} \frac{d}{dt} \frac{1}{N} \sum_{i=1}^N \langle |\mathbf{r}_i(t) - \mathbf{r}_i(0)|^2 \rangle, \tag{13.3.36}$$

where the derivative of the average *mean-square displacement* of particles is taken. The mean-square displacement in eqn. (13.3.36) is a time correlation function, as can be seen by writing

$$\langle |\mathbf{r}_i(0) - \mathbf{r}_i(t)|^2 \rangle = \langle \mathbf{r}_i^2(0) \rangle + \langle \mathbf{r}_i^2(t) \rangle - 2\langle \mathbf{r}_i(0) \cdot \mathbf{r}_i(t) \rangle. \tag{13.3.37}$$

By Liouville's theorem, $\langle \mathbf{r}_i^2(0) \rangle = \langle \mathbf{r}_i^2(t) \rangle$; the last term is the position autocorrelation function. In general, the mean-square displacement becomes a linear function of time in the long-time limit so that the diffusion constant is simply related to the slope of this linear regime. The mean-square displacement for the same heavy water model used in Fig. 13.5(b) is shown in Fig. 13.5(a). In this case, the model yields a diffusion constant of 0.055 Å^2/ps, which is smaller than the experimental value of 0.186 Å^2/ps at 298 K. Equation (13.3.36) and eqn. (13.3.35) are statistically equivalent; however, in molecular dynamics simulations, one form might yield numerically more stable results than the other.

---

[1] As an example, Alder and Wainwright showed that the decay of the velocity autocorrelation function of a hard-sphere system decays as $t^{-d/2}$, where d is the number of spatial dimensions, in a moderately dense system (Alder and Wainwright, 1967). The asymptotic behavior of the velocity autocorrelation function can be analyzed in detail theoretically using an approach known as *mode-coupling theory*, a discussion of which, however, is beyond the scope of this book (see, for example, Ernst *et al.* (1971)).

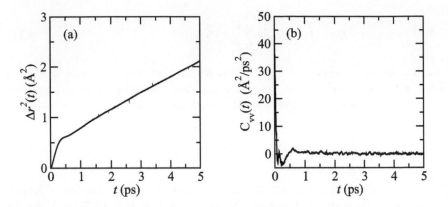

Fig. 13.5 (a) Mean-square displacement for a particular model of heavy water (Lee and Tuckerman, 2007) at 300 K. (b) Total velocity autocorrelation function for the same water model.

13.3.3 Example: The harmonic oscillator

To illustrate the concept of a time correlation function more explicitly, we calculate the position autocorrelation functions for a simple harmonic oscillator of mass m and frequency ω in the canonical ensemble. The Hamiltonian for the harmonic oscillator is

$$\mathcal{H}(x,p) = \frac{p^2}{2m} + \frac{1}{2}m\omega^2 x^2. \tag{13.3.38}$$

From Section 1.3, we know that the solution $(x_t(x,p), p_t(x,p))$ of Hamilton's equations starting from an initial condition x and p for the position and momentum is

$$x_t(x,p) = x\cos\omega t + \frac{p}{m\omega}\sin\omega t$$

$$p_t(x,p) = p\cos\omega t - m\omega x\sin\omega t. \tag{13.3.39}$$

Note that we have dropped the "0" subscript on the initial conditions, since we will need to consider each point (x,p) in phase space as initial condition in order to calculate the time correlation function. In Section 4.6, we showed that the classical canonical partition function is $Q(\beta) = 2\pi/\beta\hbar\omega$ (see eqn. (4.6.19)). The position autocorrelation function requires that eqn. (13.3.39) be integrated over all initial conditions weighted by the canonical distribution $\exp[-\beta\mathcal{H}(x,p)]$ according to the definition

$$C_{xx}(t) = \frac{1}{\hbar Q(\beta)} \int_{-\infty}^{\infty} dx \int_{-\infty}^{\infty} dp\, [xx_t(x,p)]e^{-\beta\mathcal{H}(x,p)}. \tag{13.3.40}$$

Substitution of eqns. (13.3.39) into eqn. (13.3.40) gives

$$C_{xx}(t) = (\beta\omega) \int_{-\infty}^{\infty} dx \int_{-\infty}^{\infty} dp\, x \left(x\cos\omega t + \frac{p}{m\omega}\sin\omega t \right) \exp\left[-\beta \left(\frac{p^2}{2m} + \frac{1}{2}m\omega^2 x^2 \right) \right]$$

$$= \frac{\beta\omega}{2\pi} \cos\omega t \int_{-\infty}^{\infty} dx \int_{-\infty}^{\infty} dp\, x^2 \exp\left[-\beta \left(\frac{p^2}{2m} + \frac{1}{2}m\omega^2 x^2 \right) \right]$$

$$= \frac{kT}{m\omega^2} \cos\omega t. \tag{13.3.41}$$

Because this correlation function never decays, the correlation time is infinite. In other words, a harmonic oscillator never loses memory of its initial conditions and, therefore, does not obey the Onsager regression hypothesis.

13.4 Calculating time correlation functions from molecular dynamics

The use of the Green-Kubo relations for transport coefficients requires the calculation of classical time correlation functions, which, as we will show in this section, can be computed rather easily within a molecular dynamics simulation (Berne and Harp, 1970*a*; Berne and Harp, 1970*b*; Berne, 1971). In particular, we will discuss three commonly employed approaches for generating these functions.

13.4.1 The direct method

The most straightforward and rigorous approach for computing time correlation functions from molecular dynamics calculations is based on a direct interpretation of eqn. (13.2.30). In this scheme, a set of configurations is sampled from the equilibrium distribution $f_0(\mathcal{H}(x))$. This can be achieved either by a molecular dynamics or a Monte Carlo simulation. If molecular dynamics is used to carry out the sampling, appropriate thermostatting and/or barostatting techniques should be employed to generate the desired equilibrium distribution. For Monte Carlo, the Metropolis or hybrid Monte Carlo methods of Chapter 7 (see Sections 7.3 and 7.4) can be used for a canonical distribution; these can be supplemented, if necessary, with volume sampling to generate an isothermal-isobaric distribution. An advantage of molecular dynamics is that well-equilibrated coordinates *and* velocities are generated and can be retained for subsequent calculation of the time correlation function. Given an adequate sampling of $f_0(\mathcal{H}(x))$, each of the configurations obtained is used as an initial condition to Hamilton's equations in order to generate a dynamical trajectory of a pre-specified length. The time correlation function is then computed by performing the average required by eqn. (13.2.30) over all trajectories at each time step.

Suppose we have sampled K configurations from $f_0(\mathcal{H}(x))$, which we will denote $x^{(\lambda)}$, $\lambda = 1, ..., K$. Note that x stands for a complete phase-space vector of dN coordinates and dN momenta in d dimensions. If each $x^{(\lambda)}$ is used to generate a trajectory of M steps, then we will have, overall, K trajectories consisting of configurations at discrete time points $t = 0, \Delta t, 2\Delta t, ..., M\Delta t$, denoted $\{x_0^{(\lambda)}, x_{\Delta t}^{(\lambda)}, ..., x_{M\Delta t}^{(\lambda)}\}$, where $x_0^{(\lambda)}$ is the sampled configuration $x^{(\lambda)}$. Then, the autocorrelation function $C_{AB}(t)$ between two properties A and B with corresponding phase-space functions $a(\mathbf{x})$ and $b(\mathbf{x})$ at time $t = n\Delta t$ is given by

$$C_{AB}(n\Delta t) = \frac{1}{K} \sum_{\lambda=1}^{K} a(\mathbf{x}^{(\lambda)}) b(\mathbf{x}_{n\Delta t}^{(\lambda)}). \qquad (13.4.1)$$

Thus, by letting n run from 0 to M, the time correlation function at M time points will be generated. The calculational procedure is illustrated in Fig. 13.6(a). The length of each trajectory is determined by the decay or correlation time of the particular time correlation function under consideration.

While eqn. (13.4.1) is easy to implement, its convergence usually requires a large number of initial configurations to be sampled from $f_0(\mathcal{H}(\mathbf{x}))$ and subsequent generation of a large number of trajectories. The inefficiency of this method lies in the fact that each trajectory gives just a single contribution to each time point in the correlation function. Our next approach circumvents this problem using a single trajectory.

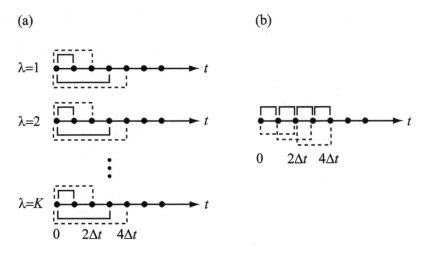

Fig. 13.6 Pictorial representations of (a) the calculation of the time correlation function $C_{AB}(n\Delta t)$ at times $\Delta t, 2\Delta t, 3\Delta t$, and $4\Delta t$ via eqn. (13.4.1), and (b) the calculation of $C_{AB}(n\Delta t)$ at points Δt and $2\Delta t$ via eqn. (13.4.3). In each panel, terms connected by a square bracket are multiplied. All similar brackets are then averaged.

13.4.2 Time correlation functions from a single trajectory

The use of a single trajectory to compute a time correlation function relies on two assumptions: 1) the system under study is large enough that the thermodynamic limit approximately applies, so that the microcanonical distribution is approximately equivalent to the canonical distribution; 2) solutions to Hamilton's equations are ergodic enough to generate an adequate sampling of $f_0(\mathcal{H}(\mathbf{x}))$. The first assumption implies that microcanonical temperature fluctuations are negligibly small (recall that such fluctuations decrease as $1/\sqrt{N}$, where N is the number of particles), while the second ensures that the trajectory can serve as a means both of sampling the equilibrium distribution and generating the dynamics of the system. If both conditions hold, then the calculation of time correlation functions can be simplified considerably.

Consider rewriting eqn. (13.2.30) in a form that exploits the equivalence of the time and phase-space averages of an ergodic system (see Section 3.7):

$$C_{AB}(\tau) = \int dx\, f_0(\mathcal{H}(x)) a(x) b(x_\tau(x)) = \lim_{\mathcal{T}\to\infty} \frac{1}{\mathcal{T}} \int_0^{\mathcal{T}} dt\, a(x_t) b(x_{t+\tau}). \quad (13.4.2)$$

Equation (13.4.2) seems to embody a contradiction. First, it implies that each configuration x_t generated by solving Hamilton's equations is an independent sampling of $f_0(\mathcal{H}(x))$. At the same time, however, it exploits the unique dependence of a trajectory on its initial conditions, which implies that any point x_t of the trajectory can be uniquely determined from any other point in the trajectory. Thus, the point $x_{t+\tau}$ evolves uniquely from x_t for $\tau > 0$. These two conditions might appear incompatible, for how can the point x_t be both an independent sampling from $f_0(\mathcal{H}(x))$ and a source for any other point $x_{t+\tau}$ of the trajectory? Once again, the existence of a finite correlation time resolves the paradox. If the time correlation function eventually decays to zero on a time scale specified by the correlation time and the total time $\mathcal{T}$ of a finite-time trajectory is much larger than the correlation time, then we can imagine breaking a trajectory into segments of length similar to the correlation time. Each segment can then be regarded as an independent "sampling" of the correlation function because over each segment, the correlation function decays to zero.

Suppose a molecular dynamics trajectory is long compared to the correlation time and consists of M points, $x_{m\Delta t}$, $m = 1, ..., M$. We can break the trajectory up into segments each having K points, where $K \ll M$ and $K\Delta t$ is comparable to the correlation time, and apply the procedure of eqn. (13.4.1) to each segment. For example, we could choose $x_{\Delta t}, ..., x_{K\Delta t}$ to be the first segment, $x_{(K+1)\Delta t}, ..., x_{2K\Delta t}$ to be the next, *etc.*, and apply eqn. (13.4.1) on these segments. Note, however, that we could just as well start with $x_{\Delta t}$ and assume $x_{\Delta t}, ..., x_{(K+1)\Delta t}$ is the first segment, $x_{(K+2)\Delta t}, ..., x_{(2K+1)\Delta t}$ is the next, *etc.* By segmenting the trajectory in as many different ways as possible, each point of the trajectory serves both as an independent sampling of $f_0(\mathcal{H}(x))$ and as a time point in the correlation function. In fact, when eqn. (13.4.2) is written in discrete form for a finite-time trajectory

$$C_{AB}(n\Delta t) = \frac{1}{M-n} \sum_{m=1}^{M-n} a(x_{m\Delta t}) b(x_{(m+n)\Delta t}), \qquad n = 0, ..., K, \quad (13.4.3)$$

it is clear that eqn. (13.4.3) automatically exploits the idea of dividing the trajectory in all possible ways. Each point of the trajectory, $x_{m\Delta t}$, serves as an "initial condition" from which the point $x_{(m+n)\Delta t}$ a time $n\Delta t$ later is determined, *i.e.*, as an independent sampling of $f_0(\mathcal{H}(x))$, and as a point generated by an "initial condition" at some earlier time in the trajectory. This approach is illustrated in Fig. 13.6(b). Indeed, as n increases, the number of time intervals that fit into the trajectory decreases, and hence, the statistics degrade. This is why it is imperative to ensure that the trajectory is long compared to the correlation time when using eqn. (13.4.3).

13.4.3 The fast Fourier transform method

For correlation functions with very long decay times, eqn. (13.4.3) requires a very long trajectory and could potentially need to be evaluated at a large number of points. Consequently, the computational overhead of eqn. (13.4.3) could be quite high, increasing roughly as M^2 for a trajectory of M time steps. The third method we will discuss is a highly efficient Fourier transform-based method that can take advantage of fast Fourier transform algorithms (Futrelle and McGinty, 1971).

In order to derive the method, we start by noting that Hamilton's equations are invariant with respect to a change in the origin of time, so that we can shift the time origin in eqn. (13.4.2) from $t = 0$ to $t = -\mathcal{T}/2$. Such a shift gives an equivalent definition of the time correlation function:

$$C_{AB}(\tau) = \lim_{\mathcal{T}\to\infty} \frac{1}{\mathcal{T}} \int_{-\mathcal{T}/2}^{\mathcal{T}/2} dt \, a(x_t)b(x_{t+\tau}). \tag{13.4.4}$$

Since $\mathcal{T}$ formally is taken to be infinite, we may write $a(x_t)$ and $b(x_t)$ in terms of their Fourier transforms:

$$\tilde{a}(\omega) = \frac{1}{\sqrt{2\pi}} \int_{-\infty}^{\infty} dt \, e^{-i\omega t} a(x_t)$$

$$\tilde{b}(\omega) = \frac{1}{\sqrt{2\pi}} \int_{-\infty}^{\infty} dt \, e^{-i\omega t} b(x_t). \tag{13.4.5}$$

Now, consider the product $\tilde{a}(\omega)\tilde{b}^*(\omega)$:

$$\tilde{a}^*(\omega)\tilde{b}(\omega) = \frac{1}{2\pi} \int_{-\infty}^{\infty} dt \int_{-\infty}^{\infty} dt' \, e^{i\omega t'} e^{-i\omega t} \, a(x_{t'})b(x_t)$$

$$= \frac{1}{2\pi} \int_{-\infty}^{\infty} dt \int_{-\infty}^{\infty} dt' \, e^{-i\omega(t-t')} \, a(x_{t'})b(x_t). \tag{13.4.6}$$

If a change of variables $s = t - t'$ is made for the t integral, we find

$$\tilde{a}^*(\omega)\tilde{b}(\omega) = \frac{1}{2\pi} \int_{-\infty}^{\infty} ds \int_{-\infty}^{\infty} dt' \, e^{-i\omega s} a(x_{t'})b(x_{t'+s})$$

$$= \frac{1}{2\pi} \int_{-\infty}^{\infty} ds \, e^{-i\omega s} \int_{-\infty}^{\infty} dt' \, a(x_{t'})b(x_{t'+s}). \tag{13.4.7}$$

Multiplying both sides by $e^{i\omega s'}$ and integrating over ω using the fact that

$$\int_{-\infty}^{\infty} d\omega \, e^{i\omega s'} e^{-i\omega s} = 2\pi\delta(s - s'), \tag{13.4.8}$$

we obtain

$$\int_{-\infty}^{\infty} dt \, a(x_t)b(x_{t+s}) = \int_{-\infty}^{\infty} d\omega \, e^{i\omega s} \tilde{a}^*(\omega)\tilde{b}(\omega). \tag{13.4.9}$$

Thus, for very large $\mathcal{T}$, we have, to a good approximation,

$$C_{AB}(t) = \frac{1}{\mathcal{T}} \int_{-\infty}^{\infty} d\omega\, e^{i\omega t} \tilde{a}^*(\omega) \tilde{b}(\omega). \tag{13.4.10}$$

In practice, since the time interval is actually finite and time is discrete, eqn. (13.4.10) can be evaluated using discrete fast Fourier transforms (FFTs). FFTs are nothing more than canned routines capable of evaluating the transforms in eqn. (13.4.5) as discrete sums of the form

$$\tilde{a}_k = \sum_{j=0}^{M-1} a(\mathrm{x}_{j\Delta t}) e^{-2\pi i j k/M}$$

$$\tilde{b}_k = \sum_{j=0}^{M-1} b(\mathrm{x}_{j\Delta t}) e^{-2\pi i j k/M}, \tag{13.4.11}$$

corresponding to discrete frequencies and times $\omega_k = 2\pi k/M\Delta t$ and $t_j = j\Delta t$, respectively, in $\mathcal{O}(M \ln M)$ operations.

The efficiency of the Fourier transform method lies in the fact that the correlation function can be computed by performing just three FFTs: Two FFTs are needed to transform a and b into the frequency domain, and a third one is needed to transform the product $\tilde{a}\tilde{b}$ back to the time domain via

$$C_{AB}(t_j) = \frac{1}{M} \sum_{k=0}^{M-1} \tilde{a}_k \tilde{b}_k e^{2\pi i j k/M}. \tag{13.4.12}$$

Moreover, since the calculation of vibrational spectra requires the Fourier transform of $C_{AB}(t)$, the final FFT in eqn. (13.4.12) can be eliminated. Finally, if an autocorrelation function $C_{AA}(t)$ is sought, then only one FFT is needed to transform a into the frequency domain, and one additional inverse FFT is required to obtain the real-time autocorrelation function. Again, if only $\tilde{C}_{AA}(\omega)$ is needed, the only one FFT is needed. Since FFTs can be evaluated in $M \ln M$ operations, the scaling with the number of trajectory points is considerably better than eqn. (13.4.3).

13.5 The nonequilibrium molecular dynamics approach

Although equilibrium time correlation functions are useful in the calculation of transport properties, their connection to experiments that employ external driving forces is obscured by the fact that the external perturbation is absent in eqn. (13.2.29). This means we can determine transport coefficients without actually observing the behavior of the system under the action of a driving force. A more intuitive approach would attempt to model the experimental conditions via direct solution of eqns. (13.2.1). Early molecular dynamics calculations based on this idea were carried out by Gosling *et al.* (1973), who employed a spatially periodic shearing force. Ciccotti *et al.* (1976, 1979) showed that a variety of transport properties could be calculated employing this approach. Lees and Edward (1972) introduced an approach by which simulations at constant shear rate could be performed using periodic boundary conditions. Finally, Ashurst and Hoover (1975) and later Edberg *et al.* (1987) introduced a general approach whereby the technique of Lees and Edwards could be coupled to thermostatting mechanisms. This methodology is known as *nonequilibrium molecular dynamics* and is described in detail in a series of reviews (Hoover, 1983; Ciccotti *et al.*, 1992; Evans and Morriss, 1980; Mundy *et al.*, 2000).

Not unexpectedly, molecular dynamics simulations based directly on the perturbed equations (13.2.1) involve certain subtleties. To see what some of these might be, consider again the example in Section 13.3.1 of flow under the action of a shearing force. In order to perform a molecular dynamics simulation based on eqns. (13.3.6) and (13.3.7), we could imagine placing N particles between movable plates, pulling the plates in opposite directions, and using the trajectory to compute $\langle \mathcal{P}_{xy} \rangle_t$; eqn. (13.3.4) would then be used to obtain the shear viscosity. However, let us take a closer look at the influence of the plates. Because the plates are physical boundaries, they create a strong inhomogeneity in the system. Moreover, the interactions between the plates and the fluid particles are generally repulsive, fluid particles are "pushed" away from the plates, thereby creating a void layer at each plate surface. In addition, these repulsive interactions set up a layer of high particle density adjacent to each void layer. To a lesser extent, these high-density layers repel or "push" particles out of their vicinity and into secondary layers of high but slightly smaller density. This effect propagates across several layers until it dissipates into the bulk. However, in realistic applications, the effect can persist for tens of angstroms before dissipating, so that very large system sizes are needed in order to render this boundary effect negligible, leaving enough bulk fluid to obtain reliable bulk properties.

As we discussed in Section 3.14.2, the effects of physical boundaries in equilibrium molecular dynamics calculations can be eliminated by employing periodic boundary conditions. Thus, it is interesting to ask if the idea of periodic boundary conditions

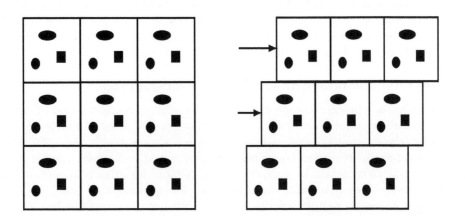

Fig. 13.7 (*Left*) Standard periodic boundary conditions. (*Right*) Lees-Edwards boundary conditions.

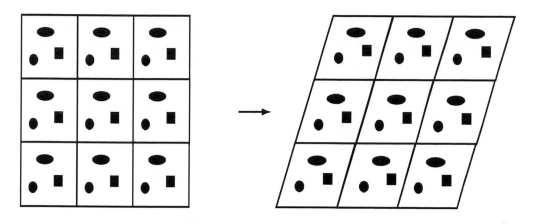

Fig. 13.8 Periodic boundary conditions under the box evolution of eqn. (13.5.1).

can be adapted for systems undergoing shear flow as a means of eliminating the physical plates. It is not immediately obvious how this can be accomplished, however, as the plates are the very source of the driving force. One option is to replace the usual periodic array employed in equilibrium molecular dynamics calculations (see Fig. 13.7, left) with an array in which layers stacked along the y-axis move in the x-direction with a speed equal to $y_l \gamma$, where y_l is the position of the lth layer along the y-axis. This scheme is depicted in Fig. 13.7, right. Such time-dependent boundary conditions are known as *Lees-Edwards boundary conditions* (Lees and Edwards, 1972). As the right panel in Fig. 13.7 suggests, it is the application of the Lees-Edwards boundary conditions that drives the flow and establishes the linear velocity profile required for planar Couette flow. The combination of eqns. (13.3.6) and (13.3.7) with the Lees-Edwards boundary conditions is known as the *SLLOD algorithm*.[2] A serious disadvantage of Lees-Edwards boundary conditions is its implementation for systems of charged particles employing Ewald summation techniques (see Appendix C), which are based in reciprocal space. In such systems, no clear definition of the reciprocal-space vectors exists under Lees-Edwards boundary conditions (Mundy *et al.*, 2000).

An alternative to the Lees-Edwards approach drives the flow by evolving the box matrix **h** in time so as to "shear" the fluid in a particular direction (see Fig. 13.8). For example, if we begin with a cubic box for which $\mathbf{h} = \mathrm{diag}(L, L, L)$, then a shear flow profile can be established by allowing **h** to evolve in time according to

$$\mathbf{h}(t) = \begin{pmatrix} L & \gamma t L & 0 \\ 0 & L & 0 \\ 0 & 0 & L \end{pmatrix}. \tag{13.5.1}$$

---

[2]Note that "SLLOD" is "DOLLS" spelled backwards. For an explanation of the amusing origin of the "SLLOD" moniker, see Evans and Morriss (1984). Suffice it to say, here, that SLLOD is not actually an acronym.

This form of the box matrix allows the simulation cell to shear along the x-direction without changing the volume (since $\det\mathbf{h}(t) = \det\mathbf{h}(0)$). As a practical matter, the box must not be allowed to "roll over" indefinitely but needs to be reset periodically using a property of the equations of motion known as *modular invariance*, which we will discuss in the next subsection. Note that when using either Lees-Edwards boundary conditions or eqn. (13.5.1), eqn. (13.3.8) is no longer conserved because the inter-particle distances become time dependent, which in turn causes the potential to become time dependent.

Another subtlety associated with the nonequilibrium molecular dynamics approach concerns the time scales that can be accessed in a typical simulation. The driving forces used in actual experiments are orders of magnitude smaller than those that can be routinely employed in calculations. For planar Couette flow, this means that a typical experimental shear rate might be $\gamma = 10^2$ s^{-1} while the simulation might require $\gamma = 10^9$ s^{-1}. Under such enormously high shear rates, it is conceivable that the behavior of the system could differ significantly from the experiment. Consequently, a careful extrapolation to the experimental limit from simulations performed at several different shear rates is generally needed in nonequilibrium molecular dynamics calculations. Clearly, this problem does not arise in the Green-Kubo approach because of the absence of explicit driving forces.

Finally, when using nonequilibrium molecular dynamics, we must ensure that conditions of linear response theory are valid in the molecular dynamics calculation. This means that the coefficient of shear viscosity η must be independent of the choice of shear rate γ. Since the extrapolation procedure requires a series of simulations at different shear rates, these simulations should be used to check that the average $\langle \mathcal{P}_{xy} \rangle_t$ varies linearly with γ. If it does not, the shear rate is too high and must be decreased.

13.5.1   Reversible integration of the SLLOD equations of motion

The developing reversible numerical integration methods for the general driven system in eqns. (13.2.1) depends on the particular form for the functions $C_i(q, p)$ and $D_i(q, p)$ appearing in them; each problem needs to be treated as a separate case. While eqns. (13.3.19) for the calculation of the diffusion constant are quite straightforward to integrate, for example, the SLLOD equations for planar Couette flow are somewhat more complicated. Hence, we will consider this latter case as an example of how to construct an integration algorithm for a nontrivial nonequilibrium molecular dynamics problem.

There are two issues that need to be considered in the SLLOD scheme. First, the equations of motion are non-Hamiltonian and involve nontrivial driving terms. Second, we must either impose Lees-Edwards boundary conditions or employ the time-dependent box matrix of eqn. (13.5.1). When performing nonequilibrium molecular dynamics simulations, the use of thermostats is often imperative, as the coupling to the external field causes the system to heat continuously, and a steady state can never be reached without some kind of temperature control mechanism. Thus, if the system is coupled to a Nosé-Hoover chain thermostat, the simulation would be based on the following equations of motion:

$$\dot{\mathbf{r}}_i = \frac{\mathbf{p}_i}{m_i} + \gamma y_i \hat{\mathbf{e}}_x$$

$$\dot{\mathbf{p}}_i = \mathbf{F}_i - \gamma p_{y_i} \hat{\mathbf{e}}_x - \frac{p_{\eta_1}}{Q_1} \mathbf{p}_i$$

$$\dot{\eta}_j = \frac{p_{\eta_j}}{Q_j} \qquad\qquad j = 1, ..., M$$

$$\dot{p}_{\eta_1} = \left[\sum_{i=1}^{N} \frac{\mathbf{p}_i^2}{m_i} - dNkT \right] - \frac{p_{\eta_2}}{Q_2} p_{\eta_1}$$

$$\dot{p}_{\eta_j} = \left[\frac{p_{\eta_{j-1}}^2}{Q_{j-1}} - kT \right] - \frac{p_{\eta_{j+1}}}{Q_{j+1}} p_{\eta_j} \qquad\qquad j = 2, ..., M - 1$$

$$\dot{p}_{\eta_M} = \left[\frac{p_{\eta_{M-1}}^2}{Q_{M-1}} - kT \right]. \tag{13.5.2}$$

In order to devise a numerical integration scheme for eqns. (13.5.2), we begin by writing the Liouville operator for the system in a form similar to eqn. (4.12.4):

$$iL = iL_{\mathrm{mNHC}} + iL_1 + iL_2, \tag{13.5.3}$$

where the three terms are defined to be

$$iL_1 = \sum_{i=1}^{N} \left[\frac{\mathbf{p}_i}{m_i} \cdot \frac{\partial}{\partial \mathbf{r}_i} + \gamma y_i \frac{\partial}{\partial x_i} \right]$$

$$iL_2 = \sum_{i=1}^{N} \mathbf{F}_i \cdot \frac{\partial}{\partial \mathbf{p}_i}$$

$$iL_{\mathrm{mNHC}} = iL_{\mathrm{NHC}} - \gamma \sum_{i=1}^{N} p_{y_i} \frac{\partial}{\partial p_{x_i}}. \tag{13.5.4}$$

Here, iL_{NHC} is the Nosé-Hoover chain Liouville operator of eqn. (4.12.5). The classical propagator $\exp(iL\Delta t)$ for a single time step is now factorized in a manner analogous to eqn. (4.12.8):

$$e^{iL\Delta t} = e^{iL_{\mathrm{mNHC}}\Delta t/2} \left[e^{iL_2\Delta t/2} e^{iL_1\Delta t} e^{iL_2\Delta t/2} \right] e^{iL_{\mathrm{NHC}}\Delta t/2} + \mathcal{O}\left(\Delta t^3 \right). \tag{13.5.5}$$

Consider, first, the three operators in the brackets. If $\gamma = 0$, these operators just produce a step of velocity Verlet integration in eqns. (3.8.7) and (3.8.9). However, with $\gamma \neq 0$, the action of the operators is only slightly more complicated. For each particle, the action of the operators can be deduced by solving the three coupled differential equations

$$\dot{x}_i = \frac{p_{x_i}}{m_i} + \gamma y_i, \qquad \dot{y}_i = \frac{p_{y_i}}{m_i}, \qquad \dot{z}_i = \frac{p_{z_i}}{m_i} \tag{13.5.6}$$

subject to initial conditions $x_i(0), y_i(0), z_i(0)$ with constant p_{x_i}, p_{y_i}, and p_{z_i}. The solution for y_i and z_i is simply

$$y_i(t) = y_i(0) + \frac{p_{y_i}}{m_i} t$$

$$z_i(t) = z_i(0) + \frac{p_{z_i}}{m_i}t. \tag{13.5.7}$$

Substituting eqn. (13.5.7) for $y_i(t)$ into eqn. (13.5.6) for x_i gives

$$\dot{x}_i = \frac{p_{x_i}}{m_i} + \gamma\left[y_i(0) + \frac{p_{y_i}}{m_it}\right]$$

$$x_i(t) = x_i(0) + \left[\frac{p_{x_i}}{m_i} + \gamma y_i(0)\right]t + \frac{t^2}{2}\gamma\frac{p_{y_i}}{m_i}. \tag{13.5.8}$$

The evolution for x_i can be understood by noting that the displacement of x_i in time is elongated by an amount that depends on the position $y_i(0)$, the shear rate, and the momentum p_{y_i}. It is this y_i-dependent elongation that gives rise to the linear flow profile. Setting $t = \Delta t$ and using the fact that $\exp(iL_2\Delta t/2)$ is a simple translation, we can express the action of the operator in brackets in eqn. (13.5.5) in pseudocode as

$$\mathbf{p}_i \longleftarrow \mathbf{p}_i + \frac{\Delta t}{2}\mathbf{F}_i$$

$$x_i \longleftarrow x_i + \Delta t\left[\frac{p_{x_i}}{m_i} + \gamma y_i\right] + \frac{\Delta t^2}{2m_i}\gamma p_{y_i}$$

$$y_i \longleftarrow y_i + \Delta t\frac{p_{y_i}}{m_i}$$

$$z_i \longleftarrow z_i + \Delta t\frac{p_{z_i}}{m_i}$$

Update Forces

$$\mathbf{p}_i \longleftarrow \mathbf{p}_i + \frac{\Delta t}{2}\mathbf{F}_i. \tag{13.5.9}$$

The action of the operator $\exp(iL_{\mathrm{mNHC}}\Delta t/2)$ is similar to the standard Nosé-Hoover chain operator $\exp(iL_{\mathrm{NHC}}\Delta t/2)$ discussed in Section 4.12 (cf. eqn. (4.12.19)). The presence of the γ-dependent operator requires only a slight modification. The operator

$$\exp\left\{-\frac{\delta_\alpha}{2}\frac{p_{\eta_1}}{Q_1}\mathbf{p}_i \cdot \frac{\partial}{\partial\mathbf{p}_i}\right\}$$

in eqn. (4.12.19) is replaced by

$$\exp\left\{-\frac{\delta_\alpha}{2}\left[\frac{p_{\eta_1}}{Q_1}\mathbf{p}_i \cdot \frac{\partial}{\partial\mathbf{p}_i} + \gamma p_{y_i}\frac{\partial}{\partial p_{x_i}}\right]\right\}$$

due to the γ-dependent term in iL_{mNHC}. The action of this operator can be derived by solving the three coupled differential equations

$$\dot{p}_{x_i} = -\frac{p_{\eta_1}}{Q_1}p_{x_i} - \gamma p_{y_i}, \qquad \dot{p}_{y_i} = -\frac{p_{\eta_1}}{Q_1}p_{y_i}, \qquad \dot{p}_{z_i} = -\frac{p_{\eta_1}}{Q_1}p_{z_i}. \tag{13.5.10}$$

The solutions for p_{y_i} and p_{z_i} are

$$p_{y_i}(t) = p_{y_i}(0)e^{-p_{\eta_1}t/Q_1}$$

$$p_{z_i}(t) = p_{z_i}(0)\mathrm{e}^{-p_{\eta_1}t/Q_1}. \tag{13.5.11}$$

Substituting the solution for $p_{y_i}(t)$ into the equation for p_{x_i} gives

$$\dot{p}_{x_i} = -\frac{p_{\eta_1}}{Q_1}p_{x_i} - \gamma p_{y_i}(0)\mathrm{e}^{-p_{\eta_1}t/Q_1}. \tag{13.5.12}$$

This equation can be solved using $\exp[p_{\eta_1}t/Q_1]$ as an integrating factor, which yields the solution

$$p_{x_i}(t) = [p_{x_i}(0) - \gamma t p_{y_i}(0)]\,\mathrm{e}^{-p_{\eta_1}t/Q_1}. \tag{13.5.13}$$

Evaluating the solutions for $p_{x_i}(t)$, $p_{y_i}(t)$, and $p_{z_i}(t)$ at $t = \delta_\alpha/2$ then gives the modification to the Suzuki-Yoshida Nosé-Hoover chain evolution necessary for treating planar Couette flow.

As shown in Appendix C, the application of periodic boundary conditions in a cubic box of side L proceeds as follows: Given a pair of particles with indices i and j, we first calculate the vector difference $\mathbf{r}_{ij} = \mathbf{r}_i - \mathbf{r}_j$ and then apply the minimum-image convention

$$
\begin{aligned}
x_{ij} &\longleftarrow x_{ij} - L \times \mathrm{NINT}\,(x_{ij}/L) \\
y_{ij} &\longleftarrow y_{ij} - L \times \mathrm{NINT}\,(y_{ij}/L) \\
z_{ij} &\longleftarrow z_{ij} - L \times \mathrm{NINT}\,(z_{ij}/L).
\end{aligned} \tag{13.5.14}
$$

Here, $\mathrm{NINT}(x)$ is the nearest integer function. As Fig. 13.7 (right) suggests, the application of periodic boundary conditions in the Lees-Edwards scheme requires a modification in the minimum-image procedure along the x-direction. For a shear rate γ, a row of boxes will be displaced with respect to a neighboring row by an amount $\gamma t L$ in time t. Hence, we must replace eqn. (13.5.14) by

$$
\begin{aligned}
y'_{ij} &= L \times \mathrm{NINT}\,(y_{ij}/L) \\
x_{ij} &\longleftarrow x_{ij} - L * \mathrm{NINT}\,\left[(x_{ij} - \gamma t y'_{ij})/L\right] \\
y_{ij} &\longleftarrow y_{ij} - y'_{ij} \\
z_{ij} &\longleftarrow z_{ij} - L \times \mathrm{NINT}\,(z_{ij}/L).
\end{aligned} \tag{13.5.15}
$$

Note that when $\gamma t = 1$, neighboring rows come back into register with each other, and the periodic array will produce the same set of minimum images as at $t = 0$. Consequently, we could reset $\gamma t = 0$ at this point and continue the simulation.

As noted previously, Lees-Edwards boundary conditions are not useful for systems with long range interactions using Ewald summation techniques, as there is no well-defined box matrix, hence no simple definition of the reciprocal-space vectors. In such systems, the use of a time-dependent box matrix is preferable. In Appendix C, it is shown that for a given box matrix $\mathbf{h}$, the application of periodic boundary conditions involves the four steps:

$$
\begin{aligned}
\mathbf{s}_i &= \mathbf{h}^{-1}\mathbf{r}_i \\
\mathbf{s}_{ij} &= \mathbf{s}_i - \mathbf{s}_j
\end{aligned}
$$

$$\mathbf{s}_{ij} \longleftarrow \mathbf{s}_{ij} - \text{NINT}(\mathbf{s}_{ij}) \quad \text{(Minimum-image convention)}$$
$$\mathbf{r}_{ij} = \mathbf{h}\mathbf{s}_{ij}, \tag{13.5.16}$$

where the third line is a reassignment of $\mathbf{s}_{ij}$ based on the minimum-image convention. In the time-dependent box-matrix method, the above prescription is performed using $\mathbf{h}(n\Delta t)$ at each time step. However, as eqn. (13.5.1) suggests, eventually the box will become long and skinny in the x-direction, causing the system to distort severely. In order to avoid this scenario, we may use a property of the minimum-image convention known as *modular invariance* (Mundy *et al.*, 2000). Note that the inverse of the matrix in eqn. (13.5.1) is

$$\mathbf{h}^{-1}(t) = \begin{pmatrix} 1/L & -\gamma t/L & 0 \\ 0 & 1/L & 0 \\ 0 & 0 & 1/L \end{pmatrix}. \tag{13.5.17}$$

Applying the prescription in eqn. (13.5.16) using eqns. (13.5.1) and (13.5.17) gives the three components of $\mathbf{r}_{ij}$ for a cubic box as

$$y'_{ij} = L \times \text{NINT} \left(y_{ij}/L\right)$$
$$x_{ij} \longleftarrow x_{ij} - L * \text{NINT} \left[\left(x_{ij} - \gamma t y_{ij}\right)/L\right] + \gamma t y'_{ij}$$
$$y_{ij} \longleftarrow y_{ij} - y'_{ij}$$
$$z_{ij} \longleftarrow z_{ij} - L \times \text{NINT} \left(z_{ij}/L\right). \tag{13.5.18}$$

In much the same way as occurs for Lees-Edwards boundary conditions, the same set of images is obtained from eqns. (13.5.16) when $\gamma t = 1$ as when $\gamma t = 0$. Thus, as in the Lees-Edwards scheme, when $\gamma t = 1$ we can reset the time-dependent element h_{21} back to 0, thereby preventing the simulation cell from becoming overly distorted. Note that a similar modular invariance condition exists if h_{21} is allowed to vary between $-L/2$ and $L/2$ rather than between 0 and L. Thus, by resetting h_{21} to $-L/2$ when $\gamma t = 1/2$, the box can be kept closer to cubic, thereby allowing use of a larger cutoff radius on the short-range interactions without requiring an increase in system size. When the diagonal elements of $\mathbf{h}$ are not all identical, it can be shown that the equivalent reset conditions for the two modular invariant forms occur when $\gamma t h_{11}/h_{22} = 1$ and when $\gamma t h_{11}/h_{22} = 1/2$.

13.5.2 Other types of flows

Although $\mathbf{u}(\mathbf{r})$ could describe any type of velocity flow field and eqns. (13.3.6) and (13.3.7) can describe both the linear and nonlinear regimes, the formal development of nonlinear response theory is beyond the scope of this book. Therefore, we will restrict ourselves to examples for which $\mathbf{u}(\mathbf{r})$ depends on linearly on $\mathbf{r}$. This means that the so-called *strain-rate tensor*, denoted $\nabla \mathbf{u}$, is a constant dyad. We will also assume hydrodynamic incompressibility for which $\nabla \cdot \mathbf{u}(\mathbf{r}) = 0$. For example, the velocity profile $\mathbf{u}(\mathbf{r}) = (\gamma y, 0, 0)$ that describes planar Couette flow has the associated strain-rate tensor

$$\nabla \mathbf{u} = \begin{pmatrix} 0 & 0 & 0 \\ \gamma & 0 & 0 \\ 0 & 0 & 0 \end{pmatrix}. \tag{13.5.19}$$

Note that the equation of motion for the matrix **h** is generally given by

$$\dot{\mathbf{h}} = (\nabla \mathbf{u})^{\mathsf{T}} \mathbf{h}, \tag{13.5.20}$$

which reproduces the evolution in eqn. (13.5.1). An example of a different flow satisfying the above requirements is planar elongational flow described by the strain-rate tensor

$$\nabla \mathbf{u} = \begin{pmatrix} \xi & 0 & 0 \\ 0 & \xi(b+1)/2 & 0 \\ 0 & 0 & \xi(b-1)/2 \end{pmatrix} \tag{13.5.21}$$

(Ciccotti *et al.*, 1992), where ξ is the elongation rate and b is a parameter such that $0 \leq b \leq 1$ that describes the type of planar elongational flow.

For a general incompressible flow with constant strain-rate tensor, we seek a set of equations of motion that have a well-defined conserved energy in the absence of time-dependent boundary conditions (Tuckerman *et al.*, 1997). These equations of motion take the form

$$\dot{\mathbf{r}}_i = \frac{\mathbf{p}_i}{m_i} + \mathbf{r}_i \cdot \nabla \mathbf{u}$$

$$\dot{\mathbf{p}}_i = \mathbf{F}_i - \mathbf{p}_i \cdot \nabla \mathbf{u} - m_i \mathbf{r}_i \cdot \nabla \mathbf{u} \cdot \nabla \mathbf{u} - \frac{p_{\eta_1}}{Q_1} \mathbf{p}_i$$

$$\dot{\zeta} = \sum_{i=1}^{N} \mathbf{r}_i \cdot \nabla \mathbf{u} \cdot \mathbf{p}_i \frac{p_{\eta_1}}{Q_1}$$

$$\dot{\eta}_j = \frac{p_{\eta_j}}{Q_j} \qquad\qquad j = 1, ..., M$$

$$\dot{p}_{\eta_1} = \left[\sum_{i=1}^{N} \frac{\mathbf{p}_i^2}{m_i} - 3NkT \right] - \frac{p_{\eta_2}}{Q_2} p_{\eta_1}$$

$$\dot{p}_{\eta_j} = \left[\frac{p_{\eta_{j-1}}^2}{Q_{j-1}} - kT \right] - \frac{p_{\eta_{j+1}}}{Q_{j+1}} p_{\eta_j} \qquad\qquad j = 2, ..., M-1$$

$$\dot{p}_{\eta_M} = \left[\frac{p_{\eta_{M-1}}^2}{Q_{M-1}} - kT \right], \tag{13.5.22}$$

where a Nosé–Hoover chain has been coupled to the system. In the absence of time-dependent boundary conditions, eqns. (13.5.22) have the conserved energy

$$\mathcal{H}' = \sum_{i=1}^{N} \frac{(\mathbf{p}_i + m_i \mathbf{r}_i \cdot \nabla \mathbf{u})^2}{2m_i} + U(\mathbf{r}) + \sum_{j=1}^{M} \frac{p_{\eta_j}^2}{2Q_j} + 3NkT\eta_1 + kT \sum_{j=2}^{M} \eta_j + \zeta. \tag{13.5.23}$$

An interesting problem to which nonequilibrium molecular dynamics with fixed boundaries can be applied is the determination of hydrodynamic boundary conditions for shear flow over a stationary corrugated surface (Tuckerman *et al.*, 1997; Mundy *et al.*, 2000). In general, the flow field $\mathbf{v}(\mathbf{r}, t)$ in a given geometry can be computed using the Navier-Stokes equation

$$\frac{\partial \mathbf{v}}{\partial t} + (\mathbf{v} \cdot \nabla)\mathbf{v} = \nabla \mathbf{P} + \eta \nabla^2 \mathbf{v}, \tag{13.5.24}$$

where $\mathbf{P}$ is the pressure tensor and η is the coefficient of shear viscosity. The most general boundary condition on the velocity profile at the surface (assuming the geometry of Fig. 13.9) is

$$\left. \frac{\partial v_x(\mathbf{r}, t)}{\partial y} \right|_{y=y_{\text{surf}}} = \frac{1}{\delta_{\text{surf}}} \left. v_x(\mathbf{r}, t) \right|_{y=y_{\text{surf}}} \tag{13.5.25}$$

(Tuckerman *et al.*, 1997), where δ_{surf} and y_{surf} are known as the slip length and hydrodynamic thickness, respectively. When $\delta_{\text{surf}} = \infty$, the right side of eqn. (13.5.25) is zero, which corresponds to "slip" boundary conditions, and when $\delta_{\text{surf}} = 0$, the right side is infinite, corresponding to the "no-slip" conditions. In general, δ_{surf} is related to the coefficient of shear viscosity via the relation $\delta_{\text{surf}} = \eta \lambda_{\text{surf}}$, where λ_{surf} is the friction coefficient of the surface. Thus, in order to use eqn. (13.5.25), one needs to determine both λ_{surf} and y_{surf}, assuming that η is already known (or has been determined by a nonequilibrium simulation with Lees-Edwards boundary conditions or

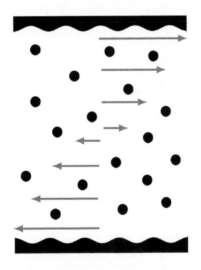

Fig. 13.9 Two-dimensional representation of a fluid confined between corrugated plates.

a time-dependent box matrix approach). These quantities can be obtained by relating them to a nonequilibrium average of the force F_x on the fluid due to the surface:

$$\langle F_x \rangle_{\text{nonequil}} = -S\lambda_{\text{surf}} v_x(y_{\text{surf}}) = -S\lambda_{\text{surf}} \gamma(y_{\text{surf}} - y_0), \tag{13.5.26}$$

where S is the area of surface in contact with the fluid, γ is the shear rate, and y_0 is the location along the y-axis where the drift velocity due to the external field is

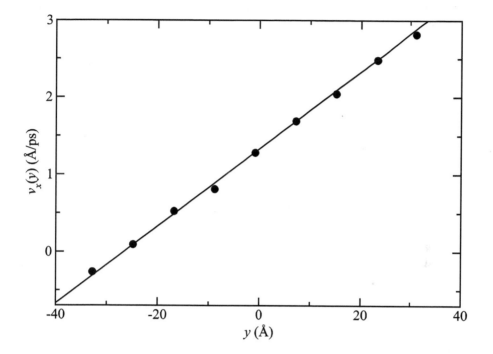

Fig. 13.10 Velocity profile at regularly spaced slabs in the y direction of a soft-sphere fluid confined between corrugated plates (Fig. 13.9) at $T = 480$ K.

0. By performing two simulations with two different values of y_0, one obtains two values of $\langle F_x \rangle_{\text{nonequil}}$, which gives two equations in the two unknowns λ_{surf} and y_{surf}. From these equations, the boundary condition can be determined via eqn. (13.5.25). In Fig. 13.10, we illustrate one such simulation by showing that a linear velocity profile can be achieved when a fluid confined between corrugated plates in the absence of moving boundaries. In this nonequilibrium molecular dynamics simulation, the fluid is described by a pair potential of the form $u(r) = \epsilon(\sigma/r)^{12}$, known as a *soft-sphere* potential. In the present simulations, $\epsilon = 480$ K, $\sigma = 3.405$ Å, and the temperature and density are $T = 480$ K, $\rho = 0.0162$ Å^{-3}. The particles interact with the corrugated walls via a potential of the form

$$U_{\text{wf}}(x, y) = \epsilon \left\{ \left[\frac{\sigma}{y - f^{(l)}(x)} \right]^{12} + \left[\frac{\sigma}{L_y - y - f^{(u)}(x)} \right]^{12} \right\}, \qquad (13.5.27)$$

where L_y is the length of the box in the y direction, and $f^{(l)}(x) = f^{(u)}(x) = a \cos(kx)$ characterizes the corrugation of the lower and upper walls, respectively. In Fig. 13.10, a corrugation amplitude of $a = 0.02\sigma$ is used with a corrugation period of 1.0 Å, and the zero of the shear field occurs at $y_0 = -26.6$ Å. The shear rate is $\gamma = 0.05$ ps^{-1}. The massive thermostatting scheme of Section 4.11 is used in order to stabilize the linear profile. Because moving boundaries are absent, the quality of the simulation can be monitored by using the conservation law in eqn. (13.5.23).

13.6   Problems

13.1. Consider two models of a velocity autocorrelation function:
$$C_1(t) = \langle v^2 \rangle e^{-\gamma t}, \qquad C_2(t) = \langle v^2 \rangle e^{-\gamma t} \cos(\alpha t).$$

a. Calculate the diffusion constants D_1 and D_2 for each model. Which one is larger?

b. Comparing the two diffusion constants, what two physical situations do these two models describe?

13.2. The classical isotropic isothermal-isobaric (NPT) ensemble is particularly useful for determining the *bulk viscosity* of a substance via Green-Kubo theory.

a. Show that the linear response formula does not change if the initial distribution is chosen to be $f_0(\mathcal{H}(\mathbf{r},\mathbf{p}), V)$, i.e., the isothermal-isobaric distribution function $\exp(-\beta(\mathcal{H}(\mathbf{r},\mathbf{p}) + PV))/\Delta(N,P,T)$.

b. Next, consider coupling a system to an external compression field described by the equations of motion
$$\dot{r}_{i,\alpha} = \frac{p_{i,\alpha}}{m_i} + \sum_\beta r_{i,\beta} M_{\beta\alpha}$$
$$\dot{p}_{i,\alpha} = F_{i,\alpha} - \sum_\beta p_{i,\beta} M_{\beta\alpha},$$

where α and β index the three spatial directions x, y, and z. Show that the equations of motion satisfy the incompressibility condition.

c. Consider the specific choice
$$M_{\alpha\beta} = \frac{1}{3}\gamma \delta_{\alpha\beta},$$

where γ is the compression rate. The coefficient of bulk viscosity η_V is given by a generalization of Newton's law of viscosity:
$$\langle V \rangle \eta_V = - \lim_{t \to \infty} \frac{\langle P(t)V(t) \rangle}{\gamma},$$

where $\langle \cdots \rangle$ on the right represents an average over the equilibrium isotropic NPT distribution function and $\langle \cdots \rangle_{\eta_V}$ on the left is the full nonequilibrium average. Using the linear response formula to evaluate $\langle P(t)V(t) \rangle$, derive the appropriate Green-Kubo expression for η_V.

13.3. Consider a particular realization of eqns. (13.2.1) with the following choice for $\mathbf{C}_i(\mathbf{r}, \mathbf{p})$ and $\mathbf{D}_i(\mathbf{r}, \mathbf{p})$:

$$\mathbf{C}_i(\mathbf{r}, \mathbf{p}) = 0$$

$$\mathbf{D}_i(\mathbf{r}, \mathbf{p}) = \varepsilon_i \hat{\mathbf{e}}_z + \sum_{j \neq i} \mathbf{f}_{ij} \left(\mathbf{r}_{ij} \cdot \hat{\mathbf{e}}_z \right),$$

where ε_i is the energy of atom i, $\hat{\mathbf{e}}_z$ is the unit vector along the z-direction, $\mathbf{f}_{ij} = \mathbf{F}_i - \mathbf{F}_j$ is the force between atoms i and j, and $\mathbf{r}_{ij} = \mathbf{r}_i - \mathbf{r}_j$ is the relative vector between atoms i and j. Write down the driven equations of motion, and derive the dissipative flux $j(\mathbf{x})$ for this choice of $\mathbf{C}_i$ and $\mathbf{D}_i$. Assuming that $F_e(t) = 1$ and that the dissipative flux also corresponds to the observable $a(\mathbf{x})$ in the linear response formula, derive a Green-Kubo relation for the thermal conductivity λ given by the nonequilibrium average

$$\lambda = \lim_{t \to \infty} \frac{V}{kT^2} \langle j \rangle_t.$$

*13.4. Show that the Einstein relation in eqn. (13.3.36) can be derived from the Green-Kubo relation in eqn. (13.3.35).

13.5. Verify that eqn. (13.5.23) is conserved by eqn. (13.5.22) in the absence of time-dependent boundary conditions.

*13.6. In Problem 5.11, we considered a particle moving through a periodic potential. Now suppose we add a linear potential $-fq$ to this system, so that the particle is driven by a constant force f. The total potential that results is known as Galton's staircase

$$U(q) = V_0 \cos(kq) - fq.$$

a. Write a program to integrate the driven equations of motion under conditions of constant energy, and use the program to show that the system never reaches a steady state by computing the average $\langle \dot{q} \rangle_t$ in the long time limit.

b. Next try coupling the system to a Nosé-Hoover chain thermostat (see Section 4.11) at temperature $kT = 1$. Does the system now reach a steady state? If so, calculate the average $\langle \dot{q} \rangle_t$ for different values of f and find a regime for which the dependence of the average on f is linear, and estimate the value of the diffusion constant D. Take $m = 1$, $k = 1$, and $V_0 = 1$.

**c. Finally, couple the system to the thermostat of Problem 4.1, using $M = 2$. Algorithms for integrating these equations are given by Liu and Tuckerman (2000) and by Ezra (2007). Repeat the analysis of part b, and estimate the diffusion constant for different values of V_0.

13.7. Verify that the cell matrix evolution in eqn. (13.5.20) reproduces eqn. (13.5.1) for planar Couette flow starting with $\mathbf{h}(0) = \text{diag}(L, L, L)$.

13.8. Consider a fluid confined between corrugated plates as in Fig. 13.9 but with the fluid particles obeying the equations of motion

$$\dot{\mathbf{r}}_i = \frac{\mathbf{p}_i}{m}$$

$$\dot{\mathbf{p}}_i = \mathbf{F}_i + F_e \hat{\mathbf{e}}_x,$$

where F_e is a constant and $\mathbf{F}_i$ is the force on particle i due to the other fluid particles and the corrugated wall. Describe the velocity profile you would expect to develop in a steady state, and give the mathematical form of this profile.

13.9. Prove that $\langle \mathcal{P}_{xy} \rangle = 0$ in the canonical ensemble.

13.10. An N-particle system with coordinate $\mathbf{r} \equiv \mathbf{r}_1, ..., \mathbf{r}_N$ and momenta $\mathbf{p} \equiv \mathbf{p}_1, ..., \mathbf{p}_N$ has a Hamiltonian of the form

$$\mathcal{H}(\mathbf{r}, \mathbf{p}) = \mathcal{H}_0(\mathbf{r}, \mathbf{p}) + \sum_{i=1}^{N} \mathbf{r}_i^\mathsf{T} (\nabla \mathbf{u}) \mathbf{p}_i$$

where $\nabla \mathbf{u}$ is a 3×3 constant matrix obtained from the derivative of the linear flow profile $\mathbf{u}(\mathbf{r})$, and

$$\mathcal{H}_0(\mathbf{r}, \mathbf{p}) = \sum_{i=1}^{N} \frac{\mathbf{p}_i^2}{2m_i} + U(\mathbf{r}_1, ..., \mathbf{r}_N).$$

a. Derive the equations of motion from this Hamiltonian.

b. Show that if $\nabla \mathbf{u}$ is the flow-profile gradient matrix for shear flow given by

$$\nabla \mathbf{u} = \begin{pmatrix} 0 & 0 & 0 \\ \gamma & 0 & 0 \\ 0 & 0 & 0 \end{pmatrix},$$

where γ is the shear rate, then the equation of motion is $\dot{\mathbf{p}}_i = \mathbf{F}_i - \gamma(\mathbf{p}_i \cdot \hat{\mathbf{e}}_x)\hat{\mathbf{e}}_y$, which is not the posited form in eqn. (13.3.7).

c. How would the equations of motion obtained in part a need to be modified so that eqn. (13.3.7) is obtained when $\nabla \mathbf{u}$ is given by the form in part b?

13.11. A system of N identical particles with coordinates $\mathbf{r}_1, ..., \mathbf{r}_N \equiv \mathbf{r}$ and momenta $\mathbf{p}_1, ..., \mathbf{p}_N \equiv \mathbf{p}$ interacts via a pair potential

$$U(\mathbf{r}_1, ..., \mathbf{r}_N) = \sum_{i=1}^{N-1} \sum_{j=i+1}^{N} u(|\mathbf{r}_i - \mathbf{r}_j|).$$

Hamilton's equations of motion are

$$\dot{\mathbf{r}}_i = \frac{\mathbf{p}_i}{m}, \qquad \dot{\mathbf{p}}_i = \mathbf{F}_i$$

where the force $\mathbf{F}_i = -\partial U / \partial \mathbf{r}_i$, and m is the mass of each particle.

a. The center of mass velocity of the system is

$$\mathbf{V} = \frac{1}{N} \sum_{i=1}^{N} \dot{\mathbf{r}}_i$$

Show that $\mathbf{V}$ is conserved over a trajectory $(\mathbf{r}(t), \mathbf{p}(t))$ satisfying these equations of motion. In your derivation, you might find it useful to remember Newton's third law of motion.

b. Suppose $\mathbf{V}$ is a constant vector $\mathbf{a} \neq 0$ over the trajectory. Let $\dot{\mathbf{r}}_i'(t)$ be the velocity of particle i in an equivalent system for which $\mathbf{a} = 0$. Then, the velocity when $\mathbf{a} \neq 0$ is $\dot{\mathbf{r}}_i(t) = \dot{\mathbf{r}}_i'(t) + \mathbf{a}$. Show that the velocity autocorrelation function

$$C_{vv}(t) = \frac{1}{3N} \sum_{i=1}^{N} \langle \dot{\mathbf{r}}_i(0) \cdot \dot{\mathbf{r}}_i(t) \rangle$$

can never decay to 0 even if $\langle \dot{\mathbf{r}}_i'(0) \cdot \dot{\mathbf{r}}_i'(t) \rangle$ decays to 0. Show, in fact, that if $\langle \dot{\mathbf{r}}_i'(0) \cdot \dot{\mathbf{r}}_i'(t) \rangle$ decays to 0, $C_{vv}(t)$ decays to a constant.

c. Show that even if $\langle |\mathbf{r}_i'(0) - \mathbf{r}_i'(t)|^2 \rangle$ becomes linear in time as $t \to \infty$, where, again, $\dot{\mathbf{r}}_i(t) = \dot{\mathbf{r}}_i'(t) + \mathbf{a}$, the mean-square displacement curve

$$\Delta R^2(t) = \frac{1}{N} \sum_{i=1}^{N} \langle |\mathbf{r}_i(0) - \mathbf{r}_i(t)|^2 \rangle$$

of the system with $\mathbf{a} \neq 0$ becomes a quadratic function of t as $t \to \infty$.

13.12. Recall that the general driven equations of motion for a system of $3N$ generalized coordinates $q_1, ..., q_{3N} \equiv q$, momenta $p_1, ..., p_{3N} \equiv p$, and Hamiltonian $\mathcal{H}(q, p) \equiv \mathcal{H}(\mathbf{x})$ are

$$\dot{q}_\alpha = \frac{\partial \mathcal{H}}{\partial p_\alpha} + C_\alpha(q, p) F_e(t)$$

$$\dot{p}_\alpha = -\frac{\partial \mathcal{H}}{\partial q_\alpha} + D_\alpha(q, p) F_e(t),$$

$\alpha = 1, ..., 3N$.

a. Evaluate $d\mathcal{H}/dt$ over a trajectory x_t, expressing your answer in terms of the dissipative flux $j(\mathrm{x}_t)$ and the driving function $F_\mathrm{e}(t)$.

b. Can the system ever reach equilibrium? If no, explain why not. If yes, give the conditions under which it might reach equilibrium.

c. Prove that eqn. (13.3.8) is conserved by eqns. (13.3.6) and (13.3.7).

d. Suppose we were to treat $\mathcal{H}'$ in eqn. (13.3.8) as a proper Hamiltonian. Determine the equations of motion produced by $\mathcal{H}'$ and explain any differences with the equations of motion in part c.

e. If the equations you derived in part d are used in the linear response theory formula to determine the coefficient of shear viscosity from the nonequilibrium average of the pressure tensor estimator $\langle \mathcal{P}_{xy} \rangle_t$, do you obtain the correct Green-Kubo relation? If not, then under what conditions would you obtain the correct Green-Kubo relation?

13.13. Consider a one-dimensional harmonic oscillator with coordinate x and momentum p for which the Hamiltonian is

$$\mathcal{H}(q,p) = \frac{p^2}{2m} + \frac{1}{2}m\omega^2 x^2.$$

Assuming a canonical distribution of initial conditions, calculate the following time correlation functions:

a. $\langle x(0)x(t) \rangle$

b. $\langle p(0)p(t) \rangle$

c. $\langle x(0)p(t) \rangle$

d. $\langle p(0)x(t) \rangle$

e. $\langle x^2(0)x^2(t) \rangle$

f. $\langle x^2(0)p^2(t) \rangle$

14
Quantum time-dependent statistical mechanics

14.1 Time-dependent systems in quantum mechanics

In this chapter, we will explore how the physical properties of a quantum system described by a Hamiltonian $\hat{\mathcal{H}}_0$ can be probed by applying a small external time-dependent perturbation. As we showed in Section 10.4, if the energy levels and energy eigenfunctions of $\hat{\mathcal{H}}_0$ are known, then using the rules of quantum statistical mechanics, all of the thermodynamic and equilibrium properties can be computed from eqns. (10.4.4) and (10.4.5). This fact emphasizes the importance of techniques that can provide information about the eigenvalue structure of $\hat{\mathcal{H}}_0$. The essence of the experimental technique known as *spectroscopy* is to employ an external electromagnetic field to induce transitions between the different eigenstates of $\hat{\mathcal{H}}_0$; from the frequencies of photons needed to induce these transitions, information about different parts of the eigenvalue spectrum can be gleaned for different electromagnetic frequency ranges (infrared, visible, ultraviolet, *etc.*) As we will see in the discussion to follow, the rules of quantum statistical mechanics help us both to interpret such experiments and to construct approximate computational procedures for calculating quantum spectral and transport properties of a condensed-phase system.

An external electromagnetic field is described by its electric and magnetic field components $\mathbf{E}(\mathbf{r}, t)$ and $\mathbf{B}(\mathbf{r}, t)$, respectively. The term "electromagnetic" arises from the notion that the electric and magnetic fields can be *unified* into a single classical field theory, an idea that was demonstrated by James Clerk Maxwell (1831–1879) between the years 1864 and 1873. Maxwell's theory of electromagnetism is embodied in a set of field equations known as *Maxwell's equations*, which describe how the electric and magnetic fields are coupled in such a unified theory. In the absence of external sources, *i.e.*, for free fields, Maxwell's equations specify how the divergence and curl of the electric and magnetic fields are related to their time rates of change fields.[1] Once the divergence and curl of a vector field are specified, then these, together with a knowledge of the time derivatives of the fields, are sufficient to determine the

---

[1] The "divergence" of a vector field is a measure of the extent to which the field expands or contracts at a point while the "curl" tells us the degree to which the field circulates around a point; these modes of behavior are about all a vector field can do. The importance of the divergence and curl of a vector field can be understood using a theorem known as Helmholtz's theorem, which states that if $C(\mathbf{r})$ and $\mathbf{D}(\mathbf{r})$ are, respectively, smooth scalar and vector functions that decay faster than $1/|\mathbf{r}|^2$ as $|\mathbf{r}| \to \infty$, then there exists a unique vector field $\mathbf{F}$ such that $\nabla \cdot \mathbf{F} = C(\mathbf{r})$ and $\nabla \times \mathbf{F} = \mathbf{D}(\mathbf{r})$. Consequently, the divergence and curl of $\mathbf{F}$ are sufficient to determine $\mathbf{F}(\mathbf{r})$, and conversely, $\mathbf{F}$ can be decomposed in terms of its divergence and curl.

spatial and time dependence of the vector fields. The elegance and beauty of Maxwell's theory lies in the simple connection it establishes between the electric and magnetic fields. For free fields, Maxwell's equations take the form (in cgs units)

$$\nabla \cdot \mathbf{E} = 0, \qquad \nabla \times \mathbf{E} = -\frac{1}{c}\frac{\partial \mathbf{B}}{\partial t}$$

$$\nabla \cdot \mathbf{B} = 0, \qquad \nabla \times \mathbf{B} = \frac{1}{c}\frac{\partial \mathbf{E}}{\partial t}, \tag{14.1.1}$$

where c is the speed of light in vacuum. While the magnetic field is always divergence-free (there are no magnetic monopoles), the electric field has zero divergence only in the absence of sources. The rates of change of the magnetic and electric fields determine, respectively, how the electric and magnetic fields "curl" to form closed loops.

Notice that the free-field Maxwell equations constitute a set of eight homogeneous partial differential equations for the six field components, $E_x(\mathbf{r},t)$, $E_y(\mathbf{r},t)$, $E_z(\mathbf{r},t)$, $B_x(\mathbf{r},t)$, $B_y(\mathbf{r},t)$, $B_z(\mathbf{r},t)$. Hence, the set is overdetermined, and there is some redundancy among the field equations. This redundancy arises because of a property exhibited by this and other types of field theories known as *gauge invariance* or *gauge freedom*.

In order to manifest the gauge freedom and remove the redundancy, it is convenient to work with related fields $\mathbf{A}(\mathbf{r},t)$ and $\phi(\mathbf{r},t)$ known as the *vector* and *scalar* potentials, respectively. These are related to the electric and magnetic fields by the transformation

$$\mathbf{B} = \nabla \times \mathbf{A}, \qquad \mathbf{E} = -\nabla\phi - \frac{1}{c}\frac{\partial \mathbf{A}}{\partial t}. \tag{14.1.2}$$

The relation $\mathbf{B} = \nabla \times \mathbf{A}$ follows from the fact that any divergence-free field can always be expressed as the curl of another field since $\nabla \cdot (\nabla \times \mathbf{A}) = 0$. The relation for $\mathbf{E}$ in eqn. (14.1.2) arises from the fact that if $\mathbf{B} = 0$, then $\nabla \times \mathbf{E} = 0$, so that $\mathbf{E}$ can be expressed as a gradient $\mathbf{E} = -\nabla\phi$. The second term in the equation for $\mathbf{E}$ is included when a nonzero magnetic field is present. Although these relations uniquely define the electric and magnetic fields, they possess some ambiguity in how the vector and scalar potentials $\mathbf{A}(\mathbf{r},t)$ and $\phi(\mathbf{r},t)$ are defined. Specifically, if new potentials $\mathbf{A}'(\mathbf{r},t)$ and $\phi'(\mathbf{r},t)$ are constructed from the original potentials via

$$\mathbf{A}'(\mathbf{r},t) = \mathbf{A}(\mathbf{r},t) + \nabla\chi(\mathbf{r},t)$$

$$\phi'(\mathbf{r},t) = \phi(\mathbf{r},t) - \frac{1}{c}\frac{\partial}{\partial t}\chi(\mathbf{r},t), \tag{14.1.3}$$

where $\chi(\mathbf{r},t)$ is an arbitrary scalar field, then the same electric and magnetic fields will result when $\mathbf{A}'$ and ϕ' are substituted into eqns. (14.1.2). The transformations in eqn. (14.1.3) are known as *gauge transformations*, and the freedom to choose $\chi(\mathbf{r},t)$ arbitrarily is the manifestation of the gauge invariance of the electric and magnetic fields mentioned above. Moreover, if the definitions in eqns. (14.1.2) are substituted into the free-field Maxwell equations, then the vector potential is seen to satisfy the classical *wave equation*

$$\nabla^2 \mathbf{A}(\mathbf{r}, t) - \frac{1}{c^2} \frac{\partial^2}{\partial t^2} \mathbf{A}(\mathbf{r}, t) = 0, \tag{14.1.4}$$

provided the gauge function $\chi(\mathbf{r}, t)$ is also chosen to be a solution of Laplace's equation $\nabla^2 \chi(\mathbf{r}, t) = 0$. This choice, known as the *Coulomb gauge*, is tantamount to requiring that $\mathbf{A}(\mathbf{r}, t)$ satisfy

$$\nabla \cdot \mathbf{A}(\mathbf{r}, t) = 0. \tag{14.1.5}$$

In the Coulomb gauge, the scalar potential satisfies Laplace's equation $\nabla^2 \phi = 0$, which, in free space with no charges present, means that ϕ is, at most, a linear function of $\mathbf{r}$ and independent of time. If we require that $\phi \to 0$ as $|\mathbf{r}| \to \infty$, then the only possible choice is $\phi = 0$, which we can take with no loss of generality.[2]

The appropriate solution to the wave equation for $\mathbf{A}(\mathbf{r}, t)$ for an electromagnetic field in a vacuum describe freely propagating waves of frequency ω, wavelength λ, and wave vector $\mathbf{k}$ ($|\mathbf{k}| = 2\pi/\lambda$) prescribing the direction of propagation is

$$\mathbf{A}(\mathbf{r}, t) = \mathbf{A}_0 \cos\left(\mathbf{k} \cdot \mathbf{r} - \omega t + \varphi_0\right). \tag{14.1.6}$$

Here, $\omega = c|\mathbf{k}|$, φ_0 is an arbitrary phase, and $\mathbf{A}_0$ is the amplitude of the wave. Since $\nabla \cdot \mathbf{A} = 0$, it follows that $\mathbf{k} \cdot \mathbf{A} = \mathbf{k} \cdot \mathbf{A}_0 = 0$, and $\mathbf{A}$ is perpendicular to the direction of propagation. From eqns. (14.1.2), the electric and magnetic fields that result are

$$\mathbf{E}(\mathbf{r}, t) = \frac{\omega}{c} \mathbf{A}_0 \sin\left(\mathbf{k} \cdot \mathbf{r} - \omega t + \varphi_0\right) \equiv \mathbf{E}_0 \sin\left(\mathbf{k} \cdot \mathbf{r} - \omega t + \varphi_0\right)$$

$$\mathbf{B}(\mathbf{r}, t) = -\mathbf{k} \times \mathbf{A}_0 \sin\left(\mathbf{k} \cdot \mathbf{r} - \omega t + \varphi_0\right) \equiv \mathbf{B}_0 \sin\left(\mathbf{k} \cdot \mathbf{r} - \omega t + \varphi_0\right). \tag{14.1.7}$$

Note that, since $\mathbf{k} \times \mathbf{A}$ is perpendicular to $\mathbf{k}$ and $\mathbf{A}_0$, it follows that $\mathbf{E}_0 \perp \mathbf{B}_0 \perp \mathbf{k}$. A snapshot in time of the free electromagnetic wave described by eqns. (14.1.7) is shown in Fig. 14.1.

In general, when an electromagnetic field interacts with matter, the field must be included in the Hamiltonian of the physical system, which complicates the mathematical treatment considerably. Therefore, in the present discussion, we will work in an approximation in which the field can be treated as an external perturbation whose degrees of freedom do not need to be explicitly included in the quantum description of the system, *i.e.*, they do not need to be included in the state vector of the system. In this approximation, the Hamiltonian for a system of N particles with charges $q_1, .., q_N$

---

[2]These choices are particular to the free-field theory. For time-dependent fields in the presence of sources, a more appropriate gauge choice is the *Lorentz gauge*:

$$\nabla \cdot \mathbf{A}(\mathbf{r}, t) + \frac{1}{c} \frac{\partial}{\partial t} \phi(\mathbf{r}, t) = 0.$$

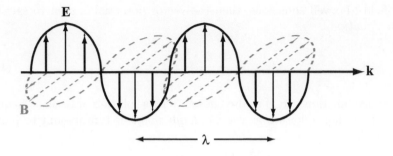

Fig. 14.1 Sketch of an electromagnetic wave described by eqn. (14.1.7).

and masses $m_1, ..., m_N$, including the external electromagnetic field, is specified in terms of the vector and scalar potentials and is given by

$$\hat{\mathcal{H}}(t) = \sum_{i=1}^{N} \frac{1}{2m_i} \left(\hat{\mathbf{p}}_i - \frac{q_i}{c} \mathbf{A}(\hat{\mathbf{r}}_i, t) \right)^2 + \sum_{i=1}^{N} q_i \phi(\hat{\mathbf{r}}_i, t) + U(\hat{\mathbf{r}}_1, ..., \hat{\mathbf{r}}_N). \qquad (14.1.8)$$

When the square in the kinetic energy is expanded out, the Hamiltonian takes the general form

$$\hat{\mathcal{H}}(t) = \hat{\mathcal{H}}_0 + \hat{\mathcal{H}}_1(t), \qquad (14.1.9)$$

where $\hat{\mathcal{H}}_0$ is the pure system Hamiltonian in the absence of the field

$$\hat{\mathcal{H}}_0 = \sum_{i=1}^{N} \frac{\hat{\mathbf{p}}_i^2}{2m_i} + U(\hat{\mathbf{r}}_1, ..., \hat{\mathbf{r}}_N), \qquad (14.1.10)$$

and $\hat{\mathcal{H}}_1(t)$ involves the coupling to the field

$$\hat{\mathcal{H}}_1(t) = -\sum_{i=1}^{N} \frac{q_i}{2m_i c} \left[\hat{\mathbf{p}}_i \cdot \mathbf{A}(\hat{\mathbf{r}}_i, t) + \mathbf{A}(\hat{\mathbf{r}}_i, t) \cdot \hat{\mathbf{p}}_i \right]$$

$$+ \sum_{i=1}^{N} \frac{q_i^2}{2m_i c^2} \mathbf{A}^2(\hat{\mathbf{r}}_i, t) + \sum_{i=1}^{N} q_i \phi(\hat{\mathbf{r}}_i, t). \qquad (14.1.11)$$

For a Hamiltonian of the form given in eqn. (14.1.9), the eigenstates of $\hat{\mathcal{H}}_0$, which satisfy

$$\hat{\mathcal{H}}_0 |E_k\rangle = E_k |E_k\rangle, \qquad (14.1.12)$$

are no longer eigenstates of $\hat{\mathcal{H}}(t)$, which means that they are not stationary states. Thus, the effect of $\hat{\mathcal{H}}_1(t)$ is to induce transitions among the eigenstates of $\hat{\mathcal{H}}_0$ as functions of time. If the system absorbs energy from the field, it can be excited from an initial state $|E_i\rangle$ with energy E_i to a final state $|E_f\rangle$ with energy E_f as depicted in Fig. 14.2. When the system returns to its initial state, the energy emitted is detected, providing information about the eigenvalue spectrum of $\hat{\mathcal{H}}_0$. This is one facet of the experimental technique known as *spectroscopy*.

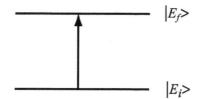

Fig. 14.2 A transition from an initial eigenstate of $\hat{\mathcal{H}}_0$ with energy E_i to a final state with energy E_f due to the external field coupling.

The remainder of this chapter will be devoted to analyzing the process in Fig. 14.2 both as an isolated event and in an ensemble of quantum systems. We will examine the behavior of quantum systems subject to time-dependent perturbations of both a general nature and specific to an external electromagnetic field-coupling, thereby providing an introduction to quantum time correlation functions and the field of linear spectroscopy. Finally, we will discuss numerical approaches for approximating quantum correlation functions. We begin with a discussion of the time-dependent Schrödinger equation when $\hat{\mathcal{H}}_1(t)$ is a weak perturbation.

14.2 Time-dependent perturbation theory in quantum mechanics

14.2.1 The interaction picture

For a system described by a Hamiltonian $\hat{\mathcal{H}}(t) = \hat{\mathcal{H}}_0 + \hat{\mathcal{H}}_1(t)$, the unperturbed Hamiltonian $\hat{\mathcal{H}}_0$ is taken to describe a physical system of interest, such as a gas, liquid, solid, or solution. $\hat{\mathcal{H}}_1(t)$ represents an arbitrary time-dependent perturbation that induces transitions between the eigenstates of $\hat{\mathcal{H}}_0$.

The state vector of the system $|\Psi(t)\rangle$ evolves in time from an initial state vector $|\Psi(t_0)\rangle$ according to the time-dependent Schrödinger equation

$$\hat{\mathcal{H}}(t)|\Psi(t)\rangle = \left(\hat{\mathcal{H}}_0 + \hat{\mathcal{H}}_1(t)\right)|\Psi(t)\rangle = i\hbar\frac{\partial}{\partial t}|\Psi(t)\rangle. \tag{14.2.1}$$

Although it might seem that obtaining an appropriate propagator from eqn. (14.2.1) is straightforward, the presence of operators on the left side of the equation, together with the fact that $[\hat{\mathcal{H}}_0, \hat{\mathcal{H}}_1(t)] \neq 0$, in general, renders this task nontrivial. However, if we view $\hat{\mathcal{H}}_1(t)$ as a weak perturbation, then we can develop a perturbative approach to the solution of eqn. (14.2.1). We begin by noting that eqn. (14.2.1) can be cast in a form more amenable to a perturbative treatment by transforming the state vector from $|\Psi(t)\rangle$ to $|\Phi(t)\rangle$ via

$$|\Psi(t)\rangle = e^{-i\hat{\mathcal{H}}_0(t-t_0)/\hbar}|\Phi(t)\rangle. \tag{14.2.2}$$

In Section 9.2.6, we introduced the concept of a picture in quantum mechanics and discussed the difference between the Schrödinger and Heisenberg pictures. The state vector $|\Phi(t)\rangle$ in eqn. (14.2.2) represents yet another quantum mechanical picture called the *interaction picture*. The interaction picture can be considered as "intermediate" between the Schrödinger and Heisenberg pictures. Recall that in the Schrödinger picture, operators are static and the state evolves in time, while the opposite is true in the Heisenberg picture. In the interaction picture, both the state vector and the operators evolve in time.

The transformation of an operator $\hat{A}$ from the Schrödinger picture to the interaction picture is given by

$$\hat{A}_I(t) = e^{i\hat{\mathcal{H}}_0(t-t_0)/\hbar}\hat{A}e^{-i\hat{\mathcal{H}}_0(t-t_0)/\hbar}, \tag{14.2.3}$$

which is equivalent to an equation of motion of the form

$$\frac{\mathrm{d}\hat{A}_I(t)}{\mathrm{d}t} = \frac{1}{i\hbar}[\hat{A}_I(t), \hat{\mathcal{H}}_0]. \tag{14.2.4}$$

Note the similarity between this transformation and that of eqn. (9.2.58) between the Schrödinger and Heisenberg pictures. Equation (14.2.4) indicates that the time evolution of operators in the interaction picture is determined solely by the unperturbed Hamiltonian $\hat{\mathcal{H}}_0$.

Equation (14.2.2) represents a transformation of the state vector between the Schrödinger and interaction pictures. The time-evolution equation for the state $|\Phi(t)\rangle$ can be derived by substituting eqn. (14.2.2) into eqn. (14.2.1), which yields

$$\left(\hat{\mathcal{H}}_0 + \hat{\mathcal{H}}_1(t)\right)e^{-i\hat{\mathcal{H}}_0(t-t_0)/\hbar}|\Phi(t)\rangle = i\hbar\frac{\partial}{\partial t}e^{-i\hat{\mathcal{H}}_0(t-t_0)/\hbar}|\Phi(t)\rangle$$

$$\left(\hat{\mathcal{H}}_0 + \hat{\mathcal{H}}_1(t)\right)e^{-i\hat{\mathcal{H}}_0(t-t_0)/\hbar}|\Phi(t)\rangle = \hat{\mathcal{H}}_0 e^{-i\hat{\mathcal{H}}_0(t-t_0)/\hbar}|\Phi(t)\rangle + e^{-i\hat{\mathcal{H}}_0(t-t_0)/\hbar}i\hbar\frac{\partial}{\partial t}|\Phi(t)\rangle$$

$$\hat{\mathcal{H}}_1(t)e^{-i\hat{\mathcal{H}}_0(t-t_0)/\hbar}|\Phi(t)\rangle = e^{-i\hat{\mathcal{H}}_0(t-t_0)/\hbar}i\hbar\frac{\partial}{\partial t}|\Phi(t)\rangle. \tag{14.2.5}$$

Multiplying on the left by $e^{i\hat{\mathcal{H}}_0(t-t_0)/\hbar}$ yields

$$e^{i\hat{\mathcal{H}}_0(t-t_0)/\hbar}\hat{\mathcal{H}}_1(t)e^{-i\hat{\mathcal{H}}_0(t-t_0)/\hbar}|\Phi(t)\rangle = i\hbar\frac{\partial}{\partial t}|\Phi(t)\rangle. \tag{14.2.6}$$

According to eqn. (14.2.3), the $\exp[i\hat{\mathcal{H}}_0(t-t_0)/\hbar]\hat{\mathcal{H}}_1(t)\exp[-i\hat{\mathcal{H}}_0(t-t_0)/\hbar]$ is the interaction-picture representation of the perturbation Hamiltonian, which we will denote as $\hat{\mathcal{H}}_I(t)$. The time evolution of the state vector in the interaction picture is, therefore, given by a Schrödinger equation of the form

$$\hat{\mathcal{H}}_I(t)|\Phi(t)\rangle = i\hbar\frac{\partial}{\partial t}|\Phi(t)\rangle. \tag{14.2.7}$$

Equation (14.2.7) shows that this time evolution is determined entirely by the interaction-picture representation of the perturbation, $\hat{\mathcal{H}}_I(t)$. According to eqn. (14.2.2), the initial condition to eqn. (14.2.7) is $|\Phi(t_0)\rangle = |\Psi(t_0)\rangle$. In the next subsection, we will develop an iterative solution to eqn. (14.2.7), which will reveal the detailed structure of the propagator for time-dependent systems.

14.2.2 Iterative solution for the interaction-picture state vector

The solution to eqn. (14.2.7) can be expressed in terms of a unitary operator $\hat{U}_I(t;t_0)$, which is the interaction-picture propagator. The initial state $|\Phi(t_0)\rangle$ evolves in time according to

$$|\Phi(t)\rangle = \hat{U}_I(t;t_0)|\Phi(t_0)\rangle = \hat{U}_I(t;t_0)|\Psi(t_0)\rangle. \tag{14.2.8}$$

Substitution of eqn. (14.2.8) into eqn. (14.2.7) yields an evolution equation for the propagator $\hat{U}_I(t;t_0)$:

$$\hat{\mathcal{H}}_I(t)\hat{U}_I(t;t_0) = i\hbar\frac{\partial}{\partial t}\hat{U}_I(t;t_0). \tag{14.2.9}$$

Equation (14.2.9) has the initial condition $\hat{U}_I(t_0;t_0) = \hat{I}$. In developing a solution to eqn. (14.2.9), we assume that $\hat{\mathcal{H}}_I(t)$ is a small perturbation so that the solution can be constructed in terms of a power series in $\hat{\mathcal{H}}_I(t)$. Such a solution is generated by rewriting eqn. (14.2.9) as an integral equation:

$$\hat{U}_I(t;t_0) = \hat{U}_I(t_0;t_0) - \frac{i}{\hbar}\int_{t_0}^{t} dt'\, \hat{\mathcal{H}}_I(t')\hat{U}_I(t';t_0)$$

$$= \hat{I} - \frac{i}{\hbar}\int_{t_0}^{t} dt'\, \hat{\mathcal{H}}_I(t')\hat{U}_I(t';t_0). \tag{14.2.10}$$

We can easily verify that eqn. (14.2.10) is the solution for $\hat{U}_I(t;t_0)$. Taking the time derivative of both sides of eqn. (14.2.10) gives

$$i\hbar\frac{\partial}{\partial t}\hat{U}_I(t;t_0) = -i\hbar\frac{i}{\hbar}\frac{\partial}{\partial t}\int_{t_0}^{t} dt'\, \hat{\mathcal{H}}_I(t')\hat{U}_I(t';t_0)$$

$$= \hat{\mathcal{H}}_I(t)\hat{U}_I(t;t_0). \tag{14.2.11}$$

Equation (14.2.10) allows a perturbation series solution to be developed systematically. We start with a zeroth-order solution by setting $\hat{\mathcal{H}}_I(t) = 0$ in eqn. (14.2.10), which gives the trivial result

$$\hat{U}_I^{(0)}(t;t_0) = \hat{I}. \tag{14.2.12}$$

This solution is now fed back into the right side of eqn. (14.2.10) to develop a first-order solution:

$$\hat{U}_I^{(1)}(t;t_0) = \hat{I} - \frac{i}{\hbar}\int_{t_0}^{t} dt'\, \hat{\mathcal{H}}_I(t')\hat{U}_I^{(0)}(t';t_0)$$

$$= \hat{I} - \frac{i}{\hbar}\int_{t_0}^{t} dt'\, \hat{\mathcal{H}}_I(t'). \tag{14.2.13}$$

This first-order solution is now substituted back into the right side of eqn. (14.2.10) to generate a second-order solution

$$\hat{U}_I^{(2)}(t;t_0) = \hat{I} - \frac{i}{\hbar} \int_{t_0}^{t} dt' \, \hat{\mathcal{H}}_I(t') \hat{U}_I^{(1)}(t';t_0)$$

$$= \hat{I} - \frac{i}{\hbar} \int_{t_0}^{t} dt' \, \hat{\mathcal{H}}_I(t') + \left(\frac{i}{\hbar}\right)^2 \int_{t_0}^{t} dt' \int_{t_0}^{t'} dt'' \, \hat{\mathcal{H}}_I(t')\hat{\mathcal{H}}_I(t''), \qquad (14.2.14)$$

and so forth. Thus we see that the kth-order solution is generated from the $(k-1)$th-order solution according to the recursion formula:

$$\hat{U}_I^{(k)}(t;t_0) = \hat{I} - \frac{i}{\hbar} \int_{t_0}^{t} dt' \, \hat{\mathcal{H}}_I(t') \hat{U}_I^{(k-1)}(t';t_0). \qquad (14.2.15)$$

Setting $k=3$ in eqn. (14.2.15), for example, yields the third-order solution

$$\hat{U}_I^{(3)}(t;t_0) = \hat{I} - \frac{i}{\hbar} \int_{t_0}^{t} dt' \, \hat{\mathcal{H}}_I(t') + \left(\frac{i}{\hbar}\right)^2 \int_{t_0}^{t} dt' \int_{t_0}^{t'} dt'' \, \hat{\mathcal{H}}_I(t')\hat{\mathcal{H}}_I(t'')$$

$$- \left(\frac{i}{\hbar}\right)^3 \int_{t_0}^{t} dt' \int_{t_0}^{t'} dt'' \int_{t_0}^{t''} dt''' \, \hat{\mathcal{H}}_I(t')\hat{\mathcal{H}}_I(t'')\hat{\mathcal{H}}_I(t'''). \qquad (14.2.16)$$

By taking the limit $k \to \infty$ in eqn. (14.2.15) and summing over all orders, we obtain the exact solution for $\hat{U}_I(t;t_0)$ as a series:

$$\hat{U}_I(t;t_0) = \sum_{k=0}^{\infty} \left(-\frac{i}{\hbar}\right)^k \int_{t_0}^{t} dt^{(1)} \cdots \int_{t_0}^{t^{(k-1)}} dt^{(k)} \, \hat{\mathcal{H}}_I(t^{(1)}) \cdots \hat{\mathcal{H}}_I(t^{(k)}). \qquad (14.2.17)$$

The propagator $\hat{U}_I(t;t_0)$, approximated at any order in perturbation theory, can be used to approximate the time evolution of the state vector $|\Phi(t)\rangle$ in the interaction picture. In general, this evolution is

$$|\Phi(t)\rangle = \hat{U}_I(t;t_0)|\Phi(t_0)\rangle, \qquad (14.2.18)$$

and from this expression, the time evolution of the original state vector $|\Psi(t)\rangle$ in the Schrödinger picture can be determined as

$$|\Psi(t)\rangle = e^{-i\hat{\mathcal{H}}_0(t-t_0)/\hbar}|\Phi(t)\rangle = e^{-i\hat{\mathcal{H}}_0(t-t_0)/\hbar}\hat{U}_I(t;t_0)|\Phi(t_0)\rangle$$

$$= e^{-i\hat{\mathcal{H}}_0(t-t_0)/\hbar}\hat{U}_I(t;t_0)|\Psi(t_0)\rangle$$

$$\equiv \hat{U}(t;t_0)|\Psi(t_0)\rangle. \qquad (14.2.19)$$

Here, we have used the fact that $|\Phi(t_0)\rangle = |\Psi(t_0)\rangle$. In the last line, the full propagator in the Schrödinger picture is identified as

$$\hat{U}(t;t_0) = e^{-i\hat{\mathcal{H}}_0(t-t_0)/\hbar}\hat{U}_I(t;t_0) = \hat{U}_0(t;t_0)\hat{U}_I(t;t_0). \qquad (14.2.20)$$

From eqn. (14.2.20), the structure of the full propagator is revealed. Let us use eqn. (14.2.20) to generate the first few orders in the propagator. Substituting eqn. (14.2.12) into eqn. (14.2.20) yields the lowest order contribution to $\hat{U}(t; t_0)$:

$$\hat{U}^{(0)}(t; t_0) = e^{-i\hat{\mathcal{H}}_0(t-t_0)/\hbar} = \hat{U}_0(t; t_0). \tag{14.2.21}$$

Thus, at zeroth order, eqn. (14.2.21) implies that the system is propagated using the unperturbed propagator $\hat{U}_0(t; t_0)$ as though the perturbation did not exist. At first order, we obtain

$$\hat{U}^{(1)}(t; t_0) = e^{-i\hat{\mathcal{H}}_0(t-t_0)/\hbar} - \frac{i}{\hbar} e^{-i\hat{\mathcal{H}}_0(t-t_0)/\hbar} \int_{t_0}^{t} dt' \; \hat{\mathcal{H}}_I(t')$$

$$= e^{-i\hat{\mathcal{H}}_0(t-t_0)/\hbar} - \frac{i}{\hbar} e^{-i\hat{\mathcal{H}}_0(t-t_0)/\hbar} \int_{t_0}^{t} dt' e^{i\hat{\mathcal{H}}_0(t'-t_0)/\hbar} \hat{\mathcal{H}}_1(t') e^{-i\hat{\mathcal{H}}_0(t'-t_0)/\hbar}$$

$$= e^{-i\hat{\mathcal{H}}_0(t-t_0)/\hbar} - \frac{i}{\hbar} \int_{t_0}^{t} dt' \; e^{-i\hat{\mathcal{H}}_0(t-t')/\hbar} \hat{\mathcal{H}}_1(t') e^{-i\hat{\mathcal{H}}_0(t'-t_0)/\hbar}$$

$$= \hat{U}_0(t; t_0) - \frac{i}{\hbar} \int_{t_0}^{t} dt' \; \hat{U}_0(t; t') \hat{\mathcal{H}}_1(t') \hat{U}_0(t'; t_0), \tag{14.2.22}$$

where in the second line, the definition

$$\hat{\mathcal{H}}_I(t) = e^{i\hat{\mathcal{H}}_0(t-t_0)/\hbar} \hat{\mathcal{H}}_1(t) e^{-i\hat{\mathcal{H}}_0(t-t_0)/\hbar} \tag{14.2.23}$$

has been used. Equation (14.2.22) indicates that at first order, the propagator is composed of two terms. The first term is simply the unperturbed propagation from t_0 to t. In the second term, the system undergoes unperturbed propagation from t_0 to t'. At t', the perturbation $\hat{\mathcal{H}}_1(t')$ acts, and finally, from t' to t, the propagation is unperturbed. In addition, we must integrate over all possible intermediate times t' at which the perturbation is applied.

In a similar manner, it can be shown that up to second-order, the full propagator is given by

$$\hat{U}^{(2)}(t; t_0) = \hat{U}_0(t; t_0) - \frac{i}{\hbar} \int_{t_0}^{t} dt' \; \hat{U}_0(t; t') \hat{\mathcal{H}}_1(t') \hat{U}_0(t'; t_0)$$

$$+ \left(\frac{i}{\hbar}\right)^2 \int_{t_0}^{t} dt' \int_{t_0}^{t'} dt'' \; \hat{U}_0(t; t') \hat{\mathcal{H}}_1(t') \hat{U}_0(t'; t'') \hat{\mathcal{H}}_1(t'') \hat{U}_0(t''; t_0). \tag{14.2.24}$$

At second-order, the new (last) term involves unperturbed propagation from t_0 to t'', action of $\hat{\mathcal{H}}_1(t'')$ at t'', unperturbed propagation from t'' to t', action of $\hat{\mathcal{H}}_1(t')$ at t', and finally unperturbed propagation from t' to t. Again, we must integrate over intermediate times t' and t'' at which the perturbation is applied. A pictorial representation of the full propagator is given in Figure 14.3. The picture on the left side of the equal sign in Fig. 14.3 indicates that the perturbation causes the system to

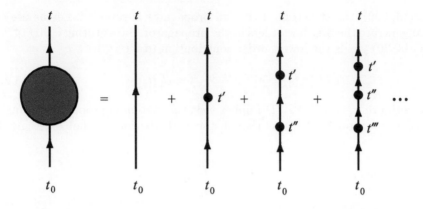

Fig. 14.3 Pictorial representation of the perturbation expansion of the time-dependent propagator.

undergo some undetermined dynamical process between t_0 and t. On the right side of the equal sign, the process is broken down in terms of the action of the perturbation $\hat{\mathcal{H}}_1$ at specific intermediate times (which must be integrated over), indicated in the figure by the dots. At the kth order, the perturbation Hamiltonian $\hat{\mathcal{H}}_1$ acts on the system at k instances in time. The limits of integration indicate that these time instances are ordered chronologically.

The specific ordering of the times at which $\hat{\mathcal{H}}_1$ acts on the unperturbed system raises an important point. In each term in the expansion for $\hat{U}_I(t; t_0)$, the order in which the operators $\hat{\mathcal{H}}_I(t')$, $\hat{\mathcal{H}}_I(t'')$,... are multiplied is critical. The reason for this is that the commutator $[\hat{\mathcal{H}}_I(t), \hat{\mathcal{H}}_I(t')]$ does not vanish if $t \neq t'$. Thus, to remove any ambiguity when specifying the order of the operators $\hat{\mathcal{H}}_I(t')$, $\hat{\mathcal{H}}_I(t'')$,... in a time series, we introduce the *time-ordering operator*, T. The action of T on a product of time-dependent operators $\hat{A}(t_1)\hat{B}(t_2)\hat{C}(t_3)\cdots\hat{D}(t_n)$ reorders the operators in the product chronologically in time from earliest to latest proceeding from right to left in the product. This means that operators earliest in time act before operators at later times. For example, the action of T on two operators $\hat{A}(t_1)$ and $\hat{B}(t_2)$ is

$$T\left[\hat{A}(t_1)\hat{B}(t_2)\right] = \begin{cases} \hat{A}(t_1)\hat{B}(t_2) & t_2 < t_1 \\ \\ \hat{B}(t_2)\hat{A}(t_1) & t_1 < t_2 \end{cases} \tag{14.2.25}$$

Let us now apply the time-ordering operator T to the second-order term. We first write the double integral

$$I_2(t_0, t) = \int_{t_0}^t dt' \int_{t_0}^{t'} dt'' \; \hat{\mathcal{H}}_I(t')\hat{\mathcal{H}}_I(t'') \tag{14.2.26}$$

as a sum of two terms generated by interchanging the dummy variables t' and t'':

$$I_2(t_0, t) = \frac{1}{2}\left[\int_{t_0}^t dt' \int_{t_0}^{t'} dt'' \hat{\mathcal{H}}_I(t')\hat{\mathcal{H}}_I(t'') + \int_{t_0}^t dt'' \int_{t_0}^{t''} dt' \hat{\mathcal{H}}_I(t'')\hat{\mathcal{H}}_I(t')\right]. \tag{14.2.27}$$

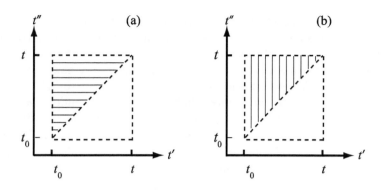

Fig. 14.4 The two integration regions in eqn. (14.2.27): (a) The region $t'' \in [t_0, t]$, $t' \in [t_0, t'']$. (b) The region $t' \in [t_0, t]$, $t'' \in [t', t]$.

Figure 14.4(a) illustrates the integration region $t'' \in [t_0, t]$, $t' \in [t_0, t'']$ in the t'-t'' plane, which is covered in the second term on the right side of eqn. (14.2.27). The same region can be covered by choosing $t' \in [t_0, t]$ and $t'' \in [t', t]$, as illustrated Fig. 14.4(b). With this choice, eqn. (14.2.27) becomes

$$I(t_0, t) = \frac{1}{2} \left[\int_{t_0}^{t} dt' \int_{t_0}^{t'} dt'' \, \hat{\mathcal{H}}_I(t')\hat{\mathcal{H}}_I(t'') + \int_{t_0}^{t} dt' \int_{t'}^{t} dt'' \, \hat{\mathcal{H}}_I(t'')\hat{\mathcal{H}}_I(t') \right]. \quad (14.2.28)$$

In the first term on the right side of eqn. (14.2.28), $t'' < t'$ and $H_I(t'')$ acts first, followed by $H_I(t')$. In the second term, $t' < t''$ and $\hat{\mathcal{H}}_I(t')$ acts first, followed by $\hat{\mathcal{H}}_I(t'')$. The two terms can thus be combined with both t' and t'' lying in the interval $[t_0, t]$ by introducing the time-ordering operator:

$$\int_{t_0}^{t} dt' \int_{t_0}^{t'} dt'' \, \hat{\mathcal{H}}_I(t')\hat{\mathcal{H}}_I(t'') = \frac{1}{2} \int_{t_0}^{t} dt' \int_{t_0}^{t} dt'' \, T\left[\hat{\mathcal{H}}_I(t')\hat{\mathcal{H}}_I(t'') \right]. \quad (14.2.29)$$

The same analysis can be applied to each order in eqn. (14.2.17), recognizing that a product of k operators can be ordered in $k!$ ways by the time-ordering operator. Equation (14.2.17) can then be rewritten in terms of a time-ordered product

$$\hat{U}_I(t; t_0) = \sum_{k=0}^{\infty} \left(-\frac{i}{\hbar} \right)^k \frac{1}{k!}$$

$$\times \int_{t_0}^{t} dt_1 \int_{t_0}^{t} dt_2 \cdots \int_{t_0}^{t} dt_k T\left[\hat{\mathcal{H}}_I(t_1)\hat{\mathcal{H}}_I(t_2) \cdots \hat{\mathcal{H}}_I(t_k) \right]. \quad (14.2.30)$$

The sum in eqn. (14.2.30) resembles the power-series expansion of an exponential, and consequently we can write the sum symbolically as

$$\hat{U}_I(t; t_0) = T\left[\exp\left(-\frac{i}{\hbar} \int_{t_0}^{t} dt' \, \hat{\mathcal{H}}_I(t') \right) \right], \quad (14.2.31)$$

which is known as a *time-ordered exponential*. Since eqn. (14.2.31) is just a shorthand for eqn. (14.2.30), it is understood that the time-ordering operator orders the operators in each term of an expansion of the exponential.

Given the formalism of time-dependent perturbation theory, we now seek to answer the following question: If the system is initially in an eigenstate of $\hat{\mathcal{H}}_0$ with energy E_i, what is the probability as a function of time t that the system will undergo a transition to a new eigenstate of $\hat{\mathcal{H}}_0$ with energy E_f? To answer this question, we first set the initial state vector $|\Psi(t_0)\rangle$ equal to the eigenvector $|E_i\rangle$ of $\hat{\mathcal{H}}_0$. Then, the amplitude as a function of time that the system will undergo a transition to the eigenstate $|E_f\rangle$ is obtained by propagating this initial state to time t with the propagator $\hat{U}(t;t_0)$ and then taking the overlap of the resulting state with the eigenstate $|E_f\rangle$:

$$A_{fi}(t) = \langle E_f|\hat{U}(t;t_0)|E_i\rangle. \tag{14.2.32}$$

The probability is just the square modulus of this complex amplitude:

$$P_{fi}(t) = \left|\langle E_f|\hat{U}(t;t_0)|E_i\rangle\right|^2. \tag{14.2.33}$$

Consider first the amplitude at zeroth order in perturbation theory. At this order, $\hat{U}(t;t_0) = \hat{U}_0(t;t_0)$, and the amplitude is simply

$$A_{fi}^{(0)}(t) = \langle E_f|e^{-i\hat{\mathcal{H}}_0(t-t_0)/\hbar}|E_i\rangle$$

$$= e^{-iE_i(t-t_0)/\hbar}\langle E_f|E_i\rangle, \tag{14.2.34}$$

which clearly vanishes by orthogonality if $E_i \neq E_f$. Thus, at zeroth order, the only possibility is a trivial one in which no transition occurs.

The lowest nontrivial $(E_i \neq E_f)$ result occurs at first order, where the transition amplitude is given by

$$A_{fi}^{(1)}(t) = \langle E_f|\hat{U}^{(1)}(t;t_0)|E_i\rangle$$

$$= -\frac{i}{\hbar}\int_{t_0}^{t}dt'\,\langle E_f|\hat{U}_0(t;t')\hat{\mathcal{H}}_1(t')\hat{U}_0(t';t_0)|E_i\rangle$$

$$= -\frac{i}{\hbar}\int_{t_0}^{t}dt'\,\langle E_f|e^{-i\hat{\mathcal{H}}_0(t-t')/\hbar}\hat{\mathcal{H}}_1(t')e^{-i\hat{\mathcal{H}}_0(t'-t_0)/\hbar}|E_i\rangle$$

$$= -\frac{i}{\hbar}\int_{t_0}^{t}dt'\,e^{-iE_f(t-t')/\hbar}e^{-iE_i(t'-t_0)/\hbar}\langle E_f|\hat{\mathcal{H}}_1(t')|E_i\rangle$$

$$= -\frac{i}{\hbar}e^{-iE_ft/\hbar}e^{iE_it_0/\hbar}\int_{t_0}^{t}dt'\,e^{i(E_f-E_i)t'/\hbar}\langle E_f|\hat{\mathcal{H}}_1(t')|E_i\rangle. \tag{14.2.35}$$

We now define a *transition frequency* ω_{fi} as $\omega_{fi} = (E_f - E_i)/\hbar$. Taking the absolute square of the last line of eqn. (14.2.35), we obtain the probability at first order as

$$P_{fi}^{(1)}(t) = \frac{1}{\hbar^2} \left| \int_{t_0}^{t} dt' \, e^{i\omega_{fi}t'} \langle E_f | \hat{\mathcal{H}}_1(t') | E_i \rangle \right|^2 . \tag{14.2.36}$$

At first order, the probability depends on the matrix element of the perturbation between the initial and final eigenstates.

Thus far, the formalism we have derived is valid for any perturbation $\hat{\mathcal{H}}_1(t)$. The specific choice of this perturbation determines the manifold of eigenstates of $\hat{\mathcal{H}}_0$ it probes, as we will demonstrate in the next subsection.

14.2.3 Fermi's Golden Rule

In Section 14.1, we formulated the Hamiltonian of a material system coupled to an external electromagnetic field, and we derived expressions for the electromagnetic field in the absence of sources or physical boundaries as solutions of the free-field wave equation. For the remainder of this chapter, we will focus on weak fields, so that the term in eqn. (14.1.11) proportional to $\mathbf{A}^2$ can be neglected. We will also focus on a class of experiments for which the wavelength of the electromagnetic radiation is long compared to the size of the sample under investigation. Under this condition, the spatial dependence of the electromagnetic field can also be neglected, since $\cos(\mathbf{k} \cdot \mathbf{r} - \omega t + \varphi_0) = \text{Re}[\exp(i\mathbf{k} \cdot \mathbf{r} - i\omega t + \varphi_0)]$, and $\exp(i\mathbf{k} \cdot \mathbf{r}) \approx 1$ in the long-wavelength limit where $|\mathbf{k}| = 2\pi/\lambda \approx 0$. Thus, $\hat{\mathcal{H}}_1(t)$ reduces to the specific form

$$\hat{\mathcal{H}}_1(t) = -2\mathcal{V}F(\omega)\cos(\omega t) = -\mathcal{V}F(\omega) \left[e^{i\omega t} + e^{-i\omega t} \right], \tag{14.2.37}$$

where $\mathcal{V}$ is a Hermitian operator.

As noted above, we are interested in the probability that the system will undergo a transition from an eigenstate $|E_i\rangle$ to $|E_f\rangle$ of $\hat{\mathcal{H}}_0$ under the perturbation of eqn. (14.2.37). However, since the perturbation is periodic in time, the problem must be stated a little differently. If we apply the perturbation over a long time, what is the mean probability per unit time of a transition or *mean transition rate*? In order to simplify the calculation of this rate, let us consider a time interval T and choose $t_0 = -T/2$ and $t = T/2$. The mean transition rate can be expressed as

$$R_{fi}^{(1)}(T) = \frac{P_{fi}^{(1)}(T)}{T}$$

$$= \frac{|F(\omega)|^2}{T\hbar^2} \left| \int_{-T/2}^{T/2} dt \, \left[e^{i(\omega_{fi}+\omega)t} + e^{i(\omega_{fi}-\omega)t} \right] \right|^2 |\langle E_f | \mathcal{V} | E_i \rangle|^2 . \tag{14.2.38}$$

For finite T, the time integral can be carried out explicitly yielding

$$\int_{-T/2}^{T/2} dt \left[e^{i(\omega_{fi}+\omega)t} + e^{i(\omega_{fi}-\omega)t} \right] =$$

$$T \left[\frac{\sin[(\omega_{fi}+\omega)T/2]}{(\omega_{fi}+\omega)T/2} + \frac{\sin[(\omega_{fi}-\omega)T/2]}{(\omega_{fi}-\omega)T/2} \right]. \tag{14.2.39}$$

Assuming $\omega > 0$, in the limit of large T, the second term on the right in eqn. (14.2.39) dominates over the first and peaks sharply at $\omega = \omega_{fi}$. Thus, we can retain only this term and write the mean transition rate as

$$R_{fi}^{(1)}(T) = \frac{1}{\hbar^2} T |F(\omega)|^2 |\langle E_f|\mathcal{V}|E_i\rangle|^2 \frac{\sin^2(\omega_{fi}-\omega)T/2}{[(\omega_{fi}-\omega)T/2]^2}. \tag{14.2.40}$$

Regarding eqn. (14.2.40) as a function of ω, at large T, this function becomes highly peaked when $\omega = \omega_{fi}$ but drops to zero rapidly away from $\omega = \omega_{fi}$. The condition $\omega_{fi} = \omega$ is equivalent to the condition $E_f = E_i + \hbar\omega$, which is a statement of energy conservation. Since $\hbar\omega$ is the energy quantum of the electromagnetic field, also known as a *photon*, the transition can only occur if the energy of the field frequency ω is exactly "tuned" for the the transition from E_i to E_f. Hence, a monochromatic field of frequency ω can be used as a probe of the allowed transitions and hence the eigenvalue structure of $\hat{\mathcal{H}}_0$.

We now consider the $T \to \infty$ limit more carefully. Denoting the rate in this limit simply as R_{fi}, the integral in eqn. (14.2.39) in this limit becomes

$$\lim_{T\to\infty} \int_{-T/2}^{T/2} dt \left[e^{i(\omega_{fi}+\omega)t} + e^{i(\omega_{fi}-\omega)t} \right] = \int_{-\infty}^{\infty} dt \left[e^{i(\omega_{fi}+\omega)t} + e^{i(\omega_{fi}-\omega)t} \right]$$

$$= 2\pi \left[\delta(\omega_{fi}+\omega) + \delta(\omega_{fi}-\omega) \right]. \tag{14.2.41}$$

Again, for $\omega > 0$ and $\omega_{fi} > 0$, only the second δ-function is ever nonzero, so we can drop the first δ-function. Note that the second δ-function in eqn. (14.2.41) can also be written as $2\pi\hbar\delta(E_f - E_i - \hbar\omega)$. Therefore, the expression for the mean rate in this limit can be written as

$$R_{fi}(\omega) = \lim_{T\to\infty} \frac{P_{fi}^{(1)}(T)}{T}$$

$$= \lim_{T\to\infty} \frac{1}{T\hbar^2} \left| \int_{-T/2}^{T/2} dt\, e^{i(\omega_{fi}-\omega)t} \right|^2 |F(\omega)|^2 |\langle E_f|\mathcal{V}|E_i\rangle|^2$$

$$= \lim_{T\to\infty} \frac{1}{T\hbar^2} \left[\int_{-T/2}^{T/2} dt\, e^{i(\omega_{fi}-\omega)t} \right]$$

$$\times \left[\int_{-T/2}^{T/2} dt\, e^{i(\omega_{fi}-\omega)t} \right] |F(\omega)|^2 |\langle E_f|\mathcal{V}|E_i\rangle|^2, \tag{14.2.42}$$

where we have dropped the "(1)" superscript (it is understood that the result is derived from first-order perturbation theory) and indicated explicitly the dependence of R_{fi}

on the frequency ω. When the integral in the third line of eqn. (14.2.42) is replaced by the δ-function, the remaining integral becomes simply T since $\omega_{fi} = \omega$, and this T cancels the T in the denominator. For this reason, the division by T in eqn. (14.2.38) is equivalent to expressing the rate as a proper derivative $\lim_{T \to \infty} P_{fi}^{(1)}(T)/T = dP_{fi}^{(1)}/dt$. The expression for the rate now becomes

$$R_{fi}(\omega) = \frac{2\pi}{\hbar}|F(\omega)|^2 \,|\langle E_f|\mathcal{V}|E_i\rangle|^2 \,\delta(E_f - E_i - \hbar\omega), \qquad (14.2.43)$$

which is known as *Fermi's Golden Rule*. The rule states that, in first-order perturbation theory, the transition rate depends only on the square of the matrix element of the operator $\mathcal{V}$ between initial and final states and explicitly requires energy conservation via the δ-function. Fermi's Golden Rule predicts the rate of transitions from a *specific* initial state $|E_i\rangle$ to a final state $|E_f\rangle$, both of which are eigenstates of $\hat{\mathcal{H}}_0$ and which are connected via the energy conservation condition $E_f = E_i + \hbar\omega$.

14.3 Time correlation functions and frequency spectra

In this section, the Fermi Golden Rule expression will be used to analyze the output of an experiment in which a monochromatic field is applied to an ensemble of systems. If we wish to calculate the transition rate for the ensemble, we must remember that the systems in the ensemble are *not* in a single initial state $|E_i\rangle$. Rather, there is a distribution of initial states prescribed by the equilibrium density matrix $\rho(\hat{\mathcal{H}}_0)$, which satisfies the equilibrium Liouville equation $[\hat{\mathcal{H}}_0, \rho(\hat{\mathcal{H}}_0)] = 0$. Thus, in the canonical ensemble, the probability that a given ensemble member is in an eigenstate of $\hat{\mathcal{H}}_0$ with energy E_i is the density matrix eigenvalue

$$w_i = \frac{e^{-\beta E_i}}{Q(N, V, T)} = \frac{e^{-\beta E_i}}{\mathrm{Tr}\left(e^{-\beta \hat{\mathcal{H}}_0}\right)}. \qquad (14.3.1)$$

The rate we seek is the ensemble average of $R_{fi}(\omega)$ over initial states, denoted $R(\omega)$, which is given by

$$R(\omega) = \langle R_{fi}(\omega)\rangle = \sum_{i,f} R_{fi}(\omega) w_i. \qquad (14.3.2)$$

Although both initial and final states are summed in eqn. (14.3.2), we know that the sum over final states is not independent, since the only permissible final states are those connected to initial states by energy conservation. Equation (14.3.2) indicates that the contribution from each possible initial state $|E_i\rangle$ to the average rate is the probability w_i that a given member of the ensemble is initially in that state. Ultimately, we sum over those final states that can be reached from the initial state without violating energy conservation to obtain the average transition rate.

When we substitute eqn. (14.2.43) for $R_{fi}(\omega)$ into eqn. (14.3.2), the average rate becomes

$$R(\omega) = \frac{2\pi}{\hbar}|F(\omega)|^2 \sum_{i,f} w_i \,|\langle E_f|\mathcal{V}|E_i\rangle|^2 \,\delta(E_f - E_i - \hbar\omega). \qquad (14.3.3)$$

Writing the δ-function as an integral, eqn. (14.3.3) becomes

$$R(\omega) = \frac{1}{\hbar^2}|F(\omega)|^2 \int_{-\infty}^{\infty} dt \sum_{i,f} w_i e^{i(E_f - E_i - \hbar\omega)t/\hbar} |\langle E_f|\mathcal{V}|E_i\rangle|^2$$

$$= \frac{1}{\hbar^2}|F(\omega)|^2 \int_{-\infty}^{\infty} dt\, e^{-i\omega t} \sum_{i,f} w_i \langle E_i|\mathcal{V}|E_f\rangle\langle E_f|\mathcal{V}|E_i\rangle e^{iE_f t/\hbar} e^{-iE_i t/\hbar}$$

$$= \frac{1}{\hbar^2}|F(\omega)|^2 \int_{-\infty}^{\infty} dt\, e^{-i\omega t} \sum_{i,f} w_i \langle E_i|\mathcal{V}|E_f\rangle\langle E_f|e^{i\hat{\mathcal{H}}_0 t/\hbar}\mathcal{V}e^{-i\hat{\mathcal{H}}_0 t/\hbar}|E_i\rangle. \qquad (14.3.4)$$

In the last line, we have used the fact that $|E_i\rangle$ and $|E_f\rangle$ are eigenstates of $\hat{\mathcal{H}}_0$ to bring the two exponential factors into the angle brackets as the unperturbed propagator $\exp(-i\hat{\mathcal{H}}_0 t/\hbar)$ and its conjugate $\exp(i\hat{\mathcal{H}}_0 t/\hbar)$. Note, however, that the operator $\exp(i\hat{\mathcal{H}}_0 t/\hbar)\mathcal{V}\exp(-i\hat{\mathcal{H}}_0 t/\hbar) = \mathcal{V}(t)$ is just the representation of the operator $\mathcal{V}$ in the interaction picture (see eqn. (14.2.3)). Thus, the average transition rate can be expressed as

$$R(\omega) = \frac{1}{\hbar^2}|F(\omega)|^2 \int_{-\infty}^{\infty} dt\, e^{-i\omega t} \sum_{i,f} w_i \langle E_i|\mathcal{V}(0)|E_f\rangle\langle E_f|\mathcal{V}(t)|E_i\rangle, \qquad (14.3.5)$$

where the $\mathcal{V}(0)$ is the operator in the interaction picture at $t = 0$. Thus, both operators in eqn. (14.3.5) are represented within the same quantum-mechanical picture. Note that the sum over final states can now be performed using the completeness relation

$$\sum_f |E_f\rangle\langle E_f| = \hat{I} \qquad (14.3.6)$$

of the eigenstates of $\hat{\mathcal{H}}_0$. Equation (14.3.5) thus becomes

$$R(\omega) = \frac{1}{\hbar^2}|F(\omega)|^2 \int_{-\infty}^{\infty} dt\, e^{-i\omega t} \sum_i w_i \langle E_i|\mathcal{V}(0)\mathcal{V}(t)|E_i\rangle$$

$$= \frac{1}{\hbar^2}|F(\omega)|^2 \int_{-\infty}^{\infty} dt\, e^{-i\omega t} \frac{1}{Q(N,V,T)} \text{Tr}\left(e^{-\beta\hat{\mathcal{H}}_0}\mathcal{V}(0)\mathcal{V}(t)\right)$$

$$= \frac{1}{\hbar^2}|F(\omega)|^2 \int_{-\infty}^{\infty} dt\, e^{-i\omega t} \langle \mathcal{V}(0)\mathcal{V}(t)\rangle. \qquad (14.3.7)$$

The last line shows that the ensemble-averaged transition rate at frequency ω is just the Fourier transform of the quantum time correlation function $\langle \mathcal{V}(0)\mathcal{V}(t)\rangle$ (Berne, 1971).

In general, a quantum time correlation function of two operators $\hat{A}$ and $\hat{B}$ with respect to an unperturbed Hamiltonian $\hat{\mathcal{H}}_0$ is given by

$$C_{AB}(t) = \frac{\text{Tr}\left[\hat{A}(0)\hat{B}(t)e^{-\beta\hat{\mathcal{H}}_0}\right]}{\text{Tr}\left[e^{-\beta\hat{\mathcal{H}}_0}\right]}. \qquad (14.3.8)$$

Although quantum time correlation functions possess many of the same properties as their classical counterparts, we point out one crucial difference at this juncture. The operators $\mathcal{V}(0)$ and $\mathcal{V}(t)$ are individually Hermitian, but since $[\mathcal{V}(0), \mathcal{V}(t)] \neq 0$, the autocorrelation function in eqn. (14.3.7) is an expectation value of a non-Hermitian operator product $\mathcal{V}(0)\mathcal{V}(t)$. Such a non-Hermitian expectation value suggests that something fundamental is missing from the above analysis.

A little reflection reveals that the problem lies with our choice of $\omega > 0$ in eqn. (14.2.39). A complete analysis requires that we examine $\omega < 0$ as well, in which case the only first term on the right side of eqn. (14.2.40) is retained. This is tantamount to substituting $-\omega$ for ω in eqn. (14.3.3), which yields

$$R(-\omega) = \frac{2\pi}{\hbar}|F(\omega)|^2 \sum_{i,f} w_i \, |\langle E_f|\mathcal{V}|E_i\rangle|^2 \, \delta(E_f - E_i + \hbar\omega). \tag{14.3.9}$$

Unlike eqn. (14.3.3), which refers to an absorption process with $E_f = E_i + \hbar\omega$, eqn. (14.3.9) describes a process for which $E_f = E_i - \hbar\omega$, or $E_f < E_i$, which is an *emission* process. The system starts in a state with energy E_i and releases an amount of energy $\hbar\omega$ as it decays to a state with lower energy E_f. We will now show that eqn. (14.3.9) can be expressed in terms of the correlation function $\mathcal{V}(t)\mathcal{V}(0)$, which when added to eqn. (14.3.7) leads to the Hermitian combination $\mathcal{V}(t)\mathcal{V}(0) + \mathcal{V}(0)\mathcal{V}(t)$.

We begin by interchanging the summation indices i and f in eqn. (14.3.9), which gives

$$R(-\omega) = \frac{2\pi}{\hbar}|F(\omega)|^2 \sum_{i,f} w_f \, |\langle E_i|\mathcal{V}|E_f\rangle|^2 \, \delta(E_i - E_f + \hbar\omega), \tag{14.3.10}$$

where

$$w_f = \frac{e^{-\beta E_f}}{Q(N,V,T)} = \frac{e^{-\beta E_f}}{\mathrm{Tr}\left[e^{-\beta \hat{\mathcal{H}}_0}\right]}. \tag{14.3.11}$$

The interchange of summation indices in eqn. (14.3.10) causes the δ-function condition in eqn. (14.3.10) to revert to that contained in eqn. (14.3.9), namely, $E_f = E_i + \hbar\omega$. Substituting this condition into the expression for w_f gives

$$w_f = \frac{e^{-\beta(E_i+\hbar\omega)}}{Q(N,V,T)} = w_i e^{-\beta\hbar\omega}. \tag{14.3.12}$$

Since $\delta(x) = \delta(-x)$, eqn. (14.3.10) can be expressed as

$$R(-\omega) = \frac{2\pi}{\hbar}|F(\omega)|^2 e^{-\beta\hbar\omega} \sum_{i,f} w_i \, |\langle E_i|\mathcal{V}|E_f\rangle|^2 \, \delta(E_f - E_i + \hbar\omega). \tag{14.3.13}$$

Comparing eqn. (14.3.13) with eqn. (14.3.3) reveals that

$$R(-\omega) = e^{-\beta\hbar\omega} R(\omega), \tag{14.3.14}$$

which is known as the condition of *detailed balance*. According to this condition, the probability per unit time of an emission event is smaller than that of an absorption event by a factor of $\exp(-\beta\hbar\omega)$ in a canonical distribution, for which the probability of finding the system with a high initial energy E_i is smaller than that for finding the system with a smaller initial energy. Equation (14.3.14) is a consequence of the statistical distribution of initial states; in fact, the individual transition rates $R_{fi}(\omega)$ satisfy the microscopic reversibility condition $R_{fi}(\omega) = R_{if}(\omega)$. If we followed all of the individual transitions of an ensemble of systems, they would all obey microscopic reversibility. However, because we introduce a statistical distribution, we no longer retain such a detailed microscopic picture, and the ensemble averaged absorption and emission rates, $R(\omega)$ and $R(-\omega)$, do not obey the microscopic reversibility condition.

If the analysis leading from eqn. (14.3.3) to eqn. (14.3.7) is carried out on eqn. (14.3.9), the result is

$$R(-\omega) = \frac{1}{\hbar^2}|F(\omega)|^2 \int_{-\infty}^{\infty} dt\, e^{-i\omega t} \langle \mathcal{V}(t)\mathcal{V}(0)\rangle, \tag{14.3.15}$$

and since $R(-\omega) \neq R(\omega)$, it follows that the correlation functions $\langle \mathcal{V}(0)\mathcal{V}(t)\rangle$ and $\langle \mathcal{V}(t)\mathcal{V}(0)\rangle$ are not equal. This could also have been gleaned from the fact that the commutator $[\mathcal{V}(0), \mathcal{V}(t)]$ does not vanish.

We now define the net energy absorption spectrum $Q(\omega)$ as the net energy absorbed per unit time at frequency ω. Since the energy absorbed is just $\hbar\omega$, and the *net* rate is the difference between the absorption and emission rates $R(\omega) - R(-\omega)$, the energy spectrum $Q(\omega)$ is given by

$$Q(\omega) = [R(\omega) - R(-\omega)]\,\hbar\omega = \hbar\omega R(\omega)\left(1 - e^{-\beta\hbar\omega}\right). \tag{14.3.16}$$

Note, however, that since $R(-\omega) = \exp(-\beta\hbar\omega)R(\omega)$, it follows that

$$R(\omega) + R(-\omega) = \left(1 + e^{-\beta\hbar\omega}\right) R(\omega) \tag{14.3.17}$$

or

$$R(\omega) = \frac{R(\omega) + R(-\omega)}{1 + e^{-\beta\hbar\omega}}. \tag{14.3.18}$$

Using eqn. (14.3.7) and eqn. (14.3.15), we express the sum $R(\omega) + R(-\omega)$ as

$$R(\omega) + R(-\omega) = \frac{1}{\hbar^2}|F(\omega)|^2 \int_{-\infty}^{\infty} dt\, e^{-i\omega t} \langle \mathcal{V}(0)\mathcal{V}(t) + \mathcal{V}(t)\mathcal{V}(0)\rangle. \tag{14.3.19}$$

Let us now define a new operator bracket

$$[\hat{A}, \hat{B}]_+ = \hat{A}\hat{B} + \hat{B}\hat{A} \tag{14.3.20}$$

known as the *anticommutator* between $\hat{A}$ and $\hat{B}$. It is straightforward to see that the anticommutator is manifestly Hermitian. Inserting the anticommutator definition into eqn. (14.3.19), we obtain

$$R(\omega) + R(-\omega) = \frac{1}{\hbar^2}|F(\omega)|^2 \int_{-\infty}^{\infty} dt\, e^{-i\omega t} \left\langle [\mathcal{V}(0), \mathcal{V}(t)]_+\right\rangle. \tag{14.3.21}$$

Finally, substituting eqn. (14.3.21) into eqn. (14.3.18) and the result into eqn. (14.3.16), the energy spectrum becomes

$$Q(\omega) = \frac{2\omega}{\hbar} |F(\omega)|^2 \tanh(\beta\hbar\omega/2) \int_{-\infty}^{\infty} dt \, e^{-i\omega t} \left\langle \frac{1}{2} [\mathcal{V}(0), \mathcal{V}(t)]_+ \right\rangle. \qquad (14.3.22)$$

Equation (14.3.22) demonstrates that the energy spectrum $Q(\omega)$ can be expressed in terms of the ensemble average of a Hermitian operator combination $[\mathcal{V}(0), \mathcal{V}(t)]_+$. In particular, $Q(\omega)$ is directly related to the Fourier transform of a symmetric quantum time correlation function $\langle [\mathcal{V}(0), \mathcal{V}(t)]_+ \rangle$.

It is instructive to examine the classical limit of the quantum spectrum in eqn. (14.3.22). In this limit, the operators $\mathcal{V}(0)$ and $\mathcal{V}(t)$ revert to classical phase-space functions so that $\mathcal{V}(0)\mathcal{V}(t) = \mathcal{V}(t)\mathcal{V}(0)$ and $[\mathcal{V}(0), \mathcal{V}(t)]_+ \longrightarrow 2\mathcal{V}(0)\mathcal{V}(t)$. Also, as $\hbar \longrightarrow 0$, $\tanh(\beta\hbar\omega/2) \longrightarrow \beta\hbar\omega/2$. Combining these results, we find that the classical limit of the quantum spectrum is just

$$Q_{\mathrm{cl}}(\omega) = \frac{\omega^2}{kT} |F(\omega)|^2 \int_{-\infty}^{\infty} dt \, e^{-i\omega t} \langle \mathcal{V}(0)\mathcal{V}(t) \rangle_{\mathrm{cl}}, \qquad (14.3.23)$$

where the notation $\langle \mathcal{V}(0)\mathcal{V}(t) \rangle_{\mathrm{cl}}$ serves to remind us that the time correlation function is a *classical* one.

14.4 Examples of frequency spectra

From eqn. (14.3.22), it is clear that in order to calculate a spectrum, we must be able to calculate a quantum time correlation function. Unfortunately, numerical evaluation of these correlation functions is an extremely difficult computational problem, an issue we will explore in more detail in Section 14.6, where we will also describe approaches for approximating quantum time correlation functions from path integral molecular dynamics. In this section, we will use a simple, analytically solvable example, the harmonic oscillator, to illustrate the general idea of a quantum time correlation function. As discussed in Section 10.4.2, expressions for equilibrium averages and thermodynamic quantities for a harmonic oscillator form the basis of simple approximations for general anharmonic systems. We will use the result we derive here for the position autocorrelation function of a harmonic oscillator to devise a straightforward approach to approximate absorption spectra from *classical* molecular dynamics trajectories.

14.4.1 Position autocorrelation function of a harmonic oscillator

We begin by considering the position autocorrelation function $\langle \hat{x}(t)\hat{x}(0) \rangle$ and symmetrized autocorrelation function $\langle [\hat{x}(t), \hat{x}(0)]_+ \rangle$ of a simple harmonic oscillator of frequency ω_0. In order to calculate the time evolution of the position operator, we use the fact that the Schrödinger operator $\hat{x}$ can be expressed in terms of the creation and annihilation operators (or raising and lowering operators) $\hat{a}^\dagger$ and $\hat{a}$, respectively, as

$$\hat{x} = \left(\frac{\hbar}{2m\omega_0} \right)^{1/2} (\hat{a} + \hat{a}^\dagger) \qquad (14.4.1)$$

(Shankar, 1994), where the action of $\hat{a}$ and $\hat{a}^\dagger$ on an energy eigenstate of the oscillator is

$$\hat{a}|n\rangle = \sqrt{n}|n-1\rangle, \qquad \hat{a}^\dagger|n\rangle = \sqrt{n+1}|n+1\rangle. \tag{14.4.2}$$

These operators satisfy the commutation relation $[\hat{a}, \hat{a}^\dagger] = 1$. In terms of the creation and annihilation operators, the Hamiltonian for a harmonic oscillator of frequency ω_0 can be written as

$$\hat{\mathcal{H}}_0 = \left(\hat{a}^\dagger \hat{a} + \frac{1}{2}\right)\hbar\omega_0. \tag{14.4.3}$$

In the interaction picture, the operators $\hat{a}$ and $\hat{a}^\dagger$ evolve according to the equations of motion

$$\frac{\mathrm{d}\hat{a}}{\mathrm{d}t} = \frac{1}{i\hbar}[\hat{a}, \hat{\mathcal{H}}_0] = -i\omega_0\hat{a}$$

$$\frac{\mathrm{d}\hat{a}^\dagger}{\mathrm{d}t} = \frac{1}{i\hbar}[\hat{a}^\dagger, \hat{\mathcal{H}}_0] = i\omega_0\hat{a}^\dagger. \tag{14.4.4}$$

Equations (14.4.4) are readily solved to yield

$$\hat{a}(t) = \hat{a}\mathrm{e}^{-i\omega_0 t}, \qquad \hat{a}^\dagger(t) = \hat{a}^\dagger\mathrm{e}^{i\omega_0 t}. \tag{14.4.5}$$

Using eqn. (10.4.26), the correlation function $\langle\hat{x}(0)\hat{x}(t)\rangle$ can be written as

$$\langle\hat{x}(0)\hat{x}(t)\rangle = \frac{\hbar}{2m\omega_0}\frac{1 - \mathrm{e}^{-\beta\hbar\omega_0}}{\mathrm{e}^{-\beta\hbar\omega_0/2}}$$

$$\times \sum_{n=0}^{\infty} \mathrm{e}^{-(n+1/2)\beta\hbar\omega_0}\langle n|(\hat{a} + \hat{a}^\dagger)(\hat{a}\mathrm{e}^{-i\omega_0 t} + \hat{a}^\dagger\mathrm{e}^{i\omega_0 t})|n\rangle. \tag{14.4.6}$$

After some algebra, we find that

$$\langle\hat{x}(0)\hat{x}(t)\rangle = \frac{\hbar}{4m\omega_0\sinh(\beta\hbar\omega_0/2)}\left[\mathrm{e}^{i\omega_0 t}\mathrm{e}^{\beta\hbar\omega_0/2} + \mathrm{e}^{-i\omega_0 t}\mathrm{e}^{-\beta\hbar\omega_0/2}\right]. \tag{14.4.7}$$

Similarly, the correlation function $\langle\hat{x}(t)\hat{x}(0)\rangle$ can be shown to be

$$\langle\hat{x}(t)\hat{x}(0)\rangle = \frac{\hbar}{4m\omega_0\sinh(\beta\hbar\omega_0/2)}\left[\mathrm{e}^{-i\omega_0 t}\mathrm{e}^{\beta\hbar\omega_0/2} + \mathrm{e}^{i\omega_0 t}\mathrm{e}^{-\beta\hbar\omega_0/2}\right]. \tag{14.4.8}$$

When we combine eqns. (14.4.7) and (14.4.8), the symmetric correlation function is found to be

$$\frac{1}{2}\left\langle[\hat{x}(0), \hat{x}(t)]_+\right\rangle = \frac{\hbar}{2m\omega_0}\coth(\beta\hbar\omega_0/2)\cos(\omega_0 t). \tag{14.4.9}$$

A comparison of eqn. (14.4.9) and eqn. (13.3.41) reveals that the quantum and classical correlation functions are related by

$$\frac{1}{2}\left\langle[\hat{x}(0), \hat{x}(t)]_+\right\rangle = \frac{\beta\hbar\omega_0}{2}\coth(\beta\hbar\omega_0/2)\left\langle x(0)x(t)\right\rangle_{\mathrm{cl}}. \tag{14.4.10}$$

The connection established in eqn. (14.4.10) between the quantum and classical position autocorrelation functions of a harmonic oscillator can be exploited as a method for approximating the quantum position autocorrelation function of an anharmonic system using the corresponding classical autocorrelation function. The latter can be obtained directly from a molecular dynamics calculation. This approximation is known as the *harmonic approximation*, within which the quantum-mechanical prefactor $(\beta\hbar\omega_0/2)\coth(\beta\hbar\omega_0/2)$ serves to capture at least some of the true quantum character of the system (Bader and Berne, 1994; Skinner and Park, 2001). The utility of this approximation depends on how well a system can be represented as a collection of harmonic oscillators.

14.4.2 The infrared spectrum

One of the most commonly used approaches to probe the vibrational energy levels of a system is infrared spectroscopy, in which electromagnetic radiation of frequency in the near infrared part of the spectrum (10^{12} to 10^{14} Hz) is used to induce transitions between the vibrational levels. By sweeping through this frequency range, the technique records the frequencies at which the transitions occur and the intensities associated with each transition.

Infrared spectroscopy makes use of the fact that the total electric dipole moment operator of a system $\hat{\boldsymbol{\mu}}$ couples to the electric field component of an electromagnetic wave via

$$\hat{\mathcal{H}}_1(t) = -\hat{\boldsymbol{\mu}} \cdot \mathbf{E}(t). \tag{14.4.11}$$

If we orient the coordinate system such that $\mathbf{E}(t) = (0, 0, E(t))$ and recall that the wavelength of infrared radiation is long compared to a typical sample size, then $E(t) = \mathcal{E}(\omega)e^{-i\omega t}$, and the perturbation Hamiltonian $\hat{\mathcal{H}}_1(t)$ is of the form given in eqn. (14.2.37). For this perturbation, the energy spectrum is given by

$$Q(\omega) = \frac{\omega}{\hbar}|\mathcal{E}(\omega)|^2\tanh(\beta\hbar\omega/2) \int_{-\infty}^{\infty} dt\, e^{-i\omega t} \left\langle [\hat{\mu}_z(0), \hat{\mu}_z(t)]_+ \right\rangle. \tag{14.4.12}$$

Since we could have chosen any direction for the electric field $\mathbf{E}(t)$, we may compute the spectrum by averaging over the three spatial directions and obtain

$$Q(\omega) = \frac{\omega}{3\hbar}|\mathcal{E}(\omega)|^2\tanh(\beta\hbar\omega/2) \int_{-\infty}^{\infty} dt\, e^{-i\omega t} \left\langle \hat{\boldsymbol{\mu}}(t) \cdot \hat{\boldsymbol{\mu}}(0) + \hat{\boldsymbol{\mu}}(0) \cdot \hat{\boldsymbol{\mu}}(t) \right\rangle. \tag{14.4.13}$$

What is actually measured in an infrared experiment is the absorptivity $\alpha(\omega)$ from the Beer-Lambert law. The product of $\alpha(\omega)$ with the frequency-dependent index of refraction $n(\omega)$ is directly proportional to $Q(\omega)$ in eqn. (14.4.13), $Q(\omega) \propto \alpha(\omega)n(\omega)$. If the quantum dipole-moment autocorrelation function is replaced by a classical autocorrelation function, with $\boldsymbol{\mu}(t) = \sum_i q_i\mathbf{r}_i(t)$ the classical dipole moment for a system of N charges $q_1, ..., q_N$, then the approximation in eqn. (14.4.10) can be employed.

Through the use of the Kramers-Krönig relations (see eqn. (14.5.20) in Section 14.5), a straightforward computational procedure can be employed to compute $n(\omega)$ (Iftimie and Tuckerman, 2005). The examples of water and ice considered by Iftimie and Tuckerman show that $n(\omega)$ has only a weak dependence on frequency so that $\alpha(\omega)n(\omega)$ is a reasonable representation of the experimental observable.

As a specific example of an infrared spectrum, we show, in Fig. 14.5(a), computed IR spectra from *ab initio* molecular dynamics calculations of pure D_2O (Lee and Tuckerman, 2007), with a comparison to experiment (Bertie *et al.*, 1989; Zelsmann, 1995). In Fig. 14.5(b), we show computed IR spectra for 1 M and a 13 M aqueous KOD solutions (Zhu and Tuckerman, 2002). In an *ab initio* molecular dynamics calculation, a molecular dynamics trajectory is generated with forces computed from electronic structure calculations performed "on the fly" as the simulation proceeds. The *ab initio* molecular dynamics technique allows chemical bond-breaking and -forming events (which occur frequently in KOD solutions as protons are transferred from water to OD^- ions) to be treated implicitly in an unbiased manner. In each of these spectra, the quantum dipole correlation function is replaced by its classical counterpart using the harmonic approximation of eqn. (14.4.10). The simulation protocol employed in Fig. 14.5(a) leads to a small red shift in the OD vibrational band compared to experiment; however, the agreement is generally reasonable. The spectra in Fig. 14.5(b) show how a strongly red-shifted OD vibrational at 1950 cm^{-1} band diminishes with concentration and disappears in the pure D_2O case. This band can be assigned to water molecules in the first solvation shell of an OD^- ion that donate a hydrogen bond to the OD^- oxygen, forming a relatively strong hydrogen bond. The stretch mode of

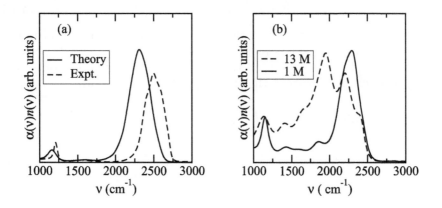

Fig. 14.5 Computed (solid line) and experimental (dashed line) IR spectra for pure D_2O (a) and KOD solutions of 1 M (solid line) and 13 M (dashed line) concentrations.

these OD groups pointing directly to the hydroxyl oxygen are strongly red-shifted. As the concentration is decreased, it is expected that this band, in particular, will exhibit diminished intensity in the infrared spectrum. At high concentration, the shoulder at ≈ 2400 cm^{-1} is due to the OD^- stretch which is only free for a fraction of the trajectory and hence is a weak signal in the IR spectrum.

14.5   Quantum linear response theory

In this section, we will show that energy spectrum can be derived directly from the ensemble density matrix and the quantum Liouville equation without explicit reference to the eigenstates of $\hat{\mathcal{H}}_0$. This approach, known as *quantum linear response theory*, is the quantum version of the classical linear response theory described in Section 13.2, and it also the basis for the calculation of quantum transport properties. Since the eigenstate approach to linear spectroscopy derived in Section 14.3 employed first-order perturbation theory (Fermi's Golden Rule), we expect to use a linearization of the quantum Liouville equation, as was done in Section 13.2, in order to establish the connection between the eigenstate and density-matrix approaches.

Recall that the quantum Liouville equation is

$$\frac{\partial \hat{\rho}(t)}{\partial t} = \frac{1}{i\hbar} \left[\hat{\mathcal{H}}(t), \hat{\rho}(t) \right]. \tag{14.5.1}$$

In order to keep the discussion as general as possible, we will consider the solution of eqn. (14.5.1) for a general class of Hamiltonians of the form

$$\hat{\mathcal{H}}(t) = \hat{\mathcal{H}}_0 - \mathcal{V}F_\mathrm{e}(t), \tag{14.5.2}$$

where $\mathcal{V}$ is a Hermitian operator and $F_\mathrm{e}(t)$ is an arbitrary function of time.

As in the classical case, we take an ansatz for $\hat{\rho}(t)$ of the form

$$\hat{\rho}(t) = \hat{\rho}_0(\hat{\mathcal{H}}_0) + \Delta\hat{\rho}(t), \tag{14.5.3}$$

where $\hat{\rho}_0(\hat{\mathcal{H}}_0)$ is the equilibrium density matrix for a system described by the unperturbed Hamiltonian $\hat{\mathcal{H}}_0$ and which therefore satisfies an equilibrium Liouville equation

$$\left[\hat{\mathcal{H}}_0, \hat{\rho}_0 \right] = 0, \qquad \frac{\partial \hat{\rho}_0}{\partial t} = 0. \tag{14.5.4}$$

Assuming that the system is in equilibrium before the perturbation is applied, the initial condition on the Liouville equation is $\hat{\rho}(t_0) = \hat{\rho}_0(\hat{\mathcal{H}}_0)$, $\Delta\hat{\rho}(t_0) = 0$. When eqn. (14.5.3) is substituted into eqn. (14.5.1) and terms involving both $\mathcal{V}$ and $\Delta\hat{\rho}$ are dropped, we obtain the following equation of motion for $\Delta\hat{\rho}(t)$:

$$\frac{\partial \Delta\hat{\rho}(t)}{\partial t} = \frac{1}{i\hbar} \left[\hat{\mathcal{H}}_0, \Delta\hat{\rho}(t) \right] - \frac{1}{i\hbar} \left[\mathcal{V}, \hat{\rho}_0 \right] F_\mathrm{e}(t). \tag{14.5.5}$$

Since this equation is a first-order inhomogeneous linear differential equation for an operator and involves a commutator, left and right integrating factors in the form of $\hat{U}_0^\dagger(t) = \exp(i\hat{\mathcal{H}}_0 t/\hbar)$ and $\hat{U}_0(t) = \exp(-i\hat{\mathcal{H}}_0 t/\hbar)$, respectively, are needed. With these integrating factors, the solution for $\Delta\hat{\rho}(t)$ becomes

$$\Delta\hat{\rho}(t) = -\frac{1}{i\hbar} \int_{t_0}^{t} ds\, \mathrm{e}^{-i\hat{\mathcal{H}}_0(t-s)/\hbar} \left[\mathcal{V}, \hat{\rho}_0 \right] \mathrm{e}^{i\hat{\mathcal{H}}_0(t-s)/\hbar} F_\mathrm{e}(s). \tag{14.5.6}$$

The ensemble average of an operator $\hat{A}$ in the time-dependent quantum ensemble is given by

$$\langle \hat{A} \rangle_t = \mathrm{Tr}\left[\hat{\rho}(t)\hat{A} \right] = \mathrm{Tr}\left[\hat{\rho}_0 \hat{A} \right] + \mathrm{Tr}\left[\Delta\hat{\rho}(t)\hat{A} \right] = \langle \hat{A} \rangle + \mathrm{Tr}\left[\Delta\hat{\rho}(t)\hat{A} \right], \tag{14.5.7}$$

where $\langle \hat{A} \rangle$ is the equilibrium ensemble average of $\hat{A}$. When eqn. (14.5.6) is substituted into eqn. (14.5.7), we obtain

$$\langle \hat{A} \rangle_t = \langle \hat{A} \rangle - \frac{1}{i\hbar} \int_{t_0}^{t} ds \, \mathrm{Tr} \left\{ \hat{A} e^{-i\hat{\mathcal{H}}_0(t-s)/\hbar} \left[\mathcal{V}, \hat{\rho}_0 \right] e^{i\hat{\mathcal{H}}_0(t-s)/\hbar} \right\} F_e(s)$$

$$= \langle \hat{A} \rangle - \frac{1}{i\hbar} \int_{t_0}^{t} ds \, \mathrm{Tr} \left\{ e^{i\hat{\mathcal{H}}_0(t-s)/\hbar} \hat{A} e^{-i\hat{\mathcal{H}}_0(t-s)/\hbar} \left[\mathcal{V}, \hat{\rho}_0 \right] \right\} F_e(s)$$

$$= \langle \hat{A} \rangle - \frac{1}{i\hbar} \int_{t_0}^{t} ds \, \mathrm{Tr} \left\{ \hat{A}(t-s) \left[\mathcal{V}(0), \hat{\rho}_0 \right] \right\} F_e(s). \tag{14.5.8}$$

In the second line, we have used the fact that the trace is invariant to cyclic permutations of the operators, $\mathrm{Tr}(\hat{A}\hat{B}\hat{C}) = \mathrm{Tr}(\hat{C}\hat{A}\hat{B}) = \mathrm{Tr}(\hat{B}\hat{C}\hat{A})$. In the last line, $\hat{A}(t-s)$ denotes the operator $\hat{A}$ in the interaction picture at time $t-s$, and $\mathcal{V}(0)$ denotes an operator in this picture at $t = 0$. Using the cyclic property of the trace again, the expression in eqn. (14.5.8) can be further simplified. Expanding the commutator in the last line of eqn. (14.5.8) yields

$$\langle \hat{A} \rangle_t = \langle \hat{A} \rangle - \frac{1}{i\hbar} \int_{t_0}^{t} ds \, \mathrm{Tr} \left[\hat{A}(t-s)\mathcal{V}(0)\hat{\rho}_0 - \hat{A}(t-s)\hat{\rho}_0 \mathcal{V}(0) \right] F_e(s)$$

$$= \langle \hat{A} \rangle - \frac{1}{i\hbar} \int_{t_0}^{t} ds \, \mathrm{Tr} \left[\hat{\rho}_0 \hat{A}(t-s)\mathcal{V}(0) - \hat{\rho}_0 \mathcal{V}(0)\hat{A}(t-s) \right] F_e(s)$$

$$= \langle \hat{A} \rangle - \frac{1}{i\hbar} \int_{t_0}^{t} ds \, \mathrm{Tr} \left[\hat{\rho}_0 \left(\hat{A}(t-s)\mathcal{V}(0) - \mathcal{V}(0)\hat{A}(t-s) \right) \right] F_e(s)$$

$$= \langle \hat{A} \rangle - \frac{1}{i\hbar} \int_{t_0}^{t} ds \, \mathrm{Tr} \left\{ \hat{\rho}_0 \left[\hat{A}(t-s), \mathcal{V}(0) \right] \right\} F_e(s). \tag{14.5.9}$$

Equation (14.5.9) is the quantum analog of the classical linear response formula given in eqn. (13.2.27) and is hence the starting point for the development of quantum Green-Kubo expressions for transport properties. Since these expressions are very similar to their classical counterparts, we will not repeat the derivations of the Green-Kubo formulae here.

The time correlation function appearing in eqn. (14.5.9) is referred to as the *after-effect function* $\Phi_{A\mathcal{V}}(t)$, which is defined as

$$\Phi_{A\mathcal{V}}(t) = \frac{i}{\hbar} \langle [\hat{A}(t), \mathcal{V}(0)] \rangle. \tag{14.5.10}$$

Note that although the operator combination $[\hat{A}(t), \mathcal{V}(0)]$ is anti-Hermitian,[3] the i prefactor in eqn. (14.5.10) fixes this: the operator $i[\hat{A}(t), \mathcal{V}(0)]$ is Hermitian. In order

---

[3] An anti-Hermitian $\hat{B}$ satisfies $\hat{B}^\dagger = -\hat{B}$.

to make contact with the treatment of Section 14.3, we set $t_0 = -\infty$ in eqn. (14.5.9) to obtain

$$\langle \hat{A} \rangle_t = \langle \hat{A} \rangle + \int_{-\infty}^{t} ds \, \Phi_{Av}(t-s) F_e(s). \tag{14.5.11}$$

According to eqn. (14.5.11), when $\mathcal{V}$ is the operator we choose to measure in the non-equilibrium ensemble, we find

$$\langle \mathcal{V} \rangle_t = \langle \mathcal{V} \rangle + \int_{-\infty}^{t} ds \, \Phi_{vv}(t-s) F_e(s), \tag{14.5.12}$$

which involves a quantum autocorrelation function of $\mathcal{V}$.

We now consider the special case of a monochromatic field of frequency ω, for which $F_e(t) = F(\omega) \exp(-i\omega t)$. Substituting this field form into eqn. (14.5.12) yields

$$\langle \mathcal{V} \rangle_t = \langle \mathcal{V} \rangle + F(\omega) \int_{-\infty}^{t} ds \, \Phi_{vv}(t-s) e^{-i\omega s}. \tag{14.5.13}$$

Because the lower limit of the time integral in eqn. (14.5.13) is $-\infty$, it ·is necessary to ensure that the potentially oscillatory integrand yields a convergent result. Thus, we multiply the integrand by a *convergence factor* $\exp(\epsilon t)$, which decays to 0 as $t \to -\infty$. After the integral is performed, the limit $\epsilon \to 0^+$ (the limit that ϵ approaches 0 from the positive side) is taken. The use of convergence factors is a formal device, the necessity of which depends on the behavior of the autocorrelation function. For nearly perfect solids and glassy systems, one would expect the decay of the correlation function to be very slow, requiring the use of the convergence factor. For ordinary liquids, the correlation function should decay rapidly to zero, obviating the need for this factor. For generality, we retain it in the present discussion. Introducing a convergence factor into eqn. (14.5.13) gives

$$\langle \mathcal{V} \rangle_t = \langle \mathcal{V} \rangle + F(\omega) \lim_{\epsilon \to 0^+} \int_{-\infty}^{t} ds \, \Phi_{vv}(t-s) e^{-i\omega s} e^{\epsilon s}. \tag{14.5.14}$$

We now change the integration variables in eqn. (14.5.14) from t to $\tau = t - s$, which yields

$$\langle \mathcal{V} \rangle_t = \langle \mathcal{V} \rangle + F(\omega) \lim_{\epsilon \to 0^+} e^{(i\omega + \epsilon)t} \int_{0}^{\infty} d\tau \, \Phi_{vv}(\tau) e^{-(i\omega + \epsilon)\tau}. \tag{14.5.15}$$

Equation (14.5.15) involves a Fourier-Laplace transform of the after-effect function at a complex variable $z = \omega - i\epsilon$. Let the function $\chi_{vv}(z)$ denote this Laplace transform (see Appendix E)

$$\chi_{vv}(z) = \int_{0}^{\infty} d\tau \, \Phi_{vv}(\tau) e^{-iz\tau}, \tag{14.5.16}$$

which is referred to as the *susceptibility*. Equation (14.5.15) can now be expressed as

$$\langle \mathcal{V} \rangle_t = \langle \mathcal{V} \rangle + F(\omega) \lim_{\epsilon \to 0^+} e^{i(\omega - i\epsilon)t} \chi_{vv}(\omega - i\epsilon). \tag{14.5.17}$$

By decomposing the susceptibility into its real and imaginary parts, we can relate it directly to the energy spectrum $Q(\omega)$. In the limit $\epsilon \to 0^+$, we obtain

$$
\chi_{vv}(\omega) = \lim_{\epsilon \to 0^+} \int_0^\infty d\tau \, e^{-\epsilon\tau} \Phi_{vv}(\tau) e^{-i\omega\tau}
$$

$$
= \lim_{\epsilon \to 0^+} \int_0^\infty d\tau \, e^{-\epsilon\tau} \Phi_{vv}(\tau) \left[\cos\omega\tau - i\sin\omega\tau \right]
$$

$$
\equiv \mathrm{Re}\left[\chi_{vv}(\omega) \right] - i\mathrm{Im}\left[\chi_{vv}(\omega) \right], \tag{14.5.18}
$$

where

$$
\mathrm{Re}\left[\chi_{vv}(\omega) \right] = \lim_{\epsilon \to 0^+} \int_0^\infty d\tau \, e^{-\epsilon\tau} \Phi_{vv}(\tau) \cos\omega\tau
$$

$$
\mathrm{Im}\left[\chi_{vv}(\omega) \right] = \lim_{\epsilon \to 0^+} \int_0^\infty d\tau \, e^{-\epsilon\tau} \Phi_{vv}(\tau) \sin\omega\tau. \tag{14.5.19}
$$

An important property of the susceptibility $\chi(z)$ is its analyticity in the complex z-plane. For any analytic function, the real and imaginary parts are not independent but satisfy a set of relations known as the *Kramers-Krönig relations*. Let $\chi'_{vv}(\omega)$ and $\chi''_{vv}(\omega)$ denote the real and imaginary parts of $\chi_{vv}(\omega)$, respectively, so that $\chi_{vv}(\omega) = \chi'_{vv}(\omega) + i\chi''_{vv}(\omega)$. The real and imaginary parts are related by

$$
\chi'_{vv}(\omega) = \frac{1}{\pi} \mathrm{P} \int_{-\infty}^\infty d\tilde{\omega} \, \frac{\tilde{\omega}\chi''_{vv}(\tilde{\omega})}{\tilde{\omega}^2 - \omega^2}
$$

$$
\chi''_{vv}(\omega) = -\frac{\omega}{\pi} \mathrm{P} \int_{-\infty}^\infty d\tilde{\omega} \, \frac{\chi'_{vv}(\tilde{\omega})}{\tilde{\omega}^2 - \omega^2}. \tag{14.5.20}
$$

Here, P indicates that the principal value of the integral is to be taken. The Kramers-Krönig relations can be expressed equivalently as

$$
\chi'_{vv}(\omega) = \frac{1}{\pi} \mathrm{P} \int_{-\infty}^\infty d\tilde{\omega} \, \frac{\chi''_{vv}(\tilde{\omega})}{\tilde{\omega} - \omega}
$$

$$
\chi''_{vv}(\omega) = -\frac{1}{\pi} \mathrm{P} \int_{-\infty}^\infty d\tilde{\omega} \, \frac{\chi'_{vv}(\tilde{\omega})}{\tilde{\omega} - \omega}, \tag{14.5.21}
$$

which are known as *Hilbert* transforms. We alluded to the use of these relations in Section 14.4 where we presented infrared spectra for water and aqueous solutions.

We will now show that the frequency spectrum of eqn. (14.3.22) can be related to the imaginary part $\chi''_{vv}(\omega)$ of the susceptibility. The spectrum of eqn. (14.3.22) is given in terms of the anticommutator of $\mathcal{V}(0)$ and $\mathcal{V}(t)$, while the susceptibility is derived from the after-effect function, which involves a commutator between $\mathcal{V}(0)$ and $\mathcal{V}(t)$. Recall, however, that the frequency spectrum is defined as

$$
Q(\omega) = \hbar\omega \left[R(\omega) - R(-\omega) \right]. \tag{14.5.22}
$$

Substituting the definitions of $R(\omega)$ and $R(-\omega)$ from eqns. (14.3.7) and (14.3.15) into this expression for $Q(\omega)$ yields

$$Q(\omega) = \hbar\omega|F(\omega)|^2 \frac{1}{\hbar^2} \int_{-\infty}^{\infty} dt\, e^{-i\omega t} \langle \mathcal{V}(0)\mathcal{V}(t) - \mathcal{V}(t)\mathcal{V}(0) \rangle$$

$$= -\omega|F(\omega)|^2 \frac{1}{\hbar} \int_{-\infty}^{\infty} dt\, e^{-i\omega t} \langle [\mathcal{V}(t), \mathcal{V}(0)] \rangle$$

$$= i\omega|F(\omega)|^2 \int_{-\infty}^{\infty} dt\, e^{-i\omega t} \Phi_{vv}(t). \tag{14.5.23}$$

Next, we divide the time integration into an integration from $-\infty$ to 0 and from 0 to ∞ so that

$$Q(\omega) = i\omega|F(\omega)|^2 \left[\int_{-\infty}^{0} dt\, e^{-i\omega t} \Phi_{vv}(t) + \int_{0}^{\infty} dt\, e^{-i\omega t} \Phi_{vv}(t) \right]$$

$$= i\omega|F(\omega)|^2 \left[\int_{0}^{\infty} dt\, e^{i\omega t} \Phi_{vv}(-t) + \int_{0}^{\infty} dt\, e^{-i\omega t} \Phi_{vv}(t) \right], \tag{14.5.24}$$

where in the first term the transformation $t \to -t$ has been made. In order to proceed, we need to analyze the time-reversal properties of the after-effect function.

Consider a general after-effect function $\Phi_{AB}(t)$:

$$\Phi_{AB}(t) = \frac{i}{\hbar} \left\langle \left[e^{i\hat{\mathcal{H}}_0 t/\hbar} \hat{A} e^{-i\hat{\mathcal{H}}_0 t/\hbar}, \hat{B} \right] \right\rangle. \tag{14.5.25}$$

Substituting $-t$ into eqn. (14.5.25) yields

$$\Phi_{AB}(-t) = \frac{i}{\hbar} \left\langle e^{-i\hat{\mathcal{H}}_0 t/\hbar} \hat{A} e^{i\hat{\mathcal{H}}_0 t/\hbar} \hat{B} - \hat{B} e^{-i\hat{\mathcal{H}}_0 t/\hbar} \hat{A} e^{i\hat{\mathcal{H}}_0 t/\hbar} \right\rangle$$

$$= \frac{i}{\hbar} \left[\text{Tr}\left(\hat{\rho}_0 e^{-i\hat{\mathcal{H}}_0 t/\hbar} \hat{A} e^{i\hat{\mathcal{H}}_0 t/\hbar} \hat{B} \right) - \text{Tr}\left(\hat{\rho}_0 \hat{B} e^{-i\hat{\mathcal{H}}_0 t/\hbar} \hat{A} e^{i\hat{\mathcal{H}}_0 t/\hbar} \right) \right]. \tag{14.5.26}$$

Because the trace is invariant under cyclic permutations of the operators and $\hat{\rho}_0$ commutes with the propagators $\exp(\pm i\hat{\mathcal{H}}_0 t/\hbar)$, we can express eqn. (14.5.26) as

$$\Phi_{AB}(-t) = \frac{i}{\hbar} \left[\text{Tr}\left(\hat{\rho}_0 e^{-i\hat{\mathcal{H}}_0 t/\hbar} \hat{A} e^{i\hat{\mathcal{H}}_0 t/\hbar} \hat{B} \right) - \text{Tr}\left(\hat{\rho}_0 \hat{B} e^{-i\hat{\mathcal{H}}_0 t/\hbar} \hat{A} e^{i\hat{\mathcal{H}}_0 t/\hbar} \right) \right]$$

$$= \frac{i}{\hbar} \left[\text{Tr}\left(e^{-i\hat{\mathcal{H}}_0 t/\hbar} \hat{\rho}_0 \hat{A} e^{i\hat{\mathcal{H}}_0 t/\hbar} \hat{B} \right) - \text{Tr}\left(\hat{B} e^{-i\hat{\mathcal{H}}_0 t/\hbar} \hat{A} e^{i\hat{\mathcal{H}}_0 t/\hbar} \hat{\rho}_0 \right) \right]$$

$$= \frac{i}{\hbar} \left[\text{Tr}\left(\hat{\rho}_0 \hat{A} e^{i\hat{\mathcal{H}}_0 t/\hbar} \hat{B} e^{-i\hat{\mathcal{H}}_0 t/\hbar} \right) - \text{Tr}\left(\hat{B} e^{-i\hat{\mathcal{H}}_0 t/\hbar} \hat{A} \hat{\rho}_0 e^{i\hat{\mathcal{H}}_0 t/\hbar} \right) \right]$$

$$= \frac{i}{\hbar} \left[\text{Tr}\left(\hat{\rho}_0 \hat{A} e^{i\hat{\mathcal{H}}_0 t/\hbar} \hat{B} e^{-i\hat{\mathcal{H}}_0 t/\hbar} \right) - \text{Tr}\left(\hat{\rho}_0 e^{i\hat{\mathcal{H}}_0 t/\hbar} \hat{B} e^{-i\hat{\mathcal{H}}_0 t/\hbar} \hat{A} \right) \right]$$

$$= \frac{i}{\hbar} \left[\mathrm{Tr}\left(\hat{\rho}_0 \hat{A}\hat{B}(t)\right) - \mathrm{Tr}\left(\hat{\rho}_0 \hat{B}(t)\hat{A}\right) \right]$$

$$= \frac{i}{\hbar} \left\langle \hat{A}\hat{B}(t) - \hat{B}(t)\hat{A} \right\rangle$$

$$= -\frac{i}{\hbar} \left\langle [\hat{B}(t), \hat{A}] \right\rangle$$

$$= -\Phi_{BA}(t). \tag{14.5.27}$$

Thus, the effect of time reversal on a general after-effect function is to reverse the order of the operators and change the overall sign. If $\hat{A} = \hat{B}$, then the after-effect function only picks up an overall change of sign upon time reversal. When eqn. (14.5.27) is introduced into eqn. (14.5.24), the energy spectrum becomes

$$Q(\omega) = i\omega|F(\omega)|^2 \left[-\int_0^\infty \mathrm{d}t \ \mathrm{e}^{i\omega t} \phi_{vv}(t) + \int_0^\infty \mathrm{d}t \ \mathrm{e}^{-i\omega t} \Phi_{vv}(t) \right]$$

$$= 2\omega|F(\omega)|^2 \int_0^\infty \mathrm{d}t \ \sin(\omega t)\phi_{vv}(t)$$

$$= 2\omega|F(\omega)|^2 \mathrm{Im}\left[\chi_{vv}(\omega)\right]. \tag{14.5.28}$$

Thus, the net absorption spectrum is related to the imaginary part of the frequency-dependent susceptibility. Note that eqns. (14.5.28) and (14.3.22) are equivalent, demonstrating that the spectrum is expressible in terms either of Hermitian or anti-Hermitian quantum time correlation functions. This derivation establishes the equivalence between the wave-function approach, leading to the Fermi Golden rule treatment of spectra, and the statistical-mechanical approach, which starts with the ensemble and its density matrix $\hat{\rho}(t)$ and makes no explicit reference to the eigenstates of $\hat{\mathcal{H}}_0$. This is significant, as the former approach is manifestly *eigenstate resolved*, meaning that it explicitly considers the transitions between eigenstates of $\hat{\mathcal{H}}_0$, which is closer to the experimental view. The latter, which builds directly from the density matrix, is closer in spirit to the path integral perspective.

14.6   Approximations to quantum time correlation functions

In this section, we will discuss the general problem of calculating quantum time correlation functions for condensed-phase systems. We will first show how to formulate the correlation function in terms of the eigenstates of $\hat{\mathcal{H}}_0$. While the eigenstate formulation is useful for analyzing the properties of time correlation functions, we have already alluded, in Section 10.1, to the computational intractability of solving the eigenvalue problem of $\hat{\mathcal{H}}_0$ for systems containing more than just a few degrees of freedom. Thus, we will also express the quantum time correlation function using the path integral formulation of quantum mechanics from Chapter 12. Although, as we will show, even the path integral representation suffers from severe numerical difficulties, it serves as a useful starting point for the development of computationally manageable approximation schemes.

Let us begin with a standard nonsymmetrized time correlation function defined by

$$C^\beta_{AB}(t) = \left\langle \hat{A}(0)\hat{B}(t) \right\rangle = \frac{1}{Q(N,V,T)} \mathrm{Tr}\left[e^{-\beta\hat{\mathcal{H}}} \hat{A} e^{i\hat{\mathcal{H}}t/\hbar} \hat{B} e^{-i\hat{\mathcal{H}}t/\hbar} \right], \qquad (14.6.1)$$

where $\hat{A}$ and $\hat{B}$ are quantum mechanical operators in the interaction picture with unperturbed Hamiltonian $\hat{\mathcal{H}}$.[4] If we evaluate the trace in the basis of the eigenvectors of $\hat{\mathcal{H}}$, then a simple formula for the quantum time correlation function results:

$$C^\beta_{AB}(t) = \frac{1}{Q(N,V,T)} \sum_n \langle E_n | e^{-\beta\hat{\mathcal{H}}} \hat{A} e^{i\hat{\mathcal{H}}t/\hbar} \hat{B} e^{-i\hat{\mathcal{H}}t/\hbar} | E_n \rangle$$

$$= \frac{1}{Q(N,V,T)} \sum_{n,m} \langle E_n | e^{-\beta\hat{\mathcal{H}}} \hat{A} | E_m \rangle \langle E_m | e^{i\hat{\mathcal{H}}t/\hbar} \hat{B} e^{-i\hat{\mathcal{H}}t/\hbar} | E_n \rangle$$

$$= \frac{1}{Q(N,V,T)} \sum_{n,m} e^{-\beta E_n} e^{i(E_m - E_n)t/\hbar} \langle E_n | \hat{A} | E_m \rangle \langle E_m | \hat{B} | E_n \rangle. \qquad (14.6.2)$$

Thus, if we are able to calculate all of the eigenvalues and eigenvectors of $\hat{\mathcal{H}}$, as well as the full set of matrix elements of $\hat{A}$ and $\hat{B}$, then the calculation of the time correlation function just requires carrying out the two sums in eqn. (14.6.2). Generally, however, this can only be done for systems having just a few degrees of freedom. In the condensed phase, for example, it is simply not possible to solve the eigenvalue problem for $\hat{\mathcal{H}}$ directly.

In the Feynman path integral formalism of Chapter 12, the eigenvalue problem is circumvented by computing thermal traces in the basis of coordinate eigenstates. We will now apply this approach to the quantum time correlation function. For simplicity, we will consider a single particle in one dimension, and we will let $\hat{A}$ and $\hat{B}$ be functions of the position operator $\hat{x}$, $\hat{A} = \hat{A}(\hat{x})$, $\hat{B} = \hat{B}(\hat{x})$. Taking the coordinate-space trace, we obtain

$$C^\beta_{AB}(t) = \frac{1}{Q(\beta)} \int \mathrm{d}x \, \langle x | e^{-\beta\hat{\mathcal{H}}} \hat{A}(\hat{x}) e^{i\hat{\mathcal{H}}t/\hbar} \hat{B}(\hat{x}) e^{-i\hat{\mathcal{H}}t/\hbar} | x \rangle \qquad (14.6.3)$$

$$= \frac{1}{Q(\beta)} \int \mathrm{d}x \, \mathrm{d}x' \, \mathrm{d}x'' \langle x | e^{-\beta\hat{\mathcal{H}}} | x' \rangle a(x') \langle x' | e^{i\hat{\mathcal{H}}t/\hbar} | x'' \rangle b(x'') \langle x'' | e^{-i\hat{\mathcal{H}}t/\hbar} | x \rangle.$$

If each of the matrix elements $\langle x | e^{-\beta\hat{\mathcal{H}}} | x' \rangle$, $\langle x' | e^{i\hat{\mathcal{H}}t/\hbar} | x'' \rangle$, and $\langle x'' | e^{-i\hat{\mathcal{H}}t/\hbar} | x' \rangle$ were expressed as path integrals, we would interpret eqn. (14.6.3) as follows: Starting at x, propagate along a real-time path to the point x'' using the propagator $\exp(-i\hat{\mathcal{H}}t/\hbar)$ and evaluate the eigenvalue $b(x'')$ of $\hat{B}$ at that point; from x'', propagate backward in real time using the propagator $\exp(i\hat{\mathcal{H}}t/\hbar)$ to the point x' and evaluate the eigenvalue $a(x')$ of $\hat{A}$; finally, propagate in imaginary time using the propagator from x' to the original starting point x. This is represented schematically in Fig. 14.6(a). Un-

---

[4]For the remainder of this chapter, we will drop the "0" subscript on the Hamiltonian, since it is assumed that $\hat{\mathcal{H}}$ represents the unperturbed Hamiltonian.

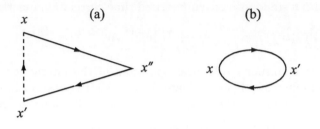

Fig. 14.6 (a) Diagram of the real- and imaginary-time paths for the correlation function in eqn. (14.6.3). (b) Same for the time correlation function in eqn. (14.6.4).

fortunately, standard Monte Carlo or molecular dynamics schemes cannot be used to compute the two real-time paths because the matrix elements of $\exp(\pm i\hat{\mathcal{H}}t/\hbar)$ are not positive-definite, and the sampling schemes of Section 7.3 break down. A possible alternative could be to devise a molecular dynamics approach with complex variables (Gausterer and Klauder, 1986; Lee, 1994; Berges *et al.*, 2007), although no known approach is stable enough to guarantee convergence of a path integral with a purely imaginary discretized action functional.

Before proceeding, we note that there are two alternative quantum time correlation functions that have important advantages over $C_{AB}^\beta(t)$. The first is a symmetrized correlation function $G_{AB}^\beta(t)$ defined by

$$G_{AB}^\beta(t) = \frac{1}{Q(N,V,T)}\mathrm{Tr}\left[\hat{A}e^{i\hat{\mathcal{H}}\tau_c^*/\hbar}\hat{B}e^{-i\hat{\mathcal{H}}\tau_c/\hbar}\right]. \tag{14.6.4}$$

Here τ_c is a complex time variable given by $\tau_c = t - i\beta\hbar/2$. Although not equal to $C_{AB}^\beta(t)$, the Fourier transform of $G_{AB}^\beta(t)$

$$\tilde{G}_{AB}^\beta(\omega) = \frac{1}{\sqrt{2\pi}}\int_{-\infty}^{\infty}\mathrm{d}t\,e^{-i\omega t}G_{AB}^\beta(t) \tag{14.6.5}$$

is related to the Fourier transform of $C_{AB}^\beta(t)$ by

$$\tilde{C}_{AB}^\beta(\omega) = e^{\beta\hbar\omega/2}\tilde{G}_{AB}^\beta(\omega), \tag{14.6.6}$$

which provides a straightforward route to the determination of a spectrum, assuming $G_{AB}^\beta(t)$ can be calculated. Equation (14.6.6) can be easily proved by performing the traces in the basis of energy eigenstates (see Problem 14.2). The advantage of $G_{AB}^\beta(t)$ over $C_{AB}^\beta(t)$ can be illustrated for a single particle in one dimension. We assume, again, that $\hat{A}$ and $\hat{B}$ are functions only of $\hat{x}$ and compute the trace in the coordinate basis, which gives

$$G_{AB}^\beta(t) = \frac{1}{Q(\beta)}\int \mathrm{d}x\,\langle x|\hat{A}(\hat{x})e^{i\hat{\mathcal{H}}\tau_c^*/\hbar}\hat{B}(\hat{x})e^{-i\hat{\mathcal{H}}\tau_c/\hbar}|x\rangle \tag{14.6.7}$$

$$= \frac{1}{Q(\beta)} \int \mathrm{d}x \, \mathrm{d}x' \langle x|e^{i\hat{\mathcal{H}}\tau_c^*/\hbar}|x'\rangle b(x') \langle x'|e^{-i\hat{\mathcal{H}}\tau_c/\hbar}|x\rangle a(x).$$

If the two matrix elements $\langle x|e^{i\hat{\mathcal{H}}\tau_c^*/\hbar}|x'\rangle$ and $\langle x'|e^{-i\hat{\mathcal{H}}\tau_c/\hbar}|x\rangle$ are represented as path integrals, the interpretation of eqn. (14.6.8) is clear. We start at x, calculate the eigenvalue $a(x)$ of $\hat{A}$, propagate along a complex time path to x', calculate the eigenvalue $b(x')$ of $\hat{B}$, and then propagate back to x along a complex conjugate time path. This process is represented schematically in Fig. 14.6(b). Since the two matrix elements in eqn. (14.6.8) are complex conjugates, and since $a(x)$ and $b(x')$ are both real, $G_{AB}^{\beta}(t)$ is, itself, a real object. More importantly, in contrast to $C_{AB}^{\beta}(t)$, the complex time paths needed to represent the two matrix elements in eqn. (14.6.8) have oscillatory phases, but they also have positive-definite weights, which tends to make them somewhat better behaved numerically. If each matrix element in eqn. (14.6.8) is discretized into paths of P points, then $G_{AB}^{\beta}(t)$ can be written as the limit of a discretized path integral of the form

$$G_{AB,P}^{\beta}(t) =$$

$$\frac{1}{Q(\beta)} \int \mathrm{d}x_1 \cdots \mathrm{d}x_{2P} \, a(x_1)b(x_{P+1})\rho(x_1, ..., x_{2P})e^{i\Phi(x_1, ..., x_{2P})} \tag{14.6.8}$$

(Krilov *et al.*, 2001), where $\rho(x_1, ..., x_{2P})$ is a positive-definite distribution given by

$$\rho(x_1, ..., x_{2P}) =$$

$$\left(\frac{mP}{2\pi|\tau_c|\hbar}\right)^P \exp\left[-\frac{mP\beta}{4|\tau_c|^2\hbar^2} \sum_{k=1}^{2P}(x_{k+1} - x_k)^2 - \frac{\beta}{2P}\sum_{k=1}^{2P}U(x_k)\right]. \tag{14.6.9}$$

Here, $x_{2P+1} = x_1$ due to the trace condition, and $\Phi(x_1, ..., x_{2P})$ is a phase factor defined by

$$\Phi(x_1, ..., x_{2P}) = \frac{mPt}{2\hbar|\tau_c|^2}\left[\sum_{k=1}^{P}(x_{k+1} - x_k)^2 - \sum_{k=P+1}^{2P}(x_{k+1} - x_k)^2\right]$$

$$- \frac{t}{\hbar P}\left[\sum_{k=2}^{P}U(x_k) - \sum_{k=P+2}^{2P}U(x_k)\right]. \tag{14.6.10}$$

In the limit $P \to \infty$, $G_{AB,P}^{\beta}(t) = G_{AB}^{\beta}(t)$. Note that the same path variables define ρ and Φ, demonstrating explicitly that the paths have a positive-definite weight as well as a phase factor. Moreover, because $G_{AB,P}^{\beta}(t)$ is real, the imaginary part of $\exp(i\Phi)$ must vanish. The fact that the paths in $G_{AB}^{\beta}(t)$ have positive-definite weights allows novel Monte-Carlo schemes to be devised for computing this function (Jadhao and Makri, 2008).

The second alternate time correlation function is the Kubo-transformed correlation function (Kubo *et al.*, 1985) defined by

$$K_{AB}^{\beta}(t) = \frac{1}{\beta Q(N,V,T)} \int_0^{\beta} d\lambda \, \text{Tr} \left[e^{-(\beta-\lambda)\hat{\mathcal{H}}} \hat{A} e^{-\lambda\hat{\mathcal{H}}} e^{i\hat{\mathcal{H}}t/\hbar} \hat{B} e^{-i\hat{\mathcal{H}}t/\hbar} \right]. \quad (14.6.11)$$

Like $G_{AB}^{\beta}(t)$, $K_{AB}^{\beta}(t)$ is also purely real. In addition, $K_{AB}^{\beta}(t)$ reduces to its classical counterpart both in the classical ($\beta \to 0$) and harmonic limits. Consequently, $K_{AB}^{\beta}(t)$ can be more readily compared to corresponding classical and harmonic time correlation functions, which can be computed straightforwardly. As with $G_{AB}^{\beta}(t)$, there is a simple relationship between the Fourier transforms of $K_{AB}^{\beta}(t)$ and $C_{AB}^{\beta}(t)$:

$$\tilde{C}_{AB}^{\beta}(\omega) = \left[\frac{\beta\hbar\omega}{1 - e^{-\beta\hbar\omega}} \right] \tilde{K}_{AB}^{\beta}(\omega). \quad (14.6.12)$$

Finally, we note that purely imaginary-time correlation functions of the form

$$\mathcal{G}_{AB}^{\beta}(\tau) = \frac{1}{Q(N,V,T)} \text{Tr} \left[e^{-\beta\hat{\mathcal{H}}} \hat{A} \, e^{-\tau\hat{\mathcal{H}}} \hat{B} e^{\tau\hat{\mathcal{H}}} \right] \quad (14.6.13)$$

can be computed straightforwardly using the numerical techniques for imaginary-time path integrals. Equation (14.6.13) results from eqn. (14.6.1) when the Wick rotation from real to imaginary time is applied (see Section 12.2 and Fig. 12.5). An example of such a correlation function is the imaginary-time mean square displacement given by

$$R^2(\tau) = \langle |\hat{x}(\tau) - \hat{x}(0)|^2 \rangle, \quad (14.6.14)$$

which for N particles in three dimensions becomes

$$R^2(\tau) = \frac{1}{N} \sum_{i=1}^{N} \left\langle [\hat{\mathbf{r}}_i(\tau) - \hat{\mathbf{r}}_i(0)]^2 \right\rangle. \quad (14.6.15)$$

Here, $\tau \in [0, \beta\hbar/2]$ ($R^2(\tau)$ is symmetric about $\tau = \beta\hbar/2$) and for a free particle, its shape is an $R^2(\tau) \propto \tau(\beta\hbar/2 - \tau)$, where the constant of proportionality depends on the number of dimensions (see Problem 14.12). This important quantity is related to the real-time velocity autocorrelation function $C_{vv}^{\beta}(t)$ (more precisely, its Fourier transform $\tilde{C}_{vv}^{\beta}(\omega)$) via a two-sided Laplace transform

$$R^2(\tau) = \frac{1}{\pi} \int_{-\infty}^{\infty} d\omega \, \frac{e^{-\beta\hbar\omega/2}}{\omega^2} \tilde{C}_{vv}^{\beta}(\omega)$$

$$\times \left\{ \cosh \left[\omega \left(\frac{\hbar\beta}{2} - \tau \right) \right] - \cosh \left(\frac{\beta\hbar\omega}{2} \right) \right\}. \quad (14.6.16)$$

Equation (14.6.16) suggests that performing the Wick rotation from real to imaginary time is a well-posed problem that requires a Fourier transform followed by a Laplace transform (see Appendix E). Unfortunately, the reverse process, transforming from imaginary time back to real time, requires an inverse Laplace transform, which is an extremely ill-posed problem numerically (see, for example, the discussion by Epstein and Schotland (2008)). This is the primary reason that the analytic continuation of imaginary-time data to real-time data is such an immense challenge (Krilov *et al.*, 2001).

Before we discuss approximation schemes for quantum time correlation functions, we need to point out that quantum effects in condensed-phase systems are sometimes squelched due to pronounced decoherence effects. In this case, off-diagonal elements of the density matrix $\exp(-\beta\hat{\mathcal{H}})$ tend to be small for large $|x - x'|$, and consequently, the sums over forward and backward real-time paths are not appreciably different. This means that there is considerable cancellation between these two sums, a fact that forms the basis of a class of approximation schemes known as *semiclassical* methods. These include the Herman-Kluk propagator (1984), the linearized semiclassical initial value representation (Miller, 2005), the linearized Feynman-Kleinert path integral method (Poulsen *et al.*, 2005; Hone *et al.*, 2008), and the forward–backward approach (Nakayama and Makri, 2005), to name just a few. Although fascinating and potentially very powerful, semiclassical approaches also carry a relatively high computational overhead, and we will not discuss them further here. Rather, we will focus on two increasingly popular approximation schemes for quantum time correlation functions that are based on the use of the imaginary-time path integral. Although these schemes are somewhat *ad hoc*, they have the advantage of being computationally inexpensive and straightforward to implement. They must, however, be used with care because there is no rigorous basis for this class of methods; because of this, we will also introduce a procedure for checking the accuracy of their results.

14.6.1 Centroid molecular dynamics

In 1993, J. Cao and G. A. Voth introduced the centroid molecular dynamics (CMD) method as an approximate technique for computing real-time quantum correlation functions. The primary object in this approach is the path centroid defined in eqn. (12.8.22). Toward the end of Section 12.8.1, we briefly discussed the centroid potential of mean force (Feynman and Kleinert, 1986); the CMD approach is rooted in this concept and in an idea put forth by Gillan (1987) for obtaining approximate quantum rate constants from the centroid density along a reaction coordinate.

CMD is based on the notion that the time evolution of the centroid on this potential of mean force surface can be used to garner approximate quantum dynamical properties of a system. In CMD, the centroid for a single particle in one dimension, denoted here as x_c, is postulated to evolve in time according to the following equations of motion:

$$\dot{x}_c = \frac{p_c}{m}, \qquad \dot{p}_c = -\frac{\mathrm{d}U_0(x_c)}{\mathrm{d}x_c} \equiv F_0(x_c) \qquad (14.6.17)$$

(Cao and Voth, 1994*a*; Cao and Voth, 1996), where m is the physical mass, p_c is a momentum conjugate to x_c, and $U_0(x_c)$ is the centroid potential of mean force given by

$$U_0(x_c) = -\frac{1}{\beta} \ln \left\{ \left(\frac{2\pi\beta\hbar^2}{m} \right)^{1/2} \oint \mathcal{D}x(\tau)\, \delta(x_0[x(\tau)] - x_c) \mathrm{e}^{-S[x(\tau)]/\hbar} \right\}, \qquad (14.6.18)$$

where $x_0[x(\tau)] = (1/\beta\hbar)\int d\tau \, x(\tau)$. In eqn. (14.6.18), $S[x(\tau)]$ is the Euclidean time action, and the δ-function restricts the functional integration to cyclic paths whose centroid position is x_c. Note that eqn. (12.8.24), in the limit $P \to \infty$, is equivalent to eqn. (14.6.18). Of course, in actual calculations, we use the discretized, finite-P version in eqn. (12.8.24). The centroid force at x_c, $F_0(x_c)$ is derived from eqn. (14.6.18) simply by spatial differentiation:

$$F_0(x_c) = -\frac{\oint \mathcal{D}x(\tau)\,\delta(x_0[x(\tau)] - x_c)\left[\frac{1}{\beta\hbar}\int_0^{\beta\hbar} d\tau' \, U'(x(\tau'))\right] e^{-S[x(\tau)]/\hbar}}{\oint \mathcal{D}x(\tau)\,\delta(x_0[x(\tau)] - x_c)\,e^{-S[x(\tau)]/\hbar}}. \quad (14.6.19)$$

In a path integral molecular dynamics or Monte Carlo calculation, the centroid force would be computed simply from

$$F_0(x_c) = -\left\langle \frac{1}{P}\sum_{k=1}^{P}\frac{\partial U}{\partial x_k}\,\delta\left(\frac{1}{P}\sum_{k=1}^{P}x_k - x_c\right)\right\rangle_f. \quad (14.6.20)$$

Although formally exact within the CMD framework, eqns. (14.6.18), (14.6.19), and eqn. (14.6.20) are of limited practical use: Their evaluation entails a full path integral calculation at each centroid configuration, which is computationally very demanding for complex systems.

In order to alleviate this computational burden, an adiabatic approximation, similar to that of Section 8.10, can be employed (Cao and Voth, 1994b; Cao and Martyna, 1996). In this approach, an ordinary imaginary-time path integral molecular dynamics calculation in the normal-mode representation of eqn. (12.8.18) is performed with two small modifications. First, the noncentroid modes are assigned masses that are significantly lighter than the centroid mass(es) so as to effect an adiabatic decoupling between the two sets of modes. According to the analysis of Section 8.10, this allows the centroid potential of mean force to be generated "on the fly" as the CMD simulation is carried out. The decoupling is achieved by introducing an adiabaticity parameter γ^2 ($0 < \gamma^2 < 1$), which is used to scale the fictitious kinetic masses of the internal modes according to $m'_k = \gamma^2 m\lambda_k$ and thereby accelerate their dynamics. Second, while the path integral molecular dynamics schemes of Section 12.8 employ thermostats on every degree of freedom, in the adiabatic CMD approach, because we require the actual dynamics of the centroids, only the noncentroid modes are coupled to thermostats.

The key assumption of CMD is that the Kubo-transformed quantum time correlation function $K_{AB}^{\beta}(t)$ of eqn. (14.6.11), for operators $\hat{A}$ and $\hat{B}$ that are functions of $\hat{x}$ can be approximated by

$$K_{AB}^{\beta}(t) \approx \frac{1}{Q(\beta)}\int dx_c dp_c \, a\,(x_c)\,b\,(x_c(t; x_c, p_c))\exp\left[-\beta\left(\frac{p_c^2}{2m} + U_0(x_c)\right)\right]. \quad (14.6.21)$$

Here, the function $b(x_c(t; x_c, p_c))$ is evaluated using the time-evolved centroid variables generated by eqns. (14.6.17), starting from $\{x_c, p_c\}$ as initial conditions. An analogous definition holds for operators $\hat{A}$ and $\hat{B}$ that are functions of momentum only. As discussed by Hernandez *et al.* (1995), eqn. (14.6.21) can be generalized for operators that are functions of both position and momentum using a procedure known as "Weyl operator ordering" (Weyl, 1927; Hillery *et al.*, 1984), which we alluded to in Section 9.2.5 (see eqn. (9.2.51)). CMD is exact in the classical limit and in the limit of a purely harmonic potential.

14.6.2 Ring polymer molecular dynamics

The method known as ring polymer molecular dynamics (RPMD), originally introduced by Craig and Manolopoulos (2004), is motivated by the primitive path integral algorithm of eqn. (12.8.5). Craig and Manolopoulos posited that the primitive equations of motion could be used to extract approximate real-time information. Indeed, like CMD, the dynamics generated by eqns. (12.8.5) possess the correct harmonic and classical limits.

The principal features that distinguish RPMD from CMD are threefold. First, the RPMD fictitious masses are chosen such that each imaginary-time slice or bead has the physical mass m. Second, RPMD uses the full chain to approximate time correlation functions. Thus, a quantum observable $\hat{A}(\hat{x})$ is assumed to evolve in time according to

$$A_P(t) = \frac{1}{P} \sum_{k=1}^{P} a(x_i(t)).$$
(14.6.22)

RPMD approximates the Kubo-transformed time correlation function $K_{AB}^{\beta}(t)$ as

$$K_{AB}^{\beta}(t) \approx \frac{1}{(2\pi\hbar)^P Q_P(\beta)} \int d^P x d^P p \, A_P(0) B_P(t) \, e^{-\beta_P \mathcal{H}_{\text{cl},P}(x,p)},$$
(14.6.23)

where $\beta_P = \beta/P$ (RPMD simulations are typically carried out at P times the actual temperature) and

$$\mathcal{H}_{\text{cl},P}(x,p) = \sum_{k=1}^{P} \frac{p_k^2}{2m} + \frac{m}{2\beta_P^2 \hbar^2} \sum_{k=1}^{P} (x_k - x_{k+1})^2 + \sum_{k=1}^{P} U(x_k)$$
(14.6.24)

with $x_{P+1} = x_1$. Note that the harmonic bead-coupling and potential energy terms are taken to be P times larger than their counterparts in eqn. (12.8.4). We adopt this convention for consistency with Craig and Manolopoulos (2004); it amounts to nothing more than a rescaling of the temperature from T to PT. For operators linear in position or momentum, the CMD and RPMD representations of observables are the same; however, they generally differ for functions that are nonlinear in these variables.

The third difference is that RPMD is purely Newtonian. The equations of motion are easily derived from eqn. (14.6.24):

$$\dot{x}_k = \frac{p_k}{m}, \qquad \dot{p}_k = -\frac{m}{\beta_P^2 \hbar^2} \left[2x_k - x_{k-1} - x_{k+1} \right] - \frac{\partial U}{\partial x_k}.$$
(14.6.25)

In this dynamics, no thermostats are used on any of the beads since all beads are treated as dynamical variables. As discussed in Section 13.2.1, however, the use of eqn. (14.6.23) assumes that the distribution $\exp[-\beta_P \mathcal{H}_{\text{cl},P}(x,p)]$ can be adequately sampled, and as Section 12.8 makes clear, this requires some care. Thus, RPMD is optimally implemented by performing a fully thermostatted path integral molecular dynamics calculation in staging or normal modes. From this trajectory, path configurations are periodically transformed back to primitive variables and saved. From these saved path configurations, independent RPMD trajectories are initiated, and these trajectories are then used to compute the approximate Kubo-transformed correlation function. Detailed discussions on the emergence of RPMD from a formally exact path integral dynamics can be found in the work of Hele *et al.* (2015*a*, 2015*b*).

14.6.3 Self-consistent quality control of time correlation functions

Since the introduction of the CMD and RPMD methods, several technical issues have been noted concerning their use in computing vibrational spectra (Witt *et al.*, 2009): Delocalization of ring polymers can cause the centroid position to be unphysical (see Section 12.7). Potentials such as harmonic bonds, which are functions of the length r of the bond, are spherical in nature, and as illustrated in Fig. 12.11, when ring polymers become delocalized over the surface, the centroid may lie inside the sphere. The effect on a vibrational spectrum, such as an infrared spectrum, computed using CMD can be a significant red-shift of the associated stretch frequency, especially at low temperatures (Witt *et al.*, 2009; Paesani and Voth, 2010). For RPMD calculations, unphysical mode frequencies of the ring polymer can interfere with the physical vibrational frequencies and cause unphysical peaks to appear in the spectrum (Witt *et al.*, 2009). As a solution to these problems, Rossi *et al.* (2014) suggested that performing an RPMD simulation with a thermostat on the internal ring-polymer modes can, with proper choice of the thermostat type and thermostat parameters, correct the spurious effects inherent in both CMD and RPMD that degrade the quality of vibrational spectra.

Technical issues such as those discussed in the preceding paragraph suggest that the quality of CMD and RPMD correlation functions is often difficult to assess. Therefore, it is important to have an internal consistency check for these predicted time correlation functions. The measure we will propose allows the inherent accuracy of the CMD or RPMD approximation to be evaluated for a given model without having to rely on experimental data as the final arbiter.

Recall that a CMD or RPMD simulation yields an approximation to the Kubo-transformed time correlation function. Consider, for example, the velocity autocorrelation function $C_{vv}^{\beta}(t)$ and its Fourier transform $\tilde{C}_{vv}^{\beta}(\omega)$. Let $\tilde{C}_{vv}^{\beta,(\text{est})}(\omega)$ denote a CMD or RPMD approximation to $\tilde{C}_{vv}^{\beta}(\omega)$. Equation (14.6.16) allows us to reconstruct the associated imaginary-time correlation function $\bar{R}^2(\tau)$ from $\tilde{C}_{vv}^{\beta,(\text{est})}(\omega)$. The approximation $\bar{R}^2(\tau)$ can then be compared directly to the numerically exact mean square displacement function $R^2(\tau)$ computed from the same simulation (see eqn. (14.6.15)). Pérez *et al.* (2009) suggested a dimensionless quantitative descriptor for the quality of an approach (CMD/RPMD) can be defined by

$$\chi^2 = \frac{1}{\beta\hbar} \int_0^{\beta\hbar} d\tau \left[\frac{\bar{R}^2(\tau) - R^2(\tau)}{R^2(\tau)} \right]^2. \tag{14.6.26}$$

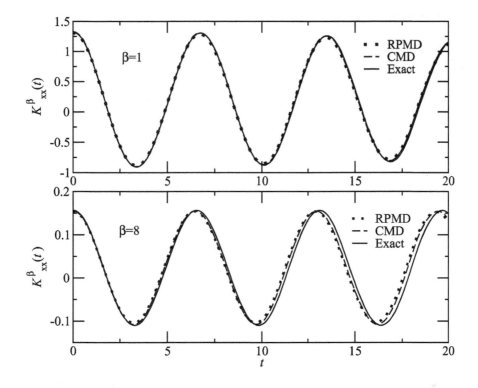

Fig. 14.7 Kubo-transformed position autocorrelation function for a mildly anharmonic potential $U(x) = x^2/2 + 0.1x^3 + 0.01x^4$ at inverse temperatures $\beta = 1$ (top) and $\beta = 8$ (bottom).

If CMD or RPMD were able to generate exact quantum time correlation functions, then χ^2 would be exactly zero. Thus, the larger χ^2, the poorer is the CMD or RPMD approximation to the true correlation function.

As illustrative examples of the CMD and RPMD schemes and the error measure in eqn. (14.6.26), we consider two one-dimensional systems with potentials given by $U(x) = x^2/2 + 0.1x^3 + 0.01x^4$ and $U(x) = x^4/4$. These potentials are simulated at inverse temperatures of $\beta = 1$ and $\beta = 8$ with $P = 8$ and $P = 32$ beads, respectively. For the CMD simulations, the adiabaticity parameter is $\gamma^2 = 0.005$. Figs. 14.7 and 14.8 show the Kubo-transformed time correlation functions $K_{xx}^{\beta}(t)$ for these two problems, respectively, comparing CMD and RPMD to the exact correlation functions, which are available for these one-dimensional examples via numerical matrix multiplication (Thirumalai *et al.*, 1983). Since the first potential is very close to harmonic, we expect CMD and RPMD to perform well compared to the exact correlation functions, which, as Fig. 14.7 shows, they do. For the strongly anharmonic potential $U(x) = x^4/4$, both methods are poor approximations to the exact correlation function. We notice, however, that the results improve at the higher temperature (lower β) for the mildly anharmonic potential, which is expected, as the higher temperature is

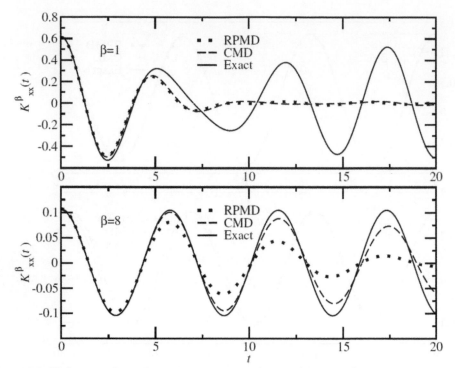

Fig. 14.8 Kubo-transformed position autocorrelation function for a quartic potential $U(x) = x^4/4$ at inverse temperatures $\beta = 1$ (top) and $\beta = 8$ (bottom).

closer to the classical limit. This trend is consistent with a study by Witt *et al.* (2009), who found significant deviations of vibrational spectra from the correct results at low temperatures. In particular, in a subsequent study by Ivanov *et al.* (2010), CMD was shown to produce severe artificial red shifts in high-frequency regions of vibrational spectra (however, see Paesani and Voth (2010)). For the quartic potential, the results are actually worse at high temperature, indicating that at low temperature (high β), the quartic potential is closer to the harmonic limit for which CMD and RPMD are exact.

The second example is a more realistic one of fluid *para*-hydrogen, described by a potential model of Silvera and Goldman (1978). Following Miller and Manolopoulos (2005) and Hone *et al.* (2006), the system is simulated at a temperature of $T = 14$ K, a density of $\rho = 0.0234$ Å^{-1}, and $N = 256$ molecules subject to periodic boundary conditions. In addition, we take $P = 32$ beads to discretize the path integral, and for CMD, the adiabaticity parameter is taken to be $\gamma^2 = 0.0444$. Figure 14.9(a) shows the Kubo-transformed velocity autocorrelation functions for this system from CMD and RPMD. The two methods appear to be in excellent agreement with each other. In fact, if these velocity autocorrelation functions are used to compute the diffusion constant using the Green-Kubo theory in eqn. (13.3.35), we obtain $D = 0.306$ Å^2/ps for CMD and $D = 0.263$ Å^2/ps for RPMD, both of which are in reasonable agreement with the

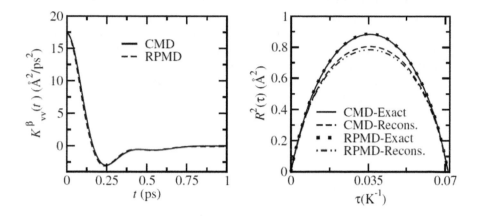

Fig. 14.9 (a) Velocity autocorrelation functions for *para*-hydrogen at $T = 14$ K for CMD and RPMD simulations. (b) Exact imaginary-time mean-square displacements and imaginary-time mean-square displacements reconstructed from the approximate CMD and RPMD real-time correlation functions in part (a).

experimental value of 0.4 Å^2/ps (Miller and Manolopoulos, 2005). Interestingly, we see that the correlation function of this condensed-phase system decays to zero in a short time, something which is not uncommon in the condensed phase at finite temperature. In Fig. 14.9(b), we show the imaginary-time mean-square displacements $R^2(\tau)$ computed directly from an imaginary-time path integral calculation and estimated from the CMD and RPMD approximate real-time correlation functions. Both approximations miss the true imaginary-time data, particularly in the peak region around $\tau = \beta\hbar/2$. The χ^2 error measure for both cases is 0.0089 for RPMD and 0.0056 for CMD. Interestingly, although Braams and Manolopoulos (2006) showed that RPMD is a more accurate approach at very short times, CMD seems to give a slightly better approximation to the true correlation function overall.

14.6.4 Approaching exact quantum time correlation functions

The CMD and RPMD methods were originally proposed as *ansätze* for approximating the Kubo-transformed time correlation function, and as such, they do not start from the exact expression for $K_{AB}^{\beta}(t)$. In this section, we will discuss a framework, termed the open-chain symmetrized correlation function (OPSCF) method, for recasting the symmetrized time correlation function $G_{AB}^{\beta}(t)$ in a way that renders it as a formally exact sampling problem from a positive-definite distribution. We first prove the existence of this positive-definite distribution and then illustrate an approach for approximating the distribution. It is worth noting that since existence of the distribution can be proved, other routes to approximating it likely exist, and we leave this as an open challenge in this area.

In the primitive coordinates, $x_1, ..., x_{2P}$, of a single particle in one dimension, the symmetrized quantum time correlation function is given by the expressions in eqns. (14.6.8) through (14.6.10). As can be discerned from the phase definition in eqn. (14.6.10), the points $x_1, ..., x_P$ represent the discretization of a path that is propagated forward in complex time via the propagator $\exp(-i\hat{\mathcal{H}}\tau_c/\hbar)$, while $x_{P+1}, ..., x_{2P}$ represent the discretization of a path that is propagated backward in complex time via the propagator $\exp(i\hat{\mathcal{H}}\tau_c^*/\hbar)$. In condensed phases, where significant quantum decoherence is often anticipated, such decoherence will cause difference between the forward and backward paths to be small. Thus, we can draw inspiration from semi-classical methods and introduce a change of variables from $x_1, ..., x_{2P}$ to a new set, denoted $r_1, ..., r_{P+1}, s_2, ..., s_P$ (Bonella *et al.*, 2010) via

$$r_1 = x_1, \qquad r_{P+1} = x_{P+1}$$

$$r_i = \frac{1}{2}\left(x_i + x_{2(P+1)-i}\right), \qquad i = 2, ..., P$$

$$s_i = x_i - x_{2(P+1)-i}, \qquad i = 2, ..., P. \qquad (14.6.27)$$

The variables r_1 and r_{P+1} are the path endpoints, $s_2, ..., s_P$ represent the difference between the forward and backward paths, and $r_2, ..., r_P$ represent the average of the forward and backward paths. Once we develop the formalism in this new set of variables, we will exploit the fact that, in many applications, $s_2, ..., s_P$ are small in order to derive one possible approximation to the symmetrized correlation function.

Before introducing any approximations, however, let us recast $G_{AB,P}^\beta(t)$ in terms of the variables $r_1, ..., r_{P+1} \equiv r$ and $s_2, ..., s_P \equiv s$ and determine what can be proved about the resulting expressions. In terms of (r, s), the expression for $G_{AB,P}^\beta(t)$ becomes

$$G_{AB,P}^\beta(t) = \frac{1}{Q(\beta)} \int dr_1 \cdots dr_{P+1}\, a(r_1)b(r_{P+1})I^\beta(r;t)$$

$$\times \exp\left[-4\alpha \sum_{i=1}^{P}(r_{i+1} - r_i)^2 - \frac{\beta}{2P}\left(U(r_1) + U(r_{P+2})\right)\right], \quad (14.6.28)$$

where

$$I^\beta(r;t) = \int ds_2 \cdots ds_P\, f^\beta(r,s;t)e^{i\chi^\beta(r,s;t)}. \qquad (14.6.29)$$

In this expression,

$$f^\beta(r,s;t) = \left(\frac{mP}{2\pi|\tau_c|\hbar}\right)^P \exp\left\{-\alpha \sum_{i=2}^{P}(s_{i+1} - s_i)^2 - \alpha\left(s_2^2 + s_P^2\right)\right.$$

$$\left. - \frac{\beta}{2P} \sum_{i=2}^{P}\left[U\left(r_i + \frac{1}{2}s_i\right) + U\left(r_i - \frac{1}{2}s_i\right)\right]\right\}, \quad (14.6.30)$$

$$\chi^{\beta}(r,s;t) = \gamma \sum_{i=2}^{P-1} (r_{i+1} - r_i)(s_{i+1} - s_i) + \gamma \left((r_2 - r_1)s_2 - (r_{P+1} - r_P)s_P \right)$$

$$- \frac{t}{P\hbar} \sum_{i=2}^{P} \left[U\left(r_i + \frac{1}{2}s_i \right) - U\left(r_i - \frac{1}{2}s_i \right) \right], \tag{14.6.31}$$

where $\alpha = mP\beta/(8|\tau_c|^2)$ and $\gamma = mPt/(\hbar|\tau_c|^2)$. Since the integral in eqn. (14.6.28) is performed over the variables $r_1, ..., r_{P+1}$, it is critical to note that $r_1 \neq r_{P+1}$, and hence, we can think of $r_1, ..., r_{P+1}$ as a discretization of an open chain derived from the average of the forward and backward complex-time paths. Amazingly, despite the presence of the complex exponential $\exp(i\chi^{\beta}(r,s;t))$ in eqn. (14.6.29), the function $I^{\beta}(r;t)$ is actually a positive-definite function of r.

In order to prove the assertion that $I^{\beta}(r;t)$ is positive-definite, note that in eqn. (14.6.28), $a(r_1) = a(x_1)$ and $b(r_{P+1}) = b(x_{P+1})$ and,, therefore, a direct comparison between eqns. (14.6.28) and (14.6.8) can be made. In particular, eqn. (14.6.8) can be rewritten in the form

$$G_{AB,P}^{\beta}(t) = \frac{1}{Q(\beta)} \int dx_1 \, dx_{P+1} \, a(x_1)b(x_{P+1}) \left| \langle x_{P+1}|e^{-i\mathcal{H}\tau_c/\hbar}|x_1\rangle \right|^2. \tag{14.6.32}$$

Note that, in recasting eqn. (14.6.8), we have renamed $x \to x_1$ and $x' \to x_{P+1}$ as dummy variables of integration to arrive at eqn. (14.6.32), and we have written $\langle x|e^{i\mathcal{H}\tau_c^*/\hbar}|x'\rangle\langle x'|e^{-i\mathcal{H}\tau_c/\hbar}|x\rangle$ as $\left| \langle x_{P+1}|e^{-i\mathcal{H}\tau_c/\hbar}|x_1\rangle \right|^2$, which is a positive-definite quantity. Therefore, this factor in the integrand of eqn. (14.6.32) is positive-definite. Now, in eqn. (14.6.28), the exponential factor is positive-definite. Since eqns. (14.6.28) and (14.6.32) are exactly the same, in order for the product

$$I^{\beta}(r;t) \exp\left[-4\alpha \sum_{i=1}^{P} (r_{i+1} - r_i)^2 - \frac{\beta}{2P} \left(U(r_1) + U(r_{P+2}) \right) \right]$$

to be positive-definite, it must follow that $I^{\beta}(r;t)$ is a positive-definite quantity as well, which completes the proof.

This result implies that *there exists an effective real-valued potential* $\tilde{U}^{\beta}(r;t)$ such that

$$I^{\beta}(r;t) \exp\left[-4\alpha \sum_{i=1}^{P} (r_{i+1} - r_i)^2 - \frac{\beta}{2P} \left(U(r_1) + U(r_{P+2}) \right) \right] = e^{-\beta \tilde{U}^{\beta}(r;t)}. \tag{14.6.33}$$

The potential $\tilde{U}^{\beta}(r;t)$ acts on an open chain consisting of $P+1$ beads whose primitive variables are $r_1, ..., r_{P+1}$. Then, since $\exp(-\beta \tilde{U}^{\beta}(r;t))$ is positive-definite, the calculation of $G_{AB,P}^{\beta}(t)$ is reduced to one of sampling open-chain configurations from the Boltzmann factor $\exp(-\beta \tilde{U}^{\beta}(r;t))$ at different points t in time and evaluating the product $a(r_1)b(r_{P+1})$ using the endpoint positions of the chain for each value of t.

It is important to note that since $|\tau_c| = [t^2 + (\beta\hbar)^2/4]^{1/2}$, as t increases, in order to maintain the accuracy of the calculation, the number of beads in the open chain must increase as well so as to keep $|\tau_c|/P$ constant.

At this point, we have established that the effective potential $\tilde{U}^\beta(r;t)$ exists, but we have not specified how it can be constructed either exactly or approximately. In the work of Cendagorta *et al.* (2018), the authors expand the terms $U(r_i \pm s_i/2)$ in eqns. (14.6.30) and (14.6.31) in a power series out to second order in s_i under the assumption alluded to above that the s_i are small. The expansion takes the form

$$U\left(r_i \pm \frac{s_i}{2}\right) \approx U(r_i)\frac{1}{2} \pm U'(r_i)s_i + \frac{1}{8}U''(r_i)s_i^2. \qquad (14.6.34)$$

When eqn. (14.6.34) is substituted into eqns. (14.6.30) and (14.6.31), the integrals over $s_2, ..., s_P$ become Gaussian integrals that can be performed analytically to yield a closed-form approximation for $\tilde{U}^\beta(r;t)$, which we know must be real despite the presence of the phase factor in eqn. (14.6.31). By working within this approximation, some of the phase information encoded in the correlation function is captured. The expression for $\tilde{U}^\beta(r;t)$ that results from this approximation (see Problem 14.14) is

$$\tilde{U}^\beta(r;t) = \frac{1}{2\beta}\left[\mathrm{K}^\mathsf{T}(r)\mathrm{M}^{-1}(r)\mathrm{K}(r) + \ln\det[\mathrm{M}(r)]\right], \qquad (14.6.35)$$

where $\mathrm{M}(r)$ is a $(P-1) \times (P-1)$ tridiagonal matrix given by

$$M_{ij}(r) = \left[4\alpha + \frac{\beta}{4P}U''(r_i)\right]\delta_{ij} - 2\alpha\left(\delta_{i,j+1} + \delta_{i+1,j}\right), \qquad (14.6.36)$$

and $\mathrm{K}(r)$ is a $(P-1)$-dimensional vector with components

$$K_i(r) = \gamma\left(2r_i - r_{i-1} - r_{i+1}\right) - \frac{t}{\beta\hbar}U'(r_i). \qquad (14.6.37)$$

An advantage of starting from the exact expression for $G^\beta_{AB}(t)$ is that correlation functions of any position-dependent operators, including those that are nonlinear in the positions, can be computed. When this is done in CMD or RPMD, there is usually a loss in accuracy. We illustrate this loss in Fig. 14.10 by computing $G^\beta_{x^2x^2}(t)$ for a harmonic oscillator $U(x) = x^2/2$ at temperature $\beta = 1$. For this example, the second-order expansion in eqn. (14.6.34) is exact, which means that the exact result must be obtained, and Fig. 14.10 shows this to be the case within the sampling error bars. It is worth noting that as t and (necessarily) P increase, sampling the configuration space of the open chain becomes more difficult, and convergence is slower.

A clear disadvantage of using eqn. (14.6.35) is that the second derivative of the potential must be computed, which, for complex systems, is something preferably to be avoided. If ample computer time is available, the need for a second-order expansion could be circumvented by rewriting integrals over $s_2, ..., s_P$ in eqn. (14.6.29) as

$$I^\beta(r;t) = \int_{-\infty}^{\infty} dy\, e^{iy} \int ds_2 \cdots ds_P\, f^\beta(r,s;t)\, \delta\left(\chi^\beta(r,s;t) - y\right). \qquad (14.6.38)$$

The most straightforward way to implement eqn. (14.6.38) is to sample a configuration $\bar{r} \equiv \bar{r}_1, ..., \bar{r}_{P+1}$ of an open chain and at this configuration perform a path integral

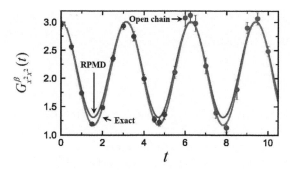

Fig. 14.10 Symmetric correlation function $G^{\beta}_{x^2 x^2}(t)$ for a harmonic oscillator $U(x) = x^2/2$ as generated using ring polymer molecular dynamics (RPMD) and the open-chain approach (symbols with error bars) compared to the exact result.

simulation in the space of the variables $s_2, ..., s_P$, which are then sampled from the distribution $f^{\beta}(\bar{r}, s; t)$ at the fixed configuration $\bar{r}$. During this simulation, a histogram of the function $\chi^{\beta}(\bar{r}, s; t)$ is collected. Once this histogram is converged, it is fed into the one-dimensional integral over y in order to obtain a value for $I^{\beta}(\bar{r}; t)$, which is then used to sample the next configuration of the open chain via, for example, a Monte Carlo algorithm. Once again, because the open chain must increase in length as t increases, we would expect that converging the distribution to an accuracy sufficient enough to integrate stably over y will become more difficult as longer times t are needed. However, when decoherence is strong, the correlation function will decay quickly, and the method based on eqn. (14.6.38) might remain stable long enough to see this rapid decay. Ultimately, the fact that the very existence of $\tilde{U}^{\beta}(r; t)$ can be proved opens the door to new approximate approaches.

14.7 Problems

14.1. a. Derive eqns. (14.4.7) and (14.4.8).

b. Show that the Fourier transforms of correlation functions $\langle \hat{x}(0)\hat{x}(t) \rangle$ and $\langle \hat{x}(t)\hat{x}(0) \rangle$ are related by

$$\frac{1}{2\pi} \int_{-\infty}^{\infty} dt \, e^{-i\omega t} \langle \hat{x}(0)\hat{x}(t) \rangle = e^{\beta \hbar \omega} \frac{1}{2\pi} \int_{-\infty}^{\infty} dt \, e^{-i\omega t} \langle \hat{x}(t)\hat{x}(0) \rangle.$$

c. Show that the Fourier transform of $\langle \hat{x}(t)\hat{x}(0) \rangle$ is related to its classical counterpart by

$$\frac{1}{2\pi} \int_{-\infty}^{\infty} dt \, e^{-i\omega t} \frac{1}{2} \langle [\hat{x}(t), \hat{x}(0)]_+ \rangle = \frac{\beta \hbar \omega}{2} \tanh(\beta \hbar \omega / 2)$$

$$\times \frac{1}{2\pi} \int_{-\infty}^{\infty} dt \, e^{-i\omega t} \langle x(t)x(0) \rangle_{\text{cl}}.$$

14.2. a. Derive eqns. (14.6.6) and (14.6.12).

    *b. Derive eqns. (14.6.8) through (14.6.10).

14.3. Derive eqn. (14.6.16).

14.4. A quantum harmonic oscillator of mass m and frequency ω is subject to a time-dependent perturbation $\hat{\mathcal{H}}_1(t) = -\alpha\hat{x}\exp(-t^2/\tau^2)$, $t \in (-\infty, \infty)$. At $t_0 = -\infty$, the oscillator is in its ground state.
   a. To the lowest nonvanishing order in perturbation theory, calculate the probability of a transition from the ground to the first excited state as $t \to \infty$.

   b. To the lowest nonvanishing order in perturbation theory, calculate the probability of a transition from the ground to the second excited state as $t \to \infty$.

14.5. Evaluate the correlation functions $G_{xx}(t)$ and $K_{xx}(t)$ for a harmonic oscillator and compare both to the classical correlation function $\langle x(0)x(t)\rangle$ for a harmonic oscillator.

14.6. The time-dependent Schrödinger equation for a single particle of mass m and charge $-e$ moving in a potential $U(\mathbf{r})$ subject to an electromagnetic field is

$$\left\{ -\frac{1}{2m}\left[-i\hbar\nabla - \frac{e}{c}\mathbf{A}(\mathbf{r},t) \right]^2 - e\phi(\mathbf{r},t) + V(\mathbf{r}) \right\}\psi(\mathbf{r},t) = i\hbar\frac{\partial}{\partial t}\psi(\mathbf{r},t).$$

Show that the Schrödinger equation is invariant under a gauge transformation

$$\mathbf{A}'(\mathbf{r},t) = \mathbf{A}(\mathbf{r},t) - \nabla\chi(\mathbf{r},t)$$

$$\phi'(\mathbf{r},t) = \phi(\mathbf{r},t) + \frac{1}{c}\frac{\partial}{\partial t}\chi(\mathbf{r},t)$$

$$\psi'(\mathbf{r},t) = e^{-ie\chi(\mathbf{r},t)/\hbar c}\psi(\mathbf{r},t).$$

14.7. Consider the free rotational motion of a rigid heteronuclear diatomic molecule of (fixed) bond length R and moment of inertia $I = \mu R^2$, where μ is the reduced mass, about an axis through its center of mass perpendicular to the internuclear bond axis. The molecule is constrained to rotate in the xy plane only. One of the atoms carries a charge q and the other a charge $-q$.
   a. Ignoring center-of-mass motion, write down the Hamiltonian $\hat{\mathcal{H}}_0$ for the molecule.

   b. Find the eigenvalues and eigenvectors of $\hat{\mathcal{H}}_0$.

c. The molecule is exposed to spatially homogeneous, monochromatic radiation with an electric field $\mathbf{E}(t)$ given by

$$\mathbf{E}(t) = E(\omega)e^{i\omega t}\hat{\mathbf{x}},$$

where $\hat{\mathbf{x}}$ is the unit vector in the x-direction. Write down the perturbation Hamiltonian $\hat{\mathcal{H}}_1$.

d. Calculate the energy spectrum $Q(\omega)$ for $\omega > 0$. Interpret your results, and in particular, explain how the allowed absorptions and emissions are manifest in your final expression. Plot the absorption part of your spectrum. Where do you expect the peak intensity to occur?

Hint: Consider using a convergence factor, $\exp(-\epsilon|t|)$, and let ϵ go to 0 at the end of the calculation.

e. Based on your results from parts a-d, plot the spectrum three-dimensional rigid rotor, for which the energy eigenvalues are $E_{lm} = \hbar^2 l(l+1)/2I$ and $m = -l, ..., l$ is the quantum number for the z-component of angular momentum. Where do you expect the peak intensity to occur in the 3-dimensional case?

14.8. Derive a discrete path integral representation for the Kubo-transformed quantum time correlation function $K_{AB}^{\beta}(t)$ defined in eqn. (14.6.11).

14.9. Consider two spin-1/2 particles at fixed points in space a distance R apart and interacting with a magnetic field $\mathbf{B} = (0, 0, B)$. The particles carry charge q and $-q$, respectively. The Hamiltonian of the system is

$$\hat{\mathcal{H}} = -\gamma\mathbf{B}\cdot\hat{\mathbf{S}} - \frac{q^2}{R},$$

where $\hat{\mathbf{S}} = \hat{\mathbf{S}}_1 + \hat{\mathbf{S}}_2$ is the total spin, and γ is the spin gyromagnetic ratio.

a. What are the allowed energy levels of this system?

b. Suppose that a time-dependent perturbation of the form

$$\hat{\mathcal{H}}_1(t) = -\gamma\mathbf{b}\cdot\hat{\mathbf{S}}e^{-t^2/\tau^2},$$

where $\mathbf{b} = (b, 0, 0)$ is applied at $t = -\infty$. At $t = -\infty$, the system is in its unperturbed ground state. To first order in perturbation theory, what is the probability, as $t \longrightarrow \infty$, that the system will make a transition from its ground state to a state with energy $-q^2/R$?

14.10. The *density of vibrational states*, also known as the *power spectrum* or *spectral density*, is the Fourier transform of the velocity autocorrelation function:

$$I(\omega) = \frac{1}{\sqrt{2\pi}} \int_{-\infty}^{\infty} dt\, e^{-i\omega t} C_{vv}(t).$$

$I(\omega)$ encodes information about the vibrational modes of a system; however, it does not provide any information about net absorption intensities. For the two model velocity autocorrelation functions in Problem 13.1, calculate the density of vibrational states and interpret them in terms of the physical situations described by these two model correlation functions.

*14.11. For the discrete correlation function $G^{\beta}_{AB,P}(t)$ defined in eqn. (14.6.8), we could analyze the importance of the phase factor $\Phi(x_1, ..., x_{2P})$ by calculating its fluctuation

$$(\delta\Phi)^2 \equiv \langle \Phi^2 \rangle - \langle \Phi \rangle^2 = \langle (\Phi - \langle \Phi \rangle)^2 \rangle$$

with respect to an equilibrium discrete path integral consisting of $2P$ imaginary-time points. Using the path integral virial theorem in eqn. (12.8.34), derive a virial estimator for the above average.

14.12. Derive analytical expressions for the imaginary-time mean-square displacement of a free particle in 1, 2, and 3 dimensions. In particular, show that in d dimensions, $R^2(\tau)$ is an inverted parabola, symmetric about the point $\tau = \beta\hbar/2$. For each number of dimensions, sketch the graph of $R^2(\tau)$ as a function of τ. Finally, determine the numerical value of $R^2(\beta\hbar/2)$ in angstroms at $T = 300$ K for an electron and for a proton.

14.13. Show that RPMD can be rewritten in terms of normal modes in such a way that the dynamical trajectories are exactly the same as those obtained using primitive variables.

14.14. Derive eqn. (14.6.35). In this derivation, you may make use of the following identity for Gaussian integrals: If $\mathbf{x}$ is an n-dimensional vector, M is an $n \times n$ matrix, and $\mathbf{v}$ is an n-dimensional vector, then

$$\int d\mathbf{x} e^{-\mathbf{x}^{\mathrm{T}} M\mathbf{x} + i\mathbf{v}\cdot\mathbf{x}} = \left[\frac{(2\pi)^n}{\det[M]} \right]^{1/2} e^{-\mathbf{v}^{\mathrm{T}} M^{-1} \mathbf{v}}.$$

14.15. Electronic excitation spectroscopy investigates the dynamics of a quantum system on two electronic surfaces denoted $U_g(x)$ and $U_e(x)$ for the ground and excited-state surfaces, respectively, with corresponding Hamiltonians $\hat{\mathcal{H}}_g$ and $\hat{\mathcal{H}}_e$. An analog of the symmetrized time correlation function $G_{eg}(t)$ for this type of process can be written as

$$G_{eg}(t) = \frac{1}{Q} \mathrm{Re} \left\{ \mathrm{Tr} \left[\hat{A} e^{i\hat{\mathcal{H}}_e \tau_c^*/\hbar} \hat{B} e^{-i\hat{\mathcal{H}}_g \tau_c/\hbar} \right] \right\}.$$

Show that an open-chain approach based on a second-order expansion of the potentials about $s_i = 0$ yields a method in which the potential energy functions appearing in eqns. (14.6.36) and (14.6.37) are replaced by the average potential $(U_{\mathrm{g}}+U_{\mathrm{e}})/2$, indicating motion on an average potential surface (reminiscent of an approach to quantum dynamics known as *Ehrenfest dynamics*).

15

The Langevin and generalized Langevin equations

15.1 The general model of a system plus a bath

Many problems in chemistry, biology, and physics do not involve homogeneous systems but are concerned, rather, with a specific process that occurs in some sort of medium. Most biophysical and biochemical processes happen in an aqueous environment, and one might be interested in a specific conformational change in a protein or the bond-breaking event in a hydrolysis reaction. In this case, the water solvent and other degrees of freedom not directly involved in the reaction serve as the "medium," which is often referred to generically as a *bath*. Organic reactions are carried out in a variety of different solvents, including water, methanol, dimethyl sulfoxide, and carbon tetrachloride. For example, a common reaction such as a Diels-Alder reaction can occur in water or in a room-temperature ionic liquid. In surface physics, we might be interested in the addition of an adsorbate to a particular site on the surface. If a reaction coordinate (see Section 8.6) for the adsorption process can be identified, the remaining degrees of freedom, including the bulk below the surface, can be treated as the environment or bath. Many other examples fall into this general paradigm, and it is, therefore, useful to develop a framework for treating such problems.

In this chapter, we will develop an approach that allows the bath degrees of freedom to be eliminated from a problem, leaving only coordinates of interest to be treated explicitly. The resulting equation of motion in the reduced subspace, known as the *generalized Langevin equation* (Langevin, 1905; Langevin, 1908) after the French physicist Paul Langevin (1872–1946), can only be taken as rigorous in certain idealized limits. However, as a phenomenological theory, the generalized Langevin equation is a powerful tool for understanding of a wide variety of physical processes. These include theories of chemical reaction rates (Kramers, 1940; Grote and Hynes, 1980; Pollak *et al.*, 1989; Pollak, 1990; Pollak *et al.*, 1990) and of vibrational dephasing and energy relaxation, to be discussed in Section 15.4.

In order to introduce the basic paradigm of a subsystem interacting with a bath, consider a classical system with generalized coordinates $q_1, ..., q_{3N}$. Suppose we are interested in a simple process that can be described by a single coordinate, which we arbitrarily take to be q_1. We will call q_1 and the remaining coordinates $q_2, ..., q_{3N}$ the *system* and *bath* coordinates, respectively. Moreover, in order to make the notation clearer, we will rename q_1 as q and the remaining bath coordinates as $y_1, ..., y_n$, where $n = 3N - 1$. In order to avoid unnecessary complexity at this point, we will assume that the system coordinate q is a simple coordinate, such as a distance between two atoms or a Cartesian spatial direction (in Section 15.8, we will introduce a general framework for treating the problem that allows this restriction to be lifted). The Hamiltonian for q and its conjugate momentum p in the absence of the bath can then be written simply as

$$\mathcal{H}(q,p) = \frac{p^2}{2\mu} + V(q), \tag{15.1.1}$$

where μ is the mass associated with q and $V(q)$ is a potential energy contribution that depends only on q and, therefore, is present even without the bath. The system is coupled to the bath via a potential $U_{\text{bath}}(q, y_1, ..., y_n)$ that involves both the coupling terms between the system and the bath and terms describing the interactions among the bath degrees of freedom. The total potential is

$$U(q, y_1, ..., y_n) = V(q) + U_{\text{bath}}(q, y_1, ..., y_n). \tag{15.1.2}$$

As an example, consider a system originally formulated in Cartesian coordinates $\mathbf{r}_1, ..., \mathbf{r}_N$ described by a pair potential

$$U(\mathbf{r}_1, ..., \mathbf{r}_N) = \sum_{i=1}^{N-1} \sum_{j=i+1}^{N} u(|\mathbf{r}_i - \mathbf{r}_j|). \tag{15.1.3}$$

Suppose the distance $r = |\mathbf{r}_1 - \mathbf{r}_2|$ between atoms 1 and 2 is a coordinate of interest, which we take as the system coordinate. All other degrees of freedom are assigned as bath coordinates. Suppose, further, that atoms 1 and 2 have the same mass. We first transform to the center-of-mass and relative coordinates between atoms 1 and 2 according to

$$\mathbf{R} = \frac{1}{2}(\mathbf{r}_1 + \mathbf{r}_2), \qquad \mathbf{r} = \mathbf{r}_1 - \mathbf{r}_2, \tag{15.1.4}$$

the inverse of which is

$$\mathbf{r}_1 = \mathbf{R} + \frac{1}{2}\mathbf{r}, \qquad \mathbf{r}_2 = \mathbf{R} - \frac{1}{2}\mathbf{r}. \tag{15.1.5}$$

The potential can then be expressed as

$$U(\mathbf{r}_1, ..., \mathbf{r}_N) = u(|\mathbf{r}_1 - \mathbf{r}_2|) + \sum_{i=3}^{N} [u(|\mathbf{r}_1 - \mathbf{r}_i|) + u(|\mathbf{r}_2 - \mathbf{r}_i|)] + \sum_{i=3}^{N-1} \sum_{j=i+1}^{N} u(|\mathbf{r}_i - \mathbf{r}_j|)$$

$$= u(r) + \sum_{i=3}^{N} \left[u\left(\left| \mathbf{R} + \frac{1}{2}r\mathbf{n} - \mathbf{r}_i \right| \right) + u\left(\left| \mathbf{R} - \frac{1}{2}r\mathbf{n} - \mathbf{r}_i \right| \right) \right]$$

$$+ \sum_{i=3}^{N-1} \sum_{j=i+1}^{N} u(|\mathbf{r}_i - \mathbf{r}_j|), \tag{15.1.6}$$

where $\mathbf{n} = (\mathbf{r}_1 - \mathbf{r}_2)/|\mathbf{r}_1 - \mathbf{r}_2| = \mathbf{r}/r$ is the unit vector along the relative coordinate direction. Equation (15.1.6) is of the same form as eqn. (15.1.2), in which the first term is equivalent to $V(q)$, the term in brackets represents the interaction between the system and the bath, and the final term is a pure bath–bath interaction.

Suppose the bath potential U_{bath} can be reasonably approximated by an expansion up to second-order about a minimum characterized by values $\bar{q}, \bar{y}_1, ..., \bar{y}_n$ of the generalized coordinates. The condition for U_{bath} to have a minimum at these values is

$$\left. \frac{\partial U_{\text{bath}}}{\partial q_\alpha} \right|_{\{q=\bar{q}, y=\bar{y}\}} = 0, \tag{15.1.7}$$

where all coordinates are set equal to their values at the minimum. Performing the expansion up to second-order gives

$$U_{\text{bath}}(q, y_1, ..., y_n) \approx U_{\text{bath}}(\bar{q}, \bar{y}_1, ..., \bar{y}_n) + \sum_\alpha \left. \frac{\partial U_{\text{bath}}}{\partial q_\alpha} \right|_{\{q=\bar{q}, y=\bar{y}\}} (q_\alpha - \bar{q}_\alpha)$$

$$+ \frac{1}{2} \sum_{\alpha,\beta} (q_\alpha - \bar{q}_\alpha) \left[\left. \frac{\partial^2 U_{\text{bath}}}{\partial q_\alpha \partial q_\beta} \right|_{\{q=\bar{q}, y=\bar{y}\}} \right] (q_\beta - \bar{q}_\beta). \tag{15.1.8}$$

The second term in eqn. (15.1.8) vanishes by virtue of the condition in eqn. (15.1.7). The first term is a constant that can be made to vanish by shifting the absolute zero of the potential (which is, anyway, arbitrary). Thus, the bath potential reduces, in this approximation, to

$$U_{\text{bath}}(q, y_1, ..., y_n) = \frac{1}{2} \sum_{\alpha=1}^{n+1} \sum_{\beta=1}^{n+1} \tilde{q}_\alpha H_{\alpha\beta} \tilde{q}_\beta, \tag{15.1.9}$$

where $H_{\alpha\beta} = \partial^2 U_{\text{bath}} / \partial q_\alpha \partial q_\beta|_{q=\bar{q}, \{y=\bar{y}\}}$ and $\tilde{q}_\alpha = q_\alpha - \bar{q}_\alpha$ are the displacements of the generalized coordinates from their values at the minimum of the potential. Note that since we have already identified the purely q-dependent term in eqn. (15.1.6), the H_{11} arising from the expansion of the bath potential can be taken to be zero or absorbed into the q-dependent function $V(q)$. Since our treatment from this point on will refer to the displacement coordinates, we will drop the tildes and let q_α refer to the displacement of a coordinate from its value at the minimum. Separating the particular coordinate q from the other coordinates gives a potential of the form

$$U_{\text{bath}}(q, y_1, ..., y_n) = \sum_\alpha C_\alpha q y_\alpha + \frac{1}{2} \sum_{\alpha=1}^{n} \sum_{\beta=1}^{n} y_\alpha \tilde{H}_{\alpha\beta} y_\beta, \tag{15.1.10}$$

where $C_\alpha = H_{1\alpha} = H_{\alpha 1}$ and $\tilde{H}_{\alpha\beta}$ is the $n \times n$ block of $H_{\alpha\beta}$ coupling only the coordinates $y_1, ..., y_n$. The potential, though quadratic, is still somewhat complicated because all of the coordinates are coupled through the matrix $H_{\alpha\beta}$. Thus, in order to simplify the potential, we introduce a linear transformation of the coordinates $y_1, ..., y_n$ to $x_1, ..., x_n$ via

$$y_\alpha = \sum_{\beta=1}^{n} R_{\alpha\beta} x_\beta, \tag{15.1.11}$$

where $R_{\alpha\beta}$ is an orthogonal matrix that diagonalizes the symmetric matrix $\tilde{H}_{\alpha\beta}$ via $\tilde{H}_{\text{diag}} = R^{\mathsf{T}} \tilde{H} R$, where R^{T} is the transpose of R and $\tilde{H}_{\text{diag}}$ contains the eigenvalues of $\tilde{H}$

on its diagonal. Letting k_α denote these eigenvalues and introducing the transformation into eqn. (15.1.10), we obtain

$$U_{\text{bath}}(q, x_1, ..., x_n) = \sum_\alpha g_\alpha q x_\alpha + \frac{1}{2} \sum_\alpha k_\alpha x_\alpha^2, \qquad (15.1.12)$$

where $g_\alpha = \sum_\beta C_\beta R_{\beta\alpha}$. The potential energy in eqn. (15.1.12) is known as a *harmonic bath* potential; it also contains a bilinear coupling to the coordinate q. We will henceforth refer to the coordinate q as the "system coordinate." In order to construct the full Hamiltonian in the harmonic bath approximation, we introduce a set of momenta $p_1, ..., p_n$, assumed to be conjugate to the coordinates $x_1, ..., x_n$, and a set of bath masses $m_1, ..., m_n$. The full Hamiltonian for the system coordinate coupled to a harmonic bath can be written as

$$\mathcal{H} = \frac{p^2}{2\mu} + V(q) + \sum_{\alpha=1}^{n} \left[\frac{p_\alpha^2}{2m_\alpha} + \frac{1}{2} m_\alpha \omega_\alpha^2 x_\alpha^2 \right] + q \sum_{\alpha=1}^{n} g_\alpha x_\alpha, \qquad (15.1.13)$$

where the spring constants k_α have been replaced by the bath frequencies $\omega_1, ..., \omega_n$ using $k_\alpha = m_\alpha \omega_\alpha^2$ so that the motion of the free bath oscillators is independent of the choice of masses m_α and depends only on the frequencies. We must not forget that eqn. (15.1.13) represents a highly idealized situation in which the possible curvilinear nature of the generalized coordinates is neglected in favor of a very simple model of the bath (Deutsch and Silbey, 1971; Caldeira and Leggett, 1983).

A real bath is often characterized by a continuous distribution of frequencies $I(\omega)$ called the *spectral density* or *density of states* (see Problem 14.10). $I(\omega)$ is obtained by taking the Fourier transform of the velocity autocorrelation function.[1] The physical picture embodied in the harmonic-bath Hamiltonian is one in which a real bath is replaced by an ideal bath under the assumption that the motion of the real bath is dominated by small displacements from an equilibrium point described by discrete frequencies $\omega_1, ..., \omega_n$. This replacement is tantamount to expressing $I(\omega)$ as a sum of harmonic-oscillator spectral density functions. It is important to note that the harmonic bath does not allow for diffusion of bath particles. In general, a set of frequencies, $\omega_1, .., \omega_n$, effective masses $m_1, ..., m_n$, and coupling constants to the system $g_1, ..., g_n$ need to be determined in order to reproduce at least some of the properties of the real bath. The extent to which this can be done, however, depends on the particular nature of the original bath. For the purposes of the subsequent discussion, we will assume that a reasonable choice can be made for these parameters and proceed to work out the classical dynamics of the harmonic-bath Hamiltonian.

15.2   Derivation of the generalized Langevin equation

We begin by deriving the classical equations of motion generated by eqn. (15.1.13). Applying Hamilton's equations, these become

$$\dot{q} = \frac{\partial \mathcal{H}}{\partial p} = \frac{p}{\mu}$$

---

[1]The density of states encodes the information about the vibrational modes of the bath; however, it does not provide any information about absorption intensities.

$$\dot{p} = -\frac{\partial \mathcal{H}}{\partial q} = -\frac{\mathrm{d}V}{\mathrm{d}q} - \sum_\alpha g_\alpha x_\alpha$$

$$\dot{x}_\alpha = \frac{\partial \mathcal{H}}{\partial p_\alpha} = \frac{p_\alpha}{m_\alpha}$$

$$\dot{p}_\alpha = -\frac{\partial \mathcal{H}}{\partial x_\alpha} = -m_\alpha \omega_\alpha^2 x_\alpha - g_\alpha q, \tag{15.2.1}$$

which can be written as the following set of coupled second-order differential equations:

$$\mu \ddot{q} = -\frac{\mathrm{d}V}{\mathrm{d}q} - \sum_\alpha g_\alpha x_\alpha$$

$$m_\alpha \ddot{x}_\alpha = -m_\alpha \omega_\alpha^2 x_\alpha - g_\alpha q. \tag{15.2.2}$$

Equations (15.2.2) must be solved subject to a set of initial conditions

$$\{q(0), \dot{q}(0), x_1(0), ..., x_n(0), \dot{x}_1(0), ..., \dot{x}_n(0)\}.$$

The second equation for the bath coordinates can be solved *in terms of the system coordinate q* by Laplace transformation, assuming that the system coordinate q acts as a kind of driving term. The Laplace transform of a function $f(t)$, alluded to briefly in Section 14.6, is one of several types of integral transforms defined to be

$$\tilde{f}(s) = \int_0^\infty \mathrm{d}t \, \mathrm{e}^{-st} f(t). \tag{15.2.3}$$

As we will now show, Laplace transforms are particularly useful for solving linear differential equations. A more detailed discussion of Laplace transforms is given in Appendix E. From eqn. (15.2.3), it can be shown straightforwardly that the Laplace transforms of $\mathrm{d}f/\mathrm{d}t$ and $\mathrm{d}^2 f/\mathrm{d}t^2$ are given, respectively, by

$$\int_0^\infty \mathrm{d}t \, \mathrm{e}^{-st} \frac{\mathrm{d}f}{\mathrm{d}t} = s\tilde{f}(s) - f(0)$$

$$\int_0^\infty \mathrm{d}t \, \mathrm{e}^{-st} \frac{\mathrm{d}^2 f}{\mathrm{d}t^2} = s^2 \tilde{f}(s) - f'(0) - sf(0). \tag{15.2.4}$$

Finally, the Laplace transform of a convolution of two functions $f(t)$ and $g(t)$ can be shown to be

$$\int_0^\infty \mathrm{d}t \, \mathrm{e}^{-st} \int_0^t \mathrm{d}\tau f(\tau) g(t - \tau) = \tilde{f}(s) \tilde{g}(s). \tag{15.2.5}$$

Equations (15.2.4) and (15.2.5), together with eqn. (E.2), are sufficient for us to solve eqns. (15.2.2).

Taking the Laplace transform of both sides of the second line in eqn. (15.2.2) yields

$$s^2\tilde{x}_\alpha(s) - \dot{x}_\alpha(0) - sx_\alpha(0) + \omega_\alpha^2\tilde{x}_\alpha(s) = -\frac{g_\alpha}{m_\alpha}\tilde{q}(s). \qquad (15.2.6)$$

The use of the Laplace transform has the effect of turning a differential equation into an algebraic equation for $\tilde{x}_\alpha(s)$. Solving this equation for $\tilde{x}_\alpha(s)$ gives

$$\tilde{x}_\alpha(s) = \frac{s}{s^2 + \omega_\alpha^2}x_\alpha(0) + \frac{1}{s^2 + \omega_\alpha^2}\dot{x}_\alpha(0) - \frac{g_\alpha}{m_\alpha}\frac{\tilde{q}(s)}{s^2 + \omega_\alpha^2}. \qquad (15.2.7)$$

We now obtain the solution to the differential equation by computing the inverse transform $\tilde{x}_\alpha(s)$ in eqn. (15.2.7). Applying the inverse Laplace transform relations in Appendix E (eqn. (E.2)), and recognizing that the last term in eqn. (15.2.7) is the product of two Laplace transforms and can be inverted using the convolution theorem, we find that the solution for $x_\alpha(t)$ is

$$x_\alpha(t) = x_\alpha(0)\cos\omega_\alpha t + \frac{1}{\omega_\alpha}\dot{x}_\alpha(0)\sin\omega_\alpha t - \frac{g_\alpha}{m_\alpha\omega_\alpha}\int_0^t d\tau \ \sin\omega_\alpha(t-\tau)q(\tau). \quad (15.2.8)$$

For reasons that will be clear shortly, we integrate the convolution term by parts to express it in the form

$$\int_0^t d\tau \ \sin\omega_\alpha(t-\tau)q(\tau) = \frac{1}{\omega_\alpha}[q(t) - q(0)\cos\omega_\alpha t]$$

$$-\frac{1}{\omega_\alpha}\int_0^t d\tau \ \cos\omega_\alpha(t-\tau)\dot{q}(\tau). \qquad (15.2.9)$$

Substituting eqn. (15.2.9) and eqn. (15.2.8) into the first line of eqn. (15.2.2) yields the equation of motion for q:

$$\mu\ddot{q} = -\frac{dV}{dq} - \sum_\alpha g_\alpha x_\alpha(t)$$

$$= -\frac{dV}{dq} - \sum_\alpha g_\alpha\left[x_\alpha(0)\cos\omega_\alpha t + \frac{p_\alpha(0)}{m_\alpha\omega_\alpha}\sin\omega_\alpha t + \frac{g_\alpha}{m_\alpha\omega_\alpha^2}q(0)\cos\omega_\alpha t\right]$$

$$-\sum_\alpha\frac{g_\alpha^2}{m_\alpha\omega_\alpha^2}\int_0^t d\tau \ \dot{q}(\tau)\cos\omega_\alpha(t-\tau) + \sum_\alpha\frac{g_\alpha^2}{m_\alpha\omega_\alpha^2}q(t). \qquad (15.2.10)$$

Equation (15.2.10) is in the form of an integro-differential equation for the system coordinate that depends explicitly on the bath dynamics. Although the dynamics of each bath coordinate are relatively simple, the collective effect of the bath on the system coordinate can be nontrivial, particularly if the initial conditions of the bath are randomly chosen, the distribution of frequencies is broad, and the frequencies are not all commensurate. Indeed, the bath might appear to affect the system coordinate in a random and unpredictable manner, especially if the number of bath degrees of freedom is large. This is just what we might expect for a real bath. Thus, in order to motivate this physical picture, the following quantities are introduced:

$$R(t) = -\sum_\alpha g_\alpha \left[\left(x_\alpha(0) + \frac{g_\alpha}{m_\alpha \omega_\alpha^2} q(0) \right) \cos \omega_\alpha t + \frac{p_\alpha(0)}{m_\alpha \omega_\alpha} \sin \omega_\alpha t \right], \qquad (15.2.11)$$

$$\zeta(t) = \sum_\alpha \frac{g_\alpha^2}{m_\alpha \omega_\alpha^2} \cos \omega_\alpha t, \qquad (15.2.12)$$

$$W(q) = V(q) - \sum_\alpha \frac{g_\alpha^2}{2 m_\alpha \omega_\alpha^2} q^2. \qquad (15.2.13)$$

In terms of these quantities, the equation of motion for the system coordinate reads

$$\mu \ddot{q} = -\frac{dW}{dq} - \int_0^t d\tau \, \dot{q}(\tau) \zeta(t - \tau) + R(t). \qquad (15.2.14)$$

Equation (15.2.14) is known as the *generalized Langevin equation* (GLE). The quantity $\zeta(t)$ in the GLE is called the *dynamic friction kernel*, $R(t)$ is called the *random force*, and $W(q)$ is identified as the *potential of mean force* acting on the system coordinate. Despite the simplifications of the bath inherent in eqn. (15.2.14), the GLE can yield considerable physical insight without requiring large-scale simulations. Before discussing predictions of the GLE, we will examine each of the terms in eqn. (15.2.14) and provide a physical interpretation of them.

15.2.1 The potential of mean force

Potentials of mean force were first discussed in Chapter 8 (see eqns. (8.6.6) and (8.6.7)). For a true harmonic bath, the potential of mean force is given by the simple expression in eqn. (15.2.13); however, as a phenomenological theory, the GLE assumes that the potential of mean force has been generated by some other means (using techniques from Chapter 8, for example the blue moon ensemble of Section 8.7 or the umbrella sampling approach of Section 8.8) and attempts to model the dynamics of the system coordinate on this surface using the friction kernel and random force to represent the influence of the bath. The use of the potential of mean force in the GLE assumes a quasi-adiabatic separation between the system and bath motions. However, considering the GLE's phenomenological viewpoint, it is also possible to use the bare potential $V(q)$ and use the GLE to model the dynamics on this surface instead. Such a model can be derived from a slightly modified version of the harmonic-bath Hamiltonian:

$$\mathcal{H} = \frac{p^2}{2\mu} + V(q) + \sum_{\alpha=1}^n \left[\frac{p_\alpha^2}{2 m_\alpha} + \frac{1}{2} \sum_{\alpha=1}^n m_\alpha \omega_\alpha^2 \left(x_\alpha + \frac{g_\alpha}{m_\alpha \omega_\alpha^2} q \right)^2 \right]. \qquad (15.2.15)$$

15.2.2 The random force

The question that immediately arises concerning the random force in eqn. (15.2.14) is why it is called "random" in the first place. After all, eqn. (15.2.11) defines a perfectly deterministic quantity. To understand why $R(t)$ can be treated as a random process, we note that a real bath, which contains a macroscopically large number of degrees of freedom, will affect the system in what appears to be a random manner, despite the fact that its time evolution is completely determined by the classical equations of motion. Recall, however, that the basic idea of ensemble theory is to disregard the detailed motion of every degree of freedom in a macroscopically large system and to replace this level of detail by an ensemble average. It is in this spirit that we replace the $R(t)$, defined microscopically in eqn. (15.2.11), with a truly random process defined by a particular time sequence of random numbers and a set of related time correlation functions that must be satisfied by this sequence.

We first note that the time correlation functions $\langle q(0)R(t)\rangle$ and $\langle \dot{q}(0)R(t)\rangle$ are identically zero for all time. To see this, consider first the correlation function

$$\langle \dot{q}(0)R(t)\rangle = \left\langle \frac{p(0)}{\mu}R(t)\right\rangle$$

$$= -\frac{1}{Q}\int \mathrm{d}p\,\mathrm{d}q\,\exp\left\{-\beta\left[\frac{p^2}{2\mu}+V(q)\right]\right\}$$

$$\times \int \prod_{\alpha=1}^{n}\mathrm{d}x_\alpha\,\mathrm{d}p_\alpha\,\exp\left\{-\beta\left[\sum_{\alpha=1}^{n}\left(\frac{p_\alpha^2}{2m_\alpha}+\frac{1}{2}\sum_{\alpha=1}^{n}m_\alpha\omega_\alpha^2 x_\alpha^2\right)+q\sum_{\alpha=1}^{n}g_\alpha x_\alpha\right]\right\}$$

$$\times \frac{p}{\mu}\sum_\alpha g_\alpha\left[\left(x_\alpha+\frac{g_\alpha}{m_\alpha\omega_\alpha^2}q\right)\cos\omega_\alpha t+\frac{p_\alpha}{m_\alpha\omega_\alpha}\sin\omega_\alpha t\right], \qquad (15.2.16)$$

where the average is taken over a canonical ensemble and Q is the partition function for the harmonic-bath Hamiltonian. Since $R(t)$ does not depend on the system momentum p, the integral over p is of the form $\int_{-\infty}^{\infty}\mathrm{d}p\,p\exp(-\beta p^2/2\mu)=0$, and the entire integral vanishes. It is left as an exercise to show that the correlation function $\langle q(0)R(t)\rangle = 0$ (see Problem 15.1). The vanishing of the correlation functions $\langle q(0)R(t)\rangle$ and $\langle \dot{q}(0)R(t)\rangle$ is precisely what we would expect from a random bath force, and hence we require that these correlation functions vanish for any model random process. Finally, the same manipulations employed above can be used to derive the autocorrelation function $\langle R(0)R(t)\rangle$ with the result

$$\langle R(0)R(t)\rangle = \frac{1}{\beta}\sum_\alpha \frac{g_\alpha^2}{m_\alpha\omega_\alpha^2}\cos\omega_\alpha t = kT\zeta(t), \qquad (15.2.17)$$

which shows that the random force and the dynamic friction kernel are related (see Problem 15.1). Equation (15.2.17) is known as the *second fluctuation dissipation theorem* (Kubo *et al.*, 1985). Once again, we require that any model random process we choose satisfy this theorem.

If the deterministic definition of $R(t)$ in eqn. (15.2.11) is to be replaced by a model random process, how should such a process be described mathematically? There are various ways to construct random time sequences that give the correct time correlation functions, depending on the physics of the problem. For instance, the influence of a relatively high-density bath, which affects the system via only soft collisions due to low amplitude thermal fluctuations, is different from a low-density, high-temperature bath that influences the system through mostly strong, impulsive collisions. Here, we construct a commonly used model, known as a *Gaussian random process*, for the former type of bath. Since for most potentials, the GLE must be integrated numerically, we seek a discrete description of $R(t)$ that acts at M discrete time points $0, \Delta t, 2\Delta t, ..., (M-1)\Delta t$. At the kth point of a Gaussian random process, $R_k \equiv R(k\Delta t)$ can be expressed as the sum of Fourier sine and cosine series as

$$R_k = \sum_{j=0}^{M-1} \left[a_j \sin\left(\frac{2\pi jk}{M}\right) + b_j \cos\left(\frac{2\pi jk}{M}\right) \right], \qquad (15.2.18)$$

where the coefficients a_j and b_j are random numbers sampled from a Gaussian distribution of the form

$$P(a_0, ..., a_{M-1}, b_0, ..., b_{M-1}) = \prod_{k=0}^{M-1} \frac{1}{2\pi\sigma_k^2} e^{-(a_k^2 + b_k^2)/2\sigma_k^2}. \qquad (15.2.19)$$

For the random force to satisfy eqn. (15.2.17) at each time point, the width, σ_k, of the distribution must be chosen according to

$$\sigma_k^2 = \frac{1}{\beta M} \sum_{j=0}^{M-1} \zeta(j\Delta t) \cos\left(\frac{2\pi jk}{M}\right), \qquad (15.2.20)$$

which can be easily evaluated using fast Fourier transform techniques. Since the random process in eqn. (15.2.18) is periodic with period M, it clearly cannot be used for more than a single period. This means that the number of points M in the trajectory must be long enough to capture the dynamical behavior sought.

15.2.3   The dynamic friction kernel

The convolution integral term in eqn. (15.2.14)

$$\int_0^t \mathrm{d}\tau \, \dot{q}(\tau)\zeta(t-\tau)$$

is called the *memory integral* because it depends, in principle, on the entire history of the evolution of q. Physically, this term expresses the fact that the bath requires a finite time to respond to any fluctuation in the motion of the system and that this lag affects how the bath subsequently influences the motion of the system. Thus, the force that the bath exerts on the system at any point in time depends on the prior motion of the system coordinate q. The memory of the system coordinate dynamics retained by the bath is encoded in the *memory kernel* or *dynamic friction kernel* $\zeta(t)$. Note that $\zeta(t)$ has units of mass·(time)$^{-2}$. Since the dynamic friction kernel is actually an autocorrelation function of the random force, it follows that the correlation time of the random force determines the decay time of the memory kernel. The finite correlation time of the memory kernel indicates that the bath, in reality, retains memory of the system motion for a finite time t_{mem}. One might expect, therefore, that the memory integral could be replaced, to a very good approximation, by an integral over a finite interval $[t - t_{\text{mem}}, t]$:

$$\int_0^t \mathrm{d}\tau \; \dot{q}(\tau)\zeta(t - \tau) \approx \int_{t - t_{\text{mem}}}^t \mathrm{d}\tau \; \dot{q}(\tau)\zeta(t - \tau). \tag{15.2.21}$$

Such an approximation proves very convenient in numerical simulations based on the generalized Langevin equation, as it permits the memory integral to be truncated, thereby reducing the computational overhead needed to evaluate it.

We now consider a few interesting limiting cases of the friction kernel. Suppose, for example, that the bath is able to respond infinitely quickly to the motion of the system. This would occur when the system mass μ is very large compared to the bath masses, $\mu \gg m_\alpha$. In such a case, the bath retains essentially no memory of the system motion, and the memory kernel reduces to a simple δ-function in time:

$$\zeta(t) = \lim_{\epsilon \to 0^+} \zeta_0 \delta(t - \epsilon). \tag{15.2.22}$$

The introduction of the parameter ϵ ensures that the entire δ-function is integrated over. Alternatively, we can recognize that for $\epsilon = 0$, only "half" of the δ-function is included in the interval $t \in [0, \infty)$, since $\delta(t)$ is an even function of time, and therefore, we could also define $\zeta(t)$ as $2\zeta_0\delta(t)$. Substituting eqn. (15.2.22) into eqn. (15.2.14) and taking the limit gives an equation of motion for q of the form

$$\mu\ddot{q} = -\frac{\mathrm{d}W}{\mathrm{d}q} - \lim_{\epsilon \to 0^+} \zeta_0 \int_0^t \mathrm{d}\tau \; \dot{q}(\tau)\delta(t - \epsilon - \tau) + R(t)$$

$$= -\frac{\mathrm{d}W}{\mathrm{d}q} - \lim_{\epsilon \to 0^+} \zeta_0 \dot{q}(t - \epsilon) + R(t)$$

$$= -\frac{\mathrm{d}W}{\mathrm{d}q} - \zeta_0 \dot{q}(t) + R(t), \tag{15.2.23}$$

where all quantities on the right are evaluated at time t. Equation (15.2.23) is known as the *Langevin equation* (LE), and it should be clear that the LE is ultimately a special case of the GLE. The LE describes the motion of a system in a potential $W(q)$ subject to an ordinary dissipative friction force as well as a random force $R(t)$. Langevin originally employed eqn. (15.2.23) as a model for Brownian motion, where the mass disparity clearly holds (Langevin, 1908). In the limit of very large friction ζ_0, the inertial term $\mu\ddot{q}$ can be neglected, resulting an equation of the form

$$\zeta_0 \dot{q} = -\frac{\mathrm{d}W}{\mathrm{d}Q} + R(t) \tag{15.2.24}$$

which is known as the *overdamped* Langevin equation. The overdamped form of the equation is used, for example, in simple theories of polymers such as the Rouse model (Rouse, 1953) (see Section 1.7 for the Hamiltonian of the Rouse model). The most common use of the LE or the overdamped LE is as a thermostatting method for generating a canonical distribution (see Section 15.5). The quantity ζ_0 is known as the *static friction coefficient*, defined generally as

$$\zeta_0 = \int_0^\infty \mathrm{d}t \ \zeta(t). \tag{15.2.25}$$

Note that the random force $R(t)$ is now completely uncorrelated, as it is required to satisfy

$$\langle R(0)R(t) \rangle = 2kT\zeta_0\delta(t). \tag{15.2.26}$$

In addition, note that ζ_0 has units of mass·(time)$^{-1}$.

The second limiting case we will consider is a sluggish bath that responds very slowly to changes in the system coordinate. For such a bath, we can take $\zeta(t)$ approximately constant over a long time interval, *i.e.*, $\zeta(t) \approx \zeta(0) \equiv \zeta$, for times that are short compared to the actual response time of the bath. In this case, the memory integral can be approximated as

$$\int_0^t \mathrm{d}\tau \ \dot{q}(\tau)\zeta(t-\tau) \approx \zeta \int_0^t \mathrm{d}\tau \ \dot{q}(\tau) = \zeta(q(t) - q(0)), \tag{15.2.27}$$

and eqn. (15.2.14) becomes

$$\mu\ddot{q} = -\frac{\mathrm{d}}{\mathrm{d}q}\left(W(q) + \frac{1}{2}\zeta(q - q(0))^2 \right) + R(t). \tag{15.2.28}$$

Here, the effect of friction is now manifest as an extra harmonic term in the potential $W(q)$, and all terms on the right are, again, evaluated at time t. This harmonic term in $W(q)$ has the effect of trapping the system in certain regions of configuration space, an effect known as *dynamic caging*. Figure 15.1 illustrates how the caging potential $\zeta[q - q(0)]^2/2$ can potentially trap the particle at what would otherwise be a point of unstable equilibrium. An example of this is a dilute mixture of small, light particles in a bath of large, heavy particles. In spatial regions where a heavy particle cluster forms a slowly moving spatial "cage," the light particles can become trapped. Only rare fluctuations in the bath open up this rigid structure, allowing the light particles to escape. After such an escape, however, the light particles can become trapped again in another cage newly formed elsewhere for a comparable time interval. Not unexpectedly, dynamic caging can cause a significant decrease in the rate of light-particle diffusion.

15.3   Analytically solvable examples

In the next few subsections, a number of simple yet illustrative examples of both Langevin and generalized Langevin dynamics will be examined in detail. In particular, we will study the free Brownian particle and compute its diffusion constant and then consider the free particle in a more general bath with memory. Finally, we will consider the harmonic oscillator and derive well-known relations for the vibrational and energy relaxation times.

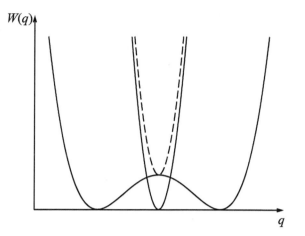

Fig. 15.1 Example of the dynamic caging phenomenon. $W(q)$ is taken to be the double-well potential. The potential $\zeta(q - q_0)^2/2$ is the single-minimum solid line, and the dashed line shows the potential shifted to the top of the barrier region.

15.3.1 The free Brownian particle

A particle diffusing in a dissipative bath with no external forces is known as a *free Brownian particle*. The dynamics is described by eqn. (15.2.23) with $W(q) = 0$:

$$\mu\ddot{q} = -\zeta_0\dot{q} + R(t). \tag{15.3.1}$$

Since only $\ddot{q}$ and $\dot{q}$ appear in the equation of motion, we can rewrite eqn. (15.3.1) in terms of the velocity $v = \dot{q}$

$$\mu\dot{v} = -\zeta_0 v + R(t). \tag{15.3.2}$$

Equation (15.3.2) can be treated as an inhomogeneous first-order equation that can be solved in terms of $R(t)$. In order to derive the solution for a given initial value $v(0)$, we take the Laplace transform of both sides, which yields

$$\mu(s\tilde{v}(s) - v(0)) = -\zeta_0\tilde{v}(s) + \tilde{R}(s). \tag{15.3.3}$$

Defining $\gamma_0 = \zeta_0/\mu$ and $f(t) = R(t)/\mu$ and solving for $\tilde{v}(s)$ gives

$$\tilde{v}(s) = \frac{v(0)}{s + \gamma_0} + \frac{\tilde{f}(s)}{s + \gamma_0}. \tag{15.3.4}$$

The function $1/(s + \gamma_0)$ has a single pole at $s = -\gamma_0$. Hence, the inverse Laplace transform (see Appendix E) yields the solution for $v(t)$ as

$$v(t) = v(0)e^{-\gamma_0 t} + \int_0^t d\tau\, f(\tau)e^{-\gamma_0(t-\tau)}. \tag{15.3.5}$$

From eqn. (15.3.5), it is clear that the solution for a free Brownian particle has two components: a transient component dependent on $v(0)$ that decays at large t, and a steady-state term involving a convolution of the random force with $\exp(-\gamma_0 t)$. Thus, the system quickly loses memory of its initial condition, and the dynamics for long times is determined by the bath, as we would expect for a random walk process such as Brownian motion. There are some subtleties associated with integrals over the (stochastic) random force $f(\tau)$, and these will be discussed in Section 15.5. Fortunately, in the examples considered here, we are only interested in correlation functions for which the random force integrals drop out, as we will see, and the problem of evaluating such integrals becomes moot.

If we wish to compute the diffusion constant of the Brownian particle, we can use eqn. (13.3.34) and calculate the velocity autocorrelation function $\langle v(0)v(t)\rangle$. From eqn. (15.3.5), the velocity correlation is obtained by multiplying both sides by $v(0)$ and averaging over a canonical distribution of the initial conditions at temperature T, $\exp(-\mu v(0)^2/kT)/Q$:

$$\langle v(0)v(t)\rangle = \Big\langle (v(0))^2 \Big\rangle e^{-\gamma_0 t} + \int_0^t d\tau\, \langle v(0)f(\tau)\rangle e^{-\gamma_0(t-\tau)}. \tag{15.3.6}$$

The second term in eqn. (15.3.6) vanishes because $\langle v(0)f(\tau)\rangle = \langle v(0)R(\tau)/\mu\rangle = 0$. Thus, interestingly, the velocity autocorrelation function is determined by the transient term, hence, the short-time dynamics. Performing the average $\langle (v(0))^2\rangle$

$$\Big\langle (v(0))^2 \Big\rangle = \frac{\int_{-\infty}^{\infty} dv\, v^2 e^{-\mu v^2/2kT}}{\int_{-\infty}^{\infty} dv\, e^{-\mu v^2/2kT}} = \frac{kT}{\mu} \tag{15.3.7}$$

yields the velocity autocorrelation function as

$$\langle v(0)v(t)\rangle = \frac{kT}{\mu} e^{-\gamma_0 t}. \tag{15.3.8}$$

Finally, from eqn. (13.3.34), we find

$$D = \int_0^{\infty} dt\, \langle v(0)v(t)\rangle = \frac{kT}{\mu} \int_0^{\infty} dt\, e^{-\gamma_0 t} = \frac{kT}{\mu\gamma_0} = \frac{kT}{\zeta_0}, \tag{15.3.9}$$

which has the expected units of length^2·(time)$^{-1}$. Note that as $\zeta_0 \to \infty$, the bath becomes infinitely dissipative and the diffusion constant goes to zero. Note that this simple picture of diffusion cannot capture the long-time algebraic decay of the velocity autocorrelation function mentioned in Section 13.3.

15.3.2   Free particle in a bath with memory

If the bath has memory, then the dynamics of the particle is given by the GLE, which, for a free particle, reads:

$$\mu\ddot{q} = -\int_0^t d\tau\, \dot{q}(\tau)\zeta(t-\tau) + R(t). \tag{15.3.10}$$

As a concrete example, suppose the dynamic friction kernel is given by an exponential function

$$\zeta(t) = \lambda A e^{-\lambda|t|}, \tag{15.3.11}$$

which could describe the long-time decay of a realistic friction kernel. Although the cusp at $t = 0$ is problematic for the short-time behavior of a typical friction kernel, the exponential friction kernel is, nevertheless, a convenient and simple model that can be solved analytically and has been studied in some detail in the literature (Berne *et al.*, 1966). Once again, let us introduce the velocity $v = \dot{q}$. For the exponential friction kernel of eqn. (15.3.11), the GLE then reads

$$\mu \dot{v} = -\lambda A \int_0^t d\tau\, v(\tau) e^{-\lambda(t-\tau)} + R(t), \tag{15.3.12}$$

where we are restricting the time domain to $t > 0$. Let us introduce the quantities $a = A/\mu$ and $f(t) = R(t)/\mu$. The Laplace transform can turn this integro-differential equation into a simple algebraic equation. Taking the Laplace transform of both sides of eqn. (15.3.12), and solving for $\tilde{v}(s)$, we obtain

$$\tilde{v}(s) = \frac{v(0)(s+\lambda)}{s^2 + s\lambda + \lambda a} + \frac{\tilde{f}(s)(s+\lambda)}{s^2 + s\lambda + \lambda a}, \tag{15.3.13}$$

where the fact that the memory integral is a convolution has been used to give its Laplace transform as a product of Laplace transforms of $v(t)$ and $\zeta(t)$. For Laplace inversion, the poles of the the function $(s+\lambda)/(s^2 + s\lambda + \lambda a)$ are needed. These occur where $s^2 + s\lambda + \lambda a = 0$, which yields two poles $s_\pm$ given by

$$s_\pm = -\frac{\lambda}{2} \pm \frac{\sqrt{\lambda^2 - 4\lambda a}}{2}. \tag{15.3.14}$$

The poles will be purely real if $\lambda \geq 4a$ and complex if $\lambda < 4a$. Performing the Laplace inversion gives the solution in the form

$$v(t) = v(0)\left[\frac{(s_+ + \lambda)e^{s_+ t}}{(s_+ - s_-)} + \frac{(s_- + \lambda)e^{s_- t}}{(s_- - s_+)}\right]$$

$$+ \int_0^t d\tau\, f(t-\tau)\left[\frac{(s_+ + \lambda)e^{s_+ \tau}}{(s_+ - s_-)} + \frac{(s_- + \lambda)e^{s_- \tau}}{(s_- - s_+)}\right]. \tag{15.3.15}$$

Since $\langle v(0)f(t)\rangle = 0$, the velocity autocorrelation function becomes

$$\langle v(t)v(0)\rangle = \left\langle (v(0))^2 \right\rangle e^{-\lambda t/2}\left[\cos \Omega t + \frac{\lambda}{2\Omega}\sin \Omega t\right], \tag{15.3.16}$$

where $\Omega = \sqrt{\lambda a - \lambda^2/4}$ for complex roots and

$$\langle v(t)v(0)\rangle = \left\langle (v(0))^2 \right\rangle e^{-\lambda t/2}\left[\cosh \alpha t + \frac{\lambda}{2\alpha}\sinh \alpha t\right], \tag{15.3.17}$$

where $\alpha = \sqrt{\lambda^2/4 - \lambda a}$. For both cases, the diffusion constant obtained from eqn. (13.3.34) is

$$D = \frac{kT}{A}. \tag{15.3.18}$$

Since $\int_0^\infty \zeta(t)dt = A$, eqn. (15.3.18) is consistent with eqn. (15.3.9) for the free Brownian particle. As $A \to \infty$, the bath becomes highly dissipative and $D \to 0$. Again, the overall decay is exponential, which means that the long-time algebraic decay of characteristic of such an autocorrelation function is not properly described.

15.3.3   The harmonic oscillator in a bath with memory

As a final example of a GLE model, consider a harmonic reaction coordinate described by a bare potential $V(q) = \mu \omega^2 q^2/2$. According to eqn. (15.2.13), the potential of mean force $W(q)$ is also a harmonic potential but with a different frequency given by

$$\tilde{\omega}^2 = \omega^2 - 2 \sum_\alpha \frac{g_\alpha^2}{\mu m_\alpha \omega_\alpha^2}, \tag{15.3.19}$$

so that

$$W(q) = \frac{1}{2} \mu \tilde{\omega}^2 q^2. \tag{15.3.20}$$

The quantity $\tilde{\omega}$ is known as the *renormalized frequency* We will examine the case in which the frequency of the oscillator is high compared to the bath frequencies, a condition that exists when the coupling between the system and the bath is weak. In the limit of high $\tilde{\omega}$, the term $-2 \sum_\alpha g_\alpha^2/\mu m_\alpha \omega_\alpha^2$ will be a small perturbation to ω^2. For a general friction kernel $\zeta(t)$, the GLE reads

$$\ddot{q} = -\tilde{\omega}^2 q - \int_0^t d\tau \, \dot{q}(\tau) \gamma(t - \tau) + f(t), \tag{15.3.21}$$

where $\gamma(t) = \zeta(t)/\mu$ and $f(t) = R(t)/\mu$. Equation (15.3.21) must be solved subject to initial conditions $q(0)$ and $\dot{q}(0)$. Taking the Laplace transform of both sides and solving for $\tilde{q}(s)$ yields

$$\tilde{q}(s) = \frac{(s + \tilde{\gamma}(s))}{\Delta(s)} q(0) + \frac{\dot{q}(0)}{\Delta(s)} + \frac{\tilde{f}(s)}{\Delta(s)}, \tag{15.3.22}$$

where

$$\Delta(s) = s^2 + \tilde{\omega}^2 + s \tilde{\gamma}(s). \tag{15.3.23}$$

In order to perform the Laplace inversion, the poles of each of the terms on the right side of eqn. (15.3.22) are needed. These are given by the zeroes of $\Delta(s)$. That is, we seek solutions of

$$s^2 + \tilde{\omega}^2 + s \tilde{\gamma}(s) = 0. \tag{15.3.24}$$

Even if we do not know the explicit form of $\tilde{\gamma}(s)$, when $\tilde{\omega}$ is large compared to the bath frequencies, it is possible to solve eqn. (15.3.24) perturbatively. We do this by positing a solution to eqn. (15.3.24) for s of the form

$$s = s_0 + s_1 + s_2 + \cdots \tag{15.3.25}$$

as an ansatz (Tuckerman and Berne, 1993). Substituting eqn. (15.3.25) into eqn. (15.3.24) gives

$$(s_0 + s_1 + s_2 + \cdots)^2 + \tilde{\omega}^2 + (s_0 + s_1 + s_2 + \cdots) \tilde{\gamma}(s_0 + s_1 + s_2 \cdots) = 0. \tag{15.3.26}$$

Assuming $\tilde{\omega}^2 >> s\tilde{\gamma}(s)$ at the root, we can solve this equation to lowest order by neglecting the $s\tilde{\gamma}(s)$ term, which gives

$$s_0^2 + \tilde{\omega}^2 = 0, \qquad\qquad s_0 = \pm i\tilde{\omega}. \tag{15.3.27}$$

Next, working to first order in the perturbation, we have

$$s_0^2 + 2s_0 s_1 + \tilde{\omega}^2 + s_0\tilde{\gamma}(s_0) = 0, \tag{15.3.28}$$

where it has been assumed that $s_0\tilde{\gamma}(s_0)$ is of the same order as s_1. Using the fact that $s_0^2 = -\tilde{\omega}^2$ and solving for s_1, we obtain

$$s_1 = -\frac{1}{2}\tilde{\gamma}(\pm i\tilde{\omega}), \tag{15.3.29}$$

which requires the evaluation of $\tilde{\gamma}(s)$ at $s = \pm i\tilde{\omega}$. Note that

$$\tilde{\gamma}(\pm i\tilde{\omega}) = \int_0^\infty dt\, \gamma(t)e^{\mp i\tilde{\omega}t}, \tag{15.3.30}$$

which contains both real and imaginary parts. Defining

$$\tilde{\gamma}(\pm i\tilde{\omega}) = \gamma'(\tilde{\omega}) \mp i\gamma''(\tilde{\omega}), \tag{15.3.31}$$

there are, to first order, two roots of $\Delta(s)$, which are given by

$$s_+ = i\left(\tilde{\omega} + \frac{1}{2}\gamma''(\tilde{\omega})\right) - \frac{1}{2}\gamma'(\tilde{\omega}) \equiv i\Omega - \frac{1}{2}\gamma'(\tilde{\omega})$$

$$s_- = -i\left(\tilde{\omega} + \frac{1}{2}\gamma''(\tilde{\omega})\right) - \frac{1}{2}\gamma'(\tilde{\omega}) \equiv -i\Omega - \frac{1}{2}\gamma'(\tilde{\omega}). \tag{15.3.32}$$

Substituting the roots in eqn. (15.3.32) into eqn. (15.3.22) yields the solution

$$q(t) = q(0)e^{-\gamma'(\tilde{\omega})t/2}\left[\cos\Omega t + \frac{\gamma'(\tilde{\omega})}{2\Omega}\sin\Omega t\right] + \frac{\dot{q}(0)}{\Omega}e^{-\gamma'(\tilde{\omega})t/2}\sin\Omega t$$

$$+ \frac{1}{\Omega}\int_0^t d\tau\, f(t-\tau)e^{-\gamma'(\tilde{\omega})\tau/2}\sin\Omega\tau$$

$$\dot{q}(t) = \dot{q}(0)\left[\cos\Omega t - \frac{\gamma'(\tilde{\omega})}{2\Omega}\sin\Omega t\right]e^{-\gamma'(\tilde{\omega})t/2} - \Omega q(0)e^{-\gamma'(\tilde{\omega})t/2}\sin\Omega t$$

$$+ \frac{1}{\Omega}\left[f(0)e^{-\gamma'(\tilde{\omega})t/2}\sin\Omega t + \int_0^t d\tau\, f'(t-\tau)e^{-\gamma'(\tilde{\omega})\tau/2}\sin\Omega\tau\right], \tag{15.3.33}$$

where we have used the fact that $\gamma''(\tilde{\omega}) \ll \Omega$, and we have neglected any terms nonlinear in $\gamma'(\tilde{\omega})$. Since $\langle q(0)f(t)\rangle = 0$ and $\langle \dot{q}(0)f(t)\rangle = 0$, the velocity and position autocorrelation functions become, respectively

$$C_{vv}(t) = \langle \dot{q}^2(0)\rangle\, e^{-\gamma'(\tilde{\omega})t/2}\left[\cos\Omega t - \frac{\gamma'(\tilde{\omega})}{2\Omega}\sin\Omega t\right]$$

$$C_{qq}(t) = \langle q^2(0) \rangle \, e^{-\gamma'(\tilde{\omega})t/2} \left[\cos \Omega t + \frac{\gamma'(\tilde{\omega})}{2\Omega} \sin \Omega t \right]. \qquad (15.3.34)$$

As eqn. (15.3.34) shows, the decay time of both correlation functions is $[\gamma'(\tilde{\omega})t/2]^{-1}$, which is denoted T_2 and is called the *vibrational dephasing time*. (We will explore vibrational and energy relaxation phenomena as an application of the GLE in greater detail in Section 15.4.) According to eqn. (13.3.41), the velocity autocorrelation function of a harmonic oscillator in isolation is proportional to $\cos \omega t$, where ω is the bare frequency of the oscillator. In this case, the correlation function does not decay because the system retains infinite memory of its initial condition. However, when coupled to a bath, the oscillator exchanges energy with the bath particles via collision events and, as a result, loses memory of its initial state on a time scale T_2. If there is a very large disparity of frequencies between the oscillator and the bath, the coupling between them will be weak and T_2 will be long, whereas if the oscillator frequency lies near or within the spectral density of the bath, vibrational energy exchange will occur readily and T_2 will be short. Thus, T_2 is an indicator of the strength of the coupling between the oscillator and the bath. As the frequency of the oscillator is increased, the coupling between the oscillator and the bath becomes weaker, and $\gamma'(\tilde{\omega})$ decreases. According to eqn. (15.3.34), this means that the correlation functions $C_{qq}(t)$ and $C_{vv}(t)$ decay more slowly, and the number of oscillations that can cycle through on the time scale T_2 grows. Two examples of the velocity autocorrelation function $C_{vv}(t)$ are shown in Fig. 15.2. In this example, the values of $\tilde{\omega}$, $\gamma'(\tilde{\omega})$, and $\gamma''(\tilde{\omega})$ correspond to a harmonic diatomic molecule of atomic type A coupled to a bath of A atoms interacting with each other and with the molecule via a Lennard-Jones potential at reduced temperature $\hat{T} = T/\epsilon = 2.5$ and reduced density $\hat{\rho} = \rho\sigma^3 = 1.05$. The frequencies $\omega = 60$ and $\omega = 90$ are expressed in Lennard-Jones reduced frequency units $\sqrt{\epsilon/(m\sigma^2)}$. A method for calculating the friction kernel for a high-frequency oscillator weakly coupled to a bath will be discussed in Section 15.8.

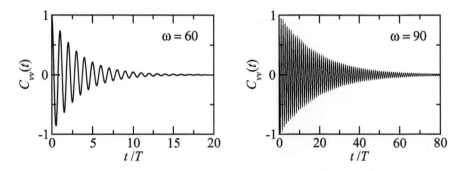

Fig. 15.2 Velocity autocorrelation function of the bond length of a harmonic diatomic coupled to a Lennard-Jones bath as described in the text. The bond frequencies $\omega = 60$ and $\omega = 90$ are expressed in units of $\sqrt{\epsilon/(m\sigma^2)}$.

15.3.4 Exact solution for an exponential memory kernel

In this section, we will consider the exact, analytical solution to eqn. (15.3.21) with an exponential memory kernel of the form

$$\gamma(t) = \gamma_0 e^{-\alpha|t|}. \tag{15.3.35}$$

Our goal in carrying out this analysis is not to obtain the detailed mathematical solution, which requires a long derivation that does not contain much physical content. Rather, we will provide enough of the derivation that the relevant physics can be readily extracted. It is important to note that the kernel in eqn. (15.3.35) is unphysical at $t = 0$ because of the cusp that exists there. Our reasons for considering such a kernel are threefold: (1) such a kernel models interesting physical behavior for $t > 0$; (2) this physical behavior can be exploited to design useful memory kernels for molecular simulation, which we will discuss later in this chapter; (3) eqn. (15.3.21) can be solved analytically.

As we have done in the previous examples, we will transform the GLE into Laplace space and examine the solutions in terms of the Laplace variable s. For the exponential memory kernel in eqn. (15.3.35), if $t \geq 0$, the absolute value of t becomes irrelevant, and $\tilde{\gamma}(s) = \gamma_0/(s + \alpha)$, hence, eqn. (15.3.22) becomes

$$\tilde{q}(s) = \frac{s^2 + \alpha s + \gamma_0}{D(s)} q(0) + \frac{s + \alpha}{D(s)} \dot{q}(0) + \frac{\tilde{f}(s)}{D(s)}, \tag{15.3.36}$$

where

$$D(s) = s^3 + \alpha s^2 + \left(\tilde{\omega}^2 + \gamma_0\right) s + \alpha \tilde{\omega}^2. \tag{15.3.37}$$

We noted in eqn. (15.3.24) that solving the GLE requires finding the poles of each term on the right of eqn. (15.3.36), which are given by the zeroes of $D(s)$. Since $D(s)$ is a cubic polynomial, there will be three roots, which we can find using the cubic formula. Of the three roots, at least one must be real, giving rise to a pure exponential decay. Of the remaining two, these can either be complex conjugates or purely real, and it is interesting to examine which conditions lead to complex roots and which lead to real roots. The general form of these three roots is

$$s_1 = A + B - \frac{\alpha}{3}$$

$$s_2 = -\frac{1}{2}(A + B) - \frac{\alpha}{3} + i\frac{\sqrt{3}}{2}(A - B)$$

$$s_3 = -\frac{1}{2}(A + B) - \frac{\alpha}{3} - i\frac{\sqrt{3}}{2}(A - B), \tag{15.3.38}$$

where

$$A = \left(-\frac{g}{2} + \sqrt{\delta}\right)^{1/3}$$

$$B = \left(-\frac{g}{2} - \sqrt{\delta}\right)^{1/3} \tag{15.3.39}$$

and

$$g = \frac{2\alpha}{3}\left(\frac{\alpha^2}{9} + \tilde{\omega}^2 - \frac{\gamma_0}{2}\right)$$

$$\delta = \frac{g^2}{4} + \frac{1}{27}\left(\tilde{\omega}^2 + \gamma_0 - \frac{\alpha^2}{3}\right)^3. \tag{15.3.40}$$

Here, δ is called the *discriminant* of the cubic. If $\delta > 0$, then two of the roots are complex (conjugates), and if $\delta < 0$, all of the roots are real. Since $g^2/4$ is a positive-definite quantity, the only condition that can determine if δ is positive or negative is whether $\tilde{\omega}^2 + \gamma_0 - \alpha^2/3$ is positive or negative. If this quantity is positive, then $\delta > 0$ and two of the roots are complex. If it is negative, then δ could be negative if $|\tilde{\omega}^2 + \gamma_0 - \alpha^2/3|^3/27 > g^2/4$, which, since $\gamma_0 > 0$, can only be true if the decay rate α of the memory kernel is large compared to the frequency $\tilde{\omega}$. However, because of the $g^2/4$ term, the prefactor γ_0 needs to be moderately large as well. This limit is clearly overdamped, as the frequency of the oscillator $\tilde{\omega}$ is low in a high-friction environment with little memory. If $\tilde{\omega}$ is high compared to γ_0 and the decay rate, then $\delta > 0$. However, $\delta > 0$ when $\tilde{\omega}^2 + \gamma_0 > \alpha^2/3$, which occurs when α is small compared to $\sqrt{\tilde{\omega}^2 + \gamma_0}$ or when the memory kernel decays slowly. This represents the dynamic caging limit discussed previously (cf. eqn. (15.2.28)). Thus, in carrying out the analysis in this section, we see that physical insight can be gained simply by examining the behavior of the poles of eqn. (15.3.36). Of course, if a complete time-dependent solution is needed, it can be obtained by inverting the Laplace transform using the three roots in eqns. (15.3.38) (see Problem 15.11). In Section 15.6, we will introduce simple algorithms for integrating the GLE with exponential memory.

15.4 Vibrational dephasing and energy relaxation in simple fluids

An application of the GLE that is of particular interest in chemical physics is the study of vibrational and energy relaxation phenomena. As we noted in Section 15.3.3, quantifying energy exchange between the system and the bath provides direct information about the strength of the system–bath coupling. For a harmonic oscillator coupled to a bath with memory, it was shown that, when the frequency of the oscillator is high compared to the spectral density of the bath, the vibrational relaxation time T_2 satisfies

$$\frac{1}{T_2} = \frac{\zeta'(\tilde{\omega})}{2\mu}. \tag{15.4.1}$$

T_2 is a measure of the decay time of the velocity and position autocorrelation functions. In addition to T_2, there is another relevant time scale, denoted T_1, which measures the rate of energy relaxation of the system. In this section, we will show how the GLE can be used to develop classical relations between T_1 and T_2 for both harmonic and anharmonic oscillators coupled to a bath. The times T_1 and T_2 are generally measured experimentally using nuclear-magnetic resonance techniques and, therefore, relate to quantum processes. However, we will see that the GLE can nevertheless provide useful insights into the physical nature of these two time scales.

Using the solutions of the GLE, it is also possible to show that the cross-correlation functions

$$C_{vq}(t) = \langle v(0)q(t) \rangle$$
$$C_{qv}(t) = \langle q(0)v(t) \rangle \tag{15.4.2}$$

have the same decay time. For this discussion, we will find it convenient to introduce a change of nomenclature and work with normalized correlation functions:

$$C_{ab}(t) = \frac{\langle a(0)b(t) \rangle}{\langle a^2 \rangle}. \tag{15.4.3}$$

In terms of the four normalized correlations functions, $C_{qq}(t)$, $C_{vv}(t)$, $C_{qv}(t)$, and $C_{vq}(t)$, the solutions of eqn. (15.3.21) can be expressed as

$$q(t) = q(0)C_{qq}(t) + \dot{q}(0)C_{vq}(t) + \int_0^t d\tau \; f(t-\tau)C_{vq}(\tau)$$

$$\dot{q}(t) = \dot{q}(0)C_{vv}(t) + q(0)C_{qv}(t) + \int_0^t d\tau \; f(t-\tau)C_{vv}(\tau) \tag{15.4.4}$$

(see Problem 15.4). Moreover, if the internal energy of the oscillator

$$\varepsilon(t) = \frac{1}{2}\mu\dot{q}^2(t) + \frac{1}{2}\mu\tilde{\omega}^2 q^2(t) \tag{15.4.5}$$

is calculated using the solutions in eqn. (15.4.4), then, as was shown by Tuckerman and Berne (1993), the autocorrelation function of $\varepsilon(t)$ is

$$C_{\varepsilon\varepsilon}(t) = \frac{1}{2}C_{vv}^2(t) + \frac{1}{2}C_{qq}^2(t) + \frac{1}{\tilde{\omega}^2}C_{qv}^2(t) \tag{15.4.6}$$

(see Problem 15.4). Since each of the correlation functions appearing in eqn. (15.4.6) has an exponential decay envelope of the form $\exp(-\zeta'(\tilde{\omega})t/2\mu)$, it follows that $C_{\varepsilon\varepsilon}(t)$ will decay as $\exp(-\zeta'(\tilde{\omega})t/\mu)$. This time scale corresponds to T_1 and is given simply by

$$\frac{1}{T_1} = \frac{\zeta'(\tilde{\omega})}{\mu}. \tag{15.4.7}$$

Equations (15.4.1) and (15.4.7) are especially useful in analyzing the dynamics of solvation. They have also found use more broadly, for example, in analyzing certain classes of polarizable force field models (Rupakheti *et al.*, 2020). A comparison of eqns. (15.4.7) and (15.4.1) reveals the prediction of the classical GLE approach namely that the vibrational dephasing and energy relaxation times for a harmonic oscillator coupled to a bath are related by

$$\frac{1}{T_2} = \frac{1}{2T_1}. \tag{15.4.8}$$

Equation (15.4.8) is true only for purely harmonic systems. However, real bonds always involve some degree of anharmonicity, which changes the relation between T_1 and T_2. The more general expression of this relation is

$$\frac{1}{T_2} = \frac{1}{2T_1} + \frac{1}{T_2^*}, \tag{15.4.9}$$

where T_2^* is a pure dephasing time. Now suppose we add to the harmonic potential $\mu\tilde{\omega}^2 q^2/2$ a small cubic term of the form $gq^3/6$ so that the potential of mean force $W(q)$ becomes

$$W(q) = \frac{1}{2}\mu\tilde{\omega}^2 q^2 + \frac{1}{6}gq^3. \tag{15.4.10}$$

Theoretical treatments of such a cubic anharmonicity have been presented by Oxtoby (1979), Levine *et al.* (1988), Tuckerman and Berne (1993), and Bader and Berne (1994), all of which lead to explicit expressions for the pure dephasing time. We note, however, that only direct solution of the full GLE, albeit an approximate one, yields $1/T_2$ in the form of eqn. (15.4.9), as we will now show.

The GLE corresponding to the potential in eqn. (15.4.10) reads

$$\ddot{q} = -\tilde{\omega}^2 q - \frac{g}{2\mu}q^2 - \int_0^t d\tau\, \dot{q}(\tau)\gamma(t-\tau) + f(t). \tag{15.4.11}$$

As long as the excursions of q in the cubic potential do not stray too far from the neighborhood of $q = 0$, the motion of q remains bound between definite turning points. However, as the energy of the oscillator fluctuations, the time required to move between the turning points varies. In other words, the period of the motion, and hence the frequency, varies as a function of the energy. Therefore, we seek a perturbative solution of eqn. (15.4.11), in which the anharmonicity is treated as an effect that causes the vibrational frequency to fluctuate in time. Equation (15.4.11) is then replaced, to lowest order in perturbation theory, by an equation of the form

$$\ddot{q} = -\omega^2(t)q - \int_0^t d\tau\, \dot{q}(\tau)\gamma(t-\tau) + f(t), \tag{15.4.12}$$

where $\omega(t) = \tilde{\omega} + \delta\omega(t)$ and $\delta\omega(t) = gf(t)/2\mu\tilde{\omega}^3$. By studying the autocorrelation function $C_{qq}(t)$ within perturbation theory, it was shown (Tuckerman and Berne, 1993) that $C_{qq}(t)$ is an oscillatory function with an exponential decay envelope. Thus, the general approximate form of $C_{qq}(t)$ is

$$C_{qq}^{(\pm)}(t) = C_{qq}^{(\pm,0)}(t)\left\langle e^{i\int_0^t d\tau \delta\omega(\tau)}\right\rangle \approx C_{qq}^{(\pm,0)}(t)e^{-\int_0^t d\tau(t-\tau)\langle\delta\omega(0)\delta\omega(\tau)\rangle}, \tag{15.4.13}$$

where $C_{qq}^{(\pm,0)}(t)$ are purely harmonic autocorrelation functions similar to those in eqn. (15.3.34). The exponential decay term in eqn. (15.4.13) is the result of a cumulant expansion applied to $\langle \exp(i \int_0^t d\tau \delta\omega(\tau) \rangle$ (see eqn. (4.8.22)). Combining the decay of $C_{qq}^{(\pm,0)}(t)$ with the long-time behavior of the integral in eqn. (15.4.13) leads to following expression for the vibrational dephasing time:

$$\frac{1}{T_2} = \frac{\zeta'(\tilde{\omega})}{2\mu} + \frac{g^2 kT}{4\mu^3 \tilde{\omega}^6} \tilde{\gamma}(0), \tag{15.4.14}$$

where $\tilde{\gamma}(0)$ is the Laplace transform of $\gamma(t)$ at $s = 0$, which is also the static friction coefficient. The second term in eqn. (15.4.14) is a consequence of the anharmonicity. Since $1/T_2^*$ is a pure dephasing time, $1/T_1$ is still $\zeta'(\tilde{\omega})/\mu$ to the same order in perturbation theory. Hence, eqn. (15.4.14) implies that $1/T_2 \geq 1/2T_1$, where equality holds for $g = 0$. This inequality between T_1 and T_2 is usually true for anharmonic systems. An analysis by Skinner and coworkers using a higher order in perturbation theory suggested possible violations of this inequality under special circumstances (Budimir and Skinner, 1987; Laird and Skinner, 1991). Further analysis of such violations and potential difficulties with their detection were subsequently discussed by Reichman and Silbey (1996).

15.5 Molecular dynamics with the Langevin equation

Because the Langevin and generalized Langevin equations replace a large number of bath degrees of freedom with the much simpler memory integral and random force terms, simulations based on these equations are convenient and often very useful. They have a much lower computational overhead than a full bath calculation and can, therefore, access much longer time scales. The Langevin equation can also be employed as a simple and efficient thermostatting method for generating the canonical distribution, which is one of the most common applications. The generalized Langevin equation can also be used as a thermostatting method; however, the need to input a dynamic friction kernel $\zeta(t)$ renders the GLE less convenient for this purpose. Under certain conditions, a friction kernel can be generated from a molecular dynamics simulation (Straub *et al.*, 1988; Berne *et al.*, 1990), but some subtleties arise, which we will discuss in Section 15.8. When such a friction kernel is available, the GLE can, to a good approximation, yield the same dynamical properties as the full molecular dynamics calculation. Because of this important property, the GLE has been employed in the development of low-dimensional or "coarse-grained" models derived from fully atomistic potential functions. Employing the GLE helps to ensure that the coarse-grained model can more faithfully reproduce the dynamics of the more detailed model from which it is obtained (Izvekov and Voth, 2006). It has also been employed in the development of molecular dynamics methods (Ceriotti *et al.*, 2010*a*) that mimic quantum effects (Ceriotti *et al.*, 2009; Ganeshan *et al.*, 2013), accelerate path-integral molecular dynamics calculations (Ceriotti *et al.*, 2010*b*; Ceriotti *et al.*, 2011), and circumvent resonances in multiple time-step methods (Morrone *et al.*, 2011).

15.5.1 Numerical integration of the Langevin equation

The Langevin equation can be used in place of the methods discussed in Chapter 4 as a means of generating a canonical distribution. In fact, it is generally preferred by mathematicians, as its stochastic nature means ergodicity of the Langevin equation can be more easily proved for certain cases, while ergodicity of deterministic methods such as Nosé-Hoover chains has never been established. In this section, we will derive a numerical integrator for the Langevin equation, following an approach introduced by Leimkuhler and Matthews (2012), using the Liouville operator formalism employed throughout this book.

We begin by writing the Langevin equation for a single coordinate q and conjugate momentum p in the form

$$\dot{q} = \frac{p}{\mu}$$

$$\dot{p} = F(q) - \gamma p + \sigma \eta(t), \tag{15.5.1}$$

where $\gamma = \zeta_0/\mu$, $\eta(t) = R(t)/\sqrt{2kT\gamma\mu}$, and $\sigma = \sqrt{2kT\gamma\mu}$. Since $\langle R(0)R(t)\rangle = kT\zeta_0\delta(t)$, it follows that $\langle \eta(0)\eta(t)\rangle = \delta(t)$.[2] The Liouville operator corresponding to eqns. (15.5.1) is

$$iL = \frac{p}{\mu}\frac{\partial}{\partial q} + F(q)\frac{\partial}{\partial p} + iL_{\text{OU}} \equiv iL_1 + iL_2 + iL_{\text{OU}}, \tag{15.5.2}$$

where iL_{OU}, known as the *Ornstein-Uhlenbeck* operator, is determined by the second and third terms on the right side of the $\dot{p}$ equation in eqns. (15.5.1). The operator iL_{OU} produces free Brownian motion (see Section 15.3.1).

In order to derive an integrator for the Langevin equation, we apply the Trotter theorem with the following operator factorization scheme for a small time step Δt (see Section 4.12):

$$e^{iL\Delta t} = e^{iL_2\Delta t/2}e^{iL_1\Delta t/2}e^{iL_{\text{OU}}\Delta t}e^{iL_1\Delta t/2}e^{iL_2\Delta t/2}. \tag{15.5.3}$$

The placement of the OU operator between half-step position updates was shown by Leimkuhler and Matthews (2012) and later by Zhang *et al.* (2017, 2019) to have the lowest error of all possible second-order factorizations.[3] Without the middle factor $\exp(iL_{\text{OU}}\Delta t)$, the factorization in eqn. (15.5.3) would be equivalent to the velocity Verlet method and would produce the algorithm in eqn. (3.10.36). (The action of the operators $\exp(iL_1\Delta t/2)$ and $\exp(iL_2\Delta t/2)$ was discussed in Section 3.10.) The new operator in this scheme is $\exp(iL_{\text{OU}}\Delta t)$. At the end of this section, we will show that the action of this operator can be worked out analytically and that it produces the following evolution:

$$e^{iL_{\text{OU}}\Delta t}q = q$$

---

[2]Because $R(t)$ and $\eta(t)$ are stochastic processes that are neither continuous nor anywhere differentiable, it is strictly correct to write eqns. (15.5.1) as equations for finite differentials, as

$$dq = \frac{p}{\mu}dt$$

$$dp = F(q)dt - \gamma p dt + \sigma dw(t),$$

where $\eta(t) = dw/dt$, with $w(t)$ known as the *Wiener process*.

[3]The integrator in eqn. (15.5.3) is often referred to colloquially as the "BAOAB" integrator, an acronym coined by Leimkuhler and Matthews who wrote eqn. (15.5.2) as $iL = iL_{\text{A}} + iL_{\text{B}} + iL_{\text{O}}$. The Trotter factorization pattern in eqn. (15.5.3), therefore, follows the letter pattern $\text{B} \cdot \text{A} \cdot \text{O} \cdot \text{A} \cdot \text{B}$.

$$e^{iL_{OU}\Delta t}p = pe^{-\gamma\Delta t} + \sigma\sqrt{\frac{1 - e^{-2\gamma\Delta t}}{2\gamma}}r, \tag{15.5.4}$$

where r is a Gaussian random number of zero mean and unit width (see Section 3.8.3). If we take eqn. (15.5.4) as given and apply the direct translation technique, then the pseudocode in eqn. (3.10.36) would be modified to read as follows:

$p = p + 0.5 * \Delta t * F$

$q = q + \Delta t * p/2\mu$

$p = p * \exp(-\gamma * \Delta t) + r * \sigma * \text{sqrt}((1 - \exp(-2 * \gamma * \Delta t))/(2 * \gamma))$

$q = q + \Delta t * p/2\mu$

Recalculate the force

$p = p + 0.5 * \Delta t * F.$ \hfill (15.5.5)

As an application of the algorithm in eqn. (15.5.5), we calculate the trajectory, phase space, and position distribution functions of a harmonic oscillator $W(q) = \mu\omega^2 q^2/2$ with $\mu = 1$, $\omega = 1$, and $kT = 1$ for $\gamma = 0.5$ and $\gamma = 8$; the results are shown in Fig. 15.3. The Langevin equation is integrated for 10^8 steps with a time step of $\Delta t = 0.01$. Figure 15.3 shows how the trajectory changes between $\gamma = 0.5$ and

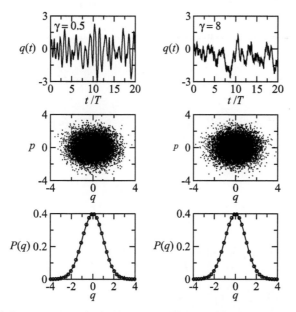

Fig. 15.3 (*Top*) Trajectories of a harmonic oscillator with $\mu = 1$, $\omega = 1$, and $kT = 1$ coupled to a bath via the Langevin equation for $\gamma = 0.5$ (*left*) and $\gamma = 8$ (*right*). Here, $T = 2\pi/\omega$ is the period of the oscillator. (*Middle*) Phase space Poincaré sections. (*Bottom*) Position probability distribution functions.

$\gamma = 8$. Despite different values of the damping constant, the computed distribution functions agree with the analytical distributions. Note that values too large or too small for γ lead to distortions in the probability distribution. In the present example, values of γ less than 10^{-3} or greater than 100 lead to such distortions.

We now present an approach for deriving eqn. (15.5.4). Recall that the operator $\exp(iL_{\text{OU}}t)$ produces free Brownian motion, which was discussed in Section 15.3.1. In principle, the operator should produce the evolution given in eqn. (15.3.5). That is, if we set $t = \Delta t$ and express eqn. (15.3.5) in terms of momentum, we would write

$$e^{iL_{\text{OU}}\Delta t}p = pe^{-\gamma\Delta t} + \sigma \int_0^{\Delta t} d\tau\, \eta(\tau)e^{-\gamma(\Delta t - \tau)}. \tag{15.5.6}$$

Although the integral in eqn. (15.5.6) over the stochastic process $\eta(t)$ is defined, evaluating it is not straightforward and requires a mathematical framework known as the stochastic *Itô calculus* (named for Japanese mathematician Kiyosi Itô, 1915–2008). Since a presentation of the Itô calculus lies beyond the scope of this book, we will present a more heuristic derivation that nevertheless draws on its ideas.

Since the Langevin equation assumes a Gaussian random process, we take

$$p(\Delta t) = pe^{-\gamma\Delta t} + r\sigma(\Delta t), \tag{15.5.7}$$

where r is a Gaussian random number of zero mean and unit width and $\sigma(\Delta t)$ is a time-step dependent width to be determined such that eqn. (15.5.7) is consistent with eqn. (15.5.6) in a statistical sense. By construction, $\sigma(\Delta t)$ is determined from $\langle p^2(\Delta t)\rangle - \langle p(\Delta t)\rangle^2 = \sigma^2(\Delta t)$. Since $\eta(t)$ satisfies $\langle \eta(t)\rangle = 0$ and $\langle \eta(t)\eta(t')\rangle = \delta(t - t')$, let us use eqn. (15.5.6) to compute the averages $\langle p^2(\Delta t)\rangle$ and $\langle p(\Delta t)\rangle$, assuming a large ensemble of trajectories all starting from the same initial momentum p. We then obtain

$$\langle p(\Delta t)\rangle = pe^{-\gamma\Delta t} + \sigma \int_0^{\Delta t} d\tau\, e^{-\gamma(\Delta t - \tau)}\langle \eta(\tau)\rangle = pe^{-\gamma\Delta t} \tag{15.5.8}$$

and

$$\langle p^2(\Delta t)\rangle = p^2 e^{-2\gamma\Delta t} + 2\sigma e^{-\gamma\Delta t} \int_0^{\Delta t} d\tau\, e^{-\gamma(\Delta t - \tau)}p\langle \eta(\tau)\rangle$$

$$+ \sigma^2 \int_0^{\Delta t} d\tau \int_0^{\Delta t} d\tau'\, \langle \eta(\tau)\eta(\tau')\rangle e^{-2\gamma\Delta t}e^{\gamma(\tau + \tau')}$$

$$= p^2 e^{-2\gamma\Delta t} + \sigma^2 e^{-2\gamma\Delta t} \int_0^{\Delta t} d\tau \int_0^{\Delta t} d\tau'\, \delta(\tau - \tau')e^{\gamma(\tau + \tau')}$$

$$= p^2 e^{-2\gamma\Delta t} + \sigma^2 e^{-2\gamma\Delta t} \int_0^{\Delta t} d\tau\, e^{2\gamma\tau}$$

$$= p^2 e^{-2\gamma\Delta t} + \frac{\sigma^2}{2\gamma}\left(1 - e^{-2\gamma\Delta t}\right). \tag{15.5.9}$$

Subtracting the square of eqn. (15.5.8) from the last line of eqn. (15.5.9), we obtain

$$\sigma^2(\Delta t) = \frac{\sigma^2}{2\gamma}\left(1 - e^{-2\gamma\Delta t}\right)$$

$$\sigma(\Delta t) = \sigma \sqrt{\frac{1 - e^{-2\gamma\Delta t}}{2\gamma}}, \tag{15.5.10}$$

which, when substituted into eqn. (15.5.7), yields the evolution scheme in eqn. (15.5.4).

In order to employ Langevin dynamics as a thermostat in a molecular dynamics calculation of N particles in three dimensions, each degree of freedom is subject to a friction and a random force. If the same random force is applied to each degree of freedom in a given time step, this would be equivalent to a gentle global thermostat, similar to global Nosé-Hoover chains discussed in Section 4.11. If a more aggressive thermostat, analogous to a "massive" Nosé-Hoover chain, is desired, then each degree of freedom would be subject to its own random force in each step of the simulation, requiring sampling $3N$ Gaussian random numbers in each step.

Finally, consider the overdamped Langevin equation in the form

$$\dot{q} = \frac{1}{\zeta_0} F(q) + \tilde{\sigma}\eta(t), \tag{15.5.11}$$

where $\tilde{\sigma} = \sqrt{2kT/\zeta_0}$. The analogous integrator for the overdamped Langevin equation advances the coordinate q according to Leimkuhler and Matthews (2012)

$$q(t + \Delta t) = q(t) + \frac{\Delta t}{\zeta_0} F(q(t)) + \frac{1}{2}\tilde{\sigma}\sqrt{\Delta t}\left[r_t + r_{t+\Delta t}\right], \tag{15.5.12}$$

where r_t and $r_{t+\Delta t}$ are the Gaussian random numbers from the previous step and the current step, respectively.

15.5.2 The stochastic Nosé-Hoover thermostat

The Langevin equation (15.5.1) has friction and noise applied directly to the physical degree of freedom, leading to a scheme for sampling a canonical distribution. However, it is also possible to combine the stochastic nature of the Langevin equation with the conventional Nosé-Hoover approach of Section 4.9.2 to produce a mild, stochastic canonical sampling scheme that is also capable of approximately preserving dynamical quantities normally generated within the microcanonical ensemble. In order to generate robust canonical sampling, an approach such as this should function as Nosé-Hoover chains do (see Section 4.11) by curing sampling failures due to conservation laws that restrict the accessible phase space while providing an option for less aggressive sampling. An approach that accomplishes this is the *Nosé-Hoover Langevin* scheme of Leimkuhler *et al.* (2009). In their scheme, the Nosé-Hoover chains of Section 4.11 are replaced by a friction and noise term added to a simple Nosé-Hoover thermostat, giving a set of equations of motion for a single coordinate q and momentum p that take the form

$$\dot{q} = \frac{p}{m}$$

$$\dot{p} = F(q) - \frac{\pi}{Q}p$$

$$\dot{\pi} = G(p) - \gamma\pi + \sigma\eta(t), \tag{15.5.13}$$

where $\sigma = \sqrt{2\gamma kTQ}$ and $G(p) = p^2/m - kT$. Note, here, that π is a thermostat momentum variable to which the friction and the stochastic noise term are applied. Since there is no conserved energy associated with these equations of motion, there is no need for a thermostat "coordinate", as was used in the Nosé-Hoover (4.9.2) and Nosé-Hoover chain (4.11) equations. We will not delve into the mechanics of eqns. (15.5.13) except to say that, as with the Nosé-Hoover chain equations, the Nosé-Hoover Langevin approach is able to generate a correct canonical distribution for a simple harmonic oscillator (see Figs. 4.10 and 4.12). Interested readers are referred to Leimkuhler *et al.* (2009) for discussions about sampling distributions and generating dynamical observables using eqns. (15.5.13). Our purpose in introducing these equations here is to make use of a variant of them in the next section, where we discuss the use of a stochastic approach to create a resonance-free multiple time-step molecular dynamics algorithm analogous to that introduced in Section 4.14.

15.5.3   The stochastic isokinetic thermostat

In the previous subsection, we suggested that Nosé-Hoover chains could be replaced by a single stochastic Nosé-Hoover thermostat known as the Nosé-Hoover Langevin scheme. At the same time, in Section 4.14, it was shown that Nosé-Hoover chains can be combined with a set of isokinetic constraints as a means of controlling resonances in multiple time-step molecular dynamics calculations. In this section, we build on these ideas to derive a stochastic isokinetic thermostat also capable of controlling these resonance artifacts. In Section 4.11, we argued that Nosé-Hoover chains could fix the sampling problems associated with the simple Nosé-Hoover thermostat, yet ergodicity for Nosé-Hoover chains cannot be proved. Stochastic approaches, by contrast, render such proofs possible, which is why they are preferred from a mathematical standpoint. Consider eqns. (4.14.8), which couple a physical degree of freedom q and momentum p to a Nosé-Hoover chain via a set of isokinetic constraints. In the stochastic version of this algorithm, we replace the full Nosé-Hoover chain with a Nosé-Hoover Langevin thermostat, leading to equations of motion of the form

$$\dot{q} = \frac{p}{m}$$

$$\dot{p} = F - \alpha p$$

$$\dot{p}_{1,j} = -\frac{p_{2,j}}{Q_2} p_{1,j} - \alpha p_{1,j}, \qquad j = 1, ..., L$$

$$\dot{p}_{2,j} = G(p_{1,j}) - \gamma p_{2,j} + \sigma\eta_j(t), \tag{15.5.14}$$

where $G(p_{1,j}) = p_{1,j}^2/Q_1 - kT$ and $\sigma = \sqrt{2kT\gamma/Q_2}$. As in Section 4.14, the isokinetic constraint condition that defines the coupling takes the form

$$\frac{p^2}{m} + \frac{L}{L+1} \sum_{j=1}^{L} \frac{p_{1,j}^2}{Q_1} = LkT, \tag{15.5.15}$$

which means that the Lagrange multiplier is

$$\alpha = \frac{Fp/m - L/(L+1)\sum_{j=1}^{L} p_{1,j}^2 p_{2,j}/Q_1 Q_1}{p^2/m + L/(L+1)\sum_{j=1}^{L} p_{1,j}^2/Q_1}. \tag{15.5.16}$$

The distribution generated by eqns. (15.5.14) is canonical in the coordinates, as was shown in Section 4.14.

Let us now assume that the force F in eqns. (15.5.14) has fast (f) and slow (s) contributions such that $F = F_f + F_s$. Introducing velocities $v = p/m$ and $v_{k,j} = p_{k,j}/Q_k$, we construct the Liouville operator with terms grouped as follows:

$$iL = iL_q + iL_v^{(f)} + iL_v^{(s)} + iL_N + iL_{OU}, \tag{15.5.17}$$

where

$$iL_q = v\frac{\partial}{\partial q}$$

$$iL_v^{(s)} = \frac{F_s}{m}\left(1 - \frac{mv^2}{LkT}\right)\frac{\partial}{\partial v} - \sum_{j=1}^{L} \frac{F_s v}{LkT} v_{1,j}\frac{\partial}{\partial v_{1,j}}$$

$$iL_v^{(f)} = \frac{F_f}{m}\left(1 - \frac{mv^2}{LkT}\right)\frac{\partial}{\partial v} - \sum_{j=1}^{L} \frac{F_f v}{LkT} v_{1,j}\frac{\partial}{\partial v_{1,j}}$$

$$iL_N = \frac{Q_1}{(L+1)kT}\left[\left(\sum_{j=1}^{L} v_{2,j}v_{1,j}^2\right)\frac{\partial}{\partial v} + \sum_{j=1}^{L}\left(v_{2,j}v_{1,j}^2 - v_{2,j}\right)\frac{\partial}{\partial v_{1,j}}\right]$$

$$+ \sum_{j=1}^{L} + \sum_{j=1}^{L} \frac{G(v_{1,j})}{Q_2}\frac{\partial}{\partial v_{2,j}}. \tag{15.5.18}$$

These operators are not only useful for deriving an integrator; they can be used to construct a proof of ergodicity of the equations of motion as was shown by Leimkuhler *et al.* (2013) using a set of conditions developed by L. Hörmander (known as *Hörmander conditions*) (Hörmander, 1967). We will omit the details here—suffice it to say that the conditions require calculation of many nested commutators of the operators (Leimkuhler *et al.*, 2013). For the purposes of deriving numerical solution algorithms, the decomposition in eqn. (15.5.18) allows us to formulate "end" and "middle" (see Section 4.12) multiple time-step propagators of the form

$$\exp(iL\Delta t) = \exp\left(iL_N\Delta t/2\right)\exp\left(iL_v^{(s)}\Delta t/2\right)$$

$$\times \left[\exp\left(iL_v^{(f)}\delta t/2\right)\exp\left(iL_q\delta t/2\right)\exp\left(iL_{\rm OU}\delta t\right)\exp\left(iL_q\delta t/2\right)\exp\left(iL_v^{(f)}\delta t/2\right)\right]^n$$

$$\times \exp\left(iL_v^{(s)}\Delta t/2\right)\exp\left(iL_N\Delta t/2\right) \tag{15.5.19}$$

and

$$\exp(iL\Delta t) = \exp\left(iL_v^{(s)}\Delta t/2\right)\left[\exp\left(iL_v^{(f)}\delta t/2\right)\exp\left(iL_q\delta t/2\right)\right.$$

$$\times \exp\left(iL_N\delta t/2\right)\exp\left(iL_{\rm OU}\delta t\right)\exp\left(iL_N\delta t/2\right)$$

$$\times \left.\exp\left(iL_q\delta t/2\right)\exp\left(iL_v^{(f)}\delta t/2\right)\right]^n\exp\left(iL_v^{(s)}\Delta t/2\right), \tag{15.5.20}$$

respectively, where, as usual, $\delta t = \Delta t/n$. Note that the integrator in eqn. (15.5.19) is an XO-RESPA algorithm while eqn. (15.5.20) is formulated as an XI-RESPA method (see Section 4.12).

Equations (15.5.19) and (15.5.20) involve operators some of which we have already encountered in our discussions of numerical solvers for equations of motion. The operator $\exp\left(iL_v^{(\alpha)}h/2\right)$, where $\alpha = s, f$, with $h = \Delta t$ or δt, changes both the velocity v and the thermostat elements $v_{1,j}$. The solution procedure for this operator is given in eqns. (4.13.16) through (4.13.20), which, in this instance, yield solutions

$$v(h/2) = \frac{v(0) + (F/m)s(h/2)}{\dot{s}(h/2)}$$

$$v_{1,j}(h/2) = \frac{v_{1,j}(0)}{\dot{s}(h/2)}, \tag{15.5.21}$$

where $v(0)$ and $v_{1,j}(0)$ denote the values these velocities have before the operator is applied. Here,

$$s(t) = \frac{1}{\sqrt{b}}\sinh\left(\sqrt{b}t\right) + \frac{a}{b}\left(\cosh\left(\sqrt{b}t\right) - 1\right), \tag{15.5.22}$$

with $a = Fv(0)/(L\beta^{-1})$ and $b = F^2m/(L\beta^{-1})$. The other new operator is $\exp(iL_Nh/2)$, which also changes v and $v_{1,j}$. For this operator, we make use of the Suzuki-Yoshida factorization approach of eqn. (4.12.16),

$$e^{iL_Nh/2} = \prod_{\alpha=1}^{n_{\rm sy}}\left[e^{iL_Nw_\alpha h/2m}\right]^m, \tag{15.5.23}$$

where w_α are the Suzuki-Yoshida weights of Section 4.12. We further factorize $\exp(iL_Nw_\alpha h/m)$ by decomposing iL_N as $iL_N = iL_{N,1} + iL_{N,2}$, where

$$iL_{N,1} = \frac{Q_1}{(L+1)kT}\left[\left(\sum_{j=1}^{L}v_{2,j}v_{1,j}^2\right)\frac{\partial}{\partial v} + \sum_{j=1}^{L}\left(v_{2,j}v_{1,j}^2 - v_{2,j}\right)\frac{\partial}{\partial v_{1,j}}\right]$$

$$iL_{N,2} = \sum_{j=1}^{L} + \sum_{j=1}^{L} \frac{G(v_{1,j})}{Q_2} \frac{\partial}{\partial v_{2,j}}. \tag{15.5.24}$$

We then factorize $\exp(iL_N w_\alpha h/2m)$ as

$$e^{iL_N w_\alpha h/2m} = e^{iL_{N,2} w_\alpha h/4m} e^{iL_{N,1} w_\alpha h/2m} e^{iL_{N,2} w_\alpha h/4m}. \tag{15.5.25}$$

The operator $\exp(iL_{N,2} w_\alpha h/4m)$ simply produces a translation of $v_{2,j}$. The operator $\exp(iL_{N,1} w_\alpha h/2m)$, however, is more complex and one we have not yet met. Its action is obtained by solving the pair of coupled first-order differential equations

$$\dot{v} = \frac{Q_1 \beta}{(L+1)} \left(\sum_{j=1}^{L} v_{2,j} v_{1,j}^2 \right) v$$

$$\dot{v}_{1,j} = \frac{Q_1 \beta}{(L+1)} \left(\sum_{j=1}^{L} v_{2,j} v_{1,j}^2 \right) v_{1,j} - v_{2,j} v_{1,j} \tag{15.5.26}$$

subject to initial conditions $v(0), v_{1,1}(0), ..., v_{1,L}(0)$, which denote the values of these variables just before the operator is applied. Because these equations are homogeneous, they can be solved by a simple separation. In particular, if we assume a solution for $v_{1,j}(t)$ of the form $u_{1,j}(t) = v_{1,j}(t) \exp(v_{2,j} t)$, then for any pair $u_{1,j}$ and $u_{1,k}$, it can be shown that $\dot{u}_{1,j}/\dot{u}_{1,k} = u_{1,j}/u_{1,k}$, and from this, some simple algebra reveals that

$$v_{1,j}(t) = v_{1,j}(0) H(t) e^{-v_{2,j} t}. \tag{15.5.27}$$

where

$$H(t) = \sqrt{\frac{L\beta^{-1}}{mv^2(0) + \frac{L}{L+1} \sum_{j=1}^{L} Q_1 v_{1,j}^2(0) e^{-2v_{2,j} t}}}. \tag{15.5.28}$$

When these are substituted into the equation for $v(t)$, we find that

$$v(t) = v(0) H(t). \tag{15.5.29}$$

The action of the operator is then obtained by evaluating eqns. (15.5.27) and (15.5.29) at $t = w_\alpha h/2m$ for each w_α in the Suzuki-Yoshida propagation.

The performance of the stochastic isokinetic method with multiple time-stepping is similar to that presented in Sections 4.14 and 4.15, and its ability to allow very large time steps has also been demonstrated for different potential energy models (Margul and Tuckerman, 2016) and for enhanced sampling techniques (see Sections 8.10 and 8.11.1) (Chen and Tuckerman, 2018). Here, we demonstrate that eqns. (15.5.14) can generate correct solvation free energies with very large time steps using the thermodynamic integration techniques discussed in Chapter 8. The solvation free energy of a molecule is a sensitive quantity that constitutes a challenging test of large time-step algorithms. Here, we consider the four polar molecules methanol, phenol, 1,4-dioxane,

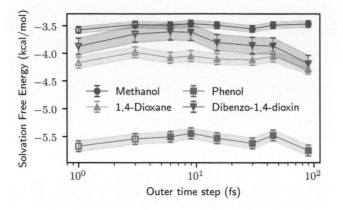

Fig. 15.4 Solvation free energies for methanol (circles), phenol (squares), 1,4-dioxane (triangles up), and dibenzo-1,4-dioxin (triangles down) as a function of the outer time step using eqns. (15.5.14) integrated using a three time-step multiple time-step algorithm (reproduced with permission from Abreu and Tuckerman, *J. Chem. Theor. Comput.* **16**, 7314 (2020), copyright American Chemical Society).

and dibenzo-1,4-dioxin solvated in a bath of 499 water molecules (Abreu and Tuckerman, 2020). Parameters for these molecules are specified by the generalized Amber force field (Wang *et al.*, 2004), while the water molecules are treated using the SPC-Fw model (Wu *et al.*, 2006). In these simulations, van der Waals and Coulomb interactions are switched separately, with the Lennard-Jones van der Waals potential being switched on using the so-called softcore method (see Section 8.2), wherein the pair potential is given by

$$u_{\text{soft}}(r, \lambda) = 4\lambda\epsilon \left[\left(\frac{\sigma^6}{r^6 + \frac{1-\lambda}{2}\sigma^6} \right)^2 - \frac{\sigma^6}{r^6 + \frac{1-\lambda}{2}\sigma^6} \right] \tag{15.5.30}$$

with $\lambda \in [0, 1]$. The electrostatic potential uses a simple linear scaling $u_{ij}(r, \lambda) = \lambda q_i q_j / r_{ij}$. The simulations employ 21 λ points and 11 λ points for the van der Waals and electrostatic potentials, respectively. A three time-step RESPA scheme is used with 0.5 fs for bonded interactions and 3.0 fs for short-range non-bonded interactions within a cutoff of 8 Å. Figure 15.4 shows the hydration free energies for the four molecules as a function of the outer time step for values from 6.0 fs to 90 fs. The figure demonstrates the power of circumventing resonances, as we see only negligible variations in the results at all time steps (Abreu and Tuckerman, 2020).

15.6 Designing memory kernels for specific tasks

Up to this point in the chapter, we have seen that the Langevin and generalized Langevin equations have many uses in both analytical models of dissipative baths and in the development of efficient molecular dynamics algorithms. Over the decades, these equations have been employed to construct rate theories (Kramers, 1940; Grote and Hynes, 1980; Tucker *et al.*, 1991; Berezhkovskii *et al.*, 1992; Pollak and Talkner, 1993) and theories of vibrational and energy relaxation (Tuckerman and Berne, 1993) (see Section 15.4) and dynamical friction (Berne *et al.*, 1990). Generalized Langevin equations have also been employed to create bath models for molecular dynamics simulations (Ceriotti *et al.*, 2010*a*) that mimic quantum effects (Ceriotti *et al.*, 2009; Ganeshan *et al.*, 2013), reduce the bead number in path-integral calculations (Ceriotti *et al.*, 2010*b*; Ceriotti *et al.*, 2011), and stave off the (now familiar) resonances that plague multiple time-step methods (Morrone *et al.*, 2011). In this section, we will develop a general framework for "designer" memory kernels. We will then consider the example of a kernel capable of mollifying resonances in multiple time-step molecular dynamics.

Consider the generalized Langevin equation for a coordinate q and momentum p with associated mass m, which we cast in the form

$$\dot{q} = \frac{p}{m}$$

$$\dot{p} = F(q) - \int_0^t d\tau \, p(\tau)\zeta(t-\tau) + R(t), \tag{15.6.31}$$

for $t \geq 0$ with the dynamic friction kernel given by a generalization of the exponential and δ-function forms in eqns. (15.3.11) and (15.2.22)

$$\zeta(t) = \zeta_0 \delta(t) - \tilde{\boldsymbol{\gamma}}^{\mathsf{T}} e^{-|t|\boldsymbol{\Gamma}} \boldsymbol{\gamma}, \tag{15.6.32}$$

which is defined for $t \in (-\infty, \infty)$. Here, $\boldsymbol{\gamma}$ is a column vector having n components, $\tilde{\boldsymbol{\gamma}}^{\mathsf{T}}$ is an n-component row vector, and $\boldsymbol{\Gamma}$ is an $n \times n$ matrix. For this particular form of friction kernel (see Section 15.3.4 for an analytical treatment of a generalized Langevin equation with exponential memory), the generalized Langevin equation can be written in the form of a standard Langevin equation in an extended phase space with n additional variables $s_1, ..., s_n \equiv \mathbf{s}$. This extended phase-space formulation of the Langevin equation takes the form

$$\dot{q} = \frac{p}{m}$$

$$\begin{pmatrix} \dot{p} \\ \dot{\mathbf{s}} \end{pmatrix} = \begin{pmatrix} F(q) \\ \mathbf{0} \end{pmatrix} - \mathbf{Z} \begin{pmatrix} p \\ \mathbf{s} \end{pmatrix} + \mathbf{B}\boldsymbol{\xi}(t), \tag{15.6.33}$$

where $\boldsymbol{\xi}(t)$ is a vector of Gaussian random noise processes, each having zero mean and unit width, $\mathbf{Z}$ is the $(n+1) \times (n+1)$ matrix

$$\mathbf{Z} = \begin{pmatrix} \zeta_0 & \tilde{\boldsymbol{\gamma}}^{\mathsf{T}} \\ \boldsymbol{\gamma} & \boldsymbol{\Gamma} \end{pmatrix}, \tag{15.6.34}$$

known as the damping matrix or *drift matrix*, and $\mathbf{B}$ is an $(n+1) \times (n+1)$ matrix known as the *diffusion matrix*. The equivalence between eqns. (15.6.33) and (15.6.31) with memory kernel given by eqn. (15.6.32) can be seen by solving eqn. (15.6.33) for $\mathbf{s}(t)$ in terms of $p(t)$ and then substituting the result into eqn. (15.6.31) (see Problem 15.12). The second fluctuation dissipation theorem (cf. eqn. (15.2.17)) allows us to derive a relation between the matrices $\mathbf{B}$ and $\boldsymbol{\zeta}$ from the autocorrelation function of the random force. This connection takes the form of a matrix equation

$$mkT\left(\mathbf{Z} + \mathbf{Z}^{\mathsf{T}}\right) = \mathbf{B}^{\mathsf{T}}\mathbf{B}. \tag{15.6.35}$$

Equation (15.6.35) implies that $\mathbf{Z} + \mathbf{Z}^{\mathsf{T}}$, which is the symmetric part of $\mathbf{Z}$, must be positive-definite. Since eqn. (15.6.33) is an ordinary Langevin equation in the extended phase space, it can be integrated using the methodology of Section 15.5.1, yet it encodes the memory of eqn. (15.6.32), which can be chosen *a priori* to model a colored-noise bath with features that influence a simulation in a desired manner. In other words, eqn. (15.6.33) allows us to design memory kernels for performing specific types of simulations while retaining the simplicity of the Langevin equation.

Because the elements of $\mathbf{Z}$ are contained in the general exponential memory kernel of eqn. (15.6.32), designing a kernel for a specific task is tantamount to customizing the drift matrix $\mathbf{Z}$. From the framework presented thus far, we can glean some important restrictions on possible choices of $\mathbf{Z}$. From eqn. (15.6.32), it is clear that the Z_{11} element is simply the white-noise term and that the time dependence, hence the memory component, is encoded in the matrix $\boldsymbol{\Gamma}$. Given this, it is important to note that the eqn. (15.6.33) will only be able to recover an equilibrium ensemble as $t \to \infty$ if $K(t)$ has a finite decay time, and this, therefore, requires that the real part of the eigenvalues of $\boldsymbol{\Gamma}$ be positive. Choosing drift matrices that satisfy requirements such as these can be nontrivial, and each choice should be carefully tested for compliance with eqn. (15.6.35) and its ability to produce stable dynamics.

With these caveats in mind, let us consider how we might design a kernel to mollify resonances in multiple time-scale molecular dynamics simulations. Recall that resonances occur from the influence of high-frequency motion (integrated using the smallest time step) on intermediate and long time scales (integrated using larger time steps). Thus, the key feature that a bath model should have is strong coupling to the highest frequencies, leading to strong damping of these frequencies, and weaker coupling and small perturbation of lower frequencies. These considerations suggest that our starting point is identifying the frequency dependence of the memory kernel. That is, we should start with the Fourier transform of $K(t)$:

$$\hat{K}(\omega) = \frac{1}{\sqrt{2\pi}} \int_{-\infty}^{\infty} K(t)e^{-i\omega t}\,\mathrm{d}t. \tag{15.6.36}$$

If we now require that the friction couple strongly to high frequencies and weakly to low frequencies, we can intuit what the general shape of $\hat{K}(\omega)$ should be, and we show this general behavior in Fig. 15.5. Since $\hat{K}(\omega)$ determines the strength of the coupling at a given value of ω, it can be seen that at the extremes, where $|\omega|$ is large, $\hat{K}(\omega)$ is large as well, while near $\omega \approx 0$, $\hat{K}(\omega) \approx 0$. There are many functional forms that could, in principle, produce the behavior in Fig. 15.5; however, we need a functional form that is consistent with the Fourier transform of eqn. (15.6.32), and since eqn.

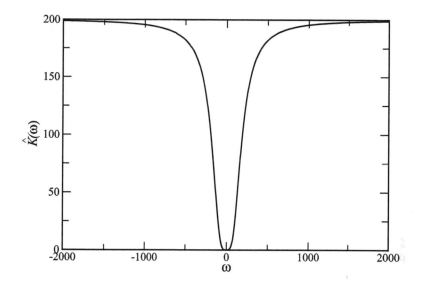

Fig. 15.5 Desired behavior of the Fourier transform of the memory kernel for mollifying resonances.

(15.6.32) is a sum of simple exponentials, $\hat{K}(\omega)$ should be composed of Lorentzian functions. For example, consider $\hat{K}(\omega)$ in the form

$$\hat{K}(\omega) = \sqrt{2\pi}\zeta_0 - \frac{\zeta_0}{2}\left[\frac{\tilde{\omega}^2}{\frac{3}{4}\tilde{\omega}^2 + \left(\omega - \frac{\tilde{\omega}}{2}\right)^2} + \frac{\tilde{\omega}^2}{\frac{3}{4}\tilde{\omega}^2 + \left(\omega + \frac{\tilde{\omega}}{2}\right)^2}\right]. \tag{15.6.37}$$

This kernel is a sum of two Lorentzian functions centered at $\omega = \pm\tilde{\omega}/2$ subtracted from a constant $\sqrt{2\pi}\zeta_0$. The memory kernel $K(t)$ to which this corresponds is

$$K(t) = \zeta_0 - \sqrt{\frac{2\pi}{3}}\tilde{\omega}\zeta_0 e^{-\sqrt{3}\tilde{\omega}|t|/2}\cos\left(\tilde{\omega}t/2\right). \tag{15.6.38}$$

Given that $K(t)$ has two exponentials (which becomes clear when $\cos(\tilde{\omega}t/2)$ is written as $(\exp(i\tilde{\omega}t/2) + \exp(-i\tilde{\omega}t/2))/2$), we then need $n = 2$ extended variables s_1 and s_2, which means that the drift matrix $\boldsymbol{\Gamma}$ is a 3×3 matrix. The 2×2 submatrix $\boldsymbol{\Gamma}$ must have eigenvalues $(\tilde{\omega}\sqrt{3}/2) \pm (i\tilde{\omega}/2)$, and a 2×2 matrix that gives us these eigenvalues is

$$\boldsymbol{\Gamma} = \begin{pmatrix} \tilde{\omega}\sqrt{3} & \tilde{\omega} \\ -\tilde{\omega} & 0 \end{pmatrix}. \tag{15.6.39}$$

Finally, the vectors $\tilde{\boldsymbol{\gamma}}^{\mathsf{T}}$ and $\boldsymbol{\gamma}$ have components proportional to $(\zeta_0\tilde{\omega})^{1/2}$, i.e.,

$$\tilde{\boldsymbol{\gamma}} = (\zeta_0\tilde{\omega})^{1/2}\begin{pmatrix} \alpha \\ \beta \end{pmatrix}, \qquad \boldsymbol{\gamma} = (\zeta_0\tilde{\omega})^{1/2}\begin{pmatrix} \alpha \\ -\beta \end{pmatrix}, \tag{15.6.40}$$

where α and β are simple numerical constants.

As an example, we consider a flexible model of water (Wu *et al.*, 2006), for which the colored-noise Langevin scheme is used to compute the diffusion constant employing a three time-step multiple time-stepping scheme. Since diffusion is dominated by low-frequency motion, the strong coupling between high-frequency modes to the bath should allow a large outer time step without affecting the diffusion constant. In this scheme, a time step of 0.5 fs is used for bonded interactions, non-bonded interactions within 9 Å are integrated with a 2 fs time step, and the outer time step is varied. The correct diffusion constant for the model is 0.25 Å^2/ps. When the colored-noise scheme is used with outer time steps of 12 fs, 16 fs, and 20 fs, the resulting diffusion constants are 0.24 Å^2/ps, 0.20 Å^2/ps, and 0.16 Å^2/ps, respectively (Morrone *et al.*, 2011). By contrast, when a simple white-noise Langevin scheme is used with a friction value of 1 ps^{-1}, the resulting diffusion constants for outer time steps of 12 fs, 16 fs, and 20 fs are 0.095 Å^2/ps, 0.044 Å^2/ps, and 0.019 Å^2/ps, respectively. This example illustrates the efficacy of the colored-noise method at mollifying resonances, allowing an outer time step of 12 fs to be achieved with minimal perturbation to the rate of diffusion.

15.7 Sampling stochastic transition paths

The numerical integration algorithm of Section 15.5.1 can be used in conjunction with the transition path sampling approach of Section 7.7 to sample a transition path ensemble of stochastic paths from a region $\mathcal{A}$ of phase space to another region $\mathcal{B}$. As noted in Section 7.7, the shooting algorithm is an effective method for generating trial moves from a path $Y(t)$ to a new path $X(t)$. Here, we describe a simple variant of the shooting algorithm for paths satisfying the Langevin equation.

In fact, the shooting algorithm can be applied almost unchanged from that described in Section 7.7. However, a few differences need to be pointed out. First, the random force in the Langevin equation is not deterministic, which means that a rule such as that given in eqn. (7.7.2) for a numerical solver such as velocity Verlet cannot be used for Langevin dynamics. Rather, we need to account for the fact that a distribution of $q(t + \Delta t)$ values can be generated from $q(t)$ due to the random force. The solver in eqn. (15.5.5) can be expressed compactly as

$$x_{(k+1)\Delta t} = x_{k\Delta t} + \delta x_d + \delta x_r, \qquad (15.7.41)$$

where the displacement δx_d is purely deterministic, and δx_r is due to the random force. If we take the random force to be a Gaussian random variable, then we can state the rule for generating trial moves in phase space from $x_{k\Delta t}$ to $x_{(k+1)\Delta t}$ as

$$T(x_{(k+1)\Delta t}|x_{k\Delta t}) = w(\delta x_r), \qquad (15.7.42)$$

where $w(x)$ is a Gaussian distribution of width determined by the friction (Chandrasekhar, 1943). In the high friction limit, eqn. (15.7.42) can be shown to be

$$T\left(q((k+1)\Delta t)|q(k\Delta t)\right) =$$

$$\sqrt{\frac{\mu\gamma}{4\pi kT\Delta t}}\exp\left[-\frac{\mu\gamma}{4kT\Delta t}\left(q((k+1)\Delta t)-q(k\Delta t)\right)^2\right] \qquad (15.7.43)$$

for a single degree of freedom q (Dellago *et al.*, 2002). Note that as $\Delta t \to 0$, the Gaussian distribution tends to a Dirac δ-function, as expected (see eqn. (A.5) in Appendix A).

The second difference in the shooting algorithm is in the choice of the shooting point. Because the trajectories are stochastic, we are free to choose a shooting point to lie on the old trajectory $Y(t)$ without modification because a trajectory launched from this point will be different from $Y(t)$. Thus, given a stochastic transition path $Y(t)$ and a randomly chosen point $y_{j\Delta t}$ on this path, we can take the rule for generating the new shooting point $x_{j\Delta t}$ to be

$$\tau(x_{j\Delta t}|y_{j\Delta t}) = \delta(x_{j\Delta t} - y_{j\Delta t}). \qquad (15.7.44)$$

Third, because the Langevin equation acts as a thermostatting mechanism, the distributions of initial conditions $f(x_0)$ and $f(y_0)$ will be canonical by construction. Thus, if canonical sampling is sought, then there is no need to apply the acceptance rule $\min[1, f(x_0)/f(y_0)]$ for each trial path move. Putting this fact together with eqn. (15.7.44) and the Gaussian form of eqn. (15.7.42), which is symmetric, gives a particularly simple acceptance criterion from eqn. (7.7.12)

$$\Lambda[X(t)|Y(t)] = h_{\mathcal{A}}(x_0)h_{\mathcal{B}}(x_{n\Delta t}). \qquad (15.7.45)$$

Thus, as long as the new path is a proper transition path from $\mathcal{A}$ to $\mathcal{B}$, it is accepted with probability 1. We can now summarize the steps of the shooting algorithm for stochastic paths as follows:

1. Choose an index j randomly on the old trajectory $Y(t)$ and take the shooting point $y_{j\Delta t}$ to be the shooting point $x_{j\Delta t}$.

2. Integrate the equations of motion backwards in time from the shooting point to the initial condition x_0 using a stochastic propagation scheme such as that of eqn. (15.5.5).

3. If the initial condition x_0 is not in the phase-space region $\mathcal{A}$, reject the trial move; otherwise, accept it.

4. Integrate the equations of motion forward in time to generate the final point $x_{n\Delta t}$ using a stochastic propagation scheme such as that of eqn. (15.5.5).

5. If $x_{n\Delta t} \in \mathcal{B}$, accept the trial move, and reject it otherwise.

6. If the path is rejected at steps 3 or 5, then the old trajectory $Y(t)$ is counted again in the calculation of averages over the transition path ensemble. Otherwise, invert the momenta along the backward portion of the path to yield the complete transition path $X(t)$ and replace the old trajectory $Y(t)$ by the new trajectory $X(t)$.

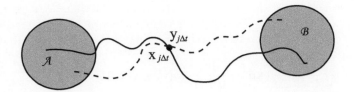

Fig. 15.6 The shooting algorithm for stochastic paths (see also Fig. 7.7)

The shooting algorithm for stochastic paths is illustrated in Fig. 15.6. One final difference between the present shooting algorithm and that for deterministic molecular dynamics is that for stochastic trajectories, it is not necessary to generate both forward and backward segments from every shooting point. A stochastic path of higher statistical weight in the transition path ensemble can be obtained by integrating only backward in time and retaining the forward part of the old trajectory or vice versa. Of course, when this is done, the difference between old and new paths is smaller and sampling becomes less efficient. However, some fraction of the shooting moves can be of this type in order to give a higher average acceptance rate.

15.8   Mori-Zwanzig theory

Our original derivation in Section 15.2 of the generalized Langevin equation was based on the introduction of a harmonic bath as a model for a true bath. While conceptually simple, such a derivation naturally raises the question of whether a GLE can be derived in a more general way for an arbitrary bath. The *Mori-Zwanzig theory* (Mori, 1965; Zwanzig, 1973) achieves this and gives us deeper physical insight into the quantities that appear in the GLE (Deutsch and Silbey, 1971; Berne, 1971; Berne and Pecora, 1976).

The Mori-Zwanzig theory begins with the full classical Hamiltonian and effectively "integrates out" the bath degrees of freedom by using a formalism known as the *projection operator* method (Kubo *et al.*, 1985). In this approach, we divide the full set of degrees of freedom into the system and the bath, as was done for the harmonic bath Hamiltonian. In the phase space, we consider the two axes corresponding to the system coordinate q and its conjugate momentum p, and the remaining $6N - 2$ axes orthogonal to the system. In order to make this phase-space picture concrete, let us introduce a two-component system vector

$$\mathbf{A} = \begin{pmatrix} q \\ p \end{pmatrix}. \tag{15.8.1}$$

Geometrically, recall that the projection of a vector $\mathbf{b}$ along the direction of another vector $\mathbf{a}$ is given by the formula

$$P\mathbf{b} = \left(\mathbf{b} \cdot \frac{\mathbf{a}}{|\mathbf{a}|} \right) \frac{\mathbf{a}}{|\mathbf{a}|}. \tag{15.8.2}$$

Here, P is an operator that gives the component of $\mathbf{b}$ along the direction of $\mathbf{a}$, with $\mathbf{a}/|\mathbf{a}|$ the unit vector along the direction of $\mathbf{a}$. An analog of this formula is used to construct projection operators in phase space parallel and perpendicular to the vector **A**. The projection operator must also eliminate or integrate out the bath degrees of freedom. Thus, we define the operator $\mathcal{P}$ that both projects along the direction of **A** and integrates out the bath according to

$$\mathcal{P} = \langle ...\mathbf{A}^\dagger \rangle \langle \mathbf{A}\mathbf{A}^\dagger \rangle^{-1} \mathbf{A}, \tag{15.8.3}$$

where the quantity on which $\mathcal{P}$ acts replaces the dots, and $\langle ... \rangle$ denotes an average over a canonical ensemble. $\mathbf{A}^\dagger$ is the Hermitian conjugate of **A**. The use of Hermitian conjugates is introduced because of the close analogy between the Hilbert space formalism of quantum mechanics and the classical phase-space propagator formalism we will employ in the present derivation (see Section 3.10). The operator that projects along the direction orthogonal to **A** is denoted $\mathcal{Q}$ and is simply $\mathcal{I} - \mathcal{P}$, where $\mathcal{I}$ is the phase-space identity operator. The operators $\mathcal{P}$ and $\mathcal{Q}$ can be shown to be Hermitian operators. Note that this definition of the projection operator is somewhat more general than the simple geometric projector of eqn. (15.8.2) in that the quantities $\langle ...\mathbf{A}^\dagger \rangle$ and $\langle \mathbf{A}\mathbf{A}^\dagger \rangle^{-1}$ are matrices. These are multiplied together and then allowed to act on **A**, ultimately producing another two-component vector. As expected for projection operators, the actions of $\mathcal{P}$ and $\mathcal{Q}$ on the vector **A** are

$$\mathcal{P}\mathbf{A} = \mathbf{A}, \qquad \mathcal{Q}\mathbf{A} = 0. \tag{15.8.4}$$

Since $\mathcal{P}\mathbf{A} = \mathbf{A}$, it follows that $\mathcal{P}^2\mathbf{A} = \mathcal{P}\mathbf{A} = \mathbf{A}$, and $\mathcal{Q}^2\mathbf{A} = -\mathcal{Q}\mathcal{Q}\mathbf{A} = 0 = \mathcal{Q}\mathbf{A}$, which also means that $\mathcal{P}$ and $\mathcal{Q}$ satisfy

$$\mathcal{P}^2 = \mathcal{P}, \qquad \mathcal{Q}^2 = \mathcal{Q}. \tag{15.8.5}$$

This condition is known as *idempotency*, and the operators $\mathcal{P}$ and $\mathcal{Q}$ are referred to as *idempotent* operators.

The projection operators $\mathcal{P}$ and $\mathcal{Q}$ can be used to analyze the dynamics of the system variables **A**. Recall that the time evolution of any quantity in the phase space is determined by the action of the classical propagator $\exp(iLt)$. The vector **A**, therefore, evolves according to

$$\mathbf{A}(t) = \mathrm{e}^{iLt}\mathbf{A}(0). \tag{15.8.6}$$

Differentiating both sides of this relation with respect to time yields

$$\frac{\mathrm{d}\mathbf{A}}{\mathrm{d}t} = \mathrm{e}^{iLt}iL\mathbf{A}(0), \tag{15.8.7}$$

where iL is the classical Liouville operator of Section 3.10 We now use the projection operators to separate this evolution equation for $\mathbf{A}(t)$ into components along $\mathbf{A}(0)$ and orthogonal to $\mathbf{A}(0)$. This is done by inserting the identity operator $\mathcal{I}$ into eqn. (15.8.7) and using the fact that $\mathcal{P} + \mathcal{Q} = \mathcal{I}$, which yields

$$\frac{\mathrm{d}\mathbf{A}}{\mathrm{d}t} = \mathrm{e}^{iLt}(\mathcal{P} + \mathcal{Q})iL\mathbf{A}(0) = \mathrm{e}^{iLt}\mathcal{P}iL\mathbf{A}(0) + \mathrm{e}^{iLt}\mathcal{Q}iL\mathbf{A}(0). \tag{15.8.8}$$

The first term can be evaluated by introducing eqn. (15.8.3) into eqn. (15.8.8):

$$\mathrm{e}^{iLt}\mathcal{P}iL\mathbf{A}(0) = \mathrm{e}^{iLt}\langle iL\mathbf{A}\mathbf{A}^\dagger \rangle \langle \mathbf{A}\mathbf{A}^\dagger \rangle^{-1} \mathbf{A}(0). \tag{15.8.9}$$

The integrations implied by the angular brackets in eqn. (15.8.9) are performed over an ensemble distribution of initial conditions $\mathbf{A}(0)$. The propagator $\exp(iLt)$ can be pulled across the ensemble averages, since the quantity $\langle i L \mathbf{A} \mathbf{A}^\dagger \rangle \langle \mathbf{A} \mathbf{A}^\dagger \rangle^{-1}$ is a matrix independent of the phase-space variables. Thus,

$$e^{iLt} \mathcal{P} i L \mathbf{A}(0) = \langle i L \mathbf{A} \mathbf{A}^\dagger \rangle \langle \mathbf{A} \mathbf{A}^\dagger \rangle^{-1} e^{iLt} \mathbf{A}(0)$$

$$= \langle i L \mathbf{A} \mathbf{A}^\dagger \rangle \langle \mathbf{A} \mathbf{A}^\dagger \rangle^{-1} \mathbf{A}(t) \equiv i \boldsymbol{\Omega} \mathbf{A}(t),$$

where $\boldsymbol{\Omega}$ is a force-constant matrix given by

$$\boldsymbol{\Omega} = \langle L \mathbf{A} \mathbf{A}^\dagger \rangle \langle \mathbf{A} \mathbf{A}^\dagger \rangle^{-1}. \tag{15.8.10}$$

Note that because of the application of the operators $\exp(iLt)$ and $\mathcal{P}$, the first term in eqn. (15.8.8) is effectively linear in $\mathbf{A}(t)$.

In order to evaluate the second term, we start with the trivial identity

$$e^{iLt} = e^{\mathcal{Q} i L t} + e^{iLt} - e^{\mathcal{Q} i L t}. \tag{15.8.11}$$

The operator difference $\exp(iLt) - \exp(\mathcal{Q} i L t)$ appearing in this identity can be evaluated as follows. We first take the Laplace transform of the $\exp(iLt) - \exp(\mathcal{Q} i L t)$:

$$\int_0^\infty dt \, e^{-st} \left[e^{iLt} - e^{\mathcal{Q} i L t} \right] = (s - iL)^{-1} - (s - \mathcal{Q} i L)^{-1}. \tag{15.8.12}$$

Equation (15.8.12) involves the difference of operator inverses. In general, given a generic operator difference of the form $\mathcal{O}_1^{-1} - \mathcal{O}_2^{-1}$, we multiply the first term by the identity operator expressed as $\mathcal{I} = \mathcal{O}_2 \mathcal{O}_2^{-1}$ and the second term by the identity operator expressed as $\mathcal{I} = \mathcal{O}_1^{-1} \mathcal{O}_1$ to yield

$$\mathcal{O}_1^{-1} - \mathcal{O}_2^{-1} = \mathcal{O}_1^{-1} \left(\mathcal{O}_2 - \mathcal{O}_1 \right) \mathcal{O}_2^{-1}. \tag{15.8.13}$$

Applying eqn. (15.8.13) to the difference in eqn. (15.8.12) gives

$$(s - iL)^{-1} - (s - \mathcal{Q} i L)^{-1} = (s - iL)^{-1} \left(s - \mathcal{Q} i L - s + iL \right) (s - \mathcal{Q} i L)^{-1}$$

$$= (s - iL)^{-1} \left(\mathcal{I} - \mathcal{Q} \right) i L \left(s - \mathcal{Q} i L \right)^{-1}$$

$$= (s - iL)^{-1} \mathcal{P} i L \left(s - \mathcal{Q} i L \right)^{-1}, \tag{15.8.14}$$

so that

$$(s - iL)^{-1} = (s - \mathcal{Q} i L)^{-1} + (s - iL)^{-1} \mathcal{P} i L \left(s - \mathcal{Q} i L \right)^{-1}. \tag{15.8.15}$$

Inverting the Laplace transform of both sides, we obtain

$$e^{iLt} = e^{\mathcal{Q} i L t} + \int_0^t d\tau \, e^{iL(t-\tau)} \mathcal{P} i L e^{\mathcal{Q} i L \tau}. \tag{15.8.16}$$

The second term in eqn. (15.8.8) can be evaluated by multiplying eqn. (15.8.16) on the right by $\mathcal{Q}iL\mathbf{A}(0)$ to give

$$e^{iLt}\mathcal{Q}iL\mathbf{A}(0) = e^{\mathcal{Q}iLt}\mathcal{Q}iL\mathbf{A}(0) + \int_0^t d\tau\ e^{iL(t-\tau)}\mathcal{P}iLe^{\mathcal{Q}iL\tau}\mathcal{Q}iL\mathbf{A}(0). \qquad (15.8.17)$$

In the equation of motion $d\mathbf{A}/dt = iL\mathbf{A}$ for $\mathbf{A}(t)$, the vector $iL\mathbf{A}$ drives the evolution. We can therefore think of $iL\mathbf{A}$ as a kind of general force driving the evolution of $\mathbf{A}$, with $iL\mathbf{A}(0)$ being the initial value of this force. Indeed, the second component of this vector is the initial physical force since $\dot{p} = F$ from Newton's second law. The action of $\mathcal{Q}$ on $iL\mathbf{A}(0)$ projects the initial force onto a direction orthogonal to $\mathbf{A}$. The evolution operator $\exp(\mathcal{Q}iLt)$ acts as a classical propagator of a dynamics in which the forces are orthogonal to $\mathbf{A}$. Therefore, $\mathbf{F}(t) \equiv \exp(\mathcal{Q}iLt)\mathcal{Q}iL\mathbf{A}(0)$ is the time evolution of the projected force in this orthogonal subspace. Of course, the propagator $\exp(\mathcal{Q}iLt)$ and the true evolution operator $\exp(iLt)$ do not produce the same time evolution. The dynamics generated by $\exp(\mathcal{Q}iLt)$ is generally not conservative and, therefore, not straightforward to evaluate. However, we will see shortly that physically interesting approximations are available for this dynamics in the case of a high frequency oscillator.

In order to complete the derivation of the GLE, we introduce the projected force $\mathbf{F}(t)$ into eqn. (15.8.17) to obtain

$$e^{iLt}\mathcal{Q}iL\mathbf{A}(0) = \mathbf{F}(t) + \int_0^t d\tau\ e^{iL(t-\tau)}\mathcal{P}iLe^{\mathcal{Q}iL\tau}\mathcal{Q}iL\mathbf{A}(0)$$

$$= \mathbf{F}(t) + \int_0^t d\tau\ e^{iL(t-\tau)}\mathcal{P}iL\mathbf{F}(\tau)$$

$$= \mathbf{F}(t) + \int_0^t d\tau\ e^{iL(t-\tau)}\langle iL\mathbf{F}(\tau)\mathbf{A}^\dagger\rangle\langle\mathbf{A}\mathbf{A}^\dagger\rangle^{-1}\mathbf{A}(0)$$

$$= \mathbf{F}(t) + \int_0^t d\tau\ \langle iL\mathbf{F}(\tau)\mathbf{A}^\dagger\rangle\langle\mathbf{A}\mathbf{A}^\dagger\rangle^{-1}e^{iL(t-\tau)}\mathbf{A}(0)$$

$$= \mathbf{F}(t) + \int_0^t d\tau\ \langle iL\mathbf{F}(\tau)\mathbf{A}^\dagger\rangle\langle\mathbf{A}\mathbf{A}^\dagger\rangle^{-1}\mathbf{A}(t-\tau). \qquad (15.8.18)$$

Since $\mathbf{F}(t)$ is orthogonal to $\mathbf{A}(t)$, it follows that

$$\mathcal{Q}\mathbf{F}(t) = \mathbf{F}(t). \qquad (15.8.19)$$

Equation (15.8.19) can be used to simplify the ensemble average appearing in eqn. (15.8.18). We first express the ensemble average of $iL\mathbf{F}(\tau)\mathbf{A}^\dagger$ as

$$\langle iL\mathbf{F}(\tau)\mathbf{A}^\dagger\rangle = \langle iL\mathcal{Q}\mathbf{F}(\tau)\mathbf{A}^\dagger\rangle. \qquad (15.8.20)$$

We next transfer the operator iL to $\mathbf{A}^\dagger$ by taking the Hermitian conjugate of iL. Recalling that L, itself, is Hermitian, we only need to change i to $-i$ so that

$$\langle iL\mathbf{F}(\tau)\mathbf{A}^\dagger\rangle = -\langle \mathcal{Q}\mathbf{F}(\tau)(iL\mathbf{A})^\dagger\rangle. \tag{15.8.21}$$

Using eqn. (15.8.19) and the fact that $\mathcal{Q}$ is Hermitian, we can write eqn. (15.8.21) as

$$\langle iL\mathbf{F}(\tau)\mathbf{A}^\dagger\rangle = -\langle \mathcal{Q}^2\mathbf{F}(\tau)(iL\mathbf{A})^\dagger\rangle = -\langle \mathcal{Q}\mathbf{F}(\tau)(\mathcal{Q}iL\mathbf{A})^\dagger\rangle = -\langle \mathbf{F}(\tau)\mathbf{F}^\dagger(0)\rangle. \tag{15.8.22}$$

Therefore, eqn. (15.8.18) becomes

$$e^{iLt}\mathcal{Q}iL\mathbf{A} = \mathbf{F}(t) - \int_0^t d\tau \, \langle \mathbf{F}(\tau)\mathbf{F}^\dagger(0)\rangle\langle \mathbf{AA}^\dagger\rangle^{-1}\mathbf{A}(t-\tau). \tag{15.8.23}$$

Finally, combining eqn. (15.8.23) with eqns. (15.8.10) and (15.8.8) gives an equation of motion for $\mathbf{A}$ in which bath degrees of freedom have been effectively eliminated:

$$\frac{d\mathbf{A}}{dt} = i\mathbf{\Omega}\mathbf{A} - \int_0^t d\tau \, \langle \mathbf{F}(\tau)\mathbf{F}^\dagger(0)\rangle\langle \mathbf{AA}^\dagger\rangle^{-1}\mathbf{A}(t-\tau) + \mathbf{F}(t). \tag{15.8.24}$$

Equation (15.8.24) takes the form of a GLE for a harmonic potential of mean force if the autocorrelation function appearing in the integral is identified with the dynamic friction kernel

$$\mathbf{K}(t) = \langle \mathbf{F}(t)\mathbf{F}^\dagger(0)\rangle\langle \mathbf{AA}^\dagger\rangle^{-1}. \tag{15.8.25}$$

The quantity $\mathbf{K}(t)$ is called the *memory function* or *memory kernel*. Note that $\mathbf{K}(t)$ is a matrix. Substituting eqn. (15.8.25) into eqn. (15.8.24) gives the generalized Langevin equation for a general bath

$$\frac{d\mathbf{A}}{dt} = i\mathbf{\Omega}\mathbf{A}(t) - \int_0^t d\tau \, \mathbf{K}(\tau)\mathbf{A}(t-\tau) + \mathbf{F}(t). \tag{15.8.26}$$

Although eqn. (15.8.26) is formally exact, the problem of determining $\mathbf{F}(t)$ and $\mathbf{K}(t)$ is generally more difficult than simply simulating the full system because of the need to generate the orthogonal dynamics of $\exp(\mathcal{Q}iLt)$ (Darve *et al.*, 2009). Taken as a phenomenological theory, however, eqn. (15.8.26) implies that if the potential of mean force is harmonic, and a memory function can be obtained that faithfully represents the dynamics of the full bath, then the GLE will yield accurate dynamical properties of a system.

If we wish to use eqn. (15.8.26) to generate dynamical properties in a low-dimensional subspace of the original system, then several subtleties arise. Consider a one-dimensional harmonic oscillator with coordinate x, momentum p, reduced mass μ, frequency ω, and potential minimum at x_0. For this problem, the force-constant matrix can be shown to be

$$i\mathbf{\Omega} = \begin{pmatrix} 0 & 1/\mu \\ -\mu\tilde{\omega}^2 & 0 \end{pmatrix} \tag{15.8.27}$$

(see Problem 15.9), where $\tilde{\omega}$ is the renormalized frequency, which can be computed using

$$\tilde{\omega}^2 = \frac{kT}{\mu\langle(x-\langle x\rangle)^2\rangle}. \tag{15.8.28}$$

More importantly, the memory kernel $\mathbf{K}(t)$ is not a simple autocorrelation function. A closer look at eqn. (15.8.25) makes clear that the required autocorrelation function is

$$\mathbf{K}(t) = \langle (e^{\mathcal{Q}iLt}\mathbf{F})\, \mathbf{F}^\dagger \rangle \langle \mathbf{A}\mathbf{A}^\dagger \rangle^{-1}, \tag{15.8.29}$$

which requires the orthogonal dynamics generated by $\exp(\mathcal{Q}iLt)$. For this example, it can be shown that

$$\mathbf{K}(t) = \begin{pmatrix} 0 & 0 \\ 0 & \zeta(t)/\mu \end{pmatrix} \tag{15.8.30}$$

and that

$$\mathbf{F}(t) = \begin{pmatrix} 0 \\ \dot{p} + \mu\tilde{\omega}^2 q \end{pmatrix}, \tag{15.8.31}$$

where $q = x - \langle x \rangle$. If we denote the nonzero component of $\mathbf{F}(t)$ as δf, we can express the exact friction kernel as

$$\frac{\zeta(t)}{\mu} = \frac{\langle \delta f e^{\mathcal{Q}iLt} \delta f \rangle}{\langle p^2 \rangle}, \tag{15.8.32}$$

which is nontrivial to evaluate. The standard autocorrelation function

$$\frac{\phi(t)}{\mu} = \frac{\langle \delta f e^{iLt} \delta f \rangle}{\langle p^2 \rangle} \tag{15.8.33}$$

is not equal to the friction kernel. It was shown by Berne *et al.* (1990) that the Laplace transforms of $\zeta(t)$ and $\phi(t)$ are related by

$$\frac{\tilde{\zeta}(s)}{\mu} = \frac{[\tilde{\phi}(s)/\mu]}{1 - \left\{ [s/(s^2 + \tilde{\omega}^2)][\tilde{\phi}(s)/\mu] \right\}}, \tag{15.8.34}$$

from which we can see that $\tilde{\zeta}(s)$ and $\tilde{\phi}(s)$ are equal only in the limit that $s \to 0$. They are also equal when $\tilde{\omega} \to \infty$, and hence, the standard correlation function $\phi(t)$ is a reasonable approximation to $\zeta(t)$ *only* in the high-frequency limit. In fact, Berne *et al.* showed formally that $\zeta(t)$ can be calculated in this limit from the correlation function

$$\frac{\zeta(t)}{\mu} = \frac{\langle \delta f e^{i\bar{L}t} \delta f \rangle}{\langle p^2 \rangle}, \tag{15.8.35}$$

where $i\bar{L}$ is the Liouville operator for a system in which the oscillator coordinate x is fixed at $x = x_0$. Equation (15.8.35) can be used, for example, to approximate the friction on a harmonic diatomic molecule by replacing it with a rigid diatomic in which the distance between the two atoms is constrained to the value of the equilibrium bond length. Importantly, as the frequency of the oscillator decreases, this rigid-bond approximation breaks down, and care is needed to determine when $\phi(t)$ is a good approximation to $\zeta(t)$.

15.9   Problems

15.1.   a. Use the definition of $R(t)$ for the harmonic-bath model in eqn. (15.2.11) to show that $\langle q(0)R(t)\rangle = 0$ and to derive eqn. (15.2.17).

b. When $R(t)$ is taken to be a true random process, it can be expressed as a Fourier series evaluated at M time points $0, \Delta t, 2\Delta t, ..., (M-1)\Delta t$ as

$$R(k\Delta t) = \sum_{j=0}^{M-1} \left[a_j \cos\left(\frac{2\pi jk}{M}\right) + b_j \sin\left(\frac{2\pi jk}{M}\right) \right],$$

where the coefficients are random numbers sampled from a Gaussian distribution

$$P(a_0, ..., a_{M-1}, b_0, ..., b_{M-1}) = \prod_{k=0}^{M-1} \frac{1}{2\pi\sigma_k^2} e^{-(a_k^2 + b_k^2)/2\sigma_k^2}$$

and

$$\sigma_k^2 = \frac{kT}{M} \sum_{j=0}^{M-1} \zeta(j\Delta t) \cos\left(\frac{2\pi jk}{M}\right).$$

Using these definitions, prove that

$$\langle R(0)R(k\Delta t)\rangle = kT\zeta(k\Delta t),$$

where the average is performed over all possible realizations of $R(k\Delta t)$.

Hint: Think carefully about what distinguishes one realization of the random force from another and define the average accordingly.

15.2.   Consider a single harmonic degree of freedom q, having a mass m, that obeys the generalized Langevin equation

$$\ddot{q} = -\omega^2 q - \int_0^t d\tau \dot{q}(\tau)\gamma(t-\tau) + f(t).$$

Here ω is the frequency associated with the motion of q, $\gamma(t) = \zeta(t)/m$, and $f(t) = R(t)/m$, where $R(t)$ and $\zeta(t)$ are the random force and friction kernel, respectively. As a crude model of a rapidly decaying friction kernel, consider a $\gamma(t)$ that is constant over a very short time t_0 and then drops suddenly to 0:

$$\gamma(t) = \gamma_0 \theta(t_0 - t).$$

Here γ_0 is a constant, $\theta(x)$ is the Heaviside step function, $t \geq 0$, and $t_0 \geq 0$.

a. Show that if t_0 is small enough that $\exp(-st_0) \approx 1 - st_0$ for all relevant values of s, then the presence of memory in this system causes the normalized velocity autocorrelation function $C_{vv}(t)$ to oscillate with a frequency less than ω and to decay slowly in time. Determine the decay constant and oscillation frequency of $C_{vv}(t)$.

b. Suppose that we now let t_0 become very large. Show that the velocity autocorrelation no longer decays but oscillates with a frequency different from ω. What is the oscillation frequency of $C_{vv}(t)$?

c. Explain the physical origin of the behavior of the velocity correlation function in the two limits considered in parts a and b in terms of the response of the bath to the system.

15.3. A simple model for electron transfer is defined by a quantum Hamiltonian of the form

$$\hat{\mathcal{H}} = \frac{\varepsilon}{2}\hat{\sigma}_z + \frac{\Delta}{2}\hat{\sigma}_x,$$

where ϵ and Δ are constants and $\hat{\sigma}_x$, $\hat{\sigma}_y$, and $\hat{\sigma}_z$ are the Pauli matrices given by

$$\hat{\sigma}_x = \begin{pmatrix} 0 & 1 \\ 1 & 0 \end{pmatrix} \qquad \hat{\sigma}_y = \begin{pmatrix} 0 & -i \\ i & 0 \end{pmatrix} \qquad \hat{\sigma}_z = \begin{pmatrix} 1 & 0 \\ 0 & -1 \end{pmatrix}.$$

The model assumes that there are only two states for the electron and is thus a simplification of the true electron-transfer problem.

a. In the Heisenberg picture, the operators that describe the observables of a system evolve in time according to the equation of motion

$$\frac{\mathrm{d}\hat{A}}{\mathrm{d}t} = \frac{1}{i\hbar}[\hat{A}, \hat{\mathcal{H}}],$$

where $\hat{A}$ is an arbitrary operator. Write down Heisenberg's equations for $\hat{\sigma}_x$, $\hat{\sigma}_y$, and $\hat{\sigma}_z$.

b. Compute the autocorrelation function $\langle \hat{\sigma}_y(0)\hat{\sigma}_y(t)\rangle$ assuming an initial canonical distribution.

c. In order to mimic the effect of an environment, the above two-level system is often coupled to a bath of quantum-mechanical harmonic oscillators for which the Hamiltonian is given by

$$\hat{\mathcal{H}} = \frac{\varepsilon}{2}\hat{\sigma}_z + \frac{\Delta}{2}\hat{\sigma}_x + \sum_\alpha \hbar\omega_\alpha \left[\hat{a}_\alpha^\dagger \hat{a}_\alpha + \frac{1}{2}\right] + \frac{\hbar}{2}\hat{\sigma}_z \sum_\alpha g_\alpha \left(\hat{a}_\alpha^\dagger + \hat{a}_\alpha\right),$$

where α is an index that runs over all of the bath modes, ω_α are the bath frequencies, $\hat{a}_\alpha$ and $\hat{a}_\alpha^\dagger$ are the bath annihilation and creation (lowering and raising) operators, respectively, and g_α are coupling constants. For this Hamiltonian, write down the Heisenberg equations of motion for *all* operators, including the Pauli matrices of the system and the creation and annihilation operators of the bath.

    d. By solving the Heisenberg equations for the bath operators, develop generalized Langevin type equations for the spin operators $\hat{\sigma}_x$ and $\hat{\sigma}_y$.

15.4. a. Derive eqn. (15.4.4).

    *b. Derive eqn. (15.4.6).

*15.5. Consider a single harmonic degree of freedom q, having a mass m, that obeys the generalized Langevin equation

$$\ddot{q} = -\omega^2 q - \int_0^t d\tau \, \dot{q}(\tau)\gamma(t-\tau) + f(t).$$

Here ω is the frequency associated with the motion of q, $\gamma(t) = \zeta(t)/m$, and $f(t) = R(t)/m$, where $R(t)$ and $\zeta(t)$ are the random force and friction kernel, respectively. Suppose $q(0) = 0$.

  a. Show that the velocity autocorrelation function $C_{vv}(t)$ obeys the following integro-differential equation:

$$\frac{d}{dt}C_{vv}(t) = -\int_0^t d\tau \, K(t-\tau)C_{vv}(\tau),$$

where $K(t) = \omega^2 + \gamma(t)$. This equation is one in a class of integro-differential equations known as *Volterra* equations.

  b. Devise a numerical algorithm for extracting the memory function and friction kernel given a velocity autocorrelation function obtained from a molecular dynamics calculation. Note that this requires inversion of the Volterra equation. Discuss any numerical difficulties you expect to arise in the implementation of your algorithm.

  c. Now consider a simple continuous model for the friction kernel $\gamma(t) = \lambda A e^{-\lambda t}$. In what time range would you expect this model to break down physically and why?

  d. Solve the Volterra equation for the velocity autocorrelation function using the simple exponential friction kernel model in part c. Discuss the influence of the parameters A and λ on your solution. In addition, examine the free particle case by taking the limit $\omega \to 0$.

  e. Finally, discretize your velocity autocorrelation function into time steps of size Δt for a given choice of A and λ, and use this discretized $C_{vv}(t)$ to test the algorithm you developed in part b. How well can you recover the exponential friction kernel? How robust is your algorithm to the addition of a little random noise to your discretized $C_{vv}(t)$?

15.6. For a particle obeying eqn. (15.3.21), show that the density of vibrational states is related to the friction kernel by

$$I(\omega) = \frac{\omega^2 \gamma'(\omega)}{[\omega^2 - \tilde{\omega}^2 - \omega\gamma''(\omega)]^2 + [\omega\gamma'(\omega)]^2},$$

where $\gamma'(\omega)$ and $\gamma''(\omega)$ are the real and imaginary parts of $\tilde{\gamma}(i\omega)$ and $I(\omega)$ is the Fourier transform of the velocity autocorrelation function

$$I(\omega) = \frac{1}{2\pi} \int_{-\infty}^{\infty} dt \, \langle \dot{q}(0)\dot{q}(t) \rangle e^{-i\omega t}.$$

In general, the connection between the autocorrelation function of a random process and a spectral density is known as the *Wiener-Khintchine theorem* (Wiener, 1930; Khintchine, 1934; Kubo *et al.*, 1985). Using eqn. (15.2.12), derive the spectral density for a harmonic bath corresponding to $\langle R(0)R(t) \rangle$.

15.7. Write a program to integrate the Langevin equation with a Gaussian random force for a harmonic oscillator with mass $m = 1$, frequency $\omega = 1$, temperature $kT = 1$, and friction $\gamma = 1$, using the integrator of Section 15.5. Verify that the correct momentum and position distribution functions of the canonical ensemble are obtained (see Problem 4.9).

15.8. Consider the adiabatic free-energy dynamics approach of Section 8.10.

    a. Reformulate this technique using a set of coupled Langevin equations with two different temperatures T_q and T for the first n coordinates and remaining $3N - n$ coordinates, respectively.

    b. Write a program to integrate your equations for the example in eqn. (8.10.27) using the parameters given following eqn. (8.10.28) and verify that you are able to generate the analytical free-energy profile in eqn. (8.10.28).

15.9. a. Derive eqns. (15.8.27), (15.8.30), and (15.8.31).

    *b. Derive eqn. (15.8.34).

*15.10. A solution contains a very low concentration of solute molecules (denoted A) and solvent molecules (denoted B). Let the number of solute molecules be N_A with positions $\mathbf{r}_1(t), ..., \mathbf{r}_{N_A}(t)$. We can introduce a phase-space function for the solute concentration $c(\mathbf{r}, t)$ at any point $\mathbf{r}$ in space as

$$c(\mathbf{r}, t) = \sum_{i=1}^{N_A} \delta(\mathbf{r} - \mathbf{r}_i(t)).$$

    a. Assuming the solution is in a cubic periodic box of length L, show that the spatial Fourier transform $\tilde{c}_{\mathbf{k}}(t)$ of $c(\mathbf{r}, t)$ is

$$\tilde{c}_{\mathbf{k}}(t) = \sum_{i=1}^{N_A} e^{-i\mathbf{k} \cdot \mathbf{r}_i(t)},$$

where $\mathbf{k} = 2\pi\mathbf{n}/L$, where $\mathbf{n}$ is a vector of integers.

b. Use the Mori-Zwanzig theory to derive a generalized Langevin equation for $\tilde{c}_\mathbf{k}(t)$ and give the explicit expressions for all terms in the equation.

c. Now consider the correlation function

$$C(\mathbf{k},t) = \frac{\langle \tilde{c}_{-\mathbf{k}}(0)\tilde{c}_\mathbf{k}(t)\rangle}{\langle \tilde{c}_{-\mathbf{k}}\tilde{c}_\mathbf{k}\rangle}.$$

Starting with your generalized Langevin equation of part b, derive an integro-differential equation satisfied by $C(\mathbf{k},t)$.

d. Show that the memory kernel in your equation is at least second-order in the wave vector $\mathbf{k}$.

e. Show that $\exp(\mathcal{Q}iLt) \rightarrow \exp(iLt)$ as $|\mathbf{k}| \rightarrow \infty$.

f. Suppose the memory kernel decays rapidly in time. In this limit, show that the correlation function satisfies an equation of the form

$$\frac{\partial}{\partial t}C(\mathbf{k},t) = -\mathbf{k}\cdot\mathbf{D}\cdot\mathbf{k}C(\mathbf{k},t),$$

where $\mathbf{D}$ is the diffusion tensor. Give an expression for $\mathbf{D}$ in terms of a velocity autocorrelation function.

15.11. Use the roots in eqn. (15.3.38) to invert the Laplace transform in eqn. (15.3.36) and obtain solutions for $q(t)$ and the position correlation function $C_{qq}(t) = \langle q(0)q(t)\rangle$ for a harmonic oscillator with an exponential memory kernel. Examine the different cases in which the roots are all real and two are complex conjugates of each other. What happens to your solutions when the discriminant $\delta = 0$?

15.12. Consider eqn. (15.6.33) without the random force term

$$\dot{q} = \frac{p}{m}$$

$$\begin{pmatrix} \dot{p} \\ \dot{s} \end{pmatrix} = \begin{pmatrix} F(q) \\ 0 \end{pmatrix} - \mathbf{Z}\begin{pmatrix} p \\ s \end{pmatrix}.$$

By solving the equation for $\mathbf{s}(t)$ in terms of $p(t)$, show that eqn. (15.6.31) without $R(t)$ results if $K(t)$ is given by eqn. (15.6.32).

15.13. Derive eqn. (15.6.35).

15.14. Write a program to integrate the overdamped Langevin equation with a Gaussian random force for a harmonic oscillator with mass $m = 1$, frequency $\omega = 1$, temperature $kT = 1$, and friction $\gamma = 1$, using the integrator in eqn. (15.5.12). Verify that the correct momentum and position distribution functions of the canonical ensemble are obtained (see Problem 4.9).

15.15. The technical difference between the d-AFED (Abrams and Tuckerman, 2008) and TAMD (Maragliano and Vanden-Eijnden, 2006) methods is that the former is based on molecular dynamics while the latter is built on the overdamped Langevin equation. That is, the equations of motion of TAMD for a system of N particles and n collective variables $f_\alpha(\mathbf{r})$ are

$$\zeta_0 \dot{\mathbf{r}}_i = \mathbf{F}_i - \sum_{\alpha=1}^{n} \kappa_\alpha \left(f_\alpha(\mathbf{r}) - s_\alpha \right) \frac{\partial f_\alpha}{\partial \mathbf{r}_i} + \sigma \mathbf{W}_i(t)$$

$$\zeta_{0,s} \dot{s}_\alpha = \kappa_\alpha \left(f_\alpha(\mathbf{r}) - s_\alpha \right) + \sigma_s w_\alpha(t) \qquad (15.9.36)$$

(see eqns. (8.10.6)), where ζ_0 and $\zeta_{0,s}$ are the friction coefficients for the physical and extended phase-space variables, respectively, $\mathbf{W}_i(t)$ and $w_\alpha(t)$ are vectors of Gaussian random processes of lengths $3N$ and n, respectively, all having zero mean and unit width, $\sigma = \sqrt{2kT\zeta_0}$, and $\sigma_s = \sqrt{2kT_s\zeta_{0,s}}$. Here, T and T_s are the temperatures of the physical and extended phase-space variables (see Section 8.10 for details). Modify your program from exercise 15.14 for the example in eqn. (8.10.27) using the parameters given following eqn. (8.10.28) and the integrator in eqn. (15.5.12). Verify that you are able to generate the analytical free-energy profile in eqn. (8.10.28).

15.16. *Active matter* is composed of entities that consume energy, allowing them to move, exert forces, and exhibit complex motion. Simulations of active matter particles employ a device in which the energy consumption is modeled as an additional random process in an overdamped Langevin equation model.

a. For a single particle in one dimension subject to a potential $U(x)$, the active-matter equations take the form

$$\gamma \dot{x} = -U'(x) + \alpha f + \sqrt{D_x} \eta$$
$$\dot{f} = -\beta f + \sqrt{D_f} \xi.$$

Here, α, β, D_x, and D_f are constants, and η and ξ are Gaussian random processes of zero–mean and correlations $\langle \eta(0)\eta(t) \rangle = \delta(t)$, and $\langle \xi(0)\xi(t) \rangle = \delta(t)$. From these equations, derive expressions for the correlation functions $\langle x(0)x(t) \rangle$ and $\langle f(0)f(t) \rangle$ if $U(x) = kx^2/2$.

b. In a system of N particles in *two* dimensions, which describes a system of colloidal disks, the equations of motion take the form

$$\dot{\mathbf{r}}_i = v\hat{\mathbf{n}}_i + \mu \sum_{j \neq i} \mathbf{F}_{ij} + \sqrt{2D_T} \boldsymbol{\eta}_i$$

$$\dot{\theta}_i = \sqrt{2D_R}\xi_i$$

where D_T and D_R are translational and rotational diffusion constants, $\mu = D_T/kT$, $\mathbf{F}_{ij}$ are pairwise forces, and $\hat{\mathbf{n}}_i = (\cos\theta_i \sin\theta_i)$. If $\mathbf{F}_{ij}$ comes from a pair potential of the form $U(r_{ij}) = k(2a - r_{ij})^2/2$ for $r_{ij} < 2a$ and $U(r_{ij}) = 0$ for $r_{ij} \geq 2a$, describe qualitatively the motion generated by these equations.

16
Discrete models and critical phenomena

16.1 Phase transitions and critical points

In Section 4.8, we studied the liquid–gas phase transition associated with the van der Waals equation of state. We demonstrated the existence of a critical isotherm and derived the thermodynamic state variables of the liquid–gas critical point. We also showed that the behavior of certain thermodynamic properties is determined by a set of power laws characterized by exponents known as *critical exponents*, and we alluded to the phenomenon of *universality*, which provides a rationalization for the observation that large classes of physically very different systems possess the same critical exponents. In this chapter, we will explore the behavior of systems near their critical points in greater detail.

To begin our discussion, consider a uniform gas of identical particles in a container of volume V. The interactions between the particles are weak, and their motion is primarily driven by the kinetic-energy (free particle) contribution to the Hamiltonian. Collisions, which are overwhelmingly dominated by two-body interactions, are infrequent. If we assume that the collisions are approximately elastic, then each colliding particle merely changes its direction of motion. The fact that an interaction event determines when and where a particle's next collision will occur is not particularly important because the frequency of collision events is small. In this case, we speak of a lack of *correlation* between collision events.

If the gas is now compressed at a given temperature so that both the pressure and the density are increased, the interactions between the particles become more important, at least locally. Formation of small, short-lived clusters might occur, due to cooperative interactions that have a greater influence on the system than merely changing the direction of a particle's motion. Collision events now exhibit short-range correlations with each other, which leads to the formation of such local structures.

If the system is compressed even further, a change of phase or *phase transition* occurs in which the gas becomes a liquid. Although phase transitions are an everyday phenomenon, their underlying microscopic details are fascinating and merit further comment. First, the macroscopic manifestation of a gas-to-liquid phase transition is a discontinuous change in the volume. At the microscopic level, the interparticle interactions give rise to long-range correlations—cooperative effects that cause the gas particles to condense, forming well-defined solvation structures quantifiable through spatial correlation functions such as those in Figs. 4.2 and 4.3.

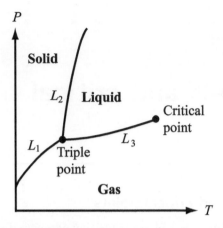

Fig. 16.1 Representative phase diagram of a simple, one-component system. L_1, L_2, and L_3 are the sublimation, melting, and boiling curves, respectively.

Further compression leads to the formation of locally ordered structures that resemble the solid. As the compression continues, long-range order sets in, and a liquid-to-solid phase transition occurs. Such a transition is accompanied by a discontinuous change in the density, although the change is not as dramatic as in the gas-to-liquid case. Other macroscopic observables known as *order parameters* (see Section 16.4) change as well. In order to map out the specific values of pressure and temperature at which the different phases exist, a *phase diagram* is used. A typical phase diagram for a simple, one-component system is shown in Fig. 16.1. In the phase diagram, the lines that separate different phases are called *coexistence curves*. Among these, there is the *melting curve* (L_2) between the liquid and solid phases, the *sublimation curve* (L_1) between the solid and gas phases, and the *boiling curve* (L_3) between the liquid and gas phases. The point at which all three curves meet is called the *triple point*.

In the above discussion, where a constant temperature is assumed, the specific value of the temperature determines whether or not a gas-to-liquid phase transition can occur. If the temperature is too high, then the system cannot exist as a liquid at any pressure. The temperature at which a gas–liquid phase transition just starts is called the *critical temperature*, denoted T_c. The existence of a critical temperature is the reason that the boiling curve in Fig. 16.1 terminates at a definite point, whereas the melting curve, in principle, does not. The point at which the boiling curve terminates is called the *critical point*.

Consider next isotherms of the equation of state for a simple fluid, which are illustrated in Fig. 16.2. For temperatures above T_c, no phase transition occurs, and the isotherms are continuous. In the phase diagram, the region to the right of the critical point is known as the *supercritical fluid* region where the system exhibits both gas-like and liquid-like properties. For temperatures below T_c, one sees a discontinuous change in the volume, signifying the transition from gas to liquid. When a phase transition is characterized by a discontinuous change in an associated thermodynamic observable, the transition is referred to as a *first-order phase transition*. In Fig. 16.2, there is one point labeled C at which the phase transition is characterized by a continuous volume

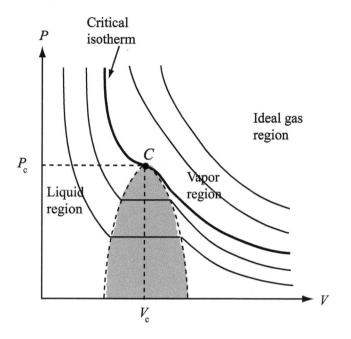

P

Critical
isotherm

Ideal gas
region

P_c

C

Liquid
region

Vapor
region

V

V_c

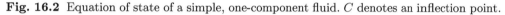

Fig. 16.2 Equation of state of a simple, one-component fluid. C denotes an inflection point.

change. This point, which is an inflection point along the isotherm, corresponds to the critical point on the phase diagram. The isotherm that contains the point C is called the *critical isotherm*. The phase transition that occurs at this point is an example of a *second-order* phase transition. For a one-component system, the critical point is the *only* point at which a second-order phase transition is possible. A two-component system, for example, could have lines of second-order phase transitions, called critical lines. As discussed in Section 4.8, C is a point of zero curvature, meaning that $\partial P/\partial \rho$ and $\partial^2 P/\partial \rho^2$ both vanish at C.

16.2 The critical exponents α, β, γ, and δ

The liquid–gas critical point is characterized by a number of important properties. First, certain thermodynamic variables are observed to diverge as the temperature T approaches the critical temperature T_c; the divergence obeys a power-law form in $|T - T_c|^{-1}$. Other thermodynamic variables are found to exhibit a nondivergent power-law dependence as the critical point is approached in either $|T - T_c|$ or $|\rho - \rho_c|$, where ρ_c is the critical density—the density at which the inflection point C occurs in Fig. 16.2. The exponents that govern the aforementioned power laws are called *critical exponents*. A second important property is the fact that large classes of systems, known as *universality classes*, possess *the same critical exponents*. This phenomenon of universality indicates that near a critical point, the details of the local interactions between specific pairs or clusters of particles become less important than long-range cooperative effects, which are largely insensitive to the particulars of an interaction potential. Thus, it is possible to obtain information about all of the systems in a universality class by studying its physically simplest members.

We start by introducing a set of critical exponents, known as the *primary exponents*, and the properties they characterize. We will illustrate these exponents first using the gas–liquid critical point we have been discussing; however, we will see very shortly see how these definitions carry over to a different type of physical system in the same universality class. The first exponent pertains to the behavior of the constant-volume (or constant-pressure) heat capacity at the critical pressure and density as the critical temperature T_c is approached from above. Recall that the heat capacity at constant volume is $C_V = (\partial E/\partial T)_V = k\beta^2(\partial^2 \ln Q(N,V,T)/\partial \beta^2)_V$. As $T \to T_c$ from above, C_V is observed to diverge with the power-law form

$$C_V \sim |T - T_c|^{-\alpha}. \tag{16.2.1}$$

The critical exponent α, therefore, characterizes the divergence in C_V as $T \to T_c$.

The second exponent γ pertains to the divergence of the isothermal compressibility κ_T, defined as $\kappa_T = -(1/V)(\partial V/\partial P)_T$ (see eqn. (4.8.41)). At the critical pressure and volume, as $T \to T_c$ from above, the isothermal compressibility is observed to diverge following the power law

$$\kappa_T \sim |T - T_c|^{-\gamma}. \tag{16.2.2}$$

The third exponent characterizes the shape of the critical isotherm near the inflection point where the density and pressure approach their critical values ρ_c and P_c. In particular, it is seen that for values of P and ρ near their critical values at $T = T_c$,

$$P - P_c \sim (\rho - \rho_c)^\delta \text{sign}(\rho - \rho_c), \tag{16.2.3}$$

where $\text{sign}(\rho - \rho_c) = (\rho - \rho_c)/|\rho - \rho_c|$ is just the sign of $(\rho - \rho_c)$.

Finally, the fourth exponent refers to the dependence of the difference $\rho_L - \rho_G$ on temperature as T_c is approached from below. Here ρ_L and ρ_G refer to the liquid and gas density values when the discontinuous change occurs. Is it observed that this difference obeys the power law

$$\rho_L - \rho_G \sim |T_c - T|^\beta. \tag{16.2.4}$$

α, β, γ, δ are the primary critical exponents.

16.3 Magnetic systems and the Ising model

The gas–liquid critical point we have been discussing is not the simplest system in its universality class due to the complexity of the interactions and the complicated ensemble distribution they generate. The problem would be simplified considerably if we could restrict the particles to specific points in space, specifically the points of a regular lattice, and use variables that take on discrete values rather than the continuous Cartesian positions $\mathbf{r}_1, ..., \mathbf{r}_N$ that characterize liquids and gases. The only requirements that we place on the simplified system is that it possess a critical point and that it belong to the same universality class.

Fortunately, it is possible to fulfill these requirements by considering the phenomenon of magnetization (the formation of magnetic ordering in a ferromagnetic system). We are interested in the transition from a disordered to an ordered state that occurs either by the application of an external field or spontaneously as the temperature is reduced. The models that are often invoked to describe this process are of a type known as *lattice models*. Since we know that magnetization involves the collective alignment of the quantum spins of particles along a particular spatial axis, the basic lattice model consists of N interacting particles having spins $\hat{\mathbf{S}}_1, ...\hat{\mathbf{S}}_N$ placed at the points of a regular lattice, as illustrated in Fig. 16.3. For simplicity, we consider the

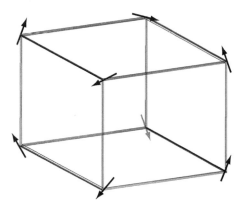

Fig. 16.3 Small part of a cubic spin lattice.

particles to be spin-1/2, so that $\hat{\mathbf{S}}_i = (\hbar/2)\hat{\boldsymbol{\sigma}}_i$, where $\hat{\boldsymbol{\sigma}}$ is the vector of Pauli matrices $\hat{\boldsymbol{\sigma}} = (\hat{\sigma}_x, \hat{\sigma}_y, \hat{\sigma}_z)$ (see Problem 16.3). When diagonalized, each of the Pauli matrices has eigenvalues ± 1. We consider only pairwise interactions among the spins described by a general spin-spin coupling tensor $\mathbf{J}_{ij}$ so that the Hamiltonian for the system in the absence of an external magnetic field is

$$\hat{\mathcal{H}}_0 = -\frac{1}{2} \sum_{i,j} \hat{\boldsymbol{\sigma}}_i \cdot \mathbf{J}_{ij} \cdot \hat{\boldsymbol{\sigma}}_j. \tag{16.3.1}$$

The model in eqn. (16.3.1) is known as the *Heisenberg model*. The '0' subscript reminds us that eqn. (16.3.1) prescribes a field-free Hamiltonian. When a magnetic field $\mathbf{B}$ is applied to the system, the total Hamiltonian takes the form $\hat{\mathcal{H}} = \hat{\mathcal{H}}_0 + \hat{\mathcal{H}}_1$, where $\hat{\mathcal{H}}_1$ is a perturbation due to the field and is given by

$$\hat{\mathcal{H}}_1 = -\sum_i \gamma \mathbf{B} \cdot \hat{\mathbf{S}}_i = -\sum_i \mathbf{h} \cdot \hat{\boldsymbol{\sigma}}_i, \tag{16.3.2}$$

where γ is the spin gyromagnetic ratio and $\mathbf{h} = \gamma \hbar \mathbf{B}/2$ (see Problems 9.2 and 10.3).

As we will discuss in the next section, the Heisenberg and liquid–gas models do not belong to the same universality class. Therefore, we need to consider a simplified version of the model in eqns. (16.3.1) and (16.3.2), which ultimately renders it easier to solve. The simplification we will make is to restrict the spins to lie along only one spatial axis, specifically, the z-axis. With this choice, the spin matrices become $\hat{\boldsymbol{\sigma}}_i = (0, 0, \hat{\sigma}_{zi})$. In addition, since $\hat{\sigma}_z$ is diagonal, we can reduce each spin matrix $\hat{\sigma}_{zi}$ to a scalar variable σ_{zi} that can take values of ± 1 only. At this point, since there is only one relevant spatial axis, we can henceforth drop the "z" label on the spin variables. Finally, we will make the assumption that the spin-spin couplings, which are now simple numbers J_{ij}, couple only nearest neighboring spins. Thus, the Hamiltonian for this idealized situation becomes

$$\mathcal{H} = -\frac{1}{2} \sum_{<i,j>} J_{ij} \sigma_i \sigma_j - h \sum_i \sigma_i, \qquad (16.3.3)$$

where the notation $\sum_{<i,j>}$ signifies that only nearest-neighbor interactions are included in the first term. The $h = 0$ limit of eqn. (16.3.3) gives the unperturbed Hamiltonian

$$\mathcal{H}_0 = -\frac{1}{2} \sum_{<i,j>} J_{ij} \sigma_i \sigma_j. \qquad (16.3.4)$$

The model described by eqn. (16.3.3) is known as the *Ising model*, named after the German physicist Ernst Ising (1900–1998). We have dropped the hat from $\mathcal{H}$ in eqns. (16.3.3) and (16.3.4), signifying that the Ising model is a type of generic discrete model obtainable as idealizations of actual classical or quantum Hamiltonians (see Problems 10.11 and 16.5 for examples of the former). It is worth noting that the Ising model is just one type of discrete spin model. Another discrete spin model, known as the *Potts model*, is specified by a Hamiltonian of the form

$$\mathcal{H} = -\sum_{<i,j>} J_{ij} \delta_{\sigma_i \sigma_j} - h \sum_i \sigma_i, \qquad (16.3.5)$$

where δ_{kl} is the usual Kroenecker delta (see Problem 16.2). A more general type of Potts model replaces the $J_{ij} \delta_{\sigma_i \sigma_j}$ term in the Hamiltonian with a term of the form $J^{ij}_{\sigma_i \sigma_j}$, allowing for a broader range of couplings between spins at sites i and j on the lattice. In fact, such generalizations of both Potts and Ising type Hamiltonians allow these models to be applied to a wide range of discrete systems beyond simple spin lattices, examples of which will be described briefly at the end of this chapter. Although we will focus throughout this chapter on techniques for solving Ising models, the mathematical similarity between the Ising and Potts Hamiltonians means that the methods we will describe can be easily applied to the solution of Potts models.

For the gas–liquid phase transition, where the density ρ was used to distinguish one phase from another, we seek a variable that can distinguish a disordered phase from an ordered, magnetized phase. To this end, we introduce the average total magnetization

$$M = \left\langle \sum_{i=1}^N \sigma_i \right\rangle, \qquad (16.3.6)$$

which is an extensive thermodynamic observable. The equivalent intensive observable is the average magnetization per spin $m = M/N$. In a perfectly ordered state, $m = \pm 1$ and $M = \pm N$, depending on whether the spins are aligned along the positive or

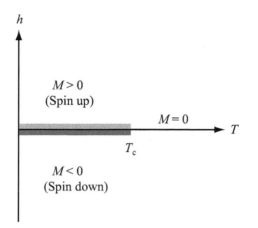

Fig. 16.4 Phase diagram of the Ising model.

negative z-direction. The applied magnetic field h plays the same role as the applied external pressure P in the gas–liquid case and therefore the phase diagram is a plot of the phases in the h-T plane, as shown in Fig. 16.4. As the figure suggests, the uniaxial nature of the Ising model leads to a single line, $h = 0$, along which a first-order phase transition between spin-up and spin-down ordered states can occur. All thermodynamic functions are smooth functions of h and T everywhere else in the phase diagram. The $h = 0$ coexistence line terminates at a critical point, the only point at which the phase transition becomes second-order. For $T > T_c$, the system is in a disordered state for $h = 0$. It could still order in the presence of a finite applied field and, therefore, is paramagnetic. For $T < T_c$, as $h \rightarrow 0$, a finite magnetization persists down to $h = 0$. If $h \rightarrow 0^+$, then the magnetization will be positive, and if $h \rightarrow 0^-$, it will be negative. As this analysis implies, at $h = 0$, the magnetization can be either positive or negative, and indeed, the Ising model exhibits a two-phase coexistence at $h = 0$. Since a plot of m vs. h for $T < T_c$ is an isotherm of the equation of state, such an isotherm shows a discontinuous change in m (see Fig. 16.5). For $T > T_c$, the magnetization vanishes as $h \rightarrow 0$. The critical isotherm shown in Fig. 16.5 separates these two modes of behavior. At $h = 0$, the isotherm has zero curvature, meaning that $\partial h/\partial m = 0$ and $\partial^2 h/\partial m^2 = 0$.

Table 16.1 draws an analogy between the gas–liquid and magnetic cases. It lists the basic thermodynamic variables and thermodynamic relations among these variables, which will be needed throughout our discussion. Accordingly, the primary critical exponents are defined as follows. At $h = 0$ and $m = 0$ (the values of the magnetic field and magnetization at the critical point), as $T \rightarrow T_c$ from above, the heat capacity at constant magnetization C_M diverges as

$$C_M \sim |T - T_c|^{-\alpha}. \tag{16.3.7}$$

Similarly, at $h = 0$ and $m = 0$, as $T \rightarrow T_c$ from above, the susceptibility diverges as

$$\chi \sim |T - T_c|^{-\gamma}. \tag{16.3.8}$$

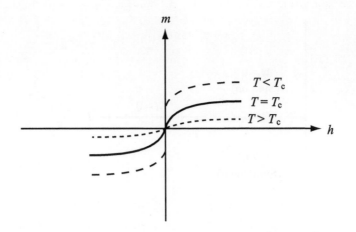

Fig. 16.5 Equation of state of the Ising model.

Table 16.1 Comparison of thermodynamic variables and relations between gas–liquid and magnetic systems.

Quantity	Gas–Liquid	Magnetic	Quantity
Pressure	P	h	Magnetic field
Volume	V	$-M = -Nm$	Magnetization
Isothermal compressibility	$\kappa_T = -(1/V)(\partial P/\partial V)$	$\chi = \partial m/\partial h$	Magnetic susceptibility
Helmholtz free energy	$A(N, V, T)$	$A(N, M, T)$	Helmholtz free energy
Gibbs free energy	$G(N, P, T)$	$G(N, h, T)$	Gibbs free energy
Pressure relation	$P = -\partial A/\partial V$	$h = \partial A/\partial M$	Magnetic field relation
Volume relation	$V = \partial G/\partial P$	$M = -\partial G/\partial h$	Magnetization relation
Const. volume heat capacity	$C_V = -T\left(\partial^2 A/\partial T^2\right)_V$	$C_M = -T\left(\partial^2 A/\partial T^2\right)_M$	Const. magnetization heat capacity
Const. pressure heat capacity	$C_P = -T\left(\partial^2 G/\partial T^2\right)_P$	$C_h = -T\left(\partial^2 G/\partial T^2\right)_h$	Const. field heat capacity

Along the critical isotherm, the behavior of the equation of state near the inflection point is

$$h \sim |m|^{\delta} \text{sign}(m). \qquad (16.3.9)$$

Finally, as $T \to T_c$ from below, the discontinuity in the magnetization depends on temperature according to the power-law

$$m \sim |T_c - T|^{\beta}. \qquad (16.3.10)$$

In this way, a perfect analogy is established between the magnetic and gas–liquid systems. Before proceeding to analyze the magnetic system model, however, we first clarify the concept of universality and provide a definition of universality classes.

16.4 Universality classes

In a perfectly ordered magnetic state, the magnetization per spin m can take one of two values, $m = 1$ or $m = -1$, depending on the direction in which the spins point. In the former case, $\sigma_1 = 1, ..., \sigma_N = 1$, while in the latter $\sigma_1 = -1, ..., \sigma_N = -1$. If we perform a variable transformation

$$\sigma_i' = -\sigma_i, \qquad (16.4.1)$$

which is simply a spin-flip transformation, the magnetization in a perfectly ordered state changes sign. Note that the spin-flip transformation has the same effect as performing a *parity transformation*, in which we let the spatial coordinate $z \to -z$.[1]

Consider next the effect of the transformation in eqn. (16.4.1) on the unperturbed Hamiltonian $\mathcal{H}_0$ of our idealized magnetic model. The unperturbed ($h = 0$) Hamiltonian is

$$\mathcal{H}_0 = -\frac{1}{2} \sum_{<i,j>} J_{ij} \sigma_i \sigma_j. \qquad (16.4.2)$$

If the spins in eqn. (16.4.2) are transformed according to eqn. (16.4.1), the Hamiltonian becomes

$$\mathcal{H}_0' = -\frac{1}{2} \sum_{<i,j>} J_{ij} \sigma_i' \sigma_j', \qquad (16.4.3)$$

which has exactly the same form as eqn. (16.4.2). Thus, the transformation in eqn. (16.4.1) preserves the form of the Hamiltonian. The Hamiltonian $\mathcal{H}_0$ is said to be *invariant* under a spin-flip transformation. This is not unexpected since the spin-flip transformation is equivalent to a parity transformation, which is merely a different choice of coordinates, and physical results should not depend on this choice. Thus, the Hamiltonian $\mathcal{H}_0$ exhibits *parity invariance*.

Readers having some familiarity with the concepts of group theory will recognize that the spin-flip transformation, together with the trivial identity transformation $\sigma_i' = \sigma_i$, form a complete group of transformations with elements $\{1, -1\}$, a group known as Z_2. The Hamiltonian $\mathcal{H}_0$ is invariant under both of the operations of this

---

[1] The general parity transformation is a complete reflection of all three spatial axes, $\mathbf{r} \to -\mathbf{r}$.

group. The magnetization of an ordered state, on the other hand, is not. Based on these notions, we introduce the concept of an *order parameter*, which is needed to define universality classes.

Suppose the unperturbed Hamiltonian $\mathcal{H}_0$ of a system is invariant with respect to all of the transformations of a group $\mathcal{G}$. If two phases can be distinguished by a specific thermodynamic average $\langle \phi \rangle$ (either a classical phase-space average or a quantum trace) that is not invariant under one or more of the transformations of $\mathcal{G}$, then $\langle \phi \rangle$ is called an *order parameter* for the system. Because the magnetization m is not invariant under one of the transformations in Z_2, it can serve as an order parameter for the Ising model with $\mathcal{H}_0$ given by eqn. (16.3.4).

The systems in a universality class are characterized by two parameters: (1) the number d of spatial dimensions in which the system exists, and (2) the dimension n of the order parameter. All systems possessing the same values of d and n belong to the same universality class. Thus, comparing the gas–liquid system and the $h = 0$ Ising model defined by eqn. (16.3.4), it should be clear why these two systems belong to the same universality class. In both cases, the number of spatial dimensions is $d = 3$. (In fact, these models can be defined in any number of dimensions.) In addition, the dimension of the order parameter for each system is $n = 1$, since for both systems, the order parameter (ρ or m) is a simple scalar quantity. Consequently, the idealized magnetic model can be used to determine the critical exponents of the $d = 3$, $n = 1$ universality class. By contrast, in the Heisenberg model of eqn. (16.4.2), the spins can point in any spatial direction, and the order parameter is the magnetization vector

$$\mathbf{M} = \left\langle \sum_{i=1}^{N} \boldsymbol{\sigma}_i \right\rangle, \tag{16.4.4}$$

which has dimension $n = 3$, corresponding to the three components of $\mathbf{M}$. Such a system could be used to determine the exponents of the $d = 3$, $n = 3$ universality class.

Having established the concept of a universality class, we will now proceed to analyze the Ising model in order to gain an understanding of systems in the $d = 3$, $n = 1$ universality class near their critical points.

16.5 Mean-field theory

We begin our treatment of the Ising model by invoking an approximation scheme known as the *mean-field theory*. In this approach, spatial correlations are neglected, and each particle is assumed to experience an "average" or "mean" field due to the other particles in the system. Before examining the magnetic system, let us first note that we previously encountered this approximation in our discussion of the van der Waals equation in Section 4.8. We derived the van der Waals equation of state from a perturbation expansion of the potential energy of a system, and we obtained the approximate Helmholtz free energy of eqn. (4.8.25):

$$A = -\frac{1}{\beta} \ln \left(\frac{Z_N^{(0)}}{N! \lambda^{3N}} \right) + \langle U_1 \rangle_0 - \frac{\beta}{2} \left(\langle U_1^2 \rangle_0 - \langle U_1 \rangle_0^2 \right) + \cdots,$$

where Z_N^0 is the configurational partition function due to the unperturbed potential U_0. Note that the second term in the free energy is just the average of the perturbation U_1, while the third term is the fluctuation in this potential $\langle (U_1 - \langle U_1 \rangle)^2 \rangle$. In the derivation of the van der Waals equation, the fluctuation term was completely neglected. Furthermore, the unperturbed configurational partition function was taken to be that of an ideal gas in a reduced volume. Thus, all of the interactions between particles were assumed to arise from U_1, and the approximation of retaining only the first two terms in the free-energy expression amounted to replacing U_1 with its mean value $\langle U_1 \rangle$ in the configurational partition function, *i.e.*,

$$Z_N = Z_N^{(0)} \langle e^{-\beta U_1} \rangle \approx Z_N^{(0)} e^{-\beta \langle U_1 \rangle} \tag{16.5.1}$$

(cf. eqn. (4.8.4)). Note that the replacement in eqn. (16.5.1) is tantamount to approximating $\langle \exp(-\beta U_1) \rangle$ by its first-order cumulant. The mean-field theory approximation can recover the first two terms in the free energy. Recall that the van der Waals equation, despite its crudeness, predicts a gas-to-liquid phase transition as well as a critical point. The four primary exponents were found in Section 4.8 to be $\alpha = 0$, $\beta = 1/2$, $\gamma = 1$, and $\delta = 3$ within the mean-field approximation. In our discussion of the van der Waals equation, we referred to the fact that the isotherms are unrealistic for $T < T_c$ owing to regions where both P and V increase simultaneously. In Fig. 4.8, the correction to a $T < T_c$ isotherm, which appears as the thin solid straight line, is necessary for the calculation of the exponent β. The location of the thin solid line is determined using a procedure known as the *Maxwell construction*, which states that the areas enclosed above and below the thin line and the isotherm must be equal. Once the isotherms for $T < T_c$ are corrected in this manner, then the exponent β can be calculated (see Problem 16.3).

In order to apply the mean-field approximation to the Ising model, we assume that the system is spatially isotropic. That is, for the spin-spin coupling J_{ij}, we assume $\sum_j J_{ij}$ is independent of the lattice location i. Since the sum in eqn. (16.3.4) is performed over nearest neighbors of i, under the assumption of isotropy, $\sum_j J_{ij} = z\tilde{J}$, where $\tilde{J}$ is a constant and z is the number of nearest neighbors of each spin ($z = 2$ in one dimension, $z = 4$ on a two-dimensional square lattice, $z = 6$ on a three-dimensional simple cubic lattice, $z = 8$ on a three-dimensional body-centered cubic lattice, *etc.*). Absorbing the factor z into the constant $\tilde{J}$, we define $J = z\tilde{J}$.

Next, we consider the Hamiltonian in the presence of an applied magnetic field h:

$$\mathcal{H} = -\frac{1}{2} \sum_{<i,j>} J_{ij} \sigma_i \sigma_j - h \sum_i \sigma_i. \tag{16.5.2}$$

The partition function is given by

$$\Delta(N, h, T) = \sum_{\sigma_1 = \pm 1} \sum_{\sigma_2 = \pm 1} \cdots \sum_{\sigma_N = \pm 1} \exp \left\{ \beta \left[\frac{1}{2} \sum_{<i,j>} J_{ij} \sigma_i \sigma_j + h \sum_i \sigma_i \right] \right\}. \tag{16.5.3}$$

To date, it has not been possible to obtain a closed-form expression for this sum in three dimensions. Thus, to simplify the problem, we write the spin-spin product $\sigma_i \sigma_j$ in terms of the difference of each spin from the magnetization per spin $m = (1/N)\langle \sum_i \sigma_i \rangle$:

$$\sigma_i\sigma_j = (\sigma_i - m + m)(\sigma_j - m + m)$$
$$= m^2 + m(\sigma_i - m) + m(\sigma_j - m) + (\sigma_i - m)(\sigma_j - m). \tag{16.5.4}$$

Since $m \sim \langle\sigma\rangle$, the last term in eqn. (16.5.4) is a fluctuation term, which is neglected in the mean-field approximation. If this term is dropped, then

$$\frac{1}{2}\sum_{<i,j>} J_{ij}\sigma_i\sigma_j \approx \frac{1}{2}\sum_{<i,j>} J_{ij}\left[-m^2 + m(\sigma_i + \sigma_j)\right]$$
$$= -\frac{1}{2}m^2 NJ + Jm\sum_i \sigma_i, \tag{16.5.5}$$

where the assumption of spatial isotropy has been used. Thus, the Hamiltonian reduces to

$$\mathcal{H} = -\frac{1}{2}\sum_{<i,j>} J_{ij}\sigma_i\sigma_j - h\sum_i \sigma_i \approx \frac{1}{2}NJm^2 - (Jm + h)\sum_i \sigma_i, \tag{16.5.6}$$

and the partition function becomes

$$\Delta(N, h, T) \approx \sum_{\sigma_1 = \pm 1}\sum_{\sigma_2 = \pm 1}\cdots\sum_{\sigma_N = \pm 1}\exp\left\{-\beta\left[\frac{1}{2}NJm^2 - (Jm + h)\sum_i \sigma_i\right]\right\}$$

$$= e^{-\beta NJm^2/2}\sum_{\sigma_1 = \pm 1}\sum_{\sigma_2 = \pm 1}\cdots\sum_{\sigma_N = \pm 1}\exp\left[\beta(Jm + h)\sum_i \sigma_i\right]$$

$$= e^{-\beta NJm^2/2}\sum_{\sigma_1 = \pm 1}\exp\left[\beta(Jm + h)\sigma_1\right] \times \cdots \times \sum_{\sigma_N = \pm 1}\exp\left[\beta(Jm + h)\sigma_N\right]$$

$$= e^{-\beta NJm^2/2}\left(\sum_{\sigma = \pm 1} e^{\beta(Jm+h)\sigma}\right)^N$$

$$= e^{-\beta NJm^2/2}\left(e^{\beta(Jm+h)} + e^{-\beta(Jm+h)}\right)^N$$

$$= e^{-\beta NJm^2/2}\left(2\cosh\beta(Jm + h)\right)^N. \tag{16.5.7}$$

From eqn. (16.5.7), the Gibbs free energy $G(N, h, T)$ can be calculated according to

$$G(N, h, T) = -\frac{1}{\beta}\ln\Delta(N, h, T) = \frac{1}{2}NJm^2 - \frac{N}{\beta}\ln\left[2\cosh\beta(Jm + h)\right]. \tag{16.5.8}$$

The average magnetization is $M = -(\partial G/\partial h)$, which means that the average magnetization per spin $m = M/N$ can be expressed as $m = -(\partial g/\partial h)$, where $g(h, T) = G(N, h, T)/N$ is the Gibbs free energy per spin:

$$g(h, T) = \frac{1}{2}Jm^2 - \frac{1}{\beta}\ln\left[2\cosh\beta(Jm + h)\right]. \tag{16.5.9}$$

Thus, the average magnetization per spin is given by

$$m = -\frac{\partial g}{\partial h} = \tanh\beta(Jm + h). \tag{16.5.10}$$

Notice, however, that since m was introduced into the Hamiltonian, the result of this derivative is an implicit relation for m that takes the form of a transcendental equation.

We now ask if an ordered phase exists at zero field. Setting $h = 0$ in eqn. (16.5.10), the transcendental equation becomes $m = \tanh(\beta Jm)$. Of course, $m = 0$ is a trivial solution to this equation; however, we seek solutions for finite m, which we can obtain by solving the equation graphically. That is, we plot the two functions $f_1(m) = m$ and $f_2(m) = \tanh(\beta Jm)$ on the same graph and then look for points at which the two curves intersect for different values of $kT = 1/\beta$. The plot is shown in Fig. 16.6. We

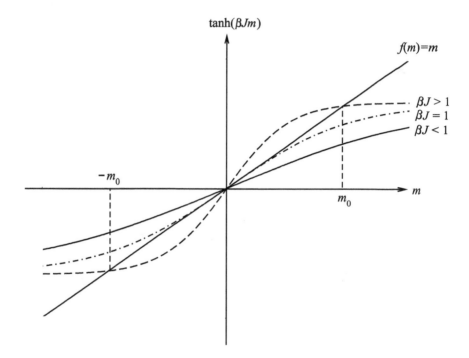

Fig. 16.6 Graphical solution of the transcendental equation $\tanh(\beta Jm) = m$.

see that depending on the value of T, the curves intersect at either three points or one point. Excluding the trivial case $m = 0$ we see that for small enough T ($\beta J > 1$), there are two other solutions, which we label m_0 and $-m_0$. These solutions correspond to magnetizations at zero field aligned along the positive and negative z-axis, respectively. When the temperature is too high, that is, when $\beta J < 1$, only the $m = 0$ solution exists. Since there is no magnetization at zero field, there is no spontaneous ordering. The case $\beta J = 1$ just separates these two regimes and corresponds, therefore, to a critical isotherm. The condition $\beta J = 1$ can be used to determine the critical temperature:

$$\beta J = \frac{J}{kT} = 1 \quad \Rightarrow \quad kT_{\mathrm{c}} = J. \tag{16.5.11}$$

In order to clarify further the behavior of the system near the critical point, consider expanding the free energy about $m = 0$ at zero field for temperatures near the critical temperature. If the expansion is carried out up to quadratic order in m, we obtain

$$g(0, T) \approx c_1 + J(1 - \beta J)m^2 + c_2 m^4 = a(T), \tag{16.5.12}$$

where c_1 and c_2 are constants with $c_2 > 0$. Note that at zero field, the Gibbs free energy per spin becomes the Helmholtz free energy per spin $a(T)$. If $\beta J > 1$, the sign of the quadratic term is negative, and a plot of the free energy as a function of m is shown in Fig. 16.7(a). We can see from the figure that the free energy has two minima at $m = \pm m_0$ and a maximum at $m = 0$, indicating that the ordered states, predicted by solving for the magnetization m, are thermodynamically stable while the disordered state with $m = 0$ is thermodynamically unstable. For $\beta J < 1$, the sign of

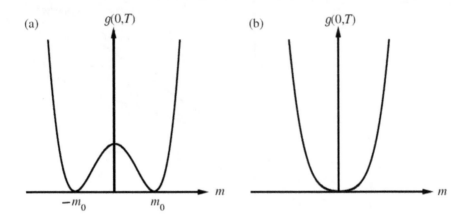

Fig. 16.7 Free energy of eqn. (16.5.12), $g(0, T)$. (a) $T < T_{\mathrm{c}}$. (b) $T > T_{\mathrm{c}}$.

the quadratic term is positive, and the free-energy plot, shown in Fig. 16.7(b), possesses a single minimum at $m = 0$, indicating that there are no solutions corresponding to ordered states.

We now turn to the calculation of the critical exponents for the Ising model within the mean-field theory. From the free-energy plot in Fig. 16.7(a), we can obtain the exponent β directly. Recall that β describes how the discontinuity associated with the first-order phase transition for $T < T_{\mathrm{c}}$ depends on temperature as $T \to T_{\mathrm{c}}$, and for this, we need to know how m_0 depends on T for $T < T_{\mathrm{c}}$. The dependence of m_0 on T is determined by the condition that $g(0, T)$ be a minimum at $m = m_0$:

$$\left. \frac{\partial g(0, T)}{\partial m} \right|_{m=m_0} = 0, \tag{16.5.13}$$

or

$$2J(1 - \beta J)m_0 + 4c_2 m_0^3 = 0$$

$$\frac{2J}{T}\left(T - \frac{J}{k}\right) + 4c_2 m_0^2 = 0$$

$$\frac{2J}{T}(T - T_c) + 4c_2 m_0^2 = 0$$

$$m_0 \sim (T_c - T)^{1/2}, \tag{16.5.14}$$

from which it is clear that $\beta = 1/2$.

In order to determine δ, the mean-field equation of state is needed, which is provided by eqn. (16.5.10). Solving eqn. (16.5.10) for h yields

$$h = kT\tanh^{-1}(m) - mJ. \tag{16.5.15}$$

We are interested in the behavior of eqn. (16.5.15) along the critical isotherm, where $m_0 \to 0$ near the inflection point. Using the expansion $\tanh^{-1}(x) \approx x + x^3/3 + \cdots$ gives

$$h \approx kT\left[m + \frac{m^3}{3}\right] - mJ$$

$$= mk\left(T - \frac{J}{k}\right) + \frac{kT}{3}m^3$$

$$= mk(T - T_c) + \frac{kT}{3}m^3. \tag{16.5.16}$$

Thus, along the critical isotherm, $T = T_c$, we find that $h \sim m^3$, which implies that $\delta = 3$.

In order to calculate γ, we examine the susceptibility χ as $T \to T_c$ from above. By definition,

$$\chi = \frac{\partial m}{\partial h} = \frac{1}{\partial h/\partial m}. \tag{16.5.17}$$

From eqn. (16.5.16),

$$\frac{\partial h}{\partial m} = k(T - T_c) + kTm^2. \tag{16.5.18}$$

For $T > T_c$, $m = 0$, hence $\partial h/\partial m \sim (T - T_c)$. Therefore, $\chi \sim (T - T_c)^{-1}$, from which it is clear that $\gamma = 1$.

Finally, the exponent α is determined by the behavior of the heat capacity C_h as $T \to T_c$ from above. Since C_h is derived from the Gibbs free energy, consider the limit of eqn. (16.5.8) for $T > T_c$, as $m \to 0$ at $h = 0$:

$$G(N, h, T) = -NkT\ln 2. \tag{16.5.19}$$

From this expression, it follows that $C_h = 0$; since, there is no divergence in C_h, we conclude that $\alpha = 0$.

In summary, we find that the mean-field exponents for the magnetic model are $\alpha = 0$, $\beta = 1/2$, $\gamma = 1$, and $\delta = 3$, which are exactly the exponents we obtained for the liquid–gas critical point using the van der Waals equation. Thus, within the mean-field theory approximation, two very different physical models yield the same critical exponents, thus providing a concrete illustration of the universality concept. As noted in Section 4.8, the experimental values of these exponents are $\alpha = 0.1$, $\beta = 0.34$, $\gamma = 1.35$, and $\delta = 4.2$, which shows that mean-field theory is not quantitatively accurate. Qualitatively, however, mean-field theory reveals many important features of critical-point behavior (even if it misses the divergence in the heat capacity) and is, therefore, a useful first approach.

In order to move beyond mean-field theory, we require an approach capable of accounting for the neglected spatial correlations. We will first examine the Ising model in one and two dimensions, where the model can be solved exactly. Following this, we will present an introduction to scaling theory and the renormalization group methodology.

16.6   Ising model in one dimension

Solving the Ising model in one dimension is a relatively straightforward exercise. As we will show, however, the one-dimensional Ising model shows no ordered phases. Why study it then? First, there are classes of problems that can be mapped onto one-dimensional Ising-like models, such as the conformational equilibria of a linear polymer (see Problems 10.9 and 16.5). Second, the mathematical techniques employed to solve the problem are applicable to other types of problems. Third, even the one-dimensional spin system must become ordered at $T = 0$, and therefore, understanding the behavior of the system as $T \to 0$ will be important in our treatment of spin systems via renormalization group methods.

From eqn. (16.5.2), the Hamiltonian for the one-dimensional Ising model is

$$\mathcal{H} = -J \sum_{i=1}^{N} \sigma_i \sigma_{i+1} - h \sum_{i=1}^{N} \sigma_i. \tag{16.6.1}$$

In order to complete the specification of the model, a boundary condition is also needed. Since the variable σ_{N+1} appears in eqn. (16.6.1), it is convenient to impose periodic boundary conditions, which leads to the condition $\sigma_{N+1} = \sigma_1$. The one-dimensional periodic chain is illustrated in Fig. 16.8. Because of the periodicity, the Hamiltonian can be written in a more symmetric manner as

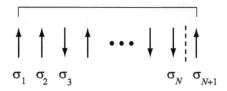

$$\sigma_1 \quad \sigma_2 \quad \sigma_3 \qquad\qquad \sigma_N \quad \sigma_{N+1}$$

Fig. 16.8 One-dimensional Ising system subject to periodic boundary conditions.

$$\mathcal{H} = -J\sum_{i=1}^{N}\sigma_i\sigma_{i+1} - \frac{h}{2}\sum_{i=1}^{N}(\sigma_i + \sigma_{i+1}). \tag{16.6.2}$$

The partition function corresponding to the Hamiltonian in eqn. (16.6.2) is

$$\Delta(N,h,T) = \sum_{\sigma_1=\pm 1}\cdots\sum_{\sigma_N=\pm 1}\exp\left[\beta J\sum_{i=1}^{N}\sigma_i\sigma_{i+1} + \frac{\beta h}{2}\sum_{i=1}^{N}(\sigma_i + \sigma_{i+1})\right]. \tag{16.6.3}$$

Since each spin sum has two terms, the total number of terms represented by the spin sums is 2^N. A powerful method for evaluating the partition function is referred to as the *transfer matrix* method, first introduced by Kramers and Wannier (1941 a, 1941 b). This method recognizes that the partition function can be expressed as a large product of matrices. Consider the matrix P, whose elements are given by

$$\langle\sigma|\mathrm{P}|\sigma'\rangle = e^{\beta J\sigma\sigma' + \beta h(\sigma+\sigma')/2}. \tag{16.6.4}$$

Since σ and σ' can be only 1 or -1, P is a 2×2 matrix with elements given by

$$\begin{aligned}
\langle 1|\mathrm{P}|1\rangle &= e^{\beta(J+h)} \\
\langle -1|\mathrm{P}|-1\rangle &= e^{\beta(J-h)} \\
\langle 1|\mathrm{P}|-1\rangle &= \langle -1|\mathrm{P}|1\rangle = e^{-\beta J}.
\end{aligned} \tag{16.6.5}$$

Written as a matrix, P appears as

$$\mathrm{P} = \begin{pmatrix} e^{\beta(J+h)} & e^{-\beta J} \\ e^{-\beta J} & e^{\beta(J-h)} \end{pmatrix}. \tag{16.6.6}$$

In terms of P, the partition function can be expressed as

$$\Delta(N,h,T) = \sum_{\sigma_1}\cdots\sum_{\sigma_N}\langle\sigma_1|\mathrm{P}|\sigma_2\rangle\langle\sigma_2|\mathrm{P}|\sigma_3\rangle\cdots\langle\sigma_{N-1}|\mathrm{P}|\sigma_N\rangle\langle\sigma_N|\mathrm{P}|\sigma_1\rangle. \tag{16.6.7}$$

Because the partition function is now a matrix product of N factors of P, each sandwiched between spin states with a spin in common, P is known as the *transfer matrix*. Using the completeness of the spin eigenvectors, each factor of the form $\sum_{\sigma_k}|\sigma_k\rangle\langle\sigma_k|$ appearing in eqn. (16.6.7) is an identity operator I, and the sum over N spins can be collapsed to a sum over just one spin σ_1:

$$\Delta(N,h,T) = \sum_{\sigma_1}\langle\sigma_1|\mathrm{P}^N|\sigma_1\rangle$$

$$= \mathrm{Tr}\left(\mathrm{P}^N\right). \tag{16.6.8}$$

Interestingly, in deriving eqn. (16.6.8), we performed the opposite set of operations used in eqns. (12.2.9) and (12.2.11) to derive the Feynman path integral. In the latter, an operator product was expanded by the *introduction* of the identity between factors of the operator.

The simplest way to calculate the trace is to diagonalize P, which yields two eigenvalues λ_1 and λ_2, in terms of which the trace is simply $\lambda_1^N + \lambda_2^N$. The eigenvalues of P are solutions of $\det(P - \lambda I) = 0$, which gives the eigenvalues

$$\lambda = e^{\beta J} \left[\cosh(\beta h) \pm \sqrt{\sinh^2(\beta h) + e^{-4\beta J}}\right]. \tag{16.6.9}$$

We denote these as values $\lambda_\pm$ (instead of $\lambda_{1,2}$), where $\lambda_\pm$ corresponds to the choice of +/- in eqn. (16.6.9). Thus, the partition function becomes

$$\Delta(N, h, T) = \text{Tr}\left[P^N\right] = \lambda_+^N + \lambda_-^N. \tag{16.6.10}$$

Although eqn. (16.6.10) is exact, since $\lambda_+ > \lambda_-$, it follows that for $N \to \infty$, $\lambda_+^N \gg \lambda_-^N$ so that the partition function is accurately approximated using the single eigenvalue λ_+. Thus, $\Delta(N, h, T) \approx \lambda_+^N$, and the free energy per spin is simply

$$g(h, T) = -kT \ln \lambda_+$$
$$= -J - kT \ln \left[\cosh(\beta h) + \sqrt{\sinh^2(\beta h) + e^{-4\beta J}}\right]. \tag{16.6.11}$$

From eqn. (16.6.11), the magnetization per spin can be computed as

$$m = \left(\frac{\partial g}{\partial h}\right) = \frac{\sinh(\beta h) + \sinh(\beta h)\cosh(\beta h)/\sqrt{\sinh^2(\beta h) + e^{-4\beta J}}}{\cosh(\beta h) + \sqrt{\sinh^2(\beta h) + e^{-4\beta J}}}. \tag{16.6.12}$$

As $h \to 0$, the magnetization vanishes, since $\cosh(\beta h) \to 1$ and $\sinh(\beta h) \to 0$. Thus, there is no magnetization at any finite temperature in one dimension, and hence, no nontrivial critical point. Note, however, that as $T \to 0$ ($\beta \to \infty$), the factors $\exp(-4\beta J)$ vanish, and $m \to \pm 1$ as $h \to 0^\pm$. This indicates that an ordered state does exist at absolute zero of temperature. The fact that m tends toward different limits depending on the approach of $h \to 0$ from the positive or negative side indicates that $T = 0$ can be thought of as a critical point, albeit an unphysical one. Indeed, such a result is expected since the entropy vanishes at absolute zero, and consequently an ordered state must exist at $T = 0$. Though unphysical, we will find this critical point useful for illustrative purposes later in our discussion of the renormalization group.

16.7 Ising model in two dimensions

In contrast to the one-dimensional Ising model, which can be solved with a few lines of algebra, the two-dimensional Ising model is a highly nontrivial problem that was first worked out exactly by Lars Onsager (1903–1976) in 1944 (Onsager, 1944). Extensive discussions of the solution of the two-dimensional Ising model can be found in the books by K. Huang (1963) and by R. K. Pathria (1972). Here, we shall give the basic idea behind two approaches to the problem and then present the solution in its final form.

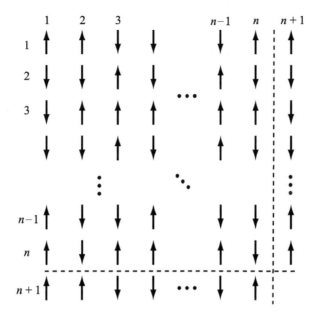

Fig. 16.9 Two-dimensional Ising system subject to periodic boundary conditions.

Transfer matrix approach: The first method follows the transfer matrix approach employed in the previous section for the one-dimensional Ising model. Consider the simple square lattice of spins depicted in Fig. 16.9, in which each row and each column contains n spins, so that $N = n^2$. If i indexes the rows and j indexes the columns, then the Hamiltonian, taking into account the restriction to nearest-neighbor interactions only, can be written as

$$\mathcal{H} = -J\sum_{i=1}^{n}\sum_{j=1}^{n}[\sigma_{i,j}\sigma_{i+1,j} + \sigma_{i,j}\sigma_{i,j+1}] - h\sum_{i=1}^{n}\sum_{j=1}^{n}\sigma_{i,j}. \tag{16.7.1}$$

As in the one-dimensional case, we impose periodic boundary conditions on the square lattice so that the spins satisfy $\sigma_{n+1,j} = \sigma_{1,j}$ and $\sigma_{i,n+1} = \sigma_{i,1}$. The partition function can now be expressed as

$$\Delta(N, h, T) = \sum_{\sigma_{1,1}=\pm1}\cdots\sum_{\sigma_{n,1}=\pm1}\sum_{\sigma_{1,2}=\pm1}\cdots\sum_{\sigma_{n,2}=\pm1}\cdots$$

$$\sum_{\sigma_{1,n}=\pm1}\cdots\sum_{\sigma_{n,n}=\pm1}\exp\left\{\beta J\sum_{i,j=1}^{n}[\sigma_{i,j}\sigma_{i+1,j} + \sigma_{i,j}\sigma_{i,j+1}] + \beta h\sum_{i,j=1}^{n}\sigma_{i,j}\right\}. \tag{16.7.2}$$

As there are $N = n^2$ spin sums, each having two terms, the total number of terms represented by the spin sums is $2^N = 2^{n^2}$.

The form of the Hamiltonian and partition function in eqns. (16.7.1) and (16.7.2) suggests that a matrix multiplication analogous to eqn. (16.6.7) involves entire columns of spins and that the elements of the transfer matrix should be determined by the columns rather than by the single spins of the one-dimensional case. (Note that we could also have used rows of spins and written eqn. (16.7.2) "row-wise" rather than "column-wise.") Let us we define a full column of spins by a variable μ_j

$$\mu_j = \{\sigma_{1,j}, \sigma_{2,j}, ..., \sigma_{n,j}\}\,. \tag{16.7.3}$$

The Hamiltonian can then be conveniently represented in terms of interactions between full columns of spins. We first introduce the two functions

$$E(\mu_j, \mu_k) = J \sum_{i=1}^{n} \sigma_{i,j}\sigma_{i,k}$$

$$\mathcal{E}(\mu_j) = J \sum_{i=1}^{n} \sigma_{i,j}\sigma_{i+1,j} + h \sum_{i,j} \sigma_{i,j} \tag{16.7.4}$$

in terms of which the Hamiltonian becomes

$$\mathcal{H} = -\sum_{j=1}^{n} \left[E(\mu_j, \mu_{j+1}) + \mathcal{E}(\mu_j)\right], \tag{16.7.5}$$

and the partition function can be expressed as

$$\Delta(N, h, T) = \sum_{\mu_1}\sum_{\mu_2}\cdots\sum_{\mu_n} \exp\left\{\beta \sum_{j=1}^{n} \left[E(\mu_j, \mu_{j+1}) + \mathcal{E}(\mu_j)\right]\right\}. \tag{16.7.6}$$

Although eqn. (16.7.6) now resembles the partition function of a one-dimensional Ising model, each sum over μ_j now represents 2^n terms. We can, nevertheless, define a $2^n \times 2^n$ transfer matrix P with elements

$$\langle\mu|P|\mu'\rangle = \exp\left\{\beta\left[E(\mu, \mu') + \frac{1}{2}\left(\mathcal{E}(\mu) + \mathcal{E}(\mu')\right)\right]\right\}, \tag{16.7.7}$$

so that the partition function becomes

$$\Delta(N, h, T) = \sum_{\mu_1}\sum_{\mu_2}\cdots\sum_{\mu_n}\langle\mu_1|P|\mu_2\rangle\langle\mu_2|P|\mu_3\rangle\cdots\langle\mu_n|P|\mu_1\rangle = \mathrm{Tr}\,[P^n]. \tag{16.7.8}$$

The partition function can now be computed from the 2^n eigenvalues of P as

$$\Delta(N, h, T) = \lambda_1^n + \lambda_2^n + \cdots\lambda_{2^n}^n. \tag{16.7.9}$$

As in the one-dimensional case, however, as $N \to \infty$, $n \to \infty$, and the contribution from the largest eigenvalue will dominate. Thus, to a very good approximation, $\Delta(N, h, T) \approx \lambda_{\mathrm{max}}^n$, and the problem of computing the partition function becomes one of simply finding the largest eigenvalue of P.

A detailed mathematical discussion of how the largest eigenvalue of P can be found is given by Huang (1963), which we will not replicate here. We simply quote the final result for the Gibbs free energy per spin at zero field in the thermodynamic limit:

$$g(0,T) = -kT \ln \left[2\cosh(2\beta J) \right] - \frac{kT}{2\pi} \int_0^\pi d\phi \ln \frac{1}{2} \left(1 + \sqrt{1 - K^2 \sin^2 \phi} \right) \quad (16.7.10)$$

(Onsager, 1944; Kaufmann, 1949), where $K = 2/[\cosh(2\beta J)\coth(2\beta J)]$. The integral is the result of taking the thermodynamic limit. In his 1952 paper, C. N. Yang obtained an exact expression for the magnetization at zero field and showed that when $T < T_c$, where T_c is given by

$$2\tanh^2(2J/kT_c) = 1, \qquad kT_c \approx 2.269185J, \qquad (16.7.11)$$

the magnetization is nonzero, indicating that spontaneous magnetization occurs in two dimensions. The magnetization is

$$m = \begin{cases} 0 & T > T_c \\ \left\{ 1 - [\sinh(2\beta J)]^{-4} \right\}^{1/8} & T < T_c \end{cases}. \qquad (16.7.12)$$

In addition to the existence of a spontaneously ordered phase, the heat capacity C_h diverges as $T \to T_c$. The expression for the heat capacity at $h = 0$ near $T = T_c$ is

$$\frac{C_h(T)}{k} = \frac{2}{\pi} \left(\frac{2J}{kT_c} \right)^2 \left[-\ln \left| 1 - \frac{T}{T_c} \right| + \ln \left(\frac{kT_c}{2J} \right) - \left(1 + \frac{\pi}{4} \right) \right], \qquad (16.7.13)$$

which diverges logarithmically. A graph of C_h vs. T is shown in Fig. 16.10. The logarithmic divergence emerges because the model is solved in two rather than three dimensions; in the latter, we would expect a power-law divergence. The other critical exponents can be derived for the two-dimensional Ising model and are $\alpha = 0$ (logarithmic divergence), $\beta = 1/8$, $\gamma = 7/4$, and $\delta = 15$. These are the exact exponents for the $d = 2$, $n = 1$ universality class. To date, the three-dimensional Ising model remains an unsolved problem. Should an exact solution emerge, the exponents for the $d = 3$, $n = 1$ universality class would be known.

Graph-theoretic approach: The second approach is a combinatorial one that leads directly to the partition of the two-dimensional Ising model in the zero-field limit. We begin by introducing the shorthand notation $K = \beta J$, and again we assume periodic boundary conditions. The partition function in the zero-field limit can be written as

$$Q(N,T) = \sum_{\sigma_{1,1}=\pm 1} \sum_{\sigma_{2,1}=\pm 1} \cdots \sum_{\sigma_{n,1}=\pm 1} \sum_{\sigma_{1,2}=\pm 1} \sum_{\sigma_{2,2}=\pm 1} \cdots \sum_{\sigma_{n,2}=\pm 1} \cdots \qquad (16.7.14)$$

$$\times \sum_{\sigma_{1,n}=\pm 1} \sum_{\sigma_{2,n}=\pm 1} \cdots \sum_{\sigma_{n,n}=\pm 1} \exp \left\{ K \sum_{i,j=1}^n \left[\sigma_{i,j}\sigma_{i+1,j} + \sigma_{i,j}\sigma_{i,j+1} \right] \right\}.$$

The combinatorial approach starts with an identity derived from the fact that the product of spins $\sigma_{i,j}\sigma_{i',j'} = \pm 1$ so that $\exp[K\sigma_{i,j}\sigma_{i',j'}] = \exp[\pm K] = \cosh(K) \pm \sinh(K)$. From these relations, it follows that

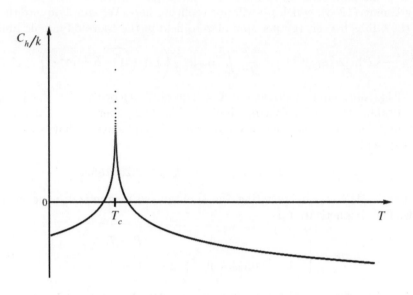

Fig. 16.10 Heat capacity of the two-dimensional Ising model (see eqn. (16.7.13)).

$$\exp\left[K\sigma_{i,j}\sigma_{i',j'}\right] = \cosh(K) + \sigma_{i,j}\sigma_{i',j'}\sinh(K)$$

$$= \cosh(K)\left[1 + \sigma_{i,j}\sigma_{i',j'}\tanh(K)\right]$$

$$= \cosh(K)\left[1 + v\sigma_{i,j}\sigma_{i',j'}\right], \qquad (16.7.15)$$

where $v = \tanh(K)$. Converting exponentiated sums into products, the partition function can be written as

$$Q(N,T) = \sum_{\{\sigma\}=\pm 1} \exp\left\{K\sum_{i,j}\left[\sigma_{i,j}\sigma_{i+1,j} + \sigma_{i,j}\sigma_{i,j+1}\right]\right\}$$

$$= \sum_{\{\sigma\}=\pm 1} \exp\left\{K\sum_{i,j}\sigma_{i,j}\sigma_{i+1,j}\right\}\exp\left\{K\sum_{i,j}\sigma_{i,j}\sigma_{i,j+1}\right\}$$

$$= \sum_{\{\sigma\}=\pm 1}\left[\prod_{i,j}e^{K\sigma_{i,j}\sigma_{i+1,j}}\right]\left[\prod_{i,j}e^{K\sigma_{i,j}\sigma_{i,j+1}}\right]$$

$$= \sum_{\{\sigma\}=\pm 1}\left[\prod_{i,j}\cosh(K)\left(1 + v\sigma_{i,j}\sigma_{i+1,j}\right)\right]\left[\prod_{i,j}\cosh(K)\left(1 + v\sigma_{i,j}\sigma_{i,j+1}\right)\right]$$

$$= [\cosh(K)]^{\nu} \sum_{\{\sigma\}=\pm 1} \prod_{i,j} [1 + v\sigma_{i,j}\sigma_{i+1,j}] [1 + v\sigma_{i,j}\sigma_{i,j+1}], \qquad (16.7.16)$$

where ν is the total number of nearest neighbors on the two-dimensional lattice and the notation $\sum_{\{\sigma\}=\pm 1}$ indicates that all spins are summed over.

Now consider just the first few terms of the product $(i, j = 1, 2)$, which contribute the following factors to the partition function:

$$(1 + v\sigma_{1,1}\sigma_{2,1})(1 + v\sigma_{1,2}\sigma_{2,2})(1 + v\sigma_{1,1}\sigma_{1,2})(1 + v\sigma_{2,1}\sigma_{2,2}).$$

Multiplying this expression out gives

$$1 + v \left[\sigma_{1,1}\sigma_{2,1} + \sigma_{1,2}\sigma_{2,2} + \sigma_{1,1}\sigma_{1,2} + \sigma_{2,1}\sigma_{2,2}\right]$$

$$+ v^2 \left[\sigma_{1,1}\sigma_{2,1}\sigma_{1,2}\sigma_{2,2} + \sigma_{1,1}\sigma_{2,1}\sigma_{1,1}\sigma_{1,2} + \sigma_{1,1}\sigma_{2,1}\sigma_{2,1}\sigma_{2,2}\right.$$
$$\left. + \sigma_{1,2}\sigma_{2,2}\sigma_{1,1}\sigma_{1,2} + \sigma_{1,2}\sigma_{2,2}\sigma_{2,1}\sigma_{2,2} + \sigma_{1,1}\sigma_{1,2}\sigma_{2,1}\sigma_{2,2}\right]$$

$$+ v^3 \left[\sigma_{1,1}\sigma_{2,1}\sigma_{1,2}\sigma_{2,2}\sigma_{1,1}\sigma_{1,2} + \sigma_{1,1}\sigma_{2,1}\sigma_{2,1}\sigma_{2,2}\sigma_{2,2}\sigma_{1,2}\right.$$
$$\left. + \sigma_{1,1}\sigma_{2,1}\sigma_{2,1}\sigma_{2,2}\sigma_{1,1}\sigma_{1,2} + \sigma_{1,1}\sigma_{1,2}\sigma_{1,2}\sigma_{2,2}\sigma_{2,2}\sigma_{2,1}\right]$$

$$+ v^4 \left[\sigma_{1,1}\sigma_{2,1}\sigma_{2,1}\sigma_{2,2}\sigma_{2,2}\sigma_{1,2}\sigma_{1,2}\sigma_{1,1}\right]. \qquad (16.7.17)$$

As the power of v increases, the number of spin factors increases. Thus, in order to keep track of the "bookkeeping," we introduce a graphical notation. Each index pair (i, j) on the spin variables $\sigma_{i,j}$ corresponds to a point on the lattice. We identify these points as the vertices of one or more graphs that can be drawn on the lattice. We also identify a spin product $\sigma_{i,j}\sigma_{i',j'}$ with an edge joining the vertices (i, j) and (i', j'). Thus, in eqn. (16.7.17), there are four vertices, $(1,1), (1,2), (2,1), (2,2)$, corresponding to four points on the lattice, and at the nth power of v, $n = 0, ..., 4$, there are n edges joining the vertices. Figure 16.11 shows the complete set of graphs corresponding to the terms in eqn. (16.7.17).

We next ask what each graph contributes to the overall partition function. The first graph, which contains no edges, obviously contributes exactly 1, and when this is summed over N spins, each of which can take on two values, we obtain a contribution of 2^N. The graphs that arise from the v^1 term contain a single edge. Consider, for example, the product $\sigma_{1,1}\sigma_{2,1}$, which must be summed over $\sigma_{1,1}$ and $\sigma_{2,1}$. The spin sum produces four terms corresponding to $(\sigma_{1,1}, \sigma_{2,1}) = (1,1), (-1,1), (1,-1), (-1,-1)$. When the spin products are taken, two of these terms will be 1 and the other two -1, and the sum yields 0 overall. The same is true for each of the remaining spin products in the v^1 term. After some reflection, we see that all of the terms proportional to v^2 and v^3 also sum to 0. However, the v^4 term, represented by a closed graph in which each vertex is included in two edges, does not vanish, since each spin variable appears twice. The contribution from this single graph (see v^4 term in Fig. 16.11), when the sum over all N spins is carried out, is $v^4 2^N$. Now, on a lattice of N spins, it is possible

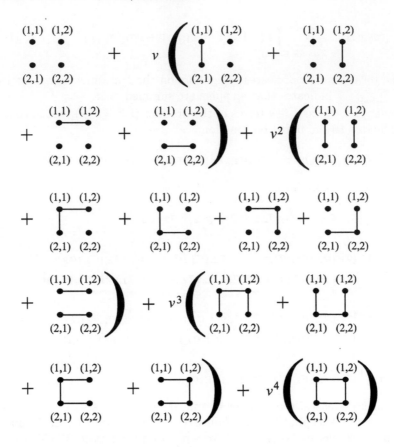

Fig. 16.11 Graphical representation of eqn. (16.7.17).

to draw $N-1$ such graphs containing four vertices. Thus, the total contribution from graphs containing four vertices is $(N-1)v^4 2^N$.

The analysis in the preceding paragraph suggests that the problem of evaluating the partition function becomes one of counting the number $n(r)$ of closed graphs that can be drawn on the lattice containing r edges and then summing the result over r, where $r = 0, 4, 6, 8, \dots$. Figure 16.12 shows some examples of graphs that occur when $r = 6$ and when $r = 8$. Once $n(r)$ is known, the partition function can be shown to take the following form:

$$Q(N,T) = 2^N \left[\cosh(K)\right]^\nu \sum_r n(r) v^r. \qquad (16.7.18)$$

We conclude this brief discussion by illustrating how eqn. (16.7.18) can be applied to the one-dimensional Ising model on a periodic lattice. The partition function was given by eqn. (16.6.10), which for $h = 0$ becomes

$$Q(N,T) = 2^N \left[\cosh^N(K) + \sinh^N(K)\right]. \qquad (16.7.19)$$

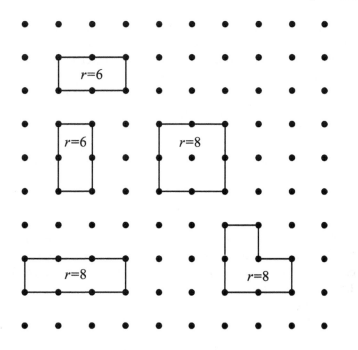

Fig. 16.12 Examples of graphs that contribute to the partition function of the two-dimensional Ising model for $r = 6$ and $r = 8$.

We now show that the graph-theoretic approach can be used to derive eqn. (16.7.19). First, note that on a one-dimensional periodic lattice, only two graphs contribute: the graph with $r = 0$ and the graph with $r = N$, which is the only closed graph that can be drawn for $r > 0$, *i.e.*, one that involves each vertex in two edges. Thus, $n(N) = 1$, and the number of nearest neighbors on the periodic lattice is $\nu = N$. Putting these facts together gives the partition function as

$$Q(N, T) = 2^N \left[\cosh(K)\right]^N \left[1 + v^N\right], \tag{16.7.20}$$

which simplifies to

$$Q(N, T) = 2^N \left[\cosh^N(K) + \sinh^N(K)\right] \tag{16.7.21}$$

in agreement with eqn. (16.7.19).

16.8 Spin correlations and their critical exponents

In Section 4.7.1, we considered spatial correlation functions in a liquid. Interestingly, it is possible to define an analogous quantity for the Ising model. Consider the following spin-spin correlation function at zero field:

$$\langle \sigma_i \sigma_j \rangle = \frac{1}{Q(N, T)} \sum_{\sigma_1} \cdots \sum_{\sigma_N} \sigma_i \sigma_j e^{-\beta \mathcal{H}}, \tag{16.8.1}$$

where $Q(N, T)$ is the canonical partition function. If σ_i and σ_j occupy lattice sites at positions $\mathbf{r}_i$ and $\mathbf{r}_j$, respectively, then at large spatial separation $r = |\mathbf{r}_i - \mathbf{r}_j|$, the correlation function should depend only on r. Heuristically, $G(r)$ is assumed to decay exponentially according to

$$G(r) \equiv \langle \sigma_i \sigma_j \rangle - \langle \sigma_i \rangle \langle \sigma_j \rangle \sim \frac{e^{-r/\xi}}{r^{d-2+\eta}} \qquad (16.8.2)$$

for $T > T_c$ (Ma, 1976). The quantity ξ is called the *correlation length*. As a critical point is approached from above, long-range order sets in, and we expect ξ to diverge as $T \to T_c^+$. This divergence is characterized by an exponent ν such that

$$\xi \sim |T - T_c|^{-\nu}. \qquad (16.8.3)$$

As $T \to T_c^+$, $\xi \to \infty$, and the exponential numerator in $G(r)$ becomes 1. In this case, $G(r)$ decays in a manner characteristic of a system with long-range order, *i.e.*, as a small inverse power of r. The exponent η appearing in the expression for $G(r)$ characterizes this decay at $T = T_c$.

The exponents ν and η cannot be determined from mean-field theory, as the mean-field approximation neglects all spatial correlations. In order to calculate these exponents, fluctuations must be restored at some level. One method that treats correlations explicitly is a field-theoretic approach known as the Landau-Ginzberg theory (Huang, 1963; Ma, 1976). This theory uses a continuous spin field to define a free-energy functional and provides a prescription for deriving the spatial correlation functions from the external field dependence of the partition function via functional differentiation. Owing to its mathematical complexity, a detailed discussion of this theory is beyond the scope of this book; instead, in the next section, we will focus on an elegant approach that is motivated by a few simple physical considerations derived from the long-range behavior spin-spin correlations.

16.9   Introduction to the renormalization group

The *renormalization group* (RG) theory is based on ideas first introduced by L. P. Kadanoff (1966) and K. G. Wilson (1971) and posits that near a critical point, where long-range correlations dominate, the system possesses self-similarity at any scale. It then proposes a series of coarse-graining operations that leave the system invariant, from which the ordered phases can be correctly identified.[2] The RG framework also offers an explanation of universality, provides a framework for calculating the critical exponents (Wilson and Fisher, 1972; Bonanno and Zappalà, 2001), and through a hypothesis known as the *scaling hypothesis*, generates sets of relations called *scaling relations* (Widom, 1965; Cardy, 1996) among the critical exponents. Although we will only explore here how the RG approach applies to the study of magnetic systems,

---

[2]The term "renormalization group" has little to do with group theory in the usual mathematical sense. Although the RG does employ a series of transformations based on the physics of a system near its critical point, the RG transformations are not unique and do not form a mathematical group. Hence, references to "the" renormalization group are also misleading, but as this usage has become the common parlance, we will continue this usage here.

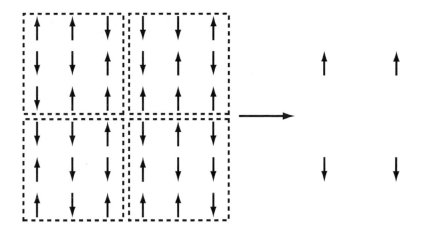

Fig. 16.13 Example of the block spin transformation on a 6×6 square lattice. The lattice on the right shows the four spins that result from applying the transformation to each 3×3 block.

the technique is very general and has been employed in problems ranging from fluid dynamics to quantum chemistry (see, for example, Baer and Head-Gordon (1998)). The proceeding discussion of the RG will be based loosely on the treatment given by Cardy (1996).

In order to illustrate the RG procedure, let us consider the example of a square spin lattice shown in Fig. 16.13. In the left half of the figure, the lattice is separated into 3×3 blocks. We now consider defining a new spin lattice from the old by applying a coarse-graining procedure that replaces each 3×3 spin block with a single spin. Of course, we need a rule for constructing this new spin lattice, so let us consider the following simple algorithm: (1) count the number of up and down spins in each block; (2) if the majority of the spins in the 3×3 block are up, replace the block by a single up spin, otherwise replace it by a single down spin. For the example on the left in Fig. 16.13, the new lattice obtained by applying this procedure is shown on the right in the figure. Such a transformation is called a *block spin transformation* (Kadanoff, 1966). Near a critical point, the system will exhibit long-range ordering, hence the coarse-graining procedure should yield a new spin lattice that is statistically equivalent to the old one; the spin lattice is then said to possess *scale invariance*.

Given the new spin lattice generated by the block spin transformation, we now wish to determine the Hamiltonian of this lattice. Since the new lattice must be statistically equivalent to the original one, the natural route to the transformed Hamiltonian is through the partition function. Thus, we consider the zero-field ($h = 0$) partition function of the original spin lattice using the Hamiltonian $\mathcal{H}_0$ in eqn. (16.3.4) for the Ising model as the starting point:

$$Q(N,T) = \sum_{\sigma_1} \cdots \sum_{\sigma_N} e^{-\beta\mathcal{H}_0(\sigma_1,\ldots,\sigma_N)} \equiv \mathrm{Tr}_\sigma e^{-\beta\mathcal{H}_0(\sigma_1,\ldots,\sigma_N)}. \qquad (16.9.1)$$

The transformation function $T(\sigma'; \sigma_1, ..., \sigma_9)$ that yields the single spin σ' for each 3×3 block of 9 spin variables can be expressed mathematically as follows:

$$T(\sigma'; \sigma_1, ..., \sigma_9) = \begin{cases} 1 & \sigma' \sum_{i=1}^{9} \sigma_i > 0 \\ 0 & \text{otherwise} \end{cases}. \tag{16.9.2}$$

This function ensures that when the spin sum over the original lattice is performed, only those terms that conform to the rule of the block spin transformation are nonzero. That is, the only nonzero terms are those for which σ' and $\sum_{i=1}^{9} \sigma_i$ have the same sign. When eqn. (16.9.2) is inserted into eqn. (16.9.1), the function $T(\sigma'; \sigma_1, ..., \sigma_9)$ projects out those configurations that are consistent with the block spin transformation rule, while the sum over the old spin variables $\sigma_1, ..., \sigma_N$ leaves a function of only the new spin variables $\{\sigma'_1, ..., \sigma'_{N'}\}$. Note that $T(\sigma'; \sigma_1, ..., \sigma_9)$ satisfies the property

$$\sum_{\sigma'=\pm 1} T(\sigma'; \sigma_1, ..., \sigma_9) = 1, \tag{16.9.3}$$

which means simply that only one of the two values of σ' can satisfy the block spin transformation rule. The new spin variables $\{\sigma'\}$ can now be used to define a new partition function. To see how this is done, let the Hamiltonian of the new lattice be defined according to

$$e^{-\beta \mathcal{H}'_0(\{\sigma'\})} = \text{Tr}_\sigma \left[\prod_{\text{blocks}} T(\sigma'; \sigma_1, ..., \sigma_9) \right] e^{-\beta \mathcal{H}_0(\{\sigma\})}, \tag{16.9.4}$$

which follows from eqn. (16.9.3). Summing both sides of eqn. (16.9.4) over the relevant spin variables yields

$$\text{Tr}_{\sigma'} e^{-\beta \mathcal{H}'_0(\{\sigma'\})} = \text{Tr}_\sigma e^{-\beta \mathcal{H}_0(\{\sigma\})}. \tag{16.9.5}$$

Equation (16.9.5) states that the partition function is preserved by the block spin transformation and, consequently, so are the equilibrium properties.

If the block spin transformation is devised in such a way that the functional form of the Hamiltonian is preserved, then the transformation can be iterated repeatedly on each new lattice generated by the transformation: each iteration will generate a system that is statistically equivalent to the original. Importantly, in a truly ordered state, each iteration will produce precisely the same lattice in the thermodynamic limit, thus signifying the existence of a critical point. If the functional *form* of the Hamiltonian is maintained, then only its parameters (*e.g.*, the strength of the spin-spin coupling) are affected by the transformation, and thus, we can regard the transformation as one that acts on these parameters. If the original Hamiltonian contains parameters $K_1, K_2, ..., \equiv \mathbf{K}$ (for example, the coupling J in the Ising model), then the transformation yields a Hamiltonian with a new set of parameters $\mathbf{K}' = (K'_1, K'_2, ...)$ that are functions of the old parameters

$$\mathbf{K}' = \mathbf{R}(\mathbf{K}). \tag{16.9.6}$$

The vector function $\mathbf{R}$ defines the transformation. These equations are called the *renormalization group equations* or *renormalization group transformations*. By iterating the RG equations, it is possible to determine if a system has an ordered phase and for what parameter values the ordered phase occurs. In an ordered phase, each iteration of the RG equations yields the same lattice with exactly the same Hamiltonian. Requiring that the Hamiltonian itself remain unchanged under an RG transformation is stronger than simply requiring that the functional form of the Hamiltonian be preserved. When the Hamiltonian is unchanged by the RG transformation, then the parameters $\mathbf{K}'$ obtained via eqn. (16.9.6) are unaltered, implying that

$$\mathbf{K} = \mathbf{R}(\mathbf{K}). \tag{16.9.7}$$

A point $\mathbf{K}$ in parameter space that satisfies eqn. (16.9.7) is called a *fixed point* of the RG transformation. Equation (16.9.7) indicates that the Hamiltonian of an ordered phase emerges from a fixed point of the RG equations.

16.9.1 RG example: The one-dimensional Ising model

In the zero-field limit, the Hamiltonian for the one-dimensional Ising model is

$$\mathcal{H}_0(\{\sigma\}) = -J \sum_{i=1}^{N} \sigma_i \sigma_{i+1}. \tag{16.9.8}$$

Let us define a dimensionless Hamiltonian $\Theta_0 = \beta \mathcal{H}_0$ and a dimensionless coupling constant $K = \beta J$ so that $\Theta_0(\{\sigma\}) = -K \sum_{i=1}^{N} \sigma_i \sigma_{i+1}$. With these definitions, the partition function becomes

$$Q(N,T) = \mathrm{Tr}_\sigma e^{-\Theta_0(\{\sigma\})}. \tag{16.9.9}$$

Consider the simple block spin transformation illustrated in Fig. 16.14. The figure shows the one-dimensional spin lattice with two different indexing schemes: The upper scheme is a straight numbering of the nine spins in the figure, while the lower scheme numbers the spins in each block. As Fig. 16.14 indicates, the block spin transformation employed in this example replaces each block of three spins with a single spin determined solely by the spin at the center of the block. Thus, for the left block, the

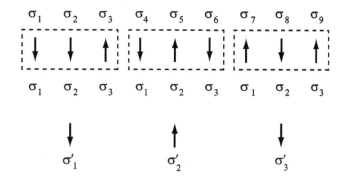

Fig. 16.14 Example of the block spin transformation applied to the one-dimensional Ising model. The three spins that result are shown below.

new spin $\sigma_1' = \sigma_2$, for the middle block, $\sigma_2' = \sigma_5$, $\sigma_3' = \sigma_8$, and so forth. Though not particularly democratic, this block spin transformation should be reasonable at low temperature where local ordering is expected and the middle spin is likely to cause neighboring spins to align with it. The transformation function $T(\sigma'; \sigma_1, \sigma_2, \sigma_3)$ for this example can expressed mathematically simply as

$$T(\sigma'; \sigma_1, \sigma_2, \sigma_3) = \delta_{\sigma'\sigma_2}. \tag{16.9.10}$$

The new spin lattice is shown below the original lattice in Fig. 16.14.

The transformation function in eqn. (16.9.10) is now used to compute the new Hamiltonian Θ_0' according to

$$e^{-\Theta_0'(\{\sigma'\})} = \sum_{\sigma_1} \sum_{\sigma_2} \sum_{\sigma_3} \cdots \sum_{\sigma_N} \left(\delta_{\sigma_1'\sigma_2} \delta_{\sigma_2'\sigma_5} \cdots \right) e^{K\sigma_1\sigma_2} e^{K\sigma_2\sigma_3} e^{K\sigma_3\sigma_4} e^{K\sigma_4\sigma_5} \cdots$$

$$= \sum_{\sigma_1} \sum_{\sigma_3} \sum_{\sigma_4} \sum_{\sigma_6} \cdots e^{K\sigma_1\sigma_1'} e^{K\sigma_1'\sigma_3} e^{K\sigma_3\sigma_4} e^{K\sigma_4\sigma_2'} \cdots. \tag{16.9.11}$$

Equation (16.9.11) encodes the information we need to determine the new coupling parameter K'. We will use the rule that when the sums over σ_3 and σ_4 are performed, the new interaction between σ_1' and σ_2' gives the contribution $\exp(K'\sigma_1'\sigma_2')$ to the partition function. If this rule is satisfied, then the functional form of the Hamiltonian will be preserved. The sum over σ_3 and σ_4 that must then be performed in eqn. (16.9.11) is

$$\sum_{\sigma_3} \sum_{\sigma_4} \exp[K\sigma_1'\sigma_3] \exp[K\sigma_3\sigma_4] \exp[K\sigma_4\sigma_2'].$$

Note that the spin product $\sigma_3\sigma_4$ has two possible values, $\sigma_3\sigma_4 = \pm 1$, which allows us to employ a convenient identity:

$$e^{\pm\theta} = \cosh\theta \pm \sinh\theta = \cosh\theta \left[1 \pm \tanh\theta\right]. \tag{16.9.12}$$

Equation (16.9.12) allows us to express $\exp(K\sigma_3\sigma_4)$ as

$$e^{K\sigma_3\sigma_4} = \cosh K \left[1 + \sigma_3\sigma_4\tanh K\right].$$

If we define $x = \tanh K$, the product of the three exponentials becomes

$$e^{K\sigma_1'\sigma_3} e^{K\sigma_3\sigma_4} e^{K\sigma_4\sigma_2'} = \cosh^3 K(1 + \sigma_1'\sigma_3 x)(1 + \sigma_3\sigma_4 x)(1 + \sigma_4\sigma_2' x)$$

$$= \cosh^3 K(1 + \sigma_1'\sigma_3 x + \sigma_3\sigma_4 x + \sigma_4\sigma_2' x$$
$$+ \sigma_1'\sigma_3^2\sigma_4 x^2 + \sigma_1'\sigma_3\sigma_4\sigma_2' x^2 + \sigma_3\sigma_4^2\sigma_2' x^2$$
$$+ \sigma_1'\sigma_3^2\sigma_4^2\sigma_2' x^3). \tag{16.9.13}$$

When summed over σ_3 and σ_4, most terms in eqn. (16.9.13) cancel, yielding

$$\sum_{\sigma_3} \sum_{\sigma_4} e^{K\sigma_1'\sigma_3} e^{K\sigma_3\sigma_4} e^{K\sigma_4\sigma_2'} = 2\cosh^3 K \left[1 + \sigma_1'\sigma_2' x^3\right] \equiv \cosh K' \left[1 + \sigma_1'\sigma_2' x'\right].$$

$$\tag{16.9.14}$$

In eqn. (16.9.14), we have expressed the interaction in its original form but with a new coupling constant K'. In order for the interaction term $[1 + \sigma_1'\sigma_2' x']$ to match the original interaction $[1 + \sigma_1'\sigma_2' x^3]$, we require $x' = x^3$ or

$$\tanh K' = \tanh^3 K$$

$$K' = \tanh^{-1}\left[\tanh^3 K\right].\tag{16.9.15}$$

Equation (16.9.15) defines the RG transformation as

$$R(K) = \tanh^{-1}\left[\tanh^3(K)\right].\tag{16.9.16}$$

We must remember, however, that eqn. (16.9.16) is particular to the block spin transformation in eqn. (16.9.10). From eqns. (16.9.14) and (16.9.15), we obtain the new Hamiltonian as

$$\Theta'_0(\{\sigma'\}) = N'g(K) - K'\sum_{i=1}^{N'}\sigma'_i\sigma'_{i+1},\tag{16.9.17}$$

where the spin-independent function $g(K)$ is given by

$$g(K) = -\frac{1}{3}\ln\left[\frac{\cosh^3 K}{\cosh K'}\right] - \frac{2}{3}\ln 2.\tag{16.9.18}$$

and N' is a constant. Thus, apart from the $N'g(K)$ term in eqn. (16.9.17), the new Hamiltonian has the same functional form as the original Hamiltonian, but it is a function of the new spin variables and coupling constant K'.

The block spin transformation of eqn. (16.9.10) could be applied again to the Hamiltonian in eqn. (16.9.17), leading to a new Hamiltonian Θ''_0 in terms of new spin variables $\sigma''_1, ..., \sigma''_{N''}$ and a new coupling constant K''. It is a straightforward exercise to show that the coupling constant K'' would be related to K' by eqn. (16.9.15). Repeated application of the block spin transformation is, therefore, equivalent to iteration of the RG equation. Since the coupling constant K depends on temperature through $K = J/kT$, this iterative procedure can determine if, for some temperature, an ordered phase exists. Recall that an ordered phase corresponds to the fixed point condition in eqn. (16.9.7). From eqn. (16.9.16), this condition for the present example becomes

$$K = \tanh^{-1}\left[\tanh^3 K\right].\tag{16.9.19}$$

In terms of $x = \tanh K$, the fixed point condition is simply $x = x^3$. Since $K \geq 0$, the only possible solutions to this fixed point equation are $x = 0$ and $x = 1$.

To understand the physical content of these solutions, consider the RG equation away from the fixed point: $x' = x^3$. Since $K = J/kT$, at high T, $K \to 0$ and $x = \tanh K \to 0^+$. At low temperature, $K \to \infty$ and $x \to 1^-$. If we view the RG equation as an iteration or recursion of the form

$$x_{n+1} = x_n^3,\tag{16.9.20}$$

and we start the recursion at $x_0 = 1$, then each successive iteration will yield 1. However, for any value $x = 1 - \epsilon$ less than 1 (here, $\epsilon > 0$), eqn. (16.9.20) eventually iterates to 0. These two scenarios are illustrated in Fig. 16.15. The iteration of eqn. (16.9.20) generates a *renormalization group flow* through the one-dimensional coupling-constant space. The fixed point at $x = 1$ is called an *unstable fixed point* because any value of x_0 other than 1, when iterated through the RG equation, flows away from this point

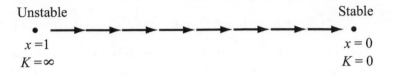

Fig. 16.15 RG flow for the one-dimensional Ising model.

toward the *stable fixed point* at $x = 0$. As the stable fixed point is approached, the coupling constant decreases until it reaches $K = 0$, corresponding to infinite temperature! At the unstable fixed point $(x = 1)$, $K = \infty$ and $T = 0$. The absence of a fixed point for any finite, nonzero value of temperature tells us that there can be no ordered phase (and hence no critical point) in one dimension. Note, however, that in one dimension, perfect ordering exists at $T = 0$. Although this is not a physically meaningful ordered phase ($T = 0$ can never be achieved), this result suggests that ordered phases and critical points are associated with the unstable fixed points of the RG equations. Recall that we also obtained ordering at $T = 0$ from the exact analytical solution of the one-dimensional Ising model in Section 16.6.

Let us make one additional observation about the $T = 0$ unstable fixed point by analyzing the behavior of the correlation length ξ at $T = 0$. ξ has units of length, but if we choose to measure it in units of the lattice spacing, then it can only depend on the coupling constant K or $x = \tanh K$, i.e., $\xi = \xi(x)$. Under the block spin transformation of eqn. (16.9.10), as Fig. 16.14 indicates, the lattice spacing increases by a factor of 3 as a result of coarse graining. Thus, in units of the lattice spacing, ξ must decrease by a factor of 3 in order to maintain the same physical distance. Thus,

$$\xi(x') = \frac{1}{3}\xi(x). \tag{16.9.21}$$

More generally, if we had taken our blocks to have b spins, eqn. (16.9.21) suggests that ξ should transform as

$$\xi(x') = \frac{1}{b}\xi(x). \tag{16.9.22}$$

In addition, the RG equation would become $x' = x^b$. We now seek a functional form for $\xi(x)$ that satisfies eqn. (16.9.22). In fact, only one functional form is possible, namely

$$\xi(x) \sim \frac{1}{\ln x}. \tag{16.9.23}$$

This can be shown straightforwardly as follows:

$$\xi(x') = \xi(x^b) \sim \frac{1}{\ln x^b} = \frac{1}{b\ln x} = \frac{1}{b}\xi(x). \tag{16.9.24}$$

Therefore, the correlation length $\xi(K) \sim 1/\ln(\tanh K) \longrightarrow \infty$ as $T \longrightarrow 0$, so that at $T = 0$ the correlation length is infinite, which is another indication that an ordered phase exists.

Finally, we examine the behavior of the RG equation at very low T where K is large. Note that eqn. (16.9.15) can be written as

$$\tanh K' = \tanh^3 K$$

$$= \tanh K \tanh^2 K$$

$$= \tanh K \left[\frac{\cosh(2K) - 1}{\cosh(2K) + 1} \right]. \qquad (16.9.25)$$

The term in brackets is very close to 1 when K is large. Thus, when K is large, eqn. (16.9.25) can be expressed as $K' \sim K$, which is a linearized version of the RG equation. On an arbitrary spin lattice, interactions between blocks are predominantly mediated by interactions between spins along the boundaries of the blocks (see Fig. 16.16 for an illustration of this in two dimensions). In one dimension, this interaction involves a single spin pair, and thus we expect a block spin transformation in one dimension to yield a coupling constant of the same order as the original coupling constant at low T where there is significant alignment between the blocks.

16.10   Fixed points of the renormalization group equations in greater than one dimension

Figure 16.16 shows a two-dimensional spin lattice and the interactions between two blocks, which are mediated by the boundary spins. In more than one dimension, these interactions are mediated by more than a single spin pair. For the case of the 3×3 blocks shown in the figure, there are three boundary spin pairs mediating the interaction between blocks. Consequently, the result of a block spin transformation should yield,

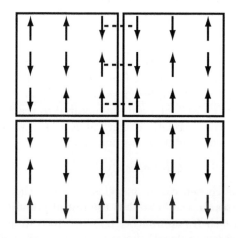

Fig. 16.16 Interactions between blocks of a square spin lattice.

at low T, a coupling constant K' roughly three times as large as the original coupling constant K, i.e., $K' \sim 3K$. In a three-dimensional lattice, using $3 \times 3 \times 3$ blocks, interactions between blocks would be mediated by $3^2 = 9$ spin pairs. Generally, in d dimensions using blocks of b^d spins, the RG equation at low T should behave as

$$K' \sim b^{d-1}K. \tag{16.10.1}$$

The number b is called the *length scaling factor*. Equation (16.10.1) implies that for $d > 1$, $K' > K$ at low T. Thus, iteration of the RG equation at low temperature yields an RG flow towards $K = \infty$, and the fixed point at $T = 0$ becomes stable (in one dimension, this fixed point was unstable). However, we know that at high temperature, the system must be in a disordered state, and hence the fixed point at $T = \infty$ must remain a stable fixed point, as it was in one dimension. These two observations suggest that for $d > 1$, there must be a third fixed point with coupling constant $\tilde{x}$ between $T = 0$ and $T = \infty$. Moreover, an iteration initiated with $x_0 = \tilde{x} + \epsilon$ ($\epsilon > 0$) must iterate to $x = 1$ where $T = 0$, and an iteration initiated from $x_0 = \tilde{x} - \epsilon$ must iterate to $x = 0$ where $T = \infty$. Hence, this fixed point is unstable and is, therefore, a critical point with $\tilde{K} = K_c$. To the extent that an RG flow in more than one dimension can be represented as a one-dimensional process, the flow diagram would appear as in Fig. 16.17. Since this unstable fixed point corresponds to a finite, nonzero temperature

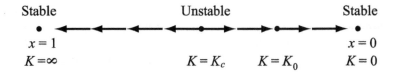

Fig. 16.17 Renormalization group flow in more than one dimension. The figure shows the iteration to each stable fixed point starting from the unstable fixed point and an arbitrary point $K = K_0$.

T_c, it is a physical critical point.

This claim is further supported by the evolution of the correlation length under the RG flow. Recall that for a length scaling factor b, the correlation length transforms as $\xi(K') = \xi(K)/b$ or $\xi(K) = b\xi(K')$. Suppose that we start at a point K near K_c and that $n(K)$ iterations of the RG equation are required to reach a value K_0 between $K = 0$ and $K = K_c$. If ξ_0 is the correlation length at $K = K_0$, which should be a finite number of order 1, then by eqn. (16.9.22) we find that

$$\xi(K) = \xi_0 b^{n(K)}. \tag{16.10.2}$$

As the starting point K is chosen closer and closer to K_c, the number of iterations needed to reach K_0 increases. In the limit that the initial point $K \to K_c$, the number of iterations is needed to reach K_0 approaches infinity. According to eqn. (16.10.2), as K approaches K_c, the correlation length becomes infinite as expected in an ordered phase. Thus, the new unstable fixed point must correspond to a critical point.

From this understanding of the correlation length behavior, we can analyze the exponent ν using the RG equation near the unstable fixed point. When $K = K_c$, the fixed point condition requires that $K_c = R(K_c)$. Near the fixed point, we can expand the RG equation to give

$$K' \approx R(K_c) + (K - K_c)R'(K_c) + \cdots. \tag{16.10.3}$$

Let us write $R'(K_c)$ as $b^{\ln R'(K_c)/\ln b}$ and define an exponent $y = \ln R'(K_c)/\ln b$. Using this exponent, eqn. (16.10.3) becomes

$$K' \approx K_c + b^y(K - K_c). \tag{16.10.4}$$

Near the critical point, ξ diverges according to

$$\xi \sim |T - T_c|^{-\nu} \sim \left| \frac{1}{K} - \frac{1}{K_c} \right|^{-\nu} \sim \left| \frac{K - K_c}{K} \right|^{-\nu} \sim \left| \frac{K - K_c}{K_c} \right|^{-\nu}. \tag{16.10.5}$$

Thus, $\xi \sim |K - K_c|^{-\nu}$. However, since $\xi(K) = b\xi(K')$, it follows that

$$|K - K_c|^{-\nu} \sim b|K' - K_c|^{-\nu} = b|b^y(K - K_c)|^{-\nu}, \tag{16.10.6}$$

which is only possible if

$$\nu = \frac{1}{y}. \tag{16.10.7}$$

Equation (16.10.7) illustrates the general result that critical exponents are related to derivatives of the RG transformation.

16.11 General linearized renormalization group theory

Our discussion in the previous section illustrates the power of the linearized RG equations. We now generalize this approach to a Hamiltonian Θ_0 with parameters $K_1, K_2, K_3, ..., \equiv \mathbf{K}$. Equation (16.9.6) for the RG transformation can be linearized about an unstable fixed point at $\mathbf{K}^*$ according to

$$K'_a \approx K^*_a + \sum_b T_{ab}(K_b - K^*_b), \tag{16.11.1}$$

where

$$T_{ab} = \left. \frac{\partial R_a}{\partial K_b} \right|_{\mathbf{K}=\mathbf{K}^*}. \tag{16.11.2}$$

Note that the matrix T is not required to be symmetric. Consequently, we define a left eigenvalue equation for T according to

$$\sum_a \phi^i_a T_{ab} = \lambda^i \phi^i_b \tag{16.11.3}$$

and a *scaling variable* u_i as

$$u_i = \sum_a \phi_a^i (K_a - K_a^*).$$
(16.11.4)

The term "scaling variable" arises from the fact that u_i transforms multiplicatively near a fixed point under the linearized RG flow:

$$
\begin{aligned}
u_i' &= \sum_a \phi_a^i (K_a' - K_a^*) \\
&= \sum_a \sum_b \phi_a^i T_{ab} (K_b - K_b^*) \\
&= \sum_b \lambda^i \phi_b^i (K_b - K_b^*) \\
&= \lambda^i u_i.
\end{aligned}
$$
(16.11.5)

Suppose the eigenvalues λ^i are real. Since $u_i' = \lambda^i u_i$, u_i will increase if $\lambda^i > 1$ and decrease if $\lambda^i < 1$. Redefining the eigenvalues λ^i as

$$\lambda^i = b^{y_i},$$
(16.11.6)

we see that

$$u_i' = b^{y_i} u_i.$$
(16.11.7)

By convention, the quantities $\{y_i\}$ are referred to as the RG eigenvalues.

From the discussion in the preceding paragraph, three cases can be identified for the RG eigenvalues:

1. If $y_i > 0$, the scaling variable u_i is called *relevant* because repeated iteration of the RG transformation drives it away from its fixed point value at $u_i = 0$.

2. If $y_i < 0$, the scaling variable u_i is called *irrelevant* because repeated iteration of the RG transformation drives it toward 0.

3. $y_i = 0$. The scaling variable u_i is referred to as *marginal* because we cannot determine from the linearized RG equations whether u_i will iterate towards or away from the fixed point.

Typically, scaling variables are either relevant or irrelevant; marginality is rare. The number of relevant scaling variables corresponds to the number of experimentally tunable parameters such as P and T in a fluid system or T and h in a magnetic system. For the former, the relevant variables are called the *thermal* and *magnetic* scaling variables, respectively. The thermal and magnetic scaling variables have corresponding RG eigenvalues y_t and y_h. An analysis of the scaling properties of the singular part $\tilde{g}(h, T)$ of the Gibbs free energy $g(h, T)$, which obeys $\tilde{g}(h, T) = b^{-d} \tilde{g}(b^{y_h} h, b^{y_t}, T)$, leads to the following relations for the primary critical exponents:

$$\alpha = 2 - \frac{d}{y_t}, \qquad \beta = \frac{d - y_h}{y_t}, \qquad \gamma = \frac{2y_h - d}{y_t}, \qquad \delta = \frac{y_h}{d - y_h}.$$
(16.11.8)

These relations are obtained by differentiating $\tilde{g}(h, T)$ to obtain the heat capacity, magnetization, and magnetic susceptibility. From eqns. (16.11.8), the relations $\alpha +$

$2\beta + \gamma = 2$ and $\alpha + \beta(1 + \delta) = 2$, which are examples of scaling relations, can be easily derived. Two other scaling relations can be derived from the scaling behavior of the spin-spin correlation function $\tilde{G}(r) = b^{-2(d-y_h)}G(r/b, b^{y_t}t)$, where $t = (T - T_c)/T$. These are $\alpha = 2 - d\nu$ and $\gamma = \nu(2 - \eta)$. Because of such scaling relations, we do not need to determine all of the critical exponents individually. For the Ising model, we see that there are four such scaling relations, indicating that only two of the exponents, ν and η, of the six total are independent. Because a subset of the critical exponents still needs to be determined by some method, numerical simulations play an important role in the implementation of the RG, and techniques such as the Wang-Landau and $\mathrm{M(RT)}^2$ schemes carried out on a lattice are useful approaches that can be employed (see Problems 16.8 and 16.9).

16.12 Understanding universality from the linearized renormalization group theory

In the linearized RG theory, at a fixed point, all scaling variables are zero, regardless of whether they are relevant, irrelevant, or marginal. Let us assume for the present discussion that there are no marginal scaling variables. From the definitions of relevant and irrelevant scaling variables, we can propose a formal procedure for locating fixed points. Begin with the space spanned by the full set of eigenvectors of T, and project out the relevant subspace by setting all the relevant scaling variables to zero by hand. The remaining subspace is spanned by the irrelevant eigenvectors of T, which defines a hypersurface in the full coupling constant space. This surface is called the *critical*

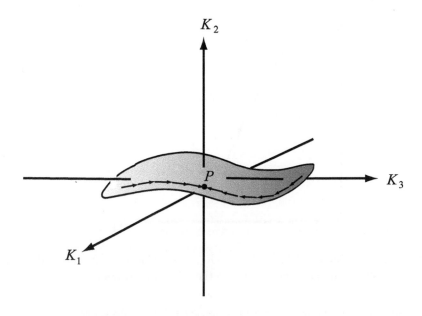

Fig. 16.18 A renormalization group trajectory.

hypersurface. Any point on the critical hypersurface belongs to the irrelevant subspace and iterates to zero under successive RG transformations. This procedure defines a trajectory on the hypersurface that leads to a fixed point, as illustrated in Fig. 16.18. This fixed point, called the *critical fixed point*, is stable with respect to irrelevant scaling variables and unstable with respect to relevant scaling variables.

In order to understand the importance of the critical fixed point, consider a simple model in which there is one relevant and one irrelevant scaling variable. Let these be denoted as u_1 and u_2, respectively, and let these variables have corresponding couplings K_1 and K_2. In an Ising model, K_1 might represent the reduced nearest-neighbor coupling, and K_2 might represent a next-nearest-neighbor coupling. Relevant variables also include experimentally tunable parameters such as temperature and magnetic field. The reason u_1 is relevant and u_2 is irrelevant is that there must be a nearest-neighbor coupling for the existence of a critical point and ordered phase at $h = 0$, but magnetization can occur even if there is no next-nearest-neighbor coupling. According to the procedure of the preceding paragraph, the condition

$$u_1(K_1, K_2) = 0$$

defines the critical surface, which in this case, is a one-dimensional curve in the $K_1 K_2$ plane as illustrated in Fig. 16.19. Here, the black curve represents the critical "surface" (curve), and the point at which the arrows meet is the critical fixed point. The full coupling constant space represents the space of all physical systems containing nearest-neighbor and next-nearest-neighbor couplings. If we wish to consider the sub-

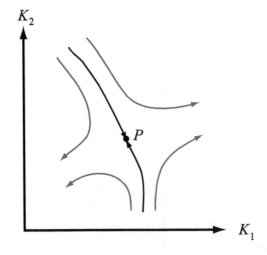

Fig. 16.19 Example curves defined by $u_1(K_1, K_2)$. The critical curve defined by $u_1(K_1, K_2) = 0$ is shown in black and iterates to the critical fixed point P.

set of systems with no next-nearest-neighbor coupling ($K_2 = 0$), the point at which the line $K_2 = 0$ intersects the critical surface defines the critical value K_{1c} and the corresponding critical temperature, and it is an unstable fixed point of an RG transformation with $K_2 = 0$. Similarly, if we consider a model for which $K_2 \neq 0$, then the point at which this line intersects the critical surface determines the critical value of K_1 for such a model. In fact, for any of these models, K_{1c} lies on the critical surface and iterates toward the critical fixed point under the full RG transformation. Thus, we have an effective definition of a universality class: All models characterized by the same critical fixed point belong to the same universality class and share the same critical properties.

16.13   Other uses of discrete models

Discrete models of the type discussed in this chapter can be applied in areas other than magnetic systems, rendering them powerful tools for extracting complex patterns of behavior within a relatively simple framework. Problem 16.5 represents one approach for studying the conformational states of a simple polymer known as a *rotational isomeric states model*. In this model, the *trans* and two *gauche* states of each dihedral angle in, for example, a simple alkane, are treated as discrete. These discrete states are then used to construct an energy model similar to a generalized Potts model (cf. eqn. (16.3.5)). From such a model, the probability associated with any set of dihedral angles can be determined and properties such as the average end-to-end distance can be estimated and compared to more advanced sampling methods (Zhu *et al.*, 2002). Ising models have been used to describe short-range order and long-range disorder in molecular crystals (Welberry, 1979), in which the "spins" represent different discrete orientations of a molecule in the crystal lattice. In the biomolecular field, Potts and Ising models have found an important application in connecting chemical composition to structure, function, and fitness. Specifically, a protein or nucleic acid can be considered as a sequence S consisting of a string $\{\sigma\}$ of letter codes corresponding to amino or nucleic acids at each of M positions in the sequence S (Levy *et al.*, 2017; Haldane and Levy, 2019; Wilburn and Eddy, 2020). The field and coupling parameters can be fit to databases of sequences, after which the model can predict conformational and fitness landscapes of mutations. A variant of the Potts model, known as the Cellular Potts model (Graner and Glazier, 1992; Savill and Hogeweg, 1997; Szabó and Merks, 2013; Guisoni *et al.*, 2018) considers the spins as different types of cells with different pairwise adhesion energies, and such a model allows for the study of tissue morphogenesis and errant phenomena such as tumor growth under active cell motion. These are just a few areas in which discrete models have been adapted for application to complex problems well beyond spin lattices. In essence, any phenomenon that can be mapped onto a set of discrete states with local interactions between them is likely to fit into the paradigm of this class of models. The primary challenge then becomes the availability (or lack) of observed data to determine parameters of sufficient quality that the chosen model becomes subsequently predictive.

16.14   Problems

16.1. Consider a block spin transformation of the one-dimensional Ising model with $h \neq 0$, in which every other spin is summed over. Such a procedure is also called a *decimation* procedure.

 a. Write down the transformation operator T for this transformation, and show that the transformation leads to a value of $b = 2$.

 b. Derive the RG equation for this transformation, and find the fixed points.

 c. Sketch the RG flow in the (x, y) plane. What is the nature of the fixed points, and what do they imply about the existence of a critical point?

16.2. The class of models known as Potts models allow, in general, N discrete states for each spin variable on the lattice. The one-dimensional three-state Potts model is defined as follows: each site i is a 'spin' which may take on three values, 1, 2, or 3. The Hamiltonian is given by

$$H = -J \sum_i \delta_{s_i, s_{i+1}}.$$

Use the same decimation procedure as in Problem 16.1, derive the RG equation, and find the fixed points. Does this model have a critical point?

*16.3. The Maxwell construction in the van der Waals theory attempts to fix the unphysical behavior of the theory resulting from $\partial P / \partial V > 0$ for $T < T_c$. The procedure is shown schematically in Fig. 16.20. The construction introduces a *tie line* for the discontinuous change in volume resulting from the first-order gas–liquid phase transition. The location of the tie line must be chosen so that the two areas enclosed by the tie line and the PV curve above and below the tie line are equal ($a_1 = a_2$). Show that the Maxwell construction together with the van der Waals equation of state in eqn. (4.8.36) leads to a mean-field value of the critical exponent β equal to $1/2$.

16.4. For the spin-$\frac{1}{2}$ Ising model in one dimension with $h = 0$, recall that the partition function could be expressed in the form

$$\Delta = \mathrm{Tr}\left(\mathrm{P}^N\right).$$

Consider an RG transformation, called the *pair cell* transformation, in which Δ is reexpressed as

$$\Delta = \mathrm{Tr}\left[\left(\mathrm{P}^2\right)^{N/2}\right].$$

The transfer matrix is redefined by $\mathrm{P}' = \mathrm{P}^2$. Find the RG equation corresponding to this transformation and show that it leads to the expected stable fixed point.

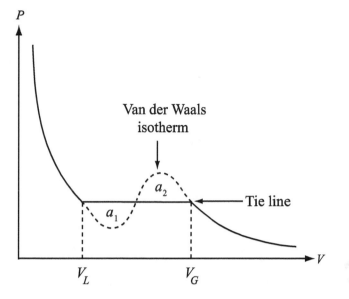

Fig. 16.20 Maxwell construction.

Hint: Try redefining the coupling constant by $u = e^K$ and show that P′ can be put in the same form as P, *i.e.*, $P'(u) = c(u)P(u')$, and that c can be defined implicitly in terms of u'.

*16.5. A simple model of a long polymer chain consists of the following assumptions:

   i. The conformational energy E of the chain is determined solely from its backbone dihedral angles.

  ii. Each dihedral angle can assume three possible values denoted t for "trans" and g^+ and g^- for the two "gauche" conformations. However, the present model is discrete in the sense that t, g^+, and g^- are the only values the dihedral angles may assume.

 iii. Each conformation has an intrinsic energy and is also influenced by the conformations of nearest neighbor dihedral angles only. If the polymer has N atomic sites, then there are $N-3$ dihedral angles numbered $\phi_1, ..., \phi_{N-3}$ by convention. The total energy $E(\phi_1, ..., \phi_{N-3})$ can be written as

$$E(\phi_1, ..., \phi_{N-3}) = \sum_{i=1}^{N-3} \varepsilon_1(\phi_i) + \sum_{i=2}^{N-3} \varepsilon_2(\phi_{i-1}, \phi_i),$$

where each ϕ_i has values t, g^+, or g^-.

 iv. The two energy functions ε_1 and ε_2 are assumed to have the following values:

$$\varepsilon_1(t) = 0$$

$$\varepsilon_1(g^+) = \varepsilon(g^-) = \varepsilon$$

$$\varepsilon_2(g^+, g^-) = \varepsilon_2(g^-, g^+) = \infty$$

$$\varepsilon_2(\phi_{i-1}, \phi_i) = 0 \quad \text{for all other combinations.}$$

a. Calculate the canonical partition function for this system. You may express your answer in terms of $\sigma \equiv \exp(-\beta\varepsilon)$.

b. Show that, in the limit $N \to \infty$, the partition function behaves as

$$\lim_{N\to\infty} \frac{1}{N} \ln Q = \ln \chi,$$

where

$$\chi = \frac{1}{2}\left[(1+\sigma) + \sqrt{1 + 6\sigma + \sigma^2}\right].$$

c. What is the probability, for large N, that all angles will be in the trans conformation?

d. What is the probability, for large N, that the angles will alternate trans, gauche, trans, gauche, ...?

16.6. Consider the Ising Hamiltonian in eqn. (16.3.3) for which each spin has z neighbors on the lattice. Each spin variable σ_i can take on three values, $-1, 0, 1$. Using mean-field theory, find the transcendental equation for the magnetization, and determine the critical temperature of this model. What are the critical exponents?

16.7. In 1978, H. J. Maris and L. J. Kadanoff introduced an RG procedure for the two-dimensional zero-field Ising model (Maris and Kadanoff, 1978; Chandler, 1987). This problem explores this procedure step by step, leading to the RG equation.

a. Consider the following labeling of spins on the two-dimensional periodic lattice shown in Fig. 16.21(a).
The first step is to sum over half of the spins on the lattice by partitioning the summand of the partition function in such a way that each spin to be summed over appears in only one Boltzmann factor. Show that the resulting partition function, for one choice of the spins summed over, corresponds to the spin lattice shown below and takes the form

$$Q = \sum_{\text{remaining spins}} \cdots \left[e^{K(\sigma_1+\sigma_2+\sigma_3+\sigma_4)} + e^{-K(\sigma_1+\sigma_2+\sigma_3+\sigma_4)}\right]$$

$$\times \left[e^{K(\sigma_2+\sigma_3+\sigma_7+\sigma_8)} + e^{-K(\sigma_2+\sigma_3+\sigma_7+\sigma_8)}\right]\cdots,$$

where $K = \beta J$ (see Fig. 16.21(b)).

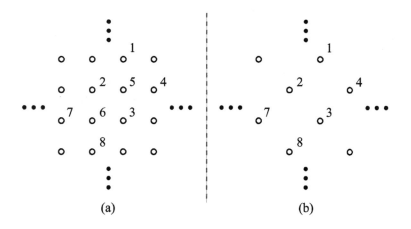

Fig. 16.21 (a) Labeling of spins on a two-dimensional lattice. (b) Spin lattice after decimation.

b. Consider trying to write one of the terms in brackets in the above expression as

$$e^{K(\sigma_1+\sigma_2+\sigma_3+\sigma_4)} + e^{-K(\sigma_1+\sigma_2+\sigma_3+\sigma_4)} = g(K)e^{K'(\sigma_1\sigma_2+\sigma_2\sigma_3+\sigma_3\sigma_4+\sigma_4\sigma_1)},$$

where the new coupling constant K' and the function $g(K)$ are to be determined by requiring that this equation be satisfied by the four nonequivalent choices of the spins:

$$\sigma_1 = \sigma_2 = \sigma_3 = \sigma_4 = \pm 1$$
$$\sigma_1 = \sigma_2 = \sigma_3 = -\sigma_4 = \pm 1$$
$$\sigma_1 = \sigma_2 = -\sigma_3 = -\sigma_4 = \pm 1$$
$$\sigma_1 = -\sigma_2 = \sigma_3 = -\sigma_4 = \pm 1.$$

Show that $g(K)$ and K' cannot be determined in this way.

c. Consider instead introducing several new coupling constants, K_1, K_2, and K_3, and writing

$$e^{K(\sigma_1+\sigma_2+\sigma_3+\sigma_4)} + e^{-K(\sigma_1+\sigma_2+\sigma_3+\sigma_4)}$$
$$= g(K)e^{(1/2)K_1(\sigma_1\sigma_2+\sigma_2\sigma_3+\sigma_3\sigma_4+\sigma_4\sigma_1)+K_2(\sigma_1\sigma_3+\sigma_2\sigma_4)+K_3\sigma_1\sigma_2\sigma_3\sigma_4}.$$

By inserting the four nonequivalent choices of the spin variables from part b, find expressions for K_1, K_2, K_3, and $g(K)$ in terms of K. Interpret the resulting partition function.

d. Note that the result of part c does not lead to an exact RG procedure. Show that if K_2 and K_3 are neglected, an RG equation of the form

$$K_1 = \frac{1}{4} \ln \cosh(4K)$$

results. Does this lead to a critical point?

e. In order to improve the results, it is proposed to neglect only K_3 and treat the K_2 term approximately. Let Σ_1 and Σ_2 be the spin sums multiplying K_1 and K_2, respectively, in the partition function expression. Consider the approximation

$$K_1\Sigma_1 + K_2\Sigma_2 \approx K'(K_1, K_2)\Sigma'_{i,j}\sigma_i\sigma_j,$$

where the sum is taken only over nearest neighbors. What is the partition function that results from this approximation?

f. Define the free energy per spin as

$$a(K) = \frac{1}{N} \ln Q(K).$$

Show that the free energy satisfies an RG equation of the form

$$a(K) = \frac{1}{2} \ln g(K) + \frac{1}{2}a(K').$$

g. Show that K' can be estimated by

$$K' \approx K_1 + K_2.$$

h. Derive the RG equation that results, and show that it predicts the existence of a critical point. What is the value of K_c?

i. By expanding the free energy in a Taylor series about $K = K_c$, calculate the critical exponent α and compare your value to the exact solution of Onsager.

j. Finally, compare the critical temperature you obtain to the Onsager result:

$$\frac{J}{kT_c} = 0.44069.$$

k. Devise an analog of the Maris-Kadanoff scheme for the one-dimensional free-field Ising model by summing over every other spin on the one-dimensional spin lattice. What is the RG equation that results? Show that your one-dimensional equation yields the expected fixed point.

16.8. The Wang-Landau method of Section 7.6 can be easily adapted for spin lattices. First, since the spin variables take on discrete values, the total energy E of the spin lattice also takes on discrete values. Therefore, we can write the canonical partition function as

$$Q(\beta) = \sum_E \Omega(E) e^{-\beta E},$$

where $\Omega(E)$ is the density of states. A trial move consists of flipping a randomly chosen spin and then applying eqn. (7.6.3) to decide whether the move is accepted. Finally, the density of states at the final energy E is modified by the scaling factor using $\ln \Omega(E) \to \ln \Omega(E) + \ln f$. Write a Wang-Landau Monte Carlo code to calculate the partition function and free energy per spin at different temperatures of a 50×50 spin lattice in the absence of an external field. Take the initial state of the lattice to be a randomly chosen set of spin values, and apply periodic boundary conditions to the lattice.

16.9. Write a simple $M(RT)^2$ Monte Carlo program to sample the distribution function of a two-dimensional Ising model in the presence of an external field h. Observe the behavior of your algorithm for temperatures $T > T_c$ and $T < T_c$ at different field strengths. For each case, calculate and plot the spin-spin correlation function $\langle \sigma_i \sigma_j \rangle - \langle \sigma_i \rangle \langle \sigma_j \rangle$.

17
Introduction to machine learning in statistical mechanics

17.1   Machine learning in statistical mechanics: What and why?

The term *artificial intelligence* (AI) brings to mind the creation of computer systems capable of mimicking the decision-making and problem-solving tasks of a human mind by emulating its thought patterns. In a broad sense, *machine learning* is a pathway to AI that uses statistical models and "training" algorithms to take in data, learn insights and patterns in the data, and apply that learning to make new predictions without additional input or programming. Viewed this way, it follows that machine learning is best poised to provide reliable predictions when data are plentiful. For example, online vendors track shoppers' purchases and employ machine learning models to learn their preferences so that they can recommend to shoppers specific items they might want to purchase in the future; however, the very notion of "preference" implies a pattern in a shopper's purchases, which can only be discerned if a large number of purchases can be analyzed for such patterns. Recommendations based only on one or a few purchases may or may not meet a shopper's needs or wants.

In many applications, machine learning is employed to perform one of several important tasks; these include regression, classification, data clustering, feature extraction and engineering, and dimensionality reduction, tasks that are particularly useful in statistical mechanics. When molecular simulation approaches are applied to investigate a complex system, large amounts of data, such as time series, conformational samples, energies, free energies, distributions, and so forth, are generated and need to be analyzed. Most of the time, this data is of very high dimension and may contain complex patterns that are difficult to discern using simple statistical analysis tools. When this is the case, machine learning becomes a powerful tool for learning these patterns and predicting trends in a system that might not have been explicitly generated in a simulation. It is also very likely to happen that a simulation yields incomplete data about a system or a particular thermodynamic or dynamical process. In such instances, machine learning can often fill in some of the missing information. In an enhanced conformational sampling simulation, for example, employing a method such as d-AFED/TAMD (Section 8.10) or replica-exchange MD (Section 7.5), a sparse sampling of points is produced as the simulation sweeps over a free-energy hypersurface. The density of points increases with the number of sweeps, and at some stage, there will be enough points to fit or "train" a regression model that can subsequently predict free energies of points not visited during the simulation while also providing a smooth closed-form representation of the surface (Schneider *et al.*, 2017; Zhang *et al.*,

2018; Wang *et al.*, 2021). The number of sampled points needed to generate an accurate model may be far fewer than that needed to reach full convergence in a direct simulation. In studying complex processes for which a suitable reaction coordinate might not be known *a priori*, a classification model can be used to look for patterns in local structural information and use this information to create an order parameter capable of parameterizing a pathway from one conformational basin to another. This has been done, for example, to study a solid-solid phase transition in a metal (Rogal *et al.*, 2019), in which environmental descriptors are used as a means of classifying individual atoms as belonging to one phase or another; the classification model ultimately allows free-energy barriers and mechanisms of the phase transition can be determined. Such an approach could also be applied to the prediction of pathways of protein or nucleic acid folding (Senior *et al.*, 2019; AlQuraishi, 2019). Finally, in the field of active matter, where systems are driven to desired patterns and modes of behavior by supplying them with energy and applying external stimuli, given an appropriate amount of data and a rule-discovery algorithm, machine learning can be leveraged to identify the underlying interactions and governing equations that lead to the emergence of such complex behavior (Cichos *et al.*, 2020).

The purpose of this chapter is to introduce basic concepts in machine learning, including a selection of machine learning models particularly useful in statistical mechanics, and to show how these models can be applied to solve a range of specific statistical mechanical problems. It is important to note that machine learning methods, in and of themselves, are not likely to reveal any new physical insights. However, because they are able to predict the outcomes of computationally intensive tasks with significantly greater efficiency than without the benefit of machine learning, their *application* can lead to such new insights by allowing time and length scales to be significantly increased. Readers interested in a more comprehensive examination of the machine learning field, which has evolved into an enormous discipline, or a broader survey of the landscape of machine learning models, are referred to, for example, C. M. Bishop, *Pattern Recognition and Machine Learning* or Hastie *et al.*, *The Elements of Statistical Learning*. Additional references include *Neural Networks and Computing* by T. W. S. Chow and S. -Y. Cho, and *Understanding Machine Learning* by S. Shalev-Shwartz and S. Ben-David. Mathematical underpinnings of machine learning are described in D. Simovici's *Mathematical Analysis for Machine Learning and Data Mining*.

17.2 Three key probability distributions

In our discussion of algorithms for training machine learning models, we will make use of several standard probability distribution functions. These are the Gaussian, Bernoulli, and categorical distributions, each of which we discuss in this section. The Gaussian, or normal, distribution is a familiar and widely used function that models the distribution of continuous variables. Thinking back to Section 7.2, the central limit theorem guarantees the convergence of a Monte Carlo calculation by establishing that estimators of averages approach a Gaussian distribution about the true average. Given a continuous random variable x, the Gaussian distribution takes the form

$$P_{\mathrm{G}}(x; \mu, \sigma) = \frac{1}{(2\pi\sigma^2)^{1/2}} e^{-(x-\mu)^2/2\sigma^2},$$ (17.2.1)

where μ and σ^2 are the mean and variance of the distribution. Here, μ determines the location of the peak of the distribution and σ determines its width, meaning that

$$\langle (x - \mu)^2 \rangle = \int_{-\infty}^{\infty} (x - \mu)^2 \, P_{\mathrm{G}}(x; \mu, \sigma) \, \mathrm{d}x = \sigma^2.$$ (17.2.2)

The number of parameters needed to specify a Gaussian distribution of one variable is clearly two. If, instead of a single random variable, we have an n-dimensional vector $\mathbf{x}$ of random variables and a vector of mean values $\boldsymbol{\mu}$, the general form of the Gaussian distribution is

$$P_{\mathrm{G}}(\mathbf{x}; \boldsymbol{\mu}, \Sigma) = \frac{1}{(2\pi)^{n/2}} \frac{1}{[\det(\Sigma)]^{1/2}} e^{-(\mathbf{x}-\boldsymbol{\mu})^{\mathsf{T}} \Sigma^{-1} (\mathbf{x}-\boldsymbol{\mu})/2},$$ (17.2.3)

where Σ is an $n \times n$ matrix known as the *covariance matrix*, related to averages of the distribution by

$$\langle (\mathbf{x} - \boldsymbol{\mu})(\mathbf{x} - \boldsymbol{\mu})^{\mathsf{T}} \rangle = \Sigma.$$ (17.2.4)

Equation (17.2.4) makes clear that Σ is a symmetric matrix, which, therefore, has real eigenvalues. For eqn. (17.2.3) to be well defined, the eigenvalues of Σ must also be positive-definite. In any case, the number of parameters needed to specify a Gaussian distribution of $\mathbf{x}$ is $n(n + 1)/2 + n = n(n + 3)/2$. However, from eqn. (17.2.4), we see that if the components of $\mathbf{x}$ are independent random variables, then Σ becomes a diagonal matrix, $\Sigma = \mathrm{diag}(\sigma_1^2 \; \sigma_2^2 \; \cdots \; \sigma_n^2)$, such that

$$\langle (\mathbf{x} - \boldsymbol{\mu})(\mathbf{x} - \boldsymbol{\mu})^{\mathsf{T}} \rangle_{ij} = \sigma_i^2 \delta_{ij},$$ (17.2.5)

which leaves us with $2n$ parameters to specify the distribution. Finally, if the components of $\mathbf{x}$ are isotropic, then all of the σ_i are the same, so that Σ reduces to $\Sigma = \sigma^2 \mathbf{I}$, where $\mathbf{I}$ is the $n \times n$ identity matrix, which leaves just $n + 1$ parameters needed to specify the distribution.

The second distribution we will employ in our discussion of machine learning concerns random variables that can take on discrete values only. For starters, suppose x is a binary random variable that can take on the values 0 or 1. An example is a coin toss where we assign tails = 0 and heads = 1. Let ν be the probability that $x = 1$ so that $1 - \nu$ is the probability that $x = 0$. Then, the probability distribution of x is a special case of a binomial distribution known as a *Bernoulli distribution*, which is given by

$$P_{\mathrm{B}}(x; \nu) = \nu^x \, (1 - \nu)^{1-x}.$$ (17.2.6)

With the convention used here, it can be easily shown that $\langle x \rangle = \nu$ and $\langle x^2 \rangle - \langle x \rangle^2 = \nu(1 - \nu)$. The Bernoulli distribution requires that just one parameter, ν, be specified.

Suppose, next, that x can take on n values, which, for simplicity, we take to be the integers $1, 2, 3, ..., n$. An example is the roll of a six-sided die. The third distribution, called the *categorical distribution*, generalizes the Bernoulli distribution to treat such cases where $n > 2$. If the probabilities that x takes on each of n values are $\nu_1, ..., \nu_n \equiv \boldsymbol{\nu}$, with $\nu_1 + \nu_2 + \cdots + \nu_n = 1$, then the categorical distribution takes the form

$$P_C(x; \boldsymbol{\nu}) = \prod_{i=1}^{n} \nu_i^{[x=i]}. \qquad (17.2.7)$$

Here, the quantity $[x = i]$ evaluates to 1 if $x = i$ and to 0 if $x \neq i$. It is straightforward to show that $\langle x \rangle = \sum_{i=1}^{n} i\nu_i$. An alternative formulation of the categorical distribution introduces an n-dimensional vector $\mathbf{x}$ having just one component with the value 1 and all other components with the value 0. There will be n such vectors, and if ν_i is the probability associated with the vector $\mathbf{x}_i$ whose single nonzero component is x_i, then the categorical distribution can be written as

$$P_C(\mathbf{x}; \boldsymbol{\nu}) = \prod_{i=1}^{n} \nu_i^{x_i}. \qquad (17.2.8)$$

With this definition, it can be shown that $\langle \mathbf{x} \rangle = \boldsymbol{\nu}$. In either formulation, the categorical distribution reduces to the Bernoulli distribution for binary random variables.

17.3   Simple linear regression as a case study

In order to introduce the basic concepts of machine learning for a regression problem, we will use the "toy" example of simple linear regression, but we will discuss it within the framework and nomenclature of machine learning. Suppose we have a set of data points $(x_1, y_1), (x_2, y_2), ..., (x_n, y_n)$ that we assume lie on a line $y_i = w_0 + w_1 x_i$, where w_0 and w_1 are the intercept and slope of the line, respectively. If the points come from a measurement of some kind and have associated measurement error, they might not satisfy a perfect linear relation. Therefore, our task is to find an optimal linear model that *best* fits the given data. To this end, we take as a data model the linear form

$$y(x, \mathbf{w}) = w_0 + w_1 x, \qquad (17.3.1)$$

where $\mathbf{w} = (w_0, w_1)$ is a two-dimensional vector of parameters whose specific values determine the optimal model. Figure 17.1 illustrates how the data might be distributed around the assumed model in eqn. (17.3.1). To determine $\mathbf{w}$, we seek an objective function, expressing the deviation of the input data from the target model in eqn. (17.3.1); this objective function, when minimized with respect to $\mathbf{w}$, yields an optimal realization of the model capable of predicting any new value of y given an input x.

Optimization of the linear model is achieved by a least-squares minimization procedure, in which the average distance between y_i and the prediction of y_i by the model in eqn. (17.3.1), *i.e.*, $y(x_i, \mathbf{w})$, is minimized with respect to $\mathbf{w}$. This procedure defines an objective function, which takes the form

$$E(\mathbf{w}) = \frac{1}{N} \sum_{i=1}^{N} |y_i - y(x_i, \mathbf{w})|^2. \qquad (17.3.2)$$

The function $E(\mathbf{w})$ in eqn. (17.3.2) has various names throughout the machine learning literature; depending on the source, it is referred to as the *error function*, *cost function*,

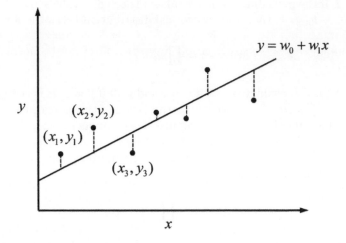

Fig. 17.1 Illustration of the process of linear regression: Seven points are distributed about the line $y = w_0 + w_1 x$, shown as a solid line. Dashed lines represent distances from each point (x_i, y_i) to the line $y = w_0 + w_1 x$ at $x = x_i$.

or *loss function*. Here, we will refer to it as the loss function, which is the most commonly used terminology. The optimal value of $\mathbf{w}$ is that which minimizes $E(\mathbf{w})$, *i.e.*, it is the solution of the minimization problem

$$\frac{\partial E}{\partial \mathbf{w}} = 0. \qquad (17.3.3)$$

In fact, the solution of the minimization problem is ultimately independent of the $1/N$ prefactor in eqn. (17.3.2), so the choice of this prefactor, which makes $E(\mathbf{w})$ an average distance, is arbitrary. If eqn. (17.3.1) is substituted into eqn. (17.3.2) to give

$$E(w_0, w_1) = \frac{1}{N} \sum_{i=1}^{N} |y_i - w_0 - w_1 x_i|^2, \qquad (17.3.4)$$

it becomes clear that eqn. (17.3.3), in this simple case, yields two equations $\partial E/\partial w_0 = 0$ and $\partial E/\partial w_1 = 0$ in the two unknowns w_0 and w_1. From eqn. (17.3.4), these two conditions yield the coupled equations

$$\langle y \rangle - w_0 - w_1 \langle x \rangle = 0$$
$$\langle xy \rangle - w_0 \langle x \rangle - w_1 \langle x^2 \rangle = 0, \qquad (17.3.5)$$

where $\langle \cdots \rangle$ indicates an average over the N data points. The solution to these equations is the familiar result

$$w_1 = \frac{\langle xy \rangle - \langle x \rangle \langle y \rangle}{\langle x^2 \rangle - \langle x \rangle^2}$$

$$w_0 = \langle y \rangle - w_1 \langle x \rangle. \tag{17.3.6}$$

At this point, we note that the minimization of the loss function in eqn. (17.3.2) is tantamount to maximizing a Gaussian probability distribution of the form

$$P(\mathbf{y}; \mathbf{w}, \sigma^2) = \frac{1}{(2\pi\sigma^2)^{N/2}} \exp\left[-\frac{1}{2\sigma^2} \sum_{i=1}^{N} |y_i - y(x_i, \mathbf{w})|^2\right], \tag{17.3.7}$$

since

$$E(\mathbf{w}) = -\frac{2\sigma^2}{N} \ln P_{\mathrm{G}}(\mathbf{y}; \mathbf{w}, \sigma^2) - \sigma^2 \ln\left(2\pi\sigma^2\right), \tag{17.3.8}$$

where $\mathbf{y} = (y_1, y_2, ..., y_N)$ is a vector of the N y-values. Therefore, the optimal parameters of our data model are those that maximize a Gaussian probability distribution between the input data and the chosen model of that data. Once the parameter vector $\mathbf{w}_{\mathrm{min}}$ that minimizes the loss function is determined, eqn. (17.3.2) can be used to compute a *root-mean-square error* (RMSE) ϵ_{RMSE} via $\epsilon_{\mathrm{RMSE}} = \sqrt{E(\mathbf{w}_{\mathrm{min}})}$. However, a more commonly used error is the *mean absolute error* (MAE), based on an L_1 norm and given by

$$\epsilon_{\mathrm{MAE}} = \frac{1}{N} \sum_{i=1}^{N} |y_i - y(x_i, \mathbf{w}_{\mathrm{min}})|. \tag{17.3.9}$$

In order to establish a robust test of the quality of a data model, we can divide the available data into two subsets: a *training set* of size N_{train} and a *test set* of size N_{test}. The size of the test set is assumed to remain fixed while the number of points in the training set is allowed to vary. Thus, when $N_{\mathrm{train}} + N_{\mathrm{test}} < N$, the rest of the data are held in reserve to augment the size of the training set. Initially, the training set size can be a small fraction of the total available data and is used to minimize the loss function and determine optimal parameters. Then, the test set is used to evaluate the accuracy of the model via an RMSE or MAE, evaluated over the points in the test set. If the error over the test set is too large, more points can be added to the training set from the reserve set and the process repeated until the magnitude of the error is deemed acceptable. What is meant by an "acceptable" error is typically one that is lower than the intrinsic error in the original data. For example, if a data set is within chemical accuracy of 1 kcal/mol, then the prediction error in the data model should be lower than 1 kcal/mol. A plot of the RMSE or MAE versus training set size, known as a *learning curve*, will reveal how well the model learns from the training data. Once an acceptable error is reached, the model is considered to be *trained*, and it can subsequently be used to predict new y values from input x values that are new to the model.

In our simple linear regression example, we assumed that x is a scalar variable. Suppose, instead, that we have N points of the form $(y_1, \mathbf{x}_1)$, $(y_2, \mathbf{x}_2)$,...,$(y_N, \mathbf{x}_N)$, where $\mathbf{x}_i$ is an n-dimensional vector and the N points, therefore, exist in an $(n + 1)$-dimensional space. If we again assume a linear relation between y and $\mathbf{x}$, then the generalization of the linear data model in eqn. (17.3.1) becomes

$$y(\mathbf{x}, \mathbf{w}) = w_0 + \mathbf{w} \cdot \mathbf{x}, \tag{17.3.10}$$

where $\mathbf{w}$ is an n-dimensional vector of parameters $\mathbf{w} = (w_1, ..., w_n)$. As with the scalar linear regression model, an analytical solution for the $n+1$ parameters $(w_0, \mathbf{w})$ can be obtained by direct minimization of the loss function in eqn. (17.3.2). The analytical solution, in particular, is expressible in terms of the inverse of the covariance matrix $C \equiv \langle \mathbf{x}\mathbf{x}^T \rangle - \langle \mathbf{x} \rangle \langle \mathbf{x} \rangle^T$ of the data vector $\mathbf{x}$ (see Problem 17.3).

As a final point, we note that, depending on the nature of the input data, certain points may overly influence the regression. In particular, if some points are outliers with respect to the posited model, a small subset of parameters might be favored as the learning procedure attempts to fit these points. This is generally undesirable and can lead to overfitting of outlier data. A control measure in the learning procedure can be introduced that prevents undue bias of certain data points. Consider modifying the loss function so that it reads

$$E(w_0, \mathbf{w}, \lambda) = \frac{1}{N} \sum_{i=1}^{N} (y_i - w_0 - \mathbf{w} \cdot \mathbf{x}_i)^2 + \frac{\lambda}{2} \mathbf{w} \cdot \mathbf{w}. \qquad (17.3.11)$$

The last term $(\lambda/2)\mathbf{w} \cdot \mathbf{w}$ in eqn. (17.3.11) is known as a *regularizer* or *ridge term* and involves a new parameter λ, which needs to be included in the optimization of the model. However, if we were to include λ into the set $(w_0, \mathbf{w})$ of optimizable parameters, then the optimization problem would become a nonlinear one, which is more difficult to solve analytically. Therefore, regularization parameters, such as λ, are typically chosen *a priori* before the optimization is performed. Such a parameter is known as a *hyperparameter*. The choice of λ is governed by its ability to lower the overall error across the test set compared to what the error would be if $\lambda = 0$. Regularization terms need not be restricted to quadratic forms. Other choices include a linear regularizer of the form $\lambda'|\mathbf{w}|$, known as a *lasso term*, or a combination of linear and quadratic regularizers, $\lambda|\mathbf{w}|^2/2 + \lambda'|\mathbf{w}|$, known as an *elastic net term*. The elastic net regularizer requires determination of two hyperparameters, λ and λ'. Depending on the nature of the data, a lasso or elastic net regularizer might be a better discriminator of the most relevant parameters $(w_0, \mathbf{w})$ weighting the input values $\mathbf{x}$.

Choosing hyperparameters in a machine learning model can be accomplished using a robust scheme known as *k-fold cross validation*. In this approach, the training data are divided into k subsets of equal size; of these subsets, $(k-1)$ are used for hyperparameter searching, and the remaining subset is used to validate the choice. This process is repeated such that each of the k subsets acts as the validation set and the validation error is retained for each of the k searches. Ultimately, the hyperparameters that give the lowest error can be selected, or the hyperparameters can be averaged over the k searches (in which case, the associated error will be an average error over the k individual errors). As a final assessment, the quality of the choice of hyperparameters should be evaluated against the test dataset. The use of k-fold cross validation ensures that the hyperparameter search is not biased toward a single validation set.

17.4 Kernel methods

Linear regression methods such as the n-dimensional generalization in eqn. (17.3.10) are special cases of a more general approach to regression problems known as *kernel regression.* In kernel regression methods, the vector $\mathbf{x}$ in eqn. (17.3.10) is replaced by a vector function $\boldsymbol{\Phi} : \mathbb{R}^n \to \mathbb{R}^{n_f}$, where n_f is a feature space of dimension $n_f \geq n$. More specifically, n_f is the number of parameters needed to define the data model. For linear regression, it is easy to see that $n_f = n$ and $\boldsymbol{\Phi}(\mathbf{x}) = \mathbf{x}$. The more general expression for the data model is

$$y(\mathbf{x}, \mathbf{w}) = w_0 + \mathbf{w} \cdot \boldsymbol{\Phi}(\mathbf{x}), \tag{17.4.1}$$

where now $\dim(\mathbf{w}) = n_f$ and $\boldsymbol{\Phi}(\mathbf{x})$ is a nonlinear function of $\mathbf{x}$. There is some vagueness to the idea of a feature space of dimension n_f, but the good news is that neither n_f nor $\boldsymbol{\Phi}(\mathbf{x})$ need to be known explicitly, as we will show shortly. The power of eqn. (17.4.1) is that it serves as a framework for obtaining different types of kernel methods. In this context, a kernel or *kernel matrix* is an $N \times N$ matrix defined by

$$K_{ij} \equiv \boldsymbol{\Phi}(\mathbf{x}_i) \cdot \boldsymbol{\Phi}(\mathbf{x}_j). \tag{17.4.2}$$

In addition, the parameters $\mathbf{w}$ are never explicitly defined but are treated as functions of the training data, which take the general form $\mathbf{w} = \sum_{i=1}^{N} \mathbf{g}(\mathbf{x}_i)$. This is called a *support-vector expansion.*

A framework that allows us to exploit these ideas is constrained optimization of the data model. That is, we define a set of N quantities ε_i, which we will require to equal $y_i - y(\mathbf{x}_i, \mathbf{w})$ via the constrained optimization procedure. To this end, we start with the quadratic loss function

$$E(\mathbf{w}) = \frac{1}{2\lambda} \sum_{i=1}^{N} |\varepsilon_i|^2 + \frac{1}{2} \mathbf{w} \cdot \mathbf{w}, \tag{17.4.3}$$

which we minimize subject to the constraint that $\varepsilon_i = y_i - y(\mathbf{x}_i, \mathbf{w})$. Here, λ is a hyperparameter similar to a regularization parameter. Since $i = 1, ..., N$, we have N constraints, which we can enforce by introducing N Lagrange multipliers α_i and an extended loss function

$$\tilde{E}(\mathbf{w}) = \frac{1}{2\lambda} \sum_{i=1}^{N} |\varepsilon_i|^2 + \frac{1}{2} \mathbf{w} \cdot \mathbf{w} - \sum_{i=1}^{N} \alpha_i \left(\varepsilon_i - y_i + w_0 + \mathbf{w} \cdot \boldsymbol{\Phi}(\mathbf{x}_i) \right). \tag{17.4.4}$$

In eqn. (17.4.4), ε_i, w_0, and $\mathbf{w}$ are all treated as optimization parameters. The minimization conditions then become

$$\frac{\partial \tilde{E}}{\partial w_0} = -\sum_{i=1}^{N} \alpha_i = 0$$

$$\frac{\partial \tilde{E}}{\partial \mathbf{w}} = \mathbf{w} - \sum_{i=1}^{N} \alpha_i \boldsymbol{\Phi}(\mathbf{x}_i) = 0$$

$$\frac{\partial \tilde{E}}{\partial \varepsilon_i} = \frac{1}{\lambda} \varepsilon_i - \alpha_i = 0. \tag{17.4.5}$$

From these equations, we see that $\sum_{i=1}^{N} \alpha_i = 0$, $\mathbf{w} = \sum_{i=1}^{N} \alpha_i \mathbf{\Phi}(\mathbf{x}_i)$, and $\varepsilon_i = \lambda \alpha_i$. Substituting these into the constraint condition $\varepsilon_i - y_i + w_0 + \mathbf{w} \cdot \mathbf{\Phi}(\mathbf{x}_i) = 0$, we obtain an expression for each y_j value

$$y_j = \sum_{i=1}^{N} \alpha_i \mathbf{\Phi}(\mathbf{x}_i) \cdot \mathbf{\Phi}(\mathbf{x}_j) + w_0 + \lambda \alpha_i. \tag{17.4.6}$$

Equations (17.4.6) can be written in matrix form as

$$\begin{pmatrix} 0 \\ \mathbf{y} \end{pmatrix} = \begin{pmatrix} 0 & \mathbf{1}^{\mathsf{T}} \\ \mathbf{1} & \mathrm{K} + \lambda \mathrm{I} \end{pmatrix} \begin{pmatrix} w_0 \\ \boldsymbol{\alpha} \end{pmatrix}, \tag{17.4.7}$$

where K is the $N \times N$ kernel matrix, $\mathbf{y}$ is an N-dimensional vector $(y_1 \ y_2 \ \cdots \ y_N)^{\mathsf{T}}$, $\mathbf{1}^{\mathsf{T}}$ is an N-dimensional row vector of 1s, $\boldsymbol{\alpha}$ is the vector $(\alpha_1 \ \alpha_2 \ \cdots \ \alpha_N)^{\mathsf{T}}$, and I is the $N \times N$ identity matrix.

A number of different regression methods emerge from eqn. (17.4.6). For example, if we set $w_0 = 0$, we obtain the *kernel ridge-regression* model for a multidimensional function $y(\mathbf{x})$. In this model, the Lagrange multipliers $\alpha_1, ..., \alpha_N$ become the optimization parameters, and the solution of eqn. (17.4.7) for the vector $\boldsymbol{\alpha}$ is

$$\boldsymbol{\alpha} = (\mathrm{K} + \lambda \mathrm{I})^{-1} \mathbf{y}. \tag{17.4.8}$$

The data model corresponding to this solution takes the form

$$y(\mathbf{x}, \boldsymbol{\alpha}) = \sum_{i=1}^{N} \alpha_i K(\mathbf{x}_i, \mathbf{x}), \tag{17.4.9}$$

where the kernel function $K(\mathbf{x}_i, \mathbf{x}) \equiv \mathbf{\Phi}(\mathbf{x}_i) \cdot \mathbf{\Phi}(\mathbf{x})$ and gives the kernel matrix K_{ij} via $K_{ij} = K(\mathbf{x}_i, \mathbf{x}_j)$. In fact, eqn. (17.4.8) can be derived by substituting the kernel ridge-regression data model into the least-squares loss function

$$E(\boldsymbol{\alpha}) = \sum_{i=1}^{N} (y_i - y(\mathbf{x}_i, \boldsymbol{\alpha}))^2 + \lambda \boldsymbol{\alpha}^{\mathsf{T}} \mathrm{K} \boldsymbol{\alpha}, \tag{17.4.10}$$

where we see that λ becomes the ridge parameter. On the other hand, if we retain the parameter w_0 and rename it α_0 for notational uniformity, then the corresponding data model, known as a *least-squares support-vector machine* model, takes the form

$$y(\mathbf{x}, \boldsymbol{\alpha}) = \sum_{i=1}^{N} \alpha_i K(\mathbf{x}_i, \mathbf{x}) + \alpha_0. \tag{17.4.11}$$

Optimizing the corresponding loss function leads to the solution

$$\begin{pmatrix} \alpha_0 \\ \boldsymbol{\alpha} \end{pmatrix} = \begin{pmatrix} 0 & \mathbf{1}^{\mathsf{T}} \\ \mathbf{1} & \mathrm{K} + \lambda \mathrm{I} \end{pmatrix}^{-1} \begin{pmatrix} 0 \\ \mathbf{y} \end{pmatrix}. \tag{17.4.12}$$

Note that in the kernel ridge and support-vector machine models presented here, the function $\mathbf{\Phi}(\mathbf{x})$ has disappeared. As we noted immediately below eqn. (17.4.1), we do not need to know the form of $\mathbf{\Phi}(\mathbf{x})$. Rather, we can bypass specification of $\mathbf{\Phi}(\mathbf{x})$ and introduce a kernel function $K(\mathbf{x}_i, \mathbf{x})$ directly. This replacement is referred to as the "kernel trick", which allows flexibility in how we specify the kernel function. When leveraging the kernel trick, a common choice is a Gaussian kernel $K(\mathbf{x}_i, \mathbf{x})$

$$K(\mathbf{x}_i, \mathbf{x}) = \mathrm{e}^{-|\mathbf{x}_i - \mathbf{x}|^2/2\sigma^2}. \tag{17.4.13}$$

The Gaussian kernel option illustrates the advantage of the kernel trick, as eqn. (17.4.13) cannot be derived from a dot product of the form $\mathbf{\Phi}(\mathbf{x})$ and $\mathbf{\Phi}(\mathbf{x}_i)$. When a Gaussian kernel is employed within the kernel ridge-regression model, for example, the method is known as *Gaussian kernel ridge regression*. In this model, σ becomes a hyperparameter to be chosen along with the ridge parameter λ. Other kernel functions could be envisioned; however, the Gaussian kernel ridge regression model is both simple and widely applicable.

A downside of kernel methods is that very large data sets (large values of N) require kernel matrices of size N^2, which can lead to significant memory issues when the matrix needs to be stored and inverted. This issue raises the question of whether more compact and flexible data models might be possible, a topic that will be addressed in the next section where neural network models are discussed.

Before leaving this section, we illustrate how kernel methods might be used in a statistical mechanical application. Suppose we have performed an enhanced sampling calculation to generate an n-dimensional free-energy surface $A(s_1, ..., s_n) \equiv A(\mathbf{s})$ using one of the techniques for generating high-dimensional free-energy surfaces such as were discussed in Sections 8.10 and 8.11. The d-AEFD/TAMD method, for example, generates global sweeps across the free-energy landscape generating a scattering of N points, $\mathbf{s}_i$ and corresponding free-energy values $A_i \equiv A(\mathbf{s}_i)$. In the early phase of a run, the N points are sparsely distributed over the surface as illustrated in the left panel of Fig. 17.2. The point distribution might not be dense enough to reveal features of the surface to the naked eye. However, the points $(\mathbf{s}_i, A_i)$ can be used to train one of the regression models described in this section in order to fill in details of the surface not easily identifiable by inspection, allowing its features to be discerned with greater clarity. This is illustrated in the right panel of Fig. 17.2. As the calculation is carried out further, more regions are sampled, the density of points $(\mathbf{s}_i, A_i)$ increases, and the model can be updated with additional training data. This is known as *active learning*. Active learning allows the features of the surface predicted by the kernel model to become sharper as the amount of training data increases. If a kernel ridge regression model is used, then at any stage in the simulation, the explicit representation of the free-energy surface $A(\mathbf{s})$ is

$$A(\mathbf{s}) = \sum_{i=1}^{N} \alpha_i K(\mathbf{s}_i, \mathbf{s}) + \alpha_0, \tag{17.4.14}$$

where $\mathbf{s}_i$, $i = 1, ..., N$ are the N training points generated in the simulation and $K(\mathbf{s}_i, \mathbf{s})$ is the kernel function expressed in terms of the extended variables $\mathbf{s}$ that parameterize

Fig. 17.2 Generation of a free-energy surface via machine learning. (*Left panel*) Points sampled in a short d-AFED/TAMD simulation of the alanine dipeptide. (*Right panel*) Free-energy surface generated using a kernel ridge regression model. Dark blue are regions of low free energy, red indicates regions of high free energy (reproduced with permission from Cendagorta et al., *J. Phys. Chem. B* **124**, 3647 (2020), copyright American Chemical Society).

the marginal probability distribution and the free-energy surface. Equation (17.4.14) allows the free energy at *any* point **s** to be evaluated. If the kernel employed is a Gaussian kernel, then eqn. (17.4.14) becomes

$$A(\mathbf{s}) = \sum_{i=1}^{N} \alpha_i e^{-|\mathbf{s}_i - \mathbf{s}|^2 / 2\sigma^2}. \qquad (17.4.15)$$

Training is performed by optimizing the loss function with a ridge term, as in eqn. (17.4.10). We defer discussion of learning curves and training protocols for this type of application until Section 17.7, where we will compare kernel methods to other types of machine learning models.

17.5 Neural networks

By far, among the most powerful machine learning models are neural networks. With their enormous flexibility in allowable architectures and alluring mathematical structure, neural networks can be constructed for a broad range of learning tasks, including speech and text recognition, image recognition and classification, customer preference analysis, risk management, materials and chemical compound design and property characterization, and computer simulation processing and enhancement. Here, we will not endeavor to describe all possible network architectures and types, as the list is simply too long; rather, we aim to introduce general concepts of neural networks and provide examples of simple architectures that are useful in statistical mechanical applications.

Neural network models are mathematical constructs, derived from neuronal patterns observed in the human brain, that take in input data and perform a series of transformations—nonlinear or linear—on the data in order to produce a desired output. One of the earliest examples of a neuron-based computational model is described in the work of Warren S. McCulloch and Walter Pitts (1943) who proposed a "logical calculus" of neuronal activity, relating the behavior of certain networks to specific psychological conditions based on assumptions about the inputs[1]. Fourteen years after the work of McCulloch and Pitts, an important theorem would be established that provided a mathematical foundation for many widely used modern neural network architectures. This theorem is the *Kolmogorov superposition theorem* (Kolmogorov, 1957) after the mathematician Andrey N. Kolmogorov (1903-1987). Following a number of subsequent refinements (Sprecher, 1965; Fridman, 1967; Lorentz, 2005), Kolmogorov's theorem asserts that a continuous scalar function $y(x_1, ..., x_n)$ of n variables $x_i \in [0, 1]$ can be expressed in terms of $2n+1$ monotonically increasing functions $\phi_q(x)$ of a single variable $(\phi : [0, 1] \rightarrow [0, 1])$ and a single continuous function $g(x)$ $(g : \mathbb{R} \rightarrow \mathbb{R})$ in the following way:

$$y(x_1, ..., x_n) = \sum_{q=1}^{2n+1} g \left(\sum_{p=1}^{n} \lambda_p \phi_q(x_p) \right), \tag{17.5.1}$$

where the coefficients $\lambda_1, ..., \lambda_n > 0$. While a detailed proof of Kolmogorov's theorem is beyond the scope of this book (a sketch of the proof is outlined in the steps of Problem 17.17), an existence proof involves establishing that the function $g(x)$ exists; this can be done by constructing a series of approximants to $y(x_1, ..., x_n)$ based on the superposition principle and then showing that this series converges exactly to $y(x_1, ..., x_n)$ (Kahane, 1975). In 1987, R. Hecht-Nielsen connected Kolmogorov's theorem to a type of neural network known as the *feed-forward network* (Hecht-Nielsen, 1987) to be discussed in this section. However, the feed-forward network construction we will describe is based on a corollary to the Kolmogorov theorem, introduced in 1991 by Vera Kurkova (1991). Kurkova's corollary is a more flexible formulation of Kolmogorov's theorem that allows the number of inner functions to be greater than $2n + 1$ while still guaranteeing that $y(x_1, ..., x_n)$ can be approximated to arbitrary accuracy. Kurkova's restatement of Kolmogorov's theorem is

$$y(x_1, ..., x_n) = \sum_{q=1}^{m} g_q \left(\sum_{p=1}^{n} \psi_{qp}(x_p) \right), \tag{17.5.2}$$

where $\psi_{qp}(x)$ are continuous monotonically increasing functions on $[0, 1]$, $g_q(x)$ is a continuous function, and $m > 0$ is an integer. The structure of a feed-forward neural network emerges by repeated application of eqn. (17.5.2) as we will now demonstrate.

In order to see how the mathematical structure of a neural network emerges from eqn. (17.5.2), note that the argument of g_q is, itself, a function of $x_1, ..., x_n$ for each value of q. We denote this function as $h_q(x_1, ..., x_n)$. Applying Kurkova's representation to h_q, we obtain

$$h_q(x_1, ..., x_n) \equiv \sum_{p=1}^{n} \psi_{qp}(x_p) = \sum_{s=1}^{m'} \gamma_{qs} \left(\sum_{r=1}^{n} \chi_{sr}(x_r) \right), \tag{17.5.3}$$

---

[1]In fact, in 1943, there was already considerable activity in the biophysics community to establish a mathematical framework for neuronal networks. The novelty of the work of McCulloch and Pitts, in addition to involving a collaboration between a neurophysiologist and logician, is its use of logic and computation as a way to understand neural activity. For a deeper look at the work of McCulloch and Pitts, see the historical and contextual analysis of G. Piccinini (2004).

where $\gamma_{qs}(x)$ is a continuous function analogous to $g_q(x)$. Let us now choose $\chi_{sr}(x_r)$ to be

$$\chi_{sr}(x_r) = w_{sr}^{(0)} x_r + a_{sr}, \tag{17.5.4}$$

where $w_{sr}^{(0)}$ and a_{sr} are constants. This function increases monotonically, as required, provided $w_{sr}^{(0)} > 0$. With this choice, we see that

$$\sum_{r=1}^{n} \chi_{sr}(x_r) = \sum_{r=1}^{n} w_{sr}^{(0)} x_r + \sum_{r=1}^{n} a_{sr} \equiv \sum_{r=1}^{n} w_{sr}^{(0)} x_r + w_{s0}^{(0)}. \tag{17.5.5}$$

We then set

$$\gamma_{qs}(x) = w_{qs}^{(1)} h(x) + b_{qs}, \tag{17.5.6}$$

where $h(x)$ is a continuous function, about which we will have more to say later in this section. Substituting eqn. (17.5.6) into eqn. (17.5.3), we obtain

$$\sum_{s=1}^{m'} \gamma_{qs}(x) = \sum_{s=1}^{m'} w_{qs}^{(1)} h(x) + \sum_{s=1}^{m'} b_{qs} \equiv \sum_{s=1}^{m'} w_{qs}^{(1)} h(x) + w_{a0}^{(1)}. \tag{17.5.7}$$

Finally, we set $g_q(x) = w_q^{(2)} h(x) + c_q$. If we now substitute eqns. (17.5.3) through (17.5.7) into eqn. (17.5.2), we obtain

$$y(x_1, ..., x_n) = \sum_{q=1}^{m} w_q^{(2)} h \left(\sum_{s=1}^{m'} w_{qs}^{(1)} h \left(\sum_{r=1}^{n} w_{sr}^{(0)} x_r + w_{s0}^{(0)} \right) + w_{q0}^{(1)} \right) + w^{(2)}, \tag{17.5.8}$$

where $w^{(2)} = \sum_{q=1}^{m} c_q$. If we iterate the Kurkova theorem once again, we obtain

$$y(x_1, ..., x_n) =$$

$$\sum_{q=1}^{m} w_q^{(3)} h \left(\sum_{s=1}^{m'} w_{qs}^{(2)} h \left(\sum_{t=1}^{m''} w_{st}^{(1)} h \left(\sum_{r=1}^{n} w_{tr}^{(0)} x_r + w_{t0}^{(0)} \right) + w_{s0}^{(1)} \right) + w_{q0}^{(2)} \right) + w^{(3)}. \tag{17.5.9}$$

Equations (17.5.8) and (17.5.9) are the mathematical representations of *two-hidden-layer* and *three-hidden-layer* feed-forward neural networks, respectively. The term "hidden layer" will be explained shortly. The parameters $w_{p0}^{(l)}$ are known as *biases*. The operational interpretation of the feed-forward neural network in eqn. (17.5.9) is as follows: We input a set of values for the n variables $x_1, ..., x_n$. These are then transformed according to the linear function $w_{tr}^{(0)} x_r + w_{t0}^{(0)}$. The result is fed into the nonlinear function $w_{st}^{(1)} h(x) + w_{s0}^{(1)}$, and the result of this transformation is fed into the function $w_{qs}^{(2)} h(x) + w_{q0}^{(2)}$. This result is fed into the function $w_q^{(3)} h(x) + w^{(3)}$ to produce the output $y(x_1, ..., x_n)$. The procedure is illustrated in Fig. 17.3. In this figure, the network is shown as a mathematical graph in which the set of nodes in the layer at the left represents the inputs $x_1, ..., x_n$. These nodes are then fully connected to a second set of nodes that represents the first transformation using the function $h(x)$. These nodes

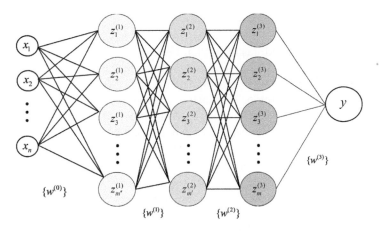

Fig. 17.3 Feed-forward neural network with three hidden layers represented as a mathematical graph. Quantities $z_\sigma^{(j)}$ are defined in eqn. (17.5.10). Each set of parameters $\{w^{(K)}\}$ determines weights of connections between hidden layers (shown with grey shading).

are fully connected to the next set, and so on, until the final output node represents the output $y(x_1, ..., x_n)$. The quantities $z_\sigma^{(j)}$ in Fig. 17.3 are defined as follows:

$$z_t^{(1)} = h\left(\sum_{r=1}^{n} w_{tr}^{(0)} x_r + w_{t0}^{(0)}\right)$$

$$z_s^{(2)} = h\left(\sum_{t=1}^{m'} w_{st}^{(1)} z_t^{(1)}\right)$$

$$z_q^{(3)} = h\left(\sum_{s=1}^{m''} w_{qs}^{(2)} z_t^{(2)}\right). \tag{17.5.10}$$

The resemblance of the graph in Fig. 17.3 to the connections between neurons in the brain has led to the use of the neural network moniker for the models in eqns. (17.5.8) and (17.5.9). Since neurons are activated when presented with input stimuli, the functions $h(x)$ are known as *activation functions*. The edges, or connections, in the graph denote the various parameters $w_{ij}^{(K)}$. These parameters are determined in the training phase by optimizing a loss function. What happens to the data in the layers that contain the activation functions $h(x)$ is, in some sense, hidden from the user of the feed-forward network, as these transformations are performed automatically by the network. For this reason, these layers are called *hidden layers*. The numbers m'', m', and m determine the numbers of nodes in the first, second, and third hidden layers, respectively. The input layer must contain n nodes for each input value, and, in the example illustrated in Fig. 17.3, the output layer contains just one node that contains the function $y(x_1, ..., x_n)$. Feed-forward neural networks with a single hidden layer are generally referred to as *single-layer perceptron* models. Network structures with more than one hidden layer are referred to as *deep neural networks*. Although deep networks usually outperform single-layer perceptron models, how deep they need to be depends on the nature of the machine learning problem, and these networks may range in depth from two to hundreds of hidden layers of different architectures.

Importantly, since the activation functions $h(x)$ are chosen *a priori*, representations of $y(x_1, ..., x_n)$ like those in eqns. (17.5.8) and (17.5.9) are no longer exact, as would be guaranteed by the Kolmogorov or Kurkova theorems, because an $h(x)$ chosen this way does not necessarily correspond to the correct choice of the functions g_q, γ_{qs},... for a given $y(x_1, ..., x_n) \equiv y(\mathbf{x})$. Therefore, it is necessary to optimize the parameters $w_{ij}^{(K)}$ of the network via the training process. For this reason, the representation of $y(\mathbf{x})$ as a neural network should actually be expressed as $y(\mathbf{x}, \mathbf{w})$ to indicate that the representation depends on the full set of parameters $\mathbf{w}$. Given N training points $(\mathbf{x}_1, y_1), ..., (\mathbf{x}_N, y_n)$, we must optimize a loss function of the form

$$E(\mathbf{w}) = \frac{1}{N} \sum_{i=1}^{N} |y_i - y(\mathbf{x}_i, \mathbf{w})|^2 \tag{17.5.11}$$

by solving $\partial E / \partial \mathbf{w} = 0$ in order to determine the optimal parameter set $\mathbf{w}$. The mathematical complexity of neural networks causes the calculation of derivatives of the network with respect to $\mathbf{w}$ to be nontrivial, a point to which we will return later in this section. First, we discuss the choice of the activation function $h(x)$.

Activation functions serve several purposes in neural networks. First, they add non-linearity into the data transformations that are performed; the corresponding weights determine how much a particular neuron contributes to the process of learning patterns in the data. Perhaps more importantly, activation functions need to effect a reshaping of the data between input and output layers such that redundancies in the data are eliminated, allowing true patterns to emerge. If we consider the input data as having an intrinsic dimension, which could be lower than the dimension of the space in which it is specified or *embedded*, then it has been suggested (Ansuini *et al.*, 2019) that as a data set propagates through the layers of a neural network, transformations performed by these layers reduce this intrinsic dimension so that the data manifold is "compressed" into a space that contains its essential, or non-redundant, features. Although there is no systematic procedure for selecting an activation function for a particular problem, sigmoid-shaped functions such as

$$h(x) = \frac{1}{1 + e^{-x}}, \qquad h(x) = \tanh(x) \tag{17.5.12}$$

are widely used. Sigmoid functions such as those in eqn. (17.5.12) effect a significant compression of the data input to them, leading to a small change in the output for a large change in the input. The result is a very small gradient at a large number of nodes, a problem known as the *vanishing gradient* problem. This problem can ameliorated by the choice of a *rectified linear unit* or ReLU function as the activation function. The ReLU function is defined as $h(x) = \max(0, x)$; a slight variant that does not return 0 is the *leaky ReLU*, defined as $h(x) = bx$ for $x < 0$ and x for $x > 0$. Here, $0 < b < 1$ is a small positive number such as 0.01. However, the ReLU and leaky ReLU functions are not differentiable at $x = 0$, which could be a problem during training when the gradient of the network is needed. Therefore, alternatives to the ReLU function that are everywhere differentiable are the softplus function

$$h(x) = \ln\left(1 + e^x\right) \tag{17.5.13}$$

and the exponential linear unit, or ELU, function

$$h(x) = \begin{cases} x & x > 0 \\ \alpha(e^x - 1) & x \le 0 \end{cases} \tag{17.5.14}$$

where α is a constant. Another differentiable activation function that does not suffer from vanishing gradients is the so-called "swish" activation function, defined by

$$h(x) = \frac{x}{1 + e^{-x}}. \tag{17.5.15}$$

Unlike the other activation functions present here, the swish function is not monotonically increasing. Although this violates the condition of monotonicity in the Kolmogorov and Kurkova theorems, empirical evidence suggests that this condition is likely sufficient but not necessary. Consequently, it is also possible to take $h(x)$ to be a Gaussian or a Lorentzian function. There is also no requirement that the same activation function be used in every layer of a neural network. Since different layers can serve different purposes as concerns learning patterns from the data, there could be advantages in using different activation functions in different layers, an issue we address in Section 17.5.1 when we discuss classification problems. Note that it is possible to tune the shape of a chosen activation function by replacing any of the $h(x)$ functions defined above with $h(ax)$, where the constant a becomes another hyperparameter that would need to be chosen at the outset, before training. The activation functions we have introduced here are shown in Fig. 17.4.

With so many possible choices for activation functions with little guidance on how to choose an optimal function for a given layer in a neural network, one might ask whether the data, itself, could dictate the selection for a given learning problem. This is, indeed, possible, via an approach known as self learning of activation functions. Self learning of activation functions can be achieved by expanding $h(x)$ in terms of a set of M basis functions as

$$h(x; \boldsymbol{\beta}) = \sum_{k=1}^{M} \beta_k \phi_k(x) \tag{17.5.16}$$

and including the set of M coefficients $\beta_1, ..., \beta_M \equiv \boldsymbol{\beta}$, together with the parameters $\mathbf{w}$, as parameters to be learning in the training phase. The neural network is then represented as $y(\mathbf{x}, \mathbf{w}, \boldsymbol{\beta})$, and optimization of the loss function requires the two conditions $\partial E(\mathbf{w}, \boldsymbol{\beta})/\partial \mathbf{w} = 0$ and $\partial E(\mathbf{w}, \boldsymbol{\beta})/\partial \boldsymbol{\beta} = 0$. Examples of possible choices of $\phi_k(x)$ are simple polynomials $\phi_k(x) = x^{k-1}$ (Goyal *et al.*, 2019) or sinc functions, $\phi_k(x) = \sin(x - kd)/[\pi(x - kd)]$, where d defines a grid spacing for x.

From a technical standpoint, the most complex operation when employing neural networks is the calculation of the derivatives needed to perform the parameter optimization. Complexity arises from the deep nesting of layers between input and output.

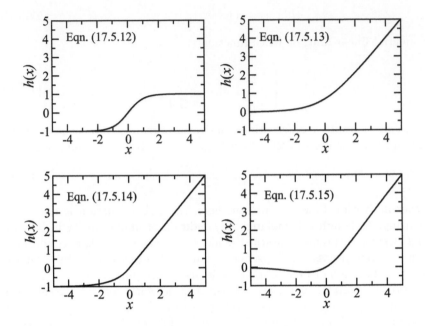

Fig. 17.4 Activation functions in eqns. (17.5.12) (tanh(x)) (upper left panel), (17.5.13) (upper right panel), (17.5.14) (lower left panel), and (17.5.15) (lower right panel).

Because of this nesting, long products arising from the application of the chain rule result when the derivatives are performed. To illustrate the structure of these products, consider a simple nested function $g(w) = h(h(h(wx_0)))$, where x_0 is a constant. We can think of this function as representing a toy network having three hidden layers with one node in each layer. From the chain rule, the derivative of g with respect to w is $g'(w) = [h'(h(h(x_0w)))][h'(h(x_0w))][h'(x_0w)]x_0$. From the pictorial representation in Fig. 17.3, if this product is read from left to right, the first term in square brackets is the derivative of the outermost layer, which produces the output result $g(w)$, the second term is the derivative of the layer just to the left of the previous layer, the third term is the next layer to the left, and finally, the last term "x_0" is the derivative of the input layer. Thus, we see that the product is a propagation backward through the layers of the network from the rightmost (output) layer back to the leftmost (input) layer. Hence, the approach for computing derivatives of the nested functions that comprise a feed-forward network via the chain rule is called *back propagation*.

Of course, computing the derivative of the loss function in eqn. (17.5.11) with a complete feed-forward network, although straightforward in principle, requires considerable bookkeeping to account for all of the terms that arise when the chain rule is applied. Suppose the network $y(\mathbf{x}, \mathbf{w})$ has K hidden layers with $m^{(i)}$ nodes in the ith layer. In order to derive the rules of back propagation, let us define a recursive variable

$$
z_r^{(k)}(\mathbf{x}) = \begin{cases} \displaystyle\sum_{a=1}^{m^{(k-1)}} h\left(z_a^{(k-1)}\right) w_{ar}^{(k-1)} + w_{0r}^{(k-1)}, & n = 2, ..., K \\[2.5em] \displaystyle\sum_{a=1}^{n} x_a w_{ar}^{(0)} + w_{0r}^{(0)}, & n = 1. \end{cases} \tag{17.5.17}
$$

Here, k indexes the hidden layer and $w_{ar}^{(k-1)}$ is the weight parameter that connects the ath node in layer $k-1$ to the rth node in layer k. The output layer of the network can be written compactly as

$$
y(\mathbf{x}, \mathbf{w}) = \sum_{a=1}^{m^{(K)}} h\left(z_a^{(K)}(\mathbf{x})\right) w_a^{(K)} + w_0^{(K)}. \tag{17.5.18}
$$

With the recursion in eqn. (17.5.17), the derivatives can also be defined recursively as a backward propagation through the layers of the network, from output back to input. Thus, the derivative of eqn. (17.5.11) with respect to $w_{rs}^{(k)}$ can be written recursively as

$$
\frac{\partial E(\mathbf{w})}{\partial w_{rs}^{(k)}} = \sum_{i=1}^{N} \sum_{a=1}^{m^{(k+1)}} \frac{\partial E(\mathbf{w})}{\partial z_a^{(k+1)}(\mathbf{x}_i)} \frac{\partial z_a^{(k+1)}(\mathbf{x}_i)}{\partial w_{rs}^{(k)}}
$$

$$
= \begin{cases} \displaystyle\sum_{i=1}^{N} \frac{\partial E(\mathbf{w})}{\partial z_s^{(k+1)}(\mathbf{x}_i)} h\left(z_i^{(k)}(\mathbf{x}_i)\right), & 0 < k \le K \\[2.5em] \displaystyle\sum_{i=1}^{N} \frac{\partial E(\mathbf{w})}{\partial z_s^{(k+1)}(\mathbf{x}_i)} x_{i,r}, & k = 0, \end{cases} \tag{17.5.19}
$$

where $x_{i,r}$ is the rth component of the ith input data point. The derivatives in eqn. (17.5.19) are expressed as

$$
\frac{\partial E(\mathbf{w})}{\partial z^{(K+1)}(\mathbf{x}_i)} \equiv \frac{\partial E(\mathbf{w})}{\partial y_i} = \frac{1}{N} \left(y(\mathbf{x}_i, \mathbf{w}) - y_i\right)
$$

$$
\frac{\partial E(\mathbf{w})}{\partial z_s^{(k)}(\mathbf{x}_i)} = \begin{cases} \displaystyle\sum_{a=1}^{m^{(k+1)}} \frac{\partial E(\mathbf{w})}{\partial z_a^{(k+1)}(\mathbf{x}_i)} w_{sa}^{(k)} h'\left(z_s^{(k)}(\mathbf{x}_i)\right), & 1 \le k \le K \\[2.5em] \displaystyle\sum_{a=1}^{m^{(k+1)}} \frac{\partial E(\mathbf{w})}{\partial z_a^{(k+1)}(\mathbf{x}_i)} w_{sa}^{(k)}, & k = 0. \end{cases} \tag{17.5.20}
$$

As noted previously, the gradient $\mathbf{G}(\mathbf{w}) = \partial E / \partial \mathbf{w}$ is needed to optimize the neural network, which requires solving $\mathbf{G}(\mathbf{w}) = 0$ for $\mathbf{w}$. However, since a neural network is a highly nonlinear function of $\mathbf{w}$, the optimization cannot be performed analytically as it can when using kernel methods. Therefore, a numerical optimization algorithm is needed. The simplest such algorithm is the *steepest descent* or *gradient descent* method. In the gradient descent algorithm, $\mathbf{w}$ is regarded as a function of an "evolution" or time-like parameter τ, and we solve the first-order differential equation $\mathrm{d}\mathbf{w}/\mathrm{d}\tau = -\mathbf{G}(\mathbf{w})$ numerically by discretizing τ into values $\tau_0, \tau_1, \tau_2, ...$ The gradient descent algorithm is then implemented by iterating

$$\mathbf{w}(\tau_{n+1}) = \mathbf{w}(\tau_n) - \eta \mathbf{G}(\mathbf{w}(\tau_n)) \tag{17.5.21}$$

until the gradient is approximately zero. The parameter η determines the step size and is known as the *learning rate* of the algorithm, which typically needs to be small in much the same way that the time step Δt in a molecular dynamics calculation must be for numerical stability. We see from eqn. (17.5.21) that the gradient descent algorithm requires the full gradient $\mathbf{G}(\mathbf{w})$ at each step, and this, in turn, needs the full set of training points. The gradient descent algorithm can be used as specified in eqn. (17.5.21) for optimization problems involving small training sets. Because eqn. (17.5.21) optimizes the full parameter set $\mathbf{w}$, it is known as a *batch optimization* approach. Gradient descent methods are often slow to converge because a small value of η is needed for stable optimization. Efficiency can be improved in batch schemes by employing more sophisticated methods such as conjugate gradient or quasi-Newton algorithms. These are standard numerical approaches and will not be discussed here. It is important to note, however, that because machine learning problems often involve very large training data sets containing hundreds of thousands or even millions of points in some situations, batch methods will become inefficient because of the cost of evaluating the full gradient vector $\mathbf{G}(\mathbf{w})$. Fortunately, it is possible to streamline the optimization problem so that only subsets of the training data are needed for each step of the iteration (LeCun *et al.*, 1989).

Note that the loss function in eqn. (17.5.11) is expressible as a sum over each observation, *i.e.*,

$$E(\mathbf{w}) = \sum_{i=1}^{N} e_i(\mathbf{w}) \tag{17.5.22}$$

and the gradient can be similarly expressed as

$$\mathbf{G}(\mathbf{w}) = \sum_{i=1}^{N} \mathbf{g}_i(\mathbf{w}). \tag{17.5.23}$$

Therefore, in the most extreme subdivision of the training data into individual observations, the gradient descent algorithm could be performed on each term $e_i(\mathbf{w})$ according to

$$\mathbf{w}(\tau_{n+1}) = \mathbf{w}(\tau_n) - \eta \mathbf{g}_i(\mathbf{w}(\tau_n)). \tag{17.5.24}$$

The update is iterated by cycling through the data, either sequentially or by choosing points at random with replacement, until the full data set is exhausted. Such an algorithm is known as *stochastic gradient descent*. Stochastic gradient descent must be iterated both through the evolution of $\mathbf{w}$ and through the data set until all of the gradients $\mathbf{g}_i$ are approximately 0. Of course, it is not necessary to work at such a fine-grained level. Rather, individual observations can be grouped in subsets, possibly of different sizes, so as to reach a compromise between the loss of evaluating the gradients and the number of iterations needed to reach full convergence. Other approaches make use of Langevin type equations (see Chapter 15) to search the parameter space for optimal solutions (Leimkuhler *et al.*, 2019).

For some applications, data on the gradient of the function $y(\mathbf{x})$ might be more readily available than on the function itself, in which case training on the gradient of $y(\mathbf{x})$ would need to be performed. The N training points would be expressed as $(\mathbf{x}_1, \nabla y(\mathbf{x}_1)), ..., (\mathbf{x}_N, \nabla y(\mathbf{x}_N)) \equiv (\mathbf{x}_1, \nabla y_1), ..., (\mathbf{x}_N, \nabla y_N)$, and although we still seek a neural network model $y(\mathbf{x}, \mathbf{w})$ for the function $y(\mathbf{x})$ itself, the training would be based on optimizing the gradient-based loss function

$$E_{\mathrm{G}}(\mathbf{w}) = \frac{1}{N} \sum_{i=1}^{N} |\nabla y_i - \nabla y(\mathbf{x}_i, \mathbf{w})|^2. \qquad (17.5.25)$$

Note that in eqn. (17.5.25), the analytical gradient of the neural network with respect to $\mathbf{x}_i$ is needed, which requires that the mathematical form of the network be everywhere differentiable. When gradient training is used, the equations for back propagation change somewhat, as illustrated in Problem 17.8.

In Section 17.4, we highlighted the example of using regression-based machine learning, specifically kernel-based learning, to fill in missing points on a free-energy surface generated by an enhanced sampling technique such as d-AFED/TAMD, which generates a scattering of points over the surface with each full sweep. Neural networks can be used in much the same way as kernel methods to perform this task (Schneider *et al.*, 2017; Zhang *et al.*, 2018; Wang *et al.*, 2021). In this case, the representation of a free-energy surface $A(s_1, ..., s_n) \equiv A(\mathbf{s})$ as a general feed-forward neural network having K hidden layers would be

$$A(\mathbf{s}, \mathbf{w}) =$$
$$\sum_{j_K=1}^{m_K} w_{j_K}^{(K)} h \left(\cdots \sum_{j_2=1}^{m_2} w_{j_3 j_2}^{(2)} h \left(\sum_{j_1=1}^{m_1} w_{j_2 j_1}^{(1)} h \left(\sum_{\alpha=1}^{n} w_{j_1 \alpha}^{(0)} s_\alpha + w_{j_1 0}^{(0)} \right) + w_{j_2 0}^{(1)} \right) + w_{j_3 0}^{(2)} \right)$$
$$\cdots + w_0^{(K)}. \qquad (17.5.26)$$

Clearly, a general feed-forward network allows for considerable flexibility in the design of the architecture. For learning high-dimensional free-energy surfaces from enhanced sampling, optimal architectures prove to be those where the early layers, those closest to the input layer, have larger numbers of nodes than layers closer to the output. That is, tapering the network such that $m_1 \geq m_2 \geq m_3 \cdots \geq m_K$ tends to lead to optimal network performance. This notion of tapering is the inspiration for a type of network known as an *autoencoder*, in which a tapered network—the *encoder*—is used to compress high-dimensional data into a lower dimensional representation or manifold, and an expanding network—the *decoder*—is used to reconstruct the original data as accurately as possible; tapering comes from the encoder phase of the autoencoder architecture. Interestingly, it has been suggested (Zhang *et al.*, 2018; Wang *et al.*, 2021) that a neural network representation of $A(\mathbf{s})$, as in eqn. (17.5.26), can be employed as a bias, similar to the metadynamics bias in the unified free-energy dynamics approach of eqns. (8.11.13). As the network becomes increasingly knowledgeable of the free-energy surface, it becomes highly effective at helping the system escape free-energy minima. Later in this chapter (see Section 17.7), we present several concrete examples of free-energy reconstruction using neural networks.

17.5.1 Neural networks for classification

Until now, our discussion of machine learning methods has focused on regression problems. Another important use of machine learning techniques is the classification of an input vector $\mathbf{x}$ into one of C classes c_k, $k = 1, ..., C$. As an example, we might wish to assign $\mathbf{x}$ to one of two options, such as "recommend" or "do not recommend" to an online shopper, where $\mathbf{x}$ could denote an item available for purchase. In the statistical mechanics of a phase transition, $\mathbf{x}$ could represent an atom or molecule that needs to be assigned to one phase or another based on its local environment. These are binary classification problems for which $C = 2$. However, it is easy to imagine situations where an input needs to be assigned to a larger number of classes. In handwritten digit recognition, an input image vector $\mathbf{x}$ would be classified as one of ten possible digits $0, ..., 9$, and in handwritten letter recognition, the number of classes would be that in the corresponding alphabet. In our phase classification example, if an atomic or molecular solid had p polymorphs, then atoms or molecules could be classified as belonging to one of these p solid phases.

Just as machine learning models for regression have an associated error in their ability to predict new values of a function, classification also has residual errors, and it is generally not possible for even the best models to achieve 100% classification accuracy. Consequently, it is worth examining classification probabilities associated with machine learning models. Using notation from Chapter 7, let us define $P(c_k|\mathbf{x})$ as the conditional probability that a given $\mathbf{x}$ will be assigned to class c_k by a given machine learning model. We note that $P(c_k|\mathbf{x}) \in [0, 1]$ and $\sum_{k=1}^{C} P(c_k|\mathbf{x}) = 1$. It should be clear that $\mathbf{x}$ will be assigned to class c_k with greater likelihood if $P(c_k|\mathbf{x}) > P(c_j|\mathbf{x})$ for all $j \neq k$. In addition to the conditional probability $P(c_k|\mathbf{x})$, we define $p(\mathbf{x})$ as the probability distribution of input vectors $\mathbf{x}$, $P(c_k)$ as the probability of the occurrence of class c_k in the machine learning model, and $p(\mathbf{x}|c_k)$ as the conditional probability distribution of inputs $\mathbf{x}$ given the class c_k. The detailed balance condition in eqn. (7.3.23) allows us to relate these probabilities via

$$P(c_k|\mathbf{x})p(\mathbf{x}) = p(\mathbf{x}|c_k)P(c_k). \tag{17.5.27}$$

Rearranging eqn. (17.5.27), we can write $P(c_k|\mathbf{x})$ as

$$P(c_k|\mathbf{x}) = \frac{p(\mathbf{x}|c_k)P(c_k)}{p(\mathbf{x})}, \tag{17.5.28}$$

a result known as *Bayes' theorem* after the English statistician, philosopher, and Presbyterian minister Thomas Bayes (1701–1761). The importance of Bayes' theorem is that it determines a key component of neural network architectures for classification problems. The term $P(c_k|\mathbf{x})$ in Bayes' theorem is known as the *posterior distribution*; the conditional probability $p(\mathbf{x}|c_k)$ is referred to as the *likelihood function*, expressing a likelihood for obtaining the data $\mathbf{x}$ given parameters, *i.e.*, class designations c_k; finally, $p(c_k)$ is known as the *prior distribution*, which allows us to input known information about the class parameters c_k if such information is known. Noting that $p(\mathbf{x})$ is a marginal distribution obtained from the sum rule

$$p(\mathbf{x}) = \sum_{k=1}^{C} p(\mathbf{x}|c_k)P(c_k), \qquad (17.5.29)$$

we see that

$$P(c_k|\mathbf{x}) = \frac{p(\mathbf{x}|c_k)P(c_k)}{\sum_{k=1}^{C} p(\mathbf{x}|c_k)P(c_k)}, \qquad (17.5.30)$$

so that $p(\mathbf{x})$ is a normalization for the product $p(\mathbf{x}|c_k)P(c_k)$ in the numerator of the theorem.

In the neural networks we derived for regression problems, we were able to express the output layer as a linear sum of the activation functions for the penultimate hidden layer as a consequence of Kurkova's theorem. For classification problems, however, we cannot assume this is possible, and consequently we must express the output layer using a final activation function. Thus, eqn. (17.5.8) for a neural network with three hidden layers would become

$$y_k(\mathbf{x}) = H\left(\sum_{q=1}^{m} w_q^{(2)} h\left(\sum_{s=1}^{m'} w_{qs}^{(1)} h\left(\sum_{r=1}^{n} w_{sr}^{(0)} x_r + w_{s0}^{(0)}\right) + w_{q0}^{(1)}\right) + w^{(2)}\right), \quad (17.5.31)$$

with a similar modification for the three-hidden-layer network in eqn. (17.5.9). Here, $H(x)$ is the outer activation function whose form we need to determine. Note that $y_k(\mathbf{x})$, $k = 1, ..., C$, which replaced the continuous function $y(\mathbf{x})$ in eqns. (17.5.8) and (17.5.9), is now interpreted as a numerical label for membership in the kth class. If we interpret $y_k(\mathbf{x})$ as the conditional probability $P(c_k|\mathbf{x})$, then it is clear that $y_k(\mathbf{x}) \in [0, 1]$ with $\sum_{k=1}^{C} y_k(\mathbf{x}) = 1$.

Bayes' theorem can now be used to determine the form of $H(x)$. The facts that $H(x)$ determines the output y_k and that $y_k \in [0, 1]$ already restrict the type of activation function $H(x)$ can be. What Bayes' theorem accomplishes is a precise specification of the particular functional form of $H(x)$. Let us begin by considering a binary classification with two classes c_1 and c_2, and let $\mathbf{z} = \mathbf{z}(\mathbf{x})$ denote the vector that results from transforming $\mathbf{x}$ through all of the hidden layers of a classification neural network. The output activation function $H(x)$ determines the conditional probability $P(c_k|\mathbf{z})$ that a particular class c_k is assigned to $\mathbf{z}$. We start by specifying a form for the distribution $p(\mathbf{z}|c_k)$, which we might take to be an exponential construct

$$p(\mathbf{z}|c_k) = \exp\left[F(\boldsymbol{\theta}_k) + B(\mathbf{z}, \boldsymbol{\phi}) + \boldsymbol{\theta}_k \cdot \mathbf{z}\right], \qquad (17.5.32)$$

where $F(\boldsymbol{\theta}_k)$ is a function of a set of parameters $\boldsymbol{\theta}_k$ that vary with the class k, $\boldsymbol{\phi}$ is a set of universal parameters, and $B(\mathbf{z}, \boldsymbol{\phi})$ is a function of $\mathbf{z}$. This form is sufficiently general to encompass the most commonly employed distribution functions such as the Gaussian and Bernoulli (see Section 17.2), binomial, Poisson, and various other distributions. For binary classification, using Bayes' theorem, we can write the probability for one of the two classes, say c_1, as

$$P(c_1|\mathbf{z}) = \frac{p(\mathbf{z}|c_1)P(c_1)}{p(\mathbf{z}|c_1)P(c_1) + p(\mathbf{z}|c_2)P(c_2)}$$

$$= \frac{1}{1 + \frac{p(\mathbf{z}|c_2)P(c_2)}{p(\mathbf{z}|c_1)P(c_1)}}$$

$$= \frac{1}{1 + e^{-a}}, \tag{17.5.33}$$

where

$$a = \ln\left(\frac{p(\mathbf{z}|c_1)P(c_1)}{p(\mathbf{z}|c_2)P(c_2)}\right), \tag{17.5.34}$$

which is a linear function of $\mathbf{z}$ of the form

$$a = \mathbf{w} \cdot \mathbf{z} + w_0$$

$$w_0 = F(\boldsymbol{\theta}_1) - F(\boldsymbol{\theta}_2) + \ln\left(\frac{P(c_1)}{P(c_2)}\right)$$

$$\mathbf{w} = \boldsymbol{\theta}_1 - \boldsymbol{\theta}_2. \tag{17.5.35}$$

We only need to determine $P(c_1|\mathbf{z})$, as we can determine $P(c_2|\mathbf{z})$ from $P(c_2|\mathbf{z}) = 1 - P(c_1|\mathbf{z})$. When a is written this way, we see that the argument of the activation function takes the expected form of a weighted linear combination of components of $\mathbf{z}$ with the bias w_0. This analysis tells us that $H(x)$ should be chosen as the logistic sigmoid function $H(x) = 1/(1 + \exp(-x))$ (eqn (17.5.12)). By the same analysis, if there are $C > 2$ classes, then the Bayes' theorem along with eqn. (17.5.32) leads to

$$P(c_k|\mathbf{z}) = \frac{p(\mathbf{z}|c_k)P(c_k)}{\sum_{l=1}^{C} p(\mathbf{z}|c_l)P(c_l)}$$

$$= \frac{e^{-a_k}}{\sum_{l=1}^{C} e^{-a_l}}, \tag{17.5.36}$$

where

$$a_k = \mathbf{w} \cdot \mathbf{z} + w_{k0} \tag{17.5.37}$$

with

$$\mathbf{w}_k = \boldsymbol{\theta}_k, \qquad w_{k0} = F(\boldsymbol{\theta}_k) + \ln P(c_k) \tag{17.5.38}$$

(see Problem 17.12). These conditions require that $H(x)$ be chosen as the *softmax* or *normalized exponential* function

$$H(x; \beta) = \frac{e^{-\beta_k x}}{\sum_{l=1}^{C} e^{-\beta_l x}} \tag{17.5.39}$$

which depends on a vector $\boldsymbol{\beta}$ of parameters.

We now turn to the determination of the correct loss function for classification. Just as the Gaussian distribution determined the loss function for regression, the Bernoulli and categorical distributions in Section 17.2 determine the form of the loss function for classification. Once again, we first consider the binary classification problem with two classes c_1 and c_2. If $y_1(\mathbf{x}, \mathbf{w})$ is the neural network used to determine membership in class c_1, with $y_2(\mathbf{x}, \mathbf{w}) = 1 - y_1(\mathbf{x}, \mathbf{w})$, then we use eqn. (17.3.8), relating the loss function to the negative logarithm of a probability distribution, to derive $E(\mathbf{w})$ from the negative logarithm of the Bernoulli distribution. Referring to the distribution in eqn. (17.2.6), we interpret x as N known input classifications $y^{(i)}$ corresponding to N input vectors $\mathbf{x}_i$. Then, since $-\ln P_{\mathrm{B}}(x) = -[x \ln \nu + (1-x) \ln(1-\nu)]$, the correct loss function $E(\mathbf{w})$ for binary classification becomes

$$E(\mathbf{w}) = -\frac{1}{N} \sum_{i=1}^{N} \left[y_1^{(i)} \ln y_1(\mathbf{x}_i, \mathbf{w}) + y_2^{(i)} \ln y_2(\mathbf{x}_i, \mathbf{w}) \right]$$

$$= -\frac{1}{N} \sum_{i=1}^{N} \left[y_1^{(i)} \ln y_1(\mathbf{x}_i, \mathbf{w}) + \left(1 - y_1^{(i)}\right) \ln \left(1 - y_1(\mathbf{x}_i, \mathbf{w})\right) \right], \quad (17.5.40)$$

which is known as a *cross-entropy* loss function. Training then proceeds by minimizing $E(\mathbf{w})$ with respect to the network parameters $\mathbf{w}$, just as in the regression problem.

When there are more than just two classes ($C > 2$), we use the categorical distribution in eqn. (17.2.7) to derive an appropriate loss function $E(\mathbf{w})$. Taking the negative log of this distribution and substituting in the input training data $(\mathbf{x}_i, y_k^{(i)})$ and the neural networks $y_k(\mathbf{x}, \mathbf{w})$, we obtain the multi-state cross-entropy loss function

$$E(\mathbf{w}) = -\frac{1}{N} \sum_{i=1}^{N} \sum_{k=1}^{C} y_k^{(i)} \ln y_k(\mathbf{x}_i, \mathbf{w}) \quad (17.5.41)$$

with $y_k^{(i)} \in [0, 1]$, and $\sum_{k=1}^{C} y_k^{(i)} = 1$.

17.5.2 Convolution layers in neural networks

We have discussed only one type of data transformation in neural networks thus far, specifically, activation of simple linear combinations in feed-forward networks. However, neural networks can incorporate other types of transformations in their effort to enhance or blur certain input features. For example, image classification via neural networks can be achieved by drawing out features capable of distinguishing one image from another. This kind of feature enhancement (or its opposite, obfuscation) can be achieved by means of filters applied to an input data stream. Filters are generally applied by convolving them with the data using a discrete version of a convolution operation, such as we encountered in Chapter 15 in our discussion of the generalized Langevin equation.

Suppose the input to a neural network is not a vector but a matrix x of dimension $n_r \times n_c$. Such a matrix could hold information about the pixels in an image, for example. Let F be a matrix of dimension $N_r \times N_c$, which we refer to as a *filter*. Then, we define the two-dimensional (2D) convolution of x with F as a matrix X of size $n_r - N_r + 1 \times n_c - N_c + 1$ given by

$$X_{IJ} \equiv (x \circ F)_{IJ} = \sum_{i=0}^{N_r-1} \sum_{j=0}^{N_c-1} x_{I+i,J+j} F_{ij}. \tag{17.5.42}$$

Convolutions can be similarly defined for 1D arrays, 3D arrays, and, generally, tensors of any dimension, depending on the number of indices needed to describe the input data. As an example of a convolution, consider an input matrix x and filter F specified as

$$x = \begin{pmatrix} 1 & 2 & 3 & 1 \\ 4 & 5 & 6 & 1 \\ 7 & 8 & 9 & 1 \end{pmatrix}, \qquad F = \begin{pmatrix} 1 & 1 \\ 1 & 1 \end{pmatrix},$$

then

$$x \circ F = \begin{pmatrix} 12 & 16 & 11 \\ 24 & 28 & 17 \end{pmatrix}.$$

In neural networks, convolution layers perform operations such as that in eqn. (17.5.42), and it is the filter matrix that must be learned via training. In addition, since convolutions are linear transformations, it is common to finalize the transformation of the layer by running X_{IJ} through an activation function to give a new matrix $Z_{IJ} = h(X_{IJ} + b)$, where b is a bias. Finally, it is also possible to train multiple filters in a convolution layer by adding an additional index to the filter. Multiple filters are used to extract multiple features from the input data. For the 2D convolution in eqn. (17.5.42), multiple filters would be included by modifying the definition to read

$$X_{IJk} = \sum_{i=0}^{N_r-1} \sum_{j=0}^{N_c-1} x_{I+i,J+j} F_{ijk}, \tag{17.5.43}$$

where $k = 1, ..., N_f$ indexes the number of desired filters.

When convolution layers are used in networks that are also partially feed-forward networks, it is necessary to feed the output of a convolution layer into an activation layer whose input is a one-dimensional vector. This is done via an intermediate *flattening* layer in which a multidimensional layer Z_{IJ} is converted into a one-dimensional array shaped for input into an activation layer.

17.6 Weighted neighbor methods

In this section, we will briefly describe two additional machine learning models that fall into a class of techniques known as *weighted neighbor* methods, specifically *k-nearest neighbors* and *random forests*. The idea behind weighted neighbor methods is to predict unknown values of a function y at a point $\mathbf{x}$ using only the nearest neighbors of $\mathbf{x}$ within the training set. This class of methods derives from the notion of regression or decision trees, depending on the desired task, in which we begin by partitioning the n-dimensional space $\mathbb{R}^n$ into $\mathcal{M}$ regions R_i, chosen such that the

(a)

(b)

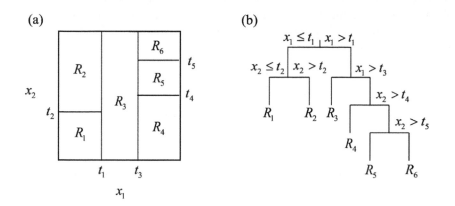

Fig. 17.5 Illustration of a splitting scheme for two-dimensional data (a) with an associated decision tree graph (b).

function $y(x_1, ..., x_n) \equiv y(\mathbf{x})$ has a constant or nearly constant value w_i within each region. This subdivision allows us to create a model for $y(\mathbf{x})$ given by

$$y(\mathbf{x}, \mathbf{w}) = \sum_{i=1}^{\mathcal{M}} w_i h(\mathbf{x} \in R_i), \qquad (17.6.1)$$

where $h(\mathbf{x} \in R)$ is an indicator function that is 1 if $\mathbf{x} \in R$ and 0 otherwise. $w_1, ..., w_{\mathcal{M}} \equiv \mathbf{w}$ are parameters representing the value of y in each region. If we minimize a least-squares loss function in eqn. (17.3.2) over the training data $(\mathbf{x}_1, y_1), ..., (\mathbf{x}_N, y_N)$, the result is

$$w_k = \frac{\frac{1}{N} \sum_{i=1}^{N} y_i h(\mathbf{x}_i \in R_k)}{\frac{1}{N} \sum_{i=1}^{N} h(\mathbf{x}_i \in R_k)}, \qquad (17.6.2)$$

which is just the average value of the target function in R_k over the training set. Unfortunately, determining the regions R_k is nontrivial, as obtaining an optimized splitting of $\mathbb{R}$ increases in complexity with the size of the training set and the dimension of $\mathbf{x}$. Figure 17.5(a) illustrates the region-splitting procedure for two-dimensional data. The definitions of the regions R_i can be gleaned from the figure; for example, R_1 is the region for which $x_1 < t_1$ and $x_2 < t_2$, R_2 is the region for which $x_1 < t_1$ and $x_2 > t_2$, and so forth. Note that the splitting procedure can be represented in a graph structure known as a *decision tree*. We will return to this decision tree graph shortly when we discuss ensemble methods. First, we introduce an approximate, yet tractable, protocol for approaching this splitting problem.

Equation (17.6.2) allows us to construct a model for $y(\mathbf{x})$ in a local neighborhood of the point $\mathbf{x}$. The approximation takes the form

$$y(\mathbf{x}) \approx \sum_{i=1}^{K} W(\mathbf{x}, \mathbf{x}_i) y_i, \qquad (17.6.3)$$

where $W(\mathbf{x}, \mathbf{x}_i)$ is a non-negative weight for the ith training point within a cluster of K neighbors of the point $\mathbf{x}^2$. Each machine learning model of this form will have a different set of associated weights. The following choice for $W(\mathbf{x}, \mathbf{x}_i)$ defines the *K-nearest neighbors* model:

$$W(\mathbf{x}, \mathbf{x}_i) = \begin{cases} \dfrac{1}{Kd_i} & i = 1, ..., K \\ \\ 0 & \text{otherwise.} \end{cases} \qquad (17.6.4)$$

Here, d_i is an additional parameter whose value depends on the specific K-nearest neighbor algorithm employed. For example, we would set $d_i = 1$ if we could assume that all neighboring points carried equal weight in determining $y(\mathbf{x})$ in the neighborhood of $\mathbf{x}$. This parameter could also be based on a distance metric that weights closer neighbors more heavily than more distant neighbors. In practice, K is a hyperparameter chosen to result in the lowest error in a cross-validation procedure. The K-nearest neighbors model results in an approximation $y_{K-\mathrm{NN}}(\mathbf{x})$ for $y(\mathbf{x})$ that takes the form

$$y_{k-\mathrm{NN}}(\mathbf{x}) = \sum_{i=1}^{K} \frac{y_i}{Kd_i}. \qquad (17.6.5)$$

Clearly, values of $y(\mathbf{x})$ can only be accurately predicted in regions for which the model has been trained.

Figure 17.5(b) depicts the decision tree corresponding to the splitting of $\mathbb{R}^2$ in Fig. 17.5(a). In general, splitting is performed according to a set of rules defined by logical functions that partition data as evenly as possible into the different regions R_i. As Fig. 17.5(b) illustrates, a decision tree consists of a root node and internal nodes set by the splitting rules, which ultimately dictate the path from the root node to a set of terminal or decision nodes, also referred to as *leaves*. In regression problems, splitting rules are determined by minimization of the relative errors (or variances) at each split until the tree grows to a pre-specified cutoff or until the data can no longer be split. The weight associated with each point in the training set is given by

$$\tilde{W}(\mathbf{x}, \mathbf{x}_i) = \begin{cases} \dfrac{1}{K} & i = 1, ..., K \\ \\ 0 & \text{otherwise.} \end{cases} \qquad (17.6.6)$$

The distinction between $W(\mathbf{x}, \mathbf{x}_i)$ and $\tilde{W}(\mathbf{x}, \mathbf{x}_i)$ will become clear by the end of this paragraph. Note that $\tilde{W}(\mathbf{x}, \mathbf{x}_i)$ also represents the weight of each point for a single decision tree. Here, K is the number of points within the same leaf at the target point $\mathbf{x}$. Unlike K-nearest neighbors, the number of neighbors within each leaf can vary among leaves in a tree. The difficulty with the use of a single decision tree in

---

[2]Weighted neighbor methods define a directed, weighted graph structure on a data set, in which nodes are represented by the data points $\{(\mathbf{x}_i, y_i)\}$ and edges are directed from point i to point j, assuming that j is among the K neighbors of i. The weight of each edge is given by the weight function W connecting points i and j.

applying eqn. (17.6.1) for regression is that one tree has a tendency to overfit the training data. An approach by which this overfitting problem can be avoided when using decision trees is to divide the training data into random subsets, an approach known as *bootstrap aggregation* or *bagging*, and to create a decision tree for each subset. Each decision tree is created with a different set of splitting rules, and the collection of all decision trees forms an ensemble known as a *random forest*. In order to determine the weights for a random forest, we average the weights in eqn. (17.6.6) over all trees in the forest (ensemble). Thus, if there are m trees in the forest, then $W(\mathbf{x}, \mathbf{x}_i) = (1/m) \sum_{j=1}^{m} \tilde{W}_j(\mathbf{x}, \mathbf{x}_i)$, where $\tilde{W}_j$ is the weight given in eqn. (17.6.6) for the jth decision tree. These average weights are then used to produce a model $y_{\mathrm{RF}}(\mathbf{x})$ using eqn. (17.6.3). A random forest generates an ensemble of decision trees, which reduces the overall variance in predictions of $y(\mathbf{x})$ without overfitting the training data.

17.7 Demonstrating machine learning in free-energy simulations

In this section, we examine the performance of the different machine learning models in this chapter for their ability to represent free-energy surfaces accurately, generate observables from these surfaces, and drive rare-event simulations. Our first demonstration involves the regression of high-dimensional free-energy surfaces produced by enhanced sampling calculations. The second demonstration illustrates the use of classification neural networks to design collective variables for use in enhanced-sampling simulations, applied here to a solid-solid phase transition in a bulk metallic system.

17.7.1 Regression of free-energy surfaces

As noted at the end of Section 17.4, regression of high-dimensional free-energy surfaces is relatively straightforward within methods such as d-AFED/TAMD or replica-exchange Monte Carlo (or replica-exchange molecular dynamics). These methods scatter points over the free-energy surface that can be used to train a machine learning model. The longer the simulation runs, the more points will be generated and the better the training will be. Since the goal of leveraging a machine learning model is to represent the function $A(s_1, ..., s_n) \equiv A(\mathbf{s})$, the input needed to train the model is the M values of $\mathbf{s}$, $\mathbf{s}_i$, $i = 1, ..., M$ and corresponding free-energy values $A(\mathbf{s}_k)$. Once trained, the model provides a compact, smooth, closed-form representation $A(\mathbf{s})$ of the free-energy surface that can be evaluated at any desired point $\mathbf{s}$. This representation can then be analyzed for its landmark points (minima and saddle points) (Chen *et al.*, 2015), fed back into a simulation as a bias to accelerate it further (Zhang *et al.*, 2018; Wang *et al.*, 2021), and employed to generate observable properties of interest via evaluation of integrals over Boltzmann factors $\exp(-\beta A(\mathbf{s}))$ (Cendagorta *et al.*, 2020). In particular, if $a(\mathbf{r})$ is a coordinate-dependent function, then we can obtain a canonical average $\langle a \rangle$ of $a(\mathbf{r})$ from an enhanced-sampling simulation as follows:

$$\langle a \rangle = \frac{\int \mathrm{d}\mathbf{s}\, \langle a \rangle_{\mathbf{r}}(\mathbf{s}) \mathrm{e}^{-\beta A(\mathbf{s})}}{\int \mathrm{d}\mathbf{s}\, \mathrm{e}^{-\beta A(\mathbf{s})}}, \tag{17.7.1}$$

where $\langle a \rangle_{\mathbf{r}}$ is given by

$$\langle a \rangle_{\mathbf{r}} = \frac{\int d\mathbf{r} \, a(\mathbf{r}) \, e^{-\beta U(\mathbf{r})} \prod_{\alpha=1}^{n} \delta \left(f_{\alpha}(\mathbf{r}) - s_{\alpha} \right)}{\int d\mathbf{r} \, e^{-\beta U(\mathbf{r})} \prod_{\alpha=1}^{n} \delta \left(f_{\alpha}(\mathbf{r}) - s_{\alpha} \right)} \qquad (17.7.2)$$

(cf. eqn. (8.6.6)). Here, $f_{\alpha}(\mathbf{r})$ is a set of collective variables and $U(\mathbf{r})$ is the potential energy of the system. Although enhanced-sampling simulations deliver sampled values of $A(\mathbf{s})$, in order to apply eqns. (17.7.1) and (17.7.2), we need an analytical representation of the free-energy surface, which is what the machine learning model provides. Thus, in applying eqn. (17.7.1), the function $A(\mathbf{s})$ is replaced by the machine-learned model, which we denote as $A_{\mathrm{ML}}(\mathbf{s})$, and averages are computed from the integral using either molecular dynamics or a Monte Carlo algorithm. If an observable of interest is a function only of collective variables, then eqn. (17.7.2) is not needed. Although we will only consider classical free-energy surfaces in this section, machine learning models can be applied equally well to quantum free-energy surfaces generated from path-integral simulations (see Section 12.7).

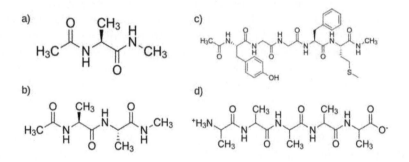

Fig. 17.6 Four molecules employed to test machine learning regression of free-energy landscapes: (a) alanine dipeptide, (b) alanine tripeptide, (c) met-enkaphalin oligopeptide (amino acid sequence Tyr-Gly-Gly-Phe-Met), (d) zwitterionic alanine pentapeptide (reproduced with permission from Cendagorta *et al.*, *J. Phys. Chem. B* **124**, 3647 (2020), copyright American Chemical Society).

For this comparative study, we will focus on a set of small peptides, commonly used as benchmark cases, and the corresponding conformational free-energy landscapes as a function of their backbone Ramachandran dihedral (ϕ, ψ) angles, which are used as collective variables. The four systems are: the alanine dipeptide, the alanine tripeptide, and the oligopeptide met-enkaphalin (amino acid sequence Tyr-Gly-Gly-Phe-Met), which are studied in vacuum; and the alanine pentapeptide, which is studied in zwitterionic form in aqueous solution. These molecules are pictured in Fig. 17.6. For the alanine dipeptide, there are just two Ramachandran angles, for the alanine tripeptide, the number of angles used is four. For met-enkephalin, ten angles are needed, and for the solvated alanine pentapeptide, the inner three residues and corresponding six Ramachandran angles are selected, as these are the same as have been used in experimental studies (Feng *et al.*, 2016). The gas-phase simulations are performed using the

CHARM22 force field (MacKerell *et al.*, 1998) while the solvated alanine pentapeptide is simulated using the OPLS-AA force field (Jorgensen *et al.*, 1996). All of the training data for the machine learning models are generated from d-AFED/TAMD simulations (see Section 8.10). The simulation parameters are set as follows: for the di- and tripeptides, $T_s = 1500$ K and $\mu_\alpha = 168.0$ amu·Å^2/rad^2; for met-enkephalin, $T_s = 400$ K and $\mu_\alpha = 2.8$ amu·Å^2/rad^2; for the alanine pentapeptide, $T_s = 1000$ K and $\mu_\alpha = 168.0$ amu·Å^2/rad^2. In all simulations, the harmonic coupling between the collective variables and the coarse-grained variables is 2.78×10^3 kcal/mol·rad^2.

Training and test data. For the alanine dipeptide, 9×10^4 values of (s_1, s_2) and $A(s_1, s_2)$ are generated as a training set. These values cover the entire free-energy surface, which is partitioned into a 300 by 300 grid of evenly spaced bins. For the alanine tripeptide, a training set of 2×10^5 points on the four-dimensional free-energy surface is randomly selected from the d-AFED/TAMD trajectory, and free-energy values $A(s_1, s_2, s_3, s_4)$ are obtained from a Gaussian fit to the histogram corresponding to these points (Chen *et al.*, 2012). For met-enkephalin, the 1081 minima and 1431 index-1 saddle points and their corresponding free-energy values identified by Chen *et al.* on the ten-dimensional surface (2015) are employed as the training set. Finally, for the aqueous alanine pentapeptide, a 1 ms simulation is performed and 10^6 free-energy points are randomly selected from a Gaussian fit to the histogram as the training set. For all systems, an additional 50,000 points are randomly generated in separate d-AFED/TAMD runs, and these points are used as a test set. The training and test sets are carefully checked to ensure there is no overlap between these two sets. For complete consistency, all machine learning models used in the comparison are trained on the same training sets for each system and tested using the same test set.

Machine learning model details. Each of the machine learning models employed in this comparative study involve hyperparameters that must be chosen. For Gaussian-based kernel methods, there are two hyperparameters, specifically, the Gaussian width σ and the regularization or ridge parameter λ. The random forest and K-nearest neighbor models have a number of hyperparameters. For K-nearest neighbors, if we choose $d_i = 1$ (as we do here), then the only hyperparameter that needs to be determined is the value of K. For the random-forest model, the key parameters are the number of trees in the ensemble or forest and the number of input variables to place into each random subset. For feed-forward neural networks, the hyperparameters are the number of hidden layers and the number of nodes in each layer. In this comparison, a ten-fold cross validation is used to perform the hyperparameter search. The resulting neural network for the alanine di- and tripeptides consists of two hidden layers with 20 nodes in each layer for the dipeptide and 40 nodes in each layer for the tripeptide. For met-enkephalin, three hidden layers are employed with 100, 50, and 50 nodes in each of the three layers, respectively. For the aqueous alanine pentapeptide, three hidden layers are employed with 60, 30, and 30 nodes in each layer, respectively. For the kernel, K-nearest neighbors and random-forest models, the resulting hyperparameters depend on the size of the training set, and as the learning curves are generated as a function of the training set size, the number of parameters determined is quite large and can be found in tables in the supporting information document accompanying the work of Cendagorta *et al.* (2020); the interested reader is encouraged to study these tables. It is worth noting that for larger training set sizes used with random forests, the number of trees is more than 200 for the di- and tripeptides and approximately 50 for met-enkephalin and the alanine pentapeptide. The learning curves are performed with respect to the test set using the L_2 error formula

$$L_2 = \sqrt{\frac{1}{N_{\text{test}}} \sum_{j=1}^{N_{\text{test}}} (A_{ML}(\mathbf{s}_j) - A_{\text{test}}(\mathbf{s}_j))^2}, \qquad (17.7.3)$$

where N_{test} is the number of points in the test set (here, 50,000) and $A_{\text{test}}(\mathbf{s}_j)$ is the (known) free energy at the jth test point.

Generating observables. In order to test the ability of the training machine learning models to generate observables from eqn. (17.7.1), we select different types of observables for each system. For the alanine tripeptide, we study the following "observable":

$$O(\{\phi, \psi\}) = \sqrt{\frac{1}{2n} \sum_{i=1}^{n} \left[\left(\phi_i - \phi_i^{(\text{min})}\right)^2 + \left(\psi_i - \psi_i^{(\text{min})}\right)^2 \right]}, \qquad (17.7.4)$$

where n is the number of Ramachandran angle pairs used to generate the free-energy surface ($n = 2$ for the tripeptide). The angles $\phi_i^{(\text{min})}$ and $\psi_i^{(\text{min})}$ are the angles at the global minimum of the free-energy surface. Although this is not a physical observable, it is a sensitive test of the ability of the machine learning model to generate an observable that depends on the full set of collective variables. For met-enkephalin, we compute the average of the $H_N H_\alpha$ nuclear magnetic resonance (NMR) J-couplings, which characterizes the indirect interaction between the nuclear spins of the C_α hydrogen and the amide hydrogen. These J-couplings can be computed using the Karplus equation (Karplus, 1959)

$$J(\phi) = A \cos^2(\phi - \phi_0)^2 + B \cos(\phi - \phi_1) + C, \qquad (17.7.5)$$

where ϕ is the Ramachandran angle, $A = 7.09$ Hz, $B = 1.42$ Hz, $C = 1.55$ Hz, and the constant angles ϕ_0 and ϕ_1 are both $60°$. $J(\phi)$ is computed for each amino acid residue in the oligopeptide. Finally, for the alanine pentapeptide, we focus on the propensities for different secondary structural motifs, specifically, α helix, β sheet, and the left-handed polyproline II helix (ppII). These are defined by simple indicators that are functions of ϕ and ψ and define specific regions in the ϕ-ψ plane for each alanine residue. The definitions are as follows:

$$\alpha : -160° < \phi < -20° \text{ and } -120° < \psi < 50°$$
$$\beta : -180° < \phi < -90° \text{ and } 50° < \psi < 180° \text{ or}$$
$$-180° < \phi < -90° \text{ and } -180° < \psi < -120° \text{ or}$$
$$160° < \phi < 180° \text{ and } 110° < \psi < 180°$$
$$\text{ppII} : -90° < \phi < -20° \text{ and } 50° < \psi < 180° \text{ or}$$

$$-90° < \phi < -20° \text{ and } -180° < \psi < 120°. \tag{17.7.6}$$

For the OPLS-AA force field used here, the populations of α, β, and ppII are 14%, 48%, and 37%, respectively. The remaining 1% of structures are characterized simply as random coil.

Results. All data for Figures 17.7, 17.8, and 17.9 are taken from Cendagorta, *et al.* (2020). Figure 17.7 depicts the learning curves for the four systems. For the simple alanine dipeptide system, we see that all models learn at approximately the same rate as a function of the number of training points. However, the kernel methods—kernel ridge regression and the least-squares support-vector machine—achieve the lowest overall error. The errors associated with the neural network and weighted–neighbor methods are roughly the same. In all cases, however, the error is only a fraction of a kcal/mol. Turning to the alanine tripeptide, for which the dimensionality of the free-energy surface is increased to four, we see that the neural network outperforms the weighted neighbor methods, while the kernel methods still achieve the lowest error overall. The weighted neighbor methods learn at a slower rate and are not able to achieve as low an error as the other models. For the ten-dimensional restricted free-energy surface of met-enkephalin, we see that the two kernel methods now perform rather differently, with the support-vector machine outperforming kernel ridge regression, while the neural network performs similarly to the support-vector machine. Note, also, that the learning rates are different for the different methods, and that the weighted-neighbor methods reach errors just below 1 kcal/mol, although it appears that with more training data, they might be able to achieve a lower error. Finally, for the alanine pentapeptide, for which we have generated the full six-dimensional free-energy surface, the neural network, kernel ridge regression, and support-vector machine models reach roughly the same error, which is lower than 1 kcal/mol; however, the neural network reaches this error with fewer training points. Once again, we see that the weighted neighbor methods underperform compared to the kernel and neural network models.

Our comparison shows that dimensionality and sample set influence the performance of different machine learning models in capturing the full free-energy surface. By contrast, accurate calculation of observables depends on how well the low free-energy regions are described by the machine learning model due to the $\exp(-\beta A(\mathbf{s}))$ factor in the integrand of eqn. (17.7.1). As we will now demonstrate, this weighting changes the comparison and highlights which methods exhibit the best performance in representing these regions.

The protocol for calculating the ensemble averages in eqn. (17.7.1) is to replace $A(\mathbf{s})$ with the representation $A_{\mathrm{ML}}(\mathbf{s})$ of the free-energy surface associated with a particular machine learning model and then perform the averages using a Metropolis Monte Carlo algorithm (see Section 7.3.3) in which trial moves of $\mathbf{s}$ are generated from a uniform distribution and the change in $A_{\mathrm{ML}}(\mathbf{s})$ is used to determine whether the trial move is accepted. In Fig. 17.8, we show the convergence of the observables in eqns. (17.7.4) for the alanine tripeptide, (17.7.5) for met-enkephalin, and (17.7.6) for the alanine-pentapeptide as a function of the training set size. For the RMSD observable in eqn. (17.7.4), we see that all models perform well for large training sets, with the neural network and least-squares support-vector machine outperforming the others in

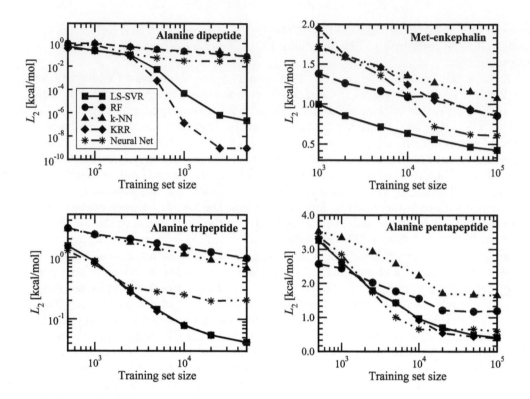

Fig. 17.7 Free-energy surface learning curves for the four peptide molecules studied.

a manner consistent with the learning curve in Fig. 17.7. Interestingly, for smaller training set sizes, we see that the random-forest method performs marginally better than the neural network and the least-square support-vector machine, suggesting that the random forest is learning the low free-energy regions with fewer data points than the kernel and neural network models. For the conformational populations of the alanine pentapeptide, we see that the neural network generates the most accurate averages across the three populations, consistent with the learning curve in Fig. 17.8. For met-enkephalin and the calculation of the average J-couplings for each of the five amino acid residues, we see from Fig. 17.8 that the neural network exhibits the lowest overall error in generating converged averages, outperforming the least-square support-vector machine. This is somewhat surprising given that the latter achieved better overall accuracy of the global free-energy surface, as reflected in the learning curve. More surprisingly, perhaps, are the accurate averages generated by the random forest for both small and large training set sizes for all residues except Phe.

For insight into the performance of the various methods for met-enkephalin, we show, in Fig. 17.9, a scatter plot of 5000 randomly selected points from the test set on models trained using 10^5 training points. The plot shows the difference between free-

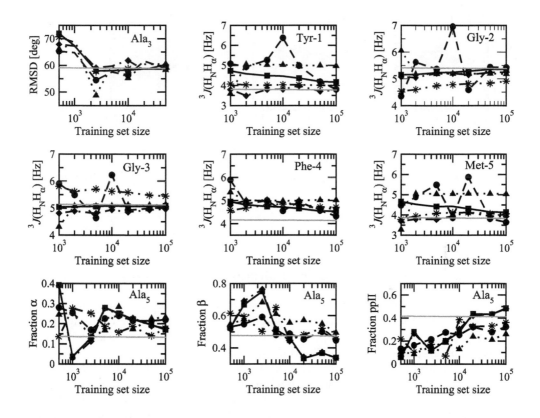

Fig. 17.8 Dependence of observables on training set size: root-mean square deviation of Ramachandran angles from the global minimum (cf. eqn. (17.7.4)) for the alanine tripeptide, NMR J-couplings for each of the five residues in met-enkephalin (cf. eqn. (17.7.5)), and conformational populations (α, β, PPII) of the alanine pentapeptide (cf. eqn. (17.7.6)). Horizontal lines indicate the fully converged values of each observable. The line types and symbols correspond to the legend given in Fig. 17.7.

energy values predicted by the least-squares support-vector machine, neural network, and random-forest machine learning models and direct simulation of the free-energy surface. The figure shows that the neural network and support-vector machine models predict the low free-energy regions accurately but then systematically underpredict larger free-energy values. This is particularly true for the support-vector machine model. By contrast, the random-forest model has roughly the same error across the full range of free-energy values shown in the figure; however, the differences are symmetric about zero, suggesting that the accuracy of the random-forest model may be due to a fortuitous error cancellation. The conclusion of this comparative study is that the most accurate predictions of the free-energy surfaces and observables using eqn. (17.7.1) is the feed-forward neural network model with the qualification that other

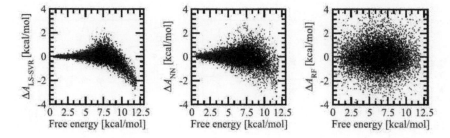

Fig. 17.9 Difference in free energy between the least-squares support-vector machine, neural network, and random-forest machine learning models and direct simulation of the free-energy surface for met-enkephalin.

models perform well in specific cases.

17.7.2   Collective variables from classification neural networks

In this section, we discuss leveraging classification neural networks for the design of collective variables that can describe rare-event processes for use in the enhanced sampling methods described in Chapters 7 and 8. We will apply classification to study a solid-solid phase transition in a bulk atomic crystal.

Suppose the crystal has p solid phases. If a sample of this bulk material contains some amount of thermal disorder, we seek to employ machine learning to classify this sample as one of the p phases. Beyond this, if there are regions in the sample where multiple phases coexist, the machine learning model should be able to identify all of these phases in such a region. A machine learning model trained to perform these classification tasks could be used to design a collective variable capable of driving transitions between different phases. In order to devise such a classification neural network, we require suitable descriptors as input functions. These descriptors need to represent the local environment of each atom in the system, which will depend on distances between an atom and its nearest neighbors as well as angles between the vectors joining the atom to its neighbors. Descriptor functions that capture these features should satisfy a number of criteria: first, they must be invariant with respect to rotations, translations, and exchanges between atoms of the same chemical element; second, they need to be smooth, differentiable functions of the atomic coordinates; and third, they should be short-ranged in order to capture only nearest neighbors. Ideally, we prefer to work with a small number of relatively simple functions.

One possible choice of descriptors is a set of functions known as *symmetry functions*, originally introduced by Behler and Parrinello (2007) (see, also, Behler (2011)), for the development of neural network potential energy functions (see Appendix C) and suggested by Geiger and Dellago (2013) as useful descriptors of atomic environments. These functions, being evaluated within a spherical region around an atom, start with a simple cutoff function $f_c(r)$. Some choices of this function are a Fermi function

$$f_c(r) = \begin{cases} \dfrac{1}{1 + e^{[\alpha_c(r - r_c + \varepsilon_c)]}} & r < r_c \\ \\ 0 & \text{otherwise} \end{cases} \tag{17.7.7}$$

or a shifted, scaled cosine function (Rogal *et al.*, 2019):

$$f_c(r) = \begin{cases} 1 & r \le r_{\min} \\ \\ \dfrac{1}{2} \left\{ \cos\left[\left(\dfrac{r - r_{\min}}{r_c - r_{\min}} \right) \pi \right] + 1 \right\} & r_{\min} < r \le r_c \\ \\ 0 & r > r_c \end{cases} \tag{17.7.8}$$

Here, r_c is a cutoff radius that defines the spherical region within which neighbors are considered. From an appropriately defined cutoff function, we build up a series of symmetry functions that capture different features of the environment around an atom at position $\mathbf{r}_i$. If the system has N atoms at positions $\mathbf{r}_1, ..., \mathbf{r}_N \equiv \mathbf{r}$, then in terms of these positions, the simplest such function is

$$G_1^{(i)}(\mathbf{r}) = \sum_{j \neq i} f_c\left(|\mathbf{r}_{ij}|\right), \tag{17.7.9}$$

where $\mathbf{r}_{ij} = \mathbf{r}_i - \mathbf{r}_j$. The sum in eqn. (17.7.9) is, in principle, taken over all j; however, because of the short-range nature of $f_c(r)$, the sum only involves neighbors of atom i within the cutoff radius r_c. Moreover, because G_1 is defined purely in terms of $f_c(r)$, these neighbors are given roughly equal weight. Other symmetry functions give different weights to these neighbors. For example, the function

$$G_2^{(i)}(\mathbf{r}) = \sum_{j \neq i} e^{-\eta(|\mathbf{r}_{ij}| - R_s)^2} f_c\left(|\mathbf{r}_{ij}|\right) \tag{17.7.10}$$

weights neighbors whose distances from $\mathbf{r}_i$ are close to the distance parameter R_s more than those whose distances are sigificantly different from R_s. The inverse width parameter η determines how quickly this weight decays to zero. Different choices of R_s and η define different G_2 symmetry function choices. In practice, we might use a range of values of R_s and η to capture different features of the local environment. Another such symmetry function employs a cosine weighting, *i.e.*,

$$G_3^{(i)}(\mathbf{r}) = \sum_{j \neq i} \cos\left(\kappa |\mathbf{r}_{ij}|\right) f_c\left(|\mathbf{r}_{ij}|\right). \tag{17.7.11}$$

Here, the parameter κ modulates the periodicity of the cosine function such that neighbors whose distances from atom i satisfy $\kappa |\mathbf{r}_{ij}| = (2n - 1)\pi/2$ will have large positive or negative weights, depending on the value of n, and zero weight if $\kappa |\mathbf{r}_{ij}| = n\pi$. The symmetry functions G_1, G_2, and G_3 depend only on the distances between neighbors of atom i. Other symmetry functions incorporate angular dependence between the vectors $\mathbf{r}_{ij}$ and $\mathbf{r}_{ik}$. An example of such a symmetry function is

$$G_4^{(i)}(\mathbf{r}) = \frac{1}{2^\xi} \sum_{j \neq i} \sum_{k \neq i} (1 + \lambda \cos\theta_{ijk})^\xi \, e^{-\eta(|\mathbf{r}_{ij}|^2 + |\mathbf{r}_{ik}|^2 + |\mathbf{r}_{jk}|^2)}$$
$$\times f_c(|\mathbf{r}_{ij}|) \, f_c(|\mathbf{r}_{ik}|) \, f_c(|\mathbf{r}_{jk}|) \qquad (17.7.12)$$

and the closely related

$$G_5^{(i)}(\mathbf{r}) = \frac{1}{2^\xi} \sum_{j \neq i} \sum_{k \neq i} (1 + \lambda \cos\theta_{ijk})^\xi \, e^{-\eta(|\mathbf{r}_{ij}|^2 + |\mathbf{r}_{ik}|^2)} f_c(|\mathbf{r}_{ij}|) \, f_c(|\mathbf{r}_{ik}|), \qquad (17.7.13)$$

which does not restrict the distance between neighbors j and k of i. In eqns. (17.7.12) and (17.7.13), the parameter λ is either 1 or -1, while the parameter ξ modulates the angular resolution. Apart from the symmetry functions, other useful descriptors capable of capturing angular information in the local environment are the Steinhardt bond-order parameters (Steinhardt *et al.*, 1983). These are defined as

$$G_{q_l}^{(i)}(\mathbf{r}) = \sqrt{\frac{4\pi}{2l+1} \sum_{m=-l}^{l} \left| q_{lm}^{(i)}(\mathbf{r}) \right|^2} \qquad (17.7.14)$$

where

$$q_{lm}^{(i)}(\mathbf{r}) = \frac{\sum_{j \neq i} Y_{lm}(\theta_{ij}, \phi_{ij}) \, f_c(|\mathbf{r}_{ij}|)}{\sum_{j \neq i} f_c(|\mathbf{r}_{ij}|)}. \qquad (17.7.15)$$

Here, $Y_{lm}(\theta, \phi)$ is a spherical harmonic, and θ_{ij} and ϕ_{ij} are the polar and azimuthal angles of the vector $\mathbf{r}_{ij}$. The combination of symmetry functions and spherical harmonics reduces the number of descriptors needed to describe local environments in an atomic crystal.

We will now apply these descriptors to the specific case of the transformation between the metastable A15 phase in solid molybdenum to the stable BCC (body-centered cubic) phase. A snapshot showing the coexistence between these two phases in a single simulation cell is shown in Fig. 17.10. The transition occurs via the migration of the interface between the two phases to the right, which transforms each layer of the A15 phase (on the right) to the BCC phase (on the left).

In order to classify both pure and mixed phases in solid molybdenum, we only need eleven radial symmetry functions of the G_2 and G_3 type and three Steinhardt parameters corresponding to $l = 6$, 7, and 8. The parameters R_s range between 2.8 Å and 6.0 Å with η fixed at 20 Å^{-2} for G_2 while κ ranges from 3.5 Å^{-1} to 7.0 Å^{-1} for G_3. The cutoff function in eqn. (17.7.8) is employed with $r_{min} = 3.8$ Å, and $r_c = 4.0$ Å. With these, we can distinguish four solid phases, A15, BCC, FCC (face-centered cubic), and HCP (hexagonal close-packed) that can exist in the system, as well as disordered or "liquid-like" phases and mixtures of these various phases.

We now proceed to describe the training procedure of the classification neural network. Because the descriptors take in raw atomic coordinates and transform them into translationally and rotationally invariant local environment variables, the only input data we need for training are system configurations, which can be generated

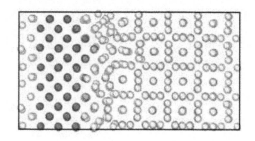

Fig. 17.10 Snapshot of a simulation cell of a system of molybdenum atoms with an interface between stable BCC phase (left region) and metastable A15 phase (right region) (reproduced with permission from Rogal *et al.*, *Phys. Rev. Lett.* **123**, 254701 (2019), copyright American Physical Society).

from molecular dynamics or Monte Carlo calculations. Training proceeds by using molecular dynamics simulations at temperatures of 300 K, 450 K, 600 K, 1000 K, 3000 K, and 4000 K to generate 176,000 pure atomic environments as well as interfacial configurations. An additional set of 125,000 atomic environments is generated as a test set. The learning curve in Fig. 17.11 shows that bulk environments are correctly classified with better than 99% accuracy, while interfacial environments containing A15, BCC, and liquid phases are correctly classified with better than 93% accuracy. The output of the neural network is a five-component classification probability vector $\mathbf{q}_i(\mathbf{r})$ for each atom.

Once complete training of the network is achieved, a collective variable capable of driving the transition is constructed. We start by defining a global classifier vector

$$\mathbf{Q}(\mathbf{r}) = \frac{1}{N} \sum_{i=1}^{N} \mathbf{q}_i(\mathbf{r}). \qquad (17.7.16)$$

The global classifier serves as a reporter on the extent to which the entire system is in one phase or the other. In Fig. 17.10, the value of $Q^{\text{bcc}} = 0.20$ while $Q^{\text{A15}} = 0.52$.

In order to drive the transition, the collective variable we employ is expressed as a path in the vector space in which the global classifier $\mathbf{Q}$ exists. The reason for working in this space is that it avoids the need to choose physical configurations between the A15 and BCC phases in order to construct a physical path (Branduardi *et al.*, 2007). Such a physical path could be biased by preconceived notions of how the transition should occur. Working in classifier space allows the neural network to decide what configurations, including pure and mixed phases, exist during the transition, which is likely to be quite complex and involve multiple ordered and disordered local environments. Thus, let $\mathbf{Q}_1, ..., \mathbf{Q}_P$ be a set of P nodal points along a putative path between the phases. This putative path exists in the two-dimensional space $(Q^{\text{bcc}}, Q^{\text{A15}})$ constructed from the BCC and A15 components of the $\mathbf{Q}$ vector. In this particular example, we start with an interface already present in the system such

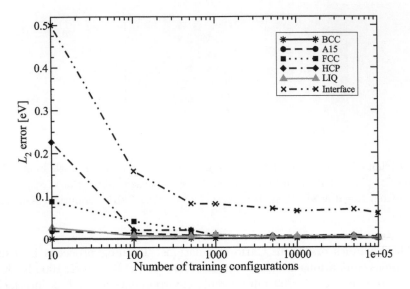

Fig. 17.11 Learning curves for each of the pure phases (BCC, A15, FCC, HCP, liquid/disordered phases) and mixed phases/interfaces.

that $\mathbf{Q}_1 = (Q_1^{\text{bcc}}, Q_1^{\text{A15}}) = (0.2, 0.5)$ and create a path with $P = 10$ points, where $\mathbf{Q}_{10} = (Q_{10}^{\text{bcc}}, Q_{10}^{\text{A15}}) = (0.65, 0.05)$. The path collective variable, inspired by the physical path form of Branduardi *et al.* (2007), allows for fluctuations around these nodal points and takes the form

$$f(\mathbf{Q}(\mathbf{r})) = \frac{1}{P-1} \frac{\sum_{k=1}^{P}(k-1)\exp\left[-\lambda|\mathbf{Q}(\mathbf{r}) - \mathbf{Q}_k|^2\right]}{\sum_{k=1}^{P}\exp\left[-\lambda|\mathbf{Q}(\mathbf{r}) - \mathbf{Q}_k|^2\right]}, \tag{17.7.17}$$

where λ is a parameter roughly determined by the inverse square distance between consecutive nodal points. An illustration of this path collective variable is given in Fig. 17.12(a). We see that $f(\mathbf{Q}(\mathbf{r})$ increases smoothly from 0 to 1 as the fraction of the BCC phase increases and that of the A15 phase decreases. Use of eqn. (17.7.17) alone can lead to large fluctuations around the nodal points, and, therefore, it is often useful to add a second collective variable that restricts these excursions. In classifier space, this collective variable takes the form

$$z(\mathbf{Q}(\mathbf{r})) = -\frac{1}{\lambda}\ln\left(\sum_{k=1}^{P}\exp\left[-\lambda|\mathbf{Q}(\mathbf{r}) - \mathbf{Q}_k|^2\right]\right). \tag{17.7.18}$$

This collective variable is illustrated in Fig. 17.12(b). The function in eqn. (17.7.18) can be used either as an additional collective variable in an enhanced-sampling simulation or to construct a restraining potential (Cuendet *et al.*, 2018; Rogal *et al.*, 2019)

$$U_r(\mathbf{r}) = \frac{1}{2}\kappa_z\left(z(\mathbf{Q}(\mathbf{r}))\right)^2, \tag{17.7.19}$$

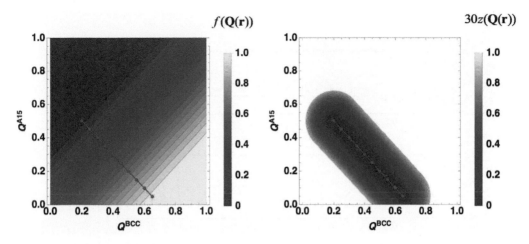

Fig. 17.12 (*Left*) Path collective variable $f(\mathbf{Q}(\mathbf{r}))$ in the Q^{bcc}-Q^{A15} plane. Red points indicate the path along the nodal points while lines are foliations of the path. (*Right*) Same for $z(\mathbf{Q}(\mathbf{r}))$ (reproduced with permission from Rogal *et al.*, *Phys. Rev. Lett.* **123**, 254701 (2019), copyright American Physical Society).

where the parameter κ_z determines the tightness of the restraint. If such a restraint is used in a simulation, then the bias must be removed, which requires reweighting with a factor $\exp(\beta U_r(\mathbf{r}))$, in order to obtain final results.

The path collective variable in eqn. (17.7.17) can now be used in an enhanced sampling simulation, such as d-AFED/TAMD or metadynamics, in order to drive the structural phase transition via migration of an interface created between the two phases. Note that using a neural network in this way incorporates machine learning directly into the enhanced sampling procedure rather than merely using it as a post-processing tool. In particular, the neural network performs a classification "on the fly" as each configuration generated by the simulation is fed into it, and it immediately outputs a classification for each atom in the system at that instant in the simulation from which $Q(\mathbf{r})$ and $f(Q(\mathbf{r}))$ can be determined. For use in molecular dynamics simulations, it is critical that the neural network employed be everywhere smoothly differentiable, which restricts the choice of activation functions.

If an enhanced sampling simulation is performed in a canonical ensemble at 300 K with fixed volume, as is shown in Fig. 17.13 for (a) d-AFED/TAMD and (b) metadynamics simulations (Rogal *et al.*, 2019), then the metastability of the A15 phase is not revealed. The reason is that the two phases have different lattice parameters, and one simulation box size cannot accommodate both phases. Nevertheless, there is a clear free-energy barrier revealed in both profiles, which is approximately 0.5 eV $\approx$ 48.2 kJ/mol and which agrees with previous independent computational studies performed on the same system (Duncan *et al.*, 2016). This free-energy barrier corresponds to the thermodynamic loss of converting each layer in the A15 crystal to the BCC structure under the constant-volume conditions. If we switch from the canonical to the isothermal-isobaric ensemble at 1 atm, then, as is revealed in Fig. 17.13(c), the

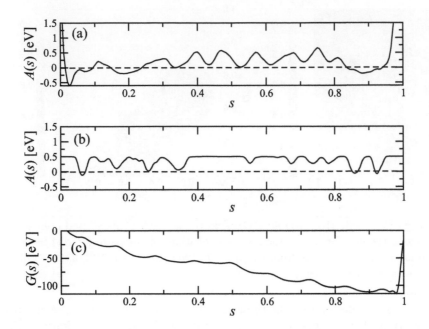

Fig. 17.13 (a) Free-energy profile at 300 K and constant volume from d-AFED/TAMD using $f(\mathbf{Q}(\mathbf{r}))$ as the collective variable. (b) Same for metadynamics. (c) Free-energy profile at 300 K and 1 atm pressure showing the metastability of the A15 phase relative to BCC (reproduced with permission from Rogal *et al.*, *Phys. Rev. Lett.* **123**, 254701 (2019), copyright American Physical Society).

free-energy profile acquires a negative slope, indicating that it is thermodynamically "downhill" from the A15 to the BCC phase, which manifestly reveals the metastability of the A15 phase relative to the BCC phase. However, the 0.5 eV barrier for each layer transition is retained, giving the profile a kind of staircase-like character[3].

17.7.3 Reaction coordinates from regression neural networks

In Section 8.12, we alluded to the challenge of determining a proper reaction coordinate to describe a particular process. We introduced the committor distribution $p_{\mathcal{B}}(\mathbf{r})$ between two stable states $\mathcal{A}$ and $\mathcal{B}$ in configuration space and the relationship between the committor and a putative reaction coordinate $q(\mathbf{r})$ capable of fully characterizing the transition from $\mathcal{A}$ to $\mathcal{B}$. Problem 8.13 asked the reader to rationalize a model of the dependence of $p_{\mathcal{B}}(\mathbf{r})$ on $q(\mathbf{r})$ (Peters *et al.*, 2007):

---

[3]The staircase-like profile is sometimes referred to as a "Galton staircase" after Sir Francis Galton (1822–1911), inventor of the Galton board, which is used to demonstrate normal distributions. The Galton staircase can be modeled by the functional form $A(s) = A_0 \cos(\alpha s) - \lambda s$. As shown by Liu and Tuckerman (2000), this type of function is a particularly challenging one for deterministic thermostatting techniques.

$$\pi_{\mathcal{B}}(q(\mathbf{r})) = \frac{1 + \tanh(q(\mathbf{r}))}{2}. \qquad (17.7.20)$$

We also presented an algorithm for computing a committor distribution, which, though useful, does not provide a closed-form expression for $p_{\mathcal{B}}(\mathbf{r})$, either directly or indirectly, because a closed-form for $q(\mathbf{r})$ that can be inserted into eqn. (17.7.20) is not specified. By employing machine learning, we can provide such a compact representation of both $q(\mathbf{r})$ and the committor distribution. There are various ways this can be achieved, but we will assume, here, that a reaction coordinate $q(\mathbf{r})$ can be obtained from a large set of possibly redundant collective variables, the space of which has been well sampled by an enhanced sampling algorithm and conformational basins on the corresponding free-energy hypersurface have been identified. What we seek is a reaction coordinate capable of describing the mechanism whereby the system transitions from one free-energy basin to another on this surface. Let the set of n collective variables be denoted $f_\alpha(\mathbf{r})$, $\alpha = 1, ..., n$. We then propose a machine learning model that expresses $q(\mathbf{r})$ in terms of $f_1(\mathbf{r}), ..., f_n(\mathbf{r})$. For example, as was suggested by Mori *et al.* (2020), a possible model for $q(\mathbf{r})$ is a simple linear combination of the collective variables

$$q(\mathbf{r}, \mathbf{w}) = w_0 + \sum_{\alpha=1}^{n} w_\alpha f_\alpha(\mathbf{r}). \qquad (17.7.21)$$

Apart from this simple linear model, any of the machine learning models, such as a feed-forward neural network or a kernel model, could be employed to represent $q(\mathbf{r})$.

Once a model is chosen, we train it by optimizing the parameters $\mathbf{w}$. Mori *et al.* suggest the use of a binary classification scheme to achieve the required training. Within such a scheme, the machine learning model $q(\mathbf{r}, \mathbf{w})$ is substituted into eqn. (17.7.20) and the loss function (cf. eqn. (17.5.40))

$$E(\mathbf{w}) = -\sum_{k=1}^{M} p_{\mathcal{B}}^*(\mathbf{r}^{(k)}) \ln \pi_{\mathcal{B}}(q(\mathbf{r}^{(k)}, \mathbf{w}))$$

$$-\sum_{k=1}^{M} \left(1 - p_{\mathcal{B}}^*(\mathbf{r}^{(k)})\right) \ln \left(1 - \pi_{\mathcal{B}}(q(\mathbf{r}^{(k)}, \mathbf{w}))\right), \qquad (17.7.22)$$

where M is the number of training points, is used to perform the optimization. In eqn. (17.7.22), we interpret $\mathbf{r}^{(k)}$ as a point in configuration space from which a trajectory is initiated that can either end in state $\mathcal{A}$ or state $\mathcal{B}$. If the trajectory ends in $\mathcal{A}$, then the target committor value $p_{\mathcal{B}}^*(\mathbf{r}^{(k)}) = 0$, and if it ends in $\mathcal{B}$, then $p_{\mathcal{B}}^*(\mathbf{r}^{(k)}) = 1$. One way to generate the trajectories needed to obtain the training data is to use the techniques in Section 7.7, such as aimless shooting. It is also helpful to add a regularization term into the loss function in eqn. (17.7.22) in order to avoid overfitting.

An alternative scheme for predicting reaction coordinates is via regression learning with a least-squares loss function. Suppose we have generated enough trajectories from a point $\mathbf{r}^{(k)}$ to obtain a converged committor distribution value $p_{\mathcal{B}}(\mathbf{r}^{(k)})$ corresponding to $\mathbf{r}^{(k)}$. Then, we can obtain a value $q^{(k)}$ for the reaction coordinate corresponding to $\mathbf{r}^{(k)}$ by inverting eqn. (17.7.20), as

$$q^{(k)} = \tanh^{-1}\left(2p_{\mathcal{B}}(\mathbf{r}^{(k)}) - 1\right). \tag{17.7.23}$$

Given a model $q(\mathbf{r}, \mathbf{w})$ for the reaction coordinate, we then train the machine learning model using these committor values via the least-squares loss function

$$E(\mathbf{w}) = \frac{1}{2M}\sum_{k=1}^{M}\left|q(\mathbf{r}^{(k)}, \mathbf{w}) - q^{(k)}\right|^2 + \text{Reg}(\mathbf{w}). \tag{17.7.24}$$

Here, $\text{Reg}(\mathbf{w})$ is a regularization term, which could be a standard ridge form, a lasso form, or an elastic net form. An optimal choice of the regularization term will depend on the choice of the machine learning model $q(\mathbf{r}, \mathbf{w})$. Finally, one could represent the committor distribution $p_{\mathcal{B}}(\mathbf{r})$ as a linear combination of collective variables as

$$p_{\mathcal{B}}(\mathbf{r}, \boldsymbol{\omega}) = \omega_0 + \sum_{\alpha=1}^{n}\omega_\alpha f_\alpha(\mathbf{r}). \tag{17.7.25}$$

and train the coefficients $\boldsymbol{\omega}$ on target committor values $p_{\mathcal{B}}^*(\mathbf{r}^{(k)})$ using either a cross-entropy or least-squares loss function.

17.8   Clustering algorithms

The machine learning models and procedures we have discussed thus far constitute examples of *supervised* learning strategies. Data in supervised learning approaches are are termed "labeled data", meaning that they are specified as inputs $\mathbf{x}_i$ and corresponding outputs y_i. In this final section, we introduce *unsupervised* learning, which is designed to handle unlabeled data. In particular, we present two examples from a class of algorithms known as *clustering* methods, which classify unlabeled data into categories, or "clusters", based on similarities between data points. Clustering reveals features that separate collections of data points from each other. Various strategies exist for performing clustering and assigning points to different groups based on particular attributes in the data. In this section, we describe two such methods: The popular K-means approach (MacQueen, 1967; Kanungo *et al.*, 2002) and the density-peaks scheme.

K-means clustering. An important feature of any cluster is its "centroid", which is the location of the average or mean over all of the data points in the cluster. The term K-means refers to a strategy whereby n data points $\mathbf{x}_1, ..., \mathbf{x}_n$ are sorted into K clusters with centroids $\boldsymbol{\mu}_1, ..., \boldsymbol{\mu}_K$. The K-means clustering algorithm proceeds first by choosing the centroids $\boldsymbol{\mu}_1, ..., \boldsymbol{\mu}_K$ randomly and then assigning the n data points $\mathbf{x}_1, ..., \mathbf{x}_n$ to each of the K clusters based on their distance to each centroid. Thus, for each data point $\mathbf{x}_i$, we compute the K distances $d_{i\gamma} = |\mathbf{x}_i - \boldsymbol{\mu}_\gamma|$, where $\gamma = 1, ..., K$ indexes the K centroids and corresponding clusters. The value of γ for which $d_{i\gamma}$ is minimal determines the cluster membership of $\mathbf{x}_i$:

$$\text{Cluster index of } \mathbf{x}_i = \arg\min_{\gamma} d_{i\gamma}. \tag{17.8.1}$$

Note that different clusters will have different numbers m_γ of points, such that $m_1 + m_2 + \cdots + m_K = n$. When all n points have been assigned to clusters in this way,

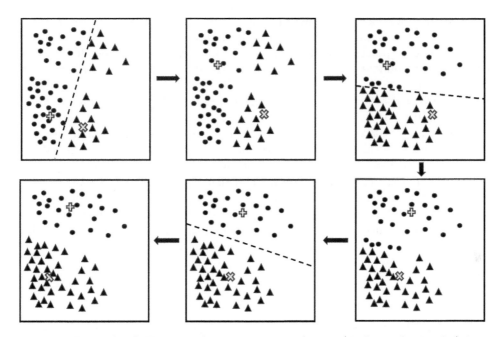

Fig. 17.14 Example of K-means clustering on two clusters (circles and triangles) in two dimensions. Following the arrows, an initial choice of centroids, shown as the "+" and "×" symbols for the two clusters, respectively, is followed by three reassignments leading to a converged assignment of cluster membership. Dashed lines divide the two clusters. In each iteration, some circles and triangles interchange according to new cluster membership assignments.

a new set of cluster centroids $\boldsymbol{\mu}_1, ..., \boldsymbol{\mu}_K$ is generated by computing the average over the data points in each cluster. Once the new centroids have been determined, the n data points are reassigned to clusters by computing new distances $d_{i\gamma}$ and using eqn. (17.8.1) to determine new cluster membership. Most likely, the assignments will change, and, consequently, the numbers $m_1, ..., m_K$ of data points in each cluster will change as well. The procedure is repeated as many times as needed until the cluster assignments no longer change. The K-means procedure is illustrated in Fig. 17.14 for a two-dimensional data set with two clusters.

$\quad K$-means clustering is both efficient and straightforward to implement. However, the parameter K must be determined *a priori*. This can be done by running the algorithm for different values of K, and for each K, calculating the average distance of all data points to their assigned cluster centroids. When plotted as a function of K, this average distance should fall off suddenly at some value of K; this value is the optimal one. Additionally, K-means clustering tends to assign outlier data points inaccurately because such points are difficult to assign to clusters and can pull centroid positions away from regions of high data population. As a final note, the K-means method can lead to inaccurate assignment of any data point that sits at a boundary between two clusters; this problem can be treated using *fuzzy clustering* approaches (Gustafson and Kessel, 1978; Bezdek *et al.*, 1984; Corsini *et al.*, 2005; Tzanov *et al.*, 2014), and although we will not describe these methods in detail here, the basic idea of fuzzy clustering is to assign points to multiple clusters with a weight for membership in each cluster. Points at boundary regions are often assigned to each cluster with weights close to 0.5 in each.

Density-based clustering. As the previous paragraph makes clear, K-means clustering assigns cluster membership based on a distance of data points to the cluster centroids, with an iterative algorithm to refine the cluster centroid locations. An alternative approach considers the density of data points near a cluster centroid by assuming that the density peaks at cluster centroids, which are surrounded by neighboring data points with a lower local density, and that the centroids are far from other points with high local density. The approach we will discuss here was introduced by Rodriguez and Laio (2014) and employs two quantities for each data point $\mathbf{x}_i$: its local density ρ_i and its distance d_i from points of higher density. Let $r_{ij} = |\mathbf{x}_i - \mathbf{x}_j|$ be the distance between data points $\mathbf{x}_i$ and $\mathbf{x}_j$. Then, the local density ρ_i is defined as

$$\rho_i = \sum_{j=1}^{n} \theta\left(r_c - r_{ij}\right), \qquad (17.8.2)$$

where $\theta(z)$ is the Heaviside step function. The quantity r_c is a cutoff distance. According to eqn. (17.8.2), ρ_i simply counts the number of points that are within a distance r_c of $\mathbf{x}_i$. In order to obtain d_i, we use the definition

$$d_i = \min_{\{j \text{ s.t. } \rho_j > \rho_i\}} r_{ij}. \qquad (17.8.3)$$

That is, we compute the smallest distance between the point $\mathbf{x}_i$ and any other point of higher density. If $\mathbf{x}_i$ is already the point of highest density, then we can compute $d_i = \max_j r_{ij}$. Cluster centroids are now recognized as points $\mathbf{x}_i$ for which d_i is anomalously large. The algorithm is illustrated in Fig. 17.15. Once the cluster centroids are determined, cluster membership of each remaining point is determined by assigning it to the same cluster as its nearest neighbor of higher density. Thus, the assignment begins with the cluster centroids, themselves, as these are points of maximum local density.

17.9 Intrinsic dimension of a data manifold

We conclude this chapter by showing how the neighbors of a data point can be used to estimate the dimensionality of a data manifold. Data sets used in learning protocols often lie in a low-dimensional space embedded in a higher dimensional one, and the lower dimension of the data set might not be obvious under such embedding. An example of such an embedding might be a Swiss roll embedded in a rectangular box (see Fig. 17.16). This lower dimensionality is known as the *intrinsic dimension* of the data manifold or space that contains the data points. Although the embedding dimension of the box in Fig. 17.16 is three, the intrinsic dimension of the Swiss-roll data set is two.

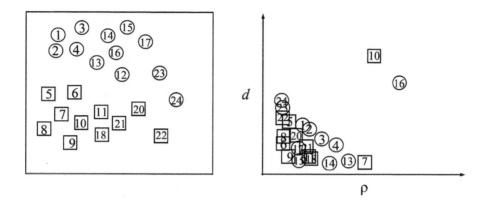

Fig. 17.15 Illustration of density-based clustering. Two clusters are shown in the left panel as circles and squares. Values of d and ρ are plotted in the right panel. The two centroids emerge as points of anomalously high d values (see right panel).

It was suggested in Section 17.5 that the change in intrinsic dimension through the layers of a neural network can be an important metric for evaluating the performance of the network. Facco *et al.* (2017) introduced an approach for estimating this intrinsic dimension using only the nearest and second nearest neighbors of each point on the data manifold. Given a set of n data points $\mathbf{x}_1, ..., \mathbf{x}_n$, let $r_1, ..., r_k$ be the k nearest neighbors of a point $\mathbf{x}_i$ in the data set. If these neighbors are arranged in ascending order such that $r_1 < r_2 < r_3 \cdots < r_k$, then r_1 and r_2 correspond to the nearest and second-nearest neighbors of $\mathbf{x}_i$, respectively. Introduce the ratio $\mu = r_2/r_1$. Since $r_2 > r_1$, it follows that $\mu \in [1, \infty)$. If the ratio μ is computed for every point in the data set, then n values of μ, $\mu_1, ..., \mu_n$ will be obtained, and a histogram of μ values can be

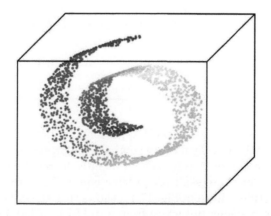

Fig. 17.16 A Swiss roll data manifold embedded in a rectangular box.

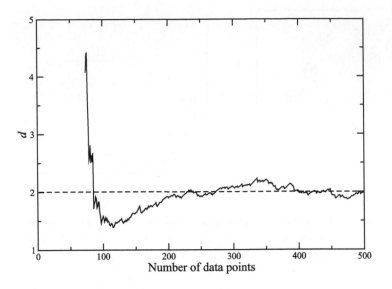

Fig. 17.17 Intrinsic dimension as a function of the size of the data set for the alanine dipeptide. The data set is taken from a d-AFED/TAMD run that uses the two Ramachandran dihedral angles as collective variables.

generated. This histogram represents a probability distribution $f(\mu)$ whose analytical form can be shown to be

$$f(\mu) = d\mu^{-d-1}\theta(\mu - 1), \qquad (17.9.1)$$

where d is the intrinsic dimension of the data set. If we now integrate $f(\mu)$, we obtain the cumulative probability $P(\mu)$ of μ values

$$P(\mu) = \int_{-\infty}^{\mu} f(y)\, dy = d\int_{1}^{\mu} y^{-d-1}\, dy = \left(1 - \mu^{-d}\right)\theta(\mu - 1). \qquad (17.9.2)$$

Equation (17.9.2) is now solved for the intrinsic dimension d to give

$$d = -\frac{\ln\left(1 - P(\mu)\right)}{\ln \mu}. \qquad (17.9.3)$$

Equation 17.9.3 prescribes a straightforward approach for calculating the intrinsic dimension of a data manifold: We only need to compute the probability $P(\mu)$ of different values of μ and feed these μ and $P(\mu)$ into eqn. (17.9.3); the result will be the intrinsic dimension d of the data set. The result of applying eqn. (17.9.3) to the sampled data set for the alanine dipeptide, shown in Fig. 17.17, shows that the intrinsic dimension of data used to obtain the free-energy surface from a d-AFED/TAMD simulation using the Ramachandran angles as collective variables (see Fig. 8.5) is two, as expected. Since this is a data set from an enhanced sampling simulation, the plot is somewhat noisier than what we would expect to observe for the synthetic data in Facco *et al.* (2017); however, a trend toward the value of two is clear. If the dimensionality of the data were not known *a priori*, this type of analysis would be capable of revealing it.

17.10   Problems

17.1. Verify the following properties for a Bernoulli distribution:

$$\sum_{x=0}^{1} P_{\mathrm{B}}(x; \nu) = 1$$

$$\langle x \rangle = \nu$$

$$\langle x^2 \rangle - \langle x \rangle^2 = \nu(1 - \nu).$$

17.2. Recall that the Shannon entropy $S[f]$ for a probability distribution $f(\mathbf{x})$ is

$$S[f] = -\int d\mathbf{x}\, f(\mathbf{x}) \ln f(\mathbf{x}).$$

a. By performing a maximization of $S[f]$ over all distributions $f(\mathbf{x})$ subject to the three constraints

$$\int d\mathbf{x}\, f(\mathbf{x}) = 1$$

$$\int d\mathbf{x}\, \mathbf{x} f(\mathbf{x}) = \boldsymbol{\mu}$$

$$\int d\mathbf{x}\, (\mathbf{x} - \boldsymbol{\mu})(\mathbf{x} - \boldsymbol{\mu})^{\mathsf{T}} f(\mathbf{x}) = \boldsymbol{\Sigma},$$

show that the resulting distribution is the Gaussian distribution of eqn. (17.2.3).

b. Show that the entropy of the Gaussian distribution in eqn. (17.2.3) is given by

$$S[P_{\mathrm{G}}] = \frac{1}{2} \ln \det(\boldsymbol{\Sigma}) + \frac{n}{2}(1 + \ln(2\pi)).$$

17.3. a. Derive eqns. (17.3.5).

b. For the linear regression model in eqn. (17.3.10), find the analytical solution for w_0 and $\mathbf{w}$ by optimizing the loss function in eqn. (17.3.11). Examine your solution in the limit that $\lambda = 0$.

17.4. Starting from the data model in eqn. (17.4.9), optimize the loss function in eqn. (17.4.10) and show that the analytical solution is given by eqn. (17.4.8).

17.5. For each of the diagrams shown above, write explicit expressions for the corresponding feed-forward neural networks, based either on eqn. (17.5.9) or on eqn. (17.5.31), depending on which general form applies.

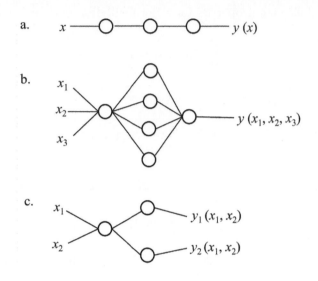

a. $x \longrightarrow y(x)$

b. $x_1, x_2, x_3 \longrightarrow y(x_1, x_2, x_3)$

c. $x_1, x_2 \longrightarrow y_1(x_1, x_2)$, $y_2(x_1, x_2)$

Fig. 17.18 Diagrams for problem 17.5.

17.6. Generalize eqn. (17.5.9) to a feed-forward neural network having M hidden layers with $m_1, ..., m_K$ nodes in each hidden layer.

*17.7. Derive eqns. (17.5.19) and (17.5.20).

*17.8. Gradient training of a neural network requires optimization of the loss function given in eqn. (17.5.25). Derive the back propagation scheme for this loss function.

17.9. Consider the following application of Bayes' theorem to a regression neural network. Let eqn. (17.3.7) be used to construct a likelihood function for the network in terms of its loss function. We now define a prior distribution

$$p(\mathbf{w}) = \left(\frac{\lambda}{2\pi}\right)^{n/2} e^{-\lambda \mathbf{w} \cdot \mathbf{w}},$$

where $\dim(\mathbf{w}) = n$. If we additionally define a new loss function as the logarithm of the posterior probability distribution, show that the new loss function is given by eqn. (17.5.11) but with an additional ridge regularization term.

17.10. The following ten data points are assumed to lie approximately along the curve $y(x) = \sin(2\pi x)$: (0,0.30), (1/9, 0.86), (2/9, 1.00), (1/3, 0.98), (4/9, 0.10), (5/9, 0.06), (2/3, −0.90), (7/9, −0.40), (8/9, −0.50), (1, 0.29) The points are plotted on the accompanying figure along with the function $y(x) = \sin(2\pi x)$. Write a program to train a regression neural network with a single hidden layer on these ten points. In particular, train three such networks, the first with one node in the hidden layer, the second with three nodes in the hidden layer, and the third with ten nodes in the hidden layer. For this problem, the use of the gradient descent algorithm in eqn. (17.5.21) is relatively straightforward; however, more advanced readers might wish to try the stochastic gradient descent approach. Plot the three functions that result from each trained network along with the ten training points. Is there an optimal number of nodes in the hidden layer? What happens if the number of nodes in the hidden layer is too large?

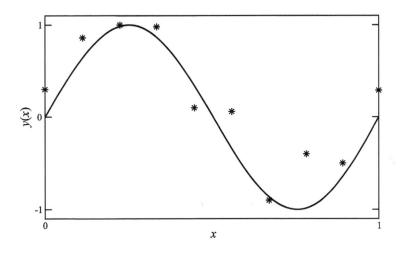

Fig. 17.19 Plot for problem 17.10.

17.11. Consider the matrix x and filter F given below

$$ x = \begin{pmatrix} 1 & 3 & 5 & 2 \\ 5 & 7 & 9 & 2 \\ 2 & 4 & 5 & 1 \end{pmatrix}, \qquad F = \begin{pmatrix} 1 & 2 \\ 2 & -1 \end{pmatrix}. $$

Determine the convolution matrix $X = x \circ F$.

17.12. Consider a classification neural network with $C > 2$ classes. Derive eqns. (17.5.36) through (17.5.38) and show that the final activation function $H(x)$ should be the softmax function given in eqn. (17.5.39). If the neural network has no hidden layers, to what explicit form does the learning model simplify?

17.13. Determine how the back propagation approach needs to be modified for a classification neural network with $C > 2$ classes. **Hint:** Do not forget to take into account the constraint that $\sum_{k=1}^{C} y_k^{(i)} = 1$.

17.14. The phase classification example in Section 17.7.2 employs molecular dynamics based enhanced sampling in order to generate the free-energy profiles in Fig. 17.13. The forces needed by these methods require derivatives of the form $\partial f(\mathbf{Q}(\mathbf{r}))/\partial \mathbf{r}_i$ on atom i. These derivatives are computed using the chain rule, which means that products of the form

$$\frac{\partial f}{\partial \mathbf{Q}} \cdot \frac{\partial \mathbf{Q}}{\partial G_k} \frac{\partial G_k}{\partial \mathbf{r}_i}$$

need to be computed. Here, G_k is one of the symmetry functions discussed in Section 17.7.2. Discuss qualitatively how each term is most efficiently evaluated. Some of the approaches in Appendix C might be useful for your discussion. Derive an explicit expression for the force on atom i for a neural network that contains one hidden layer with M nodes, whose output classifies p phases, and whose input consists only of G_2 descriptors with a single value of η and n different values of R_s.

17.15. Let $P(x)$ and $Q(x)$ be two normalized probability distributions. The relative Shannon entropy between P and Q with respect to P is known as the Kullback-Leibler (KL) divergence and is defined by

$$\text{KL}(P||Q) = -\int dx \, P(x) \ln \left[\frac{Q(x)}{P(x)} \right].$$

The KL divergence is a measure of a statistical "distance" between P and Q.
 a. If $P(x)$ is a Gaussian distribution of mean μ and width σ, and $Q(x)$ is a one-dimensional Gaussian distribution of mean ν and width λ, calculate the KL divergence between P and Q.
 *b. Repeat for multivariate Gaussian distributions of a vector $\mathbf{x}$ of dimension n, assuming that $P(\mathbf{x})$ has a mean vector $\boldsymbol{\mu}$ and covariance matrix $\boldsymbol{\Sigma}$ and that $Q(\mathbf{x})$ has a mean vector $\boldsymbol{\nu}$ and covariance matrix $\boldsymbol{\Lambda}$.

17.16. Derive a probability density corresponding to the K-nearest neighbors learning model, and show that this distribution cannot be normalized. That is, show that the integral of this distribution over all space is divergent.

**17.17. Define the L_2 norm of a function $f(x_1, ..., x_n)$ of n variables over an n-dimensional volume Ω as

$$||f(x_1, ..., x_n)|| = \left[\int_\Omega dx_1 \cdots dx_n \ (f(x_1, ..., x_n))^2 \right]^{1/2}.$$

Let $x_1, ..., x_n$ be variables such that $x_i \in [0,1]$, $i = 1, ..., n$, and let $\phi_i(x)$ be monotonically increasing functions $\phi_i : [0,1] \to [0,1]$. Let ϵ and δ, with $0 < \epsilon < 1$ and $0 < \delta < 1$, be ordinary numbers. Finally, let $\gamma_1(x)$ be a function such that $\gamma_1 : \mathbb{R} \to \mathbb{R}$, and suppose we can choose γ_1 such that $||\gamma_1|| \le ||f||$,

$$\left\| f(x_1, ..., x_n) - \sum_{q=1}^{2n+1} \gamma_1 \left(\sum_{i=1}^{n} \lambda_i \phi_q(x_i) \right) \right\| \le (1 - \epsilon)||f||,$$

and $||\gamma_1|| = \delta||f||$. Let us now define a series of functions $\gamma_j : \mathbb{R} \to \mathbb{R}$ and $h_j : [0,1]^n \to \mathbb{R}$ such that

$$h_j(x_1, ..., x_n) = \sum_{q=1}^{2n+1} \gamma_j \left(\sum_{i=1}^{n} \lambda_i \phi_q(x_i) \right).$$

With these definitions, note that $||f - h_1|| = (1 - \epsilon)||f||$ and $||\gamma_1|| = \delta||f||$.

  a. Show that this series of functions leads to an approximation to $f(x_1, ..., x_n)$ such that

$$\left\| f - \sum_{j=1}^{r} h_j \right\| \le (1 - \epsilon)^r \, ||f||$$

  and

$$||\gamma_r|| = \delta \, (1 - \epsilon)^{r-1} \, ||f||.$$

  b. Now let $r \to \infty$. Show that

$$\lim_{r \to \infty} \left\| f - \sum_{j=1}^{r} h_j \right\| \le (1 - \epsilon)^r \, ||f|| = 0$$

  and

$$\lim_{r \to \infty} ||\gamma_r|| = 0.$$

  c. Finally, show that

$$f(x_1, ..., x_n) = \sum_{q=1}^{2n+1} \sum_{j=1}^{\infty} \gamma_j \left(\sum_{i=1}^{n} \lambda_i \phi_q(x_i) \right) \equiv \sum_{q=1}^{2n+1} g \left(\sum_{i=1}^{n} \lambda_i \phi_q(x_i) \right)$$

  as in eqn. (17.5.1).

  d. Can this procedure provide guidance on how to construct a feed-forward neural network for regression of a function? Explain.

17.18. In Problem 8.12, we considered the effect of an invertible transformation on the partition function of a complex system. Machine learning models can be trained to learn such transformations. Let $\mathbf{x} \in \mathbb{R}^n$ be a vector, and let $\mathbf{f} : \mathbb{R}^n \rightarrow \mathbb{R}^n$ be an invertible, smooth mapping with inverse $\mathbf{f}^{-1}$. Let $\mathbf{z} = \mathbf{f}(\mathbf{x})$ and $\mathbf{x} = \mathbf{f}^{-1}(\mathbf{z})$, and let $P(\mathbf{x})$ be a probability distribution function of $\mathbf{x}$.

   a. Show that under the transformation $\mathbf{z} = \mathbf{f}(\mathbf{x})$, the probability distribution $P(\mathbf{x})$ transforms as

$$\tilde{P}(\mathbf{z}) = P(\mathbf{x}) \left| \det\left(\frac{\partial \mathbf{f}^{-1}}{\partial \mathbf{z}} \right) \right| = P(\mathbf{x}) \left| \det\left(\frac{\partial \mathbf{f}}{\partial \mathbf{x}} \right) \right|,$$

   where $|\det(\partial \mathbf{f}/\partial \mathbf{x})|$ is the determinant of the transformation.

   b. Suppose we now have a sequence of invertible transformations from $\mathbf{z}_0 \equiv \mathbf{x}$ to $\mathbf{z}_K$: $\mathbf{z}_1 = \mathbf{f}_1(\mathbf{z}_0)$, $\mathbf{z}_2 = f_2(\mathbf{z}_1),...,\mathbf{z}_K = \mathbf{f}_K(\mathbf{z}_{K-1})$. If $P_0(\mathbf{z}_0)$ is a probability distribution of $\mathbf{z}_0$ and $P_K(\mathbf{z}_K)$ is a distribution of $\mathbf{z}_K$, show that

$$\ln P_K(\mathbf{z}_K) = \ln P_0(\mathbf{z}_0) - \sum_{k=1}^{K} \ln \left| \det\left(\frac{\partial \mathbf{f}_k}{\partial \mathbf{z}_{k-1}} \right) \right|.$$

   c. The sequence of random variables $\{\mathbf{z}_k - \mathbf{f}_K(\mathbf{z}_{k-1})\}$ is known as a *flow*, and the sequence of distributions $\{P_k(\mathbf{z}_k)\}$ is known as a *normalizing flow*. Based on the discussion of neural networks given in Section 17.5, discuss how a normalizing flow can be formulated and learned as a feed-forward neural network model.

Appendix A
Properties of the Dirac delta-function

The Dirac delta(δ)-function is informally defined as having infinite height, zero width, and unit area. When centered around $x = 0$, it can be heuristically represented as

$$\delta(x) = \begin{cases} \infty & x = 0 \\ 0 & x \neq 0 \end{cases}.$$ (A.1)

The δ-function has two important properties. First, its integral is unity:

$$\int_{-\infty}^{\infty} dx \ \delta(x) = 1.$$ (A.2)

Second, the integral of a δ-function times any arbitrary function $f(x)$ is

$$\int_{-\infty}^{\infty} dx \ \delta(x) f(x) = f(0).$$ (A.3)

Equation (A.1) is not rigorous and, therefore, is not especially useful when we wish to derive the properties of the δ-function. For this reason, we replace eqn. (A.1) with a definition involving a limit of a sequence of functions $\delta_\sigma(x)$ known as δ sequences:

$$\delta(x) = \lim_{\sigma \to 0} \delta_\sigma(x).$$ (A.4)

The choice of the δ-sequence is not unique. Several possible choices are:
1. Normalized Gaussian function

$$\delta_\sigma(x) = \frac{1}{\sqrt{2\pi\sigma^2}} e^{-x^2/2\sigma^2},$$ (A.5)

2. Fourier integral

$$\delta_\sigma(x) = \frac{1}{\pi} \frac{\sigma}{\sigma^2 + x^2} = \frac{1}{2\pi} \int_{-\infty}^{\infty} dk \ e^{ikx - |\sigma|x},$$ (A.6)

3. Scaled *sinc* function

$$\delta_\sigma(x) = \frac{1}{\pi\sigma} \text{sinc}\left(\frac{x}{\sigma}\right) = \frac{1}{\pi x} \sin\left(\frac{x}{\sigma}\right).$$ (A.7)

Let us employ the normalized Gaussian sequence in eqn. (A.5) to prove eqns. (A.2) and (A.3). Equation (A.2) follows immediately from the fact that eqn. (A.5) is properly normalized so that

$$\int_{-\infty}^{\infty} dx\, \delta(x) = \lim_{\sigma \to 0} \int_{-\infty}^{\infty} dx\, \delta_\sigma(x)$$

$$= \lim_{\sigma \to 0} \frac{1}{\sqrt{2\pi\sigma^2}} \int_{-\infty}^{\infty} dx\, e^{-x^2/2\sigma^2}$$

$$= 1. \tag{A.8}$$

To prove eqn. (A.3), we need to perform the integral

$$\int_{-\infty}^{\infty} dx\, \delta(x) f(x) = \lim_{\sigma \to 0} \int_{-\infty}^{\infty} dx\, \delta_\sigma(x) f(x)$$

$$= \lim_{\sigma \to 0} \frac{1}{\sqrt{2\pi\sigma^2}} \int_{-\infty}^{\infty} dx\, e^{-x^2/2\sigma^2} f(x). \tag{A.9}$$

This integral can be carried out by introducing a Taylor series expansion for $f(x)$ about $x = 0$ into eqn. (A.9):

$$\frac{1}{\sqrt{2\pi\sigma^2}} \int_{-\infty}^{\infty} dx\, e^{-x^2/2\sigma^2} f(x) = \frac{1}{\sqrt{2\pi\sigma^2}} \int_{-\infty}^{\infty} dx\, e^{-x^2/2\sigma^2} \sum_{n=0}^{\infty} \frac{f^{(n)}(0)}{n!} x^n$$

$$= \frac{1}{\sqrt{2\pi\sigma^2}} \sum_{n=0}^{\infty} \frac{f^{(n)}(0)}{n!} \int_{-\infty}^{\infty} dx\, x^n e^{-x^2/2\sigma^2}, \tag{A.10}$$

where $f^{(n)}(0)$ is the nth derivative of $f(x)$ evaluated at $x = 0$. If n is odd, the integral vanishes because the product of an odd and an even function yields equal areas of opposite sign for $x > 0$ and $x < 0$. For n even, the integrals are just the moments of the Gaussian distribution. These moments have the general form

$$\frac{1}{\sqrt{2\pi\sigma^2}} \int_{-\infty}^{\infty} dx\, x^n e^{-x^2/2\sigma^2} = (n-1)!!\sigma^n, \tag{A.11}$$

where $(n-1)!! = 1 \cdot 3 \cdot 5 \cdots (n-1)$ and $-1!! \equiv 1$. Since we need the limit $\sigma \to 0$, the only term that does not vanish in this limit is the $n = 0$ term. Thus, we find

$$\lim_{\sigma \to 0} \int_{-\infty}^{\infty} dx\, \delta_\sigma(x) f(x) = f(0). \tag{A.12}$$

In a similar manner, it is also possible to prove the more general property

$$\int_{-\infty}^{\infty} dx\, \delta(x-a) f(x) = f(a), \tag{A.13}$$

which is left as an exercise for the reader.

Consider, next, a δ-function of the form $\delta(ax)$. This can be simplified according to

$$\delta(ax) = \frac{1}{|a|}\delta(x), \tag{A.14}$$

which also implies that $\delta(-x) = \delta(x)$. Similarly, a δ-function of the form $\delta(x^2 - a^2)$ can be rewritten equivalently as

$$\delta\left(x^2 - a^2\right) = \frac{1}{2|a|}\left[\delta(x-a) + \delta(x+a)\right]. \tag{A.15}$$

More generally, given a δ-function $\delta(g(x))$ where $g(x)$ is a function with n zeroes at points, $\bar{x}_1, ..., \bar{x}_n$, such that $g'(\bar{x}_i) \neq 0$, the δ-function can be simplified according to

$$\delta(g(x)) = \sum_{i=1}^{n} \frac{\delta(x - \bar{x}_i)}{|g'(\bar{x}_i)|}. \tag{A.16}$$

It is easy to see that eqns. (A.14) and (A.15) are special cases of eqn. (A.16). Equation (A.16) can also be proved using δ-sequences by performing a change of variables $y = g(x)$ and employing similar arguments to those used in proving eqn. (A.3).

Appendix B
Calculus of functionals

Recall that an ordinary scalar function $f(x_1, ..., x_n) \equiv f(\mathbf{x})$ of a vector $\mathbf{x} \in \Omega$, where Ω is an n-dimensional space, specifies a rule for mapping a point $\mathbf{x}$ in Ω to a single number. By contrast, a *functional* $F[f]$ specifies a rule for mapping a function $f(\mathbf{x})$ onto a single number. Note the difference in notation: A functional is specified with square brackets "[ ]" to specify the argument whereas an ordinary function $f(\mathbf{x})$ uses parentheses "( )". A functional $F[f]$ can depend on the entire range of f or just a subset of this full range. Some examples of functionals are listed below.

1. A simple functional can take the form of a definite integral over Ω of $f(\mathbf{x})$:

$$F[f] = \int_\Omega d\mathbf{x}\, f(\mathbf{x}). \tag{B.1}$$

An integral is a natural mathematical construct for expressing a functional, as evaluation of an integral requires all values of $f(\mathbf{x})$. That is, in order to compute an integral, the full range of f over the entire space Ω is needed. A functional need not be a simple integral of $f(\mathbf{x})$. If $g(y)$ is integrable over Ω, then a more general form of a functional is

$$F[f] = \int_\Omega d\mathbf{x}\, g(f(\mathbf{x})). \tag{B.2}$$

In addition, the integral could contain some other fixed function $w(\mathbf{x})$, *i.e.*,

$$F[f] = \int_\Omega d\mathbf{x}\, w(\mathbf{x}) g(f(\mathbf{x})). \tag{B.3}$$

Note, finally, that if $\omega \subset \Omega$, then a functional that involves only part of the range of f can also be defined:

$$F[f] = \int_\omega d\mathbf{x}\, w(\mathbf{x}) g(f(\mathbf{x})). \tag{B.4}$$

Examples in the book of functionals expressed as single integrals are the modified Shannon entropy functional in eqn. (4.4.4), in which $\mathbf{x}$ is the phase-space vector, or the Thomas-Fermi kinetic energy functional in eqn. (11.5.46), in which $\mathbf{x}$ is the Cartesian spatial vector $\mathbf{r}$. Because functionals of this type are single integrals, they are known as *local functionals*.

2. Functionals can also take the form of multiple integrals involving $f(\mathbf{x})$ at different points. An example is a double integral

$$F[f] = \int_\Omega d\mathbf{x} \int_\Omega d\mathbf{x}'\, g(f(\mathbf{x}))h(f(\mathbf{x}'))w(\mathbf{x}, \mathbf{x}'), \qquad (\text{B.5})$$

where $w(\mathbf{x}, \mathbf{x}')$ is a fixed function. Functionals of this type can involve any number of integrals. The exchange energy functional in eqn. (11.5.53), for example, contains two integrals. Because functionals of this type involve multiple integrals of $f(\mathbf{x})$ at different points, they are known as *nonlocal functionals*.

3. Functionals can involve any number of derivatives of $f(\mathbf{x})$, such as the gradient $\nabla f(\mathbf{x})$, the Laplacian $\nabla^2 f(\mathbf{x})$, and/or higher–order derivatives. Local functionals involving derivatives of f take the general form

$$F[f] = \int_\Omega d\mathbf{x}\, w(\mathbf{x})g(f(\mathbf{x}), \nabla f(\mathbf{x}), \nabla^2 f(\mathbf{x}), ...). \qquad (\text{B.6})$$

Examples include the action functional in eqns. (1.8.1) and (12.4.5), the Euclidean action integral in eqn. (12.4.11), and generalized gradient functional approximations to exchange and correlation in density functional theory. Note that such functionals can also be nonlocal, as in:

$$F[f] = \int_\Omega d\mathbf{x} \int_\Omega d\mathbf{x}'\, g(f(\mathbf{x}), \nabla f(\mathbf{x}), \nabla^2 f(\mathbf{x}), ..., f(\mathbf{x}'), \nabla f(\mathbf{x}'), \nabla^2 f(\mathbf{x}'), ...). \qquad (\text{B.7})$$

4. In some instances, a functional may involve values of $f(\mathbf{x})$ at a set of M discrete points $\mathbf{x}_1, ..., \mathbf{x}_M$, *i.e.*,

$$F[f] = g(f(\mathbf{x}_1), f(\mathbf{x}_2), ..., f(\mathbf{x}_M)). \qquad (\text{B.8})$$

In order to create a calculus of functionals, we first need to define the concept of a derivative of a functional, also known as a *functional derivative*. Functional derivatives require the concept of a small change in a functional, for which we use the notation of a variation δF defined as

$$\delta F = F[f + \delta f] - F[f]. \qquad (\text{B.9})$$

The variation δF, therefore, requires changing $f(\mathbf{x})$ by a small amount $\delta f(\mathbf{x})$. However, unlike with simple functions, where a small change $\mathbf{x} + \delta \mathbf{x}$ in a point can be understood as a displacement of the point $\mathbf{x}$ by $\delta \mathbf{x}$, the displacement $\delta f(\mathbf{x})$ of a function is, itself, another function! This brings some arbitrariness to the process of defining the variation δF, and makes the process of taking a limit $\delta f(\mathbf{x}) \to 0$, which is ultimately needed to define a derivative, complex. The easiest way to proceed in computing a derivative of a functional with respect to its function argument is to introduce an arbitrary function $\eta(\mathbf{x})$ and a multiplicative parameter ε such that $\delta f(\mathbf{x}) = \varepsilon \eta(\mathbf{x})$. In this way, the derivative can be defined through an integral relation

$$\int_\Omega d\mathbf{x} \, \frac{\delta F}{\delta f(\mathbf{x})} \eta(\mathbf{x}) \equiv \lim_{\varepsilon \to 0} \left(\frac{F[f + \varepsilon \eta] - F[f]}{\varepsilon} \right) = \left(\frac{d}{d\varepsilon} F[f + \varepsilon \eta] \right) \bigg|_{\varepsilon=0}. \tag{B.10}$$

The quantity $\delta F/\delta f(\mathbf{x})$ is known as the *functional derivative* of $F[f]$. Since $\eta(\mathbf{x})$ is arbitrary, we can take $\eta(\mathbf{x}) = \delta^{(n)}(\mathbf{x} - \mathbf{x}_0)$, in which case, evaluation of the integral on the left side of eqn. (B.10) yields the functional derivative of $F[f]$, $\delta F/\delta f(\mathbf{x}_0)$ at the point $\mathbf{x}_0$. However, doing so is not really necessary, as the ε derivative on the right side of eqn. (B.10) will yield an integral expression involving $\eta(\mathbf{x})$ explicitly, so that what remains in the integrand can be deduced as $\delta F/\delta f(\mathbf{x})$. As an example, consider the functional

$$F[f] = \int_\Omega d\mathbf{x} \, (f(\mathbf{x}))^p, \tag{B.11}$$

where p is an arbitrary power. Applying eqn. (B.10), we simply need to evaluate

$$\frac{d}{d\varepsilon} F[f + \varepsilon \eta] = \frac{d}{d\varepsilon} \int_\Omega d\mathbf{x} \, (f(\mathbf{x}) + \varepsilon \eta(\mathbf{x}))^p$$

$$= \int_\Omega d\mathbf{x} \, \frac{\partial}{\partial \varepsilon} (f(\mathbf{x}) + \varepsilon \eta(\mathbf{x}))^p$$

$$= \int_\Omega d\mathbf{x} \, p \, (f(\mathbf{x}) + \varepsilon \eta(\mathbf{x}))^{p-1} \eta(\mathbf{x}). \tag{B.12}$$

Now, setting $\varepsilon = 0$, we obtain

$$\left(\frac{d}{d\varepsilon} F[f + \varepsilon \eta] \right)_{\varepsilon=0} = \int_\Omega d\mathbf{x} \, p \, (f(\mathbf{x}))^{p-1} \eta(\mathbf{x}), \tag{B.13}$$

from which we see that

$$\frac{\delta F}{\delta f(\mathbf{x})} = p \, (f(\mathbf{x}))^{p-1}. \tag{B.14}$$

Note that, in this case, because functional derivatives are taken at specific point values $f(\mathbf{x})$ of the function, the derivative of a local functional removes the integral, leaving an expression that is evaluated at a specific point $\mathbf{x}$.

As a second example, consider the functional in eqn. (B.5), with $g(f) = h(f) = f$ and with $w(\mathbf{x}, \mathbf{x}')$ being a symmetric function, meaning $w(\mathbf{x}, \mathbf{x}') = w(\mathbf{x}', \mathbf{x})$. It is left as a exercise for the reader to show that application of eqn. (B.10) leads to the result

$$\frac{\delta F}{\delta f(\mathbf{x})} = 2 \int_\Omega d\mathbf{x}' \, f(\mathbf{x}') w(\mathbf{x}, \mathbf{x}'). \tag{B.15}$$

As a third example, consider the functional

$$F[f] = \int_\Omega d\mathbf{x} \, g(f(\mathbf{x}), \nabla f(\mathbf{x})). \tag{B.16}$$

Applying eqn. (B.10) by taking a ε derivative, we find that

$$\frac{\mathrm{d}}{\mathrm{d}\varepsilon}F[f + \varepsilon\eta] = \int_\Omega \mathrm{d}\mathbf{x}\, g(f(\mathbf{x}) + \varepsilon\eta(\mathbf{x}), \nabla f(\mathbf{x}) + \varepsilon\nabla\eta(\mathbf{x})), \qquad (\text{B.17})$$

which involves $\eta(\mathbf{x})$ and its gradient $\nabla\eta(\mathbf{x})$. Taking the derivative with respect to ε and setting $\varepsilon = 0$, we obtain

$$\left(\frac{\mathrm{d}}{\mathrm{d}\varepsilon}F[f + \varepsilon\eta]\right)\bigg|_{\varepsilon=0} = \int_\Omega \mathrm{d}\mathbf{x}\, \left[\frac{\partial g}{\partial f(\mathbf{x})}\eta(\mathbf{x}) + \frac{\partial g}{\partial\nabla f(\mathbf{x})}\nabla\eta(\mathbf{x})\right]. \qquad (\text{B.18})$$

Although the first term is in the form required by eqn. (B.10), the second term, which involves $\nabla\eta(\mathbf{x})$, is not. However, if we assume that $f(\mathbf{x})$ vanishes at the boundary $\partial\Omega$ of Ω, then we can integrate the second term by parts, dropping the boundary term, and we obtain

$$\left(\frac{\mathrm{d}}{\mathrm{d}\varepsilon}F[f + \varepsilon\eta]\right)\bigg|_{\varepsilon=0} = \int_\Omega \mathrm{d}\mathbf{x}\, \left[\frac{\partial g}{\partial f(\mathbf{x})} - \nabla\cdot\frac{\partial g}{\partial\nabla f(\mathbf{x})}\right]\eta(\mathbf{x}), \qquad (\text{B.19})$$

from which we see that

$$\frac{\delta F}{\delta f(\mathbf{x})} = \frac{\partial g}{\partial f(\mathbf{x})} - \nabla\cdot\left(\frac{\partial g}{\partial\nabla f(\mathbf{x})}\right). \qquad (\text{B.20})$$

It is left as an exercise to the reader to show that if $F[f]$ is

$$F[f] = \int_\Omega \mathrm{d}\mathbf{x}\, g(f(\mathbf{x}), \nabla f(\mathbf{x}), \nabla^2 f(\mathbf{x})), \qquad (\text{B.21})$$

then

$$\frac{\delta F}{\delta f(\mathbf{x})} = \frac{\partial g}{\partial f(\mathbf{x})} - \nabla\cdot\left(\frac{\partial g}{\partial\nabla f(\mathbf{x})}\right) + \nabla^2\frac{\partial g}{\partial\nabla^2 f(\mathbf{x})}. \qquad (\text{B.22})$$

As a further exercise, the reader should verify the following result: Let $f(x)$ be a function of a single variable x, and consider a local functional of the form

$$F[f] = \int \mathrm{d}x\, g(f(x), f'(x), f''(x), f'''(x), ..., f^{(n)}(x)). \qquad (\text{B.23})$$

The functional derivative of $F[f]$ is then

$$\frac{\delta F}{\delta f(x)} = \frac{\partial g}{\partial f(x)} - \frac{\mathrm{d}}{\mathrm{d}x}\frac{\partial g}{\partial f'(x)} + \frac{\mathrm{d}^2}{\mathrm{d}x^2}\frac{\partial g}{\partial f''(x)} + \cdots + (-1)^n\frac{\mathrm{d}^n}{\mathrm{d}x^n}\frac{\partial g}{\partial f^{(n)}(x)}, \qquad (\text{B.24})$$

where $f^{(n)}(x)$ is the nth derivative of $f(x)$.

According to eqn. (B.10), all that is required to define a functional derivative is a a simple parameter derivative with respect to ε; therefore, it is straightforward to define higher order functional derivatives. For example, a second functional derivative $\delta^2 F/\delta f(\mathbf{x})\delta f(\mathbf{x}')$ is defined by

$$\int_\Omega \mathrm{d}\mathbf{x}\,\mathrm{d}\mathbf{x}'\,\frac{\delta^2 F}{\delta f(\mathbf{x})\delta f(\mathbf{x}')}\eta(\mathbf{x})\eta(\mathbf{x}') = \left(\frac{\mathrm{d}^2}{\mathrm{d}\varepsilon^2}F[f+\varepsilon\eta]\right)\bigg|_{\varepsilon=0}, \qquad (\text{B}.25)$$

with analogous definitions of third, fourth,..., nth functional derivatives of $F[f]$, i.e.,

$$\int_\Omega \mathrm{d}\mathbf{x}_1\cdots\mathrm{d}\mathbf{x}_n\,\frac{\delta^n F}{\delta f(\mathbf{x}_1)\cdots\delta f(\mathbf{x}_n)}\eta(\mathbf{x}_1)\cdots\eta(\mathbf{x}_n) = \left(\frac{\mathrm{d}^n}{\mathrm{d}\varepsilon^n}F[f+\varepsilon\eta]\right)\bigg|_{\varepsilon=0}. \quad (\text{B}.26)$$

With this general definition, we can also define a functional Taylor series for a finite displacement $\Delta f(\mathbf{x})$ of $f(\mathbf{x})$:

$$F[f+\Delta f] = F[f] + \sum_{n=1}^\infty \frac{1}{n!}\int_\Omega \mathrm{d}\mathbf{x}_1\cdots\mathrm{d}\mathbf{x}_n\frac{\delta^n F}{\delta f(\mathbf{x}_1)\cdots\delta f(\mathbf{x}_n)}\Delta f(\mathbf{x}_1)\cdots\Delta f(\mathbf{x}_n).$$
$$(\text{B}.27)$$

To conclude this appendix, we briefly discuss the concept of functional integration. Section 12.4 in Chapter 12 provides a fairly detailed discussion of functional integration in the context of the Feynman path integral. Functional integration of a functional $F[f]$ is tantamount to the limit of a "sum" over all possible functions $f(\mathbf{x})$ via a process in which f is systematically varied. As was done for the functional derivative, we start by defining functional integration in discretized form. In order to keep the discussion simple and avoid the complexities of multiple integration, we will consider functionals $F[f]$ of functions $f(x)$ of a single variable $x \in [a, b]$. Moreover, let us consider a simple local functional of the form

$$F[f] = \int_a^b g(f(x))\,\mathrm{d}x. \qquad (\text{B}.28)$$

Let us further assume that the functional integral of $F[f]$ must include all functions satisfying fixed endpoint conditions $f(a) = f_a$ and $f(b) = f_b$. To construct such a functional integral, we first discretize the integral in eqn. (B.28) using, for example, a trapezoidal rule integration:

$$\int_a^b g(f(x))\,\mathrm{d}x = \frac{1}{2}\left(g(f_a)+g(f_b)\right) + \lim_{M\to\infty}\frac{b-a}{M}\sum_{k=1}^{M-1}g(f(x_k)), \qquad (\text{B}.29)$$

where the interval $[a, b]$ has been discretized into $M+1$ points x_k such that $x_0 = a$, and $x_M = b$. Ultimately, the exact result is recovered in the limit that $M \to \infty$. Equation (B.29) now replaces the integral in eqn. (B.28) as

$$F[f] = \frac{1}{2}\left(g(f_a)+g(f_b)\right) + \lim_{M\to\infty}\frac{b-a}{M}\sum_{k=1}^{M-1}g(f(x_k)). \qquad (\text{B}.30)$$

Let us define $f(x_k) \equiv f_k$. Integration over all functions $f(x)$ satisfying the endpoint condition $f(x_0) = f(a) = f_a$ and $f(x_M) = f(b) = f_b$ now requires that we perform an integration over all f_k, $k = 0, ..., M$, subject to the condition that f_0 and f_M are fixed. We indicate this (following the notation of Chapter 12) as

$$\int_{f(a)=f_a}^{f(b)=f_b} \mathcal{D}f \, F[f] = \int_{f(a)=f_a}^{f(b)=f_b} \mathcal{D}f \int_a^b g(f(x)) \, \mathrm{d}x$$

$$\equiv \lim_{M \to \infty} \frac{b-a}{M} \int \prod_{k=0}^M \mathrm{d}f_k \left[\frac{g(f_a) + g(f_b)}{2} + \sum_{k=1}^{M-1} g(f_k) \right]_{f_0 = f_a, f_M = f_b} \tag{B.31}$$

Equation (B.31) provides a definition of functional integration for local functionals. As discussed in Chapter 12, the integral in eqn. (B.31) is performed by carrying out the integration on the right side as a function of the number of points M and then taking the limit $M \to \infty$ of the result. If $F[f]$ includes derivatives $f'(x)$, $f''(x)$,..., these are evaluated numerically, e.g., $f'(x) \to (f_k - f_{k-1})/\Delta$, where $\Delta = (b-a)/M$, and so forth. As illustrated in Chapter 12, there are approaches for performing the integration directly in terms of the continuous functions themselves; however, there are very few examples for which this can be done, these typically being restricted to quadratic functionals. In general, the integration needs to be carried out in discrete form for finite M numerically using the right side of eqn. (B.31). As the numerical calculation is repeated for increasing M, the result will converge asymptotically to the infinite-M limit, which is then taken as the result. The processes of discrete and continuous functional integration based on eqn. (B.31) are illustrated in Figs. 12.6 and 12.8.

Appendix C
Evaluation of energies and forces

In any Monte Carlo or molecular dynamics calculation of a many-body system, the most time-consuming step is the evaluation of energies and, when needed, forces. In this appendix, we will discuss the efficient evaluation of energies and forces for simple nonpolarizable force fields under periodic boundary conditions

Let us begin by considering a system described by a typical nonpolarizable force field as in eqn. (3.11.1). The bonding and bending terms are straightforward and computationally inexpensive to evaluate; dihedral angles can be calculated from the angle between the planes defined by atoms 1, 2, and 3 and atoms 2, 3, and 4. By far, the most expensive part of the calculation is the evaluation of the nonbonded (nb) forces and energies given by the Lennard-Jones and Coulomb terms:

$$U_{\text{nb}}(\mathbf{r}_1, ..., \mathbf{r}_N) = \sum_{i > j \in \text{nb}} \left\{ 4\epsilon_{ij} \left[\left(\frac{\sigma_{ij}}{r_{ij}} \right)^{12} - \left(\frac{\sigma_{ij}}{r_{ij}} \right)^6 \right] + \frac{q_i q_j}{r_{ij}} \right\}. \tag{C.1}$$

As written, the evaluation of these energies and their associated forces requires $\mathcal{O}(N^2)$ operations, a calculation that quickly becomes intractable as N becomes very large. Simulations of even small proteins in solution typically require at least 10^4 to 10^5 atoms, while simulations in materials science, such as crack formation and propagation, can involve up to 10^9 atoms. Many interesting physical and chemical processes such as these require long time scales to be accessed, and in order to reach these time scales, a very large number of energy and force evaluations is needed. Such calculations would clearly not be possible if the quadratic scaling could not be ameliorated. The goal, therefore, is to evaluate the terms in eqn. (C.1) with $\mathcal{O}(N)$, or at worst, $\mathcal{O}(N \ln N)$, scaling.

The first thing we see about eqn. (C.1) is that the Lennard-Jones and Coulomb terms have substantially different length scales. The former is relatively short range and could possibly be truncated as a means of improving the scaling, but this is not true of the Coulomb interaction, which is very long range and would, therefore, suffer from severe truncation artifacts (see, for example, Patra *et al.* (2003)). Let us begin, therefore, by examining the Coulomb term more closely. In order to reduce the computational overhead of the long-range Coulomb interaction, we only need to recognize that any function we might characterize as long ranged in real space becomes short ranged in reciprocal or Fourier space. Moreover, since we are dealing with a periodic system, a Fourier representation is entirely appropriate. Therefore, we can tame the Coulomb interaction by dividing it into a contribution that is short ranged in real space and one that is short ranged in reciprocal space. Given two rapidly convergent terms, the Coulomb interaction should be easier to evaluate and exhibit better scaling.

We begin by introducing a simple identity that pertains to a function known as the error function $\mathrm{erf}(x)$ defined by

$$\mathrm{erf}(x) = \frac{2}{\sqrt{\pi}} \int_0^x dt\, e^{-t^2}. \tag{C.2}$$

In the limit $x \to \infty$, $\mathrm{erf}(x) \to 1$. Note also that $\mathrm{erf}(0) = 0$. The error function has a complement $\mathrm{erfc}(x)$ defined by

$$\mathrm{erfc}(x) = 1 - \mathrm{erf}(x) = \frac{2}{\sqrt{\pi}} \int_x^\infty dt\, e^{-t^2}. \tag{C.3}$$

Both functions are defined for $x \geq 0$. Note that as $x \to \infty$, $\mathrm{erfc}(x) \to 0$. From these definitions, it is clear that $\mathrm{erf}(x) + \mathrm{erfc}(x) = 1$. This identity can now be used to divide up the Coulomb interaction into short-range and long-range components. Consider writing $1/r$ as

$$\frac{1}{r} = \frac{\mathrm{erfc}(\alpha r)}{r} + \frac{\mathrm{erf}(\alpha r)}{r}, \tag{C.4}$$

where the parameter α has units of inverse length. Since $\mathrm{erfc}(\alpha r)$ decays rapidly as r increases, the first term in eqn. (C.4) is short ranged. The parameter α can be used to tune the range over which $\mathrm{erfc}(\alpha r)/r$ is nonnegligible. The second term in eqn. (C.4) is long ranged and behaves asymptotically as $1/r$. Introducing eqn. (C.4) into eqn. (C.1), we can write the nonbonded forces in terms of short- and long-ranged components as

$$U_{\mathrm{nb}}(\mathbf{r}_1, ..., \mathbf{r}_N) = U_{\mathrm{short}}(\mathbf{r}_1, ..., \mathbf{r}_N) + U_{\mathrm{long}}(\mathbf{r}_1, ..., \mathbf{r}_N)$$

$$U_{\mathrm{short}}(\mathbf{r}_1, ..., \mathbf{r}_N) = \sum_{\mathbf{S}} \sum_{i > j \in \mathrm{nb}} \left\{ 4\epsilon_{ij} \left[\left(\frac{\sigma_{ij}}{r_{ij,\mathbf{S}}} \right)^{12} - \left(\frac{\sigma_{ij}}{r_{ij,\mathbf{S}}} \right)^6 \right] + \frac{q_i q_j \mathrm{erfc}(\alpha r_{ij,\mathbf{S}})}{r_{ij,\mathbf{S}}} \right\}$$

$$U_{\mathrm{long}}(\mathbf{r}_1, ..., \mathbf{r}_N) = \sum_{\mathbf{S}} \sum_{i > j \in \mathrm{nb}} \frac{q_i q_j \mathrm{erf}(\alpha r_{ij,\mathbf{S}})}{r_{ij,\mathbf{S}}}, \tag{C.5}$$

where $r_{ij,\mathbf{S}} = |\mathbf{r}_i - \mathbf{r}_j + \mathbf{S}|$, with $\mathbf{S} = \mathbf{m}L$ for a periodic cubic box of side L, and $\mathbf{S} = \mathbf{hm}$ for a general box matrix $\mathbf{h}$. Here $\mathbf{m}$ is a vector of integers. Note that if a system is sufficiently large, then the sums over the lattice vectors $\mathbf{S}$ reduce to the single term $\mathbf{m} = (0, 0, 0)$. As we will soon see, the use of the potential-energy contributions in eqn. (C.5) has a distinct advantage over eqn. (C.1).

Let us consider first evaluating the short-range forces in eqn. (C.5). For notational convenience, we denote this term compactly as $U_{\mathrm{short}} = \sum_{\mathbf{S}} \sum_{i > j} u_{\mathrm{short}}(r_{ij,\mathbf{S}})$, where $u_{\mathrm{short}}(r)$ consists of the Lennard-Jones and complementary error function terms. Because the range of these interactions is finite, we can reduce the computational overhead needed to evaluate them by introducing a cutoff radius r_c, beyond which we assume that the forces are negligible. The cutoff r_c is important for determining α.

Experience has shown that a good balance between the short- and long-range components is achieved if $\alpha = 3.5/r_c$. If r_c is chosen to be half the length L of a cubic simulation box, $r_c = L/2$, then the value $\alpha = 7/L$ should be used. For large systems, however, a typical value of r_c is in the range of 10 to 12 Å, which is roughly a factor of 3 larger than a typical value of σ_{ij}. In most simulations, it is the case that at $r = r_c$, $u_{\text{short}}(r)$ is not exactly zero, in which case the potential energy exhibits a small jump discontinuity as the distance between two particles passes through $r = r_c$. This discontinuity leads to a degradation in energy conservation. One approach for circumventing this problem is simply to shift the potential $u_{\text{short}}(r)$ by an amount $u_{\text{short}}(r_c)$, so that the short-range interaction between two particles i and j becomes

$$\tilde{u}_{\text{short}}(r_{ij}) = \begin{cases} u_{\text{short}}(r_{ij}) - u_{\text{short}}(r_c) & \text{if } r_{ij} < r_c \\ 0 & \text{otherwise} \end{cases} \tag{C.6}$$

However, there is a second problem with the use of cutoffs, which is that the forces also exhibit a discontinuity at $r = r_c$; a simple shift does not remove this discontinuity. A more robust truncation protocol that ensures continuous energies and forces is to smoothly switch off $u_{\text{short}}(r)$ to zero via a switching function $S(r)$. When a switch is used, the potential becomes $\tilde{u}_{\text{short}}(r) = u_{\text{short}}(r)S(r)$. An example of such a function $S(r)$ was given in eqn. (3.14.4). Switching functions become particularly important for simulations in the isothermal-isobaric ensemble, as the pressure estimator in eqn. (4.7.57) is especially sensitive to discontinuities in the force (Martyna *et al.*, 1999).

An important point concerning truncation of the potential is its effect on energies and pressures. In fact, when the potential is assumed to be zero beyond r_c, many weak interactions among particle pairs will not be included. However, the contribution of these neglected interactions can be estimated using eqns. (4.7.46) and (4.7.69). Instead of integrating these expressions from 0 to ∞, if the lower limit is taken to be r_c, then the integrals can be performed in the approximation that for $r > r_c$, $g(r) \approx 1$. For example, the corrections to the energy per particle and pressure for a Lennard-Jones potential in this approximation would be

$$\Delta u = 2\pi\rho \int_{r_c}^{\infty} dr \, r^2 u(r)$$

$$= 8\epsilon\pi\rho \int_{r_c}^{\infty} dr \, r^2 \left[\left(\frac{\sigma}{r}\right)^{12} - \left(\frac{\sigma}{r}\right)^{6} \right]$$

$$= \frac{8\pi\epsilon\rho\sigma^3}{3} \left[\frac{1}{3} \left(\frac{\sigma}{r_c}\right)^{9} - \left(\frac{\sigma}{r_c}\right)^{3} \right], \tag{C.7}$$

$$\Delta P = -\frac{2\pi\rho^2}{3} \int_{r_c}^{\infty} dr \, r^3 u'(r)$$

$$= 16\pi\rho^2\epsilon \int_{r_c}^{\infty} dr \, r^2 \left[2 \left(\frac{\sigma}{r}\right)^{12} - \left(\frac{\sigma}{r}\right)^{6} \right]$$

$$= \frac{16\pi\rho^2\epsilon\sigma^3}{3} \left[\frac{2}{3} \left(\frac{\sigma}{r_c} \right)^9 - \left(\frac{\sigma}{r_c} \right)^3 \right]. \tag{C.8}$$

Equations (C.7) and (C.8) are called the *standard long-range corrections* (Allen and Tildesley, 1989; Frenkel and Smit, 2002). They should be included in order to obtain accurate thermodynamics averages from simulations using truncated potentials.

When periodic boundary conditions are employed (see left panel of Fig. 13.7), interparticle distances are computed using the *minimum image convention*. The rule of this convention is that in the infinite periodic array, particle i interacts with the periodic image of particle j to which it is closest. For a system with an arbitrary box matrix $\mathbf{h}$, the three components of the vector difference $\mathbf{r}_{ij} = \mathbf{r}_i - \mathbf{r}_j$ are computed as follows for a system with an arbitrary box matrix $\mathbf{h}$:

$$
\begin{aligned}
\mathbf{s}_i &= \mathbf{h}^{-1}\mathbf{r}_i \\
\mathbf{s}_{ij} &= \mathbf{s}_i - \mathbf{s}_j \\
\mathbf{s}_{ij} &\longleftarrow \mathbf{s}_{ij} - \mathrm{NINT}(\mathbf{s}_{ij}) \qquad \text{Minimum image convention} \\
\mathbf{r}_{ij} &= \mathbf{h}\mathbf{s}_{ij},
\end{aligned}
\tag{C.9}
$$

where $\mathrm{NINT}(x)$ is the nearest-integer function. The first line transforms the coordinates into the scaled coordinates of eqn. (5.8.2); the second line gives the raw interparticle vector differences; the third line applies the minimum image convention to these differences; the last line transforms the differences back to the original Cartesian variables.

A cutoff radius alone is insufficient to reduce the computational overhead associated with the short-range forces since it is still necessary to look over all N^2 pairs to determine if the condition $r_{ij} < r_c$ is satisfied for a given particle pair. However, since the cutoff radius ensures that a given particle will only interact with a finite number of other particles, we can build a list of all particles j with which a particle i interacts. Such a list is known as a *Verlet neighbor list* (Verlet, 1967). Since particles typically diffuse, the neighbor list must be updated from time to time during the simulation. The average frequency of updates is controlled by a parameter δ known as the *skin length*. Figure C.1 illustrates the construction of a neighbor list. For each particle i, we identify all particles that lie within a distance $r_c + \delta$ of i and store the indices of these neighboring particles in an array. Once the list is built, we simply loop over its elements and calculate all of the forces and energies of the pairs. Because some of the pairs in the list are in the distance range $r_c < r_{ij} < r_c + \delta$, some of the energies and forces will be zero. The smaller δ is chosen, the fewer such interactions there will be, which saves computational time. However, the larger δ is chosen, the more infrequently the list needs to be updated. Thus, the value of δ should be optimized at the start of any simulation.

At the start of a simulation, the initial positions $\mathbf{r}_1(0), ..., \mathbf{r}_N(0)$ are saved and an initial Verlet list is generated. At each time step $k\Delta t$, $k = 1, 2, 3, ...$, the displacement Δ_i of each particle from its initial position $\Delta_i = |\mathbf{r}_i(k\Delta t) - \mathbf{r}_i(0)|$ is computed and the maximum over all of these displacements $\Delta_{\max}$ is subsequently determined. If at the nth step $\Delta_{\max} > \delta/2$, then the interactions contained in the list will no longer

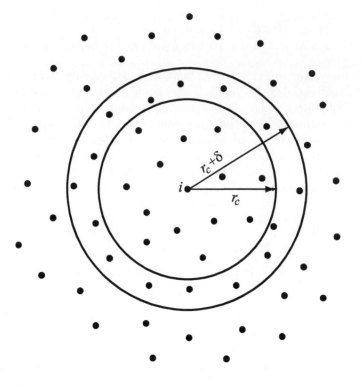

Fig. C.1 Building a Verlet list of the neighbors of particle i.

be correct, which means that the neighbor list needs to be recomputed. The saved positions $\mathbf{r}_1(0), ..., \mathbf{r}_N(0)$ are overwritten by $\mathbf{r}_1(n\Delta t), ..., \mathbf{r}_N(n\Delta t)$, and the Verlet list is regenerated using these positions. At each time step subsequent to the list update, the displacement test is performed using $\Delta_i = |\mathbf{r}_i((k+n)\Delta t) - \mathbf{r}_i(n\Delta t)|$. If $\Delta_{\max} > \delta/2$, then the neighbor list must be generated again. In this way, we can determine automatically when the neighbor list must be recalculated.

It is worth mentioning in passing that an alternative to the Verlet list is the *link list* or *cell list* (Hockney and Eastwood, 1981; Allen and Tildesley, 1989; Frenkel and Smit, 2002). This method consists of dividing the system into cells of size equal to or slightly larger than the cutoff radius r_c. Thus, each particle interacts only with particles in the same cell or in nearest-neighbor cells. A link list is generally more efficient than a Verlet list in large systems, as the calculation of the latter scales as $\mathcal{O}(N^2)$ while the former scales as $\mathcal{O}(N)$. However, the logic to write such a list if more involved. A very useful intermediate approach uses a Verlet list for computing the forces and energies and a link list to create the Verlet list. This improves considerably the efficiency of generating the Verlet list. Finally, fast algorithms for computing short range forces on massively parallel architectures have been developed (Plimpton, 1995) and have proved highly successful, although we will not discuss such algorithms here.

We now turn to the evaluation of the long-range energy and forces in eqn. (C.5). Because $\mathrm{erf}(\alpha r)/r$ behaves as $1/r$ for large r, the long-range energy and forces cannot be evaluated efficiently in real space. However, because a long-ranged function in real space is short ranged in Fourier or reciprocal space, if the latter is used, the scaling can be improved, thereby reducing the computational overhead. Moreover, a Fourier expansion of $\text{erf}(\alpha r)/r$ is consistent with the periodic boundary conditions typically used in Monte Carlo and molecular dynamics calculations. For simplicity, we will consider the case of a cubic simulation cell of length L and volume $V = L^3$. As we saw in Section 11.2, the reciprocal space of such a cell is composed of all reciprocal-space vectors $\mathbf{g} = 2\pi\mathbf{n}/L$, where $\mathbf{n}$ is a vector of integers. Using the Poisson summation rule, we can expand the error function term in a Fourier series as

$$\sum_{\mathbf{S}} \frac{\text{erf}(\alpha|\mathbf{r} + \mathbf{S}|)}{|\mathbf{r} + \mathbf{S}|} = \frac{1}{V} \sum_{\mathbf{g}} C_{\mathbf{g}} e^{i\mathbf{g}\cdot\mathbf{r}}. \tag{C.10}$$

In eqn. (C.10), the expansion coefficients are

$$C_{\mathbf{g}} = \sum_{\mathbf{S}} \int_{D(V)} d\mathbf{r} \, \frac{\text{erf}(\alpha|\mathbf{r} + \mathbf{S}|)}{|\mathbf{r} + \mathbf{S}|} e^{-i\mathbf{g}\cdot\mathbf{r}}$$

$$= \int_{\text{all space}} \frac{\text{erf}(\alpha r)}{r} e^{-i\mathbf{g}\cdot\mathbf{r}}$$

$$= \frac{4\pi}{|\mathbf{g}|^2} e^{-|\mathbf{g}|^2/4\alpha^2}. \tag{C.11}$$

The second line in eqn. (C.11) again follows from the Poisson summation formula, which allows the sum over $\mathbf{S}$ to be eliminated in favor of an integral over all space. Note that the coefficient corresponding to $\mathbf{n} = (0,0,0)$ is not defined, hence this term must be excluded from the sum in eqn. (C.10). Moreover, it is not possible, in practice, to perform a sum over an infinite number of reciprocal-space vectors as required by eqn. (C.10). However, we note that the factor $\exp(-|\mathbf{g}|^2/4\alpha^2)$ decays to zero very quickly as $|\mathbf{g}| \to \infty$, which means that the sum over reciprocal-space vectors can be truncated and restricted only to a very small part of reciprocal space. We can readily see, therefore, the advantage of working in Fourier space when faced with the calculation of long-range forces! Because $C_{\mathbf{g}}$ only depends on the magnitude $|\mathbf{g}|$, the most natural way to truncate the Fourier sum is to restrict the sum to all reciprocal-space vectors $\mathbf{g}$ with magnitudes $|\mathbf{g}| \leq g_{\max}$, where $g_{\max}$ is chosen such that $\exp(-g_{\max}^2/4\alpha^2)$ is negligible. The Fourier sum is consequently restricted to $\mathbf{g}$-vectors that lie within a sphere in reciprocal space of radius $g_{\max}$. In addition, because the coefficients $C_{\mathbf{g}}$ are real and depend only on $|\mathbf{g}|$, they satisfy $C_{-\mathbf{g}} = C_{\mathbf{g}}$ and therefore we only need to keep half the $\mathbf{g}$-vectors in the sphere. For example, we could choose the hemisphere $\mathcal{S}$ for which g_x is positive or 0. When the sum is performed over half of reciprocal space, we simply need to multiply the result by 2.

Taking into account the truncation of the Fourier sum and the restriction of the $\mathbf{g}$-vectors to a single hemisphere, the long-range potential energy can be expressed as

$$U_{\text{long}}(\mathbf{r}_1, ..., \mathbf{r}_N) = \frac{2}{V} \sum_{i=1}^{N-1} \sum_{j=i+1}^{N} q_i q_j \sum_{\mathbf{g} \in \mathcal{S}} \frac{4\pi}{|\mathbf{g}|^2} e^{-|\mathbf{g}|^2/4\alpha^2} e^{i\mathbf{g}\cdot(\mathbf{r}_i - \mathbf{r}_j)}. \tag{C.12}$$

As written, the sum over particles in eqn. (C.12) scales as $\mathcal{O}(N^2)$. However, the computational overhead of eqn. (C.12) can be reduced substantially with a few clever tricks. First, consider writing eqn. (C.12) as

$$U_{\text{long}}(\mathbf{r}_1, ..., \mathbf{r}_N) = \frac{1}{V} \sum_{i=1}^{N} \sum_{j=1, j \neq i}^{N} q_i q_j \sum_{\mathbf{g} \in \mathcal{S}} \frac{4\pi}{|\mathbf{g}|^2} e^{-|\mathbf{g}|^2/4\alpha^2} e^{i\mathbf{g} \cdot (\mathbf{r}_i - \mathbf{r}_j)}. \tag{C.13}$$

Next, we add and subtract the term with $i = j$ to yield

$$U_{\text{long}}(\mathbf{r}_1, ..., \mathbf{r}_N) = \frac{1}{V} \sum_{i,j=1}^{N} q_i q_j \sum_{\mathbf{g} \in \mathcal{S}} \frac{4\pi}{|\mathbf{g}|^2} e^{-|\mathbf{g}|^2/4\alpha^2} e^{i\mathbf{g} \cdot (\mathbf{r}_i - \mathbf{r}_j)}$$

$$- \frac{1}{V} \sum_{i=1}^{N} q_i^2 \sum_{\mathbf{g} \in \mathcal{S}} \frac{4\pi}{|\mathbf{g}|^2} e^{-|\mathbf{g}|^2/4\alpha^2}. \tag{C.14}$$

When this is done, the first term can be written as the square magnitude of a single sum:

$$U_{\text{long}}(\mathbf{r}_1, ..., \mathbf{r}_N) = \frac{1}{V} \sum_{\mathbf{g} \in \mathcal{S}} \frac{4\pi}{|\mathbf{g}|^2} e^{-|\mathbf{g}|^2/4\alpha^2} \left| \sum_{i=1}^{N} q_i e^{i\mathbf{g} \cdot \mathbf{r}_i} \right|^2$$

$$- \frac{1}{V} \sum_{i=1}^{N} q_i^2 \sum_{\mathbf{g} \in \mathcal{S}} \frac{4\pi}{|\mathbf{g}|^2} e^{-|\mathbf{g}|^2/4\alpha^2}. \tag{C.15}$$

Note the presence of the structure factor $S(\mathbf{g}) = \sum_i q_i \exp(i\mathbf{g} \cdot \mathbf{r}_i)$ in the first term of eqn. (C.15). The reciprocal-space sum in the second term of eqn. (C.15) can either be evaluated once in the beginning and stored, or it can also be performed analytically if we extend the sum to all of reciprocal space. For the latter, we first write

$$U_{\text{long}}(\mathbf{r}_1, ..., \mathbf{r}_N) = \frac{1}{V} \sum_{\mathbf{g} \in \mathcal{S}} \frac{4\pi}{|\mathbf{g}|^2} e^{-|\mathbf{g}|^2/4\alpha^2} |S(\mathbf{g})|^2$$

$$- \frac{1}{2V} \sum_{i=1}^{N} q_i^2 \sum_{\mathbf{g} \neq (0,0,0)} \frac{4\pi}{|\mathbf{g}|^2} e^{-|\mathbf{g}|^2/4\alpha^2}. \tag{C.16}$$

We now note that

$$\frac{1}{V} \sum_{\mathbf{g}} \frac{4\pi}{|\mathbf{g}|^2} e^{-|\mathbf{g}|^2/4\alpha^2} = \lim_{r \to 0} \frac{\text{erf}(\alpha r)}{r}. \tag{C.17}$$

Since r and $\text{erf}(\alpha r)$ are both zero at $r = 0$, the limit in eqn. (C.17) can be performed using L'Hôpital's rule, so that

$$\lim_{r \to 0} \frac{\text{erf}(\alpha r)}{r} = \lim_{r \to 0} \frac{2}{r\sqrt{\pi}} \int_0^{\alpha r} e^{-t^2} \, dt = \lim_{r \to 0} \frac{2\alpha}{\sqrt{\pi}} e^{-\alpha^2 r^2} = \frac{2\alpha}{\sqrt{\pi}}. \tag{C.18}$$

From this limit, eqn. (C.17) becomes

$$U_{\text{long}}(\mathbf{r}_1, ..., \mathbf{r}_N) = \frac{1}{V} \sum_{\mathbf{g} \in \mathcal{S}} \frac{4\pi}{|\mathbf{g}|^2} e^{-|\mathbf{g}|^2/4\alpha^2} |S(\mathbf{g})|^2 - \frac{\alpha}{\sqrt{\pi}} \sum_{i=1}^{N} q_i^2. \qquad \text{(C.19)}$$

Equation (C.19) is known as the *Ewald sum* for the long-range part of the Coulomb interaction (Ewald, 1921). The second term in eqn. (C.19) is known as the self-interaction correction since it cancels out the $i = j$ term, which describes a long-range interaction of a particle with itself. The error that results when the second reciprocal-space sum in eqn. (C.15) is extended over all of Fourier space can be compensated for by multiplying the last term by $\text{erfc}(g_{\text{max}}/2\alpha)$.

Although the sum in eqn. (C.19) is more efficient to evaluate than the double sum in eqn. (C.12), two important technical problems remain. First, the $|S(\mathbf{g})|^2$ factor in eqn. (C.19) leads to an interaction between *all* charged particles. Unfortunately, if a system contains molecules, then charged particles involved in common bond, bend, or torsional interactions must not also have a charge–charge interaction, as these are built into the intramolecular potential energy function. Therefore, these unwanted Coulomb interactions need to be excluded from the Ewald sum. One way to solve this problem is to add these unwanted terms with the opposite sign in real space so that they become new contributions to the intramolecular potential and approximately cancel their reciprocal-space counterparts. If this is done, eqn. (C.19) becomes

$$U_{\text{long}}(\mathbf{r}_1, ..., \mathbf{r}_N) = \frac{1}{V} \sum_{\mathbf{g} \in \mathcal{S}} \frac{4\pi}{|\mathbf{g}|^2} e^{-|\mathbf{g}|^2/4\alpha^2} |S(\mathbf{g})|^2$$

$$- \sum_{i,j \in \text{intra}} \frac{q_i q_j \text{erf}(\alpha r_{ij})}{r_{ij}} - \frac{\alpha}{\sqrt{\pi}} \sum_{i=1}^{N} q_i^2, \qquad \text{(C.20)}$$

where "intra" denotes the full set of bonded interactions. The second term in eqn. (C.20) can, therefore, be incorporated into the bonded terms in eqn. (3.11.1). Equation (C.20) is straightforward to implement, but because we are attempting to achieve the cancellation in real space rather than in Fourier space, the cancellation is imperfect. Indeed, Procacci *et al.* (1998) showed that better cancellation is achieved if the excluded interactions are corrected using a reciprocal-space expression that accounts for reciprocal-space truncation. In particular, these authors showed that the excluded interactions can be computed more precisely if an additional correction is added, which is given by

$$U_{\text{corr}}(\mathbf{r}_1, ..., \mathbf{r}_N) = \sum_{i,j \in \text{intra}} q_i q_j \chi(r_{ij}, g_{\text{max}}) + \frac{\alpha}{\sqrt{\pi}} \text{erfc}(g_{\text{max}}/2\alpha) \sum_{i=1}^{N} q_i^2, \qquad \text{(C.21)}$$

where

$$\chi(r, g_{\text{max}}) = \frac{2}{\pi} \int_{g_{\text{max}}}^{\infty} dg\ e^{-g^2/4\alpha^2} \frac{\sin(gr)}{gr}. \qquad \text{(C.22)}$$

An important technical point about the form of the Ewald sum is that it causes the potential to acquire an explicit volume dependence. Thus, when calculating the pressure or performing simulations in the NPT ensemble, it becomes necessary to use the pressure estimators in eqns. (4.7.58) and (5.8.29).

The second technical problem associated with the Ewald sum is the high computational overhead associated with the calculation of the structure factor $S(\mathbf{g}) = \sum_i q_i \exp(i\mathbf{g} \cdot \mathbf{r}_i)$ for large systems. As L increases, the number of $\mathbf{g}$-vectors satisfying the criterion $|\mathbf{g}| \leq g_{\max}$ becomes quite large, and the particle sum in $S(\mathbf{g})$, which also increases, must be carried out at each $\mathbf{g}$-vector. The *smooth particle-mesh Ewald* (SPME) method, an important advance introduced by Essmann *et al.* (1995), substantially reduces the cost of the Ewald sum and improves the scaling with system size. The SPME approach can be viewed as a kind of "smearing" of the charges, which are located at arbitrary spatial points $\mathbf{r}_i$, over a finite set of points on a regular rectangular lattice. In practice, this "smearing" is realized mathematically by employing an interpolation formula to express the exponential $\exp(i\mathbf{g} \cdot \mathbf{r}_i)$ in terms of the exponential factors that depend on the coordinates of the grid points rather than the arbitrary particle coordinate $\mathbf{r}_i$. Essmann *et al.* achieved this interpolation using a set of spline functions known as the *cardinal B-spline* functions. The functions, denoted $M_n(x)$, are defined as follows: $M_2(x) = 1 - |x - 1|$ for $0 \leq x \leq 2$ and $M_2(x) = 0$ for $x < 0$ and $x > 2$. For $n > 2$, $M_n(x)$ are defined via the recursion relations

$$M_n(x) = \frac{x}{n-1} M_{n-1}(x) + \frac{n-x}{n-1} M_{n-1}(x-1). \tag{C.23}$$

The cardinal B-spline functions have several important properties. First, they have compact support, meaning that $M_n(x)$ is zero outside the interval $0 \leq x \leq n$. Second, $M_n(x)$ is $n - 2$ times continuously differentiable. Third, the derivative $\mathrm{d}M_n/\mathrm{d}x$ can be obtained from $M_{n-1}(x)$ using another recurrence relation

$$\frac{\mathrm{d}}{\mathrm{d}x} M_n(x) = M_{n-1}(x) - M_{n-1}(x-1). \tag{C.24}$$

Two other useful properties are:

$$M_n(x) = M_n(n - x)$$

$$\sum_{j=-\infty}^{\infty} M_n(x - j) = 1. \tag{C.25}$$

A plot of the first few cardinal B-spline functions is shown in Fig. C.2.

We now introduce the scaled particle coordinates $\mathbf{s}_i = V^{-1/3}\mathbf{r}_i$ (which reduce to $\mathbf{s}_i = \mathbf{r}_i/L$ for a cubic box), so that $\exp(i\mathbf{g} \cdot \mathbf{r}_i) = \exp(2\pi i \mathbf{n} \cdot \mathbf{s}_i)$. Let the lattice be cubic with N_l points along each direction and define new coordinates $\mathbf{u}_i = N_l \mathbf{s}_i$, such that $u_{i,\alpha} \in [0, N_l]$, $\alpha = x, y, z$. Then, $\exp(2\pi i \mathbf{n} \cdot \mathbf{s}_i) = \exp(2\pi i \mathbf{n} \cdot \mathbf{u}_i/N_l)$. We can then approximate the exponential $\exp(2\pi i n_\alpha u_{i,\alpha}/N_l)$ using a cardinal B-spline interpolation formula as

$$e^{2\pi i n_\alpha u_{i,\alpha}/N_l} \approx b_n(n_\alpha) \sum_{k=-\infty}^{\infty} M_n(u_{i,\alpha} - k) e^{2\pi i n_\alpha k/N_l}, \tag{C.26}$$

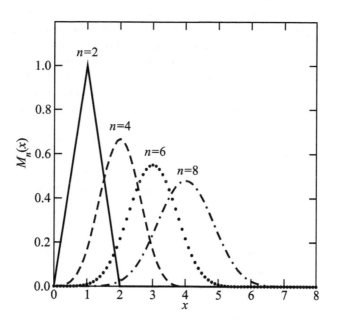

Fig. C.2 Cardinal B-spline functions for $n = 2, 4, 6, 8$.

where

$$b_n(\nu) = e^{2\pi i(n-1)\nu/N_l} \sum_{k=0}^{n-2} M_n(k+1)e^{2\pi i\nu k/N_l}. \tag{C.27}$$

(Remember that n and n_α are *different* indices!) The sum in eqn. (C.26) is not actually infinite because $M_n(x)$ has compact support. The number of nonzero terms, and hence the accuracy of the interpolation, increases with n. Using eqn. (C.26), the structure factor can then be evaluated using a (discrete) fast Fourier transform of the form

$$S(\mathbf{n}) = b_n(n_x)b_n(n_y)b_n(n_z)\tilde{Q}(\mathbf{n}), \tag{C.28}$$

where

$$\tilde{Q}(\mathbf{n}) = \sum_{k_x=0}^{N_l-1} \sum_{k_y=0}^{N_l-1} \sum_{k_z=0}^{N_l-1} e^{2\pi i n_x k_x/N_l} e^{2\pi i n_y k_y/N_l} e^{2\pi i n_z k_z/N_l} Q(\mathbf{k}), \tag{C.29}$$

and

$$Q(\mathbf{k}) = \sum_{i=1}^{N} q_i \sum_{j_1,j_2,j_3} M_n(u_{x,i} - k_x - j_1 N_l)M_n(u_{y,i} - k_y - j_2 N_l)M_n(u_{z,i} - k_z - j_3 N_l). \tag{C.30}$$

Because the computational overhead of a fast Fourier transform of length N_l scales as $\mathcal{O}(N_l \log N_l)$, and $N_l \propto N$, it follows that using the smooth particle-mesh Ewald technique to evaluate the structure factor reduces the cost of an Ewald sum to $\mathcal{O}(N \log N)$, which becomes a significant computational savings in large systems.

Finally, we note that Ewald summation need not be restricted to periodic systems. In fact, Martyna, Tuckerman, and coworkers (1999, 2004a, 2002) showed that the Ewald summation formalism could be easily modified in a unified manner for use in systems with zero, one, or two periodic dimensions, corresponding to clusters, wires, and surfaces, respectively.

Force fields employing simple functions, such as harmonic bonds, harmonic bends, cosine-series torsions, and nonbonded interactions like those in eqn. (C.1), are crude approximations of the true physics of interactions, which include charge polarization, charge transfer, and chemical events involving bond-breaking and forming. More sophisticated force fields, not discussed in this appendix, attempt to include chemical events by including polarization through fluctuating charges and dipole moments in the system or via models that attach a light, negatively charged particle to each atom by a spring (the so-called Drude model). Specialized reactive force fields have also been developed to describe specific chemical events. A considerable literature exists on force fields, and an entire book could be written about this subject alone. It is not our intention to review the literature on force fields in a book on statistical mechanics. However, we, nevertheless, conclude this appendix with a brief discussion of a new class of force fields based on the machine learning models of Chapter 17. The advantage of a machine-learning based interaction model is that much of the physics, such as polarization, manybody effects, and chemical reactivity, missing from the fixed-charge models discussed thus far can be captured within a single universal form, such as a neural network or Gaussian kernel model provided that these missing physical effects are present in the data on which the model is trained.

As discussed in Chapter 17, a machine learning model needs a set of descriptors for the input information. For the development of a force field, such a descriptor could provide information about the local environment of each atom in the system, which involves other atoms with which a given atom interacts most strongly. As noted in Section 17.7.2, these descriptors must be invariant to translations and rotations of the system. Because of this requirement, the symmetry functions given in eqns. (17.7.7) through (17.7.13) are particularly discriminating of these local atomic environments and have been employed as descriptors to create force fields in terms of multi-layer feed-forward neural networks (Behler and Parrinello, 2007; Behler, 2011; Behler, 2015; Behler, 2021) (see Section 17.5). If these atomic descriptors are employed, then the construction of a machine learning force field must also be based on contributions from individual atoms. Moreover, since these descriptors are "nearsighted" via inclusion of the cutoff function $f_c(r_{ij})$ in the symmetry functions, the machine learning model will learn short-range interactions, including both bonded and short-range nonbonded contributions. Let us denote the corresponding sum of bonded and short-ranged potentials as $U_s(\mathbf{r}_1, ..., \mathbf{r}_N) = U_{\text{bonded}}(\mathbf{r}_1, ..., \mathbf{r}_N) + U_{\text{short}}(\mathbf{r}_1, ..., \mathbf{r}_N)$. The use of atom-centered descriptors suggests that there should be a corresponding decomposition of U_s in terms of contributions from each atom, that is,

$$U_s = \sum_{i=1}^{N} u_{s,i}. \tag{C.31}$$

If the long-range term U_{long} is now added to U_s, the full potential becomes $U = U_s + U_{\text{long}}$. The utility of eqn. (C.31) is that each term $u_{s,i}$ is represented as a machine learning model. In the work of Behler, *et al.*, a feed-forward network is employed, a schematic of which is illustrated in Fig. C.3. The fact that $u_{s,i}$ is short ranged is important for employing machine learning in that it restricts the dimensionality of the function that must be learned, since the number of neighbors of atom i accounted for in the symmetry functions is relatively small (typically fewer then 100) compared to those included in the long-range contributions, and this renders tractable the use of a model such as a neural network. Nevertheless, the number of neighbors is sufficient to include manybody effects. The analytical function to which this diagram corresponds is given in eqn. (17.5.9), which is needed in order to compute the derivatives needed for the forces.

The training of a machine learning potential requires input of coordinates and corresponding energies. If desired, gradient training using energies and forces, together with coordinates, is also possible and can reduce the amount of training data needed while improving the accuracy of the model if it is to be used in a molecular dynamics calculation. Ultimately, the physics contained within the machine learning force field

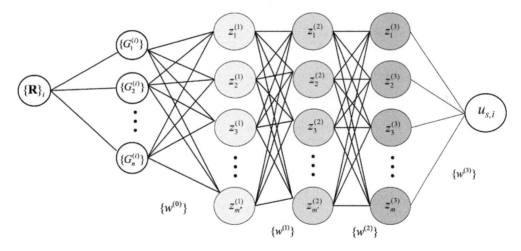

Fig. C.3 A feed-forward neural network with three hidden layers that are used to generate an atomic contribution $u_{s,i}$ to the short-range potential energy in eqn. (C.31). The quantities $z_\sigma^{(j)}$ are defined in eqn. (17.5.10). Each set of parameters $\{w^{(K)}\}$ determines the weights of the connections between hidden layers. The input layer is a set of symmetry functions of different types for each atom i (see eqns. (17.7.9) through (17.7.13)), which are generated from the set of coordinates $\{\mathbf{R}\}_i$ of atom i and all of its neighbors within the cutoff prescribed by the symmetry functions.

will match that contained in the training data. For example, if the training data is produced by *ab initio* quantum chemical calculations at the level of density functional theory, then the accuracy of the machine learning model will reflect the level of the density functional theory calculations employed. If, in addition, the training data includes bond-breaking and forming events, then the machine learning model will be able to account for these as well. An important question concerns the number of configurations or, equivalently, atomic environments needed to produce a potential free of "holes", which occur when the system visits configurations whose energies and forces cannot be predicted by the model because of insufficient training data at these configurations. In order to have a feeling for this, consider the neural network potential employed by Cendagorta *et al.* (2021) in their study of structure-II hydrogen clathrate (see Fig. 12.17). Although this potential was only needed to describe one solid phase, in order that enough environments were included, the training data consisted of six types of structures. "Clathrate structures" included a structure-II clathrate unit cell with a range of H_2 occupancies of the small and large cages; "H_2" structures consisted of simulation boxes of pure H_2 molecules; "$H_2O + H_2$" structures consisted of simulation boxes with varying concentrations of hydrogen and water molecules; "Large cage" structures consisted of an isolated large cage from the structure-II clathrate; "Small cage" structures consisted of an isolated small cage from the structure-II clathrate; "H_2O" structures consisted of simulation boxes of pure water. Configurations within each of these six structure classes were generated either randomly or by performing high-temperature (> 300 K) molecular dynamics calculations so as to generate a sufficient number of different environments to obtain a potential free of holes. The numbers of structures in each of the six classes were, respectively, 2,432, 533, 716, 539, 319, and 1,440 for a total of 5,979 different environments. The RMSE error in the test set was 0.93 meV/atom, and at the conditions of the simulations performed by Cendagorta *et al.* (2021), no holes were observed. The process of generating this potential exemplifies the large variety of environments needed to produce a reliable model free of holes.

Appendix D
Proof of the Trotter theorem

The Trotter theorem figures prominently throughout the book, in both the development of numerical solvers for ordinary differential equations and in the derivation of the Feynman path integral. However, in eqn. (3.10.18), we presented the theorem without proof. Therefore, in this appendix, we outline the proof of the theorem following a technique presented by Schulman (1981).

Let $\hat{P}$ and $\hat{Q}$ be linear operators on a general normed vector space $\mathcal{V}$, also known as a *Banach space*, and let $\psi \in \mathcal{V}$. The Trotter theorem is equivalent to the statement that there exists a linear operator $\hat{R}$ on $\mathcal{V}$ such that the difference

$$\hat{R}^t \psi - \lim_{n \to \infty} \left(\hat{P}^{t/n} \hat{Q}^{t/n} \right) \psi = 0, \tag{D.1}$$

where $0 \le t < \infty$. Before proceeding, it is useful to introduce the following definition: A *contraction semigroup* on $\mathcal{V}$ is a family of bounded linear operators $\hat{P}^t$, $0 \le t < \infty$, which are defined everywhere on $\mathcal{V}$ and constitute a mapping $\mathcal{V} \to \mathcal{V}$ such that the following statements are true:

$$\hat{P}^0 = 1, \qquad \hat{P}^t \hat{P}^s = \hat{P}^{t+s}, \qquad t \ge 0, \ s \le \infty, \tag{D.2}$$

$$\lim_{t \to \infty} \hat{P}^t \psi = \psi, \qquad ||\hat{P}^t|| \le 1. \tag{D.3}$$

Here, the norm $||\hat{P}^t||$ is defined to be

$$||\hat{P}^t|| = \inf_{\beta \in B} \left\{ \beta \mid ||\hat{P}^t \phi|| \le \beta ||\phi|| \ \forall \phi \in \mathcal{V}, \ ||\phi|| \le 1 \right\}. \tag{D.4}$$

Let $\hat{A}$, $\hat{B}$, and $\hat{A} + \hat{B}$ be infinitesimal generators of the contraction semigroups $\hat{P}^t$, $\hat{Q}^t$, and $\hat{R}^t$, respectively. This means, for example, that the action of $\hat{A}$ on a vector ψ is

$$\hat{A}\psi = \lim_{t \to 0} \frac{1}{t} \left(\hat{P}^t \psi - \psi \right). \tag{D.5}$$

Next, let h be a positive real number. It is straightforward to verify the following identity for the contraction semigroups:

$$\left(\hat{P}^h \hat{Q}^h - 1 \right) \psi = \left(\hat{P}^h - 1 \right) \psi + \hat{P}^h \left(\hat{Q}^h - 1 \right) \psi. \tag{D.6}$$

Using the infinitesimal generators allows us to write

$$\left(\hat{P}^h\hat{Q}^h - 1\right)\psi = h\left(\hat{A} + \hat{B}\right)\psi + \mathcal{O}(h), \tag{D.7}$$

where $\mathcal{O}(h)$ denotes any vector ϕ such that

$$\lim_{h\to 0} \frac{||\phi||}{h} = 0. \tag{D.8}$$

In other words, $||x||$ goes to 0 faster than h does. Note that we can also write

$$\left(\hat{R}^h - 1\right)\psi = h\left(\hat{A} + \hat{B}\right)\psi + \mathcal{O}(h). \tag{D.9}$$

Consequently,

$$\left(\hat{P}^h\hat{Q}^h - \hat{R}^h\right)\psi = \mathcal{O}(h). \tag{D.10}$$

Now let $h = t/n$. We need to show that

$$\left|\left|\left[\left(\hat{P}^h\hat{Q}^h\right)^n - \hat{R}^{hn}\right]\psi\right|\right| \to 0 \tag{D.11}$$

as $n \to \infty$.

To see how this limit can be demonstrated, consider first the case $n = 2$. It is straightforward to show that

$$\left(\hat{P}^h\hat{Q}^h\right)^2 - \hat{R}^{2h} = \left(\hat{P}^h\hat{Q}^h - \hat{R}^h\right)\hat{R}^h + \hat{P}^h\hat{Q}^h\left(\hat{P}^h\hat{Q}^h - \hat{R}^h\right). \tag{D.12}$$

Likewise, for $n = 3$, a little algebra reveals that

$$\left(\hat{P}^h\hat{Q}^h\right)^3 - \hat{R}^{3h} = \left(\hat{P}^h\hat{Q}^h - \hat{R}^h\right)\hat{R}^{2h} + \hat{P}^h\hat{Q}^h\left(\hat{P}^h\hat{Q}^h - \hat{R}^h\right)\hat{R}^h$$
$$+ \left(\hat{P}^h\hat{Q}^h\right)^2\left(\hat{P}^h\hat{Q}^h - \hat{R}^h\right). \tag{D.13}$$

Generally, therefore,

$$\left(\hat{P}^h\hat{Q}^h\right)^n - \hat{R}^{nh} = \left(\hat{P}^h\hat{Q}^h - \hat{R}^h\right)\hat{R}^{(n-1)h} + \hat{P}^h\hat{Q}^h\left(\hat{P}^h\hat{Q}^h - \hat{R}^h\right)\hat{R}^{(n-2)h} + \cdots$$
$$+ \left(\hat{P}^h\hat{Q}^h\right)^{n-1}\left(\hat{P}^h\hat{Q}^h - \hat{R}^h\right). \tag{D.14}$$

We now let the operators on the left and right sides of eqn. (D.14) act on ψ and take the norm. This yields

$$\left|\left|\left[\left(\hat{P}^h\hat{Q}^h\right)^n - \hat{R}^{nh}\right]\psi\right|\right| = \left|\left|\left(\hat{P}^h\hat{Q}^h - \hat{R}^h\right)\hat{R}^{(n-1)h}\psi\right.\right.$$

$$\left.\left. + \hat{P}^h\hat{Q}^h\left(\hat{P}^h\hat{Q}^h - \hat{R}^h\right)\hat{R}^{(n-2)h}\psi\right.\right.$$

$$+ \cdots + \left(\hat{P}^h \hat{Q}^h \right)^{n-1} \left(\hat{P}^h \hat{Q}^h - \hat{R}^h \right) \psi \Big\|. \quad \text{(D.15)}$$

Recall that for ordinary vectors **a**, **b**, and **c**, such that **a** = **b**+**c**, the triangle inequality $|\mathbf{a}| \leq |\mathbf{b}| + |\mathbf{c}|$ holds. Similarly, we have

$$\left\| \left[\left(\hat{P}^h \hat{Q}^h \right)^n - \hat{R}^{nh} \right] \psi \right\| \leq \left\| \left(\hat{P}^h \hat{Q}^h - \hat{R}^h \right) \hat{R}^{(n-1)h} \psi \right\|$$
$$+ \left\| \hat{P}^h \hat{Q}^h \left(\hat{P}^h \hat{Q}^h - \hat{R}^h \right) \hat{R}^{(n-2)h} \psi \right\| + \cdots. \quad \text{(D.16)}$$

On the right, there are n terms, all of order $\mathcal{O}(h)$. Thus, the right side varies as $n\mathcal{O}(h) = n\mathcal{O}(t/n)$. As $n \to \infty$, $n\mathcal{O}(t/n) \to 0$. Hence,

$$\lim_{n \to \infty} \left\| \left[\left(\hat{P}^h \hat{Q}^h \right)^n - \hat{R}^{nh} \right] \psi \right\| \to 0, \quad \text{(D.17)}$$

which implies eqn. (D.1).

Appendix E
Laplace transforms

The Laplace transform of a function $f(t)$ is just one of a general class of integral transforms of the form

$$\tilde{f}(s) = \int_0^\infty dt\, K(st) f(t),$$
(E.1)

where $K(x)$ is a kernel that is typically a smooth and rapidly decaying function of x. The Laplace transform corresponds to the choice $K(x) = e^{-x}$.

Interestingly, even if the integral of $f(t)$ over the interval $t \in [0, \infty)$ does not exist, the Laplace transform $\tilde{f}(s)$ nevertheless can. If there exists a positive constant s_0 such that $|e^{-s_0 t} f(t)| \leq M$ for M finite, then $\tilde{f}(s)$ exists for $s > s_0$. By contrast, a function such as $f(t) = \exp(t^2)$ does not have a Laplace transform.

Laplace transforms of elementary functions are generally straightforward to evaluate. Some examples are given below:

$$f(t) = t^n, \qquad\qquad \tilde{f}(s) = \frac{n!}{s^{n+1}},$$

$$f(t) = e^{-at}, \qquad\qquad \tilde{f}(s) = \frac{1}{s+a},$$

$$f(t) = \cos(\omega t), \qquad\qquad \tilde{f}(s) = \frac{s}{s^2 + \omega^2},$$

$$f(t) = \sin(\omega t), \qquad\qquad \tilde{f}(s) = \frac{\omega}{s^2 + \omega^2},$$

$$f(t) = \cosh(\alpha t), \qquad\qquad \tilde{f}(s) = \frac{s}{s^2 - \alpha^2},$$

$$f(t) = \sinh(\alpha t), \qquad\qquad \tilde{f}(s) = \frac{\alpha}{s^2 - \alpha^2}.$$
(E.2)

It is left as an exercise to the reader to verify each of these relations. In addition to these elementary transforms another useful result is the Laplace transform of the convolution $f(t)$ between two functions $g(t)$ and $h(t)$. Recall that this convolution is defined as

$$f(t) = \int_0^t d\tau\, g(\tau) h(t - \tau).$$
(E.3)

The Laplace transform of $f(t)$ is $\tilde{f}(s) = \tilde{g}(s)\tilde{h}(s)$, which is known as the *convolution theorem*. The proof of the convolution theorem proceeds by writing

$$\tilde{g}(s)\tilde{h}(s) = \lim_{a\to\infty} \int_0^a dx \, e^{-sx} g(x) \int_0^{a-x} dy \, e^{-sy} h(y). \tag{E.4}$$

The use of the upper limit $a - x$ rather than simply a in the y integral is permissible because of the $\lim_{a\to\infty}$ and the fact that the integrals are assumed to decay rapidly. The use of a triangular integration region rather than a square one simplifies the algebra. We then introduce the change of variables $x = t - z$, $y = z$ into the integral, which leads to

$$\tilde{g}(s)\tilde{h}(s) = \lim_{a\to\infty} \int_0^a dt \, e^{-st} \int_0^t dz \, g(t - z)h(z). \tag{E.5}$$

When $a \to \infty$, eqn. (E.5) is the Laplace transform of a convolution.

Laplace transforms are particularly useful for solving ordinary linear differential and linear integro-differential equations. In particular, an ordinary linear differential equation for a function $f(t)$ can be converted into a simple algebraic equation for $\tilde{f}(s)$. This conversion is accomplished by expressing derivatives as simple algebraic expressions via Laplace transformation. Consider first the Laplace transform of the function $g(t) = f'(t) = df/dt$, which is given by

$$\tilde{g}(s) = \int_0^\infty dt \, e^{-st} \frac{df}{dt}. \tag{E.6}$$

An integration by parts shows that

$$\tilde{g}(s) = e^{-st} f(t) \Big|_0^\infty + s \int_0^\infty dt \, e^{-st} f(t)$$

$$= s\tilde{f}(s) - f(0), \tag{E.7}$$

where it is assumed that $\lim_{a\to\infty} e^{-sa} f(a) = 0$. Similarly, if $h(t) = f''(t) = d^2 f/dt^2$, then by the same analysis

$$\tilde{h}(s) = s^2 \tilde{f}(s) - sf(0) - f'(0). \tag{E.8}$$

In general, if $F(t) = d^n f/dt^n$, then

$$\tilde{F}(s) = s^n \tilde{f}(s) - s^{n-1} f(0) - s^{n-2} f'(0) - \cdots - f^{(n-1)}(0). \tag{E.9}$$

In Section 15.2, we showed how these relations can be used to convert a linear second-order differential equation for $f(t)$ into an algebraic equation for $\tilde{f}(s)$, which can then be easily solved. However, once an expression for $\tilde{f}(s)$ is found, obtaining $f(t)$ requires performing an inverse Laplace transform, and the inversion of Laplace transforms is considerably less straightforward than the forward Laplace transform.

Of course, the simplest way to invert a Laplace transform is to look up the given form for $\tilde{f}(s)$ in a table of Laplace transforms and find the corresponding form for $f(t)$. However, if $\tilde{f}(s)$ cannot be found in the table, then an explicit inversion must be performed. Unfortunately, as we noted earlier, $f(t)$ can diverge exponentially and its Laplace transform will still exist, a fact that makes Laplace inversion rather tricky.

From a numerical standpoint, a forward Laplace transform can be perform straight-forwardly, while a numerical Laplace inversion is a highly ill-posed problem mathematically (Epstein and Schotland, 2008).

To see how the form of the inverse Laplace transform arises, let us assume that $f(t)$ exhibits an exponential divergence $e^{\gamma t}$, and let us define a function $g(t)$ by $f(t) = e^{\gamma t} g(t)$. Since the Laplace transform restricts t to the interval $[0, \infty)$, we will assume $g(t) = 0$ for $t < 0$. $g(t)$ is a well-behaved function, and therefore we can define it via its Fourier transform

$$g(t) = \frac{1}{\sqrt{2\pi}} \int_{-\infty}^{\infty} d\omega \, e^{i\omega t} \hat{g}(\omega). \tag{E.10}$$

The Fourier transform $\hat{g}(\omega)$ can, itself, be expressed in terms of $g(t)$ as

$$\hat{g}(\omega) = \frac{1}{\sqrt{2\pi}} \int_{0}^{\infty} dt \, g(t) e^{-i\omega t}. \tag{E.11}$$

Substituting eqn. (E.11) into eqn. (E.10) yields

$$g(t) = \frac{1}{2\pi} \int_{-\infty}^{\infty} d\omega \, e^{i\omega t} \int_{0}^{\infty} du \, g(u) e^{-i\omega u}. \tag{E.12}$$

Since $g(t) = e^{-\gamma t} f(t)$, eqn. (E.12) implies that

$$f(t) = \frac{e^{\gamma t}}{2\pi} \int_{-\infty}^{\infty} d\omega \, e^{i\omega t} \int_{0}^{\infty} du \, f(u) e^{-\gamma u} e^{-i\omega u}. \tag{E.13}$$

If we let $s = \gamma + i\omega$, then eqn. (E.13) becomes

$$f(t) = \frac{1}{2\pi i} \int_{\gamma - i\infty}^{\gamma + i\infty} ds \, e^{st} \int_{0}^{\infty} du \, f(u) e^{-us}$$

$$= \frac{1}{2\pi i} \int_{\gamma - i\infty}^{\gamma + i\infty} ds \, e^{st} \tilde{f}(s). \tag{E.14}$$

Equation (E.14) defines the Laplace inversion problem as an integral over a complex variable s, which can be performed using techniques of complex integration and the calculus of residues. For $t > 0$, eqn. (E.14) can be rewritten as a contour integral using a contour of the type shown in Fig. E.1 whose leading edge is parallel to the imaginary s axis and is chosen far enough to the right to enclose all of the poles of $\tilde{f}(s)$. The contour is then closed in the left half of the complex plane. Denoting this contour as B for the *Bromwich contour*, we can rewrite eqn. (E.14) as

$$f(t) = \frac{1}{2\pi i} \oint_{B} ds \, e^{st} \tilde{f}(s). \tag{E.15}$$

As an example, consider inverting the Laplace transform for $\tilde{f}(s) = \alpha/(s^2 - \alpha^2)$, which we know from eqn. (E.2) to be the Laplace transform of $f(t) = \sinh(\alpha t)$. We set up the contour integration as

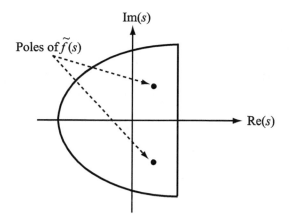

Fig. E.1 Bromwich contour, which contains all of the poles of $\tilde{f}(s)$. Laplace inversion employs such a contour extended to infinity in all directions.

$$f(t) = \frac{1}{2\pi i} \oint_B ds \; \frac{\alpha e^{st}}{(s-\alpha)(s+\alpha)}. \tag{E.16}$$

The integrand in eqn. (E.16) has two first-order poles at $s = \alpha$ and $s = -\alpha$. Thus, we need to choose the leading edge of the contour to lie to the right of the point $s = \alpha$ on the real s-axis. Once we do this and apply the residue theorem to each of the poles, we obtain

$$f(t) = \frac{\alpha}{2\pi i} 2\pi i \left[\frac{e^{\alpha t}}{2\alpha} - \frac{e^{-\alpha t}}{2\alpha} \right]$$

$$= \frac{1}{2} \left[e^{\alpha t} - e^{-\alpha t} \right]$$

$$= \sinh(\alpha t). \tag{E.17}$$

In Section 15.2, we showed how to solve ordinary linear differential equations using Laplace transforms. Now let us use the result obtained in eqn. (E.17) in another example that involves the application of the Laplace transform to the solution of an integral equation. Consider the integral equation

$$x(t) = t + a^2 \int_0^t d\tau \; (t-\tau) x(\tau), \tag{E.18}$$

which we must solve for $x(t)$ in terms of the constant a. We can make use of the fact that the integral appearing in eqn. (E.18) is a convolution, and its Laplace transform will simply be $\tilde{x}(s)/s^2$, since the Laplace transform of $f(t) = t$ is $\tilde{f}(s) = 1/s^2$. Thus, taking the Laplace transform of both sides of eqn. (E.18), we obtain the algebraic equation

$$\tilde{x}(s) = \frac{1}{s^2} + a^2 \frac{\tilde{x}(s)}{s^2}. \tag{E.19}$$

Solving this for $\tilde{x}(s)$, we obtain

$$\tilde{x}(s) = \frac{1}{s^2 - a^2}, \tag{E.20}$$

and, as we have already shown, the inverse Laplace transform of $\tilde{x}(s)$, which gives us the solution $x(t)$ is

$$x(t) = \frac{1}{a}\sinh(at). \tag{E.21}$$

References

Abrams, C. F. and Vanden-Eijnden, E. (2010). Large-scale conformational sampling of proteins using temperature-accelerated molecular dynamics. *Proc. Natl. Acad. Sci. U.S.A.*, **107**, 4961.

Abrams, J. B., Rosso, L., and Tuckerman, M. E. (2006). Efficient and precise solvation free energies via alchemical adiabatic molecular dynamics. *J. Chem. Phys.*, **125**, 074115.

Abrams, J. B. and Tuckerman, M. E. (2008). Efficient and direct generation of multidimensional free energy surfaces via adiabatic dynamics without coordinate transformations. *J. Phys. Chem. B*, **112**, 15742.

Abreu, C. R. A. and Tuckerman, M. E. (2020). Molecular dynamics with very large time steps for the calculation of solvation free energies. *J. Chem. Theory Comput.*, **16**, 7314.

Agarwal, A. and Delle Site, L. (2016). Grand-canonical adaptive resolution centroid molecular dynamics: Implementation and application. *Comp. Phys. Comm.*, **206**, 26–34.

Alavi, S., Ripmeester, J. A., and Klug, D. D. (2005). Molecular dynamics study of structure II hydrogen clathrates. *J. Chem. Phys.*, **123**, 024507.

Alder, B. J. and Wainwright, T. E. (1957). Phase transition for a hard sphere system. *J. Chem. Phys.*, **27**, 1208.

Alder, B. J. and Wainwright, T. E. (1959). Studies in molecular dynamics. 1. General method. *J. Chem. Phys.*, **31**, 459.

Alder, B. J. and Wainwright, T. E. (1967). Velocity autocorrelations for hard spheres. *Phys. Rev. Lett.*, **18**, 988.

Allen, M. P. and Tildesley, D. J. (1989). *Computer Simulations of Liquids*. Claredon Press, Oxford.

AlQuraishi, M. (2019). AlphaFold at CASP13. *Bioinform.*, **35**, 4862.

Andersen, H. C. (1980). Molecular dynamics at constant pressure and/or temperature. *J. Chem. Phys.*, **72**, 2384.

Andersen, H. C. (1983). RATTLE—A velocity version of the SHAKE algorithm for molecular dynamics calculations. *J. Comput. Phys.*, **52**, 24.

Anderson, M. H., Ensher, J. R., Matthews, M. R., Wieman, C. E., and Cornell, E. A. (1995). Observation of Bose-Einstein condensation in a dilute vapor. *Science*, **269**, 198.

Ansuini, A., Laio, A., Macke, J. H., and Zoccolan, D. (2019). Intrinsic dimension of data representations in deep neural networks. In *32nd Conference on Neural Information Processing Systems (NeurIPS 2019)* (ed. H. Wallach, H. Larochelle, A. Beygelzimer, F. d'Alché-Buc, E. Fox, and R. Garnett), Volume 32. Curran Associates, Inc.

Ashurst, W. T. and Hoover, W. G. (1975). Dense-fluid shear viscosity via nonequilibrium molecular dynamics. *Phys. Rev. A*, **11**, 658.

Awasthi, S. and Nair, N. N. (2017). Exploring high dimensional free energy landscapes: Temperature accelerated sliced sampling. *J. Chem. Phys.*, **146**, 094108.

Bader, J. S. and Berne, B. J. (1994). Quantum and classical relaxation rates from classical simulations. *J. Chem. Phys.*, **100**, 8359.

Baer, R. and Head-Gordon, M. (1998). Electronic structure of large systems: Coping with small gaps using the energy renormalization group method. *J. Chem. Phys.*, **109**, 10159.

Barducci, A., Bussi, G., and Parrinello, M. (2008). Well-tempered metadynamics: A smoothly converging and tunable free-energy method. *Phys. Rev. Lett.*, **100**, 020603.

Beck, T. L. (2011). Hydration free energies by energetic partitioning of the potential distribution theorem. *J. Stat. Phys.*, **145**, 335.

Beck, T. L., Paulaitis, M. E., and Pratt, L. R. (2006). *The Potential Distribution Theorem and Models of Molecular Solutions*. Cambridge University Press, Cambridge.

Behler, J. (2011). Atom-centered symmetry functions for constructing high-dimensional neural network potentials. *J. Chem. Phys.*, **134**, 074106.

Behler, J. (2015). Constructing high-dimensional neural network potentials: A tutorial review. *Intl. J. Quant. Chem.*, **115**, 1032.

Behler, J. (2021). Four generations of high-dimensional neural network potentials. *Chem. Rev.*, **121**, 10037.

Behler, J. and Parrinello, M. (2007). Generalized neural-network representation of high-dimensional potential-energy surfaces. *Phys. Rev. Lett.*, **98**, 146401.

Berendsen, H. J. C., Postma, J. P. M., van Gunsteren, W. F., DiNola, A., and Haak, J. R. (1984). Molecular dynamics with coupling to an external bath. *J. Chem. Phys.*, **81**, 3684.

Berezhkovskii, A. M., Pollak, E., and Zitserman, V. Y. (1992). Activated rate-processes - generalization of the Kramers-Grote-Hynes and Langer theories. *J. Chem. Phys.*, **97**, 2422.

Berges, J., Borsanyi, S., Sexty, D., and Stamatescu, I. O. (2007). Lattice simulations of real-time quantum fields. *Phys. Rev. D*, **75**, 045007.

Berne, B. J. (1971). Time correlation functions in condensed media. In *Physical Chemistry, Vol. VIIIB* (ed. H. Eyring, D. Henderson, and W. Jost), p. 539. Academic Press, New York.

Berne, B. J., Boon, J. P., and Rice, S. A. (1966). On calculation of autocorrelation functions of dynamical variables. *J. Chem. Phys.*, **45**, 1086.

Berne, B. J. and Harp, G. D. (1970a). On the calculation of time correlation functions. *Adv. Chem. Phys.*, **17**, 63.

Berne, B. J. and Harp, G. D. (1970b). Time correlation functions, memory functions and molecular dynamics. *Phys. Rev. A*, **2**, 975.

Berne, B. J., Pechukas, P., and Harp, G. D. (1968). Molecular reorientation in liquids and gases. *J. Chem. Phys.*, **49**, 3125.

Berne, B. J. and Pecora, R. (1976). *Dynamic Light Scattering*. Wiley, New York.

Berne, B. J., Tuckerman, M. E., Straub, J. E., and Bug, A. L. R. (1990). Dynamic friction on rigid and flexible bonds. *J. Chem. Phys.*, **93**, 5084.

Bertie, J. E., Ahmed, M. K., and Eysel, H. H. (1989). Infrared intensities of liquids. 5. Optical and dielectric-constants, integrated-intensities, and dipole-moment derivatives of H_2O and D_2O at 22°C. *J. Phys. Chem.*, **93**, 2210.

Beutler, T. C., Mark, A. E., van Schaik, R. C., Gerber, P. R., and van Gunsteren, W. F. (1994). Avoiding singularities and numerical instabilities in free energy calculations based on molecular simulations. *Chem. Phys. Lett.*, **222**, 529.

Bezdek, J. C., Ehrlich, R., and Full, W. (1984). FCM: The fuzzy c-means clustering algorithm. *Computers and Geosci.*, **10**, 191.

Bianucci, M., Mannella, M., and Grigolini, P. (1996). Linear response of Hamiltonian chaotic systems as a function of the number of degrees of freedom. *Phys. Rev. Lett.*, **77**, 1258.

Binnig, G., Quate, C. F., and Gerber, C. (1986). Atomic force microscope. *Phys. Rev. Lett.*, **56**, 930.

Bishop, C. M. (2006). *Pattern Recognition and Machine Learning*. Springer, Singapore.

Bolhuis, P. G., Chandler, D., Dellago, C., and Geissler, P. L. (2002). Transition path sampling: Throwing ropes over rough mountain passes, in the dark. *Ann. Rev. Phys. Chem.*, **53**, 291.

Bonanno, A. and Zappalà, D. (2001). Towards an accurate determination of critical exponents with the renormalization group flow equations. *Phys. Lett. B*, **504**, 181.

Bond, S. D., Leimkuhler, B. J., and Laird, B. B. (1999). The Nosé-Poincaré method for constant temperature molecular dynamics. *J. Comp. Phys.*, **151**, 114.

Bonella, S., Monteferrante, M., Pierleoni, C., and Ciccotti, G. (2010). Path integral based calculations of symmetrized time correlation functions. I. *J. Chem. Phys.*, **133**, 164104.

Bonomi, M., Branduardi, D., Bussi, G., Camilloni, C., Provasi, D., Raiteri, P., Donadio, D., Marinelli, F., Pietrucci, F., Broglia, R. A., and Parrinello, M. (2009). PLUMED: A portable plugin for free-energy calculations with molecular dynamics. *Comp. Phys. Comm.*, **180**, 1961.

Booth, G. H., Thom, A. J. W., and Alavi, A. (2009). Fermion Monte Carlo without fixed nodes: A game of life, death, and annihilation in Slater determinant space. *J. Chem. Phys.*, **131**, 054106.

Braams, B. J. and Manolopoulos, D. E. (2006). On the short-time limit of ring polymer molecular dynamics. *J. Chem. Phys.*, **125**, 124105.

Branduardi, D., Gervasio, F. L., and Parrinello, M. (2007). From a to b in free energy space. *J. Chem. Phys.*, **126**, 054103.

Brown, F. R., Butler, F. P., Chen, H., Christ, N. H., Dong, Z. H., Schaffer, W., Unger, L. I., and Vaccarino, A. (1991). Hadron masses in QCD with 2 flavors of dynamic fermions at beta = 5.7. *Phys. Rev. Lett.*, **67**, 1062.

Brown, F. R. and Christ, N. H. (1988). Parallel supercomputers for lattice gauge theory. *Science*, **239**, 1393.

Budimir, J. and Skinner, J. L. (1987). On the relationship between T_1 and T_2 for stochastic relaxation models. *J. Stat. Phys.*, **49**, 1029.

Bulgac, A. and Kusnezov, D. (1990). Canonical ensemble averages from pseudomi-crocanonical dynamics. *Phys. Rev. A*, **42**, 5045.

Bussi, G., Gervasio, F., Laio, A., and Parrinello, M. (2006). Free-energy landscape for beta hairpin folding from combined parallel tempering and metadynamics. *J. Am. Chem. Soc.*, **128**, 13435.

Caldeira, A. O. and Leggett, A. J. (1983). Quantum tunnelling in a dissipative system. *Ann. Phys.*, **149**, 374.

Calvo, F. (2002). Sampling along reaction coordinates with the Wang-Landau method. *Mol. Phys.*, **100**, 3421.

Cao, J. and Berne, B. J. (1989). On energy estimators in path-integral Monte Carlo simulations—dependence of accuracy on algorithm. *J. Chem. Phys.*, **91**, 6359.

Cao, J. and Berne, B. J. (1993). A Born-Oppenheimer approximation for path-integrals with an application to electron solvation in polarizable fluids. *J. Chem. Phys.*, **99**, 2902.

Cao, J. and Martyna, G. J. (1996). Adiabatic path integral molecular dynamics methods. II. Algorithms. *J. Chem. Phys.*, **104**, 2028.

Cao, J. and Voth, G. A. (1993). A new perspective on quantum time correlation functions. *J. Chem. Phys.*, **99**, 10070.

Cao, J. and Voth, G. A. (1994*a*). The formulation of quantum statistical mechanics based on the Feynman path centroid density. II. Dynamical properties. *J. Chem. Phys.*, **100**, 5106–5117.

Cao, J. and Voth, G. A. (1994*b*). The formulation of quantum statistical mechanics based on the Feynman path centroid density. IV. Algorithms for centroid molecular dynamics. *J. Chem. Phys.*, **101**, 6168.

Cao, J. and Voth, G. A. (1996). Semiclassical approximations to quantum dynamical time correlation functions. *J. Chem. Phys.*, **104**, 273.

Car, R. and Parrinello, M. (1985). Unified approach for molecular dynamics and density-functional theory. *Phys. Rev. Lett.*, **55**, 2471.

Cardy, J. (1996). *Scaling and Renormalization in Statistical Physics*. Cambridge University Press, Cambridge.

Carter, E. A., Ciccotti, G., Hynes, J. T., and Kapral, R. (1989). Constrained reaction coordinate dynamics for the simulation of rare events. *Chem. Phys. Lett.*, **156**, 472.

Causo, M. S., Ciccotti, G., Bonella, S., and Builleumier, R. (2006). An adiabatic linearized path-integral approach for quantum time-correlation functions II: A cumulant expansion method for improving convergence. *J. Chem. Phys.*, **110**, 16026.

Cendagorta, J. R., Bačić, Z., and Tuckerman, M. E. (2018). An open-chain imaginary-time path-integral sampling approach to the calculation of approximate symmetrized quantum time correlation functions. *J. Chem. Phys.*, **148**, 102340.

Cendagorta, J. R., Powers, A., Hele, T. J. H., Marsalek, O., Bačić, Z., and Tuckerman, M. E. (2016). Competing quantum effects in the free energy profiles and diffusion rates of hydrogen and deuterium molecules through clathrate hydrates. *Phys. Chem. Chem. Phys.*, **18**, 32169.

Cendagorta, J. R., Shen, H., Bačić, Z., and Tuckerman, M. E. (2021). Enhanced sampling path integral methods using neural network potential energy surfaces with application to diffusion in hydrogen hydrates. *Adv. Theor. Simulat.*, **4**, 2000258.

Cendagorta, J. R., Tolpin, J., Schneider, E., Topper, R. Q., and Tuckerman, M. E. (2020). Comparison of the performance of machine learning models in representing high-dimensional free energy surfaces and generating observables. *J. Phys. Chem. B*, **124**, 3647.

Ceperley, D. M. (1995). Path integrals in the theory of condensed helium. *Rev. Mod. Phys.*, **67**, 279.

Ceperley, D. M. and Pollock, E. L. (1984). Simulation of quantum many-body systems by path-integral methods. *Physical Review B*, **30**, 2555.

Ceriotti, M., Bussi, G., and Parrinello, M. (2009). Nuclear quantum effects in solids using a colored-noise thermostat. *Phys. Rev. Lett.*, **103**, 030603.

Ceriotti, M., Bussi, G., and Parrinello, M. (2010*a*). Colored-noise thermostats à la carte. *J. Chem. Theory Comput.*, **6**, 1170.

Ceriotti, M., Manolopoulos, D. E., and Parrinello, M. (2011). Accelerating the convergence of path integral dynamics with a generalized Langevin equation. *J. Chem. Phys.*, **134**, 084104.

Ceriotti, M., Parrinello, M., Markland, T. E., and Manolopoulos, D. E. (2010*b*). Efficient stochastic thermostatting of path integral molecular dynamics. *J. Chem. Phys.*, **133**, 124104.

Chandler, D. (1987). *Introduction to Modern Statistical Mechanics*. Oxford University Press, New York.

Chandler, D. and Wolynes, P. G. (1981). Exploiting the isomorphism between quantum-theory and classical statistical-mechanics of polyatomic fluids. *J. Chem. Phys.*, **74**, 4078.

Chandrasekhar, S. (1943). Stochastic problems in physics and astronomy. *Rev. Mod. Phys.*, **15**, 1.

Chen, B., Ivanov, I., Klein, M. L., and Parrinello, M. (2003). Hydrogen bonding in water. *Phys Rev. Lett.*, **91**, 215503.

Chen, M., Cuendet, M. A., and Tuckerman, M. E. (2012). Heating and flooding: A unified approach for rapid generation of free energy surfaces. *J. Chem. Phys.*, **137**, 024102.

Chen, M., Yu, T. -Q., and Tuckerman, M. E. (2015). Locating landmarks on high-dimensional free energy surfaces. *Proc. Natl. Acad. Sci. U.S.A.*, **112**, 3235.

Chen, P. -Y. and Tuckerman, M. E. (2018). Molecular dynamics based enhanced sampling of collective variables with very large time steps. *J. Chem. Phys.*, **148**, 024106.

Chin, S. A. (1997). Symplectic integrators from composite operator factorizations. *Phys. Lett. A*, **226**, 344.

Chipot, C. and Pohorille, A., eds. (2007). *Free Energy Calculations*. Springer–Verlag, Berlin.

Chow, T. W. S. and Cho, S. Y. (2007). *Neural Networks and Computing*. Imperial College Press, London.

Ciccotti, G. and Delle Site, L. (2019). The physics of open systems for the simulation of complex molecular environments in soft matter. *Soft Matt.*, **15**, 2114.

Ciccotti, G., Jacucci, G., and McDonald, I. R. (1976). Transport properties of molten alkali-halides. *Phys. Rev. A*, **13**, 426.

Ciccotti, G., Jacucci, G., and McDonald, I. R. (1979). Thought experiments by molecular dynamics. *J. Stat. Phys.*, **21**, 1.

Ciccotti, G., Pierleoni, C., and Ryckaert, J. P. (1992). Theoretical foundation and rheological application of nonequilibrium molecular dynamics. In *Microscopic Simulations of Complex Hydrodynamic Phenomena* (ed. M. Marechal and B. L. Holian), p. 25. Plenum Press, New York.

Cichos, F., Gustavsson, K., Mehlig, B., and Volpe, G. (2020). Machine learning for active matter. *Nature Mach. Intell.*, **2**, 94–103.

Coker, D. F., Berne, B. J., and Thirumalai, D. (1987). Path integral Monte-Carlo studies of the behavior of excess electrons in simple fluids. *J. Chem. Phys.*, **86**, 5689.

Corsini, R., Lazzerini, B., and Marcelloni, F. (2005). Relational clustering algorithm based on the fuzzy C-means algorithm. *Intl. J. Soft Comput.*, **2005**, 439.

Cortes-Huerto, R., Praprotnik, M., Kremer, K., and Delle Site, L. (2021). From adaptive resolution to molecular dynamics of open systems. *Euro. Phys. J. B*, **94**, 189.

Craig, I. R. and Manolopoulos, D. E. (2004). Quantum statistics and classical mechanics: Real time correlation functions from ring polymer molecular dynamics. *J. Chem. Phys.*, **121**, 3368.

Craig, I. R. and Manolopoulos, D. E. (2005). A refined ring polymer molecular dynamics theory of chemical reaction rates. *J. Chem. Phys.*, **123**, 034102.

Crooks, G. E. (1998). Nonequilibrium measurements of free energy differences for microscopically reversible Markovian systems. *J. Stat. Phys.*, **90**, 1481.

Crooks, G. E. (1999). Entropy production fluctuation theorem and the nonequilibrium work relation for free energy differences. *Phys. Rev. E*, **60**, 2721.

Cuendet, M. A. (2006). Statistical mechanical derivation of Jarzynski's identity for thermostatted non-Hamiltonian dynamics. *Phys. Rev. Lett.*, **96**, 120602.

Cuendet, M. A., Margul, D. T., Schneider, E., Vogt-Maranto, L., and Tuckerman, M. E. (2018). Endpoint-restricted adiabatic free energy dynamics approach for the exploration of biomolecular conformational equilibria. *J. Chem. Phys.*, **149**, 072316.

Cuendet, M. A. and Tuckerman, M. E. (2014). Free energy reconstruction from metadynamics or adiabatic free energy dynamics simulations. *J. Chem. Theory Comput.*, **10**, 2975.

Cynn, H., Yoo, C. S., Baer, B., Iota-Herbei, V., McMahan, A. K., Nicol, M., and Carlson, S. (2001). Martensitic FCC-to-HCP transformation observed in xenon at high pressure. *Phys. Rev. Lett.*, **86**, 4552.

Dama, J. F., Parrinello, M., and Voth, G. A. (2014). Well-tempered metadynamics converges asymptotically. *Phys. Rev. Lett.*, **112**.

Darve, E. (2007). Thermodynamic integration using constrainted and unconstrained dynamics. In *Free Energy Calculations* (ed. C. Chipot and A. Pohorille), pp. 119–170. Springer-Verlag, Berlin.

Darve, E. and Pohorille, A. (2001). Calculating free energies using average force. *J. Chem. Phys.*, **115**, 9169.

Darve, E., Rodriguez-Gomez, D., and Pohorille, A. (2008). Adaptive biasing force method for scalar and vector free energy calculations. *J. Chem. Phys.*, **128**, 144120.

Darve, E., Solomon, J., and Kia, A. (2009). Computing generalized Langevin equations and generalized Fokker–Planck equations. *Proc. Natl. Acad. Sci. U.S.A.*, **106**, 10884.

Davisson, C. and Germer, L. H. (1927). Reflection of electrons by a crystal of nickel. *Nature*, **119**, 558.

De Leon, N., Mehta, M. A., and Topper, R. Q. (1991). Cylindrical manifolds in phase space as mediators of chemical reaction dynamics and kinetics I. Theory. *J. Chem. Phys.*, **94**, 8310.

Delbuono, G. S., Rossky, P. J., and Schnitker, J. (1991). Model dependence of quantum isotope effects in liquid water. *J. Chem. Phys.*, **95**, 3728.

Dellago, C., Bolhuis, P. G., Csajka, F. S., and Chandler, D. (1998). Transition path sampling and the calculation of rate constants. *J. Chem. Phys.*, **108**, 1964.

Dellago, C., Bolhuis, P. G., and Geissler, P. L. (2002). Transition path sampling. *Adv. Chem. Phys.*, **123**, 1.

Delle Site, L. and Praprotnik, M. (2017). Molecular systems with open boundaries: Theory and simulation. *Phys. Rep.*, **693**, 1.

Deng, Y. Q. and Roux, B. (2009). Computations of standard binding free energies with molecular dynamics simulations. *J. Phys. Chem. B*, **113**, 2234.

Deutsch, J. M. and Silbey, R. (1971). Exact generalized Langevin equation for a particle in a harmonic lattice. *Phys. Rev.*, **3**, 2049.

Duane, S., Kennedy, A. D., Pendleton, B. J., and Roweth, D. (1987). Hybrid Monte Carlo. *Phys. Lett. B*, **195**, 216.

Duncan, J., Harjunmaa, A., Terrell, R., Drautz, R., Henkelman, G., and Rogal, J. (2016). Collective atomic displacements during complex phase boundary migration in solid-solid phase transformations. *Phys. Rev. Lett.*, **116**.

Edberg, R., Morriss, G. P., and Evans, D. J. (1987). Rheology of n-alkanes by nonequilibrium molecular dynamics. *J. Chem. Phys.*, **86**, 4555.

Engle, R. D., Skeel, R. D., and Dress, M. (2005). Monitoring energy drift with shadow Hamiltonians. *J. Comp. Phys.*, **206**, 432.

Epstein, C. L. and Schotland, J. (2008). The bad truth about Laplace's transform. *SIAM Rev.*, **50**, 504.

Ernst, M. H., Hague, E. H., and van Leeuwen, J. M. (1971). Asymptotic time behavior of correlation functions 1. Kinetic terms. *Phys. Rev. A*, **4**, 2055.

Essmann, U., Perera, L., Berkowitz, M. L., Darden, T., Lee, H., and Pedersen, L. G. (1995). A smooth particle-mesh Ewald method. *J. Chem. Phys.*, **103**, 8577.

Evans, D. J. and Morriss, G. P. (1980). *Statistical Mechanics of Nonequilibrium Liquids*. Harcourt Brace Javanovich, London.

Evans, D. J. and Morriss, G. P. (1984). Non-Newtonian molecular dynamics. *Comput. Phys. Rep.*, **I**, 299.

Ewald, P. P. (1921). Die Berechnung optischer und electrostatischer Gitterpotentiale. *Ann. Phys.*, **64**, 253.

Ezra, G. S. (2004). On the statistical mechanics of non-Hamiltonian systems: The generalized Liouville equation, entropy, and time-dependent metrics. *J. Math. Chem.*, **35**, 29.

Ezra, G. S. (2007). Reversible measure-preserving integrators for non-Hamiltonian systems. *J. Chem. Phys.*, **125**, 034104.

Facco, E., d'Errico, M., Rodriguez, A., and Laio, A. (2017). Estimating the intrinsic dimension of datasets by a minimal neighborhood information. *Sci. Rep.*, **7**.

Fanourgakis, G. S., Schenter, G. K., and Xantheas, S. S. (2006). A quantitative account of quantum effects in liquid water. *J. Chem. Phys.*, **125**, 141102.

Fanourgakis, G. S. and Xantheas, S. S. (2006). The flexible, polarizable thole-type interaction potential for water (TTM2-F) revisited. *J. Phys. Chem. A*, **110**, 4100.

Feng, Y., Huang, J., Kim, S., Shim, J. H., MacKerell, A. D., and Ge, N. H. (2016). Structure of penta-alanine investigated by two-dimensional infrared spectroscopy and molecular dynamics simulations. *J. Phys. Chem. B*, **120**, 5325.

Fermi, E., Pasta, J., and Ulam, S. (1955). Studies in nonlinear problems. I. *Los Alamos Report*, **LA 1940**.

Fernandez, J. M. and Li, H. (2004). Force-clamp spectroscopy minotors the folding trajectory of a single protein. *Science*, **303**, 1674.

Feynman, R. P. (1948). Space-time approach to non-relativistic quantum mechanics. *Rev. Mod. Phys.*, **20**, 367.

Feynman, R. P. (1998). *Statistical Mechanics: A Set of Lectures*. Addison-Wesley, Reading.

Feynman, R. P. and Hibbs, A. R. (1965). *Quantum Mechanics and Path Integrals*. McGraw-Hill, New York.

Feynman, R. P. and Kleinert, H. (1986). Effective classical partition functions. *Phys. Rev. A*, **34**, 5080.

Flügge, S. (1994). *Practical Quantum Mechanics 2nd edn*. Springer, Berlin.

Fredrickson, G. H. (2006). *The Equilibrium Theory of Inhomogeneous Polymers*. Oxford University Press, Oxford.

Frenkel, D. and Smit, B. (2002). *Understanding Molecular Simulation*. Academic Press, San Diego.

Fridman, B. L. (1967). An improvement in the smoothness of the functions in Kolmogorov's theorem on superpositions. *Dokl. Akad. Nauk USSR*, **117**, 1019.

Futrelle, R. P. and McGinty, D. J. (1971). Calculation of spectra and correlation-functions from molecular dynamics data using fast Fourier-transform. *Chem. Phys. Lett.*, **12**, 285.

Ganeshan, S., Ramírez, R., and Fernández-Serra, M. V. (2013). Simulation of quantum zero-point effects in water using a frequency-dependent thermostat. *Phys. Rev. B*, **87**, 134207.

Gans, J. and Shalloway, D. (2000). Shadow mass and the relationship between velocity and momentum in symplectic numerical integration. *Phys. Rev. E*, **61**, 4587.

Gauss, K. F. (1829). Über ein allgemeines Grundgesetz der Mechanik. *Reine Angew. Math.*, **IV**, 232.

Gausterer, H. and Klauder, J. R. (1986). Complex Langevin equations and their applications to quantum statistical and lattice field models. *Phys. Rev. D*, **33**, 3678.

Geiger, P. and Dellago, C. (2013). Neural networks for local structure detection in polymorphic systems. *J. Chem. Phys.*, **139**, 164105.

Geissler, P. L., Dellago, C., and Chandler, D. (1999). Kinetic pathways of ion pair dissociation in water. *J. Phys. Chem. B*, **103**, 3706.

Geissler, P. L., Dellago, C., Chandler, D., Hutter, J., and Parrinello, M. (2001). Autoionization in liquid water. *Science*, **291**, 291.

Geva, E., Shi, Q., and Voth, G. A. (2001). Quantum-mechanical reaction rate constants from centroid molecular dynamics simulations. *J. Chem. Phys.*, **115**, 9209.

Gillan, M. J. (1987). Quantum-classical crossover of the transition rate in the damped double well. *J. Phys. C.*, **20**, 3621.

Gilson, M. K., Given, J. A., Bush, B. L., and McCammon, J. A. (1997). The statistical thermodynamic basis for computation of binding affinities. *Biophys. J.*, **72**, 1047.

Glaesemann, K. R. and Fried, L. E. (2002). Improved heat capacity estimator for path integral simulations. *J. Chem. Phys.*, **117**, 3020.

Goldstein, H. (1980). *Classical Mechanics*. Addison-Wesley, Reading.

Gosling, E. M., McDonald, I. R., and Singer, K. (1973). Calculation by molecular dynamics of shear viscosity of a simple fluid. *Mol. Phys.*, **26**, 1475.

Goyal, M., Goyal, R., and Lall, B. (2019). Learning activation functions: A new paradigm for understanding neural networks. *Proc. Machine Learning Res.*, **101**, 1.

Graner, F. and Glazier, J. A. (1992). Simulation of biological cell sorting using a two-dimensional extended Potts model. *Phys. Rev. Lett.*, **69**, 2013.

Grote, R. F. and Hynes, J. T. (1980). The stable states picture of chemical-reactions 2. Rate constants for condensed and gas-phase reaction models . *J. Chem. Phys.*, **73**, 2715.

Guisoni, N., Mazzitello, K. I., and Diambra, L. (2018). Modeling active cell movement with the Potts model. *Front. Phys.*, **6**.

Gustafson, D. E. and Kessel, W. C. (1978). Fuzzy clustering with a fuzzy covariance matrix. In *Proceedings of the 1978 IEEE Conference on Decision and Control including the 17th Symposium on Adaptive Processes*. IEEE Control Systems Society.

Habershon, S., Markland, T. E., and Manolopoulos, D. E. (2009). Competing quantum effects in the dynamics of a flexible water model. *J. Chem. Phys.*, **131**, 024501.

Haldane, A. and Levy, R. M. (2019). Influence of multiple-sequence-alignment depth on Potts statistical models of protein covariation. *Phys. Rev. E*, **99**.

Hall, R. W. and Berne, B. J. (1984). Nonergodicity in path integral molecular dynamics. *J. Chem. Phys.*, **81**, 3641.

Harp, G. D. and Berne, B. J. (1968). Linear- and angular-momentum autocorrelation functions in diatomic liquids. *J. Chem. Phys.*, **49**, 1249.

Harp, G. D. and Berne, B. J. (1970). Time-correlation functions, memory functions and molecular dynamics. *Phys. Rev. A*, **2**, 975.

Hastie, T., Tibshirani, R., and Friedman, J. (2017). *The Elements of Statistical Learning*. Springer, New York.

Hayes, R. L., Paddison, S. J., and Tuckerman, M. E. (2009). Proton transport in triflic acid hydrates studied via path integral Car-Parrinello molecular dynamics. *J. Phys. Chem. B*, **113**, 16574.

Hayes, R. L., Paddison, S. J., and Tuckerman, M. E. (2011). Proton transport in triflic acid pentahydrate studied via *ab initio* path integral molecular dynamics. *J. Phys. Chem. A*, **115**, 6112–6124.

Hecht-Nielsen, R. (1987). Kolmogorov's mapping neural network existence theorem.

In *Proceedings of the International Conference on Neural Networks, volume 3* (ed. I. of Electrical and E. Engineers), p. 11. IEEE Press, New York.

Heidari, M., Kremer, K., Golestanian, R., Potestio, R., and Cortes-Huerto, R. (2020). Open-boundary Hamiltonian adaptive resolution. From grand canonical to non-equilibrium molecular dynamics simulations. *J. Chem. Phys.* (19), 194104.

Hele, T. J. H., Willatt, M. J., Muolo, A., and Althorpe, S. C. (2015*a*). Boltzmann-conserving classical dynamics in quantum time-correlation functions: "Matsubara dynamics". *J. Chem. Phys.*, **142**(13), 134103.

Hele, T. J. H., Willatt, M. J., Muolo, A., and Althorpe, S. C. (2015*b*). Communication: Relation of centroid molecular dynamics and ring-polymer molecular dynamics to exact quantum dynamics. *J. Chem. Phys.*, **142**(19), 191101.

Henderson, R. L. (1974). A uniqueness theorem for fluid pair correlation functions. *Phys. Lett. A*, **49**, 197.

Herman, M. F., Bruskin, E. J., and Berne, B. J. (1982). On path-integral Monte Carlo simulations. *J. Chem. Phys.*, **76**, 5150.

Herman, M. F. and Kluk, E. (1984). A semiclassical justification for the use of non-spreading wavepackets in dynamics calculations. *Chem. Phys.*, **91**, 27.

Hernandez, R., Cao, J. S., and Voth, G. A. (1995). On the Feynman path centroid density as a phase-space distribution in quantum-statistical mechanics. *J. Chem. Phys.*, **103**, 5018.

Hillery, M., O'Connel, R. F., Scully, M. O., and Wigner, E. P. (1984). Distribution functions in physics: Fundamentals. *Phys. Rep.*, **106**, 121.

Hirshberg, B., Rizzi, V., and Parrinello, M. (2019). Path integral molecular dynamics for bosons. *Proc. Natl. Acad. Sci. U.S.A.*, **116**, 21445.

Hirshberg, B., Rizzi, V., and Parrinello, M. (2020). Path integral molecular dynamics for fermions: Alleviating the sign problem with the Bogliubov inequality. *J. Chem. Phys.*, **152**, 171102.

Hockney, R. W. and Eastwood, J. W. (1981). *Computer Simulation Using Particles*. McGraw-Hill, New York.

Hohenberg, P. and Kohn, W. (1964). Inhomogeneous electron gas. *Phys. Rev.*, **136**, B864.

Hone, T. D., Poulsen, J. A., Rossky, P. J., and Manolopoulos, D. E. (2008). Comparison of approximate quantum simulation methods applied to normal liquid helium at 4 K. *J. Phys. Chem. B*, **112**, 294.

Hone, T. D., Rossky, P. J., and Voth, G. A. (2006). A comparative study of imaginary time path integral based methods for quantum dynamics. *J. Chem. Phys.*, **124**, 154103.

Hong, R. S., Chan, E. J., Vogt-Maranto, L., Mattei, A., Sheikh, A. Y., and Tuckerman, M. E. (2021). Insights into the polymorphic structures and enantiotropic layer-slip transition in Paracetamol form III from enhanced molecular dynamics. *Cryst. Growth & Design*, **21**, 886.

Hoover, W. G. (1983). Nonequilibrium molecular dynamics. *Ann. Rev. Phys. Chem.*, **34**, 103.

Hoover, W. G. (1985). Canonical dynamics: Equilibrium phase-space distributions. *Phys. Rev. A*, **31**, 1695.

Hörmander, L. (1967). Hypoelliptic second-order differential equations. *Acta Math.*, **119**, 147.

Huang, C. and Carter, E. A. (2010). Nonlocal orbital-free kinetic energy density functional for semiconductors. *Phys. Rev. B*, **81**, 045206.

Huang, K. (1963). *Statistical Mechanics* (2nd edn). John Wiley & Sons, New York.

Hummer, G. and Szabo, A. (2001). Free energy reconstruction from nonequilibrium single-molecule pulling experiments. *Proc. Natl. Acad. Sci. U.S.A.*, **98**, 3658.

Hwang, J. K., Chu, Z. T., Yadav, A., and Warshel, A. (1991). Simulations of quantum-mechanical corrections for rate constants of hydride-transfer reactions in enzymes and solutions. *J. Phys. Chem.*, **95**, 8445.

Hwang, J. K. and Warshel, A. (1996). How important are quantum mechanical nuclear motions in enzyme catalysis? *J. Am. Chem. Soc.*, **118**, 11745.

Iftimie, R., Minary, P., and Tuckerman, M. E. (2005). *Ab initio* molecular dynamics: Concepts, recent developments and future trends. *Proc. Natl. Acad. Sci. U.S.A.*, **102**, 6654.

Iftimie, R. and Tuckerman, M. E. (2005). Decomposing total IR spectra of aqueous systems into solute and solvent contributions: A computational approach using maximally localized Wannier orbitals. *J. Chem. Phys.*, **122**, 214508.

Ivanov, S. D., Witt, A., Shiga, M., and Marx, D. (2010). On artificial frequency shifts in infrared spectra obtained from centroid molecular dynamics: Quantum liquid water. *J. Chem. Phys.*, **132**, 031101.

Izaguirre, J. A. and Hampton, S. S. (2004). Shadow hybrid Monte Carlo: An efficient propagator in phase space of macromolecules. *J. Comp. Phys.*, **200**, 581.

Izvekov, S. and Voth, G. A. (2006). Modeling real dynamics in the coarse-grained representation of condensed phase systems. *J. Chem. Phys.*, **125**, 151101.

Jadhao, V. and Makri, N. (2008). Iterative Monte Carlo for quantum dynamics. *J. Chem. Phys.*, **129**, 161102.

Janežič, D., Praprotnik, M., and Merzel, F. (2005). Molecular dynamics integration and molecular vibrational theory. I. New symplectic integrators. *J. Chem. Phys.*, **122**, 174101.

Jang, S., Jang, S., and Voth, G. A. (2001). Applications of higher order composite factorization schemes in imaginary time path integral simulations. *J. Chem. Phys.*, **115**, 7832.

Janke, W. (ed) (2010). *Rugged Free Energy Landscapes*. Springer–Verlag, Berlin.

Jarzynski, C. (1997). Nonequilibrium equality for free energy differences. *Phys. Rev. Lett.*, **78**, 2690.

Jarzynski, C. (2004). Nonequilibrium work theorem for a system strongly coupled to a thermal environment. *J. Stat. Mech.: Theor. and Expt.*, P09005.

Jorgensen, W. L., Maxwell, D. S., and Tirado-Rives, J. (1996). Development and testing of the OPLS all-atom force field on conformational energetics and properties of organic liquids. *J. Am. Chem. Soc.*, **118**, 11225.

Kadanoff, L. P. (1966). Scaling laws for Ising models near T_c. *Physics*, **2**, 263.

Kahane, J. -P. (1975). Sur le théorème de superposition de Kolmogorov. *J. Approx. Theory*, **13**, 229.

Kalos, M. H. and Whitlock, P. A. (1986). *Monte Carlo Methods. Volume 1: Basics.*

John Wiley & Sons, New York.

Kanungo, T., Mount, D. M., Netanyahu, N. S., Piatko, C. D., Silverman, R., and Wu, A. Y. (2002). An efficient k-means clustering algorithm: Analysis and implementation. *IEEE Trans. Pattern Anal. Mach. Intell.*, **24**, 881.

Kapil, V., Behler, J., and Ceriotti, M. (2016). High order path integrals made easy. *J. Chem. Phys.*, **145**, 234103.

Karplus, M. (1959). Contact electron-spin coupling of nuclear magnetic moments. *J. Chem. Phys.*, **30**, 11.

Kästner, J. and Thiel, W. (2005). Bridging the gap between thermodynamic integration and umbrella sampling provides a novel analysis method: Umbrella integration. *J. Chem. Phys.*, **123**, 144104.

Kaufmann, B. (1949). Crystal Statistics. II. Partition function evaluated by spinor analysis. *Phys. Rev.*, **76**, 1232.

Khintchine, A. I. (1934). Correlation theories of stationary stochastic processes. *Math. Ann.*, **109**, 604.

Kirkwood, J. G. (1935). Statistical mechanics of fluid mixtures. *J. Chem. Phys.*, **3**, 300.

Kladko, K. and Fulde, P. (1998). On the properties of cumulant expansions. *Intl. J. Quant. Chem.*, **66**, 377.

Kohn, W. and Sham, L. J. (1965). Self-consistent equations including exchange and correlation effects. *Phys. Rev.*, **140**, A1138.

Kolmogorov, A. N. (1957). On the representation of continuous functions of many variables by superpositions of continuous functions of one variable and addition. *Dokl. Akad. Nauk USSR*, **14**, 953.

Kone, A. and Kofke, D. A. (2005). Selection of temperature intervals for parallel tempering simulations. *J. Chem. Phys.*, **122**, 206101.

Korol, R., Bou-Rabee, N., and Miller III, T. F. (2019). Cayley modification for strongly stable path-integral and ring-polymer molecular dynamics. *J. Chem. Phys.*, **151**, 124103.

Korol, R., Rosa-Raíces, J. L., Bou-Rabee, N., and Miller, III, T. F. (2020). Dimension-free path-integral molecular dynamics without preconditioning. *J. Chem. Phys.*, **152**, 104102.

Kraeutler, V., van Gunsteren, W. F., and Huenenberger, P. H. (2001). A fast SHAKE algorithm to solve distance constraint equations for small molecules in molecular dynamics simulations. *J. Comput. Chem.*, **22**, 501.

Kramers, H. A. (1940). Brownian motion in a field of force and the diffusion model of chemical reactions . *Physica*, **7**, 284.

Kramers, H. A. and Wannier, G. H. (1941*a*). Statistics of the two-dimensional Ferromagnet. Part I. *Phys. Rev.*, **60**, 252.

Kramers, H. A. and Wannier, G. H. (1941*b*). Statistics of the two-dimensional Ferromagnet. Part II. *Phys. Rev.*, **60**, 263.

Kreis, K., Kremer, K., Potestio, R., and Tuckerman, M. E. (2017). From classical to quantum and back: Hamiltonian adaptive resolution path integral, ring polymer, and centroid molecular dynamics. *J. Chem. Phys.*, **147**, 244104.

Kreis, K., Tuckerman, M. E., Donadio, D., Kremer, K., and Potestio, R. (2016). From classical to quantum and back: A Hamiltonian scheme for adaptive multiresolution classical/path-integral simulations. *J. Chem. Theory Comput.*, **12**, 3030.

Krilov, G., Sim, E., and Berne, B. J. (2001). Quantum time correlation functions from complex time Monte Carlo simulations: A maximum entropy approach. *J. Chem. Phys.*, **114**, 1075.

Kubo, R., Toda, M., and Hashitsume, N. (1985). *Statistical Physics II*. Springer-Verlag, Berlin.

Kumar, S., Bouzida, D., Swendsen, R. H., Kollman, P. A., and Rosenberg, J. M. (1992). The weighted histogram analysis method for free-energy calculations on biomolecules. 1. The method. *J. Comp. Chem.*, **13**, 1011.

Kurkova, V. (1991). Kolmogorov's theorem and multilayer neural networks. *Neural Comput.*, **3**, 617–622.

Laio, A. and Parrinello, M. (2002). Escaping the free-energy minima. *Proc. Natl. Acad. Sci. U.S.A.*, **99**, 12562.

Laio, A., Rodriguez-Fortea, A., Gervasio, F. L., Ceccarelli, M., and Parrinello, M. (2005). Assessing the accuracy of metadynamics. *J. Phys. Chem. B*, **109**, 6714.

Laird, B. B. and Skinner, J. L. (1991). T_2 can be greater than $2T_1$ even at finite temperature. *J. Chem. Phys.*, **94**, 4405.

Langevin, P. (1905). Une formule fondamentale de théorie cinétique. *Ann. Chim. Phys.*, **5**, 245.

Langevin, P. (1908). Sur la théorie du mouvement Brownien. *CR Hebd. séances Acad. Sci.*, **146**, 530.

LeCun, Y., Boser, B., Denker, J. S., Henderson, D., Howard, R. E., Hubbard, W., and Jackel, L. D. (1989). Backpropagation applied to handwritten zip code recognition. *Neural Comp.*, **1**, 541.

Lee, H. -S. and Tuckerman, M. E. (2006). Structure of liquid water at ambient temperature from *ab initio* molecular dynamics performed in the complete basis set limit. *J. Chem. Phys.*, **125**, 154507.

Lee, H. -S. and Tuckerman, M. E. (2007). Dynamical properties of liquid water from *ab initio* molecular dynamics performed in the complete basis set limit. *J. Chem. Phys.*, **126**, 164501.

Lee, S. (1994). The convergence of complex Langevin simulations. *Nucl. Phys. B*, **413**, 827.

Lees, A. W. and Edwards, F. (1972). The computer study of transport processes under extreme conditions. *J. Phys. C*, **5**, 1921.

Leimkuhler, B., Margul, D. T., and Tuckerman, M. E. (2013). Stochastic, resonance-free multiple time-step algorithms for molecular dynamics with very large time steps. *Mol. Phys.*, **111**, 3579.

Leimkuhler, B. and Matthews, C. (2012). Rational construction of stochastic numerical methods for molecular sampling. *Appl. Math. Res. eXpress*, **2013**, 34.

Leimkuhler, B., Matthews, C., and Vlaar, T. (2019). Partitioned integrators for thermodynamic parameterization of neural networks. *Found. Data Sci.*, **1**, 457.

Leimkuhler, B., Noorizadeh, E., and Theil, F. (2009). A gentle stochastic thermostat for molecular dynamics. *J. Stat. Phys.*, **135**, 261.

Lennard-Jones, J. E. (1924). On the determination of molecular force fields. *Proc.*

Royal Soc. London A, **106**, 463.

Levine, A. M., Shapiro, M., and Pollak, E. (1988). Hamiltonian theory for vibrational dephasing rates of small molecules in liquids. *J. Chem. Phys.*, **88**, 1959.

Levy, R. M, Haldane, A., and Flynn, W. F. (2017). Potts Hamiltonian models of protein co-variation, free energy landscapes, and evolutionary fitness. *Curr. Opin. Struct. Biol.*, **43**, 55.

Lin, L., Morrone, J. A., Car, R., and Parrinello, M. (2011). Displaced path integral formulation for the momentum distribution of quantum particles. *Phys. Rev. Lett.*, **105**, 110602.

Liphardt, J., Dumont, S., Smith, S. B., Tinoco, I., and Bustamante, C. (2002). Equilibrium information from nonequilibrium measurements in an experimental test of Jarzynski's equality. *Science*, **296**, 1832.

Liu, P., Kim, B., Friesner, R. A., and Berne, B. J. (2005). Replica exchange with solute tempering: A method for sampling biological systems in explicit water. *Proc. Natl. Acad. Sci. U.S.A.*, **102**, 13749–13754.

Liu, Y. and Tuckerman, M. E. (2000). Generalized Gaussian moment thermostatting: A new continuous dynamical approach to the canonical ensemble. *J. Chem. Phys.*, **112**, 1685.

Liu, Z. H. and Berne, B. J. (1993). Electron solvation in methane and ethane. *J. Chem. Phys.*, **99**, 9054.

Lorentz, G. G. (2005). *Approximation of Functions*. AMS Chelsea Publishing, Providence.

Ma, Q., Izaguirre, J. A., and Skeel, R. D. (2003). Verlet-I/r-RESPA/Impulse is limited by nonlinear instabilities. *SIAM J. Sci. Comput.*, **24**, 1951.

Ma, S. K. (1976). *Modern Theory of Critical Phenomena*. Benjamin/Cummings, Reading.

MacKerell, A., Jr., Bashford, D., Bellott, M., Dumbrack, R. L., Evanseck, J. D., Field, M. J., Fischer, S., Guo, H., Ha, S., Joseph-McCarthy, D., Kcuhnir, L., Kuczera, K., Lau, F. T. K., Mattos, C., Michnick, S., Ngo, T., Nguyen, D. T., Prodhom, B., Reiher, W. E., III, Roux, B., Schlenkrich, M., Smith, J. C., Stote, R., Straub, J., Watanabe, M., Wiorkiewicz-Kuczera, J., Yin, D., and Karplus, M. (1998). All-atom empirical potential for molecular modeling and dynamics studies of proteins. *J. Phys. Chem. B*, **102**, 3586.

MacQueen, J. B. (1967). Some methods for classification and analysis of multivariate observations. In *Proceedings of the Fifth Berkeley Symposiumon Mathematical Statistics and Probability* (ed. L. M. LeCam and J. Neyman), Volume 1, p. 281. University of California Press, Berkeley.

Manousiouthakis, V. I. and Deam, M. W. (1999). Strict detailed balance is unnecessary in Monte Carlo simulation. *J. Chem. Phys.*, **110**, 2753.

Maragliano, L. and Vanden-Eijnden, E. (2006). A temperature accelerated method for sampling free energy and determining reaction paths in rare event simulations. *Chem. Phys. Lett.*, **426**, 168.

Maragliano, L. and Vanden-Eijnden, E. (2008). Single-sweep methods for free energy calculations. *J. Chem. Phys.*, **128**, 184110.

Marchi, M., Sprik, M., and Klein, M. L. (1990). Calculation of the molar volume of electron solvation in liquid ammonia. *J. Phys. Chem.*, **94**, 431.

Margul, D. and Tuckerman, M. E. (2016). A stochastic, resonance-free multiple time-step algorithm for polarizable models that permits very large time steps. *J. Chem. Theory Comput.*, **12**, 2170.

Marinari, E. and Parisi, G. (1992). Simulated tempering: A new Monte Carlo scheme. *Europhys. Lett.*, **19**, 451.

Maris, H. J. and Kadanoff, L. J. (1978). Teaching the renormalization group. *Am. J. Phys.*, **46**, 652.

Markland, T. E. and Manolopoulos, D. E. (2008*a*). A refined ring polymer contraction scheme for systems with electrostatic interactions. *Chem. Phys. Lett.*, **464**, 256.

Markland, T. E. and Manolopoulos, D. E. (2008*b*). An efficient ring polymer contraction scheme for imaginary time path integral simulations. *J. Chem. Phys.*, **129**, 024105.

Marsalek, O. and Markland, T. E. (2016). *Ab initio* molecular dynamics with nuclear quantum effects at classical cost: Ring polymer contraction for density functional theory. *J. Chem. Phys.*, **144**, 054112.

Martyna, G. J., Deng, Z. H., and Klein, M. L. (1993). Quantum simulation studies of singlet and triplet bipolarons in liquid ammonia. *J. Chem. Phys.*, **98**, 555.

Martyna, G. J., Hughes, A., and Tuckerman, M. E. (1999). Molecular dynamics algorithms for path integrals at constant pressure. *J. Chem. Phys.*, **110**, 3275.

Martyna, G. J., Klein, M. L., and Tuckerman, M. E. (1992). Nosé-Hoover chains: The canonical ensemble via continuous dynamics. *J. Chem. Phys.*, **97**, 2635.

Martyna, G. J., Tobias, D. J., and Klein, M. L. (1994). Constant pressure molecular dynamics algorithms. *J. Chem. Phys.*, **101**, 4177.

Martyna, G. J. and Tuckerman, M. E. (1999). A reciprocal space based method for treating long range interactions in *ab initio* and force-field-based calculations in clusters. *J. Chem. Phys.*, **110**, 2810.

Martyna, G. J., Tuckerman, M. E., Tobias, D. J., and Klein, M. L. (1996). Explicit reversible integrators for extended systems dynamics. *Mol. Phys.*, **87**, 1117.

Marx, D., Chandra, A., and Tuckerman, M. E. (2010). Aqueous basic solutions: Hydroxide solvation, structural diffusion and comparison to the hydrated proton. *Chem. Rev.*, **110**, 2174.

Marx, D. and Hutter, J. (2009). *Ab Initio Molecular Dynamics*. Cambridge University Press, Cambridge.

Marx, D. and Parrinello, M. (1994). *Ab initio* path-integral molecular dynamics. *Z. Phys. B*, **95**, 143.

Marx, D. and Parrinello, M. (1996). *Ab initio* path integral molecular dynamics: Basic ideas. *J. Chem. Phys.*, **104**, 4077.

Marx, D., Tuckerman, M. E., Hutter, J., and Parrinello, M. (1999*a*). The nature of the hydrated excess proton in water. *Nature*, **367**, 601.

Marx, D., Tuckerman, M. E., and Martyna, G. J. (1999*b*). Quantum dynamics via adiabatic *ab initio* centroid molecular dynamics. *Comp. Phys. Comm.*, **118**, 166.

McCammon, J. A., Gelin, B. R., and Karplus, M. (1977). Dynamics of folded proteins. *Nature*, **267**, 585.

McCammon, J. A., Gelin, B. R., Karplus, M., and Wolynes, P. G. (1976). Hinge-bending mode in lysozyme. *Nature*, **262**, 325.

McCulloch, W. S. and Pitts, W. (1943). A logical calculus of the idea immanent in nervous activity. *Bull. Math. Biophys.*, **5**, 115.

McGreevy, R. L. and Pusztai, L. (1988). Reverse Monte Carlo simulation: A new technique for the determination of disordered structures. *Mol. Simulat.*, **1**, 359.

McQuarrie, D. A. (2000). *Statistical Mechanics*. University Science Books, Sausalito.

Medders, G. R., Babin, V., and Paesani, F. (2014). Development of a "first-principles" water potential with flexible monomers. III. Liquid phase properties. *J. Chem. Theory Comput.*, **10**, 2906.

Merli, P. G., Missiroli, G. F., and Pozzi, G. (1976). Statistical aspect of electron interference phenomena. *Am. J. Phys.*, **44**, 306.

Metropolis, N., Rosenbluth, A. W., Rosenbluth, M. N., Teller, A. H., and Teller, E. (1953). Equation of state calculations by fast computing machines. *J. Chem. Phys.*, **21**, 1087.

Miller, T. F., III, Eleftheriou, M., Pattnaik, P., Ndirango, A., Newns, D., and Martyna, G. J. (2002). Symplectic quaternion scheme for biophysical molecular dynamics. *J. Chem. Phys.*, **116**, 8649.

Miller, T. F., III and Manolopoulos, D. E. (2005). Quantum diffusion in liquid para-hydrogen from ring-polymer molecular dynamics. *J. Chem. Phys.*, **122**, 184503.

Miller, W. H. (2005). Quantum dynamics of complex molecular systems. *Proc. Natl. Acad. Sci. U.S.A.*, **102**, 6660.

Minary, P., Martyna, G. J., and Tuckerman, M. E. (2003). Algorithms and novel applications based on the isokinetic ensemble. I. Biophysical and path integral molecular dynamics. *J. Chem. Phys.*, **118**, 2510.

Minary, P., Morrone, J. A., Yarne, D. A., Tuckerman, M. E., and Martyna, G. J. (2004*a*). Long range interactions on wires: A reciprocal space based formalism. *J. Chem. Phys.*, **121**, 11949.

Minary, P. and Tuckerman, M. E. (2004). Reaction pathway of the [4+2] Diels-Alder adduct formation on Si(100)-2×1. *J. Am. Chem. Soc.*, **126**, 13920.

Minary, P., Tuckerman, M. E., and Martyna, G. J. (2004*b*). Long-time molecular dynamics for enhanced conformational sampling in biomolecular systems. *Phys. Rev. Lett.*, **93**, 150201.

Minary, P., Tuckerman, M. E., and Martyna, G. J. (2007). Dynamical spatial warping: A novel method for the conformational sampling of biophysical structure. *SIAM J. Sci. Comput.*, **30**, 2055.

Minary, P., Tuckerman, M. E., Pihakari, K. A., and Martyna, G. J. (2002). A new reciprocal space based treatment of long range interactions on surfaces. *J. Chem. Phys.*, **116**, 5351.

Miura, S. and Okazaki, S. (2000). Path integral molecular dynamics for Bose-Einstein and Fermi-Dirac statistics. *J. Chem. Phys.*, **112**, 10116.

Mjalli, F. S. and Naser, J. (2015). Viscosity model for choline chloride-based deep eutectic solvents. *Asia-Pacific J. Chem. Eng.*, **10**(2), 273–281.

Mori, H. (1965). Transport collective motion and Brownian motion. *Prog. Theor. Phys.*, **33**, 423.

Mori, Y., Okazaki, K., Mori, T., Kim, K., and Matubayasi, N. (2020). Learning reaction coordinates via cross-entropy minimization: Application to alanine dipeptide. *J. Chem. Phys.*, **153**, 054115.

Morrone, J. A. and Car, R. (2008). Nuclear quantum effects in water. *Phys Rev. Lett.*, **101**, 017801.

Morrone, J. A., Markland, T. E., Ceriotti, M., and Berne, B. J. (2011). Efficient multiple time scale molecular dynamics: Using colored noise thermostats to stabilize resonances. *J. Chem. Phys.*, **134**, 014103.

Morrone, J. A., Srinivasan, V., Sebastiani, D., and Car, R. (2007). Proton momentum distribution in water: An open path integral molecular dynamics study. *J. Chem. Phys.*, **126**, 234504.

Mukamel, S. (1995). *Principles of Nonlinear Optical Spectroscopy*. Oxford University Press, New York.

Mundy, C. J., Balasubramanian, S., Bagchi, K., Martyna, G. J., and Klein, M. L. (2000). Nonequilibrium molecular dynamics. *Rev. Comp. Chem.*, **14**, 291.

Nakayama, A. and Makri, N. (2005). Simulation of dynamical properties of normal and superfluid helium. *Proc. Natl. Acad. Sci. U.S.A.*, **102**, 4230.

Noe, F., Olsson, S., Kohler, J., and Wu, H. (2019). Boltzmann generators: Sampling equilibrium states of many-body systems with deep learning. *Science*, **365**, 6457.

Nosé, S. (1984). A unified formulation of the constant-temperature molecular dynamics methods. *J. Chem. Phys.*, **81**, 511.

Nosé, S. and Klein, M. L. (1983). Constant-pressure molecular dynamics for molecular systems. *Mol. Phys.*, **50**, 1055.

Oberhofer, H., Dellago, C., and Geissler, P. L. (2005). Biased sampling of nonequilibrium trajectories: Can fast switching simulations outperform conventional free energy calculation methods? *J. Phys. Chem. B*, **109**, 6902.

Olender, R. and Elber, R. (1996). Classical trajectories with a very large time step: Formalism and numerical examples. *J. Chem. Phys.*, **105**, 9299.

Onsager, L. (1944). Crystal statistics. I. A two-dimensional model with an order–disorder transition. *Phys. Rev.*, **65**, 117.

Oxtoby, D. W. (1979). Hydrodynamic theory for vibrational dephasing in liquids. *J. Chem. Phys.*, **70**, 2605.

Paesani, F., Iuchi, S., and Voth, G. A. (2007). Quantum effects in liquid water from an *ab initio*-based polarizable force field. *J. Chem. Phys.*, **127**, 074506.

Paesani, F. and Voth, G. A. (2009). The properties of water: Insights from quantum simulations. *J. Phys. Chem. B*, **113**, 5702.

Paesani, F. and Voth, G. A. (2010). A quantitative assessment of the accuracy of centroid molecular dynamics for the calculation of the infrared spectrum of liquid water. *J. Chem. Phys.*, **132**, 014105.

Paesani, F., Zhang, W., Case, D. A., Cheathem, T. E., and Voth, G. A. (2006). An accurate and simple quantum model for liquid water. *J. Chem. Phys.*, **125**, 184507.

Park, S., Khalili-Araghi, F., and Schulten, K. (2003). Free energy calculation from steered molecular dynamics simulations using Jarzynski's equality. *J. Chem. Phys.*, **119**, 3559.

Park, S. and Schulten, K. (2004). Calculating potentials of mean force from steered molecular dynamics simulations. *J. Chem. Phys.*, **120**, 5946.

Parr, R. G. and Yang, W. T. (1989). *Density Functional Theory of Atoms and Molecules.* Clarendon Press, Oxford.

Parrinello, M. and Rahman, A. (1980). Crystal-structure and pair potentials: A molecular dynamics study. *Phys. Rev. Lett.*, **45**, 1196.

Parrinello, M. and Rahman, A. (1984). Study of an F center in molten KCl. *J. Chem. Phys.*, **80**, 860.

Passerone, D. and Parrinello, M. (2001). Action-derived molecular dynamics in the study of rare events. *Phys. Rev. Lett.*, **87**, 108302.

Pathria, R. K. (1972). *Statistical Mechanics* (2nd edn). Elsevier Butterworth Heinemann, Amsterdam.

Patra, M., Karttunen, M., Hyvonen, M. T., Falck, E., Lindqvist, P., and Vattulainen, I. (2003). Molecular dynamics simulations of lipid bilayers: Major artifacts due to truncating electrostatic interactions. *Biophys. J.*, **84**, 3636.

Paul, S., Nair, N. N., and Vashisth, H. (2019). Phase space and collective variable based simulation methods for studies of rare events. *Mol. Simulat.*, **45**, 1273.

Peierls, R. (1938). On a minimum property of the free energy. *Phys. Rev.*, **54**, 918.

Pérez, A. and Tuckerman, M. E. (2011). Improving the convergence of closed and open path integral molecular dynamics via higher order Trotter factorization schemes. *J. Chem. Phys.*, **135**, 064104.

Pérez, A., Tuckerman, M. E., and Müser, M. H. (2009). A comparative study of the centroid and ring-polymer molecular dynamics methods for approximating quantum time correlation functions from path integrals. *J. Chem. Phys.*, **130**, 184105.

Peters, B. (2006). Using the histogram test to quantify reaction coordinate error. *J. Chem. Phys.*, **125**, 241101.

Peters, B., Beckham, G. T., and Trout, B. L. (2007). Extensions to the likelihood maximization approach for finding reaction coordinates. *J. Chem. Phys.*, **127**, 034109.

Peters, B. and Trout, B. L. (2006). Obtaining reaction coordinates by likelihood maximization. *J. Chem. Phys.*, **125**, 054108.

Piaggi, P. M. and Parrinello, M. (2018). Predicting polymorphism in molecular crystals using orientational entropy. *Proc. Natl. Acad. Sci. U.S.A.*, **115**, 10251.

Piana, S. and Laio, A. (2007). A bias-exchange approach to protein folding. *J. Phys. Chem. B*, **111**, 4553.

Piccinini, G. (2004). The first computational theory of mind and brain: A close look at McCulloch and Pitts's "Logical Calculus of Ideas Immanent in Nervous Activity". *Synthese*, **141**, 175.

Plimpton, S. (1995). Fast parallel algorithms for short-range molecular dynamics. *J. Comp. Phys.*, **117**, 1.

Pollak, E. (1990). Variational transition-state theory for activated rate processes. *J. Chem. Phys.*, **93**, 1116.

Pollak, E., Grabert, H., and Hanggi, P. (1989). Theory of activated rate-processes for arbitrary frequency-dependent friction: Solution of the turnover problem. *J. Chem. Phys.*, **91**, 4073.

Pollak, E. and Talkner, P. (1993). Activated rate-processes: Finite-barrier expansion for the rate in the spatial-diffusion limit . *Phys. Rev. E*, **47**, 922.

Pollak, E., Tucker, S. C., and Berne, B. J. (1990). Variational transition-state theory for reaction-rates in dissipative systems. *Phys. Rev. Lett.*, **65**, 1399.

Potestio, R., Fritsch, S., Español, P., Delgado-Buscalioni, R., Kremer, K., Everaers, R., and Donadio, D. (2013). Hamiltonian adaptive resolution simulation for molecular liquids. *Phys. Rev. Lett.*, **110**, 108301.

Poulsen, J. A., Nyman, G., and Rossky, P. J. (2005). Static and dynamic quantum effects in molecular liquids: A linearized path integral description of water. *Proc. Natl. Acad. Sci. U.S.A.*, **102**, 6709.

Pratt, L. R. and Laviolette, R. A. (1998). Quasi-chemical theories of associated liquids. *Mol. Phys.*, **94**, 909.

Procacci, P., Marchi, M., and Martyna, G. J. (1998). Electrostatic calculations and multiple time scales in molecular dynamics simulation of flexible molecular systems. *J. Chem. Phys.*, **108**, 8799.

Quigley, D. and Rodger, P. M. (2008). Metadynamics simulations of ice nucleation and growth. *J. Chem. Phys.*, **128**, 154518.

Quigley, D. and Rodger, P. M. (2009). A metadynamics-based approach to sampling crystallisation events. *Mol. Simulat.*, **35**, 613.

Rahman, A. (1964). Correlations in motion of atoms in liquid argon. *Phys. Rev.*, **136**, A405.

Rahman, A. and Stillinger, F. H. (1971). Molecular dynamics study of liquid water. *J. Chem. Phys.*, **55**, 3336.

Ramshaw, J. D. (2002). Remarks on non-Hamiltonian statistical mechanics. *Europhys. Lett.*, **59**, 319.

Rathore, N., Chopra, M., and de Pablo, J. J. (2005). Optimal allocation of replicas in parallel tempering simulations. *J. Chem. Phys.*, **122**, 024111.

Rathore, N., Knotts IV, T. A., and de Pablo, J. J. (2003). Density-of-states simulations of proteins. *J. Chem. Phys.*, **118**, 4285.

Reichman, D. R. and Silbey, R. J. (1996). On the relaxation of a two-level system: Beyond the weak-coupling approximation. *J. Chem. Phys.*, **104**, 1506.

Reiss, C. A., van Mechelen, J. B., Goubitz, K., and Peschar, R. (2018). Reassessment of Paracetamol orthorhombic form III and determination of a novel low-temperature monoclinic form III-m from powder diffraction data. *Acta Cryst. C*, **74**, 392.

Ricci, M. A., Nardone, M., Ricci, F. P., Andreani, C., and Soper, A. K. (1995). Microscopic structure of low temperature liquid ammonia: A neutron diffraction study. *J. Chem. Phys.*, **102**, 7650.

Rodriguez, A. and Laio, A. (2014). Clustering by fast search and find of density peaks. *Science*, **344**, 1492.

Rogal, J., Schneider, E., and Tuckerman, M. E. (2019). Neural-network-based path collective variables for enhanced sampling of phase transformations. *Phys. Rev. Lett.*, **123**, 245701.

Rosa-Raíces, J. L., Sun, J., Bou-Rabee, N., and Miller, T. F. (2021). A generalized class of strongly stable and dimension-free t-RPMD integrators. *J. Chem. Phys.*, **154**, 024106.

Ross, M. and McMahan, A. K. (1980). Condensed xenon at high pressure. *Phys. Rev. B*, **21**, 1658.

Rossi, M., Ceriotti, M., and Manolopoulos, D. E. (2014). How to remove the spurious resonances from ring polymer molecular dynamics. *J. Chem. Phys.*, **140**, 234116.

Rosso, L., Abrams, J. B., and Tuckerman, M. E. (2005). Mapping the backbone dihedral free-energy surfaces in small peptides in solution using adiabatic free-energy dynamics. *J. Phys. Chem. B*, **109**, 4162.

Rosso, L., Minary, P., Zhu, Z. W., and Tuckerman, M. E. (2002). On the use of the adiabatic molecular dynamics technique in the calculation of free energy profiles. *J. Chem. Phys.*, **116**, 4389.

Rouse, P. E. (1953). A theory of the linear viscoelastic properties of dilute solutions of coiling polymers. *J. Chem. Phys.*, **21**, 1272.

Rupakheti, C., Lamoureux, G., MacKerell, A. D., and Roux, B. (2020). Statistical mechanics of polarizable force fields based on classical drude oscillators with dynamical propagation by the dual-thermostat extended Lagrangian. *J. Chem. Phys.*, **153**, 114108.

Ryckaert, J. P. and Ciccotti, G. (1983). Introduction of Andersen demon in the molecular dynamics of systems with constraints. *J. Chem. Phys.*, **78**, 7368.

Ryckaert, J. P., Ciccotti, G., and Berendsen, H. J. C. (1977). Numerical-integration of Cartesian equations of motion of a system with constraints: Molecular-dynamics of n-alkanes. *J. Comput. Phys.*, **23**, 327.

Sandu, A. and Schlick, T. (1999). Masking resonance artifacts in force-splitting methods for biomolecular simulations by extrapolative Langevin dynamics. *J. Comput. Phys.*, **151**, 74.

Sardanashvily, G. (2002*a*). The Lyapunov stability of first-order dynamic equations with respect to time-dependent Riemannian metrics. *arXiv.nlin.CD/0201060v1*.

Sardanashvily, G. (2002*b*). The Lyapunov stability of first-order dynamic equations with respect to time-dependent Riemannian metrics: An example. *arXiv.nlin.CD/0203031v1*.

Savill, N. J. and Hogeweg, P. (1997). Modelling morphogenesis: From single cells to crawling slugs. *J. Theor. Biol.*, **184**, 229.

Scharf, D., Martyna, G. J., and Klein, M. L. (1993). Path-integral Monte Carlo study of a lithium impurity in para-hydrogen: Clusters and the bulk liquid. *J. Chem. Phys.*, **99**, 8997.

Schlick, T., Mandzuik, M., Skeel, R. D., and Srinivas, K. (1998). Nonlinear resonance artifacts in molecular dynamics simulations. *J. Comput. Phys.*, **140**, 1.

Schneider, E., Dai, L., Topper, R. Q., Drechsel-Grau, C., and Tuckerman, M. E. (2017). Stochastic neural network approach for learning high-dimensional free energy surfaces. *Phys. Rev. Lett.*, **119**, 150601.

Schoell-Paschinger, E. and Dellago, C. (2006). A proof of Jarzynski's nonequilibrium work theorem for dynamical systems that conserve the canonical distribution. *J. Chem. Phys.*, **125**, 054105.

Schulman, L. S. (1981). *Techniques and Applications of Path Integration*. Wiley-Interscience, New York.

Senior, A. W., Evans, R., Jumper, J., Kirkpatrick, J., Sifre, L., Green, T., Qin, C., Žídek, A., Nelson, A. W. R., Bridgland, A., Penedones, H., Petersen, S., Simonyan, K., Crossan, S., Kohli, P., Jones, D. T., Silver, D., Kavukcuoglu, K., and Hassabis,

D. (2019). Protein structure prediction using multiple deep neural networks in the 13th critical assessment of protein structure prediction (CASP13). *Proteins: Struct., Funct., Bioinform.*, **87**, 1141.

Sergi, A. (2003). Non-Hamiltonian equilibrium statistical mechanics. *Phys. Rev. E*, **67**, 021101.

Shalev-Shwartz, S. and Ben-David, S. (2014). *Understanding Machine Learning*. Cambridge University Press, New York.

Shankar, R. (1994). *Principles of Quantum Mechanics* (2nd edn). Plenum Press, New York.

Shtukenberg, A. G., Tan, M., Vogt-Maranto, L., Chan, E. J., Xu, W., Yang, J., Tuckerman, M. E., Hu, C. T., and Kahr, B. (2019). Melt crystallization for Paracetamol polymorphism. *Cryst. Growth & Design*, **19**, 4070.

Silvera, I. F. and Goldman, V. V. (1978). The isotropic intermolecular potential for H_2 and D_2 in the solid and gas phases. *J. Chem. Phys.*, **69**, 4209.

Simovici, D. (2018). *Mathematical Analysis for Machine Learning and Data Mining*. World Scientific, Singapore.

Skeel, R. D. and Hardy, D. J. (2001). Practical construction of modified Hamiltonians. *SIAM J. Sci. Comput.*, **23**, 1172.

Skinner, J. L. and Park, K. (2001). Calculating vibrational energy relaxation rates from classical molecular dynamics simulations: Quantum correction factors for processes involving vibration-vibration energy transfer. *J. Phys. Chem. B*, **105**, 6716.

Smith, E. L., Abbott, A. P., and Ryder, K. S. (2014). Deep eutectic solvents (DESs) and their applications. *Chem. Rev.*, **114**(21), 11060–11082.

Song, H., Vogt-Maranto, L., Wiscons, R., Matzger, A. J., and Tuckerman, M. E. (2020). Generating cocrystal polymorphs with information entropy driven by molecular dynamics-based enhanced sampling. *J. Phys. Chem. Lett.*, **11**, 9751.

Soper, A. K. (1996). Empirical potential Monte Carlo simulation of fluid structure. *Chem. Phys.*, **202**, 295.

Sprecher, D. (1965). On the structure of continuous functions of several variables. *Trans. Amer. Math. Soc.*, **115**, 340.

Sprik, M. and Ciccotti, G. (1998). Free energy from constrained molecular dynamics. *J. Chem. Phys.*, **109**, 7737.

Sprik, M., Impey, R. W., and Klein, M. L. (1985). Study of electron solvation in liquid ammonia using quantum path integral Monte-Carlo calculations. *J. Chem. Phys.*, **83**, 5802.

Sprik, M., Impey, R. W., and Klein, M. L. (1986). Study of electron solvation in polar solvents using path integral Monte Carlo calculations. *J. Stat. Phys.*, **43**, 967.

Steinczinger, Z. and Pusztai, L. (2012). An independent, general method for checking consistency between diffraction data and partial radial distribution functions derived from them: The example of liquid water. *Condens. Matt. Phys.*, **15**, 23606.

Steinhardt, P. J., Nelson, D. R., and Ronchetti, M. (1983). Bond-orientational order in liquids and glasses. *Phys. Rev. B*, **28**, 784.

Stillinger, F. H. and Rahman, A. (1972). Molecular dynamics study of temperature effects on water structure and kinetics. *J. Chem. Phys.*, **57**, 1281.

Stillinger, F. H. and Rahman, A. (1974). Improved simulation of liquid water by molecular dynamics. *J. Chem. Phys.*, **60**, 1545.

Strang, G. (1968). On the construction of and comparison of difference schemes. *SIAM J. Numer. Anal.*, **5**, 506.

Straub, J. E., Borkovec, M., and Berne, B. J. (1988). Molecular dynamics study of an isomerizing diatomic in a Lennard-Jones fluid. *J. Chem. Phys.*, **89**, 4833.

Suzuki, M. (1991*a*). Decomposition formulas of exponential operators and Lie exponentials with some applications to quantum mechanics and statistical physics. *J. Math. Phys.*, **32**, 400.

Suzuki, M. (1991*b*). General theory of fractal path-integrals with applications to many-body theories and statistical physics. *J. Math. Phys.*, **32**, 400.

Suzuki, M. (1992). General nonsymmetric higher-order decomposition of exponential operators and symplectic integrators. *J. Phys. Soc. Japan*, **61**, 3015.

Suzuki, M. (1995). Hybrid exponential product formulas for unbounded operators with possible applications to Monte Carlo simulations. *Phys. Lett. A*, **201**, 425.

Sweet, C. R., Petrone, P., Pande, V. S., and Izaguirre, J. A. (2008). Normal mode partitioning of Langevin dynamics for biomolecules. *J. Chem. Phys.*, **128**, 145101.

Swope, W. C., Andersen, H. C., Berens, P. H., and Wilson, K. R. (1982). A computer-simulation method for the calculation of equilibrium-constants for the formation of physical clusters of molecules: Application to small water clusters. *J. Chem. Phys.*, **76**, 637.

Szabó, A. and Merks, R. M. H. (2013). Cellular Potts modeling of tumor growth, tumor invasion, and tumor evolution. *Front. Oncol.*, **3**.

Takahashi, M. and Imada, M. (1984). Monte Carlo calculations of quantum systems. II. Higher order correction. *J. Phys. Soc. Japan*, **53**, 3765.

Tarasov, V. E. (2004). Phase-space metric for non-Hamiltonian systems. *J. Phys. A*, **38**, 2145.

Tesi, M. C., van Rensburg, E. J. J., Orlandini, E., and Whittington, S. G. (1996). Monte Carlo study of the interacting self-avoiding walk model in three dimensions. *J. Stat. Phys.*, **82**, 155.

Thirumalai, D., Bruskin, E. J., and Berne, B. J. (1983). An iterative scheme for the evaluation of discretized path integrals. *J. Chem. Phys.*, **79**, 5063.

Tobias, D. J., Martyna, G. J., and Klein, M. L. (1993). Molecular dynamics simulations of a protein in the canonical ensemble. *J. Phys. Chem.*, **97**, 12959.

Tolman, R. C. (1918). A general theory of energy partition with applications to quantum theory. *Phys. Rev.*, **11**, 261.

Torrie, G. M. and Valleau, J. P. (1974). Monte-Carlo free-energy estimates using non-Boltzmann sampling: Application to subcritical Lennard-Jones fluid. *Chem. Phys. Lett.*, **28**, 578.

Torrie, G. M. and Valleau, J. P. (1977). Non-physical sampling distributions in Monte-Carlo free-energy estimation: Umbrella sampling. *J. Comput. Phys.*, **23**, 187.

Toxvaerd, S. (1994). Hamiltonians for discrete dynamics. *Phys. Rev. E*, **50**, 2274.

Trotter, H. F. (1959). On the produce of semi-groups of operators. *Proc. Amer. Math. Soc.*, **10**, 545.

Trout, B. L. and Parrinello, M. (1998). The dissociation mechanism of H_2O in water studied by first-principles molecular dynamics. *Chem. Phys. Lett.*, **288**, 343.

Tucker, S. C., Tuckerman, M. E., Berne, B. J., and Pollak, E. (1991). Comparison of rate theories for generalized Langevin dynamics. *J. Chem. Phys.*, **95**, 5809.

Tuckerman, M. E. (2002). *Ab initio* molecular dynamics: Basic concepts, current trends and novel applications. *J. Phys. Condens. Matt.*, **14**, R1297.

Tuckerman, M. E. (2019). Machine learning transforms how microstates are sampled. *Science*, **365**, 982.

Tuckerman, M. E., Alejandre, J., Lopez-Rendon, R., Jochim, A. L., and Martyna, G. J. (2006). A Liouville-operator derived measure-preserving integrator for molecular dynamics simulations in the isothermal-isobaric ensemble. *J. Phys. A*, **39**, 5629.

Tuckerman, M. E. and Berne, B. J. (1993). Vibrational relaxation in simple fluids: Comparison of theory and simulation. *J. Chem. Phys.*, **98**, 7301.

Tuckerman, M. E., Liu, Y., Ciccotti, G., and Martyna, G. J. (2001). Non-Hamiltonian molecular dynamics: Generalizing Hamiltonian phase space principles to non-Hamiltonian systems. *J. Chem. Phys.*, **115**, 1678.

Tuckerman, M. E., Martyna, G. J., and Berne, B. J. (1992). Reversible multiple time scale molecular dynamics. *J. Chem. Phys.*, **97**, 1990.

Tuckerman, M. E., Martyna, G. J., Klein, M. L., and Berne, B. J. (1993). Efficient molecular dynamics and hybrid Monte Carlo algorithms for path integrals. *J. Chem. Phys.*, **99**, 2796.

Tuckerman, M. E. and Marx, D. (2001). Heavy-atom skeleton quantization and proton tunneling in "intermediate-barrier" hydrogen bonds. *Phys. Rev. Lett.*, **86**, 4946.

Tuckerman, M. E., Marx, D., Klein, M. L., and Parrinello, M. (1996). Efficient and general algorithms for path integral Car-Parrinello molecular dynamics. *J. Chem. Phys.*, **104**, 5579.

Tuckerman, M. E., Marx, D., and Parrinello, M. (2002). The nature and transport mechanism of hydrated hydroxide ions in aqueous solution. *Nature*, **417**, 925.

Tuckerman, M. E., Mundy, C. J., Balasubramanian, S., and Klein, M. L. (1997). Modified nonequilibrium molecular dynamics for fluid flows with energy conservation. *J. Chem. Phys.*, **106**, 5615.

Tuckerman, M. E., Mundy, C. J., and Martyna, G. J. (1999). On the classical statistical mechanics of non-Hamiltonian systems. *Europhys. Lett.*, **45**, 149.

Tzanov, A. T., Cuendet, M. A., and Tuckerman, M. E. (2014). How accurately do current force fields predict experimental peptide conformations? An adiabatic free energy dynamics study. *J. Phys. Chem. B*, **118**, 6539.

Valsson, O. and Parrinello, M. (2014). Variational approach to enhanced sampling and free energy calculations. *Phys. Rev. Lett.*, **113**, 090601.

VandeVondele, J. and Rothlisberger, U. (2002). Canonical adiabatic free energy sampling (CAFES): A novel method for the exploration of free energy surfaces. *J. Phys. Chem. B*, **106**, 203.

Verlet, L. (1967). Computer experiments on classical fluids. I. Thermodynamical properties of Lennard-Jones molecules. *Phys. Rev.*, **159**, 98.

Voth, G. A. (1993). Feynman path integral formulation of quantum mechanical transition state theory. *J. Phys. Chem.*, **97**, 8365.

Voth, G. A., Chandler, D., and Miller, W. H. (1989). Rigorous formulation of quantum transition state theory and its dynamical corrections. *J. Chem. Phys.*, **91**, 7749.

Waldram, J. R. (1985). *Theory of Thermodynamics*. Cambridge University Press, New York.

Wallqvist, A., Thirumalai, D., and Berne, B. J. (1987). Path integral Monte-Carlo study of the hydrated electron. *J. Chem. Phys.*, **86**, 6404.

Wang, D., Wang, Y., Chang, J., Zhang, L., Wang, H., and E, W. (2021). Efficient sampling of high-dimensional free energy landscapes using adaptive reinforced dynamics. *Nature Comp.Sci.*, **2**, 20.

Wang, F. and Landau, D. P. (2001). Determinating the density of states for classical statistical models: A random walk algorithm to produce a flat histogram. *Phys. Rev. E*, **64**, 056101.

Wang, J., Wolf, R. M., Caldwell, J. W., Kollman, P. A., and Case, D. A. (2004). Development and testing of a general amber force field. *J. Comp. Chem.*, **25**, 1157.

Wang, Q., Johnson, J. K., and Broughton, J. Q. (1997). Path integral grand canonical Monte Carlo. *J. Chem. Phys.*, **107**, 5108.

Watanabe, M. and Reinhardt, W. P. (1990). Direct dynamic calculation of entropy and free-energy by adiabatic switching. *Phys. Rev. Lett.*, **65**, 3301.

Weber, H. J. and Arfken, G. B. (2005). *Mathematical Methods for Physicists*. Academic Press, Burlington.

Weingarten, D. (1982). Monte Carlo evaluation of hadron masses in lattice gauge theories with fermions. *Phys. Lett. B*, **109**, 57.

Weingarten, D. and Petcher, D. N. (1981). Monte Carlo integration for lattice gauge theories with fermions. *Phys. Lett. B*, **99**, 333.

Welberry, T. R. (1979). Crystal growth-disorder models and Ising models. *Mol. Cryst. Liq. Cryst.*, **50**, 21.

Weyl, H. (1927). Quantenmechanik und Gruppentheorie. *Z. Phys.*, **46**, 1.

Widom, B. (1963). Some topics in the theory of fluids. *J. Chem. Phys.*, **39**, 2808.

Widom, B. (1965). Equation of state in the neighborhood of the critical point. *J. Chem. Phys.*, **43**, 3898.

Widom, B. (1982). Potential-distribution theory and the statistical mechanics of fluids. *J. Phys. Chem.*, **86**, 869.

Wiener, N. (1930). Generalized harmonic analysis. *Acta Math.*, **55**, 117.

Wilburn, G. W. and Eddy, S. R. (2020). Remote homology search with hidden Potts models. *PLOS Comp. Biol.*, **16**, e1008085.

Wilson, K. G. (1971). Renormalization group and critical phenomena. 1. Renormalization group and Kadanoff scaling picture. *Phys. Rev. B*, **4**, 3174.

Wilson, K. G. and Fisher, M. E. (1972). Critical exponents in 3.99 dimensions. *Phys. Rev. Lett.*, **28**, 240.

Witt, A., Ivanov, S. D., Shiga, M., Forbert, H., and Marx, D. (2009). On the applicability of centroid and ring-polymer path integral molecular dynamics for vibrational spectroscopy. *J. Chem. Phys.*, **130**, 194510.

Wu, Y., Tepper, H. L., and Voth, G. A. (2006). Flexible simple point-charge water model with improved liquid-state properties. *J. Chem. Phys.*, **124**, 024503.

Yan, Q. L., Faller, R., and de Pablo, J. J. (2002). Density-of-states Monte Carlo method for simulation of fluids. *J. Chem. Phys.*, **116**, 8745.

Yang, C. N. (1952). The spontaneous magnetization of a two-dimensional Ising model. *Phys. Rev.*, **85**, 808.

Yang, J., Erriah, B., Hu, C. T., Reiter, E., Zhu, X., López-Mejías, V., Carmona-Sepúlveda, I. P., Ward, M. D., and Kahr, B. (2020). A deltamethrin crystal polymorph for more effective malaria control. *Proc. Natl. Acad. Sci. U.S.A.*, **117**, 26633.

Yang, J., Hu, C. T., Zhu, X., Zhu, Q., Ward, M. D., and Kahr, B. (2017). DDT polymorphism and the lethality of crystal forms. *Angew. Chem. Intl. Ed.*, **56**, 10165.

Yoshida, H. (1990). Construction of higher-order symplectic integrators. *Phys. Lett. A*, **150**, 262.

Yu, T. -Q., Chen, P. -Y., Chen, M., Samanta, A., Vanden-Eijnden, E., and Tuckerman, M. E. (2014). Order-parameter-aided temperature-accelerated sampling for the exploration of crystal polymorphism and solid-liquid phase transitions. *J. Chem. Phys.*, **140**, 214109.

Yu, T. -Q., Martyna, G. J., and Tuckerman, M. E. (2010). Measure-preserving integrators for molecular dynamics in the isothermal-isobaric ensemble derived from the Liouville operator. *Chem. Phys.*, **370**, 294.

Yu, T. -Q. and Tuckerman, M. E. (2011). Temperature-accelerated method for exploring polymorphism in molecular crystals based on free energy. *Phys. Rev. Lett.*, **107**, 015701.

Zelsmann, H. R. (1995). Temperature dependence of the optical constants for liquid H_2O and D_2O in the far IR region. *J. Mol. Struct.*, **350**, 95.

Zhang, F. (1997). Operator-splitting integrators for constant-temperature molecular dynamics. *J. Chem. Phys.*, **106**, 6102.

Zhang, L., Wang, H., and E, W. (2018). Reinforced dynamics for enhanced sampling in large atomic and molecular systems. *J. Chem. Phys.*, **148**, 124113.

Zhang, Y., Poe, D., Heroux, L., Squire, H., Doherty, B. W., Long, Z., Dadmun, M., Gurkan, B., Tuckerman, M. E., and Maginn, E. J. (2020). Liquid structure and transport properties of the deep eutectic solvent ethaline. *J. Phys. Chem. B*, **124**(25), 5251–5264.

Zhang, Z. J., Liu, X. Z. J., Chen, Z., Zheng, H., Yan, K., and Liu, J. (2017). Unified thermostat scheme for efficient configurational sampling for classical/quantum canonical ensembles via molecular dynamics. *J. Chem. Phys.*, **147**, 034109.

Zhang, Z. J., Liu, X. Z. J., Yan, K. Y., Tuckerman, M. E., and Liu, J. (2019). Unified efficient thermostat scheme for the canonical ensemble with holonomic or isokinetic constraints via molecular dynamics. *J. Phys. Chem. A*, **123**, 6056.

Zhu, X., Hu, C. T., Erriah, B., Vogt-Maranto, L., Yang, J., Yang, Y., Qiu, M., Fellah, N., Tuckerman, M. E., Ward, M. D., and Kahr, B. (2021). Imidacloprid crystal polymorphs for disease vector control and pollinator protection. *J. Am. Chem. Soc.*, **143**, 17144.

Zhu, Z. W. and Tuckerman, M. E. (2002). *Ab initio* molecular dynamics investigation of the concentration dependence of charged defect transport in basic solutions via calculation of the infrared spectrum. *J. Phys. Chem. B*, **106**, 8009.

Zhu, Z. W., Tuckerman, M. E., Samuelson, S. O., and Martyna, G. J. (2002). Using novel variable transformations to enhance conformational sampling in molecular dynamics. *Phys. Rev. Lett.*, **88**, 100201.

Zwanzig, R. W. (1954). High-temperature equation of state by a perturbation method. 1. Nonpolar gases. *J. Chem. Phys.*, **22**, 1420.

Zwanzig, R. W. (1973). Nonlinear generalized Langevin equations. *J. Stat. Phys.*, **9**, 215.

Index